Table of Atomic Masses Based on Carbon-12

Name	Symbol	Atomic Number	Atomic Mass	Name	Symbol	Atomic Number	Mass
Actinium	Ac	89	227.028	Meitnerium	Mt	109	(268)
Aluminum	Al	13	26.9815	Mendelevium	Md	101	(258)
Americium	Am	95	(243)	Mercury	Hg	80	200.59
Antimony	Sb	51	121.760	Molybdenum	Mo	42	95.94
Argon	Ar	18	39.948	Neodymium	Nd	60	144.24
Arsenic	As	33	74.9216	Neon	Ne	10	20.1797
Astatine	At	85	(210)	Neptunium	Np	93	237.048
Barium	Ba	56	137.327	Nickel	Ni	28	58.6934
Berkelium	Bk	97	(247)	Niobium	Nb	41	92.9064
Beryllium	Be	4	9.01218	Nitrogen	N	7	14.0067
Bismuth	Bi	83	208.980	Nobelium	No	102	(259)
Bohrium	Bh	107	(267)	Osmium	Os	76	190.23
Boron	B	5	10.811	Oxygen	O	8	15.9994
Bromine	Br	35	79.904	Palladium	Pd	46	106.42
Cadmium	Cd	48	112.411	Phosphorus	P	15	30.9738
Calcium	Ca	20	40.078	Platinum	Pt	78	195.078
Californium	Cf	98	(251)	Plutonium	Pu	94	(244)
Carbon	C	6	12.0107	Polonium	Po	84	(209)
Cerium	Ce	58	140.116	Potassium	K	19	39.0983
Cesium	Cs	55	132.905	Praseodymium	Pr	59	140.908
Chlorine	Cl	17	35.4527	Promethium	Pm	61	(145)
Chromium	Cr	24	51.9961	Protactinium	Pa	91	231.036
Cobalt	Co	27	58.9332	Radium	Ra	88	226.025
Copper	Cu	29	63.546	Radon	Rn	86	(222)
Curium	Cm	96	(247)	Rhenium	Re	75	186.207
Darmstadtium	Ds	110	(281)	Rhodium	Rh	45	102.906
Dubnium	Db	105	(262)	Rubidium	Rb	37	85.4678
Dysprosium	Dy	66	162.50	Ruthenium	Ru	44	101.07
Einsteinium	Es	99	(252)	Rutherfordium	Rf	104	(261)
Erbium	Er	68	167.26	Samarium	Sm	62	150.36
Europium	Eu	63	151.964	Scandium	Sc	21	44.9559
Fermium	Fm	100	(257)	Seaborgium	Sg	106	(266)
Fluorine	F	9	18.9984	Selenium	Se	34	78.96
Francium	Fr	87	(223)	Silicon	Si	14	28.0855
Gadolinium	Gd	64	157.25	Silver	Ag	47	107.868
Gallium	Ga	31	69.723	Sodium	Na	11	22.9898
Germanium	Ge	32	72.61	Strontium	Sr	38	87.62
Gold	Au	79	196.967	Sulfur	S	16	32.066
Hafnium	Hf	72	178.49	Tantalum	Ta	73	180.948
Hassium	Hs	108	(269)	Technetium	Tc	43	(98)
Helium	He	2	4.00260	Tellurium	Te	52	127.60
Holmium	Ho	67	164.930	Terbium	Tb	65	158.925
Hydrogen	H	1	1.00794	Thallium	Tl	81	204.383
Indium	In	49	114.818	Thorium	Th	90	232.038
Iodine	I	53	126.904	Thulium	Tm	69	168.934
Iridium	Ir	77	192.217	Tin	Sn	50	118.710
Iron	Fe	26	55.845	Titanium	Ti	22	47.867
Krypton	Kr	36	83.80	Tungsten	W	74	183.84
Lanthanum	La	57	138.906	Uranium	U	92	238.029
Lawrencium	Lr	103	(262)	Vanadium	V	23	50.9415
Lead	Pb	82	207.2	Xenon	Xe	54	131.29
Lithium	Li	3	6.941	Ytterbium	Yb	70	173.04
Lutetium	Lu	71	174.967	Yttrium	Y	39	88.9059
Magnesium	Mg	12	24.3050	Zinc	Zn	30	65.39
Manganese	Mn	25	54.9380	Zirconium	Zr	40	91.224

Atomic masses in this table are relative to carbon-12 and limited to six significant figures, although some atomic masses are known more precisely. For certain radioactive elements the numbers listed (in parentheses) are the mass numbers of the most stable isotopes.

ELEVENTH EDITION

Chemistry for Changing Times

John W. Hill
University of Wisconsin–River Falls

Doris K. Kolb

With Special Contributions by
Terry W. McCreary
Murray State University

Upper Saddle River, NJ 07458

Library of Congress Cataloging-in-Publication Data

Hill, John William
    Chemistry for changing times.—11th ed. / John W. Hill, Doris K. Kolb;
  with special contributions by Terry McCreary.
      p. cm.
    Includes bibliographical references and index.
    ISBN 0-13-228084-1
    1. Chemistry—Textbooks.   I. Kolb, Doris K.   II. McCreary, Terry.   III. Title.
  QD33.2.H54 2007
    540—dc22   2006015210

Editor in Chief, Science: *Daniel Kaveney*
Executive Editor: *Kent Porter Hamann*
Editorial Assistant: *Joya Carlton*
Assistant Editor: *Jennifer Hart*
Production Editor: *Shari Toron*
Manager, Composition: *Allyson Graesser*
Desktop Administration: *Joanne Del Ben*
Electronic Page Makeup: *Joanne Del Ben, Jackie Ambrosius*
Executive Managing Editor: *Kathleen Schiaparelli*
Assistant Managing Editor: *Beth Sweeten*
Senior Managing Editor, Science Media: *Nicole Jackson*
Media Editor, Physical Sciences: *David Alick*
Manufacturing Manager: *Alexis Heydt-Long*
Manufacturing Buyer: *Alan Fischer*
Creative Director: *Juan López*
Art Director: *Jonathan Boylan*
Interior Designer: *Dianne Densberger*
Cover Designer: *Kenny Beck*
Cover Image Credit: *Freeman Patterson / Masterfile*
Director of Creative Services: *Paul Belfanti*
Senior Managing Editor, Art Production and Management: *Patricia Burns*
Manager, Production Technologies: *Matthew Haas*
Managing Editor, Art Management: *Abigail Bass*
Art Production Editor: *Rhonda Aversa*
Illustrations: *Precision Graphics*
Director, Image Resource Center: *Melinda Reo*
Manager, Rights and Permissions: *Zina Arabia*
Manager, Visual Research: *Beth Brenzel/Elaine Soares*
Manager, Cover Visual Research & Permissions: *Karen Sanatar*
Image Permission Coordinator: *Nancy Seise*
Photo Researcher: *Truitt & Marshall*
Proofreader: *Michael Rossa*

© 2007, 2004, 2001, 1998 Pearson Education, Inc.
Pearson Prentice Hall
Pearson Education, Inc.
Upper Saddle River, NJ 07458

Printed in the United States of America

10   9   8   7   6   5   4   3   2   1

ISBN 0-13-228084-1

Pearson Education LTD., *London*
Pearson Education Australia PTY, Limited, *Sydney*
Pearson Education Singapore, Pte. Ltd.
Pearson Education North Asia Ltd, *Hong Kong*
Pearson Education Canada, Ltd, *Toronto*
Pearson Educación de *Mexico*, S.A. de C.V.
Pearson Education—Japan, *Tokyo*
Pearson Education Malaysia, Pte. Ltd.

Doris K. Kolb

(1927–2005)

This eleventh edition of *Chemistry for Changing Times* is dedicated to the memory of Doris K. Kolb, who died of pancreatic cancer on December 20, 2005. Doris has been my esteemed coauthor since the seventh edition.

Doris Jean Kasey was born on August 4, 1927 in Louisville, Kentucky. She received a B.S. degree with a major in chemistry from the University of Louisville and M.S. and Ph.D. degrees from The Ohio State University. She married Kenneth E. Kolb in 1948.

A distinguished scientist, Doris Kolb was the first female Ph.D. employed by Standard Oil of Indiana. She did research that led to a better understanding of the browning of fruits, and she helped establish fundamental standards for pure fatty acids. While at Standard Oil, she helped pioneer a live television program, *Spotlight on Research*, which aired on WTTW in Chicago.

Doris taught at Corning Community College in New York and at Illinois Central College. She was the first recipient of ICC's Outstanding Teacher Award. While at ICC, she developed several courses, mostly for nursing students. Doris joined the faculty at Bradley University in 1986 and taught there until 2005. Her excellence in teaching was further recognized with the prestigious Catalyst Award of the Chemical Manufacturers Association.

Doris was exceptionally active in the American Chemical Society (ACS), holding many offices at the local and national level. Much of her service to the ACS was in the area of chemical education. Most notably, she served for 16 years on the Board of Publications for the *Journal of Chemical Education*, including eight years as chair. She also served as Chair of the ACS Division of Chemical Education. She was a Feature Editor for the *Journal*'s column, "Overhead Projector Demonstrations."

Doris authored more than 60 papers, several book chapters, and was coauthor of two books, including this one. She presented more than 50 talks at ACS meetings and organized numerous symposia. In September 2005, she and Ken were recipients of the Division of Chemical Education's Outstanding Service Award.

Doris was a leader in community affairs. She helped establish Planned Parenthood in Peoria and served as its first executive director. She was active with Planned Parenthood for the rest of her life and was honored with the Margaret Sanger Award in 1975 and the Betty Osborne Award in 1991. Doris was also a pioneer in recognizing the effect of smoking on health and helped found Group Against Smokers' Pollution (GASP). She was also active in the Peoria Women's Club, the Peoria Garden Club, the Peoria Symphony Guild, the Peoria Fine Arts Society, and the PEO Sisterhood. She was a member of the Universalist Unitarian Church.

Doris was a poet of some note. Her main interest was humorous poetry. She not only added a touch of fun to *Chemistry for Changing Times*, but she gave readings in many venues and taught independent learning classes in humorous poetry. The last class she taught was a class in poetry in the fall of 2005.

To me, Doris and Ken were friends and helpful supporters long before Doris joined the author team. She has provided much to the spirit and flavor of the book. Doris's contributions to *Chemistry for Changing Times*—and indeed to all of chemistry and chemical education—will live on for many years to come, not only in her publications, but in the hearts and minds of her many students, colleagues, and friends. Let us dedicate our lives, as Doris did hers, to making this world a better place.

John W. Hill

Brief Contents

GREEN CHEMISTRY

The eleventh edition of *Chemistry for Changing Times* is pleased to present the Green Chemistry essays listed below. Each essay includes media activities. Once students access the Companion Website with GradeTracker, the static investigations and exercises become a dynamic and exciting discovery activity for them. The topics have been carefully chosen to introduce students to the concepts of Green Chemistry—a new approach to designing chemicals and chemical transformations that are beneficial for human health and the environment. The Green Chemistry essays in this edition highlight cutting-edge research by chemists, molecular scientists, and engineers to explore the fundamental science and practical applications of chemistry that is "benign by design." These examples emphasize the responsibility of chemists for the consequences of the new materials they create and the importance of building a sustainable chemical enterprise.

Contents

1 Chemistry: A Science for All Seasons   1

2 Atoms: Are They for Real?   36

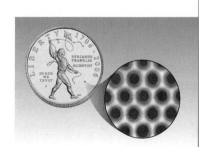

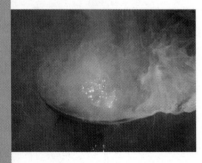

6   Chemical Accounting: Mass and Volume Relationships   154

7   Acids and Bases: Please Pass the Protons   190

8   Oxidation and Reduction: Burn and Unburn   212

9 Organic Chemistry: The Infinite Variety of Carbon Compounds 238

10 Polymers: Giants Among Molecules 272

11   Chemistry of Earth: Metals and Minerals   300

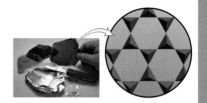

12   Air: The Breath of Life   322

16 Food: Molecular Gastronomy 458

17 Household Chemicals: Helps and Hazards 512

20 Poisons: Chemical Toxicology 630

Preface

Chemistry for Changing Times is now in its eleventh edition. Times have changed immensely since the first edition appeared in 1972 and are now changing more rapidly than ever—especially in the vital areas of biochemistry, the environment, energy, drugs, and health and nutrition. The book is changing accordingly. We have thoroughly updated the material in the appropriate chapters. A major change is that we have introduced the concept of green chemistry throughout the book, but especially in the end-of-chapter Green Chemistry exercises (see page v–vi).

A Special Kind of Course

Our knowledge base has expanded enormously since that first edition, never more so than in the last few years. We have faced tough choices in deciding what to include and what to leave out. We now live in what has been called the "information age." Unfortunately, information is not knowledge; the information may or may not be valid. Our focus, more than ever, is on helping students evaluate information. May we all someday gain the gift of wisdom.

A major premise is that a chemistry course for students who are not majoring in science should be quite different from the course we offer our science majors. It must present basic chemical concepts with intellectual honesty, but it need not—probably should not—focus on esoteric theories or rigorous mathematics. It should include lots of modern everyday applications. The textbook should be appealing to look at, easy to understand, and interesting to read.

Three-fourths of the legislation considered by the U.S. Congress involves questions having to do with science or technology, yet only rarely does a scientist or engineer enter politics. Most of the people who make important decisions regarding our health and our environment are not trained in science, but it is critical that these decision makers be scientifically literate. In the judicial system, decisions often depend on scientific evidence, but judges and jurors frequently have little education in the sciences. A chemistry course for students who are not science majors should emphasize practical applications of chemistry to problems involving such things as environmental pollution, radioactivity, energy sources, and human health. The students who take our liberal arts chemistry courses include future teachers, business leaders, lawyers, legislators, accountants, artists, journalists, jurors, and judges.

Objectives

Our main objectives in a chemistry course for students who are not majoring in science are as follows:

- To attract lots of students from a variety of disciplines. If students do not enroll in the course, we can't teach them.
- To help students learn so that they may become productive, creative, ethical, and engaged citizens.
- To use topics of current interest to illustrate chemical principles. We want students to appreciate the importance of chemistry in the real world.
- To relate chemical problems to the everyday lives of our students. Chemical problems become more significant to students when they can see a personal connection.

- To instill in students an appreciation for chemistry as an open-ended learning experience. We hope that our students will develop a curiosity about science and will want to continue learning throughout their lives.
- To acquaint students with scientific methods. We want students to be able to read about science and technology with some degree of critical judgment. This is especially important because many of the scientific problems discussed are complex and controversial.
- To help students become literate in science. We want our students to develop a comfortable knowledge of science so that they find news articles relating to science interesting rather than intimidating.

New Features in the Eleventh Edition

In preparing this new edition, we have responded to suggestions from users and reviewers of the tenth edition, and we have used our own writing and teaching experience. The text is fully revised and updated to reflect the latest scientific developments in a fast-changing world.

The organization of the text makes it easier for the instructor to skip sections or (in some cases) whole chapters. At most institutions, the course for nonscience majors brings together a group of students quite heterogeneous both in their science backgrounds and in their academic interests. A major challenge to the instructor is to find the balance between these needs and interests. As authors we have tried to create a text that is flexible and that can be used in a variety of ways.

- The first eight chapters deal with some of the basic chemistry, including atoms and elements, structures of ionic and molecular compounds, acids and bases, and oxidation-reduction reactions. Most of the numerical manipulations are concentrated in Chapter 6.
- Chapters 9 and 10 introduce organic chemistry and polymers.
- Chapters 11 through 14 (which deal with Earth, air, water, and energy, respectively) focus on environmental issues. Environmentalism permeates the entire book, especially in the incorporation of the concept of green chemistry in the end-of-chapter Green Chemistry exercises.
- Chapter 15 introduces biochemistry and provides a background for the some of the material in subsequent chapters.
- The final five chapters delve into topics related to health and personal care (food, household chemicals, fitness and health, drugs, and poisons, respectively) in some detail.

Clearly, for a one-semester course, choices will have to be made. In some places, the text refers to later chapters where further details of a particular topic are to be found, but it is fairly easy for an instructor to select the parts most important and useful to his or her students. If you have specific questions about skipping sections or chapters, please contact John Hill at jwhill602@comcast.net.

Improvements in Pedagogy

The following changes have been made to strengthen and improve the pedagogy in this edition.

- We have revised about 25% of the end-of-chapter problems and increased their variety. In some of the problems we have referred back to figures within the text, to aid the visual learner.
- We have added several more worked-out Examples and their accompanying Exercises and we have changed many of the others to improve pedagogy.
- We have added to the Critical Thinking Exercises in each chapter. These exercises require the student to apply information and learning from the chapter in both concrete and abstract fashion.

- Each chapter has a category of Collaborative Group Projects as a part of the end-of-chapter exercises. This will make it easy for instructors who want to encourage collaborative work and to make group assignments. In this way, the students' learning of chemistry can be extended far beyond the textbook.
- We have added voice balloons to figures to point out important features. We also use voice balloons in text displays and problem solving to guide the student through the learning process and thus improve the pedagogy.
- The chapter summaries have been organized into a new format, presented by sections, with key terms highlighted in red for easy recognition. Figures and photographs have been included in the summaries to aid the visual learner in revisiting important concepts.
- We have updated the References and Readings at the end of each chapter.
- Keeping in mind that today's student is more visually oriented than ever, we have made extensive changes in the photographs, figures, graphs, and other illustrations.

Applications

Focusing on the importance of providing interesting, relevant applications, we have added several new box features:

- Nanoworld (Chapter 1)
- Cell Phones and Microwaves and Power Lines, Oh My! (Chapter 4)
- A Compound by Any Other Name Would Smell As Sweet . . . (Chapter 5)
- Photochromic Glass (Chapter 8)
- A Closed Ecosystem? (Chapter 12)
- Energy Return on Energy Invested (Chapter 14)
- Infectious Prions: Deadly Protein (Chapter 15)
- Enzymes and Green Chemistry (Chapter 15)
- It's a Drug! No, It's a Food! No, It's . . . a Dietary Supplement! (Chapter 16)
- Chemistry and Athletic Performance (Chapter 18)
- Chemistry of Sports Materials (Chapter 18)
- Cisplatin: The Platinum Standard of Cancer Treatment (Chapter 19)
- Some Chemistry of Love, Trust, and Sexual Fidelity (Chapter 19)
- Renaissance Poisoners and Chemistry and Counterterrorism (Chapter 20)

Visualization

New color photographs and diagrams have been added. Visual material adds greatly to the general appeal of a textbook. Color diagrams can also be highly instructive, and colorful photographs relating to descriptive chemistry do much to enhance the learning process. We have added more illustrations that use both microscopic (molecular) and macroscopic (visual) views to help students visualize chemical phenomena. Some of the figure captions feature questions to focus attention on the concept illustrated in the figure.

Readability

Over the years, students have told us that they have found this textbook easy to read. The language is simple, and the style is conversational. Explanations are clear and easy to understand. The friendly tone of the book has been maintained in this edition. Since the format and the amount of open space on a page also contribute to readability, we have made conscious improvements in the design of this edition. For example, many of the margin notes have been incorporated directly into the text to ensure that pages don't appear to be crowded.

Units of Measurement

The United States continues to use the traditional English system for many kinds of measurement even though the metric system has long been used internationally. A modern version of the metric system, the Système International (SI), is now widely used, especially by scientists. So what units should be used in a text for liberal arts students? In presenting chemical principles, we use primarily metric units. In other parts of the book we use those units that the students are most likely to encounter elsewhere in the same context.

Chemical Structures

The structures of many complicated molecules are presented in the text, especially in the later chapters. These structures are presented mainly to emphasize that they are actually known and to illustrate the fact that substances with similar properties often have similar structures. Students should not feel that they must learn all these structures, but they should take the time to look at them. We hope that they will come to recognize familiar features in these molecules.

Chapter Summaries and Glossary

The chapter summaries have been organized into a new format with key terms highlighted in red. The Glossary (Appendix B) gives definitions of terms that appear in boldface throughout the text. These terms include all key terms highlighted in the chapter summaries.

Questions and Problems

Worked-out Examples and accompanying Exercises are given within almost all of the chapters. Each Example carefully guides the student through the process for solving a particular type of problem. It is then followed by one or more Exercises that allow the student to check his or her comprehension right away. Most Examples are now followed by two Exercises, labeled A and B. The goal in an A Exercise is to apply to a similar situation the method outlined in the Example. In a B Exercise, students often must combine that method with other ideas previously learned. Many of the B Exercises provide a context closer to that in which chemical knowledge is applied, and they thus serve as a bridge between the worked Examples and the more challenging problems at the end of the chapter. The A and B Exercises provide a simple way for the instructor to assign homework that is closely related to the Examples. Answers to all the in-chapter Exercises are given in Appendix C.

The end-of-chapter exercises include

- Review Questions that for the most part simply ask for a recall of material in the chapter.
- A set of matched-pair Problems, arranged according to subject matter from each chapter, with answers to the odd-numbered problems given in Appendix C.
- Additional Problems that are not grouped by type. Some of these are more challenging than the matched-pair Problems and often require a synthesis of ideas from more than one chapter. Other Additional Problems pursue an idea further than is done in the text or introduce new ideas.

Answers to many Review Questions and Additional Problems are also given in Appendix C.

References and Suggested Readings

An updated list of recommended books and articles appears at the end of each chapter. A student whose interest has been sparked by a topic can delve more deeply into the subject in the library. Instructors might also find these lists useful.

Supplementary Materials

The most important learning aid is the teacher. In order to make the instructor's job easier and enrich the education of students, we have provided a variety of supplementary materials.

Print Resources for Students

Student Study Guide (0-13-227113-3). Prepared by Richard Jones of Sinclair Community College. This book assists students through the text material and contains learning objectives, chapter outlines, key terms, and additional problems along with self-tests and answers.

Chemical Investigations for Changing Times, Eleventh Edition (0-13-1755005). Prepared by C. Alton Hassell and Paula Marshall. Contains 56 laboratory experiments and is specifically referenced to *Chemistry for Changing Times*.

Study Card (0-13-239660-2). Prepared by Stacy Brown. This is a concise, quick reference card covering key topics in each chapter of the text. Presented in an easy-to-read format, the *Eleventh Edition Study Card* enables students to master concepts, theories, and facts for exams as well as easily review information on the go.

Print Resources for Instructors

Instructor's Resource Manual (0-13-227115-X). Prepared by Paul Karr and David Pietz of Wayne State College. This useful guide describes all the different resources available to instructors and shows how to integrate them into your course. Organized by chapter, this manual offers lecture outlines, answers and solutions to all questions and problems that are not answered by the authors in Appendix C, suggested in-class demonstrations recommended by Doris Kolb, and other suggested resources. The lecture outline is also available in an electronic format.

Instructor's Manual for Chemical Investigations for Changing Times (0-13-227116-8). Prepared by Paula Marshall and C. Alton Hassell. This laboratory manual reference includes notes for experiments, safety regulations, procedural instructions, and specifications for equipment and supplies.

Transparencies (0-13-199002-0). Selected by Terry McCreary and John W. Hill. This set contains 150 full-color acetates.

Test Item File (0-13-227114-1). Prepared by Rill Ann Reuter, Winona State University. The *Test Item File* now contains over 2400 test questions that are referenced to the text.

Media Resources for Students

Chemistry for Changing Times Companion Website with GradeTracker (0-13-243412-1) (http://www.prenhall.com/hill)

- **Key Concepts:** A collection of learning goals and key terms to help students identify what they should know, understand, and be able to do after reading the chapter. Definitions of key terms can be accessed through the online glossary.

- **Review Activities:** A variety of assessment opportunities to help students check their understanding of key concepts with the choice to view helpful hints and receive instant feedback on selected answers. Instructors have the option to assign any of these online self-grading activities for homework credit or simply allow students to work at their own pace toward mastery.

- **Media Enhancements:** Selected review questions are enhanced with media—short movies, animations, and 3D molecules—to help students visualize key concepts. Follow the media icons in the textbook to the Companion Website.

- **Application and Critical Thinking Activities:** Collections of thought-provoking questions designed to involve students in scenarios and independent and collaborative research opportunities that focus on current chemistry issues. Pearson's *Research Navigator*™, included as a *Companion Website* offering, connects students to four exclusive databases of source material, including the EBSCO Academic Journal and Abstract Database, New York Times Search by Subject Archive, "Best of the Web" Link Library, and Financial Times Article Archive and Company Financials, helping students quickly and efficiently make the most of their research time.

- **Green Chemistry:** Online versions of the Green Chemistry activities found in the textbook designed to promote awareness. Students are prompted to communicate their findings in a variety of reporting strategies.

Media Resources for Instructors

Instructor's Resource Center on CD/DVD (0-13-199003-9). Prepared by John Singer and James Noblet. This fully searchable and integrated collection of resources includes everything you need organized in one easy-to-access place. It is designed to help you make efficient and effective use of your lecture preparation time as well as to enhance your classroom presentations and assessment efforts. This package features nearly all of the art from the text, including tables; three pre-built PowerPoint presentations; PDF files of the art for high-resolution printing; all the interactive and dynamic media objects from the *Companion Website*; the *Instructor's Resource Manual*, and a set of "clicker" questions for use with Classroom Response Systems. This CD/DVD set also features a search engine tool that enables you to find relevant resources via a number of different parameters, such as key terms, learning objectives, figure numbers, and resource type. Also included is the TestGen test-generation software and a TestGen version of the Test Item File that enables you to create and tailor exams to your needs, or to create online quizzes for delivery in WebCT, Blackboard, or CourseCompass.

Course Management Options. Prentice Hall offers prebuilt courses in a variety of Course Management systems, each of which lets you easily post your syllabus, communicate with students online or offline, administer quizzes, and record student results and track their progress. **OneKey's online course management content** is all you need to plan and administer your course and includes the best teaching and learning resources all in one place. Conveniently organized by textbook chapter, these compiled resources help you save time and help your students reinforce and apply what they have learned in class. Available resources include Summary with Key Terms, Tools, Research Navigator Web research center, Math Toolkit, Practice Problems, and the Test Item File. Resources from the *Instructor's Resource Center* on CD/DVD are also included. OneKey is available for our nationally hosted CourseCompass course management system. If desired, WebCT and Blackboard cartridges containing only the Test Item File are also available for download. Visit http://www.prenhall.com/cms

CourseCompass™—the easiest way to get your course online! Three clicks and you're up. The course includes media resources from the Instructor's Resource Center on CD/DVD as well as assessment items from the companion website and the Test Item File. Visit www.prenhall.com/demo for details on how to communicate with your students, customize content to meet your course needs, create quizzes and test, track grades, and many more online options.

Blackboard®—for campuses that use the user-friendly system Blackboard, consider a prebuilt course that includes media resources from the Instructor's Resource Center on CD/DVD as well as assessment items from the Companion Website with GradeTracker and the Test Item File. Visit www.prenhall.com/demo for details on how to communicate with your students, customize content to meet your course needs, create quizzes and tests, track grades, and utilize many more online options.

WebCT®—for campuses that use the sophisticated course management tools of WebCT, the prebuilt course offers everything mentioned previously as well as the ability to create algorithmic questions using WebCT's calculation format.

OneKey content—Along with all the material from the Companion Website with GradeTracker (www.prenhall.com/hill), OneKey includes

- Resources from the Instructor's Resource Center on CD/DVD
- Test bank questions, converted from our TestGen test item file

Acknowledgments

Terry W. McCreary of Murray State University, a distinguished teacher and author, has made many contributions, including preparing the art manuscript, revising the chapter summaries into the new format, reviewing the content of each chapter, and providing invaluable help with new Examples, Exercises, Problems, and all other aspects of the text.

Through the last three decades we have greatly benefited from hundreds of helpful reviews. It would take far too many pages to list all of those reviewers here. Many of you have contributed to the flavor of the book and helped us minimize our errors. Please know that your contributions are deeply appreciated. For the eleventh edition, we are grateful for challenging reviews from

Iffat Ali, *Lakeland Community College*

Stacy Brown, *The Citadel*

Patrick Buick, *Florida Atlantic University*

Susan Collier, *SUNY at Brockport*

Darwin Dahl, *Western Kentucky University*

Jeannine Eddleton, *Virginia Tech*

Wavell Fogleman, *Plymouth State University*

Jennifer Garlitz, *Bowling Green State*

Linda Hobart, *Finger Lakes Community College*

Ramon Lopez de la Vega, *Florida International University*

Joseph Maloy, *Seton Hall University*

Shane Phillips, *California State University, Stanislaus*

Danaé Quirk Dorr, *Minnesota State University–Mankato*

Rill Ann Reuter, *Winona State University*

Bruce Richardson, *Highline Community College*

Elsa Santos, *Colorado State University*

Steven Summers, *Seminole Community College*

Christopher Truitt, *Texas Tech*

Kendra Twomey, *Massasoit Community College*

Martin Zysmilich, *George Washington University*

We also appreciate the many people who have called or written or e-mailed with corrections and other helpful suggestions. Cynthia S. Hill prepared much of the original material on biochemistry, food, and health and fitness. For this edition, we are especially indebted to Ron Fedie and Arlin Gyberg of Augsburg College for several excellent ideas for the revision.

Four of the verses that appear in this volume were first published in the *Journal of Chemical Education*. We acknowledge, with thanks, the permission to reprint them here. Doris Kolb wrote those verses plus all of the others.

We also want to thank our colleagues at the University of Wisconsin–River Falls and Bradley University for all their help and support.

We also owe a debt of gratitude to the many creative people at Prentice Hall who have contributed their talents to this edition. Kent Porter Hamann, our chemistry editor, has provided valuable guidance throughout the project. As Chemistry Editorial Assistant, Joya Carlton contributed organizational and editorial skills and kept the entire team up to date on developments. We are grateful to Shari Toron, Production Editor, who excelled in the critical role of keeping the project on schedule; to Assistant Editor Jennifer Hart, who provided careful management of the supplements for the text; to Kathleen Schiaparelli, Beth Sweeten, and Joanne Del Ben in production and Art Director Jonathan Boylan for their diligence and patience in bringing all the parts together to yield a finished work; and to our Media Editor, David Alick, who has brought new facets to the media set accompanying our text. We are indebted to our copy editor Patricia Daly, whose expertise helped improve the consistency of the text; and to proofreader Michael Rossa and accuracy checker Rill Ann Reuter, whose sharp eyes caught many of our errors and typos. We also salute our photo researcher, Jerry Marshall of Truitt & Marshall, who vetted hundreds of images in the search for quality photographic illustrations.

We owe a very special kind of thanks to our wonderful spouses, Ina and Ken. Ina has done typing, library research, and so many other things. Ken has done chapter reviews, made suggestions, and given invaluable help for many editions. Most of all, we are grateful to both of them for their enduring love and their boundless patience. Terry W. McCreary would like to thank his wife, Geniece, and their children, Corinne and Yvette, for their unflagging support, understanding, and love.

Finally, we also thank all those many students whose enthusiasm has made teaching such a joy. It is gratifying to have students learn what you are trying to teach them, but it is a supreme pleasure to find that they want to learn even more. Finally, we want to thank all of you who have made so many helpful suggestions. We welcome and appreciate all your comments, corrections, and criticisms.

John W. Hill
jwhill602@comcast.net

Doris K. Kolb

Terry W. McCreary
terry.mccreary@murraystate.edu

To the Student

Welcome to Our Chemical World!

Chemistry is fun. Through this book, we would like to share with you some of the excitement of chemistry and some of the joy of learning about it. You do not need to exclude chemistry from your learning experiences. Learning chemistry will enrich your life—now and long after this course is over—through a better understanding of the natural world, the technological questions now confronting us, and the choices we must face as citizens within a scientific and technological society.

Learning chemistry involves thinking logically, critically, and creatively. Skills gained in this course can be exceptionally useful in many aspects of your life. You will learn how to use the language of chemistry: symbols, formulas, and equations. More important, you will learn how to obtain meaning from information. The most important thing you will learn is how to learn. Memorized material will quickly fade into oblivion unless it is arranged on a framework of understanding.

Chemistry Directly Affects Our Lives

How does the human body work? How does aspirin cure headaches, reduce fevers, and perhaps lessen the chance of a heart attack or stroke? Is ozone a good thing or a threat to our health? Are iron supplement pills poisonous? Is global warming real? If so, did humans contribute to it, and what are some of the possible consequences? Why do most weight-loss diets seem to work in the short run but fail in the long run? Why do our moods swing from happy to sad? Can a chemical test on urine predict possible suicide attempts? How does penicillin kill bacteria without harming our healthy body cells? Chemists have found answers to questions such as these and continue to seek the knowledge that will unlock still other secrets of our universe. As these mysteries are resolved, the direction of our lives often changes—sometimes dramatically. We live in a chemical world—a world of drugs, biocides, food additives, fertilizers, fuels, detergents, cosmetics, and plastics. We live in a world with toxic wastes, polluted air and water, and dwindling petroleum reserves. Knowledge of chemistry will help you better understand the benefits and hazards of this world and will enable you to make intelligent decisions in the future.

Chemical Dependency

We are all chemically dependent. Even in the womb we depend on a constant supply of oxygen, water, glucose, and a multitude of other chemicals.

Our bodies are intricate chemical factories. They are durable but delicate systems. Innumerable chemical reactions that allow our bodies to function properly are constantly taking place within us. Thinking, learning, exercising, feeling happy or sad, putting on too much weight or not gaining enough, and virtually all life processes are made possible by these chemical reactions. Everything that we ingest is part of a complex process that determines whether our bodies work effectively or not. The consumption of some substances can initiate chemical reactions that will stop body functions. Other substances, if consumed, can cause permanent handicaps, and still others can make living less comfortable. A proper balance of the right foods provides the chemicals and generates the reactions we need in order to function at our best. The knowledge of chemistry that you will soon be gaining will help you better understand how your body works so that you will be able to take proper care of it.

Changing Times

We live in a world of increasingly rapid change. It has been said that the only constant is change itself. At present, we are facing some of the greatest problems that humans have ever encountered, and the dilemmas with which we are now confronted seem to have no perfect solutions. We are sometimes forced to make a best choice among only bad alternatives, and our decisions often provide only temporary solutions to our problems. Nevertheless, if we are to choose properly, we must understand what our choices are. Mistakes can be costly, and they cannot always be rectified. It is easy to pollute, but cleaning up pollution once it is there is enormously expensive. We can best avoid mistakes by collecting as much information as possible and evaluating it carefully before making critical decisions. Science is a means of gathering and evaluating information, and chemistry is central to all the sciences.

Chemistry and the Human Condition

Above all else, our hope is that you will learn that the study of chemistry need not be dull and difficult. Rather, it can enrich your life in so many ways—through a better understanding of your body, your mind, your environment, and the world in which you live. After all, the search to understand the universe is an essential part of what it means to be human.

Highlights of the 11ᵗʰ Edition

Chemistry for Changing Times, the most successful book in liberal arts chemistry, defined the course in its first edition. With each subsequent revision, this text has reflected the changing times and the changing needs of the market.

Visually appealing, understandable, and interesting to read, the goal of the eleventh edition of *Chemistry for Changing Times* is to help inform students as scientifically literate consumers and decision-makers. The authors present basic chemical concepts with abundant everyday applications, personalizing the chemistry experience for today's students. In this way, the text focuses students on evaluating information about real-life issues instead of memorizing rigorous theory and mathematics. Important in this new edition is the use of green chemistry as a theme to show the positive impact of chemistry on the future.

Green Chemistry

The concept of **green chemistry** is used throughout the book and appears prominently in the **Green Chemistry** essays at the end of the chapters that include media exercises. ▼

Nanotechnology

Rich Gurney, Simmons College

A flurry of discovery of new physical phenomena and properties followed the discovery of new elements. The isolation and purification of bulk uranium led to many discoveries involving radioactivity, and today radiological procedures have become a mainstay of medical diagnosis. As stated in Chapter 3, the arrangement of various parts of atoms determines the bulk properties of different kinds of matter for large collections of atoms. You may be surprised to learn that the properties of a collection of atoms are also influenced by the size of the collection. Nearly everyone can list the physical properties of the element gold. The yellow, malleable metal is ubiquitous and pervasive in nearly all cultures and civilizations. However, a solution of gold *nanoparticles*—submicroscopic in size—is a brilliant red, blue, or gold color, depending on the size of the nanoparticles.

The way in which the atoms are connected can also have a profound effect on the physical properties of the

collection. Graphite, one allotrope of carbon, is utilize pencils because it is soft. Graphite rubbed on pa leaves a trail on the page. Carbon nanotubes, an allotr with a different connectivity (Chapter 10), are m stronger than steel. The discovery of many new phys properties of carbon nanotubes has made them bot extremely useful and very valuable material.

Nanoscience and nanotechnology both operate w in the nanoscopic world, a world whose typical bou aries are defined by one-billionth to one-millionth meter. Nanometers are really, really small. There are *billion* nanometers in one meter. A nanoparticle is o three to five atoms wide, making the nanoparticle ab 40,000 times smaller than the width of an average hur hair. While the discipline of nanotechnology is relativ new, examples of nanotechnology can be found throu out history. By varying the sizes of gold nanoparticles tisans were able to produce beautifully stained, g windows as far back as medieval times and colorf glazed ceramics during the Ming dynasty. Today, g nanoparticles are being tested for use in therapies neurodegenerative diseases that involve protein aggr tion, such as Alzheimer's and Parkinson's.

▲ The beads fluoresce in different colors, but they are made of the same plastic and contain the same nanoparticles—tiny particles of a cadmium-selenium compound. The different colors are the result of different sizes of nanoparticles used.

▲ Carbon nanotubes (magnified 35,000×) show promise as a superstrong material.

WEB INVESTIGATIONS

Investigation 1
Real-World Applications of Nanotechnology
Nanotechnology is already having a profound impact on the lives of many people around the world as many consumer products include nanomaterials. You may already have some nanoscale items in your home. Nanoparticle-containing bandages will be arriving in your neighborhood pharmacy before you know it. Do a keyword search for "Nano Materials Consumer Products" using a major search engine to investigate the consumer products on the market that already include nanomaterials. Create a list of consumer products that you or your family may have already encountered. What functions do the nanoparticles serve in these consumer products? Search the Web site of your favorite newspaper to find reports on consumer products containing nanoscale materials. What impact have these consumer products had on our society?

Investigation 2
Green Nanotechnology
Nanotechnology involves the measurement, observation, manipulation, and production of materials usually between 1 and 100 nanometers. If these materials are so small, why should our society be overly concerned with ensuring the field of nanotechnology develops in a "Green" way? How will the manufacture of nanomaterials impact our society if this burgeoning field does not embrace the 12 principles of green chemistry and engi-

neering? Based on your web research of green nanotechnology, which of the 12 principles of green chemistry have the most relevance to this field?

Investigation 3
Microchip Manufacture and Green Solvents
Microprocessors and digital memory chips are both examples of integrated circuits or microchips. These devices can be found in everything from laptop computers to microwave ovens, automobiles to electric toothbrushes. Historically, the fabrication of microchips, via nanotechnology, has required the use of hazardous chemicals and the consumption of large amounts of purified water. SC Fluids Inc., working in collaboration with Los Alamos National Laboratories, designed a greener alternative to the nanofabrication that replaces the hazardous volatile organic solvents with a supercritical carbon dioxide (CO_2) resist remover, SCORR. SC Fluids, Inc. was awarded the Presidential Green Chemistry Challenge Small Business Award in 2002 for this development. Research the term "SC Fluids Scorr" online and determine which of the 12 principles of green chemistry are addressed by this technology. Why would a microchip manufacturing company be interested in adopting the **SCORR** technology? Why is carbon dioxide a greener solvent as compared to volatile organic compounds? In Chapter 17, you will learn about another commercial application of supercritical carbon dioxide—dry cleaning.

COMMUNICATE YOUR RESULTS

Exercise 1
Get the Word Out
Many products of nanotechnology exist on the consumer market, but most people are not aware of their existence or benefits. Choose one item from the list of products that you discovered in Investigation 1 that incorporate nanotechnology, and research the product to uncover the benefits. Create a one-page pamphlet that you could give to your friends and family members to educate them about nanotechnology in the consumer product. If the information is available, discuss how green chemistry has impacted the creation, development, or manufacture of the product.

Exercise 2
Educate Your Community
While the burgeoning field of nanotechnology is embracing green chemistry and green technology, many members of our global community have not yet appreciated the implications of adopting such methods. Craft a one-page letter to the editor of your local newspaper discussing the importance and need for adopting green practices specifically with respect to the field of nanotechnology.

Critical Thinking

Critical Thinking Exercises. Critical thinking is introduced in Chapter 1 and carried throughout the text. At the end of every chapter is an expanded set of Critical Thinking Exercises that encourage students to think critically about and evaluate the most up-to-date, relevant issues. These exercises require the student to apply information and learning from the chapter in both concrete and abstract ways. ▶

Critical Thinking Exercises

Apply knowledge that you have gained in this chapter and one or more of the FLaReS principles (Chapter 1) to evaluate the following statements or claims.

6.1 Suppose that someone has published a paper claiming a new value for Avogadro's number. The author says that he has made some very careful laboratory measurements and his calculations indicate that the true value for the Avogadro constant is 3.01875×10^{23}. Is this claim credible in your opinion? What questions would you ask the person about his claim?

6.2 A chemistry teacher asked his students, "What is the mass, in grams, of a mole of bromine?" One student said "80"; another said "160"; and several others gave answers of 79, 81, 158, and 162. The teacher stated that all of these answers were correct. Do you believe his statement?

6.3 Some automobile tire stores claim that filling your car tires with pure, dry nitrogen is much better than using plain air. They make the following claims: (1) The pressure inside nitrogen-filled tires does not rise or fall with temperature changes. (2) Nitrogen leaks out of tires much more slowly than air because the nitrogen molecules are bigger. (3) Nitrogen is not very reactive, and moisture and oxygen in air cause corrosion that shortens tire life by 25 to 30%. Use information you have gained in this chapter and from other sources as necessary to evaluate these claims.

6.4 A Web site on fireworks provides directions for preparing potassium nitrate, KNO_3 (molar mass 101 g/mol), using potassium carbonate (138 g/mol) and ammonium nitrate (80 g/mol), according to the following equation:

$$K_2CO_3 + 2\,NH_4NO_3 \longrightarrow 2\,KNO_3 + CO_2 + H_2O + NH_3$$

The directions claim that one should "mix one kilogram of potassium carbonate with two kilograms of ammonium nitrate. The carbon dioxide and ammonia come off as gases, and the water can be evaporated, leaving two kilograms of pure potassium nitrate." Use information from this chapter to evaluate this claim.

6.5 A battery manufacturer claims that lithium batteries deliver the same power as batteries using nickel, zinc, or lead, but the lithium batteries are much lighter because lithium has a lower atomic mass than does nickel, zinc, or lead.

COLLABORATIVE GROUP PROJECTS

Prepare a PowerPoint, poster, or other presentation (as directed by your instructor) for presentation to the class.

94. List five chemical activities you have engaged in today. Compare your list with that of the other members of your group.

95. Prepare a brief biographical report on one of the following and share it with your group.
    a. Francis Bacon
    b. Rachel Carson
    c. Thomas Malthus
    d. George Washington Carver

    (The following problem is best done in a group of four students.)

96. Make copies of the following form. Student 1 should write a word from the list in the first column of the form and its definition in the second column. Then she or he should fold the first column under to hide the word and pass the sheet to Student 2, who uses the definition to determine what word was defined and place that word in the third column. Student 2 then folds the second column under to hide it and passes the sheet to Student 3, who writes a def-

inition for the word in the third column and then folds the third column under and passes the form to Student 4. Finally, Student 4 writes the word corresponding to the definition given by Student 3.

Compare the word in the last column with that in the first column. Discuss any differences in the two definitions. If the word in the last column differs from that in the first column, determine what went wrong in the process.
    a. hypothesis   b. theory
    c. mixture   d. substance

Text Entry	Student 1		Student 2	Student 3		Student 4
Word	Definition		Word	Definition		Word

◀ **NEW Collaborative Group Projects.** These end-of-chapter exercises, which extend student learning of chemistry beyond the text, are available for instructors who want to engage students in collaborative work with group assignments.

Conceptual Problem Solving

Conceptual Examples guide students through the process of learning and understanding important chemical concepts.

- Each Example shows a title indicating the skill being covered.
- Solutions are expanded with more explanation to guide the students through solving the problem.
- **A and B Exercises** are in almost all of the Examples. The **A** Exercise is entirely parallel to the Example; the **B** Exercise requires the student to incorporate information from earlier material. The dual Exercises help the students synthesize their learning into a coherent whole rather than just learning isolated facts.
- **Voice balloons** show students the logic of the problem-solving process. ▶

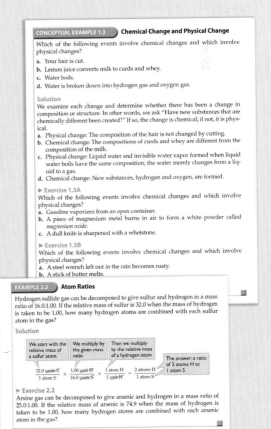

CONCEPTUAL EXAMPLE 1.3   Chemical Change and Physical Change

Which of the following events involve chemical changes and which involve physical changes?
a. Your hair is cut.
b. Lemon juice converts milk to curds and whey.
c. Water boils.
d. Water is broken down into hydrogen gas and oxygen gas.

Solution

We examine each change and determine whether there has been a change in composition or structure. In other words, we ask "Have new substances that are chemically different been created?" If so, the change is chemical; if not, it is physical.
a. Physical change: The composition of the hair is not changed by cutting.
b. Chemical change: The compositions of curds and whey are different from the composition of the milk.
c. Physical change: Liquid water and invisible water vapor formed when liquid water boils have the same composition; the water merely changes from a liquid to a gas.
d. Chemical change: New substances, hydrogen and oxygen, are formed.

▶ Exercise 1.3A
Which of the following events involve chemical changes and which involve physical changes?
a. Gasoline vaporizes from an open container.
b. A piece of magnesium metal burns in air to form a white powder called *magnesium oxide*.
c. A dull knife is sharpened with a whetstone.

▶ Exercise 1.3B
Which of the following events involve chemical changes and which involve physical changes?
a. A steel wrench left out in the rain becomes rusty.
b. A stick of butter melts.

EXAMPLE 2.2   Atom Ratios

Hydrogen sulfide gas can be decomposed to give sulfur and hydrogen in a mass ratio of 16.0:1.00. If the relative mass of sulfur is 32.0 when the mass of hydrogen is taken to be 1.00, how many hydrogen atoms are combined with each sulfur atom in the gas?

Solution

We start with the relative mass of a sulfur atom | We multiply by the given mass ratio | Then we multiply by the relative mass of a hydrogen atom | The answer: a ratio of 2 atoms H to 1 atom S

$$\frac{32.0 \text{ units S}}{1 \text{ atom S}} \times \frac{1.00 \text{ unit H}}{16.0 \text{ units S}} \times \frac{1 \text{ atom H}}{1 \text{ unit H}} = \frac{2 \text{ atoms H}}{1 \text{ atom S}}$$

▶ Exercise 2.2
Arsine gas can be decomposed to give arsenic and hydrogen in a mass ratio of 25.0:1.00. If the relative mass of arsenic is 74.9 when the mass of hydrogen is taken to be 1.00, how many hydrogen atoms are combined with each arsenic atom in the gas?

Visualization

NEW **Illustrations** using both microscopic (molecular) and macroscopic (visual) views help students visualize chemical phenomena. ▼

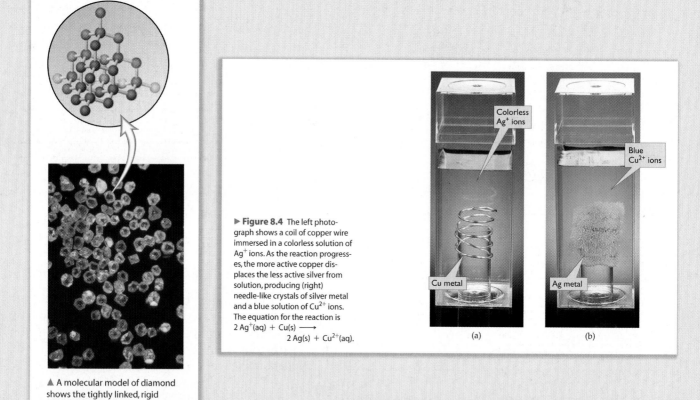

▶ **Figure 8.4** The left photograph shows a coil of copper wire immersed in a colorless solution of Ag^+ ions. As the reaction progresses, the more active copper displaces the less active silver from solution, producing (right) needle-like crystals of silver metal and a blue solution of Cu^{2+} ions. The equation for the reaction is
$$2\,Ag^+(aq) + Cu(s) \longrightarrow 2\,Ag(s) + Cu^{2+}(aq).$$

Colorless Ag^+ ions

Blue Cu^{2+} ions

Cu metal

Ag metal

(a)

(b)

▲ A molecular model of diamond shows the tightly linked, rigid structure that explains why diamonds are so hard.

NEW **Questions** shown at the end of some of the figure captions direct the student to the things that are particularly important to visualize and to expand on the concept illustrated in the figure or photograph. ▼

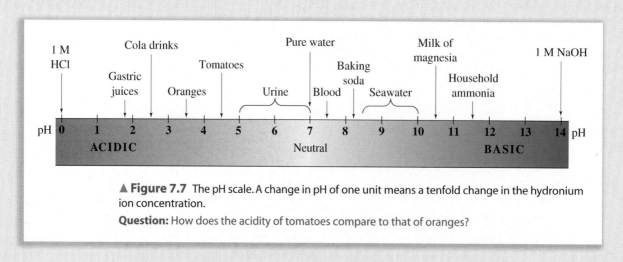

1 M HCl

Cola drinks

Gastric juices

Tomatoes

Oranges

Pure water

Urine

Blood

Baking soda

Seawater

Milk of magnesia

Household ammonia

1 M NaOH

pH 0   1   2   3   4   5   6   7   8   9   10   11   12   13   14 pH

ACIDIC   Neutral   BASIC

▲ **Figure 7.7** The pH scale. A change in pH of one unit means a tenfold change in the hydronium ion concentration.

Question: How does the acidity of tomatoes compare to that of oranges?

SUMMARY

Section 3.1—Davy, Faraday, and others showed that matter is electrical in nature. They were able to decompose compounds into elements by electrolysis or passing electricity through molten salts. Electrodes are the pieces of metal or carbon that carry electricity through the electrolyte or solution/compound undergoing electrolysis. The electrolyte contains ions—charged atoms or groups of atoms. The anode is the positive electrode and anions or negatively charged ions move toward it. The cathode is the negative electrode and cations or positively charged ions move toward it. Experiments with cathode ray tubes showed that matter contained negatively charged particles, which were called electrons. By deflecting cathode rays with a magnet, Thomson was able to determine the mass-to-charge ratio for the electron. Goldstein's experiment, using gas discharge tubes with perforated cathodes, showed that matter also contained positively charged particles. Millikan's oil drop experiment measured the charge on the electron, so its mass could then be calculated. That mass was found to be smaller than that of the lightest atom.

Sections 3.2–3.3—In his studies of cathode rays, Roentgen accidentally discovered X-rays, which are a highly penetrating form of radiation used in medical diagnosis. Becquerel accidentally discovered another type of radiation that comes from certain unstable elements. Marie (Sklodowska) Curie named this new discovery radioactivity and studied it extensively, discovering two elements in the process and winning two Nobel Prizes. Radioactivity was soon found to be three different types of radiation: Alpha particles have four times the mass of a hydrogen atom and a positive charge twice that of an electron. Beta particles are energetic electrons. Gamma rays are a form of energy like X-rays but more penetrating.

Section 3.4—Rutherford's later experiments with alpha particles and gold foil showed that most of the alpha particles emitted toward the foil passed through it. A few were deflected or, occasionally, bounced almost directly back to the source. This indicated that all the positive charge and most of the mass of an atom must be in a tiny core, which he called the nucleus.

Alpha radiation

Section 3.5—Rutherford called the unit of positive charge the proton, with about the mass of a hydrogen atom and with a charge equal in size but opposite in sign to the electron. In 1932 Chadwick discovered the neutron, a nuclear particle

as massive as a proton but with no charge. The number of protons in an atom is the atomic number. Atoms of the same element have the same number of protons but may have different numbers of neutrons; such atoms are called isotopes. Different isotopes of an element are mostly identical chemically. Protons and neutrons collectively are called nucleons, and the number of nucleons is called the mass number or nucleon number. The difference between the nucleon number and the atomic number is the number of neutrons. The general symbol for an isotope of element X is written $_{Z}^{A}X$, where A is the nucleon number and Z is the atomic number.

Section 3.6—Light from the sun or from an incandescent lamp is a continuous spectrum, containing all colors. Light from a gas discharge tube is a line spectrum, containing only certain colors. Bohr explained line spectra by proposing that an electron in an atom could have only certain discrete energy levels, with these levels differing by a quantum or discrete unit of energy. An atom drops to the lowest energy state or ground state after being in a higher-energy state or excited state. Line spectra occur from these changes in energy levels giving off specific wavelengths or colors of light. Bohr also deduced that the energy levels of an atom could handle at most $2n^2$ electrons, where n is the energy level or shell number. A description of the shells occupied by electrons of an atom is one way of giving the atom's electron configuration or arrangement of electrons.

Section 3.7—de Broglie theorized that electrons have wave properties. Schrödinger developed equations that described each electron's location in terms of an orbital, the volume of space that an electron usually occupies. Each orbital holds at most two electrons. Orbitals in the same shell and with the same energy make up a subshell, each of which is designated by a letter. An s orbital is spherical and a p orbital is dumbbell shaped. The d and f orbitals have more complex shapes. The first main shell can hold only one s orbital; the second level can hold one s and three p orbitals; and the third level can hold one s, three p, and five d orbitals. In writing an electron configuration using subshell notation (quantum mechanical notation), the shell number and subshell letter are given, followed by a superscript giving the number of electrons in that subshell. The order of the shells and subshells can be remembered with a chart or by using the periodic table.

Sections 3.8—Elements in the periodic table are arranged in columns or groups, and in rows or periods. Electrons in the outermost shell of an atom are called valence electrons and

---

◄ Chapter Summaries. The chapter summaries are organized in a new format, presented by sections, with key terms highlighted in red for easy recognition. Figures and photographs are included in the summaries to aid the visual learner in revisiting important concepts.

Applications that Focus on the Environment

NEW Applications Added essays include many interesting, relevant, and environmental applications. ▼

A Closed Ecosystem?

If you think for a moment about the oxygen and nitrogen cycles, you may realize that a *closed ecosystem* should be possible. In such a system, the plant and animal life would be balanced. The O_2 given off by the plants would sustain the animal life, and the CO_2 given off by the animals would be sufficient to keep the plants alive. Other animal wastes would supply the nitrogen needed by the plants, and the animals would partially consume the plants or one another for food. Trace elements would need to be present in sufficient quantity. Sunlight would provide the energy necessary to "run" the ecosystem; nothing else goes in and nothing comes out.

Earth is itself a closed ecosystem. The planet absorbs sunlight energy, but no significant amount of matter enters or leaves. But can such a system be constructed artificially? The answer is of great interest in a number of fields, including space exploration. A long space voyage would require that huge amounts of resources be carried along—at enormous expense. The required resources could be minimized by carrying along plants to provide oxygen and consumer carbon dioxide and human wastes.

Tiny ecosystems are available commercially as desktop novelties. These systems have a limited lifetime of a few years. *Biosphere II* was a large-scale experiment involving humans and a semiclosed ecosystem.

▲ The Ecosphere® is a sealed, self-contained ecosystem in which plants and animals can live for a year or more, with sunlight as the only "input."

The experiment was cut short because of several problems, including rising levels of nitrogen oxides, water pollution, unforeseen low levels of carbon dioxide, and other problems that could only be examined in such an experiment. Although additional studies and experiments are slated, Earth remains the only long-term successful closed ecosystem.

Student Media Resources

Chemistry for Changing Times excels at identifying the connections between our world and the chemistry that surrounds us. In this edition, John Hill and Doris Kolb continue their legacy of making chemistry exciting and interesting for thousands of students.

Chemistry for Changing Times offers a full package of textbook and Web resources. The Companion Website with GradeTracker (*http://www.prenhall.com/hill*) is designed to complement the text through the use of multiple review, practice, exploration, and assessment activities. Instructors can choose to assign any of the self-grading activities for credit or simply allow students to work at their own pace toward mastery.

Companion Website with GradeTracker

NEW Key Concepts summary statements identify what all students should Know, Understand, and Be Able To do after reading each chapter. The accompanying online glossary means a quick review is only a click away! ▶

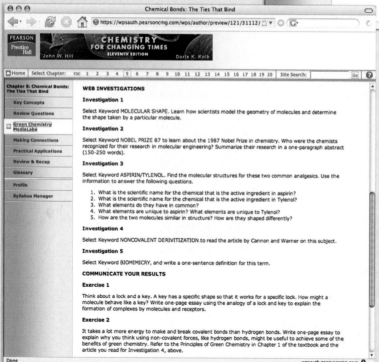

Green Chemistry

◀ **NEW** Online versions of the Green Chemistry activities found in the textbook promote awareness through online research and encourage students to present findings using a variety of strategies.

NEW Review questions help students assess their understanding of key concepts, offering helpful hints and instant feedback. ▶

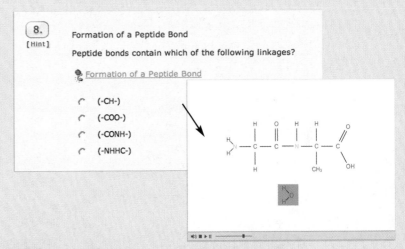

6. **CORRECT** The experiment of what scientist determined that an atom has a tiny, very dense nucleus with electrons occupying most of the space of the atom?

Your Answer: Rutherford

The results of Rutherford's experiment were quite surprising.

7. **INCORRECT** How many electrons can exist in the $n = 3$ shell?

Your Answer: 8
Correct Answer: 18

The $n = 2$ shell can only hold 8 electrons.

8. **CORRECT** In an oxygen-18 isotope, there are 8 protons and 10 neutrons.

Your Answer: True

Oxygen has an atomic number of 8. In an isotope, the number of neutrons varies, but the number of protons remains the same.

[Hint]

Review & Recap
Hint for Question 7

Remember that the quantum number n is equal to the
of shells and is equal to the number of subshells (s, p,
Do you remember the Pauli Principle?

NEW Critical thinking and application activities help bridge the gap between the textbook and real-world chemistry issues and encourage individual and collaborative research.

Many of the Review Questions have been enhanced with media—animations, movies, and 3D molecules—to help students visualize the concepts presented in the text. Follow the media icons (🎬) from the text to the Companion Website with GradeTracker. ▶

8. **[Hint]** Formation of a Peptide Bond

Peptide bonds contain which of the following linkages?

🎬 Formation of a Peptide Bond

- ◌ (-CH-)
- ◌ (-COO-)
- ◌ (-CONH-)
- ◌ (-NHHC-)

Instructor Media Resources

Classroom Presentation Tools

💿 **Instructor's Resource on CD/DVD**
(0-13-199003-9)

This searchable, integrated book-specific lecture resource features almost all the art from the text, including tables; printable, high-resolution PDF files of all included art; several prebuilt PowerPoint presentations for each chapter, as well as Classroom Response System ("Clicker Questions") slides; the interactive animations, movies, and 3D molecules from the Companion Website; Worked Examples; a Test Item File and TestGen test-generation software; and fully editable lecture outlines from the Instructor's Resource Manual—all in one convenient-to-use resource.

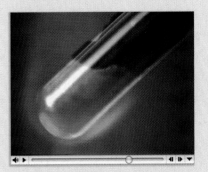

Online Homework/Course Management Options

Prentice Hall offers three content cartridges for online, text-specific course management systems depending on your preferred platform. Hundreds of text-specific problems are provided.

Visit *www.prenhall.com/demo* for details on how to communicate with your students online, customize content to meet your course needs, create online quizzes and tests, track grades, and much more.

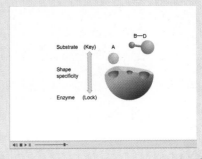

CHAPTER 1

Chemistry

Chemistry is everywhere, not just in a laboratory. Chemistry occurs in soil and rocks, in waters, in clouds, and in us.

A Science for All Seasons

Look around you. Everything you see is made of chemicals. Chemistry is involved with the food we eat, the air we breathe, the clothes we wear, the medicines we take, the vehicles we ride in, and the buildings we live and work in.

Everything we *do* also involves chemistry. Whenever we eat a sandwich, bathe, drive a car, listen to music, or ride a bicycle, we use chemistry. Even when we are asleep, chemical reactions go on constantly throughout our bodies.

Most developments in health and medicine involve a lot of chemistry. The astounding advances in biotechnology—decoding the human genome, developing new drugs, improving nutrition, and much more—have a huge chemical component. Understanding environmental problems requires a knowledge of chemistry. The great worldwide issues of global warming and ozone depletion involve chemistry.

The news media often carry reports about toxic chemicals and cancer-causing chemicals polluting our air, water, and food. Half a century ago the media rarely mentioned such health risks. Rather, they emphasized the wonders of chemical technology: miracle drugs, amazing plastics, fertilizers, detergents, and more. Has chemistry changed over the years, or just our perception of it?

Certain chemicals do indeed cause problems, but many others are extremely helpful. Some chemicals kill bacteria that cause dreadful diseases; some relieve our pain and suffering; some increase our food production; some provide fuel for our heating, cooling, lighting, and transportation; and some provide materials for building our machines, making our clothing, and constructing our houses. Chemistry has provided ordinary people with luxuries that were not available even to the mightiest of kings in ages past. Chemicals are highly important to our lives—so much so that life itself would be impossible without chemicals.

So what is chemistry anyway? According to the usual definition, **chemistry** is the study of matter and the changes it undergoes. *Matter* is anything that has *mass*, which means that if you can weigh it, it is matter. Even though you cannot see air, it has mass; therefore, air is matter. Matter is the stuff of which all material things are made. The science of chemistry deals with every kind of matter, from the tiniest parts of atoms to the most complex materials in living plants and animals. Thus, the word *chemical* may sound ominous, but it is nothing more than a name for matter—any and all matter. Gold, water, salt, sugar, coffee, ice cream, a computer, a pencil—all are chemicals or are made entirely of chemicals.

▲ "I don't use chemicals in *my* garden."

▲ "Mom won't let me drink that stuff. It has chemicals in it."

1

▲ Aristotle (384–322 B.C.E.), Greek philosopher and tutor of Alexander the Great, believed that we could understand nature through logic. The idea of experimental science did not triumph over Aristotelian logic until about C.E. 1500.

We use the more universal designations C.E. and B.C.E. rather than the perhaps more familiar A.D. and B.C. Common Era (C.E.) is used when referring to the past 2000 years. Before the Common Era (B.C.E.) is used for more ancient times.

▼ A late sixteenth-century painting, titled *The Alchemist*, by Giovanni Stradano, showing an alchemist in his laboratory. (Source: Giovanni Stradano, The Alchemist. Studiolo, Palazzo Vecchio, Florence, Italy. Scala/Art Resource, New York)

Then there are those *changes* that matter undergoes. Sometimes matter changes on its own, as when a tool left out in the wet grass gets rusty. But often we change matter to make it more useful, as when we light a candle or cook an egg. Most changes in matter are accompanied by changes in energy. For example, when we burn gasoline, the reaction gives off energy that we can use to propel an automobile or a lawn mower.

Your own body is an incredibly marvelous chemical factory. It takes the food you eat and turns it into skin, bones, blood, and muscle, while also generating energy for all your many activities. Your body is an amazing chemical plant that operates continuously 24 hours a day for as long as you live. Chemistry not only affects your own individual life every moment, but it also transforms society as a whole. Chemistry shapes our civilization.

1.1 SCIENCE AND TECHNOLOGY: THE ROOTS OF KNOWLEDGE

Chemistry is a *science*, but what is science? Let's examine the roots of science. Our study of the material universe has two facets: the *technological* (or *factual*) and the *philosophical* (or *theoretical*).

Technology, the application of knowledge for practical purposes, arose long before science and originated in prehistoric times. Early people used fire to bring about chemical changes. For example, they cooked food, baked pottery, and smelted ores to produce metals such as copper. They made beer and wine by fermentation and obtained dyes and drugs from plant materials. These things—and many others—were accomplished without an understanding of the scientific principles involved.

The Greek philosophers, about 2500 years ago, were perhaps the first to formulate *theories*—explanations of the behavior of matter. They generally did not test their theories by experimentation, however. Nevertheless, their view of nature—attributed mainly to Aristotle—dominated natural philosophy for 2000 years.

The experimental roots of chemistry are planted in **alchemy**, a mystical mixture of chemistry and magic that flourished in Europe during the Middle Ages, from about C.E. 500 to 1500. Alchemists searched for a "philosophers' stone" that would turn cheaper metals into gold, and they sought an elixir that would confer immortality on those exposed to it. Alchemists never achieved these goals, but they discovered many new chemical substances and perfected techniques such as distillation and extraction that are still used today. Modern chemists inherited from the alchemists an abiding interest in human health and the quality of life.

Technology also developed rapidly during the Middle Ages in Europe even though it was not aided by the Aristotelian philosophy that prevailed. The beginnings of modern science were more recent, however, and coincided with the emergence of the experimental method. What we now call science grew out of **natural philosophy**—that is, out of philosophical speculation about nature. Science had its true beginnings in the seventeenth century, when astronomers, physicists, and physiologists began to rely on experimentation.

Modern Alchemy

Modern chemists can transform matter in ways that would astound the alchemists. They can change ordinary salt into lye and laundry bleach. They can convert sand into transistors and computer chips. And they can turn crude oil into plastics, fibers, pesticides, drugs, detergents, and a host of other products. Many products of modern chemistry are much more valuable than the glittering metal that the alchemists vainly sought. For example, a relatively new form of carbon, called *nanotubes* because of their size and shape, is worth many times more than gold. Nanotubes have thousands of potential uses. A pound of purified nanotubes costs about $200,000. A pound of gold costs only about $5000.

1.2 THE BACONIAN DREAM AND THE CARSONIAN NIGHTMARE

Francis Bacon was a philosopher who practiced law and served as a judge. Although not a scientist, he was interested in science and argued that it should be experimental. His dream was that science could solve the world's problems and enrich human life with new inventions, thereby increasing happiness and prosperity.

By the middle of the twentieth century, science and technology appeared to have made the Baconian dream come true. Many dread diseases, such as smallpox, polio, and plague, had been almost eliminated. The use of fertilizers, pesticides, and scientific animal and plant breeding had increased and enriched our food supply. New materials provided us better clothing and shelter, swifter transportation, and nearly instantaneous communication. Nuclear energy seemed to promise unlimited power for our every need. Science and technology had done much toward creating our "modern" world.

The Baconian dream has lost much of its luster in recent decades. People have learned that the products of science are not an unmitigated good. Some have predicted that science might bring not wealth and happiness, but death and destruction.

Rachel Carson, a biologist, was among the more noteworthy critics of modern technology. Her poetic and polemic book *Silent Spring* was published in 1962. The book's main theme is that through our use of chemicals to control insects, we are threatening the destruction of all life, including ourselves. People in the pesticide industry and their allies roundly denounced Carson as a "propagandist," while other scientists rallied to her support. By the late 1960s, however, massive fish kills, the threatened extinction of several species of birds, and the disappearance of fish from rivers, lakes, and areas of the ocean that had long been productive had caused many scientists to move into Carson's camp. Popular support for Carson's views was overwhelming.

Carson was not the first prophet of doom. As early as 1798, Thomas Malthus had predicted in his "Essay upon the Principles of Population," that the rapid increase in world population would outpace the increase in food supply and result in widespread famine. During the nineteenth century and much of the twentieth century, however, technological developments enabled food production to keep up with population growth, at least in developed countries.

Today the picture has changed. In spite of declining birth rates in the developed countries, population growth in much of the rest of the world still threatens to overtake even the most optimistic projections of food production. Some scientists project a dismal future for our world; others confidently predict that science and technology, properly applied, will save us from disaster.

▲ Francis Bacon (1561–1626), English philosopher and Lord Chancellor to King James I.

▲ Rachel Carson (1907–1964) at Woods Hole, Massachusetts, in 1951.

1.3 SCIENCE: TESTABLE, REPRODUCIBLE, EXPLANATORY, PREDICTIVE, AND TENTATIVE

What *is* science? If scientists disagree about what is and what will be, is science merely a guessing game in which one guess is as good as another? Science is difficult to define precisely, but we will try to describe it.

Scientific Hypotheses

Science is an accumulation of knowledge about nature and our physical world and the theories that we use to explain that knowledge. Science is based on observations and experimental tests of our assumptions. Science is not simply a collection of unalterable facts. We cannot force nature to fit our preconceived ideas. Science is good at correcting errors, but it is not especially good at establishing truths. Science is an unfinished work. The things we have learned from science fill millions of books and scholarly journals, but what we know pales in comparison to what we don't yet know.

Example of a "fact" that may be discarded: It was long thought that exercise caused muscles to tire by a build-up of lactic acid. This acid was also thought to cause soreness after exercise. Recent findings indicate that lactic acid actually *delays* muscle tiredness. The cause of muscle tiring and soreness is now thought to be complex, perhaps related in part to an excess of potassium ions outside the muscle cells.

Scientists collect data by making careful observations. Data reported by a scientist must also be observable by other scientists; the data must be *reproducible*. Careful observations and measurements are essential, but scientific work is not fully accepted until it has been verified by other scientists.

Scientists develop *testable* **hypotheses** (guesses) as tentative explanations of observed data and test these hypotheses by designing and performing experiments. This is the main thing that distinguishes science from the arts and humanities: The tenets of science are *testable*. In the humanities, people often still argue about some of the same questions that were being debated thousands of years ago: What is truth? What is beauty? These arguments persist because the answers cannot be tested and confirmed objectively. However, experiments can be devised to answer most scientific questions. Ideas can be tested and thereby either verified or rejected. As a result, scientists have established a firm foundation of knowledge so that each new generation can build on the past.

Five characteristics of science are that it is

1. *testable*,
2. *reproducible*,
3. *explanatory*, and
4. *predictive*, but always
5. *tentative*.

What Science Is Not

News media generally try to be fair, presenting both sides of an issue. Science is not fair; not all ideas are considered to be equal. Only ideas that have survived experimental testing or that can be tested by experiment are considered valid. Ideas that are beautiful, elegant, or even sacrosanct can be destroyed by experimental data.

Science is not a democratic process. Majority rule does not determine what is sound science. Science does not accept notions that are proven false by experiment.

Scientific Laws

Large amounts of scientific data can sometimes be summarized in brief statements called **scientific laws**. For example, Robert Boyle (1627–1691), an Englishman, conducted many experiments on gases. In each experiment, he found the volume of the gas to decrease when the pressure applied to the gas was increased. Many scientific laws can be stated mathematically. For example, Boyle's law can be written as $PV = k$, where P is the pressure on a gas, V is its volume, and k is a constant number. If P is doubled, V will be cut in half. Scientific laws are *universal*; under the stated conditions they hold everywhere in the observable universe.

Experimental observations are just the beginning of the intellectual processes of science. There are many different paths to scientific discovery, and there is no general set of rules. Science is not just a straightforward logical process for cranking out discoveries.

A Possible Scientific Process

Observation reported
↓
Observation confirmed by others
↓
Hypothesis suggested
↓
Experiments designed to test hypothesis
↓
Experiments unsuccessful
↓
Hypothesis rejected
↓
New hypothesis offered
↓
New experiments tried
↓
Experiments successful
↓
Experiments repeated and results confirmed
↓
Theory formulated
↓
Many further experiments
↓
and so on

Scientific Theories

Science is a body of knowledge. Scientists organize this knowledge on a framework of detailed explanations called **theories**. The theories represent the best current explanations for various phenomena, but they are always *tentative*. In the future, a theory may have to be modified or even discarded in the light of new observations. Science is a body of knowledge that is rapidly growing and always changing.

Theories provide organization for scientific knowledge, but they also are useful for their *predictive* value. Predictions based on theories are then tested by further experiments. Theories that make successful predictions generally are widely accepted by the scientific community. A theory developed in one area is often found to apply in others.

Scientific Models

Scientists often use models to help explain complicated phenomena. A **scientific model** uses tangible items or pictures to represent invisible processes. For example, the invisible particles of a gas can be visualized as billiard balls or marbles, or as

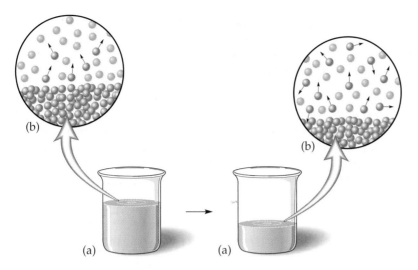

= Water molecule

= Air (nitrogen or oxygen) molecule

◀ Figure 1.1 The evaporation of water. (a) When a container of water is left standing open to the air, the water slowly disappears. (b) Scientists explain evaporation in terms of the motion of molecules.

In assessing data, it is important to note that a *correlation* between two items isn't necessarily evidence that one *causes* the other. For example, many people have allergies in the fall when goldenrod is in bloom. However, research has shown that the main cause of these allergies is ragweed pollen. There is a correlation between the goldenrod blooming and autumnal allergies, but goldenrod pollen is not the cause. Ragweed happens to bloom at the same time.

dots or circles on paper. We know that when a glass of water is left standing for a period of time, the water disappears through a process called evaporation (Figure 1.1). Scientists use a theory, termed the kinetic–molecular theory, based on tiny, invisible particles called *molecules*, which are in constant motion. To explain evaporation, the kinetic–molecular model pictures the liquid water as molecules. In the bulk of the liquid, these molecules are held together by forces of attraction. The molecules collide with one another like billiard balls on a playing table. Sometimes a "hard break" of billiard balls causes one ball to fly off the table. Likewise, some of the molecules of a liquid gain enough energy through collisions to break the attraction to their neighbors, escape from the liquid, and disperse among the widely spaced air molecules. The water in the glass gradually disappears. It is much more rewarding to understand evaporation through the use of a model than merely to have a name for it.

Molecular Modeling

Molecular models are three-dimensional representations of molecules. They can be put together using simple balls to represent atoms and sticks to connect the balls. The models can also be made with various space-filling model kits, or they can be computer generated. Using a computer, one can draw various kinds of molecules and then rotate them on the screen to get a three-dimensional perspective.

Some molecular modeling programs can compute the structures of complicated molecules, even indicating the distribution of electrons. These computational programs can be used to look at individual molecules or to study their interactions with each other. Computer modeling is especially useful in the study of complex biological molecules and drug molecules.

1.4 THE LIMITATIONS OF SCIENCE

We sometimes hear people say that we could solve all our problems if we would only attack them using the scientific method. We have seen already that there is no single scientific method, but why can't the procedures of the scientist be applied to social, political, ethical, and economic problems? Why do scientists disagree over environmental, social, and political issues?

Often the disagreement results from the inability to control *variables*. A **variable** is something that can change over the course of an experiment. If, for example, we wanted to study in the laboratory how the volume of a gas varies with changes in

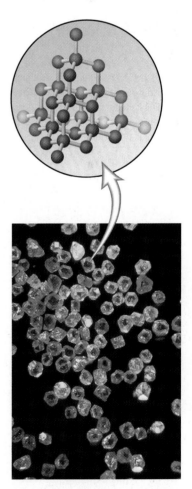

▲ A molecular model of diamond shows the tightly linked, rigid structure that explains why diamonds are so hard.

Historically, scientists have tried to vary only one quantity at a time, holding most factors constant. Today, with computers to do the complicated mathematics, scientists can often vary several factors at the same time and still determine the effect of each variable.

pressure, we would hold constant such factors as temperature and the amount and kind of gas. If, on the other hand, we wanted to determine the health effect of low levels of a particular pollutant on a human population, we would find it difficult, if not impossible, to control such variables as the people's diets, habits, and exposure to other substances.

We can make observations, formulate hypotheses, and conduct experiments, even if most of the variables are not subject to control. Interpretation of the results, however, is much more difficult and more subject to disagreement. Nonscientists sometimes use some of the methods and language of scientists. Artists *experiment* with new techniques and new materials. Playwrights and novelists *observe* life as it is before trying to express its essence in their writings. Scientific ideas and methods pervade almost every aspect of society.

1.5 SCIENCE AND TECHNOLOGY: RISKS AND BENEFITS

Science and technology are interrelated. In everyday life, people often fail to distinguish between the two. A distinction can be useful, however. *Technology* is the sum total of the processes by which humans modify the materials of nature to better satisfy their needs and wants. These processes need not be based on scientific principles. For example, prehistoric people used technology; they smelted ores to produce metals such as copper and iron without understanding the chemistry involved.

Most people recognize that society has benefited from science and technology, but there are risks associated with technological advances. How can we determine when the benefits outweigh the risks? One approach, called **risk–benefit analysis**, involves the estimation of a desirability quotient (DQ).

$$DQ = \frac{\text{Benefits}}{\text{Risks}}$$

If benefits are large and risks are small, the DQ is large.
If benefits are small and risks are large, the DQ is small.
If both benefits and risks are large, the DQ is uncertain.
If both benefits and risks are small, the DQ is uncertain.

A *benefit* is anything that promotes well-being or has a positive effect. Benefits may be economic, social, or psychological. A *risk* is any hazard that leads to loss or injury. Some of the risks in modern technology have led to disease, death, economic loss, and environmental deterioration. Risks may involve one individual, a group, or society as a whole.

Most people recognize the automobile as a technological development that benefits us greatly. But driving a car involves risk, including the risk of injury or death in a traffic accident. Most people consider the benefits of driving a car to outweigh the risks.

Weighing the benefits and risks connected with a product is more difficult when one considers a group of people. For example, pasteurized milk is a safe, clean, nutritious beverage for many people of northern European descent. However, a few people in this group can't tolerate lactose, the sugar in milk. And some are allergic to milk proteins. For these groups of people, drinking milk can be harmful. Because milk allergies and lactose intolerance are relatively uncommon among people of northern European descent, the benefits are large and the risks are small, resulting in a large DQ. Skim or low-fat milk is generally beneficial for this group. However, adults in much of the rest of the world are lactose intolerant and would find that milk has a small DQ. Thus, milk is not always suitable for use in programs to relieve malnutrition.

In 1958, the drug thalidomide was introduced in Germany. Many pregnant women took the drug to prevent morning sickness. But soon thalidomide was shown to involve an enormous risk. Many malformed infants were born to women who had taken the drug during pregnancy. For thalidomide, then, there are small benefits, huge risks, and a very small DQ. Thalidomide was eventually judged to present unacceptable risks and was banned. However, there is now a renewed interest in thalidomide. Its use in treating a debilitating skin condition of leprosy was

▲ For most people of northern European ancestry, milk is a wholesome food; its benefits far outweigh its risks. Other people have high rates of lactose intolerance among adults, and the desirability quotient for milk is much smaller.

approved in September 1998. Current research is evaluating the use of thalidomide in treating Kaposi's sarcoma (a complication of AIDS), as well as for graft-versus-host reactions in organ transplants and autoimmune diseases, for ulcers in AIDS, and for tuberculosis.

The artificial sweetener aspartame is the most highly studied of all food additives. There is little evidence that use of an artificial sweetener helps one to lose weight. (Aspartame may provide some benefit to diabetics, however, because sugar consumption presents a large risk to them.) There are anecdotal reports of problems with the sweetener, but these have not been confirmed in controlled studies. To most people, the risk involved in using aspartame is small. This leads to an uncertain DQ—and, it seems, to endless debate over the safety of aspartame.

Other technologies provide large benefits and present large risks. For these technologies, too, the DQ is uncertain. An example is the conversion of coal to liquid fuels. Most people find liquid fuels to provide large benefits in the areas of transportation, home heating, and industry. The risks associated with coal conversion are also large, however. These include the risks to mine workers, air and water pollution, and the exposure of conversion plant workers to toxic chemicals. The result, again, is an uncertain DQ and political controversy.

There are yet other problems in risk–benefit analysis. Some technologies benefit one group of people while presenting a risk to another. For example, it may be economically advantageous to a community to spend as little as possible on a sewage treatment plant and to dump raw wastes into a nearby stream. These wastes might present a hazard to downstream communities, however. Difficult political decisions are needed in such cases.

Other technologies provide current benefits but present future risks. For example, although nuclear power now provides useful electricity, wastes from nuclear power plants, if improperly stored, might present hazards for centuries. Thus, the use of nuclear power is controversial.

Science and technology obviously involve *both* risks and benefits. The determination of benefits is almost entirely a social judgment; risk assessment also involves social decisions, but scientific investigation can help considerably in risk evaluation.

According to the National Safety Council, the chemical industry ranks at or near the top each year in worker safety among 42 basic industries. The number of incidents of occupational illness and injury in the chemical industry is only one-third of the average rate for all U.S. industries. This is in sharp contrast to the popular belief that chemicals are so dangerous. (Some are, of course, but even they can be used safely with proper precautions.)

CONCEPTUAL EXAMPLE 1.1   **Risk–Benefit Analysis**

Some medical doctors think heroin is more effective for the relief of severe pain than other medicines. However, heroin is highly addictive and often renders the user unable to function in society. Do a risk–benefit analysis for the use of heroin in treating the pain of **(a)** a young athlete's broken leg and **(b)** a terminally ill cancer patient.

Solution

a. The heroin would provide the benefit of pain relief, but its use for such purposes has been judged to be too risky by the U.S. Food and Drug Administration. The DQ is low.

b. The heroin would provide the benefit of pain relief. The risk of addiction in a dying person is irrelevant. Heroin is used this way in Great Britain, but it is banned for any purpose in the United States. The DQ is uncertain. (Both answers involve judgments that are not clearly scientific; people can differ in their assessments of each.)

▶ **Exercise 1.1A**

Chloramphenicol is a powerful antibacterial drug that often destroys bacteria unaffected by other drugs. It is highly dangerous to some individuals, however, causing fatal aplastic anemia in about 1 in 30,000 people. Do a risk–benefit analysis for the use of chloramphenicol in **(a)** sick farm animals, from which people might consume milk or meat with residues of the drug, and **(b)** a person with Rocky Mountain spotted fever faced with a high probability of death or permanent disability.

▶ **Exercise 1.1B**

A four-year clinical trial (6800 participants) of the drug tamoxifen in healthy women at high risk showed a 49% lower rate of breast cancer than a group of 6800 women taking a placebo. Tamoxifen has a low rate of serious side effects, including potentially fatal blood clots, uterine cancer, hot flashes, loss of libido, and cataracts. Do a risk–benefit analysis for the use of tamoxifen in **(a)** all women and **(b)** women at high risk.

▶ **Exercise 1.1C**

Rotavirus is the most common cause of severe diarrhea among children. This virus causes hospitalization of approximately 55,000 children each year in the United States and the death of over 600,000 children annually worldwide. A vaccine can prevent most cases. There is a strong association between the vaccine and bowel obstruction in some infants. Do a risk–benefit analysis for the use of the vaccine in **(a)** the United States and **(b)** worldwide.

Cost–Benefit Analysis and Health Care

When buying a new home, car, or major appliance, people usually consider the cost of the purchase against the benefits it will provide. They are much less likely to make such a cost–benefit analysis when entering a hospital or scheduling a surgical procedure. Health care insurers, however, do this for their clients all the time.

Some people complain because they must get approval from their health insurance company before they check into a hospital or schedule a medical treatment. Insurance companies may argue that they must have such approval rights because some treatments are still experimental and certain procedures are too unreliable to justify their high cost. Because the insurers bear the financial risks, but their clients receive any health benefits that result, these two groups have different viewpoints. It is not surprising that they might disagree about the DQ for a particular type of health care.

1.6 CHEMISTRY: ITS CENTRAL ROLE

Science is a unified whole. The various areas of science interact and support one another. Accordingly, chemistry not only is useful in itself but is also fundamental to other scientific disciplines. The application of chemical principles has revolutionized biology and medicine, has provided materials for powerful computers used in mathematics, and has profoundly influenced other fields such as psychology. The social goals of better health and more and better food, housing, and clothing are dependent to a large extent on the knowledge and techniques of chemists. Many modern materials have been developed by chemists and are used by engineers, and even more amazing materials are on the way. Recycling of basic materials—paper, glass, and metals—involves chemical processes. Chemistry is indeed a central science (Figure 1.2). There is scarcely a single area of our daily lives that is not affected by chemistry.

Chemistry is also important to the *economy* of industrial nations. In the United States, chemical process industries make about 70,000 consumer products, including pharmaceutical and personal care products, agricultural products, plastics, coatings, soaps, and detergents. The U.S. chemical industry employs about a million people in about 10,000 plants. It is the nation's fifth largest industry with sales of about $450 billion per year. The exports of the chemical industry help to keep the U.S. international trade deficit from being even worse than it is. The U.S. chemical industry has had a trade surplus most years since 1981.

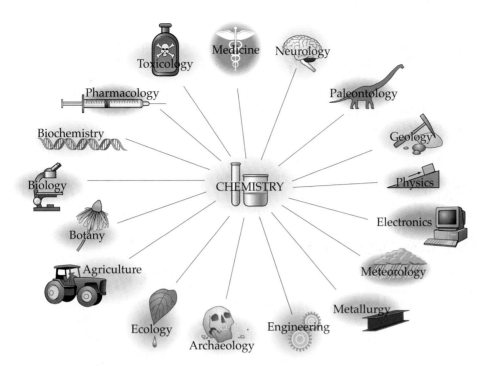

◀ **Figure 1.2** Chemistry has a central role among the sciences.

1.7 SOLVING SOCIETY'S PROBLEMS: SCIENTIFIC RESEARCH

Chemistry is a powerful force in shaping society today. Chemical research not only plays a pivotal role in other sciences, but it also has a profound influence on society as a whole. Chemists, like other scientists, do research in one of two categories, *applied research* or *basic research*. The two often overlap, and it isn't always possible to label a particular project as one or the other.

Applied Research

Most chemists work in the area of applied research. They test polluted soil, air, and water. They analyze foods, fuels, cosmetics, detergents, and drugs. They synthesize new substances for use as drugs or pesticides and formulate plastics for new applications. These activities are examples of **applied research**—work oriented toward the solution of a particular problem in industry or the environment.

Among the most monumental accomplishments in the realm of applied research were those of George Washington Carver. Born in slavery, Carver attended Simpson College and later graduated from Iowa State University. A botanist and agricultural chemist, Carver taught and did research at Tuskegee Institute. He developed over 300 products from peanuts, from hand cleaner to fruit punch to insulating board. He made other new products from sweet potatoes, pecans, and clay. Carver also taught southern farmers to rotate crops and to use legumes to replenish the nitrogen removed from the soil by cotton crops. Carver's work helped to revitalize the economy of the South.

Another example of applied research is the synthesis of an antifungal drug. Many drugs are effective against bacteria, but few are useful and safe against fungal infections in humans. Seeking an effective fungal antibiotic, Rachel Brown (1898–1980) and Elizabeth Hazen (1885–1975) of the New York State Department of Health discovered nystatin, a compound effective against such fungal infections as thrush and vaginitis. Nystatin also has been used in veterinary medicine, in agriculture (to protect fruit), and in the art world (to prevent the growth of mold on art treasures in Florence, Italy, after the objects were damaged by floods). Brown and Hazen set out to find a fungicide and did so—an excellent example of applied research.

▲ George Washington Carver (1860–1943), research scientist, in his laboratory at Tuskegee Institute.

Basic Research: The Search for Knowledge

Many chemists are involved in **basic research**, the search for knowledge for its own sake. Some chemists work out the fine points of atomic and molecular structure. Others measure the intricate energy changes that accompany complex chemical reactions. Chemists synthesize new compounds and determine their properties. This type of investigation is called basic research. Done for the sheer joy of unraveling the secrets of nature and discovering order in our universe, basic research is characterized by the absence of any predictable, marketable product.

Findings from basic research often *are* applied at some point. This may be the hope but is not the primary goal of the researcher. In fact, most of our modern technology is based on results obtained in basic research. Without this base of factual information, technological innovation would be haphazard and slow.

Applied research is carried out mainly by industries seeking a competitive edge with a novel, better, or more salable product. Its ultimate aim is usually profit for the stockholders. Basic research is conducted mainly at universities and research institutes. Most of its support comes from federal and state governments and foundations, although some larger industries also support it.

An example of basic research later applied to improving human welfare is the work of Gertrude Elion (1918–1999) and George Hitchings (1905–1998), who studied compounds called purines in an attempt to understand their role in the chemistry of the cell. Their basic research at Burroughs Wellcome Research Laboratories in North Carolina led to the discovery of a number of valuable new drugs for carrying out successful organ transplants and for treating various diseases such as gout, malaria, herpes, and cancer. Elion and Hitchings shared the 1988 Nobel Prize in Physiology and Medicine, and in 1991 Elion became the first woman to be inducted into the National Inventors Hall of Fame.

▲ Gertrude Elion, a basic research chemist, won the Nobel Prize in 1988 because of a chance discovery.

Importance of Basic Research

There are two compelling reasons why society must support basic science. One is substantial: The theoretical physics of yesterday is the nuclear defense of today; the obscure synthetic chemistry of yesterday is curing disease today. The other reason is cultural. The essence of our civilization is to explore and analyze the nature of man and his surroundings. As proclaimed in the Bible in the Book of Proverbs: "Where there is no vision, the people perish."

Arthur Kornberg (1918–), American Biochemist,
Nobel Prize in Physiology and Medicine, 1959

Spinning Nuclei

An example of basic research later applied to improving human welfare is the work of two physicists, Otto Stern (1888–1969) and Isidor Isaac Rabi (1898–1988). In 1930 they determined that certain atomic nuclei have a property called *spin*, and that in spinning, the nuclei act as tiny magnets. For their basic research in measuring these minuscule magnetic fields and thoroughly studying this phenomenon, Stern, a German, and Rabi, an American, won the Nobel Prize in physics in 1943 and 1944, respectively. In the 1940s teams led by Edward M. Purcell (1912–1997) and Felix Bloch (1905–1983) used the fact that nuclear spin is influenced by its chemical environment (that is, by the atoms around the nucleus) to work out the structure of complicated molecules. This technique, called *nuclear magnetic resonance* (NMR), has become a major tool of chemists in determining molecular structure. Purcell and Bloch shared the Nobel Prize in physics in 1952 for this research. In the 1960s scientists applied the principles of NMR to scans of the human body. This technique, called *magnetic resonance imaging* (MRI), has replaced many exploratory surgical operations with a noninvasive procedure and made other surgery processes much more precise.

1.8 CHEMISTRY: A STUDY OF MATTER AND ITS CHANGES

Earlier we defined chemistry as the study of matter and the changes it undergoes. Because the entire physical universe is made up of nothing more than matter and energy, the field of chemistry extends from atoms to stars, from rocks to living organisms. Now let's look at matter a little more closely.

Matter is the stuff that makes up all material things; it is anything that occupies space and has mass. Matter has *mass*; you can weigh it. Wood, sand, water, air, and people have mass and are therefore matter. **Mass** is a measure of the quantity of matter that an object contains. The greater the mass of an object, the more difficult it is to change its velocity. You can easily deflect a tennis ball coming toward you at 30 meters per second (m/s), but you would have difficulty stopping a cannonball of the same size moving at the same speed. A cannonball has more mass than a tennis ball of equal size.

The mass of an object does not vary with location. An astronaut has the same mass on the moon as on Earth. In contrast, **weight** measures a force. On Earth, it measures the force of attraction between our planet and the mass in question. On the moon, where gravity is one-sixth that on Earth, an astronaut weighs only one-sixth as much as on Earth (Figure 1.3). Weight varies with gravity; mass does not.

CONCEPTUAL EXAMPLE 1.2   **Mass and Weight**

On the planet Mercury gravity is 0.376 times that on Earth. **(a)** What would be the mass on Mercury of a person who has a mass of 62.5 kilograms (kg) on Earth? **(b)** What would be the weight on Mercury of a person who weighs 124 pounds (lb) on Earth?

Solution

a. The person's mass would be the same (62.5 kg) as on Earth; the quantity of matter has not changed.
b. The person would weigh only $0.376 \times 124 \text{ lb} = 46.6 \text{ lb}$; the force of attraction between planet and person is only 0.376 times that on Earth.

▶ **Exercise 1.2A**
At the surface of Venus, the force of gravity is 0.903 times that on Earth's surface. **(a)** What would be the mass of a standard 1.00-kg object on Venus? **(b)** A man who weighs 198 lb on Earth would weigh how much on the surface of Venus?

▶ **Exercise 1.2B**
On Jupiter, at the boundary between the gaseous atmosphere and the liquid that makes up the bulk of the planet, the force of gravity is 2.34 times that on Earth. **(a)** What would be the mass of a 52.5-kg woman at that location on Jupiter? **(b)** A man who weighs 212 lb on Earth would weigh how much on Jupiter?

◀ **Figure 1.3** Astronaut John W. Young leaps from the lunar surface, where gravity pulls at him with only one-sixth the force on Earth.

Question: If an astronaut has a mass of 72 kg, what would be his mass on the moon? If an astronaut weighs 180 lb on Earth, what would he weigh on the moon?

Physical and Chemical Properties

We can use our knowledge of chemistry to change matter to make it more useful. Chemists can change crude oil into gasoline, plastics, pesticides, drugs, detergents, and thousands of other products. Changes in matter are accompanied by changes in energy. Often we change matter to extract part of its energy. For example, we burn gasoline to get energy to propel our automobiles.

To distinguish between samples of matter, we can compare their properties (Figure 1.4). A **physical property** of a substance is a physical characteristic or behavior, such as color, odor, and hardness (Table 1.1). A **chemical property** describes how a substance reacts with other types of matter—how its atomic building blocks can change (Table 1.2).

Chemical properties are inherent in a substance; we observe those properties through chemical change.

▶ **Figure 1.4** A comparison of the physical properties of two elements. Copper, obtained as pellets (upper left), can be hammered into thin foil or drawn into wire (lower left). When hammered, lumps of sulfur (lower right) crumble into a fine powder (upper right).

Question: What additional physical properties of copper and sulfur are apparent from the photograph?

TABLE 1.1	Some Examples of Physical Properties
Property	**Examples**
Temperature	0 °C for ice water, 100 °C for boiling water.
Mass	A nickel weighs 5 g; a penny weighs 2.5 g.
Structure	Ice is crystalline; glass is amorphous.
Color	Sulfur is yellow; bromine is reddish-brown.
Taste	Acids are sour; bases are bitter.
Odor	Benzyl acetate smells like jasmine; hydrogen sulfide smells like rotten eggs.
Boiling point	Water boils at 100 °C; ethyl alcohol boils at 78.5 °C.
Freezing point	Water freezes at 0 °C; methane freezes at −182 °C.
Heat capacity	Water has a high heat capacity; iron has a low heat capacity.
Hardness	Diamond is exceptionally hard; sodium metal is soft.
Conductivity	Copper conducts electricity; diamond does not. Aluminum is a good conductor of heat; glass is a poor heat conductor.
Solubility	Ethyl alcohol dissolves in water; gasoline does not.
Density	1.00 g/mL for water; 19.3 g/cm^3 for gold.

TABLE 1.2	Some Examples of Chemical Properties
Substance	**Typical Chemical Property**
Iron	gets rusty (combines with oxygen to form iron oxide).
Carbon	burns (combines with oxygen to form carbon dioxide).
Silver	tarnishes (combines with sulfur to form silver sulfide).
Nitroglycerin	explodes (decomposes to produce a mixture of gases).
Carbon monoxide	is toxic (combines with hemoglobin, causing anoxia).
Neon	is inert (does not react with anything).

A **physical change** involves an alteration in the physical appearance of matter without changing its chemical identity or composition. An ice cube can melt to form a liquid, but it is still water. Melting is a physical change, and the temperature at which it occurs—the melting point—is a physical property. A **chemical change** involves a change in the chemical identity of matter into other substances that are chemically different. In exhibiting a chemical property, matter undergoes a chemical change; the substance(s) in the original matter is replaced by one or more new substances. Iron metal reacts with oxygen from the air to form rust (iron oxide); when sulfur burns in air, sulfur, which is made up of one type of atom, and oxygen (from air), which is made up of another type of atom, combine to form sulfur dioxide, which is comprised of molecules that have sulfur and oxygen atoms in the ratio 1:2. (A *molecule* is a group of atoms bound together as a single unit. More about atoms and molecules later.)

It is difficult at times to determine whether a change is physical or chemical, but we can decide on the basis of what happens to the composition or structure of the matter involved. *Composition* refers to the types of atoms that are present and their relative proportions, and *structure* to the arrangement of those atoms in particular assemblages or in space. A chemical change results in a change in composition or structure, whereas a physical change does not.

CONCEPTUAL EXAMPLE 1.3   **Chemical Change and Physical Change**

Which of the following events involve chemical changes and which involve physical changes?

a. Your hair is cut.
b. Lemon juice converts milk to curds and whey.
c. Water boils.
d. Water is broken down into hydrogen gas and oxygen gas.

Solution
We examine each change and determine whether there has been a change in composition or structure. In other words, we ask "Have new substances that are chemically different been created?" If so, the change is chemical; if not, it is physical.
a. Physical change: The composition of the hair is not changed by cutting.
b. Chemical change: The compositions of curds and whey are different from the composition of the milk.
c. Physical change: Liquid water and invisible water vapor formed when liquid water boils have the same composition; the water merely changes from a liquid to a gas.
d. Chemical change: New substances, hydrogen and oxygen, are formed.

▶ **Exercise 1.3A**
Which of the following events involve chemical changes and which involve physical changes?
a. Gasoline vaporizes from an open container.
b. A piece of magnesium metal burns in air to form a white powder called *magnesium oxide*.
c. A dull knife is sharpened with a whetstone.

▶ **Exercise 1.3B**
Which of the following events involve chemical changes and which involve physical changes?
a. A steel wrench left out in the rain becomes rusty.
b. A stick of butter melts.
c. A wooden log is burned.
d. A piece of wood is ground up into sawdust.

1.9 CLASSIFICATION OF MATTER

Matter can be classified in many ways. We will examine three ways of doing so in this section. First we look at the physical forms or *states of matter*.

The States of Matter

There are three familiar states of matter: solid, liquid, and gas (Figure 1.5). They can be classified by bulk properties (a *macro* view) or by arrangement of the particles that comprise them (a "*molecular*" or *micro* view). A **solid** object ordinarily maintains its shape and volume regardless of its location. A **liquid** occupies a definite volume but assumes the shape of the occupied portion of its container. If you have a 355-milliliter (mL) soft drink, you have 355 mL whether the soft drink is in a can, in a bottle, or, through a mishap, on the floor—which demonstrates another property of liquids. Unlike solids, liquids flow readily. A **gas** maintains neither shape nor volume. It expands to fill completely whatever container it occupies. Gases flow and are easily compressed. For example, enough air for many minutes of breathing can be compressed into a steel tank for underwater diving.

Bulk properties of *solids*, *liquids*, and *gases* are explained using the kinetic–molecular theory. In solids, the particles are close together and in fixed positions. In liquids, the particles are close together, but they are free to move about. In gases, the particles are far apart and are in rapid random motion. We discuss the states of matter in more detail in Chapter 5.

Substances and Mixtures

Matter can be either pure or mixed (Figure 1.6). Pure matter is considered to be a substance. A **substance** has a definite, or fixed, composition that does not vary from one sample to another. Pure gold (24-carat gold) consists entirely of gold atoms; it is a substance. All samples of pure water are comprised of molecules consisting of two hydrogen atoms and one oxygen atom; water is a substance.

The composition of a **mixture** of two or more substances is variable. The substances retain their identities. They do not change chemically; they simply mix. Mixtures can be separated by physical means. Mixtures can be either *homogeneous* or *heterogeneous*. Homogeneous mixtures are also called *solutions*. A saline solution—a solution of salt in water—is a homogeneous mixture. The proportions of salt and water can vary from one solution to another, but the water is still water and

Phases of Water

Classifications of Matter

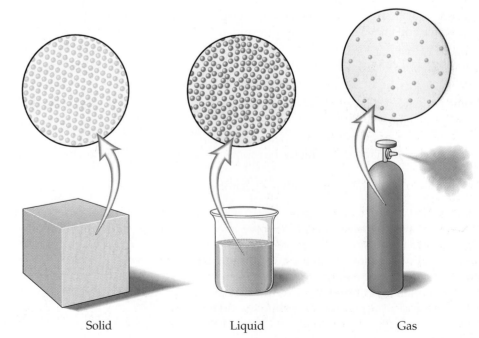

▶ **Figure 1.5** The kinetic–molecular theory can be used to interpret (or explain) the bulk properties of *solids, liquids,* and *gases.* In solids, the particles are close together and in fixed positions. In liquids, the particles are close together, but they are free to move about. In gases, the particles are far apart and are in rapid random motion.

Question: Based on this figure, why does a quart of water vapor weigh so much less than a quart of liquid water?

Solid   Liquid   Gas

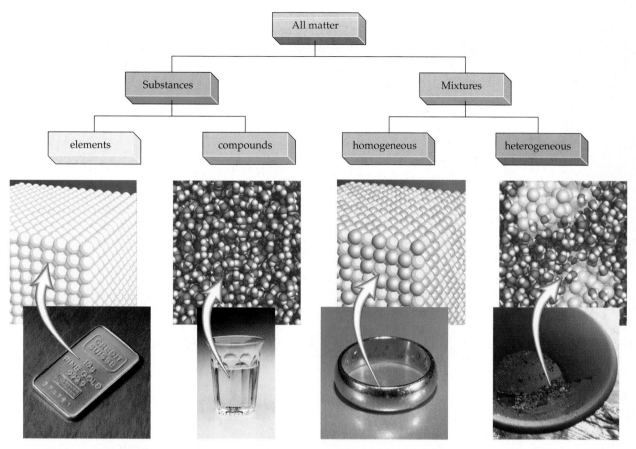

▲ **Figure 1.6** A scheme for classifying matter. The "molecular-level" views are of gold—an *element*; water—a *compound*; 12-carat gold—a *homogeneous mixture* of silver and gold; and a miner's pan containing a *heterogeneous mixture* of gold flakes in water.

the salt remains salt. The two substances can be separated by physical means. For example, the water can be boiled away, leaving the salt behind.

Sand and water form a heterogeneous mixture; the appearance is not the same throughout. The grains of sand are different from the liquid water.

Elements and Compounds

A *substance* is either an element or a compound. An **element** is one of the fundamental substances from which all material things are constructed. Elements cannot be broken down into simpler substances by any chemical process. A **compound** is a substance made up of two or more elements chemically combined. Compounds have a fixed composition. For example, water has fixed proportions, by mass, of hydrogen (H) and oxygen (O). At present there are 114 known elements. A list of these elements is provided on the inside front cover. Oxygen, carbon, sulfur, aluminum, and iron are familiar elements. Aluminum oxide ("sand" on sandpaper), carbon dioxide, and iron sulfide ("fool's gold") are compounds.

Because elements are so fundamental to our study of chemistry, we find it useful to refer to them in a shorthand form. Each element can be represented by a **chemical symbol** made up of one or two letters derived from the name of the element. Symbols for all the elements are listed in the Table of Atomic Masses (inside front cover). The first letter of the symbol is always capitalized; the second is always lowercase. (It makes a difference. For example, Co is the symbol for cobalt, an element, but CO is the formula for carbon monoxide, a compound.)

Symbols are the alphabet of chemistry. Most are based on the English names of the elements, but a few are based on Latin names (Table 1.3).

 Mixtures and Compounds

 Periodic Table

Of the 114 elements now known, 25 are too unstable to exist in nature. In this book we deal with only about one-third of the known elements in any detail.

A chemical symbol in a formula stands for one atom of the element. If more than one atom is to be indicated in a formula, a subscript number is used after the symbol. For example, the formula H_2 represents two atoms of hydrogen, and the formula CH_4 stands for one atom of carbon and four atoms of hydrogen.

TABLE 1.3 Some Elements with Symbols Derived from Latin Names

Usual English Name	Latin Name	Symbol	Spanish Name[2]	French Name[2]
Copper	Cuprum	Cu	Cobre	Cuivre
Gold	Aurum	Au	Oro	Or
Iron	Ferrum	Fe	Hierro	Fer
Lead	Plumbum	Pb	Plomo	Plomb
Mercury	Hydrargyrum	Hg	Mercurio	Mercure
Potassium	Kalium[1]	K	Potasio	Potassium
Silver	Argentum	Ag	Plata	Argent
Sodium	Natrium[1]	Na	Sodio	Sodium
Tin	Stannum	Sn	Estaño	Étain

[1]The elements potassium and sodium were unknown in ancient times. The names kalium and natrium shown here are the names that were used for the compounds potassium carbonate and sodium carbonate.

[2]Element names in languages other than English are often close to the Latin names in Table 1.3. For example, lead is *plomo* in Spanish and *plomb* in French (compare to plumbum), silver is *argent* in French, iron is *fer* in French and *hierro* in Spanish (compare to ferrum), tin is *estaño* in Spanish (compare to stannum), and gold is *oro* in Spanish and *or* in French (close to the Latin aurum). The closeness is even more apparent in pronunciation than in spelling.

CONCEPTUAL EXAMPLE 1.4   **Elements and Compounds**

Which of the following represent elements and which represent compounds?

C Ca HI BN In HBr

Solution

C, Ca, and In represent elements (each is a single symbol). HI, BN, and HBr are composed of two symbols each and represent compounds.

▶ **Exercise 1.4A**

Which of the following represent elements and which represent compounds?

He CuO No NO KI Os

▶ **Exercise 1.4B**

How many *different* elements are represented in the entire list of Exercise 1.4A?

Atoms and Molecules

An **atom** is the smallest characteristic part of an element. Each element is composed of atoms of a particular kind. For example, the element copper is made up of copper atoms, and gold is made up of gold atoms. All copper atoms are alike in a fundamental way and are different from gold atoms. The smallest characteristic part of most compounds is a molecule. A **molecule** is a group of atoms bound together as a unit. Each molecule of a given compound has the same atoms in the same proportions as all the other molecules of the compound. For example, all water molecules have two hydrogen (H) atoms and one oxygen (O) atom, as indicated by the formula (H_2O). We will discuss atoms in some detail in Chapter 2, and much of the focus of many of the remaining chapters is on molecules.

1.10 THE MEASUREMENT OF MATTER

Accurate measurements of such quantities as mass, volume, time, and temperature are essential to the compilation of dependable scientific data. These data are of critical importance in all science-related fields. Measurements of temperature and blood pressure are routinely made in medicine, and modern medical diagnosis de-

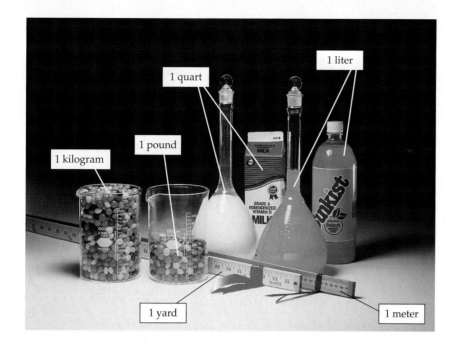

◄ **Figure 1.7** Comparisons of metric and customary units of measure. The red ribbon is 1 in. wide and is tied around a stick 1 yd long. The green ribbon is 1 cm wide and is tied around a stick 1 m long (1 m = 1.0936 yd; 1 cm = 0.3937 in.).

Question: From this figure, approximately how many pounds are in a kilogram? Which two units are closer in size, the inch and centimeter or the quart and the liter?

pends on a whole battery of other measurements, including careful chemical analyses of blood and urine.

The measurement system agreed on since 1960 is the *International System of Units* or **SI units** (from the French *Système Internationale*), a modernized version of the metric system established in France in 1791. Most countries use metric measures in everyday life, but in the United States SI units are used mainly in science laboratories. However, metric measures are increasingly being used in commerce, especially in businesses with an international component. The contents of most bottled beverages are now given in metric units, and metric measurements are also common in sporting events. Figure 1.7 compares some metric and customary units.

Because SI units are based on the decimal system, it is easy to convert from one unit to another. All measured quantities can be expressed in terms of the seven base units listed in Table 1.4. We use the first six in this text.

TABLE 1.4 **The Seven SI Base Units**		
Physical Quantity	**Name of Unit**	**Symbol of Unit**
Length	meter*	m
Mass	kilogram	kg
Time	second	s
Temperature	kelvin	K
Amount of substance	mole	mol
Electric current	ampere	A
Luminous intensity	candela	cd

*Spelled *metre* in most countries.

Exponential Numbers: Powers of Ten

Scientists deal with objects smaller than atoms and as large as the universe. We usually use exponential numbers to describe the sizes of such objects. An electron has a diameter of about 10^{-15} meter (m) and a mass of about 10^{-30} kilogram (kg). At the other extreme, a galaxy typically measures about 10^{23} m across and has a mass of about 10^{41} kg. It is difficult even to imagine numbers so small or so large. The accompanying figure offers some perspectives on size.

Measurements using the basic SI units are often of awkward magnitude. As we have just seen, we can use exponential numbers. (See Appendix A for a detailed discussion of exponential numbers.) However, in many cases it is more convenient to use prefixes (Table 1.5) to indicate units larger and smaller than the base unit. The following examples show how prefixes and powers of ten are interconverted.

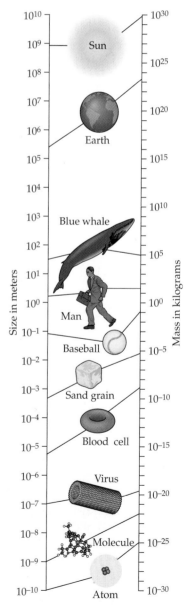

▲ Comparison of very large and very small objects is much easier with exponential notation.

An essential part of the SI system is the use of exponential (powers of ten) notation for numbers. If you are not already familiar with this notation, you will find a discussion of this topic in Appendix A.

TABLE 1.5 Approved Numerical Prefixes*

Exponential Expression	Decimal Equivalent	Prefix	Pronounced	Symbol
10^{12}	1,000,000,000,000	tera-	TER-uh	T
10^9	1,000,000,000	giga-	GIG-uh	G
10^6	1,000,000	mega-	MEG-uh	M
10^3	1,000	kilo-	KIL-oh	k
10^2	100	hecto-	HEK-toe	h
10	10	deka-	DEK-uh	da
10^{-1}	0.1	deci-	DES-ee	d
10^{-2}	0.01	centi-	SEN-tee	c
10^{-3}	0.001	milli-	MIL-ee	m
10^{-6}	0.000,001	micro-	MY-kro	μ
10^{-9}	0.000,000,001	nano-	NAN-oh	n
10^{-12}	0.000,000,000,001	pico-	PEE-koh	p
10^{-15}	0.000,000,000,000,001	femto-	FEM-toe	f

*The most commonly used prefixes are shown in color.

EXAMPLE 1.5   Prefixes and Powers of Ten

Convert each of the following measurements to a unit that replaces the power of ten by a prefix.

a. 2.89×10^{-3} g   **b.** 4.30×10^3 m

Solution

Our goal is to replace each power of ten with the appropriate prefix from Table 1.5. For example, $10^{-3} = 0.001$, corresponding to milli(unit). (It doesn't matter what the unit is; here we are dealing only with the prefixes.)

a. 10^{-3} corresponds to the prefix *milli-*; 2.89 mg; that is, $10^{-3} \times$ (unit) = milli(unit).
b. 10^3 corresponds to the prefix *kilo-*; 4.30 km; that is, $10^3 \times$ (unit) = kilo(unit).

▶ **Exercise 1.5**

Convert each of the following measurements to a unit that replaces the power of ten by a prefix.

a. 7.24×10^3 g   **b.** 4.29×10^{-6} m   **c.** 7.91×10^{-3} s
d. 2.29×10^{-2} g   **e.** 7.90×10^6 m

EXAMPLE 1.6   Prefixes and Powers of Ten

Use exponential notation to express each of the following measurements in terms of an SI base unit.

a. 4.12 cm   **b.** 947 μs   **c.** 3.17 nm

Solution

a. Our goal is to find the power of ten that relates the given unit to the SI base unit. That is, centi(base unit) = $10^{-2} \times$ (base unit)

$$4.12 \text{ centimeter} = 4.12 \times 10^{-2} \text{ m}$$

b. To change microsecond to the base unit second, we replace the prefix *micro-* by 10^{-6}. To obtain an answer in the conventional exponential form, we also need to replace the coefficient 947 by 9.47×10^2. The result of these two changes is

$$947 \ \mu s = 947 \times 10^{-6} \text{ s} = 9.47 \times 10^2 \times 10^{-6} \text{ s} = 9.47 \times 10^{-4} \text{ s}$$

c. To change nanometer to the base unit meter, we replace the prefix *nano-* by 10^{-9}. The answer in exponential form is 3.17×10^{-9} m.

▶ **Exercise 1.6A**

Use exponential notation to express each of the following measurements in terms of an SI base unit.

a. 7.45 nm **b.** 5.25 ms
c. 1.415 km **d.** 2.06 mm

▶ **Exercise 1.6B**

Use exponential notation to express each of the following measurements in terms of an SI base unit.

a. 284 nm **b.** 119 ms
c. 754 km **d.** 6.19×10^6 mm

Mass

The SI base quantity of mass is the **kilogram (kg)**, about 2.2 pounds (lb). This base quantity is unusual in that it already has a prefix. A more convenient mass unit for most laboratory work is the gram (g).

$$1 \text{ kg} = 10^3 \text{ g} = 1000 \text{ g} \qquad \text{or} \qquad 1 \text{ g} = 0.001 \text{ kg} = 10^{-3} \text{ kg}$$

The milligram (mg) is a suitable unit for small quantities of materials, such as some drug dosages.

$$1 \text{ mg} = 10^{-3} \text{ g} = 0.001 \text{ g}$$

Chemists can now detect masses in the microgram (μg), nanogram (ng), picogram (pg), and even smaller ranges.

Length, Area, and Volume

The SI base unit of length is the **meter (m)**, a unit about 10% longer than 1 yard (yd). The kilometer (km) is used to measure distances along a highway.

$$1 \text{ km} = 1000 \text{ m}$$

In the laboratory, we usually find lengths smaller than the meter to be more convenient. For example, we use the centimeter (cm), which is about the width of a typical calculator button—and the millimeter (mm), which is about the thickness of the cardboard backing in a notepad.

$$1 \text{ cm} = 0.01 \text{ m} \qquad 1 \text{ mm} = 0.001 \text{ m}$$

For measurements at the atomic and molecular level, we use the micrometer (μm), the nanometer (nm), and the picometer (pm). For example, a chlorophyll molecule is about 0.1 μm or 100 nm long, and the diameter of a sodium atom is 372 pm.

The units for area and volume are derived from the base unit of length. The SI unit of area is the square meter (m^2), but we often find square centimeters (cm^2) or square millimeters (mm^2) more convenient for laboratory work.

$$1 \text{ cm}^2 = (10^{-2} \text{ m})^2 = 10^{-4} \text{ m}^2 \qquad 1 \text{ mm}^2 = (10^{-3} \text{ m})^2 = 10^{-6} \text{ m}^2$$

Similarly, the SI unit of volume is the cubic meter (m^3), but two units more likely to be used in the laboratory are the cubic centimeter (cm^3 or cc) and the cubic decimeter (dm^3). A cubic centimeter is about the volume of a sugar cube, and a cubic decimeter is slightly larger than 1 quart (qt).

$$1 \text{ cm}^3 = (10^{-2} \text{ m})^3 = 10^{-6} \text{ m}^3 \qquad 1 \text{ dm}^3 = (10^{-1} \text{ m})^3 = 10^{-3} \text{ m}^3$$

The cubic decimeter is commonly called a liter. A **liter (L)** is 1 cubic decimeter or 1000 cubic centimeters.

$$1 \text{ L} = 1 \text{ dm}^3 = 1000 \text{ cm}^3$$

The milliliter (mL) and/or cubic centimeter are frequently used in laboratories.

$$1 \text{ mL} = 1 \text{ cm}^3$$

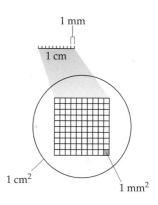

▲ Area units such as mm^2 and cm^2 are derived from units of length.

Question: How many mm are in 1 cm? How many mm^2 are in 1 cm^2?

The units liter, deciliter, and milliliter are commonly used for volumes of liquids and gases; the units cubic centimeter and cubic decimeter are more often used for solids.

Mass does not vary with temperature, but volume does, and so does density. We can assume the density of water to be 1.00 g/mL near room temperature.

Time

The SI base unit for measuring intervals of time is the second (s). Extremely short time periods are expressed through the usual SI prefixes: milliseconds (ms), microseconds (μs), nanoseconds (ns), and picoseconds (ps).

$$1 \text{ ms} = 10^{-3} \text{ s} \qquad 1 \text{ μs} = 10^{-6} \text{ s} \qquad 1 \text{ ns} = 10^{-9} \text{ s} \qquad 1 \text{ ps} = 10^{-12} \text{ s}$$

Long time intervals, in contrast, are usually expressed in traditional, non-SI units: minutes (min), hours (h), days (d), and years (y).

$$1 \text{ min} = 60 \text{ s} \qquad 1 \text{ h} = 60 \text{ min} \qquad 1 \text{ d} = 24 \text{ h} \qquad 1 \text{ y} = 365 \text{ d}$$

Problem Solving: Estimation

Many chemistry problems require calculations that yield numerical answers. When you use a calculator, it will always give you an answer—but the answer may not be correct. You may have punched a wrong number or used the wrong function. You can learn to make *estimations* of answers that enable you to know whether your answer is reasonable. At times an estimated answer is good enough, as the following Example and Exercises illustrate. This ability to estimate answers can be important in everyday life as well as in chemistry. Note that estimation does *not* require a detailed calculation. Only a rough calculation—or none at all—is required.

EXAMPLE 1.7   **Mass, Length, Area, Volume**

Without doing a detailed calculation, determine which of the following is a reasonable **(a)** mass (weight) and **(b)** height for a typical two-year-old child.

Mass:	10 mg	10 g	10 kg	100 g
Height:	85 mm	85 cm	850 cm	8.5 m

Solution
a. In customary units, a two-year-old child should weigh about 20–25 lb. Knowing that 1 kg is about 2.2 lb, the only reasonable answer is 10 kg.
b. In customary units, a two-year-old child should be about 30–36 in. tall. Knowing that 1 cm is a little less than 0.5 in., the answer must be a little *more* than twice 30–36, or a bit more than 70–72 cm. Thus the only reasonable answer is 85 cm.

▶ **Exercise 1.7A**
Without doing a detailed calculation, determine which of the following is a reasonable area for the front cover of your textbook.

$$500 \text{ mm}^2 \qquad 50 \text{ cm}^2 \qquad 500 \text{ cm}^2 \qquad 50 \text{ m}^2$$

▶ **Exercise 1.7B**
Without doing a detailed calculation, determine which of the following is a reasonable volume for your textbook.

$$1600 \text{ mm}^3 \qquad 16 \text{ cm}^3 \qquad 1600 \text{ cm}^3 \qquad 1.6 \text{ m}^3$$

Problem Solving: Unit Conversions

It is easy to convert from one metric unit to another. We use the unit conversion method of problem solving. If you are not familiar with that method, you should study Appendix A before proceeding.

EXAMPLE 1.8 **Unit Conversions**

Convert **(a)** 1.83 kg to grams and **(b)** 729 μL to milliliters.

Solution

a. We start with the given quantity 1.83 kg and use the equivalence 1 kg = 1000 g to form a conversion factor (Appendix A) that allows us to cancel the unit kg and end with the unit g.

$$1.83 \text{ kg} \times \frac{1000 \text{ g}}{1 \text{ kg}} = 1830 \text{ g}$$

b. Here we start with the given quantity 729 μL and use the equivalences 10^6 μL = 1 L and 10^3 mL = 1 L to form conversion factors that allows us to cancel the unit μL and end with the unit mL.

$$729 \text{ μL} \times \frac{1 \text{ L}}{10^6 \text{ μL}} \times \frac{1000 \text{ mL}}{1 \text{ L}} = 0.729 \text{ mL}$$

▶ **Exercise 1.8**

Convert **(a)** 7.5 m to millimeters, **(b)** 2056 mL to liters, **(c)** 2.06 g to micrograms, and **(d)** 0.738 cm to millimeters.

Nanoworld

For over two centuries, chemists have been able to rearrange atoms to make molecules. These molecules generally have dimensions in the *picometer* (10^{-12} m) range. This ability has led to revolutions in the design of drugs, plastics, and many other materials. Over the last several decades, scientists have made vast strides in handling materials in the *micrometer* (10^{-6} m) range. The revolution in microelectronic devices—computers, cellular phones, and so on—was spurred by the ability of scientists to produce computer chips by photolithography on a micrometer scale.

Recently the news has focused on *nanotechnology*. Just what are the pundits talking about? The prefix *nano-* means one-billionth (10^{-9}). With nanotechnology it is possible to bridge the gap between picometer-sized molecules and micrometer-sized electronics. By tailoring the structure of materials in the range from about 1 nm to 100 nm, a scientist can systematically change the properties of materials. A nanometer-sized object contains just a few hundred to a few thousand atoms or molecules, and such objects often have entirely new properties. For example, bulk gold is yellow, but a gold ring made up of nanometer-sized particles appears red. Carbon *nanotubes* are made of the same element as the graphite in pencil lead, but nanotubes are much, much stronger. Scientists can now control matter on every scale, from nanometers to meters, giving us great new power in materials design. We will examine some of the many practical applications of nanotechnology in subsequent chapters.

Bulk gold = Yellow

Nanogold = Red

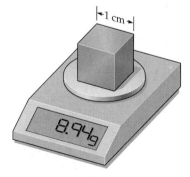

◄1 cm►

▲ One cubic centimeter of copper weighs 8.94 g, so the density of copper is 8.94 g/cm³.

Question: What would be the mass of 2 cm³ of copper? What would be the density of 2 cm³ of copper?

1.11 DENSITY

In everyday life, we might speak of lead as "heavy" or aluminum as "light," but such descriptions are imprecise at best. Scientists use the term *density* to describe this important property. The **density**, *d*, of a substance is the quantity of mass, *m*, per unit of volume, *V*.

$$d = \frac{m}{V}$$

For substances that don't mix, such as oil and water, the concept of density allows us to predict which will float on the other. It isn't just the mass of the materials, but the

▶ **Figure 1.8** At room temperature, the density of water is 1.00 g/mL. A coin sinks in water (left), but a cork floats on water (right).

Question: Is the density of the coin less than, equal to, or greater than 1.00 g/mL? Is the density of the cork less than, equal to, or greater than 1.00 g/mL?

Scientists and others are often faced with the task of converting from one unit to another. Unit conversions are discussed in some detail in Appendix A, where you will also find tables of common conversion factors.

mass per unit volume—the density—that determines the result. Density is one of the properties that helps us to know that in an oil-and-vinegar salad dressing, oil (lower density) floats on water (higher density). Figure 1.8 gives more examples.

We can rearrange the equation for density to give

$$m = d \times V \quad \text{and} \quad V = \frac{m}{d}$$

These equations are useful for calculations. Densities of some common substances are listed in Table 1.6. They are reported, as is customary, in grams per milliliter (g/mL) for liquids and grams per cubic centimeter (g/cm³) for solids. Values listed in the table are used in some of the following Examples and Exercises and in some of the end-of-chapter problems.

As with mass, length, area, and volume (page 20), it is often sufficient to estimate answers to problems involving densities. For example, an estimated answer is good enough to determine relative volumes of materials or whether or not a material will float or sink in water. These ideas are illustrated in the following example and exercise. Note again that estimation does *not* require a detailed calculation.

TABLE 1.6   Densities of Some Common Substances at Specified Temperatures

Substance[*]	Density	Temperature
Solids		
Copper (Cu)	8.94 g/cm³	25 °C
Gold (Au)	19.3 g/cm³	25 °C
Magnesium (Mg)	1.738 g/cm³	20 °C
Water (ice) (H_2O)	0.917 g/cm³	0 °C
Liquids		
Ethyl alcohol (CH_3CH_2OH)	0.789 g/mL	20 °C
Hexane ($CH_3CH_2CH_2CH_2CH_2CH_3$)	0.660 g/mL	20 °C
Mercury (Hg)	13.534 g/mL	25 °C
Urine (a mixture)	1.003–1.030 g/mL	25 °C
Water (H_2O)	0.997 g/mL	0 °C
	1.000 g/mL	4 °C

[*]Formulas are provided for possible future reference.

EXAMPLE 1.9   Mass, Volume, and Density

Answer the following without doing a detailed calculation. **(a)** Which has the greater volume, a 50.0-g block of copper or a 50.0-g block of gold? **(b)** Which has the greater mass, 225 mL of ethyl alcohol or 225 mL of hexane?

Solution

a. From Table 1.6 we see that the density of gold (19.3 g/cm^3) is greater than that of copper (8.94 g/cm^3). It takes a larger block of copper to have a mass of 50.0 g than of gold. A 50.0-g block of copper has a greater volume than a 50.0-g block of gold.

b. From Table 1.6 we see that the density of ethyl alcohol (0.789 g/mL) is greater than that of hexane (0.660 g/mL). Because it is more dense (that is, it has more mass in each unit of volume), 225 mL of ethyl alcohol has a greater mass than does 225 mL of hexane.

▶ **Exercise 1.9A**

Without doing a detailed calculation, determine which has the greater volume, a 500.0-g block of ice or a 500.0-g block of magnesium.

▶ **Exercise 1.9B**

Wood does not dissolve in water and will float on water if its density is less than that of water. Padouk (d = 0.86 g/cm^3) and ebony (d = 1.2 g/cm^3) are tropical woods. Which will float on water and which will sink in water?

When we need a numerical answer, we can use the relationship of mass and volume to density to form conversion factors. Then we use the unit conversion method of problem solving.

EXAMPLE 1.10   Density from Mass and Volume

What is the density of iron if 156 g of iron occupies a volume of 20.0 cm^3?

Solution

The given quantities are

$$m = 156\ g \quad \text{and} \quad V = 20.0\ cm^3$$

We can use the equation that defines density.

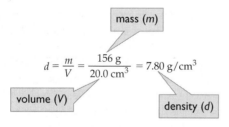

$$d = \frac{m}{V} = \frac{156\ g}{20.0\ cm^3} = 7.80\ g/cm^3$$

▶ **Exercise 1.10A**

What is the density, in grams per milliliter, of a salt solution if 52.5 mL has a mass of 58.5 g?

▶ **Exercise 1.10B**

What is the density, in grams per cubic centimeter, of a metal alloy if a cube that measures 2.00 cm on an edge has a mass of 89.2 g?

Water, ethyl alcohol, and ethylene glycol (the principal ingredient in many antifreeze solutions) all are colorless liquids. They can be distinguished by measuring their densities. At 20 °C, the density of water is 0.998 g/mL, that of ethanol is 0.789 g/mL, and that of ethylene glycol is 1.114 g/mL.

EXAMPLE 1.11 **Mass from Density and Volume**

What is the mass of 1.00 L of gasoline if its density is 0.703 g/mL?

Solution

We can express density as a ratio, 0.703 g/1.00 mL, and use it as a conversion factor. We also need the factor 1000 mL/1 L to convert liters to milliliters.

$$m = d \times V = \frac{0.703 \text{ g}}{1 \text{ mL}} \times 1 \text{ L} \times \frac{1000 \text{ mL}}{1 \text{ L}} = 703 \text{ g}$$

▶ **Exercise 1.11A**

What is the mass of 200.0 mL of glycerol, which has a density of 1.264 g/mL at 20 °C?

▶ **Exercise 1.11B**

What is the mass of a lead brick that is 2.50 cm × 3.50 cm × 8.25 cm on an edge? (The density of lead at 20 °C is 11.34 g/cm^3.)

EXAMPLE 1.12 **Volume from Mass and Density**

What volume is occupied by 461 g of mercury? (See Table 1.6.)

Solution

Here we can express the inverse of density as a ratio, 1.00 mL/13.6 g, and use it as a conversion factor.

$$V = 461 \text{ g} \times \frac{1 \text{ mL}}{13.534 \text{ g}} = 34.1 \text{ mL}$$

▶ **Exercise 1.12A**

What volume is occupied by a 845-g piece of magnesium? (See Table 1.6.)

▶ **Exercise 1.12B**

What volume is occupied by 225 g of hexane? (See Table 1.6.)

1.12 ENERGY: HEAT AND TEMPERATURE

The physical and chemical changes that matter undergoes are almost always accompanied by changes in energy. **Energy** is required to make something happen that wouldn't happen by itself; it is the ability to change matter, either physically or chemically.

Two important concepts in science that are related, and sometimes confused, are *temperature* and *heat*. When two objects at different temperatures are brought together, heat flows from the warmer to the cooler object until both are at the same temperature. **Heat** is energy on the move, the energy that flows from a warmer object to a cooler one. **Temperature** is a measure of how hot or cold an object is. Temperature tells us the direction in which heat will flow. Heat flows from more energetic (higher-temperature) to less energetic (lower-temperature) atoms or molecules. For example, if you touch a hot test tube, heat will flow from the tube to your hand. If the tube is hot enough, your hand will be burned.

The SI base unit of temperature is the **kelvin (K)**. For laboratory work, we often use the more familiar **Celsius scale**. On this temperature scale, the freezing point of water is 0 degrees Celsius (°C) and the boiling point is 100 °C. The interval between these two reference points is divided into 100 equal parts, each a *degree Celsius*. The Kelvin scale is called an *absolute scale* because its zero point is the coldest temperature possible, or absolute zero. (This fact was determined by theoretical considerations and has been confirmed by experiment, as we will see in Chapter 6.) The zero point on the Kelvin scale, 0 K, is equal to -273.15 °C, often rounded to -273 °C. A kelvin is the same size as a degree Celsius, and so the freezing point of water on the Kelvin scale is 273 K. The Kelvin scale has no negative temperatures, and we don't use a degree sign with the K. To convert from degrees Celsius to kelvins, simply add 273.15 to the Celsius temperature.

$$K = °C + 273.15$$

EXAMPLE 1.13 **Temperature Conversions**

Ether boils at 36 °C. What is the boiling point of ether on the Kelvin scale?

Solution

$$K = °C + 273.15$$
$$K = 36 + 273.15 = 309 \text{ K}$$

▶ **Exercise 1.13A**
What is the boiling point of water (100 °C) expressed in kelvins?

▶ **Exercise 1.13B**
Express a temperature of -78 °C in kelvins.

The Fahrenheit temperature scale is widely used in the United States. Figure 1.9 compares the three temperature scales. Note that on the Fahrenheit scale, the freezing point of water is 32 °F and the boiling point is 212 °F, so that a 10-degree temperature interval on the Celsius scale equals an 18-degree interval on the Fahrenheit scale. Conversion between Fahrenheit and Celsius temperatures is discussed in Appendix A.

The SI-derived unit of energy is the **joule (J)**, but the **calorie (cal)** is the more familiar unit in everyday life.

$$1 \text{ cal} = 4.184 \text{ J}$$

A calorie is the amount of heat required to raise the temperature of 1 g of water 1 °C.

The "calorie" used for measuring the energy content of foods is actually a **kilocalorie (kcal)**.

$$1000 \text{ cal} = 1 \text{ kcal} = 4184 \text{ J}$$

Because it varies slightly with temperature, a calorie is defined more precisely as the quantity of heat needed to raise the temperature of 1 g of water from 14.5 to 15.5 °C.

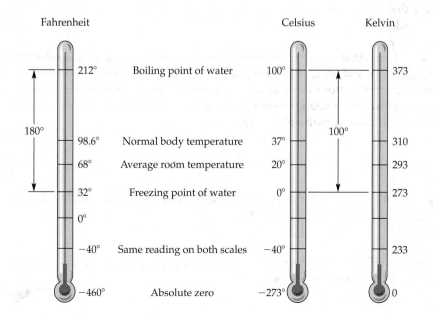

Temperature

◀ **Figure 1.9** A comparison of the Fahrenheit, Celsius, and Kelvin temperature scales.

A dieter might be aware that a banana split contains 1500 "calories." If the same dieter realized that this was really 1,500,000 calories, giving up a banana split might be easier! The concept of heat is discussed further in Appendix A.

EXAMPLE 1.14 **Energy Conversions**

When 1.00 g of gasoline burns, it yields about 10.3 kcal of energy. What is the quantity of energy in kilojoules (kJ)?

Solution

$$10.3 \text{ kcal} \times \frac{1000 \text{ cal}}{1 \text{ kcal}} \times \frac{4.184 \text{ J}}{1 \text{ cal}} \times \frac{1 \text{ kJ}}{1000 \text{ J}} = 43.1 \text{ kJ}$$

▶ **Exercise 1.14A**

A European woman on average consumes food with an energy content of 7525 kJ per day. What is her daily intake in kilocalories (food calories)?

▶ **Exercise 1.14B**

It takes about 12 kJ of energy to melt a single ice cube. How many kilocalories does it take to melt three ice cubes?

Body Temperature, Hypothermia, and Hyperthermia (Fever)

Carl Wunderlich (1815–1877), a German physician, first recognized fever as a symptom of disease. He averaged thousands of human temperature measurements and reported the value as 37 °C. When this value was converted to the Fahrenheit scale, it somehow (improperly) acquired an extra significant figure, and 98.6 °F became widely (and incorrectly) known as the *normal body temperature*. In recent years millions of measurements have revealed that *average* normal body temperature is actually 98.2 °F and that body temperatures in healthy people range from 97.7 °F to 99.5 °F.

The healthy human body maintains a fairly constant temperature. Heat is a by-product of metabolism, and we must constantly get rid of some of it. Because heat always flows spontaneously from a hot object to a cold one, we can get rid of excess heat by simple conduction only if the surrounding temperature is less than 37 °C. Above that temperature, the body depends on air movement, blood vessel dilation, and perspiration to keep its temperature normal. If the outside temperature is above 40 °C, the U.S. National Weather Service usually issues a heat advisory because the body can *gain* heat from the environment. The body temperature then rises above the normal range, a condition known as *hyperthermia*. Certain diseases also cause abnormally high body temperatures or *fevers*. Mild fevers (up to 102 °F) are usually not dangerous and may even help the body fight off an infection. Prolonged high fevers (104 °F or more in adults) can be fatal.

When exposed to prolonged cold, the body can lose too much heat to the environment. The body temperature drops below the normal range, a condition known as *hypothermia*. A drop of only 2 or 3 °F leads to shivering, a condition in which muscles contract in an attempt to generate more heat. Prolonged or severe hypothermia (body temperature below 93 °F) can lead to unconsciousness and death.

1.13 CRITICAL THINKING

One of the hallmarks of science is the ability to think critically—to evaluate statements and claims in a rational, objective fashion. This ability can be learned, and it will serve you well in everyday life as well as in science courses. Critical thinking is important for workers in all types of professions and employees in all kinds of industries. We will use an approach adapted from one developed by James Lett,[1] first outlining the approach and then working through some simple examples.

---

[1]Lett used the acronym FiLCHeRS (ignore the vowels) as a mnemonic for six rules: *f*alsifiability, *l*ogic, *c*omprehensiveness, *h*onesty, *r*eplicability, and *s*ufficiency. See Reference 9 for a full description. See Shermer's "Baloney Detection" approach (Reference 11) for another approach to critical thinking.

You can use the acronym **FLaReS** (ignore the vowels) to remember four rules used to test a claim: *f*alsifiability, *l*ogic, *r*eplicability, and *s*ufficiency. We will first list the rules and describe them briefly, and then we will illustrate them with examples.

- **Falsifiability:** Can any conceivable evidence show the claim to be false? It must be possible to think of evidence that would prove the claim false. Falsifiability is an essential component of the scientific method. A hypothesis that cannot be falsified is of no value. Science cannot prove anything true in an absolute sense, although it can provide overwhelming evidence. Science can prove something false.

- **Logic:** Any argument offered as evidence in support of any claim must be sound. An argument is sound if its conclusion follows inevitably from its premises and if its premises are true. It is unsound if a premise is false, and it is unsound if there is a single exception in which the conclusion does not necessarily follow from the premises.

- **Replicability:** If the evidence for any claim is based on an experimental result, it is necessary for the evidence to be replicable in subsequent experiments or trials. Scientific research is almost always reviewed by other qualified scientists. The peer-reviewed research is then published in a form that enables others to repeat the experiment. Sometimes bad science slips through the peer-review process, but such science eventually fails the test of replicability. For example, in May 2005 the journal *Science* published research by Korean biomedical scientist Hwang Woo-suk in which he claimed to have tailored embryonic stem cells so that every patient could receive custom treatment. Others were unable to reproduce Hwang's findings, and evidence emerged that he had intentionally fabricated results. *Science* retracted the article in January 2006. Research results published in media that are not peer reviewed have little or no standing in the scientific community.

- **Sufficiency:** The evidence offered in support of any claim must be adequate to establish the truth of that claim, with these stipulations: (1) The burden of evidence for any claim rests on the claimant, (2) extraordinary claims demand extraordinary evidence, and (3) evidence based on authority and/or testimony is never adequate.

If a claim passes all four FLaReS tests, then it *might* be true. On the other hand, it could still be proven false. However, if a claim fails even one of the FLaReS tests it is likely to be false.

Critical Thinking Examples

1.1 A psychic claims he can bend a spoon using only the powers of his mind. However, he says he can do so only when the conditions are right; there must be no one with negative energy present. Apply the FLaReS tests to evaluate the psychic's claim.

Solution

1. Is the claim falsifiable? No. If the psychic fails, he can always claim that someone present had negative energy.

2. Is the claim logical? No. What kind of matter or energy could move from the psychic's mind to the spoon with enough force to bend it? Just what is "negative energy"?

3. Is the claim reproducible? No. The psychic can do it only when "conditions are right." (Actually, one such psychic was caught cheating; he was seen bending the spoon with his hands. Now his proponents claim he cheats only sometimes!)

4. Is the claim sufficient? No. Any such claim is extraordinary; it would require extraordinary evidence. The burden of proof for any such claim rests on the claimant, and only flimsy evidence is provided.

1.2 Some people claim that repetitive biorhythm cycles that date from a person's birth date can predict airplane crashes. They say that more crashes occur when the pilot, copilot, and/or navigator are experiencing critically low points in their intellectual, emotional, and/or physical cycles. Several studies show no such correlation. Apply the FLaReS tests to evaluate this claim.

Solution

1. Is the claim falsifiable? Yes. Crash data and birth dates of the crew members can be analyzed to see whether or not there is a correlation.
2. Is the claim logical? No. Hundreds of thousands of people are born each day. It is highly unlikely that all of them would share the same up-and-down cycles day by day for a lifetime.
3. Is the claim reproducible? No. It has not held up in other studies.
4. Is the claim sufficient? No. It does not address the real causes of airplane crashes.

1.3 Many "psychics" claim to be able to predict the future. These predictions are often made near the end of one year for the next year. Look up several such predictions and apply the FLaReS tests to evaluate the claim.

Solution

1. Is the claim falsifiable? Yes. You can write down a claim and then check it yourself at the end of the year.
2. Is the claim logical? No. If psychics could really predict the future, they could readily win lotteries, ward off all kinds of calamities by providing advance warning, and so on.
3. Is the claim reproducible? No. Psychics usually make broad, general claims. They later make claim-specific references to "hits" while ignoring "misses." (Some past predictions: World War III will begin in 1958; Kennedy will *not* be elected in 1960; Fidel Castro will die in 1969; George H. W. Bush will be elected in 1992. None of the well-known psychics, including James Van Praagh, John Edward, Sylvia Browne, or the Jamison sisters, Terry and Linda, predicted the terrorist attacks on the United States on September 11, 2001.)
4. Is the claim sufficient? No. Any such claim is extraordinary; it would require extraordinary evidence. The burden of proof for any such claim rests on the claimant, and only flimsy evidence is provided.

Critical Thinking Exercises

Apply knowledge that you have gained in this chapter and one or more of the FLaReS principles to evaluate the following statements or claims. You may also find Shermer's "Baloney Detection" articles (Reference 11) to be helpful.

1.1 An alternative health practitioner claims that a nuclear power plant releases radiation at a level so low that it cannot be measured, but that this radiation is harmful to the thyroid gland. He sells a thyroid extract that he claims can fix the problem.

1.2 A doctor claims that she can cure a patient of arthritis by simply massaging the affected joints. If she has the patient's complete trust, she can cure the arthritis within a year. Several of her patients have testified that the doctor has cured their arthritis.

1.3 Some people claim that crystals have special powers. Crystal therapists claim that they can use quartz to restore balance and harmony to a person's spiritual energy.

1.4 A sixth-grade student has a beautiful tigereye stone that her grandfather gave her. When she holds it in her hand, she can think more clearly.

She claims that the stone really works because one day when she left the stone at home, she made the worst grade she had ever received on an exam. She believes that her stone has magic power. What do you think?

1.5 A woman claims that she has memorized the New Testament. She offers to quote any chapter of any book entirely from memory.

1.6 Some people claim *ear coning* cleans your ears and mind. You lie on your side and then place a narrow, cylindrical cone of wax into your ear canal to form a tight seal. The open end of the cone is set on fire. The negative pressure created is supposed to remove undesirable earwax and realign and cleanse energy flows, thus sharpening mental functioning, vision, hearing, smell, taste, and color perception. Ear coning has been in use for many centuries; however, a study published in the refereed journal *Laryngoscope* showed that ear candles do not produce negative pressure and do not remove wax. Further investigations revealed 21 ear injuries from ear coning.

SUMMARY

Sections 1.1–1.3—**Chemistry** is the study of matter and the changes it undergoes. The roots of chemistry lie in **natural philosophy**—philosophical speculation about nature—and in **alchemy**, a mystical investigation practiced in the Middle Ages. **Technology** is the practical applica-

tion of knowledge, while **science** is an accumulation of knowledge about nature and our physical world based on observations and experimental tests of our assumptions. Science is *testable, reproducible, explanatory, predictive*, and always *tentative*. In one common scientific method, a confirmed observation about nature may lead to a **hypothesis**—a tentative explanation of observations. If a hypothesis stands up to testing and further experimentation, it may become a **theory**—the best current explanation for a phenomenon. A theory is always tentative and can be rejected if it does not continue to stand up to further testing. A valid theory can be used to predict new scientific facts. A **scientific law** is a brief statement that summarizes large amounts of data. A **scientific model** uses tangible items or pictures to represent invisible processes and help explain complex phenomena.

Sections 1.4–1.6—Scientists disagree over social and political issues partly because of the inability to control **variables**, those things that can change during an experiment. Science and technology have provided many benefits for our world, but they have also introduced new risks. A **risk–benefit analysis** can help us decide which outweighs the other. Chemistry is fundamental to other scientific disciplines and plays a central role in science.

Section 1.7—Chemists are usually involved in some kind of research. The purpose of **applied research** is to make particular kinds of useful products, while **basic research** is carried out simply to obtain new knowledge or to answer fundamental questions. Basic research often may result in a useful product or process.

Section 1.8—**Matter** is anything that has mass and takes up space. **Mass** is a measure of the amount of matter in an object, while **weight** represents the gravitational force of attraction for an object.

Physical properties of matter can be observed without making new substances. When **chemical properties** are observed, new substances are formed. A **physical change** does not entail a change in chemical composition. A **chemical change** does involve such a change.

Section 1.9—Matter can be classified according to its physical state. A **solid** has definite shape and volume; a **liquid** has

definite volume but takes the shape of its container; and a **gas** takes both the shape and the volume of its container. Matter also can be classified according to composition. A pure **substance** always has the same composition, no matter how it is made or found. A **mixture** may have different compositions depending on how it is prepared. Pure substances are either elements or compounds. An **element** is

composed of atoms all of one type, an **atom** being the smallest particle possible of an element. There are only about a hundred elements, each of which is represented by a **chemical symbol**, which is made up of one or two letters derived from the element's name. A **compound** is made of two or more elements, chemically combined in fixed proportions. Many compounds exist as **molecules**, groups of atoms bound together as a unit.

Section 1.10—Scientific measurements are made using **SI units**, an agreed-upon standard version of the metric system. Base units are the **meter (m)** for length, the **kilogram (kg)** for mass, and the **kelvin (K)** for temperature. Although it is not

truly SI, the **liter (L)** is a common unit of volume. Prefixes make basic units larger or smaller by factors of ten.

Section 1.11—**Density** can be thought of as "how heavy something is for its size." It is the amount of mass per unit volume:

$$d = \frac{m}{V}$$

Density can be used as a conversion factor. The density of water is almost exactly 1 g/mL.

Section 1.12—When matter undergoes a physical or chemical change, there is also a change in **energy**, the ability to change matter physically or chemically. Energy is either given off or absorbed in each process. **Heat** is energy flow from a hot object to a cold one. **Temperature** is a measure of how hot or cold an object is. The **kelvin (K)** is the SI unit of temperature, but the **Celsius scale (°C)** is more commonly used. On the Celsius scale water boils at 100 °C and freezes at 0 °C. The SI unit of heat is the **joule (J)**. The **calorie (cal)** is a more familiar unit, the amount of heat needed to raise the temperature of 1 g of water by 1 °C. The "food calorie" is actually a **kilocalorie (kcal)** or 1000 calories.

Section 1.13—Claims may be tested with critical thinking. The acronym **FLaReS** can be useful in such exercises. The letters FLRS represent rules used to test a claim: *f*alsifiability, *l*ogic, *r*eplicability, and *s*ufficiency.

REVIEW QUESTIONS

1. Define *chemistry*. What is a chemical?

2. What is matter? How is it related to mass?

3. State five distinguishing characteristics of science. Which characteristic best serves to distinguish science from other disciplines?

4. Why were the ancient Greek philosophers like Aristotle not successful as scientists?

5. What is alchemy? How is it related to chemistry?

6. What is natural philosophy? How is it related to science?

7. What did Francis Bacon envision for us as a result of science?

8. What is the main theme of Rachel Carson's *Silent Spring*?

9. Why have Thomas Malthus's predictions not been fulfilled in developed countries?

10. What is a scientific hypothesis? How are hypotheses tested?

11. What is a scientific law?

12. What is a theory?

13. Why can't scientific methods always be used to solve social, political, ethical, and economic problems?

14. How does technology differ from science?

15. What is risk–benefit analysis?

16. What sorts of judgments go into (a) the evaluation of benefits, and (b) the evaluation of risks?

17. What is a DQ? What does a large DQ mean? Why is it often difficult to estimate a DQ?

18. Explain the difference between mass and weight.

19. Distinguish between chemical and physical properties.

20. How do substances and mixtures differ?

21. How do gases, liquids, and solids differ in their properties?

22. What are the names and symbols of the basic SI units for mass, length, and temperature? What derived units are used more often in the laboratory?

23. Following is an incomplete table of SI prefixes, their symbols, and their meanings. Fill in the blank cells. The first row is completed as an example.

Prefix	Symbol	Definition
Tera	T	10^{12}
	M	
Centi		
	μ	
Milli		
		10^{-1}
	K	
Nano		

24. Give an example of an object that weighs (a) roughly 1 g; (b) roughly 1 kg.

25. Give an example of an object that has a dimension that is (a) roughly 1 mm; (b) roughly 1 m; (c) roughly 1 cm.

26. What is the SI-derived unit for volume? What volume units are more often used in the laboratory?

27. What is energy?

28. What is applied research? What is basic research? Give an example of each.

PROBLEMS

A word of advice: You cannot learn to work problems by reading them or watching your instructor work them, just as you cannot become a piano player solely by reading about piano-playing skills or attending a performance. Working through problems will help you improve your understanding of the ideas presented in the chapter and to practice your estimation skills and your ability to synthesize concepts. Plan to work through the great majority of these problems.

Risk–Benefit Analysis

29. Penicillin kills bacteria, thus saving the lives of thousands of people who otherwise might die of infectious diseases. Penicillin causes allergic reactions in some people; in extreme cases the allergic reaction can lead to death if the resulting condition is not treated. Do a risk–benefit analysis of the use of penicillin for society as a whole.

30. Do a risk–benefit analysis of the use of penicillin for a person who is allergic to it. (See Problem 29.)

31. Synthetic food colors make food more attractive and increase sales. A few such dyes are suspected carcinogens (cancer inducers). Who derives most of the benefits from the use of food colors? Who assumes most of the risk associated with use of these dyes?

32. An artificial sweetener is 4000 times as sweet as table sugar but exhibits possible toxic side effects. Do a risk–benefit analysis of the sweetener.

33. Nitrogen mustard is extremely toxic—a form of "mustard gas" was used in World War I—but it is effective in treating some forms of skin cancer. Do a risk–benefit analysis of nitrogen mustard.

34. Some researchers think that a glass or two of red wine provides some protection against heart disease. Excessive consumption of alcoholic beverages causes a host of medical problems. Do a risk–benefit analysis of red wine consumption for **(a)** the population as a whole and for **(b)** an alcoholic.

Matter

35. Which of the following are examples of matter?
    a. natural gas   b. love
    c. iron   d. automobile exhaust

36. Which of the following are examples of matter?
    a. the human body   b. air   c. an idea
    d. blue light   e. a checking-account balance

Mass and Weight

37. Which of the following is a reasonable mass for a typical adult student?

    70 mg   70 g   700 g   70 kg

38. A formerly chubby but now trim person completes a successful diet. Has the person's weight changed? Has the person's mass changed? Explain.

39. Two samples are weighed under identical conditions in a laboratory. Sample A weighs 1.00 lb and Sample B weighs 2.00 lb. Does Sample B have twice the mass of Sample A?

40. Sample A on the moon has exactly the same mass as Sample B on Earth. Do the two samples weigh the same? Explain.

Length, Area, and Volume

41. Which of the following is a reasonable volume for a tea cup?

    25 mL   250 mL   2.5 L   25 L

42. Which of the following is a reasonable height for a typical adult student?

    17 mm   170 mm   170 cm   17 m

43. Earth's oceans contain $3.50 \times 10^8 \, mi^3$ of water and cover an area of $1.40 \times 10^8 \, mi^2$. What is the volume of ocean water in cubic kilometers?

44. What is the area of the oceans in square kilometers? See Problem 43.

Physical and Chemical Properties and Change

45. Which of the following describes a physical change, and which describes a chemical change?
    a. Sheep are sheared and the wool is spun into yarn.
    b. Silkworms feed on mulberry leaves and produce silk.
    c. Milk that has been left outside a refrigerator overnight turns sour.
    d. A cake is baked from a mixture of flour, baking powder, sugar, eggs, shortening, and milk.

46. Which of the following describes a physical change, and which a chemical change?
    a. Because a lawn is watered and fertilized, the grass grows thicker.
    b. An overgrown lawn is manicured by mowing it with a lawn mower.
    c. Ice cubes form when a tray filled with water is placed in a freezer.
    d. An egg is fried in a skillet.

47. Identify the following as physical or chemical properties.
    a. Metals reflect light.
    b. Gasoline and oil burn in a weed-trimmer engine.
    c. Water has a density of 1.0 g/mL.
    d. Nitrogen is a gas a room temperature.

48. Identify the following as physical or chemical properties.
    a. Solid iron melts at a temperature of 1535 °C.
    b. Solid sulfur is yellow.
    c. Natural gas burns with a blue flame.
    d. Diamond is extremely hard.

Substances and Mixtures

49. Identify each of the following as a substance or a mixture.
    a. carbon dioxide   b. oxygen
    c. smog   d. a carrot
    e. blueberry pancakes   f. adhesive tape

50. Identify each of the following as a substance or a mixture.
    a. helium gas used to fill a balloon
    b. the juice squeezed from an orange
    c. distilled water
    d. carbon dioxide gas

51. Which of the following mixtures are homogeneous and which are heterogeneous?
    a. gasoline
    b. Italian salad dressing
    c. raisin pudding
    d. an intravenous glucose solution

52. Which of the following mixtures are homogeneous and which are heterogeneous?
    a. maple syrup
    b. distilled water
    c. liquid oxygen
    d. chicken noodle soup

53. Every sample of the sugar glucose (collected anywhere on Earth) consists of 8 parts (by mass) oxygen, 6 parts carbon, and 1 part hydrogen. Is glucose a substance or a mixture? Explain.

54. An advertisement for shampoo says, "Pure shampoo, with nothing artificial added." Is this shampoo a substance or a mixture? Explain.

Elements and Compounds

55. Which of the following represent elements and which represent compounds?
    a. H   b. He
    c. HF   d. Hf

56. Which of the following represent elements and which represent compounds?
    a. C   b. CO
    c. Cf   d. CF_4

57. Without consulting tables, write symbols for each of the following.
    a. carbon   b. chlorine   c. iron

58. Without consulting tables, write symbols for each of the following.
    a. calcium   b. potassium   c. plutonium

59. Without consulting tables, name each of the following.
    a. H   b. O   c. Na

60. Without consulting tables, name each of the following.
    a. N   b. Pb   c. U

61. In his 1789 textbook, *Traité élementaire de Chimie*, Antoine Lavoisier (Chapter 2) listed 33 known elements, one of which was baryte. Which of the following observations best shows that baryte cannot be an element?
    (a) Baryte is insoluble in water.
    (b) Baryte melts at 1580 °C.
    (c) Baryte has a density of $4.48 \, g/cm^3$.
    (d) Baryte is formed in hydrothermal veins and around hot springs.
    (e) Baryte is formed as a solid when sulfuric acid and barium hydroxide are mixed.
    (f) Baryte is formed as a sole product when a particular metal is burned in oxygen.

62. In 1774 Joseph Priestley isolated a gas that he called "dephlogisticated air." Which of the following observations best shows that dephlogisticated air is an element?
   (a) Dephlogisticated air combines with charcoal to form "fixed air."
   (b) Dephlogisticated air combines with "inflammable air" to form water.
   (c) Dephlogisticated air combines with a metal to form a solid called a "calx."
   (d) Dephlogisticated air has never been separated into simpler substances.
   (e) Dephlogisticated air and "mephitic air" are the main components of the atmosphere.

The Metric System: Measurement and Unit Conversion

63. Change the unit used to report each of the following measurements by replacing the power of ten with an appropriate SI prefix.
   a. 8.01×10^{-6} g   **b.** 7.9×10^{-3} L   **c.** 1.05×10^{3} m

64. Use exponential notation to express each of the following measurements in terms of an SI base unit.
   a. 45 mg   **b.** 125 ns   **c.** $10.7 \, \mu$L

65. Carry out the following conversions.
   a. 37.4 mL to L   **b.** 1.55×10^{2} km to m
   c. 0.198 g to mg   **d.** 1.19 m^2 to cm^2
   e. 78 μs to ms

66. Carry out the following conversions.
   a. 546 mm to m   **b.** 65 ns to μs
   c. 87.6 mg to kg   **d.** 46.3 dm^3 to L
   e. 181 pm to μm

67. For each of the following, indicate which is the larger unit.
   a. mm or cm   **b.** kg or g   **c.** dL or μL

68. For each of the following, indicate which is the larger unit.
   a. L or cm^3 **b.** dm^3 or mL   **c.** μs or ps

69. How many milliliters are there in 1.00 cm^3? In 15.3 cm^3?

70. How many millimeters are there in 1.00 cm? In 1.83 m?

Density

(You may need data from Table 1.6 for some of these problems.)

71. What is the density, in grams per milliliter, of (a) a salt solution if 75.0 mL has a mass of 87.5 g? (b) 2.75 L of the liquid glycerol, which has a mass of 3465 g?

72. What is the density, in grams per milliliter, of (a) a sulfuric acid solution if 5.00 mL has a mass of 7.52 g? (b) a 10.0-cm^3 block of plastic with a mass of 9.23 g?

73. What is the mass, in grams, of (a) 125 mL of castor oil, a laxative, which has a density of 0.962 g/mL? (b) 477 mL of blood plasma, $d = 1.027$ g/mL?

74. What is the mass, in grams, of (a) 30.0 mL of the liquid propylene glycol, a moisturizing agent for foods, which has a density of 1.036 g/mL at 25 °C? (b) 1.000 L of mercury at 25 °C?

75. What is the volume of (a) a 475-g piece of copper (in cubic centimeters)? (b) a 253-g sample of mercury (in milliliters)?

76. What is the volume of (a) 227 g of hexane (in milliliters)? (b) a 454-g block of ice (in cubic centimeters)?

Energy: Heat and Temperature

77. Convert 37 °C to kelvins.

78. Temperature-dependent scientific data often is recorded at 298 K. What is that value in degrees Celsius?

79. The label on a 100-mL container of orange juice packaged in New Zealand reads in part, "energy … 161 kJ." What is that value in kilocalories (food "calories")?

80. To vaporize 1.00 g of sweat (water) from your skin, the water must absorb 584 calories. What is that value in kilojoules?

ADDITIONAL PROBLEMS

81. The brass weight and the pillow (left photo) have the same mass. Which has the greater density? Explain.

82. Arrange the following in order of increasing length (shortest first): (1) a 1.21-m chain, (2) a 75-in. board, (3) a 3-ft 5-in. rattlesnake, (4) a yardstick.

83. Arrange the following in order of increasing mass (lightest first): (1) a 5-lb bag of potatoes, (2) a 1.65-kg cabbage, (3) 2500 g sugar.

84. One of the women (right photo) has a mass of 38.5 kg and a height of 1.51 m. Which one is it likely to be?

85. Some metal chips having a volume of 3.29 cm^3 are placed on a piece of paper and weighed. The combined mass is found to be 18.43 g. The paper itself weighs 1.21 g. Calculate the density of the metal.

86. A glass container weighs 48.462 g. A sample of 4.00 mL of antifreeze solution is added, and the container plus the antifreeze weigh 54.51 g. Calculate the density of the antifreeze solution.

87. A rectangular block of balsa wood ($d = 0.11$ g/cm^3) is 7.6 cm $\times$ 7.6 cm $\times$ 94 cm. What is the mass of the block in grams?

88. A rectangular block of gold-colored material measures 3.00 cm $\times$ 1.25 cm $\times$ 1.50 cm and has a mass of 28.12 g. Can the material be gold? Explain.

89. A 5.79-mg piece of gold is hammered into gold leaf of uniform thickness with an area of 44.6 cm². What is the thickness of the gold leaf?

90. A box with a square base measuring 0.80 m on each side and having a height of 1.20 m is filled with 3.2 kg of expanded polystyrene packing material. What is the bulk density, in grams per cubic centimeter, of the packing material? (The bulk density includes the air between the pieces of polystyrene foam.)

91. What is the mass, in metric tons, of a cube of gold that is 36.1 cm on each side? (1 metric ton = 1000 kg)

92. A 10.5-inch (26.7-cm) iron skillet has a mass of 7.00 lb (3180 g). How many cubic centimeters of iron does it contain?

93. Refer to Problems 43 and 44. What is the average depth of Earth's oceans in kilometers?

COLLABORATIVE GROUP PROJECTS

Prepare a PowerPoint, poster, or other presentation (as directed by your instructor) for presentation to the class.

94. List five chemical activities you have engaged in today. Compare your list with that of the other members of your group.

95. Prepare a brief biographical report on one of the following and share it with your group.
   a. Francis Bacon
   b. Rachel Carson
   c. Thomas Malthus
   d. George Washington Carver

(The following problem is best done in a group of four students.)

96. Make copies of the following form. Student 1 should write a word from the list in the first column of the form and its definition in the second column. Then she or he should fold the first column under to hide the word and pass the sheet to Student 2, who uses the definition to determine what word was defined and place that word in the third column. Student 2 then folds the second column under to hide it and passes the sheet to Student 3, who writes a def-

inition for the word in the third column and then folds the third column under and passes the form to Student 4. Finally, Student 4 writes the word corresponding to the definition given by Student 3.

Compare the word in the last column with that in the first column. Discuss any differences in the two definitions. If the word in the last column differs from that in the first column, determine what went wrong in the process.
   a. hypothesis   b. theory
   c. mixture   d. substance

Text Entry	Student 1	Student 2	Student 3	Student 4
Word	Definition	Word	Definition	Word

REFERENCES AND READINGS

1. Ashworth, William J. "Metrology and the State: Science, Revenue, and Commerce." *Science*, 19 November 2004, pp. 1314–1317. A history of measurement in Great Britain.

2. Ball, Philip. *The Ingredients: A Guided Tour of the Elements.* New York: Oxford University Press, 2003.

3. Bate, Roger. *What Risk?* Stoneham, MA: Butterworth-Heinemann, 1999.

4. Carson, Rachel. *Silent Spring.* Boston: Houghton Mifflin, 1962. The classic book on dangers to the environment.

5. Derry, Gregory N. *What Science Is and How It Works.* Princeton, NJ: Princeton University Press, 1999.

6. Emsley, John. *The Elements*, 3rd ed. Oxford, UK: Oxford University Press, 1998.

7. Guttman, Burton S. "The Real Method of Scientific Discovery." *Skeptical Inquirer*, January/February 2004, pp. 45–47.

8. Hill, John W., Ralph H. Petrucci, Terry W. McCreary, and Scott S. Perry. *General Chemistry*, 4th ed. Upper Saddle River, NJ: Prentice Hall, 2005. Chapter 1 provides a more extensive treatment of some of the topics in this chapter.

9. Lett, James. "A Field Guide to Critical Thinking." *The Skeptical Inquirer*, Winter 1990, pp. 153–160.

10. Sagan, Carl. *The Demon-Haunted World: Science as a Candle in the Dark.* New York: Random House, 1996.

11. Shermer, Michael. "Baloney Detection: How to Draw Boundaries between Science and Pseudoscience," *Scientific American*, Part I. November 2001, p. 36, and Part II. December 2001, p. 34.

12. Todd, Stuart. "Alchemy: Secret Chemistry?" *Chemistry and Industry*, December 20, 2004, pp. 10–12.

13. Tyson, Neil de Grasse. "On Being Dense." *Natural History*, January 1996, pp. 66–67.

14. Tyson, Neil de Grasse. "The Long and Short of It." *Natural History*, April 2005, pp. 24–28.

15. Watson, Bruce. "Sounding the Alarm." *Smithsonian*, September 2002, pp. 115–117. Looking back on *Silent Spring* after 40 years.

16. "What We Don't Know." *Science*, 1 July 2005, pp. 75–102. A special section on the scientific puzzles that drive reasearch. See especially, "How Hot Will the Greenhouse Be?" p. 100, and "What Can Replace Cheap Oil—and When?" p. 101.

GREEN CHEMISTRY

Introduction to Green Chemistry

Paul Anastas, ACS Green Chemistry Institute

In the following Web Investigations, you will learn more about the concepts, principles, and recent advances in green chemistry. These should help to frame your perspective on the chemistry you will be learning through-out this course. You will see many more examples and explore many of these concepts in the Green Chemistry activities in future chapters.

WEB INVESTIGATIONS

Investigation 1
The Green Chemistry Institute
What is green chemistry? Visit the American Chemical Society's website and search for "Green Chemistry Institute." Follow the link on the Web page to "What are Green Chemistry and Green Engineering." The definition of *green chemistry* discusses the design of chemical products and processes. What are three chemical products you use every day?

Investigation 2
Pollution Prevention
What is pollution prevention? Pollution prevention is defined as "practices that reduce or eliminate the creation of pollutants." Perform a keyword search for "pollution prevention" using your favorite search engine. Describe how one pollution prevention program works to minimize or prevent pollution.

Investigation 3
Nobel Prize
Perhaps the highest honor a chemist can receive is the Nobel Prize in chemistry. When the 2005 Nobel Prize in chemistry was awarded, the Nobel Committee described the winning work as "a great step forward for green chemistry." Investigate the winners of this award online. What are the three main reasons that the Nobel Committee said this work is a good example of green chemistry?

▲ The medal for the Nobel Prize in chemistry is the highest honor a chemist can receive.

▲ Richard R. Schrock, left, receives the Nobel Prize in chemistry from King Carl XVI Gustaf of Sweden, December 10, 2005. This prize was shared with two other scientists, Robert H. Grubbs and Yves Chauvin.

Investigation 4
Chemicals in the Environment

The Toxic Release Inventory (TRI) was established by Congress in 1986 with the Emergency Planning and Community Right-to-Know Act. According to the Environmental Protection Agency, what is the Toxic Release Inventory? From their website you can find out more about TRI releases in your hometown or where you live now. By entering your home or current zip code, answer the following questions for your area of interest: What potential hazardous waste sites exist that are part of a Superfund site? What facilities have reported hazardous waste activities? What detailed hazardous waste information for large-quantity generators exists?

COMMUNICATE YOUR RESULTS

Exercise 1
Getting the Word Out

Compare the results of a keyword search for "green chemistry" or "sustainable chemistry" in your favorite search engine, the website of a prominent newspaper or television network, and a website dedicated to the study of chemistry such as chemistry.org. Based on how much—or how little—you have noticed in the media, do you think that coverage of the Green Chemistry Initiative is adequate? Compose a letter to a newspaper or television or radio network, explaining what green chemistry is and urging expanded coverage.

Exercise 2
An Ounce of Prevention . . .

There is an old saying that "an ounce of prevention is worth a pound of cure." This is related to the first principle of green chemistry, which states that "It is better to prevent waste than to treat or clean up waste after it has been created." Write a 250-word essay that explains the benefits of preventing pollution rather than treating and disposing of waste. These benefits may involve the economy, the environment, worker safety, compliance with regulations, public relations, marketing, and so on.

COOPERATIVE EXERCISE

The World Wide Web is vast, with a rapidly increasing abundance of information. Working in groups can help! After you have become acquainted with the concepts of green chemistry, form small groups of "reporters." Return to the Environmental Protection Agency's website to read about the Presidential Green Chemistry Award winners, and look closely at a winner of one of these awards. Write a brief report on the company and on why it won the award. Compose a letter of commendation to the company.

CHAPTER 2

Atoms

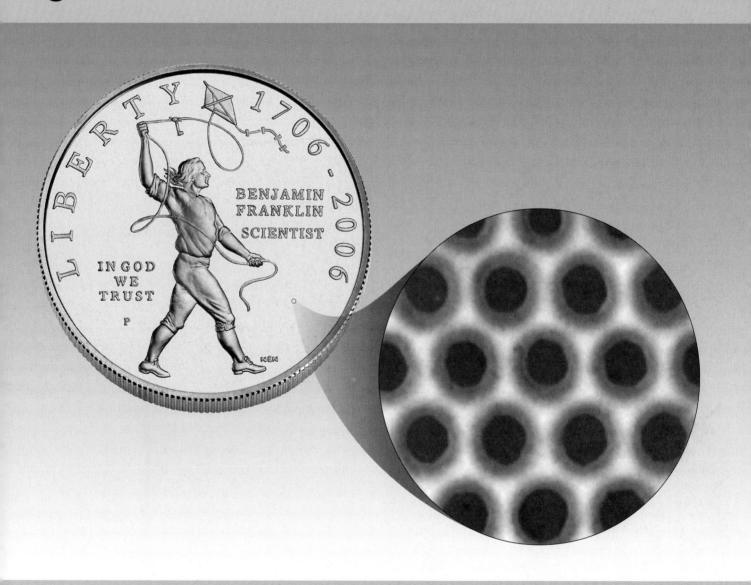

False color image of the atoms on the surface of the element silver (Ag). Each bright spot indicates a silver atom. The image was made by a technique called scanning tunneling microscopy. The hypothesis that all matter is composed of atoms is over 2000 years old, but it was only in the latter part of the twentieth century that scientists were able to obtain images of atoms.

Are They for Real?

We hear something about atoms almost every day. Some say we live in the Atomic Age. The terms *atomic power*, *atomic energy*, and *atomic bomb* are a part of our ordinary vocabulary. But just what are atoms?

Every material thing in the world is made up of atoms, tiny particles that are much, much too small to see. The smallest speck of matter that can be detected by the human eye is made up of many billions of atoms. There are more than 10^{22} atoms in a penny. That is 10,000,000,000,000,000,000,000 atoms. Imagine the atoms in a penny being enlarged until they were barely visible, like tiny grains of sand. The atoms in a single penny would make enough "sand" to cover the entire state of Texas to a depth of several feet. Comparing an atom to a penny is like comparing a grain of sand to a Texas-size sandbox.

Why should we care about something as tiny as an atom? Because our world is made up of atoms, and atoms are a part of all we do. *Everything* is made of atoms, including you and me. And because chemistry is the study of the behavior of matter, chemistry studies the behavior of atoms.

Atoms are not all alike. Each element has its own kind of atoms. On Earth, about 90 elements occur in nature. About two dozen more have been synthesized by scientists. As far as we know, the entire universe is made up of these same elements. An atom is the smallest particle that is characteristic of a given element.

2.1 ATOMS: THE GREEK IDEA

A pool of water can be separated into drops, and then each drop can be split into smaller and smaller drops. Suppose you could keep splitting these drops into still smaller ones even after they became much too small to see. Would you ever reach a point at which the tiny drop could no longer be separated into smaller droplets of water?

The Greek philosopher Leucippus, who lived in the fifth century B.C.E., and his pupil Democritus (ca. 460–ca. 370 B.C.E.) might well have discussed this question as they strolled along the beach of the Aegean Sea. Based only on intuition, Leucippus thought that there must ultimately be tiny particles of water that could not be subdivided. After all, from a distance the sand on the beach looked continuous, but closer inspection showed it to be made up of tiny grains (Figure 2.1).

Democritus expanded on Leucippus's idea. He called the particles *atomos* (meaning "cannot be cut"), from which we derive the modern name *atom* for the

▲ This 1983 postage stamp from Greece has an image of Democritus.

Question: Sand looks continuous when you look at a beach from a distance. Is it really continuous? Water looks continuous, even when viewed up close. Is it really continuous? Is a cloud continuous? Is air?

▲ **Figure 2.2** Democritus imagined that "atoms" of water might be smooth, round balls and that atoms of fire could have sharp edges.

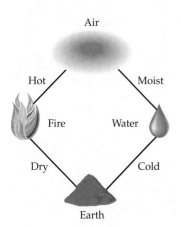

▲ **Figure 2.3** The Greek view of matter was that there were only four elements connected by four "principles."

tiny unit particle of an element. Democritus thought that each kind of atom was distinct in shape and size (Figure 2.2). He thought real substances were mixtures of various kinds of atoms. The Greeks at that time believed that there were four basic elements: earth, air, fire, and water. The relationships among these elements and the four "principles"—hot, moist, dry, and cold—are shown in Figure 2.3.

Four centuries later, the Roman poet Lucretius (ca. 95–ca. 55 B.C.E.) wrote a long didactic poem (a poem meant to teach), "On the Nature of Things," in which he presented strong arguments for the atomic nature of matter. Unfortunately, a few centuries earlier, Aristotle (ca. 384–ca. 322 B.C.E.), considered the greatest of the ancient Greek philosophers, had declared that matter was continuous, not atomistic, and the people of that time had no way to determine which view was correct. To most of them, the continuous view of matter seemed more logical and reasonable, and so Aristotle's view prevailed for 2000 years, even though it was wrong.

2.2 LAVOISIER: THE LAW OF CONSERVATION OF MASS

The eighteenth century saw the triumph of careful observation and measurement. Antoine Laurent Lavoisier (1743–1794) perhaps did more than anyone to establish chemistry as a quantitative science. He found that when a chemical reaction was carried out in a closed system, the total mass of the system was not changed. Perhaps the most important chemical reaction that Lavoisier performed was decomposition of the red oxide of mercury to form metallic mercury and a gas he named oxygen. Karl Wilhelm Scheele (1742–1786), a Swedish apothecary, and Joseph Priestley (1733–1804), a Unitarian minister who later fled England and eventually settled in America, had carried out the same reaction earlier, but Lavoisier was the first to weigh all the substances present before and after the reaction. He was also the first to interpret the reaction correctly.

Lavoisier carried out many quantitative experiments. He found that when coal was burned, it united with oxygen to form carbon dioxide. He experimented with animals, observing that when a guinea pig breathed, oxygen was consumed and carbon dioxide was formed. Lavoisier therefore concluded that respiration was related to combustion. In each of these reactions, he found that matter was *conserved*—its amount remained constant.

Lavoisier summarized his findings in a scientific law. The **law of conservation of mass** states that matter is neither created nor destroyed during a chemical change (Figure 2.4). The total mass of the reaction products is always equal to the total mass of the reactants (starting materials).

Scientists had by this time abandoned the Greek idea of the four elements and were almost universally using Robert Boyle's operational definition put forth over a century before. In his book *The Sceptical Chymist* (published in 1661), Boyle said that a supposed *element* must be tested to see if it really was simple. If a substance could be broken down into simpler substances, it was not an element. The simpler substances might be elements and would be so regarded until such time (if it ever came) as they in turn could be broken down into still simpler substances. On the other hand, two or more elements might combine to form a complex substance called a *compound*.

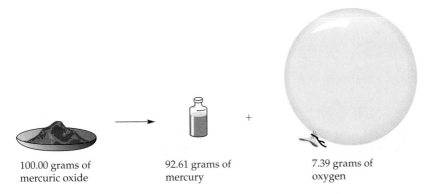

100.00 grams of mercuric oxide

92.61 grams of mercury

7.39 grams of oxygen

◀ **Figure 2.4** Although mercuric oxide (a red solid) has none of the properties of mercury (a silver liquid) or oxygen (a colorless gas), when 100.00 g of mercuric oxide is decomposed by heating, the products are 92.61 g of mercury and 7.39 g of oxygen. Properties are completely changed in this reaction, but there is no change in mass.

Using Boyle's definition, Lavoisier included a table of elements in his book *Elementary Treatise on Chemistry*. The table included some substances we now know to be compounds. (It also included light and heat, which he called "caloric.") Lavoisier was the first to use systematic names for chemical elements. He is often called the "father of modern chemistry," and his book is usually regarded as the first chemistry textbook.

The law of conservation of mass is the basis for many chemical calculations. For example, we can calculate the mass of iron ore needed to produce a ton of iron metal. (Such calculations are discussed in detail in Chapter 6.) This law is not just a matter of academic interest. It states that we cannot create materials from nothing; we can make new materials only by changing the way atoms are combined. Nor can we get rid of wastes by the destruction of matter. We must put wastes somewhere. However, through chemical reactions, we can change some kinds of potentially hazardous wastes to less harmful forms. Such transformations of matter from one form to another are what chemistry is all about.

▲ Antoine Lavoisier and his wife, Marie, portrayed in a painting by Jacques Louis David in 1788.

2.3 PROUST: THE LAW OF DEFINITE PROPORTIONS

By the end of the eighteenth century, Lavoisier and other scientists noted that many substances were composed of two or more elements. Each compound had the same elements in the same proportions, regardless of where it came from or who prepared it. The painstaking work of Joseph Louis Proust (1754–1826) convinced most chemists of the general validity of these observations. In one set of experiments, for example, Proust found that basic copper carbonate, whether prepared in the laboratory or obtained from natural sources, was always composed of 57.48% by mass copper, 5.43% carbon, 0.91% hydrogen, and 36.18% oxygen (Figure 2.5). To summarize these and many other experiments, Proust in 1799 formulated a new scientific law. The **law of definite proportions** states that a compound always contains the same elements in certain definite proportions and in no other combinations. (This generalization is also sometimes called the *law of constant composition*.)

Proust, like Lavoisier, was a member of the French nobility. He was working in Spain, temporarily safe from the ravages of the French Revolution. His laboratory was destroyed and he was reduced to poverty, however, when the French troops of Napoleon Bonaparte occupied Madrid in 1808.

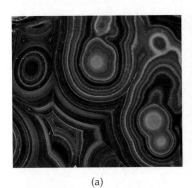

(a)

(b)

(c)

▲ **Figure 2.5** Basic copper carbonate occurs in nature as the mineral *malachite* (a). It is formed as a patina on copper roofs (b). It can also be synthesized in the laboratory (c). Regardless of its source, basic copper carbonate always has the same composition. Analysis of this compound led Proust to formulate the law of definite proportions.

▲ Jöns Jakob Berzelius (1779–1848) was the first person to prepare an extensive list of atomic weights. Published in 1828, it agrees remarkably well with most of our accepted values today.

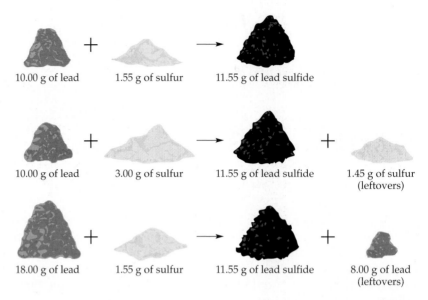

| 10.00 g of lead | + | 1.55 g of sulfur | → | 11.55 g of lead sulfide |

| 10.00 g of lead | + | 3.00 g of sulfur | → | 11.55 g of lead sulfide | + | 1.45 g of sulfur (leftovers) |

| 18.00 g of lead | + | 1.55 g of sulfur | → | 11.55 g of lead sulfide | + | 8.00 g of lead (leftovers) |

▲ **Figure 2.6** An example showing how Berzelius's experiments illustrate the law of definite proportions.

An early illustration of the law of definite proportions is found in the work of the noted Swedish chemist J. J. Berzelius, illustrated in Figure 2.6. Berzelius heated a quantity (say, 10.00 g) of lead with various amounts of sulfur to form lead sulfide. Lead is a soft, grayish metal, and sulfur is a yellow solid. Lead sulfide is a shiny, black solid. Therefore, it was easy to tell when all the lead had reacted. Excess sulfur was washed away with carbon disulfide, a solvent that dissolves sulfur but not lead sulfide. As long as he used at least 1.55 g of sulfur with 10.00 g of lead, Berzelius got exactly 11.55 g of lead sulfide. Any sulfur in excess of 1.55 g was left over, unreacted. If he used more than 10.00 g of lead with 1.55 g of sulfur, he got 11.55 g of lead sulfide, with lead left over.

The law of definite proportions is further illustrated by the electrolysis of water. In 1783 Henry Cavendish (1731–1810), a wealthy, eccentric English nobleman, found that water forms when hydrogen burns in oxygen. (It was Lavoisier, however, who correctly interpreted the experiment and who first used the names *hydrogen* and *oxygen*.) Later, in 1800, two English chemists, William Nicholson and Anthony Carlisle, decomposed water into hydrogen and oxygen gases by passing an electric current through the water (Figure 2.7). (The Italian scientist Alessandro Volta had invented the chemical battery only six weeks earlier.) The two gases are always produced in a 2:1 volume ratio. Although this is a volume ratio and Berzelius's experiment is a mass ratio, both substantiate the law of definite proportions. This scientific law led to rapid developments in chemistry and dealt a death blow to the ancient Greek idea of water as an element.

The law of definite proportions is the basis for chemical formulas (Chapter 6), such as H_2O. It also has wider meaning. Not only do compounds have constant composition, but they also have constant properties. Pure water always dissolves salt or sugar, and at normal pressure it always freezes at 0 °C and boils at 100 °C.

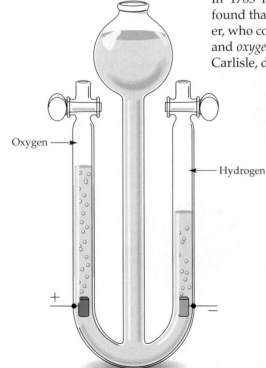

Oxygen ⟶

⟵ Hydrogen

◀ **Figure 2.7** Electrolysis of water. Hydrogen and oxygen are always produced in a volume ratio of 2:1.

Question: Under the same conditions of temperature and pressure, molecules of hydrogen occupy the same volume as an equal number of oxygen molecules. What ratio of hydrogen molecules to oxygen molecules is produced by the electrolysis of water?

2.4 JOHN DALTON AND THE ATOMIC THEORY OF MATTER

Lavoisier's law of conservation of mass and Proust's law of definite proportions were repeatedly verified by experiment. This work led to attempts to develop theories to explain these laws.

In 1803 John Dalton, an English schoolteacher, proposed a model to explain the accumulating experimental data. By this time, the composition of a number of substances was known with a fair degree of accuracy. (To avoid confusion, we will use modern values, terms, and examples rather than those actually used by Dalton.) For example, all samples of water have an oxygen to hydrogen mass ratio of 8.01:1.00. Similarly, all samples of ammonia have a nitrogen to hydrogen mass ratio of 4.68:1.00. Dalton explained these unvarying ratios by assuming that matter is made of atoms.

As Dalton refined his model, he discovered another law that his theory would have to explain. Proust had stated that a compound contains elements in certain proportions and only those proportions. Dalton's new law, called the **law of multiple proportions**, stated that elements might combine in *more* than one set of proportions, with each set corresponding to a different compound. For example, carbon combines with oxygen in a mass ratio of 1.00:2.66 (or 3.00:8.00) to form carbon dioxide, a gas familiar as a product of respiration and of the burning of coal and wood. But Dalton found that carbon also combines with oxygen in a mass ratio of 1.00:1.33 (or 3.00:4.00) to form carbon monoxide, a poisonous gas produced when a fuel is burned in the presence of a limited air supply.

Dalton then used his **atomic theory** to explain the various laws. Following are the important points of Dalton's atomic theory, with some modern modifications that we will consider later.

▲ John Dalton (1766–1848).

 Law of Multiple Proportions

Dalton's Atomic Theory

1. All matter is composed of extremely small particles called atoms.

2. All atoms of a given element are alike, but atoms of one element differ from the atoms of any other element.

3. Compounds are formed when atoms of different elements combine in fixed proportions.

4. A chemical reaction involves a *rearrangement* of atoms. No atoms are created, destroyed, or broken apart in a chemical reaction.

Modern Modifications

1. Dalton assumed atoms to be indivisible. This isn't quite true, as we will see in the next chapter.

2. Dalton assumed that all the atoms of a given element were identical in all respects, including mass. We now know this to be incorrect, as we will see on page 64.

3. Unmodified. The numbers of each kind of atom in simple compounds usually form a simple ratio. For example, the ratio of carbon atoms to oxygen atoms is 1:1 in carbon monoxide and 1:2 in carbon dioxide.

4. Unmodified for *chemical* reactions. Atoms are broken apart in *nuclear* reactions.

The ancient Greeks' ideas of atoms were mainly intuitive. However, by 1800 scientists had accumulated considerable evidence for the existence of atoms.

Explanations Using Atomic Theory

Dalton's theory clearly explains the difference between elements and compounds. *Elements* are composed of only one kind of atom. For example, a sample of the element phosphorus has only phosphorus atoms in it. (We will explain more precisely what we mean by *kind* in Section 3.5.) *Compounds* are made up of two or more kinds of atoms chemically combined in definite proportions.

Dalton set up a table of relative atomic masses based on hydrogen as 1. Many of Dalton's atomic masses were inaccurate, as we might expect because of the equipment available at that time. These relative atomic masses are usually simply called

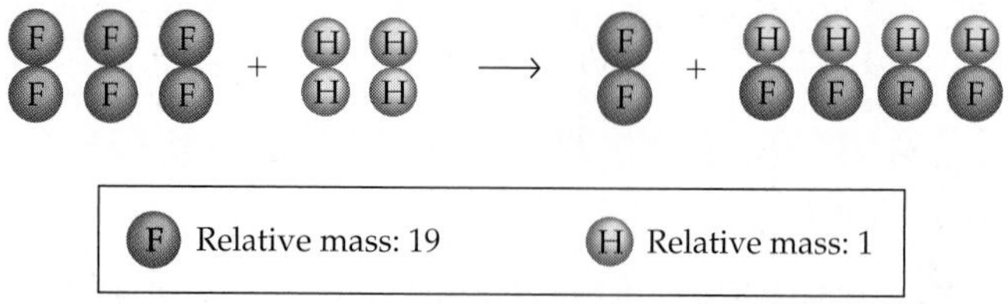

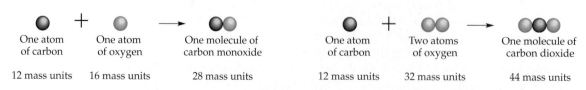

F Relative mass: 19   H Relative mass: 1

▲ **Figure 2.8** The law of definite proportions and the law of conservation of mass interpreted in terms of Dalton's atomic theory.

Question: Use Dalton's atomic theory to interpret Berzelius's experiment (Figure 2.6). Assume sulfur atoms have a relative mass of 32.066 and lead atoms a relative mass of 207.2. What is the atom ratio of lead to sulfur in lead sulfide? How does the experiment confirm the law of conservation of mass?

atomic masses. Historically, the relative masses were usually determined by comparison with a standard mass, a technique called *weighing*. For this historical reason, we often refer to these relative masses as *atomic weights*. You will find a table of atomic masses on the inside front cover of this book. We will use these modern values to show how Dalton's atomic theory explains the various laws.

To explain the law of definite proportions, Dalton's reasoning went something like this: Why should 1.0 g of hydrogen always combine with 19 g of fluorine? Why shouldn't 1.0 g of hydrogen also combine with 18 g of fluorine? Or 20 g of fluorine? Or any other mass of fluorine? If an atom of fluorine has a mass 19 times that of a hydrogen atom, the compound formed by the union of one atom of each element would have to consist of 1 part by mass of hydrogen and 19 parts by mass of fluorine. Matter must be atomic for the law of definite proportions to be valid (Figure 2.8).

Atomic theory explains the law of conservation of mass. When fluorine atoms combine with hydrogen atoms to form hydrogen fluoride, the atoms are merely rearranged. Matter is neither lost nor gained; the mass does not change.

Atomic theory also explains the law of multiple proportions. For example, 1.00 g of carbon combines with 1.33 g of oxygen to form carbon monoxide, and with 2.66 g of oxygen to form carbon dioxide. Carbon dioxide has twice the mass of oxygen per gram of carbon as does carbon monoxide. This is explained by the fact that one atom of carbon combines with *one* atom of oxygen to form carbon monoxide, and one atom of carbon combines with *two* atoms of oxygen to form carbon dioxide (Figure 2.9). Using modern values, we assign an oxygen atom a mass of 16.0 and a carbon atom a mass of 12.0. On this scale carbon monoxide is seen to be made up of one atom of carbon combined with one atom of oxygen to give a mass ratio of 12.0 parts carbon to 16.0 parts oxygen (or 3.00:4.00). Carbon dioxide is composed of one atom of carbon combined with two atoms of oxygen to give a mass ratio of 12.0 parts carbon to $2 \times 16.0 = 32.0$ parts oxygen (or 3.00:8.00). Table 2.1 shows some of the proportions in which nitrogen and oxygen combine.

Dalton also invented a set of symbols (Figure 2.10) to represent the different kinds of atoms. In fact, symbols similar to Dalton's are used in Figure 2.8 and Table 2.1. These symbols have since been replaced by modern symbols of one or two letters (inside front cover).

Isotopes

As we have noted, Dalton's second assumption has been modified. Not all atoms of an element have the same mass. Atoms of an element with different masses are called *isotopes*. For example, while most carbon atoms have a relative atomic mass of 12 (carbon-12), 1.1% of carbon atoms have a relative atomic mass of 13 (carbon-13). We discuss isotopes in more detail in Chapter 3.

One atom of carbon	One atom of oxygen	One molecule of carbon monoxide	One atom of carbon	Two atoms of oxygen	One molecule of carbon dioxide
12 mass units	16 mass units	28 mass units	12 mass units	32 mass units	44 mass units

▲ **Figure 2.9** The law of multiple proportions. A carbon atom can combine with either one or two atoms of oxygen.

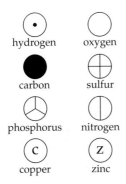

▲ **Figure 2.10** Some of Dalton's symbols for the elements.

Question: Using Dalton's symbols, draw diagrams for sulfur dioxide and sulfur trioxide. (Place the sulfur atom in the center and arrange the oxygen atoms around it.) What law does these two compounds illustrate?

TABLE 2.1	The Law of Multiple Proportions		
Compound	Representation[a]	Mass of N per 1.000 g of O	Ratio of the Masses of N[b]
Nitrous oxide		1.750 g	$(1.750 \div 0.4375) = 4.000$
Nitric oxide		0.8750 g	$(0.8750 \div 0.4375) = 2.000$
Nitrogen dioxide		0.4375 g	$(0.4375 \div 0.4375) = 1.000$

[a] ● = nitrogen atom and ○ = oxygen atom

[b] We obtain the ratio of the masses of N that combine with a given mass of O by dividing each quantity in the third column by the smallest (0.4375 g).

What a Difference an O Makes!

The law of multiple proportions is illustrated by the fact that carbon forms two oxides, carbon monoxide (CO) and carbon dioxide (CO_2). Both are colorless, odorless gases, but how very different they are! Carbon dioxide is always present in your body. It is collected from your cells and carried by your bloodstream to your lungs and then exhaled. The bubbles in "soda"-type drinks are also carbon dioxide. Most fuels produce CO_2 (and H_2O) when they burn.

At temperatures around −80 °C carbon dioxide solidifies and becomes dry ice, which we often use to keep ice cream from melting. Carbon monoxide is quite another thing. It is so highly poisonous that only 0.2% of CO in the air is enough to kill. Unfortunately, most fuels also produce some CO when they burn. This is why an indoor kerosene heater should always be well vented and a car engine should never be left running in a closed garage.

Problem Solving: Mass and Atom Ratios

We can use proportions, such as those determined by Dalton, to calculate the amount of one substance needed to combine with a given quantity of another substance. To learn how to do this, let's look at some examples.

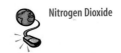
Nitrogen Dioxide

EXAMPLE 2.1   **Mass Ratios**

The gas *methane* (CH_4), the main component of the fuel natural gas, can be decomposed to give carbon (C) and hydrogen (H) in a ratio of 3.00 parts by mass of carbon to 1.00 part by mass of hydrogen. How much hydrogen can be made from 90.0 g of methane?

Solution

We can express the parts ratio in any units we choose—pounds, grams, kilograms—as long as it is the same for both elements. Using grams as the units, we see that 3.00 g of C and 1.00 g of H would mean 4.00 g CH_4 at the start. To convert g CH_4 to g H, the conversion factor we need includes 1.00 g H and 4.00 g CH_4.

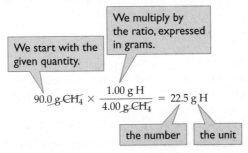

We start with the given quantity.

We multiply by the ratio, expressed in grams.

$$90.0 \text{ g } CH_4 \times \frac{1.00 \text{ g H}}{4.00 \text{ g } CH_4} = 22.5 \text{ g H}$$

the number   the unit

▶ **Exercise 2.1A**

The gas ammonia can be decomposed to give 3.00 parts by mass of hydrogen and 14.0 parts by mass of nitrogen. What mass of nitrogen is obtained if 1.27 g of ammonia is decomposed?

▶ **Exercise 2.1B**

Nitrous oxide, sometimes called "laughing gas," can be decomposed to give 7.00 parts by mass of nitrogen and 4.00 parts by mass of oxygen. What mass of nitrogen is obtained if enough nitrous oxide is decomposed to yield 36.0 g of oxygen?

EXAMPLE 2.2 **Atom Ratios**

Hydrogen sulfide gas can be decomposed to give sulfur and hydrogen in a mass ratio of 16.0:1.00. If the relative mass of sulfur is 32.0 when the mass of hydrogen is taken to be 1.00, how many hydrogen atoms are combined with each sulfur atom in the gas?

Solution

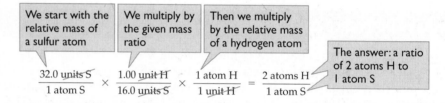

We start with the relative mass of a sulfur atom. We multiply by the given mass ratio. Then we multiply by the relative mass of a hydrogen atom. The answer: a ratio of 2 atoms H to I atom S

$$\frac{32.0 \text{ units S}}{1 \text{ atom S}} \times \frac{1.00 \text{ unit H}}{16.0 \text{ units S}} \times \frac{1 \text{ atom H}}{1 \text{ unit H}} = \frac{2 \text{ atoms H}}{1 \text{ atom S}}$$

▶ **Exercise 2.2**

Arsine gas can be decomposed to give arsenic and hydrogen in a mass ratio of 25.0:1.00. If the relative mass of arsenic is 74.9 when the mass of hydrogen is taken to be 1.00, how many hydrogen atoms are combined with each arsenic atom in the gas?

Despite some inaccuracies, Dalton's atomic theory was a great success. Why? Because it served—and still serves—to explain a large amount of experimental data. It also successfully predicted how matter would behave under a wide variety of circumstances. Dalton arrived at his atomic theory by reasoning based on experimental facts, and with modest modification it has stood the test of time and modern, highly sophisticated instrumentation. Formulation of so successful a theory was quite a triumph for a Quaker schoolteacher in 1803.

2.5 OUT OF CHAOS: THE PERIODIC TABLE

Before moving on, let's take a look at a remarkable parallel development. New elements were being discovered with surprising frequency, and by 1830, there were 55 known elements, all with different properties and with no apparent order in these properties. John Dalton had set up a table of relative atomic masses in his book *A New System of Chemical Philosophy* in 1808. Dalton's rough values were improved in subsequent years, notably by Berzelius, who published a table of atomic weights in 1828 containing 54 elements. Most of Berzelius's values agree well with modern values.

 Interactive Periodic Table

Relative Atomic Masses

Although it was impossible to determine actual masses of atoms in the 1800s, chemists were able to determine relative atomic masses by measuring the amounts of various elements that combined with a given mass of another element. Dalton's atomic masses were based on an atomic mass of 1 for hydrogen. As more accurate atomic weights were determined, this standard was replaced by one in which oxygen was assigned a value of 16.0000. The oxygen standard was used until 1961,

Reihen	Gruppe I. — R^2O	Gruppe II. — RO	Gruppe III. — R^2O^3	Gruppe IV. RH^4 RO^2	Gruppe V. RH^3 R^2O^5	Gruppe VI. RH^2 RO^3	Gruppe VII. RH R^2O^7	Gruppe VIII. — RO^4
1	H=1							
2	Li=7	Be=9,4	B=11	C=12	N=14	O=16	F=19	
3	Na=23	Mg=24	Al=27,3	Si=28	P=31	S=32	Cl=35,5	
4	K=39	Ca=40	—=44	Ti=48	V=51	Cr=52	Mn=55	Fe=56, Co=59, Ni=59, Cu=63.
5	(Cu=63)	Zn=65	—=68	—=72	As=75	Se=78	Br=80	
6	Rb=85	Sr=87	?Yt=88	Zr=90	Nb=94	Mo=96	—=100	Ru=104, Rh=104, Pd=106, Ag=108.
7	(Ag=108)	Cd=112	In=113	Sn=118	Sb=122	Te=125	J=127	
8	Cs=133	Ba=137	?Di=138	?Ce=140	—			— — — —
9	(—)		—	—	—			
10	—	—	?Er=178	?La=180	Ta=182	W=184	—	Os=195, Ir=197, Pt=198, Au=199.
11	(Au=199)	Hg=200	Tl=204	Pb=207	Bi=208	—	—	— — — —
12	—	—		Th=231	—	U=240	—	— — — —

when it was replaced by a more logical one based on an isotope of carbon, carbon-12. Adoption of this new standard caused little change in atomic masses. These relative atomic masses are usually expressed in **atomic mass units (amu)**, commonly referred to today simply as *units (u)*.

Mendeleev's Periodic Table

Various attempts were made to arrange the elements in some sort of systematic fashion. The most successful arrangement—one that soon became widely accepted by chemists—was published in 1869 by Dmitri Ivanovich Mendeleev (1834–1907), a Russian chemist. Mendeleev's **periodic table** arranged the elements primarily in order of increasing atomic mass, although in a few cases he put a slightly heavier element before a lighter one in order to place elements with similar chemical properties in the same column (Figure 2.11). For example, he put tellurium, with an atomic mass of 127.6 u, ahead of iodine, which has an atomic mass of 126.9 u. He did this in order to place tellurium in the same column as sulfur and selenium, which it resembles in chemical properties. This rearrangement also put iodine in the same column as chlorine and bromine, which it resembles.

Mendeleev left gaps in his table. This was also necessary in order to place elements in groups with similar properties. Instead of considering these blank spaces as defects, he boldly predicted the existence of elements yet undiscovered. Further, he even predicted the properties of some of the missing elements. For example, the missing elements he called eka-boron, eka-aluminum, and eka-silicon were soon discovered and named scandium, gallium, and germanium.[1] From their positions in the periodic table, Mendeleev predicted the properties of these elements with amazing success (Table 2.2). This remarkable predictive value led to wide acceptance of Mendeleev's table.

◄ **Figure 2.11** Mendeleev's original periodic table.

▲ Dmitri Mendeleev, the Russian chemist who invented the periodic table of the elements, continues to be honored in his native land. Element 101 (Md) is named mendelevium in his honor.

TABLE 2.2	Properties of Germanium: Predicted and Observed	
Property	**Predicted by Mendeleev for Eka-Silikon (1871)**	**Observed by Winkler for Germanium (1886)**
Atomic mass	72	72.6
Density (g/cm^3)	5.5	5.47
Color	Dirty gray	Grayish white
Density of oxide (g/cm^3)	EsO$_2$: 4.7	GeO$_2$: 4.703
Boiling point of chloride	EsCl$_4$: below 100 °C	GeCl$_4$: 86 °C
Density of chloride (g/cm^3)	EsCl$_4$: 1.9	GeCl$_4$: 1.887

[1]Gallium was discovered in 1875 by P. E. Lecoq de Boisbaudran, who named it for his native land, Gaul (France). Swedish chemist Lars F. Nilson discovered scandium in 1879 and named it for Scandinavia. Finally, in 1886 C. A. Winkler discovered germanium and named it for his country, Germany.

Precursors of the Periodic Table

Dobereiner's "Triads"

As early as 1816 Johann Dobereiner, a German chemist, noticed that there were several groups of three elements that were very similar (lithium, sodium, potassium; calcium, strontium, barium; sulfur, selenium, tellurium; chlorine, bromine, iodine). In each case, the middle element seemed to be halfway between the other two in atomic mass, reactivity, and other properties. Dobereiner published this in 1829.

De Chancourtois's "Telluric Helix"

In 1862 Beguyer de Chancourtois, a French geologist, arranged the elements in order of atomic mass. When he wound the list spirally around a cylinder, he found that similar elements fell along the same vertical lines.

Newlands's "Law of Octaves"

In 1863 John Newlands, an English chemist, noted that when elements were listed in order of atomic mass, every eighth element had similar properties; however, the rule seemed to break down for elements past calcium.

Meyer's System of Elements

In 1868 Lothar Meyer in Germany came up independently with an arrangement of elements similar to that of Mendeleev. Many believe that he should share the credit for the periodic table. Unfortunately, Meyer did not write up his table until December 1869, and it was not published until March 1870. Mendeleev had published his table in 1869.

The modern periodic table (inside front cover) contains 114 elements. Each element is represented by a "box" in the periodic table, and the data typically shown are

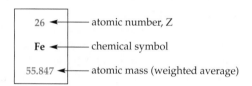

We will discuss the periodic table and its theoretical basis in Chapter 3.

2.6 ATOMS: REAL AND RELEVANT

Are atoms real? Certainly they are real as a concept, a highly useful concept at that. Scientists can even observe computer-enhanced *images* of individual atoms. These portraits reveal little detail of atoms, but they provide powerful (though still indirect) evidence that atoms exist.

Are atoms relevant? Much of modern science and technology—including the production of new materials and the technology of pollution control—is ultimately based on the concept of atoms. We have seen that atoms are conserved in chemical reactions. Thus, material things—things made of atoms—can be recycled, for the atoms are not destroyed no matter how we use them. The one way we might lose a material from a practical standpoint is to spread the atoms so thinly that it would take too much time and energy to put them back together again.

2.7 LEUCIPPUS REVISITED: MOLECULES

Now back to Leucippus and his musings by the seashore. We now know that if we keep dividing drops of water into smaller drops, we will ultimately obtain a small particle—called a *molecule*—that is still water. A **molecule** is a group of atoms, chemically bonded or connected together. Molecules are represented by chemical formulas. The symbol H represents an *atom* of hydrogen; the formula H_2 represents a *molecule* of hydrogen, which is composed of two hydrogen atoms. The formula H_2O represents a molecule of water, which is composed of two hydrogen atoms and one oxygen atom. If we divide a water molecule, we will obtain those *two atoms* of hydrogen and *one atom* of oxygen.

And if we divide these atoms ... but that is a story for another time.

Dalton regarded the atom as indivisible, as did his successors up until the discovery of radioactivity in 1895. We examine the changing concept of the atom in the next chapter.

Recycling

Consider the following two different pathways for the recycling of iron.

1. Iron ore (hematite) is mined from the ground and converted to pig iron and then into steel (an alloy of iron with carbon). The steel is used in making an automobile, which is driven for a decade and then sent to the junkyard. The junkyard compresses the automobile and sends it to a recycling plant, where the steel is recovered and ultimately used again in a new automobile. Once the iron was removed from its ore, it has been conserved in its elemental metallic form.

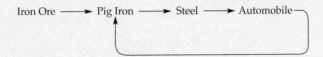

2. Pig iron (obtained from iron ore) is used to make wire for a fence. Over the years the wire gets rusty. Eventually it is sent to a junkyard and ends up in a vat of sulfuric acid where it is completely dissolved as iron sulfate. The iron sulfate is later poured down the drain so that it flows into the river and eventually winds up in the ocean. The original iron atoms have become dissolved ions that are usable by plants. Marine plants absorb and incorporate the iron, so that it is available to any nearby marine creatures that eat plants. The original iron atoms are now widely separated in space. Perhaps a few of them might even become part of the hemoglobin in your own bloodstream. The iron has been recycled, but it will never again resemble the original pig iron.

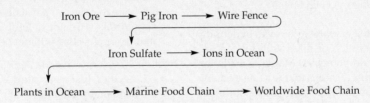

Critical Thinking Exercises

Apply knowledge that you have gained in this chapter and one or more of the FLaReS principles (Chapter 1) to evaluate the following statements or claims.

2.1 A doctor claims that by giving patients small amounts of selenium, he can cure some types of cancer. He says that selenium atoms have the ability to get inside certain kinds of cancer cells and kill them. He has a list of patients that he says he has cured with this treatment.

2.2 A health food store has a large display of bracelets made of copper metal. Some people claim that wearing a copper bracelet will protect the wearer against arthritis or rheumatoid diseases.

2.3 Mary has just learned that her red blood cell count is low, and her doctor has given her some pills that contain iron. The doctor says that the pills should raise the level of hemoglobin in her bloodstream and keep Mary from becoming anemic. Should Mary take the pills or should she seek another opinion?

SUMMARY

Section 2.1—The concept of atoms was first suggested in ancient Greece by Leucippus and Democritus. However, it was rejected for almost 2000 years in favor of Aristotle's view of matter, which declared that matter was continuous in nature.

Section 2.2—The law of conservation of mass resulted from careful experiments by Lavoisier and others, who weighed all the reactants and all the products for a number of chemical reactions and found that no change in mass occurred. Boyle said that a supposed element must be tested; if it could be broken down into simpler substances, it was not really an element.

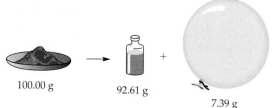

100.00 g   92.61 g   +   7.39 g

Section 2.3—The law of definite proportions (or the law of constant composition) was formulated by Proust, based on his experiments and on those of Berzelius. It states that a given compound always contains the same elements in exactly the same proportions by mass.

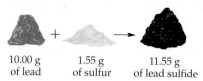

10.00 g   1.55 g   11.55 g
of lead   of sulfur   of lead sulfide

Section 2.4—In 1803 John Dalton explained the laws of definite proportions and of conservation of mass with his atomic theory, which had four main points: (1) Matter is made up of tiny particles called atoms; (2) atoms of the same element are alike; (3) compounds are formed when atoms of different elements combine in certain proportions; and (4) during

chemical reactions atoms are rearranged, not destroyed. In his studies Dalton also discovered the law of multiple proportions, which states that different elements might combine in two or more different sets of proportions, each set corresponding to a different compound. His atomic theory also explained this new law. These laws can be used to perform calculations involving the amounts of elements that combine or form in a compound or reaction. In the two centuries following, atomic theory has undergone only minor modification.

Section 2.5—Berzelius published a table of atomic weights in 1828, which agree well with modern values. In 1869 Mendeleev published his version of the periodic table, a systematic arrangement of the elements that allowed him to predict the existence and properties of other elements. The modern periodic table contains over 110 elements. For each element is listed its symbol and the average mass of an atom of that element in atomic mass units, which are very tiny units of mass.

26	atomic number, Z
Fe	chemical symbol
55.847	atomic mass (weighted average)

Section 2.6—Since atoms are conserved in chemical reactions, matter (which is made of atoms) always can be recycled. If we hope to recycle a particular kind of matter, however, we should take care not to let it spread too thinly throughout nature, or recycling will not be practical.

Section 2.7—A molecule is a group of atoms chemically bonded together. Just as an atom is the smallest unit particle of an element, the smallest unit particle of most compounds is a molecule.

REVIEW QUESTIONS

1. Distinguish between **(a)** the atomic view and the continuous view of matter and **(b)** the ancient Greek definition of an element and the modern one.

2. What was Democritus' contribution to atomic theory? Why did the idea that matter was continuous (rather than atomic) prevail for so long? What discoveries finally refuted the idea?

3. Consider the ideas of *discrete* (atomic) and *continuous* as applied to foods at the macroscopic (visible to the unaided eye) level. Which of the two designations would you use for each of the following?
   a. milk
   b. mashed potatoes
   c. tomatoes
   d. gumballs in a machine
   e. a bag of potatoes
   f. modeling clay
   g. tomato juice
   h. turkey gravy

4. Describe Lavoisier's contribution to the development of modern chemistry.

5. State the law of conservation of mass. How does it apply to a reaction between iron and sulfur to form iron sulfide?

6. How did Robert Boyle define an element?

7. State the law of definite proportions and illustrate it using the compound zinc sulfide.

8. State the law of multiple proportions. For a fixed mass of chlorine (Cl) in each of the following compounds, what is the relationship between (ratio of)
   a. masses of oxygen in ClO_2 and in ClO?
   b. masses of fluorine in ClF_3 and ClF?

9. Outline the main points of Dalton's atomic theory.

10. A photographic flashbulb weighing 0.750 g contains magnesium and air. The flash produces magnesium oxide. After cooling, the bulb weighs 0.750 g. What law does this illustrate?

11. Ammonia produced in an industrial plant contains 3 kg of hydrogen for every 14 kg of nitrogen. The ammonia dissolved in a window-cleaning preparation contains 30 g of hydrogen for every 140 g of nitrogen. What law does this illustrate?

12. In the figure, the blue spheres represent phosphorus atoms, the red ones represent oxygen atoms, and "Initial" represents a mixture. Which one of the three other rectan- gles could *not* represent that mixture after chemical reaction(s) occur? Explain briefly.

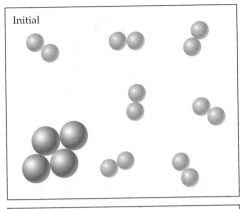

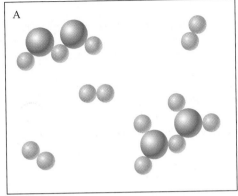

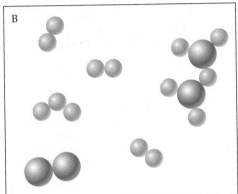

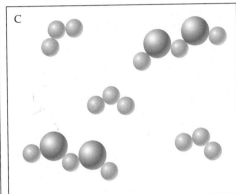

13. Consider the following set of compounds. What principle does this group illustrate?

$$N_2O \quad NO \quad NO_2 \quad N_2O_4$$

14. Use Dalton's atomic theory to explain each of the follow- ing laws and give an example that illustrates each law.
   a. conservation of mass
   b. definite proportions
   c. multiple proportions

15. When 3.00 g of carbon is burned in 8.00 g of oxygen, 11.00 g of carbon dioxide is formed. What mass of carbon diox- ide is formed when 3.00 g of carbon is burned in 50.00 g of oxygen? What law does this illustrate?

16. Heptane is always composed of 84.0% carbon and 16.0% hydrogen. What law does this illustrate?

17. The ancient Greeks thought that water was an element. In 1800 Nicholson and Carlisle decomposed water into hy- drogen and oxygen. What did their experiment prove?

18. What did each of the following contribute to the develop- ment of modern chemistry?
   a. J. J. Berzelius
   b. Henry Cavendish
   c. Joseph Proust
   d. Dmitri Mendeleev

PROBLEMS

Conservation of Mass

19. A balloon filled with helium floats near the ceiling. After several days, the balloon is deflated and lying on the floor. Have the helium atoms been destroyed? If so, how? If not, where are they?

20. A piece of iron is dissolved in a solution of hydrochloric acid. Have the iron atoms been destroyed? If so, how? If not, where are they?

21. Methane consists of carbon and hydrogen atoms. When methane is burned in a Bunsen burner, the only visible product is water vapor condensed on a cold beaker held above the flame. Have the carbon atoms of the methane been destroyed? If so, how? If not, where are they?

22. Aspirin consists of carbon, hydrogen, and oxygen atoms. A student in an organic chemistry laboratory prepares im- pure aspirin, and then dissolves it in hot water in order to purify it by recrystallization. No crystals form on cooling, and so the student pours out the solution and starts the experiment over. Have the carbon, hydrogen, and oxygen atoms of the aspirin been destroyed? If so, how? If not, where are they?

23. A student heats 1.0000 g of zinc powder with 0.2000 g of sulfur. He reports that he obtains 0.6080 g of zinc sulfide and recovers 0.5920 g of unreacted zinc. Show by calcula- tion whether or not his results obey the law of conserva- tion of mass.

24. A student heats 0.5585 g of iron with 0.3550 g of sulfur. She reports that she obtains 0.8792 g of iron sulfide and recovers 0.0433 g of unreacted sulfur. Show by calculation whether or not her results obey the law of conservation of mass.

25. When 1.00 g zinc and 0.80 g sulfur are allowed to react, all the zinc is used up, 1.50 g of zinc sulfide is formed, and some unreacted sulfur remains. What is the mass of *unreacted* sulfur (choose one)?
    (a) 0.20 g   (b) 0.30 g   (c) 0.50 g
    (d) impossible to determine from this information alone

26. A city has to come up with a plan to solve its solid waste problem. The solid wastes consist of many different kinds of materials, and the materials are comprised of many different kinds of atoms. The options for disposal include burying the wastes in a landfill, incinerating them, and dumping them at sea. Which method, if any, will get rid of the atoms that make up the waste? Which method, if any, will change the chemical form of the waste?

Definite Proportions

27. When 18.0 g of water is decomposed by electrolysis, 16.0 g of oxygen and 2.0 g of hydrogen are formed. According to the law of definite proportions, how much hydrogen is formed by the electrolysis of 630 g of water?

28. Hydrogen from the decomposition of water has been promoted as the fuel of the future (Chapter 14). How much water would have to be electrolyzed to produce 100 kg of hydrogen? (See Problem 27.)

29. With a plentiful supply of air, 3.0 parts carbon react with 8.0 parts oxygen to produce carbon dioxide. Use this mass ratio to calculate how much carbon is required to produce 960 g of carbon dioxide.

30. When 31 g of phosphorus reacts with oxygen, 71 g of an oxide of phosphorus is the product. What mass of oxygen is needed to produce 13 g of this product?

Multiple Proportions

31. A compound containing only oxygen and rubidium has 0.187 g O per gram Rb. The relative atomic masses are O = 16.0 and Rb = 85.5. What is a possible O-to-Rb mass ratio for a different oxide of rubidium (choose one)?
    (a) 8.0:85.5   (b) 16.0:85.5
    (c) 32.0:16.0   (d) 32.0:171

32. A sample of an oxide of tin with the formula SnO consists of 0.742 g of tin and 0.100 g of oxygen. A sample of a second oxide of tin consists of 0.555 g of tin and 0.150 g of oxygen. What is the formula of this second oxide?

Dalton's Atomic Theory

33. Are the following findings, expressed to the nearest atomic mass unit, in agreement with Dalton's atomic theory? Explain your answers. (a) An atom of calcium has a mass of 40 u and one of vanadium, 50 u. (b) An atom of calcium has a mass of 40 u and one of potassium, 40 u.

34. To the nearest atomic mass unit, one atom of calcium has a mass of 40 u and another calcium atom has a mass of 44 u. Do these findings support or contradict Dalton's atomic theory? Explain.

35. A neutron strikes an atom of uranium-235 and splits it into two smaller atoms. Do these findings support or contradict Dalton's atomic theory? Explain.

36. According to Dalton's atomic theory, when elements react, their atoms combine in (choose one)
    (a) a simple whole number ratio that is unique for each set of elements.
    (b) exactly a 1:1 ratio.
    (c) one or more simple whole number ratios.
    (d) pairs.
    (e) random proportions.

37. Hydrogen and oxygen combine in a mass ratio of 1:8 to form water. If every water molecule consists of two atoms of hydrogen and one atom of oxygen, the mass of an oxygen atom must be what fraction or multiple of that of a hydrogen atom?
    (a) $\frac{1}{16}$ (b) $\frac{1}{8}$ (c) 8 times   (d) 16 times

38. Hydrogen and nitrogen combine in a mass ratio of 3:14 to form ammonia. If every ammonia molecule consists of three atoms of hydrogen and one atom of nitrogen, the mass of a nitrogen atom must be what fraction or multiple of that of a hydrogen atom?
    (a) $\frac{3}{14}$ (b) 3 times   (c) $\frac{14}{3}$ (d) 14 times

Chemical Compounds

39. A colorless liquid is thought to be a pure compound. Analyses of three samples of the material yield the following results.

	Mass of Sample	Mass of Carbon	Mass of Hydrogen
Sample 1	1.000 g	0.862 g	0.138 g
Sample 2	1.549 g	1.295 g	0.254 g
Sample 3	0.988 g	0.826 g	0.162 g

Could the material be a pure compound? Explain.

40. A blue solid called azulene is thought to be a pure compound. Analyses of three samples of the material yield the following results.

	Mass of Sample	Mass of Carbon	Mass of Hydrogen
Sample 1	1.000 g	0.937 g	0.0629 g
Sample 2	0.244 g	0.229 g	0.0153 g
Sample 3	0.100 g	0.094 g	0.0063 g

Could the material be a pure compound?

ADDITIONAL PROBLEMS

41. By experiment, we find that 1.008 g of hydrogen combines with 35.453 g of chlorine to form 36.461 g of the compound hydrogen chloride. Using Dalton's atomic theory, the best explanation for this is that
    (a) hydrogen and chlorine atoms are neither created nor destroyed in the process, which means the mass of reactants is equal to the mass of the product.
    (b) hydrogen and chlorine atoms combine in a 1:35 ratio.
    (c) hydrogen and chlorine atoms combine in more than one small-whole-number ratio.
    (d) one atom of hydrogen combines with 35.453 atoms of chlorine in the reaction.
    (e) the product is a mixture because the ratio of elements is not a whole number ratio.

42. When 0.2250 g of magnesium is heated with 0.5331 g of nitrogen in a closed container, the magnesium is completely converted to 0.3114 g of magnesium nitride. What mass of unreacted nitrogen must remain?

43. When we burn a 10-kg piece of wood, only 0.05 kg of ash remains. Explain this apparent contradiction of the law of conservation of mass.

44. The gas silane can be decomposed, to yield silicon and hydrogen in a ratio of 7 parts by mass of silicon to 1 part by mass of hydrogen. If the relative mass of silicon atoms is 28 and the mass of hydrogen atoms is taken to be 1, how many hydrogen atoms are combined with each silicon atom?

45. Jan Baptista van Helmont (1579–1644), a Flemish alchemist, performed an experiment in which he planted a young willow tree in a weighed bucket of soil. After 5 years, he found that the tree had gained 75 kg, yet the soil had lost only 0.057 kg. He had added only water to the system, and so he concluded that the substance of the tree had come from water. Criticize his conclusion.

46. Two experiments were performed in which sulfur was burned completely in pure oxygen gas, producing sulfur dioxide and leaving some unreacted oxygen. In the first experiment, 0.312 g of sulfur produced 0.623 g of sulfur dioxide. In the second experiment, 1.305 g of sulfur was burned. What mass of sulfur dioxide was produced?

47. In an experiment, about 15 mL of hydrochloric acid solution was placed in a flask and approximately 10 g of sodium carbonate was placed into a balloon. The opening of the balloon was then carefully stretched over the top of the flask, taking care not to allow the sodium carbonate to fall into the acid in the flask. The flask was placed on an electronic balance, and the mass of the flask and its contents was found to be 238.0 g. The sodium carbonate was then slowly shaken into the acid. The balloon began to fill with gas. When the reaction was complete, the mass of the flask and its contents, including the gas in the balloon, was found to be 238.0 g. What law does this experiment illustrate? Explain.

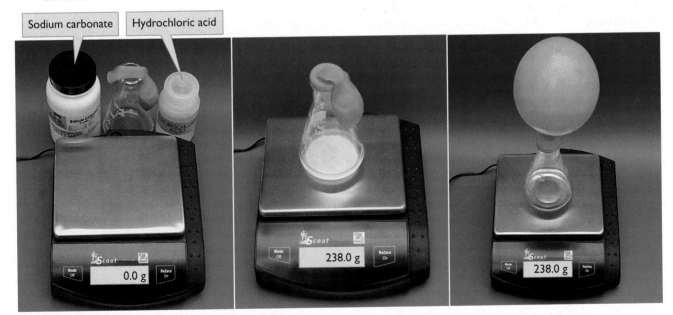

Sodium carbonate   Hydrochloric acid

48. In an experiment, 3.06 g hydrogen was allowed to react with an excess of oxygen to form 27.35 g water. In a second experiment, electric current broke down a sample of water into 1.45 g hydrogen and 11.51 g oxygen. Are these results consistent with the law of constant composition? Show why or why not.

49. Use Figure 2.4 to calculate the mass of mercury oxide that would be needed to produce 100.0 g of mercury metal.

50. Gold chloride, $AuCl_3$, is formed when gold metal is dissolved in *aqua regia*, a highly corrosive acid mixture. Consult Figure 2.11, and determine the mass ratio of gold to chlorine in gold chloride:
    a. based on Mendeleev's values from his periodic table.
    b. based on values in the modern periodic table in the inside front cover of this book.

COLLABORATIVE GROUP PROJECTS

Prepare a PowerPoint, poster, or other presentation (as directed by your instructor) for presentation to the class.

51. Prepare a brief biographical report on one of the following.
    a. Henry Cavendish
    b. Joseph Proust
    c. John Newlands
    d. Lothar Meyer
    e. Dmitri Mendeleev
    f. John Dalton
    g. Antoine Lavoisier

52. Prepare a brief report on early Greek contributions and ideas in the field of science, focusing on the work of one of the following: Aristotle, Leucippus, Democritus, Thales, Anaximander, Anaximenes, Heraclitus, Empedicles, or other Greek philosopher of the time before 300 B.C.E.

53. Write a brief essay on recycling of one of the following: metals, paper, plastics, glass, food wastes, or grass clippings. Contrast a recycling method that maintains the properties of an element with one that changes them.

REFERENCES AND READINGS

1. Cobb, Cathy, and Harold Goldwhite. *Creations of Fire: Chemistry's Lively History from Alchemy to the Atomic Age.* New York: Plenum, 1996.

2. Emsley, John. *Nature's Building Blocks: An A–Z Guide to the Elements.* New York: Oxford University Press, 2002.

3. Gordin, Michael D. *A Well-Ordered Thing: Dmitrii Mendeleev and the Shadow of the Periodic Table.* New York: Basic Books, 2004.

4. Gribbin, John. *Almost Everyone's Guide to Science.* London: Phoenix (Orion Books), 1998.

5. Hellman, Hal. *Great Feuds in Science: Ten of the Liveliest Disputes Ever.* New York: Wiley, 1998.

6. Hill, John W., Ralph H. Petrucci, Terry W. McCreary, and Scott S. Perry. *General Chemistry*, 4th edition. Upper Saddle River, NJ: Prentice Hall, 2005. Chapter 2 provides a more extensive treatment of some of the topics in this chapter.

7. Jaffe, Bernard. *Crucibles: The Story of Chemistry.* New York: Fawcett World Library, 1957.

8. Kolb, Doris. "Chemical Principles Revisited: But if Atoms Are So Tiny …" *Journal of Chemical Education*, September 1977, pp. 543–547.

9. Lloyd, G. E. R. *Methods and Problems in Greek Science.* New York: Cambridge University Press, 1991.

10. Strathern, Paul. *Mendeleyev's Dream: The Quest for the Elements.* London: Hamish Hamilton, 2000.

11. Stwertka, Albert. *Guide to the Elements.* New York: Oxford University Press, 1998.

12. Willis, Randall C. "The Poets of Chemistry." *Today's Chemist at Work*, August 2002, pp. 41–44. Goethe, Dobereiner, and Runge.

13. Young, Louise B. (ed.). *The Mystery of Matter.* New York: Oxford University Press, 1965.

GREEN CHEMISTRY

Designing Molecules

John Thompson, Lane Community College

People have been deliberately changing matter since prehistoric times. With the development of alchemy, then chemistry, we began to analyze these processes scientifically and recognize that the changes in matter were actually the chemical reactions of molecules and compounds. Chemical synthesis—the creation of a desired molecule by controlled chemical reactions—played an important role in the industrial revolution as new materials and products were made on a large scale for the first time. The goal of industrial chemistry was to maximize the amount of the desired product molecule while minimizing material and production costs. This goal, however, did not take into account environmental impacts. As pollution increased, nations responded with regulations to protect our air, water, and soil. Today regulations have increased the cost of doing business while pollution has strained ecosystems and impacted human health.

Chemists are molecule designers much as architects are designers of buildings. Green chemists understand that chemical reactions have impacts beyond the synthesis of a product. The principles of green chemistry are used as a guide to consider these larger impacts when designing a chemical synthesis. Green chemistry can play a significant role in the challenges we face as an industrial society. President Clinton created the Presidential Green Chemistry Challenge Awards in 1995 to promote the use of green chemistry for pollution prevention. The following investigations will give you the opportunity to examine some of the past Presidential Green Chemistry Challenge award winners.

WEB INVESTIGATIONS

Investigation 1
Alternative Synthetic Pathways
The BHC Company developed a new synthetic process to manufacture ibuprofen and was awarded the Alternative Synthetic Pathway Award in 1997. Search the internet to find out more about BHC Company's award. What green chemistry principles were used to improve the original ibuprofen synthesis?

Investigation 2
Alternative Solvents and Reaction Conditions
Polylactic acid (PLA) is the first polymer that is made 100% from corn. Cargill Dow LLC received the Alternative Solvents and Reaction Conditions Award in 2002. What can you find out from the Environmental Protection Agency's website about Cargill Dow LLC's award-winning work? What green chemistry principles were used in the PLA synthesis? (You may want to review the 12 principles.) What are the implications (pro and con) of using corn rather than petroleum for the production of plastics?

▲ Polylactic acid is a polymer that is made from a renewable resource, corn. It also readily degrades in sunlight.

Investigation 3
Designing Safer Chemicals

Paints and coatings have changed significantly over the years. The hazardous nature of lead paints was discovered in the mid-twentieth century. Since the 1980s, we have moved from primarily oil-based to water-based (latex) paints. Today new paints are being developed with reduced amounts of volatile organic compounds (VOCs). VOCs may contribute to indoor air pollution. If you were developing a new paint, what properties would you want that paint to have? Archer Daniels Midland Company received the Designing Safer Chemicals Award in 2005 for developing a coalescent that significantly reduces the need for VOCs in latex paint. Search online to find out more about the award given to Archer Daniels Midland. Consider whether this product is a positive improvement to the paint or if we are trading product quality for a reduction in the chemical hazard. Does this new product appear to be economically viable?

▲ Volatile organic compounds (VOCs) are found in paints, stains, thinners, and many other consumer goods. VOCs can contribute to indoor pollution and to photochemical smog.

COMMUNICATE YOUR RESULTS

Exercise 1
Evaluate a Green Improvement

Consider the new ibuprofen synthesis discussed in Investigation 1. Make a list of the improvements made by the BHC Company to the synthesis of ibuprofen and match each one to the appropriate green chemistry principle. How big of an impact do you think this green improvement had? Do you think that further green chemistry improvements could be made to this process?

Exercise 2
You're the Consultant

You have been hired by World's Best Chemicals, Inc., to help them redesign their production process for their top-selling product *Amazing Stuff*. As a strong proponent of green chemistry, you want to convince the CEO that he should "green up" this process. Make the best case you can using the 12 principles of green chemistry as your guide. Remember that the reason this process is being redesigned is to increase profitability.

Exercise 3
You're the Boss

As the director of research at a major chemical products firm, you set the direction for the next generation of products that corporation will develop. Pick a particular product that you are interested in improving, and use the 12 principles of green chemistry to set the parameters for the properties of this product and how it will be produced. Try to be as realistic as possible with your proposal, focusing on the two or three principles that will make the greatest improvement to the current product. While applying green chemistry principles reduces waste and hazard, significant leaps in product quality often take years of development. Write a memo to your staff proposing your product and the desired green improvements.

Exercise 4
Create the Ideal Future

Today humanity faces polluted air, water, and soil; acid rain that threatens the health of forests and endangers species; and the coming impacts of global warming. Chemical and chemistry-related industries have contributed to these problems, yet these industries are important for our quality of life. Consider the year 2050. What would the ideal civilization look like and what role would green chemistry and chemical industries play in this society? Write a two-page paper describing your vision for the year 2050. Include information about what changes would have to be made for your vision to be realized.

CHAPTER 3

Atomic Structure

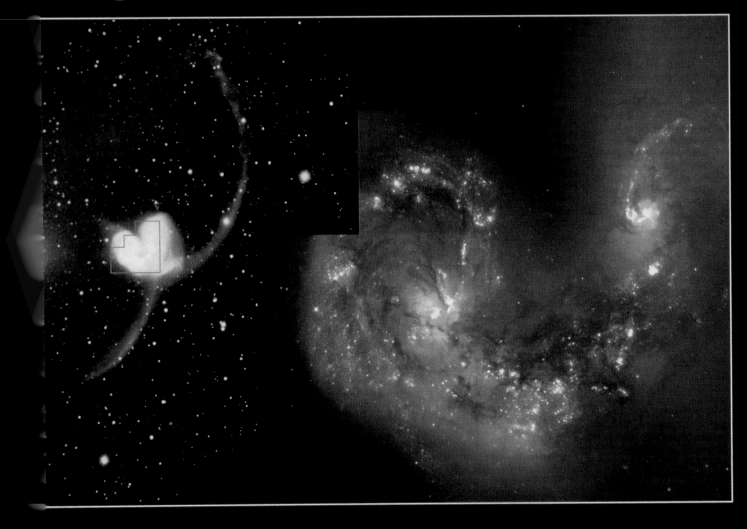

On the left is a photograph of the Antennae galaxies from an Earth-based telescope. On the right, the high resolution of the Hubble telescope shows stars being born in an intense, brilliant "fireworks show" as the two galaxies collide. The cores of the twin galaxies are the orange blobs, left and right of center. Filaments of dark dust crisscross the cores, and a wide band of chaotic dust extends between the cores. Spiral patterns of bright blue star clusters show the firestorm of star birth triggered by the collision. All that we know about objects in outer space comes from energy given off or absorbed by atoms. This energy also reveals the structure of atoms, the subject of this chapter.

Images of the Invisible

Atoms are exceedingly tiny particles, much too small to see. Yet in this chapter, we discuss their inner structure. If atoms are so small that we cannot see them, even with a microscope, how can we possibly know anything about their structures?

It turns out that we really know quite a lot about the structure of atoms. Although scientists have never examined the interior of an atom directly, they have been able to obtain a great deal of *indirect* information. By designing some clever experiments and exercising their powers of deduction, they have been able to construct an amazingly detailed model of what an atom must be like.

Atoms are much too small to be seen with an ordinary light microscope. However, since 1970 scientists have been able to obtain images of individual atoms. In order to see even rough images of atoms, they must use special kinds of instruments such as the scanning tunneling microscope (STM). In this way they obtain pictures such as the one shown on page 36. We can see outlines of atoms in such photographs, and we can tell quite a bit about how they are arranged, but these pictures tell us nothing about the inner structure of the atoms. Why do we care about the structure of particles as tiny as atoms? It is the arrangement of various parts of their atoms that determines the properties of different kinds of matter. Only by understanding atomic structure can we learn how atoms combine to make millions of different substances. With such knowledge, we can modify and synthesize materials to meet our needs more precisely. A knowledge of atomic structure is even essential to our health. Many medical diagnoses are based on chemical analyses that have been developed from our understanding of atomic structure.

Perhaps of greater interest to you is the fact that your understanding of chemistry (as well as much of biology and other sciences) depends, at least in part, on your knowledge of atomic structure. Let's start our study of atomic structure by going back to the time of John Dalton.

3.1 ELECTRICITY AND THE ATOM

Dalton, who set forth his atomic theory in 1803, regarded the atom as hard and indivisible. However, it wasn't long before evidence accumulated to show that matter has electrical properties. Indeed, the electrolytic decomposition of water by Nicholson and Carlisle in 1800 (Section 2.3) had already indicated this. Electricity played an important role in unraveling the structure of the atom.

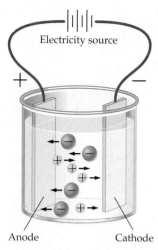

Electricity source

Anode   Cathode

▲ **Figure 3.1** An electrolysis apparatus. The electricity source (for example, a battery) directs electrons through wires from the anode to the cathode. Cations (+) are attracted to the cathode (−), and anions (−) are attracted to the anode (+). This migration of ions is the flow of electricity through the solution.

Electrolytes and Nonelectrolytes

▲ This 20-pound English bank note honors Michael Faraday.

Electrochemistry, which includes the study of electrochemical cells and electrolysis, is discussed in Chapter 8. It is essential that you learn the terms introduced in this section because they are used throughout the text.

Static electricity has been known since ancient times, but the notion of continuous electric current was born with the nineteenth century. In 1800 Alessandro Volta invented an electrochemical cell much like a modern battery. If the poles of a cell are connected by a wire, current flows through the wire. The current is sustained by chemical reactions inside the cell. Volta's invention soon was applied in many areas of science and everyday life.

Electrolysis

Soon after Volta's invention, Humphry Davy (1778–1829), a British chemist, built a powerful battery that he used to pass electricity through molten (melted) salts. Davy quickly discovered several new elements. In 1807 he liberated highly reactive potassium metal from molten potassium hydroxide. Shortly thereafter he produced sodium metal by passing electricity through molten sodium hydroxide. Within a year, Davy had also produced magnesium, strontium, barium, and calcium metals for the first time. The science of electrochemistry was born.

Davy's protégé Michael Faraday (1791–1867) greatly extended this new science. Faraday defined many of the terms we still use today, such as **electrolysis**, the splitting of compounds by electricity (Figure 3.1). Lacking in his own formal education, he consulted English classical scholar William Whewell (1794–1866), who suggested the name **electrolyte** for a compound that conducts electricity when melted or dissolved in water. In the electrolysis apparatus, **electrodes**, carbon rods or metal strips inserted into a molten compound or solution, carry the electric current. The electrode that bears a positive charge is the **anode**, and the negatively charged electrode is the **cathode**. The entities that carry the electric current through a melted compound or solution are called *ions*. An **ion** is an atom or a group of atoms bonded together that has an electric charge. An ion with a negative charge is an **anion**; it travels toward the anode. A positively charged ion is a **cation**; it moves toward the cathode.

Faraday's work established that atoms are electrical in nature, but further details of atomic structure had to wait several decades for more powerful sources of electrical voltage and the development of gas discharge tubes.

Cathode Ray Tubes

Faraday tried and failed to pass electricity through a tube that had part of the air pumped out. His vacuum was not sufficient for the voltage he had available. By 1875 tubes with better vacuum were available. William Crookes (1832–1919), an English chemist, passed an electric current through such a tube containing air at low pressure. His experiment is shown in Figure 3.2. Metal electrodes are sealed in the tube. It is connected to a vacuum pump, and most of the air is removed. A beam of current is seen as a green fluorescence, observed when the beam strikes a screen coated with zinc sulfide. This beam, which seems to leave the cathode and travel to the anode, is called a **cathode ray**.

Thomson's Experiment: Mass-to-Charge Ratio

Considerable speculation arose as to the nature of cathode rays and many experiments were undertaken. Were these rays actually beams of particles, or did they consist of a form of energy much like visible light? The answer came (as scientific answers should) from an experiment performed by the English physicist Joseph John Thomson in 1897. Thomson showed that cathode rays were deflected in an electric field (see again Figure 3.2). The beam was attracted to the positive plate and repelled by the negative plate. Thomson therefore concluded that cathode rays consisted of negatively charged particles. His experiments also showed that the particles were the same regardless of the materials from which the electrodes were made or the type of gas in the tube. He concluded that these negative particles are part of all kinds of atoms. Thomson named these negatively charged units **electrons**. *Cathode rays*, then, are beams of electrons emanating from the cathode of a gas discharge tube.

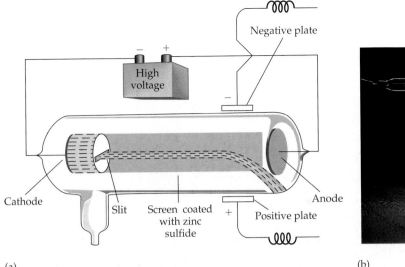

(a)

(b)

▲ **Figure 3.2** Thomson's apparatus, showing deflection of cathode rays (a beam of electrons). Cathode rays are themselves invisible but are observed through the green fluorescence produced when they strike a zinc sulfide–coated screen. The diagram (a) shows deflection of the beam in an electric field. The photograph (b) shows the deflection in a magnetic field. The magnetic field is created by the magnet to the right of and slightly behind the screen. Cathode rays travel in straight lines unless some kind of external field is applied.

Cathode rays are deflected in magnetic fields as well as in electric fields. By measuring the amount of deflection in fields of known strength, Thomson was able to calculate the *ratio* of the mass of the electron to its charge. He could not measure either the mass or the charge separately. This is like knowing that each 1-ft length of a steel beam has a mass of 25 lb. With this data alone, we cannot find either the total mass or length of a beam. Once the beam's mass or its length is known, it is easy to calculate the other from the known value and the 25 lb/1 ft ratio. Thomson was awarded the Nobel Prize in physics in 1906.

Goldstein's Experiment: Positive Particles

In 1886 German scientist Eugen Goldstein performed experiments with gas discharge tubes that had perforated cathodes (Figure 3.3). He found that although electrons were formed and sped off toward the anode as usual, positive particles were also formed and shot in the opposite direction toward the cathode. Some of these positive particles went through the holes in the cathode. In 1907 a study of the deflection of these particles in a magnetic field indicated that they were of varying mass. The lightest particles, formed when there was a little hydrogen gas in the tube, were later shown to have a mass 1837 times that of an electron.

Millikan's Oil-Drop Experiment: Electron Charge

The charge on the electron was determined in 1909 by Robert A. Millikan (1868–1953), a physicist at the University of Chicago. Millikan observed electrically charged oil drops in an electric field. A diagram of his apparatus is shown in Figure 3.4. A spray bottle is used to form tiny droplets of oil. Some acquire negative charges by picking up electrons from the friction generated as the particles rub against the opening of the spray nozzle and against each other. (The charge is static electricity, just like the charge you get from walking across a nylon carpet.) The negative droplets can acquire one or more extra electrons by this process. Charges can also be produced by irradiation with X-rays.

Some of the oil droplets pass into a chamber where they can be viewed through a microscope. The negative plate at the bottom of the chamber repels the negatively charged droplets; the positive plate attracts them. By manipulating the charge on each plate and observing the behavior of the droplets, the charge on each droplet can be determined. Millikan took the smallest possible difference in charge between two droplets to be the charge of an individual electron. For his research, he received the Nobel Prize in physics in 1923.

Millikan's Oil Drop Experiment

The greater the charge on a particle, the more it is deflected in an electric (or magnetic) field. (The greater the charge, the greater the attraction or repulsion.) The greater the mass of the particle, the less it is deflected by a force. (Consider two spheres coming toward you. It is much easier to deflect a ping-pong ball than a cannonball.)

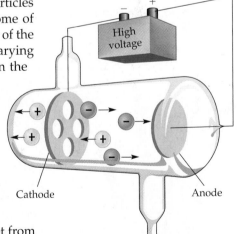

▲ **Figure 3.3** Goldstein's apparatus for the study of positive particles. Some positive ions, attracted toward the cathode, pass through the holes in the cathode. The deflection of these particles (not shown) in a magnetic field can be studied in the region to the left of the cathode.

▶ **Figure 3.4** The Millikan oil-drop experiment. Oil drops irradiated with X-rays pick up electrons and become negatively charged. Their fall due to gravity can be balanced by adjusting the voltage of the electric field. The charge on the oil drop can be determined from the applied voltage and the mass of the oil drop. The charge on each drop is that of some whole number of electrons.

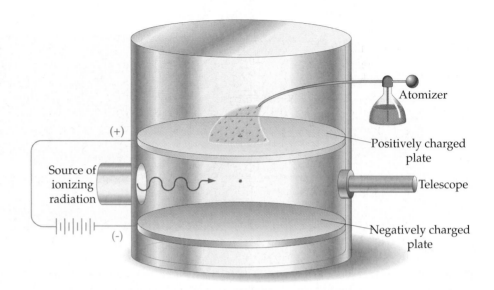

You can find out more about numbers such as 9.1×10^{-28} and 1×10^{27} in Appendix A. However, in practice we seldom use the actual mass and charge of the electron. For many purposes, the electron is considered the unit of electrical charge. The charge is shown as a superscript minus sign (meaning $^{1-}$). In indicating charges on ions, we use $^-$ to indicate a net charge of one electron, $^{2-}$ to indicate a charge of two electrons, and so on.

Serendipity is an aptitude for making fortunate discoveries by accident. Horace Walpole coined the term, alluding to the Persian fairy tale *The Three Princes of Serendip*, who made many such discoveries.

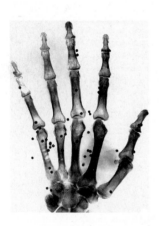

▲ X-rays were used in medicine shortly after they were discovered by Wilhelm Roentgen (1845–1923) in 1895. Michael Purpin of Columbia University made this X-ray in 1896 to aid in the surgical removal of gunshot pellets (the dark spots) from the hand of a patient.

From Millikan's value for the charge and Thomson's value for the mass-to-charge ratio, the mass of the electron was readily calculated. That mass was found to be only 9.1×10^{-28} g. It would take more than 1×10^{27} electrons to weigh one gram—that's a 1 followed by 27 zeros, or a trillion trillion trillion electrons. More important, that mass is much *smaller* than that of the lightest atom. This means that electrons are much smaller than atoms.

3.2 SERENDIPITY IN SCIENCE: X-RAYS AND RADIOACTIVITY

Let's return now to the structure of the atom and look at a little scientific serendipity. Often scientific discoveries are described as happy accidents. Have you ever wondered why these accidents always seem to happen to scientists? It is probably because scientists are trained observers. The same accident could happen right before the eyes of an untrained person and go unnoticed. Or, if noticed, its significance might not be grasped.

Roentgen: The Discovery of X-Rays

Two serendipitous discoveries the last years of the nineteenth century profoundly changed the world. In 1895 German scientist Wilhelm Conrad Roentgen (1845–1923) was working in a dark room, studying the glow produced in certain substances by cathode rays. To his surprise, he noted this glow on a chemically treated piece of paper some distance from the cathode ray tube. The paper even glowed when taken into the next room. Roentgen had discovered a new type of ray that could travel through walls. When he waved his hand between the radiation source and the glowing paper, he suddenly was able to see the bones of his own hand through the paper. He called these mysterious rays, which seemed to make his flesh disappear, **X-rays**.

Today X-rays are one of the most widely used tools in the world for medical diagnosis. Not only are they employed for examining decayed teeth, broken bones, and diseased lungs, but they are also the basis for such procedures as mammography and computerized tomography (Chapter 4). In the United States alone, payment for various radiological procedures totals more than $20 billion each year. How ironic that Roentgen himself made no profit at all from his discovery. He considered X-rays a "gift to humanity" and refused to patent any part of the discovery. However, he did receive much popular acclaim and in 1901 was awarded the first Nobel Prize in physics.

The Discovery of Radioactivity

Certain chemicals exhibit *fluorescence* after exposure to strong sunlight; they continue to glow even when taken into a dark room. In 1895 Antoine Henri Becquerel (1852–1908), a French physicist, was studying fluorescence by wrapping photographic film in black paper, placing a few crystals of the fluorescing chemical on top of the paper, and then placing the package in strong sunlight. If the glow was like ordinary light, it would not pass through the paper. On the other hand, if it was similar to X-rays, it would pass through the black paper and fog the film.

While working with a uranium compound, Becquerel made an important accidental discovery. When placed in sunlight, the compound fluoresced and fogged the film. On several cloudy days when exposure to sunlight was not possible, he prepared samples and placed them in a drawer. To his great surprise, the photographic film was fogged even though the uranium compound had not been exposed to sunlight. Further experiments showed that the radiation coming from the uranium compound was unrelated to fluorescence but was a characteristic of the element uranium.

Other scientists immediately began to study this new radiation. Becquerel had a graduate student from Poland, Marie Sklodowska, who gave the phenomenon a name: radioactivity. **Radioactivity** is the spontaneous emission of radiation from certain unstable elements. Marie later married Pierre Curie, a French physicist. Together they discovered the radioactive elements polonium and radium, and with Becquerel they shared the 1903 Nobel Prize in physics.

After her husband's death in 1906, Marie Curie continued to work with radioactive substances, winning the Nobel Prize for chemistry in 1911. For more than 50 years she was the only person ever to have received two Nobel Prizes.

▲ Marie Sklodowska Curie in her laboratory and Marie and Pierre Curie on a French postage stamp.

3.3 THREE TYPES OF RADIOACTIVITY

Scientists soon showed that three types of radiation emanated from various radioactive elements. Ernest Rutherford (1871–1937), a New Zealander who spent his career in Canada and Great Britain, chose the names alpha, beta, and gamma for the three types of radiation. When passed through a strong magnetic or electric field, the alpha form was deflected in a manner indicating that it consisted of a beam of positive particles (Figure 3.5). Later experiments showed that an **alpha particle** has a mass four times that of a hydrogen atom and a charge twice the magnitude of, but opposite in sign to, that of an electron.

The beta radiation was shown to be made up of negatively charged particles identical to those of cathode rays. Therefore, a **beta particle** is an electron.

Gamma rays are not deflected by a magnetic field. They are a form of energy, much like the X-rays used in medical work but even more penetrating. The three types of radioactivity are summarized in Table 3.1.

The discoveries of the late nineteenth century paved the way for an entirely new picture of the atom, which developed rapidly during the early years of the twentieth century.

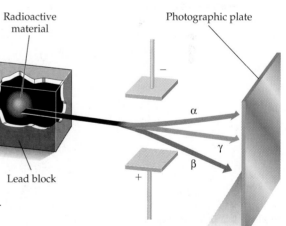

▲ **Figure 3.5** Behavior of radioactive rays in an electric field.

Separation of Alpha, Beta and Gamma Rays

Alpha particles are identical to helium ions (He^{2+}) or helium nuclei. They are helium atoms with both electrons removed.

TABLE 3.1	Types of Radioactivity		
Name	**Greek Letter**	**Mass (u)**	**Charge**
Alpha	α	4	2+
Beta	β	$\dfrac{1}{1837}$	1–
Gamma	γ	0	0

3.4 RUTHERFORD'S EXPERIMENT: THE NUCLEAR MODEL OF THE ATOM

Rutherford Experiment:
Nuclear Atom

At Rutherford's suggestion, two of his coworkers, Hans Geiger (1882–1945) a German physicist, and Ernest Marsden (1889–1970), an English undergraduate student, bombarded very thin metal foils with alpha particles from a radioactive source (Figure 3.6). In an experiment with gold foil, most of the particles behaved as Rutherford expected, going right through the foil with little or no scattering. However, a few particles were deflected sharply. Occasionally one was sent right back in the direction from which it had come! Rutherford had assumed the positive charge to be spread evenly over all the space occupied by the atom, but obviously it was not. To explain the experiment, Rutherford concluded that all the positive charge and nearly all the mass of an atom are concentrated at the center of the atom in a tiny core called the **nucleus**.

When an alpha particle, which is positively charged, approached the positively charged nucleus, it was strongly repelled and therefore sharply deflected (Figure 3.7). Because only a few alpha particles were deflected, Rutherford concluded that the nucleus must occupy only a tiny fraction of the volume of an atom. Most of the alpha particles passed right through because most of an atom is empty space. The space outside the nucleus isn't completely empty, however. It contains the negatively charged electrons. Rutherford concluded that the electrons had so little mass that they were no match for the alpha "bullets." It would be analogous to a mouse trying to stop the charge of a bull elephant.

Rutherford's nuclear theory of the atom, set forth in 1911, was revolutionary. He postulated that all the positive charge and nearly all the mass of an atom are

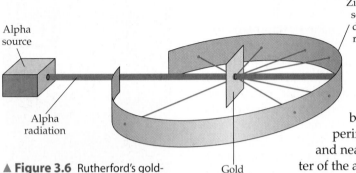

▲ **Figure 3.6** Rutherford's gold-foil experiment. Most alpha particles passed right through the gold foil, but now and then a particle was deflected.

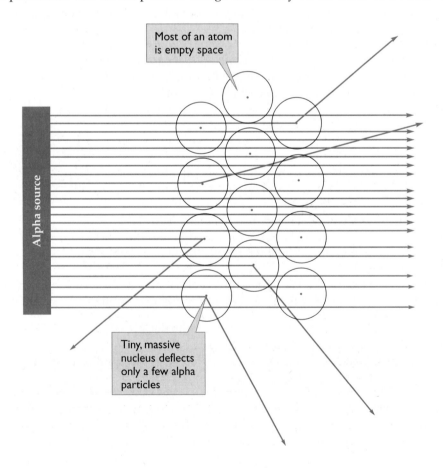

▶ **Figure 3.7** Model explaining the results of Rutherford's gold-foil experiment. Most of the alpha particles pass right through the foil because it is mainly empty space. But some alpha particles are deflected as they pass close to a dense, positively charged atomic nucleus. Once in a while an alpha particle approaches an atomic nucleus head-on and is knocked back in the direction from which it came.

concentrated in a tiny, tiny nucleus. The negatively charged electrons have almost no mass, yet they occupy nearly all the volume of an atom. To picture Rutherford's model, visualize a sphere as big as a giant indoor football stadium. The nucleus at the middle of the sphere is as small as a pea but weighs several million tons. A few flies flitting here and there throughout the sphere represent the electrons.

3.5 THE ATOMIC NUCLEUS

In 1914 Rutherford suggested that the smallest positive particle (the one formed when there is hydrogen gas in the Goldstein apparatus—see Section 3.1) is the unit of positive charge in the nucleus. This particle, called a **proton**, has a charge equal in magnitude to that of the electron and has nearly the same mass as a hydrogen atom. Rutherford's suggestion was that protons constitute the positively charged matter in all atoms. The nucleus of a hydrogen atom consists of one proton, and the nuclei of larger atoms contain greater numbers of protons.

Except for hydrogen atoms, atomic nuclei are heavier than indicated by the number of positive charges (number of protons). For example, the helium nucleus has a charge of 2+ (and therefore two protons, according to Rutherford's theory), but its mass is *four* times that of hydrogen. This excess mass puzzled scientists at first. But in 1932 English physicist James Chadwick (1891–1974) discovered a particle with about the same mass as a proton but with no electric charge. It was called a **neutron**, and its existence explains the unexpectedly high mass of the helium nucleus. Whereas the hydrogen nucleus contains only one proton of mass 1 u, the helium nucleus contains not only two protons (2 u) but also two neutrons (2 u), giving the nucleus a total mass of 4 u.

With the discovery of the neutron, the list of "building blocks" we will need for "constructing" atoms is complete. The properties of these particles are summarized in Table 3.2.

TABLE 3.2 Subatomic Particles				
Particle	**Symbol**	**Mass (u)**	**Charge**	**Location in Atom**
Proton	p^+	1	1+	Nucleus
Neutron	n	1	0	Nucleus
Electron	e^-	$\dfrac{1}{1837}$	1−	Outside nucleus

The number of protons in the nucleus of an atom of any element is the **atomic number (Z)** of that element. This number determines the kind of atom—that is, the identity of the element—and it is found on any periodic table or list of the elements. An *element*, then, is a substance in which all the atoms have the same atomic number. Dalton had said that the mass of an atom determines the element. We now know it is not the mass but the number of protons that determines the identity of an element. For example, an atom with 26 protons (one whose atomic number $Z = 26$) is an atom of iron (Fe). An atom with 50 protons ($Z = 50$) is an atom of tin (Sn). In a neutral atom (without an electric charge) the positive charge of the protons is exactly neutralized by the negative charge of the electrons. The attractive forces between the unlike charges help hold the atom together.

A proton and a neutron have almost the same mass, 1.0073 u and 1.0087 u, respectively. This is equivalent to saying that two different people weigh 100.7 kg and 100.9 kg. The difference is so small that it usually can be ignored. Thus, for many purposes, we assume the masses of the proton and the neutron to be the same, 1 u. The proton has a charge equal in magnitude but opposite in sign to that of an electron. This charge on a proton is written as 1+. The electron has a charge of 1− and a mass of 0.00055 u. The electrons in an atom contribute so little to its total mass that their mass is usually disregarded and treated as if it were 0.

Isotopes of Hydrogen

Water in which both hydrogen atoms are deuterium is called *heavy water*, often written D_2O. Heavy water boils at 101.4 °C and freezes at 3.8 °C. Its density is 1.108 g/cm^3. (The density of ordinary water is 1.000 g/cm^3.)

We will refer to A as the nucleon number in order to stress the fact that the mass number of an atom is equal to the total number of nucleons.

Isotopes

Atoms of a given element can have different numbers of neutrons in their nuclei. For example, most hydrogen atoms have a nucleus consisting of a single proton and no neutrons. However, about 1 hydrogen atom in 6700 has a neutron as well as a proton in the nucleus. This heavier hydrogen atom is called *deuterium*. Both kinds are hydrogen atoms (any atom with $Z = 1$—that is, with one proton—is a hydrogen atom). Atoms that have this sort of relationship—the same number of protons but different numbers of neutrons—are called **isotopes** (Figure 3.8). A third, rare isotope of hydrogen is *tritium*, which has two neutrons and one proton in the nucleus.

Most, but not all, elements exist in nature in isotopic forms. For example, tin (Sn) is present in nature in 10 different isotopic forms. It also has 15 radioactive isotopes that do not occur in nature. This fact also requires a modification of Dalton's original theory. He said that all atoms of the same element are alike. We now say that all atoms of the same element have the same number of protons. Different isotopes of an element have atoms with the same number of protons but with different numbers of neutrons (and therefore different masses).

Isotopes usually are of little importance in ordinary chemical reactions. All three hydrogen isotopes react with oxygen to form water. Because the isotopes differ in mass, compounds formed with different hydrogen isotopes have different physical properties, but such differences are usually slight. In nuclear reactions, however, isotopes are of utmost importance, as we shall see in the next chapter.

Symbols for Isotopes

Collectively, the two principal nuclear particles, protons and neutrons, are called **nucleons**. Isotopes are represented by symbols with subscripts and superscripts.

$$^{A}_{Z}X$$

In this general symbol, Z is the nuclear charge (atomic number or number of protons), and A is the **mass number**, or the **nucleon number** (the number of protons plus the number of neutrons). As an example, the isotope with the symbol

$$^{35}_{17}Cl$$

has 17 protons and 35 nucleons. The number of neutrons is therefore $35 - 17 = 18$.

Isotopes often are named by placing the nucleon number as a suffix to the name of the element. The three hydrogen isotopes are therefore represented as

$$^{1}_{1}H \quad ^{2}_{1}H \quad ^{3}_{1}H$$

and named as hydrogen-1, hydrogen-2, and hydrogen-3.

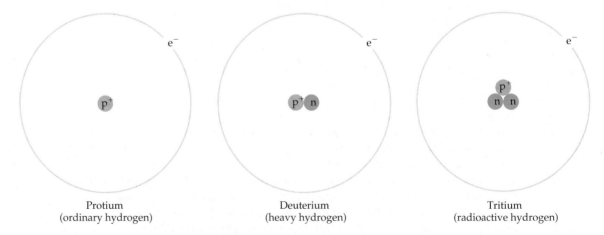

Protium
(ordinary hydrogen)

Deuterium
(heavy hydrogen)

Tritium
(radioactive hydrogen)

▲ **Figure 3.8** The three isotopes of hydrogen. Each has one proton and one electron, but they differ in the number of neutrons in the nucleus.

Question: What is the atomic number and nucleon number of each of the isotopes? What is the mass of each, in atomic mass units, to the nearest whole number?

EXAMPLE 3.1   **Number of Neutrons**

How many neutrons are there in the $^{235}_{92}U$ nucleus?

Solution

Simply subtract the atomic number Z (number of protons) from the nucleon number A (number of protons plus neutrons).

$$A - Z = \text{number of neutrons}$$
$$235 - 92 = 143$$

There are 143 neutrons in the nucleus.

▶ **Exercise 3.1A**
How many neutrons are there in the $^{131}_{53}I$ nucleus?

▶ **Exercise 3.1B**
A potassium isotope has 21 neutrons in its nucleus. What is the nucleon number and name of the isotope?

EXAMPLE 3.2   **Number of Neutrons**

How many neutrons are there in the strontium-90 nucleus?

Solution

From the periodic table on the inside front cover, we find the atomic number of strontium is 38. The nucleon number is given as 90. The number of neutrons is therefore

$$90 - 38 = 52$$

▶ **Exercise 3.2A**
How many neutrons are there in the molybdenum-90 nucleus?

▶ **Exercise 3.2B**
Use the $^{A}_{Z}X$ notation to represent the isotope of uranium having 148 neutrons.

CONCEPTUAL EXAMPLE 3.3   **Isotopes**

Refer to the following isotope symbols, in which we use the letter X as the symbol for all elements so that the symbol will not identify the elements. **(a)** Which are isotopes of the same element? **(b)** Which have the same nucleon number? **(c)** Which have the same number of neutrons?

$$^{16}_{8}X \quad ^{16}_{7}X \quad ^{14}_{7}X \quad ^{14}_{6}X \quad ^{12}_{6}X$$

Solution

a. Isotopes of the same element will have the same atomic number (subscript). Therefore, $^{16}_{7}X$ and $^{14}_{7}X$ are isotopes of nitrogen (N), and $^{14}_{6}X$ and $^{12}_{6}X$ are isotopes of carbon (C).
b. The nucleon number is the superscript, so $^{16}_{8}X$ and $^{16}_{7}X$ have the same nucleon number. The first is an isotope of oxygen, and the second an isotope of nitrogen. $^{14}_{7}X$ and $^{14}_{6}X$ also have the same nucleon number. The first is an isotope of nitrogen, and the second an isotope of carbon.
c. To determine the number of neutrons, we subtract the atomic number from the nucleon number. We find that $^{16}_{8}X$ and $^{14}_{6}X$ each have eight neutrons ($16 - 8 = 8$ and $14 - 6 = 8$, respectively).

▶ **Exercise 3.3A**
Which of the following are isotopes of the same element?

$$^{90}_{37}X \quad ^{90}_{35}X \quad ^{88}_{37}X \quad ^{88}_{38}X \quad ^{93}_{38}X$$

▶ **Exercise 3.3B**
How many different elements are represented in Exercise 3.3A?

3.6 ELECTRON ARRANGEMENT: THE BOHR MODEL

Let us now turn our attention once more to electrons. Rutherford demonstrated that atoms have a tiny, positively charged nucleus with electrons outside the nucleus. Evidence soon accumulated that the electrons were not randomly distributed but were arranged in an ordered fashion. We will examine the evidence soon. But first let's take a side trip into some colorful chemistry and physics to provide a background for our study of the electron structure of atoms.

Fireworks and Flame Tests

Chemists of the eighteenth and nineteenth centuries developed flame tests that used the colors of flames to identify several elements (Figure 3.9). Sodium salts give a persistent yellow flame, potassium salts a fleeting lavender flame, and lithium salts a brilliant red flame. Like those of fireworks, these flame colors result from the electron structures of atoms of the specific elements.

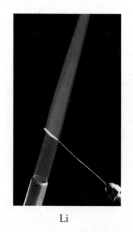
Li

Na

K

Ca

Sr

▲ **Figure 3.9** Certain chemical elements can be identified by the characteristic colors their compounds impart to flames. Five examples are shown here.

Fireworks originated earlier than flame tests, in ancient China. The brilliant colors of aerial displays still mark our celebrations of patriotic holidays. The colors of fireworks are attributable to specific elements. Brilliant reds are produced by strontium compounds, whereas barium compounds are used to produce green, sodium compounds yield yellow, and copper salts produce blue.

The colors of fireworks and flame tests are not what they seem to the unaided eye. If the light from the flame is passed through a prism, it is separated into light of several different colors.

Continuous and Line Spectra

When white light from an incandescent lamp is passed through a prism, it produces a continuous spectrum, or rainbow of colors (Figure 3.10). A similar phenomenon occurs when sunlight passes through raindrops. The different colors of light represent different wavelengths. Blue light has shorter wavelengths than red light, but there is no sharp transition in moving from one color to the next. All wavelengths

▲ Flame colors (see Figure 3.9) also are the basis for the brilliant colors of fireworks displays. Strontium compounds produce red, copper compounds produce blue, and sodium compounds produce yellow.

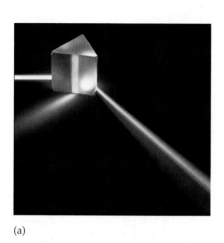

(a)   (b)

◄ **Figure 3.10** A glass prism (a) separates white light into a continuous spectrum or rainbow of colors. Sunlight passing through raindrops (b) causes a rainbow in the sky.

▲ Danish physicist Niels Bohr (1885–1962) received the 1922 Nobel Prize in physics for his planetary model of the atom with its quantized electron energy levels. Element 107 is named bohrium in his honor.

are present in a continuous spectrum. White light is simply a combination of all the various colors.

If the light from a gas discharge tube containing a particular element is passed through a prism, only narrow colored lines are observed (Figure 3.11). Each line corresponds to light of a particular wavelength. The pattern of lines emitted by an element is called its *line spectrum*. The line spectrum of an element is characteristic of that element and can be used to identify it. Not all the lines in the spectrum of an atom are visible; some lines appear as infrared or ultraviolet radiation.

The visible line spectrum of hydrogen is fairly simple, consisting of four lines in that portion of the electromagnetic spectrum. To explain the hydrogen spectrum, Danish physicist Niels Bohr (1885–1962) worked out a model for the electron structure of the hydrogen atom.

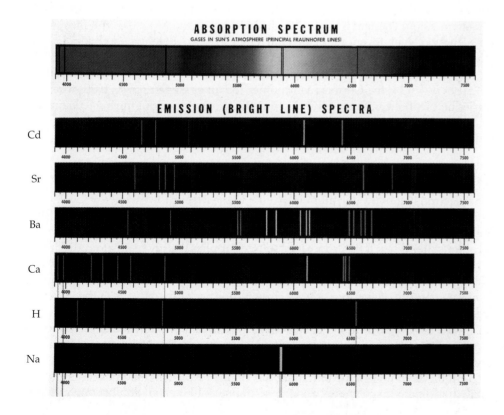

◄ **Figure 3.11** Line spectra of selected elements. Some of the components of the light emitted by excited atoms appear as colored lines. A continuous spectrum is shown at the top for comparison. The numbers are wavelengths of light given in Angstrom units ($1 \text{ Å} = 10^{-10}$ m). Each element has its own characteristic line spectrum that is different from all others and can be used to identify the element.

Bohr's Explanation of Line Spectra

Niels Bohr presented his explanation of line spectra in 1913. He suggested that electrons cannot have just any amount of energy but can have only certain specified amounts; that is to say, the energy of an electron is *quantized*. A **quantum** (plural: quanta) is a tiny unit of energy, the value of which depends on the frequency. A specified energy value for an electron is called its **energy level**.

An electron, by absorbing a quantum of energy (for example, when atoms of the element are heated), is elevated to a higher energy level (Figure 3.12). By giving up a quantum of energy, the electron can return to a lower energy level. The energy released shows up as a line spectrum. Each line has a specific wavelength corresponding to a quantum of energy. An electron moves practically instantaneously from one energy level to another, and there are no intermediate stages.

▶ **Figure 3.12** Possible electron shifts between energy levels in atoms to produce the lines found in spectra. Not all the lines are in the visible portion of the spectrum. The four colored lines correspond to the colored lines in the hydrogen spectrum.

Question: Which transition involves the greater change in energy, a drop from energy level 6 to energy level 2, or a drop from energy level 7 to energy level 6? What color of light is absorbed when an electron in a hydrogen atom is boosted from energy level 2 to energy level 3?

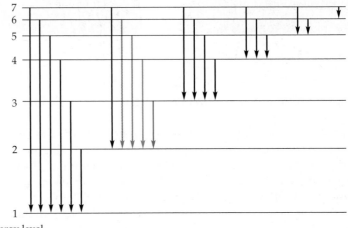

Energy level

Consider as an analogy a person on a ladder. The person can stand on the first rung, the second rung, the third rung, and so on, but is unable to stand between rungs. As the person goes from one rung to another, the potential energy (energy due to position) changes by definite amounts or quanta. As an electron moves from one energy level to another, its total energy (both potential and kinetic) also changes by quanta.

Bohr based his model of the atom on the laws of planetary motion that had been set down by the German astronomer Johannes Kepler (1571–1630) three centuries before. Bohr imagined the electrons to be orbiting about the nucleus much as planets orbit the sun (Figure 3.13). Different energy levels were pictured as different orbits. The modern picture is different, as we shall see in subsequent sections.

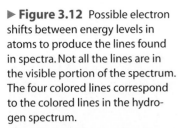

▲ **Figure 3.13** The nuclear atom, as envisioned by Bohr, has most of its mass in an extremely small nucleus. Electrons orbit about the nucleus, occupying most of the volume of the atom but contributing little to its mass.

Question: What element is represented in this drawing?

Ground States and Excited States

The electron in a hydrogen atom is usually in the first or lowest energy level. Given the choice, electrons usually remain in their lowest possible energy levels (those nearest the nucleus); atoms whose electrons are thus situated are said to be in their **ground state**. When a flame or other source supplies energy to an atom (hydrogen, for example) and an electron jumps from the lowest possible level to a higher level, the atom is said to be in an **excited state**. An atom in an excited state eventually emits a quantum of energy as the electron jumps back down to one of the lower levels and ultimately reaches the ground state.

Bohr's theory was a spectacular success in explaining the line spectrum of hydrogen. It established the important idea of energy levels in atoms. Bohr was awarded the Nobel Prize in physics in 1922 for this work.

Atoms larger than hydrogen have more than one electron, and Bohr was also able to deduce that a given energy level of an atom could contain at most only a certain number of electrons. We shall simply state Bohr's findings in this regard. The maximum number of electrons that can be in a given level is indicated by the formula

$$\text{Maximum number of electrons} = 2n^2$$

A *photon* is a quantum or "particle" of light.

where n is the energy level being considered. For the first energy level ($n = 1$), the maximum population is $2 \times 1^2 = 2 \times 1 = 2$. For the second energy level ($n = 2$), the maximum number of electrons is $2 \times 2^2 = 2 \times 4 = 8$. For the third level, the maximum is $2 \times 3^2 = 2 \times 9 = 18$. (However, the outermost level usually has no more than 8 electrons.)

The various energy levels are often called *shells*. The first energy level ($n = 1$) is the first shell, the second energy level ($n = 2$) is the second shell, and so on.

EXAMPLE 3.4   Electron Shell Capacity

What is the maximum number of electrons in the fifth shell (fifth energy level)?

Solution

The maximum number of electrons in a given shell is given by $2n^2$. For the fifth level, $n = 5$, and so we have

$$2 \times 5^2 = 2 \times 25 = 50$$

▶ **Exercise 3.4A**

What is the maximum number of electrons in the fourth shell (fourth energy level)?

▶ **Exercise 3.4B**

What is the lowest shell (value of n) that can hold 18 electrons?

Building Atoms: Main Shells

Imagine building up atoms by adding one electron to the proper shell as *each* proton is added to the nucleus, keeping in mind that electrons will go to the lowest energy level (shell) available. (We can ignore the neutrons in the nucleus; they are not involved in this process.) For hydrogen, with a nucleus of only one proton ($Z = 1$), the single electron goes into the first shell. For helium, with a nucleus having two protons ($Z = 2$), both electrons go into the first shell. According to Bohr, two electrons is the maximum population of the first shell; that level is filled in the helium atom.

With lithium ($Z = 3$), two electrons go into the first shell; the other must go into the second shell. This process of adding electrons is continued until the second shell is filled with eight electrons, as in a neon atom ($Z = 10$), which has two of its 10 electrons in the first shell and the remaining eight in the second shell.

A sodium atom ($Z = 11$) has 11 electrons. Two are in the first shell, the second shell is filled with eight electrons, and the remaining electron is in the third shell. We can indicate the **electron configuration** (or arrangement) of the first 11 elements as follows:

Element	1st shell	2nd shell	3rd shell
H	1		
He	2		
Li	2	1	
⋮	⋮	⋮	
Ne	2	8	
Na	2	8	1

Sometimes the main-shell configuration is abbreviated by simply listing the symbol for the element followed by the number of electrons in each shell, starting with the lowest energy level. The configuration for sodium is simply given as

Na   2   8   1

We could now continue to add electrons to the third shell until we get to argon. The configuration for argon is

$$\text{Ar} \quad 2 \quad 8 \quad 8$$

After argon, the shell model works best if it is enhanced with more detail. The topic of atomic structure is discussed further and in greater detail in Section 3.8. Meanwhile, Example 3.5 shows how to determine a few main-shell configurations.

EXAMPLE 3.5 **Main-Shell Electron Configurations**

What are the main-shell electron configurations for (a) fluorine and (b) aluminum?

Solution

a. Fluorine has the symbol F ($Z = 9$); it has nine electrons. Two of these electrons go into the first shell, and the remaining seven go into the second level.

$$\text{F} \quad 2 \quad 7$$

b. Aluminum ($Z = 13$) has 13 electrons. Two go into the first shell, eight go into the second, and the remaining three go into the third shell.

$$\text{Al} \quad 2 \quad 8 \quad 3$$

Notice that fluorine is in group 7A of the periodic table and that it has seven outer-shell electrons. Aluminum is in group 3A, and it has three electrons in its outermost shell.

▶ **Exercise 3.5**

Give the main-shell electron configurations for (a) beryllium and (b) magnesium. These elements are in the same group of the periodic table. What do you notice about the number of electrons in their outermost shells?

3.7 ELECTRON ARRANGEMENT: THE QUANTUM MODEL

The simple planetary Bohr model of the atom has been replaced for many purposes by more sophisticated models in which electrons are treated as waves and their locations are indicated as probabilities. The theory that the electron should have wavelike properties was first suggested in 1924 by Louis de Broglie (1892–1987), a young French physicist. Although it was hard to accept because of Thomson's evidence that electrons are particles, de Broglie's theory was experimentally verified within a few years.

Erwin Schrödinger (1887–1961), an Austrian physicist, used highly mathematical *quantum mechanics* in the 1920s to develop equations that describe the properties of electrons in atoms. The solutions to these equations express the probability of finding an electron in a given volume of space. These shaped volumes of space, called **orbitals**, replace the planetary orbits of the Bohr model.

Suppose you had a camera that could photograph electrons and you left the shutter open while an electron zipped about the nucleus. The developed picture would give a record of where the electron had been. (Doing the same thing with an electric fan would give a blurred image of the rapidly moving blades, a picture resembling a disk.) The electrons in the first shell would appear as a fuzzy ball (often referred to as a *charge cloud* or an *electron cloud* [Figure 3.14]).

Building Atoms by Orbital Filling

Schrödinger concluded that each electron orbital could contain a maximum of two electrons, and that some shells could contain more than one orbital. The first shell contains a single spherical orbital named 1s. The second shell contains four orbitals: 2s is spherical, and the other three orbitals, called 2p, are dumbbell shaped (Figure 3.15). Orbitals in the same shell with the same letter designation make up a **subshell (sublevel)**. The second shell has two subshells; the 2s subshell with only

Fortunately, we need not understand the elaborate mathematics used by Schrödinger in order to make use of some of his results.

An *s* orbital

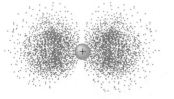

A *p* orbital

▲ **Figure 3.14** Charge-cloud representations of atomic orbitals.

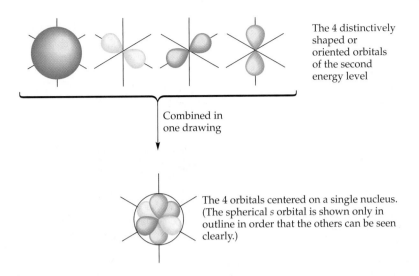

The 4 distinctively shaped or oriented orbitals of the second energy level

Combined in one drawing

The 4 orbitals centered on a single nucleus. (The spherical *s* orbital is shown only in outline in order that the others can be seen clearly.)

◀ **Figure 3.15** Electron orbitals of the second main shell. In these drawings, the nucleus of the atom is located at the intersection of the axes. The eight electrons that would be placed in the second shell of Bohr's model are distributed among these four orbitals in the current model of the atom, with two electrons per orbital.

one orbital and the 2*p* subshell with three orbitals. The third shell contains nine orbitals distributed among three subshells: a spherical 3*s* orbital in the 3*s* subshell, three dumbbell-shaped 3*p* orbitals in the 3*p* subshell, and five 3*d* orbitals with more complicated shapes in the 3*d* subshell.

In building up the electron configuration of atoms of the various elements, the lower sublevels (subshells) are filled first.

Hydrogen (Z = 1) has only one electron in the *s* orbital of the first shell. The electron configuration is

$$H \quad 1s^1$$

Helium (Z = 2) has two electrons and an electron configuration

$$He \quad 1s^2$$

Lithium (Z = 3) has three electrons—two in the first shell and one in the *s* orbital of the second shell:

$$Li \quad 1s^2 2s^1$$

Skipping to nitrogen (Z = 7), we place the seven electrons as follows:

$$N \quad 1s^2 2s^2 2p^3$$

Let's review what this notation means.

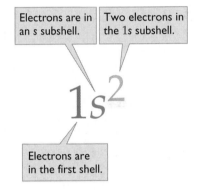

Electrons are in an *s* subshell. | Two electrons in the 1*s* subshell.

$1s^2$

Electrons are in the first shell.

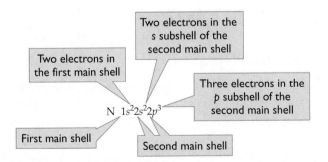

Two electrons in the *s* subshell of the second main shell

Two electrons in the first main shell

Three electrons in the *p* subshell of the second main shell

N $1s^2 2s^2 2p^3$

First main shell

Second main shell

Look next at argon (Z = 18) with its 18 electrons. The configuration is

$$Ar \quad 1s^2 2s^2 2p^6 3s^2 3p^6$$

Note that the highest occupied subshell, 3*p*, is filled. When we move to potassium (Z = 19), we find that the 4*s* sublevel fills before the 3*d* sublevel. The order of filling the various electron sublevels is shown in Figure 3.16, and Table 3.3 gives the electron configuration for the first 20 elements.

▶ **Figure 3.16** An order-of-filling chart for determining the electron configurations of atoms.

Electron Configurations

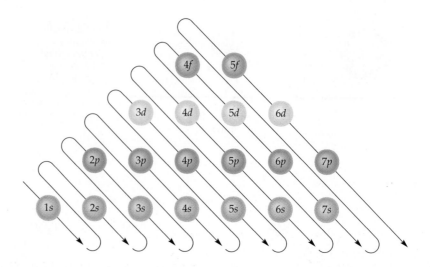

TABLE 3.3	Electron Structures for Atoms of the First 20 Elements	
Name	**Atomic Number**	**Electron Structure***
Hydrogen	1	$1s^1$
Helium	2	$1s^2$
Lithium	3	$1s^2 2s^1$
Beryllium	4	$1s^2 2s^2$
Boron	5	$1s^2 2s^2 2p^1$
Carbon	6	$1s^2 2s^2 2p^2$
Nitrogen	7	$1s^2 2s^2 2p^3$
Oxygen	8	$1s^2 2s^2 2p^4$
Fluorine	9	$1s^2 2s^2 2p^5$
Neon	10	$1s^2 2s^2 2p^6$
Sodium	11	$1s^2 2s^2 2p^6 3s^1$
Magnesium	12	$1s^2 2s^2 2p^6 3s^2$
Aluminum	13	$1s^2 2s^2 2p^6 3s^2 3p^1$
Silicon	14	$1s^2 2s^2 2p^6 3s^2 3p^2$
Phosphorus	15	$1s^2 2s^2 2p^6 3s^2 3p^3$
Sulfur	16	$1s^2 2s^2 2p^6 3s^2 3p^4$
Chlorine	17	$1s^2 2s^2 2p^6 3s^2 3p^5$
Argon	18	$1s^2 2s^2 2p^6 3s^2 3p^6$
Potassium	19	$1s^2 2s^2 2p^6 3s^2 3p^6 4s^1$
Calcium	20	$1s^2 2s^2 2p^6 3s^2 3p^6 4s^2$

*Valence electrons are shown in red.

EXAMPLE 3.6 **Subshell Notation**

Without referring to Table 3.3, use subshell notation to write out the electron configuration for **(a)** oxygen and **(b)** sulfur. What similarity of features do the electron configurations exhibit?

Solution

a. Oxygen (Z = 8) has eight electrons. Place them in the lowest unfilled energy sublevels. Two go into the $1s$ orbital and two into the $2s$ orbital. That leaves four electrons to be placed in the $2p$ subshell. The electron configuration is $1s^2 2s^2 2p^4$.

b. Sulfur (Z = 16) atoms have 16 electrons each. The electron configuration is $1s^2 2s^2 2p^6 3s^2 3p^4$. Note that the total of the superscripts is 16 and that we have not exceeded the maximum capacity for any sublevel.

Both O and S have electron configurations with four electrons in their highest energy sublevel (outermost subshell).

▶ **Exercise 3.6A**
Without referring to Table 3.3, use subshell notation to write out the electron configurations for **(a)** fluorine and **(b)** chlorine. What similarity of features do the electron configurations exhibit?

▶ **Exercise 3.6B**
Use Figure 3.16 to write the electron configurations for **(a)** titanium (Ti) and **(b)** gallium (Ga).

3.8 ELECTRON CONFIGURATIONS AND THE PERIODIC TABLE

In general, the properties of elements can be correlated with their electron configurations. (We explore bond formation using electron configurations in Chapter 5.) Because the number of electrons equals the number of protons, the periodic table tells us about electron configuration as well as atomic number.

The modern periodic table (inside front cover) has horizontal rows and vertical columns.

- Each vertical column is a **group** or *family*. Elements in a group have similar chemical properties.
- A horizontal row of the periodic table is called a **period**. The properties of elements vary periodically across a period.

In the United States, the groups are often indicated by a numeral followed by the letter A or B.

- An element in an A group is a **main group element**.
- An element in a B group is a **transition element**.

The International Union of Pure and Applied Chemistry (IUPAC) recommends numbering the groups from 1 to 18. Both systems are indicated on the periodic table on the inside front cover, but we follow the traditional U.S. method in this book.

Family Features: Outer Electron Configurations

The period in which an element appears in the periodic table tells us how many main electron shells there are in that atom. Phosphorus, for example, is in the third period, and so the phosphorus atom has three main shells. The group number (for main group elements) tells us how many electrons are in the outermost shell. An electron in the outermost shell of an atom is called a **valence electron**. From the fact that phosphorus is in group 5A, we can deduce that it has five valence electrons. Two of these are in an s orbital, and the other three are in p orbitals. We can indicate the outer electron configuration of the phosphorus atom as

$$P \quad 3s^2 3p^3$$

These valence electrons determine most of the chemistry of an atom. Since all the elements in the same group of the periodic table have the same number of valence electrons, they should have similar chemistry, and they do. Figure 3.17 relates the subshell configurations to the groups in the periodic table.

Family Groups

Elements within a group have similar properties. All the elements in group 1A have one valence electron, and all except hydrogen are very reactive metals (hydrogen is a nonmetal). All have the outer electron structure ns^1, where n denotes the number of the outermost main shell. The metals in group 1A are called **alkali metals**; they react vigorously with water to evolve hydrogen gas. There are important trends

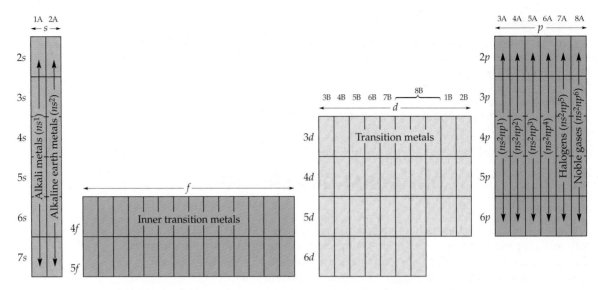

▲ **Figure 3.17** Valence shell electron configurations and the periodic table.

within a family. For example, lithium is the hardest metal in the group. Sodium is softer than lithium; potassium is softer still; and so on down the group. Lithium is also the least reactive toward water. Sodium, potassium, rubidium, and cesium are progressively more reactive. Francium is highly radioactive and extremely rare; few of its properties have been measured. Hydrogen is the odd one in group 1A. It is not an alkali metal. Rather, it is a characteristic nonmetal. Based on its properties, hydrogen probably should be put in a group of its own.

Group 2A elements are known as **alkaline earth metals**. The metals in this group have the outer electron structure ns^2. They are fairly soft and moderately reactive with water. Beryllium is an odd member of the group in that it is rather hard and does not react with water. As in other families, there are trends in properties within the group. For example, magnesium, calcium, strontium, barium, and radium are progressively more reactive toward water.

Group 7A elements, often called **halogens**, also consist of reactive elements. All have seven valence electrons in the configuration ns^2np^5. Halogens react vigorously with alkali metals to form crystalline solids (this is discussed further in Chapter 5). There are trends in the halogen family. Fluorine is most reactive toward alkali metals, chlorine is the next most reactive, and so on. Fluorine and chlorine are gases at room temperature, bromine is a liquid, and iodine is a solid. (Astatine, like francium, is highly radioactive and extremely rare; few of its properties have been determined.)

Members of group 8A, to the far right of the periodic table, have a complete set of valence electrons and therefore undergo few, if any, chemical reactions. They are called **noble gases**, known for their lack of chemical reactivity.

EXAMPLE 3.7 **Valence Shell Electron Configurations**

Write out the subshell notation for the electrons in the highest main shell for **(a)** strontium (Sr) and **(b)** arsenic (As).

Solution
a. Strontium is in group 2A and thus has two valence electrons in an s subshell. Because strontium is in the fifth period of the periodic chart, its outer main shell is $n = 5$. Its outer electron configuration is therefore $5s^2$.
b. Arsenic is in group 5A and the fourth period. Its five outer (valence) electrons are in the $n = 4$ shell and have the configuration $4s^24p^3$.

► **Exercise 3.7A**

Write out the subshell notation for the electrons in the highest main shell for **(a)** rubidium (Rb), **(b)** selenium (Se), and **(c)** germanium (Ge).

► **Exercise 3.7B**

The valence electrons in aluminum have the configuration $3s^2 3p^1$. Gallium is directly below aluminum, and indium is directly below gallium in the periodic table. Use only this information to write the configuration for the valence electrons in **(a)** gallium and **(b)** indium.

Metals and Nonmetals

Elements in the periodic table also are divided into two classes by a heavy, stepped diagonal line. Those to the left of the line are *metals*. A **metal** has a characteristic luster and generally is a good conductor of heat and electricity. Except for mercury, which is a liquid, all metals are solids at room temperature. Metals generally are *malleable*; that is, they can be hammered into thin sheets. Most also are *ductile*; they can be drawn into wires.

Elements to the right of the stepped line are *nonmetals*. A **nonmetal** element lacks metallic properties. Several nonmetals are gases (oxygen, nitrogen, fluorine, chlorine). Others are solids (carbon, sulfur, phosphorus, iodine). Bromine is the only nonmetal that is a liquid at room temperature.

Some of the elements bordering the stepped line are called *semimetals* or *metalloids*, elements that have intermediate properties. Metalloids have properties that resemble those of both metals and nonmetals. There is a lack of agreement on just which elements fit in this category.

3.9 WHICH MODEL TO CHOOSE?

For some purposes in this text, we use only main-shell configurations of atoms to picture the distribution of electrons in atoms. At other times, the electron clouds of the quantum mechanical model are more useful. Even Dalton's model sometimes proves to be the best way to describe certain phenomena (the behavior of gases, for example). The choice of model is always based on which one is most helpful in understanding a particular concept. This, after all, is the whole purpose of scientific models.

Critical Thinking Exercises

Apply knowledge that you have gained in this chapter and one or more of the FLaReS principles (Chapter 1) to evaluate the following statements or claims.

3.1 Suppose you read in the newspaper that a chemist in South America claims to have discovered a new element with an atomic mass of 34. An extremely rare element, it was found in a sample recovered from the Andes Mountains. Unfortunately, the chemist has used up all of the sample in his analyses.

3.2 Some aboriginal tribes have rain-making ceremonies in which they toss pebbles of gypsum up into the air. (Gypsum is the material used to make plaster of Paris by heating the rock to remove some of its water.) Sometimes it really does rain several days after these rain-making ceremonies.

SUMMARY

Section 3.1—Davy, Faraday, and others showed that matter is electrical in nature. They were able to decompose compounds into elements by **electrolysis** or passing electricity through molten salts. **Electrodes** are the pieces of metal or carbon that carry electricity through the **electrolyte** or solution/compound undergoing electrolysis. The electrolyte contains **ions**—charged atoms or groups of

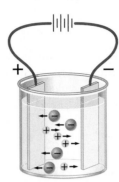

atoms. The **anode** is the positive electrode and **anions** or negatively charged ions move toward it. The **cathode** is the negative electrode and **cations** or positively charged ions move toward it. Experiments with cathode ray tubes showed that matter contained negatively charged particles, which were called **electrons**. By deflecting cathode rays with a magnet, Thomson was able to determine the mass-to-charge ratio for the electron. Goldstein's experiment, using gas discharge tubes with perforated cathodes, showed that matter also contained positively charged particles. Millikan's oil drop experiment measured the charge on the electron, so its mass could then be calculated. That mass was found to be smaller than that of the lightest atom.

Sections 3.2–3.3—In his studies of cathode rays, Roentgen accidentally discovered **X-rays**, which are a highly penetrating form of radiation used in medical diagnosis. Becquerel accidentally discovered another type of

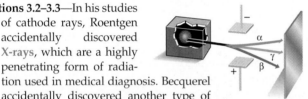

radiation that comes from certain unstable elements. Marie (Sklodowska) Curie named this new discovery **radioactivity** and studied it extensively, discovering two elements in the process and winning two Nobel Prizes. Radioactivity was soon found to be three different types of radiation: **Alpha particles** have four times the mass of a hydrogen atom and a positive charge twice that of an electron. **Beta particles** are energetic electrons. **Gamma rays** are a form of energy like X-rays but more penetrating.

Section 3.4—Rutherford's later experiments with alpha particles and gold foil showed that most of the alpha particles emitted toward the foil

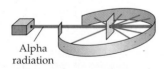

Alpha radiation

passed through it. A few were deflected or, occasionally, bounced almost directly back to the source. This indicated that all the positive charge and most of the mass of an atom must be in a tiny core, which he called the **nucleus**.

Section 3.5—Rutherford called the unit of positive charge the **proton**, with about the mass of a hydrogen atom and with a charge equal in size but opposite in sign to the electron. In 1932 Chadwick discovered the **neutron**, a nuclear particle

as massive as a proton but with no charge. The number of protons in an atom is the **atomic number**. Atoms of the same element have the same number of protons but may have different numbers of neutrons; such atoms are called **isotopes**. Different isotopes of an element are mostly identical chemically. Protons and neutrons collectively are called **nucleons**, and the number of nucleons is called the **mass number** or **nucleon number**. The difference between the nucleon number and the atomic number is the number of neutrons. The general symbol for an isotope of element X is written $^A_Z X$, where A is the nucleon number and Z is the atomic number.

Section 3.6—Light from the sun or from an incandescent lamp is a continuous spectrum, containing all colors. Light from a gas discharge tube is a line spectrum, containing only certain colors. Bohr explained line spectra by proposing that an electron in an atom could have only

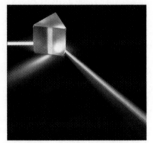

certain discrete energy levels, with these levels differing by a **quantum** or discrete unit of energy. An atom drops to the lowest energy state or **ground state** after being in a higher-energy state or **excited state**. Line spectra occur from these changes in energy levels giving off specific wavelengths or colors of light. Bohr also deduced that the energy levels of an atom could handle at most $2n^2$ electrons, where n is the energy level or **shell** number. A description of the shells occupied by electrons of an atom is one way of giving the atom's **electron configuration** or arrangement of electrons.

Section 3.7—de Broglie theorized that electrons have wave properties. Schrödinger developed equations

that described each electron's location in terms of an **orbital**, the volume of space that an electron usually occupies. Each orbital holds at most two electrons. Orbitals in the same shell and with the same energy make up a **subshell**, each of which is designated by a letter. An *s* orbital is spherical and a *p* orbital is dumbbell shaped. The *d* and *f* orbitals have more complex shapes. The first main shell can hold only one *s* orbital; the second level can hold one *s* and three *p* orbitals; and the third level can hold one *s*, three *p*, and five *d* orbitals. In writing an electron configuration using subshell notation (quantum mechanical notation), the shell number and subshell letter are given, followed by a superscript giving the number of electrons in that subshell. The order of the shells and subshells can be remembered with a chart or by using the periodic table.

Sections 3.8—Elements in the periodic table are arranged in columns or **groups**, and in rows or **periods**. Electrons in the outermost shell of an atom are called **valence electrons** and are responsible for the reactivity of that atom. Elements in a group usually have the same number of valence electrons and similarities in properties. Groups are designated by a number and by the letter A or B. The A-group elements are called **main group elements** and are on both sides of the periodic table. The B-group elements are **transition elements**, in the middle of the table. The first column of the table, or group 1A elements, is known as the **alkali metals**. Except for hydrogen they are all soft, low-melting, highly reactive metals. The second column, group 2A, is known as the **alkaline earth elements**. They are fairly soft and fairly reactive metals. Group 7A elements are the **halogens** and are reactive nonmetals. Group 8A elements, the **noble gases**, react very little or not at all. A diagonal line divides the periodic table into metals and nonmetals. **Metals** conduct electricity and heat, and are shiny, malleable, and ductile. **Nonmetals** tend to lack the properties of metals. Some elements on the diagonal line are sometimes referred to as metalloids, and have properties intermediate between metals and nonmetals.

REVIEW QUESTIONS

1. What did each of the following scientists contribute to our knowledge of the atom?
   a. Crookes
   b. Thomson
   c. Goldstein
   d. Millikan
   e. Roentgen
   f. Becquerel

2. Which is *not* a property of cathode rays? (a) They are composed of negatively charged particles. (b) They have the same characteristics from element to element. (c) They can be deflected by electric fields. (d) They are composed of particles with a mass of 1 u.

3. What is radioactivity? How did the discovery of radioactivity contradict Dalton's atomic theory?

4. Define or identify each of the following.
   a. alpha particle
   b. beta particle
   c. gamma ray
   d. deuterium
   e. tritium
   f. photon

5. What are isotopes? Give an example.

6. How are X-rays and gamma rays similar? How are they different?

7. The following table describes four atoms.

	Atom A	Atom B	Atom C	Atom D
Number of protons	10	11	11	10
Number of neutrons	11	10	11	10
Number of electrons	10	11	11	10

   Are atoms A and B isotopes? A and C? A and D? B and C?

8. What is the mass of each atom in Question 7?

9. What is the atomic nucleus? What two principal subatomic particles are found in the nucleus?

10. Discuss Rutherford's gold-foil experiment. What two things did it tell us about the structure of the atom?

11. Give the distinguishing characteristics of the proton, the neutron, and the electron.

12. Compare Dalton's model of the atom with the nuclear model of the atom.

13. In an atom, what are the extranuclear (outside the nucleus) subatomic particles?

14. If the nucleus of an atom contains ten protons, how many electrons are there in the neutral atom?

15. What is the symbol, name, and atomic mass of the element with Z = 98? You may use a periodic table and table of atomic masses.

16. What is the symbol, name, and atomic mass of the element that has 18 protons in the nucleus of its atoms?

17. What is meant by the nucleon number of an isotope?

18. Explain what is meant by the term *atomic mass*.

19. Give the nuclear symbols for the hydrogen isotopes called protium, deuterium, and tritium.

20. How did Bohr refine the model of the atom?

21. What particles travel in the orbits of the Bohr model of the atom?

22. Define the following terms.
    a. ground state
    b. excited state

23. When an electron moves from the fourth shell to the second, is energy being emitted or absorbed?

24. Which atom absorbs more energy, one in which an electron moves from the second shell to the third shell or an otherwise identical atom in which an electron moves from the first to the third shell?

25. In what shell is the (a) 3d subshell (b) 4p subshell?

26. What is the electron capacity of the (a) 3d subshell and (b) 4p subshell in an atom?

27. Use the periodic table to determine the numbers of protons in atoms of the following elements.
    a. helium
    b. sodium
    c. chlorine
    d. oxygen
    e. magnesium
    f. sulfur

28. How many electrons are there in the neutral atoms of the elements listed in Question 27?

PROBLEMS

Nuclear Symbols and Isotopes

For items 29–34, you may refer to the periodic table.

29. Give the nuclear symbol and name for **(a)** an isotope with a nucleon number of 12 and an atomic number of 5 and **(b)** an isotope with 53 protons and 72 neutrons.

30. Give the nuclear symbol and name for **(a)** an isotope with $Z = 35$ and $A = 83$ and **(b)** a neutral atom with 26 electrons and 29 neutrons.

31. Give the nuclear symbol for the following isotopes.
    a. gallium-69
    b. cobalt-60

32. Give the nuclear symbol for the following isotopes.
    a. molybdenum-99
    b. technetium-98

33. Indicate the number of protons and the number of neutrons in atoms of the following isotopes:
    a. $^{62}_{30}\text{Zn}$
    b. $^{241}_{94}\text{Pu}$

34. Indicate the number of protons and the number of neutrons in atoms of the following isotopes:
    a. $^{107}_{47}\text{Ag}$
    b. $^{81}_{36}\text{Kr}$

35. Which of the following pairs represent isotopes?
    a. $^{70}_{34}\text{X}$ and $^{70}_{33}\text{X}$
    b. $^{57}_{28}\text{X}$ and $^{66}_{28}\text{X}$

36. Which of the following pairs represent isotopes?
    a. $^{186}_{74}\text{X}$ and $^{186}_{73}\text{X}$
    b. $^{8}_{4}\text{X}$ and $^{6}_{4}\text{X}$
    c. $^{22}_{11}\text{X}$ and $^{44}_{22}\text{X}$

The Bohr Model

37. According to Bohr, what is the maximum number of electrons in the fourth shell ($n = 4$)?

38. If the third shell of an atom in the ground state contains two electrons, what is the total number of electrons in the atom?

39. What are the main-shell electron configurations for the elements listed in Question 27?

40. Neutral atoms of an element have 5 p^+ and 6 n its nucleus. A student lists the main-shell electron configuration of the atom as

    B 2 4

    The configuration is incorrect. Identify the error.

Quantum Mechanical Notation for Electron Structure

41. In the quantum mechanical notation $2s^2$, how many electrons are described? What is the general shape of the orbitals described in the notation? How many orbitals are included in the notation?

42. In the quantum mechanical notation $2p^6$, how many electrons are described? What is the general shape of the orbitals described in the notation? How many orbitals are included in the notation?

43. Neutral atoms of an element have 9 p^+ and 10 n in their nuclei and the main-shell electron configuration 2 7. Use quantum mechanical notation to describe the electron configuration of the atoms.

44. Give the electron configurations (using quantum mechanical notation) for the elements in Problem 27. You may refer to the periodic table.

45. Without referring to the periodic table, give the atomic numbers of the following elements.
    a. $1s^2 2s^2$
    b. $1s^2 2s^2 2p^3$
    c. $1s^2 2s^2 2p^6 3s^2 3p^1$

46. Identify the elements described in Problem 45. You may refer to the periodic table.

47. None of the following ground state electron configurations is reasonable. In each case, explain why.
    a. $1s^2 2s^2 3s^2$
    b. $1s^2 2s^2 2p^2 3s^1$
    c. $1s^2 2s^2 2p^6 2d^5$

48. None of the following electron configurations is reasonable. In each case, explain why.
    a. $1s^1 2s^1$
    b. $1s^2 2s^2 2p^7$
    c. $1s^2 2p^2$

49. What is the difference between the electron configurations of sulfur (S) and chlorine (Cl)?

50. What subshell contains one electron in aluminum (Al) but none in magnesium (Mg)?

51. If three electrons were added to the outermost shell of a phosphorus atom, the new electron configuration would resemble that of what element?

52. If two electrons were removed from the outermost shell of a magnesium atom, the new electron configuration would resemble that of what element?

Electron Structure and the Periodic Table

53. If a neutral atom in its ground state contains only five electrons in its outermost p sublevel, it is an atom of what group of elements?

54. If a neutral atom in its ground state has a d sublevel that contains five electrons, to what group of elements does the atom belong?

55. If an atom contains only *s* electrons in its outermost energy level, is the element a metal or a nonmetal?

56. Give the symbol for a metal that has two *p* electrons in its outermost energy level.

57. Referring only to the periodic table, indicate what similarity in electron structure is shared by fluorine (F) and chlorine (Cl). What is the difference between their electron structures?

58. Referring only to the periodic table, indicate the similarity in the electron structures of oxygen (O) and sulfur (S). What is the difference between their electron structures?

59. List two characteristic properties of metals.

60. When a crystal of the element iodine is struck by a hammer, it shatters. Without looking at the periodic table, is iodine likely to be a metal or a nonmetal? Explain.

61. Identify the following elements as metals or nonmetals. You may refer to the periodic table.
    a. sulfur
    b. chromium
    c. iodine
    d. holmium

62. Indicate the group numbers of the following families.
    a. alkali metals
    b. halogens
    c. alkaline earth metals

63. Identify the periods of the following elements. You may refer to the periodic table.
    a. chlorine
    b. osmium
    c. hydrogen
    d. lithium

64. List some of the properties of alkali metals.

65. What is the most distinguishing property of noble gases?

66. Which of the following elements are halogens?
    a. Ag
    b. At
    c. As

67. Which of the following elements are alkali metals?
    a. K
    b. Y
    c. W

68. Which of the following elements are noble gases?
    a. Fe
    b. Ne
    c. Ge
    d. He
    e. Xe

69. Which of the following elements are transition metals?
    a. Ti
    b. Tc
    c. Te

70. Which of the following elements are alkaline earth metals?
    a. Bi
    b. Ba
    c. Be
    d. Br

71. How many electrons are in the outermost shells of group 2A elements?

72. In what period of elements are electrons first introduced into the fourth shell?

ADDITIONAL PROBLEMS

73. Elements are defined on a theoretical basis as being composed of atoms that share the same atomic number. On the basis of this theory, would you think it possible for someone to discover **(a)** a new element that would fit between magnesium (atomic number 12) and aluminum (atomic number 13)? **(b)** A new element with atomic number 122?

74. What is the difference between a Bohr orbit and the orbital of wave mechanics?

75. How many subshells are there in the **(a)** second main shell ($n = 2$), and **(b)** third main shell ($n = 3$)?

76. What is the lowest numbered main shell in which **(a)** *p* orbitals are found, and **(b)** a *d* subshell can be found?

77. Before neutrons were discovered, some scientists thought that the nucleus contained electrons to balance some or all the protons. How could this model explain the fact that fluorine has an atomic number of 9 and a mass number of 19?

78. Tell in words what is meant by each of the following electron configuration notations. Which element corresponds to each configuration?
    a. $1s^2 2s^2 2p^5$ b. $1s^2 2s^2 2p^6 3s^2 3p^6 3d^{10} 4s^1$

79. What subshell(s) is(are) being filled in each of the following regions of the periodic table?
    a. groups 1A and 2A
    b. groups 3A through 7A

80. Give the period number and group number for the element whose atoms have the electron configuration
    a. $1s^2 2s^2 2p^6$
    b. $1s^2 2s^2 2p^6 3s^2 3p^2$
    c. $1s^2 2s^2 2p^6 3s^1$
    d. $1s^2 2s^2$
    e. $1s^2 2s^2 2p^3$
    f. $1s^2 2s^2 2p^6 3s^2 3p^1$

81. Consider the following statements concerning atoms and subatomic particles. Identify each statement as true or false and explain your choice.
    a. Neutrons have neither mass nor charge.
    b. Isotopes of an element have the same number of protons.
    c. Carbon-14 and nitrogen-14 have the same atomic number and the same nucleon number.

82. Without referring to any tables in the text, mark an appropriate location for each of the following in the blank periodic table provided: **(a)** the fourth period noble gas, **(b)** the third period alkali metal, **(c)** the fourth period halogen, and **(d)** a metal in the fourth period and in group 3B.

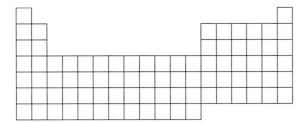

83. Refer back to Figure 3.11. A patient is found to be suffering from heavy-metal poisoning. An emission spectrum is obtained from a sample of the patient's blood, by spraying the blood into a hot flame. The brightest emission lines are found at 5520, 5540, 5890, and 5900Å. What two elements are likely to be present? Which one is more likely to be the culprit?

COLLABORATIVE GROUP PROJECTS

Prepare a PowerPoint, poster, or other presentation (as directed by your instructor) for presentation to the class.

84. Prepare a brief biographical report on one of the following.
    a. Humphry Davy   b. William Crookes
    c. Wilhelm Roentgen   d. Marie Curie
    e. Robert Millikan   f. Niels Bohr
    g. Alessandro Volta   h. Ernest Rutherford
    i. Michael Faraday   j. J. J. Thomson

85. Draw a grid containing 16 squares, in two rows of 8, representing the 16 elements in the periodic table from lithium to argon. For each of the elements, give **(a)** the main-shell electron configuration and **(b)** the quantum mechanical notation for electron configuration.

86. Prepare a brief report on one of the alkali metals or alkaline earth metals. List sources and commercial uses.

87. Using this book and other references, design a time line of important milestones in the elucidation of atomic structure. Begin with the ancient Greeks and conclude with the present.

88. Prepare a brief report on one of the halogens or noble gases. List sources and commercial uses.

REFERENCES AND READINGS

1. Abrahams, Marc (ed.). *The Best Annals of Improbable Research*. New York: Freeman, 1998. A collection of anecdotes about scientific research showing that scientists are real people and that some of them, at least, can laugh at themselves.

2. Asimov, Isaac. *Atom: Journey Across the Subatomic Cosmos*. New York: Dutton, 1991.

3. Bowden, Mary Ellen. *Chemical Achievers: The Human Face of the Chemical Sciences*. Philadelphia: Chemical Heritage Foundation, 1997.

4. Bowden, Mary Ellen. *Chemistry Is Electric!* Philadelphia: Chemical Heritage Foundation, 1997.

5. Conkling, John A. "Pyrotechnics." *Scientific American*, July 1990, pp. 96–102. Discusses the science of fireworks.

6. Farmelo, Graham. "The Discovery of X-rays." *Scientific American*, November 1995, pp. 86–91.

7. Knight, David M. *Humphry Davy: Science and Power*. Cambridge, MA: Blackwell, 1992.

8. Krebs, Robert E. *The History and Use of Our Earth's Chemical Elements: A Reference Guide*. Westport, CT: Greenwood Press, 1998.

9. Parker, Barry. *Quantum Legacy: The Discovery That Changed Our Universe*. Amherst, NY: Prometheus Books, 2002.

10. Pizzi, Richard A. "Humphry Davy, Self-Made Chemist." *Today's Chemist at Work*, April 2004, pp. 49–51.

11. Quinn, Susan. *Madame Curie: A Life*. New York: Simon & Schuster, 1995.

12. Strathern, Paul. *Mendeleyev's Dream: The Quest for the Elements*. London: Hamish Hamilton, 2000.

13. Walker, Jearl. "The Amateur Scientist: The Spectra of Street Lights Illuminate Basic Principles of Quantum Mechanics." *Scientific American*, January 1984, pp. 138–143.

14. Wolff, Peter. *Breakthroughs in Chemistry*. New York: Signet Science Library, 1967. Chapters 6–9 relate the research of Faraday, Mendeleev, Marie Curie, and Bohr.

GREEN CHEMISTRY

Nanotechnology

Rich Gurney, Simmons College

A flurry of discovery of new physical phenomena and properties followed the discovery of new elements. The isolation and purification of bulk uranium led to many discoveries involving radioactivity, and today radiological procedures have become a mainstay of medical diagnosis. As stated in Chapter 3, the arrangement of various parts of atoms determines the bulk properties of different kinds of matter for large collections of atoms. You may be surprised to learn that the properties of a collection of atoms are also influenced by the size of the collection. Nearly everyone can list the physical properties of the element gold. The yellow, malleable metal is ubiquitous and pervasive in nearly all cultures and civilizations. However, a solution of gold *nanoparticles*—submicroscopic in size—is a brilliant red, blue, or gold color, depending on the size of the nanoparticles.

The way in which the atoms are connected can also have a profound effect on the physical properties of the collection. Graphite, one allotrope of carbon, is utilized in pencils because it is soft. Graphite rubbed on paper leaves a trail on the page. Carbon nanotubes, an allotrope with a different connectivity (Chapter 10), are much stronger than steel. The discovery of many new physical properties of carbon nanotubes has made them both an extremely useful and very valuable material.

Nanoscience and nanotechnology both operate within the nanoscopic world, a world whose typical boundaries are defined by one-billionth to one-millionth of a meter. Nanometers are really, really small. There are one *billion* nanometers in one meter. A nanoparticle is only three to five atoms wide, making the nanoparticle about 40,000 times smaller than the width of an average human hair. While the discipline of nanotechnology is relatively new, examples of nanotechnology can be found throughout history. By varying the sizes of gold nanoparticles, artisans were able to produce beautifully stained, glass windows as far back as medieval times and colorfully glazed ceramics during the Ming dynasty. Today, gold nanoparticles are being tested for use in therapies for neurodegenerative diseases that involve protein aggregation, such as Alzheimer's and Parkinson's.

▲ The beads fluoresce in different colors, but they are made of the same plastic and contain the same nanoparticles—tiny particles of a cadmium–selenium compound. The different colors are the result of different sizes of nanoparticles used.

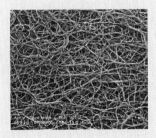

▲ Carbon nanotubes (magnified 35,000×) show promise as a superstrong material.

Investigation 1
Real-World Applications of Nanotechnology

Nanotechnology is already having a profound impact on the lives of many people around the world as many consumer products include nanomaterials. You may already have some nanoscale items in your home. Nanoparticle-containing bandages will be arriving in your neighborhood pharmacy before you know it. Do a keyword search for "Nano Materials Consumer Products" using a major search engine to investigate the consumer products on the market that already include nanomaterials. Create a list of consumer products that you or your family may have already encountered. What functions do the nanoparticles serve in these consumer products? Search the website of your favorite newspaper to find reports on consumer products containing nanoscale materials. What impact have these consumer products had on our society?

Investigation 2
Green Nanotechnology

Nanotechnology involves the measurement, observation, manipulation, and production of materials usually between 1 and 100 nanometers. If these materials are so small, why should our society be overly concerned with ensuring the field of nanotechnology develops in a "Green" way? How will the manufacture of nanomaterials impact our society if this burgeoning field does not embrace the 12 principles of green chemistry and engineering? Based on your Web research of green nanotechnology, which of the 12 principles of green chemistry have the most relevance to this field?

Investigation 3
Microchip Manufacture and Green Solvents

Microprocessors and digital memory chips are both examples of integrated circuits or microchips. These devices can be found in everything from laptop computers to microwave ovens, automobiles to electric toothbrushes. Historically, the fabrication of microchips, via nanotechnology, has required the use of hazardous chemicals and the consumption of large amounts of purified water. SC Fluids Inc., working in collaboration with Los Alamos National Laboratories, designed a greener alternative to the nanofabrication that replaces the hazardous volatile organic solvents with a supercritical carbon dioxide (CO_2) resist remover, SCORR. SC Fluids, Inc. was awarded the Presidential Green Chemistry Challenge Small Business Award in 2002 for this development. Research the term "SC Fluids Scorr" online and determine which of the 12 principles of green chemistry are addressed by this technology. Why would a microchip manufacturing company be interested in adopting the **SCORR** technology? Why is carbon dioxide a greener solvent as compared to volatile organic compounds? In Chapter 17, you will learn about another commercial application of supercritical carbon dioxide—dry cleaning.

Exercise 1
Get the Word Out

Many products of nanotechnology exist on the consumer market, but most people are not aware of their existence or benefits. Choose one item from the list of products that you discovered in Investigation 1 that incorporate nanotechnology, and research the product to uncover the benefits. Create a one-page pamphlet that you could give to your friends and family members to educate them about nanotechnology in the consumer product. If the information is available, discuss how green chemistry has impacted the creation, development, or manufacture of the product.

Exercise 2
Educate Your Community

While the burgeoning field of nanotechnology is embracing green chemistry and green technology, many members of our global community have not yet appreciated the implications of adopting such methods. Craft a one-page letter to the editor of your local newspaper discussing the importance and need for adopting green practices specifically with respect to the field of nanotechnology.

Exercise 3
E-waste and Future Green Solutions

SC Fluids, Inc. has significantly reduced the waste and energy use with its SCORR process, applying green chemistry principles 3, 4, 5, and 6 to the nanofabrication of microchips. In a two-page essay, discuss the implications on our society if microchips and all components of electronic devices were, according to green principle 10, "designed so that at the end of their function they did not persist in the environment and instead broke down into innocuous degradation products." For inspiration on the magnitude of the problem of "E-waste" or electronic waste in our society, visit the Web page of photographer Chris Jordan, who has documented the magnitude of our mass consumption of electronic devices. Estimate the weight of all electronic devices currently in your household and project the space these items will occupy in a landfill because they were not created with principle 10 in mind.

Exercise 4
Contact Congress

As an indirect consumer of microchips, a proponent of green chemistry, and a taxpayer you have an opportunity to make a difference by educating the legislative branch of government and encouraging funding to these research initiatives. Based upon your research into the SC Fluids SCORR technology and your knowledge of the 12 principles of green chemistry, project how the technology will transform our society, and write a convincing and gripping letter to your members of Congress outlining the need for additional funds to support this research.

Cooperative Exercise

Together with members of your class, contact your friends from high school who have attended other colleges and universities to form a community of students who are interested in learning more about green chemistry and nanotechnology. To facilitate the formation of your group, you must first collectively create a questionnaire to assess group members' knowledge of green chemistry and nanotechnology. You might also begin to create a document to educate your friends by compiling all of your *Educate Your Community* letters to the editors from Exercise 2, describing the benefits of green chemistry in this field. You might also collect your *Get the Word Out* one-page pamphlets from Exercise 1, to send to interested friends. Once you have formed a dedicated and interested community, determine the collective goals of the group. You might also consider issues of sustainability on your campuses and determine which colleges and universities offer courses on green chemistry, green engineering, or nanotechnology.

CHAPTER 4

Nuclear Chemistry

Nearly all energy on Earth comes from the sun, a six-billion-year-old nuclear fusion reactor. Solar nuclear reactions fuse hydrogen nuclei into helium nuclei, releasing enormous quantities of heat and light. In this chapter, we focus on nuclear changes that offer both promise and peril.

The Heart of Matter

The term *nuclear energy* conjures up images of a mighty, often fearsome force! The images of giant mushroom clouds from nuclear explosions that devastated entire cities are hallmarks of an age. Most people have heard of the accident at Three Mile Island and the catastrophic events at Chernobyl. However, there is also a life-giving side to nuclear energy: Our sun is one huge nuclear power plant, supplying the energy that warms our planet as well as the light necessary for plant growth. The twinkling light from every star we see in the night sky is produced by unimaginably powerful nuclear reactions.

Here on Earth, scientists have developed many applications of nuclear energy apart from weapons of war and power plants for electricity. Uses range from medical diagnosis and treatment of diseases like cancer, to accurate dating of archaeological artifacts, to fire safety in most modern homes.

In Chapter 3 we discussed the structure of an atom, focusing mainly on the electrons in the atom, the particles that determine its chemistry. Let us now take a closer look at that tiny speck in the center of the atom—the atomic nucleus.

If an atom is incomprehensibly small, the infinitesimal size of the atomic nucleus is completely beyond our imagination. The diameter of an atom is 100,000 times greater than the diameter of its nucleus. If an atom could be blown up in size until it was as large as your classroom, the nucleus would be about as big as the period at the end of this sentence.

Yet this tiny nucleus contains almost all the atom's mass. How incredibly dense the atomic nucleus must be! A cubic centimeter of water weighs 1 g and 1 cm^3 of gold about 19 g. A cubic centimeter of pure atomic nuclei would weigh more than 100 million metric tons!

Even more amazing than its density is the enormous amount of energy contained within each atomic nucleus. Some atomic nuclei undergo reactions that can fuel the most powerful bombs ever built or provide electricity for millions of people.

Much of what we hear today about nuclear technology is negative: nuclear accidents, the problem of nuclear waste disposal, and the global crisis of nuclear weapons proliferation. However, there is a positive side: The use of radioactive isotopes in medicine saves lives every day, and many applications of nuclear chemistry in science and industry have improved the human condition significantly. In this chapter you will learn about both the destructive and healing power of that infinitesimally small and incredibly dense heart of every atom ... the nucleus.

85

4.1 NATURAL RADIOACTIVITY

Recall from Chapter 3 that most elements occur in nature in several isotopic forms, with nuclei differing in their number of neutrons. Many of these nuclei are unstable and undergo **radioactive decay**. The nuclei that undergo such decay are called **radioisotopes**, and the process results in the production of one or more types of radiation.

Background Radiation

Humans have always been exposed to radiation. Even as you read this sentence you are being bombarded by cosmic rays, which originate from the sun and outer space. Other radiation reaches us from natural radioactive isotopes in air, water, soil, and rocks. We cannot escape radiation: It occurs in many natural processes, including those in our bodies. One naturally occurring isotope of potassium, ^{40}K, exists in all our cells. This ever-present radiation is called **background radiation.** Figure 4.1 shows that over three-fourths of the average radiation exposure comes from background radiation. Most of the remaining one-fourth comes from medical irradiation such as X-rays. Other sources, such as fallout from testing of nuclear bombs, releases from the nuclear industry, and occupational exposure, account for only a minute fraction of our total exposure.

In the past, accidents at the Three Mile Island plant near Harrisburg, Pennsylvania, in 1979 and at Chernobyl in Ukraine in 1986 did much to increase public apprehension about nuclear technology. No one was hurt at Three Mile Island, but the accident at Chernobyl was catastrophic. Dozens of people were killed outright; and hundreds of others died later of severe radiation poisoning. The long-term impact on thousands more throughout Europe exposed to increased levels of radioactivity is still uncertain.

Harmful effects arise from the interaction of radiation with living tissue. Radiation with enough energy to knock electrons from atoms and molecules, converting them into ions (electrically charged atoms or groups of atoms), is called **ionizing radiation**. Nuclear radiation and X-rays are examples. Because radiation is invisible and because it has such great potential for harm, we are very much concerned with our exposure to it.

Radiation Damage to Cells

Radiation-caused chemical changes in living cells can be highly disruptive. Ionizing radiation can devastate living cells by interfering with their normal chemical processes. Molecules can be splintered into reactive fragments called *free radicals*, which can disrupt vital cellular processes. White blood cells, the body's first line of defense against bacterial infection, are particularly vulnerable. Radiation also affects bone marrow, causing a drop in the production of red blood cells, which

In 1993 a Ukrainian Academy of Science study group led by Vladimir Chernousenko investigated the aftereffects of the Chernobyl accident. Chernousenko claims that 15,000 of those who helped clean up the accident site have died and that another 250,000 have been left as invalids. He also says that 200,000 children have experienced radiation-induced illnesses and that half of the children in Ukraine and Belarus have symptoms.

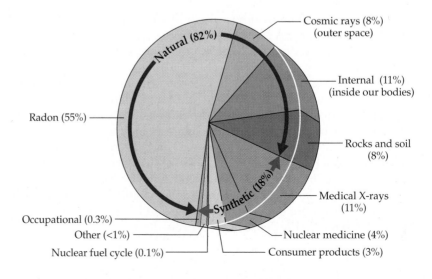

▶ **Figure 4.1** Most of our exposure to ionizing radiation comes from natural sources (blue shades). About 18% of our exposure comes from human activities (other colors).

Question: What natural source of ionizing radiation contributes most to our exposure? What makes up the largest proportion of our exposure from human activities?

results in anemia. Radiation also has been shown to induce leukemia, a cancerlike disease of the blood-forming organs.

Ionizing radiation also can cause changes in the molecules of heredity (DNA) in reproductive cells. Such changes show up as mutations in the offspring of exposed parents. Little is known of the effects of such exposure on humans. However, many of the mutations that occurred during the evolution of present species may have been caused by background radiation.

Because of the potentially devastating effects on living things, a knowledge of radiation, radioactive decay, and nuclear chemistry in general is crucial. We first describe how to write balanced nuclear equations and how to identify the various types of radiation that are emitted.

Cell Phones and Microwaves and Power Lines, Oh My!

The word *radiation* has several meanings. *Ionizing* radiation that comes from decaying atomic nuclei is highly energetic and can damage tissue, as described previously. *Electromagnetic* radiation includes light in its many forms—visible light, radio waves, television broadcasts, microwaves, ultraviolet light, military ULF (ultra low frequency), and others. A few types of electromagnetic radiation—X-rays and gamma rays—have enough energy to ionize tissue, and ultraviolet light has been strongly implicated in melanoma or skin cancer. But what about microwaves and radio waves, which don't have anywhere near that much energy? Although very large amounts of microwaves can cause burns (by heating), there is no evidence that low levels of microwaves pose any threat to human health. Likewise, three separate studies reported in 2000, 2003, and 2005 were unable to demonstrate a connection between cell phone usage and cancer. (A fourth study reported in 2003 did show that cell phone users were twice as likely to be involved in rear-end collisions in automobiles, so clearly there is some hazard associated with their use.) And according to the American Cancer Society, the largest study on the subject showed no link between cancer and electromagnetic fields from electric blankets and computer monitors. Studies continue, but it appears that the effects of most nonionizing radiation on humans are, at the very most, quite small and difficult to measure accurately.

4.2 NUCLEAR EQUATIONS

It is rather easy to write balanced equations for nuclear processes. These equations differ in two ways from the chemical equations we will discuss in Chapter 6. First, as we will see in Chapter 6, chemical equations must have the same elements on both sides of the arrow, whereas nuclear equations rarely do. Second, in ordinary chemical equations we must balance atoms; in nuclear equations we balance the *nucleons* (protons and neutrons). What this really means is that we must balance the atomic numbers (number of protons) and nucleon numbers (number of nucleons) of the starting materials and products. For this reason, we must always specify the *isotope* of each element appearing in a nuclear equation. We use nuclear symbols (Section 3.5) in writing nuclear equations because this makes the equations easier to balance.

 Radioactive Decay

Radon-222 atoms break down spontaneously, giving off alpha (α) particles as shown in Figure 4.2(a). The process is called *alpha decay*. Because alpha particles are identical to helium nuclei, this reaction can be summarized by the equation

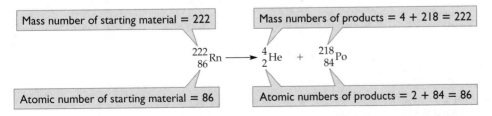

Mass number of starting material = 222

Mass numbers of products = 4 + 218 = 222

$$^{222}_{86}\text{Rn} \longrightarrow {}^{4}_{2}\text{He} + {}^{218}_{84}\text{Po}$$

Atomic number of starting material = 86

Atomic numbers of products = 2 + 84 = 86

We use the symbol $^{4}_{2}\text{He}$ for the alpha particle (rather than α) because it allows us to check the balance of mass and atomic numbers more readily. The atomic number $Z = 84$ identifies the new element as polonium (Po). Note that the nucleon number A of the starting material must equal the total of the nucleon numbers of the products. The same is true for atomic numbers.

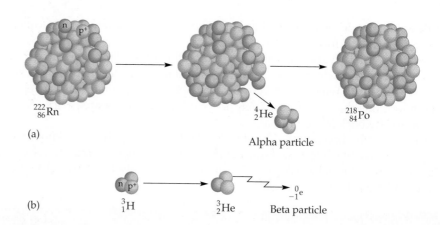

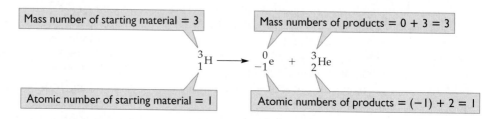

▶ **Figure 4.2** Nuclear emission of (a) an alpha particle and (b) a beta particle.

Question: What changes occur in a nucleus when it emits an alpha particle? A beta particle?

The heaviest isotope of hydrogen, hydrogen-3, often called tritium, decomposes by *beta decay*. Because a beta (β) particle is identical to an electron, this process can be written as

$$\text{Mass number of starting material} = 3 \qquad \text{Mass numbers of products} = 0 + 3 = 3$$

$$^{3}_{1}\text{H} \longrightarrow \ ^{0}_{-1}\text{e} \ + \ ^{3}_{2}\text{He}$$

$$\text{Atomic number of starting material} = 1 \qquad \text{Atomic numbers of products} = (-1) + 2 = 1$$

The atomic number $Z = 2$ identifies the product isotope as helium.

The process is a little more complicated than indicated by the preceding reaction. Within the nucleus, a neutron is converted into a proton (which remains) and an electron (which is ejected).

$$^{1}_{0}\text{n} \longrightarrow \ ^{1}_{1}\text{p} \ + \ ^{0}_{-1}\text{e}$$

With beta decay, the atomic number increases, but the nucleon number remains the same. A pictorial representation of this reaction is shown in Figure 4.2(b).

When gamma radiation (symbolized by γ) occurs, the emitted radiation has no charge and no mass. Neither the nucleon number nor atomic number of the emitting atoms is changed; the nuclei simply become less energetic. Table 4.1 compares the properties of alpha, beta, and gamma radiation.

Two other types of radioactive decay are *positron emission* and *electron capture*. These two processes have the same effect on the atomic nucleus: Both result in a decrease of 1 in atomic number with no change in nucleon number. However, they occur by different pathways (Figure 4.3).

The **positron** (β^{+}) is a particle equal in mass but opposite in charge to the electron. It is represented as $^{0}_{+1}\text{e}$. Fluorine-18 decays by positron emission.

$$^{18}_{9}\text{F} \longrightarrow \ ^{0}_{+1}\text{e} \ + \ ^{18}_{8}\text{O}$$

We can envision a proton in the nucleus changing into a neutron and a positron.

$$^{1}_{1}\text{p} \longrightarrow \ ^{1}_{0}\text{n} \ + \ ^{0}_{+1}\text{e}$$

TABLE 4.1 Common Types of Radiation in Nuclear Reactions					
Radiation	**Mass (u)**	**Charge**	**Identity**	**Velocity**[*]	**Penetrating Power**
Alpha (α)	4	2+	He^{2+}	$0.1c$	Very low
Beta (β)	0.00055	1−	e^{-}	$<0.9c$	Moderate
Gamma (γ)	0	0	High-energy photon	c	Extremely high

[*]c is the speed of light.

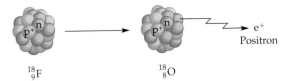

(a) Nuclear change accompanying positron emission.

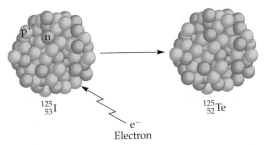

(b) Nuclear change accompanying electron capture.

◀ **Figure 4.3** Nuclear change accompanying (a) positron emission and (b) electron capture.

Question: What changes occur in a nucleus when it emits a positron? When it undergoes electron capture?

After the positron is emitted, the original nucleus has one less proton and one more neutron than it had before. The nucleon number of the product nucleus is the same, but its atomic number has been reduced by 1. The emitted positron quickly encounters an electron (there are numerous electrons in all kinds of matter), and both particles are annihilated, with the production of two gamma rays.

$$_{+1}^{0}e + _{-1}^{0}e \longrightarrow 2_{0}^{0}\gamma$$

A *photon* is a tiny bundle of energy, a "particle" of insignificant mass. Visible light and all other kinds of electromagnetic radiation are made up of photons.

Electron capture (EC) is a process in which a nucleus absorbs an electron from an inner electron shell, usually the first or second. When an electron from a higher shell drops to the level vacated by the captured electron, an X-ray is released. Once inside the nucleus, the captured electron combines with a proton to form a neutron.

$$_{1}^{1}p + _{-1}^{0}e \longrightarrow _{0}^{1}n$$

Air is about 1% argon. Most of this argon is believed to have come from the radioactive decay of potassium-40, which decays by electron capture.

Iodine-125, used in medicine to diagnose pancreatic function and intestinal fat absorption, decays by EC.

$$_{53}^{125}I + _{-1}^{0}e \longrightarrow _{52}^{125}Te$$

Notice that, unlike alpha, beta, gamma, or positron emission, the electron is a reactant (on the left side) and not a product. Conversion of a proton to a neutron (by the absorbed electron) yields a nucleus lower by 1 in atomic number but unchanged in atomic mass. Emission of a positron and absorption of an electron have the same effect on an atomic nucleus (lowering the atomic number by 1), except that positron emission is accompanied by gamma radiation and electron capture by X-radiation. Positron-emitting isotopes and those that undergo electron capture both have important medical applications (Section 4.7).

The five types of decay discussed here are summarized in Table 4.2.

 U-238 Decay Series

TABLE 4.2 Radioactive Decay and Nuclear Change					
Type of Decay	Decay Particle	Particle Mass (u)	Particle Charge	Change in Nucleon Number	Change in Atomic Number
Alpha decay	α	4	2+	Decreases by 4	Decreases by 2
Beta decay	β	0	1−	No change	Increases by 1
Gamma radiation	γ	0	0	No change	No change
Positron emission	β^+	0	1+	No change	Decreases by 1
Electron capture (EC)	e^- absorbed	0	1−	No change	Decreases by 1

EXAMPLE 4.1   **Balancing Nuclear Equations**

Write balanced nuclear equations for each of the following processes. In each case, indicate what new element is formed.

a. Plutonium-239 emits an alpha particle when it decays.
b. Protactinium-234 undergoes beta decay.
c. Carbon-11 emits a positron when it decays.
d. Carbon-11 undergoes electron capture.

Solution

a. We start by writing the symbol for plutonium-239 and a partial equation showing that one of the products is an alpha particle (helium nucleus).

$$^{239}_{94}\text{Pu} \longrightarrow {}^{4}_{2}\text{He} + ?$$

Mass and charge are conserved. The new element must have a mass of $239 - 4 = 235$ and a charge of $94 - 2 = 92$. The nuclear charge $Z = 92$ identifies the element as uranium (U).

$$^{239}_{94}\text{Pu} \longrightarrow {}^{4}_{2}\text{He} + {}^{235}_{92}\text{U}$$

b. Write the symbol for protactinium-234 and a partial equation showing that one of the products is a beta particle (electron).

$$^{234}_{91}\text{Pa} \longrightarrow {}^{0}_{-1}\text{e} + ?$$

The new element still has a nucleon number of 234. It must have a nuclear charge $Z = 92$ in order for the total charge to be the same on each side of the equation. The nuclear charge identifies the new atom as another isotope of uranium.

$$^{234}_{91}\text{Pa} \longrightarrow {}^{0}_{-1}\text{e} + {}^{234}_{92}\text{U}$$

c. Write the symbol for carbon-11 and a partial equation showing that one of the products is a positron.

$$^{11}_{6}\text{C} \longrightarrow {}^{0}_{+1}\text{e} + ?$$

To balance the equation, a particle with $A = 11 - 0 = 11$ and $Z = 6 - 1 = 5$ (boron) is required.

$$^{11}_{6}\text{C} \longrightarrow {}^{0}_{+1}\text{e} + {}^{11}_{5}\text{B}$$

d. We write the symbol for carbon-11 and a partial equation showing it capturing an electron.

$$^{11}_{6}\text{C} + {}^{0}_{-1}\text{e} \longrightarrow ?$$

To balance the equation, the product must have $A = 11 + 0 = 11$ and $Z = 6 + (-1) = 5$ (boron).

$$^{11}_{6}\text{C} + {}^{0}_{-1}\text{e} \longrightarrow {}^{11}_{5}\text{B}$$

As we mentioned previously, positron emission and electron capture result in identical changes in atomic number, and therefore the identical elements are formed. Also, as parts **(c)** and **(d)** illustrate, carbon-11 (and certain other nuclei) can undergo more than one type of radioactive decay.

▶ **Exercise 4.1**

Write balanced nuclear equations for each of the following processes. In each case, indicate what new element is formed.
a. Radium-226 decays by alpha emission.
b. Sodium-24 undergoes beta decay.
c. Gold-188 decays by positron emission.
d. Argon-37 undergoes electron capture.

We mentioned that nuclear equations differ greatly from ordinary chemical equations. Some important differences are summarized in Table 4.3. Some nuclear equations involve processes other than the five simple nuclear processes we have discussed here. Regardless, all nuclear equations must be balanced according to nucleon numbers and atomic numbers. When an unknown particle has an atomic number that does not correspond to an atom, that particle may be a subatomic particle. A list of nuclear symbols for subatomic particles is given in Table 4.4.

TABLE 4.3   Some Differences Between Chemical Reactions and Nuclear Reactions

Chemical Reactions	Nuclear Reactions
Atoms retain their identity.	Atoms usually change from one element to another.
Reactions involve only electrons, and usually only outermost electrons.	Reactions involve mainly protons and neutrons. It does not matter what the valence electrons are doing.
Reaction rates can be speeded up by raising the temperature.	Reaction rates are unaffected by changes in temperature.
Energy absorbed or given off in reactions is comparatively small.	Reactions sometimes involve enormous changes in energy.
Mass is conserved. The mass of products equals the mass of starting materials.	Huge changes in energy are accompanied by measurable changes in mass ($E = mc^2$).

TABLE 4.4   Nuclear Symbols for Subatomic Particles

Particles	Symbols	Nuclear Symbols
Proton	p	$_1^1\text{p}$ or $_1^1\text{H}$
Neutron	n	$_0^1\text{n}$
Electron	e^- or β	$_{-1}^0\text{e}$ or $_{-1}^0\beta$
Positron	e^+ or β^+	$_{+1}^0\text{e}$ or $_{+1}^0\beta$
Alpha particle	α	$_2^4\text{He}$ or $_2^4\alpha$
Beta particle	β or β^-	$_{-1}^0\text{e}$ or $_{-1}^0\beta$
Gamma ray	γ	$_0^0\gamma$

EXAMPLE 4.2   **More Nuclear Equations**

In the upper atmosphere, a nitrogen-14 nucleus absorbs a neutron. A carbon-14 nucleus and another particle are formed. What is the other particle?

Solution
We start by writing the symbols for nitrogen-14 and a neutron (from Table 4.4) on the left of an equation, and the symbol for carbon-14 on the right.

$$_7^{14}\text{N} + _0^1\text{n} \longrightarrow _6^{14}\text{C} + ?$$

The total of nucleon numbers on the left is 15, and on the right is 14, so the missing particle must have a nucleon number of $15 - 14 = 1$. The total atomic number on the left is 7 and on the right is 6, so the missing particle must have an atomic number of 1.

$$_7^{14}\text{N} + _0^1\text{n} \longrightarrow _6^{14}\text{C} + _1^1?$$

From Table 4.4, the particle with an atomic number of 1 and a nucleon number of 1 is a proton.

$$_7^{14}\text{N} + _0^1\text{n} \longrightarrow _6^{14}\text{C} + _1^1\text{p}$$

▶ **Exercise 4.2**
In one reaction that might be a future energy source, a hydrogen-2 nucleus combines with a hydrogen-3 nucleus, forming a helium-4 nucleus and another particle. What is the other particle formed?

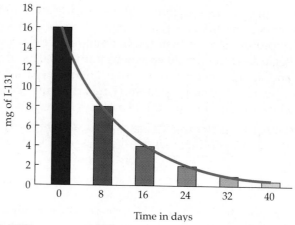

▶ **Figure 4.4** The radioactive decay of iodine-131, which has a half-life of eight days.

Question: What quantity of an original 32-mg sample of iodine-131 remains after five half-lives?

The half-life of an element can be very long (millions of years, as with uranium-238 at 4.5 billion years) or extremely short (tiny fractions of a second, as with boron-9 at 8×10^{-19} s).

We cannot say when *all* the atoms of a radioactive isotope will have decayed. For most samples, we can assume that the activity is essentially gone after about 10 half-lives. (After 10 half-lives, the activity is $\frac{1}{2} = \frac{1}{1024}$ of the original value.) Generally, we can say that $\frac{1}{1000}$ of the original activity remains.

4.3 HALF-LIFE

Thus far we have discussed radioactivity as applied to single atoms. In the laboratory, we generally deal with great numbers of atoms—numbers far larger than the number of all the people on Earth. If we could see the nucleus of an individual atom, we could tell whether or not it could undergo radioactive decay by noting its composition. Certain combinations of protons and neutrons are unstable. However, we could not determine when the atom would undergo a change. Radioactivity is a random process, generally independent of outside influences.

With large numbers of atoms, the process of radioactive decay becomes more predictable. We can measure the half-life, a property characteristic of each radioisotope. The **half-life** of a radioactive isotope is the period in which one-half of the original number of atoms undergo radioactive decay to form a new element. Suppose, for example, we had 16.00 mg of the radioactive isotope iodine-131. The half-life of iodine-131 is 8.0 days. This means that in 8.0 days, half the iodine-131, or 8.00 mg, will have decayed, and there will be 8.00 mg left. In another 8.0 days, half of the remaining 8.00 mg will have decayed. After two half-lives or 16.0 days, then, one-quarter of the original iodine-131, or 4.00 mg, will remain. Two half-lives, then, do not make a whole. The concept of half-life is illustrated by the graph in Figure 4.4.

We can calculate the fraction of the original isotope that remains after a given number of half-lives from the relationship

$$\text{Fraction remaining} = \frac{1}{2^n}$$

where n is the number of half-lives.

EXAMPLE 4.3 Half-Lives

You obtain a new sample of cobalt-60, half-life 5.271 years, with a mass of 4.00 mg. How much cobalt-60 remains after 15.813 years (three half-lives)?

Solution

The fraction remaining after three half-lives is

$$\frac{1}{2^n} = \frac{1}{2^3} = \frac{1}{2 \times 2 \times 2} = \frac{1}{8}$$

The amount of cobalt-60 remaining is $\left(\frac{1}{8}\right)(4.00 \text{ mg}) = 0.50$ mg.

▶ **Exercise 4.3A**

You have 1.224 mg of freshly prepared gold-189, half-life 30 min. How much of the gold-189 sample remains after five half-lives?

▶ **Exercise 4.3B**

The half-life of phosphorus-32 is 14.3 days. What percentage of an original sample's radioactivity remains after four half-lives?

Half-Lives

You obtain a 2.60-mg sample of mercury-190, half-life 20 min. How much of the mercury-190 sample remains after 2.0 h?

Solution

There are 120 min and thus $\left(\frac{120}{20}\right) = 6$ half-lives in 2.0 h. The fraction remaining after six half-lives is

$$\frac{1}{2^n} = \frac{1}{2^6} = \frac{1}{2 \times 2 \times 2 \times 2 \times 2 \times 2} = \frac{1}{64}$$

The amount of mercury-190 remaining is $\left(\frac{1}{64}\right)(2.60)$ mg = 0.0406 mg.

▶ **Exercise 4.4A**

A sample of 16.0 mg of nickel-57, half-life 36.0 h, is produced in a nuclear reactor. How much of the nickel-57 sample remains after 7.5 days?

▶ **Exercise 4.4B**

Technetium-99 decays to ruthenium-99 with a half-life of 210,000 y. Starting with 1.00 mg of technetium-99, how long will it take for 0.75 mg of ruthenium-99 to form?

What Makes for Nuclear Stability?

Most isotopes are radioactive, but some are stable. What are the factors that tend to make an atomic nucleus stable?

1. *Even* numbers of either protons or neutrons or, especially, of both. Of the 264 stable isotopes, 157 have *even* numbers of both protons and neutrons, and only four have odd numbers of both protons and neutrons. Elements of even Z have more stable isotopes than those of odd Z.

2. So-called "magic numbers" of either protons or neutrons. (Magic numbers are 2, 8, 20, 50, 82, and 126.)

3. An atomic number of 83 or less. All isotopes with Z > 83 are radioactive.

4. There should be no more protons than neutrons in the nucleus, and the ratio of neutrons to protons should be close to 1 if the atomic number is 20 or below. As atomic numbers get larger, the stable n/p ratio also increases, up to about 1.5. There is a zone of stability within which the n/p ratio should lie for an atom of a given atomic number.

4.4 RADIOISOTOPIC DATING

The half-lives of certain isotopes can be used to estimate the ages of rocks and archaeological artifacts. Uranium-238 decays with a half-life of 4.5 billion years. The initial products of this decay are also radioactive, and breakdown continues until an isotope of lead (lead-206) is formed. By measuring the relative amounts of uranium-238 and lead-206, chemists can estimate the age of a rock. Some of the older rocks on Earth have been found to be 3.0–4.5 billion years old. Moon rocks and meteorites have been dated at a maximum age of about 4.5 billion years. Thus, the age of Earth (and the solar system itself) is generally estimated to be about 4.5 billion years.

 Radioactive Half-Lives

Carbon-14 Dating

The dating of artifacts derived from plants or animals usually involves radioactive carbon-14. Of the carbon on Earth, about 99% is carbon-12 and 1% is carbon-13, both of which are stable isotopes. However, in the upper atmosphere carbon-14 is formed by the bombardment of ordinary nitrogen by neutrons from cosmic rays.

$$^{14}_{7}\text{N} + ^{1}_{0}\text{n} \longrightarrow ^{14}_{6}\text{C} + ^{1}_{1}\text{H}$$

This process leads to a tiny but steady-state concentration of carbon-14 in Earth's CO_2. Plants use CO_2, and animals consume plants and other animals, so living things constantly incorporate this isotope into their own cells. When they die, however, the incorporation of carbon-14 ceases, and the carbon-14 in the organisms decays—with a half-life of 5730 years—back to nitrogen-14. Thus, we need merely to measure the carbon-14 activity remaining in an artifact of plant or animal origin to determine its age. For instance, a sample that has half the carbon-14 activity of new plant material is 5730 years old; it has been dead for one half-life. Similarly, an artifact with 25% of the carbon-14 activity of new plant material is 11,460 years old; it has been dead for two half-lives.

Carbon-14 dating, as outlined here, assumes that the formation of the isotope was constant over the years. This is not quite the case. However, for the most recent 7000 years or so, carbon-14 dates have been correlated with those obtained from the annual growth rings of trees. Calibration curves have been constructed from which accurate dates can be determined. Generally, carbon-14 is reasonably accurate for dating objects up to about 50,000 years old. Objects older than 50,000 years have too little of the isotope left for accurate measurement.

Charcoal from the fires of an ancient people, dated by determining the carbon-14 activity, is used to estimate the age of other artifacts found at the same archaeological site. Recently, carbon dating was used to confirm that ancient Hebrew writing found on a stone tablet unearthed in Israel dates from the ninth century B.C.E.

The Shroud of Turin

The Shroud of Turin is a very old piece of linen cloth, about 4 m long, bearing a faint human likeness. Since about C.E. 1350 it had been alleged to be part of the burial shroud of Christ. However, carbon-14 dating studies in 1988 by three different nuclear laboratories indicated that the flax used in making the cloth was not grown until sometime between C.E. 1260 and 1390. Therefore, the cloth could not possibly have existed at the time of Christ. Unlike the Dead Sea Scrolls, which were shown by carbon-14 dating to be authentic records from a civilization that existed about 2000 years ago, the Shroud of Turin has been shown to be less than 800 years old.

Tritium Dating

Tritium, the radioactive isotope of hydrogen, can also be used for dating. Its half-life of 12.26 years makes it useful for dating items up to about 100 years old. An interesting application is the dating of brandies. These alcoholic beverages are quite expensive when aged from 10 to 50 years. Tritium dating can be used to check the veracity of advertising claims about the age of the most expensive kinds.

Many other isotopes are useful for estimating the ages of objects and materials. Several of the more important ones are listed in Table 4.5.

TABLE 4.5 Several Isotopes Useful in Radioactive Dating			
Isotope	**Half-Life (years)**	**Useful Range**	**Dating Applications**
Carbon-14	5730	500 to 50,000 years	Charcoal, organic material
Hydrogen-3 (tritium)	12.26	1 to 100 years	Aged wines
Lead-210	22	1 to 75 years	Skeletal remains
Potassium-40	1.25×10^9	10,000 years to the oldest Earth samples	Rocks, the Earth's crust, the moon's crust
Rhenium-187	4.3×10^{10}	4×10^7 years to the oldest samples in the universe	Meteorites
Uranium-238	4.51×10^9	10^7 years to the oldest Earth samples	Rocks, the Earth's crust

EXAMPLE 4.5   **Radioisotopic Dating**

An old wooden implement has carbon-14 activity one-eighth that of new wood. How old is the artifact? The half-life of carbon-14 is 5730 years.

Solution
Using the relationship

$$\text{Fraction remaining} = \frac{1}{2^n}$$

we see that one-eighth is $\frac{1}{2^n}$, where $n = 3$; that is, the fraction $\frac{1}{8}$ is $\frac{1}{2^3}$. The carbon-14 has gone through three half-lives. The wood is therefore about $3 \times 5730 = 17{,}190$ years old.

▶ **Exercise 4.5A**
How old is a piece of a fur garment that has carbon-14 activity $\frac{1}{16}$ that of living tissue? The half-life of carbon-14 is 5730 years.

▶ **Exercise 4.5B**
Strontium-90 has a half-life of 28.5 years. How long will it take for the strontium-90 now on Earth to be reduced to $\frac{1}{32}$ of its present amount?

4.5 ARTIFICIAL TRANSMUTATION

During the Middle Ages, alchemists tried to turn base metals, such as lead, into gold. However, they were trying to do it chemically and were therefore doomed to failure because chemical reactions involve only atoms' outer electrons. *Transmutation* (changing one element into another) requires altering the *nucleus*.

Thus far we have considered only natural forms of radioactivity. Other nuclear reactions can be brought about by bombardment of stable nuclei with alpha particles, neutrons, or other subatomic particles. These particles, given sufficient energy, penetrate the formerly stable nucleus and result in some form of radioactive emission. Just as in natural radioactive processes, one element is changed into another. Because the change would not have occurred naturally, the process is called *artificial transmutation*.

In 1919, a few years after his famous gold-foil experiment (Chapter 3), Ernest Rutherford reported on the bombardment of a variety of light elements with alpha particles. One such experiment, in which he bombarded nitrogen, resulted in the production of protons, as shown in this balanced nuclear equation.

$$^{14}_{7}\text{N} + {}^{4}_{2}\text{He} \longrightarrow {}^{17}_{8}\text{O} + {}^{1}_{1}\text{H}$$

(The hydrogen nucleus is simply a proton; hence the alternative symbol ${}^{1}_{1}\text{H}$ for the proton.) This provided the first empirical verification of the existence of protons in atomic nuclei, which Rutherford had first postulated in 1914.

Recall that Eugen Goldstein had produced protons in his gas discharge tube experiments in 1886 (Chapter 3). The significance of Rutherford's experiment lay in the fact that he obtained protons from the *nucleus* of an atom other than hydrogen, thus establishing their nature as constituents of nuclei. Rutherford's experiment was the first induced nuclear reaction.

▲ Ernest Rutherford (1871–1937) carried out the first nuclear bombardment experiment. Element 104 (Rf) is named rutherfordium in his honor.

EXAMPLE 4.6   **Artificial Transmutation Equations**

When potassium-39 is bombarded with neutrons, chlorine-36 is produced. What other particle is emitted?

$$^{39}_{19}\text{K} + {}^{1}_{0}\text{n} \longrightarrow {}^{36}_{17}\text{Cl} + \text{?}$$

Solution

Write a balanced nuclear equation. To balance the equation, we need four mass units and two charge units (that is, a particle with $A = 4$ and $Z = 2$—an alpha particle).

$$^{39}_{19}\text{K} + {}^{1}_{0}\text{n} \longrightarrow {}^{36}_{17}\text{Cl} + {}^{4}_{2}\text{He}$$

▶ **Exercise 4.6A**

Technetium-97 is produced by bombarding molybdenum-96 with a deuteron (hydrogen-2 nucleus). What other particle is emitted?

$$^{96}_{42}\text{Mo} + {}^{2}_{1}\text{H} \longrightarrow {}^{97}_{43}\text{Tc} + ?$$

▶ **Exercise 4.6B**

Some scientists want to form uranium-235 by bombarding another nucleus with an alpha particle. They expect a neutron to be produced along with the uranium-235 nucleus. What nucleus must be bombarded with the alpha particle?

4.6 USES OF RADIOISOTOPES

Most of the 3000 known radioisotopes are produced by artificial transmutation from stable isotopes. The value of both naturally occurring and artificial radioisotopes goes far beyond their contributions to our knowledge of chemistry.

Tracers

Scientists in a wide variety of fields use radioisotopes as **tracers** in physical, chemical, and biological systems. Isotopes of a given element, whether radioactive or not, behave nearly identically in chemical and physical processes. Because radioactive isotopes are easily detected through their decay products, it is relatively easy to trace their movement, even through a complicated system. For example, we can use radioisotopes to

▲ Scientists can trace the uptake of phosphorus by a green plant by adding a compound containing some phosphorus-32 to the applied fertilizer. When the plant is later placed on a photographic film, radiation from the phosphorus isotopes exposes the film, much as light does. This type of exposure, called a *radiograph*, shows the distribution of phosphorus in the plant.

- **Detect leaks in underground pipes.** Suppose there is a leak in the pipe, which is buried beneath a concrete floor. We could locate the leak by digging up extensive areas of the floor, or we could add a small amount of radioactive material to liquid poured into the drain and trace the flow of the liquid with a Geiger counter (an instrument that detects radioactivity). Once we locate the leak, only a small area of the floor would have to be dug up to repair the leak. A compound containing a short-lived nuclide (for example, $^{131}_{53}\text{I}$, half-life 8.04 days) is usually employed.

- **Determine frictional wear in piston rings.** The ring is subjected to neutron bombardment, which converts some of the carbon in the steel to carbon-14. Wear in the piston ring is assessed by the rate at which the radioactivity of the carbon-14 appears in the engine oil.

- **Determine the uptake of phosphorus and its distribution in plants.** This can be done by incorporating phosphorus-32, a β^- emitter with a 14.3-day half-life, into phosphate fertilizers fed to the plants. Radioisotopes are also used to study the effectiveness of weed killers, compare the nutritional value of various feeds, determine optimal insect control methods, and monitor the fate and persistence of pesticides in soil and groundwater.

One of the most successful agricultural techniques has been inducing heritable genetic alterations known as mutations. Exposing seeds or other parts of plants to neutrons or gamma rays increases the likelihood of genetic mutations. At first glance, this procedure does not seem the most promising of endeavors. However, genetic variability is vital, not only in an effort to improve varieties, but also to protect species from extinction. Emile Frison, a Belgian plant pathologist, announced in early 2003 that bananas might go extinct within 10 years due to their lack of disease resistance. The lack of genetic variability places the entire population at risk.

◀ Figure 4.5 Gamma radiation delays the decay of mushrooms. Those on the left were irradiated; the ones on the right were not.

Question: Are the mushrooms on the left now radioactive?

Radioisotopes are also used as sources to irradiate foodstuffs as a method of preservation (Figure 4.5). The radiation destroys microorganisms that cause food spoilage. Irradiated food shows little change in taste or appearance. Some people are concerned about possible harmful effects of chemical substances produced by the radiation, but there is no good evidence of harm to laboratory animals fed irradiated food, nor are there any known adverse effects in humans in countries where irradiation has been used for years. There is no residual radiation in the food after the sterilization process because gamma rays do not have nearly enough energy to change nuclei.

4.7 NUCLEAR MEDICINE

Nuclear medicine involves two distinct uses of radioisotopes: therapeutic and diagnostic. In radiation therapy, an attempt is made to treat or cure disease with radiation. The diagnostic use of radioisotopes is aimed at obtaining information about the state of a patient's health.

Radiation Therapy

Cancer is not one disease but many. Some forms are particularly susceptible to radiation therapy. The aim of radiation therapy is to destroy cancerous cells before too much damage is done to healthy tissue. Radiation is most lethal to rapidly reproducing cells, and this is precisely the characteristic of cancer cells that allows radiation therapy to be successful. Radiation is carefully aimed at cancerous tissue while minimizing the exposure of normal cells. If the cancer cells are killed by the destructive effects of the radiation, the malignancy is halted.

Patients undergoing radiation therapy often get sick from the treatment. Nausea and vomiting are the usual early symptoms of radiation sickness. Radiation therapy can also interfere with white blood cell replenishment and increase susceptibility to infection.

Diagnostic Uses of Radioisotopes

Radioisotopes are used for diagnostic purposes to provide information about the type or extent of an illness. Table 4.6 lists some radioisotopes in common use in medicine. The list is necessarily incomplete, but even this abbreviated discussion should give you an idea of their importance. The claim that nuclear science has saved many more lives than nuclear bombs have destroyed is not an idle one.

Radioactive iodine-131 is used to determine the size, shape, and activity of the thyroid gland, as well as to treat cancer located in this gland and to control a hyperactive thyroid. Small doses are used for diagnostic purposes, and large doses for treatment of thyroid cancer. After the patient drinks a solution of potassium iodide incorporating iodine-131, the body concentrates iodide in the thyroid. A detector showing the differential uptake of the isotope is used in diagnosis. The resulting *photoscan* can pinpoint the location of tumors or other abnormalities in the thyroid. In cancer treatment, radiation from therapeutic (large) doses of iodine-131 kills the

TABLE 4.6 **Some Radioisotopes and Their Medical Applications**

Isotope	Name	Half-Life*	Use
^{11}C	Carbon-11	20.39 m	Brain scans
^{51}Cr	Chromium-51	27.8 d	Blood volume determination
^{57}Co	Cobalt-57	270 d	Measuring vitamin B_{12} uptake
^{60}Co	Cobalt-60	5.271 y	Radiation cancer therapy
^{153}Gd	Gadolinium-153	242 d	Determining bone density
^{67}Ga	Gallium-67	78.1 h	Scan for lung tumors
^{131}I	Iodine-131	8.040 d	Thyroid therapy
^{192}Ir	Iridium-192	74 d	Breast cancer therapy
^{59}Fe	Iron-59	44.496 d	Detection of anemia
^{32}P	Phosphorus-32	14.3 d	Detection of skin cancer or eye tumors
^{238}Pu	Plutonium-238	86 y	Provides power for pacemakers
^{226}Ra	Radium-226	1600 y	Radiation therapy for cancer
^{75}Se	Selenium-75	120 d	Pancreas scans
^{24}Na	Sodium-24	14.659 h	Locating obstructions in blood flow
^{99m}Tc	Technetium-99m	6.0 h	Imaging of brain, liver, bone marrow, kidney, lung, or heart
^{201}Tl	Thallium-201	73 h	Detecting heart problems with treadmill stress test
^{3}H	Tritium	12.26 y	Determining total body water
^{133}Xe	Xenon-133	5.27 d	Lung imaging

*Abbreviations: y, years; d, days; h, hours; m, minutes.

PET scans using carbon-11 have shown that a schizophrenic brain metabolizes only about one-fifth as much glucose as a normal brain. These scans can also reveal metabolic changes that occur in the brain during tactile learning (learning by the sense of touch).

thyroid cells in which the radioisotope has concentrated. This occurs even in thyroid-type cells that have spread to other parts of the body.

A radioisotope widely used in medicine is gadolinium-153. This isotope is used to determine bone mineralization. Its widespread use is an indication of the large number of people, mostly women, who suffer from osteoporosis (reduction in the quantity of bone) as they grow older. Gadolinium-153 gives off two characteristic radiations, a gamma ray and an X-ray. A scanning device compares these radiations after they pass through bone. Bone densities are then determined by differences in absorption of the rays.

Technetium-99m is used in a variety of diagnostic tests (Figure 4.6). The *m* stands for "metastable," which means that this isotope gives up some energy to become a more stable version of the same isotope (same atomic number, same atomic mass). The energy it gives up is the gamma ray needed to detect the isotope.

$$^{99m}_{43}Tc \longrightarrow \,^{99}_{43}Tc + \gamma$$

Notice that the decay of technetium-99m produces no alpha or beta particles, which could cause unnecessary damage to the body. Technetium-99m also has a short

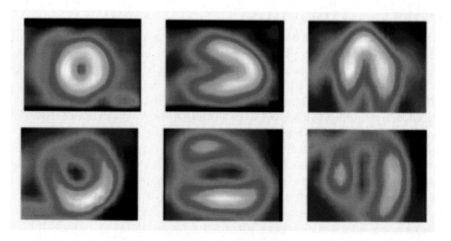

▶ **Figure 4.6** Gamma ray imaging using technetium-99m. Shown at the top are directional slices through a healthy human heart, and at the bottom, through a diseased heart. The lighter regions of the images indicate regions receiving adequate blood flow.

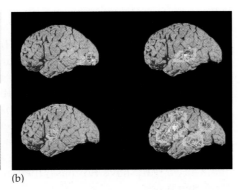

(a)   (b)

▲ **Figure 4.7** Modern computer technology used for medical diagnosis. (a) Patient in position for positron emission tomography (PET), a technique that uses radioisotopes to scan internal organs. (b) Images created by PET scanning, showing different parts of the brain involved in different functions.

half-life (6.0 h), which means that the radioactivity does not linger in the body long after the scan has been completed. With so short a half-life, use of the isotope must be carefully planned. In fact, the isotope itself is not what is purchased. Technetium-99m is formed by the decay of molybdenum-99.

$$^{99}_{42}\text{Mo} \longrightarrow {}^{99\text{m}}_{43}\text{Tc} + {}^{0}_{-1}\text{e} + \gamma$$

A container of this molybdenum isotope is obtained, and the decay product, technetium-99m, is "milked" from the container as needed.

Using modern computer technology, positron emission tomography (PET) can measure dynamic processes occurring in the body, such as blood flow or the rate at which oxygen or glucose is being metabolized. PET scans can pinpoint the area of brain damage that triggers severe epileptic seizures. Compounds incorporating positron-emitting isotopes, such as carbon-11 or oxygen-15, are inhaled or injected prior to the scan. Before the emitted positron can travel very far in the body, it encounters an electron (numerous in any ordinary matter), and two gamma rays are produced, exiting from the body in exactly opposite directions.

$$^{11}_{6}\text{C} \longrightarrow {}^{11}_{5}\text{B} + {}^{0}_{+1}\text{e}$$
$$^{0}_{+1}\text{e} + {}^{0}_{-1}\text{e} \longrightarrow 2\gamma$$

Detectors, positioned on opposite sides of the patient, record the gamma rays. An image of an area in the body is formed using computerized calculations of the points at which annihilation of the positrons and electrons occurs (Figure 4.7).

EXAMPLE 4.7   **Positron Emission Equations**

One of the isotopes used for PET scans is oxygen-15, a positron emitter. What new element is formed when oxygen-15 decays?

Solution
First write the nuclear equation

$$^{15}_{8}\text{O} \longrightarrow {}^{0}_{+1}\text{e} + ?$$

The nucleon number A does not change, but the atomic number Z becomes $8 - 1 = 7$; and so the new product is nitrogen-15.

$$^{15}_{8}\text{O} \longrightarrow {}^{0}_{+1}\text{e} + {}^{15}_{7}\text{N}$$

Recall that by "nucleons" we mean both protons and neutrons.

▶ **Exercise 4.7**
Phosphorus-30 is a positron-emitting radioisotope suitable for use in PET scans. What new element is formed when phosphorus-30 decays?

4.8 PENETRATING POWER OF RADIATION

The danger of radiation to living organisms comes from its potential for damaging cells and tissue. The ability to inflict injury comes from the penetrating power of the radiation. The two aspects of nuclear medicine just discussed (therapeutic and diagnostic) also are dependent on the penetrating powers of the various types of radiation.

All other things being equal, the more massive the particle, the less its penetrating power. Alpha particles, which are helium nuclei with a mass of 4 u, are the least penetrating of the three main types of radioactivity. Beta particles, which are identical to the almost massless electrons, are somewhat more penetrating. Gamma rays, like X-rays, truly have no mass; they are considerably more penetrating than the other two types.

But all other things are not always equal. The faster a particle moves or the more energetic the radiation is, the more penetrating power it has.

It may seem contrary to common sense that the biggest particles make the least headway. Consider that penetrating power reflects the ability of radiation to make its way through a sample of matter. It is as if you were trying to roll some rocks through a field of boulders. The alpha particle acts as if it were a boulder itself. Because of its size, it cannot get very far before it bumps into and is stopped by other boulders. The beta particle acts as if it were a small stone. It can sneak between and perhaps ricochet off boulders until it makes its way farther into the field (Figure 4.8). The gamma ray can be compared with a grain of sand that can get through the smallest openings.

The danger of a specific type of radiation to human tissue depends on the location of the source as well as on the penetrating power. If a radioactive substance is outside the body, alpha particles of low penetrating power are the least dangerous; they are stopped by the outer layer of skin. Beta particles also usually are stopped before they reach vital organs. Gamma rays readily pass through tissues, and so an external gamma source can be quite dangerous. People working with radioactive materials can protect themselves with one or both of the following actions:

- Move away from the source; intensity of radiation decreases with distance from the source.
- Use shielding. A sheet of paper can stop most alpha particles, a block of wood or a thin sheet of aluminum can stop beta particles, but it takes several meters of concrete or several centimeters of lead to stop gamma rays (Figure 4.9).

When the radioactive source is *inside* the body, as in the case of medicinal applications, the situation is reversed. The nonpenetrating alpha particles can do great damage. All such particles are trapped within the body, which must then absorb all the energy released by the particle. Alpha particles inflict all their damage in a tiny area because they do not travel far. Therefore, getting the therapeutic radioisotope close to the targeted cells is vital. Beta particles distribute their damage over a somewhat larger area because they travel farther. Tissue may recover from limited damage spread over a large area; it is less likely to survive concentrated damage. Diagnostic applications rely on the highly penetrating power of gamma rays. In most cases, the radiation created inside the body must be detected by instruments outside the body, so a minimum of absorption is desirable.

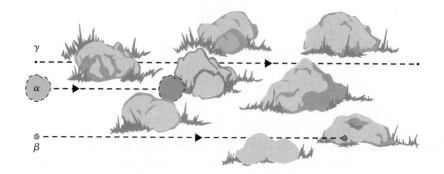

▶ **Figure 4.8** Shooting radioactive particles through matter is like rolling rocks through a field of boulders—the larger rocks stop more quickly.

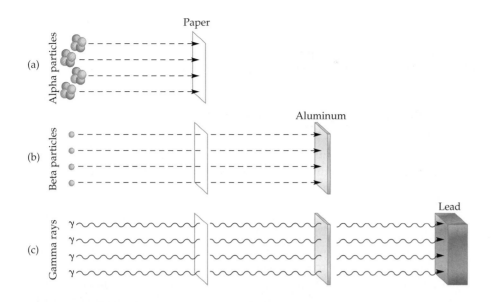

Figure labels: (a) Alpha particles / Paper; (b) Beta particles / Aluminum; (c) Gamma rays / Lead

◄ **Figure 4.9** The relative penetrating powers of alpha, beta, and gamma radiation. Alpha particles are stopped by a sheet of paper (a). Beta particles will not penetrate a sheet of aluminum (b). It takes several centimeters of lead to block gamma rays (c).

CONCEPTUAL EXAMPLE 4.8 **Radiation Hazard**

Most modern fire detectors contain a tiny amount of americium-241, a solid element that is an alpha emitter. The americium is in a chamber that includes thin aluminum shielding, and the device is usually mounted on a ceiling. Rate the radiation hazard of this device in normal use as either (a) high, (b) moderate, or (c) very low, and explain your choice.

Analysis and Conclusions

Alpha particles have very little penetrating ability—they are stopped by a sheet of paper or a layer of skin. The emitted alpha particles cannot exit the chamber, because even thin metal is more than sufficient to absorb the alpha particles. Even if alpha particles could escape, they would likely be stopped by the dead cells of your hair, or by the air after a short distance. Therefore, the radiation hazard is probably best rated as (c)—very low.

▶ **Exercise 4.8**

Radon-222 is a gas that can diffuse from the ground into homes. Like americium-241, it is an alpha emitter. Is the radiation hazard from radon-222 likely to be higher or lower than that from a smoke detector? Explain.

4.9 ENERGY FROM THE NUCLEUS

We have seen that radioactivity—quiet and invisible—can be beneficial or dangerous. A much more dramatic—and equally paradoxical—aspect of nuclear chemistry is the release of nuclear energy by either fission (splitting of heavy nuclei into smaller nuclei) or fusion (combining of light nuclei to form heavier ones).

Einstein and the Equivalence of Mass and Energy

The potential power in the nucleus was worked out by Albert Einstein, a famous and most unusual scientist. Whereas most scientists work with glassware and instruments in laboratories, Einstein worked with a pencil and a note pad. By 1905, at the age of 26, he had already worked out his special theory of relativity and had developed his famous **mass–energy equation** in which mass (m) is multiplied by the square of the speed of light (c).

$$E = mc^2$$

The equation suggests that mass and energy are just two different aspects of the same thing, and a little bit of mass can yield enormous energy. The atomic bombs

▲ Albert Einstein (1879–1955). Element 99 (Es) is named einsteinium in his honor.

that destroyed Hiroshima and Nagasaki (Section 4.10) converted less than an ounce of matter into energy.

A chemical reaction that gives off heat must lose mass in the process, but the change in mass is far too small to measure. Reaction energy must be enormous—such as the energy given off by nuclear explosions—in order for the mass loss to be measurable. (Converting a single gram of matter completely to energy would power a 100-W light bulb for over 28 million years!)

Binding Energy

As we see in the atomic bomb and in nuclear power plants, nuclear fission involves a tremendous release of energy. Where does all this energy come from? It is locked inside the atomic nucleus. When protons and neutrons combine to form atomic nuclei, a small amount of mass is converted to energy. This is the **binding energy** that holds the nucleons together in the nucleus. For example, the helium nucleus contains two protons and two neutrons. The mass of these four particles is 2×1.0073 u $+ 2 \times 1.0087$ u $= 4.0320$ u (Figure 4.10). However, the actual mass of the helium nucleus is only 4.0015 u, and the missing mass amounts to 0.0305 u. Using Einstein's equation $E = mc^2$, we can calculate (see Problem 70) a value of 28.3 million electron volts (MeV) for the binding energy of the helium nucleus (1 MeV $= 1.6022 \times 10^{-13}$ J). This is the amount of energy it would take to separate one helium nucleus into two protons and two neutrons.

When binding energy per nucleon is calculated for all the elements and plotted against nucleon number, a graph such as that in Figure 4.11 is obtained. The elements with the highest binding energies per nucleon have the most stable nuclei. They include iron and nearby elements. When uranium atoms undergo nuclear fission, they split into fragments with higher binding energies; in other words, the fission reaction converts large atoms into smaller ones with greater nuclear stability.

We can also see from Figure 4.11 that even more energy can be obtained by combining small atoms, such as hydrogen or deuterium, to form larger atoms with more stable nuclei. This kind of reaction is called **nuclear fusion**. It is what happens when a hydrogen bomb explodes, and it is also the source of the sun's energy.

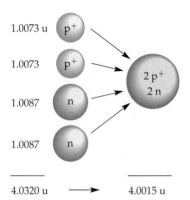

▲ **Figure 4.10** Nuclear binding energy in $^4_2$He. The mass of a helium-4 nucleus is 4.0015 u, which is 0.0305 u less than the masses of two protons and two neutrons. The missing mass is equivalent to the binding energy of the helium-4 nucleus.

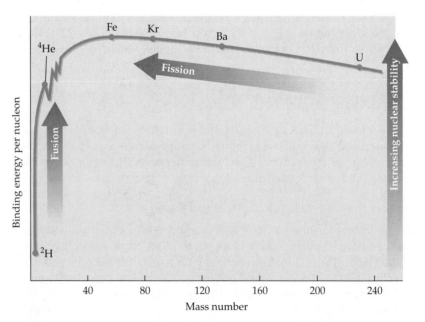

▲ **Figure 4.11** Nuclear stability is greatest near iron in the periodic table. Fission of very large atoms or fusion of very small ones results in greater nuclear stability.

Question: Which process, fission of uranium nuclei or fusion of hydrogen nuclei, releases more energy?

4.10 THE BUILDING OF THE BOMB

In 1934 the Italian scientists Enrico Fermi and Emilio Segrè (1905–1989) bombarded uranium atoms with neutrons. They were trying to make elements higher in atomic number than uranium, which then had the highest known number. To their surprise, they found four radioactive species among the products. One presumably was element 93, formed by the initial conversion of uranium-238 to uranium-239,

$$^{238}_{92}U + ^{1}_{0}n \longrightarrow ^{239}_{92}U$$

which then underwent beta decay.

$$^{239}_{92}U \longrightarrow ^{0}_{-1}e + ^{239}_{93}Np$$

They were unable to explain the remaining radioactivity.

Nuclear Fission

In repeating the Fermi–Segrè experiment in 1938, German chemists Otto Hahn (1879–1968) and Fritz Strassman (1902–1980) were perplexed to find isotopes of barium among the many reaction products. Hahn wrote to Lise Meitner, his former long-time colleague, to ask what she thought about these strange results.

Lise Meitner was an Austrian physicist who had worked with Hahn in Berlin. Because she was Jewish, she had recently fled to Sweden when the Nazis took over Austria in 1938. On hearing about Hahn's work, she noted that barium atoms were only about half the size of uranium atoms. Was it possible that the uranium nucleus might be splitting into fragments? She made some calculations that convinced her that the uranium nuclei had indeed been split apart. Her nephew, Otto Frisch (1904–1979), was visiting for the winter holidays, and they discussed this new discovery with great excitement. It was Frisch who later coined the term **nuclear fission** (Figure 4.12).

Frisch was working with Niels Bohr at the University of Copenhagen, and when he returned to Denmark he took the news about the fission reaction to Bohr, who happened to be going to the United States to attend a physics conference. The discussions in the corridors about this new reaction would be the most important talks to take place at that meeting.

Meanwhile, Enrico Fermi had just received the 1938 Nobel Prize in physics. Because Fermi's wife Laura was Jewish, and the fascist Italian dictator Mussolini was an ally of Hitler, Fermi accepted the award in Stockholm and then immediately fled with his wife and children to the United States. Thus, by 1939 the United States had received news about the German discovery of nuclear fission and had also acquired from Italy one of the world's foremost nuclear scientists.

▲ Enrico Fermi (1901–1954). Element 100 (Fm) is named fermium in his honor.

▲ Lise Meitner (1878–1968). Element 109 (Mt) is named meitnerium in her honor.

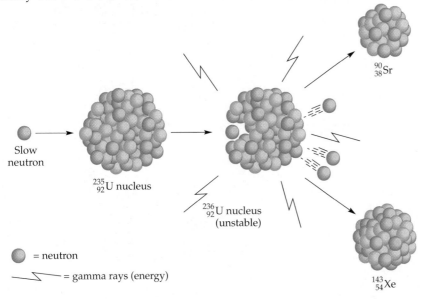

Slow neutron

$^{235}_{92}U$ nucleus

$^{236}_{92}U$ nucleus (unstable)

$^{90}_{38}Sr$

$^{143}_{54}Xe$

● = neutron

⟍⟋ = gamma rays (energy)

◀ **Figure 4.12** One possible way a uranium atom can undergo fission. The neutrons produced in the fission can split other uranium atoms, thus sustaining a chain reaction.

Question: If each of the neutrons emitted by the ^{236}U nucleus caused another ^{235}U nucleus to split in the manner depicted, how many new neutrons would be released?

Nuclear Chain Reaction

Leo Szilard (1898–1964) was one of the first scientists to realize that nuclear fission could be a practical chain reaction. Szilard had been born in Hungary and educated in Germany, but he came to the United States in 1937 as another Jewish refugee. He saw that neutrons released in the fission of one atom could trigger the fission of other uranium atoms, thus setting off a **chain reaction** (Figure 4.13). Because massive amounts of energy could be obtained from the fission of uranium, he saw that the fission process might produce a bomb with tremendous explosive force.

Aware of the destructive forces that could be produced and concerned that Germany might develop such a bomb, Szilard prevailed on Einstein to sign a letter to President Franklin D. Roosevelt indicating the importance of the discovery. It was critical that the U.S. government act quickly.

The Manhattan Project

President Roosevelt launched a highly secret research project for the study of atomic energy. Called the Manhattan Project, it eventually became a massive research effort involving more scientific brainpower than has ever been devoted to a single project. Amazingly, it was conducted under such extreme secrecy that even Vice President Harry Truman did not know about its existence until after Roosevelt's death.

Begun in 1939, the Manhattan Project included four separate research teams trying to learn how to

- sustain the nuclear fission chain reaction,
- enrich uranium to about 90% with the fissile isotope, ^{235}U,
- make plutonium-239 (another fissile isotope), and
- construct a bomb based on nuclear fission.

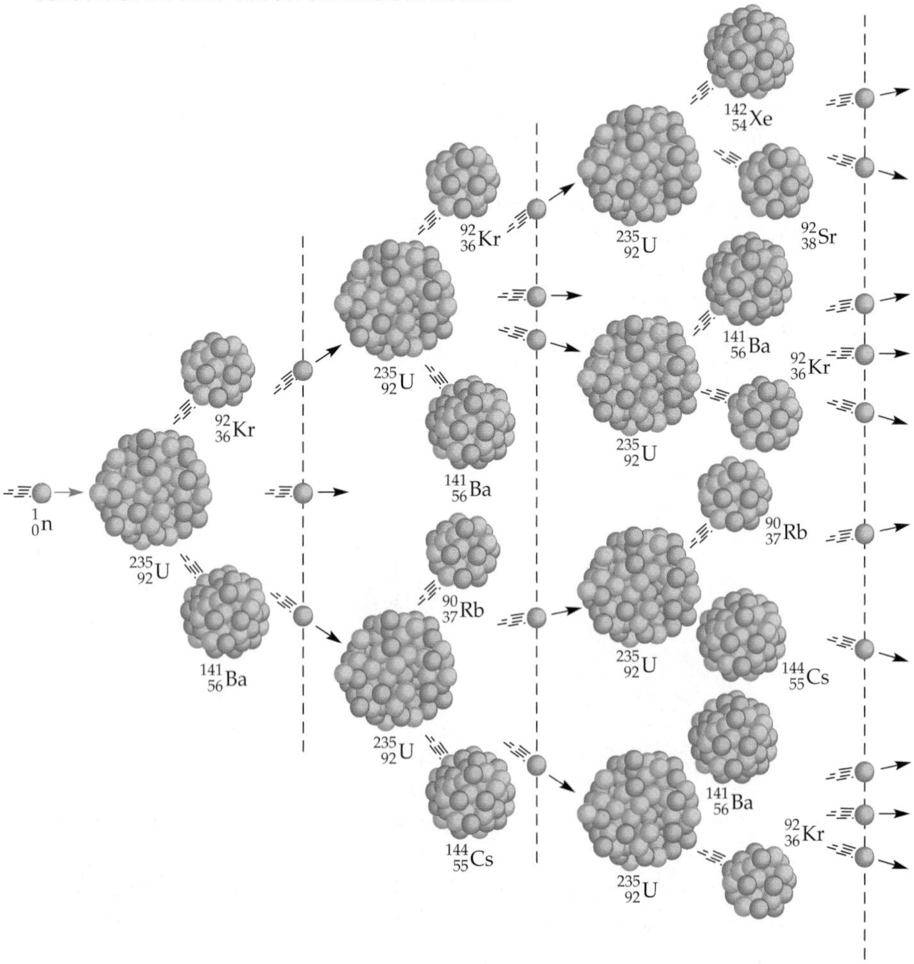

▶ **Figure 4.13** Schematic representation of a nuclear chain reaction. Neutrons released in the fission of one uranium-235 nucleus can strike other nuclei, causing them to split and release more neutrons, as well as a variety of new nuclei. For simplicity, fission fragments are not shown.

By that time it had been established that neutron bombardment could initiate the fission reaction, but there were many things about fission that were unknown. Enrico Fermi and his group, working in a lab under the bleachers at Stagg Field on the campus of the University of Chicago, worked on the fission reaction itself and how to sustain it.

They found that the neutrons used to trigger the reaction had to be slowed down in order to increase the probability that they would hit a uranium nucleus. Because graphite slows down neutrons, a large "pile" of graphite was built to house the reaction. Then the amount of uranium "fuel" was gradually increased. The major question was the **critical mass**[1] or the amount of uranium-235 needed to sustain the fission reaction. There had to be enough fissile nuclei for the neutrons released in one fission process to have a good chance of being captured by another fissile nucleus before escaping from the pile.

On December 2, 1942, Fermi and his group achieved the first sustained nuclear fission reaction. The critical mass for uranium (enriched to about 94% ^{235}U) for this reactor turned out to be about 16 kg.

Isotopic Enrichment

Natural uranium is 99.27% uranium-238, which does not undergo fission. Uranium-235, the fissile isotope, makes up only 0.72% of natural uranium. Because making a bomb required that the uranium be enriched to about 90% uranium-235, it was necessary to find a way to separate the uranium isotopes. This was the job of a top-secret research team in Oak Ridge, Tennessee.

Chemical separation was almost impossible. The separation method eventually used involved the conversion of uranium to uranium hexafluoride, UF_6, which can be vaporized. Molecules containing uranium-235 are slightly lighter in mass and therefore move slightly faster than molecules containing the uranium-238 isotope. Vapors of uranium hexafluoride were allowed to pass through a series of thousands of pinholes, and the molecules containing uranium-235 gradually outdistanced the others. Enough enriched uranium-235 was finally obtained to make a small explosive device.

The Synthesis of Plutonium

While the tedious work of separating uranium isotopes was under way at Oak Ridge, other workers, led by Glenn T. Seaborg, approached the problem of obtaining fissionable material by another route. Although uranium-238 would not fission when bombarded by neutrons, it was found that this more common isotope of uranium *would* form a new element, named neptunium (Np). This product quickly decayed to another new element, plutonium (Pu).

$$^{238}_{92}U + ^{1}_{0}n \longrightarrow ^{239}_{92}U$$

$$^{239}_{92}U \longrightarrow ^{0}_{-1}e + ^{239}_{93}Np$$

$$^{239}_{93}Np \longrightarrow ^{239}_{94}Pu + ^{0}_{-1}e$$

Plutonium-239 was found to be fissile and thus was suitable material for the making of a bomb. A series of large reactors were built near Hanford, Washington, to produce plutonium.

Bomb Construction

The actual building of the nuclear bombs was carried out at Los Alamos, New Mexico, under the direction of J. Robert Oppenheimer (1904–1967). In a top-secret laboratory at a remote site, a group of scientists planned and then constructed

▲ Glenn T. Seaborg (1912–1999). Element 106 (Sg) is named seaborgium in his honor. Seaborg is the only person to have an element named for him while still alive. He is shown here pointing to his namesake element on the periodic table of elements.

[1]The critical mass required for a nuclear explosion is not a fixed quantity. It depends on the concentration of the fissile material, its configuration, and the nature of the material surrounding the fissile isotope.

▶ **Figure 4.14** Replicas of the uranium bomb, "Little Boy" dropped on Hiroshima and the plutonium bomb "Fat Man" (far right) dropped on Nagasaki.

Question: How does a uranium bomb differ from a plutonium bomb?

▲ **Figure 4.15** The mushroom cloud over Nagasaki, following the detonation of "Fat Man," August 9, 1945.

what would become known as atomic bombs. Two different models were pursued; one based on ^{235}U, the other on ^{239}Pu.

The critical mass of uranium-235 must not be exceeded prematurely, so it was important that no single piece of fissionable material in the bomb be that large. The bomb was designed to contain a number of pieces of subcritical mass, plus a neutron source to initiate the fission reaction. Then, at the chosen time, all the pieces would be forced together by setting off a charge of TNT (trinitrotoluene), thus triggering a runaway nuclear chain reaction.

The synthesis of plutonium turned out to be easier than the isotopic separation, and by July 1945, enough fissile material had been made for three bombs to be assembled; two using plutonium, and one using uranium. The first atomic bomb (one of the plutonium devices) was tested in the desert near Alamogordo, New Mexico, on July 16, 1945. The heat from the explosion vaporized the 30-m steel tower on which the bomb was placed and melted the sand for several hectares around the site. The light produced was the brightest anyone had ever seen.

Some of the scientists were so awed by the force of the blast that they argued against its use against Japan. A few, led by Leo Szilard, suggested a demonstration of its power at an uninhabited site. But fear of a well-publicized "dud" and the desire to avoid millions of casualties in an invasion of Japan led President Harry S. Truman to order the dropping of bombs on Japanese cities. The lone uranium bomb called "Little Boy" (Figure 4.14) was dropped on Hiroshima on August 6, 1945 and caused over 100,000 casualties (Figure 4.15). Three days later, a plutonium bomb called "Fat Man" was dropped on Nagasaki with comparable results. World War II ended with the surrender of Japan on August 14, 1945.

4.11 RADIOACTIVE FALLOUT

When a nuclear explosion occurs in the open atmosphere, radioactive materials can rain down on parts of Earth thousands of miles away, days and weeks later. This is called *radioactive fallout*. The uranium atom can split in many different ways as shown. Some examples are as follows:

$$^{235}_{92}\text{U} + ^{1}_{0}\text{n} \longrightarrow ^{90}_{38}\text{Sr} + ^{143}_{54}\text{Xe} + 3\ ^{1}_{0}\text{n}$$

$$\longrightarrow ^{102}_{39}\text{Y} + ^{131}_{53}\text{I} + 3\ ^{1}_{0}\text{n}$$

$$\longrightarrow ^{95}_{37}\text{Rb} + ^{137}_{55}\text{Cs} + 4\ ^{1}_{0}\text{n}$$

The primary fission products are radioactive. They decay to daughter isotopes, many of which are also radioactive. In all, over 200 different fission products are produced, with half-lives varying from less than a second to more than a billion years. In addition, the neutrons produced in the explosion act on molecules in the atmosphere to produce carbon-14, tritium, and other radioisotopes. Fallout is therefore exceedingly complex. We consider only two of the more worrisome isotopes here.

Of all the isotopes, strontium-90 presents the greatest hazard to people. The isotope has a half-life of 28.5 years. Strontium-90 reaches us primarily through dairy products and vegetables. Because of its similarity to calcium (both are group 2A elements), strontium-90 is incorporated into bone. There it remains a source of internal radiation for many years.

Iodine-131 may present a greater threat immediately after a nuclear explosion. Its half-life is only 8 days, but it is produced in relatively large amounts. Iodine-131 is efficiently carried through the food chain. In the body it is concentrated in the thyroid gland, and it is precisely this characteristic that makes it so useful for diagnostic scanning. However, for a healthy individual, the incorporation of radioactive iodine offers no useful information, only damaging side effects.

By the late 1950s, radioactive isotopes from atmospheric testing of nuclear weapons were detected in the environment. Concern over radiation damage from nuclear fallout led to a movement to ban atmospheric testing. Many scientists were leaders in the movement. Linus Pauling, who won the Nobel Prize in chemistry in 1954 for his bonding theories and for his work in determining the structure of proteins, was a particularly articulate advocate of banning atmospheric tests. In 1963 a nuclear test ban treaty was signed by the major powers—with the exception of France and the People's Republic of China, which continued aboveground tests. Since the signing of the treaty, other countries have joined the nuclear club. Pauling, who had endured being called a Communist and a traitor because of his outspoken position, was awarded the Nobel Prize for peace in 1962.

4.12 NUCLEAR POWER PLANTS

A significant portion of today's electric power is generated by nuclear power plants. In the United States, one-fifth of all the electricity produced comes from nuclear power plants. Europeans rely even more on nuclear energy. France, for example, obtains more than 70% of its electric power from nuclear plants, while Belgium, Spain, Switzerland, and Sweden each generate about one-third of their power from nuclear reactors.

Ironically, the same nuclear reactions that occurred in the detonation of the bomb dropped on Hiroshima are used extensively today under the familiar concrete containment tower of a nuclear plant. The key difference is that the power plant employs a *slow, controlled release of energy* from the nuclear chain reaction, rather than an explosion. The slower process results from using uranium fuel that is less enriched (2.5–3.5% ^{235}U rather than the 90% or so for weapons-grade uranium).

One of the main problems with the production of nuclear power comes from the products of the nuclear reactions. As in nuclear fallout, most of the daughter nuclei produced by the fission of ^{235}U are themselves radioactive, some with very long half-lives. We present a further discussion of problems associated with nuclear waste in Chapter 14. Perhaps a more serious problem is the potential transformation of spent nuclear fuel into weapons-grade material (see the box titled Nuclear Proliferation and Dirty Bombs).

4.13 THERMONUCLEAR REACTIONS

We opened this chapter with a picture of the sun, a "nuclear reactor" vital to sustaining life on Earth. The reactions that take place in the sun are somewhat different from the ones previously discussed. They are called **thermonuclear reactions** because they require enormously high temperatures (millions of degrees) to initiate

To minimize the absorption of radioactive iodine, many people in the area of Chernobyl were given large amounts of potassium iodide. This effectively diluted the amount of radioactive iodine actually absorbed by the thyroid.

Nuclear Proliferation and Dirty Bombs

As power is generated in nuclear plants, important changes occur in the fuel. Neutron bombardment converts fissile ^{235}U to radioactive daughter products. Eventually, the concentration of ^{235}U becomes too low to sustain the nuclear chain reaction. The fuel rods must therefore be replaced about every three years. The rods then become high-level nuclear waste.

As ^{235}U undergoes fission, the nonfissile (and very concentrated) ^{238}U nuclei absorb neutrons and are converted to ^{239}Pu. (This transmutation is described in Section 4.10.) If the rods are used long enough, ^{240}Pu is also formed. However, if the fuel rods are removed after only about three months, the fissile plutonium-239 can be easily separated from the other fission products by chemical means. Therefore, less sophisticated technology is needed to produce a nuclear weapon from plutonium than from uranium. Plutonium is thus a greater source of concern for weapons proliferation, and operation of a nuclear plant can be a method of producing materials suitable for use in a nuclear weapon. In recent years, North Korea has restarted both a nuclear power plant and a plutonium separation facility with the capability of producing enough ^{239}Pu for several weapons each year (Figure 4.16). Iran also is constructing a facility to enrich uranium, which could lead to the production of nuclear bombs. These and other developing countries raise the fear of a dangerous proliferation of nuclear weapons.

Another security concern is a "dirty bomb," a device that uses a conventional explosive, such as dynamite, to disperse radioactive material that might be stolen from a hospital or other facility that uses radioactive isotopes. In most cases, the conventional bomb would do more immediate harm than the radioactive substances. At the levels most likely to be used, the dirty bomb would not contain enough radioactive material to kill people or cause severe illness.

▲ **Figure 4.16** Satellite photograph of the plutonium processing plant at Yongbyon, North Korea.

them. The intense temperatures and pressures on the sun cause nuclei to fuse and release unimaginable amounts of energy. Instead of splitting large nuclei into smaller fragments (fission), small nuclei are fused into larger ones (fusion). The principal reaction in the sun is thought to be the fusion of four hydrogen nuclei to produce one helium nucleus and two positrons.

$$4{}^{1}_{1}\text{H} \longrightarrow {}^{4}_{2}\text{He} + 2{}^{0}_{+1}\text{e}$$

The fusion of 1 g of hydrogen releases an amount of energy equivalent to the burning of nearly 20 tons of coal. Every second the sun fuses 600 tons of hydrogen, producing millions of times more energy than has been produced in the entire history of humankind. Much current research is aimed at reproducing such a reaction in the laboratory, by using ultrapowerful magnets to contain the intense heat required for ignition. Fusion technology is discussed further in Chapter 14. To date, however, fusion reactions on Earth have been limited to the uncontrolled reactions in explosion tests of hydrogen (thermonuclear) bombs and to small amounts of energy produced by very expensive experimental fusion reactors.

4.14 THE NUCLEAR AGE

With the splitting of the atom, the Chinese saying, "May you live in interesting times" seems quite appropriate. The goal of the alchemists, to change one element into another, has been achieved through the application of scientific principles.

New elements have been formed, and the periodic chart has been extended beyond uranium ($Z = 92$) to element 116. This modern alchemy produces plutonium by the ton; neptunium ($Z = 93$), americium ($Z = 95$), and curium ($Z = 96$) by the kilogram; and berkelium ($Z = 97$) and einsteinium ($Z = 99$) by the milligram. Radioactive isotopes have killed tiny cancer cells, harmful microorganisms in food, and human beings in accidents and in war.

Radioactive isotopes are used in medical diagnosis (Section 4.7) and in dating objects of unknown age (Section 4.4). Figure 4.17 suggests a number of other constructive uses of nuclear energy. We live in an age in which the extraordinary forces present in the atom have been unleashed as a true double-edged sword. The threat of nuclear war—and nuclear terrorism—has been a constant specter for the last six decades, and yet it is hard to believe that the world would be a better place if we had not discovered the secrets of the atomic nucleus.

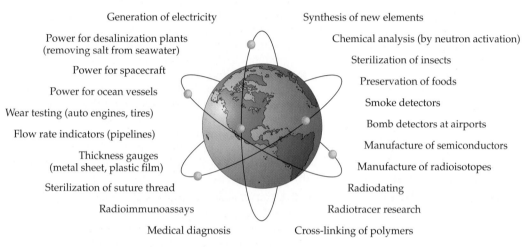

▲ **Figure 4.17** Some constructive uses of nuclear energy.

Critical Thinking Exercises

Apply knowledge that you have gained in this chapter and one or more of the FLaReS principles (Chapter 1) to evaluate the following statements or claims.

4.1 A cave in Mexico is claimed to have wonderful healing powers. People who are ill sometimes sit for hours on the benches inside this cave. Studies have shown that the cave really is unusual. Its walls contain radioactive substances that constantly give off a low level of radiation. Do you think the cave really has healing power?

4.2 One of the assumptions behind the technique of radioisotopic dating is supposition that nuclear decay processes are occurring at the same rate currently as they always have. One criticism of conclusions based on radiocarbon dating is that decay rates could have changed over time. How would you respond to such an allegation?

4.3 For more than 600 years it has been alleged that the Shroud of Turin was the burial shroud of Jesus Christ. In 1988 several laboratories carried out carbon-14 analyses indicating that the flax from which the shroud was made was grown during the period between C.E. 1260 and 1390. Recently there have been new claims that the 1988 analyses are unreliable because the shroud is coated with pollen from plants grown in the fourteenth century, and it is the age of this pollen that the scientists were actually measuring in 1988. At least one of the scientists has offered to repeat the analysis on a very carefully cleaned sample of the cloth, but further access to the shroud has been refused.

4.4 Most modern smoke detectors contain a small amount of radioactive americium, an alpha emitter. A rookie firefighter wearing a tank of breathing air refuses to enter a burning home because he claims the radioactive material in the smoke detectors escapes during a fire.

4.5 A young girl refuses to visit her grandmother after the woman has undergone radiation treatment for breast cancer. The teenager is afraid that she will catch radiation sickness from her grandmother.

SUMMARY

Section 4.1—Some isotopes of elements are unstable and their nuclei undergo changes in nucleon number, atomic number, or energy; this process is called radioactive decay. The nuclei that undergo such changes are called radioisotopes. We are exposed to naturally occurring radioisotopes and other natural ever-present radiation, called background radiation. Radiation that causes harm by dislodging electrons from living tissue and forming ions is called ionizing radiation and includes nuclear radiation and X-rays. Radiation can disrupt normal chemical processes in cells and can damage DNA, causing mutations in some cases.

Section 4.2—Nuclear equations are used to represent nuclear processes. Equations are written and balanced so that the sum of nucleon numbers on each side is the same, and the sum of atomic numbers on each side is the same. Four types of radioactive decay are alpha ($_2^4$He) decay, beta ($_{-1}^{0}$e) decay, gamma ($_0^0\gamma$) decay, and positron ($_{+1}^{0}$e) decay. Electron capture is a fifth type of decay in which a nucleus absorbs one of the atom's electrons. There are enormous differences between nuclear reactions and chemical reactions.

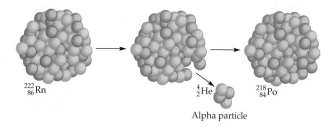

$$_{86}^{222}\text{Rn} \qquad _{2}^{4}\text{He} \qquad _{84}^{218}\text{Po}$$

Alpha particle

Sections 4.3–4.4—The half-life of a radioactive isotope is the time it takes for half of a sample to undergo radioactive decay. The fraction of a radioisotope remaining after n half-lives is given by the expression

$$\text{Fraction remaining} = \frac{1}{2^n}$$

Half-lives of certain isotopes can be used to estimate the ages of various objects. Carbon-14 dating is the best-known of the dating methods. With a half-life of 5730 years, carbon-14 dating can be used to estimate the age of once-living items up to 50,000 years old. The decay of tritium can be used to date items up to 100 years old. Other isotopes can be used to date rocks, the Earth's crust, and meteorites.

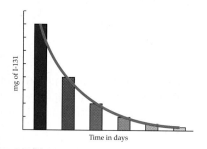

Section 4.5—Transmutation, the conversion of one element into another, cannot be carried out by chemical means, but can be accomplished by nuclear processes. Bombarding one nucleus with other energetic particles can cause transmutation. These processes can be represented with nuclear equations.

Sections 4.6–4.7—Radioisotopes have many uses. A radioisotope and a stable isotope of an element behave nearly the same, so radioisotopes can be used as tracers in physical and biological systems. A radioactive atom in a molecule labels it so that it can be followed by a radiation detector. Radioisotopes have many uses in agriculture, including the production of useful mutations. Radiation can be used to irradiate foodstuffs as a method of preservation. Radiation therapy to destroy cancer depends on the fact that radiation is more damaging to cancer cells than to healthy cells. Radioisotopes are used in diagnosis of various disorders. Iodine-131 is used for thyroid diagnoses, gadolinium-153 for bone mineralization examinations, and technetium-99m for a variety of diagnostic tests. Positron emission tomography (PET) involves radioisotopes that emit positrons, which are then annihilated in the body, producing gamma rays that can be used to build an image.

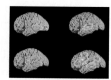

Section 4.8—Different types of radiation have different penetrating abilities. Alpha particles (helium nuclei) are relatively slow and low in penetrating power. Beta particles (electrons) are much faster and more penetrating. Gamma rays (high-energy photons) travel at the speed of light and have great penetrating power. Radiation hazard depends on the location of the source; alpha particles from a source inside the body are highly damaging. Radiation hazard can be decreased by moving away from the source, or with shielding.

Sections 4.9–4.10—Einstein's mass–energy equation, $E = mc^2$, shows that mass and energy are different aspects of the same thing. The total mass of the nucleons in an atom is greater than the actual mass of the nucleus. The missing mass is present as binding energy holding the nucleons together. Binding energy can be released either by breaking down heavy nuclei into smaller ones, a process called nuclear fission, or by joining small nuclei to form larger ones, called nuclear fusion. Fermi and Segrè bombarded uranium atoms with neutrons and found radioactive species among the products, while Hahn and Strassman found light nuclei among the reaction products. Meitner, aided by Frisch, hypothesized that the uranium was undergoing fission. Szilard saw that neutrons released in the fission of one atom could trigger the fission of other atoms, setting off a chain reaction. The nuclear fission reaction became the center of the Manhattan Project. Its goals were

(a) to achieve sustained nuclear fission and determine the **critical mass** or minimum amount of fissile material required; (b) to enrich the amount of fissile uranium-235 in ordinary uranium; (c) to synthesize plutonium-239, which is also fissile; and (d) to construct a nuclear fission bomb before the Germans were able to do so. World War II ended shortly after the dropping of atomic bombs on Hiroshima and Nagasaki.

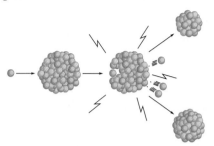

Sections 4.11–4.14—In addition to the devastation at the site of a nuclear explosion, much radioactive debris or radioactive fallout is produced. Strontium-90 and iodine-131 that are produced are particularly problematic. Partly because of fallout, a nuclear test ban treaty was signed by most of the major nations; only underground testing is not banned. Nuclear power plants use the same reaction as atomic bombs, but the reaction is much slower and is controlled because of the low concentration of fissile material. Disposal of the products of nuclear power plants is an important problem facing us, as is the potential conversion of nuclear fuel into weapons. **Thermonuclear reactions**, also known as fusion reactions, combine small nuclei to form larger ones. Nuclear fusion produces even more energy than nuclear fission. Fusion is the basis of the hydrogen bomb and the source of the sun's energy. Much research is aimed at producing controlled fusion commercially. The forces within the nucleus truly constitute a double-edged sword for civilization.

REVIEW QUESTIONS

1. Define or identify each of the following.
   a. half-life
   b. positron
   c. background radiation
   d. radioisotope
   e. fission
   f. fusion
   g. artificial transmutation
   h. binding energy

2. How does the size of the nucleus compare with that of an atom as a whole?

3. Why are isotopes important in nuclear reactions but not particularly so in most chemical reactions?

4. What is meant by the nucleon number of an isotope?

5. Give the nuclear symbols for protium, deuterium, and tritium (which are hydrogen-1, hydrogen-2, and hydrogen-3, respectively).

6. Give the nuclear symbol for the isotope with a nucleon number of 8 and an atomic number of 5.

7. Give the nuclear symbol for the isotope with $Z = 35$ and $A = 83$.

8. Give the nuclear symbol for the isotope with 53 protons and 72 neutrons.

9. Give the nuclear symbols for the following isotopes. You may refer to the periodic table.
   a. gallium-69
   b. molybdenum-98
   c. molybdenum-99
   d. technetium-98

10. Indicate the number of protons and the number of neutrons in atoms of the following isotopes:
    a. $^{62}_{30}Zn$
    b. $^{241}_{94}Pu$
    c. $^{99m}_{43}Tc$
    d. $^{81m}_{36}Kr$

11. Which of the following pairs represent isotopes?
    a. $^{70}_{34}X$ and $^{70}_{33}X$
    b. $^{57}_{28}X$ and $^{66}_{28}X$
    c. $^{186}_{74}X$ and $^{189}_{74}X$
    d. $^{8}_{2}X$ and $^{6}_{4}X$
    e. $^{22}_{11}X$ and $^{44}_{22}X$

12. In which of the following atoms are there more protons than neutrons?
    a. ^{58}Fe
    b. ^{1}H
    c. ^{3}H
    d. ^{12}C

13. The longest-lived isotope of fermium (Fm) has a nucleon number of 257. How many neutrons are there in the nucleus of this isotope?

14. The longest-lived isotope of technetium (Tc) has 54 neutrons. What is the nucleon number of this isotope?

15. The two principal isotopes of lithium are lithium-6 and lithium-7. The atomic mass of lithium is 6.9 u. Which is the predominant isotope of lithium?

16. The two principal isotopes of bromine are bromine-79 and bromine-81. Find the atomic mass of bromine on the periodic table, then fill in the blanks that follow: For every 100 atoms of bromine, roughly _____ are bromine-79 and _____ are bromine-81.

17. Give the nuclear symbols for the following subatomic particles.
    a. alpha particle
    b. beta particle
    c. neutron
    d. positron

18. What changes occur in the nucleon number and atomic number of the nucleus during emission of each the following?
    a. beta particle
    b. neutron
    c. proton

19. What changes occur in the nucleon number and atomic number of the nucleus during emission of each of the following?
    a. alpha particle
    b. gamma ray
    c. positron

20. Explain how radioisotopes can be used for therapeutic purposes.

21. Which radioisotope has been used extensively for the treatment of overactive or cancerous thyroid glands?

22. Describe the use of a radioisotope as a diagnostic tool in medicine.

23. What are some of the characteristics that make technetium-99m such a useful radioisotope for diagnostic purposes?

24. From which type of radiation would **(a)** a pair of gloves be sufficient to shield the hands: the heavy alpha particles or the massless gamma rays? **(b)** heavy lead shielding be necessary to protect a worker: alpha, beta, or gamma?

25. What form of radiation is detected in PET scans?

26. Plutonium is especially hazardous when inhaled or ingested because it emits alpha particles. Why would alpha particles cause more damage to tissue than beta particles?

27. List two ways in which workers can protect themselves from the radioactive materials with which they work.

28. Which subatomic particles are responsible for carrying on the chain reactions characteristic of nuclear fission?

29. What is the source of the greatest proportion of our exposure to artificial radiation?

30. Compare nuclear fission and nuclear fusion. Why is energy liberated in each case?

31. The compounds $^{235}UF_6$ and $^{238}UF_6$ are nearly chemically identical. How are they separated?

32. Discuss the process by which ^{239}Pu is created in a nuclear power plant.

PROBLEMS

Nuclear Equations

33. Write a balanced equation for emission of **(a)** a beta particle by silver-108, **(b)** an alpha particle by bismuth-210, and **(c)** a positron by copper-64.

34. Write a balanced equation for the **(a)** alpha decay of radium-226, **(b)** beta decay of polonium-209, and **(c)** capture of an electron by fluorine-18.

35. Complete the following equations.
    a. $^{179}_{79}Au \longrightarrow ^{175}_{77}Ir + ?$
    b. $^{23}_{10}Ne \longrightarrow ^{23}_{11}Na + ?$
    c. $^{121}_{51}Sb + ? \longrightarrow ^{121}_{52}Te + ^{1}_{0}n$

36. Complete the following equations.
    a. $^{10}_{5}B + ^{1}_{0}n \longrightarrow ^{4}_{2}He + ?$
    b. $^{12}_{6}C + ^{2}_{1}H \longrightarrow ^{13}_{6}C + ?$
    c. $^{154}_{62}Sm + ^{1}_{0}n \longrightarrow 2\,^{1}_{0}n + ?$

37. Radiological laboratories often have a container of molybdenum-99, which decays to form technetium-99m. What other particle is formed? Write an equation to show this reaction.
$$^{99}_{42}Mo \longrightarrow ^{99m}_{43}Tc + ?$$

38. Write an equation for the decay of technetium-99m to technetium-99. What is the other product of the reaction?
$$^{99m}_{43}Tc \longrightarrow ^{99}_{43}Tc + ?$$

39. When magnesium-24 is bombarded with a neutron, a proton is ejected. What new element is formed? (*Hint:* Write a balanced nuclear equation.)

40. When chlorine-37 is bombarded with a neutron, a proton is ejected. What new element is formed?

41. A radioactive isotope decays to give an alpha particle and bismuth-211. What was the original element?

42. A radioisotope decays to give an alpha particle and protactinium-233. What was the original element?

43. A nucleus of silver-109 absorbs a neutron, forming a new nucleus A. Nucleus A decays by beta emission to nucleus B. Write the complete symbol for B.

44. A proposed method of making more fissile nuclear fuel is to bombard the relatively abundant isotope thorium-232 with a neutron. The product A of this bombardment decays quickly by beta emission to nucleus B, and nucleus B decays quickly to fissile nucleus C. Write the complete symbol for C.

Half-Life

45. C. E. Bemis and colleagues at Oak Ridge National Laboratory confirmed the synthesis of element 104, the half-life of which was only 4.5 s. Only 3000 atoms of the element were created in the tests. How many atoms were left after **(a)** 4.5 s and **(b)** a total of 9.0 s?

46. Krypton-81m is used for lung ventilation studies. Its half-life is 13 s. How long does it take the activity of this isotope to reach one-quarter of its original value?

47. A 100-mg technetium-99m sample is used in a medical study. How much of the technetium-99m sample remains after 24 h? The half-life of technetium-99m is 6.0 h.

48. The half-life of molybdenum-99 is 67 h. How much time passes before a sample with an activity of 160 counts/min has decreases to 5.0 counts/min?

49. A patient is injected with a radiopharmaceutical labeled with technetium-99m, half-life 6.0 h, in preparation for a gamma ray scan in order to evaluate kidney function. If the original activity of the sample was 48 μCi, what activity remained after **(a)** 24 h and **(b)** after a total of 48 h?

50. Radium-223 has a half-life of 11.4 days. Approximately how long would it take for the radioactivity associated with a sample of ^{223}Ra to decrease to 1% of its initial value?

51. Look again at Problem 43. Nucleus A has a half-life of 31 seconds. How long will it take for all but 1/64 of nucleus A to decay?

52. Look again at Problem 44. Nucleus A has a half-life of 22 minutes. Beginning with 1.00 kg of A, how long will it take to form 875 g of B?

Radioisotopic Dating

53. Living matter has a carbon-14 activity of 16 counts/min per gram of carbon. What is the age of an artifact for which the carbon-14 activity is 8 counts/min per gram of carbon?

54. A piece of wood from an Egyptian tomb has carbon-14 activity of 980 counts per hour. A piece of new wood of the same size gave 3920 counts/h. What is the age of the wood from the tomb?

55. The ratio of carbon-14 to carbon-12 in a piece of charcoal from an archaeological excavation is found to be one-half the ratio in a sample of modern wood. Approximately how old is the site? How old would it be if the ratio were 25% of the ratio in a sample of modern wood?

56. You are offered a case of brandy supposedly bottled in the time of Napoleon (1769–1821) for a really great price. Before buying it, you insist on testing a sample of the brandy and find that it has a tritium content 12.5% of that of newly produced brandy. How long ago was the brandy bottled? Is it likely to be authentic Napoleon-era brandy?

ADDITIONAL PROBLEMS

57. Write a balanced nuclear equation for **(a)** the bombardment of $^{121}_{51}Sb$ by alpha particles to produce $^{124}_{53}I$, followed by **(b)** the radioactive decay of the $^{124}_{53}I$ by positron emission.

58. To make element 106, a 0.25-mg sample of californium-249 was used as the target. Four neutrons were emitted to yield a nucleus with 106 protons and a mass of 263 u. What was the bombarding particle?

59. One atom of element 109 with a nucleon number of 266 was produced in 1982 by bombarding a target of bismuth-209 with iron-58 nuclei for 1 week. How many neutrons were released in the process?

60. Element 109 undergoes alpha emission to form element 107, which in turn also emits an alpha particle. What are the atomic number and nucleon number of the isotope formed by these two steps? Write balanced nuclear equations for the two reactions.

61. Radium-223 nuclei usually decay by alpha emission. For every billion alpha decays, one atom emits a carbon-14 nucleus. Write a balanced nuclear equation for each type of emission.

62. A particular uranium alloy has a density of 18.75 g/cm^3. What volume is occupied by a critical mass of 49 kg of this alloy? The critical mass can be decreased to 16 kg if the alloy is surrounded by a layer of natural uranium (which acts as a neutron reflector). What is the volume of the smaller mass? Compare your answers to the volume of a baseball, a volleyball, and a basketball.

63. Plutonium has a density of 19.1 g/cm^3. What volume is occupied by a mass of 16.3 kg of plutonium? With the proper neutron reflecting coating (made of beryllium), the critical mass can be lowered to 2.5 kg! Compare this in size to the volume of a baseball, a volleyball, and a basketball.

64. What is the new nucleus formed in each of the following processes?
    a. Lead-196 goes through two successive EC processes.
    b. Bismuth-215 decays through two successive beta emissions.
    c. Protactinium-231 decays through four successive alpha emissions.
    d. Neptunium-237 undergoes a series of seven alpha and four beta decays to form a stable isotope.

65. Complete the following nuclear equations.
    a. $^{10}_{5}B + ^{1}_{0}n \longrightarrow ? + ^{1}_{1}H$
    b. $^{121}_{51}Sb + ? \longrightarrow ^{121}_{52}Te + ^{1}_{0}n$
    c. $^{59}_{27}Co + ^{1}_{0}n \longrightarrow ^{56}_{25}Mn + ?$

66. There are few technological applications for the transuranium elements ($Z > 92$). One important one is in smoke detectors, which may use the decay of a tiny amount of americium-241 to neptunium-237. What particle is emitted from that decay process?

67. In the first step of the chain reaction of a nuclear explosion, a ^{235}U nucleus absorbs a neutron. The resulting ^{236}U nucleus is unstable and can fission into a ^{92}Kr and a ^{141}Ba nucleus as shown in the figure. What are the other products of this reaction? In light of this reaction, explain how the chain reaction can continue.

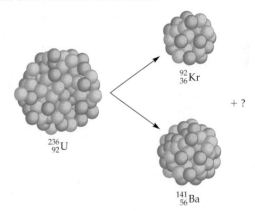

68. In 1932, James Chadwick discovered a new subatomic particle when he bombarded beryllium-9 with alpha particles. One of the products was carbon-12. What particle did Chadwick discover?

69. At a wine auction, you have the opportunity to purchase a bottle of wine presumably bottled by Thomas Jefferson around 1800. The tritium activity is about 8% that of new wine, and the half-life of tritium is 12.26 years. Can the claim of the bottle's origin be authentic? Explain.

70. In Einstein's mass–energy equation, $E = mc^2$, where mass is in kilograms and the speed of light is 3.00×10^8 m/s. The units of energy are joules (4.184 J = 1 cal; 1000 cal = 1 kcal).

a. Calculate the energy released, in calories and kilocalories, when 1 gram of matter is converted to energy.
b. A bowl of cornflakes supplies 110 kcal (110 food calories). How many bowls of cornflakes are equivalent to the energy in part (a) from one gram of matter?

71. Out of every five atoms of boron, one has a mass of 10 u and four have a mass of 11 u. What is the atomic mass of boron? Use the periodic table only to check your answer.

COLLABORATIVE GROUP PROJECTS

Prepare a PowerPoint, poster, or other presentation (as directed by your instructor) for presentation to the class.

72. Write a brief report on the impact of nuclear science on one of the following.
    a. war and peace   b. industrial progress
    c. medicine   d. agriculture
    e. human, animal, and plant genetics

73. Write a brief report on one of the following.
    a. Radiodating of archaeological objects
    b. Use of radioisotopes in medicine
    c. The many and varied uses of nuclear energy

74. Write a brief biography of one of the following scientists.
    a. Otto Hahn   b. Enrico Fermi
    c. Glenn T. Seaborg   d. J. Robert Oppenheimer
    e. Lise Meitner   f. Albert Einstein

75. Write a brief essay on the discovery of radioactivity. (You might want to take a look at the *Journal of Chemical Education*, January 1992, p. 10.)

76. The positron is a particle of antimatter. Search the Web for information about antimatter, and the positron in particular.

77. Find a website that is strongly in favor of nuclear power plants and one that is strongly opposed. Note their sponsors and analyze their viewpoints. Try to find a website with a balanced approach.

78. Write a short research report describing the operation of smoke detectors.

79. In July 1999 researchers at Lawrence Berkeley Laboratory reported the creation of the heaviest element to date, element 118. Two years later, that report was found to be fraudulent (see Reference 14). Report on the scientific ethical questions that this episode raises.

80. Find the location of the nuclear plant closest to where you live. Try to determine risks and rewards of the plant. To how many houses does it provide power? What are the environmental impacts of the plant under normal operating conditions?

81. Assemble two groups of two to four people each to debate the following resolution: The use of the atomic bomb on Hiroshima and Nagasaki was justified. Decide beforehand which team will take the affirmative and which the negative. Each team member is allowed a three- to six-minute speech followed by a cross-examination by the opposing team members. Have the rest of the class judge and give written or oral comments.

82. Prepare a presentation on the subject of "cold fusion." Other useful search-engine keywords are "Pons" and "Fleischmann." The presentation should discuss the differences between the hypothesized cold fusion and conventionally understood fusion reactions and should provide a balanced and impartial view of the subject.

REFERENCES AND READINGS

1. Armbruster, Paul, and Fritz Hessberger. "Making New Elements." *Scientific American*, September 1998, pp. 72–77. Discusses synthesis of the transuranium elements.

2. Bartell, Lawrence S. "A Chemist's Job on the Manhattan Project." *Chemical Heritage,* Summer 2002, pp. 12–13, 29–31.

3. Bowman, Sheridan. *Radiocarbon Dating (Interpreting the Past, No. 1).* Berkeley, CA: University of California Press, June 1990.

4. Brennan, Mairin B. "Positron Emission Tomography Merges Chemistry with Biological Imaging." *Chemical and Engineering News*, February 19, 1996, pp. 26–33.

5. Budavari, Susan (ed.). *The Merck Index*, 11th edition. Rahway, NJ: Merck and Co., 1989. Contains extensive tables of radioisotopes (pp. MISC 31–45) and radioisotopes used in medical diagnosis and therapy (pp. MISC 46–52).

6. Cobb, Cathy, and Harold Goldwhite. *Creations of Fire: Chemistry's Lively History from Alchemy to the Atomic Age.* New York: Plenum, 1995.

7. Freemantle, Michael. "Nuclear Power for the Future." *Chemical and Engineering News*, September 13, 2004, pp. 31–35.

8. Gove, H. E. *Relic, Icon or Hoax? Carbon Dating the Turin Shroud.* Bristol and Philadelphia: Institute of Physics Press, December 1996.

9. Herken, Gregg. *Brotherhood of the Bomb: The Tangled Lives and Loyalties of Robert Oppenheimer, Ernest Lawrence, and Edward Teller.* New York: H. Holt, 2002.

10. Hoffman, Darleane C., and Diana M. Lee. "Chemistry of the Heaviest Elements—One Atom at a Time." *Journal of Chemical Education*, March 1999, pp. 331–347.

11. Jaworowski, Zbigniew. "Radiation Risk and Ethics." *Physics Today*, September 1999, pp. 24–29.

12. Murphy, Marina. "Isotopic Techniques to Provide Fresh Evidence." *Chemistry and Industry*, October 4, 2004, p. 14.

13. Scheider, W. *A Serious but Not Ponderous Book About Nuclear Energy.* Cambridge: Cavendish Press, 2001.

14. Schwarzschild, Bertram. "Lawrence Berkeley Lab Concludes that Evidence of Element 118 Was a Fabrication." *Physics Today*, vol. 55, 2002, p. 15.

15. Seaborg, Eric. *Adventures in the Atomic Age: From Watts to Washington.* New York: Farrar Straus & Giroux, September 2001.

16. Seaborg, Glenn T. "The Positive Power of Radioisotopes." *The Skeptical Inquirer*, January–February 1995, pp. 39–40, 62.

17. Sime, Ruth Lewin. "Meitner and the Discovery of Nuclear Fission." *Scientific American*, January 1998, p. 80.

18. Wiesner, Emilie, and Frank A. Settle, Jr. "Politics, Chemistry, and the Discovery of Nuclear Fission." *Journal of Chemical Education*, July 2001, pp. 889–895.

19. Zimmer, Carl. "How Old Is It?" *National Geographic*, September 2001, pp. 78–101.

CHAPTER 5

Chemical Bonds

Elements combine by forming chemical bonds. The compounds so formed have very different properties from the original elements. Here (upper left) pieces of aluminum foil are added to liquid bromine. The two elements react slowly at first (upper right), and then more vigorously (lower left); the heat given off by the reaction causes the bromine to vaporize, forming heavy reddish-brown fumes. The product (lower right) is a white solid, aluminum bromide.

The Ties That Bind

Aluminum is a silvery solid, familiar as aluminum foil. Bromine is an orange-red liquid that gives off amber fumes. When these two elements combine, as shown on the facing page, they form aluminum bromide, a white solid. Aluminum bromide is quite different from either aluminum or bromine. When elements combine to form a compound, they do not keep their original properties. They are transformed into a completely different substance with its own characteristic properties.

Why should compounds be so unlike their component elements? The answer lies in the bonds that hold the various atoms together. We have learned a bit about the structure of the atom and examined the atomic nucleus. Now we are ready to consider chemical bonds, the ties that bind atoms together.

Chemical bonds are the forces that hold atoms together in molecules, and ions together in ionic crystals. The vast number and incredible variety of chemical compounds—tens of millions are known—result from the fact that atoms can form bonds with many different other types of atoms. In addition, chemical bonds determine the three-dimensional shape of a molecule. Some important consequences of chemical structure and bonding include

- Whether a substance is a solid, liquid, or gas at room temperature
- The strengths of materials (as well as adhesives that hold materials together) used for building bridges, houses, and many other structures
- Whether a liquid is light and volatile (like gasoline) or heavy and viscous (like motor oil)
- The taste, odor, and drug activity of chemical compounds
- The structural integrity of skin, muscles, bones, and teeth
- The toxicity of certain molecules to living organisms

Chemical bonding is related to the arrangement of electrons in compounds. In this chapter, we look at different types of chemical bonds and some of the unusual properties of compounds that result.

5.1 THE ART OF DEDUCTION: STABLE ELECTRON CONFIGURATIONS

In our discussion of the atom and its structure (Chapters 2 and 3), we followed the historical development of some of the more important atomic concepts. Some of the nuclear concepts (Chapter 4) were approached in the same way. We could continue to look at chemistry in this manner, but that would require several volumes of print—and perhaps more of your time than you care to spend. We won't abandon the historical approach entirely, but we will emphasize another important aspect of scientific endeavor: deduction.

The art of deduction works something like this.

- **Fact:** Noble gases, such as helium, neon, and argon, are inert; they undergo few, if any, chemical reactions.
- **Theory:** The inertness of noble gases results from their electron structures; each (except helium) has an octet of electrons in its outermost shell.
- **Deduction:** Other elements that can alter their electron structures to become like those of noble gases would become less reactive by doing so.

We can use an example to illustrate this deductive argument. Sodium has 11 electrons, one of which is in the third shell. Recall that electrons in the outermost shell are called **valence electrons**, while those in all the other shells are lumped together as **core electrons** (Section 3.8). If the sodium atom got rid of its valence electron, the remaining core electrons would have the same electron structure as an atom of the noble gas neon. Using main-shell configurations (Chapter 3), we can represent this as

The outermost shell is "filled" when it contains eight electrons, two *s* and six *p*. The first shell is an exception: It holds only two *s* electrons.

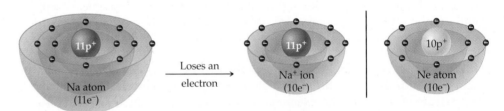

Similarly, if a chlorine atom could gain an electron, it would have the same electron structure as argon.

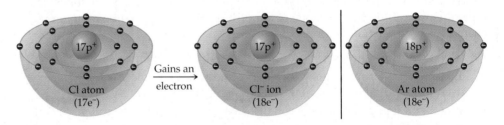

The sodium atom, having lost an electron, becomes positively charged. It has 11 protons (11+) and only 10 electrons (10−). It is written Na^+ and is called a *sodium ion*. The chlorine atom, having gained an electron, becomes negatively charged. It has 17 protons (17+) and 18 electrons (18−). It is written Cl^- and is called a *chloride ion*. Note that a positive charge, as in Na^+, indicates that one electron has been lost. Similarly, a negative charge, as in Cl^-, indicates that one electron has been gained. It is important to note that even though Cl^- and Ar are **isoelectronic** (have the same electron configuration), they are *different* chemical species. In the same way, a sodium atom does not *become* an atom of neon when it loses an electron; the sodium ion is simply isoelectronic with neon.

5.2 LEWIS (ELECTRON-DOT) SYMBOLS

In forming ions, the cores of sodium atoms and chlorine atoms do not change. It is convenient therefore to let the symbol represent the *core* of the atom (nucleus plus inner electrons). The valence electrons are then represented by dots. The equations of the preceding section then can be written as follows:

$$Na\cdot \longrightarrow Na^+ + 1\,e^-$$

and

$$\cdot\ddot{\underset{..}{C}}l: + 1\,e^- \longrightarrow :\ddot{\underset{..}{C}}l:^-$$

In these representations, the symbol of the element represents the core, and dots stand for valence electrons. These electron-dot symbols are usually called **Lewis symbols**.

Lewis Symbols and the Periodic Table

It is especially easy to write Lewis symbols for most of the main group elements. The number of valence electrons for most of these elements is equal to the group number (Table 5.1). Because of their more complicated electron configurations, elements in the central part of the periodic table (the transition metals) do not easily lend themselves to the electron dot symbolism.

▲ Electron-dot symbols are called *Lewis symbols* after G. N. Lewis, the famous American chemist (1875–1946) who invented them. Lewis also made important contributions in the fields of thermodynamics, acids and bases, and spectroscopy. His achievements in any of these areas might have merited a Nobel Prize, yet he never received that award.

TABLE 5.1	Lewis Symbols for Selected Main Group Elements						
Group 1A	Group 2A	Group 3A	Group 4A	Group 5A	Group 6A	Group 7A	Noble Gases
H·							He:
Li·	·Be·	·Ḃ·	·Ċ·	:Ṅ·	:Ö·	:F̈·	:Ṅe:
Na·	·Mg·	·Äl·	·Ṡi·	:P̈·	:S̈·	:C̈l·	: Är:
K·	·Ca·				:Se·	:Br·	:Kr:
Rb·	·Sr·				:Te·	:Ï·	:Xe:
Cs·	·Ba·						

EXAMPLE 5.1   **Lewis Symbols**

Without referring to Table 5.1, give Lewis symbols for magnesium, oxygen, and phosphorus. You may use the periodic table.

Solution

Magnesium is in group 2A, oxygen is in group 6A, and phosphorus is in group 5A. The Lewis symbols therefore have two, six, and five dots, respectively. They are

$$\cdot Mg\cdot \qquad :\ddot{O}: \qquad :\dot{P}\cdot$$

▶ **Exercise 5.1**

Without referring to Table 5.1, give Lewis symbols for each of the following elements. You may use the periodic table.
a. Ar   **b.** Ca   **c.** F   **d.** N   **e.** K   **f.** S

In writing a Lewis symbol, only the number *of dots is important. The dots need not be drawn in any specific positions, except that there should be no more than two dots on any given side of the chemical symbol (right, left, top, or bottom).*

5.3 SODIUM REACTS WITH CHLORINE: FACTS

Sodium is a highly reactive metal. It is soft enough to be cut with a knife. When freshly cut, it is bright and silvery, but it dulls rapidly because it reacts with oxygen in the air. In fact, it reacts so readily in air that it is usually stored under oil or

Chlorine gas is actually composed of Cl_2 molecules, not separate Cl atoms. Each atom of the molecule takes an electron from a sodium atom. Two sodium ions and two chloride ions are formed.

$$Cl_2 + 2\,Na \longrightarrow 2\,Cl^- + 2\,Na^+$$

Formation of Sodium Chloride

▶ **Figure 5.1** Sodium, a soft, silvery metal, reacts with chlorine, a greenish gas, to form sodium chloride (ordinary table salt), a white crystalline solid.

Question: What particles, ions or molecules, make up sodium chloride?

kerosene. Sodium reacts violently with water also, becoming so hot that it melts. A small piece forms a spherical bead after melting and races around on the surface of the water as it reacts.

Chlorine is a greenish-yellow gas. It is familiar as a disinfectant for city water supplies and swimming pools. (The actual substance added is often a compound that reacts with water to form chlorine.) Chlorine is extremely irritating to the eyes and nose. In fact, it was used as a poison gas in World War I.

If a piece of sodium is dropped into a flask containing chlorine gas, a violent reaction ensues, producing sodium chloride, beautiful white crystals that you might sprinkle on your food at the dinner table. These white crystals are ordinary table salt. Sodium chloride has very few properties in common with either sodium or chlorine (Figure 5.1).

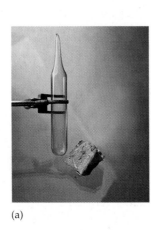

(a)　　　　　　　　　(b)　　　　　　　　　(c)

5.4 SODIUM REACTS WITH CHLORINE: THE THEORY

A sodium atom achieves a filled valence shell by losing one electron. A chlorine atom achieves a filled valence shell by adding one electron. What happens when sodium atoms come into contact with chlorine atoms? The obvious: Chlorine extracts an electron from a sodium atom.

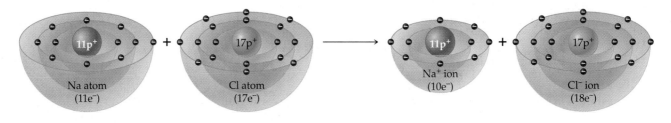

Electron loss is known as *oxidation*, and gain of electrons is called *reduction*. We shall explore this concept in Chapter 8.

In the abbreviated Lewis form, this reaction is written

$$Na\cdot + \ :\!\ddot{C}l\cdot \ \longrightarrow Na^+ + \ :\!\ddot{\underset{..}{C}l}\!:^-$$

Ionic Bonds

The Na^+ and Cl^- ions formed from sodium and chlorine atoms have opposite charges and are strongly attracted to one another. These ions arrange themselves in an orderly fashion. Each sodium ion attracts (and is attracted to) six chloride ions (top and bottom, front and back, left and right) as shown in Figure 5.2(a). The arrangement is repeated many times in all directions (Figure 5.2b). The result is a **crystal** of sodium chloride. The orderly microscopic arrangement of the ions is reflected in the macroscopic shape of each salt crystal (Figure 5.2c). Even the tiniest grain of salt has billions and billions of each type of ion. The forces holding the crystal together—the attractive forces between positive and negative ions—are called **ionic bonds.**

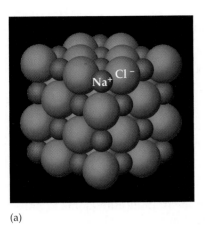 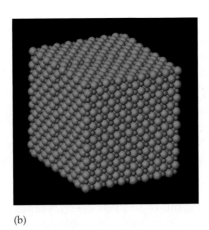 

(a)   (b)   (c)

▲ **Figure 5.2** Structure of a sodium chloride crystal. (a) Each Na^+ ion (small sphere) is surround-ed by six Cl^- ions (large spheres), and each Cl^- ion by six Na^+ ions. (b) This arrangement repeats it-self many, many times. (c) The highly ordered pattern of alternating Na^+ and Cl^- ions is observed in the macroscopic world as a crystal of sodium chloride.

Atoms and Ions: Distinctively Different

Ions are emphatically different from the atoms from which they are made, much as a whole peach (an atom) and a peach pit (a positive ion) are different from one another. The names and symbols of atoms and ions may look a lot alike, but the similarities end there (Figure 5.3). Unfortunately, the situation is con-fusing because people talk about needing iron to perk up "tired blood" and calcium for healthy teeth and bones. What they really mean is iron(II) *ions* (Fe^{2+}) and calcium *ions* (Ca^{2+}). You wouldn't think of eating iron nails to get iron. (However, some enriched cereals do indeed have powdered iron added; the iron metal is readily converted to Fe^{2+} ions in the stomach.) Nor would you eat highly reactive calcium metal.

Similarly, when you are warned to reduce your sodium intake, your doctor is not concerned that you are eating too much sodium metal—that would be ex-ceedingly unpleasant—but that your intake of Na^+

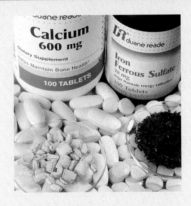

◀ **Figure 5.3** Forms of iron and calcium that are useful to the human body are the ions (usually Fe^{2+} and Ca^{2+}) taken in ionic compounds such as $FeSO_4$ and $CaCO_3$. The elemental forms (Fe and Ca) are very different from forms of the same elements in ionic compounds.

ions—usually as sodium chloride—may be too high. Although the names are similar, the atom and the ion are quite different chemically. It is important that we make careful distinctions and use precise terminology, as we shall do here.

5.5 USING LEWIS SYMBOLS: MORE IONIC COMPOUNDS

As we might expect, potassium, a metal in the same family as sodium—and there-fore similar to sodium in properties—also reacts with chlorine. The reaction yields potassium chloride (KCl).

$$K\cdot \ + \ \cdot\ddot{\underset{..}{Cl}}: \ \longrightarrow \ K^+ \ + \ :\ddot{\underset{..}{Cl}}:^-$$

Potassium also reacts with bromine, a reddish-brown liquid in the same family as chlorine—and therefore similar to chlorine in properties—to form stable white, crystalline potassium bromide (KBr).

$$K\cdot \ + \ \cdot\ddot{\underset{..}{Br}}: \ \longrightarrow \ K^+ \ + \ :\ddot{\underset{..}{Br}}:^-$$

EXAMPLE 5.2 **Electron Transfer to Form Ions**

Use Lewis symbols to show the transfer of electrons from sodium atoms to bromine atoms to form ions with noble gas configurations.

Solution

Sodium has one valence electron, and bromine has seven. Transfer of the single electron from sodium to bromine leaves each with a noble gas configuration.

$$\text{Na} \cdot \; + \; \cdot \ddot{\ddot{\text{Br}}} \text{:} \; \longrightarrow \; \text{Na}^+ \; + \; \text{:} \ddot{\ddot{\text{Br}}} \text{:}^-$$

▶ **Exercise 5.2A**

Use Lewis symbols to show the transfer of electrons from lithium atoms to fluorine atoms to form ions with noble gas configurations.

▶ **Exercise 5.2B**

Use Lewis symbols to show the transfer of electrons from rubidium atoms to iodine atoms to form ions with noble gas configurations.

Magnesium, a group 2A metal, is harder and less reactive than sodium. Magnesium reacts with oxygen, a group 6A element that is a colorless gas, to form another stable white, crystalline solid called magnesium oxide (MgO).

$$\cdot \text{Mg} \cdot \; + \; \cdot \ddot{\text{O}} \text{:} \; \longrightarrow \; \text{Mg}^{2+} \; + \; \text{:} \ddot{\ddot{\text{O}}} \text{:}^{2-}$$

Magnesium must give up two electrons and oxygen must gain two electrons for each to have the same configuration as the noble gas neon.

An atom such as oxygen, which needs two electrons to complete a noble gas configuration, may react with potassium atoms, which have only one electron each to give. In this case, two atoms of potassium are needed for each oxygen atom. The product is potassium oxide (K_2O).

$$\begin{array}{c} \text{K} \cdot \\ \text{K} \cdot \end{array} + \; \cdot \ddot{\text{O}} \text{:} \; \longrightarrow \; \begin{array}{c} \text{K}^+ \\ \text{K}^+ \end{array} + \; \text{:} \ddot{\ddot{\text{O}}} \text{:}^{2-}$$

By this process, each potassium atom achieves the argon configuration. As before, oxygen assumes the neon configuration.

> When the Mg atom becomes a Mg^{2+} ion, its second electron shell becomes its outermost shell. The Mg^{2+} ion is isoelectronic with the Ne atom and has the same stable electron structure: 2 8.

EXAMPLE 5.3 **Electron Transfer to Form Ions**

Use Lewis symbols to show the transfer of electrons from magnesium atoms to nitrogen atoms to form ions with noble gas configurations.

Solution

$$\begin{array}{c} \cdot \text{Mg} \cdot \\ \cdot \text{Mg} \cdot \\ \cdot \text{Mg} \cdot \end{array} + \; \begin{array}{c} \cdot \ddot{\text{N}} \cdot \\ \cdot \ddot{\text{N}} \cdot \end{array} \; \longrightarrow \; \begin{array}{c} \text{Mg}^{2+} \\ \text{Mg}^{2+} \\ \text{Mg}^{2+} \end{array} + \; \begin{array}{c} \text{:} \ddot{\ddot{\text{N}}} \text{:}^{3-} \\ \text{:} \ddot{\ddot{\text{N}}} \text{:}^{3-} \end{array}$$

Each of three magnesium atoms gives up two electrons (a total of six), and each of the two nitrogen atoms acquires three (a total of six). Notice that the total positive and negative charges on the products are equal (6+ and 6−). Magnesium reacts with nitrogen to yield magnesium nitride (Mg_3N_2).

▶ **Exercise 5.3**

Use Lewis symbols to show the transfer of electrons from aluminum atoms to oxygen atoms to form ions with noble gas configurations.

Generally speaking, metallic elements in groups 1A and 2A (those from the left side of the periodic table) react with nonmetallic elements in groups 6A and 7A (those from the right side) to form ionic compounds. These are stable crystalline solids.

Ions and the Octet Rule

The Octet Rule

Each atom of a metal tends to give up the electrons in its outer shell, and each atom of a nonmetal tends to take on enough electrons to complete its valence shell. The resulting ions have noble gas configurations. A set of eight valence electrons—an *octet*—is the characteristic arrangement of all noble gases except helium. When atoms react with each other, they often tend to attain this stable noble-gas electron configuration. Thus they are said to follow the **octet rule**, or the "rule of eight." (In the case of helium, a maximum of two electrons can exist in its single electron shell, and so hydrogen follows the "rule of two.")

In following the octet rule, atoms of group 1A metals give up one electron to form 1+ ions, those of group 2A metals give up two electrons to form 2+ ions, and group 3A metals give up three electrons to form 3+ ions. Group 7A nonmetal atoms take on one electron to form 1− ions, and group 6A atoms tend to pick up two electrons to form 2− ions. Atoms of B group metals can give up various numbers of electrons to form positive ions with various charges. These periodic relationships are summarized in Figure 5.4.

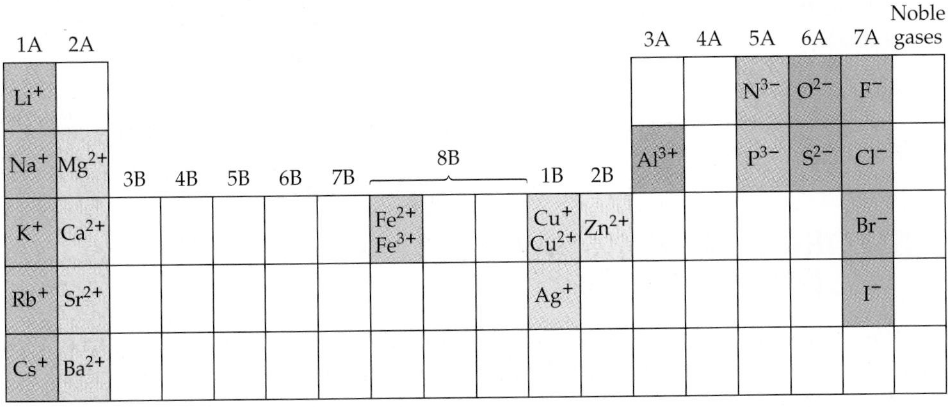

1A	2A										1B	2B	3A	4A	5A	6A	7A	Noble gases
Li^+															N^{3-}	O^{2-}	F^-	
Na^+	Mg^{2+}	3B	4B	5B	6B	7B		8B					Al^{3+}		P^{3-}	S^{2-}	Cl^-	
K^+	Ca^{2+}						Fe^{2+} Fe^{3+}			Cu^+ Cu^{2+}	Zn^{2+}						Br^-	
Rb^+	Sr^{2+}									Ag^+							I^-	
Cs^+	Ba^{2+}																	

◀ **Figure 5.4** Periodic relationships of some simple ions. The transition elements (B groups) often form ions of more than one kind, with different charges.

Question: Can you propose a formula for the simple ion formed from selenium (Se)? For the simple ion formed from germanium (Ge)?

Table 5.2 lists symbols and names for some ions formed by the gain or loss of electrons. You can calculate the charge on the negative ions in the table by subtracting 8 from the group number. For example, the charge on the oxide ion (oxygen is in group 6A) is $6 - 8 = -2$. The nitride ion (nitrogen is in group 5A) has a charge of $5 - 8 = -3$.

TABLE 5.2 Symbols and Names for Some Simple (Monatomic) Ions

Group	Element	Name of Ion	Symbol for Ion
1A	Hydrogen	Hydrogen ion	H^+
	Lithium	Lithium ion	Li^+
	Sodium	Sodium ion	Na^+
	Potassium	Potassium ion	K^+
2A	Magnesium	Magnesium ion	Mg^{2+}
	Calcium	Calcium ion	Ca^{2+}
3A	Aluminum	Aluminum ion	Al^{3+}
5A	Nitrogen	Nitride ion	N^{3-}
6A	Oxygen	Oxide ion	O^{2-}
	Sulfur	Sulfide ion	S^{2-}
7A	Fluorine	Fluoride ion	F^-
	Chlorine	Chloride ion	Cl^-
	Bromine	Bromide ion	Br^-
	Iodine	Iodide ion	I^-
1B	Copper	Copper(I) ion (cuprous ion)	Cu^+
		Copper(II) ion (cupric ion)	Cu^{2+}
	Silver	Silver ion	Ag^+
2B	Zinc	Zinc ion	Zn^{2+}
8B	Iron	Iron(II) ion (ferrous ion)	Fe^{2+}
		Iron(III) ion (ferric ion)	Fe^{3+}

EXAMPLE 5.4 **Determining Formulas by Electron Transfer**

What is the formula of the compound formed by the reaction of sodium and sulfur?

Solution

Sodium is in group 1A; the sodium atom has one valence electron. Sulfur is in group 6A; the sulfur atom has six valence electrons.

$$\text{Na·} \quad \cdot \ddot{\underset{..}{S}} \cdot$$

Sulfur needs two electrons to gain an argon configuration, but sodium has only one to give. The sulfur atom therefore must react with two sodium atoms.

$$\begin{matrix} \text{Na·} \\ \text{Na·} \end{matrix} + \cdot \ddot{\underset{..}{S}} \cdot \longrightarrow \begin{matrix} \text{Na}^+ \\ \text{Na}^+ \end{matrix} + \; :\ddot{\underset{..}{S}}:^{2-}$$

The formula of the compound, called sodium sulfide, is Na_2S.

▶ **Exercise 5.4A**

What are the formulas of the compounds formed by the reaction of **(a)** calcium with fluorine, and **(b)** lithium with oxygen?

▶ **Exercise 5.4B**

Use data in Figure 5.4 to predict the formulas of the two compounds that can be formed from iron (Fe) and chlorine.

5.6 FORMULAS AND NAMES OF BINARY IONIC COMPOUNDS

Names of simple positive ions (*cations*) are derived from those of their parent elements by the addition of the word *ion*. A sodium atom, on losing an electron, becomes a *sodium ion* (Na^+). A magnesium atom (Mg), on losing two electrons, becomes a *magnesium ion* (Mg^{2+}).

Names of simple negative ions (*anions*) are derived from those of their parent elements by changing the usual ending to *-ide* and adding the word *ion*. A chlor*ine* atom, on gaining an electron, becomes a chlor*ide ion* (Cl^-). A sul*fur* atom gains two electrons, becoming a sul*fide ion* (S^{2-}).

Simple ions of opposite charge can be combined to form **binary** (two-element) **compounds**. To get the correct formula for a binary compound, simply write each ion with its charge (positive ion to the left), then cross over the numbers (but not the plus and minus signs) and write them as subscripts. The process is best learned by practice, as is provided in the following examples and exercises and by problems at the end of the chapter.

When a metal forms more than one ion, the charges on the different ions are denoted by Roman numerals in parentheses. For example, Fe^{2+} is iron(II) ion and Fe^{3+} is iron(III) ion.

EXAMPLE 5.5 **Determining Formulas from Ionic Charges**

Give the formulas for **(a)** calcium chloride and **(b)** aluminum oxide.

Solution

a. First, write the symbols for the ions. (We write the charge on chloride ion explicitly as "1−" to illustrate the crossover method. You may omit the "1" when you are comfortable with the process.)

$$Ca^{2+} \quad Cl^{1-}$$

Then cross over the numbers as subscripts.

$$Ca^{2+} \quad\quad Cl^{1-}$$

Then rewrite the formula, dropping the charges. The formula for calcium chloride is

$$Ca_1Cl_2 \quad \text{or (dropping the "1") simply} \quad CaCl_2$$

b. Write the symbols for the ions.

$$Al^{3+} \quad O^{2-}$$

Cross over the numbers as subscripts.

Then rewrite the formula, dropping the charges. The formula for aluminum oxide is

$$Al_2O_3$$

▶ **Exercise 5.5**
Give the formulas for **(a)** potassium oxide, **(b)** calcium nitride, and **(c)** calcium sulfide.

The crossover method works because it is based on the transfer of electrons and the conservation of charge. Two Al atoms lose three electrons each (a total of six electrons lost), and three O atoms gain two electrons each (a total of six electrons gained). Electrons lost equal electrons gained. Similarly, two Al^{3+} ions have six positive charges (three each), and three O^{2-} ions have six negative charges (two each). The net charge on Al_2O_3 is 0, just as it should be.

Now you are able to translate the "English," such as aluminum oxide, into the "chemistry," Al_2O_3. You also can translate in the other direction.

EXAMPLE 5.6   **Naming Ionic Compounds**

What are the names of **(a)** MgS and **(b)** $FeCl_3$?

Solution
a. From Table 5.2 we can determine that MgS is made up of Mg^{2+} (magnesium ion) and S^{2-} (sulfide ion). The name is simply magnesium sulfide.
b. From Table 5.2 we can determine that the ions in $FeCl_3$ are

$$Fe^{3+} \quad Cl^-$$

How do we know the iron ion in $FeCl_3$ is Fe^{3+} and not Fe^{2+}? Because there are three Cl^- ions, each $1-$, the one Fe ion must be $3+$ because the compound $FeCl_3$ is neutral. The names of these ions are iron(III) ion (or ferric ion) and chloride ion. Therefore, the compound is iron(III) chloride (or, by the older system, ferric chloride).

▶ **Exercise 5.6**
What are the names of **(a)** CaF_2 and **(b)** $CuBr_2$?

A Compound by Any Other Name Would Smell As Sweet . . .

As you read labels on foodstuff and pharmaceuticals, you will see some names that are beginning to sound familiar, and some that are confusing or mysterious. For example, why do we have two names for Fe^{2+}? Some names are historical: Iron, copper, gold, silver, and other elements have been known for thousands of years. In naming compounds of these elements, an *-ous* ending came to be used for the ion of smaller charge, and an *-ic* ending for the ion of larger charge. The modern system uses Roman numerals to indicate charge. Although the new system is more logical and easier to apply, the old names persist, especially in everyday life and in some of the biomedical sciences.

3 mg	100%
horus 109 mg	11%

Boron 150 mcg

*Daily Value (DV) n

DIENTS: Dicalcium Phosphate, Potassium Chloride, se, Ascorbic Acid, Ferrous Fumarate, Calcium Carł ocopheryl Acetate, Croscarmellose Sodium, Niacin sium Stearate, Dextrin, d-Calcium Pantothenat nellose, Manganese Sulfate, Polyethylene Glycol, Si ie, Maltodextrin, Cupric Sulfate, Corn Starch, Dex

▲ Dietary supplements often use older, Latin-derived names.

5.7 COVALENT BONDS: SHARED ELECTRON PAIRS

We might expect a hydrogen atom, with its one electron, to acquire another electron and assume the configuration of the noble gas helium. In fact, hydrogen atoms do just that in the presence of atoms of a reactive metal such as lithium—that is, a metal that finds it easy to give up an electron.

$$\text{Li} \cdot \; + \; \text{H} \cdot \; \longrightarrow \; \text{Li}^+ \; + \; \text{H} \colon^-$$

But what if there are no other kinds of atoms around, only hydrogen? One atom can't gain an electron from another, for all hydrogen atoms have an equal attraction for electrons. Two hydrogen atoms can compromise, however, by *sharing a pair* of electrons.

$$\text{H} \cdot \; + \; \cdot \text{H} \; \longrightarrow \; \text{H} \colon \text{H}$$

By sharing electrons, the two hydrogen atoms form a hydrogen molecule. The bond formed by a shared pair of electrons is called a **covalent bond**.

$$\text{H} \colon \text{H}$$
$$\underset{\text{covalent bond (shared pair of electrons)}}{\uparrow}$$

Consider next the case of chlorine. A chlorine atom readily takes an extra electron from anything willing to give one up. But again, what if the only things around are other chlorine atoms? Chlorine atoms also can attain a more stable arrangement by sharing a pair of electrons.

$$: \ddot{\text{Cl}} \cdot \; + \; \cdot \ddot{\text{Cl}} : \; \longrightarrow \; : \ddot{\text{Cl}} : \ddot{\text{Cl}} :$$

The shared pair of electrons in the chlorine molecule is another example of a covalent bond; they are called a **bonding pair**. The other electrons that stay on one atom and are not shared are called *nonbonding pairs* or **lone pairs**.

$$: \ddot{\text{Cl}} : \ddot{\text{Cl}} :$$

For simplicity, the hydrogen molecule is often represented as H_2, and the chlorine molecule as Cl_2. In each case, the covalent bond between the atoms is understood. Sometimes the covalent bond is indicated by a dash, H—H and Cl—Cl. Lone pairs of electrons often are not shown. Each chlorine atom in the chlorine molecule has eight electrons around it, an arrangement like that of the noble gas argon. Thus, the atoms in a covalent bond follow the octet rule by sharing electrons, even as those in an ionic bond follow it by giving up or taking on electrons.

Multiple Bonds

In some molecules, atoms must share more than one pair of electrons in order to fulfill the octet rule. In carbon dioxide (CO_2), for example, the carbon atom shares *two* pairs of electrons with each of the two oxygen atoms.

$$: \ddot{\text{O}} : : \text{C} : : \ddot{\text{O}} :$$

Note that each atom has an octet of electrons about it as a result of this sharing. We say that the atoms are joined by a **double bond**, a covalent linkage in which the two atoms share two pairs of electrons.

Atoms also can share three pairs of electrons. In the nitrogen (N_2) molecule, for example, each nitrogen atom shares three pairs of electrons with the other.

$$: \text{N} : : : \text{N} :$$

The atoms are joined by a **triple bond**, a covalent linkage in which two atoms share three pairs of electrons. Note that each of the nitrogen atoms has an octet of electrons around it.

Covalent bonds are usually represented as dashes. The three kinds of covalent bonds are simply written as follows:

$$\text{H}\!-\!\text{Cl} \quad \text{O}\!=\!\text{C}\!=\!\text{O} \quad \text{N}\!\equiv\!\text{N}$$

Names of Covalent Compounds

Covalent or *molecular* compounds are those in which electrons are shared, not transferred. Molecular compounds generally have molecules that consist of two or more nonmetals. Many molecular compounds have common and widely used names. Examples are water (H_2O), methane (CH_4), and ammonia (NH_3). For other molecular compounds, the naming process is more systematic. The prefixes *mono-*, *di-*, *tri-*, and so on are used to indicate the number of atoms of each element in the molecule. A list of these prefixes for up to 10 atoms is given in Table 5.3. For example, the compound N_2O_4 is called *dinitrogen tetroxide*. (The *a* often is dropped from *tetra-* and other prefixes when they precede another vowel.) We often leave off the *mono-* prefix (NO_2 is nitrogen dioxide) but do include it to distinguish between two compounds of the same pair of elements (CO is carbon monoxide; CO_2 is carbon dioxide).

TABLE 5.3	Prefixes that Indicate the Number of Atoms of an Element in a Covalent Compound
Prefix	**Number of Atoms**
Mono-	1
Di-	2
Tri-	3
Tetra-	4
Penta-	5
Hexa-	6
Hepta-	7
Octa-	8
Nona-	9
Deca-	10

EXAMPLE 5.7 **Naming Covalent Compounds**

What are the names of **(a)** SCl_2 and **(b)** SF_6?

Solution
a. With one sulfur atom and two chlorine atoms, SCl_2 is sulfur dichloride.
b. With one sulfur atom and six fluorine atoms, SF_6 is sulfur hexafluoride.

▶ **Exercise 5.7A**
What are the names of **(a)** BrF_3 and **(b)** BrF_5?

▶ **Exercise 5.7B**
What are the names of **(a)** N_2O and **(b)** N_2O_5?

EXAMPLE 5.8 **Formulas of Covalent Compounds**

Give the formula for tetraphosphorus hexoxide.

Solution
The *tetra-* indicates four phosphorus atoms, and the *hex-* specifies six oxygen atoms. The formula is P_4O_6.

▶ **Exercise 5.8A**
Give the formulas for **(a)** phosphorus trichloride and **(b)** dichlorine heptoxide.

▶ **Exercise 5.8B**
Give the formulas for **(a)** nitrogen triiodide and **(b)** disulfur dichloride.

5.8 UNEQUAL SHARING: POLAR COVALENT BONDS

So far we have seen that atoms combine in two different ways. Atoms that are quite different in electron structure (from opposite sides of the periodic table) react by the complete transfer of one or more electrons from one atom to another to form an ionic bond. Atoms that are identical combine by sharing a pair of electrons to form a covalent bond. Now let's consider bond formation between atoms that are different, but not different enough to form ionic bonds.

Hydrogen Chloride

Hydrogen and chlorine react to form a colorless gas called hydrogen chloride. This reaction may be represented as

$$\text{H} \cdot \; + \; \cdot \ddot{\underset{..}{\text{Cl}}} \colon \; \longrightarrow \; \text{H} \colon \ddot{\underset{..}{\text{Cl}}} \colon$$

Hydrogen and chlorine both actually consist of diatomic molecules; the reaction is more accurately represented by the scheme

$$H:H \ + \ :\ddot{\underset{..}{Cl}}:\ddot{\underset{..}{Cl}}: \longrightarrow$$
$$2 \ H:\ddot{\underset{..}{Cl}}:$$

We use the individual atoms in order to focus on the sharing of electrons to form a covalent bond.

Ignoring the lone pair electrons and using a dash to represent the covalent bond, we can write the hydrogen chloride molecule as H—Cl. Both hydrogen and chlorine need an electron to achieve a noble gas configuration—a helium configuration for hydrogen, and an argon configuration for chlorine. They achieve these configurations by sharing a pair of electrons to form a covalent bond.

EXAMPLE 5.9 Covalent Bonds from Lewis Structures

Use Lewis structures to show the formation of a covalent bond (a) between two fluorine atoms and (b) between a fluorine atom and a hydrogen atom.

Solution

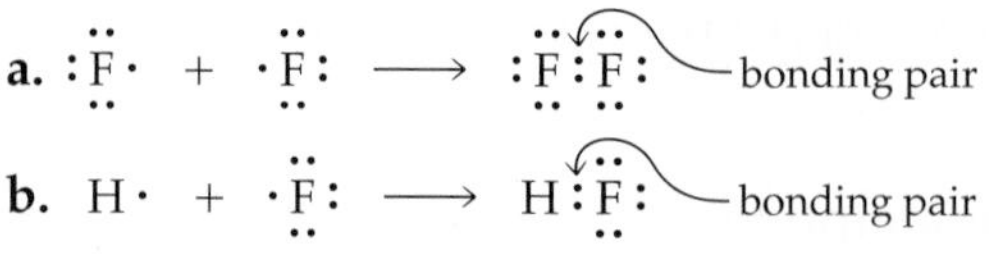

a. $:\ddot{F}\cdot \ + \ \cdot\ddot{F}: \longrightarrow :\ddot{F}\!:\!\ddot{F}:$ ‿ bonding pair

b. $H\cdot \ + \ \cdot\ddot{F}: \longrightarrow H\!:\!\ddot{F}:$ ‿ bonding pair

▶ **Exercise 5.9**

Use Lewis structures to show the formation of a covalent bond between (a) two bromine atoms, (b) between a hydrogen atom and a bromine atom, and (c) between an iodine atom and a chlorine atom.

One might reasonably ask why a hydrogen molecule and a chlorine molecule react at all. Have we not just explained that they themselves were formed to provide more stable arrangements of electrons? Yes, indeed, we did say that. But there is stable, and there is *more* stable. The chlorine molecule represents a more stable arrangement than two separate chlorine atoms. But given the opportunity, a chlorine atom selectively forms a bond with a hydrogen atom rather than with another chlorine atom.

For convenience and simplicity, the reaction of hydrogen (molecule) and chlorine (molecule) to form hydrogen chloride often is represented as

$$H_2 + Cl_2 \longrightarrow 2 \ HCl$$

The bonds between the atoms and the lone pairs on the chlorine atoms are not shown explicitly, but remember that they are there.

Each molecule of hydrogen chloride consists of one atom of hydrogen and one atom of chlorine. These unlike atoms share a pair of electrons. *Share*, however, does not necessarily mean "share equally." Chlorine atoms have a greater attraction for a shared pair of electrons than hydrogen atoms do; chlorine is said to be more *electronegative* than hydrogen.

Electronegativity

The **electronegativity** of an element is a measure of the attraction of an atom *in a molecule* for a pair of shared electrons. The atoms to the right in the periodic table are, in general, more electronegative than those to the left. The ones on the right are precisely the atoms that, in forming ions, tend to gain electrons and form negative ions. The ones on the left—metals—tend to give up electrons and become positive ions. The more electronegative an atom is, the greater its tendency to pull the electrons in the bond toward its end of the bond when it is involved in covalent bonding. Linus Pauling, the great American chemist mentioned in Chapter 4, devised a scale of relative electronegativity values by assigning fluorine, the most electronegative element, a value of 4.0. Figure 5.5 displays the electronegativity values for some of the common elements that we will encounter in this text.

Chlorine (3.0) is more electronegative than hydrogen (2.1). In the hydrogen chloride molecule, the shared electrons are held more tightly by the chlorine atom, and this results in the chlorine end of the molecule being more negative than the hydrogen end. When the electrons in a covalent bond are not equally shared, the bond is said to be *polar*. Thus, the bond in a hydrogen chloride molecule is de-

1A																		Noble gases
H 2.1	2A											3A	4A	5A	6A	7A		
Li 1.0	Be 1.5											B 2.0	C 2.5	N 3.0	O 3.5	F 4.0		
Na 0.9	Mg 1.2	3B	4B	5B	6B	7B		8B		1B	2B	Al 1.5	Si 1.8	P 2.1	S 2.5	Cl 3.0		
K 0.8	Ca 1.0													As 2.0	Se 2.4	Br 2.8		
																I 2.5		

◄ Figure 5.5 Pauling electronegativity values of several common elements.

Question: Can you estimate a value for the electronegativity of germanium (Ge)? For the electronegativity of rubidium (Rb)?

scribed as a **polar covalent bond**, whereas the bond in a hydrogen molecule or a chlorine molecule is a **nonpolar covalent bond**. A polar covalent bond is not an ionic bond. In an ionic bond, one atom completely loses an electron. In a polar covalent bond, the atom at the positive end of the bond (hydrogen in HCl) still has some share in the bonding pair of electrons (Figure 5.6). To distinguish this arrangement from that in an ionic bond, the following notation is used:

$$\overset{\delta+}{H}-\overset{\delta-}{Cl}$$

The line between the atoms represents the covalent bond, a pair of shared electrons. The $\delta+$ and $\delta-$ (read "delta plus" and "delta minus") signify which end is partially positive and which is partially negative. (The word *partially* is used to distinguish this charge from the full charge on an ion.)

We can use the electronegativity values for the atoms in a compound to predict the type of bonding. When the electronegativity difference is zero or very small (<0.5), the bond is *nonpolar covalent*, with essentially equal sharing of the electrons. When the electronegativity difference is large (>2.0), complete electron transfer occurs and an *ionic* bond is formed, as in the case of sodium and chlorine. When the electronegativity difference is between 0.5 and 2.0, *polar covalent* bonds are formed.

$$\overset{\delta+}{H}-\overset{\delta-}{\underset{\cdot\cdot}{Cl}}\!:$$

(a)   (b)

▲ Figure 5.6 Representation of the polar hydrogen chloride molecule. (a) The electron-dot formula, with the shared electron pair shown nearer the chlorine atom. The symbols $\delta+$ and $\delta-$ indicate partial positive and partial negative charges, respectively. (b) An *electrostatic potential* diagram depicting the unequal distribution of electron density in the hydrogen chloride molecule.

EXAMPLE 5.10   Covalent Bonds from Lewis Structures

Use data from Figure 5.5 to classify bonds between each of the following pairs of atoms as nonpolar covalent, polar covalent, or ionic:

a. H, H   **b.** O, H   **c.** C, H

Solution

a. Two H atoms have exactly the same electronegativity; the electronegativity difference is 0; the bond is nonpolar covalent.
b. The electronegativity difference is $3.5 - 2.1 = 1.4$; the bond is polar covalent.
c. The electronegativity difference is $2.5 - 2.1 = 0.4$; the bond is nonpolar covalent.

► Exercise 5.10A
Use data from Figure 5.5 to classify bonds between each of the following pairs of atoms as nonpolar covalent, polar covalent, or ionic:

a. H, Br   **b.** Na, O   **c.** C, C

► Exercise 5.10B
Use a periodic table to classify the following bonds as nonpolar covalent or polar covalent:

a. C—N   **b.** C—O   **c.** C=C

▲ Chlorine hogs the electron blanket, leaving hydrogen partially, but positively, exposed.

5.9 POLYATOMIC MOLECULES: WATER, AMMONIA, AND METHANE

To obtain an octet of electrons, an oxygen atom must share electrons with two hydrogen atoms, a nitrogen atom must share electrons with three hydrogen atoms, and a carbon atom must share electrons with four hydrogen atoms. In general, many nonmetals often form a number of covalent bonds equal to eight minus the group number. Oxygen, which is in group 6A, forms $8 - 6 = 2$ covalent bonds in most molecules. Nitrogen, in group 5A, forms $8 - 5 = 3$ covalent bonds in most molecules. Carbon, in group 4A, forms $8 - 4 = 4$ covalent bonds in most molecules, including the great host of organic compounds (Chapter 9). The following simple guidelines—sometimes called the HONC rules—will enable you to write formulas for many molecules.

- Hydrogen forms 1 bond.
- Oxygen forms 2 bonds.
- Nitrogen forms 3 bonds.
- Carbon forms 4 bonds.

Water

Water is one of the most familiar chemical substances. The electrolysis experiment of Nicholson and Carlisle (Chapter 2) and the fact that both hydrogen and oxygen are diatomic gases indicate that the molecular formula for water is H_2O. In order to be surrounded by an octet, oxygen shares two pairs of electrons. However, because a hydrogen atom shares only one pair of electrons, an oxygen atom must bond with two hydrogen atoms.

$$\cdot \ddot{O}\!: \ + \ 2\,H\cdot \ \longrightarrow \ H\!:\!\ddot{O}\!: \quad \text{or} \quad H\!-\!O$$
$$\qquad\qquad\qquad\qquad\qquad\ \ H \qquad\qquad\qquad |$$
$$\qquad\qquad\qquad\qquad\qquad\qquad\qquad\qquad\qquad H$$

This arrangement completes the valence shell octet in the oxygen atom, giving it the neon structure. It also completes the outer shell of the hydrogen atoms, each of which now has the helium structure. Oxygen has a higher electronegativity than does hydrogen, so the H—O bonds formed are *polar covalent*.

Ammonia

A nitrogen atom has five electrons in its valence shell. It can assume the neon configuration by sharing three pairs of electrons with *three* hydrogen atoms. The result is the compound ammonia.

$$\cdot \ddot{N}\cdot \ + \ 3\,H\cdot \ \longrightarrow \ H\!:\!\ddot{N}\!:\!H \quad \text{or} \quad H\!-\!N\!-\!H$$
$$\qquad\qquad\qquad\qquad\qquad\qquad H \qquad\qquad\qquad |$$
$$\qquad\qquad\qquad\qquad\qquad\qquad\qquad\qquad\qquad\ H$$

In ammonia, the bond arrangement is that of a tripod with a hydrogen atom at the end of each "leg" and the nitrogen atom with its unshared pair of electrons sitting at the top. (We will see why it has this shape in Section 5.13.) The electronegativity of N is 3.0, that of H is 2.1, so all three N—H bonds are *polar covalent*.

Methane

A carbon atom has four electrons in its valence shell. It can assume the neon configuration by sharing pairs of electrons with four hydrogen atoms, forming the compound methane.

$$\qquad\qquad\qquad\qquad\qquad\qquad\qquad\qquad\qquad\qquad H$$
$$\qquad\qquad\qquad\qquad\qquad H \qquad\qquad\qquad\qquad |$$
$$\cdot \dot{C}\cdot \ + \ 4\,H\cdot \ \longrightarrow \ H\!:\!\ddot{C}\!:\!H \quad \text{or} \quad H\!-\!C\!-\!H$$
$$\qquad\qquad\qquad\qquad\qquad H \qquad\qquad\qquad\qquad |$$
$$\qquad\qquad\qquad\qquad\qquad\qquad\qquad\qquad\qquad\qquad H$$

The four bonds in the methane molecule, as written, appear to be planar but actually are not—as we shall see in Section 5.13. The electronegativity difference between H and C is so small that the C—H bonds are considered nonpolar.

5.10 POLYATOMIC IONS

 Naming Polyatomic Ions

Many compounds contain both ionic and covalent bonds. Sodium hydroxide, commonly known as lye, consists of sodium ions (Na^+) and hydroxide ions (OH^-). The hydroxide ion contains an oxygen atom covalently bonded to a hydrogen atom, plus an "extra" electron. That extra electron gives hydroxide ion a negative charge and gives both the O and H atoms filled shells.

$$e^- + \cdot \ddot{O} \cdot + \cdot H \longrightarrow :\ddot{O}:H^-$$

The formula for sodium hydroxide is NaOH; for each sodium ion there is one hydroxide ion.

There are many groups of atoms that (like hydroxide ion) remain together through most chemical reactions. **Polyatomic ions** are charged particles containing two or more covalently bonded atoms. A list of common polyatomic ions is given in Table 5.4. You can use these ions, in combination with the simple ions in Table 5.2, to determine formulas for compounds that contain polyatomic ions.

TABLE 5.4 Some Common Polyatomic Ions

Charge	Name	Formula
1+	Ammonium ion	NH_4^+
	Hydronium ion	H_3O^+
1−	Hydrogen carbonate (bicarbonate) ion	HCO_3^-
	Hydrogen sulfate (bisulfate) ion	HSO_4^-
	Acetate ion	$CH_3CO_2^-$ (or $C_2H_3O_2^-$)
	Nitrite ion	NO_2^-
	Nitrate ion	NO_3^-
	Cyanide ion	CN^-
	Hydroxide ion	OH^-
	Dihydrogen phosphate ion	$H_2PO_4^-$
	Permanganate ion	MnO_4^-
2−	Carbonate ion	CO_3^{2-}
	Sulfate ion	SO_4^{2-}
	Chromate ion	CrO_4^{2-}
	Monohydrogen phosphate ion	HPO_4^{2-}
	Oxalate ion	$C_2O_4^{2-}$
	Dichromate ion	$Cr_2O_7^{2-}$
3−	Phosphate ion	PO_4^{3-}

Acetate ion

Ammonium ion

Hydrogen carbonate ion (bicarbonate ion)

Carbonate ion   Nitrite ion

EXAMPLE 5.11 **Formulas Using Polyatomic Ions**

 Building Blocks for Naming Ionic Compounds

What is the formula for ammonium sulfide?

Solution
Ammonium ion is found in Table 5.4; sulfide ion is a sulfur atom (group 6A) with two additional electrons. The ions are

$$NH_4^+ \quad S^{2-}$$

Crossing over,

we get

$$(NH_4^+)_2S_1^{2-}$$

Dropping the charges gives

$$(NH_4)_2S$$

The parentheses with a subscript 2 indicate that the entire ammonium unit is taken twice; there are two nitrogen atoms and eight ($4 \times 2 = 8$) hydrogen atoms.

▶ **Exercise 5.11A**

What are the formulas for **(a)** calcium acetate, **(b)** ammonium nitrate, and **(c)** potassium permanganate?

▶ **Exercise 5.11B**

How many **(a)** nitrogen atoms are in the formula for ammonium nitrate; and **(b)** carbon atoms are in the formula for calcium acetate?

EXAMPLE 5.12   **Naming Compounds with Polyatomic Ions**

What is the name of the compound NaCN?

Solution

The ions are Na^+, sodium ion, and CN^-, cyanide ion (found in Table 5.4). The compound is sodium cyanide.

▶ **Exercise 5.12A**

What are the names of **(a)** $CaCO_3$ and **(b)** $Mg_3(PO_4)_2$?

▶ **Exercise 5.12B**

What are the names of **(a)** K_2CrO_4 and **(b)** $(NH_4)_2Cr_2O_7$?

5.11 RULES FOR WRITING LEWIS FORMULAS

As we have seen, electrons are transferred or shared in ways that leave most atoms with octets of electrons in their outermost shells. In this section we describe how to write **Lewis formulas** for molecules. To write a Lewis formula, we first put the atoms of the molecule in their proper places, and then we place all the electrons of the molecule so that each atom has a filled shell.

The *skeletal structure* of a molecule tells us the order in which the atoms are attached to one another. Drawing a skeletal structure takes some practice. However, in the absence of experimental evidence, the following guidelines help us to devise likely skeletal structures:

- Hydrogen atoms form only single bonds; they are always at the end of a sequence of atoms. Hydrogen is often bonded to carbon, nitrogen, or oxygen.
- Oxygen tends to have two bonds, nitrogen usually has three bonds, and carbon has four bonds.
- Polyatomic molecules and ions often consist of a central atom surrounded by more electronegative atoms. (Hydrogen is an exception; it is always on the outside, even when bonded to a more electronegative element.) The central atom of a polyatomic molecule is often the *least* electronegative atom.

After choosing a skeletal structure for a polyatomic molecule or ion, we can use the following steps to write a Lewis formula:

1. Determine the total number of valence electrons. This total is the sum of the valence electrons for all the atoms in the molecule. For a polyatomic anion, also add the number of negative charges. For a polyatomic cation, subtract the number of positive charges.

Examples:

N_2O_4 has $(2 \times 5) + (4 \times 6) = 34$ valence electrons.

NO_3^- has $[(1 \times 5) + (3 \times 6)] + 1 = 24$ valence electrons.

NH_4^+ has $[(1 \times 5) + (4 \times 1)] - 1 = 8$ valence electrons.

2. Write a reasonable skeletal structure and connect bonded pairs of atoms by a dash (one electron pair).

3. Place electrons in pairs around outer atoms so that each (except hydrogen) has an octet.

4. Subtract the number of electrons assigned so far (both in bonds and as lone pairs) from the total calculated in step 1. Any electrons that remain are assigned in pairs to the central atom(s).

5. If a central atom has fewer than eight electrons after step 4, one or more multiple bonds are likely. Move one or more lone pairs from an outer atom to the space between the atoms to form a double or triple bond. A deficiency of two electrons suggests a double bond, and a shortage of four electrons indicates a triple bond or two double bonds to the central atom.

EXAMPLE 5.13   **Lewis Formulas**

Give Lewis formulas for **(a)** methanol, CH_3OH, **(b)** the BF_4^- ion, and **(c)** carbon dioxide, CO_2.

Solution

a. We start by following the preceding rules:

   1. The total number of valence electrons is $4 + (4 \times 1) + 6 = 14$.

   2. The skeletal structure must have all the H atoms on the outside. That means the C and H atoms must be bonded to each other. A reasonable skeletal structure is

$$
\begin{array}{c}
\text{H} \\
| \\
\text{H} - \text{C} - \text{O} - \text{H} \\
| \\
\text{H}
\end{array}
$$

   3. Now, we count five bonds with two electrons each, making a total of ten electrons. Thus four of the 14 valence electrons are left to be assigned. They are placed (as two lone pairs) on the oxygen atom.

$$
\begin{array}{c}
\text{H} \\
| \\
\text{H} - \text{C} - \ddot{\text{O}} - \text{H} \\
| \\
\text{H}
\end{array}
$$

   (The remaining steps are not necessary; both carbon and oxygen have octets of electrons.)

b. Again, we start by applying the preceding rules:

   1. There are $3 + (4 \times 7) + 1 = 32$ electrons.

   2. The skeletal structure is

$$
\begin{array}{c}
\text{F} \\
| \\
\text{F} - \text{B} - \text{F} \\
| \\
\text{F}
\end{array}
$$

   3. Place three lone pairs on each fluorine atom.

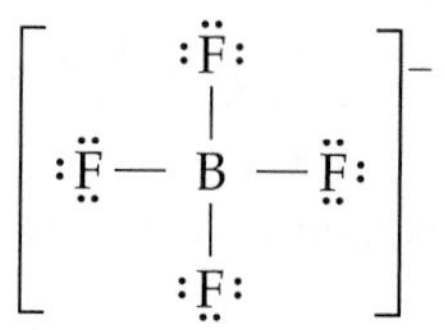

$$\left[\; :\!\ddot{F}\!-\!B\!-\!\ddot{F}\!: \;\right]^{-}$$

with $:\!\ddot{F}\!:$ above and $:\!\ddot{F}\!:$ below the boron

   4. We have assigned 32 electrons. None remain to be assigned. A negative sign is added to show that this is the structure for an anion, not a molecule.

c. Again, we start by applying the rules:

   1. There are $4 + (2 \times 6) = 16$ valence electrons.

   2. The skeletal structure is O—C—O.

   3. Place three lone pairs on each oxygen atom.

$$:\!\ddot{O}\!-\!C\!-\!\ddot{O}\!:$$

   4. We have assigned 16 electrons. None remain to be placed.

   5. The central carbon atom has only four electrons. It needs to form two double bonds in order to have an octet. Move a lone pair from each oxygen atom to the space between the atoms to form a double bond on each side of the carbon atom.

$$:\!\ddot{O}\!=\!C\!=\!\ddot{O}\!:$$

▶ **Exercise 5.13A**
Give Lewis formulas for **(a)** oxygen difluoride, OF_2, and **(b)** methyl chloride, CH_3Cl.

▶ **Exercise 5.13B**
Give Lewis formulas for **(a)** azide ion, N_3^-, and **(b)** nitryl fluoride, NO_2F (O—N—O—F skeleton).

   The rules we have used here lead to results that are summarized and illustrated for selected elements in Table 5.5. Figure 5.7 relates the number of covalent bonds that a particular element forms to its position on the periodic table.

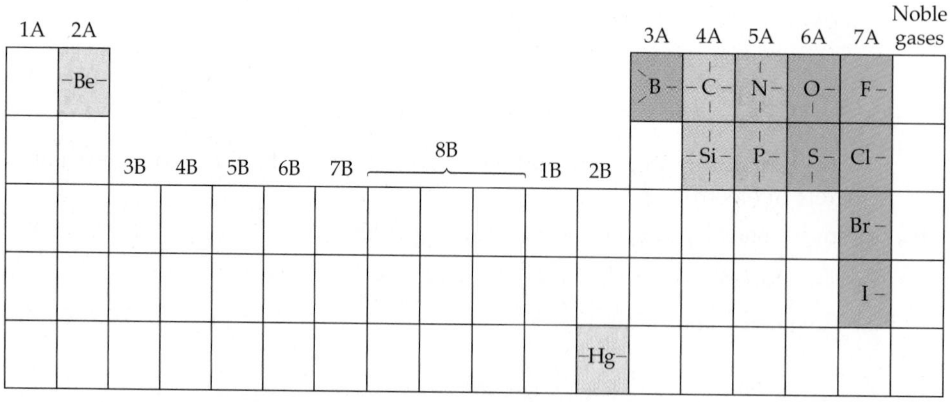

▲ **Figure 5.7** Covalent bonding of representative elements of the periodic table.

Question: What is the relationship between an atom's position on the periodic table and the number of covalent bonds it tends to form? Why are the group 1A and most of the group 2A elements omitted from the table?

TABLE 5.5 Number of Bonds Formed by Selected Elements

Electron-Dot Symbol	Bond Picture	Number of Bonds	Representative Molecules	Ball-and-Stick Models
H·	H—	1	H—H H—Cl	HCl
He:		0	He	He
·C̈·	$-\overset{\displaystyle |}{\underset{\displaystyle |}{C}}-$	4	$H-\overset{\displaystyle H}{\underset{\displaystyle H}{C}}-H$ $H-\overset{\displaystyle O}{C}-F$	CH$_4$
·N̈·	$-\overset{\displaystyle |}{N}-$	3	$H-\overset{\displaystyle |}{\underset{\displaystyle H}{N}}-H$ $H-\overset{\displaystyle N-O-H}{C}-H$	NH$_3$
·Ö:	$-O-$	2	$H-\overset{\displaystyle |}{\underset{\displaystyle H}{O}}$ $H-\overset{\displaystyle O}{C}-H$	H$_2$O
·F̈:	—F	1	H—F F—F	F$_2$
·C̈l:	—Cl	1	Cl—Cl $H-\overset{\displaystyle H}{\underset{\displaystyle H}{C}}-Cl$	CH$_3$Cl

5.12 ODD-ELECTRON MOLECULES: FREE RADICALS

Molecules with odd numbers of valence electrons obviously cannot satisfy the octet rule. Examples of such molecules are nitrogen monoxide (NO, also called nitric oxide), with $5 + 6 = 11$ valence electrons; nitrogen dioxide (NO$_2$), with 17 valence electrons; and chlorine dioxide (ClO$_2$), which has 19 outer electrons. Obviously, one of the atoms in each of these molecules has an odd number of electrons and therefore cannot have an octet.

An atom or molecule with unpaired electrons is called a **free radical**. Most free radicals are highly reactive and have only transitory existence as intermediates in chemical reactions. Every atom or molecule with an odd number of electrons must have one unpaired electron. Filled shells and subshells have all their electrons paired, with two electrons in each orbital (Section 3.8). Therefore, we need only consider valence electrons to determine whether or not an atom or molecule is a free radical. Lewis structures of NO, NO$_2$, and ClO$_2$ show one atom of each with an unpaired electron; this atom obviously does not have an octet of electrons in its outer shell.

$$:\dot{N}::\ddot{O}: \quad :\ddot{O}:\dot{N}::\ddot{O}: \quad :\ddot{O}:\dot{C}l:\ddot{O}:$$

Nitrogen oxides are major components of smog (Chapter 12). Chlorine atoms from the breakdown of chlorofluorocarbons in the stratosphere, lead to depletion of the ozone shield (Chapter 12). Some free radicals are quite stable, however, and have important functions in the body as well as in industrial processes (see Useful Applications of Free Radicals on page 136).

Useful Applications of Free Radicals

The very reactive nature of free radicals does not preclude their importance and use in a variety of natural and industrial applications. Nitric oxide (NO) has been shown to be of tremendous importance in helping cells to communicate. That work was awarded the Nobel Prize in physiology and medicine in 1998. Interestingly, one of the physiological effects of Viagra is the production of small quantities of NO in the bloodstream. Hydroxyl radicals (·OH) have been implicated in the formation of cancerous cells due to their rapid reaction with DNA. Many plastics, including polyethylene and polyvinylchloride (PVC), are made using free radicals to initiate the reactions. Liquid polyester resin is used to repair automobiles and to construct small boats and surfboards. The resin is converted to hard plastic by the addition of a free-radical catalyst called methyl ethyl ketone peroxide. Chlorine dioxide is widely used to bleach paper and other products. Another important application was its use to decontaminate the Senate office building after anthrax bacteria were discovered there in 2001 (right).

 VSEPR Movie

5.13 MOLECULAR SHAPES: THE VSEPR THEORY

We have represented molecules in two dimensions on paper, but molecules have three-dimensional shapes. These shapes are important in determining the properties of molecular substances. For example, the shapes of the molecules that make up gasoline determine their octane rating (Chapter 14), and drug molecules must have the right atoms in the right places to be effective (Chapter 19). We can use Lewis structures as part of the process of predicting molecular shapes. Figure 5.8 shows the shapes that we consider in this book.

We use the **valence shell electron pair repulsion (VSEPR) theory** to predict the arrangement of atoms about a central atom. The basis of the VSEPR theory is that electron pairs arrange themselves about a central atom in a way that minimizes repulsion between like-charged particles. This means that they will get as far apart as possible. Table 5.6 gives the geometric shapes associated with the arrangement of two, three, or four entities about a central atom.

- When *two* substituent electron pairs are as far apart as possible, they are on opposite sides of the central atom at an angle of 180°.

- *Three* groups assume a triangular arrangement about the central atom, forming angles of separation of 120°.

- *Four* groups form a tetrahedral array around the central atom, giving a separation of about 109.5°.

▶ **Figure 5.8** Shapes of molecules. In a *linear* molecule (a), all the atoms are along a line; the bond angle is 180°. A *bent* molecule (b) has an angle less than 180°. Connecting the three outer atoms of a *triangular* molecule (c) with imaginary lines produces a triangle with an atom at the center. Imaginary lines connecting all four atoms of a *pyramidal* molecule (d) form a three-sided pyramid. Connecting the four outer atoms of a *tetrahedral* molecule (e) with imaginary lines produces a tetrahedron (a four-sided figure in which each side is a triangle) with an atom at the center.

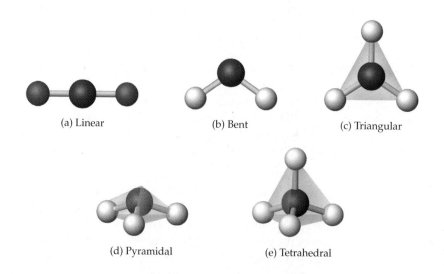

(a) Linear   (b) Bent   (c) Triangular

(d) Pyramidal   (e) Tetrahedral

TABLE 5.6   Bonding and the Shape of Molecules

Number of Bonded Atoms	Number of LP*	Number of Sets	Molecular Shape	Examples	Ball-and-Stick Models
2	0	2	Linear	$BeCl_2$ $HgCl_2$ CO_2 HCN	BeCl₂
3	0	3	Triangular	BF_3 $AlBr_3$ CH_2O	BF₃
4	0	4	Tetrahedral	CH_4 CBr_4 $SiCl_4$	CH₄
3	1	4	Pyramidal	NH_3 PCl_3	NH₃
2	2	4	Bent	H_2O H_2S SCl_2	H₂O
2	1	3	Bent	SO_2 O_3	SO₂

*LP: Lone pair(s) of electrons.

We can determine the shapes of many molecules (and polyatomic ions) by following this simple procedure:

1. Draw a Lewis structure in which a shared electron pair (bonding pair, BP) is indicated by a line. Use dots to show any unshared pairs (lone pairs, LPs) of electrons.
2. To determine shape, count the number of atoms *and LPs* attached to the *central* atom. Note that a multiple bond counts only as one *set* of electrons. Examples are

Four sets (2 atoms, 2 LPs) Four sets (3 atoms, 1 LP) Four sets (4 atoms)

Two sets (2 atoms) Two sets (2 atoms) Three sets (3 atoms) Three sets (2 atoms, 1 LP)

3. Determine the number of electron sets and draw a shape *as if* all were bonding pairs.

4. Sketch this shape, placing the electron pairs as far apart as possible (Table 5.6). If there is *no* LP, this is the shape of the molecule. If there *are* LPs, remove them, leaving the BPs exactly as they were. (This may seem strange, but it stems from the fact that *all* the sets determine the geometry, but only the arrangement of bonded atoms is considered in the shape of the molecule.)

EXAMPLE 5.14 **Shapes of Molecules**

What are the shapes of **(a)** the H_2CO molecule and **(b)** SCl_2 molecule?

Solution

a. We follow the preceding rules, starting with the Lewis structure.

1. The Lewis structure is

$$
\begin{array}{c}
H \\
| \\
C = \ddot{O} \\
| \\
H
\end{array}
$$

2. There are three sets to consider: two C—H single bonds and one C=O double bond.

3. The three sets get as far apart as possible, giving a triangular arrangement of the sets.

$$
\begin{array}{c}
H \diagdown \;\; 120° \\
\;\;\;\;\; C = \ddot{O}: \\
H \diagup
\end{array}
$$

4. All the sets are bonding pairs; the molecular shape is triangular, the same as the arrangement of the electrons.

b. Again, we follow the rules, starting with the Lewis structure.

1. The Lewis structure is

$$
\begin{array}{c}
:\ddot{S} — \ddot{C}l: \\
| \\
:\ddot{C}l:
\end{array}
$$

2. There are four sets on the sulfur atom to consider.

3. The four sets get as far apart as possible, giving a tetrahedral arrangement of the sets about the central atom.

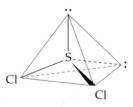

4. Two of the sets are bonding pairs and two are LP. Ignore the LP; the molecular shape is *bent*, with a bond angle of about 109.5°.

▶ **Exercise 5.14A**

What are the shapes of **(a)** the PH_3 molecule and **(b)** the nitrate ion (NO_3^-)?

▶ **Exercise 5.14B**

What are the shapes of **(a)** the carbonate ion (CO_3^{2-}) and **(b)** the hydrogen peroxide (H_2O_2) molecule?

5.14 SHAPES AND PROPERTIES: POLAR AND NONPOLAR MOLECULES

Polarity Activity

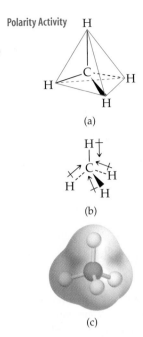

In Section 5.8 we discussed polar and nonpolar bonds. A diatomic molecule is non-polar if its bond is nonpolar, as in H_2 or Cl_2, and polar if its bond is polar, as in HCl. In addition to the symbolism in Section 5.8, where the partial charges in a polar bond are indicated by $\delta+$ and $\delta-$, we use an arrow with a plus sign at the tail end ($+\!\!\longrightarrow$) to indicate a **dipole**, that is, a molecule with a positive end and a negative end. The plus sign indicates the part of the molecule with a partial positive charge, and the head of the arrow signifies the end of the molecule with a partial negative charge.

$$\begin{array}{ccc} \text{H---H and Cl---Cl} & \overset{\delta^+}{\text{H}}\!-\!\overset{\delta^-}{\text{Cl}} & \text{or} \quad \overset{\longrightarrow}{\text{H---Cl}} \\ \text{Nonpolar} & \text{Polar} \end{array}$$

For molecules with three or more atoms, we must consider the polarity of the individual bonds as well as the molecular geometry of the molecule to determine whether the molecule as a whole is polar. A **polar molecule** has separate centers of positive and negative charge, just as a magnet has north and south poles. Many properties of compounds—such as melting point, boiling point, and solubility—depend on the polarity of their molecules.

Methane: A Tetrahedral Molecule

There are four pairs of electrons on the central carbon atom in methane. Using the VSEPR theory, we would expect a tetrahedral arrangement and bond angles of 109.5° (Figure 5.9a). The actual bond angles are 109.5°, as predicted by the theory. All four electron pairs are shared with hydrogen atoms, and therefore all four pairs occupy identical volumes. Each carbon-to-hydrogen bond is slightly polar (Figure 5.9b), but the methane molecule as a whole is symmetric. The slight bond polarities cancel out, leaving the methane molecule, as a whole, nonpolar (Figure 5.9c).

▲ **Figure 5.9** The methane molecule. In (a) black lines indicate covalent bonds; the red lines outline a tetrahedron; all bond angles are 109.5°. The slightly polar C—H bonds cancel each other (b), resulting in a nonpolar, tetrahedral molecule (c).

Question: How does the electrostatic potential diagram in (c) look different from the one in Figure 5.6?

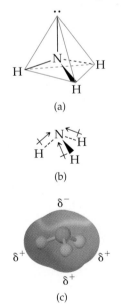

Ammonia: A Pyramidal Molecule

With ammonia, NH_3, several things change because we have only three bonds and a lone pair. The N—H bonds of ammonia are more polar than the C—H bonds of methane. More important, the molecule has a different geometry as a result of three BPs and one LP about the nitrogen atom. The VSEPR theory predicts a tetrahedral arrangement of the four sets of electrons, giving bond angles of 109.5° (Figure 5.10a). The lone pair of electrons occupies a greater volume than a bonding pair, pushing the BPs slightly closer together. The actual bond angles are slightly less, about 107°. The pyramidal geometry can be envisioned as a tripod with a hydrogen atom at the end of each leg and the nitrogen atom with its lone pair sitting at the top. Each nitrogen-to-hydrogen bond is somewhat polar (Figure 5.10b). The asymmetric structure makes the ammonia molecule polar, with a partial negative charge on the nitrogen atom and partial positive charges on the three hydrogen atoms (Figure 5.10c).

▲ **Figure 5.10** The ammonia molecule. In (a), black lines indicate covalent bonds; the red lines outline a tetrahedron. The polar N—H bonds do not cancel each other completely (b) resulting in a polar, pyramidal molecule (c).

Question: How does the electrostatic potential diagram in (c) look similar to the one in Figure 5.6?

Water: A Bent Molecule

The O—H bonds in water are even more polar than the N—H bonds in ammonia because oxygen is more electronegative than nitrogen. (Electronegativity increases from left to right in the periodic table.) Just because a molecule contains polar bonds, however, does not mean that the molecule as a whole is polar. If the atoms in the water molecule were in a straight line (that is, in a linear arrangement as in CO_2), the two polar bonds would cancel one another out and the molecule would be nonpolar.

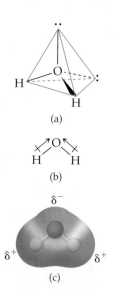

▲ **Figure 5.11** The water molecule. In (a), black lines indicate covalent bonds; the red lines outline a tetrahedron. The polar O—H bonds do not cancel each other (b), resulting in a polar, bent molecule (c).

The reverse of sublimation is *deposition*, a process in which a substance goes directly from the gaseous to the solid state.

In its physical and chemical properties, however, water acts like a polar molecule. Molecules such as water and ammonia, in which the polar bonds do not cancel out, act as dipoles; they have a positive end and negative end.

We can understand the dipole in the water molecule by using VSEPR theory. The two bonds and two LP should form a tetrahedral arrangement (Figure 5.11a). Ignoring the LP, the molecular shape has the atoms in a bent arrangement, with a bond angle of about 104.5°. As with ammonia, we explain the difference between the actual angle and the predicted angle by the fact that the two LPs occupy a greater volume than the bonding pairs. These larger sets push the smaller BPs closer together. The two polar O—H bonds do not cancel each other (Figure 5.11b), leaving a polar, bent water molecule (Figure 5.11c).

5.15 INTERMOLECULAR FORCES AND THE STATES OF MATTER

The shape, size, and polarity of molecules determine how they interact with each other. The physical state of a material (whether it is a gas, liquid, or solid) depends on the strength of the intermolecular forces that hold the molecules or ions together, relative to the thermal energy (temperature) that acts to separate them.

Solids, Liquids, and Gases

Solids are highly ordered assemblies of particles—atoms, molecules, or ions—in close contact with one another. Their motion is primarily vibration within a solid lattice. Table salt (sodium chloride) is a typical crystalline solid. In this solid, ionic bonds hold the Na^+ and Cl^- ions in position and maintain the orderly arrangement (recall Figure 5.2). In liquids, the particles are still in close contact, but they are more loosely organized and are freer to move about. This is why liquids are able to flow to conform to the shape of their container. In gases the molecules are separated by relatively great distances and are moving quite rapidly in random directions. In this state the atoms or molecules have essentially no interactions with one another. We shall look more closely at gases in the next chapter.

To get a better image of a liquid at the molecular level, think of a box of marbles being shaken continuously. The marbles move back and forth, rolling over one another. The particles of a liquid (like the marbles) are not so rigidly held in place as are particles in a solid. Even so, there must be some force attracting the particles in a liquid to one another, because those particles are in contact with one another.

Most solids can be changed to liquids; that is, they can be *melted*. The solid is heated, and the heat energy is absorbed by the particles of the solid. The energy causes the particles to vibrate in place with more and more vigor until, finally, the forces holding the particles in a particular arrangement are overcome. The solid becomes a liquid. The temperature at which this happens is the **melting point** of the solid. A high melting point is one indication that the forces holding a solid together are very strong.

A liquid can change to a gas or vapor in a process called **vaporization**. Again, one need only supply sufficient heat to achieve this change. Energy is absorbed by the liquid particles, which move faster and faster as a result. Finally, this increasingly violent motion overcomes the attractive forces holding the liquid particles in contact and the particles fly away from one another. The liquid becomes a gas. The temperature at which this happens is the **boiling point** of the liquid.

Removing energy from the sample and slowing down the particles can reverse the entire sequence of changes. Vapor changes to liquid in a process called **condensation**; liquid changes to solid in a process called *freezing*. Figure 5.12 presents a diagram of the changes in state that occur as energy is added to or removed from a sample. Some substances go directly from the solid state to the gaseous state, a process called **sublimation**. For example, ice cubes left in a freezer for a long time "shrink," and frost on the grass will disappear even if the temperature remains below zero. In each case, the ice sublimes.

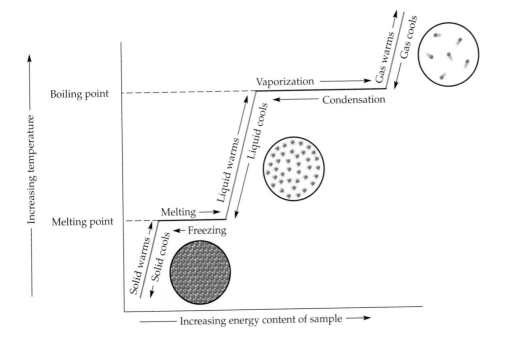

◀ **Figure 5.12** Diagram of changes in state of matter on heating or cooling.

Question: What factors determine the actual temperatures at which the changes of state occur?

Intermolecular Forces

Ionic Bonds

The quantity of energy required to accomplish a change of state depends on the strength of the forces responsible for maintaining the solid or the liquid state. The *ionic bonds* in salt crystals are very strong. Sodium chloride must be heated to about 800 °C before it melts. Generally, ionic bonds are the strongest of all the forces that hold solids and liquids together. We will now consider some other interactions that hold the particles of solids and liquids together.

Dipole Forces

Not all solids are held together by ionic bonds. Hydrogen chloride melts at −112 °C and boils at −85 °C (it is a gas at room temperature). The attractive forces between molecules are not nearly as strong as the interionic forces in NaCl crystals. We know that covalent bonds hold the hydrogen and chlorine *atoms* together to form the hydrogen chloride *molecule*, but what makes one molecule interact with another in the solid or liquid state?

Remember that the hydrogen chloride molecule is a *dipole*: It has a positive end and a negative end. Two dipoles brought close enough together attract one another. In solid HCl, the molecules line up so that the positive end of one molecule attracts the negative end of neighboring molecules. A crystalline lattice of dipoles (like HCl) might look something like Figure 5.13(a). When enough thermal energy is added, the orderly arrangement is broken and the solid melts. In the liquid, the oppositely

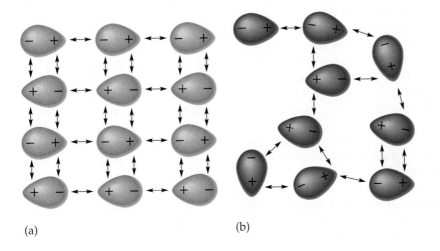

(a)   (b)

◀ **Figure 5.13** An idealized representation of dipole forces in (a) a solid and (b) a liquid. In a real liquid or solid, interactions are more complex.

charged dipoles in the liquid still attract one another, but in a more random fashion, as shown in Figure 15.3(b). In general, the **dipole forces** are much weaker than attractive forces between ions, but they are stronger than the forces between nonpolar molecules of comparable size.

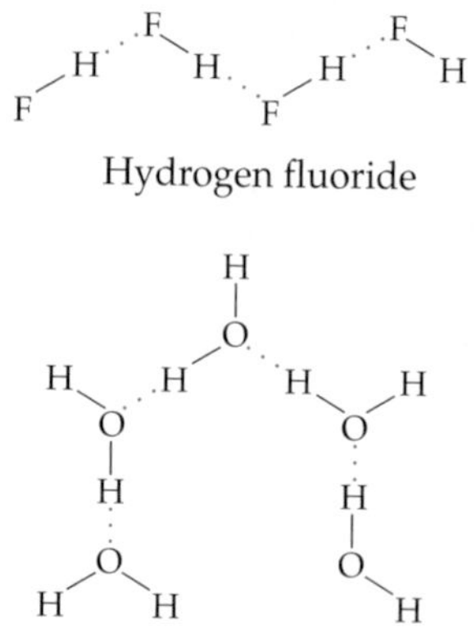

Hydrogen Bonds

Certain polar molecules exhibit stronger attractive forces than expected on the basis of ordinary dipolar interactions. These forces are strong enough to be given a special name, *hydrogen bonds*. The name is a bit misleading because it emphasizes only one component of the interaction. Not all compounds containing hydrogen exhibit this strong attractive force; in most cases, the hydrogen *must* be attached to a small, very electronegative atom such as fluorine, oxygen, or nitrogen. These atoms permit us to offer an explanation for the extra strength of hydrogen bonds as compared with other dipolar forces. Fluorine, oxygen, and nitrogen are all highly electronegative, and they are small (they are at the top of the periodic table). A hydrogen–fluorine bond, for example, is strongly polarized, with a negative fluorine end and a positive hydrogen end. Both hydrogen and fluorine are small atoms, and so the negative end of one dipole can approach very closely the positive end of a second dipole. This results in an unusually strong interaction between the hydrogen atom of one molecule and a lone pair of electrons on another. This special force between two molecules is called a **hydrogen bond**. Hydrogen bonds are often explicitly represented by *dotted* lines to emphasize their exceptional strength compared to ordinary dipolar interactions. A dotted line is used to distinguish a hydrogen bond from the much stronger covalent bond, which is represented by a *solid* line (Figure 5.14).

Water has both an unusually high melting point and an unusually high boiling point for a compound with such small molecules. These abnormal values are attributed to water's ability to form hydrogen bonds. Remember however, that during melting and boiling, only hydrogen bonds *between molecules* are overcome. The covalent bonds between the hydrogen atoms and oxygen atoms in each water molecule remain intact; there is no chemical change. In general, covalent bonds are about 40 or so times stronger than hydrogen bonds.

The unique properties of water resulting from hydrogen bonding are discussed in Chapter 13. Hydrogen bonding also plays an important role in biological molecules. The three-dimensional structures of proteins and enzymes (Chapter 15) and the genetic code embedded within each molecule of DNA are dependent on the presence and strength of hydrogen bonds.

Dispersion Forces

If one understands that positive attracts negative, it is easy enough to understand how ions or polar molecules maintain contact with one another. But how can we explain the fact that *nonpolar* compounds can exist in the liquid and solid states? Even hydrogen can exist as a liquid or a solid if the temperature is low enough (its melting point is $-259\ ^\circ C$). Some force must be holding these molecules in contact with one another in the liquid and solid states.

Up to this point we have pictured the electrons in a covalent bond as being held in place between the two atoms sharing the bond. But the electrons are *not* really static; they actually move about in the bonds. On average, the two electrons in the hydrogen molecule (or any nonpolar bond) are between and equidistant from the two nuclei. At any given instant, however, the electrons may be at one end of the molecule. At some other time, a moment later, the electrons may be at the other end of the molecule. At the instant the electrons of one molecule are at one end, the electrons in the next molecule move away from its adjacent end. Thus, at this instant there is an attractive force between the electron-rich end of one molecule and the electron-poor end of the next. These momentary, usually weak, attractive forces between molecules are called **dispersion forces** (Figure 5.15). To a large extent, dispersion forces determine the physical properties of nonpolar compounds.

▲ **Figure 5.14** Hydrogen bonding in hydrogen fluoride and in water.

Dispersion forces are weak for small, nonpolar molecules such as H_2, N_2, and CH_4. For comparable molecular sizes and shapes, intermolecular forces increase in the order dispersion forces (weakest), dipolar interactions, and hydrogen bonding (strongest).

▲ **Figure 5.15** How dispersion forces arise. In (a) the two nonpolar molecules have their electrons evenly distributed. In (b), the electrons in the lefthand molecule are, for a moment, concentrated on the left. The partial positive charge also formed attracts the electrons in the righthand molecule. This causes the situation in (c), where each molecule has a small, momentary dipole and the opposite charges attract.

Forces in Solutions

To complete our look at chemical bonding, we shall briefly examine the interactions that occur in solutions. A **solution** is an intimate, homogeneous mixture of two or more substances. By *intimate* we mean that mixing occurs down to the level of individual ions and molecules. In a solution of salt in water, for example, there are no clumps of ions floating around, but single ions randomly distributed among the water molecules. *Homogeneous* means that the mixing is thorough. All parts of the solution have the same distribution of components. For a salt solution, the saltiness is the same at the top, bottom, and middle of the solution. The substance being dissolved, and usually present in a lesser amount (salt in a salt solution), is called the **solute.** The substance doing the dissolving, and usually present in a greater amount, is the **solvent** (water in the salt–water solution) (Figure 5.16).

Ordinarily solutions form most readily when the substances involved have *similar* strengths of forces between their molecules (or atoms or ions). An old chemical adage is, "Like dissolves like." Nonpolar solutes dissolve best in nonpolar solvents. For example, oil and gasoline, both nonpolar, mix (Figure 5.17a), but oil and water do not (Figure 5.17b). Ethyl alcohol is a liquid held together by hydrogen bonds. So is water. Ethyl alcohol readily dissolves in water because the two substances can hydrogen bond with one another (Figure 5.17c). In general, a solute dissolves when attractive forces between it and the solvent overcome the attractive forces operating in the pure solute and in the pure solvent.

Why, then, does salt dissolve in water? Ionic solids are held together by strong ionic bonds. We have already indicated that high temperatures are required to melt ionic solids and break these bonds. Yet if we simply place sodium chloride in water at room temperature, the salt dissolves. And when such a solid dissolves, its bonds *are* broken. The difference between the two processes is the difference between brute force and persuasion. In the melting process, we simply put in enough energy

● = Solvent molecule
● = Solute molecule

▲ **Figure 5.16** In a solution, solute molecules (orange spheres) are randomly distributed among solvent molecules (purple spheres).

 Forces Between Molecules and Ions

◄ **Figure 5.17** (a) Lawn mowers with two-cycle engines are fueled and lubricated with a solution of nonpolar lubricating oil in nonpolar gasoline. (b) In Italian dressing, polar vinegar and nonpolar olive oil are mixed. However, the two liquids do not form a solution and separate on standing. (c) Wine is a solution of polar ethyl alcohol in polar water.

Question: What are some other examples of nonpolar–polar mixtures?

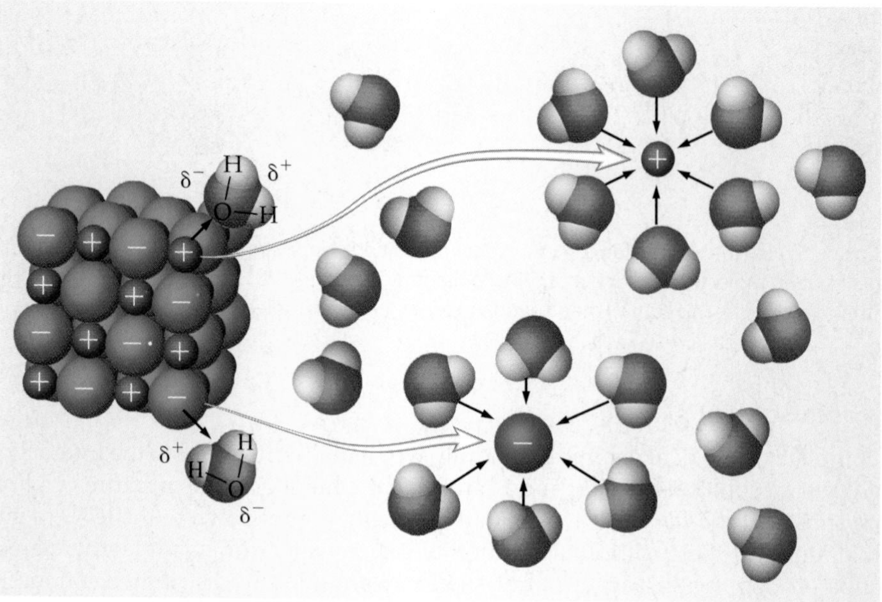

▶ **Figure 5.18** An ionic solid (sodium chloride) dissolving in a polar solvent (water). The hydrogen ends of the water molecules surround the negatively charged chloride ions, while the oxygen ends of the solvent water molecules surround the positively charged sodium ions.

(as heat) to break the crystal apart. In the dissolving process, we offer the ions an attractive alternative to the ionic interactions in the crystal.

It works this way. Water molecules surround the crystal. Those that approach a negative ion align themselves so that the positive ends of their dipoles point toward the ion. With a positive ion, the process is reversed, and the negative end of the water dipole points toward the ion. Still, the attraction between a dipole and an ion is not as strong as that between two ions. To compensate for their weaker attractive power, several molecules surround each ion, and in this way the many *ion–dipole* interactions overcome the *ion–ion* interactions (Figure 5.18).

In an ionic solid, the positive and negative ions are strongly bonded together in an orderly crystalline arrangement. In solution, cations and anions move about more or less independently, each surrounded by a cage of solvent molecules. Water—including the water in our bodies—is an excellent solvent for many ionic compounds. Water also dissolves many molecules that are polar covalent like itself. These solubility principles explain how nutrients reach the cells of our bodies (dissolved in blood, which is mostly water) and how many kinds of pollutants get into our water supplies.

5.16 A CHEMICAL VOCABULARY

Learning chemical symbolism is much like learning a foreign language. Once you have learned a basic "vocabulary," the rest is a lot easier. Initially, the task is complicated by different chemical species or definitions sounding a lot alike (sodium atoms versus sodium ions). Another complication is that we have several different representations of the *same* chemical species (Figure 5.19).

In Chapter 1, we introduced symbols for the chemical elements. Chapter 4 introduced the structure of the nucleus and symbolism to distinguish different isotopes from one another. In the first part of this chapter, we introduced chemical

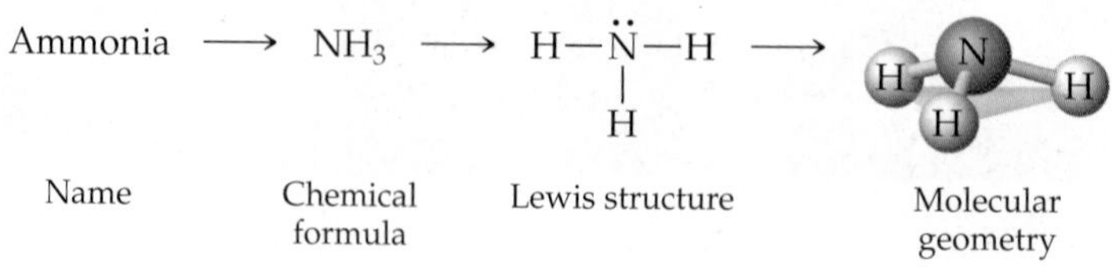

▶ **Figure 5.19** Several representations of the ammonia molecule.

names and chemical formulas. Now you also know how to write Lewis symbols using dots to represent valence electrons, and you can predict the shapes of thousands of different molecules with the VSEPR theory. In the following chapter, you will add to your toolbox the mathematics of chemistry.

Critical Thinking Exercises

Apply knowledge that you have gained in this chapter and one or more of the FLaReS principles (Chapter 1) to evaluate the following statements or claims.

5.1 Some people believe that crystals have special powers. Crystal therapists claim that they can use quartz crystals to restore balance and harmony to a person's spiritual energy. Do you think their claim is credible?

5.2 A "fuel-enhancer" device is being sold by an entrepreneur. The device contains a powerful magnet and is placed on the fuel line of an automobile. The inventor claims that the device "separates the positive and negative charges in the hydrocarbon fuel molecules, increasing their polarity and al-lowing them to react more readily with oxygen." Evaluate this claim for credibility.

5.3 Sodium chloride, NaCl, is a metal–nonmetal compound held together by ionic bonds. A scientist has studied mercury(II) chloride, $HgCl_2$, and says that the atoms are held together by covalent bonds. The scientist bases this contention largely on the fact that a water solution of the substance does not conduct an electric current, indicating that it does not contain an appreciable amount of ions. In your opinion, is the scientist correct?

5.4 Another scientist, noting that the noble gas xenon does not contain any ions, states that xenon atoms must be held together by covalent bonds. Is this statement plausible?

SUMMARY

Sections 5.1–5.2—Outermost shell electrons are called valence electrons; those in all other shells are core electrons. The noble gases are inert because they have an octet of valence electrons. Other elements become less reactive by gaining or losing electrons to attain an electron configuration that is isoelectronic with, or has the same electron configuration as, a noble gas. The number of valence electrons for most main group elements is the same as the group number. Electron-dot symbols or Lewis symbols use dots singly or in pairs to represent valence electrons of an atom or ion.

Sections 5.3–5.4—The element sodium reacts violently with elemental chlorine to give sodium chloride or table salt. In this reaction, an electron is transferred from a sodium atom to a chlorine atom.

$$\text{Na} \cdot + \cdot \ddot{\underset{..}{Cl}} \cdot \longrightarrow \text{Na}^+ + \ddot{\underset{..}{Cl}} \colon^-$$

The ions formed have opposite charges and are strongly attracted to one another; this attraction is called an ionic bond. The positive and negative ions arrange themselves in a regular array, forming a crystal of sodium chloride. The ions of an element have very different properties from the atoms of that element.

Section 5.5—Metal atoms tend to give up their valence electrons to become ions, while nonmetal atoms tend to accept (8 − group number) electrons to become ions.

$$\begin{array}{l} \text{K} \cdot \\ \text{K} \cdot \end{array} + \cdot \ddot{\underset{..}{O}} \colon \longrightarrow \begin{array}{l} \text{K}^+ \\ \text{K}^+ \end{array} + \ddot{\underset{..}{O}} \colon^{2-}$$

In either case the ions formed (except hydrogen ion) tend to have eight valence electrons; the ions are said to follow the octet rule. The symbol for an ion is given as the element symbol with a superscript to indicate the number and type (+ or −) of charge. Some elements, especially the transition metals, can form ions of different charges. The formula of an ionic compound always has the same number of positive charges as negative charges.

Section 5.6—A binary compound contains two different elements; a binary ionic compound contains cation(s) from a metal and anion(s) from a nonmetal. The formula of a binary ionic compound represents the number and type of ions in the compound. The formula is found by "crossing over" the charges of the ions.

Binary ionic compounds are named by naming the cation first, and then putting an -ide ending on the stem of the anion. Examples include calcium chloride, potassium oxide, aluminum nitride, and so on. A metal that can form differently charged cations is named with a Roman numeral to indicate the charge: Fe^{2+} is iron(II) ion and Fe^{3+} is iron(III) ion.

Section 5.7—Nonmetal atoms can bond by sharing pairs of electrons, forming a covalent bond. A single bond is one pair of shared electrons; a **double bond** is two pairs; and a **triple bond** is three pairs.

$$:\ddot{Cl}:\ddot{Cl}: \quad :\ddot{O}::C::\ddot{O}: \quad :N:::N:$$

Shared pairs of electrons are called **bonding pairs (BPs)** while unshared pairs are called **nonbonding pairs** or **lone pairs (LPs)**. Most binary covalent compounds are named by naming the first element in the formula, then the stem of the second element with an *-ide* ending, and using prefixes such as *mono-*, *di-*, *tri-*, and so on, to indicate the number of atoms.

Sections 5.8–5.9—Bonding pairs are shared equally when the two atoms are the same but may be unequally shared when the atoms are different. The **electronegativity** of an element is the atom's attraction in a molecule for a bonding pair. Electronegativity generally increases to the right and up on the periodic table, so fluorine is the most electronegative element. When electrons are shared unequally, the more electronegative element takes on a partial negative charge ($\delta-$), the other element takes on a partial positive charge ($\delta+$), and the bond is said to be a **polar covalent bond**. The greater the difference in electronegativity, the more polar is the bond. A bonding pair shared equally is a **nonpolar covalent bond**. Carbon tends to form four bonds, nitrogen three, oxygen two, and hydrogen can only form one bond. A water molecule has two BP (O—H bonds) and two LP, an ammonia molecule has three BP (N—H bonds) and one LP, and a methane molecule has four BP (C—H bonds).

Section 5.10—A **polyatomic ion** is a charged particle containing two or more covalently bonded atoms. Compounds containing polyatomic ions are named, and formulas are written, in the same fashion as compounds containing monatomic ions, except that parentheses are placed around a polyatomic ion if there is more than one of the ion.

Ammonium sulfide **Potassium carbonate** **Calcium nitrate**

$$(NH_4)_2S \qquad K_2CO_3 \qquad Ca(NO_3)_2$$

Names of polyatomic ions often end in *-ate*, but if there are two anions containing the same elements (including oxygen), the one with one fewer oxygen atom ends in *-ite*.

Sections 5.11–5.12—An **electron-dot structure** or **Lewis structure** shows the arrangement of atoms, bonds, and lone pairs in a molecule or polyatomic ion. To draw a Lewis structure we (a) draw a reasonable skeletal structure, showing the arrangement of atoms; (b) count the valence electrons; (c) connect bonded pairs with a dash (one electron pair); (d) place electron pairs around outer atoms to give each an octet; and (e) place remaining electrons on the central atom. A multiple bond(s) is used if there are not enough electrons to give each atom (except hydrogen) an octet.

$$\begin{array}{c} H \\ | \\ H-C-\ddot{O}-H \\ | \\ H \end{array} \qquad :\ddot{O}=C=\ddot{O}:$$

There are exceptions to the octet rule. Atoms and molecules with unpaired electrons are called **free radicals**. They are highly reactive and short lived. Examples of free radicals are NO and ClO_2.

Section 5.13–5.14—The shape of a molecule can be predicted with **VSEPR theory**, which assumes that sets of electrons around a central atom will get as far away from each other

as possible. Two sets of electrons are 180° apart, three sets are about 120° apart, and four sets are about 109° apart. Once these angles have been established, we determine the shape of the molecule by examining only the bonded atoms. Simple molecules may have shapes described as linear, bent, triangular, pyramidal, or tetrahedral.

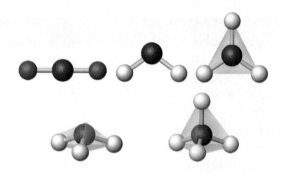

Like a polar bond, a **polar molecule** has separation between its centers of positive charge and negative charge. A molecule with nonpolar bonds is nonpolar. A molecule with polar bonds is nonpolar if its shape causes the polar bonds to "cancel." If the polar bonds do not cancel, the molecule is polar or is said to be a **dipole**. The water molecule has polar O—H bonds and is bent, and so it is polar.

Sections 5.15–5.16—When a substance is melted or vaporized, the forces that hold the particles (molecules, atoms, or ions) close to one another are overcome. Several properties are related to the strengths of those forces. The **melting point** of a solid is the temperature at which the forces holding the particles together in a regular arrangement are overcome. The reverse of the melting process, a liquid changing to a solid, is called **freezing**. **Vaporization** is the conversion of a liquid to a gas; the reverse process is called **condensation**. The temperature at which a substance vaporizes at ordinary pressure is called its **boiling point**. Some substances undergo **sublimation**, a conversion directly from the solid to the gas state. Ionic solids have very strong forces holding the ions to one another, so most ionic solids have high melting points and high boiling points.

Not all solids are held together by ionic bonds. **Dipole forces** are the net result of the attraction of the positive end of one dipole for the negative end of another dipole. A special, strong type of dipole force occurs when the positive end is a hydrogen atom attached to F, O, or N, and the negative end is a pair of electrons on an atom of F, O, or N. Such a force is called a **hydrogen bond**. Nonpolar molecules also have forces between them. These **dispersion forces** result from electron motion of neighboring atoms, which causes tiny, short-lived dipoles. Dispersion forces are generally weak.

A **solution** is a **homogeneous** mixture, a mixture in which all parts have the same composition. The phase that is dissolved is called the **solute**, and the dissolving phase is called the **solvent**. Nonpolar solutes dissolve in nonpolar solvents. Polar substances and many ionic substances dissolve in polar solvents.

REVIEW QUESTIONS

1. Which group of elements in the periodic table is characterized by an especially stable electron arrangement?

2. What is the structural difference between a sodium atom and a sodium ion?

3. How does sodium metal differ from sodium ions (in sodium chloride, for example) in properties?

4. What is the structural difference between a sodium ion and a neon atom? What is similar?

5. What are the structural differences among chlorine atoms, chlorine molecules, and chloride ions? How do their properties differ?

6. Indicate charges on simple ions formed from the following elements.
   a. group 3A   b. group 6A
   c. group 1A   d. group 7A

7. In what group of the periodic table would elements that form ions with the following charges likely be found?
   a. 2+   b. 3−   c. 1−

8. How many covalent bonds do each of the following usually form? You may refer to the periodic table.
   a. H   b. C
   c. O   d. F
   e. N   f. Br

9. Of the elements H, O, N, and C, which one(s) can readily form double bonds?

10. Of the elements H, O, N, and C, which one(s) can readily form triple bonds?

11. In what ways are liquids and solids similar? In what ways are they different?

12. List four types of interactions between particles in the liquid and solid states. Give an example of each type.

13. Define each of the following terms.
   a. melting   b. vaporization
   c. condensation   d. freezing

14. In which process is energy *absorbed* by the material undergoing the change of state?
   a. melting or freezing
   b. condensation or vaporization

15. Label each arrow with the term listed in Question 13 that correctly identifies the process presented.

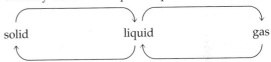

solid   liquid   gas

16. Define each of the following terms.
   a. solution   b. solute   c. solvent

PROBLEMS

Lewis Symbols for Elements

17. Give Lewis symbols for each of the following elements. You may use the periodic table.
   a. fluorine   b. calcium
   c. nitrogen   d. carbon

18. Give Lewis symbols for each of the following elements. You may use the periodic table.
   a. potassium   b. aluminum
   c. iodine   d. oxygen

Lewis Structures for Ions

19. Give Lewis structures for each of the following.
   a. I⁻   b. Sr^{2+} c. N^{3-}

20. Give Lewis structures for each of the following.
   a. Al^{3+} b. Cl^- c. S^{2-}

21. Using Lewis formulas, show the formation of an ion from an atom for each of the following.
   a. chlorine   b. magnesium

22. Using Lewis formulas, show the formation of an ion from an atom for each of the following.
   a. bromine   b. lithium

Lewis Symbols and Ionic Compound Formation

23. Use Lewis symbols to show the transfer of electrons from magnesium atoms to bromine atoms to form ions with noble gas configurations.

24. Use Lewis symbols to show the transfer of electrons from lithium atoms to sulfur atoms to form ions with noble gas configurations.

25. Use Lewis symbols to show the transfer of electrons from calcium atoms to nitrogen atoms to form ions with noble gas configurations.

26. Use Lewis symbols to show the transfer of electrons from potassium atoms to phosphorus atoms to form ions with noble gas configurations.

Lewis Structures for Ions and Ionic Compounds

27. Give Lewis symbols for the following.
   a. Al and Al^{3+} b. Br and Br^- c. O^{2-} and Ne

28. Give Lewis symbols for the following.
   a. Ca and Ca^{2+}
   b. S and S^{2-}
   c. Rb and Rb^+

29. Give electron structures (subshell notation) for the elements and ions in Problem 27.

30. Give electron structures (subshell notation) for the elements and ions in Problem 28.

31. Give Lewis formulas for each of the following ionic compounds.
   a. magnesium fluoride
   b. calcium chloride
   c. sodium oxide
   d. potassium sulfide

32. Give Lewis formulas for each of the following ionic compounds.
   a. sodium fluoride
   b. potassium chloride
   c. potassium iodide

33. Give Lewis formulas for each of the following ionic compounds.
    a. magnesium oxide
    b. aluminum nitride
    c. aluminum sulfide

34. Give Lewis formulas for each of the following ionic compounds.
    a. sodium nitride
    b. aluminum chloride
    c. calcium nitride

Names and Symbols for Simple Ions

35. Without referring to Table 5.2, name the following ions.
    a. K^+
    b. Ca^{2+}
    c. Zn^{2+}
    d. Br^-
    e. Li^+
    f. S^{2-}

36. Without referring to Table 5.2, name the following ions.
    a. Na^+
    b. Mg^{2+}
    c. Al^{3+}
    d. Cl^-
    e. O^{2-}
    f. N^{3-}

37. Without referring to Table 5.2, name the following ions.
    a. Fe^{2+}
    b. Cu^+
    c. I^-

38. Without referring to Table 5.2, name the following ions.
    a. Fe^{3+}
    b. Cu^{2+}
    c. Ag^+

39. Give symbols for the following ions.
    a. sodium ion
    b. aluminum ion
    c. oxide ion
    d. copper(II) ion

40. Give symbols for the following ions.
    a. bromide ion
    b. calcium ion
    c. potassium ion
    d. iron(II) ion

Names and Formulas for Binary Ionic Compounds

41. Name the following binary ionic compounds.
    a. KI
    b. CaF_2
    c. $MgCl_2$
    d. $FeCl_2$
    e. Na_2S
    f. CuO

42. Name the following binary ionic compounds.
    a. $MgBr_2$
    b. Li_2S
    c. Al_2O_3
    d. Fe_2O_3
    e. $CuCl_2$
    f. NaF

43. Give formulas for the following binary ionic compounds.
    a. calcium sulfide
    b. aluminum oxide
    c. iron(II) sulfide
    d. copper(I) chloride

44. Give formulas for the following binary ionic compounds.
    a. strontium bromide
    b. aluminum fluoride
    c. copper(II) iodide
    d. iron(III) sulfide

Names and Formulas for Polyatomic Ions

45. Name the following polyatomic ions.
    a. CO_3^{2-}
    b. HPO_4^{2-}
    c. MnO_4^-
    d. OH^-

46. Name the following ions.
    a. NO_3^-
    b. SO_4^{2-}
    c. $H_2PO_4^-$
    d. HCO_3^-

47. Give formulas for the following polyatomic ions.
    a. ammonium ion
    b. hydrogen sulfate ion
    c. cyanide ion
    d. nitrite ion

48. Give formulas for the following polyatomic ions.
    a. phosphate ion
    b. hydrogen carbonate ion
    c. dichromate ion
    d. oxalate ion

Names and Formulas for Ionic Compounds with Polyatomic Ions

49. Give formulas for the following.
    a. lithium hydroxide
    b. sodium carbonate
    c. calcium oxalate
    d. zinc nitrite

50. Give formulas for the following.
    a. ammonium acetate
    b. magnesium cyanide
    c. potassium monohydrogen phosphate
    d. sodium dihydrogen phosphate

51. Give formulas for the following.
    a. potassium permanganate
    b. silver hydrogen sulfate
    c. iron(III) hydroxide
    d. copper(II) sulfate

52. Give formulas for the following.
    a. ammonium chromate
    b. sodium phosphate
    c. cobalt(III) carbonate
    d. zinc monohydrogen phosphate

53. Name the following.
    a. KNO_2
    b. LiCN
    c. NH_4I
    d. $NaNO_3$
    e. $KMnO_4$
    f. $CaSO_4$

54. Name the following.
    a. $NaHSO_4$
    b. $Al(OH)_3$
    c. Na_2CO_3
    d. $KHCO_3$
    e. NH_4NO_2
    f. $Ca(HSO_4)_2$

Molecules: Covalent Bonds

55. Use Lewis symbols to show the sharing of electrons between two iodine atoms to form an iodine (I_2) molecule. Label all electron pairs as bonding pairs (BPs) or lone pairs (LPs).

56. Use Lewis symbols to show the sharing of electrons between a hydrogen atom and a fluorine atom.

57. Use Lewis symbols to show the sharing of electrons between a phosphorus atom and hydrogen atoms to form a molecule in which phosphorus has an octet of electrons.

58. Use Lewis symbols to show the sharing of electrons between a silicon atom and hydrogen atoms to form a molecule in which silicon has an octet of electrons.

59. Use Lewis symbols to show the sharing of electrons between a carbon atom and fluorine atoms to form a molecule in which each atom has an octet of electrons.

60. Use Lewis symbols to show the sharing of electrons between a nitrogen atom and chlorine atoms to form a molecule in which each atom has an octet of electrons.

Names and Formulas for Covalent Compounds

61. Give formulas for the following covalent compounds.
 a. dinitrogen tetroxide
 b. bromine trichloride
 c. nitrogen triiodide
 d. disulfur difluoride

62. Give formulas for the following covalent compounds.
 a. oxygen difluoride
 b. phosphorus trichloride
 c. chlorine trifluoride
 d. tricarbon dioxide

63. Name the following covalent compounds.
 a. CS_2 b. N_2S_4
 c. PF_5 d. S_2F_{10}

64. Name the following covalent compounds.
 a. CBr_4 b. Cl_2O_7
 c. P_4S_{10} d. I_2O_5

65. Chlorine dioxide is used to bleach flour. Give the formula for chlorine dioxide.

66. Tetraphosphorus trisulfide is used in the tips of "strike anywhere" matches. Give the formula for tetraphosphorus trisulfide.

Lewis Formulas for Covalent Compounds

67. Give Lewis formulas that follow the octet rule for the following covalent molecules.
 a. SiF_4 b. N_2H_4
 c. CH_5N d. NOH_3

68. Give Lewis formulas that follow the octet rule for the following covalent molecules.
 a. NCl_3 b. C_2H_4
 c. C_2H_2 d. CH_2O

69. Give Lewis formulas that follow the octet rule for the following covalent molecules.
 a. COF_2 b. PCl_3
 c. H_3PO_3 d. HCN

70. Give Lewis formulas that follow the octet rule for the following covalent molecules.
 a. SCl_2 b. H_2SO_4
 c. XeO_3 d. $HClO_4$

71. Give Lewis formulas that follow the octet rule for the following ions.
 a. ClO^- b. $HPO_4{}^{2-}$
 c. $ClO_2{}^-$ d. $BrO_3{}^-$

72. Give Lewis formulas that follow the octet rule for the following ions.
 a. CN^- b. $IO_4{}^-$
 c. $HSO_4{}^-$ d. $PO_4{}^{3-}$

Electronegativity: Polar Covalent Bonds

73. Classify the following covalent bonds as polar or nonpolar.
 a. H—O   b. N—Cl   c. B—F

74. Classify the following covalent bonds as polar or nonpolar.
 a. H—N   b. Be—F   c. P—Cl

75. Use the symbol ($\longleftrightarrow$) to indicate the direction of the dipole in each polar bond in Problem 73.

76. Use the symbol ($\longleftrightarrow$) to indicate the direction of the dipole in each polar bond in Problem 74.

77. Use the symbols $\delta+$ and $\delta-$ to indicate partial charges, if any, on the following bonds.
 a. Si—Cl   b. Cl—Cl   c. O—F

78. Use the symbols $\delta+$ and $\delta-$ to indicate partial charges, if any, on the following bonds.
 a. N—H   b. C—F   c. C—C

Classifying Bonds

79. Classify the bonds in the following as ionic or covalent. For bonds that are covalent, indicate whether they are polar or nonpolar.
 a. KF   b. IBr   c. MgO

80. Classify the bonds in the following as ionic or covalent. For bonds that are covalent, indicate whether they are polar or nonpolar.
 a. NO   b. CaO   c. NaBr

81. Classify the bonds in the following as ionic or covalent. For bonds that are covalent, indicate whether they are polar or nonpolar.
 a. Br_2 b. F_2 c. HCl

82. Classify the hydrogen–fluorine bond in Problem 56 as polar or nonpolar. Label the ends of the molecule with symbols that indicate polarity.

Intermolecular Forces

83. For which of the following would hydrogen bonding be an important intermolecular force?

 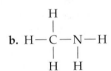
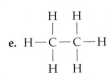

84. In which of the following are dispersion forces the only type of intermolecular force: Br_2, HF, HCl?

85. In which of the following are dipole interactions an important intermolecular force: Br_2, HCl, NaCl?

86. In which of the following are ionic bonds important: Br_2, HBr, NaBr?

VSEPR Theory: The Shapes of Molecules

87. Use VSEPR theory to predict the shape of each of the following molecules.
    a. silane (SiH_4)
    b. hydrogen selenide (H_2Se)
    c. arsine (AsH_3)

88. Use the VSEPR theory to predict the shape of each of the following molecules.
    a. beryllium chloride ($BeCl_2$)
    b. boron trichloride (BCl_3)
    c. carbon tetrafluoride (CF_4)

89. Use the VSEPR theory to predict the shape of each of the following molecules.
    a. oxygen difluoride (OF_2)
    b. silicon tetrachloride ($SiCl_4$)
    c. phosphorus trifluoride (PF_3)

90. Use the VSEPR theory to predict the shape of each of the following molecules.
    a. nitrogen trichloride (NCl_3)
    b. sulfur dichloride (SCl_2)
    c. dichlorodifluoromethane (CCl_2F_2)

Polar and Nonpolar Molecules

91. The molecule BeF_2 is linear. Is it polar or nonpolar? Explain.

92. The molecule SF_2 is bent. Is it polar or nonpolar? Explain.

93. Look again at the molecules in Problem 87. For each, are the bonds polar? What are the approximate bond angles? Is the molecule as a whole polar?

94. Look again at the molecules in Problem 88. For each, are the bonds polar? What are the approximate bond angles? Is the molecule as a whole polar?

95. Look again at the molecules in Problem 89. For each, are the bonds polar? What are the approximate bond angles? Is the molecule as a whole polar?

96. Look again at the molecules in Problem 90. For each, are the bonds polar? What are the approximate bond angles? Is the molecule as a whole polar?

Molecules That Are Exceptions to the Octet Rule

97. Which of the following atoms or molecules are free radicals?
    a. Br   b. F_2 c. CCl_3

98. Which of the following atoms or molecules are free radicals?
    a. S   b. NO_2 c. N_2O_4

99. Free radicals are one class of molecules in which atoms do not conform to the octet rule. Another exception to the octet rule involves atoms with fewer than eight electrons, as seen in elements of group 3 and in beryllium. In some covalent molecules, these atoms can have six or four electrons, respectively. Write Lewis structures for the following covalent molecules:
    a. $AlBr_3$ b. BeH_2 c. BH_3

100. Exceptions to the octet rule include molecules with atoms having more than eight valence electrons, most typically 10 or 12. Atoms heavier than Si can "expand their valence shell," meaning the "extra" electrons are accommodated in unoccupied, higher-energy atomic orbitals. Draw Lewis structures for the following molecules or ions:
    a. XeF_4 b. I_3^-
    c. SF_4 d. KrF_2

Solutions

101. The disinfectant used to clean the skin prior to an injection is actually a solution of 3 parts water and 7 parts alcohol. Which component is the solvent and which is the solute?

102. Explain why a salt dissolves in water.

103. Benzene (C_6H_6) is a nonpolar solvent. Would you expect NaCl to dissolve in benzene? Explain.

104. Motor oil is nonpolar. Would you expect it to dissolve in water? In benzene (C_6H_6)? Explain.

105. A 250-mL can of motor oil is poured into a can that contains 2.0 L of gasoline. Which component is the solute and which is the solvent?

106. Which of the following would you expect to dissolve in water and which in benzene (C_6H_6)?
    a. KBr   b. C_5H_{12}
    c. C_8H_{18} d. $CaCl_2$

ADDITIONAL PROBLEMS

107. Why does neon tend not to form chemical bonds?

108. Draw a charge-cloud picture for the H_2S molecule. Use the symbols $\delta+$ and $\delta-$ to indicate the polarity of the molecule.

109. The gas phosphine (PH_3) is used as a fumigant to protect stored grain and other durable produce from pests. Phosphine is generated in situ by adding water to aluminum phosphide or magnesium phosphide. Give formulas for the two phosphides.

110. There are two different covalent molecules with the formula C_2H_6O. Give Lewis formulas for the two molecules.

111. Solutions of iodine chloride (ICl) are used as disinfectants. Are the molecules of ICl ionic, polar covalent, or nonpolar covalent?

112. Consider the hypothetical elements X, Y, and Z with the following Lewis formulas.

    $$:\ddot{X}\cdot \quad :\ddot{Y}\cdot \quad :\ddot{Z}\cdot$$

    a. To which group in the periodic table would each element belong?
    b. Give the Lewis formula for the simplest compound of each with hydrogen.
    c. Give Lewis formulas for the ions formed when X reacts with sodium and when Y reacts with sodium.

113. Potassium is a soft, silvery metal that reacts violently with water and ignites spontaneously in air. Your doctor recommends you take a potassium supplement. Would you take potassium metal? If not, what would you take?

114. Is there any such thing as a sodium chloride molecule? Explain.

115. Use subshell notations to give electron configurations for the most stable simple ion formed by each of the following elements.
a. Ba   **b.** K   **c.** Se
d. I   **e.** N   **f.** Te

116. Why is Na^+ smaller than Na? Why is Cl^- larger than Cl?

117. The halogens (F, Cl, Br, and I) tend to form only one single bond in binary molecules. Explain.

118. A science magazine for the general public contains the statement, "Some of these hydrocarbons are very light, like methane gas—just a single carbon molecule attached to three hydrogen molecules." Evaluate the statement and correct any inaccuracies.

COLLABORATIVE GROUP PROJECTS

Prepare a PowerPoint, poster, or other presentation (as directed by your instructor) for presentation to the class. Projects 119 and 120 are best done in a group of four students.

119. Make copies of the form in Problem 96, page 33, except replace the column headings "Word" and "Definition" with "Name" and "Lewis Structure." Student 1 should write the name of an element or compound from the list below in the first column of the form and its Lewis structure in the second column. Proceed as in Problem 96, page 33, ending by comparing the name in the last column with that in the first column. Discuss any differences in the two names. If the name in the last column differs from that in the first column, determine what went wrong in the process.
a. bromide ion
b. calcium fluoride
c. phosphorus trifluoride
d. carbon disulfide

120. Make copies of the form in Problem 96, page 33, using the column headings "Name" and "Structure." Then Student 1 should write a name from the list below in the first column of the form and its structure in the second column. Proceed as in Problem 96, page 33, ending by comparing the name in the last column with that in the first column. Discuss any differences in the two names. If the name in the last column differs from that in the first column, determine what went wrong in the process.
a. ammonium nitrate
b. potassium phosphate
c. lithium carbonate
d. copper(I) chloride

121. Starting with 20 balloons, blown up to about the same size and tied, tie two balloons together. Next, tie three balloons together. Repeat this with four, five, and six balloons. (*Suggestion*: Three sets of two balloons can be twisted together to form a six-balloon set, and so on.) Show how these balloon sets can be used to explain the VSEPR theory.

122. Prepare a brief biographical report on one of the following.
a. Gilbert N. Lewis   **b.** Linus Pauling

123. There are several different definitions and scales for electronegativity. Search for information on some of them and write a brief compare-and-contrast essay on two of them.

124. Search the Web for information on the hydrogen bonding **(a)** in water, **(b)** in DNA, or **(c)** in synthetic polymers. What are some of the important results of hydrogen bonding in these systems?

REFERENCES AND READINGS

1. Atkins, P. W. *Molecules*. New York: W. H. Freeman, 1987. A friendly and colorful discussion about some common molecules and their chemical bonds.

2. Ball, Philip. *Stories of the Invisible: A Guided Tour of the Molecules*. Oxford, UK: Oxford University Press, 2001.

3. Barrett, J. *Structure and Bonding*. London: Royal Society of Chemistry, 2001.

4. Emsley, John. *Molecules at an Exhibition: Portraits of Interesting Molecules in Everyday Life*. Oxford, UK: Oxford University Press, 1998. A witty "tour" through a "museum" of various elements and compounds.

5. Gillespie, Ronald J. "Covalent and Ionic Molecules: Why Are BeF_2 and AlF_3 High Melting Point Molecules Whereas BF_3 and SiF_4 Are Gases?" *Journal of Chemical Education*, July 1998, pp. 923–925.

6. Gillespie, Ronald J. "Teaching Molecular Geometry with the VSEPR Model." *Journal of Chemical Education*, March 2004, pp. 298–304.

7. Hill, John W., Ralph H. Petrucci, Terry W. McCreary, and Scott S. Perry. *General Chemistry*, 4th edition. Upper Saddle River, NJ: Prentice Hall, 2005. Chapters 9 and 10.

8. Logan, S. R. "The Role of Lewis Structures in Teaching Covalent Bonding." *Journal of Chemical Education*, November 2001, pp. 1457–1458.

9. Normile, Dennis. "Search for Better Crystals Explores Inner, Outer Space." *Science*, December 22, 1995, pp. 1921–1922.

10. Pfennig, Brian W., and Richard L. Frock. "The Use of Molecular Modeling and VSEPR Theory in the Undergraduate Curriculum to Predict the Three-Dimensional Structure of Molecules." *Journal of Chemical Education*, 1999, p. 1018.

GREEN CHEMISTRY

Chemical Bonds Green Chemistry

John C. Warner, Center for Green Chemistry, University of Massachusetts
Kathryn Parent, Green Chemistry Institute, American Chemical Society

You've learned in this chapter that when atoms bond to form compounds, the properties of the material usually change considerably. In molecules, atoms join by sharing electrons and forming covalent bonds. As discussed in Chapter 5, these covalent bonds make the molecules adopt specific geometric shapes, with each atom held somewhat rigidly in a specific position relative to other atoms in the molecule.

In nature, the specific geometry of a molecule controls how it will react with other substances. Biological molecules are recognized by enzymes and other biological receptors, rather like a key is recognized by a lock. The shape of the molecule and receptor fit together to form complexes. Complexes result from various non-covalent forces, such as hydrogen bonding. When the molecule binds to the receptor it triggers a response, causing a physiological event to begin or end.

Medicinal chemists design drug molecules that are very similar in shape to biological molecules. The drug molecules "look" similar enough that they still bind to the enzyme or receptor, but they are a little different and behave differently once the complex is formed. The science by which molecules interact via geometric orientations of atoms is called *molecular recognition*. The 1987 Nobel Prize in chemistry was awarded for pioneering work in this area. Scientists are now learning to use these non-covalent mechanisms to control chemical and physical properties by mimicking nature.

In the following Web Investigations, you will learn more about molecular recognition, medicinal chemistry, and non-covalent derivatization—ways by which chemists can mimic materials and processes found in nature.

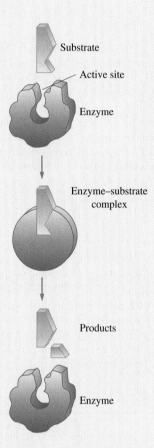

▲ Enzymes—biological catalysts—cause chemical reactions to occur based on enzyme shape and reactant shape.

WEB INVESTIGATIONS

Investigation 1
Shape and Function
You can begin your research with a keyword search for "molecular shape." Learn how scientists model the geometry of molecules and determine the shape taken by a particular molecule.

Investigation 2
The Nobel Prize
Search the Web to discover who won the 1987 Nobel Prize in chemistry. Who were the chemists recognized for their research in molecular engineering? Summarize their research in a one-paragraph abstract (150–250 words).

Investigation 3
Structure of Pain Relief
Search the Web for images of the molecular structures of aspirin and Tylenol, two common analgesics. Use the information to answer the following questions.

- What is the scientific name for the chemical that is the active ingredient in aspirin?
- What is the scientific name for the chemical that is the active ingredient in Tylenol?
- What elements do they have in common?
- What elements are unique to aspirin? What elements are unique to Tylenol?

- How are the two molecules similar in structure? How are they shaped differently?

Investigation 4
Noncovalent Interaction
Search for and read the article by Amy Cannon and John Warner entitled "Noncovalent Derivitization: green chemistry applications of crystal engineering" published in 2002.

Investigation 5
Biomimicry
Search using the keyword "biomimicry" and condense your readings into a one-sentence definition for this term.

COMMUNICATE YOUR RESULTS

Exercise 1
Shape and Function
Think about a lock and a key. A key has a specific shape so that it works for a specific lock. How might a molecule behave like a key? Write one-page essay using the analogy of a lock and key to explain the formation of complexes by molecules and receptors.

Exercise 2
Noncovalent Interaction
It takes a lot more energy to make and break covalent bonds than hydrogen bonds. Write a one-page essay to explain why you think using non-covalent forces, like hydrogen bonds, might be useful to achieve some of the benefits of green chemistry. Refer to the principles of green chemistry in Chapter 1 of the textbook and the article you read for Investigation 4, above.

Exercise 3
Biomimicry
Find three interesting case studies of biomimicry on the Web. Prepare a PowerPoint presentation on these case studies to share with your class. Your presentation should explain the natural material or process and what it inspired (or might inspire) scientists to develop.

CHAPTER 6

Chemical Accounting

When molding plaster of Paris, the ingredients must be mixed in the proper amounts and in the proper proportions. Enough material must be prepared to fill the mold. Too little water gives a weak, crumbly product. Add too much water and the soupy mixture does not solidify. In this chapter we will investigate chemical reactions and learn how the amounts of reactants and products are determined.

Mass and Volume Relationships

Much chemistry can be discussed and understood with little or no mathematics, but there are also many interesting quantitative aspects. In this chapter, we will consider some of the basic calculations used in chemistry and related fields such as biology and medicine. Chemists use many kinds of mathematics, from simple arithmetic to sophisticated calculus and complicated computer algorithms, but our calculations here will require no more than a knowledge of simple algebra.

6.1 CHEMICAL SENTENCES: EQUATIONS

Chemistry is a study of matter and the changes it undergoes and of the energy that brings about these changes or is released when these changes occur. In Chapter 5, we discussed the symbols and formulas used to represent elements and compounds. Now that we have learned the letters (symbols) and words (formulas) of our chemical language, we are ready to write sentences (chemical equations). A **chemical equation** is a shorthand way of describing chemical change using symbols and formulas to represent the elements and compounds involved in the change.

We can describe a chemical reaction in words. For example,

Carbon reacts with oxygen to form carbon dioxide.

We can also describe the same reaction with chemical symbols and formulas.

$$C + O_2 \longrightarrow CO_2$$

The plus sign (+) indicates that carbon and oxygen are added together or combined in some way. The arrow ($\longrightarrow$) is read "yield(s)", or "react(s) to produce". Substances on the left of the arrow ($C + O_2$ in this case) are **reactants** or *starting materials*. Those on the right (here, CO_2) are the **products** of the reaction.

At the submicroscopic (atomic or molecular) level, the chemical equation means that one atom of carbon (C) reacts with one molecule of oxygen (O_2) to produce one molecule of carbon dioxide (CO_2).

Sometimes we indicate the physical states of the reactants and products by writing the initial letter of the state immediately following the formula. Thus, (g) indicates a gaseous substance, (l) a liquid, and (s) a solid. The label (aq) indicates an aqueous solution—that is, a water solution. Using these labels, our equation becomes

$$C(s) + O_2(g) \longrightarrow CO_2(g)$$

Reactants and products need not be written in any particular order in a chemical equation, except that all reactants must be to the left of the arrow and all products to the right. In other words, we could also write the preceding equation as

$$O_2 + C \longrightarrow CO_2$$

155

Counting Atoms

Balancing Equations

Balancing Chemical Equations

We can represent the reaction of carbon and oxygen to form carbon dioxide quite simply, but many other chemical reactions require more thought. For example, hydrogen reacts with oxygen to form water. Using formulas, we can represent this reaction as

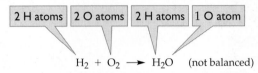

$$H_2 + O_2 \longrightarrow H_2O \quad \text{(not balanced)}$$

However, this representation shows two oxygen atoms in the reactants (as O_2), but only one in the product (in H_2O). Because matter is neither created nor destroyed in a chemical reaction (the law of conservation of matter, Chapter 2), the equation must be balanced to represent the chemical reaction correctly. That means the same number of each type of atom must appear on both sides. To balance the oxygen atoms, we need only place the coefficient 2 in front of the formula for water.

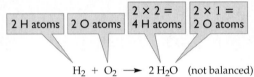

$$H_2 + O_2 \longrightarrow 2\,H_2O \quad \text{(not balanced)}$$

This coefficient means that two molecules of water are produced. As is the case with subscripts, a coefficient of 1 is understood when no other number appears. A coefficient preceding a formula multiplies *everything* in the formula. In the preceding equation, the coefficient 2 not only increases the number of oxygen atoms to two but also increases the number of hydrogen atoms to four on the product side.

But the equation is still not balanced. As we took care of the oxygen, we unbalanced the hydrogen. To balance the hydrogen, we place a coefficient 2 in front of H_2.

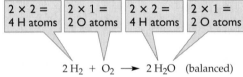

$$2\,H_2 + O_2 \longrightarrow 2\,H_2O \quad \text{(balanced)}$$

Now there are four hydrogen atoms and two oxygen atoms on each side of the equation. Atoms are conserved: The equation is balanced (Figure 6.1) and the law of conservation of mass is obeyed. Figure 6.2 illustrates two common pitfalls in the process of balancing equations, as well as the correct method. Remember not to add to or change chemical species on either side, but use coefficients to equate the numbers of atoms of each type.

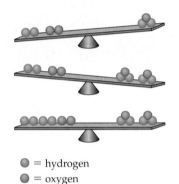

● = hydrogen
● = oxygen

▲ **Figure 6.1** To balance the equation for the reaction in which hydrogen and oxygen react to form water, the same number of each kind of atom must appear on each side (atoms are conserved). When the equation is balanced, there are four H atoms and two O atoms on each side.

Question: Why can't we balance the equation by removing one of the oxygen atoms from the left side of the top balance?

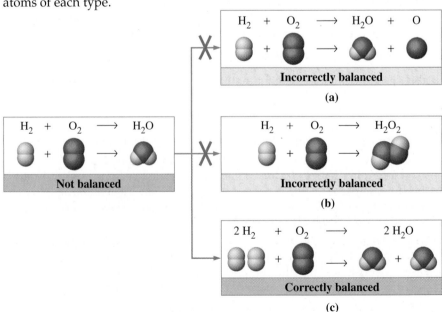

▶ **Figure 6.2** Balancing the equation for the reaction between hydrogen and oxygen to form water. (a) Incorrect. There is no atomic oxygen (O) as a product. Extraneous products cannot be introduced simply to balance an equation. (b) Incorrect. The product of the reaction is water (H_2O), not hydrogen peroxide (H_2O_2). A formula can't be changed simply to balance an equation. (c) Correct. An equation can be balanced only through the use of correct formulas and coefficients.

EXAMPLE 6.1 Balancing Equations

Balance the following representation of the chemical reaction involved when an airbag deploys.

$$NaN_3 \longrightarrow Na + N_2$$

Solution

Sodium atoms are balanced, but the nitrogen atoms are not. For this sort of problem, we will use the concept of the least common multiple. There are three nitrogen atoms on the left (reactants side) and two on the right (products side). The least common multiple of 2 and 3 is 6. Therefore, we need *three* N_2 and *two* NaN_3:

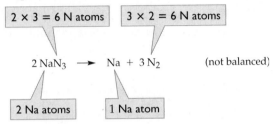

We now have *two* sodium atoms on the left. We can get two on the right by placing the coefficient 2 in front of Na.

$$2\,NaN_3 \longrightarrow 2\,Na + 3\,N_2 \quad \text{(balanced)}$$

Checking, we count two Na atoms and six N atoms on each side. The equation is balanced.

▶ **Exercise 6.1A**

The reaction between hydrogen and nitrogen to give ammonia, called the Haber process, is typically the first step in the industrial production of nitrogen fertilizers, represented as

$$H_2 + N_2 \longrightarrow NH_3$$

Balance the equation.

▶ **Exercise 6.1B**

Iron ores like Fe_2O_3 are *smelted* by reaction with carbon to produce metallic iron and carbon dioxide, represented as

$$Fe_2O_3 + C \longrightarrow CO_2 + Fe$$

Balance the equation.

Although simple equations can be balanced by trial and error, a couple of strategies often help:

1. If an element occurs in just one substance on each side of the equation, try balancing that element *first*.
2. Balance any reactants or products that exist as the free element *last*.

Perhaps the most important step in any strategy is to check an equation to ensure that it is indeed balanced. Remember that for each element, the same number of atoms of the element must appear on each side of the equation; atoms are conserved in chemical reactions.

EXAMPLE 6.2 Balancing Equations

When a fuel such as methane is burned in sufficient air, the products are carbon dioxide and water, represented as

$$CH_4 + O_2 \longrightarrow CO_2 + H_2O \quad \text{(not balanced)}$$

Balance the equation.

Solution

In this equation, oxygen appears in two different products and by itself in O_2; we leave the oxygen for last and balance the other two elements first. Carbon is already balanced, with one atom on each side of the equation. For hydrogen, the least common multiple of 2 and 4 is 4, and so we place the coefficient 2 in front of H_2O to balance hydrogen. Now we have four hydrogen atoms on each side.

$$CH_4 + O_2 \longrightarrow CO_2 + 2\,H_2O \quad \text{(not balanced)}$$

Now for the oxygen. There are four oxygen atoms on the right. If we place a 2 in front of O_2 on the left, the oxygen atoms balance.

$$CH_4 + 2\,O_2 \longrightarrow CO_2 + 2\,H_2O \quad \text{(balanced)}$$

The equation now has one C atom, four H atoms, and four O atoms on each side; it is balanced.

▶ **Exercise 6.2A**

Butane is a common fuel. Its combustion is represented as

$$C_4H_{10} + O_2 \longrightarrow CO_2 + H_2O$$

Balance the equation.

▶ **Exercise 6.2B**

Does it take more oxygen (per molecule) to burn butane (see Exercise 6.2A) than it does to burn methane? How much more CO_2 is produced?

We have made the task of balancing equations deceptively easy by considering fairly simple reactions. It is more important at this point for you to understand the principle than to be able to balance complicated equations. You should know what is meant by a balanced equation and be able to handle simple reactions. It is essential that you understand the information that is contained in these balanced equations.

6.2 VOLUME RELATIONSHIPS IN CHEMICAL EQUATIONS

So far we have looked at chemical reactions in terms of individual atoms and molecules. In the real world, however, chemists work with quantities of matter that contain billions of billions of atoms. John Dalton postulated that atoms of different elements had different masses. Therefore, equal masses of different elements would contain different numbers of atoms. Consider the analogous situation of golf balls and Ping-Pong balls. A kilogram of golf balls contains a smaller number of balls than a kilogram of Ping-Pong balls. One could determine the number of balls in each case simply by counting them. For atoms, however, such a straightforward method is not possible, because the smallest visible particle of matter contains more atoms than could be counted in 10 lifetimes.

Experiments with gases led French scientist Joseph Louis Gay-Lussac (1778–1850) to an approach to quantifying atoms. In 1809 he announced the results of some chemical reactions that he had carried out with gases and summarized these experiments in a new law: The **law of combining volumes** states that when all measurements are made at the same temperature and pressure, the volumes of gaseous reactants and products are in a small whole-number ratio. One such experiment is illustrated in Figure 6.3. When hydrogen reacts with nitrogen to form ammonia, three volumes of hydrogen combine with one volume of nitrogen to yield two volumes of ammonia. The small whole-number ratio is 3:1:2.

Gay-Lussac thought there must be some relationship between the numbers of molecules and the volumes of gaseous reactants and products. But it was Amedeo Avogadro who first explained the law of combining volumes in 1811. **Avogadro's**

▲ Amedeo Avogadro (1776–1856) and his hypothesis. The quotation reads, "Equal volumes of all gases at the same temperature and pressure contain the same number of molecules."

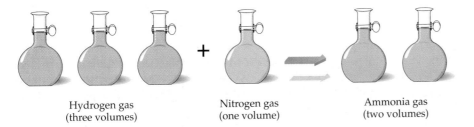

Hydrogen gas
(three volumes)

Nitrogen gas
(one volume)

Ammonia gas
(two volumes)

◀ **Figure 6.3** Gay-Lussac's law of combining volumes. Three volumes of hydrogen gas react with one volume of nitrogen gas to yield two volumes of ammonia gas.

Question: Can you sketch a similar figure for the reaction
$2 SO_2(g) + O_2(g) \longrightarrow 2 SO_3(g)$?

hypothesis, based on a shrewd interpretation of experimental facts, was that equal volumes of all gases, when measured at the same temperature and pressure, contain the same number of molecules (Figure 6.4).

The equation for the combination of hydrogen and nitrogen to form ammonia is

$$3 H_2(g) + N_2(g) \longrightarrow 2 NH_3(g)$$

The coefficients of the molecules are the same as the combining ratio of the gas volumes, 3:1:2 (Figure 6.3). The equation says that a nitrogen molecule and three hydrogen molecules react to produce two ammonia molecules. If you had 1 million nitrogen molecules, you would need 3 million hydrogen molecules to produce 2 million ammonia molecules. The equation provides the combining ratios. According to the equation, three volumes of hydrogen reacts with one volume of nitrogen to produce two volumes of ammonia because each volume of hydrogen contains the same number of molecules as that same volume of nitrogen.

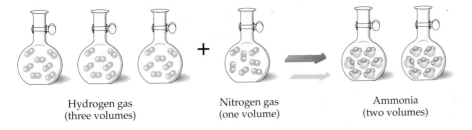

Hydrogen gas
(three volumes)

Nitrogen gas
(one volume)

Ammonia
(two volumes)

◀ **Figure 6.4** Avogadro's explanation of Gay-Lussac's law of combining volumes. Equal volumes of each of the gases contain the same number of molecules.

Question: Can you sketch a similar figure for the reaction
$2 NO(g) + O_2(g) \longrightarrow 2 NO_2(g)$?

EXAMPLE 6.3   **Volume Relationships of Gases**

What volume of oxygen is required to burn 0.556 L of propane if both gases are measured at the same temperature and pressure?

$$C_3H_8(g) + 5 O_2(g) \longrightarrow 3 CO_2(g) + 4 H_2O(g)$$

Solution
The coefficients in the equation indicate that each volume of $C_3H_8(g)$ requires 5 volumes of $O_2(g)$. Thus, we use $5 L O_2(g)/1 L C_3H_8(g)$ as the ratio to find the volume of oxygen required.

$$? L O_2(g) = 0.556 \, \cancel{L \, C_3H_8(g)} \times \frac{5 L O_2(g)}{1 \, \cancel{L \, C_3H_8(g)}} = 2.78 \, L O_2(g)$$

▶ **Exercise 6.3A**
Using the equation in Example 6.3, calculate the volume of $CO_2(g)$ produced when 0.492 L of propane is burned if the two gases are compared at the same temperature and pressure.

▶ **Exercise 6.3B**
If 10.0 L each of propane and oxygen are combined at the same temperature and pressure, which gas will be left over after reaction? What volume of that gas will remain?

6.3 AVOGADRO'S NUMBER: 6.02×10^{23}

Avogadro's hypothesis, which has been verified many times and in several ways over the years, states that equal volumes of gas at the same temperature and pressure contain equal numbers of molecules. This means that if we weigh equal volumes of several gases, the ratio of their masses should be the same as the mass ratio of the molecules themselves.

Avogadro had no way of knowing how many molecules were in a given volume of gas. Scientists since his time have determined the number of atoms in various weighed samples of substances. The numbers are extremely large, even for tiny samples. In defining atomic masses, the mass of a carbon-12 atom is defined as exactly 12 u. The number of carbon-12 atoms in a 12-g sample of carbon-12 is called **Avogadro's number** and has been determined experimentally to be 6.0221367×10^{23}. For our purposes, we usually round it off to three significant figures: 6.02×10^{23}.

Whose Number Is It, Anyway?

Avogadro died before his ideas were recognized by the scientific community. Acceptance finally came in 1860 at a scientific conference at which Stanislao Cannizzaro (1826–1910) effectively communicated Avogadro's ideas from half a century earlier. Scientists knew that the number must be enormous, but it wasn't until 1865 that reasonable estimates were made. That year Josef Loschmidt (1821–1895) measured the size of air molecules and found them to be about a "millionth of a millimeter" in diameter. This rough measurement indicated a value of 4×10^{22}, not a bad estimate for a first attempt. Later measurements, using various approaches, have shown the actual diameter of air molecules to be a bit smaller than Loschmidt had determined, and the number to be closer to 6×10^{23}. In German-speaking countries, this value is usually called the Loschmidt number, but in most places, it is called Avogadro's number in spite of the fact that Avogadro never knew how big the number was.

6.4 THE MOLE: "A DOZEN EGGS AND A MOLE OF SUGAR, PLEASE"

We buy socks by the pair (2 socks), eggs by the dozen (12 eggs), soda pop by the case (24 cans), pencils by the gross (144 pencils), and paper by the ream (500 sheets). A dozen is the same number whether we are counting a dozen melons or a dozen oranges. But a dozen oranges and a dozen melons do not weigh the same. If a melon weighs five times as much as an orange, a dozen melons will weigh five times as much as a dozen oranges.

Chemists count atoms and molecules by the *mole*. (A single carbon atom is much too small to see or weigh, but a mole of carbon atoms fills a tablespoon and weighs 12 g.) A mole of carbon and a mole of titanium each contain the same number of atoms. But a titanium atom has a mass four times that of a carbon atom, and so a mole of titanium has a mass four times that of a mole of carbon.

A **mole (mol)** is an amount of substance that contains the same number of elementary units as there are atoms in exactly 12 g of carbon-12. That number is 6.02×10^{23}, Avogadro's number. The elementary units may be atoms (such as S or Ca), molecules (such as O_2 or CO_2), ions (such as K^+ or SO_4^{2-}), or any other kind of formula unit. A mole of NaCl, for example, contains 6.02×10^{23} NaCl formula units, which means that it contains 6.02×10^{23} Na^+ ions and 6.02×10^{23} Cl^- ions.

Formula Masses

Each element has a characteristic atomic mass. Because chemical compounds are made up of two or more elements, the masses of compounds are combinations of atomic masses. For any substance, the **formula mass** is the average[1] mass of a for-

---

[1] We use the word *average* when denoting the mass of an individual molecule because molecules of a compound may have different isotopes of one or more of their constituent elements.

◀ How big is Avogadro's number?

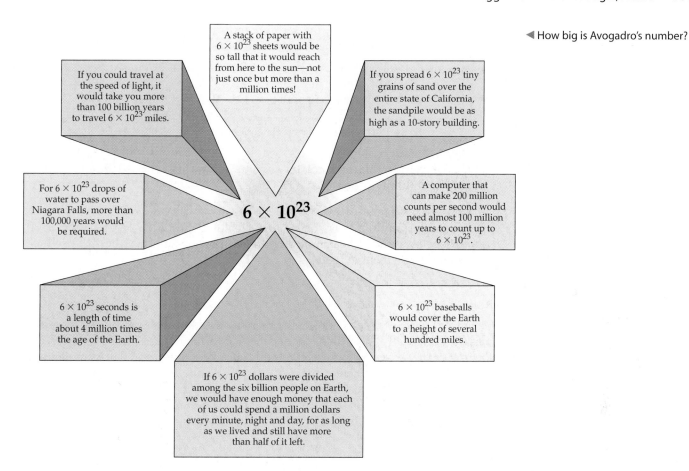

A stack of paper with 6×10^{23} sheets would be so tall that it would reach from here to the sun—not just once but more than a million times!

If you could travel at the speed of light, it would take you more than 100 billion years to travel 6×10^{23} miles.

If you spread 6×10^{23} tiny grains of sand over the entire state of California, the sandpile would be as high as a 10-story building.

For 6×10^{23} drops of water to pass over Niagara Falls, more than 100,000 years would be required.

6×10^{23}

A computer that can make 200 million counts per second would need almost 100 million years to count up to 6×10^{23}.

6×10^{23} seconds is a length of time about 4 million times the age of the Earth.

6×10^{23} baseballs would cover the Earth to a height of several hundred miles.

If 6×10^{23} dollars were divided among the six billion people on Earth, we would have enough money that each of us could spend a million dollars every minute, night and day, for as long as we lived and still have more than half of it left.

mula unit relative to that of a carbon-12 atom. Simply put, it is the sum of the masses of the atoms represented in a formula. If the formula represents a molecule, the term *molecular mass* is often used. For example, because the formula O_2 specifies two O atoms per molecule of oxygen, the formula (or molecular) mass of oxygen (O_2) is twice the atomic mass of oxygen.

$$2 \times \text{atomic mass of O} = 2 \times 16.0 \text{ u} = 32.0 \text{ u}$$

EXAMPLE 6.4 **Calculating Molecular Masses**

Calculate **(a)** the molecular mass of nitrogen dioxide (NO_2), an amber colored gas that is a constituent of smog, and **(b)** the formula mass of ammonium sulfate [$(NH_4)_2SO_4$] a fertilizer commonly used by home gardeners.

Solution

a. We start with the molecular formula: NO_2. Then, to determine the molecular mass, we need only to add the atomic mass of nitrogen to twice the atomic mass of oxygen.

$$1 \times \text{atomic mass of N} = 1 \times 14.0 \text{ u} = 14.0 \text{ u}$$
$$2 \times \text{atomic mass of O} = 2 \times 16.0 \text{ u} = \underline{32.0 \text{ u}}$$
$$\text{Formula mass of } NO_2 = 46.0 \text{ u}$$

Using a calculator, we need only write down the final answer, 46.0 u. That is, we have no need to record the numbers 14.0 and 32.0.

b. We must make certain that all the atoms in the formula unit are accounted for, which means paying particular attention to all the subscripts and parentheses in the formula. The "$(NH_4)_2$" means that both the "N" and the "H_4" must be multiplied by 2—that is, the formula indicates a total of two N atoms and eight H atoms. Combining the atomic masses, we have

$$2 \times \text{atomic mass of N} = 2 \times 14.0\ u = 28.0\ u$$
$$8 \times \text{atomic mass of H} = 8 \times 1.01\ u = 8.08\ u$$
$$1 \times \text{atomic mass of S} = 1 \times 32.0\ u = 32.0\ u$$
$$4 \times \text{atomic mass of O} = 4 \times 16.0\ u = \underline{64.0\ u}$$
$$\text{Formula mass of } (NH_4)_2SO_4 = 132.1\ u$$

▶ **Exercise 6.4A**

Calculate the formula mass of **(a)** sodium azide (NaN_3) used in automobile airbags, and **(b)** phosphoric acid (H_3PO_4).

▶ **Exercise 6.4B**

Calculate the formula mass of **(a)** *para*-dichlorobenzene ($C_6H_4Cl_2$) used as a moth repellent, and **(b)** calcium dihydrogen phosphate [$Ca(H_2PO_4)_2$], used as a mineral supplement in foods.

Molar Mass

The **molar mass** of a substance is the mass of 1 mol of that substance. The molar mass is numerically equal to the atomic mass or formula mass, but it is expressed in the unit *grams per mole* (g/mol). The atomic mass of sodium is 23.0 u; its molar mass is 23.0 g/mol. The molecular mass of carbon dioxide is 44.0 u; its molar mass is 44.0 g/mol. The formula mass of ammonium sulfate is 132.1 u; its molar mass is 132.1 g/mol. We can use these facts, together with the basic definition of the number of elementary units in a mole, to write the following relationships.

$$1\ \text{mol Na} = 23.0\ \text{g Na}$$
$$1\ \text{mol } CO_2 = 44.0\ \text{g } CO_2$$
$$1\ \text{mol } (NH_4)_2SO_4 = 132.1\ \text{g } (NH_4)_2SO_4$$

These relationships supply the conversion factors we need to make conversions between mass in grams and amount in moles, as illustrated in the following examples.

EXAMPLE 6.5 **Mole-to-Mass Conversions**

How many grams of N_2 are in 0.400 moles N_2?

Solution

The molecular mass of N_2 is $2 \times 14.0\ u = 28.0\ u$. The molar mass of N_2 is therefore 28.0 g/mol. Using the molar mass as a conversion factor (red), we have

$$?\ \text{g } N_2 = 0.400\ \text{mol } N_2 \times \frac{28.0\ \text{g } N_2}{1\ \text{mol } N_2} = 11.2\ \text{g } N_2$$

▶ **Exercise 6.5**

Calculate the mass, in grams, of **(a)** 0.0728 mol silicon, **(b)** 55.5 mol H_2O, and **(c)** 0.0728 mol $Ca(H_2PO_4)_2$.

EXAMPLE 6.6 **Mass-to-Mole Conversions**

Calculate the number of moles of Na in a 62.5-g sample of sodium metal.

Solution

The molar mass of Na is 23.0 g/mol. To convert from a mass in grams to an amount in moles, we must use the *inverse* of the molar mass as a conversion factor (1 mol Na/23.0 g Na) to get the proper cancellation of units. When we start with grams, we must have grams in the denominator of our conversion factor (red).

$$?\ \text{mol Na} = 62.5\ \text{g Na} \times \frac{1\ \text{mol Na}}{23.0\ \text{g Na}} = 2.72\ \text{mol Na}$$

▶ **Exercise 6.6**

Calculate the amount, in moles, of **(a)** 3.71 g Fe, **(b)** 165 g butane, C_4H_{10}, and **(c)** 0.100 mol $Mg(NO_3)_2$.

◀ **Figure 6.5** One mole of each of several familiar substances: from left to right on the table; sugar, salt, carbon, and copper; the balloon contains helium. Each contains Avogadro's number of formula units of the substance it contains. There are 6.02×10^{23} molecules of sugar ($C_{12}H_{22}O_{11}$), 6.02×10^{23} formula units of NaCl, 6.02×10^{23} atoms of carbon, 6.02×10^{23} atoms of copper, and 6.02×10^{23} atoms of helium in the respective samples.

Question: How many Na^+ ions and how many Cl^- ions are there in the dish of salt? If the balloon was filled with oxygen gas (O_2), how many oxygen atoms would it contain?

Figure 6.5 is a photograph showing 1.00-mol samples of several different chemical substances. Each container contains Avogadro's number of formula units of the substance. There are just as many copper atoms in the smallest dish as there are helium atoms in the balloon.

Molar Volume: 22.4 L at Standard Temperature and Pressure

A mole of gas contains Avogadro's number of molecules (atoms if it is a noble gas). Furthermore, Avogadro's number of molecules of gas occupies the same volume (at a given temperature and pressure) regardless of the size or mass of the individual molecules. The volume occupied by 1 mol of gas is the **molar volume** of a gas.

Because the volume of a gas is altered by changes in temperature or pressure (Section 6.7), a particular set of conditions has been chosen for reference purposes as "standard." Standard pressure is 1 atmosphere (atm), which is the normal pressure of the air at sea level; and standard temperature is 0 °C, which is the freezing point of water. A mole of any gas at **standard temperature and pressure (STP)** occupies a volume of about 22.4 L. This is known as the standard molar volume of a gas.

A cube that measures 28.2 cm along each edge has a volume of 22.4 L (Figure 6.6). It is just a bit smaller than a cubic foot. At a temperature of 0 °C and 1 atm pressure, a 22.4-L container holds 28.0 g of N_2, 32.0 g of O_2, 44.0 g of CO_2, and so on; it holds 1 mol of any gas.

We can readily calculate the density (g/L) of a gas at STP. We begin with the molecular mass of the gas (g/mol) and use the conversion factor 1 mol gas = 22.4 L. Because the conversion factor can be inverted, we also can calculate molar mass from density, using the same factor 1 mol gas = 22.4 L.

▶ **Figure 6.6** A cube 28.2 cm on a side contains 22.4 L. Here the molar volume of a gas is compared with a basketball, a football, and a soccer ball. At STP, the cube holds 16.0 g of CH_4, 28.0 g of N_2, 32.0 g of O_2, 39.9 g of Ar, and so on. It holds 1 mol of any gas at STP.

Question: How many $C_2H_6(g)$ molecules would the cube hold at STP?

EXAMPLE 6.7   **Density of a Gas at STP**

Calculate the density of **(a)** nitrogen gas and **(b)** methane (CH_4) gas, both at STP.

Solution

a. The molar mass of N_2 gas is 28.0 g/mol. We multiply by the conversion factor 1 mol N_2 = 22.4 L, arranged to cancel units of *moles*.

$$\frac{28.0 \text{ g } N_2}{1 \text{ mol } N_2} \times \frac{1 \text{ mol } N_2}{22.4 \text{ L } N_2} = 1.25 \text{ g/L}$$

b. The molar mass of CH_4 gas is (1×12.0) g/mol + (4×1.01) g/mol = 16.0 g/mol. Again we use the conversion factor 1 mol CH_4 = 22.4 L.

$$\frac{16.0 \text{ g } CH_4}{1 \text{ mol } CH_4} \times \frac{1 \text{ mol } CH_4}{22.4 \text{ L } CH_4} = 0.714 \text{ g/L}$$

▶ **Exercise 6.7A**

Calculate the density of He at STP.

▶ **Exercise 6.7B**

Estimate the density of air at STP (assume 78% N_2 and 22% O_2) and compare this value to the value of He you calculated in Exercise 6.7A.

EXAMPLE 6.8   **Molar Mass from Gas Densities**

The density of diethyl ether vapor is 3.30 g/L. Calculate the molar mass of diethyl ether.

Solution

This time we are given the density in g/L. We want molar mass, which has units of g/mol. Once again we have the factor 22.4 L = 1 mol diethyl ether. Clearly, we must cancel the unit of liters and obtain the unit of moles in the denominator.

$$\frac{3.30 \text{ g}}{1 \text{ L}} \times \frac{22.4 \text{ L}}{1 \text{ mol}} = 73.9 \text{ g/mol}$$

▶ **Exercise 6.8A**

The density of an unknown gas at STP is 2.30 g/L. Calculate its molar mass.

▶ **Exercise 6.8B**

An unknown gaseous compound contains only hydrogen and carbon and its density is 1.34 g/mol at STP. What is the formula for the compound?

What Is a Mole?

A mole is a particular amount
Of substance—just its formulary weight
Expressed in grams, with Avogadro's count
Of units making up the aggregate.
A mole is a specific quantity:
Its volume measures twenty-two point four
In liters, for a gas at STP.
A mole's a counting unit, nothing more.
A mole is but a single molecule
By Avogadro's number multiplied;
One entity, extremely minuscule,
A trillion trillion times intensified.
A mole is a convenient amount,
For molecules are just too small to count.

6.5 MOLE AND MASS RELATIONSHIPS IN CHEMICAL EQUATIONS

Chemists and other scientists and engineers are often confronted with questions such as, How many grams of phosphoric acid can I make from 660 g of phosphorus? or How many grams of hydrogen peroxide do I need to convert 5.82 g of lead sulfide to lead sulfate? There is no simple way to calculate grams of one substance directly from grams of another. Because chemical reactions involve atoms and molecules, quantities in reactions are calculated in moles of atoms, ions, or molecules. To do such calculations, it is necessary to convert grams of a substance to moles, as we did in Example 6.6. We also must write balanced equations to determine molar ratios.

Recall that chemical equations not only represent ratios of atoms and molecules but also give us mole ratios. For example, the equation

$$C + O_2 \longrightarrow CO_2$$

tells us that one atom of carbon reacts with one molecule (two atoms) of oxygen to form one molecule of carbon dioxide (one atom of carbon and two atoms of oxygen). The equation also indicates that 1 mol (6.02×10^{23} atoms) of carbon reacts with 1 mol (6.02×10^{23} molecules) of oxygen to yield 1 mol (6.02×10^{23} molecules) of carbon dioxide. Because the molar mass in grams of a substance is numerically equal to the formula mass of the substance in atomic mass units, the equation also tells us (indirectly) that 12.0 g (1 mol) of carbon reacts with 32.0 g (1 mol) of oxygen to yield 44.0 g (1 mol) of CO_2 (Figure 6.7).

We need not use exactly 1 mol of each reactant. The important thing is to keep the *ratio* constant. For example, in the preceding reaction, the mass ratio of oxygen to carbon is 32.0:12.0 or 8.0:3.0. We could use 8.0 g of oxygen and 3.0 g of carbon to produce 11.0 g of CO_2. In fact, to calculate the amount of oxygen needed to react with a given amount of carbon, we need only multiply the amount of carbon by the factor 32.0:12.0. The following examples illustrate these relationships.

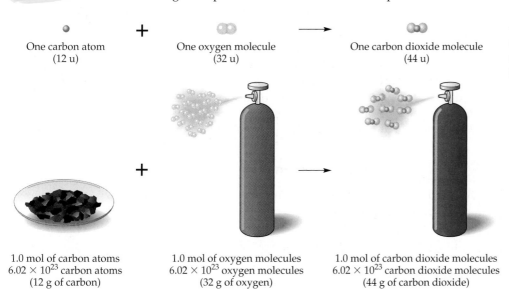

One carbon atom (12 u)   One oxygen molecule (32 u)   One carbon dioxide molecule (44 u)

1.0 mol of carbon atoms
6.02×10^{23} carbon atoms
(12 g of carbon)

1.0 mol of oxygen molecules
6.02×10^{23} oxygen molecules
(32 g of oxygen)

1.0 mol of carbon dioxide molecules
6.02×10^{23} carbon dioxide molecules
(44 g of carbon dioxide)

◀ **Figure 6.7** We cannot weigh single atoms or molecules, but we can weigh equal numbers of these tiny particles.

CONCEPTUAL EXAMPLE 6.9   **Molecular, Molar, and Mass Relationships**

Nitrogen monoxide (nitric oxide), an air pollutant discharged by internal combustion engines, combines with oxygen to form nitrogen dioxide, a yellowish-brown gas that irritates the respiratory system and eyes. The equation for this reaction is

$$2\,NO + O_2 \longrightarrow 2\,NO_2$$

State the molecular, molar, and mass relationships indicated by the equation.

Solution

The molecular and molar relationships can be obtained directly from the equation; no calculation is necessary. The mass relationship requires a little calculation:

Molecular: Two molecules of NO react with one molecule of O_2 to form two molecules of NO_2.

Molar: 2 mol of NO reacts with 1 mol of O_2 to form 2 mol of NO_2.

Mass: 60.0 g of NO (2 mol NO × 30.0 g/mol) reacts with 32.0 g (1 mol O_2 × 32.0 g/mol) of O_2 to form 92.0 g (2 mol NO_2 × 46.0 g/mol) of NO_2.

▶ **Exercise 6.9**

Hydrogen sulfide, a gas that smells like rotten eggs, burns in air to produce sulfur dioxide and water according to the equation

$$2\,H_2S + 3\,O_2 \longrightarrow 2\,SO_2 + 2\,H_2O$$

State the molecular, molar, and mass relationships indicated by this equation.

Molar Relationships in Chemical Equations

The quantitative relationship between reactants and products in a chemical reaction is called **stoichiometry**. The ratio of moles of reactants and products is given by the coefficients in a balanced chemical equation. Consider the combustion of propane, the main component of bottled gas.

$$C_3H_8 + 5\,O_2 \longrightarrow 3\,CO_2 + 4\,H_2O$$

The coefficients in the balanced equation allow us to make statements such as

- 1 mol of C_3H_8 reacts with 5 mol of O_2.
- 3 mol of CO_2 is produced for every 1 mol of C_3H_8 that reacts.
- 4 mol of H_2O is produced for every 3 mol of CO_2 produced.

We can turn these statements into conversion factors known as stoichiometric factors. (Conversion factors are explained in Appendix A.) A **stoichiometric factor** relates the amounts, in *moles*, of any two substances involved in a chemical reaction. Note that we can set up conversion factors for any two compounds involved in a reaction. Typical problems ask you to calculate how much of one compound is equivalent to a given amount of one of the other compounds. All you need to do to solve such a problem is put together a conversion factor relating the two compounds. In the examples that follow, stoichiometric factors are shown in color.

EXAMPLE 6.10 **Molar Relationships**

When 0.105 mol of propane is burned in a plentiful supply of oxygen, how many moles of oxygen is consumed?

$$C_3H_8 + 5\,O_2 \longrightarrow 3\,CO_2 + 4\,H_2O$$

Solution

The equation tells us that 5 mol O_2 is required to burn 1 mol C_3H_8. We can write

$$1 \text{ mol } C_3H_8 \backsimeq 5 \text{ mol } O_2$$

where we use the symbol $\backsimeq$ to mean "is stoichiometrically equivalent to." From this relationship we can construct conversion factors to relate moles of oxygen to moles of propane. The possible conversion factors are

$$\frac{1 \text{ mol } C_3H_8}{5 \text{ mol } O_2} \quad \text{and} \quad \frac{5 \text{ mol } O_2}{1 \text{ mol } C_3H_8}$$

Which one do we use? Only if we multiply the given quantity (0.105 mol C_3H_8) by the factor on the right do we get an answer with the asked-for units (moles of oxygen).

$$? \text{ mol O}_2 = 0.105 \text{ mol } C_3H_8 \times \frac{5 \text{ mol O}_2}{1 \text{ mol } C_3H_8} = 0.525 \text{ mol O}_2$$

▶ **Exercise 6.10**

For the combustion of propane in Example 6.10, **(a)** How many moles of carbon dioxide is formed when 0.529 mol of C_3H_8 is burned? **(b)** How many moles of water is produced when 76.2 mol of C_3H_8 is burned? **(c)** How many moles of carbon dioxide is produced when 1.010 mol of O_2 is consumed?

Mass Relationships in Chemical Equations

A chemical equation defines the stoichiometric relationship in terms of moles, but problems are seldom formulated in moles. Typically, you are given an amount of one substance in grams and asked to calculate how many grams of another substance can be made from it. Such calculations involve several steps.

1. Write a balanced chemical equation for the reaction.
2. Determine the molar masses of the substances involved in the calculation.
3. Write down the given quantity and use the molar mass to convert this quantity to moles.
4. Use the balanced chemical equation to convert moles of the given substance to moles of the desired substance.
5. Use the molar mass to convert moles of the desired substance to grams of the desired substance.

If we know the quantity of any substance in an equation, we can determine the quantities of all the other substances. The conversion process is diagrammed in Figure 6.8. It is best learned from examples and by working out exercises.

 Stoichiometry Calculation

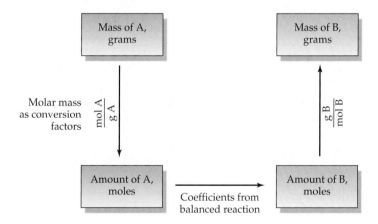

◀ **Figure 6.8** We can use a chemical equation to relate *moles* of any two substances represented in the equation. These substances may be a reactant and a product, two reactants, or two products. We cannot directly relate the mass of one substance to the mass of another substance. To obtain *mass* relationships, we must convert the mass of each substance to moles, relate moles of one substance to moles of the other through a stoichiometric factor, and then convert moles of the second substance to a mass.

EXAMPLE 6.11 **Mass Relationships**

Calculate the mass of oxygen needed to react with 10.0 g of carbon in the reaction that forms carbon dioxide.

Solution
Step 1 The balanced equation is

$$C + O_2 \longrightarrow CO_2$$

Step 2 The molar masses are $2 \times 16.0 = 32.0$ g/mol for O_2 and 12.0 g/mol for C.

Step 3 We convert the mass of the given substance, carbon, to an amount in moles.

$$? \text{ mol C} = 10.0 \text{ g } \cancel{\text{C}} \times \frac{1 \text{ mol C}}{12.0 \text{ g } \cancel{\text{C}}} = 0.833 \text{ mol C}$$

Step 4 We use coefficients from the balanced equation to establish the stoichiometric factor (red) that relates the amount of oxygen to that of carbon.

$$0.833 \cancel{\text{ mol C}} \times \frac{1 \text{ mol O}_2}{1 \cancel{\text{ mol C}}} = 0.833 \text{ mol O}_2$$

Step 5 We convert from moles of oxygen to grams of oxygen.

$$0.833 \cancel{\text{ mol O}_2} \times \frac{32.0 \text{ g O}_2}{1 \cancel{\text{ mol O}_2}} = 26.7 \text{ g O}_2$$

We can also combine the five steps into a single setup. Note that the units in the denominators of the conversion factors are chosen so that each cancels the unit in the numerator of the preceding term.

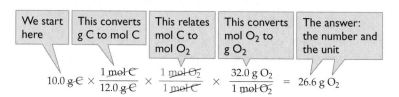

$$10.0 \text{ g } \cancel{\text{C}} \times \frac{1 \cancel{\text{ mol C}}}{12.0 \text{ g } \cancel{\text{C}}} \times \frac{1 \cancel{\text{ mol O}_2}}{1 \cancel{\text{ mol C}}} \times \frac{32.0 \text{ g O}_2}{1 \cancel{\text{ mol O}_2}} = 26.6 \text{ g O}_2$$

(The slightly different answers are due to rounding in the intermediate steps.)

▶ **Exercise 6.11A**
Calculate the mass of oxygen (O_2) needed to react with 0.334 g of nitrogen (N_2) in the reaction that forms nitrogen dioxide.

▶ **Exercise 6.11B**
Calculate the mass of carbon dioxide formed by burning 775 g of each of **(a)** methane (CH_4) and **(b)** butane (C_4H_{10}).

EXAMPLE 6.12 **Mass Relationships**

The decomposition of sodium azide (NaN_3) produces sodium metal and nitrogen gas. The gas is used to inflate automobile airbags. What mass of nitrogen, in grams, can be made from 60.0 g of sodium azide?

Solution

1. We start by writing and balancing the chemical equation, which shows that 2 mol NaN_3 produces 2 mol Na and 3 mol N_2.

$$2 \text{ NaN}_3 \longrightarrow 2 \text{ Na} + 3 \text{ N}_2$$

2. The molar mass of NaN_3 is 23.0 g/mol + (3 × 14.0) g/mol = 65.0 g/mol and the molar mass of N_2 is (2 × 14.0) g/mol = 28.0 g/mol.

3. We convert the mass of the given substance, sodium azide, to an amount in moles.

$$60.0 \text{ g } \cancel{\text{NaN}_3} \times \frac{1 \text{ mol NaN}_3}{65.0 \text{ g } \cancel{\text{NaN}_3}} = 0.923 \text{ mol NaN}_3$$

4. We use coefficients from the balanced equation to establish the stoichiometric factor that relates the amount of nitrogen gas to that of sodium azide.

$$0.923 \cancel{\text{ mol NaN}_3} \times \frac{3 \text{ mol N}_2}{2 \cancel{\text{ mol NaN}_3}} = 1.38 \text{ mol N}_2$$

The units in the denominators of the conversion factors are chosen so that each cancels the unit in the numerator of the preceding term.

5. We convert from moles of nitrogen gas to grams of nitrogen gas.

$$1.38 \; \text{mol N}_2 \times \frac{28.0 \; \text{g N}_2}{1 \; \text{mol N}_2} = 38.6 \; \text{g N}_2$$

As is usually the case, all of the steps just outlined can be combined into a single setup.

$$60.0 \; \text{g NaN}_3 \times \frac{1 \; \text{mol NaN}_3}{65.0 \; \text{g NaN}_3} \times \frac{3 \; \text{mol N}_2}{2 \; \text{mol NaN}_3} \times \frac{28.0 \; \text{g N}_2}{1 \; \text{mol N}_2} = 38.8 \; \text{g NaN}_3$$

Notice that the units of the *numerator* in one stoichiometric factor are the units in the *denominator* of the next stoichiometric factor. In this way, the correct cancellation of units occurs and the units of the final numerator are the units of your answer.

▶ **Exercise 6.12A**

Ammonia reacts with phosphoric acid (H_3PO_4) to form ammonium phosphate [$(NH_4)_3PO_4$]. What mass in grams of ammonia is needed to react completely with 74.8 g of phosphoric acid?

▶ **Exercise 6.12B**

a. The decomposition of potassium chlorate ($KClO_3$) produces potassium chloride (KCl) and O_2 gas. What mass in grams of oxygen can be made from 2.47 g of potassium chlorate?
b. Phosphorus reacts with oxygen to form tetraphosphorus decoxide. The equation is

$$P_4 + O_2 \longrightarrow P_4O_{10} \qquad \text{(not balanced)}$$

What mass in grams of tetraphosphorus decoxide can be made from 3.50 g of phosphorus?

▲ Air-bag technology often relies on the rapid production of nitrogen gas provided by the decomposition of sodium azide, NaN_3.

6.6 THE GAS LAWS

Experiments with gases were instrumental in developing the concepts of Avogadro's number and of molar ratios in reactions. Let's now look at the behavior of gases more closely, using a model known as the **kinetic–molecular theory** (Figure 6.9). The basic postulates of this theory are the following.

1. Particles of a gas (usually *molecules*, but in the case of noble gases, *atoms*) are in rapid, constant motion and move in straight lines.
2. The particles of a gas are tiny compared with the distances between them.
3. Because they are so far apart, there is very little attraction between particles of a gas.
4. Particles collide with one another. Energy is *conserved* in these collisions, although one particle can gain energy at the expense of another.
5. Temperature is a measure of the average kinetic energy of the gas molecules.

We discussed (Section 5.15) how intermolecular forces hold molecules together in solids and liquids. In gases the particles are separated by relatively great distances and are moving about at random, and so they seldom interact. The behavior of gases can be described by the gas laws, which tell how the volume of a gas changes with temperature or pressure.

Boyle's Law: Pressure and Volume

A simple gas law, discovered by Robert Boyle in 1662, describes the relationship between the pressure and volume of a gas. **Boyle's law** states that *for a given amount of gas at a constant temperature, the volume of the gas varies inversely with its pressure.* That is, in a closed container of gas, when the pressure increases, the volume decreases; when the pressure decreases, the volume increases.

Pressure Volume Relationships in the Gas Phase

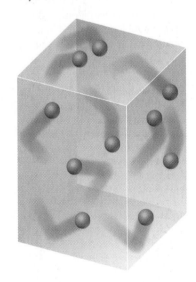

▲ **Figure 6.9** According to the kinetic–molecular theory, molecules of a gas are in constant, random motion. They move in straight lines and undergo collisions with each other and with the walls of the container.

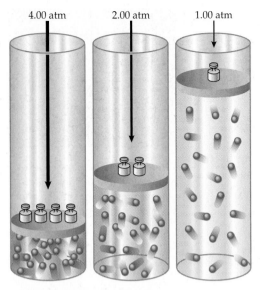

▲ **Figure 6.10** A kinetic–molecular-theory view of Boyle's law. As the pressure is reduced from 4.00 atm to 2.00 atm and then to 1.00 atm, the volume of the gas doubles and then doubles again.

Question: What would happen to the volume if the pressure were changed to 0.500 atm?

A bicycle pump illustrates Boyle's law. When you push down on the plunger, you increase the pressure of the air in the pump. The volume of the air decreases, and the higher-pressure air flows into the bicycle tire.

Think of gases as pictured in the kinetic–molecular theory. A gas exerts a particular pressure because its molecules bounce against the container walls with a certain frequency and speed (Figure 6.10). If the volume of the container is increased while the amount of gas remains fixed, the number of molecules per unit volume of gas decreases. The frequency with which molecules strike a unit area of the container walls decreases, and the gas pressure decreases. Thus, as the volume of a gas is increased, its pressure decreases.

Mathematically, for a given amount of gas at a constant temperature, Boyle's law is written

$$V \propto \frac{1}{P}$$

where the symbol $\propto$ means "is proportional to." This relationship can be changed to an equation by inserting a proportionality constant, a.

$$V = \frac{a}{p}$$

Multiplying both sides of the equation by P, we get

$$PV = a \quad \text{(at constant temperature and amount of gas)}$$

Another way to state Boyle's law, then, is that for a given amount of gas at a constant temperature, the product of the pressure and volume is a constant. This is an elegant and precise, if somewhat abstract, way of summarizing a lot of experimental data. If the product $P \times V$ is to be constant, then when V increases P must decrease, and vice versa. This relationship is demonstrated in Figure 6.11 by a pressure–volume graph.

▶ **Figure 6.11** A graphic representation of Boyle's law. As the pressure of the gas is increased, its volume decreases. When the pressure is doubled ($P_2 = 2 \times P_1$), the volume of the gas decreases to one-half its original value $\left(V_2 = \frac{1}{2} \times V_1\right)$. The pressure–volume product is a constant ($PV = a$).

Question: What would happen to the volume if the pressure were quadrupled?

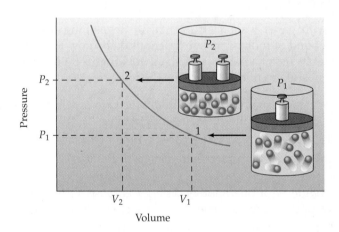

Boyle's law has a number of practical applications perhaps best illustrated by some examples. In Example 6.13, we see how to estimate an answer. Sometimes this is all we need. Even when we want a quantitative answer, however, the estimate helps us to determine whether or not our answer is reasonable.

EXAMPLE 6.13 **Boyle's Law: Pressure–Volume Relationships**

A gas is enclosed in a cylinder fitted with a piston. The volume of the gas is 2.00 L at 0.524 atm. The piston is moved to increase the gas pressure to 5.15 atm. Which of the following is a reasonable value for the volume of the gas at the greater pressure?

0.20 L   0.40 L   1.00 L   16.0 L

Solution

The pressure increase from 0.524 atm to 5.15 atm is almost tenfold. The volume should drop to about one-tenth of the initial value. We estimate a volume of 0.20 L. (The calculated value is 0.203 L.)

▶ **Exercise 6.13**

A gas is enclosed in a 10.2-L tank at 1208 mmHg. (The mmHg is a pressure unit; 760 mmHg = 1 atm.) Which of the following is a reasonable value for the pressure when the gas is transferred to a 30.0-L tank?

300 mmHg   400 mmHg   3,600 mmHg   12,000 mmHg

Example 6.14 illustrates quantitative calculations using Boyle's law. Note that in these applications any units can be used for pressure and volume as long as the same units are used throughout a calculation. As long as we use the same sample of a confined gas at a constant temperature, the product of the initial volume (V_1) times the initial pressure (P_1) is equal to the product of the final volume (V_2) times the final pressure (P_2). Thus, the following useful equation representing Boyle's law can be written.

$$V_1P_1 = V_2P_2$$

Gases are usually stored under high pressure, even though they are used at atmospheric pressure. This practice allows a large quantity of gas to be stored in a small volume.

EXAMPLE 6.14 **Boyle's Law: Pressure–Volume Relationships**

A cylinder of oxygen has a volume of 2.25 L. The pressure of the gas is 1470 pounds per square inch (psi) at 20 °C. What volume will the oxygen occupy at standard atmospheric pressure (14.7 psi) assuming no temperature change?

Solution

We find it helpful to first separate the initial from the final condition.

Initial	Final	Change
P_1 = 1470 psi	P_2 = 14.7 psi	↓ The pressure goes down; therefore,
V_1 = 2.25 L	V_2 = ?	↑ the volume goes up.

Then use the equation $V_1P_1 = V_2P_2$ and solve for the desired volume or pressure. In this case, we solve for V_2.

$$V_2 = \frac{V_1P_1}{P_2}$$

$$V_2 = \frac{2.25 \text{ L} \times 1470 \text{ psi}}{14.7 \text{ psi}} = 225 \text{ L}$$

Because the final pressure (14.7 psi) is *less than* the initial pressure (1470 psi), we expect the final volume (225 L) to be *larger than* the original volume (2.25 L), and we see that it is.

Hint: Regardless of which variable is to be determined, it is recommended that you solve the equation for the unknown *before* substituting values into the equation. Rearranging a few letters takes less time than rearranging and rewriting complex terms.

▶ **Exercise 6.14A**

A sample of air occupies 73.3 mL at 98.7 atm and 0 °C. What volume will the air occupy at 4.02 atm and 0 °C?

▶ **Exercise 6.14B**

A sample of helium occupies 535 mL at 988 mmHg and 25 °C. If the sample is transferred to a 1.05-L flask at 25 °C, what will be the gas pressure in the flask?

Charles's Law: Temperature and Volume

In 1787 the French physicist Jacques Charles (1746–1823), a pioneer hot air balloonist, studied the relationship between the volume and temperature of gases. He found that when a fixed mass of gas is cooled at constant pressure, its volume decreases. When the gas is heated, its volume increases. Temperature and volume vary directly; that is, they rise or fall together. But this law requires a bit more thought. If a quantity of gas that occupies 1.00 L is heated from 100 °C to 200 °C at constant pressure, the volume does not double but only increases to about 1.27 L. The relationship between temperature and volume is not as tidy as it may seem at first.

Zero pressure or zero volume really means *zero*; no pressure or volume can be measured. Zero degrees Celsius (0 °C) means only the freezing point of water. This zero point is arbitrarily set, much as mean sea level is set as the arbitrary zero for measuring altitudes on Earth. Temperatures below 0 °C are often encountered, as are altitudes below sea level.

Charles noted that for each degree Celsius rise in temperature, the volume of a gas increases by $\frac{1}{273}$ of its volume at 0 °C. If we plot volume against temperature, we get a straight line (Figure 6.12). We can extrapolate the line beyond the range of measured temperatures to the temperature at which the volume of the gas would become zero. This temperature is −273.15 °C. In 1848 William Thomson (Lord Kelvin) made this temperature the zero point on an absolute temperature scale now called the Kelvin scale. As noted in Chapter 1, the unit of temperature on this scale is the kelvin (K).

A modern statement of **Charles's law** is that *the volume of a fixed amount of a gas at a constant pressure is directly proportional to its absolute temperature.* Mathematically, this relationship is expressed as

$$V = bT \quad \text{or} \quad \frac{V}{T} = b$$

where b is a proportionality constant. To keep $\frac{V}{T}$ equal to a constant value, when the temperature increases, the volume must also increase. When the temperature decreases, the volume must decrease accordingly (Figure 6.13).

As long as we use the same sample of trapped gas at a constant pressure, the initial volume (V_1) divided by the initial absolute temperature (T_1) is equal to the final volume (V_2) divided by the final absolute temperature (T_2). We can use the following equation to solve problems involving Charles's law.

$$\frac{V_1}{T_1} = \frac{V_2}{T_2}$$

The kinetic–molecular model readily explains the relationship between gas volume and temperature. When we heat a gas, we supply the gas molecules with energy and they begin to move faster. These speedier molecules strike the walls of the

Any real gas liquifies—and the liquid freezes—before the gas ever reaches this zero-volume temperature. Extrapolation of the line to zero volume is therefore an exercise only for the imagination.

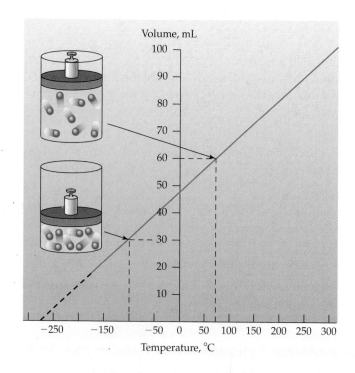

Volume, mL

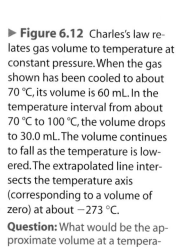

▶ **Figure 6.12** Charles's law relates gas volume to temperature at constant pressure. When the gas shown has been cooled to about 70 °C, its volume is 60 mL. In the temperature interval from about 70 °C to 100 °C, the volume drops to 30.0 mL. The volume continues to fall as the temperature is lowered. The extrapolated line intersects the temperature axis (corresponding to a volume of zero) at about −273 °C.

Question: What would be the approximate volume at a temperature of 150 °C?

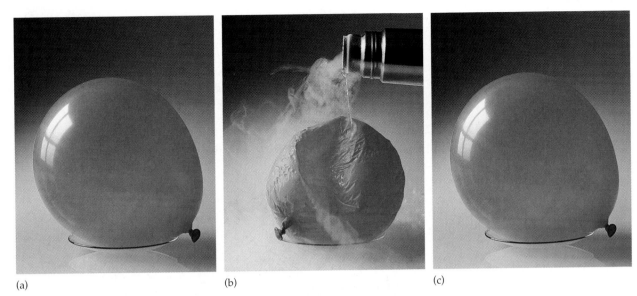

(a)   (b)   (c)

▲ **Figure 6.13** A dramatic illustration of Charles's law. (a) Liquid nitrogen (boiling point, −196 °C) cools the balloon and its contents to a temperature far below room temperature. (b) As the balloon warms back to room temperature, the volume of air increases proportionately (about fourfold).

container harder and more often. For the pressure to stay the same, the volume of the container must increase so that the increased molecular motion will be distributed over a greater space.

EXAMPLE 6.15   **Charles's Law: Temperature–Volume Relationship**

A balloon indoors, where the temperature is 27 °C, has a volume of 2.00 L. What would its volume be **(a)** in a hot room where the temperature is 47 °C, and **(b)** outdoors, where the temperature is −23 °C? (Assume no change in pressure in either case.)

Solution

First, and most important, convert all temperatures to the Kelvin scale:

$$T(K) = t(°C) + 273$$

The initial temperature (T_1) in each case is $(27 + 273) = 300$ K and the final temperatures are **(a)** $(47 + 273) = 320$ K and **(b)** $(-23 + 273) = 250$ K.

a. We start by separating the initial from the final condition.

Initial	Final	Change
$t_1 = 27 °C$	$t_2 = 47 °C$	⇑
$T_1 = 300$ K	$T_2 = 320$ K	⇑
$V_1 = 2.00$ L	$V_2 = ?$	⇑

Solving the equation

$$\frac{V_1}{T_1} = \frac{V_2}{T_2}$$

for V_2, we have

$$V_2 = \frac{V_1 T_2}{T_1}$$

$$V_2 = \frac{2.00 \text{ L} \times 320 \text{ K}}{300 \text{ K}} = 2.13 \text{ L}$$

As expected, because the temperature increases, the volume must also increase.

b. We have the same initial conditions as in **(a)**, but different final conditions.

Initial	Final	Change
$t_1 = 27\ °C$	$t_2 = -23\ °C$	⇓
$T_1 = 300\ K$	$T_2 = 250\ K$	⇓
$V_1 = 2.00\ L$	$V_2 = ?$	⇓

Again using Charles's law, we solve the equation for V_2:

$$V_2 = \frac{V_1 T_2}{T_1}$$

$$V_2 = \frac{2.00\ L \times 250\ K}{300\ K} = 1.67\ L$$

As we expected, the volume decreased because the temperature decreased.

▶ **Exercise 6.15A**

a. A sample of oxygen gas occupies a volume of 2.10 L at 25 °C. What volume will this sample occupy at 150 °C? (Assume no change in pressure.)
b. A sample of hydrogen occupies 692 L at 602 °C. If the pressure is held constant, what volume will the gas occupy after being cooled to 23 °C?

▶ **Exercise 6.15B**

At what Celsius temperature will the initial volume of oxygen in Exercise 6.15A occupy 0.750 L? (Assume no change in pressure.)

The Ideal Gas Law

Gas Laws

Boyle's law and Charles's law are useful when the temperature or pressure can be held constant. Often, however, both the temperature and the pressure change at the same time, and the amount of gas may change as well. The **ideal gas law**, expressed in the equation

$$\frac{PV}{nT} = R \quad \text{or} \quad PV = nRT$$

involves four variables (R is a constant). The number of moles is given by n. The constant R (called the *gas constant*) can be calculated from the fact that 1 mol of gas occupies 22.4 L at 273 K (0 °C) and 1 atm (Section 6.2). If P is in atmospheres, V in liters, and T in kelvins, then R has a value of

We read this R value as "0.0821 liter-atmosphere per mole-kelvin."

$$0.0821\ \frac{L \cdot atm}{mol \cdot K}$$

The ideal gas equation can be used to calculate any of the four quantities—P, V, n, or T—if the other three are known.

EXAMPLE 6.16 **Ideal Gas Law**

Use the ideal gas law to calculate **(a)** the volume occupied by 1.00 mol of nitrogen gas at 244 K and 1.00 atm pressure, and **(b)** the pressure exerted by 0.500 mol of oxygen in a 15.0-L container at 303 K.

Solution
a. We start by solving the ideal gas equation for V.

$$V = \frac{nRT}{P}$$

$$V = \frac{1.00\ mol}{1.00\ atm} \times \frac{0.0821\ L \cdot atm}{mol \cdot K} \times 244\ K = 20.0\ L$$

b. Here we solve the ideal gas equation for P.

$$P = \frac{nRT}{V}$$

$$P = \frac{0.500 \text{ mol}}{15.0 \text{ L}} \times \frac{0.0821 \text{ L} \cdot \text{atm}}{\text{mol} \cdot \text{K}} \times 303 \text{ K} = 0.829 \text{ atm}$$

▶ **Exercise 6.16A**

Determine (**a**) the pressure exerted by 0.0330 mol of oxygen in an 18.0-L container at 313 K, and (**b**) the volume occupied by 0.200 mol of nitrogen gas at 298 K and 0.980 atm.

▶ **Exercise 6.16B**

Determine the volume of nitrogen gas produced from the decomposition of 130 g sodium azide (about the amount in a typical automobile airbag) at 25 °C and 1 atm.

Hint for Exercise 6.16B: Begin with a stoichiometric calculation that converts the mass of sodium azide to moles of nitrogen gas. (See Example 6.12.)

6.7 SOLUTIONS

We briefly introduced solutions in Chapter 5. Recall that a *solution* is a homogeneous mixture of two or more substances. The substance being dissolved is the *solute*, and the substance doing the dissolving is the *solvent*. The solute is usually the component present in the lesser quantity, and the solvent is usually present in the greater quantity. There are many solvents: Hexane dissolves grease. Ethanol dissolves many drugs. Isopentyl acetate, a component of banana oil, is a solvent for the glue used in making model airplanes. Water is no doubt the most familiar solvent, dissolving as it does many common substances such as sugar, salt, and ethanol. We focus our discussion here on **aqueous solutions**, those in which water is the solvent, and we take a more quantitative look at the relationship between solute and solvent.

Solution Concentrations

We say that some substances, such as sugar and salt, are *soluble* in water. Of course, there is a limit to the quantity of sugar or salt we can dissolve in a given volume of water, but we still find it convenient to say that they are soluble in water because an appreciable quantity dissolves. Other substances, such as an iron nail or sand (silicon dioxide), we consider to be *insoluble* because the limit of solubility is near zero. Such terms as *soluble* and *insoluble* are useful, but they are imprecise and must be used with care. Two other roughly estimated but sometimes useful terms are *dilute* and *concentrated*. A **dilute solution** is one that contains a little bit of solute in lots of solvent. For example, a pinch of sugar in a liter of water is a dilute, faintly sweet, sugar solution. A **concentrated solution** is one in which lots of solute is dissolved in a relatively small quantity of solvent. A sugar solution with lots of solute in a relatively small amount of water is a concentrated, very sweet solution. The dilute solution is quite "thin"—little changed in appearance from that of pure water. The concentrated solution is thick and rather syrupy.

Scientific work generally requires more precise measurement of quantities than "a pinch of sugar in a liter of water." Further, quantitative work often requires the unit amount of substance (moles) because substances enter into chemical reactions according to *molar* ratios.

Molarity

A concentration unit that chemists often use is *molarity*. For reactions involving solutions, the amount of solute is usually measured in moles and the quantity of solution in liters or milliliters. The **molarity (M)** is the amount of solute, in moles, per liter of solution.

$$\text{Molarity (M)} = \frac{\text{moles of solute}}{\text{liters of solution}}$$

EXAMPLE 6.17   Solution Concentration: Molarity

Calculate the molarity of a solution made by dissolving 3.50 mol of NaCl in enough water to produce 2.00 L of solution.

Solution

$$\text{Molarity (M)} = \frac{\text{moles of solute}}{\text{liters of solution}} = \frac{3.50 \text{ mol NaCl}}{2.00 \text{ L solution}} = 1.75 \text{ M NaCl}$$

We read 1.75 M NaCl as "1.75 molar NaCl."

▶ **Exercise 6.17A**

Calculate the molarity of a solution that has 0.0500 mol of NH_3 in 5.75 L of solution.

▶ **Exercise 6.17B**

Calculate the molarity of a solution made by dissolving 0.750 mol of H_3PO_4 in enough water to produce 775 mL of solution.

Usually when preparing a solution, we must *weigh* the solute. Balances do not display units of moles, so we usually work with a given mass and divide by the molar mass of the substance, as illustrated in Example 6.18.

EXAMPLE 6.18   Solution Concentration: Molarity

What is the molarity of a solution in which 333 g of potassium hydrogen carbonate is dissolved in enough water to make 10.0 L of solution?

Solution

First, we must convert grams of $KHCO_3$ to moles of $KHCO_3$.

$$333 \text{ g KHCO}_3 \times \frac{1 \text{ mol KHCO}_3}{100.1 \text{ g KHCO}_3} = 3.33 \text{ mol KHCO}_3$$

Now use this value as the numerator in the defining equation for molarity. The solution volume, 10.0 L, is the denominator.

$$\text{Molarity} = \frac{3.33 \text{ mol KHCO}_3}{10.0 \text{ L solution}} = 0.333 \text{ M KHCO}_3$$

▶ **Exercise 6.18**

Calculate the molarity of each of the following solutions.
a. 18.0 mol of H_2SO_4 in 2.00 L of solution
b. 3.00 mol of KI in 2.39 L of solution
c. 0.206 mol of HF in 752 mL of solution (HF is used for etching glass.)

Frequently we need to know the *mass* of solute required to prepare a given volume of solution of a particular molarity. In such calculations we can use molarity as a conversion factor between moles of solute and liters of solution. Thus, in Example 6.19, the expression 0.15 M NaCl means 0.15 mol of NaCl per liter of solution, expressed as the conversion factor

$$\frac{0.15 \text{ mol NaCl}}{1 \text{ L solution}}$$

EXAMPLE 6.19   Solution Preparation: Molarity

How many grams of NaCl is required to prepare 0.500 L of typical over-the-counter saline solution (about 0.15 M NaCl)?

Solution

First we use the molarity as a conversion factor to calculate moles of NaCl.

$$0.500 \text{ L solution} \times \frac{0.15 \text{ mol NaCl}}{1 \text{ L solution}} = 0.075 \text{ mol NaCl}$$

Then we use the molar mass to calculate the grams of NaCl.

$$0.75 \text{ mol NaCl} \times \frac{58.4 \text{ g NaCl}}{1 \text{ mol NaCl}} = 44 \text{ g NaCl}$$

▶ **Exercise 6.19A**
What mass in grams of potassium hydroxide is required to prepare 2.00 L of 6.00 M KOH?

▶ **Exercise 6.19B**
What mass in grams of potassium hydroxide is required to prepare 100.0 mL of 1.00 M KOH?

Quite often, solutions of known molarity are available. For example, commercial concentrated hydrochloric acid is 12 M. How would you calculate the *volume* needed to get a certain number of moles of solute? We can again rearrange the definition of molarity to obtain

$$\text{Liters of solution} = \frac{\text{moles of solute}}{\text{molarity}}$$

EXAMPLE 6.20 **Moles from Molarity and Volume**

Concentrated hydrochloric acid has a concentration of 12.0 M HCl. How many milliliters of this solution would one need to get 0.425 mol of HCl?

Solution

$$\text{Liters of HCl solution} = \frac{\text{moles of solute}}{\text{molarity}} = \frac{0.425 \text{ mol HCl}}{12.0 \text{ M HCl}}$$

$$= \frac{0.425 \text{ mol HCl}}{12.0 \text{ mol HCl/L}} = 0.0354 \text{ L}$$

We would need 0.0354 L (35.4 mL) of the solution to have 0.425 mol. Remember that molarity is moles per liter of *solution*, not per liter of solvent.

▶ **Exercise 6.20A**
What volume in milliliters of 15.0 M aqueous ammonia (NH_3) solution do you need to get 0.445 mol of NH_3?

▶ **Exercise 6.20B**
What mass in grams of HNO_3 is in 500 mL of rain that has a concentration of 2.0×10^{-5} M HNO_3?

Percent Concentrations

For many practical applications, we often express solution concentrations in percentage composition. Then, if we require a precise quantity of solution, we simply measure out a mass or volume. There are different ways to express percentage, depending on whether mass or volume is being measured. If both the solute and solvent are liquids, **percent by volume** is often used because liquid volumes are so easily measured.

$$\text{Percent by volume} = \frac{\text{volume of solute}}{\text{volume of solution}} \times 100\%$$

For example, ethanol (CH_3CH_2OH) for medicinal purposes—generally known as USP (an abbreviation of *United States Pharmacopeia*, the official publication of standards for pharmaceutical products) ethanol—is 95% by volume. That is, it consists of 95 mL CH_3CH_2OH per 100 mL of aqueous solution.

▲ Rubbing alcohol is a water–isopropyl alcohol solution that is 70% isopropyl alcohol [$(CH_3)_2CHOH$] by volume.

Notice in Example 6.21 that any units of volume for solute and solvent may be used, as long as the two have the same units.

EXAMPLE 6.21 **Percent by Volume**

Two-stroke engines use a mixture of 120 mL of oil dissolved in enough gasoline to make of 4.0 liters of fuel. What is the percent by volume of oil in this mixture?

Solution

$$\text{Percent by volume} = \frac{120 \text{ mL oil}}{4000 \text{ mL solution}} \times 100\% = 3.0\%$$

▶ **Exercise 6.21A**

What is the volume percent of ethanol in a solution that has 58.0 mL water in 625 mL of an ethanol–water solution?

▶ **Exercise 6.21B**

Assume that the volumes are additive, and determine the volume percent toluene ($C_6H_5CH_3$) in a solution made by mixing 40.0 mL of toluene with 75.0 mL of benzene (C_6H_6).

EXAMPLE 6.22 **Solution Preparation: Percent by Volume**

Describe how to make 775 mL of vinegar (about a 5.0% by volume solution of acetic acid in water).

Solution

We begin by rearranging the equation for percent by volume to solve for volume of solute.

$$\text{Volume of solute} = \frac{\text{percent by volume} \times \text{volume of solution}}{100\%}$$

Substituting, we have

$$= \frac{5.0\% \times 775 \text{ mL}}{100\%} = 39 \text{ mL}$$

Take 39 mL of acetic acid and add enough water to make 775 mL of solution. Notice that we *don't* simply add 775 mL of water, because the final volume of solution must be 775 mL.

▶ **Exercise 6.22A**

Describe how to prepare 450 mL of an aqueous solution that is 70.0% isopropyl alcohol by volume.

▶ **Exercise 6.22B**

Describe how you would prepare exactly 2.00 L of an aqueous solution that is 9.77% acetic acid by volume.

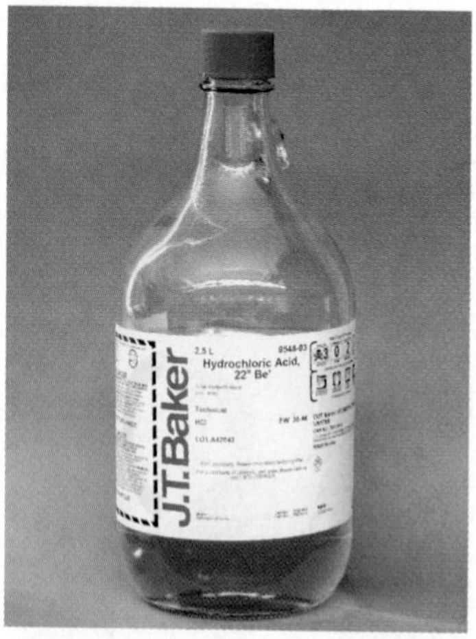

▲ "Reagent grade" concentrated hydrochloric acid is a 38% by mass solution of HCl in water.

Many commercial solutions are labeled with the concentration in **percent by mass**. For example, sulfuric acid is sold as a solution that is 35.7% H_2SO_4 for use in storage batteries, 77.7% H_2SO_4 for the manufacture of phosphate fertilizers, and 93.2% H_2SO_4 for pickling steel. Each of these figures is a percent by mass: 35.7 g of H_2SO_4 per 100 g of sulfuric acid solution, and so on.

$$\text{Percent by mass} = \frac{\text{mass of solute}}{\text{mass of solution}} \times 100\%$$

EXAMPLE 6.23   **Percent by Mass**

What is the percent by mass of a solution of 25.5 g of NaCl dissolved in 425 g (425 mL) of water?

Solution

Use these values in the above percent-by-mass equation:

$$\text{Percent by mass} = \frac{25.5 \text{ g NaCl}}{(25.5 + 425) \text{ g solution}} \times 100\% = 5.66\% \text{ NaCl}$$

▶ **Exercise 6.23A**

Hydrogen peroxide solutions for home use are 3.0% by mass solutions of H_2O_2 in water. What is the percent by mass of a solution of 9.40 g of H_2O_2 dissolved in 335 g (335 mL) of water?

▶ **Exercise 6.23B**

Sodium hydroxide (NaOH, lye) is used to make soap and is very soluble in water. What is the percent by mass of a solution that contains 1.00 kg of NaOH dissolved in 950 mL of water?

Notice in Example 6.23 that any units of mass for solute and solvent may be used, as long as the two have the same units.

EXAMPLE 6.24   **Solution Preparation: Percent by Mass**

Describe how to make 430 g of an aqueous solution that is 4.85% by mass $NaNO_3$.

Solution

We begin by rearranging the equation for percent by mass to solve for mass of solute.

$$\text{Mass of solute} = \frac{\text{percent by mass} \times \text{mass of solution}}{100\%}$$

Substituting, we have

$$= \frac{4.85\% \times 430 \text{ g}}{100\%} = 20.9 \text{ g}$$

Take 20.9 g of $NaNO_3$ and add enough water to make 430 g of solution.

▶ **Exercise 6.24A**

Describe how you would prepare 125 g of an aqueous solution that is 4.50% glucose by mass.

▶ **Exercise 6.24B**

Describe how you would prepare 1750 g of *isotonic saline*, a commonly used intravenous (IV) solution that is 0.89% sodium chloride by mass.

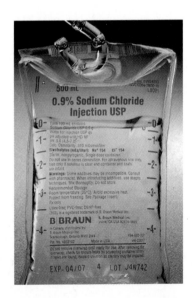

▲ Isotonic or "normal" saline used by hospitals is about 0.9% sodium chloride by mass.

Note that for percent concentrations, the mass or volume of the solute needed doesn't depend on what the solute is. A 10% by mass solution of NaOH contains 10 g of NaOH per 100 g of solution. Similarly, 10% HCl and 10% $(NH_4)_2SO_4$ and 10% $C_{110}H_{190}N_3O_2Br$ each contain 10 g of the specified solute per 100 g of solution. For *molar* solutions, however, the mass of solute in a solution of specified molarity is different for different solutes. A liter of a 0.10 M solution requires 4.0 g (0.10 mol) of NaOH, 3.7 g (0.10 mol) of HCl, 13.2 g (0.10 mol) of $(NH_4)_2SO_4$, or 166 g (0.10 mol) of $C_{110}H_{190}N_3O_2Br$.

Critical Thinking Exercises

Apply knowledge that you have gained in this chapter and one or more of the FLaReS principles (Chapter 1) to evaluate the following statements or claims.

6.1 Suppose that someone has published a paper claiming a new value for Avogadro's number. The author says that he has made some very careful laboratory measurements and his calculations indicate that the true value for the Avogadro constant is 3.01875×10^{23}. Is this claim credible in your opinion? What questions would you ask the person about his claim?

6.2 A chemistry teacher asked his students, "What is the mass, in grams, of a mole of bromine?" One student said "80"; another said "160"; and several others gave answers of 79, 81, 158, and 162. The teacher stated that all of these answers were correct. Do you believe his statement?

6.3 Some automobile tire stores claim that filling your car tires with pure, dry nitrogen is much better than using plain air. They make the following claims: (1) The pressure inside nitrogen-filled tires does not rise or fall with temperature changes. (2) Nitrogen leaks out of tires much more slowly than air because the nitrogen molecules are bigger. (3) Nitrogen is not very reactive, and moisture and oxygen in air cause corrosion that shortens tire life by 25 to 30%. Use information you have gained in this chapter and from other sources as necessary to evaluate these claims.

6.4 A website on fireworks provides directions for preparing potassium nitrate, KNO_3 (molar mass 101 g/mol), using potassium carbonate (138 g/mol) and ammonium nitrate (80 g/mol), according to the following equation:

$$K_2CO_3 + 2\,NH_4NO_3 \longrightarrow 2\,KNO_3 + CO_2 + H_2O + NH_3$$

The directions claim that one should "mix one kilogram of potassium carbonate with two kilograms of ammonium nitrate. The carbon dioxide and ammonia come off as gases, and the water can be evaporated, leaving two kilograms of pure potassium nitrate." Use information from this chapter to evaluate this claim.

6.5 A battery manufacturer claims that lithium batteries deliver the same power as batteries using nickel, zinc, or lead, but the lithium batteries are much lighter because lithium has a lower atomic mass than does nickel, zinc, or lead.

SUMMARY

Section 6.1—A **chemical equation** is shorthand for a chemical change, using symbols and formulas instead of words: $C + O_2 \longrightarrow CO_2$. Materials to the left of the arrow are starting materials or **reactants**, while those to the right of the arrow are the **products** or what we end with. Physical states may be indicated with (s), (l), or (g). Because matter is conserved, an equation must be balanced; it must have the same number and type of each atom on each side. Equations are balanced by placing coefficients in front of each reactant or product; we do *not* balance an equation by changing the formulas of reactants or products. The balanced equation $2\,NaN_3 \longrightarrow 2\,Na + 3\,N_2$ means that two molecules of NaN_3 react to give two atoms of Na and three molecules of N_2.

● = hydrogen
● = oxygen

Sections 6.2–6.3—Gay-Lussac's experiments were summarized in the **law of combining volumes**: At a given temperature and pressure, the volumes of gaseous reactants and products is in a small whole-number ratio (1:1, 2:1, 4:3, etc.). **Avogadro's hypothesis** explained this law by stating that equal volumes of all gases contain the same number of molecules at fixed temperature and pressure. **Avogadro's number** is defined as the number of ^{12}C atoms in exactly 12 grams of ^{12}C: 6.02×10^{23} atoms.

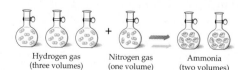

Hydrogen gas (three volumes)   +   Nitrogen gas (one volume)   →   Ammonia (two volumes)

Section 6.4—A **mole (mol)** is the amount of a substance that contains Avogadro's number of elementary units—atoms, molecules, or formula units, depending on the substance. The **formula mass** is the average mass of a formula unit of a sub-

stance, relative to that of a ^{12}C atom. If the formula represents a molecule, the term *molecular mass* often is used. The **molar mass** is the mass of one mole of a substance. Molar mass is the same number as formula mass, molecular mass, or atomic mass, but has units of grams per mole (g/mol). Molar mass is used to convert from grams to moles of a substance, and vice versa. The **molar volume** of a gas is the volume occupied by 1 mol of any gas. At 0 °C and 1 atm, known as **standard temperature and pressure (STP)**, that molar volume is 22.4 L for any gas. We can use molar volume, 1 mol gas = 22.4 L, to convert from moles of gas to volume, and vice versa.

Section 6.5—A balanced equation can be read in terms of atoms and molecules, or in terms of moles. **Stoichiometry** or mass relationships in chemical reactions can be found by using **stoichiometric factors**, which relate moles of one substance to moles of another in a chemical reaction. Stoichiometric factors can be obtained without calculation, directly from the balanced equation. To evaluate mass relationships, molar masses and stoichiometric factors are both used as conversion factors.

Section 6.6—The behavior of gases is explained using **kinetic–molecular theory**, which describes gas molecules: (1) They move rapidly, constantly, and in straight lines; (2) they are far apart; (3) there is little attraction between them; (4) when they collide, energy is conserved; (5) temperature is a measure of average kinetic energy of the gas molecules. **Boyle's law** says that the volume of a fixed amount of gas varies inversely with pressure at constant temperature, or $V_1 P_1 = V_2 P_2$. **Charles's law** says that the volume of a fixed amount of gas varies directly with absolute (Kelvin) temperature, or $V_1/T_1 = V_2/T_2$. The

ideal gas law shows the relationship among pressure (P), volume (V), number of moles (n), and absolute temperature (T): $PV = nRT$. When P = atmospheres, V = liters, n = moles of gas, and T = kelvins, the value of the constant R is 0.0821 liter · atm/(mol · K).

Section 6.7—A solution is a homogeneous mixture, consisting of a solute dissolved in a solvent. A substance is soluble in a solvent if some appreciable quantity of the substance dissolves in the solvent. Any substance that does not dissolve significantly in a solvent is insoluble in that solvent. An **aqueous solution** has water as the solvent. Concentration of a solution can be expressed in many ways. A **concentrated solution** has a relatively large amount of solute compared to solvent, and a **dilute solution** has little solute compared to the solvent. One quantitative unit of concentration is **molarity (M)**, moles of solute dissolved per liter of solution.

$$\text{Molarity (M)} = \frac{\text{moles of solute}}{\text{liters of solution}}$$

There are several types of percent concentration units. Two important ones are **percent by volume** and **percent by mass**.

$$\text{Percent by volume} = \frac{\text{volume of solute}}{\text{volume of solution}} \times 100\%$$

$$\text{Percent by mass} = \frac{\text{mass of solute}}{\text{mass of solution}} \times 100\%$$

For each concentration unit, if we know two of the three terms in the equation, we can solve for the third.

REVIEW QUESTIONS

1. Define or illustrate each of the following.
   a. formula unit
   b. formula mass
   c. mole
   d. Avogadro's number
   e. molar mass
   f. molar volume

2. Explain the difference between the atomic mass of oxygen and the formula mass of oxygen (gas).

3. What is Avogadro's hypothesis? How does it explain Gay-Lussac's law of combining volumes?

4. Consider the law of conservation of mass and explain why we must work with balanced chemical equations.

5. State Boyle's law in words and as a mathematical equation.

6. Use the kinetic–molecular theory to explain Boyle's law.

7. State Charles's law in words and as a mathematical equation. Why must an absolute temperature scale rather than the Celsius scale be used for calculations involving Charles's law?

8. Use the kinetic–molecular theory to explain Charles's law.

9. State the ideal gas law in words and in the form of a mathematical equation.

10. Define or explain and illustrate the following terms.
    a. solution
    b. solvent
    c. solute
    d. aqueous solution

11. Define or explain and illustrate the following terms.
    a. concentrated solution
    b. dilute solution
    c. soluble
    d. insoluble

12. Explain how to calculate each of the following concentrations.
    a. percent by mass
    b. percent by volume
    c. molarity

PROBLEMS

Interpreting Formula Units

13. How many hydrogen atoms are indicated in one formula unit of each of the following?
    a. $Ca(H_2PO_4)_2$
    b. NH_2OH
    c. $C_6H_5COOCH_3$

14. How many oxygen atoms are indicated in one formula unit of each of the following?
    a. $Ti(SO_3)_2$
    b. $Mg_3(PO_4)_2$
    c. $Cu(CH_3COO)_2$

15. How many atoms of each kind (Fe, C, H, and O) does the notation $2\ Fe(HOOCCOO)_3$ indicate?

16. How many atoms of each kind (N, S, H, and O) does the notation $6\ NH_4HSO_4$ indicate?

Interpreting Chemical Equations

17. Consider the following equation. **(a)** Explain its meaning at the molecular level. **(b)** Interpret it in terms of moles. **(c)** State the mass relationships conveyed by the equation.

$$4\ NH_3 + 3\ O_2 \longrightarrow 2\ N_2 + 6\ H_2O$$

18. Translate the following chemical equations into words:
    a. $Ti + O_2 \longrightarrow TiO_2$
    b. $2\ C + O_2 \longrightarrow 2\ CO$
    c. $H_2 + Cl_2 \longrightarrow 2\ HCl$

Balancing Chemical Equations

19. Indicate whether the following equations are balanced. (You need not balance the equation; just determine whether it is balanced as written.)
    a. $Ca + 2\ H_2O \longrightarrow Ca(OH)_2 + H_2$
    b. $2\ LiOH + CO_2 \longrightarrow Li_2CO_3 + H_2O$
    c. $4\ LiH + AlCl_3 \longrightarrow 2\ LiAlH_4 + 2\ LiCl$
    d. $2\ Sn + 2\ H_2SO_4 \longrightarrow 2\ SnSO_4 + SO_2 + 2\ H_2O$

20. Indicate whether the following equations are balanced as written.
    a. $10\ K + 2\ KNO_3 \longrightarrow 6\ K_2O + N_2$
    b. $2\ NH_3 + O_2 \longrightarrow 3\ H_2O + N_2$
    c. $4\ BF_3 + 3\ H_2O \longrightarrow H_3BO_3 + 3\ HBF_4$
    d. $6\ NaOH + 3\ Cl_2 \longrightarrow 5\ NaCl + NaClO_3 + 3\ H_2O$

21. Balance the following equations.
    a. $Al + O_2 \longrightarrow Al_2O_3$
    b. $H_2 + V_2O_5 \longrightarrow V_2O_3 + H_2O$
    c. $H_2O + Cl_2O_5 \longrightarrow HClO_3$

22. Balance the following equations.
    a. $HCl + MnO_2 \longrightarrow MnCl_2 + Cl_2 + H_2O$
    b. $Ca + MnO_2 \longrightarrow CaO + Mn$
    c. $C_5H_{12} + O_2 \longrightarrow CO_2 + H_2O$

23. Write balanced equations for the following processes.
    a. Iron metal reacts with oxygen gas to form rust, iron(III) oxide.
    b. Calcium carbonate (the active ingredient in most antacids) reacts with stomach acid (HCl) to make calcium chloride, water, and carbon dioxide.
    c. Heptane (C_7H_{16}) burns in oxygen to make carbon dioxide and water.

24. Write balanced equations for the following processes.
    a. Nitrogen gas and oxygen gas react to form nitric oxide, NO.
    b. Ozone (O_3) decomposes into oxygen gas.
    c. Xenon hexafluoride reacts with water to make xenon trioxide and hydrogen fluoride.

Molar Volume and Volume Relationships in Chemical Equations

25. What is the volume of each of the following gases at STP?
    a. 1.00 mole N_2
    b. 1.00 mole H_2
    c. 2.00 moles C_2H_6

26. What is the mass of one molar volume of each of the gases in Problem 25?

27. Look again at Figure 6.3. Make a similar sketch to show that when hydrogen reacts with oxygen to form steam at 100 °C, two volumes of hydrogen unite with one volume of oxygen to yield two volumes of steam.

28. Look again at Figure 6.4. Make a similar sketch to show that when hydrogen reacts with oxygen to form steam at 100 °C, each volume of gas—hydrogen, oxygen, or steam—contains the same number of molecules.

29. Consider the following equation.

$$2\ C_4H_{10}(g) + 13\ O_2(g) \longrightarrow 8\ CO_2(g) + 10\ H_2O(g)$$

    a. What volume in liters of $H_2O(g)$ is formed when 3.44 L of $C_4H_{10}(g)$ is burned? Assume both gases are measured under the same conditions.
    b. What volume in milliliters of $O_2(g)$ is required to burn 29 mL of $C_4H_{10}(g)$? Assume both gases are measured under the same conditions.

30. Consider the following equation.

$$C_2H_4(g) + 3\ O_2(g) \longrightarrow 2\ CO_2(g) + 2\ H_2O(g)$$

    a. What volume in liters of $CO_2(g)$ is formed when 125 L of $C_2H_4(g)$ is burned? Assume both gases are measured under the same conditions.
    b. What volume in liters of $O_2(g)$ is required to form 36 L of $CO_2(g)$? Assume both gases are measured under the same conditions.

Molar Volume and Gas Densities

31. Calculate the density of krypton (Kr) gas, in grams per liter, at STP.

32. Calculate the density of carbon monoxide (CO) gas, in grams per liter, at STP.

33. Calculate the molar mass of **(a)** a gas that has a density of 2.12 g/L at STP, and **(b)** an unknown liquid the vapor of which has a density of 2.97 g/L at STP.

34. Calculate the molar mass of **(a)** a gas that has a density of 1.98 g/L at STP, and **(b)** an unknown liquid for which 3.33 liters at STP is found to weigh 10.88 g.

Avogadro's Number

35. How many **(a)** oxygen *molecules* and how many **(b)** oxygen *atoms* are there in 1.00 mol of O_2?

36. How many calcium ions and how many chloride ions are there in 1.00 mol of $CaCl_2$?

37. Choose one of the following to complete the phrase accurately: One *mole* of fluorine gas (F_2),
a. has a mass of 19.0 g.
b. contains 6.02×10^{23} F atoms.
c. contains 12.04×10^{23} F atoms.
d. has a mass of 6.02×10^{23} g.

38. For calcium nitrate, **(a)** how many calcium ions and how many nitrate ions are there in 1.00 mol $Ca(NO_3)_2$? **(b)** How many nitrogen atoms and how many oxygen atoms are there in 1.00 mol $Ca(NO_3)_2$?

Formula Masses and Molar Masses

You may round all atomic masses to one decimal place.

39. Calculate the molar mass of each of the following compounds.
a. KIO_4
b. Na_2HPO_4
c. C_4H_9OH
d. $(NH_4)_2CrO_4$

40. Calculate the molar mass of each of the following compounds.
a. K_2S
b. C_6H_6
c. $Fe(NO_3)_3$
d. $Al_2(SO_3)_3$

41. Calculate the mass, in grams, of each of the following.
a. 7.57 mol $CaSO_4$
b. 0.0236 mol $CuCl_2$
c. 2.50 mol $C_{12}H_{22}O_{11}$

42. Calculate the mass, in grams, of each of the following.
a. 4.61 mol $AlCl_3$
b. 0.615 mol Cr_2O_3
c. 0.158 mol IF_5

43. Calculate the amount, in moles, of each of the following.
a. 6.63 g Sb_2S_3
b. 19.1 g MoO_3
c. 434 g $AlPO_4$
d. 11.8 g $Be(NO_3)_2$

44. Calculate the amount, in moles, of each of the following.
a. 16.3 g SF_6
b. 25.4 g $Pb(C_2H_3O_2)_2$
c. 15.6 g $CoCl_3$
d. 25.3 g $(NH_4)_2C_2O_4$

Mole and Mass Relationships in Chemical Equations

45. Consider the reaction for the combustion of ethane, a minor component of natural gas.

$$2\,C_2H_6 + 7\,O_2 \longrightarrow 4\,CO_2 + 6\,H_2O$$

a. How many moles of CO_2 is produced when 2.09 mol of ethane is burned?
b. How many moles of oxygen is required to burn 4.47 mol of ethane?

46. Consider the reaction for the combustion of octane.

$$2\,C_8H_{18} + 25\,O_2 \longrightarrow 16\,CO_2 + 18\,H_2O$$

a. How many moles of H_2O is produced when 2.81 mol of propane is burned?
b. How many moles of CO_2 is produced when 4.06 mol of oxygen is consumed?

47. What mass of **(a)** ammonia, in grams, can be made from 440 g of H_2, and **(b)** hydrogen, in grams, is needed to react completely with 892 g of N_2?

$$N_2 + H_2 \longrightarrow NH_3 \quad \text{(not balanced)}$$

48. Toluene (C_7H_8) and nitric acid (HNO_3) are used in the production of trinitrotoluene TNT ($C_7H_5N_3O_6$), an explosive.

$$C_7H_8 + HNO_3 \longrightarrow C_7H_5N_3O_6 + H_2O \quad \text{(not balanced)}$$

What mass of **(a)** nitric acid, in grams, is required to react with 454 g of C_7H_8 and **(b)** TNT can be made from 829 g of C_7H_8?

49. Phosphine (a toxic gas used as a fumigant to protect stored grain) is generated by the action of water on magnesium phosphide. The equation is

$$Mg_3P_2 + H_2O \longrightarrow PH_3 + Mg(OH)_2 \quad \text{(not balanced)}$$

What mass in grams of magnesium phosphide is needed to produce 134 g of PH_3?

50. In an oxyacetylene welding torch, acetylene (C_2H_2) burns in pure oxygen with a very hot flame.

$$C_2H_2 + O_2 \longrightarrow CO_2 + H_2O \quad \text{(not balanced)}$$

What mass of oxygen, in grams, is required to react with 52.0 g of C_2H_2?

Boyle's Law

51. A sample of helium occupies 1521 mL at 719 mmHg. Assume that the temperature is held constant and determine (a) the volume of the helium at 752 mmHg and (b) the pressure, in mmHg, if the volume is changed to 315 mL.

52. A decompression chamber used by deep-sea divers has a volume of 10.3 m^3 and operates at an internal pressure of 4.50 atm. What volume, in cubic meters, would the air in the chamber occupy if it were at 1.00 atm pressure, assuming no temperature change?

53. Oxygen used in respiratory therapy is stored at room temperature under a pressure of 150 atm in a gas cylinder with a volume of 60.0 L.
    a. What volume would the gas occupy at 0.987 atm? Assume no temperature change.
    b. If the oxygen flow to the patient is adjusted to 8.00 L/min, at room temperature and 0.987 atm, how long will the tank of gas last?

54. The pressure within a 2.25-L balloon is 1.10 atm. If the volume of the balloon increases to 7.05 L, what will be the final pressure within the balloon if the temperature does not change?

Charles's Law

55. A gas at a temperature of 100 °C occupies a volume of 154 mL. What will the volume be at a temperature of 10 °C, assuming no change in pressure?

56. A balloon is filled with helium. Its volume is 5.90 L at 26 °C. What will its volume be at 78 °C, assuming no pressure change?

57. A 567-mL sample of a gas at 305 °C and 1.20 atm is cooled at constant pressure until its volume becomes 425 mL. What is the new gas temperature?

58. A sample of gas at STP is to be heated at constant pressure until its volume triples. What is the new gas temperature?

The Ideal Gas Law

59. Will the volume of a fixed amount of gas increase, decrease, or remain unchanged with
    a. an increase in pressure at constant temperature?
    b. a decrease in temperature at constant pressure?
    c. a decrease in pressure coupled with an increase in temperature?

60. Will the pressure of a fixed amount of a gas increase, decrease, or remain unchanged with
    a. an increase in temperature at constant volume?
    b. a decrease in volume at constant temperature?
    c. an increase in temperature coupled with a decrease in volume?

61. According to the kinetic–molecular theory, (a) what change in temperature occurs if the molecules of a gas begin to move more slowly, on average, and (b) what change in pressure occurs when molecules of the gas strike the walls of the container less often?

62. For each of the following, indicate whether a given gas would have the same or different densities in the two containers. If the densities are different, in which container is the density greater?
    a. Containers A and B have the same volume and are at the same temperature, but the gas in A is at a higher pressure.
    b. Containers A and B are at the same pressure and temperature, but the volume of A is greater than that of B.
    c. Containers A and B are at the same pressure and volume, but the gas in A is at a higher temperature.

63. Calculate (a) the volume, in liters, of 1.12 mol $H_2S(g)$ at 62 °C and 1.38 atm, and (b) the pressure, in atmospheres, of 4.64 mol $CO(g)$ in a 3.96-L tank at 29 °C.

64. Calculate (a) the volume, in liters, of 0.00600 mol of a gas at 31 °C and 0.870 atm, and (b) the pressure, in atmospheres, of 0.0108 mol $CH_4(g)$ in a 0.265-L flask at 37 °C.

65. How many moles of $Kr(g)$ are there in 2.22 L of the gas at 0.918 atm and 45 °C?

66. How many grams of $CO(g)$ are there in 745 mL of the gas at 1.03 atm and 36 °C?

Molarity of Solutions

67. Calculate the molarity of each of the following solutions.
    a. 23.4 mol of HCl in 2.50 L of solution
    b. 0.0875 mol of Li_2CO_3 in 316 mL of solution

68. Calculate the molarity of each of the following solutions.
    a. 8.82 mol of H_2SO_4 in 3.75 L of solution
    b. 0.611 mol of C_2H_5OH in 96.3 mL of solution

69. How many grams of solute are needed to prepare (a) 3.50 L of 0.400 M NaOH and (b) 65.0 mL of 2.90 M $C_6H_{12}O_6$?

70. How many grams of solute are needed to prepare (a) 0.500 L of 0.167 M $K_2Cr_2O_7$ and (b) 625 mL of 0.0100 M $KMnO_4$?

71. What volume of (a) 6.00 M NaOH is required to contain 1.25 mol of NaOH and (b) 0.0250 M KH_2AsO_4 is needed to get 8.10 g of KH_2AsO_4?

72. What volume of (a) 2.50 M NaOH is required to contain 1.05 mol of NaOH and (b) 4.25 M $H_2C_2O_4$ is needed to get 2.25 g of $H_2C_2O_4$?

Percent Concentrations of Solutions

73. What is the volume percent concentration of **(a)** 35.0 mL of water in 725 mL of an ethanol–water solution, and **(b)** 78.9 mL of acetone in 1550 mL of an acetone–water solution?

74. What is the volume percent concentration of **(a)** 58.0 mL of water in 625 mL of an acetic acid–water solution, and **(b)** 79.1 mL of methanol in 755 mL of a methanol–water solution?

75. Describe how you would prepare 3375 g of an aqueous solution that is 8.2% NaCl by mass.

76. Describe how you would prepare 2.44 kg of an aqueous solution that is 16.3% KOH by mass.

77. Describe how you would prepare exactly 2.00 L of an aqueous solution that is 2.00% acetic acid by volume.

78. Describe how you would prepare exactly 500 mL of an aqueous solution that is 30.0% isopropyl alcohol by volume.

ADDITIONAL PROBLEMS

79. Which of the following correctly represents the decomposition of potassium chlorate to produce potassium chloride and oxygen gas?
    (a) $KClO_3(s) \longrightarrow KClO_3(s) + O_2(g) + O(g)$
    (b) $2\,KClO_3(s) \longrightarrow 2\,KCl(s) + 3\,O_2(g)$
    (c) $KClO_3(s) \longrightarrow KClO(s) + O_2(g)$
    (d) $KClO_3(s) \longrightarrow KCl(s) + O_3(g)$

80. Both magnesium and aluminum react with an acidic solution to produce hydrogen. Why is it that only one of the following equations correctly describes the reaction?

$$Mg(s) + 2\,H^+(aq) \longrightarrow Mg^{2+}(aq) + H_2(g)$$
$$Al(s) + 2\,H^+(aq) \longrightarrow Al^{3+}(aq) + H_2(g)$$

81. Write a balanced chemical equation to represent **(a)** the decomposition, by heating, of solid mercury(II) nitrate to produce pure liquid mercury, nitrogen dioxide gas, and oxygen gas, and **(b)** the reaction of aqueous sodium carbonate with aqueous hydrochloric acid (hydrogen chloride) to produce water, carbon dioxide gas, and aqueous sodium chloride.

82. Joseph Priestley discovered oxygen in 1774 by heating "red calx of mercury," mercury(II) oxide. The calx decomposed to its elements. The equation is

$$HgO \longrightarrow Hg + O_2 \quad \text{(not balanced)}$$

What mass of oxygen is produced by the decomposition of 10.8 g of HgO?

83. Compare 0.50 mol $H_2(g)$ and 1.0 mol He(g) at STP. Will the two gases **(a)** have the same number of atoms? **(b)** have the same number of molecules? **(c)** occupy equal volumes?

84. How many liters of hydrogen gas (at STP) are produced from the electrolysis of 1.00 L $H_2O(l)$?

85. How much of the magnetic oxide of iron (Fe_3O_4) can be made from 12.0 grams of pure iron and an excess of oxygen? The equation is

$$Fe + O_2 \longrightarrow Fe_3O_4 \quad \text{(not balanced)}$$

86. Which of the following laws is correctly described by the mathematical proportionality?
    a. Boyle's law: $V \propto P$
    b. Charles's law: $V \propto T$

87. Which of the following would *not* result in an increase in the volume of a gas?
    (a) An increase in temperature
    (b) An increase in pressure
    (c) A decrease in temperature
    (d) A threefold increase in pressure together with a twofold reduction in temperature

88. Which of the following gases has the greatest density at STP? Explain how this may be answered *without* calculating the densities.
    (a) Cl_2
    (b) SO_3
    (c) N_2O
    (d) PF_3

89. Choose the answer that correctly completes the statement: At 0 °C and 0.500 atm, 4.48 L $NH_3(g)$ (a) contains 0.20 mol NH_3; (b) has a mass of 3.40 g; (c) contains 6.02×10^{22} molecules; (d) contains 0.40 mol NH_3.

90. When heated above about 900 °C, limestone (calcium carbonate) decomposes to quicklime (calcium oxide) and carbon dioxide. What mass in grams of quicklime can be produced from 4.72×10^9 g limestone?

91. Ammonia reacts with oxygen to produce nitric acid (HNO_3) and water. What mass of nitric acid, in grams, can be made from 971 g of ammonia?

92. Hydrogen peroxide from the drug store is 3% H_2O_2 by mass dissolved in water. How many moles of H_2O_2 are in a typical 16-oz bottle (1 oz = 29.6 mL)?

93. What is the mass percent of **(a)** NaOH in a solution of 4.12 g NaOH in 100.0 g water, and **(b)** ethanol in a solution of 5.00 mL ethanol (density 0.789 g/mL) in 50.0 g water?

94. What is the volume percent of ethanol **(a)** in a solution that has 58.0 mL water in 625 mL of an ethanol–water solution, and **(b)** in a solution that has 10.00 mL acetone in 1.25 L of an acetone–water solution?

95. Laughing gas (dinitrogen monoxide, N_2O, also called nitrous oxide) can be made by heating ammonium nitrate with great care. The other product is water.
    a. Write a balanced equation for the process.
    b. Draw the Lewis structure for N_2O.
    c. How much N_2O can be made from 4.00 g of ammonium nitrate?
    d. If the water is produced as steam, how many liters will be formed at STP from 4.00 g ammonium nitrate?

96. Mining helmets once used calcium carbide (CaC_2) as a source of light. When water is dripped on the solid, calcium hydroxide and acetylene (C_2H_2) are produced, the latter of which when burned produces a bright flame. **(a)** What mass of CaC_2 is needed to provide light for 3.0 hours if 9.5 mole acetylene (at STP) is used every hour? **(b)** What volume, in liters, would the acetylene occupy at STP if it were not burned?

97. For many years the noble gases were called "inert gases," and it was thought that they formed no chemical compounds. Neil Bartlett made the first noble gas compound in 1962. Xenon hexafluoride is prepared according to the equation

$$Xe + F_2 \longrightarrow XeF_6 \quad \text{(not balanced)}$$

If you had a large excess of Xe but only 2.0 L of fluorine gas (at STP), how much XeF_6 could you produce if the reaction proceeded perfectly?

98. In what volume of solution must 31.7 g of oxalic acid ($H_2C_2O_4$) be dissolved to make a solution that is 0.0859 M in oxalic acid?

99. In tests for intoxication, blood alcohol levels are expressed as percent by volume. A blood alcohol level of 0.080% by volume means 0.080 mL ethanol per 100 mL of blood and is considered proof of intoxication. If a person's blood volume is 5.0 L, what volume of alcohol in the blood gives a blood alcohol level of 0.165% by volume?

100. Determine the molarity of **(a)** Li^+ and NO_3^- in 0.647 M $LiNO_3$, and **(b)** Ca^{2+} and I^- in 0.035 M CaI_2.

101. CuS can be converted (smelted) to copper metal by reaction with O_2 with SO_2 as the other product. **(a)** What mass of Cu could be produced from 1000 kg CuS ore? **(b)** What volume of $SO_2(g)$ (at STP) would be made from the same amount of ore?

102. In Problem 46, you were asked to determine the mass of CO_2 produced from the combustion of octane. If the reaction were given as

$$C_8H_{18} + \frac{25}{2}O_2 \longrightarrow 8\,CO_2 + 9\,H_2O$$

would your answer change? Explain.

103. Calculate the density of air (assume 22% oxygen and 78% nitrogen by volume). Calculate the density of steam (100% H_2O vapor) under the same conditions. Comment, without doing a calculation, on whether you think the density of moist air would be greater than or less than that of dry air.

104. Look again at Critical Thinking Exercise 6.4, then calculate **(a)** the correct mass of ammonium nitrate that would react with 1.00 kg of potassium carbonate, and **(b)** the correct mass of potassium nitrate that would be obtained.

COLLABORATIVE GROUP PROJECTS

Prepare a PowerPoint, poster, or other presentation (as directed by your instructor) for presentation to the class.

105. Prepare a brief biographical report on one of the following and share it with your group.
   a. Joseph Louis Gay-Lussac
   b. Amedeo Avogadro
   c. Robert Boyle
   d. Jacques Charles

106. In the view of many scientists, hydrogen holds great promise as a source of clean energy. In 2003, the U.S. government pledged $1.2 billion for research into hydrogen-powered cars. Doing your own search, explore the proposals for using hydrogen as a fuel. Write a brief summary of what you find and use the appropriate balanced equation to explain why this is considered a clean, renewable source. More discussion of hydrogen as a fuel is found in Chapter 14.

REFERENCES AND READINGS

1. Ault, Addison. "How to Say How Much: Amounts and Stoichiometry." *Journal of Chemical Education*, October 2001, pp. 1345–1347.

2. Gorin, George. "Mole, Mole per Liter, and Molar: A Primer on SI and Related Units for Chemistry Students." *Journal of Chemical Education*, January 2003, pp. 103–104.

3. Haim, Liliana, Eduardo Cortón, Santiago Kocmur, and Lydia Galagovsky. "Learning Stoichiometry with Hamburger Sandwiches."*Journal of Chemical Education*, September 2003, pp. 1021–1022.

4. Hill, John W., Ralph H. Petrucci, Terry W. McCreary, and Scott S. Perry. *General Chemistry*, 4th edition. Upper Saddle River, NJ: Prentice Hall, 2005. Chapter 3, "Stoichiometry," and Chapter 5, "Gases," provide a more extensive treatment of some of the topics in this chapter.

5. Jensen, William B. "The Origin of Stoichiometry Problems." *Journal of Chemical Education*, November 2003, p. 1248.

6. Wakeley, Dawn M., and Hans de Grys. "Developing an Intuitive Approach to Moles." *Journal of Chemical Education*, August 2000, pp. 1007–1009.

GREEN CHEMISTRY

Atom Economy

Margaret Kerr, Worcester State College

In Chapter 6, you learned how to write and balance chemical equations. You also learned how to calculate the molar mass of a substance, convert from grams to moles, and determine the amount of product formed from given amounts of reactants.

The amount of product calculated to form from the given reactants is called the *theoretical yield*. This is the maximum amount of product that can be produced by the amount of reactants that are used. When a chemical reaction is performed in the lab, the actual amount of product that is collected is called the *actual yield*. Chemists typically will refer to their experimental results in *percent yield*. This value is simply the yield of product they actually produce divided by the theoretical yield, multiplied by 100.

$$\% \text{ yield} = \frac{\text{actual yield}}{\text{theoretical yield}} \times 100\%$$

If all of the reactants turn into products, then a 100% yield would be obtained.

Reporting a 100% yield for a reaction sounds good, but the yield is only for the desired product. Anything else that is formed as part of the reaction is not counted as part of the yield. By-products are often discarded as waste and become part of the chemical waste stream. Some reactions produce significantly more by-products than they do desired products. An alternate measure of

▲ Uranium is enriched for nuclear reactors and nuclear weapons by converting it to uranium hexafluoride, UF_6. The percent yield of the process is reasonably efficient, but the atom economy is extremely poor—over 99% of the uranium is waste, stored in cylinders like these.

reaction efficiency is called *atom economy (AE)*. Atom economy expresses the number of the atoms put into the reaction mixture that actually become part of the product. Atom economy is an integral part of the 12 principles of green chemistry, introduced in Chapter 1.

An example of the difference between percent yield and atom economy can be illustrated with the chemical reaction used in airbags, from Example 6.12 on page 168. The balanced equation is

$2\,NaN_3$	$\rightarrow$	$2\,Na$	$+$	$3\,N_2$
Starting Material		By-product (waste)		Desired Product
Number of Atoms (moles) Reacted		Number of Atoms (moles) in By-product		Number of Atoms (moles) in Product
Na = 2		Na = 2		
N = 6				N = 6
Total Mass		Mass of Atoms in By-product		Mass of Atoms in Product
130.02 g		45.98 g		84.04 g

$$\% \text{ AE} = \frac{\text{mass of desired product}}{\text{mass of desired product} + \text{mass of all by-products}} \times 100\%$$

$$\% \text{ AE} = \frac{84.04 \text{ g}}{84.04 \text{ g} + 45.98 \text{ g}} \times 100\% = 64.62\%$$

Based on these calculations, when this reaction is run, only about 65% of the atoms are utilized. Thirty-five percent of the atoms are wasted and are likely to enter the waste stream.

WEB INVESTIGATIONS

Investigation 1
Atom Economy

Visit the Environmental Protection Agency's website. Follow the links to the 1998 Presidential Green Chemistry challenge award winner, Dr. Barry Trost. Describe why atom economy is such an important concept.

Investigation 2
Synthesis of Ibuprofen

Go to the Green Chemistry Institute's page from the American Chemical Society's website and search for the Greener Synthesis of Ibuprofen by the BHC company. Describe how this process utilizes the green principle of atom economy.

Investigation 3
Waste Minimization

Return to the Environmental Protection Agency's website. Follow the links to the EPA National Waste Minimization Program website. How can atom economy assist with this program?

Investigation 4
Application of Atom Economy

Locate the "Greener Education Materials" database online. Search for the online module about atom economy by Dr. Michael Cann from the University of Scranton. Go through the module and answer the questions found at the end.

COMMUNICATE YOUR RESULTS

Exercise 1
Waste Minimization

How does atom economy fit in with the statement, "Green Chemistry is the use of chemistry for pollution prevention"? The Environmental Protection Agency's website is a great place to start.

Exercise 2
Atom Economy of Pain Relievers

Calculate the percent atom economy for the reaction of salicylic acid ($C_7H_6O_3$) with acetic anhydride [$(CH_3CO)_2O$] to form aspirin (acetylsalicylic acid, $C_9H_8O_4$) and acetic acid (CH_3COOH). Acetic acid is produced as a by-product. Prepare a series of slides (PowerPoint or other) showing your calculations, comparing the percent atom economy for the synthesis of aspirin to that of the "brown" synthesis of ibuprofen and to the "green" synthesis of ibuprofen.

Exercise 3
Atom Economy and Industrial Synthesis

Use your favorite search engine to find out about the Haber process and review how ammonia is made industrially using this process. What are the primary uses for ammonia? What is the atom economy of this reaction? Based on the major uses for ammonia, do you think that high atom economy is more important for ammonia synthesis than for ibuprofen synthesis; or is a high atom economy equally important for the two? Justify your answer in a brief (one-page) report.

Exercise 4
Atom Economy and the Consumer

Do you think a product made from a more atom-economical synthesis should cost more or less than a less efficiently produced product? Would you pay more for a product made more efficiently? How much more would you pay? Or should industry be driven to atom-economic synthesis only if it results in a lower-priced product? Use the thoughts you have generated with these questions to write a one-page argument to your supervisor justifying research to develop a more atom-economical synthesis for product X.

CHAPTER 7

Acids and Bases

The chemical substances we call acids and bases are all around us, even in the food we eat. For example, the acids in lemon juice add tangy flavor to a fruit salad.

Please Pass the Protons

Have you ever tasted a lemon or a grapefruit? Or felt a burning sensation on your arm after using spray-on oven cleaner? These are but two examples of the presence of acids (citric acid in lemons and grapefruit) and bases (lye or sodium hydroxide in oven cleaner) in our daily lives. Other familiar acids are vinegar (acetic acid), vitamin C (ascorbic acid), and battery acid (sulfuric acid). Some familiar bases are drain cleaner (sodium hydroxide), baking soda (sodium bicarbonate), and ammonia.

From "acid indigestion" to "acid rain," the word *acid* is frequently in the news and in advertisements. Air and water pollution often involve acids and bases. Acid rain, for example, is a serious environmental problem. In arid areas, alkaline (basic) water is sometimes undrinkable.

Did you know that our senses recognize four tastes related to acid–base chemistry? Acids taste sour, bases taste bitter, and the compounds formed when acids react with bases (salts) taste salty. The sweet taste is more complicated. To taste sweet, a compound must have both an acidic-type group and a basic-type group, plus just the right geometry to fit the sweet-taste receptor.

In this chapter we discuss some of the chemistry of acids and bases. You use them every day. Your body processes them continuously. And you will probably hear and read about them as long as you live. What you learn here can help you gain a better understanding of these important classes of compounds.

7.1 ACIDS AND BASES: EXPERIMENTAL DEFINITIONS

Acids and bases are chemical opposites, and so their properties are quite different—often opposite. Let us begin by listing a few of their properties.

An *acid* is a compound that

- causes litmus indicator dye to turn red.
- tastes sour.
- dissolves active metals (such as zinc or iron), producing hydrogen gas.
- reacts with bases to form water and ionic compounds called salts.

A *base* is a compound that

- causes litmus indicator dye to turn blue.
- tastes bitter.

SAFETY ALERT: Although all acids taste sour and all bases are bitter, a taste test is not the best general-purpose way to determine whether a substance is an acid or a base. Some acids and bases are poisonous, and many are quite corrosive unless greatly diluted. You should NEVER TASTE LABORATORY CHEMICALS. Too many of them are toxic, and there is a risk that they might be contaminated.

▲ Sour candy owes its pucker power to citric acid and/or malic acid.

191

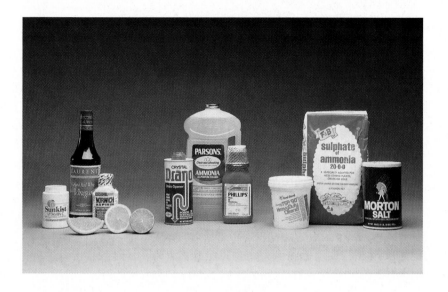

► **Figure 7.1** Some common acids (left), bases (center), and salts (right). Acids, bases, and salts are components of many familiar consumer products.

Question: How would each of the three classes of compounds affect the indicator dye litmus? (See Figure 7.2.)

- feels slippery on the skin.
- reacts with acids to form water and salts.

We can identify foods that are acidic by their sour taste. Vinegar and lemon juice are examples. Vinegar is a solution of acetic acid (about 5%) in water. Lemons, limes, and other citrus fruits contain citric acid. Lactic acid gives yogurt its tart taste, and phosphoric acid is often added to carbonated drinks to impart tartness. The bitter taste of tonic water, on the other hand, is attributable to the presence of quinine, a base. Figure 7.1 shows some common acids and bases.

A litmus test is a common way to identify a substance as an acid or a base. Litmus (Figure 7.2) is an **acid–base indicator**, one of many such compounds. If you dip a strip of neutral (violet-colored) litmus paper into an unknown solution and it turns pink, the solution is acidic. If it turns blue, the solution is basic. If the strip does not turn pink or blue, the solution is neither acidic nor basic. Many natural food colors, such as those in grape juice, red cabbage, and blueberries, are acid–base indicators. So are the colors in most flower petals.

Blue litmus turns red

Acid solution

Red litmus turns blue

Base solution

► **Figure 7.2** Strips of paper impregnated with litmus dye are often used to distinguish between acids and bases. The sample on the left turns blue litmus red and is therefore acidic. The sample on the right turns red litmus blue and is basic.

7.2 ACIDS, BASES, AND SALTS

Acids and bases have certain characteristic properties. But why do they have these properties? We use several different theories to explain these properties.

The Arrhenius Theory

Svante Arrhenius developed the first successful theory of acids and bases in 1887. According to Arrhenius's concept,

- an **acid** is a molecular substance that breaks up in aqueous solution into hydrogen ions (H^+) and anions.

(Because a hydrogen ion is a hydrogen atom from which the sole electron has been removed, H^+ ions are also called *protons*.) The acid is said to *ionize*.

In water, then, the properties of acids are those of the H^+ ion. It is the hydrogen ion that turns litmus red, tastes sour, and reacts with active metals and bases. Table 7.1 lists some common acids. Notice that each formula contains one or more hydrogen atoms. Chemists often indicate an acid by writing the formula with the H atoms first. HCl, H_2SO_4, and HNO_3 are acids; NH_3 and CH_4 are not. The formula $HC_2H_3O_2$ (acetic acid) indicates that one H atom ionizes and three do not.

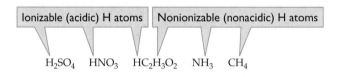

Arrhenius's concept of a base is that

- a **base** is a substance that releases hydroxide ions (OH^-) in aqueous solution.

Some bases are ionic solids that contain OH^-; sodium hydroxide (NaOH) is a familiar example. These compounds simply release hydroxide ions into water when the solid is made into an aqueous solution:

$$NaOH(s) \xrightarrow{H_2O} Na^+(aq) + OH^-(aq)$$

Other bases are molecular substances such as ammonia that ionize to produce OH^- when placed in water (see page 195).

Experimental evidence indicates that the properties of bases in water are due to OH^-. Table 7.2 lists some common bases. Most of these are ionic compounds containing positive metal ions, such as Na^+ or Ca^{2+}, and negative hydroxide ions. When these compounds dissolve in water, they all provide OH^- ions, and thus they are all bases. The properties of bases are those of hydroxide ions, just as the properties of acids are those of hydrogen ions.

▲ Swedish chemist Svante Arrhenius (1859–1927) proposed the theory that acids, bases, and salts in water are composed of ions. He also was the first to relate carbon dioxide in the atmosphere to the greenhouse effect (Chapter 12).

 Introduction to Aqueous Acids

 Nitric Acid

 Introduction to Aqueous Bases

TABLE 7.1	Some Familiar Acids		
Name	**Formula**	**Acid Strength**	**Common Uses/Notes**
Sulfuric acid	H_2SO_4	Strong	Battery acid; extremely corrosive
Nitric acid	HNO_3	Strong	Manufacture of fertilizers, explosives
Hydrochloric acid	HCl	Strong	Cleaning of metals, bricks; removing scale from boilers
Phosphoric acid	H_3PO_4	Moderate	Manufacture of fertilizers; colas; rust removers
Hydrogen sulfate ion	HSO_4^-	Moderate	Toilet bowl cleaners ($NaHSO_4$)
Lactic acid	$CH_3CHOHCOOH$	Weak	Yogurt; acidulant for soda pop
Acetic acid	CH_3COOH	Weak	Vinegar; acidulant
Carbonic acid	H_2CO_3	Weak	Unstable; formed in aqueous CO_2
Boric acid	H_3BO_3	Very weak	Antiseptic eye wash, roach poison
Hydrocyanic acid	HCN	Very weak	Plastics manufacture; extremely toxic

TABLE 7.2	Common Bases		
Name	**Formula**	**Classification**	**Common Uses/Notes**
Sodium hydroxide	NaOH	Strong	Acid neutralization; soap making; dehorning calves
Potassium hydroxide	KOH	Strong	Making liquid soaps; absorbing CO_2
Lithium hydroxide	LiOH	Strong	Alkaline storage batteries
Calcium hydroxide	$Ca(OH)_2$	Strong*	Mortar, plaster, cement; water purification
Magnesium hydroxide	$Mg(OH)_2$	Strong*	Antacid, laxative
Ammonia	NH_3	Weak	Fertilizer, household cleansers

*Although these bases are classified as strong, they are not very soluble. Calcium hydroxide is only slightly soluble in water, and magnesium hydroxide is practically insoluble.

Arrhenius further proposed that the essential reaction between an acid and a base, *neutralization*, is the combination of H^+ and OH^- to form water. The cation originally associated with the OH^- and the anion associated with the H^+ give rise to an ionic compound, a **salt**.

$$\text{An acid } + \text{ a base } \longrightarrow \text{ a salt } + \text{ water}$$

CONCEPTUAL EXAMPLE 7.1   Ionization of Acids and Bases

Write equations to show **(a)** the ionization of nitric acid (HNO_3) in water, and **(b)** the ionization of solid potassium hydroxide (KOH) in water.

Solution

a. An HNO_3 molecule ionizes to form a hydrogen ion and a nitrate ion. Because this reaction occurs in water, we can use (aq) to indicate that these substances are in aqueous solution.

$$HNO_3(aq) \longrightarrow H^+(aq) + NO_3^-(aq)$$

b. Potassium hydroxide (KOH), an ionic solid, simply dissolves in the water, forming separate $K^+(aq)$ and $OH^-(aq)$ ions.

$$KOH(s) \xrightarrow{H_2O} K^+(aq) + OH^-(aq)$$

▶ **Exercise 7.1A**

Write equations to show **(a)** the ionization of HBr (hydrobromic acid) in water, and **(b)** the ionization of solid calcium hydroxide in water.

▶ **Exercise 7.1B**

Formulas for carboxylic acids (compounds containing a —COOH group; Chapter 9) are often written with the ionizable hydrogen *last*. For example, instead of writing acetic acid as $HC_2H_3O_2$, it often is written CH_3COOH. Write an equation to show the ionization of CH_3COOH in water.

Limitations of the Arrhenius Theory

The Arrhenius theory is limited in several ways.

- A simple free proton does not exist in water solution. The H^+ ion has such a high positive charge density that it immediately seeks out a negative charge. It finds a lone pair of electrons on the O atoms of an H_2O molecule and attaches itself to form a *hydronium ion*, H_3O^+.

Water   Hydronium ion

- It does not explain the basicity of ammonia and related compounds. Ammonia seems out of place in Table 7.2 because it contains no hydroxide ions.
- It applies only to reactions in aqueous solution.

In water the H^+ ion is probably associated with several H_2O molecules—for example, four H_2O molecules in the ion $H(H_2O)_4^+$ or $H_9O_4^+$. For most purposes, however, we simply use H^+ and ignore the associated water molecules. However, we should understand that H^+ is a simplification of the real situation: When we mention protons in water, we really mean their sources (hydronium ions).

Like many scientific theories, a better one based on newer data has supplanted Arrhenius's theory.

The Brønsted–Lowry Acid–Base Theory

The shortcomings of the Arrhenius theory were largely overcome by a theory proposed independently, in 1923, by J. N. Brønsted in Denmark and T. M. Lowry in Great Britain. In the Brønsted–Lowry theory,

- an **acid** is a *proton donor*, and
- a **base** is a *proton acceptor*.

The theory describes the ionization of hydrogen chloride in this way:

$$HCl(aq) + H_2O \longrightarrow H_3O^+(aq) + Cl^-(aq)$$

The acid molecules donate hydrogen ions (protons) to the water molecules, and so the acid (HCl) acts as a proton donor.

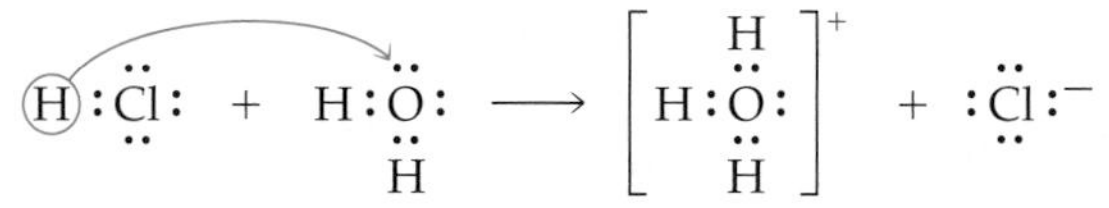

Acid
(proton donor)   Hydrochloric acid

The HCl molecule donates a proton to the water molecule, producing a hydronium ion and a chloride ion, a solution called hydrochloric acid. Other acids react similarly; they donate hydrogen ions to water to produce hydronium ions. If we let HA represent any acid, the reaction is

$$HA(aq) + H_2O \longrightarrow H_3O^+(aq) + A^-(aq)$$

Even when the solvent is something other than water, the acid acts as a proton donor, transferring H^+ ions to the solvent molecules.

CONCEPTUAL EXAMPLE 7.2   **Brønsted–Lowry Acids**

Write an equation to show the reaction of HNO_3 as a Brønsted–Lowry acid with water. What is the role of water in the reaction?

Solution

As a Brønsted–Lowry acid, HNO_3 donates a proton to water, forming a hydronium ion and a nitrate ion.

$$HNO_3(aq) + H_2O \longrightarrow H_3O^+(aq) + NO_3^-(aq)$$

The water molecule accepts a proton from HNO_3; water is a Brønsted–Lowry base in this reaction.

▶ Exercise 7.2A

Write an equation to show the reaction of HBr as a Brønsted–Lowry acid with water.

▶ Exercise 7.2B

Write an equation to show the reaction of HBr as a Brønsted–Lowry acid with methanol, CH_3OH.

Where does the OH^- come from in the ionization of bases such as ammonia (NH_3)? The Arrhenius theory is inadequate in answering this question, but the Brønsted–Lowry theory explains how ammonia acts as a base in water. Ammonia is a gas at room temperature. When it is dissolved in water, some of the ammonia molecules react as shown by the following equation.

$$NH_3(aq) + H_2O \longrightarrow NH_4^+(aq) + OH^-(aq)$$

► **Figure 7.3** A Brønsted–Lowry acid is a proton donor. A base is a proton acceptor.

Question: Can you write an equation in which the acid is represented as HA and the base as :B⁻?

Acid   Base

An ammonia molecule accepts a proton from a water molecule; NH_3 acts as a Brønsted–Lowry base. (Recall that the N atom of ammonia has a lone pair of electrons to which a proton can attach.) The water molecule acts as a proton donor—an acid. The ammonia molecule becomes an ammonium ion. When a proton leaves a water molecule, it leaves behind the electron pair that joined it to the O atom. The water molecule becomes a negatively charged hydroxide ion.

In general, then, a base is a proton acceptor (Figure 7.3). This definition includes not only hydroxide ions but also neutral molecules such as ammonia. It also includes other negative ions such as oxide (O^{2-}), carbonate (CO_3^{2-}), and bicarbonate (HCO_3^-). The idea of an acid as a proton donor and a base as a proton acceptor greatly expands our concept of acids and bases.

7.3 ACIDIC AND BASIC ANHYDRIDES

In the Brønsted–Lowry view, many metal oxides act directly as bases because the oxide ion can accept a proton. These metal oxides also react with water to form metal hydroxides, compounds that are bases in the Arrhenius sense. Similarly, many nonmetal oxides react with water to form acids.

Nonmetal Oxides: Acidic Anhydrides

Many acids are made by the reaction of nonmetal oxides with water. For example, sulfur trioxide reacts with water to form sulfuric acid.

$$SO_3 + H_2O \longrightarrow H_2SO_4$$

Similarly, carbon dioxide reacts with water to form carbonic acid.

$$CO_2 + H_2O \longrightarrow H_2CO_3$$

In general, nonmetal oxides react with water to form acids.

$$\text{Nonmetal oxide} + H_2O \longrightarrow \text{acid}$$

Nonmetal oxides that act in this way are called **acidic anhydrides**. *Anhydride* means "without water." These reactions explain why rainwater is acidic (Section 7.7).

Carbon Dioxide Behaves as an Acid in Water

Carbonic Acid

CONCEPTUAL EXAMPLE 7.3   **Acidic Anhydrides**

Give the formula for the acid formed when sulfur dioxide reacts with water.

Solution
Simply write the equation for the reaction, following the pattern in the preceding examples.

$$SO_2 + H_2O \longrightarrow H_2SO_3$$

The formula for the acid is derived by adding the two H atoms and one O atom of water to SO_2. It is H_2SO_3.

► **Exercise 7.3A**
Give the formula for the acid formed when selenium dioxide (SeO_2) reacts with water.

► **Exercise 7.3B**
Give the formula for the acid formed when dinitrogen pentoxide (N_2O_5) reacts with water. (*Hint:* Two molecules of acid are formed.)

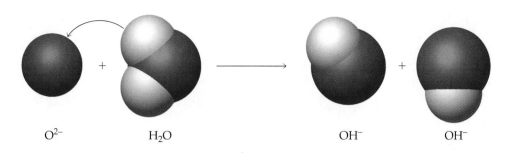

$$O^{2-} \qquad H_2O \qquad OH^- \qquad OH^-$$

◀ **Figure 7.4** Metal oxides are basic because the oxide ion reacts with water to form two hydroxide ions.

Question: Can you write an equation that shows how solid sodium oxide, Na_2O, reacts with water to form sodium hydroxide?

Metal Oxides: Basic Anhydrides

Just as acids can be made from nonmetal oxides, many common hydroxide bases can be made from metal oxides. For example, calcium oxide (lime) reacts with water to form calcium hydroxide (slaked lime).

$$CaO + H_2O \longrightarrow Ca(OH)_2$$

Another example is the reaction of lithium oxide with water to form lithium hydroxide.

$$Li_2O + H_2O \longrightarrow 2\,LiOH$$

In general, metal oxides react with water to form bases (Figure 7.4). These metal oxides are called **basic anhydrides**.

$$\text{Metal oxide} + H_2O \longrightarrow \text{base}$$

CONCEPTUAL EXAMPLE 7.4   **Basic Anhydrides**

Give the formula for the base formed by the addition of water to barium oxide (BaO).

Solution

Simply write the equation for the reaction. (Because barium ion has a 2+ charge, the O in BaO is present as an O^{2-} ion. As shown in Figure 7.4, the oxide ion reacts with water to form *two* hydroxide ions.)

$$BaO + H_2O \longrightarrow Ba(OH)_2$$

The formula for the base, barium hydroxide, is $Ba(OH)_2$.

▶ **Exercise 7.4A**

Give the formula for the base formed by the addition of water to strontium oxide (SrO).

▶ **Exercise 7.4B**

What base is formed by the addition of water to potassium oxide (K_2O)? (*Hint:* Two moles of base are formed for each mole of potassium oxide.)

7.4 STRONG AND WEAK ACIDS AND BASES

When gaseous hydrogen chloride (HCl) reacts with water, it reacts completely to form hydronium ions and chloride ions. Essentially no HCl molecules remain.

$$HCl + H_2O \longrightarrow H_3O^+ + Cl^-$$

For many purposes, we simply write the reaction as the ionization of HCl and use (aq) to indicate the involvement of water.

$$HCl(aq) \longrightarrow H^+(aq) + Cl^-(aq)$$

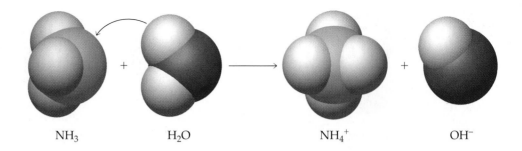

$$NH_3 \qquad H_2O \qquad NH_4^+ \qquad OH^-$$

▶ **Figure 7.5** Ammonia is a base because it accepts a proton from water. A solution of ammonia in water contains ammonium ions and hydroxide ions. Only a small fraction of the ammonia molecules react, however; most remain unchanged. Ammonia is therefore a *weak* base.

Question: Amines are related to ammonia in that one or more of the H atoms of NH_3 is replaced by a carbon containing group that we represent as R. Amines act in the same way as ammonia. Can you write an equation that shows how RNH_2 reacts with water to form R-substituted ammonium ions and hydroxide ions?

A strong acid or base is one that is almost completely ionized in solution. The word *strong* does not refer to the amount of acid or base in the solution. A solution that contains a relatively large amount of acid or base, whether strong or weak, in a given volume of solution is called a *concentrated* solution. A solution with only a little solute in that same volume of solution is a *dilute* solution.

The poisonous gas hydrogen cyanide (HCN) also ionizes in water to produce hydrogen ions and cyanide ions. But HCN reacts only to a slight extent. In a solution that has 1 mol of HCN in 1 L of water, only one HCN molecule in 40,000 reacts to produce a hydrogen ion.

$$HCN(aq) \rightleftharpoons H^+(aq) + CN^-(aq)$$

We indicate this slight ionization by using a double arrow. A short arrow points to the right with a longer arrow pointing left to indicate that most of the HCN remains intact as HCN molecules.

- An acid such as HCl that reacts completely with water is called a **strong acid**.
- An acid such as HCN that reacts only slightly with water is a **weak acid**.

There are not many strong acids. The first three acids listed in Table 7.1 (sulfuric, nitric, and hydrochloric) are the common ones. Most acids are weak acids.

Bases are also classified as strong or weak.

- A **strong base** is completely ionized in water.
- A **weak base** is only slightly ionized in water.

Perhaps the most familiar strong base is sodium hydroxide (NaOH), commonly called lye. It exists as sodium ions and hydroxide ions even in the solid state. Other strong bases include potassium hydroxide (KOH) and the hydroxides of all the other group 1A metals. Except for $Be(OH)_2$, group 2A hydroxides are also strong bases. However, $Ca(OH)_2$ is only slightly soluble in water, and $Mg(OH)_2$ is nearly insoluble. The concentration of hydroxide ions in water from either is therefore not very high.

The most familiar weak base is ammonia (NH_3). It reacts with water to a slight extent to produce ammonium ions (NH_4^+) and hydroxide ions (Figure 7.5).

$$NH_3 + H_2O \rightleftharpoons NH_4^+ + OH^-$$

In its reaction with HCl (page 195), water acts as a base (proton acceptor). In its reaction with NH_3, water acts as an acid (proton donor). A substance, such as water, that can either donate a proton or accept a proton is said to be *amphiprotic* (see also Additional Problem 67).

7.5 NEUTRALIZATION

When an acid reacts with a base, the products are water and a salt. If a solution containing hydrogen ions (an acid) is mixed with another solution containing exactly the same amount of hydroxide ions (a base), the resulting solution no longer affects litmus, and it no longer dissolves zinc or iron. Nor does it feel slippery on the skin. It is no longer either acidic or basic; it is neutral. The reaction of an acid with a base is called **neutralization** (Figure 7.6). In water it is simply the reaction of hydrogen ions with hydroxide ions to form water molecules.

$$H^+ + OH^- \longrightarrow H_2O$$

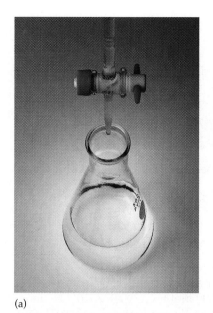
(a)

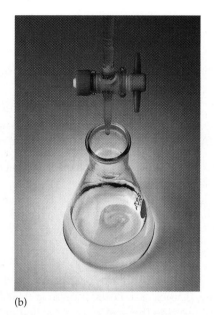
(b)

(c)

▲ **Figure 7.6** The amount of acid (or base) in a solution is determined by careful neutralization. Here a 5.00-mL sample of vinegar, some water, and a few drops of phenolphthalein (an acid–base indicator) are added to a flask (a). A solution of 0.1000 M NaOH is added slowly from a buret (a device for precise measurement of volumes of solutions) (b). As long as the acid is in excess, the solution is colorless. When the acid has been neutralized and a tiny excess of base is present, the phenolphthalein indicator turns pink (c).

Question: Can you write an equation for the reaction of acetic acid ($HC_2H_3O_2$), the acid in vinegar, with aqueous NaOH?

If sodium hydroxide is neutralized by hydrochloric acid, the products are water and sodium chloride (ordinary table salt).

A base	An acid	A salt	Water

$$NaOH(aq) \ + \ HCl(aq) \longrightarrow NaCl(aq) \ + \ H_2O$$

EXAMPLE 7.5   **Neutralization Reactions**

Write the equation for the neutralization reaction between potassium hydroxide and nitric acid.

Solution

The OH^- of the base and the H^+ of the acid combine to form water. The cation of the base (K^+) and the anion of the acid (NO_3^-) form a solution of the salt potassium nitrate (KNO_3).

$$KOH(aq) + HNO_3(aq) \longrightarrow KNO_3(aq) + H_2O$$

▶ **Exercise 7.5A**

Write the equation for the neutralization reaction between lithium hydroxide and acetic acid.

▶ **Exercise 7.5B**

Write the equation for the neutralization reaction between calcium hydroxide and hydrochloric acid. (*Hint:* Be sure to write the correct formula for reactants and the salt before attempting to balance the equation.)

7.6 THE pH SCALE

In solutions, the concentrations of ions are measured in moles per liter. Because hydrogen chloride is completely ionized in water, a solution of 1 molar hydrochloric acid (1 M HCl), for example, contains 1 mol of H^+ ions per liter of solution. In other

Acids and Bases

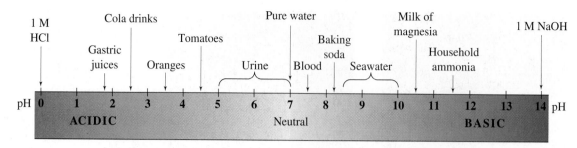

▲ **Figure 7.7** The pH scale. A change in pH of one unit means a tenfold change in the hydronium ion concentration.

Question: How does the acidity of tomatoes compare to that of oranges?

The relationship between pH and [H⁺] is perhaps easier to see when the equation is written in the following form.

$$[H^+] = 10^{-pH}$$

TABLE 7.3	Relationship Between pH and Concentration of Hydronium Ions
Concentration of H_3O^+ (mol/L)	**pH**
1×10^{-0}	0
1×10^{-1}	1
1×10^{-2}	2
1×10^{-3}	3
1×10^{-4}	4
1×10^{-5}	5
1×10^{-6}	6
1×10^{-7}	7
1×10^{-8}	8
1×10^{-9}	9
1×10^{-10}	10
1×10^{-11}	11
1×10^{-12}	12
1×10^{-13}	13
1×10^{-14}	14

words, 1 L of 1 M HCl contains 6.02×10^{23} hydrogen ions, and 0.500 L of 0.00100 M HCl contains 0.500 L × 0.00100 mol/L = 0.000500 mol H⁺ or 3.01×10^{20} H⁺ ions.

We can describe the acidity of a particular solution in moles per liter: The hydrogen ion concentration of a 0.00100 M HCl solution is 1×10^{-3} mol/L. However, exponential notation isn't terribly convenient. More often we see the acidity of this solution reported simply as pH 3.

We usually use the **pH** scale, first proposed in 1909 by the Danish biochemist S. P. L. Sorensen, to describe the degree of acidity or basicity. The pH scale lies mainly in the range 0 to 14. The neutral point on the scale is 7, with values below 7 indicating increasing acidity and those above 7 increasing basicity. Thus, pH 6 is slightly acidic, whereas pH 12 is strongly basic (Figure 7.7).

The numbers on the pH scale are directly related to the hydrogen ion concentration. We might expect that pure water would be completely in the form of H_2O molecules, but it turns out that about 1 out of every 500 million molecules is split into H⁺ and OH⁻ ions. This gives a concentration of hydrogen ions and of hydroxide ions in pure water of 0.0000001 mol/L, or 1×10^{-7} M. Can you see why 7 is the pH of pure water? It is simply the power of 10 for the molar concentration of H⁺, with the negative sign removed. (The H in pH stands for "hydrogen," and the p represents "power.") Thus, we define pH as the negative logarithm of the molar concentration of hydrogen ions (Table 7.3):

$$pH = -\log[H^+]$$

The brackets about the H⁺ indicate molar concentration.

Although pH is an acidity scale, note that its value goes down when acidity goes up. Not only is the relationship an inverse one, but it is also logarithmic. A decrease of 1 pH unit represents a tenfold increase in acidity, and when pH goes down by 2 units, acidity increases by a factor of 100. This relationship may seem strange at first, but once you understand the pH scale, you will appreciate its convenience.

Table 7.3 summarizes the relationship between hydrogen ion concentration and pH. A pH of 4 means a hydrogen ion concentration of 1×10^{-4} mol/L, or 0.0001 M. If the concentration of hydrogen ions is 0.01 M, or 1×10^{-2} M, the pH is 2. The pH values for various common solutions are listed in Table 7.4.

EXAMPLE 7.6 **pH from Hydrogen Ion Concentration**

What is the pH of a solution that has a hydrogen ion concentration of 1×10^{-5} M?

Solution

The hydrogen ion concentration is 1×10^{-5} M. The exponent is −5; the pH is therefore the negative of this exponent, or 5.

▶ Exercise 7.6A
What is the pH of a solution that has a hydrogen ion concentration of 1×10^{-11} M?

▶ Exercise 7.6B
What is the pH of a solution that is 0.0010 M HCl? (*Hint:* HCl is a strong acid.)

TABLE 7.4 The Approximate pH Values of Some Common Solutions	
Solution	pH
Hydrochloric acid (4%)	0
Gastric juice	1.6–1.8
Lemon juice	2.1
Vinegar (4%)	2.5
Soda pop	2.0–4.0
Rainwater (thunderstorm)	3.5–4.2
Milk	6.3–6.6
Urine	5.5–7.0
Rainwater*	5.6
Saliva	6.2–7.4
Pure water	7.0
Blood	7.4
Fresh egg white	7.6–8.0
Bile	7.8–8.6
Milk of magnesia	10.5
Washing soda	12.0
Sodium hydroxide (4%)	13.0

*Rainwater saturated with carbon dioxide from the atmosphere but unpolluted.

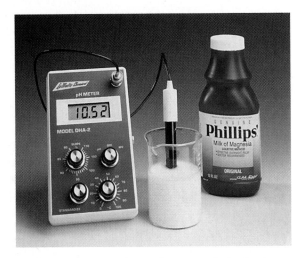

▲ A pH meter is a simple, rapid, accurate device for determining pH.

Question: Is the solution in the beaker more basic or less basic than household ammonia? (Refer to Figure 7.7.)

EXAMPLE 7.7 **Hydrogen Ion Concentration from pH**

What is the hydrogen ion concentration of a solution that has a pH of 4?

Solution
The pH value is 4. This means the exponent of 10 is -4. The hydrogen ion concentration is therefore 1×10^{-4} M.

▶ Exercise 7.7A
What is the hydrogen ion concentration of a solution that has a pH of 2?

▶ Exercise 7.7B
What is the concentration of a HNO_3 solution that has a pH of 3? (*Hint:* HNO_3 is a strong acid.)

EXAMPLE 7.8 **pH from Hydrogen Ion Concentration**

Which of the following is a reasonable pH for a solution that is 8×10^{-4} M in H^+?
(a) 2.9   (b) 3.1   (c) 4.2   (d) 4.8

Solution
The $[H^+]$ is greater than 1×10^{-4} M, and so the pH must be less than 4. That rules out (c) and (d). The $[H^+]$ is less than 10×10^{-4} (or 1×10^{-3} M), and so the pH must be greater than 3. That rules out (a). The only reasonable answer is (b), a value between 3 and 4.

(With a scientific calculator, you can get the actual pH by the following key strokes: [8][exp][4][±][log]. This gives a value of -3.1 for the log of 8×10^{-4}. Because the pH is the negative log of the H^+ concentration, the pH is 3.10.)

▶ Exercise 7.8
Which of the following is a reasonable pH for a solution that is 2×10^{-10} M in H^+?
(a) 2.0   (b) 8.7   (c) 9.7   (d) 10.2

$$pH = -\log[H^+]$$

For coffee it's 5; for tomatoes it's 4;
While household ammonia's 11 or more.
It's 7 for water, if in a pure state,
But rainwater's 6 and seawater's 8.
It's basic at 10, quite acidic at 2,
And well above 7 when litmus turns blue.
Some find it a puzzlement. Doubtless their fog
Has something to do with that negative log!

7.7 ACID RAIN

Carbon dioxide is the anhydride of carbonic acid (Section 7.3). Raindrops falling through the air absorb CO_2, which is converted to H_2CO_3. Rainwater is therefore a dilute solution of carbonic acid, a weak acid. Rain saturated with carbon dioxide has a pH of 5.6. In many areas of the world, particularly those downwind from industrial centers, rainwater is much more acidic, with a pH as low as 3 or less. Rain with a pH below 5.6 is called **acid rain**.

Acid rain is due to acidic pollutants in the air. As we shall see in Chapter 12, several air pollutants are acid anhydrides. These include sulfur dioxide (SO_2), mainly from burning high-sulfur coal in power plants and metal smelters, and nitrogen dioxide (NO_2) and nitric oxide (NO), from automobile exhaust fumes.

Some acid rain is due to natural pollutants, such as those resulting from volcano eruptions and lightning. Volcanoes give off sulfur oxides and sulfuric acid, and lightning produces nitrogen oxides and nitric acid.

Acid rain is an important environmental problem that involves both air pollution (Chapter 12) and water pollution (Chapter 13). It can have serious effects on plant and animal life.

7.8 ANTACIDS: A BASIC REMEDY

The stomach secretes hydrochloric acid to aid in the digestion of food. Sometimes overindulgence or emotional stress leads to *hyperacidity* (too much acid secreted). Hundreds of brands of antacids (Figure 7.8) are sold in the United States to treat this condition. Despite the many brand names, there are only a few different antacid ingredients, all of which are bases. Common ingredients are sodium bicarbonate, calcium carbonate, aluminum hydroxide, magnesium carbonate, and mag-

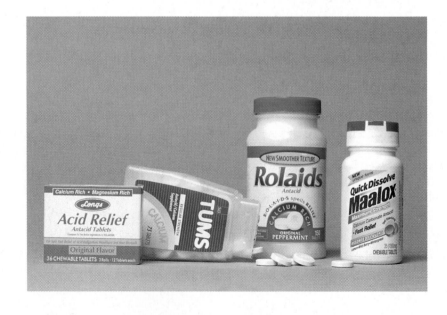

▶ **Figure 7.8** A great variety of antacids is available to consumers. All antacids are basic compounds that act by neutralizing hydrogen ions in stomach acid. Claims of "fast action" are almost meaningless because all acid–base reactions are almost instantaneous. Some tablets may dissolve a little slower than others. You can speed their action by chewing them.

nesium hydroxide. Even for a given brand name, there is often a variety of products. For example, there are more than a dozen varieties of Alka-Seltzer®.

Sodium bicarbonate ($NaHCO_3$), commonly called *baking soda*, is one of the earliest antacids and is still sometimes used. The bicarbonate ions react with the acid to form carbonic acid, which then breaks down to carbon dioxide and water.

$$HCO_3^-(aq) + H^+(aq) \longrightarrow H_2CO_3(aq)$$
$$H_2CO_3(aq) \longrightarrow CO_2(g) + H_2O(l)$$

The $CO_2(g)$ is largely responsible for the burps that result from the use of bicarbonate-containing antacids. Sodium bicarbonate is the principal antacid in most forms of Alka-Seltzer for heartburn relief. Overuse of sodium bicarbonate can make the blood too alkaline, a condition called **alkalosis**. Antacids that contain sodium ion are not recommended for people with hypertension (high blood pressure).

Calcium carbonate ($CaCO_3$) is safe in small amounts, but regular use can cause constipation. It also appears that calcium carbonate can actually result in increased acid secretion after a few hours. Tums® and many store brand antacids have calcium carbonate as the only active ingredient.

Aluminum hydroxide [$Al(OH)_3$], like calcium carbonate, can cause constipation. There is also some concern that antacids containing aluminum ions can deplete the body of essential phosphate ions. Aluminum hydroxide is the active ingredient in Amphojel®.

A suspension of magnesium hydroxide [$Mg(OH)_2$] in water is sold as "milk of magnesia." Magnesium carbonate ($MgCO_3$) is also used as an antacid. In small doses, these magnesium compounds act as antacids, but in large doses they act as laxatives.

Many antacid products have a mixture of antacids. Rolaids® and Mylanta® contain calcium carbonate and magnesium hydroxide. Maalox® liquid has aluminum hydroxide and magnesium hydroxide. These products balance the tendency of magnesium compounds to cause diarrhea with that of aluminum and calcium compounds to cause constipation.

Although antacids are generally safe for occasional use, they can interact with other medications; and anyone who has severe or repeated attacks of indigestion should consult a physician. Self-medication can sometimes be dangerous.

You can make your own aspirin-free "Alka-Seltzer™." Simply place half a teaspoon of baking soda in a glass of orange juice. (What is the acid and what is the base in this reaction?)

Drugs such as ranitidine (Zantac™), famotidine (Pepcid AC™), and cimetidine (Tagamet HB™) are not antacids. These drugs act by blocking receptors on cells in the lining of the stomach that normally bind a substance called histamine. Binding histamine causes the cells to produce acid; blocking histamine binding reduces acid production. Nexium™ (esomeprazole) acts by inhibiting *proton pumps*. Proton pumps are used by cells that line the stomach to produce stomach acid. By inhibiting the action of the proton pumps, esomeprazole reduces the production of stomach acid.

Why Doesn't "Stomach Acid" Dissolve the Stomach?

We know that strong acids are corrosive to skin. The gastric juice in your stomach is a solution containing about 0.5% hydrochloric acid. Why doesn't the acid in your stomach destroy your stomach lining? The cells that line the stomach are protected by a layer of mucus, a viscous solution of a sugar–protein complex called *mucin*, and other substances in water. The mucus serves as a physical barrier, but its role is not simply passive. Rather, the mucin acts like a sponge that soaks up bicarbonate ions from the cellular side and hydrochloric acid from within the stomach. The bicarbonate ions neutralize the acid within the mucus. When aspirin, alcohol, bacteria, or other agents damage the mucus, the exposed cells are damaged and an ulcer can form.

7.9 ACIDS AND BASES IN INDUSTRY AND IN US

Acids and bases play an important role in industry, both as products for use in many areas and as by-products that can damage the environment. Their use requires caution, and their misuse can be dangerous to human health. Acids and bases are also important participants in the biochemistry of every living thing.

Acids and Bases in Industry and at Home

Sulfuric acid is by far the leading chemical product in the United States. Nearly 40 billion kg is produced each year, most of it for making fertilizers and other industrial chemicals. Around the home we use sulfuric acid in automobile batteries and in some special kinds of drain cleaners.

SAFETY ALERT: Concentrated acids and bases can cause severe burns. They must be handled with great care. In working with them, always follow directions carefully. Wear safety goggles to protect your eyes and protective clothing to protect your skin and clothes.

▲ Treating the soil with lime makes it "sweeter" (less acidic).

Hydrochloric acid is used in industry to remove rust from metal, in construction to remove excess mortar from bricks and etch concrete for painting, and in the home to remove lime deposits from fixtures and toilet bowls. The product used in the home is often called *muriatic acid*, an old name for hydrochloric acid. Concentrated solutions (about 38% HCl) cause severe burns, but dilute solutions can be used safely in the home if handled carefully. Annual U.S. production of hydrochloric acid is more than 4 billion kg.

Lime (CaO) is the cheapest and most widely used commercial base. It is made by heating limestone ($CaCO_3$) to drive off CO_2.

$$CaCO_3(s) + heat \longrightarrow CaO(s) + CO_2(g)$$

Annual U.S. production of calcium oxide is about 22 billion kg. Much of the lime is "slaked" by adding water, forming calcium hydroxide [$Ca(OH)_2$], which is generally safer to handle than lime. Slaked lime is used to make mortar and cement and also to "sweeten" acidic soil.

Sodium hydroxide (commonly known as *lye*) is the strong base most often used in the home. It is employed as an oven cleaner in products such as Easy Off®, to open clogged drains in products such as Drano®, and in both commercial and home-made soaps. Annual U.S. production of sodium hydroxide is about 9 billion kg.

Ammonia is produced in huge volume, mainly for use as fertilizer. Annual U.S. production is nearly 11 billion kg. Ammonia is used around the home in a variety of cleaning products (Chapter 17).

Acids and Bases in Health and Disease

When they are misused, acids and bases can be damaging to human health. Concentrated strong acids and bases are corrosive poisons (Chapter 20) that can cause serious chemical burns. Once the chemical agents are removed, the injuries are similar to burns caused by heat, and they are often treated in the same way. Besides being a strong acid, sulfuric acid is also a powerful dehydrating agent that can react with water in the cells.

Strong acids and bases, even in dilute solutions, break down or *denature* the protein molecules in living cells, much as cooking does. Generally, the fragments are not able to carry out the functions of the original proteins. In cases of severe exposure, this fragmentation continues until the tissue has been completely destroyed.

Acids and bases affect human health in more subtle ways. A delicate balance must be maintained between acids and bases in the blood, body fluids, and cells. If the acidity of the blood changes too much, the blood loses its capacity to carry oxygen. In living cells, proteins function properly only at an optimum pH. If the pH changes too much in either direction, the proteins can't carry out their usual functions. Fortunately, the body has a complex but efficient mechanism for maintaining a proper acid–base balance. (Consult Reference 4 for an explanation of this mechanism.)

SUMMARY

Sections 7.1–7.2—Acids taste sour, turn litmus red, and react with active metals to form hydrogen. Bases taste bitter, turn litmus blue, and feel slippery to the skin. Acids and bases react to form salts and water. An **acid–base indicator** such as litmus has different colors in acid and in base and is used to determine whether something is acidic or basic. According to Arrhenius's definition, an **acid** produces hydrogen ions (H^+, protons) in water, and a **base** produces hydroxide ions (OH^-). Neutralization is the combination of H^+ and OH^- to form water. The remaining cations and anions give rise to an ionic **salt**. In the Brønsted–Lowry theory, an acid is a proton donor and a base is a proton acceptor. When a Brønsted–Lowry acid dissolves in water, the H_2O molecules pick up H^+ to form hydronium ions (H_3O^+). A Brønsted–Lowry base releases or forms OH^- in water. The Brønsted–Lowry theory is more general than the Arrhenius theory.

Acid   Base

Critical Thinking Exercises

Apply knowledge that you have gained in this chapter and one or more of the FLaReS principles (Chapter 1) to evaluate the following statements or claims.

7.1 A television advertisement claimed that the antacid Maalox neutralizes stomach acid faster and therefore relieves heartburn faster than Pepcid AC®, a drug that inhibits the release of stomach acid. To illustrate this claim, two flasks of acid were shown. In one, Maalox rapidly neutralized the acid. In the other, Pepcid AC did not neutralize the acid. Did this visual demonstration validate the claim made in the advertisement?

7.2 Canadians in the province of Ontario claim that industrial plants that burn coal in the United States are polluting the air in Canada and causing damage to their buildings, trees, fish, and other wildlife. Is their claim reasonable?

7.3 In arguing a case, a witness makes the following statement: "Although runoff from our plant did appear to contaminate a stream, the pH of the stream before the contamination was 6.4, and after contamination it was 5.4. So the stream is now only slightly more acidic than before." Evaluate the statement.

7.4 A Jamaican recipe for fish uses the juice of several limes, but no heat is used to cook the fish. The directions state that the lime juice in effect cooks the fish. Is this claim reasonable?

7.5 An advertisement claims that vinegar in a glass cleaning product will remove the spots left on glass by tap water. The spots are largely calcium carbonate deposits. Is the claim reasonable?

Sections 7.3–7.5—Some nonmetal oxides (such as CO_2 and SO_3) are **acidic anhydrides** in that they react with water to form acids. Some metal oxides (such as Li_2O and CaO) are **basic anhydrides**; they react with water to form bases. A **strong acid** is one that reacts completely with water to form H^+ and an anion. A **weak acid** reacts only slightly with water, and most of the acid exists as intact molecules. Common strong acids include sulfuric, hydrochloric, and nitric acids. Likewise, a **strong base** is completely ionized in water, and a **weak base** is only slightly ionized. Sodium hydroxide and potassium hydroxide are two common strong bases. The reaction between an acid and base is called **neutralization**. In aqueous solution, it is the combination of H^+ and OH^- to form water. The remaining cations and anions give rise to an ionic salt.

Section 7.6—The **pH** scale is an acidity and basicity scale, defined as

$$pH = -\log[H^+]$$

where $[H^+]$ is "molar concentration of H^+." A pH of 7 ($[H^+] = 1 \times 10^{-7}$ M) is neutral, pH values lower than 7 represent increasing concentrations of H^+ and are increasingly acidic, and pH values greater than 7 represent decreasing concentrations of H^+ and are increasingly basic. A change in pH of one unit represents a tenfold change in $[H^+]$.

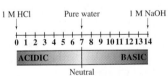

Sections 7.7–7.8—**Acid rain** is rain with a pH less than 5.6. Acid rain arises from sulfur oxides and nitrogen oxides from natural sources as well as from industry and automobile exhaust fumes. Acid rain can have serious effects on plant and animal life. An antacid is a base such as sodium bicarbonate, magnesium hydroxide, aluminum hydroxide, or calcium carbonate that is taken to relieve hyperacidity. Overuse of some antacids can make the blood too alkaline (basic), a condition called **alkalosis.**

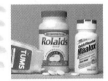

Section 7.9—Sulfuric acid is the number one chemical product in the United States, used for making fertilizers and other industrial chemicals. Hydrochloric acid (muriatic acid) is used for rust removal and etching mortar and concrete. Lime or calcium oxide is made from limestone and is the cheapest and most widely used base. Lime is an ingredient in mortar and cement and is used in agriculture. Sodium hydroxide is used to make many industrial products as well as soap. Ammonia is a weak base produced mostly as a fertilizer. Concentrated strong acids and bases are corrosive poisons that can cause serious burns. Living organisms have an optimum pH for their continued good health.

REVIEW QUESTIONS

1. Define and illustrate the following terms.
   a. acid
   b. base
   c. salt

2. Describe the effect on litmus and the action on iron or zinc of a solution that has been neutralized.

3. List four general properties each of **(a)** acidic solutions and **(b)** basic solutions.

4. Can a substance be a Brønsted–Lowry acid if it does not contain H atoms? Are there any characteristic atoms that must be present in a Brønsted–Lowry base?

5. Give the formulas and the names of two strong acids and two weak acids.

6. Give the formulas and the names of two strong bases and one weak base.

7. Strong acids and weak acids both have properties characteristic of hydrogen ions. How do strong acids and weak acids differ?

8. What is meant by the proton as used in acid–base chemistry? How does it differ from the proton of nuclear chemistry (Chapter 4)?

9. What is an acidic anhydride? A basic anhydride?

10. Describe the neutralization of an acid or base.

11. Magnesium hydroxide is completely ionic, even in the solid state, yet it can be taken internally as an antacid. Explain why it does not cause injury as sodium hydroxide would.

12. Name some of the active ingredients in antacids. What is the medical use of antacids?

13. What is alkalosis? What antacid ingredient might cause alkalosis if taken in excess?

14. What is the leading chemical product of U.S. industry?

15. What are the effects of strong acids and strong bases on the skin?

16. According to the Arrhenius theory, all acids have one element in common. What is that element? Are all compounds containing that element acids? Explain.

PROBLEMS

Acids and Bases: The Arrhenius Theory

17. What ion is responsible for the properties of acidic solutions (in water)?

18. What ion is responsible for the properties of basic solutions (in water)?

19. Write an equation that represents the action of perchloric acid ($HClO_4$) as an Arrhenius acid.

20. Write an equation that represents the action of hydrogen sulfate ion (HSO_4^-) as an Arrhenius acid. (Be sure to include the correct charges for ions.)

Acids and Bases: The Brønsted–Lowry Acid–Base Theory

21. Give the Brønsted–Lowry definition of an acid. Write an equation that illustrates the definition.

22. Give the Brønsted–Lowry definition of a base. Write an equation that illustrates the definition.

23. Use the definitions of *acid* and *base* to identify the first compound in each equation as an acid or a base. (*Hint:* What is produced by the reaction?)
    a. $C_5H_5N + H_2O \longrightarrow C_5H_5NH^+ + OH^-$
    b. $C_6H_5OH + H_2O \longrightarrow C_6H_5O^- + H_3O^+$
    c. $CH_3COCOOH + H_2O \longrightarrow CH_3COCOO^- + H_3O^+$

24. Use the definitions of *acid* and *base* to identify the first compound in each equation as an acid or a base.
    a. $CH_3CH_2SH + H_2O \longrightarrow CH_3CH_2S^- + H_3O^+$
    b. $CH_3NH_2 + H_2O \longrightarrow CH_3NH_3^+ + OH^-$
    c. $C_6H_5SO_2NH_2 + H_2O \longrightarrow C_6H_5SO_2NH^- + H_3O^+$

25. Write the equation that shows hydrogen chloride gas reacting as a Brønsted–Lowry acid in water. What is the name of the acid formed?

26. Write the equation that shows hydrogen bromide gas reacting as a Brønsted–Lowry acid in water. Based on the answer to Problem 25, suggest a name for the acid formed.

27. Write the equation that shows how ammonia acts as a Brønsted–Lowry base in water.

28. Hydroxylamine ($HONH_2$) is not an Arrhenius base even though it has OH, but it *is* a Brønsted–Lowry base. Write an equation that shows how hydroxylamine acts as a Brønsted–Lowry base in water.

Acids and Bases: Names and Formulas

29. Give formulas for the following acids and bases.
    a. hydrochloric acid
    b. strontium hydroxide
    c. potassium hydroxide
    d. boric acid

30. Give formulas for the following acids and bases.
    a. rubidium hydroxide
    b. aluminum hydroxide
    c. hydrocyanic acid
    d. nitric acid

31. Name the following and classify each as an acid or a base.
    a. HNO_3
    b. $CsOH$
    c. H_2CO_3

32. Name the following and classify each as an acid or a base.
    a. $Mg(OH)_2$
    b. NH_3
    c. CH_3COOH

33. Often, an acid ending in *-ic acid* will have a corresponding acid ending with *-ous acid*. The *-ous* acid has one less oxygen atom than the *-ic* acid. With this information, write formulas for **(a)** nitrous acid and **(b)** sulfurous acid.

34. Use the information in Problem 33 to write a formula for phosphorous acid.

Acidic and Basic Anhydrides

35. Give the formula for the compound formed when **(a)** sulfur trioxide reacts with water and **(b)** magnesium oxide reacts with water. In each case, is the product an acid or a base?

36. Give the formula for the compound formed when **(a)** potassium oxide reacts with water and **(b)** carbon dioxide reacts with water. In each case, is the product an acid or a base?

Strong and Weak Acids and Bases

37. Thallium hydroxide (TlOH) is ionic in the solid state and is quite soluble in water. Classify TlOH as a strong acid, weak acid, weak base, or strong base.

38. Hydrogen iodide (HI) gas reacts completely with water to form hydronium ions and iodide ions. Classify HI as a strong acid, weak acid, weak base, or strong base.

39. Hydrogen sulfide (H_2S) gas reacts slightly with water to form relatively few hydronium ions and hydrogen sulfide ions (HS^-). Classify H_2S as a strong acid, weak acid, weak base, or strong base.

40. Methylamine (CH_3NH_2) gas reacts slightly with water to form relatively few hydroxide ions and methylammonium ions ($CH_3NH_3^+$). Classify CH_3NH_2 as a strong acid, weak acid, weak base, or strong base.

41. Identify each of the following substances as either a strong acid, a weak acid, a strong base, a weak base, or a salt.
    a. H_3PO_4
    b. LiOH
    c. NH_4NO_3
    d. HCl

42. Identify each of the following substances as either a strong acid, a weak acid, a strong base, a weak base, or a salt.
    a. Na_2SO_4
    b. KOH
    c. NH_4I
    d. $BaCl_2$

43. Which of the following aqueous solutions has the highest concentration of H^+ ion? Which has the lowest?
    (a) 0.10 M HCl
    (b) 0.10 M NH_3
    (c) 0.10 M CH_3COOH

44. Which of the following aqueous solutions has the highest concentration of H^+ ion? Which has the lowest?
    (a) 0.10 M HNO_3
    (b) 0.10 M H_2CO_3
    (c) 0.10 M NaOH

Ionization of Acids and Bases

45. Write equations showing the ionization of the following as Arrhenius acids or bases.
    a. HI
    b. LiOH
    c. $HClO_2$

46. Write equations showing the ionization of the following as Arrhenius acids or bases.
    a. HNO_2
    b. $Ba(OH)_2$
    c. HBr

47. Write equations showing the ionization of the following as Brønsted–Lowry acids.
    a. $HClO_2(aq)$
    b. $HNO_2(aq)$
    c. $HCN(aq)$

48. Write equations to show the ionization of the following as Brønsted–Lowry acids in water.
    a. HBr
    b. HIO_3
    c. $CH_3CHOHCOOH$

Neutralization

49. Write equations for the reaction of **(a)** potassium hydroxide with hydrochloric acid and **(b)** lithium hydroxide with nitric acid.

50. Write equations for the reaction of **(a)** 1 mol calcium hydroxide with 2 mol hydrochloric acid and **(b)** 1 mol sulfuric acid with 2 mol potassium hydroxide.

51. Write the equation for the reaction of 1 mol phosphoric acid with 3 mol sodium hydroxide.

52. Write the equation for the reaction of 1 mol sulfuric acid with 1 mol calcium hydroxide.

The pH Scale

53. Indicate whether each of the following pH values represents an acidic, basic, or neutral solution.
    a. 4
    b. 7
    c. 3.5
    d. 9

54. Lime juice is quite sour. Which of the following is a reasonable pH for lime juice?
    (a) 2
    (b) 7
    (c) 9
    (d) 12

55. What is the pH of a solution that has a hydrogen ion concentration of 1.0×10^{-5} M?

56. What is the pH of a solution that has a hydrogen ion concentration of 1.0×10^{-1} M?

57. What is the hydrogen ion concentration of a solution that has a pH of 12?

58. What is the hydrogen ion concentration of a solution that has a pH of 4?

59. Which of the following is a reasonable pH for a solution that is 2×10^{-4} M in H^+?
    (a) 2.4   (b) 3.7
    (c) 4.0   (d) 4.7

60. Which of the following is a reasonable pH for a solution that is 7×10^{-9} M in H^+?
    (a) 7.1   (b) 7.9
    (c) 8.1   (d) 9.7

Antacids

61. Mylanta liquid has 200 mg $Al(OH)_3$ and 200 mg $Mg(OH)_2$ per teaspoonful. Write the equations for the neutralization of stomach acid [HCl(aq)] by each of these substances.

62. What is the Brønsted–Lowry base in each of the following compounds used as ingredients in antacids?
    a. $NaHCO_3$
    b. $Mg(OH)_2$
    c. $MgCO_3$
    d. $CaCO_3$

ADDITIONAL PROBLEMS

63. According to the Arrhenius theory, are all compounds containing OH groups bases? Explain.

64. Lime deposits on brass faucets are mostly $CaCO_3$. The deposits can be removed by soaking the faucet in hydrochloric acid. Write an equation for the reaction that occurs.

65. Strontium iodide can be made by the reaction of solid strontium carbonate ($SrCO_3$) with hydroiodic acid (HI). Write the equation for this reaction.

66. A paste of sodium hydrogen carbonate (sodium bicarbonate) and water can be used to relieve the pain of an ant bite. The irritant in the ant bite is an organic acid called formic acid (HCOOH). Write an equation for the reaction that occurs.

67. Like water, the hydrogen phosphate ion (HPO_4^{2-}) is amphiprotic. That is, it can act either as a Brønsted–Lowry acid or as a Brønsted–Lowry base. Write equations that illustrate both these reactions.

68. A term akin to pH, called pOH, is related to [OH^-] just as pH is related to [H^+]. What is the pOH **(a)** of a solution that has a hydroxide ion concentration of 1.0×10^{-2} M? **(b)** Of a 0.001 M KOH solution?

69. The pOH of a solution (Problem 68) is related to the pH of the solution by the relationship pH + pOH = 14. What is the pH **(a)** of a solution that has a pOH of 3? **(b)** Of a 0.01-M NaOH solution?

70. Three varieties of Tums have calcium carbonate as the only active ingredient: Regular Tums have 500 mg; Tums E-X, 750 mg; and Tums ULTRA, 1000 mg. How many regular Tums would you have to take to get the same quantity of calcium carbonate as you would get with two Tums E-X? With two Tums ULTRA tablets?

71. Milk of magnesia has 400 mg of $Mg(OH)_2$ per teaspoon. Calculate the mass of stomach acid that can be neutralized by 1.00 teaspoon of milk of magnesia, assuming the stomach acid is 0.50% HCl by mass.

COLLABORATIVE GROUP PROJECTS

Prepare a PowerPoint, poster, or other presentation (as directed by your instructor) for presentation to the class.

72. Prepare a brief report on one of the following acids or bases. List sources (including local sources for the acid or base, if available) and commercial uses.
a. ammonia
b. hydrochloric acid
c. phosphoric acid
d. nitric acid
e. sodium hydroxide
f. sulfuric acid

73. Examine the labels of at least five antacid preparations. Make a list of the ingredients in each. Look up the properties (medical use, side effects, toxicity, and so on) of each ingredient in a reference book such as *The Merck Index* (Reference 1).

74. Examine the labels of at least five toilet bowl cleaners and five drain cleaners. Make a list of the ingredients in each. Look up the formulas and properties of each ingredient in a reference book such as *The Merck Index* (Reference 1). Which ingredients are acids? Which are bases?

75. Examine the labels of at least three different substances or mixtures sold to adjust the pH of pool water, and list the acidic or basic ingredients of each, and their properties (from *The Merck Index* or a similar reference). Also explain how pool water is tested for pH, and the desired pH for pool water.

REFERENCES AND READINGS

1. Budavari, S., M. J. O'Neill, and A. Smith. *The Merck Index: An Encyclopedia of Chemicals, Drugs, and Biologicals*, 13th edition. Rahway, NJ: Merck & Co., 2001.

2. Gorman, Jessica. "Tums of the Sea." *Science News*, August 7, 2002, pp. 104–105. Dissolved CO_2 is neutralized by shells floating in the sea.

3. "Heartburn: Picking the Right Remedy." *Consumer Reports*, September 2002, pp. 42–45.

4. Hill, John W., Stuart J. Baum, and Rhonda J. Scott-Ennis. *Chemistry and Life*, 6th edition. Upper Saddle River, NJ: Prentice Hall, 2000. Chapters 9 and 10.

5. Jensen, William B. "Acids and Bases: Ancient Concepts in Modern Science." *ChemMatters*, April 1983, pp. 14–15. Part of a special issue on acid–base chemistry.

6. Jensen, William B. "The Symbol for pH." *Journal of Chemical Education*, January 2004, p. 21.

7. Kolb, Doris. "The pH Concept." *Journal of Chemical Education*, January 1979, pp. 49–53.

8. Lowenstein, Jerome. *Acids and Bases*. New York: Oxford University Press, 1993. A discussion of acid–base physiology.

9. Preston, Richard A. *Acid–Base, Fluids, and Electrolytes Made Ridiculously Simple*. Miami, FL: Medmaster, 2002.

GREEN CHEMISTRY

Soap: Waste Not, Want Not

Irvin Levy, Gordon College

Acid–base chemistry is extremely important in the actual practice of chemistry. Many important chemical reactions with economic value involve acids and bases. One of the most commonly used bases is sodium hydroxide, commonly called "lye." This potent alkaline material is used in the synthesis of many useful products. Lye is used in one of the earliest practiced chemical reactions: the preparation of soap.

Soap is produced when fats or oils are mixed with a base. Fats and oils are comprised of mixtures of large organic molecules called triglycerides. When treated with base, the triglyceride molecules are transformed into four separate molecular fragments as shown below.

$$CH_3-(CH_2)_n-\overset{\overset{\text{O}}{\|}}{C}-OCH_2$$
$$CH_3-(CH_2)_n-\overset{\overset{\text{O}}{\|}}{C}-OCH_2 \quad + \quad 3\,NaOH \longrightarrow$$
$$CH_3-(CH_2)_n-\overset{\overset{\text{O}}{\|}}{C}-OCH_2$$

Triglyceride
(n = 12-14)

$$CH_3-(CH_2)_n-\overset{\overset{\text{O}}{\|}}{C}-O^- \quad Na^+ \qquad HOCH_2$$
$$CH_3-(CH_2)_n-\overset{\overset{\text{O}}{\|}}{C}-O^- \quad Na^+ \qquad HOCH$$
$$CH_3-(CH_2)_n-\overset{\overset{\text{O}}{\|}}{C}-O^- \quad Na^+ \qquad HOCH_2$$

Soap   Glycerol

In most processes that involve chemical reactions, there are desired products and unwanted by-products (waste). For example, in the reaction above, the soap is produced along with glycerol. In traditional methods of soap-making, the glycerol is left in with the soap to act as a moisturizer. On an industrial scale it is possible to remove the glycerol from the soap, using it for other purposes. In both of these cases, the "waste" from the soap-making process is found to have a useful purpose, making it unnecessary for it to end up in a waste container. Not all chemical reactions are as convenient as this one; other reactions produce hazardous by-products for which no useful purpose is known.

In the Web Investigations you will examine several ways that "waste" from one process can be used to feed another process, as well as methods used in industry to measure the waste produced in an industrial process. In Communicate Your Results you will examine the production of soap with respect to the 12 principles of green chemistry to ascertain whether soap production can fall into the "benign by design" paradigm of green chemistry. You will also consider practical ways that leftover materials from one process could be redirected as raw materials for another.

WEB INVESTIGATION

Investigation 1
The Chemistry of Soap
Do a Web search using the keyword "saponification" to learn more about the chemistry of soap preparation. When making soap, it is important that the amount of base added is determined with care. One must use enough base to convert the fats and oils to soap (you wouldn't get very clean washing with bacon grease!) but

not so much base that the harsh lye remains in the soap, causing skin irritation. Check out some of the formulations for soap by searching using the keywords "soap recipes." The keywords "lye calculator" should show you some of the tools available for soapmakers. Finally, "soap manufacturing" will provide information about the industrial process of soap-making.

Investigation 2
Turning Waste into Value

Fats and oils have some nutritive value, but the majority of fats and oils used in the soap-making process are not widely used as foods. Return to several of the "soap recipe" sites and make a list of some of the fats and oils used for making soap. In modern times soap is usually made by adding sodium hydroxide to a mixture of fats. In earlier eras, one could not simply purchase sodium hydroxide. Where did the alkali come from in those days? A search using the terms "colonial soap" should provide for some background on the production of alkaline materials from sources that would otherwise be discarded as waste.

Investigation 3
E-Factor as a Measure of Waste Production

With most endeavors it is useful to have some objective way to measure success. When thinking about the waste produced by a process, one measure of success is called E-factor. The search "waste" and "E-factor" should help you to learn how this measurement is calculated.

COMMUNICATE YOUR RESULTS

Exercise 1
Soap and Green Chemistry

Consider the production of soap in relation to the 12 principles of green chemistry. Which of the principles most directly apply? Prepare a poster demonstrating this relationship.

Exercise 2
The Manufacture of Soap

Using the results from your Web Investigation, describe how glycerol is removed from raw soap industrially. Describe some of the uses that have been found for the glycerol. Suppose that glycerol was a useless by-product instead of a valuable commodity. How would that affect the E-factor of the soap manufacturing process?

Exercise 3
Finding Value in "Waste"

Interview employees from the dining services at your school to estimate the amount of bacon grease (lard) produced in a normal week. Using the information you have obtained in the Web Investigation, design a soap recipe with lard as the major source of fat. Use the lye calculator to determine the amount of base needed to produce your soap. Produce a table of materials needed in order to produce 1.0 kilogram of your soap. Estimate the amount of soap that can be made in a semester using the bacon grease as the raw material.

Writing in the format of a proposal, persuade the school to allow a student organization on campus to use the grease to make and market soap at the school's campus store or local craft fairs. The E-factor should be discussed as part of your persuasive argument.

▲ Most restaurants produce a large amount of waste grease. Green chemistry is finding ways to use such wastes.

CHAPTER 8

Oxidation and Reduction

The defensive spray of the African bombardier beetle is formed when the beetle is alarmed and mixes the contents of two storage chambers. One chamber contains hydroquinones and hydrogen peroxide; the other contains enzymes. To activate the spray, the beetle mixes the contents of the two compartments. The enzymes speed the breakdown of hydrogen peroxide to oxygen and water and the oxidation of hydroquinones to highly irritating quinones (see Problem 54). The reaction is exothermic, heating the spray to 100 °C. The oxygen also acts as a propellant, causing the mixture to be expelled with an audible "pop." Reduction–oxidation reactions are vital parts of all living and many nonliving systems on Earth.

Burn and Unburn

From a simple campfire to the most advanced electric battery, from the trees consuming carbon dioxide and making oxygen in a rainforest to the people hurrying along a city street, we depend on an important group of reactions called *reduction–oxidation* (or *redox*) reactions. These reactions are extremely diverse: Charcoal burns, iron rusts, bleach removes stains, the food we eat is converted to energy for our brain and muscles, and film is developed.

Oxidation and reduction always occur together. They are opposite aspects of a single process, the redox reaction. You can't have one without the other (Figure 8.1). When one substance is oxidized, another is reduced. However, it is sometimes convenient to discuss only a part of the process—the oxidation part or the reduction part.

▲ **Figure 8.1** Oxidation and reduction always occur together. Pictured here on the left is ammonium dichromate. In the reaction (center), the ammonium ion (NH_4^+) is oxidized and the dichromate ion ($Cr_2O_7^{2-}$) is reduced. Considerable heat and light are evolved. The equation for the reaction is

$$(NH_4)_2Cr_2O_7 \longrightarrow Cr_2O_3 + N_2 + 4\,H_2O$$

The water is driven off as vapor, and the nitrogen gas escapes, leaving pure Cr_2O_3 as the visible product (right).

213

Reduced forms of matter—food, coal, and gasoline—are high in energy. Oxidized forms—carbon dioxide and water—are low in energy. The energy in foods and fossil fuels is released when these materials are oxidized. Let's examine the processes of oxidation and reduction in some detail. By doing so, we can better understand the chemical reactions that keep us alive and enable us to maintain our civilization.

Oxidation—Reduction Reactions
Part 1

Oxidation—Reduction Reactions
Part 2

8.1 OXIDATION AND REDUCTION: THREE VIEWS

The term *oxidation* stems from the early recognition of oxygen's involvement in oxide formation; *reduction* then meant removal of oxygen from an oxide. When oxygen combines with other elements or compounds, the process is called **oxidation**. The substances that combine with oxygen are said to have been *oxidized*. Originally the term *oxidation* was limited to reactions involving combination with oxygen. Then as chemists came to realize that combination with chlorine (or bromine or other active nonmetals) was not all that different from reaction with oxygen, they broadened the definition of oxidation.

Reduction is the opposite of oxidation. When hydrogen burns, it combines with oxygen to form water.

$$2 H_2 + O_2 \longrightarrow 2 H_2O$$

The hydrogen is oxidized in this reaction, but at the same time the oxygen is reduced. Whenever oxidation occurs, reduction must occur also. Oxidation and reduction always happen at the same time and in exactly equivalent amounts.

Because oxidation and reduction are chemical opposites and constant companions, it is convenient to link their definitions together. We can view oxidation and reduction in at least three different ways (Figure 8.2).

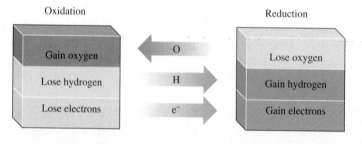

▶ **Figure 8.2** Three different views of oxidation and reduction.

1. *Oxidation* is a gain of oxygen atoms.
   Reduction is a loss of oxygen atoms.

At high temperatures (such as those in automobile engines), nitrogen, which is normally quite unreactive, combines with oxygen to form nitric oxide.

$$N_2 + O_2 \longrightarrow 2 NO$$

Nitrogen gains oxygen atoms; there are no O atoms in the N_2 molecule and one O atom in each of the NO molecules. Therefore, nitrogen is oxidized.

Now consider what happens when methane is burned to form carbon dioxide and water.

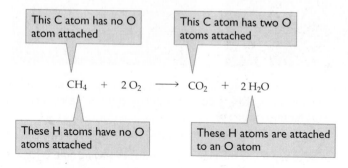

Both carbon and hydrogen gain oxygen atoms, and so both elements are oxidized. When lead dioxide is heated at high temperatures, it decomposes as follows.

$$2\,PbO_2 \longrightarrow 2\,PbO + O_2$$

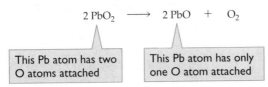

This Pb atom has two O atoms attached

This Pb atom has only one O atom attached

The lead dioxide loses oxygen, and so it is reduced.

CONCEPTUAL EXAMPLE 8.1 **Redox—Gain or Loss of Oxygen Atoms**

In each of the following reactions, is the reactant undergoing oxidation or reduction? (These are not complete chemical equations.)

a. $Pb \longrightarrow PbO_2$ 　　　　b. $SnO_2 \longrightarrow SnO$

c. $KClO_3 \longrightarrow KCl$ 　　　d. $Cu_2O \longrightarrow 2\,CuO$

Solution

a. Lead gains oxygen atoms (it has none on the left and two on the right); it is oxidized.
b. Tin loses an oxygen atom (it has two on the left and only one on the right); it is reduced.
c. There are three oxygen atoms on the left and none on the right. The compound loses oxygen; it is reduced.
d. The two copper atoms on the left share a single oxygen atom; they have half an oxygen atom each. On the right, each copper atom has an oxygen atom all its own. Cu has gained oxygen; it is oxidized.

▶ **Exercise 8.1**
In each of the following reactions, is the reactant undergoing oxidation or reduction? (These are not complete chemical equations.)
a. $3\,Fe \longrightarrow Fe_3O_4$ 　　　b. $NO \longrightarrow NO_2$
c. $Cr_2O_3 \longrightarrow CrO_3$ 　　　d. $C_3H_6O \longrightarrow C_3H_6O_2$

A second view of oxidation and reduction involves hydrogen atoms.

2. *Oxidation* is a loss of hydrogen atoms.
   Reduction is a gain of hydrogen atoms.

Look once more at the burning of methane:

$$CH_4 + 2\,O_2 \longrightarrow CO_2 + 2\,H_2O$$

The oxygen gains hydrogen to form water; the oxygen is reduced. (From our first definition we see that the carbon and hydrogen of CH_4 gain oxygen; CH_4 is oxidized.)

Methyl alcohol (CH_3OH), when passed over hot copper gauze, forms formaldehyde and hydrogen gas.

$$CH_3OH \longrightarrow CH_2O + H_2$$

The C and O atoms have 4 H atoms attached

The C and O atoms have only two H atoms attached

Because the methyl alcohol loses hydrogen, it is oxidized in this reaction.
Methyl alcohol can be made by reaction of carbon monoxide with hydrogen.

$$CO + 2\,H_2 \longrightarrow CH_3OH$$

Because the carbon monoxide gains hydrogen atoms, it is reduced.
Biochemists often find the gain or loss of hydrogen atoms a useful way to look at oxidation–reduction processes. For example, a substance called NAD^+ has the

formula $C_{21}H_{27}N_7O_{14}P_2$. In a variety of biochemical redox reactions, NAD^+ is changed to NADH, which has the formula $C_{21}H_{29}N_7O_{14}P_2$. These molecules are rather complex, but it is easy to see that NAD^+ gains two hydrogen atoms in changing to NADH; NAD^+ is oxidized.

CONCEPTUAL EXAMPLE 8.2 **Redox—Gain or Loss of Hydrogen Atoms**

In each of the following reactions, is the reactant undergoing oxidation or reduction? (These are not complete chemical equations.)

a. $C_2H_6O \longrightarrow C_2H_4O$ 　　　　**b.** $C_2H_2 \longrightarrow C_2H_6$

Solution

a. There are six hydrogen atoms in the compound on the left and only four in the one on the right. The compound loses hydrogen atoms; it is oxidized.

b. There are two hydrogen atoms in the compound on the left and six in the one on the right. The compound gains hydrogen atoms; it is reduced.

▶ **Exercise 8.2**

In each of the following reactions, is the reactant undergoing oxidation or reduction? (These are not complete chemical equations.)

a. $C_6H_6 \longrightarrow C_6H_{12}$ 　　　　**b.** $C_3H_6O \longrightarrow C_3H_4O$

A third view of oxidation and reduction involves gain or loss of electrons.

3. *Oxidation* is a loss of electrons.
　　Reduction is a gain of electrons.

When magnesium metal reacts with chlorine, magnesium ions and chloride ions are formed.

$$Mg + Cl_2 \longrightarrow Mg^{2+} + 2\,Cl^-$$

Because the magnesium atom loses electrons, it is oxidized; and because the chlorine atoms gain electrons, they are reduced.

It is easy to see that when magnesium atoms become Mg^{2+} ions they lose electrons, and that when chlorine atoms become Cl^- ions they must gain electrons. But which is oxidation and which is reduction? Perhaps Figure 8.3 can help. The charge on a simple ion is often referred to as its *oxidation number*. An increase in oxidation number (increase in positive charge) is oxidation; a decrease in oxidation number is reduction. The charge on a Mg atom is zero; thus conversion to Mg^{2+} is an increase in oxidation number (oxidation). For Cl atoms the charge is also zero, and a change to Cl^- is therefore a decrease in oxidation number (reduction).

This view of oxidation and reduction as a gain or loss of electrons is especially useful in electrochemistry (Section 8.3).

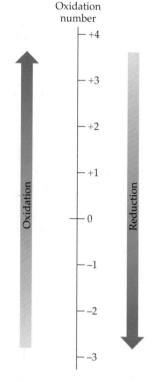

Oxidation
number

+4

+3

+2

+1

0

−1

−2

−3

▲ **Figure 8.3** An increase in oxidation number means a loss of electrons and is therefore oxidation. A decrease in oxidation number means a gain of electrons and is therefore reduction.

Question: Manganese goes from oxidation number +4 in MnO_2 to +7 in MnO_4^-. Is it oxidized or reduced? Sulfur goes from oxidation number 0 in S_8 to −2 in S^{2-}. Is it oxidized or reduced?

CONCEPTUAL EXAMPLE 8.3 **Redox—Gain or Loss of Electrons**

In each of the following reactions, is the reactant undergoing oxidation or reduction? (These are not complete chemical equations.)

a. $Zn \longrightarrow Zn^{2+}$ 　　　　**b.** $Fe^{3+} \longrightarrow Fe^{2+}$
c. $S^{2-} \longrightarrow S$ 　　　　**d.** $AgNO_3 \longrightarrow Ag$

Solution

a. In forming a 2+ ion, a Zn atom loses two electrons. Its oxidation number increases from 0 to +2. Zinc is oxidized.

b. To go from a 3+ ion to a 2+ ion, iron gains an electron. Its oxidation number decreases from +3 to +2. Iron is reduced.

c. To go from a 2− ion to an atom with no charge, sulfur loses two electrons. Its oxidation number increases. Sulfur is oxidized.

d. To answer this question, you must recognize that $AgNO_3$ is an ionic compound with Ag^+ and NO_3^- ions. In going from Ag^+ to Ag, silver gains an electron. Its oxidation number decreases. Silver is reduced.

▶ **Exercise 8.3**
In each of the following reactions, is the reactant undergoing oxidation or reduction? (These are not complete chemical equations.)
a. $Cu^{2+} \longrightarrow Cu$
b. $MnO_4^{2-} \longrightarrow MnO_4^-$
c. $Sn^{2+} \longrightarrow Sn^{4+}$
d. $Cu \longrightarrow CuSO_4$

Why do we have different ways to look at oxidation and reduction? Oxidation as a gain of oxygen is historical and specific; but the definition in terms of electrons applies more broadly. Which one should we use? Usually we use the clearest or most convenient definition. For the combustion of carbon

$$C + O_2 \longrightarrow CO_2$$

it is most convenient to see that carbon is oxidized by gaining oxygen atoms. Similarly, for the reaction

$$CH_2O + H_2 \longrightarrow CH_4O$$

it is easy to see that the reactant gains hydrogen atoms and is thereby reduced. Finally, in the case of the reaction

$$3\,Sn^{2+} + 2\,Bi^{3+} \longrightarrow 3\,Sn^{4+} + 2\,Bi$$

it is clear that tin is oxidized because it loses electrons, increasing in oxidation number from +2 to +4. Similarly, we see that bismuth is reduced because it gains electrons, decreasing in oxidation number from +3 to 0.

Two mnemonics:

Leo the lion.
LEO says GER
Loss of
   Electrons is
      Oxidation.
Gain of
   Electrons is
      Reduction.

OIL RIG
Oxidation
   Is
      Loss of electrons.
Reduction
   Is
      Gain of electrons.

8.2 OXIDIZING AND REDUCING AGENTS

Oxidation and reduction occur together. However, in looking at oxidation–reduction reactions, we often find it convenient to focus on the role played by a particular reactant. For example, in the reaction

$$CuO + H_2 \longrightarrow Cu + H_2O$$

we see that copper oxide is reduced and hydrogen is oxidized. We can also see that if one substance is oxidized, the other must *cause* it to be oxidized. In the preceding example, CuO causes H_2 to be oxidized. Therefore, CuO is called the **oxidizing agent**. Conversely, H_2 causes CuO to be reduced, and so H_2 is the **reducing agent**. Each oxidation–reduction reaction has an oxidizing agent and a reducing agent among the reactants. The reducing agent is the substance being oxidized; the oxidizing agent is the substance being reduced.

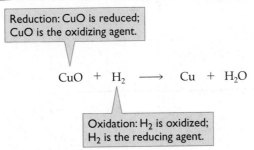

Reduction: CuO is reduced; CuO is the oxidizing agent.

$$CuO + H_2 \longrightarrow Cu + H_2O$$

Oxidation: H_2 is oxidized; H_2 is the reducing agent.

CONCEPTUAL EXAMPLE 8.4   **Oxidizing and Reducing Agents**

Identify the oxidizing agents and reducing agents in the following reactions.
a. $2\,C + O_2 \longrightarrow 2\,CO$
b. $N_2 + 3\,H_2 \longrightarrow 2\,NH_3$
c. $SnO + H_2 \longrightarrow Sn + H_2O$
d. $Mg + Cl_2 \longrightarrow Mg^{2+} + 2\,Cl^-$

Solution

We can determine the answers by one or more of the preceding methods.

a. C gains oxygen and is oxidized, and so it must be the reducing agent. O_2 is therefore the oxidizing agent.

b. N_2 gains hydrogen and is reduced, and so it is the oxidizing agent. H_2 therefore is the reducing agent.

c. SnO loses oxygen and is reduced, and so it is the oxidizing agent. H_2 is therefore the reducing agent.

d. Mg loses electrons and is oxidized, and so it is the reducing agent. Cl_2 is therefore the oxidizing agent.

▶ **Exercise 8.4**

Identify the oxidizing agents and reducing agents in the following reactions.

a. $Se + O_2 \longrightarrow SeO_2$

b. $CH_3CN + 2\,H_2 \longrightarrow CH_3CH_2NH_2$

c. $V_2O_5 + 2\,H_2 \longrightarrow V_2O_3 + 2\,H_2O$

d. $2\,K + Br_2 \longrightarrow 2\,K^+ + 2\,Br^-$

8.3 ELECTROCHEMISTRY: CELLS AND BATTERIES

Voltaic Cells I: The Copper-Zinc Cell

We saw in Chapter 3 that electricity can produce chemical change, a process called *electrolysis*. For example, an electric current through molten sodium chloride produces sodium metal and chlorine gas. Here we see the reverse process, in which chemical change produces electricity.

An electric current in a wire is simply a flow of electrons. Oxidation–reduction reactions in which electrons are transferred from one substance to another can be used to produce electricity. This is what happens in dry cell and storage batteries.

When a coil of copper wire is placed in a solution of silver nitrate (Figure 8.4a), the copper atoms give up their outer electrons to the silver ions. (We can omit the nitrate ions from the equation because they do not change.) The copper metal dissolves, going into solution as copper(II) ions that color the solution blue. The silver ions come out of solution as beautiful needles of silver metal (Figure 8.4b). The copper is oxidized; the silver ions are reduced.

Because the reaction simply involves transfer of electrons, it can occur even when the silver ions are separated from the copper metal. By placing the reactants in separate compartments and connecting them with a wire, the electrons will flow

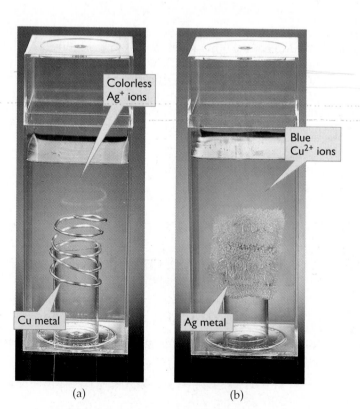

Colorless Ag^+ ions

Blue Cu^{2+} ions

Cu metal

Ag metal

(a)   (b)

▶ **Figure 8.4** The left photograph shows a coil of copper wire immersed in a colorless solution of Ag^+ ions. As the reaction progresses, the more active copper displaces the less active silver from solution, producing (right) needle-like crystals of silver metal and a blue solution of Cu^{2+} ions. The equation for the reaction is

$2\,Ag^+(aq) + Cu(s) \longrightarrow$
$\qquad 2\,Ag(s) + Cu^{2+}(aq)$.

through the wire to get from the copper metal to the silver ions. This flow of electrons constitutes an electric current, and it can be used to run a motor or light a lamp.

In the **electrochemical cell** pictured in Figure 8.5, there are two separate compartments. One contains copper metal in a blue solution of copper(II) sulfate, and the other contains silver metal in a colorless solution of silver nitrate. Copper atoms give up electrons much more readily than silver atoms, and so electrons flow away from the copper and toward the silver. The copper metal slowly dissolves as copper atoms give up electrons to form copper(II) ions. The electrons flow through the wire to the silver, where silver ions pick them up to become silver atoms.

As time goes by, the copper bar slowly disappears and the silver bar gets bigger. The blue solution becomes darker blue as Cu atoms are converted to Cu^{2+} ions. Those Cu^{2+} ions give the left compartment a positive charge, so (negative) nitrate ions move from the right compartment, through the porous partition, into the copper sulfate solution. (That is why the partition is porous. If the nitrate ions were unable to move through the barrier, the cell would not work.) For each copper atom that gives up two electrons, two silver ions each pick up one electron, and two nitrate ions move from the right compartment to the left compartment.

The two pieces of metal where electrons are transferred are called **electrodes**. The electrode where oxidation occurs is called the **anode**. The one where reduction occurs is the **cathode**. In our cell, copper gives up electrons, it is oxidized, and the copper bar is therefore the anode. Silver ions gain electrons and are reduced, so the silver bar is the cathode.

Electrochemical reactions are often represented as two *half-reactions*. The following representation shows the two half-reactions for the copper–silver cell and how they are added to give the overall cell reaction.

Oxidation:	$Cu(s) \longrightarrow Cu^{2+}(aq) + 2e^-$
Reduction:	$2\,Ag^+(aq) + 2e^- \longrightarrow 2\,Ag(s)$
Overall reaction:	$Cu(s) + 2\,Ag^+(aq) \longrightarrow Cu^{2+}(aq) + 2\,Ag(s)$

Note that the electrons cancel when the two half-reactions are added.

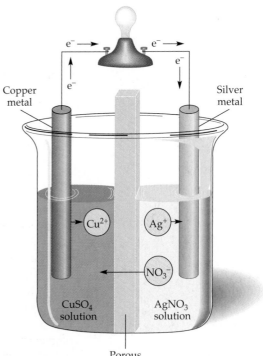

▲ **Figure 8.5** A simple electrochemical cell. The half-reactions are
Oxidation (anode):
  $Cu(s) \longrightarrow Cu^{2+}(aq) + 2\,e^-$
Reduction (cathode):
  $Ag^+(aq) + e^- \longrightarrow Ag(s)$

Question: To balance the equation for the overall reaction we need two Ag^+ and two Ag(s). Why?

Note that *anode* and *oxidation* both begin with vowels, whereas *cathode* and *reduction* begin with consonants.

CONCEPTUAL EXAMPLE 8.5   **Oxidation and Reduction Half-Reactions**

Represent the following reaction as two half-reactions and label them as an oxidation half-reaction and a reduction half-reaction.

$$Mg + Cl_2 \longrightarrow Mg^{2+} + 2\,Cl^-$$

Solution

Magnesium is oxidized from Mg to Mg^{2+}, a process that involves loss of two electrons from the Mg atom. The oxidation half-reaction is therefore

Oxidation:	$Mg \longrightarrow Mg^{2+} + 2\,e^-$

The reduction half-reaction involves chlorine. Each of the two Cl atoms in the Cl_2 molecule must gain an electron to form a Cl^- ion. The reduction half-reaction is therefore

Reduction:	$Cl_2 + 2\,e^- \longrightarrow 2\,Cl^-$

▶ **Exercise 8.5**

Represent the following reaction as two half-reactions and label them as an oxidation half-reaction and a reduction half-reaction.

$$2\,Al + 3\,Br_2 \longrightarrow 2\,Al^{3+} + 6\,Br^-$$

EXAMPLE 8.6   **Balancing Redox Equations**

Balance the following half-reactions and combine them to give a balanced overall reaction.

$$Sn^{2+} \longrightarrow Sn^{4+}$$

$$Bi^{3+} \longrightarrow Bi$$

Solution

Atoms are balanced in both, but electric charge is not. To balance charge in the first half-reaction, we add two electrons on the right side.

$$Sn^{2+} \longrightarrow Sn^{4+} + 2\,e^-$$

The second half-reaction requires three electrons on the left side.

$$Bi^{3+} + 3\,e^- \longrightarrow Bi$$

Before we can combine the two, however, we must set electron loss equal to electron gain. (Electrons lost by the substance being oxidized must be gained by the substance being reduced.) To do this, we multiply the first half-reaction by 3 and the second by 2.

$$3 \times (Sn^{2+} \longrightarrow Sn^{4+} + 2\,e^-) = 3\,Sn^{2+} \longrightarrow 3\,Sn^{4+} + 6\,e^-$$
$$\underline{2 \times (Bi^{3+} + 3\,e^- \longrightarrow Bi) = 2\,Bi^{3+} + 6\,e^- \longrightarrow 2\,Bi}$$
$$3\,Sn^{2+} + 2\,Bi^{3+} \longrightarrow 3\,Sn^{4+} + 2\,Bi$$

Note that both atoms and charges balance in the overall reaction.

▶ **Exercise 8.6A**

Balance the following half-reactions and combine them to give a balanced overall reaction.

$$Fe \longrightarrow Fe^{3+}$$

$$Mg^{2+} \longrightarrow Mg$$

▶ **Exercise 8.6B**

Balance the following half-reactions and combine them to give a balanced overall reaction.

$$Pb \longrightarrow Pb^{2+}$$

$$Ag(NH_3)_2^+ \longrightarrow Ag + 2\,NH_3$$

Dry Cells

The familiar *dry cell* (Figure 8.6) is used in flashlights and many other small portable devices. It has a zinc anode—in this case, the container itself. A carbon rod in the center of the cell is the cathode. The space between the cathode and the anode contains a moist paste of graphite powder (carbon), manganese dioxide (MnO_2), and ammonium chloride (NH_4Cl). The anode reaction is the oxidation of the zinc cylinder to zinc ions. The cathode reaction involves reduction of manganese dioxide. A simplified version of the overall reaction is

$$Zn + 2\,MnO_2 + H_2O \longrightarrow Zn^{2+} + Mn_2O_3 + 2\,OH^-$$

Alkaline cells are similar, but most contain moist manganese dioxide and potassium hydroxide. They are more expensive than ordinary zinc–carbon cells, but they last longer—both in storage and in use.

Lead Storage Batteries

Although we often refer to dry cells as batteries, a **battery** is actually a series of electrochemical cells. The 12-volt (V) storage battery used in automobiles, for example, is a series of six 2-V cells. Each cell (Figure 8.7) contains a pair of electrodes, one lead and the other lead dioxide, in a chamber filled with sulfuric acid. Lead–acid

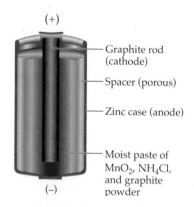

(+)

Graphite rod (cathode)

Spacer (porous)

Zinc case (anode)

Moist paste of MnO_2, NH_4Cl, and graphite powder

(−)

▲ **Figure 8.6** Cross section of a zinc–carbon cell. The half-reactions are
Oxidation (anode):
$$Zn(s) \longrightarrow Zn^{2+}(aq) + 2\,e^-$$
Reduction (cathode):
$$2\,MnO_2(s) + H_2O + 2\,e^- \longrightarrow$$
$$Mn_2O_3(s) + 2\,OH^-(aq)$$

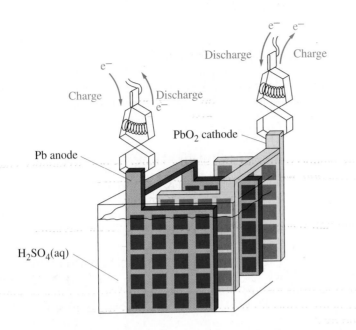

◀ **Figure 8.7** One cell of a lead–acid battery has two anode plates and two cathode plates. Six such cells make up the common 12-V car battery. The half-reactions are
Oxidation (anode):
$$Pb(s) + SO_4^{2-}(aq) \longrightarrow PbSO_4(s) + 2\,e^-$$
Reduction (cathode):
$$PbO_2(s) + 4\,H^+(aq) + SO_4^{2-}(aq) + 2\,e^- \longrightarrow PbSO_4(s) + 2\,H_2O$$

batteries have been known for 150 years. They are used in power tools and in certain appliances as well as in cars. Three hundred million of them are made per year.

An important feature of the lead storage battery is that it can be recharged. It discharges as it supplies electricity when you turn on the ignition to start a car or when the motor is off and the lights are on. But it is recharged when the car is moving and an electric current is supplied to the battery by the mechanical action of the car. The net reaction during discharge is

$$Pb + PbO_2 + 2\,H_2SO_4 \longrightarrow 2\,PbSO_4 + 2\,H_2O$$

The reaction during recharge is just the reverse.

$$2\,PbSO_4 + 2\,H_2O \longrightarrow Pb + PbO_2 + 2\,H_2SO_4$$

Lead storage batteries are durable, but they are heavy and contain corrosive sulfuric acid.

Other Batteries

Much of current battery technology involves the use of lithium, which has an extraordinarily low density as well as a fairly high voltage. Lithium cells make today's lightweight laptop computers possible. Lithium–SO_2 cells are used in submarines and rocket vehicles (such as the Jupiter probe); a lithium–iodine cell is used in pacemakers; and lithium–FeS_2 batteries are used in cameras, radios, and compact disc players.

The rechargeable Ni–Cad cell (Cd anode, NiO cathode) has long been popular for portable radios and cordless appliances. Its biggest competitor is the nickel–metal hydride cell, which replaces cadmium with a hydrogen-absorbing nickel alloy such as $ZrNi_2$ or $LaNi_5$.

The small "button" cells used in hearing aids and hand calculators formerly contained mercury (zinc anode, HgO cathode), but they are gradually being phased out and replaced with zinc–air cells. Tiny silver oxide cells (zinc anode, Ag_2O cathode) are used mainly in watches and cameras.

Fuel Cells

An interesting kind of battery is the fuel cell. When fossil fuels, our major energy source, are burned to generate electricity, only about 35–40% of their energy of combustion is actually harnessed. In a **fuel cell**, the fuel is oxidized at the anode and oxygen is reduced at the cathode with 70–75% efficiency. As we shall see in Chapter 14, most present-day fuel cells use hydrogen as fuel with platinum, nickel, or rhodium electrodes, in a solution of potassium hydroxide.

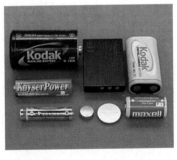

▲ Today consumers see a wide range of cells and batteries, many of which are tailored for specific applications.

8.4 CORROSION

A redox process of particular economic importance is the corrosion of metals. A 2002 study estimated that in the United States alone corrosion cost $276 billion a year. Perhaps 20% of all the iron and steel production in the United States each year goes to replace corroded items. Let's look first at the corrosion of iron.

The Rusting of Iron

In moist air, iron is oxidized, particularly at a nick or scratch.

$$Fe \longrightarrow Fe^{2+} + 2\,e^-$$

As iron is oxidized, oxygen is reduced.

$$O_2 + 2\,H_2O + 4\,e^- \longrightarrow 4\,OH^-$$

The net result, initially, is the formation of insoluble iron(II) hydroxide, which appears dark green or black.

$$2\,Fe + O_2 + 2\,H_2O \longrightarrow 2\,Fe(OH)_2$$

This product is usually further oxidized to iron(III) hydroxide.

$$4\,Fe(OH)_2 + O_2 + 2\,H_2O \longrightarrow 4\,Fe(OH)_3$$

Iron(III) hydroxide [$Fe(OH)_3$], sometimes written as $Fe_2O_3 \cdot 3\,H_2O$, is the familiar iron rust.

Oxidation and reduction often occur at separate points on the metal's surface. Electrons are transferred through the iron metal. The circuit is completed by an electrolyte in aqueous solution. In the Snow Belt, this solution is often the slush from road salt and melting snow. The metal is pitted in an anodic area, where iron is oxidized to Fe^{2+}. These ions migrate to the cathodic area, where they react with hydroxide ions formed by the reduction of oxygen.

$$Fe^{2+} + 2\,OH^- \longrightarrow Fe(OH)_2$$

As indicated previously, this iron(II) hydroxide is then oxidized to $Fe(OH)_3$, or rust. This process is diagrammed in Figure 8.8. Notice that the anodic area is protected from oxygen by a water film, whereas the cathodic area is exposed to air.

Protection of Aluminum

Aluminum is more reactive than iron, and yet corrosion is not a serious problem with aluminum. We use aluminum foil, we cook with aluminum pots, and we buy beverages in aluminum cans. Even after many years, they have not corroded. How is this possible?

The aluminum surface reacts with oxygen in the air to form a thin layer of oxide. However, instead of being porous and flaky like iron oxide, aluminum oxide is hard, tough, and adheres strongly, protecting the metal from further oxidation. Nonetheless, corrosion can sometimes be a problem with aluminum. Certain substances, such as salt, can interfere with the protective oxide coating on aluminum, allowing the metal to oxidize. This problem has caused mag wheels on automobiles to crack, and some planes with aluminum landing gear have had wheels shear off.

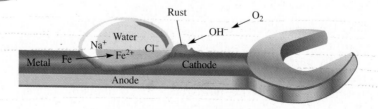

▶ **Figure 8.8** The corrosion of iron requires water, oxygen, and an electrolyte.

Silver Tarnish

The tarnish on silver results from oxidation of the silver surface by hydrogen sulfide (H_2S) in the air or from food. It produces a film of black silver sulfide (Ag_2S) on the metal surface. You can use silver polish to remove the tarnish, but in doing so, you also lose part of the silver. An alternative method involves the use of aluminum metal to reduce the silver ions back to silver metal.

$$3\,Ag^+ + Al \longrightarrow 3\,Ag + Al^{3+}$$

This reaction also requires an electrolyte, and sodium bicarbonate ($NaHCO_3$) is usually used. The tarnished silver is placed in contact with aluminum metal (such as foil) and covered with a solution of sodium bicarbonate. A precious metal is conserved at the expense of a cheaper one.

▲ Silver tarnish (Ag_2S) is readily removed from silver metal upon contact with aluminum metal in a baking-soda solution.

Question: To what is the Ag_2S converted? To what is the Al converted?

8.5 EXPLOSIVE REACTIONS

Chemical explosions (that is, those not based on nuclear fission or fusion) are usually the result of redox reactions. Explosions happen when a chemical reaction occurs rapidly and with a considerable increase in volume. They often involve compounds of nitrogen, such as nitroglycerine (the active ingredient in dynamite), ammonium nitrate (a common fertilizer), or trinitrotoluene (TNT). These compounds can decompose readily, yielding nitrogen gas as one of the products.

Explosive mixtures are used for mining, earthmoving projects, and the demolition of buildings, as well as in making bombs. Trucks loaded with ammonium nitrate (NH_4NO_3) mixed with fuel oil have been used in terrorist attacks around the world. This mixture, sometimes called ANFO (ammonium nitrate–fuel oil), is an explosive with enormous destructive power. The ammonium nitrate is both oxidized and reduced. Ammonium ion is the reducing agent, whereas nitrate ion is the oxidizing agent. The fuel oil provides additional oxidizable material. Using $C_{17}H_{36}$ as the formula for a typical molecule in fuel oil, we can write an equation for the explosive reaction as follows.

$$52\,NH_4NO_3(s) + C_{17}H_{36}(l) \longrightarrow 52\,N_2(g) + 17\,CO_2(g) + 122\,H_2O(g)$$

Notice that this reaction includes 53 mol of starting material, most of it solid and the rest in the liquid state, and 191 mol of products, all gases. A reaction of solid and/or liquid reactants that generates gaseous products involves a huge volume increase and a possibility of explosion. In this case, the reaction is rapid and the volume increase is enormous.

Ammonium nitrate is a widely used fertilizer, and fuel oil is used in many home furnaces. The fact that both these materials are so accessible to potential terrorists is cause for serious concern.

8.6 OXYGEN: AN ABUNDANT AND ESSENTIAL OXIDIZING AGENT

Oxygen itself is the most common oxidizing agent. Making up one-fifth of the air, it oxidizes the wood in our campfires and the gasoline in our automobiles. It even "burns" the food we eat to give us the energy to move and to think.

Certainly one of the most important elements on Earth, oxygen is one of about two dozen elements essential to life. It is found in most compounds that are important to living organisms. Foodstuffs—carbohydrates, fats, and proteins—all contain oxygen. The human body is approximately 65% water (by mass). Because water is 89% oxygen by mass and many other compounds in your body also contain oxygen, almost two-thirds of your body mass is oxygen.

Oxygen comprises about half of the accessible portion of Earth by mass. It occurs in all three subdivisions of the outermost structure of Earth. Oxygen occurs as

The structure and chemistry of Earth are discussed in Chapter 11, the atmosphere in Chapter 12, and water in Chapter 13.

Physical activity under conditions in which plenty of oxygen is available is called *aerobic* exercise. When the available oxygen is insufficient, the exercise is called *anaerobic*, and it often leads to weakness and pain resulting from "oxygen debt." (See Chapter 18.)

▲ Fuels burn more rapidly in pure oxygen than in air. Huge quantities of liquid oxygen are used to burn the fuels that blast rockets into orbit.

O_2 molecules in the atmosphere (the gaseous mass surrounding Earth). In the hydrosphere (the oceans, seas, rivers, and lakes), oxygen is combined with hydrogen in water. In the lithosphere (the outer solid portion), oxygen is combined with silicon (sand is SiO_2), aluminum (in clays), and other elements.

The atmosphere is about 21% elemental oxygen (O_2) by volume. [The rest is mainly nitrogen (N_2), which is rather unreactive.] The free, uncombined oxygen in the air is taken into our lungs, passes into our bloodstreams, is carried to our body tissues, and reacts with the food we eat. This is the process that provides us with all our energy. Fuels such as natural gas, gasoline, and coal also need oxygen to burn and release their stored energy. Oxidation, or combustion, of these fossil fuels currently supplies about 86% of the energy that turns the wheels of civilization.

Pure oxygen is obtained by liquefying air and then letting the nitrogen and argon boil off. Nitrogen boils at $-196\,°C$, argon at $-186\,°C$, and oxygen at $-183\,°C$. About 20 billion kg of oxygen is produced annually in the United States, but most of it is used directly by industry, much of it by steel plants. About 1% is compressed into tanks for use in welding, hospital respirators, and other purposes.

Not everything that oxygen does is desirable. In addition to causing corrosion (Section 8.4), it promotes food spoilage and wood decay.

Reactions with Other Elements

Many oxidation reactions involving atmospheric oxygen are quite complicated, but we can gain some understanding by looking at some of the simpler chemical reactions of oxygen. For example, as we have seen, oxygen combines with many metals to form metal oxides, and with nonmetals to form nonmetal oxides.

EXAMPLE 8.7 **Writing Equations: Reaction of Oxygen with Other Elements**

Magnesium combines readily with oxygen when ignited in air. Write the equation for this reaction.

Solution

Magnesium is a group 2A metal. Oxygen occurs as diatomic molecules (O_2). The two react to form MgO. The reaction is

$$Mg(s) + O_2(g) \longrightarrow MgO(s) \quad \text{(not balanced)}$$

To balance the equation, we need 2 MgO and 2 Mg.

$$2\,Mg(s) + O_2(g) \longrightarrow 2\,MgO(s)$$

▶ **Exercise 8.7A**
Zinc burns in air to form zinc oxide (ZnO). Write the equation for the reaction.

▶ **Exercise 8.7B**
Selenium (Se) burns in air to form selenium dioxide. Write the equation for the reaction.

Reactions with Compounds

Oxygen reacts with many compounds, oxidizing one or more of the elements in the compound. As we have seen, combustion of a fuel is oxidation and produces oxides. There are many other examples. Hydrogen sulfide, a gaseous compound with a rotten-egg odor, burns, producing oxides of hydrogen (water) and of sulfur (sulfur dioxide).

$$2\,H_2S + 3\,O_2 \longrightarrow 2\,H_2O + 2\,SO_2$$

EXAMPLE 8.8   Writing Equations: Reaction of Oxygen with Compounds

Carbon disulfide, a highly flammable liquid, combines readily with oxygen, burning with a blue flame. What products are formed? Write the equation.

Solution

Carbon disulfide is CS_2. The products are oxides of carbon (carbon dioxide) and of sulfur (sulfur dioxide). The balanced equation is

$$CS_2 + 3\,O_2 \longrightarrow CO_2 + 2\,SO_2$$

▶ **Exercise 8.8A**

When heated in air, lead sulfide (PbS) combines with oxygen to form lead(II) oxide (PbO) and sulfur dioxide. Write a balanced equation for the reaction.

▶ **Exercise 8.8B**

Write a balanced equation for the combustion (reaction with oxygen) of ethanol (C_2H_5OH).

Ozone

In addition to diatomic O_2, the normal form of oxygen, the element also has a tri-atomic form (O_3) called *ozone*. Ozone is a powerful oxidizing agent and a harmful air pollutant. It can be extremely irritating to both plants and animals and is especially destructive to rubber. On the other hand, a layer of ozone in the upper stratosphere serves as a shield that protects life on Earth against ultraviolet radiation from the sun (Chapter 12). Ozone clearly illustrates that the same substance can be extremely beneficial or quite harmful, depending on where it happens to be.

Ozone

8.7 OTHER COMMON OXIDIZING AGENTS

Many oxidizing agents—sometimes called *oxidants*—are important in the laboratory, in industry, and in the home. They are used as antiseptics, disinfectants, and bleaches and play a role in many chemical syntheses.

Hydrogen peroxide (H_2O_2) is a common oxidizing agent that has the advantage of being converted to water in most reactions. Pure hydrogen peroxide is a syrupy liquid. It is available (in laboratories) as a dangerous 30% solution that has powerful oxidizing power, or as a 3% solution sold in stores for various uses around the home. When combined with a specially designed catalyst, hydrogen peroxide can be used to oxidize colored compounds and water impurities. The green chemistry MediaLab at the end of this chapter will introduce the use of hydrogen peroxide oxidation in laundry applications and water disinfection.

An oxidizing agent often used in the laboratory is potassium dichromate $(K_2Cr_2O_7)$. It is an orange material that turns green when it is reduced to chromium(III) compounds. One of the compounds that potassium dichromate oxidizes is ethyl alcohol. The old Breathalyzer test for intoxication made use of a dichromate solution. The exhaled breath of the person being tested mixed with the acidic dichromate, and the degree of color change indicated the level of alcohol.

Many antiseptics—compounds applied to living tissue to kill microorganisms or prevent their growth—are mild oxidizing agents. For example, a 3% solution of hydrogen peroxide is often used to treat minor cuts, and tincture of iodine has long been a household antiseptic.

Ointments for treating acne often contain 5–10% benzoyl peroxide, a powerful antiseptic and also a skin irritant. It causes old skin to slough off and be replaced by new, fresher-looking skin. When used on areas exposed to sunlight, however, benzoyl peroxide may promote skin cancer.

Oxidizing agents are also used as disinfectants. A good example is chlorine, which is used to kill disease-causing microorganisms in drinking water. In recent years, concern has been raised over disinfection by-products that form when

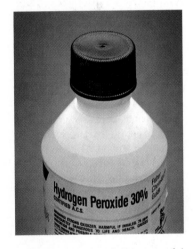
▲ Hydrogen peroxide is a powerful oxidizing agent. Dilute solutions are used as disinfectant and to bleach hair.

A *tincture* is a solution of one or more active ingredients in ethyl alcohol.

SAFETY ALERT Strong oxidizing agents such as 30% aqueous hydrogen peroxide can cause severe burns. In general, strong oxidizing agents are corrosive and many are highly toxic. They can also initiate a fire or add to the severity of a fire when brought into contact with combustible materials.

chlorine is used to disinfect water containing certain impurities. While chlorine kills harmful microorganisms, it can also oxidize smaller molecules to form harmful by-products, such as trichloromethane. Chemists are working to find alternate methods for effectively disinfecting water. The MediaLab at the end of this chapter will lead you through some aspects of the current research and debate among scientists. Swimming pools are often chlorinated with calcium hypochlorite [$Ca(OCl)_2$]. Because calcium hypochlorite is alkaline, it also raises the pH of the water. (When a pool becomes too alkaline, the pH is lowered by adding hydrochloric acid. Swimming pools are usually maintained at pH 7.2–7.8.)

Bleaches are oxidizing agents, too. A **bleach** removes unwanted color from fabrics or other material. Nearly any oxidizing agent could do the job, but some might be unsafe, or harmful to fabrics, or perhaps too expensive.

Laundry bleaches (Chapter 17) are usually sodium hypochlorite (NaOCl) as an aqueous solution (in products such as Purex® and Clorox®) or calcium hypochlorite [$Ca(OCl)_2$], known as bleaching powder. The powder is usually preferred for large industrial operations, such as the whitening of paper or fabrics. Nonchlorine bleaches contain sodium percarbonate (a combination of Na_2CO_3 and H_2O_2) or sodium perborate (a combination of $NaBO_2$ and H_2O_2).

For lightening hair color, bleaches are usually 6 or 12% solutions of hydrogen peroxide, which oxidizes the dark pigment (melanin) in the hair to colorless products (Chapter 17).

Stain removal is more complicated than bleaching. A few stain removers are oxidizing agents, but some are reducing agents, some are solvents or detergents, and some have quite different action. Stains often require rather specific stain removers.

8.8 SOME REDUCING AGENTS OF INTEREST

In every reaction involving oxidation the oxidizing agent is reduced, and the substance undergoing oxidation acts as a reducing agent. Let's now consider reactions in which the purpose of the reaction is reduction.

Most metals occur in nature as compounds. In order to prepare the free metals, the compounds must be reduced. Metals are often freed from their ores with coal or *coke* (elemental carbon obtained by heating coal to drive off volatile matter). Tin(IV) oxide is one of the many ores that can be reduced with coal or coke.

$$SnO_2 + C \longrightarrow Sn + CO_2$$

Sometimes a metal can be obtained by heating its ore with a more active metal. Chromium oxide, for example, can be reduced by heating it with aluminum.

$$Cr_2O_3 + 2\,Al \longrightarrow Al_2O_3 + 2\,Cr$$

Reduction in Photography

Perhaps a more familiar reducing agent, by use if not by name, is the developer used in black-and-white photography. Photographic film is coated with a silver salt (Ag^+Br^-). Silver ions that have been exposed to light react with the developer, a reducing agent (such as the organic compound hydroquinone), to form metallic silver.

$$C_6H_4(OH)_2 + 2\,Ag^+ \longrightarrow C_6H_4O_2 + 2\,Ag + 2\,H^+$$

Hydroquinone · Silver metal

The silver ions not exposed to light are not reduced by the developer. The film is then treated with "hypo," a solution of sodium thiosulfate ($Na_2S_2O_3$), which washes out unexposed silver bromide to form the negative. This leaves the negative dark where the metallic silver was deposited (where it was originally exposed to light) and transparent where light did not strike it. Light is then shone through the negative onto light-sensitive paper to make the positive print. Figure 8.9 shows positive and negative prints.

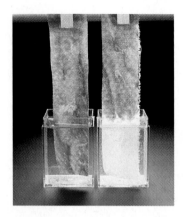

▲ Pure water (left) has little effect on a dried tomato sauce stain. Sodium hypochlorite bleach (right) removes the stain by oxidizing the colored tomato pigments to colorless products.

Solutions of potassium permanganate ($KMnO_4$), a common laboratory oxidizing agent, are good disinfectants. The solutions are a deep purplish pink. Highly diluted solutions, a pale pink (called "pinky water" in India), are used in developing countries to disinfect foods. Their use is limited in that the permanganate ion is reduced to brown manganese dioxide, leaving a brown stain on the treated surface.

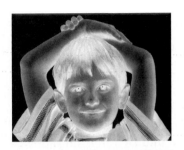 

◀ **Figure 8.9** A photographic negative (left) and a positive print.

Antioxidants

In food chemistry, certain reducing agents are called **antioxidants**. Ascorbic acid (vitamin C) can prevent the browning of fruit (such as sliced apples or pears) by inhibiting air oxidation. Whereas vitamin C is water soluble, tocopherol (vitamin E) and beta-carotene (vitamin A precursor) are fat-soluble antioxidants. All these vitamins are believed to retard various oxidation reactions that are potentially damaging to vital components of living cells (Chapter 16).

Hydrogen as a Reducing Agent

Hydrogen is an excellent reducing agent that can free many metals from their ores, but it is generally used to produce more expensive metals, such as tungsten (W).

$$WO_3 + 3\,H_2 \longrightarrow W + 3\,H_2O$$

Hydrogen can be used to reduce many kinds of chemical compounds. Ethylene, for example, can be reduced to ethane.

$$C_2H_4 + H_2 \longrightarrow C_2H_6$$

The reaction requires a **catalyst**, a substance that increases the rate of a chemical reaction without itself being used up. Nickel is used in this case.

Hydrogen also reduces nitrogen, from the air, in the industrial production of ammonia.

$$N_2 + 3\,H_2 \longrightarrow 2\,NH_3$$

Ammonia is the source of most nitrogen fertilizers in modern agriculture. This reaction employs an iron catalyst.

A stream of pure hydrogen burns quietly in air with an almost colorless flame; but when a mixture of hydrogen and oxygen is ignited by a spark or a flame, an explosion results. The product in both cases is water.

$$2\,H_2 + O_2 \longrightarrow 2\,H_2O$$

Certain metals, such as platinum and palladium, have an unusual affinity for hydrogen. They absorb large volumes of the gas. Palladium can absorb up to 900 times its own volume of hydrogen. It is interesting to note that hydrogen and oxygen can be mixed at room temperature with no perceptible reaction. But if a piece of platinum gauze is added, the gases react violently at room temperature. The platinum acts as a catalyst. It lowers the *activation energy* for the reaction. (The **activation energy** for a chemical reaction is the minimum energy needed to get the reaction started.) The heat from the initial reaction heats up the platinum, making it glow; it then ignites the hydrogen–oxygen mixture, causing an explosion.

Nickel, platinum, and palladium are often used as catalysts for reactions involving hydrogen. These metals have the greatest catalytic activity when they are finely divided and have lots of active surface area. Hydrogen adsorbed on the surface of these metals is more reactive than ordinary hydrogen gas. Catalysts are an important area of research in green chemistry. Chemists are able to develop more energy efficient and sustainable processes by designing catalysts to enable chemical reactions to occur at lower temperatures or pressures. Using catalysts can also allow a reaction to occur using milder chemicals, such as oxygen or hydrogen peroxide instead of chlorine to oxidize stains in laundry or harmful microorganisms in drinking water.

A nickel catalyst is used in converting unsaturated fats to saturated fats (Chapter 16). Unsaturated fats ordinarily react very slowly with hydrogen, but in the presence of nickel metal, the reaction proceeds readily. The nickel increases the rate of the reaction, but it does not increase the amount of product formed.

8.9 A CLOSER LOOK AT HYDROGEN

Hydrogen is an especially vital element because it and oxygen are the components of water. By mass, hydrogen makes up only about 0.9% of the outer, accessible portion of Earth. However, because of its low atomic mass, it ranks rather high in abundance by number of atoms. If we had a random sample of 10,000 atoms from the outer, accessible portion of Earth, more than half (5330) would be oxygen, 1590 would be silicon, and 1510 would be hydrogen.

Unlike oxygen, hydrogen is seldom found as a free, uncombined element on Earth. Most of it is combined with oxygen in water. Some is combined with carbon in petroleum and natural gas, which are mixtures of hydrocarbons. Nearly all compounds derived from plants and animals contain combined hydrogen.

Because hydrogen has the lowest atomic number (1) and the smallest atoms, it is the first element in the periodic table. However, it is difficult to know exactly where to place it. Hydrogen is usually shown as the first member of group 1A because it has the same valence electronic structure (ns^1) as alkali metals. Like the atoms of elements in group 1A, a hydrogen atom can lose an electron to form a 1+ ion, but unlike the other group 1A elements, hydrogen is not a metal. Like the atoms of group 7A elements, a hydrogen atom can pick up an electron to become a 1− ion, but unlike the other group 7A elements, hydrogen is not a halogen. We place hydrogen in group 1A in the periodic table on the inside front cover, but you should keep in mind that it is a unique element, in a class by itself.

Small amounts of elemental hydrogen can be made for laboratory use by reacting zinc with hydrochloric acid.

$$Zn(s) + 2\,HCl(aq) \longrightarrow ZnCl_2(aq) + H_2(g)$$

Because hydrogen does not dissolve in water, it can be collected by water displacement (Figure 8.10). Commercial quantities of hydrogen are obtained as by-products of petroleum refining or by the reaction of natural gas with steam. About 8 billion kg

Hydrogen ranks low in abundance by mass on Earth. If we look beyond our home planet, however, hydrogen becomes much more significant. The sun, for example, is made up largely of hydrogen. The planet Jupiter is also mainly hydrogen. In fact, hydrogen is by far the most abundant element in the universe.

▶ **Figure 8.10** Hydrogen gas in the laboratory is prepared by the reaction of zinc with hydrochloric acid. The gas bubbles from the reaction flask and is trapped in the inverted bottle. Initially, the bottle was filled with water, but the hydrogen pushes the water out as it collects in the bottle.

of hydrogen is produced each year in the United States. At present, hydrogen is used mainly to make ammonia and methanol. Its possible use as a fuel of the future is discussed in Chapter 14.

Hydrogen is a colorless, odorless gas and the lightest of all substances. Its density is only one-fourteenth that of air, and for this reason it was once used to fill lighter-than-air craft (Figure 8.11). Unfortunately, hydrogen can be ignited by a spark, which is what occurred in 1937 when the German airship *Hindenburg* was destroyed in a disastrous fire and explosion as it was landing in Lakehurst, New Jersey. The use of hydrogen in airships was discontinued after that, and the dirigible industry never recovered. Today the few airships that are still in service are filled with nonflammable helium, but they are used mainly in advertising.

▲ **Figure 8.11** Hydrogen is the most buoyant gas, but it is highly flammable. The disastrous fire in the hydrogen-filled German zeppelin *Hindenburg* led to the replacement of hydrogen by nonflammable helium, which buoys the Fuji blimp.

Question: What is the equation for the reaction pictured on the left? Would a similar reaction occur if lightning struck the blimp pictured on the right? Explain.

Photochromic Glass

Eyeglasses with photochromic lenses eliminate the need for sunglasses because the lenses darken when exposed to bright light. This response to light is the result of oxidation-reduction reactions. Ordinary glass is a complex matrix of silicates (Chapter 11) that is transparent to visible light. Photochromic lenses have silver chloride (AgCl) and copper(I) chloride (CuCl) crystals uniformly embedded in the glass. Silver chloride is susceptible to oxidation and reduction by light. First the light displaces an electron from a chloride ion:

$$Cl^- \xrightarrow{\text{oxidation}} Cl + e^-$$

The electron then reduces a silver ion to a silver atom:

$$Ag^+ + e^- \xrightarrow{\text{reduction}} Ag$$

Clusters of silver atoms block the transmittance of light, causing the lenses to darken. This process occurs almost instantly. The degree of darkening depends on the intensity of the light.

To be useful in eyeglasses, the photochromic process must be reversible. The darkening process is reversed by the copper(I) chloride. When the lenses are removed from light, the chlorine atoms formed by the exposure to light are reduced by the copper(I) ions that are oxidized to copper(II) ions:

$$Cl + Cu^+ \longrightarrow Cu^{2+} + Cl^-$$

The copper(II) ions then oxidize the silver atoms.

$$Cu^{2+} + Ag \longrightarrow Cu^+ + Ag^+$$

The net effect is that the silver and chloride atoms are converted to their original oxidized and reduced states, and the lenses become transparent once more.

▲ Photochromic glass darkens in the presence of light.

8.10 OXIDATION, REDUCTION, AND LIVING THINGS

Perhaps the most important oxidation–reduction processes are the ones that maintain life on this planet. We obtain energy for all our physical and mental activities by metabolizing food through respiration. The process has many steps, but eventually the food we eat is converted mainly into carbon dioxide, water, and energy.

Bread and many of the other foods we eat are largely made up of carbohydrates (Chapter 16). If we represent carbohydrates with the simple example glucose ($C_6H_{12}O_6$), we can write the overall equation for their metabolism as follows.

$$C_6H_{12}O_6 + 6\,O_2 \longrightarrow 6\,CO_2 + 6\,H_2O + energy$$

This process constantly occurs in animals, including humans. The carbohydrate is oxidized in the process.

Meanwhile, plants need carbon dioxide and water, from which they produce carbohydrates. The energy needed comes from the sun, and the process is called **photosynthesis** (Figure 8.12). The chemical equation is

$$6\,CO_2 + 6\,H_2O + energy \longrightarrow C_6H_{12}O_6 + 6\,O_2$$

Notice that this process in plant cells is exactly the reverse of the process going on inside animals. In food metabolism in animals we focus on an oxidation process. In photosynthesis we focus on a reduction process.

The carbohydrates produced by photosynthesis are the ultimate source of all our food because fish, fowl, and other animals either eat plants or eat other animals that eat plants. Note that the photosynthesis process not only makes carbohydrates but also yields free elementary oxygen (O_2). In other words, photosynthesis does not just provide all the food we eat, it also provides all the oxygen we breathe.

There are many oxidation reactions that occur in nature (with oxygen being reduced in the process). The net photosynthesis reaction is unique in that it is a natural reduction of carbon dioxide (with oxygen being oxidized). Many reactions in nature use oxygen. Photosynthesis is the only natural process that produces it.

▶ **Figure 8.12** Photosynthesis occurs in green plants. The chlorophyll pigments that catalyze the photosynthesis process give the green color to much of the land area of Earth.

Critical Thinking Exercises

Apply knowledge that you have gained in this chapter and one or more of the FLaReS principles (Chapter 1) to evaluate the following statements or claims.

8.1 Over the years some people have claimed that someone has invented an automobile engine that burns water instead of gasoline. They say that we have not heard about this wonderful invention because the oil companies have bought the rights to the engine from the inventor so that they can keep it from the public and people will continue to burn gasoline in their cars. Do you find this claim believable?

8.2 A friend tells you that if you take a sugar cube and hold it in a flame, it will melt but will not burn; but he claims that if you dip the sugar cube into cigarette ashes before you place it in the flame, it will burn. He explains that cigarette ashes contain something that is a catalyst for the burning of sugar. Do you think your friend's claim is valid?

8.3 A Web site claims that electricity can remove rust. The procedure given is as follows: Connect a wire to the object to be derusted, and immerse in a bath of sodium bicarbonate (baking soda) solution. Immerse a second wire in the bath. Connect the wires to a battery so that the object to be derusted is the cathode. The site claims that hydrogen gas is produced at the cathode and oxygen at the anode; and the hydrogen somehow changes the rust back to iron metal. Evaluate this claim.

8.4 A friend claims that an automobile can be kept rust free simply by connecting the negative pole of the car's battery to the car's body. He reasons that this action makes the car the cathode, and oxidation does not occur at the cathode. He claims that for years he has kept his car from rusting by doing this. Evaluate this claim.

8.5 A salesman claims that your tap water is contaminated. As evidence, he connects a battery to a pair of "special electrodes" and dips the electrodes in a glass of your tap water. Within a few minutes, white and brown "gunk" appears in the water. The salesman sells a special purifying filter that he claims will remove the contaminants and leave the water pure and clear. Evaluate this claim.

SUMMARY

Sections 8.1–8.2—Oxidation and reduction are processes that occur together. They may be defined in three ways. **Oxidation** occurs when a substance gains oxygen atoms, or loses hydrogen atoms, or loses electrons. **Reduction** is the opposite of oxidation; it occurs when a substance loses oxygen atoms, or gains hydrogen atoms, or gains electrons. We usually use the definition that is most convenient for the situation. In a redox or oxidation–reduction reaction, the substance that causes oxidation is the **oxidizing agent**. The substance that causes reduction is the **reducing agent**. In a redox reaction the oxidizing agent is reduced, and the reducing agent is oxidized.

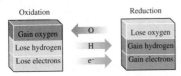

Section 8.3—A redox reaction involves transfer of electrons. That transfer can be made to take place in an **electrochemical cell**, which has two solid **electrodes** where electrons are transferred. The electrode where oxidation occurs is the **anode**, and the one at which reduction occurs is the **cathode**. Electrons are transferred from anode to cathode through a wire, which is an electric current. When the oxidation half-reaction is added to the reduction half-reaction, the electrons in the half-reactions cancel and we get the overall reaction. Knowing this, we can balance many redox reactions.

A dry cell (flashlight battery) is an electrochemical cell with a zinc case, a central carbon rod, and containing a paste of manganese dioxide. As the cell is used, the zinc is oxidized and the MnO_2 is reduced. Alkaline cells are similar but contain potassium hydroxide. A **battery** is a series of cells connected. An automotive lead storage battery contains six cells. Each cell contains lead and lead dioxide plates in sulfuric acid, and the battery can be recharged. A **fuel cell** generates electricity by oxidizing a fuel in oxygen in a special cell.

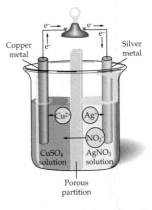

Sections 8.4–8.5—Corrosion of metal is ordinarily an undesirable redox reaction. When iron corrodes, the iron is first oxidized to Fe^{2+}, and oxygen gas is reduced to hydroxide ion. The $Fe(OH)_2$ that forms is further oxidized to $Fe(OH)_3$ or common rust. Oxidation and reduction often occur in separate places on the metal surface, and an electrolyte (such as salt) on the surface can speed the corrosion reaction. Aluminum corrodes, but the oxide layer is tough and adherent and protects the underlying metal. Silver corrodes in the presence of sulfur compounds, forming black silver sulfide. The sulfide can be removed electrochemically. An explosive

reaction is a redox reaction that occurs rapidly and with a considerable increase in volume (of gases).

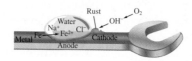

Sections 8.6–8.7—Oxygen is an oxidizing agent that is essential to life. It makes up about one-fifth of air and about two-thirds of body mass. Our food is "burned" in oxygen as we breathe, and fuels such as coal use oxygen to burn and release most of the energy used in civilization. Pure oxygen is obtained from liquefied air, and most of it is used in industry. Oxygen reacts with metals and nonmetals to form the corresponding oxides. It reacts with many compounds to form oxides of the elements in the compound. Ozone, O_3, is a form of oxygen that is an air pollutant in the biosphere but a shield from ultraviolet radiation in the upper atmosphere. Other oxidizing agents include hydrogen peroxide, potassium dichromate, benzoyl peroxide, and chlorine. A bleach is an oxidizing agent that removes unwanted color from fabric or material.

Sections 8.8–8.9—An antioxidant is a reducing agent that retards damaging oxidation reactions in living cells. Vitamins C and E and beta-carotene are antioxidants. Common reducing agents include carbon, hydrogen, and active metals. Carbon in the form of coal or coke is often used as a reducing agent to obtain metals from their oxides. Active metals such as aluminum also are used. Hydrogen can free many metals from their ores. Some reactions of hydrogen require a catalyst that speeds up the reaction without itself being used up. A catalyst lowers the activation energy—the minimum energy needed to start the reaction. Formation of ammonia from hydrogen and nitrogen uses a catalyst.

Hydrogen is a vital element because it is a component of water and of living tissue. Hydrogen's lowest atomic number and smallest atoms make it unique in its behavior among the elements. Much of the hydrogen used today is produced from petroleum refining or from reactions of natural gas and is used to make ammonia and methanol. Hydrogen was once used for lighter-than-air craft but its flammability caused it to be replaced by helium.

Section 8.10—Oxidation of glucose and other substances produce the energy needed in the animal kingdom. The reduction of carbon dioxide in photosynthesis is the most important reduction process on Earth. We could not exist without it; it provides the food we eat and the oxygen we breathe.

REVIEW QUESTIONS

1. Explain how we view oxidation and reduction in terms of the following.
   a. oxygen atoms gained or lost
   b. hydrogen atoms gained or lost
   c. electrons gained or lost

2. What happens to the oxidation number of one of its elements when a compound is oxidized, and when it is reduced?

3. Considering the following reaction, which statement is true?

   $$Cu(s) + 4\,H^+ + SO_4^{2-}(aq) \longrightarrow$$
   $$Cu^{2+}(aq) + 2\,H_2O(l) + SO_2(g)$$

   (a) Cu is oxidized.
   (b) SO_2 is the oxidizing agent.
   (c) H^+ is reduced.
   (d) SO_4^{2-} is the reducing agent.

4. What is an electrochemical cell? An electrochemical battery?

5. What is the purpose of a porous plate between the two electrode compartments in an electrochemical cell?

6. What is a half-reaction? How are half-reactions combined to give an overall redox reaction?

7. How does an alkaline cell differ from a regular carbon–zinc dry cell?

8. From what material is the case of a carbon–zinc dry cell made? What purpose does this material serve? What happens to it as the cell discharges?

9. Describe what happens when a lead storage battery discharges.

10. What happens when a lead storage battery is charged?

11. Describe what happens when iron corrodes. How does road salt speed this process?

12. Why does aluminum corrode more slowly than iron, even though aluminum is more reactive than iron?

13. How does silver tarnish? How can tarnish be removed without the loss of silver?

14. List four common oxidizing agents.

15. List three common reducing agents.

16. Name some oxidizing agents used as antiseptics and disinfectants.

17. Relate the chemistry of photosynthesis to the chemistry that provides energy for your heartbeat.

18. Describe how a bleaching agent, such as hypochlorite (ClO^-), works.

PROBLEMS

Recognizing Oxidation and Reduction

19. Each of the following "equations" shows only part of a chemical reaction. Indicate whether the reactant shown is being oxidized or reduced. Explain.
    a. $C_2H_4O \longrightarrow C_2H_4O_2$
    b. $H_2O_2 \longrightarrow H_2O$
    c. $C_6H_4(OH)_2 \longrightarrow C_6H_4O_2$
    d. $Sn^{4+} \longrightarrow Sn^{2+}$
    e. $Cl_2 \longrightarrow 2\,Cl^-$

20. In which of the following partial reactions is the reactant undergoing oxidation? Explain.
    a. $C_2H_4O \longrightarrow C_2H_6O$
    b. $WO_3 \longrightarrow W$
    c. $Fe^{3+} \longrightarrow Fe^{2+}$
    d. $C_{27}H_{33}N_9O_{15}P_2 \longrightarrow C_{27}H_{35}N_9O_{15}P_2$
    e. $2\,H^+ \longrightarrow H_2$

21. Write balanced oxidation half-reaction equations of the type $Na \longrightarrow Na^+ + e^-$ for the following metals.
    a. K   b. Ca   c. Zn

22. Write balanced reduction half-reaction equations of the type $Cl_2 + 2\,e^- \longrightarrow 2\,Cl^-$ for the following nonmetals.
    a. O_2 b. I_2 c. N_2

Oxidizing Agents and Reducing Agents

23. Identify the oxidizing agent and the reducing agent in each reaction.
    a. $3\,C + Fe_2O_3 \longrightarrow 3\,CO + 2\,Fe$
    b. $P_4 + 5\,O_2 \longrightarrow P_4O_{10}$
    c. $C + H_2O \longrightarrow CO + H_2$
    d. $H_2SO_4 + Zn \longrightarrow ZnSO_4 + H_2$

24. Identify the oxidizing agent and the reducing agent in each reaction.
    a. $CuS + H_2 \longrightarrow Cu + H_2S$
    b. $4\,K + CCl_4 \longrightarrow C + 4\,KCl$
    c. $C_3H_4 + 2\,H_2 \longrightarrow C_3H_8$
    d. $Fe^{3+} + Ce^{3+} \longrightarrow Fe^{2+} + Ce^{4+}$

25. Look again at Figure 8.1. What is the oxidizing agent?

26. What is the reducing agent in Figure 8.1?

Half-Reactions

27. Separate the following redox reactions into half-reactions and label each half as oxidation or reduction.
    a. $Fe(s) + 2\,H^+(aq) \longrightarrow Fe^{2+}(aq) + H_2(g)$
    b. $2\,Al(s) + 3\,Cr^{2+}(aq) \longrightarrow 3\,Cr(s) + 2\,Al^{3+}(aq)$

28. Separate the following redox reactions into half-reactions and label each half as oxidation or reduction.
    a. $2\,Al(s) + 6\,H^+(aq) \longrightarrow 2\,Al^{3+}(aq) + 3\,H_2(g)$
    b. $2\,Cu^+(aq) + Mg(s) \longrightarrow Mg^{2+}(aq) + 2\,Cu(s)$

29. Label each of the following half-reactions as oxidation or reduction and then combine them to obtain a balanced overall redox reaction.
    a. $2\,I^- \longrightarrow I_2 + 2\,e^-$ and $Cl_2 + 2\,e^- \longrightarrow 2\,Cl^-$
    b. $HNO_3 + H^+ + e^- \longrightarrow NO_2 + H_2O$ and $SO_2 + 2\,H_2O \longrightarrow H_2SO_4 + 2\,H^+ + 2\,e^-$

30. Label each of the following half-reactions as oxidation or reduction and then combine them to obtain a balanced overall redox reaction.
    a. $2\,H_2O_2 \longrightarrow 2\,O_2 + 4\,H^+ + 4\,e^-$ and $Fe^{3+} + e^- \longrightarrow Fe^{2+}$
    b. $WO_3 + 6\,H^+ + 6\,e^- \longrightarrow W + 3\,H_2O$ and $C_2H_6O \longrightarrow C_2H_4O + 2\,H^+ + 2\,e^-$

Oxidation and Reduction: Chemical Reactions

31. In the following reactions, which substance is oxidized? Which is the oxidizing agent?
    a. $H_2CO + H_2O_2 \longrightarrow H_2CO_2 + H_2O$
    b. $5\,C_2H_6O + 4\,MnO_4^- + 12\,H^+ \longrightarrow$
       $5\,C_2H_4O_2 + 4\,Mn^{2+} + 11\,H_2O$

32. In the following reactions, which element is oxidized and which is reduced?
    a. $2\,HNO_3 + SO_2 \longrightarrow H_2SO_4 + 2\,NO_2$
    b. $2\,CrO_3 + 6\,HI \longrightarrow Cr_2O_3 + 3\,I_2 + 3\,H_2O$

33. Acetylene (C_2H_2) reacts with hydrogen to form ethane (C_2H_6). Is the acetylene oxidized or reduced? Explain.

34. Unsaturated vegetable oils react with hydrogen to form saturated fats. A typical reaction is

    $$C_{57}H_{104}O_6 + 3\,H_2 \longrightarrow C_{57}H_{110}O_6$$

    Is the unsaturated oil oxidized or reduced? Explain.

35. Molybdenum metal, used in special kinds of steel, can be manufactured by the reaction of its oxide with hydrogen.

    $$MoO_3 + 3\,H_2 \longrightarrow Mo + 3\,H_2O$$

    Which substance is reduced? Which is the reducing agent?

36. To test for iodide ions (for example, in iodized salt), a solution is treated with chlorine to liberate iodine.

    $$2\,I^- + Cl_2 \longrightarrow I_2 + 2\,Cl^-$$

    Which substance is oxidized? Which is reduced?

37. When the water pump failed in the nuclear reactor at Three Mile Island in 1979, zirconium metal reacted with very hot water to produce hydrogen gas.

    $$Zr + 2\,H_2O \longrightarrow ZrO_2 + 2\,H_2$$

    What substance was oxidized in the reaction? What was the oxidizing agent?

38. Unripe grapes are exceptionally sour because of a high concentration of tartaric acid $(C_4H_6O_6)$. As the grapes ripen, this compound is converted to glucose $(C_6H_{12}O_6)$. Is the tartaric acid oxidized or reduced?

39. Vitamin C (ascorbic acid) is thought to protect our stomachs from the carcinogenic effect of nitrite ions (NO_2^-) by converting the ions to nitric oxide (NO). Is the nitrite ion oxidized or reduced? Is ascorbic acid an oxidizing agent or a reducing agent?

40. In the reaction in Problem 39, ascorbic acid ($C_6H_8O_6$) is converted to dehydroascorbic acid ($C_6H_6O_6$). Is ascorbic acid oxidized or reduced in this reaction?

Combination with Oxygen

41. Give formulas for the products formed when each of the following substances react with oxygen (O_2).
   a. S   **b.** H_2
   c. CH_4 **d.** C_3H_8

42. Give formulas for the products formed when each of the following substances react with oxygen (O_2). (There may be more than one correct answer.)
   a. C   **b.** N_2
   c. CS_2 **d.** $C_6H_{12}O_6$

ADDITIONAL PROBLEMS

43. The dye indigo (used to color blue jeans) is formed by exposure of indoxyl to air.

$$2\,C_8H_7ON + O_2 \longrightarrow C_{16}H_{10}N_2O_2 + 2\,H_2O$$
$$\text{Indoxyl} \qquad\qquad \text{Indigo}$$

What substance is oxidized? What is the oxidizing agent?

44. Why do signs in hospitals warn against smoking? State a reason other than the long-term health risks.

45. Why do some mechanics lightly coat their tools with grease or oil before storing them?

46. When aluminum wire is added to a blue solution of copper(II) chloride, the blue solution turns colorless and reddish-brown copper metal comes out of solution. Write an equation for the reaction. (Chloride ion is not involved in the reaction.)

47. Hydrogen has such a strong affinity for oxygen that it can remove oxygen atoms from many metal oxides to yield the free metal. For example, when hydrogen is passed over heated copper(II) oxide, metallic copper and water are formed. Write an equation for the reaction.

48. When lead-based paints are exposed to air containing hydrogen sulfide, they turn black because the Pb^{2+} ions react with the H_2S to form black lead sulfide (PbS). **(a)** Write the equation for the reaction. Hydrogen peroxide can be used to lighten the paints by oxidizing the black sulfide (S^{2-}) to white sulfates (SO_4^{2-}). **(b)** Write the equation for this reaction.

49. To oxidize 1.0 kg of fat, our bodies require about 2000 L of oxygen. A good diet contains about 80 g of fat per day. What volume (at STP) of oxygen is required to oxidize that fat?

50. The oxidizing agent we use to obtain energy from food is oxygen (from the air). If you breathe 15 times a minute (at rest), taking in and exhaling 0.5 L of air with each breath, what volume of air do you breathe each day? Air is 21% oxygen by volume. What volume of oxygen do you breathe each day?

51. The overall photosynthesis reaction is given on page 230. Which substance is oxidized? Which is the oxidizing agent? Which substance is reduced? Which is the reducing agent?

52. The photosynthesis reaction (Problem 51) can be expressed as the net result of two processes, one of which requires light and is called the *light reaction*. This reaction may be written as

$$12\,H_2O \xrightarrow{\text{light}} 6\,O_2 + 24\,H^+ + 24\,e^-$$

   a. Is the light reaction an oxidation or reduction?
   b. Write the equation for the other half-reaction, called the *dark reaction*, for photosynthesis.

53. Indicate whether the first-named substance in each change undergoes an oxidation, a reduction, or neither. Explain your reasoning.
   a. A violet solution of V^{2+}(aq) is converted to a green solution of V^{3+}(aq).
   b. Nitrogen dioxide converts to dinitrogen tetroxide when cooled.
   c. Carbon monoxide reacts with hydrogen to form methane.

54. Look again at the photograph on page 212. When the beetle mixes the contents of the two compartments, several reactions occur. Write equations for the following:
   a. Hydrogen peroxide breaks down to oxygen and water.
   b. *para*-Hydroquinone (HOC_6H_4OH) is oxidized by hydrogen peroxide to *para*-benzoquinone (O=C_6H_4=O).

55. As a rule, metallic elements act as reducing agents, not as oxidizing agents. Explain. *(Hint: Consider the charges on metal ions.)* Do nonmetals act only as oxidizing agents and not as reducing agents? Explain your answer.

56. Consider the following reaction of calcium hydride (CaH_2) with molten sodium metal. Identify the species being oxidized and the species being reduced according to **(a)** the second definition shown in Figure 8.2 and **(b)** the third definition in Figure 8.2. What difficulty arises?

$$CaH_2(s) + 2\,Na(l) \longrightarrow 2\,NaH(s) + Ca(l)$$

57. If a stream of hydrogen gas is ignited in air, the flame can then be immersed in a jar of chlorine gas and it will still burn. Write the balanced equation for the reaction that occurs in the jar. What is the oxidizing agent in the reaction?

58. Refer back to Figure 8.8 and Section 8.4. A steel shovel is pushed partway into moist soil. After a few days, the part of the shovel that was underground is found to be coated with a green-black film, while the part of the steel above the ground is coated with orange-red material. Explain why the different parts of the shovel are coated with different colors of material.

COLLABORATIVE GROUP PROJECTS

Prepare a PowerPoint, poster, or other presentation (as directed by your instructor) for presentation to the class.

59. Prepare a brief report on one of the following types of electrochemical cells or batteries and share it with your group. If possible, give the half-reactions, the overall reaction, and uses.
    a. lithium–SO_2 cell
    b. lithium–iodine cell
    c. lithium–FeS_2 battery
    d. Ni–Cd cell
    e. silver oxide cell
    f. nickel–metal hydride cell

60. Prepare a brief report on one of the following methods of protecting steel from corrosion and share it with your group. Give any pertinent chemical reactions involved in the process and tell how it is related to oxidation and reduction.
    a. galvanization
    b. coating with tin
    c. cathodic protection

61. Prepare a brief report on the restoration of the Statue of Liberty for its centennial in 1986. What role did oxidation–reduction play in the need for restoration?

62. Go to one or more of the websites about fuel cells. Write out the reactions that occur in the different types in development. Prepare a brief report focusing on one of the following:
    a. a cost–benefit analysis for their use in automobiles versus the use of gasoline, electricity via batteries, or natural gas
    b. safety considerations of using a hydrogen–oxygen fuel cell
    c. political–economic factors that may be hindering or promoting fuel cell research

REFERENCES AND READINGS

1. Bowden, M. E. *Chemistry Is Electric!* Philadelphia, PA: Chemical Heritage, 1997.

2. Dell, R. M., and D. A. J. Rand. *Understanding Batteries.* London: Royal Society of Chemistry, 2001.

3. "Car Batteries that Last." *Consumer Reports*, October 2002, pp. 29–31.

4. Cox, Amy L., and James R. Cox. "Determining Oxidation–Reduction on a Simple Number Line." *Journal of Chemical Education*, August 2002, pp. 965–967.

5. Elsworth, John F. "Entertaining Chemistry—Two Colorful Reactions." *Journal of Chemical Education*, April 2000, 484–485.

6. Ennis, John L. "Photography at Its Genesis." *Chemical and Engineering News*, December 18, 1989, pp. 26–42.

7. Hill, John W., Ralph H. Petrucci, Terry W. McCreary, and Scott S. Perry. *General Chemistry*, 4th edition. Upper Saddle River, NJ: Prentice Hall, 2005. Chapter 4, "Chemical Reactions in Aqueous Solutions," and Chapter 18, "Electrochemistry."

8. Kolb, Doris. "The Chemical Equation: Part II—Oxidation–Reduction Reactions." *Journal of Chemical Education*, May 1978, pp. 326–331.

9. "Picking the Right Battery." *Consumer Reports*, December 2002, p. 55.

10. Smith, Michael J., and Colin A. Vincent. "Structure and Content of Some Primary Batteries." *Journal of Chemical Education*, April 2001, pp. 519–521.

11. "Stain Removers: Which Are Best?" *Consumer Reports*, March 2000, pp. 52–53. Provides some effective home remedies.

GREEN CHEMISTRY

TAML® Peroxide Oxidation

Kathryn Parent, ACS Green Chemistry Institute

Whether it is a cozy fire in your fireplace, a long-lasting battery in your laptop computer, photosynthesis in your houseplants, or bleach in your washing machine, chemistry plays an important role in our everyday lives. In Chapter 8, you are learning about the coupled chemical reactions called reduction and oxidation (redox).

Professor Terry Collins, at Carnegie Mellon University, designed a molecule he calls TAML® to help hydrogen peroxide oxidize other molecules. In Chapter 8 you are also learning how palladium acts as a catalyst for the reaction between hydrogen and oxygen. The TAML® activator is also a catalyst; it speeds the reaction of hydrogen peroxide with a wide variety of other molecules. In the following Web Investigation, you will explore the application of this elegant redox chemistry to two different applications from your everyday life: stain removal and water disinfection.

WEB INVESTIGATIONS

Investigation 1
Peroxide Bleach

Visit the website of the American Chemical Society (ACS) and search the site for the article in the April 2004 issue of *ChemMatters* magazine, "Building a Better Bleach," pages 17–19. What is the active ingredient in household bleach? What is the environmental concern for releasing bleach into the waste stream? What does the acronym **TAML** stand for? What naturally occurring metal atom does the TAML® molecule shown contain? What important biological molecule (a protein found in your bloodstream) contains the same metal atom?

▶ Bleach removes stains by a redox reaction. Although most bleaches are chlorine bleaches, alternative bleaches are being developed.

Investigation 2

TAML® Peroxide Oxidation

Visit the website of the Institute for Green Oxidation Chemistry to learn more about the TAML® activator that Collins and his research group designed. What are the four research areas at the Institute for Green Oxidation Chemistry? Which of the four research areas does Collins consider the single most important long-term development goal for the Institute?

Investigation 3

Water Disinfection

Visit the Chlorine Chemistry Council Web page. Make a list of the advantages of using chlorine to disinfect water. Now visit the American Water Works Association Web page. What are some of the concerns with using chlorine to disinfect water?

Visit the EPA website. Browse the site's educational resources to find the document entitled "Water on Tap." What are the five treatment steps in the process shown in the Water Treatment Plant diagram. What are *disinfection by-products* (*DBPs*)? Trihalomethanes are one type of DBP. Look at the table in Appendix A: National Primary Drinking Water Standards. What are the potential health effects of exposure to trihalomethanes? List three other DBPs listed in the table in Appendix A.

Return to the ACS Web page and find the article "Many Faces of Chlorine" in the October 18, 2004 issue of *Chemical & Engineering News*. What uses of chlorine are cause for concern, according to Professor Collins?

COMMUNICATE YOUR RESULTS

Exercise 1

Bleach and You

Survey your classmates and tabulate a list of the different detergents and bleaching agents used to do laundry and the reasons for using them. Visit a local store that carries laundry supplies and survey the off-the-shelf detergents and bleaches that are currently available. Classify the different options into main categories, and generate a tabulated report listing brand, active ingredient, state (solid/liquid), and price per unit (grams, ounces, pounds or gallons). Based on your research, develop three overarching characteristics you would look for in a good bleach. Which one will you purchase next time you buy a detergent or bleach? Why?

Exercise 2

TAML® Applications

Besides laundry, list four other practical applications for the TAML® activators and hydrogen peroxide. Choose one of the applications and do further research to learn more about it. Develop a 5–10-minute (7–15 slides) PowerPoint presentation summarizing the application you selected and give an oral report to your classmates.

Exercise 3

Chlorine Versus TAML®

Write a two-page report, summarizing the advantages and disadvantages of chlorine bleach and TAML® activators with peroxide to disinfect water. Conclude with a recommendation for use of one or the other.

CHAPTER 9

Organic Chemistry

The compound cinnamaldehyde, a molecular model of which is shown here, is a component of cinnamon flavors and aromas. Cinnamon is obtained from the bark of the cinnamon tree, *Cinnamomum zeylanicum*. Cinnamaldehyde is organic in the original sense because it is obtained from a tree. It is also organic in the modern sense in that it is a compound of the element carbon. Numerous organic chemicals are obtained from plants and animals, but many of the millions of organic compounds known today are synthetic. See Problem 67, page 268.

The Infinite Variety of Carbon Compounds

The definition of organic chemistry has changed over the years. The changes serve as a good example of the dynamic character of science and of how scientific concepts change in response to experimental evidence. Until the nineteenth century, chemists believed that *organic* chemicals originated only in tissues of living *organisms* and required a "vital force" for their production. All chemicals not manufactured by living tissue were regarded as *inorganic*. Some chemists even believed that organic and inorganic chemicals followed different laws. This all changed in 1828, when Friedrich Wöhler synthesized the organic compound urea, which is found in urine, from ammonium cyanate, an inorganic compound. This important event led other chemists to attempt synthesis of organic chemicals from inorganic ones and changed the very definition of organic chemicals.

Organic chemistry is now defined as the chemistry of carbon-containing compounds. Most of these compounds do come from living things or from things that were once living, but this is not necessarily the case. Perhaps the most remarkable thing about organic compounds is that there are so many of them. Of the tens of millions of known chemical compounds, over 95% are compounds of carbon.

The word *organic* has several different meanings. Organic fertilizer is organic in the original sense; it is derived from living organisms. Organic foods are those grown without synthetic pesticides or fertilizers. Organic chemistry is simply the chemistry of carbon compounds.

9.1 THE UNIQUE CARBON ATOM

Carbon atoms are unique in their ability to bond to each other so strongly that they can form long chains. Silicon and a few other elements can form chains, but only short ones; carbon chains often contain thousands of carbon atoms.

Carbon chains can also have branches or form rings of various sizes. Add to this the fact that carbon atoms also bond strongly to other elements, such as hydrogen, oxygen, and nitrogen, and that these atoms can be arranged in many different ways, and it soon becomes obvious why there are so many carbon compounds.

In addition to the millions of carbon compounds already known, new ones are being discovered every day. Carbon can form an almost infinite number of molecules of various shapes, sizes, and compositions.

Organic compounds always contain carbon, and almost all also contain hydrogen. Many organic compounds contain oxygen and nitrogen, and quite a few contain sulfur and halogens. Nearly all the common elements are found in at least a few organic compounds.

239

We use thousands of carbon compounds every day without even realizing it because they are silently carrying out important chemical reactions within our bodies. Many of these carbon compounds are so vital that we literally could not live without them.

Hydrocarbons

Methane, Ethane, Propane

The simplest organic compounds are hydrocarbons. A **hydrocarbon** contains only hydrogen and carbon. There are several kinds of hydrocarbons, classified according to the type of bonding between carbon atoms. We will consider several of these in turn in the following sections.

9.2 ALKANES

Each carbon atom forms four bonds, and each hydrogen atom forms only one bond, and so the simplest hydrocarbon molecule that is possible is CH_4. Called methane, CH_4 is the main component of natural gas. It has the structure

$$H-\underset{\underset{H}{|}}{\overset{\overset{H}{|}}{C}}-H$$

Methane is the first member of a group of related compounds called **alkanes**, hydrocarbons that contain only single bonds. Alkanes can have from one to several hundred or more carbon atoms. The next member of the series is ethane,

$$H-\underset{\underset{H}{|}}{\overset{\overset{H}{|}}{C}}-\underset{\underset{H}{|}}{\overset{\overset{H}{|}}{C}}-H$$

Ethane is a minor constituent of natural gas. It is seldom encountered as a pure compound in everyday life, but many compounds derived from it are common.

We saw in Chapter 5 that the methane molecule is tetrahedral, in accord with VSEPR theory. In fact, the tetrahedral shape results whenever a carbon atom is connected to four other atoms, be they hydrogen, carbon, or other elements. Figure 9.1 shows models of methane and ethane. The ball-and-stick models show the bond angles best, but the space-filling models more accurately reflect the shapes of the molecules. Ordinarily we use simple structural formulas such as the one shown previously for ethane because they are much easier to draw. A **structural formula**

(a)

(b)

▶ **Figure 9.1** Ball-and-stick (a) and space-filling (b) models of methane (left) and ethane (right).

shows which atoms are bonded to each other, but it does not attempt to show the actual shape of the molecule.

Alkanes often are called **saturated hydrocarbons** because each carbon atom is bonded to the maximum number of hydrogen atoms. In constructing a formula, we connect the carbon atoms to each other through single bonds; then we add enough hydrogen atoms to give each carbon atom four bonds. All alkanes have two more than twice as many hydrogen atoms as carbon atoms. That is, alkanes can be represented by a general formula C_nH_{2n+2} in which n is the number of carbon atoms.

The three-carbon alkane is propane. Models of propane are shown in Figure 9.2. To draw its structural formula, we place three carbon atoms in a row.

$$C—C—C$$

Then we add enough hydrogen atoms (eight in this case) to give each carbon atom a total of four bonds. The structural formula of propane is therefore

$$\begin{array}{ccccccc} & H & & H & & H & \\ & | & & | & & | & \\ H - & C & - & C & - & C & - H \\ & | & & | & & | & \\ & H & & H & & H & \end{array}$$

Condensed Structural Formulas

The complete structural formulas that we have used so far show all the carbon and hydrogen atoms and how they are attached to one another. But these formulas take up a lot of space, and they are quite a bit of trouble to draw or to type. For these reasons, chemists usually prefer to use **condensed structural formulas**. Condensed structures show how many hydrogen atoms are attached to each carbon atom without showing the bonds to each hydrogen atom. For example, ethane and propane are written $CH_3—CH_3$ and $CH_3—CH_2—CH_3$. These formulas can be simplified even further by omitting some (or all) of the bond lines, resulting in CH_3CH_3 and $CH_3CH_2CH_3$.

Homologous Series

Note that methane, ethane, and propane form a pattern. We can build alkanes of any length simply by tacking carbon atoms together in long chains and adding sufficient hydrogen atoms to give each carbon atom four bonds. Even the naming of these compounds follows a pattern (Table 9.1), with a stem name indicating the number of carbon atoms. For compounds of five carbon atoms or more, each stem is derived from the Greek or Latin name for the number. The compound names end in -*ane*, signifying that the compounds are *alkanes*. Table 9.2 gives condensed structural formulas and names for continuous-chain (unbranched) alkanes up to 10 carbon atoms in length.

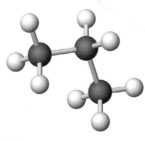

(a)

(b)

▲ **Figure 9.2** Ball-and-stick (a) and space-filling (b) models of propane.

TABLE 9.1	Word Stems Indicating the Number of Carbon Atoms in Organic Molecules
Stem	**Number**
Meth-	One
Eth-	Two
Prop-	Three
But-	Four
Pent-	Five
Hex-	Six
Hept-	Seven
Oct-	Eight
Non-	Nine
Dec-	Ten

TABLE 9.2 The First Ten Continuous-Chain Alkanes			
Name	**Molecular Formula**	**Condensed Structural Formula**	**Number of Possible Isomers**
Methane	CH_4	CH_4	—
Ethane	C_2H_6	CH_3CH_3	—
Propane	C_3H_8	$CH_3CH_2CH_3$	—
Butane	C_4H_{10}	$CH_3CH_2CH_2CH_3$	2
Pentane	C_5H_{12}	$CH_3CH_2CH_2CH_2CH_3$	3
Hexane	C_6H_{14}	$CH_3CH_2CH_2CH_2CH_2CH_3$	5
Heptane	C_7H_{16}	$CH_3CH_2CH_2CH_2CH_2CH_2CH_3$	9
Octane	C_8H_{18}	$CH_3CH_2CH_2CH_2CH_2CH_2CH_2CH_3$	18
Nonane	C_9H_{20}	$CH_3CH_2CH_2CH_2CH_2CH_2CH_2CH_2CH_3$	35
Decane	$C_{10}H_{22}$	$CH_3CH_2CH_2CH_2CH_2CH_2CH_2CH_2CH_2CH_3$	75

Notice that the molecular formula for each alkane in Table 9.2 differs from the one preceding it by precisely one carbon atom and two hydrogen atoms—that is, by a CH_2 unit. Such a series of compounds has properties that vary in a regular and predictable manner. This principle, called *homology*, gives order to organic chemistry in much the same way that the periodic table gives organization to the chemistry of the elements. Instead of studying the chemistry of a bewildering array of individual carbon compounds, organic chemists study a few members of a **homologous series** from which they can deduce the properties of other compounds in the series.

We need not have stopped at 10 carbon atoms as we did in Table 9.2. One could hook together 100 or 1000 or 1 million carbon atoms. We can make an infinite number of alkanes simply by lengthening the chain. But lengthening the chain is not the only option. With four carbon atoms, chain branching is also possible.

Isomerism

When we extend the carbon chain to four atoms and add enough hydrogen atoms to give each carbon atom four bonds, we get $CH_3CH_2CH_2CH_3$. This formula represents butane, a compound that boils at about 0 °C. A second compound, which has a boiling point of −12 °C, has the same molecular formula, C_4H_{10}, as butane. The structural formula of the second compound, however, is not the same as that of butane. Instead of having four carbon atoms connected in a continuous chain, this new compound has a continuous chain of only three carbon atoms. The fourth carbon is branched off the middle carbon of the three-carbon chain. To show the two structures more clearly, let's use complete structural formulas, showing the attachment of all the hydrogen atoms.

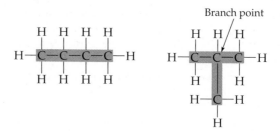

Compounds that have the same molecular formula but different structural formulas are called **isomers**. Because it is an isomer of butane, the branched four-carbon alkane is called isobutane. (As we shall see in Section 9.7, isobutane is also named 2-methylpropane because it has a methyl group attached to the second carbon of propane.) Condensed structural formulas for the two isomeric butanes are written as follows.

$$CH_3CH_2CH_2CH_3 \qquad \begin{matrix} CH_3CHCH_3 \\ | \\ CH_3 \end{matrix}$$

Butane   Isobutane

Figure 9.3 shows ball-and-stick models of the two compounds.

The number of isomers increases rapidly with the number of carbon atoms (Table 9.2). There are three pentanes, five hexanes, nine heptanes, and so on. Isomerism is common in carbon compounds and provides another reason for the existence of millions of organic compounds.

Propane and the butanes are familiar fuels. Although they are gases at ordinary temperatures and under normal atmospheric pressure, they are liquefied under pressure and are usually supplied in tanks as liquefied petroleum gas (LPG). Gasoline is a mixture of hydrocarbons, mostly alkanes, with 5–12 carbon atoms.

Not all the possible isomers of the larger molecules have been isolated. Indeed, the task rapidly becomes more and more prohibitive as you proceed up the series. There are, for example, over 4 billion possible isomers with the molecular formula $C_{30}H_{62}$.

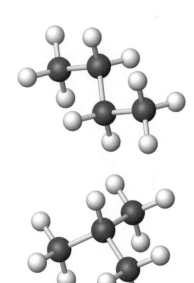

▲ **Figure 9.3** Ball-and-stick models of butane (top) and isobutane (bottom).

EXAMPLE 9.1   **Hydrocarbon Formulas**

Without referring to Table 9.2, give the molecular formula, the complete structural formula, and the condensed structural formula for heptane.

Solution
The stem *hept-* means seven carbon atoms, and the ending *-ane* indicates an alkane. For the complete structural formula, we can write out a string of seven carbon atoms.

$$C-C-C-C-C-C-C$$

Then we attach enough hydrogen atoms to the carbon atoms to give each carbon four bonds. This requires three hydrogen atoms on each end carbon and two each on the others.

$$\begin{array}{ccccccc} & H & H & H & H & H & H & H \\ & | & | & | & | & | & | & | \\ H- & C- & C- & C- & C- & C- & C- & C-H \\ & | & | & | & | & | & | & | \\ & H & H & H & H & H & H & H \end{array}$$

For the condensed form, simply write each carbon atom's set of hydrogen atoms next to the carbon.

$$CH_3CH_2CH_2CH_2CH_2CH_2CH_3$$

For the molecular formula, we can simply count the carbon and hydrogen atoms to arrive at C_7H_{16}. Alternatively, we could use the general formula C_nH_{2n+2} with $n = 7$ to get C_7H_{16}.

▶ **Exercise 9.1**
Give the molecular, complete structural, and condensed structural formulas for **(a)** hexane and **(b)** octane.

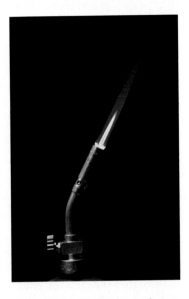

▲ A propane torch. Propane burns in air with a hot flame.

▲ Gasoline is a mixture containing dozens of hydrocarbons, many of which are alkanes.

Properties of Alkanes

Note in Table 9.3 that after the first few members of the alkane series, the rest show about a 20–30 °C increase in boiling point with each added CH_2 group. Note also that at room temperature  alkanes with from 1 to 4 carbon atoms per molecule are

TABLE 9.3   Physical Properties and Uses/Occurrences of Selected Alkanes

Name	Molecular Formula	Melting Point (°C)	Boiling Point (°C)	Density at 20 °C (g/mL)	Use/Occurrence
Methane	CH_4	−183	−162	(Gas)	Natural gas (main component); fuel
Ethane	C_2H_6	−172	−89	(Gas)	Natural gas (minor component); production of chemicals
Propane	C_3H_8	−188	−42	(Gas)	LPG (bottled gas; fuel)
Butane	C_4H_{10}	−138	0	(Gas)	LPG (fuel; lighter fuel)
Pentane	C_5H_{12}	−130	36	0.626	Gasoline component (fuel)
Hexane	C_6H_{14}	−95	69	0.659	Gasoline component (fuel); extraction solvent (food oils)
Heptane	C_7H_{16}	−91	98	0.684	Gasoline component (fuel)
Octane	C_8H_{18}	−57	126	0.703	Gasoline component (fuel)
Decane	$C_{10}H_{22}$	−30	174	0.730	Gasoline component (fuel)
Dodecane	$C_{12}H_{26}$	−10	216	0.749	Gasoline component (fuel)
Tetradecane	$C_{14}H_{30}$	6	254	0.763	Diesel fuel component
Hexadecane	$C_{16}H_{34}$	18	280	0.775	Diesel fuel component
Octadecane	$C_{18}H_{38}$	28	316	(Solid)	Paraffin wax component
Eicosane	$C_{20}H_{42}$	37	343	(Solid)	Paraffin wax component

Recall that water has a density of 1.00 g/mL at room temperature. All the alkanes listed in Table 9.3 have densities less than that.

SAFETY ALERT: Hydrocarbons and most other organic substances are flammable. Volatile ones may form explosive mixtures with air. Hydrocarbons are the main ingredients in gasoline, motor oils, fuel gases (natural gas and bottled gas), and fuel oils. They are also found on labels as "petroleum distillates" and "mineral spirits" in products such as floor cleaners, furniture polish, paints, paint thinners, wood stains and varnishes, and car waxes and polishes. Some of these products also contain flammable alcohols, esters, and ketones. Although products containing flammable substances can be quite dangerous, they are safe to use with proper precautions. Use only as directed and in devices designed for their use. Keep all such products—gasoline, paint thinners, petroleum solvents, and so on—away from open flames.

Skin lotions and creams are discussed in Chapter 17.

 Cycloalkanes

gases, those with 5 to about 16 carbon atoms per molecule are liquids, and those with more than 16 carbon atoms per molecule are solids. The densities of liquid and solid alkanes are less than that of water.

Alkanes are nonpolar molecules and are essentially insoluble in water; hence they float on top of water. They dissolve many organic substances of low polarity—such as fats, oils, and waxes.

Alkanes undergo few chemical reactions, but one important chemical property is that they burn, producing a lot of heat. Alkanes have many uses, but they mainly serve as fuels.

The physiologic properties of alkanes vary. Methane appears to be inert physiologically. We probably could breathe a mixture of 80% methane and 20% oxygen without ill effect. This mixture would be flammable, however, and no fire or spark of any kind could be permitted in such an atmosphere. Breathing an atmosphere of pure methane (the "gas" of a gas-operated stove) can lead to death—not because of the presence of methane but because of the absence of oxygen, a condition called *asphyxia*. Light liquid alkanes, such as those in gasoline, dissolve and wash away body oils when spilled on the skin. Repeated contact may cause *dermatitis*. (Other components of gasoline are less innocuous.) If swallowed, most alkanes do little harm in the stomach; in fact, *mineral oil* is a mixture of long-chain liquid alkanes that has been used for many years as a laxative. However, in the lungs, alkanes cause chemical pneumonia by dissolving fatlike molecules from the cell membranes in the alveoli, allowing the lungs to fill with fluid. Heavier liquid alkanes, when applied to the skin, act as emollients (skin softeners). Petroleum jelly (Vaseline™ is one brand) is a semisolid mixture of hydrocarbons that can be applied as an emollient or simply as a protective film.

9.3 CYCLIC HYDROCARBONS: RINGS AND THINGS

The hydrocarbons we have encountered so far (alkanes) have been composed of open-ended chains of carbon atoms. Carbon atoms can also connect to form closed rings. The simplest possible ring-containing hydrocarbon, or **cyclic hydrocarbon**, has the molecular formula C_3H_6.

This compound is called cyclopropane (Figure 9.4).

Names of cycloalkanes (cyclic hydrocarbons containing only single bonds) are formed by adding the prefix *cyclo-* to the name of the open-chain compound with the same number of carbon atoms as are in the ring.

Chemists often use geometric figures to represent cyclic compounds (Figure 9.5). For example, a triangle is used to represent the cyclopropane ring, and a hexagon to represent cyclohexane.

▶ **Figure 9.4** Ball-and-stick model of cyclopropane. Cyclopropane is a potent, quick-acting anesthetic with few undesirable side effects. It is no longer used in surgery, however, because it forms an explosive mixture with air at nearly all concentrations.

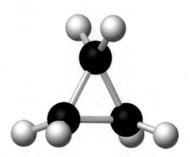

$$CH_2$$
$$CH_2 \quad CH_2$$

$$CH_2$$
$$CH_2—CH_2$$

$$CH_2 \quad CH_2$$
$$CH_2 \quad CH_2$$
$$CH_2$$

$$CH_2$$
$$CH_2 \quad CH$$
$$\parallel$$
$$CH_2 \quad CH$$
$$CH_2$$

Cyclopropane   Cyclohexane   Cyclohexene

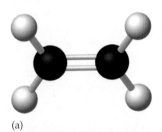

◀ **Figure 9.5** Structural formulas and symbolic representations of some cyclic hydrocarbons.

EXAMPLE 9.2   **Structural Formulas of Cyclic Hydrocarbons**

Give the structural formula for cyclobutane. What geometric figure is used to represent cyclobutane?

Solution
Cyclobutane has four carbon atoms arranged in cyclic fashion.

$$\begin{array}{c} C—C \\ |\quad\ | \\ C—C \end{array}$$

Each carbon atom needs two hydrogen atoms to complete its set of four bonds.

$$\begin{array}{c} CH_2—CH_2 \\ |\qquad\ | \\ CH_2—CH_2 \end{array}$$

Cyclobutane is represented by a square.

▶ **Exercise 9.2A**
Give the structural formula and geometric representation for cyclopentane.

▶ **Exercise 9.2B**
What is the general formula for a cycloalkane?

9.4 UNSATURATED HYDROCARBONS: ALKENES AND ALKYNES

Two carbon atoms can share more than one pair of electrons. In ethylene (C_2H_4), the two carbon atoms share two pairs of electrons and are therefore joined by a double bond (Figure 9.6).

$$\begin{array}{cc} H \quad\quad H \\ \ddot{C}::\ddot{C} \\ H \quad\quad H \end{array} \quad or \quad \begin{array}{cc} H \qquad H \\ \diagdown\quad\diagup \\ C=C \\ \diagup\quad\diagdown \\ H \qquad H \end{array} \quad or \quad CH_2{=}CH_2$$

Ethylene (also called ethene) is the simplest member of the alkene family. An **alkene** is a hydrocarbon that contains one or more carbon-to-carbon double bonds. Alkenes with one double bond have the general formula C_nH_{2n}, where n is the number of carbon atoms.

   Ethylene is the most important commercial organic chemical. Annual U.S. production is over 20 billion kg. More than half goes into the manufacture of polyethylene, one of the most familiar plastics. Another 15% or so is converted to ethylene glycol, the major component of many formulations of antifreeze used in automobile radiators.

(a)

(b)

▲ **Figure 9.6** Ball-and-stick (a) and space-filling (b) models of ethylene.

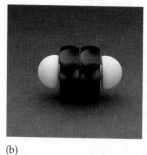

▶ **Figure 9.7** Ball-and-stick (a) and space-filling (b) models of acetylene.

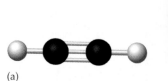

(a)   (b)

In acetylene (C_2H_2), the two carbon atoms share three pairs of electrons; the carbon atoms are joined by a triple bond (Figure 9.7).

$$H \colon C \colon\colon\colon C \colon H \quad \text{or} \quad H-C\equiv C-H$$

Acetylene (also called ethyne) is the simplest member of the alkyne family. An **alkyne** is a hydrocarbon that contains one or more carbon-to-carbon triple bonds. Alkynes with one triple bond have the general formula C_nH_{2n-2}, where n is the number of carbon atoms.

Acetylene is used in oxyacetylene torches for cutting and welding metals. Such torches can produce very high temperatures. Acetylene is also converted to a variety of other chemical products.

Collectively, alkenes and alkynes are called *unsaturated hydrocarbons*. An **unsaturated hydrocarbon** is so called because it can add more hydrogen atoms. Recall that a *saturated hydrocarbon* (alkane) has the maximum number of hydrogen atoms attached to each carbon atom; it has no double or triple bonds.

▲ Alkenes occur widely in nature. Ripening fruits and vegetables give off ethylene, which triggers further ripening. Food processors artificially introduce ethylene to hasten the normal ripening process: 1 kg of tomatoes can be ripened by exposure to as little as 0.1 mg of ethylene for 24 hr. Unfortunately, the tomatoes don't taste much like those that ripen on the vine. The red color of tomatoes is due to lycopene, a hydrocarbon with many double bonds (a *polyene*).

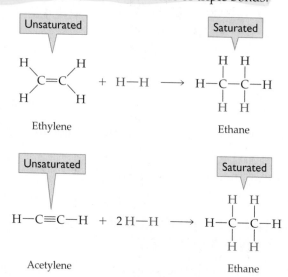

Double bonds are common in various biochemicals important to life. Unsaturated fats are a well-known example.

EXAMPLE 9.3 **Molecular Formulas of Hydrocarbons**

What is the molecular formula for 1-octene? (The "1-" is a part of a systematic name that indicates the location of the double bond; it follows the first carbon atom of the chain.)

Solution

The stem *oct-* indicates eight carbon atoms, the ending *-ene* tells us that the compound is an alkene. Using the general formula C_nH_{2n} for an alkene, with $n = 8$, we see that an alkene with eight carbon atoms has the molecular formula C_8H_{16}.

Alkenes and alkynes typically undergo addition reactions in which all the atoms of the reactants are incorporated into a single product. In the reactions here, ethylene (C_2H_4) adds hydrogen (H_2) to form ethane (C_2H_6), and acetylene (C_2H_2) adds 2 H_2 to form ethane (C_2H_6).

▶ **Exercise 9.3** C_6H_{12} C_7H_{12}

What are the molecular formulas for **(a)** 3-hexene and **(b)** 1-heptyne?

Properties of Alkenes and Alkynes

The physical properties of alkenes and alkynes are quite similar to those of the corresponding alkanes. Those with 2–4 carbon atoms per molecule are gases at room temperature, those with 5–18 carbon atoms are liquids, and most of those with more than 18 carbon atoms are solids. Like alkanes, alkenes and alkynes are insoluble in water and they float on water.

Like alkanes, both alkenes and alkynes burn. However, these unsaturated hydrocarbons undergo many more chemical reactions than alkanes. An alkene or alkyne can undergo an **addition reaction** across the double or triple bond. The addition of hydrogen to double and triple bonds is shown on page 246. Chlorine, bromine, water, and many other kinds of molecules also add to double and triple bonds.

One of the most unusual features of alkene (and alkyne) molecules is that they can add to each other to form large molecules called *polymers*. These interesting molecules are discussed in Chapter 10.

9.5 AROMATIC HYDROCARBONS: BENZENE AND RELATIVES

Another type of hydrocarbon is represented by benzene. Discovered by Michael Faraday in 1825, benzene has a molecular formula of C_6H_6. The structure of benzene puzzled chemists for decades. The formula seems to indicate an unsaturated compound, but benzene does not react as if it contains any double or triple bonds. It does not readily undergo addition reactions the way unsaturated compounds usually do.

Finally, in 1865 August Kekulé proposed a structure with a ring of six carbon atoms, each attached to one hydrogen atom.

The two structures shown appear to contain double bonds, but in fact they do not. Both structures represent the same molecule, and the actual structure of this molecule is a hybrid of these two structures. The benzene molecule actually has six identical carbon-to-carbon bonds that are neither single bonds nor double bonds but something in between. In other words, the three pairs of electrons that would form the three double bonds are not tied down in one location but are spread around the ring (Figure 9.8). Today we usually represent the benzene ring with a circle inside a hexagon.

The hexagon represents the ring of six carbon atoms, and the inscribed circle the ring of six unassigned electrons. Because its ring of electrons resists being disrupted, the benzene molecule is an exceptionally stable structure.

Benzene and similar compounds are called *aromatic hydrocarbons*. This is because quite a few of the first benzenelike substances to be discovered had strong aromas. Even though many benzene derivatives have turned out to be odorless, the name has stuck. Today an **aromatic compound** is any compound that contains a benzene ring or has certain properties similar to those of benzene.

An unsaturated hydrocarbon tends to burn with a smoky flame, especially in limited oxygen. Saturated hydrocarbons usually burn with a clean yellow flame. Natural gas, propane, and butane all burn almost soot free, while paint thinner, gasoline, and kerosene (containing unsaturated hydrocarbons) burn with sooty flames.

Testing for Unsaturated Hydrocarbons with Bromine

Benzene

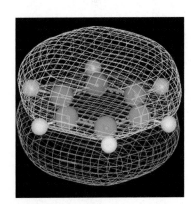

▲ **Figure 9.8** A computer-generated model of the benzene molecule. The six unassigned electrons occupy the yellow and blue areas below and above the plane of the carbon and hydrogen atoms.

A circle in a cyclic structure simply indicates an aromatic compound. It does not always mean six electrons. Naphthalene (Figure 9.9), for example, has a circle in each ring to indicate that it is an aromatic hydrocarbon. The total of unassigned electrons, however, is only 10. Some chemists still prefer to represent naphthalene (and other aromatic compounds) with alternate double and single bonds.

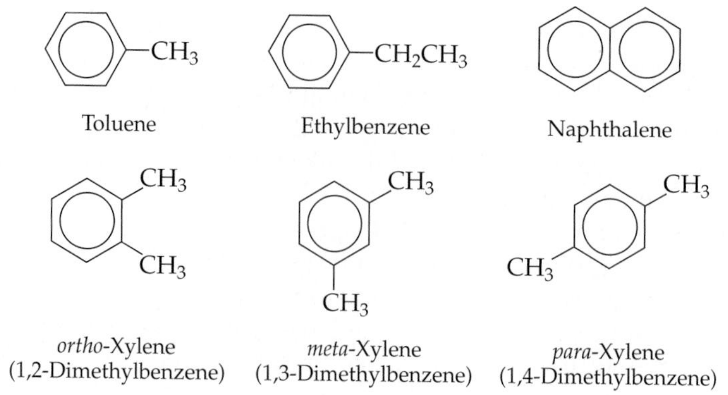

▶ **Figure 9.9** Some aromatic hydrocarbons. Like benzene, toluene and the three xylenes are components of gasoline and also serve as solvents. All these compounds are intermediates in the synthesis of polymers (Chapter 10) and other organic chemicals.

Properties of Aromatic Hydrocarbons

Structures of some common aromatic hydrocarbons are shown in Figure 9.9. Benzene, toluene, and xylenes are all liquids that float on water. They are used mainly as solvents and fuels, but they are also used to make other benzene derivatives. Their vapors can act as narcotics when inhaled. Because benzene may cause leukemia after long exposure, its use has been restricted. Naphthalene is a volatile, white, crystalline solid used as an insecticide, especially as a moth repellant.

9.6 CHLORINATED HYDROCARBONS: MANY USES, SOME HAZARDS

Many other organic compounds are considered to be derived from hydrocarbons by replacing one or more hydrogen atoms by another atom or group of atoms. For example, replacement of hydrogen atoms by chlorine atoms gives chlorinated hydrocarbons. When chlorine gas (Cl_2) is mixed with methane (CH_4) in the presence of ultraviolet (UV) light, a reaction takes place at a very rapid (even explosive) rate. The result is a mixture of products, some of which may be familiar to you (see Example 9.4 and Exercise 9.4). We can write an equation for the reaction in which one hydrogen atom of methane is replaced by a chlorine atom to give methyl chloride (chloromethane).

$$CH_4(g) + Cl_2(g) \longrightarrow CH_3Cl(g) + HCl(g)$$

EXAMPLE 9.4   **Formulas of Chlorinated Hydrocarbons**

What is the formula for dichloromethane (also known as methylene chloride)?

Solution

The *dichloro* indicates two chlorine atoms. The *methane* part of the name indicates that the compound is derived from methane. We conclude that two hydrogen atoms of methane (CH_4) have been replaced by chlorine atoms; the formula for dichloromethane is therefore CH_2Cl_2.

▶ **Exercise 9.4**

What is the formula for **(a)** trichloromethane (also known as chloroform) and **(b)** tetrachloromethane (carbon tetrachloride)?

Methyl chloride is used mainly in making silicone polymers (Chapter 10). Methylene chloride (dichloromethane) is a solvent used, for example, as a paint remover. Chloroform (trichloromethane), also a solvent, was used as an anesthetic in earlier times, but such use is now considered dangerous. The dosage required for effective anesthesia is too close to a lethal dose. Carbon tetrachloride (tetrachloromethane) has been used as a dry-cleaning solvent and in fire extinguishers, but it is no longer recommended for either use. Exposure to carbon tetrachloride (or most other chlorinated hydrocarbons) can cause severe damage to the liver. The use of a carbon tetrachloride fire extinguisher in conjunction with water to put out a fire can be deadly. Carbon tetrachloride reacts with water at elevated temperatures to form phosgene ($COCl_2$), an extremely poisonous gas that was used during World War I.

A variety of more complicated chlorinated hydrocarbons are of considerable interest. DDT (dichlorodiphenyltrichloroethane) and other chlorinated hydrocarbons used as insecticides are discussed in Chapter 16. In Chapter 10, we study polychlorinated biphenyls (PCBs). For now, let's just say that many chlorinated hydrocarbons have similar properties. Most are only slightly polar, and they do not dissolve in water, which is highly polar. Instead, they dissolve (and dissolve in) fats, oils, greases, and other substances of low polarity. This is why certain chlorinated hydrocarbons make good dry-cleaning solvents; they remove grease and oily stains from fabrics. This is also why DDT and PCBs cause problems for fish and birds and perhaps for people; the toxic substances are concentrated in fatty animal tissues rather than being easily excreted in (aqueous) urine.

Chlorofluorocarbons and Fluorocarbons

Carbon compounds containing fluorine as well as chlorine are called *chlorofluorocarbons (CFCs)*. CFCs have been used as the dispersing gases in aerosol cans, for making foamed plastics, and as refrigerants. Three common CFCs are best known by their industrial code designations:

$CFCl_3$	CF_2Cl_2	CF_2ClCF_2Cl
CFC 11	CFC 12	CFC 114

At room temperature, CFCs are gases or liquids with low boiling points. They are essentially insoluble in water and inert toward most other substances. These properties make them ideal propellants for use in aerosol cans for deodorant, hair spray, and food products. Unfortunately, the inertness of these compounds allows them to persist in the environment. They diffuse into the stratosphere, where they undergo chemical reactions that lead to depletion of the ozone layer that protects Earth from harmful UV radiation. We look at this problem and some attempts to solve it in Chapter 12.

Fluorinated compounds have found some interesting uses. Some have been used as blood extenders. Oxygen is quite soluble in certain *perfluorocarbons*. (The prefix *per-* means that all hydrogen atoms have been replaced, in this case by fluorine atoms.) These compounds can therefore serve as temporary substitutes for hemoglobin, the oxygen-carrying protein in blood. Perfluoro compounds have been used to treat premature babies, whose lungs are often underdeveloped.

Polytetrafluoroethylene (PTFE or Teflon®, Chapter 10) is a perfluorinated polymer. It has many interesting applications because of its resistance to corrosive chemicals and to high temperatures. It is especially noted for its unusual nonstick properties.

▲ Perfluorinated greases are used in highly corrosive or reactive environments that would cause most hydrocarbon greases to burst into flame.

Oxygen-Containing Organic Compounds

Many organic compounds contain oxygen as well as carbon and hydrogen. In Sections 9.8 through 9.13, we introduce several important families of oxygen-containing organic compounds. First, however, we introduce another fundamental concept of organic chemistry in Section 9.7.

9.7 THE FUNCTIONAL GROUP

Double and triple carbon–carbon bonds and halogen substituents are examples of **functional groups**, groups of atoms that give a family of organic compounds its characteristic chemical and physical properties. Table 9.4 lists a number of common functional groups found in organic molecules. Each of these groups is the basis for its own homologous series.

In many simple molecules, a functional group is attached to a hydrocarbon stem called an *alkyl group*. An **alkyl group** is derived from an alkane by removing a

TABLE 9.4 Selected Organic Functional Groups

Name of Class	Functional Group[a]	General Formula of Class
Alkane	None	R—H
Alkene	$-\overset{\vert}{C}=\overset{\vert}{C}-$	$R_2C=CR_2$
Alkyne	$-C\equiv C-$	$RC\equiv CR$
Alcohol	$-\overset{\vert}{\underset{\vert}{C}}-OH$	R—OH
Ether	$-\overset{\vert}{\underset{\vert}{C}}-O-\overset{\vert}{\underset{\vert}{C}}-$	R—O—R'
Aldehyde	$-\overset{O}{\overset{\Vert}{C}}-H$	$R-\overset{O}{\overset{\Vert}{C}}-H$
Ketone	$-\overset{O}{\overset{\Vert}{C}}-$	$R-\overset{O}{\overset{\Vert}{C}}-R'$
Carboxylic acid	$-\overset{O}{\overset{\Vert}{C}}-OH$	$R-\overset{O}{\overset{\Vert}{C}}-OH$
Ester	$-\overset{O}{\overset{\Vert}{C}}-O-\overset{\vert}{\underset{\vert}{C}}-$	$R-\overset{O}{\overset{\Vert}{C}}-O-R'$
Amine	$-\overset{\vert}{\underset{\vert}{C}}-\overset{\vert}{N}-$	$R-\overset{}{\underset{H}{N}}-H$ $R-\overset{}{\underset{H}{N}}-R'$ $R-\overset{}{\underset{R''}{N}}-R'$
Amide	$-\overset{O}{\overset{\Vert}{C}}-\overset{}{\underset{\vert}{N}}-$	$R-\overset{O}{\overset{\Vert}{C}}-\overset{}{\underset{H}{N}}-H$ $R-\overset{O}{\overset{\Vert}{C}}-\overset{}{\underset{H}{N}}-R'$ $R-\overset{O}{\overset{\Vert}{C}}-\overset{}{\underset{R''}{N}}-R'$

[a]Neutral functional groups are shown in green, acidic groups in red, and basic groups in blue.

hydrogen atom. The methyl group (CH_3—), for example, is derived from methane (CH_4), and the ethyl group (CH_3CH_2—) from ethane (CH_3CH_3). Propane yields two different alkyl groups, depending on whether the hydrogen atom is missing from the end or from the middle carbon atom.

$$CH_3CH_2CH_2- \qquad CH_3CHCH_3$$

Propyl group Isopropyl group

Table 9.5 lists the four alkyl groups derived from the two butanes.

Often the letter R is used to stand for alkyl groups in general. Thus, ROH is a general formula for alcohols, and RCl represents any alkyl chloride.

9.8 THE ALCOHOL FAMILY

When a hydroxyl group (OH) is substituted for any hydrogen atom in an alkane (RH), the molecule becomes an **alcohol** (ROH). Like many organic compounds, alcohols can be called by their common names or by the systematic names of IUPAC.

TABLE 9.5 Common Alkyl Groups

Name		Structural Formula	Condensed Structural Formula
Derived from Propane	Propyl		$CH_3CH_2CH_2-$
	Isopropyl		CH_3CHCH_3
Derived from Butane	Butyl		$CH_3CH_2CH_2CH_2-$
	Secondary butyl (*sec*-butyl)		$CH_3CHCH_2CH_3$
Derived from Isobutane	Isobutyl		CH_3CHCH_2- with CH_3
	Tertiary butyl (*tert*-butyl)		CH_3-C-CH_3 with CH_3

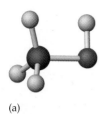

(a)

(b)

▲ **Figure 9.10** Ball-and-stick (a) and space-filling (b) models of the methanol molecule.

1-propanol

(a)

(b)

▲ **Figure 9.11** Ball-and-stick (a) and space-filling (b) models of the ethanol molecule.

Is it poison? seems to be a simple question, but it is a difficult one to answer. Toxicity depends on the nature of the substance, the amount, and the route by which it is taken into the body. Toxicity is discussed in detail in Chapter 20.

The IUPAC names for alcohols are based on those of alkanes, with the ending changed from *-e* to *-ol*. If more than one location is possible, a number is used to indicate the location of the OH group. Methanol, ethanol, and 1-propanol are the first three members of the homologous series of alcohols. (The 1 in 1-propanol indicates that OH is attached to an end carbon atom of the three-carbon chain. An isomeric compound, 2-propanol, has an OH group on the second carbon atom of the chain.)

$$CH_3—OH \qquad CH_3CH_2—OH \qquad CH_3CH_2CH_2—OH$$

Methanol   Ethanol   1–Propanol

Common names for these alcohols are methyl alcohol, ethyl alcohol, and propyl alcohol. The IUPAC names are the ones used in most scientific literature.

Methyl Alcohol (Methanol)

The simplest alcohol is methanol or methyl alcohol (Figure 9.10). Methanol is sometimes called wood alcohol because it was once made from wood. The modern industrial process makes methanol from carbon monoxide and hydrogen at high temperature and pressure in the presence of a catalyst.

$$CO(g) + 2\,H_2(g) \longrightarrow CH_3OH(l)$$

Methanol is important as a solvent and as a chemical intermediate. It is also an additive and possible complete replacement for gasoline in automobiles.

Ethyl Alcohol (Ethanol)

The next member of the homologous series of alcohols is ethyl alcohol (CH_3CH_2OH), also called ethanol or grain alcohol (Figure 9.11). Ethanol is made by fermentation of grain (or other starchy or sugary materials). If the sugar is glucose, the reaction is

$$C_6H_{12}O_6(aq) \xrightarrow{\text{yeast}} 2\,CH_3CH_2OH(aq) + 2\,CO_2(g)$$

All ethyl alcohol for beverages and for automotive fuel is made in this way. Ethanol for industrial use only is made by reacting ethylene with water.

$$CH_2=CH_2(g) + H_2O(l) \xrightarrow{H^+} CH_3CH_2OH(l)$$

This industrial alcohol is identical to that made by fermentation and is generally cheaper, but by law it cannot be used in alcoholic beverages. Since it carries no excise tax, the law requires that noxious substances be added to the alcohol to prevent people from drinking it. The resulting *denatured alcohol* is not fit to drink. This is the kind of alcohol commonly found on the shelves in chemical laboratories.

TABLE 9.6   **Approximate Relationship Among Drinks Consumed, Blood-Alcohol Level, and Behavior***

Number of Drinks[†]	Blood-Alcohol Level (percent by volume)	Behavior[‡]
2	0.05	Mild sedation; tranquility
4	0.10	Lack of coordination
6	0.15	Obvious intoxication
10	0.30	Unconsciousness
20	0.50	Possible death

*Data are for a 70-kg (154-lb) moderate drinker.

[†]Rapidly consumed 30-mL (1-oz) "shots" of 90-proof whiskey, 360-mL (12-oz) bottles of beer, or 150-mL (5-oz) glasses of wine.

[‡]An inexperienced drinker would be affected more strongly, or more quickly, than one who is ordinarily a moderate drinker. Conversely, an experienced heavy drinker would be affected less.

Gasoline in many parts of the United States contains up to 10% ethyl alcohol from fermentation. Because the United States has a large corn (maize) surplus, there is a generous government subsidy for gasoline producers who make their ethanol this way. As a result, many factories now make ethanol by fermentation of corn.

Toxicity of Alcohols

Although ethyl alcohol is an ingredient in wine, beer, and other alcoholic beverages, alcohols in general are rather toxic. Methanol, for example, is oxidized in the body to formaldehyde (HCHO). Drinking as little as 1 oz (about 30 mL or 2 tablespoonfuls) can cause blindness and even death. Many poisonings each year result from mistaking methanol for its less toxic relative ethanol.

Ethanol is not as poisonous as methanol, but it is still toxic. One pint (about 500 mL) of pure ethyl alcohol, rapidly ingested, will kill most people. Of course, even strong alcoholic beverages seldom contain more than 45% ethanol (90 proof).

Generally, ethanol acts as a mild depressant; it slows down both physical and mental activity. Table 9.6 lists the effects of various doses. Although ethanol generally is a depressant, in small amounts it seems to act as a stimulant, perhaps by relaxing tensions and relieving inhibitions.

Excessive ingestion of ethanol over a long period of time can alter brain cell function, cause nerve damage, and shorten life span by contributing to diseases of the liver, cardiovascular system, and practically every other organ of the body. In addition, about half of fatal automobile accidents involve at least one drinking driver. Babies born to alcoholic mothers often are small, deformed, and mentally retarded. Some investigators believe that this *fetal alcohol syndrome* can occur even if mothers drink only moderately. Ethanol, by far the most abused drug in the United States, is discussed more fully in Chapter 19.

Rubbing alcohol is a 70% solution of isopropyl alcohol. Because isopropyl alcohol is also more toxic than ethyl alcohol, it is not surprising that people become ill, and sometimes die, after drinking rubbing alcohol.

Multifunctional Alcohols

Several alcohols have more than one hydroxyl group. Examples are ethylene glycol, propylene glycol, and glycerol.

<table>
<tr><td>

H   H
| |
H—C—C—H
| |
OH OH

Ethylene glycol

</td><td>

H   H   H
| | |
H—C—C—C—H
| | |
H   OH OH

Propylene glycol

</td><td>

H   H   H
| | |
H—C—C—C—H
| | |
OH OH OH

Glycerol

</td></tr>
</table>

Ethylene glycol is the main ingredient in many permanent antifreeze mixtures. Its high boiling point keeps it from boiling away in an automobile radiator. Ethylene glycol is a syrupy liquid with a sweet taste, but it is quite toxic. It is oxidized in the liver to oxalic acid.

$$\underset{\text{Ethylene glycol}}{\overset{\displaystyle \text{OH \quad OH}}{\underset{\displaystyle \text{CH}_2\text{—CH}_2}{|\quad\quad|}}} \xrightarrow{\text{liver enzymes}} \underset{\text{Oxalic acid}}{\overset{\displaystyle \text{O \quad O}}{\text{HO—C—C—OH}}}$$

Oxalic acid forms crystals of its calcium salt, calcium oxalate (CaC_2O_4), which can damage the kidneys, leading to kidney failure and death. Propylene glycol is now marketed as a safer permanent antifreeze. It is a high-boiling antifreeze and it is not poisonous.

Glycerol (or glycerin) is a sweet, syrupy liquid made as a by-product from fats during soap manufacture (Chapter 17). It is used in lotions to keep the skin soft and as a food additive to keep cakes moist. Its reaction with nitric acid makes nitroglycerin,

The *proof* of an alcoholic beverage is merely twice the percentage of alcohol by volume. The term has its origin in an old seventeenth-century English method for testing whiskey. Dealers were often tempted to increase profits by adding water to the booze. A qualitative method for testing the whiskey was to pour some of it on gunpowder and ignite it. If the gunpowder ignited after the alcohol had burned away, this was considered "proof" that the whiskey did not contain too much water.

▲ Warning labels on alcoholic beverages alert consumers to the hazards of excessive consumption.

$$\underset{\substack{\text{Isopropyl alcohol}\\\text{(2-propanol)}}}{\overset{\displaystyle \text{CH}_3\text{CHCH}_3}{\underset{\displaystyle \text{OH}}{|}}}$$

the explosive material in dynamite. Incidentally, nitroglycerin is also important as a vasodilator, a medication taken by heart patients to relieve angina pain.

EXAMPLE 9.5   **Structural Formulas of Alcohols**

Write the structural formula for *tert*-butyl alcohol, sometimes used as an octane booster in gasoline.

Solution

An alcohol can be considered an alkyl group joined to a hydroxyl group. From Table 9.5, we see that the *tert*-butyl group is

$$CH_3-\underset{\underset{CH_3}{|}}{\overset{\overset{CH_3}{|}}{C}}-$$

Connecting this group to an OH group gives *tert*-butyl alcohol.

$$CH_3-\underset{\underset{CH_3}{|}}{\overset{\overset{CH_3}{|}}{C}}-OH$$

▶ **Exercise 9.5**

Write the structural formulas for **(a)** butyl alcohol and **(b)** *sec*-butyl alcohol.

9.9 PHENOLS

When a hydroxyl group is attached to a benzene ring, the compound is called a **phenol**. Although it may appear to be an alcohol, it actually is not. The benzene ring greatly alters the properties of the hydroxyl group. In fact, phenol is a weak acid (sometimes called carbolic acid), and it is highly poisonous.

Phenol was the first antiseptic used in an operating room—by Joseph Lister in 1867. Up until that time, surgery was not antiseptic, and many patients died from infections following surgical operations. Although phenol has a strong germicidal action, it is far from an ideal antiseptic because it causes severe skin burns and it kills healthy cells along with harmful microorganisms.

Phenol

Phenol is still sometimes employed as a disinfectant for floors and furniture, but other phenolic compounds are now used as antiseptics. Hexylresorcinol, for example, is a more powerful germicide than phenol, and it is less damaging to the skin and has fewer other side effects.

A benzene ring with hydroxyl groups at both ends of the ring is called hydroquinone. As we noted in Chapter 8, it is commonly used as a photographic developer.

9.10 ETHERS

Compounds with two alkyl groups attached to the same oxygen atom are called **ethers**. The general formula is ROR (or ROR' because the alkyl groups need not be alike). Best known is diethyl ether ($CH_3CH_2OCH_2CH_3$) often called simply *ether*.

Diethyl ether was once used as an anesthetic but today is used mainly as a solvent; it dissolves many organic substances that are insoluble in water. It boils at 36 °C, and so it evaporates readily, making it easy to recover dissolved materials.

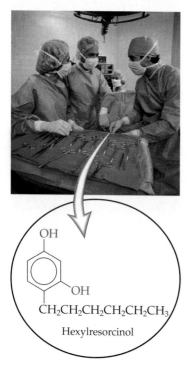

Hexylresorcinol

▲ Phenolic compounds help to ensure antiseptic conditions in hospital operating rooms.

Hydroquinone

 Diethyl ether

Although diethyl ether has little chemical reactivity, it is highly flammable, and great care must be used to avoid sparks or flames when it is in use.

Another problem with ethers is that over time they can react slowly with oxygen to form unstable peroxides, which may decompose explosively. Beware of previously opened containers of ether, especially old ones.

The ether produced in the largest amount commercially is a cyclic compound called ethylene oxide. Its two carbon atoms and an oxygen atom form a three-membered ring. Ethylene oxide is a toxic gas, much of which is used to make ethylene glycol.

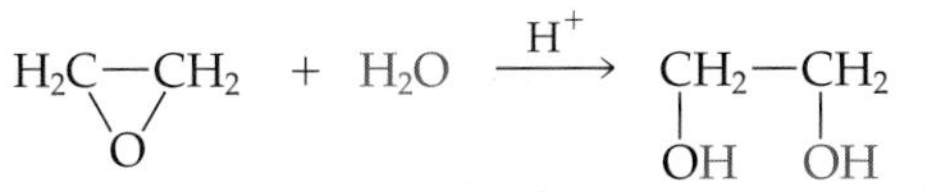

$$H_2C\text{---}CH_2 + H_2O \xrightarrow{H^+} \underset{\underset{OH}{|}}{CH_2}\text{---}\underset{\underset{OH}{|}}{CH_2}$$

Ethylene oxide   Water   Ethylene glycol

Ethylene oxide is also used to sterilize medical instruments and as an intermediate in the synthesis of nonionic surfactants (Chapter 17). Most of the ethylene glycol produced is used in making polyester fibers (Chapter 10) and antifreeze.

> ### CONCEPTUAL EXAMPLE 9.6   The Functional Group

Classify each of the following as an alcohol, an ether, or a phenol.

HO—⬡—Cl $CH_3CH_2OCH_3$ $\underset{\underset{OH}{|}}{CH_3CHCH_2CH_3}$

(a)   (b)   (c)

⬡—CH_2OH (cyclic ether)

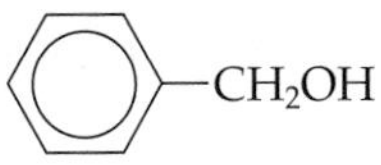

(d)    (e)

Solution

a. The functional group, an OH, is attached directly to the benzene ring; the compound is a phenol.
b. The O atom is between two alkyl groups; the compound is an ether.
c. The OH group is attached to an alkyl group; the compound is an alcohol.
d. An alcohol; the OH is not attached directly to the benzene ring.
e. An ether; the O atom is between two C atoms.

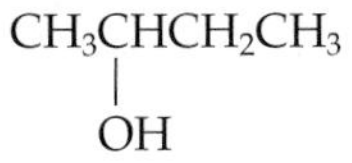

▶ **Exercise 9.6**

Classify the following as alcohols, ethers, or phenols.

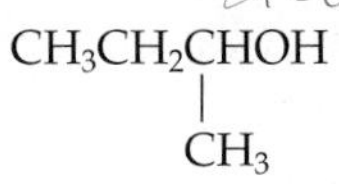

alcohol   ether   ether

$\underset{\underset{CH_3}{|}}{CH_3CH_2CHOH}$ CH_3O—⬡ $\underset{\underset{CH_3}{|}}{CH_3CH_2CHOCH_3}$

(a) 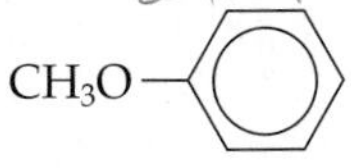   (b) 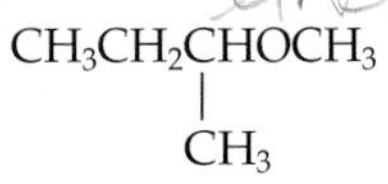   (c)

HO—⬡ phenol   (cyclic) ether
          |
        Br

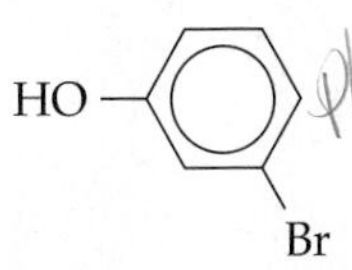

(d) 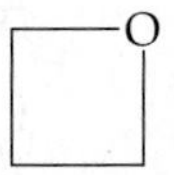   (e)

EXAMPLE 9.7 **Formulas for Ethers**

Give the formula for isopropyl methyl ether.

Solution

Isopropyl methyl ether has an oxygen atom joined to an isopropyl group (three carbons joined to oxygen by the middle carbon) and a methyl group. The formula is

CH₃CHOCH₃ ← Methyl group
Isopropyl group ↙ |
                  CH₃

▶ **Exercise 9.7A**

Give the formula for methyl propyl ether.

▶ **Exercise 9.7B**

Give the formula for ethyl *tert*-butyl ether.

9.11 ALDEHYDES AND KETONES

Two families of organic compounds that share the same functional group are **aldehydes** and **ketones**. Both families contain the **carbonyl group** ($C=O$) but aldehydes have a hydrogen atom attached to the carbonyl carbon, whereas ketones have the carbonyl carbon attached to two other carbon atoms.

$$\begin{array}{ccc} O & O & O \\ \| & \| & \| \\ -C- & R-C-H & R-C-R' \end{array}$$

A carbonyl group   An aldehyde   A ketone

To simplify typing, these structures are often written on one line with the $C=O$ understood:

$-C(=O)-$ or $-CO-$ $R-CH(=O)$ or $R-CHO$ $R-C(=O)-R'$ or $R-CO-R'$

A carbonyl group   An aldehyde   A ketone

Models of three familiar carbonyl compounds are shown in Figure 9.12.

▶ **Figure 9.12** Ball-and-stick (a) and space-filling (b) models of formaldehyde (left), acetaldehyde (center), and acetone (right).

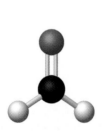 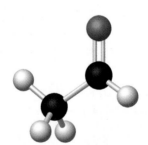 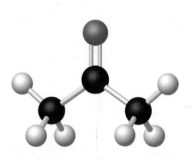

(a)

(b)

9.11 Aldehydes and Ketones   257

Some Common Aldehydes

The simplest aldehyde is formaldehyde (HCHO). It is a gas at room temperature but is readily soluble in water. As a 40% solution called formalin, it is used as a preservative for biological specimens and in embalming fluid.

Formaldehyde is used in making certain plastics (Chapter 10). It also is used to disinfect homes, ships, and warehouses. Commercially, formaldehyde is made by the oxidation of methanol. This is the same net reaction that occurs in the human body when methanol is ingested and that accounts for the high toxicity of methyl alcohol.

$$CH_3OH \xrightarrow{\text{oxidation}} \underset{\text{Formaldehyde}}{H-\overset{\overset{\text{O}}{\|}}{C}-H}$$

Methanol   Formaldehyde

The next member of the homologous series of aldehydes is acetaldehyde (ethanal), formed by the oxidation of ethanol.

$$CH_3CH_2OH \xrightarrow{\text{oxidation}} CH_3-\overset{\overset{\text{O}}{\|}}{C}-H$$

Ethanol   Acetaldehyde

The next two members of the aldehyde series are propionaldehyde (propanal) and butyraldehyde (butanal). Both have strong, unpleasant odors. Benzaldehyde has an aldehyde group attached to a benzene ring. Also called (synthetic) oil of almond, benzaldehyde is used in perfumery and flavoring. It is the flavor ingredient in maraschino cherries.

$$\underset{\text{Propionaldehyde}}{CH_3CH_2-\overset{\overset{\text{O}}{\|}}{C}-H} \qquad \underset{\text{Butyraldehyde}}{CH_3CH_2CH_2-\overset{\overset{\text{O}}{\|}}{C}-H} \qquad \underset{\text{Benzaldehyde}}{C_6H_5-\overset{\overset{\text{O}}{\|}}{C}-H}$$

Some Common Ketones

The simplest ketone is acetone, made by the oxidation of isopropyl alcohol.

$$\underset{\text{Isopropyl alcohol}}{CH_3-\underset{\underset{\text{OH}}{|}}{CH}-CH_3} \xrightarrow{\text{oxidation}} \underset{\text{Acetone}}{CH_3-\overset{\overset{\text{O}}{\|}}{C}-CH_3}$$

Acetone is a common solvent for such organic materials as fats, rubbers, plastics, and varnishes. It also finds use in paint and varnish removers. It is a common ingredient in some fingernail polish removers.

Two other familiar ketones are ethyl methyl ketone and isobutyl methyl ketone, which, like acetone, are frequently used as solvents.

$$\underset{\text{Ethyl methyl ketone}}{CH_3CH_2-\overset{\overset{\text{O}}{\|}}{C}-CH_3} \qquad \underset{\text{Isobutyl methyl ketone}}{CH_3\underset{\underset{\text{CH}_3}{|}}{CH}CH_2-\overset{\overset{\text{O}}{\|}}{C}-CH_3}$$

Systematic names for aldehydes are based on those of alkanes, with the ending changed from -e to -al. Thus, by the IUPAC system formaldehyde is named methanal. IUPAC names are seldom used for simple aldehydes because of possible confusion with the corresponding alcohols. For example, methanal is easily confused with methanol, either orally or (especially) when handwritten.

Organic chemists often write equations that show only the organic reactants and products. Inorganic substances are omitted for the sake of simplicity.

By the IUPAC system, acetone is named propanone. The systematic names for ketones are based on those of alkanes, with the ending changed from -e to -one. When necessary, a number is used to indicate the location of the carbonyl group. For example, $CH_3CH_2COCH_2CH_2CH_3$ is 3-hexanone.

CONCEPTUAL EXAMPLE 9.8   Aldehyde or Ketone?

Identify each of the following compounds as an aldehyde or a ketone.

(a) Ketone $CH_3CH_2CH_2CH_2-\overset{\overset{\text{O}}{\|}}{C}-H$ (b) aldehyde   (c) aldehyde $C_6H_5-\overset{\overset{\text{O}}{\|}}{C}-CH_3$

Solution

a. A hydrogen atom is attached to the carbonyl carbon atom; the compound is an aldehyde.

b. The carbonyl group is between two other (ring) carbon atoms; the compound is a ketone.

c. A ketone. (Remember that the corner of the hexagon stands for a carbon atom.)

▶ **Exercise 9.8**

Identify each of the following compounds as an aldehyde or a ketone.

Ketone

$$CH_3CH_2CH_2CCH_3$$
(with O double bonded to C)

aldehyde

CHO (on ring structure)

aldehyde

$$H_3C-\bigcirc-CHO$$

(a)   (b)   (c)

A ketone is a carbon chain
With carbonyl inside.
If carbonyl is on the end,
Then it's an aldehyde.

9.12 CARBOXYLIC ACIDS

The functional group of organic acids is called the **carboxyl group**, and the acids are called **carboxylic acids**.

$$-\overset{O}{\overset{\|}{C}}-OH \qquad R-\overset{O}{\overset{\|}{C}}-OH$$

A carboxyl group   A carboxylic acid

Formic Acid

Like aldehydes and ketones, these structures are often written on one line:

$$-C(=O)OH \text{ or } -COOH \qquad R-C(=O)OH \text{ or } R-COOH$$

A carboxyl group   A carboxylic acid

The systematic (IUPAC) names for carboxylic acids are based on those of alkanes, with the ending changed from *-e* to *-oic acid*. For example, $CH_3CH_2CH_2CH_2COOH$ is pentanoic acid.

The simplest carboxylic acid is formic acid (HCOOH), also called methanoic acid. It was first obtained by the destructive distillation of ants (the Latin word *formica* means "ant"). The sting of an ant smarts because the ant injects formic acid as it stings. The stings of wasps and bees also contain formic acid (as well as other poisonous compounds).

$$H-\overset{O}{\overset{\|}{C}}-OH$$

Formic acid

Acetic acid (ethanoic acid, CH_3COOH) can be made by the aerobic fermentation of a mixture of cider and honey. This produces a solution (vinegar) containing about 4–10% acetic acid plus a number of other compounds that give vinegar its flavor. Acetic acid is probably the most familiar weak acid used in academic and industrial chemistry laboratories.

The third member of the homologous series of acids, propionic acid (propanoic acid, CH_3CH_2COOH), is seldom encountered in everyday life. The fourth member is more familiar, at least by its odor. If you've ever smelled rancid butter, you know the odor of butyric acid (butanoic acid, $CH_3CH_2CH_2COOH$). It is one of the most foul-smelling substances imaginable. Butyric acid can be isolated from butterfat or synthesized in the laboratory. It is one of the ingredients of body odor, and extremely small quantities of this and other chemicals enable bloodhounds to track fugitives.

The acid with a carboxyl group attached directly to a benzene ring is called benzoic acid.

▲ Acetic acid is a familiar weak acid in chemistry laboratories. It is also the principal active ingredient in vinegar.

$$\bigcirc-\overset{O}{\overset{\|}{C}}-OH$$

Benzoic acid

Carboxylic acid salts—calcium propionate, sodium benzoate, and others—are widely used as food additives to prevent mold (Chapter 16).

EXAMPLE 9.9 **Structural Formulas of Oxygen-Containing Organic Compounds**

Give the structural formula for each of the following compounds.

a. propionaldehyde
b. acetic acid
c. ethyl methyl ketone

Solution

a. Propionaldehyde has three carbon atoms with an aldehyde function.

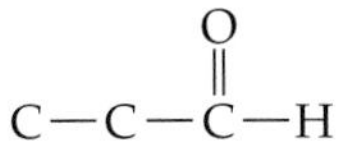

$$C-C-\overset{\overset{\displaystyle O}{\|}}{C}-H$$

Adding the proper number of hydrogen atoms to the other two carbon atoms gives the structure

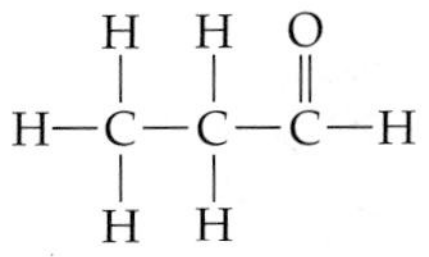

$$H-\overset{\overset{\displaystyle H}{|}}{\underset{\underset{\displaystyle H}{|}}{C}}-\overset{\overset{\displaystyle H}{|}}{\underset{\underset{\displaystyle H}{|}}{C}}-\overset{\overset{\displaystyle O}{\|}}{C}-H \quad \text{or} \quad CH_3CH_2CHO$$

b. Acetic acid has two carbon atoms with a carboxylic acid function.

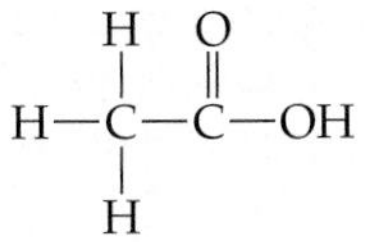

$$H-\overset{\overset{\displaystyle H}{|}}{\underset{\underset{\displaystyle H}{|}}{C}}-\overset{\overset{\displaystyle O}{\|}}{C}-OH \quad \text{or} \quad CH_3COOH$$

c. Ethyl methyl ketone has a ketone function between an ethyl and a methyl group.

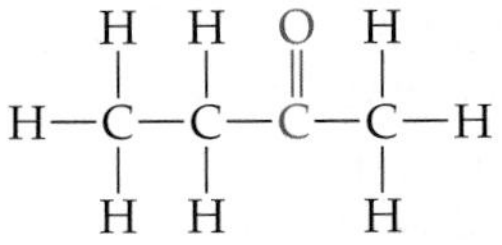

$$H-\overset{\overset{\displaystyle H}{|}}{\underset{\underset{\displaystyle H}{|}}{C}}-\overset{\overset{\displaystyle H}{|}}{\underset{\underset{\displaystyle H}{|}}{C}}-\overset{\overset{\displaystyle O}{\|}}{C}-\overset{\overset{\displaystyle H}{|}}{\underset{\underset{\displaystyle H}{|}}{C}}-H \quad \text{or} \quad CH_3CH_2COCH_3$$

▶ **Exercise 9.9A**
Give the structural formula for each of the following compounds.
a. butyric acid $CH_3CH_2CH_2COOH$
b. acetaldehyde CH_3CHO
c. diethyl ketone $CH_3CH_2COCH_2CH_3$

▶ **Exercise 9.9B**
Give the structural formula for each of the following compounds.
a. hexanoic acid $CH_3CH_2CH_2CH_2CH_2COOH$
b. 3-octanone $CH_3CH_2COCH_2CH_2CH_2CH_2CH_3$
c. heptanal $CH_3CH_2CH_2CH_2CH_2CH_2CHO$

9.13 ESTERS: THE SWEET SMELL OF RCOOR′

Esters are derived from carboxylic acids and alcohols or phenols. The general reaction involves splitting out a molecule of water.

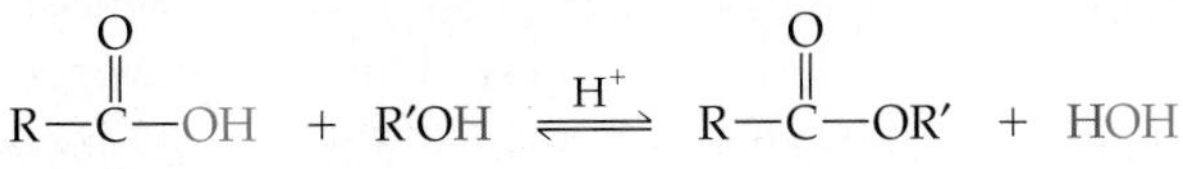

$$R-\overset{\overset{\displaystyle O}{\|}}{C}-OH \;+\; R'OH \;\overset{H^+}{\rightleftharpoons}\; R-\overset{\overset{\displaystyle O}{\|}}{C}-OR' \;+\; HOH$$

An acid   An alcohol   An ester

TABLE 9.7 Ester Flavors and Fragrances

Ester	Formula	Flavor/Fragrance
Methyl butyrate	$CH_3CH_2CH_2COOCH_3$	Apple
Ethyl butyrate	$CH_3CH_2CH_2COOCH_2CH_3$	Pineapple
Propyl acetate	$CH_3COOCH_2CH_2CH_3$	Pear
Pentyl acetate	$CH_3COOCH_2CH_2CH_2CH_2CH_3$	Banana
Pentyl butyrate	$CH_3CH_2CH_2COOCH_2CH_2CH_2CH_2CH_3$	Apricot
Octyl acetate	$CH_3COOCH_2CH_2CH_2CH_2CH_2CH_2CH_2CH_3$	Orange
Methyl benzoate	$C_6H_5COOCH_3$	Ripe kiwifruit
Ethyl formate	$HCOOCH_2CH_3$	Rum
Methyl salicylate	$o\text{-}HOC_6H_4COOCH_3$	Wintergreen
Benzyl acetate	$CH_3COOCH_2C_6H_5$	Jasmine

The name of an ester ends in *-ate* and is formed by naming the part from the alcohol first and the part from the carboxylic acid last. For example, the ester derived from butyric acid and methyl alcohol is methyl butyrate.

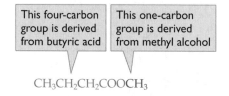

This four-carbon group is derived from butyric acid

This one-carbon group is derived from methyl alcohol

$$CH_3CH_2CH_2COOCH_3$$

Although carboxylic acids often have strongly unpleasant odors, the esters derived from them are usually quite fragrant, especially when dilute. Many esters have fruity odors and tastes. Some examples are given in Table 9.7. Esters are widely used as flavorings in cakes, candies, and other foods and as ingredients in perfumes.

Acids and Esters

Unless the hydrocarbon chain is reasonably long,
A carboxylic acid's apt to smell both foul and strong.
An ester, on the other hand, is so extremely sweet
It can be used in perfumes and delicious things to eat.

Salicylates: Pain Relievers Based on Salicylic Acid

Salicylic acid is both a carboxylic acid and a phenol. We can use it to illustrate some of the reactions of these two families of compounds.

Salicylic acid

Since ancient times, various peoples around the world had used willow bark to treat fevers. Edward Stone, an English clergyman, reported to the Royal Society in 1763 that an extract of willow bark was useful in reducing fever. Salicylic acid was first isolated from willow bark in 1860. Soon after its isolation, salicylic acid was used in medicine as an *antipyretic* (fever reducer) and as an *analgesic* (pain reliever). However, it is sour and irritating when taken orally. Chemists sought to use chemical reactions to modify its structure to reduce these undesirable properties while retaining or even enhancing its desirable properties. These reactions are summarized in Figure 9.13. The first such modification was simply to neutralize the acid (Reaction 1 in Figure 9.13). The resulting salt, sodium salicylate, was first used in 1875. It was less unpleasant to swallow than the acid but was still highly irritating to the stomach.

By 1886 chemists had produced another derivative (Reaction 2), the phenyl ester of salicylic acid, called phenyl salicylate or salol. Salol was less unpleasant to swallow, and it passed largely unchanged through the stomach. In the small intestine, salol was hydrolyzed to the desired salicylic acid, but phenol was formed as a by-product. A large dose could produce phenol poisoning.

Acetylsalicylic acid, an ester of the phenol group of salicylic acid with acetic acid, was first produced in 1853. It is usually made by reacting salicylic acid with acetic anhydride, the acid anhydride of acetic acid (Reaction 3). The German Baeyer Company introduced acetylsalicylic acid as a medicine in 1899 under the trade name Aspirin. It soon became the best-selling drug in the world.

Another derivative, methyl salicylate, is made by reacting the carboxyl group of salicylic acid with methanol, producing a different ester (Reaction 4). Methyl salicylate, called oil of wintergreen, is used as a flavoring agent. It also finds use in rub-on analgesics. When applied to the skin, it causes a mild warming sensation, providing some relief for sore muscles.

Aspirin and its use as a drug are discussed in more detail in Chapter 19.

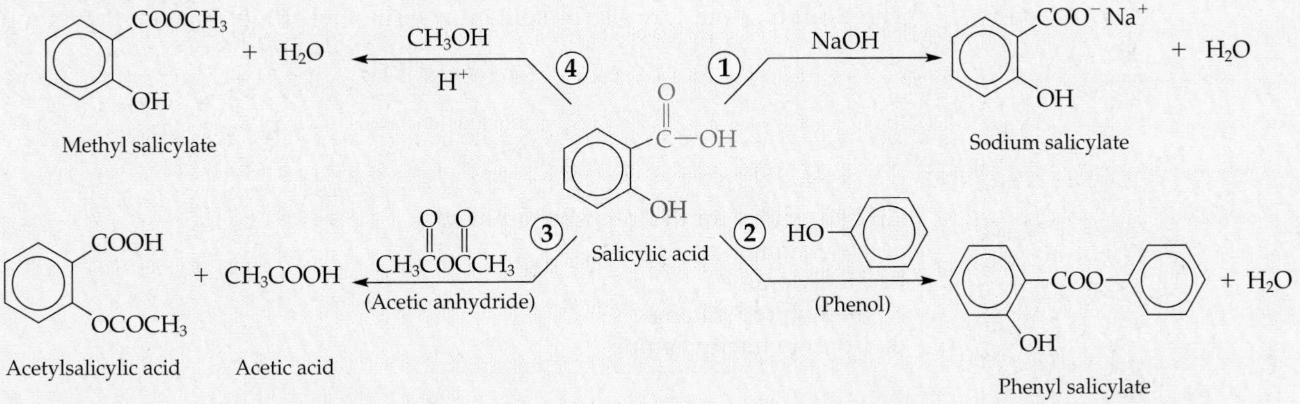

▲ **Figure 9.13** Some reactions of salicylic acid. In Reactions 1, 2, and 4, salicylic acid reacts as a carboxylic acid; the reactions occur at the carboxyl group. In Reaction 3, salicylic acid reacts as a phenol; the reaction takes place at the hydroxyl group.

Question: Identify and name the functional groups present in (a) salicylic acid, (b) methyl salicylate, (c) acetylsalicylic acid, and (d) phenyl salicylate.

Nitrogen-Containing Organic Compounds

Many organic substances of interest to us in the chapters that follow contain nitrogen. It is the fourth most common element in organic compounds after carbon, hydrogen, and oxygen. We will look at a few simple examples in the following sections.

9.14 AMINES AND AMIDES

The two families we introduce in this section, amines and amides, provide a vital background for the material ahead.

Amines

Amines contain the elements carbon, hydrogen, and nitrogen. An **amine** is derived from ammonia by replacing one, two, or three of the hydrogen atoms by one, two, or three alkyl or aromatic groups:

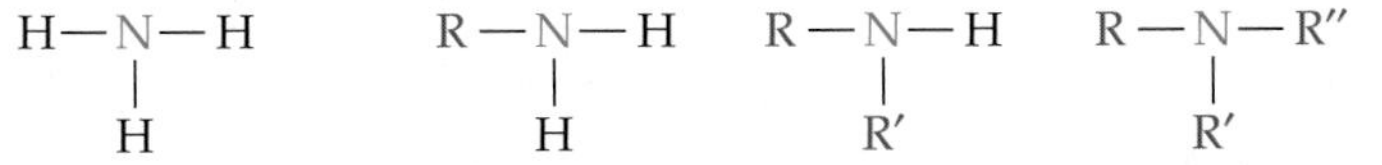

The simplest amine is methylamine (CH_3NH_2). The next higher homolog is ethylamine ($CH_3CH_2NH_2$). However, with two carbon atoms we can have isomers: ethylamine and dimethylamine (CH_3NHCH_3) both have the molecular formula C_2H_7N. With three carbon atoms, there are several possibilities, including trimethylamine [$(CH_3)_3N$].

CONCEPTUAL EXAMPLE 9.10   **Amine Isomers: Structures and Names**

Give structures and names for the other three-carbon amines.

Solution

Three carbon atoms can be in one alkyl group, and there are two such propyl groups.

$$CH_3CH_2CH_2NH_2 \qquad \begin{array}{c} CH_3CHCH_3 \\ | \\ NH_2 \end{array}$$

Propylamine   Isopropylamine

Three carbon atoms can also be split into one methyl group and one ethyl group.

$$CH_3CH_2NHCH_3$$

Ethylmethylamine

▶ **Exercise 9.10**

Give structures for the following amines.

a. butylamine
b. diethylamine
c. methylpropylamine
d. isopropylmethylamine

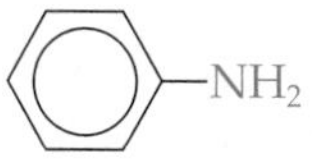

Aniline

▶ **Figure 9.14** Some amines of interest. Amphetamine (a) is a stimulant drug (Chapter 19). Cadaverine (b) has the odor of decaying flesh. 1,6-Hexanediamine (c) is used in the synthesis of nylon (Chapter 10). Pyridoxamine (d) is a B vitamin (Chapter 18).

Question: 1,6-Hexanediamine is the IUPAC name for a compound also known by the common name hexamethylenediamine. Cadaverine is a common name. What is the IUPAC name for cadaverine?

The amine with an NH_2 group attached directly to a benzene ring has the special name *aniline*. Like many other aromatic amines, aniline is used in making dyes. Aromatic amines tend to be toxic, and some are strongly carcinogenic.

Simple amines are similar to ammonia in odor, basicity, and other properties. It is the higher amines that are the most interesting, though. Figure 9.14 shows a variety of these. Notice that each structure contains an **amino (NH_2) group**.

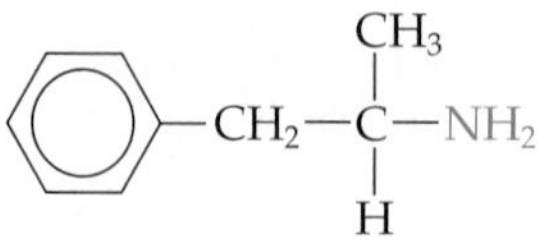

Amphetamine

$$H_2NCH_2CH_2CH_2CH_2CH_2CH_2NH_2$$

1,6-Hexanediamine

$$H_2NCH_2CH_2CH_2CH_2CH_2NH_2$$

Cadaverine

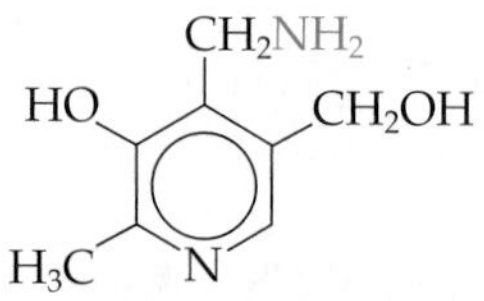

Pyridoxamine

Among the most important kinds of organic molecules are amino acids. As the name implies, these compounds have carboxylic acid and amine functional groups. Amino acids are the building blocks from which proteins are constructed. The simplest amino acid, H_2NCH_2COOH, is glycine. Amino acids and proteins are considered in detail in Chapter 15.

Amides

Another important group of nitrogen-containing compounds is *amides*, which contain oxygen bonded to the same carbon as nitrogen. Thus, an **amide** has the nitrogen atom attached directly to a carbonyl group.

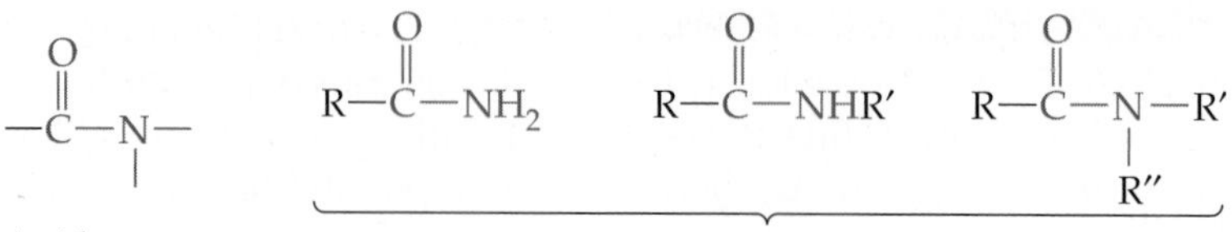

Amide group   Amides

Like those of other compounds that have C=O groups, the formulas of amides are often written on one line as

$$RCONH_2 \qquad RCONHR' \qquad RCONR'R''$$

Note that urea (H_2NCONH_2), the compound that helped change the understanding of organic chemistry (page 239), is an amide.

Complex amides are of much greater interest than the simple ones considered here. Your body contains many kinds of proteins, all held together by amide linkages (Chapter 15). Nylon, silk, and wool molecules also contain hundreds of amide functional groups.

Names for simple amides are derived from those of the corresponding carboxylic acids. For example, $HCONH_2$ is formamide (IUPAC, methanamide) and CH_3CONH_2 is acetamide (IUPAC, ethanamide). Alternatively, we can say amide names are based on those of alkanes, with the ending changed from -e to -amide. For example, $CH_3CH_2CH_2CH_2CONH_2$ is pentanamide.

CONCEPTUAL EXAMPLE 9.11 **Amine or Amide?**

Which of the following are amides and which are amines? Identify the functional groups.

a. $CH_3CH_2CH_2NH_2$ b. CH_3CONH_2
c. $CH_3CH_2NHCH_3$ d. $CH_3COCH_2CH_2NH_2$

Solution
a. An amine; the NH_2 is an amine function; there is no C=O group.
b. An amide; the $CONH_2$ is the amide function.
c. An amine; the NH is an amine function.
d. An amine; the NH_2 is an amine function; there is a C=O group, but the NH_2 is not attached to it.

▶ Exercise 9.11
Which of the following are amides and which are amines? Identify the functional groups.
a. CH_3NHCH_3 b. $CH_3CONHCH_3$
c. $CH_3CH_2N(CH_3)_2$ d. $CH_3NHCH_2CONH_2$

9.15 HETEROCYCLIC COMPOUNDS: ALKALOIDS AND OTHERS

Cyclic hydrocarbons feature rings of carbon atoms. Now let's look at some ring compounds that have atoms other than carbon within the ring. These **heterocyclic compounds** usually have one or more nitrogen, oxygen, or sulfur atoms.

CONCEPTUAL EXAMPLE 9.12 **Heterocyclic Compounds**

Which of the following structures represent heterocyclic compounds?

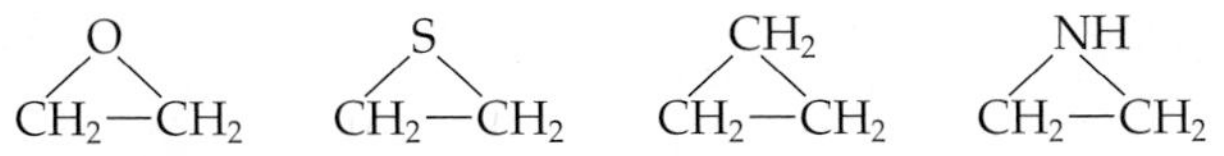

(a)   (b)   (c)   (d)

Solution
Compounds **a**, **b**, and **d** have oxygen, sulfur, and nitrogen atoms, respectively, in a ring structure; these represent heterocyclic compounds.

▶ Exercise 9.12
Which of the following structures represent heterocyclic compounds?

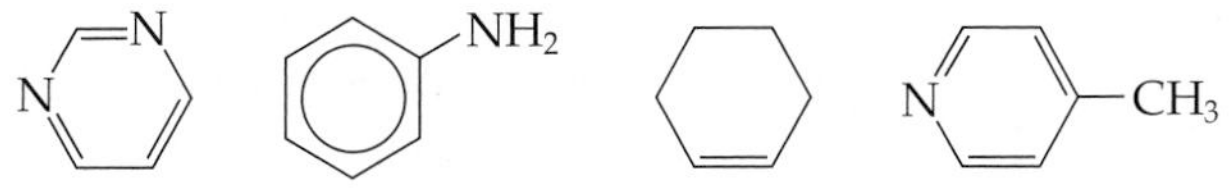

(a)   (b)   (c)   (d)

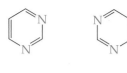

Pyrimidine   Purine

Many amines, particularly heterocyclic ones, occur naturally in plants. Like other amines, these compounds are basic. They are called **alkaloids**, which means "like alkalis." Among the familiar alkaloids are morphine, caffeine, nicotine, and cocaine. The actions of these compounds as drugs are considered in Chapter 19. Of more immediate interest are pyrimidine, which has two nitrogen atoms in a six-membered ring, and purine, which has four nitrogen atoms in two rings that share a common side. Compounds related to pyrimidine and purine are constituents of nucleic acids (Chapter 15).

Critical Thinking Exercises

Apply knowledge that you have gained in this chapter and one or more of the FLaReS principles (Chapter 1) to evaluate the following statements or claims.

9.1 A television advertisement claims that gasoline with added ethanol burns cleaner than gasoline without added ethanol.

9.2 A news feature states that scientists have found that the odor of a certain ester improves workplace performance.

9.3 Alcohols, including ethanol, are toxic. An environmental activist states that all toxic chemicals should be banned from the home.

9.4 An advertisement refers to organic calcium carbonate as being superior to other forms of calcium carbonate because it contains a life force.

9.5 A Web page claims that acetylsalicylic acid (aspirin), polyester plastics, and polystyrene plastics all are carcinogens (cancer-causing agents). The reasoning given is that all three contain benzene rings in their structure, and benzene is a carcinogen.

9.6 A method is suggested for treating methanol poisoning. The method involves giving the patient a great deal of ethanol, though not enough to cause ethanol poisoning. The reasoning used is that the high concentration of ethanol will cause more ethanol to be oxidized to acetaldehyde, and less methanol to be oxidized to formaldehyde in the body. With less formaldehyde produced, the toxic effects of the methanol should be reduced.

SUMMARY

Sections 9.1–9.2—Organic chemistry is the study of carbon compounds. More than 95% of all known compounds contain carbon. Carbon is unique in its ability to form long chains, rings, and branches. **Hydrocarbons** contain only carbon and hydrogen. **Alkanes** are hydrocarbons that contain only single bonds; they have names ending in *-ane*. Alkanes are **saturated hydrocarbons** because each carbon atom is bonded to the maximum number of hydrogen atoms. A **structural formula** (above, top) shows which atoms are bonded to one another. A **condensed structural formula** omits the C—H bond lines and is easier to write. Condensed structural formulas for the first three alkanes are methane (CH_4), ethane (CH_3CH_3), and propane ($CH_3CH_2CH_3$). The rest of the straight-chain alkanes can be generated by adding a CH_2 unit to the previous molecule.

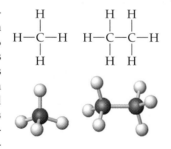

Such a **homologous series** of compounds has properties that vary in a regular and predictable manner.

With more than three carbon atoms it is possible to have **isomers**, compounds with the same molecular formulas but different structures. Alkanes are nonpolar, insoluble in water, and undergo few chemical reactions. They are used primarily as fuels.

Sections 9.3–9.4—**Cyclic hydrocarbons** have one or more closed rings. Geometric figures often are used to represent cyclic compounds: a triangle for cyclopropane, a hexagon for cyclohexane, etc. An **alkene** is a hydrocarbon with at least one carbon–carbon double bond. Ethylene ($CH_2{=}CH_2$) is the simplest alkene and the most important one commercially. It is used to make polyethylene plastic and ethylene glycol. An **alkyne** is a hydrocarbon with at least one carbon–carbon triple bond. Acetylene

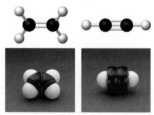

(HC≡CH) is the simplest alkyne, used in welding torches and as a starting material for other chemical products. Alkenes and alkynes are **unsaturated hydrocarbons** to which more hydrogen atoms can be added, and they undergo **addition reactions** in which a small molecule adds to the double or triple bond. They also can add to one another to form polymers.

Sections 9.5–9.6—Benzene (C_6H_6) is drawn as though it has double bonds, but its ring of carbon atoms has six pairs of bonding electrons between carbon atoms, plus six unassigned electrons in a ring.   Benzene and similar compounds are called **aromatic compounds** because of this structure, which is exceptionally stable. Aromatic hydrocarbons are used as solvents, fuels, and to make other benzene derivatives.

Chlorinated hydrocarbons are derived from hydrocarbons by replacing a hydrogen atom(s) with a chlorine atom(s). Chloroform (former anesthetic), carbon tetrachloride (former cleaning solvent), and DDT (former insecticide) are chlorinated compounds. Chlorofluorocarbons contain both chlorine and fluorine. They are used as aerosol propellants and as refrigerants, but their use has been curtailed because of their contribution to ozone layer destruction. Perfluorinated compounds, where all hydrogens have been replaced with fluorine, have important uses in plastics and in medicine.

Section 9.7—A **functional group** is a group of atoms that confers characteristic properties on a family of organic compounds. Compounds with the same functional group undergo similar reactions and have similar properties. The functional group is often attached to an **alkyl group (R—)**, an alkane with a hydrogen atom removed.

Sections 9.8–9.9—A hydroxyl group (—OH) on an alkyl group produces a molecule called an **alcohol (ROH)**. Methanol (CH_3OH), ethanol (CH_3CH_2OH), and isopropanol [$(CH_3)_2CHOH$] are well-known and widely used alcohols. Alcohols with more than one —OH group include ethylene glycol, used in antifreeze, and glycerol, a food additive and lotion additive.

A **phenol** is a compound with an —OH group on a benzene ring. Phenols are slightly acidic, and some are used as antiseptics.

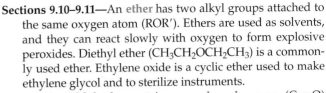

Hexylresorcinol

Sections 9.10–9.11—An **ether** has two alkyl groups attached to the same oxygen atom (ROR′). Ethers are used as solvents, and they can react slowly with oxygen to form explosive peroxides. Diethyl ether ($CH_3CH_2OCH_2CH_3$) is a commonly used ether. Ethylene oxide is a cyclic ether used to make ethylene glycol and to sterilize instruments.

An **aldehyde** contains a **carbonyl group** (C=O) with a hydrogen atom attached to the carbonyl carbon. A **ketone** has two other carbon atoms attached to the carbonyl carbon. Formaldehyde is used in making plastics and as a disinfectant. Benzaldehyde is a flavoring ingredient. Acetone, the simplest ketone, is a widely used solvent.

Sections 9.12–9.13—A **carboxylic acid** has a **carboxyl group** (—COOH) as its functional group. Formic acid (HCOOH) is the acid in ant, bee, and wasp stings. Acetic acid (CH_3COOH) is the acid in vinegar. Butyric acid is the ingredient that gives rancid butter its odor. Carboxylic acid salts are used as preservatives.

The structure of an **ester** (RCOOR′) is similar to a carboxylic acid, with an alkyl group replacing the hydrogen atom of the carboxyl group. Esters are made from a carboxylic acid and an alcohol or phenol. They often have fruity or flowery odors and are used as flavorings and in perfumes.

Section 9.14—An **amine** is a hydrocarbon derivative of ammonia in which one or more hydrogen atoms are replaced by alkyl or aromatic groups. Many amines simply contain an alkyl group and an **amino group** (—NH₂). Like ammonia, amines are basic and often have strong odors. Amino acids, the building blocks of proteins, contain both amine and carboxylic acid functional groups. An **amide** contains a carbonyl group whose carbon atom is attached to a nitrogen atom. Proteins, nylon, silk, and wool contain amide groups. A **heterocyclic compound** has a cyclic-hydrocarbon structure with one or more nitrogen, sulfur, or oxygen atoms in the ring. **Alkaloids** are amines, especially heterocyclic amines, that occur naturally in plants, and include morphine, caffeine, nicotine, and cocaine. Heterocyclic structures are constituents of DNA and RNA.

REVIEW QUESTIONS

1. What is organic chemistry?

2. List three characteristics of the carbon atom that make possible the existence of millions of organic compounds.

3. Define, illustrate, or give an example of each of the following terms.
   a. hydrocarbon   b. alkyne
   c. alkane   d. alkene

4. What is a homologous series? Give an example.

5. What are isomers? How can you tell whether or not two compounds are isomers?

6. What is a saturated hydrocarbon? What is an unsaturated hydrocarbon? Give examples of two types of the latter.

7. What is the meaning of the circle inside the hexagon in the modern representation of the structure of benzene?

8. What is an aromatic hydrocarbon? How can you recognize an aromatic compound from its structure?

9. Which alkanes are gases at room temperature? Which are liquids? Which are solids? State your answers in terms of the number of carbon atoms per molecule.

10. Compare the densities of liquid alkanes with that of water. When you add hexane to water in a beaker, what do you expect to observe?

11. What are the chemical names for the alcohols known by the following familiar names?
    a. grain alcohol   b. rubbing alcohol   c. wood alcohol

12. What are some of the long-term effects of excessive ethanol consumption?

13. Give an important historical use for diethyl ether. What is its main use today?

14. Give an important use for phenols.

15. What structural feature distinguishes aldehydes from ketones?

16. How do carboxylic acids and esters differ in odor? In chemical structure?

17. For what family is each of the following the general formula?
    a. ROH   b. RCOR′   c. RCOOR′
    d. ROR′   e. RCOOH   f. RCHO

18. What is an alkaloid? Name three common alkaloids.

PROBLEMS

Organic and Inorganic

19. Classify the following compounds as organic or inorganic.
    a. C_6H_{10} b. CH_3NH_2
    c. $HClO_3$ d. $NaNH_2$

20. Classify the following compounds as organic or inorganic.
    a. $CaCl_2$ b. CH_2Cl_2
    c. $C_6H_{12}O_2$ d. $Co(NH_3)_6Cl_2$

Names and Formulas of Hydrocarbons

21. How many carbon atoms are there in each of the following?
    a. butane   b. cyclooctane
    c. heptane   d. 2-pentene

22. How many carbon atoms are there in each of the following?
    a. propane   b. cyclopentane
    c. ethylene   d. nonane

23. Name the following hydrocarbons.
    a. $CH_3CH_2CH_3$ b. $H-C\equiv C-H$ c. $CH_2=CH_2$

24. Name the following hydrocarbons.

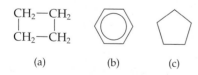

      (a)   (b)   (c)

25. Give the molecular formulas and structural formulas for the following hydrocarbons.
    a. pentane
    b. heptane

26. Give the structural formulas of four-carbon alkanes (C_4H_{10}). Identify butane and isobutane.

27. Give structures for the following alkyl groups.
    a. ethyl   b. isopropyl

28. Give structures for the following alkyl groups.
    a. *tert*-butyl   b. propyl

29. Name the following alkyl groups.
    a. CH_3- b. CH_3CH_2CH-
    $\qquad\qquad\qquad\qquad\qquad\quad |$
    $\qquad\qquad\qquad\qquad\qquad\;\; CH_3$

30. Name the following alkyl groups.
    a. $CH_3CH_2CH_2CH_2-$ b. CH_3CHCH_2-
    $\qquad\qquad\qquad\qquad\qquad\qquad\qquad\quad |$
    $\qquad\qquad\qquad\qquad\qquad\qquad\qquad\; CH_3$

Names and Formulas: Alcohols and Phenols

31. Name the following compounds.
    a. CH_3CH_2OH
    b. $CH_3CH_2CH_2CH_2OH$

32. Name the following compounds.
    a. $CH_3CH_2CHCH_3$ b. CH_3CHCH_2OH
    $\qquad\quad |$ $|$
    $\qquad\quad OH$ CH_3

33. Give the structural formula for each of the following alcohols.
    a. methyl alcohol
    b. propyl alcohol

34. Give the structural formula for each of the following.
    a. hexyl alcohol
    b. *tert*-butyl alcohol

35. Give the structure for phenol.

36. How do phenols differ from alcohols? How are they similar?

Names and Formulas: Ethers

37. Give the structure for each of the following.
    a. diethyl ether   b. butyl methyl ether

38. Give the structure for **(a)** methyl propyl ether, an anesthetic known as Neothyl, and **(b)** methyl *tert*-butyl ether, a gasoline antiknock agent that reduces the emission of carbon monoxide in automotive exhaust gas.

Names and Formulas: Aldehydes and Ketones

39. Give the structure of each of the following compounds.
    a. acetaldehyde   b. formaldehyde

40. Give the structure of each of the following compounds.
    a. butyraldehyde   b. propionaldehyde

41. Name these compounds.

    a.
    $$CH_3CH_2\overset{\displaystyle O}{\overset{\|}{C}}{-}H$$
    b.
    $$CH_3\overset{\displaystyle O}{\overset{\|}{C}}CH_2CH_2CH_3$$

42. Name these compounds.

    a.
    $$CH_3CH_2\overset{\displaystyle O}{\overset{\|}{C}}CH_2CH_2CH_3$$
    b.
    (benzaldehyde structure) $\overset{\displaystyle O}{\overset{\|}{C}}{-}H$

Names and Formulas: Carboxylic Acids

43. Name these compounds.
    a. HCOOH   b. CH_3CH_2COOH

44. Name the following compounds by the IUPAC system.
    a. $CH_3CH_2CH_2CH_2COOH$
    b. $CH_3CH_2CH_2CH_2CH_2CH_2CH_2COOH$

45. Give the condensed structural formula for each of the following.
    a. heptanoic acid   b. decanoic acid

46. Give the condensed structural formula for each of the following.
    a. hexanoic acid   b. butanoic acid

47. Give the structural formula for each of the following.
    a. acetic acid   b. pentanoic acid

48. Give the structural formula for each of the following.
    a. butyric acid   b. nonanoic acid

Names and Formulas: Esters

49. Give the structural formula for each of the following.
    a. ethyl acetate   b. methyl butyrate

50. Give the structural formula for each of the following.
    a. ethyl butyrate   b. methyl acetate

Names and Formulas: Nitrogen-Containing Compounds

51. Give the structural formula for each of the following.
    a. ethylamine   b. dimethylamine

52. Give the structural formula for each of the following.
    a. methylamine   b. isopropylamine

53. Name the following compounds.
    a. $CH_3CH_2CH_2NH_2$
    b. $CH_3CH_2NHCH_2CH_3$

54. Name the following compounds.
    a. $CH_3CH_2NHCH_3$ b. (benzene ring)$-NH_2$

Isomers and Homologs

55. Indicate whether the structures in each set represent the same compound or isomers.

    a. CH_3CH_3 and $\overset{\displaystyle CH_3}{\overset{|}{}}$ CH_3

    b. $CH_3\overset{\displaystyle CH_2}{\underset{CH_3}{\overset{|}{CH_2}}}$ and $CH_3CH_2CH_3$

    c. $CH_3\underset{CH_3}{\overset{|}{C}}HCH_2CH_2CH_3$ and $CH_3CH_2\underset{CH_3}{\overset{|}{C}}HCH_2CH_3$

56. Indicate whether the structures in each set represent the same compound or isomers.

    a. $CH_3\underset{CH_3}{\overset{|}{C}}HCH_2OH$ and $CH_3\underset{OH}{\overset{|}{C}}HCH_2CH_3$

    b. $CH_3\underset{NH_2}{\overset{|}{C}}HCH_2CH_3$ and $CH_3CH_2\underset{CH_3}{\overset{|}{C}}HNH_2$

57. Classify the following pairs as homologs, identical, isomers, or none of these.
    a. $CH_3CH_2CH_3$ and $CH_3CH_2CH_2CH_3$
    b. (cyclobutane) $\begin{array}{c}CH_2 \\ CH_2 \quad CH_2 \\ CH_2{-}CH_2\end{array}$ and $CH_3CH_2CH_2CH_2CH_3$

58. Classify the following pairs as homologs, identical, isomers, or none of these.
    a. $CH_3\underset{CH_3}{\overset{|}{C}}HCH_2CH_3$ and $\overset{\displaystyle CH_3}{\underset{CH_3CHCH_2CH_3}{\overset{|}{}}}$
    b. $CH_3{-}\underset{CH_3}{\overset{|}{C}}H{-}CH{=}CH_2$
    and
    $\begin{array}{c}CH_2{-}CH_2{-}CH_2 \\ CH_2{-}CH_2\end{array}$

Classification of Hydrocarbons

59. Indicate whether each of the following compounds is saturated or unsaturated. Classify each as an alkane, alkene, or alkyne.
    a. $CH_3\underset{CH_3}{\overset{|}{C}}{=}CH_2$ b. $CH_3{-}\overset{\displaystyle CH_3}{\underset{CH_3}{\overset{|}{C}}}{-}CH_3$

60. Indicate whether each of the following compounds is saturated or unsaturated. Classify each as an alkane, alkene, or alkyne.

a. $CH_3C{\equiv}CCH_3$

b.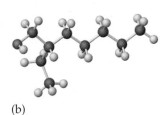

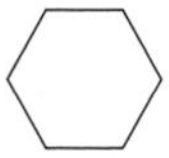

Functional Groups

61. Give the structure of the **(a)** carbonyl functional group and **(b)** carboxyl group.

62. Give the structure of the **(a)** amide functional group and **(b)** amino group.

63. Classify each of the following as an alcohol, amine, amide, ketone, aldehyde, ester, carboxylic acid, or ether. Identify the functional groups in each.

a.
$$\overset{O}{\overset{\|}{CH_3CH_2COCH_3}}$$

b.
$$\overset{O}{\overset{\|}{CH_3CH_2CH}}$$

c. $CH_3CH_2CH_2NH_2$

d. $CH_3CH_2OCH_3$

e.
$$\overset{O}{\overset{\|}{CH_3CH_2CCH_2CH_3}}$$

f.
$$\overset{O}{\overset{\|}{CH_3CH_2COH}}$$

64. Classify each of the following as an alcohol, amine, amide, ketone, aldehyde, ester, carboxylic acid, or ether. Identify the functional groups in each.

a. $HCOOH$

b. $CH_3CH_2COOCH_3$

c. $CH_3CH_2CH_2CH_2OH$

d. $CH_3CH_2CONHCH_2CH_2CH_3$

e. $CH_3CH_2CH_2COOH$

f. $HCOOCH_2CH_2CH_3$

65. Which of the following represent heterocyclic compounds? Classify each also as an amine, ether, or (cyclo)alkane.

a.
$$\begin{array}{c} CH_2{-}CH_2 \\ |\qquad\quad| \\ CH_2{-}CH_2 \end{array}$$

b.
$$\begin{array}{c} CH_2{-}NH \\ |\qquad\quad| \\ CH_2{-}CH_2 \end{array}$$

66. Which of the following represent heterocyclic compounds? Classify each also as an amine, ether, or (cyclo)alkane.

a.
$$\begin{array}{c} CH_2{-}O \\ |\qquad\quad| \\ CH_2{-}CH_2 \end{array}$$

b.
$$\begin{array}{c} CH_2{-}CH{-}NH_2 \\ |\qquad\quad| \\ CH_2{-}CH_2 \end{array}$$

ADDITIONAL PROBLEMS

67. A molecular model of cinnamaldehyde is shown on page 238. The color code for the atoms is carbon (black), hydrogen (white), and oxygen (red). What is the molecular formula of cinnamaldehyde?

68. Using the following color code for atoms—carbon (black), hydrogen (white), and oxygen (red), nitrogen (blue), and chlorine (green)—give the structural formula and molecular formula for the compounds whose models are shown here.

(a)

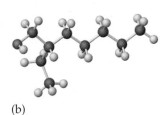

(b)

(c)

(d)

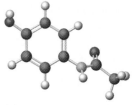

(e)

69. On page 247 we noted that alkenes and alkynes can add hydrogen to form alkanes. The reactions, called hydrogenations, are carried out using H_2 gas and a catalyst such as nickel or platinum. Write equations, using structural formulas, for the complete hydrogenation of each of the following.

a.
$$CH_3C{\equiv}CCHCH_2CH_2CH_2CH_3$$
$$\qquad\qquad | $$
$$\qquad\qquad CH_3$$

b.
$$CH_2{=}CCH_2CH_2CH_3$$
$$\qquad | $$
$$\qquad CH_3$$

70. On page 252 we noted that alkenes can add water to form alcohols. The reactions, called hydrations, are carried out using H_2O gas and an acid (H^+) catalyst. Write equations, using structural formulas, for the hydration of each of the following.

a. $CH_3CH{=}CHCH_3$

b.

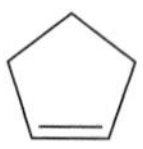

71. On page 257 we noted that alcohols can be oxidized to aldehydes and ketones. Give the structure of the alcohol that can be oxidized to each of the following.

a. $CH_3CH_2CH_2CHO$

b. $CH_3CH_2COCH_2CH_3$

c. $CH_3COCH(CH_3)_2$

d. C_6H_5CHO

72. Thymol is used as an antimold agent in preserving books.

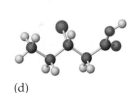

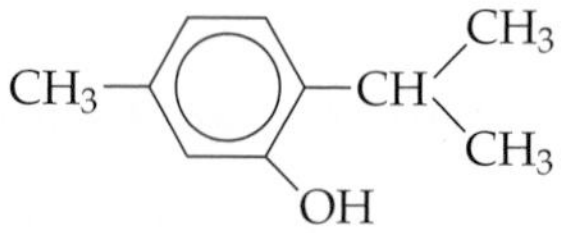

To what family does thymol belong? Name the alkyl groups on the benzene ring.

73. Consider the following set of compounds. What principle does the series illustrate?

$$CH_3OH \qquad CH_3CH_2OH$$
$$CH_3CH_2CH_2OH \quad CH_3CH_2CH_2CH_2OH$$

74. Consider the following set of compounds. What principle does the series illustrate?

$$CH_3CH_2CH_2CH_2OH \quad CH_3CH_2OCH_2CH_3$$

$$\underset{\underset{CH_3}{|}}{CH_3CHCH_2OH} \qquad \underset{\underset{CH_3}{|}}{CH_3CHOCH_3}$$

75. Methanol is a possible replacement for gasoline. The complete combustion of methanol forms carbon dioxide and water.

$$CH_3OH + O_2 \longrightarrow CO_2 + H_2O$$

Balance the equation. What mass of carbon dioxide is formed by the complete combustion of 775 g of methanol?

76. Water adds to ethylene to form ethyl alcohol.

$$CH_2{=}CH_2 + H_2O \xrightarrow{H^+} CH_3CH_2OH$$

Balance the equation. What mass of ethyl alcohol is formed by the addition of water to 445 g of ethylene?

COLLABORATIVE GROUP PROJECTS

Prepare a PowerPoint, poster, or other presentation (as directed by your instructor) for presentation to the class.

77. The theory that organic chemistry was the chemistry of living organisms, which was overturned by Wöhler's discovery in 1828, was known as vitalism. Using key words such as *vitalism*, *vital force*, and *life force*, search the Internet for information about the vital force theory. What was the effect of this philosophy on areas outside chemistry?

78. Many organic molecules contain more than one functional group. Look up structural formulas for each of the following in this text (use the index) or in a reference work such as *The Merck Index*. Identify and name the functional groups in each.
   a. butesin   b. estrone   c. tyrosine
   d. morphine   e. eugenol   f. methyl anthranilate

79. Prepare a brief report on one of the alcohols with three or more carbon atoms per molecule and share it with your group. List sources and commercial uses.

80. Prepare a brief report on one of the carboxylic acids with three or more carbon atoms per molecule and share it with your group. List sources and commercial uses.

REFERENCES AND READINGS

1. Asimov, Isaac. *A Short History of Chemistry*. Garden City, NY: Doubleday Anchor Books, 1965. Chapter 6 provides a brief history of organic chemistry.

2. Benfey, O. T. "August Kekulé and the Birth of the Structural Theory of Organic Chemistry in 1858." *Journal of Chemical Education*, January 1958, pp. 21–25.

3. Brock, William H. *The Chemical Tree*. New York: W. W. Norton, 2000.

4. Bruice, Paula Yurkanis. *Organic Chemistry*, 4th edition. Upper Saddle River, NJ: Prentice Hall, 2003.

5. Buckingham, John. *Chasing the Molecule*. Phoenix Mill, Thrupp, Stroud, UK: Sutton Publishing, 2004.

6. Hill, John W., Stuart J. Baum, and Rhonda J. Scott-Ennis. *Chemistry and Life*, 6th edition. Upper Saddle River, NJ: Prentice Hall, 2000. Chapters 13–17 provide more detail on organic chemistry.

7. Julian, Maureen M. "What Compound Was Discovered as a Result of an Insurance Claim?" *Journal of Chemical Education*, October 1981, p. 793.

GREEN CHEMISTRY

The Art of Organic Synthesis: Green Chemists Find a Better Way

Thomas E. Goodwin, Hendrix College

At one time or another, most of us have taken medicine, either purchased "over the counter" or obtained by prescription. Many of the drugs that we take are *synthetic* organic chemicals (as opposed to compounds that occur in nature). These synthetic compounds are first prepared by an organic chemist carrying out research at a pharmaceutical company or in a research lab at a college or university. (You will learn more about drugs in Chapter 19.) Medicinal chemists who prepare these organic compounds synthesize them by carrying out chemical transformations (reactions) in a rational way, to prepare larger and more complicated organic compounds from simpler ones.

You have seen a few examples in this chapter of the many types of reactions they use. For example, the oxidation of an alcohol is shown on page 257, the synthesis of an ester on page 259, and several reactions of salicylic acid in Figure 9.13. The traditional methods for such transformations involve heating a solution of the reacting compounds (the solutes) to produce the desired products. This process obviously requires a solvent and an energy source. A green chemist might look at this tradition and think: Could I use less energy for a shorter time? Could I find a more environmentally friendly solvent? Could I eliminate the solvent altogether?

In this Web Investigation, you will learn that microwaves are not just for cooking; they can bring about greener organic transformations. You will see that organic chemists are finding ways to replace nonrenewable, petroleum-derived solvents. You will find that in some cases the solvent can be eliminated altogether, thus removing what is often the component of lowest atom economy in the reaction mixture. (You may review the concepts of *atom economy* and *E-factor* in the Green Chemistry activities for Chapters 6 and 7, respectively.)

WEB INVESTIGATIONS

Investigation 1
Microwaved Organic Reactions in a Snap*
Search wikipedia, explore.com or a comparable online encyclopedia to refresh your memory on microwave radiation. Now use your favorite search engine to perform a search using "microwave synthesis" as the keywords. What do you think is the most significant advantage of microwaves in chemical synthesis and analysis?

Investigation 2
Moving Away from Petroleum-Based Solvents
Visit the Scripps Research Institute Web page and search the site for "organic reactions on water." Read the article titled "Solubility Doesn't Matter—Organic Reactions in Aqueous Suspensions" by Jason Bardi. Describe the advantages of running organic reactions "on water." Now do a Web search for "supercritical water oxidation." How does supercritical water differ from liquid or gaseous

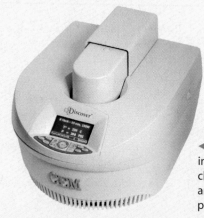

◀ Microwave reactors can increase the rate of many chemical reactions for analyses and for many other purposes.

water? Determine whether anyone has received an award in green chemistry that involves supercritical water.

Investigation 3
Solventless Organic Reactions
Search the Web for "dry media reactions." How do these reactions differ from those in Investigation 1? Why are these types of reactions desirable?

*CAUTION: You must never attempt to carry out a chemical reaction in your home microwave oven. That could be quite dangerous!

Exercise 1
Microwave Laboratory Safety

No activity in life is completely risk free, and the use of microwave radiation to accelerate chemical reactions is no exception. Research safety recommendations for microwave use in chemical laboratories. Write a one- to two-page paper on some of the safety precautions and hazards that are discussed.

Exercise 2
Commercially Available Microwave Reactors

Several companies sell their own versions of specialized laboratory equipment for microwave acceleration of organic reactions. Use a Web browser to search for microwave equipment using two or three of these company names coupled with the word *microwave*: Anton-Paar, Ashwin-Ushas, Biotage, CEM, Milestone. Use the information from their websites to write a short essay comparing and contrasting their products.

Exercise 3
Intermolecular Forces and Solubility

In Chapter 3 of this textbook, you learned about intermolecular forces among molecules and ions, and how they affect physical properties like solubility. Review that material, then consider the Latin phrase *corpora non agunt nisi soluta* (substances do not interact with each other if they are not dissolved). In a one-page paper, use your knowledge of intermolecular forces to discuss why using water as a solvent in a reaction when the starting materials are petroleum-based organic chemicals might be expected to present a problem.

Exercise 4
Twelve Principles of Green Chemistry

Use a Web browser to search for the 12 principles of green chemistry. Review these principles and select the principles most relevant to this Green Chemistry activity. Discuss your choices with one to two of your classmates and describe how the principles you selected are related to the central themes of this activity.

CHAPTER 10

Polymers

Polymers, in the form of films, fibers, and molded objects, pervade nearly every aspect of our lives. Many goods and containers are made of high-density polyethylene; pigments often are added for aesthetics and to protect the contents from light. Polymers are made by joining small molecules (monomers) together to form giant molecules (polymers). In the synthesis of polyethylene shown here in a computer-generated representation (insert), the yellow dots indicate a new bond forming as an ethylene molecule (upper left) is added to the growing polymer chain.

Giants Among Molecules

Look around you. Polymers are everywhere. Around the home, carpets, curtains, upholstery, towels, sheets, floor tile, books, furniture, and most toys and containers (not to mention such things as telephones, toothbrushes, and piano keys) are made of polymers. Clothes are made of polymers. In cars, the dashboard, seats, tires, steering wheel, floor mats, ceiling, and many parts that you cannot see are made of polymers. Much of the food you eat contains polymers, and many important molecules in your body are polymers. You couldn't live without them. Some of the polymers that pervade our lives come from nature, but many are synthetic, made in chemical plants in an attempt to improve on nature in some way.

10.1 POLYMERIZATION: MAKING BIG ONES OUT OF LITTLE ONES

Polymers are composed of *macromolecules* (from the Greek *makros*, meaning "large" or "long"). Macromolecules may not seem large to the human eye (in fact, many of these giant molecules are invisible), but when compared with other molecules, they are enormous.

A **polymer** (from the Greek *poly*, meaning "many," and *meros*, meaning "parts") is made from much smaller molecules called *monomers* (from the Greek *monos*, meaning "one"). Sometimes thousands of monomer units combine to make one polymer molecule. A **monomer** is a small-molecule building block from which a polymer is made. The process by which monomers are converted to polymers is called *polymerization*. A polymer is as different from its monomer as a long strand of spaghetti is from tiny specks of flour. For example, polyethylene, the familiar waxy material used to make plastic bags, is made from the monomer ethylene, a gas.

10.2 NATURAL POLYMERS

Polymers have served humanity for centuries in starches and proteins used for food; in wood used for shelter; and in wool, cotton, and silk used for clothing. Starch is a polymer made up of glucose ($C_6H_{12}O_6$) units. (Glucose is a simple sugar.) Cotton is made of cellulose, also a glucose polymer; and wood is largely cellulose as well. Proteins are polymers made up of amino acid monomers. Wool and silk are two of the thousands of different kinds of proteins found in nature.

273

Living things could not exist without polymers. Each plant and animal requires many different specific types of polymers. Probably the most amazing natural polymers are nucleic acids, which carry the coded genetic information that makes each individual unique. The polymers found in nature are discussed in Chapter 15. In this chapter we focus mainly on macromolecules made in the laboratory.

10.3 CELLULOID: BILLIARD BALLS AND COLLARS

The oldest attempts to improve on nature simply involved chemical modification of natural macromolecules. The synthetic material **celluloid**, as its name implies, was derived from natural cellulose (from cotton and wood, for example). When cellulose is treated with nitric acid, a derivative called cellulose nitrate is formed. In response to a contest to find a substitute for ivory for use in billiard balls, American inventor John Wesley Hyatt (1837–1920) found a way to soften cellulose nitrate by treating it with ethyl alcohol and camphor. The softened material could be molded into smooth, hard balls. Thus, Hyatt brought the game of billiards within the economic reach of more people—and possibly saved a few elephants.

Celluloid was also used in movie film and for stiff collars (so they didn't require laundering and repeated starching). Because of its dangerous flammability (cellulose nitrate is also used as smokeless gunpowder), celluloid was removed from the market when safer substitutes became available. Today, movie film is made from cellulose acetate, another semisynthetic modification of cellulose. And high, stiff collars on men's shirts are out of fashion.

It didn't take long for the chemical industry to recognize the potential of synthetics. Scientists found ways to make macromolecules from small molecules rather than simply modifying large ones. The first such truly synthetic polymers were phenol–formaldehyde resins, first made in 1909. These complex polymers are discussed later in this chapter. Let's look at some simpler ones first.

10.4 POLYETHYLENE: FROM THE BATTLE OF BRITAIN TO BREAD BAGS

The prevalent plastic polyethylene is the simplest and least expensive synthetic polymer. It is familiar today in the plastic bags used for packaging fruit and vegetables, in garment bags for dry-cleaned clothing, in garbage-can liners, and in many other items. Polyethylene is made from ethylene ($CH_2{=}CH_2$), an unsaturated hydrocarbon (Chapter 9) produced in large quantities from the cracking of petroleum.

With pressure and heat and in the presence of a catalyst, ethylene monomers join together in long chains.

Using condensed formulas, this becomes

$$\cdots + CH_2{=}CH_2 + CH_2{=}CH_2 + CH_2{=}CH_2 + CH_2{=}CH_2 + \cdots \longrightarrow {\sim}CH_2CH_2CH_2CH_2CH_2CH_2CH_2CH_2{\sim}$$

These equations can be tedious to draw, and so we often use abbreviated forms like these:

Polyethylene

A century ago, celluloid was widely used as a substitute for more expensive substances such as ivory, amber, and tortoiseshell. The movie industry was once known as the "celluloid industry," and even today a Web search using the key word *celluloid* yields far more hits about movies than about the polymer.

The ellipses ($\cdots$) and tildes ($\sim$) are like et ceteras; they indicate that the number of monomers and the polymer structure are extended for many units in each direction.

The molecular fragment enclosed within the brackets is called the *repeat unit* of the polymer. In the formula for the polymer product, the repeat unit is placed within brackets with bonds extending to both sides. The subscript n indicates that this unit is repeated many times in the full polymer structure.

The simplicity of the abbreviated formula facilitates certain comparisons between the monomer and the polymer. Note that the monomer ethylene contains a double bond and polyethylene does not. The double bond of the reactant contains two pairs of electrons. One of these pairs is used to connect one monomer unit to the next in the polymer (indicated by the lines sticking out to the sides in the repeat unit). This leaves only a single pair of electrons—a single bond—between the two carbon atoms of the repeat unit. Note that each repeat unit in the polymer has the same composition (C_2H_4) as the monomer.

Molecular models provide three-dimensional representations. Figure 10.1 presents models of a tiny part of a very long molecule, which can vary in number of carbon atoms from a few hundred to several thousand.

Polyethylene was invented shortly before the start of World War II. It proved to be tough and flexible, an excellent electric insulator, and able to withstand both high and low temperatures. Before long, it was used for insulating cables in radar, a top-secret invention that helped British pilots detect enemy aircraft before the aircraft could be spotted visually. Without polyethylene, the British could not have had effective radar, and without radar, the Battle of Britain might have been lost. The invention of this simple plastic helped to change the course of history.

Today, there are three principal kinds of polyethylene. *High-density polyethylenes* (*HDPEs*) have mostly linear molecules that pack closely together and can assume a fairly ordered, crystalline structure. HDPEs therefore are rather rigid and have good tensile strength. They are used for such items as threaded bottle caps, toys, bottles, and gallon milk jugs.

Low-density polyethylenes (*LDPEs*), on the other hand, have a lot of side chains branching off the polymer molecules. The branches prevent the molecules from packing closely together and assuming a crystalline structure. LDPEs are waxy, bendable plastics that are lower melting than high-density polyethylenes. Objects

LDPE has densities ranging from 0.910 to 0.940 g/cm³, and HDPE from 0.941 to 0.960 g/cm³.

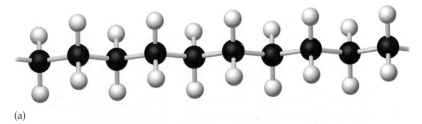

(a)

(b)

◀ **Figure 10.1** Ball-and-stick (a) and space-filling (b) models of a short segment of a polyethylene molecule.

▲ **Figure 10.2** Two bottles, both made of polyethylene, were heated in the same oven for the same length of time.

Question: Which of these bottles is made of HDPE and which of LDPE? Explain.

Polyethylene with molecular masses of three to six million, called ultra-high-molecular-weight polyethylene (UHMWPE) can be used to make fibers which are so strong they replaced Kevlar (see Problem 40) for use in bullet-proof vests. Large sheets of UHMWPE can be used instead of ice for skating rinks.

made of HDPE hold their shape in boiling water, whereas those made of LDPE are severely deformed (Figure 10.2). LDPEs are used to make plastic bags, plastic film, squeeze bottles, electric wire insulation, and many common household products where flexibility is important.

The third type of polyethylene, called *linear low-density polyethylenes* (LLDPEs), is actually a **copolymer**, a polymer formed from two or more different monomers. LLDPEs are made by polymerizing ethylene with a branched-chain alkene such as 4-methyl-1-pentene.

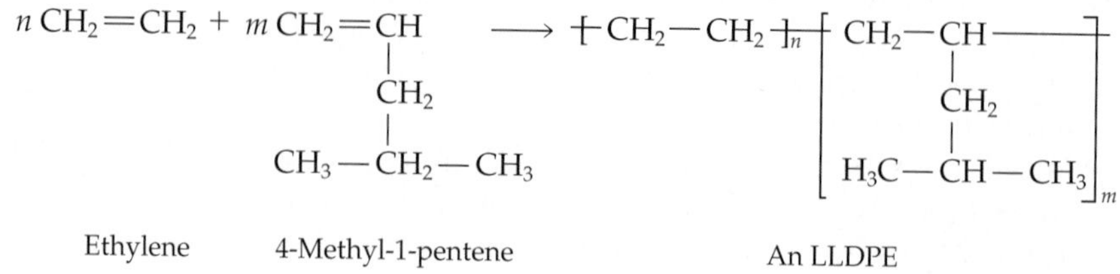

$$n\ CH_2{=}CH_2 + m\ CH_2{=}CH \longrightarrow \left[CH_2{-}CH_2\right]_n \left[\begin{array}{c} CH_2{-}CH{-} \\ | \\ CH_2 \\ | \\ H_3C{-}CH{-}CH_3 \end{array}\right]_m$$

Ethylene 4-Methyl-1-pentene An LLDPE

(monomer side: CH_2, CH_3—CH_2—CH_3)

LLDPEs are used to make such things as plastic films for use as landfill liners, trash cans, tubing, and automotive parts.

Fullerenes: Buckyballs and Nanotubes

We saw in Chapter 9 that carbon atoms can form long chains, branched chains, and rings. Here we see in polymer molecules just how long some of these carbon chains can be. Carbon atoms, with their four bonds, can form still other structures. In 1985 scientists discovered a variety of molecules formed exclusively of carbon atoms. A particularly prominent one had a molecular mass of 720 u, corresponding to the formula C_{60}. The molecule is a roughly spherical collection of hexagons and pentagons very much like a soccer ball. Because C_{60} resembles the geodesic-domed structures that architect R. Buckminster Fuller pioneered, the scientists named it "buckminsterfullerene" in Fuller's honor. The general name *fullerenes* is now used for C_{60} and similar molecules with formulas such as C_{70}, C_{74}, and C_{82}. These substances are often colloquially called "buckyballs."

Later, scientists discovered tube-shaped carbon molecules called *nanotubes*. We can visualize a nanotube as a fullerene that has been stretched out into a hollow cylinder by the insertion of many, many more C atoms. We can also picture them as a two-dimensional array of hexagonal rings of carbon atoms, rather like ordinary "chicken wire." The "wire" is then rolled into a cylinder and capped at each end by half a C_{60} molecule. These nanotubes have unusual mechanical and electrical properties that are of great interest in current research.

▶ Ball-and-stick models of C_{60}, a "buckyball" (a), and a carbon nanotube (b). Nanotubes differ in length, as indicated here by a break in the structure. They can be either single walled (shown) or multi-walled (tubes inside tubes).

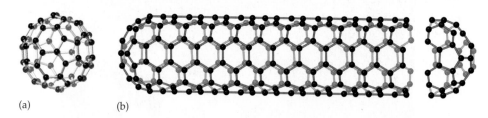

(a)   (b)

Thermoplastic and Thermosetting Polymers

Polyethylene is one of a variety of thermoplastic polymers. A **thermoplastic polymer** can be softened by heat and pressure and then reshaped. It can be repeatedly melted down and remolded. Thermoplastics can be reshaped because their molecules can slide past one another when heat and pressure are applied. Total production of thermoplastic polymers in the United States in 2003 was about 35 billion kg, of which almost 16 billion kg was polyethylene. [Production figures for chemicals are reported each year in *Chemical and Engineering News* (see Reference 6).]

Not all polymers can be readily melted. About 10% of U.S. production is made up of *thermosetting polymers*, which harden permanently when formed. They cannot be softened by heat and remolded; instead, strong heating causes them to discolor and decompose. The permanence of thermosetting plastics is due to cross-linking of the polymer chains. We look at some thermosetting polymers later in this chapter.

More than 221 billion kg of plastics and rubber was manufactured globally in 2003. Of this, 176 billion kg was processed into plastics products. About 19 billion kg of rubber was made into tires and other rubber goods. The remainder, almost 30 billion kg of polymers, was used to make fibers, paints, adhesives, and coatings.

10.5 ADDITION POLYMERIZATION: ONE + ONE + ONE + · · · GIVES ONE!

There are two general types of polymerization reactions: addition polymerization and condensation polymerization. In **addition polymerization** (also called *chain-reaction polymerization*), the monomer molecules add to one another in such a way that the polymeric product contains all the atoms of the starting monomers. The polymerization of ethylene to form polyethylene is an example. In polyethylene, as we noted in Section 10.4, the two carbon atoms and the four hydrogen atoms of each monomer molecule are incorporated into the polymer structure. In *condensation polymerization* (Section 10.7), a portion of the monomer molecule is not incorporated in the final polymer but is split out as the polymer is formed.

Polypropylene

Most of the many familiar addition polymers are made from derivatives of ethylene in which one or more of the hydrogen atoms are replaced by another atom or group. Replacing one of the hydrogen atoms with a methyl group gives the monomer propylene (propene). Polypropylene looks like polyethylene, except that there are methyl groups (CH_3) attached to every other carbon atom.

Polypropylene

$$\sim CH_2-\underset{\underset{CH_3}{|}}{CH}-CH_2-\underset{\underset{CH_3}{|}}{CH}-CH_2-\underset{\underset{CH_3}{|}}{CH}-CH_2-\underset{\underset{CH_3}{|}}{CH}\sim \quad or \quad \left[CH_2-\underset{\underset{CH_3}{|}}{CH} \right]_n$$

Polypropylene

The chain of carbon atoms is called the polymer *backbone*. Groups such as the CH_3 of polypropylene are called *pendant groups*.

Polypropylene is a tough plastic material that resists moisture, oils, and solvents. It is molded into hard-shell luggage, battery cases, and various kinds of appliance parts. It is also used to make packaging material, fibers for textiles such as upholstery fabrics and indoor–outdoor carpets, and ropes that float. Because of its high melting point (121°C), polypropylene objects can be sterilized with steam.

Polystyrene

Replacing one of the hydrogen atoms in ethylene with a benzene ring gives a monomer called styrene, with the formula $C_6H_5CH=CH_2$, where C_6H_5 represents the benzene ring. Polymerization of styrene produces polystyrene, which has benzene rings as pendant groups.

Polystyrene

$$CH_2=CH \qquad \sim CH_2CH-CH_2CH-CH_2CH-CH_2CH\sim$$

Styrene   Polystyrene

▲ Styrofoam insulation saves energy by reducing the transfer of heat from a warm house to the outside in winter or from the hot outside to the cooled inside in summer.

Polystyrene is the plastic used to make transparent "throwaway" drinking cups. With color and filler added, it is the material of thousands of inexpensive toys and household items. When a gas is blown into polystyrene liquid, it foams and hardens into the familiar solid Styrofoam used for ice chests and disposable coffee cups. The polymer can easily be formed into shapes as packing material for shipping instruments and appliances, and it is widely used for home insulation.

► Polyvinyl chloride polymers can be coated onto copper wire for insulation, made into colorful resilient flooring, or formed into many other familiar consumer products.

Vinyl Polymers

Would you like a tough synthetic material that looks like leather at a fraction of the cost? Perhaps a clear, rigid material from which unbreakable bottles could be made? Do you need an attractive, long-lasting floor covering? Or lightweight, rust-proof, easy-to-construct plumbing? Polyvinyl chloride (PVC) has all these properties—and more.

Replacing one of the hydrogen atoms of ethylene with a chlorine atom gives vinyl chloride (CH_2=CHCl), a compound that is a gas at room temperature. Polymerization of vinyl chloride yields the tough thermoplastic material PVC. A segment of the PVC molecule is illustrated below.

$$\sim CH_2CH-CH_2CH-CH_2CH-CH_2CH\sim$$
$$\quad\ |\qquad\quad |\qquad\quad |\qquad\quad |$$
$$\quad\ Cl\qquad\ Cl\qquad\ Cl\qquad\ Cl$$

Polyvinyl chloride

PVC is readily formed into various shapes. The clear, transparent polymer is used in plastic wrap and clear plastic bottles. Adding color and other ingredients to vinyl plastics yields artificial leather. Most floor tile and shower curtains are also made from vinyl plastics, and they are widely used to simulate wood in home siding panels and window frames. About 40% of the PVC produced is molded into pipes.

The monomer from which vinyl plastics are made is a carcinogen. Several people who worked closely with vinyl chloride gas later developed a kind of cancer known as angiosarcoma. (Carcinogens are discussed in Chapter 20.)

PTFE: The Nonstick Coating

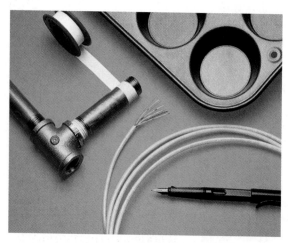

In 1938, young Roy Plunkett, a chemist at DuPont, was working with the gas tetrafluoroethylene (CF_2=CF_2). He opened the valve on a tank of the gas—and nothing came out. Rather than discarding the tank, he decided to investigate. The tank was found to be filled with a waxy, white solid. He attempted to analyze the solid but ran into a problem: It simply wouldn't dissolve, even in hot concentrated acids. Plunkett had discovered the polymer of tetrafluoroethylene, called polytetrafluoroethylene (PTFE), and best known by its trade name, Teflon®.

$$\sim CF_2-CF_2-CF_2-CF_2-CF_2-CF_2-CF_2-CF_2\sim$$

Teflon

Because its C—F bonds are exceptionally strong and resistant to heat and chemicals, PTFE is a tough, unreactive, nonflammable material. It is used to make electric insulation, bearings, and gaskets. It is also widely used to coat surfaces of cookware to give them non-sticking properties.

▲ The plumber's tape, the coating on the muffin pan, and the insulation on the wire are all Teflon.

CONCEPTUAL EXAMPLE 10.1   **Repeat Units in Polymers**

What is the repeat unit in polyvinylidene chloride? A segment of the polymer is represented as

$$\begin{array}{cccccc} & H & Cl & H & Cl & H & Cl \\ & | & | & | & | & | & | \\ \sim C & - C & - C & - C & - C & - C \sim \\ & | & | & | & | & | & | \\ & H & Cl & H & Cl & H & Cl \end{array}$$

Solution
The repeat unit is

$$\begin{array}{cc} H & Cl \\ | & | \\ C = C \\ | & | \\ H & Cl \end{array}$$

Joining three of these units forms the segment shown; joining hundreds of the units would form a molecule of the polymer.

▶ **Exercise 10.1**
What is the repeat unit in polyacrylonitrile? A segment of the polymer is represented as

$$\sim CH_2CHCH_2CHCH_2CHCH_2CHCH_2CHCH_2CH\sim$$
$$||||||$$
$$CNCNCNCNCNCN$$

▲ Granular polymer resins are the basic stock for many molded polymer goods.

EXAMPLE 10.2   **Structure of Polymers**

Give the structure of the polymer made from vinyl fluoride ($CH_2{=}CHF$). Show at least four repeat units.

Solution
The carbon atoms become bonded in a chain with only single bonds between the carbon atoms. The fluorine atom is a substituent on the chain. (Two of the electrons in the double bond of the monomer are used to join the units.) The polymer is

$$\sim CH_2CHCH_2CHCH_2CHCH_2CH\sim$$
$$||||$$
$$FFFF$$

▶ **Exercise 10.2A**
Give the structure of the polymer made from methyl vinyl ether ($CH_2{=}CHOCH_3$). Show at least four repeat units.

▶ **Exercise 10.2B**
Give the structure of the polymer made from vinyl acetate ($CH_2{=}CHOCOCH_3$). Show at least four repeat units.

▲ A giant bubble of tough, transparent plastic film emerges from a die of an extruding machine. The film is used in packaging, consumer products, and food services.

In everyday life, many polymers are called plastics. In chemistry, a *plastic* material is one that can be made to flow under heat and pressure. The material can then be shaped in a mold or in other ways. Plastic products often are made from powders. In *compression molding*, heat and pressure are applied directly to the polymer powder in the mold cavity. In *transfer molding*, the powder is softened by heating outside the mold and then poured into molds to harden.

There also are several methods of molding molten polymers. In *injection molding*, the plastic is melted in a heating chamber and then forced by a plunger into cold molds to set. In another method, *extrusion molding*, the melted polymer is extruded through a die in continuous form to be cut into lengths or coiled. Bottles and similar hollow objects often are *blow-molded*; a "bubble" of molten polymer is blown up like a balloon inside a hollow mold.

Table 10.1 lists some of the more important addition polymers, along with a few of their uses.

TABLE 10.1   Some Addition Polymers

Monomer	Polymer	Polymer Name	Some Uses
$CH_2\!=\!CH_2$		Polyethylene	Plastic bags, bottles, toys, electrical insulation
$CH_2\!=\!CH\!-\!CH_3$		Polypropylene	Indoor-outdoor carpeting, bottles, luggage
$CH_2\!=\!CH\!-\!\bigcirc$		Polystyrene	Simulated wood furniture, insulation, cups, toys, packing materials
$CH_2\!=\!CH\!-\!Cl$		Polyvinyl chloride (PVC)	Plastic wrap, simulated leather, plumbing, garden hoses, floor tile
$CH_2\!=\!CCl_2$		Polyvinylidene chloride (Saran)	Food wrap, seatcovers
$CF_2\!=\!CF_2$		Polytetrafluoroethylene (Teflon)	Nonstick coating for cooking utensils, electrical insulation
$CH_2\!=\!CH\!-\!C\!\equiv\!N$		Polyacrylonitrile (Acrilan, Creslan, Dynel)	Yarns, wigs, paints
$CH_2\!=\!CH\!-\!OCOCH_3$		Polyvinyl acetate	Adhesives, textile coatings, chewing gum resin, paints
$CH_2\!=\!C(CH_3)COOCH_3$		Polymethyl methacrylate (Lucite, Plexiglas)	Glass substitute, bowling balls

Conducting Polymers: Polyacetylene

Acetylene (H—C≡C—H) has a triple bond instead of a double bond, but it can still undergo addition polymerization, forming polyacetylene (Figure 10.3). Notice that, unlike polyethylene, which has a carbon chain containing only single bonds, every other bond in polyacetylene is a double bond.

$$\sim CH{=}CH{-}CH{=}CH{-}CH{=}CH{-}CH{=}CH{-}CH{=}CH{-}CH{=}CH\sim$$

The alternating double and single bonds form a *conjugated* system. They make it easy for electrons to travel along the chain, and so this polymer is able to conduct electricity. (Most plastics are electric insulators!) Polyacetylene and similar conjugated polymers can be used as lightweight substitutes for metal. In fact, the plastic even looks like metal, having a silvery luster.

Polyacetylene, the first conducting polymer, was discovered in 1970. Since then, a number of other polymers with electric conductivity have been made.

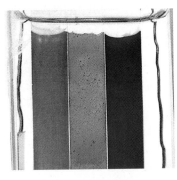

▲ New conducting polymers have special conjugated systems that appear red, green, or blue when electricity is passed through them.

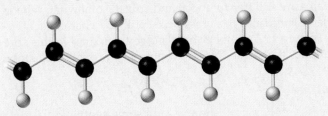

▲ **Figure 10.3** A ball-and-stick model of polyacetylene.

Question: Write an equation for the formation of polyacetylene similar to that we wrote for polyethylene on page 274, in which the repeat polymer unit is placed within brackets.

10.6 RUBBER AND OTHER ELASTOMERS

Although rubber is a natural polymer, it was the basis for much of the development of the synthetic polymer industry. During World War II, Japanese occupation of Malaysia and Indonesia cut off most of the Allies' supply of natural rubber. The search for synthetic substitutes resulted in much more than just a replacement for natural rubber. The plastics industry, to a large extent, developed out of the search for synthetic rubber.

Natural rubber can be broken down into a simple hydrocarbon called isoprene. Isoprene is a volatile liquid, whereas rubber is a semisolid, elastic material. Chemists can make polyisoprene, a substance identical to natural rubber, except that the isoprene comes from petroleum refineries rather than from the cells of rubber trees.

$$n\,CH_2{=}\underset{\underset{CH_3}{|}}{C}{-}CH{=}CH_2 \longrightarrow {+}CH_2{-}\underset{\underset{CH_3}{|}}{C}{=}CH{-}CH_2{+}_n$$

Isoprene   Polyisoprene (rubber)

Chemists have also developed several synthetic rubbers and devised ways to modify these various polymers to change their properties.

Vulcanization: Cross-Linking

The long-chain molecules that make up rubber can be coiled and twisted and intertwined with one another. When rubber is stretched, its coiled molecules are straightened. Natural rubber is soft and tacky when hot. It can be made harder by reaction with sulfur. This process, called **vulcanization**, cross-links the hydrocarbon chains with sulfur atoms (Figure 10.4). Charles Goodyear discovered vulcanization and was issued U.S. Patent 3633 in 1844.

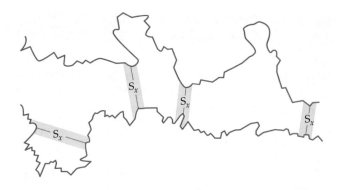

► **Figure 10.4** Vulcanized rubber has hydrocarbon chains (represented here by red lines) cross-linked by sulfur atoms. The subscript x indicates a small, indefinite number, usually not more than 4.

Its three-dimensional cross-linked structure makes vulcanized rubber a harder, stronger substance that is suitable for automobile tires. Surprisingly, cross-linking also improves the elasticity of rubber. With just the right degree of cross-linking, the individual chains are still free to uncoil and stretch somewhat. When stretched vulcanized rubber is released, the cross-links pull the chains back to their original arrangement (Figure 10.5). Rubber bands owe their snap to this sort of molecular structure. Materials that act in this stretchable way are called **elastomers**.

Synthetic Rubber

Natural rubber is a polymer of isoprene, and some synthetic elastomers are closely related. For example, polybutadiene is made from the monomer butadiene ($CH_2\!=\!CH\!-\!CH\!=\!CH_2$), which differs from isoprene only in that it lacks a methyl group on the second carbon atom. Polybutadiene is made rather easily from the monomer.

$$n\ CH_2\!=\!CH\!-\!CH\!=\!CH_2 \longrightarrow \ -\!\!\left[CH_2\!-\!CH\!=\!CH\!-\!CH_2\right]_n\!\!-$$

However, it has only fair tensile strength and poor resistance to gasoline and oils. These properties limit its value for automobile tires, the main use of elastomers.

Another synthetic elastomer, polychloroprene (Neoprene), is made from a monomer similar to isoprene, but with a chlorine in place of the methyl group on isoprene.

$$n\ CH_2\!=\!\underset{\underset{Cl}{|}}{C}\!-\!CH\!=\!CH_2 \longrightarrow \ -\!\!\left[CH_2\!-\!\underset{\underset{Cl}{|}}{C}\!=\!CH\!-\!CH_2\right]_n\!\!-$$

Neoprene is more resistant to oil and gasoline than other elastomers are. It is used to make gasoline hoses and similar items used at automobile service stations.

(a)	(b)
(c)	⊙ = Carbon
	⊙ = Sulfur

▲ **Figure 10.5** Vulcanization of rubber cross-links the molecular chains. (a) In unvulcanized rubber, the chains slip past one another when the rubber is stretched. (b) Vulcanization involves the addition of sulfur cross-linkages between the chains. (c) When vulcanized rubber is stretched, the sulfur cross-linkages prevent the chains from slipping past one another. Vulcanized rubber is stronger than unvulcanized rubber.

Styrene-butadiene rubber (SBR) is a copolymer of styrene (about 25%) and butadiene (about 75%). A segment of an SBR molecule might look something like this.

$$\sim\!CH_2CH\!=\!CHCH_2\!-\!CH_2CH\!-\!CH_2CH\!=\!CHCH_2\!-\!CH_2CH\!=\!CHCH_2\!\sim$$

| Butadiene unit | Styrene unit | Butadiene unit | Butadiene unit |

SBR is more resistant to oxidation and abrasion than natural rubber, but its mechanical properties are less satisfactory.

Like those of natural rubber, SBR molecules contain double bonds and can be cross-linked by vulcanization. SBR accounts for about a third of the total production for the U.S. of elastomers and is used mainly for making tires.

Polymers in Paints

A surprising use for elastomers is in paints and other coatings. The substance in a paint that hardens to form a continuous surface coating, often called the *binder* or *resin*, is a polymer, usually an elastomer. Paint made with elastomers is more resistant to cracking over time. Various kinds of polymers can be used as binders, depending on the specific qualities desired in the paint. Over the last few decades, the popularity of latex paint has soared. Organic solvents are not needed; the latex binder is dispersed in water, so that the brushes and rollers are easily cleaned in soap and water. This is a good example of green chemistry, because the hazardous organic solvents historically used in paints are replaced with water.

10.7 CONDENSATION POLYMERS: SPLITTING OUT WATER

The polymers considered so far are all addition polymers. All the atoms of the monomer molecules are incorporated into the polymer molecules. In a condensation polymer, part of the monomer molecule is not incorporated in the final polymer. During **condensation polymerization**, also called *step-reaction polymerization*, small molecules, such as water, ammonia, or HCl, are split out as by-products.

Nylon and Other Polyamides

As an example, let's consider the formation of nylon. (There are several different nylons, each prepared from a different monomer or set of monomers, but all share certain common structural features.) The monomer in one type of nylon, called nylon 6, is a six-carbon carboxylic acid with an amino group on the sixth carbon atom, 6-aminohexanoic acid ($HOOCCH_2CH_2CH_2CH_2CH_2NH_2$).

In the polymerization reaction, a carboxyl group of one monomer molecule forms an amide bond with the amine group of another.

$$\cdots + HO-\overset{\overset{\displaystyle O}{\|}}{C}CH_2(CH_2)_3CH_2\overset{\overset{\displaystyle H}{|}}{N}-H + HO-\overset{\overset{\displaystyle O}{\|}}{C}CH_2(CH_2)_3CH_2\overset{\overset{\displaystyle H}{|}}{N}-H + \cdots \longrightarrow$$

$$\sim\!\overset{\overset{\displaystyle O}{\|}}{C}CH_2(CH_2)_3CH_2\overset{\overset{\displaystyle H}{|}}{N}-\overset{\overset{\displaystyle O}{\|}}{C}CH_2(CH_2)_3\overset{\overset{\displaystyle H}{|}}{N}\!\sim + n\,H_2O$$

Amide linkage

A polymer related to SBR is poly(styrene-butadiene-styrene), or SBS. Called a *block copolymer*, SBS has molecules made up of three segments: one end has a chain of polystyrene repeat units; the middle, a long chain of polybutadiene repeat units; and the other end another chain of polystyrene repeat units. SBS is a hard rubber used for such things as shoe soles and tire treads where durability is important.

▲ Synthetic polymers serve as binders in paints. Pigments in paint provide color or opacity. Titanium dioxide (TiO_2), a white solid, is the pigment most widely used.

 Synthesis of Nylon 6,10

Water molecules are formed as a by-product. This formation of a nonpolymeric by-product distinguishes condensation polymerization from addition polymerization. Note that the formula of a repeat unit is not the same as that of the monomer.

Because the linkages holding the polymer together are amide bonds, nylon 6 is a **polyamide**. Another nylon is made by the condensation of two different monomers, 1,6-hexanediamine ($H_2NCH_2CH_2CH_2CH_2CH_2CH_2NH_2$) and adipic acid ($HOOCCH_2CH_2CH_2CH_2COOH$). Each monomer has six carbon atoms; the polymer is called nylon 66.

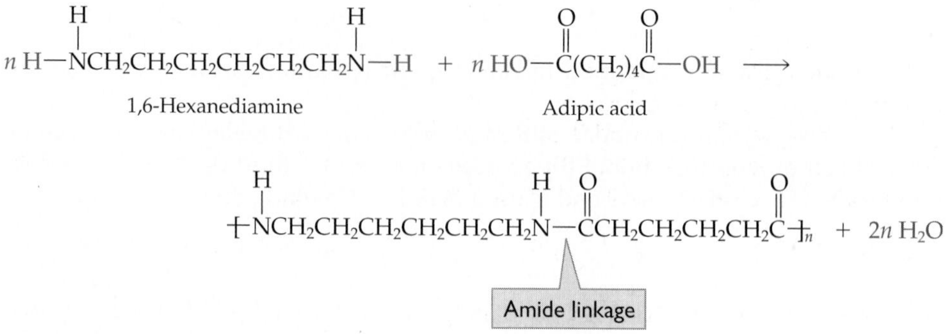

$$n \; H-NCH_2CH_2CH_2CH_2CH_2CH_2N-H \;+\; n \; HO-C(CH_2)_4C-OH \longrightarrow$$

1,6-Hexanediamine   Adipic acid

$$+NCH_2CH_2CH_2CH_2CH_2CH_2N-CCH_2CH_2CH_2CH_2C+_n \;+\; 2n \; H_2O$$

Amide linkage

Polymers are often represented by line-angle formulas (Chapter 9). For example, nylon 66 is represented as

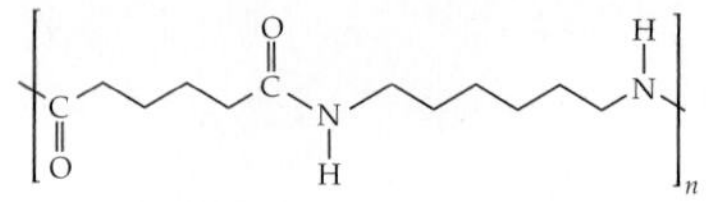

This was the original nylon polymer discovered in 1937 by DuPont chemist Wallace Carothers. Note that one monomer has two amino groups and the other has two carboxyl groups, but the product is still a polyamide, quite similar to nylon 6. Silk and wool, which are protein fibers, are natural polyamides.

Although nylon can be molded into various shapes, most nylon is made into fibers. Some is spun into fine thread to be woven into silklike fabrics, and some is made into yarn that is much like wool. Carpeting, which was once made primarily from wool, is now made largely from nylon.

Polyethylene Terephthalate and Other Polyesters

A **polyester** is a condensation polymer made from molecules with alcohol and carboxylic acid functional groups. The most common polyester is made from ethylene glycol and terephthalic acid. It is called polyethylene terephthalate (PET).

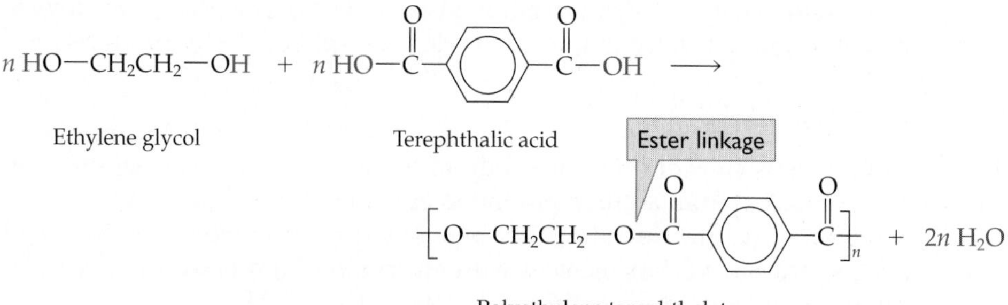

$$n \; HO-CH_2CH_2-OH \;+\; n \; HO-C-\bigcirc-C-OH \longrightarrow$$

Ethylene glycol   Terephthalic acid   Ester linkage

$$+O-CH_2CH_2-O-C-\bigcirc-C+_n \;+\; 2n \; H_2O$$

Polyethylene terephthalate

The hydroxyl groups in ethylene glycol react with the carboxylic acid groups in terephthalic acid to produce long chains held together by many ester linkages.

PET can be molded into bottles for beverages and other liquids. It can also be formed as a film, used for the magnetically coated tape in audio- and videocassettes. Polyester molecules make excellent fibers that are widely used in wash-and-wear clothing.

Phenol–Formaldehyde and Related Resins

Let us now go back to Bakelite, the original synthetic polymer. Bakelite, a phenol–formaldehyde resin, was first synthesized by Leo Baekeland, who received U.S. Patent 942,699 for the process in 1909.

Phenol–formaldehyde resins are formed by splitting out water molecules, the hydrogen atoms coming from the benzene ring and the oxygen atoms from the aldehyde. The reaction proceeds stepwise, with formaldehyde adding first to the 2- or 4-position of the phenol molecule.

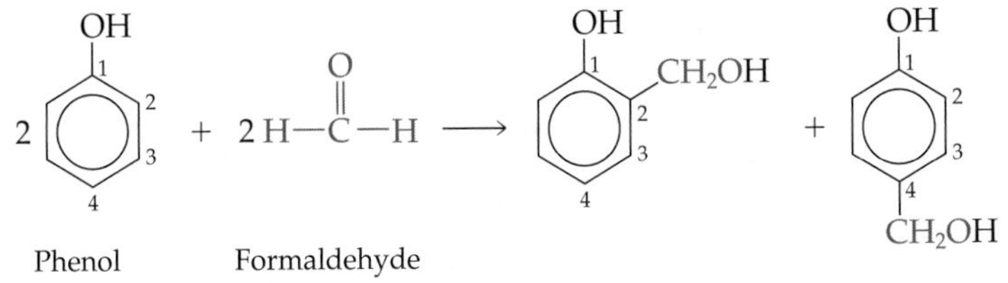

The substituted molecules then react by splitting out water. (Remember that there are hydrogen atoms at all the unsubstituted corners of a benzene ring.) The hookup of molecules continues until an extensive network is achieved.

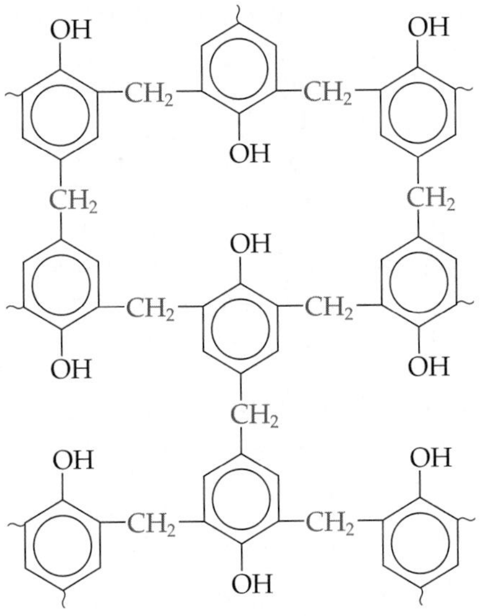

Phenol–formaldehyde resin

Melamine

Water is driven off by heat as the polymer sets. The structure of the polymer is extremely complex, a three-dimensional network somewhat like the framework of a giant building. Note that the phenolic rings are joined together by CH_2 units from the formaldehyde. These polymers are **thermosetting resins**; they cannot be melted and remolded. Instead they decompose when heated to high temperatures.

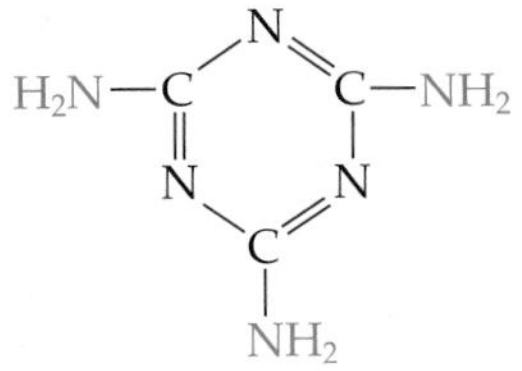

Formaldehyde is also condensed with urea [$H_2N(C{=}O)NH_2$] to make urea–formaldehyde resins and with melamine to form melamine–formaldehyde resins. (Melamine is formed by condensation of three molecules of urea.)

Both resins, like phenol–formaldehyde polymers, are thermosetting. The polymers are complex three-dimensional networks formed by the splitting out of H_2O from formaldehyde ($H_2C{=}O$) molecules and amino ($-NH_2$) groups. Urea–formaldehyde resins are used to bind wood chips together in panels of particle board. Melamine–formaldehyde resins are used in plastic (Melmac) dinnerware and Formica countertops.

Other Condensation Polymers

There are many other kinds of condensation polymers, but we will mention only a few.

Polycarbonates are tough, "clear as glass" polymers strong enough to be used in bulletproof windows. They are also used in protective helmets, safety glasses, and even in dental crowns. One polycarbonate, commonly called Lexan, has the following repeat unit.

$$\sim O-\overset{\displaystyle O}{\overset{\displaystyle \|}{C}}-O-\!\!\!\bigcirc\!\!\!-\overset{\displaystyle CH_3}{\underset{\displaystyle CH_3}{C}}-\!\!\!\bigcirc\!\!\!-\sim$$

Polyurethanes are similar to nylon in structure. The repeat unit in a common polyurethane is

$$\sim\overset{\displaystyle O}{\overset{\displaystyle \|}{C}}-NH-CH_2CH_2CH_2CH_2CH_2CH_2-NH-\overset{\displaystyle O}{\overset{\displaystyle \|}{C}}-O-CH_2CH_2CH_2CH_2-O\sim$$

Polyurethanes may be elastomers or tough and rigid, depending on the monomers used. They are popular in foamed padding ("foam rubber") in cushions, mattresses, and padded furniture. They are also used for skate wheels, in running shoes, and in protective gear for sports activities.

Epoxy resins make excellent surface coatings, and they are powerful adhesives. The following is a typical repeat unit.

$$\sim O-\!\!\!\bigcirc\!\!\!-\overset{\displaystyle CH_3}{\underset{\displaystyle CH_3}{C}}-\!\!\!\bigcirc\!\!\!-O-CH_2\underset{\displaystyle OH}{CH}CH_2\sim$$

Epoxy adhesives usually have two components that are mixed just before they are used. The polymer chains become cross-linked, and the bonding is extremely strong.

Composite Materials

Composite materials are made up of high-strength fibers (of glass, graphite, synthetic polymers, or ceramics) held together by a polymeric matrix, usually a thermosetting condensation polymer. The fiber reinforcement provides the support, and the surrounding plastic protects the fibers from breaking.

The most commonly used composite materials thus far have been polyester resins reinforced with glass fibers. They are widely used in boat hulls, molded chairs, automobile panels, and in sports gear, such as tennis rackets. Some composite materials have the strength and rigidity of steel at a fraction of the weight.

▶ Various sports cars, including the Dodge Viper ACR (American Club Racer) shown here, have plastic composite bodies.

Silicones

Not all polymers are based on chains of carbon atoms. A silicone (polysiloxane) is a good example of a different type of polymer. A **silicone** is a polymer based on a series of alternating silicon and oxygen atoms.

$$\begin{array}{ccccccccccccc}
& R & & R & & R & & R & & R & & R & & R \\
& | & & | & & | & & | & & | & & | & & | \\
\sim Si & - O - & Si & - O - & Si & - O - & Si & - O - & Si & - O - & Si & - O - & Si \sim \\
& | & & | & & | & & | & & | & & | & & | \\
& R & & R & & R & & R & & R & & R & & R
\end{array}$$

(In simple silicones, R represents a hydrocarbon group, such as methyl, ethyl, or butyl.)

Silicones can be linear, cyclic, or cross-linked networks. They are heat-stable and resistant to most chemicals, and they are excellent waterproofing materials. Depending on chain length and amount of cross-linking, silicones can be oils or greases, rubbery compounds, or solid resins. Silicone oils are used as hydraulic fluids and lubricants, whereas other silicones are used in making such products as sealants, auto polish, shoe polish, and waterproof sheeting. Fabrics for raincoats and umbrellas are frequently treated with silicone.

An interesting silicone toy is Silly Putty. It can be molded like clay or rolled up and bounced like a ball. On standing, it flows like a liquid.

Perhaps the most remarkable silicones of all are the ones used for synthetic human body parts. Kinds of silicone replacements range from finger joints to eye sockets. Artificial ears and noses are also made from silicone polymers. They can even be specially colored to match the surrounding skin.

For many years, silicone gel was used for breast implants. In most cases, these implants have remained perfectly stable over the years. However, because of a manufacturing error, some batches of gel were not sufficiently cured. As a result, the gel implants later disintegrated, causing leakage into body tissues and leading to health problems in some cases. Silicone implants have become a topic of much controversy and litigation. Today only saline-filled implants are approved for use in the United States; they have a silicone–elastomer shell.

Silicon is in the same group (4A) as carbon and, like carbon, is tetravalent and able to form chains. However, carbon can form chains of just carbon atoms (as in polyethylene), whereas silicone polymers have chains of alternating silicon and oxygen atoms.

▲ Cookware made of silicone is colorful, flexible, and non-stick.

EXAMPLE 10.3   **Condensed Structural Formulas for Polymers**

Write the condensed structural formula for the polymer formed from dimethylsilanol, $(CH_3)_2Si_2(OH)_2$.

Solution

Let's start by writing out the structural formula of the monomer.

$$\begin{array}{c}
CH_3 \\
| \\
HO - Si - OH \\
| \\
CH_3
\end{array}$$

Because there are no double bonds in this molecule, we do not expect addition polymerization. Rather, we expect a condensation reaction in which an OH group of one molecule and an H atom of another combine to form a molecule of water. Moreover, because there are two OH groups per molecule, each monomer can form bonds with the neighbors on both sides. This is a key requirement for polymerization. We can represent the reaction as follows.

$$\begin{array}{c}
CH_3 \\
| \\
HO - Si - OH \\
| \\
CH_3
\end{array}
+
\begin{array}{c}
CH_3 \\
| \\
HO - Si - OH \\
| \\
CH_3
\end{array}
+
\begin{array}{c}
CH_3 \\
| \\
HO - Si - OH \\
| \\
CH_3
\end{array}
+ \cdots
\longrightarrow
\left[
\begin{array}{c}
CH_3 \\
| \\
Si - O \\
| \\
CH_3
\end{array}
\right]_n
+ \, n\,H_2O$$

> ▶ **Exercise 10.3**
>
> **(a)** Write the structural formula for the polymer formed from 3-hydroxy-propanoic acid ($HOCH_2CH_2COOH$), showing at least four repeat units.
> **(b)** Write a condensed structural formula in which the repeat unit is shown in brackets.

10.8 PROPERTIES OF POLYMERS

Polymers differ from substances with small molecules in three main ways. First, the long chains can be entangled with one another; the polymer molecules form a tangled mass much like a dish of spaghetti. Especially at low temperatures, it is difficult to untangle the polymer molecules. This lends strength to polymers in plastics, elastomers, and other materials.

Second, although intermolecular forces affect polymers just as they do small molecules, these forces are greatly multiplied in large molecules. The larger the molecules are, the greater the intermolecular forces between them. Even when ordinarily weak dispersion forces are the only operative intermolecular forces, they can strongly bind polymer chains together. This too makes for strong polymeric materials. For example, polyethylene is nonpolar with only dispersion forces between molecules, but (as we indicated in the marginal note on page 276) ultra-high-molecular-weight polyethylene forms fibers so strong they can be used in bullet-proof vests.

Third, large polymer molecules move more slowly than small molecules do. A group of small molecules (monomers) can move around faster and more randomly when independent than they can when joined together in a long chain (polymer). The slower speed of molecules makes a polymeric material different from one made of small molecules. For example, a polymer dissolved in a solvent will form a solution that is a lot more viscous than the pure solvent.

Crystalline and Amorphous Polymers

Some polymers are highly crystalline, and their molecules line up neatly to form long fibers of great strength. Other polymers are largely amorphous, composed of randomly oriented molecules that get tangled up with one another (Figure 10.6). Crystalline polymers tend to make good synthetic fibers, whereas amorphous polymers make good elastomers.

Sometimes the same polymer is crystalline in one region and amorphous in another. For example, scientists have designed spandex fibers (used for stretch fabrics [Lycra] in ski pants, exercise clothing, and swimsuits) so as to combine the tensile strength of crystalline fibers with the elasticity of amorphous rubber. Two molecular structures are combined in one polymer chain, with blocks of crystalline character alternating with amorphous blocks. The amorphous block is soft and rubbery; the crystalline part is quite rigid. The resulting polymer exhibits both sets of properties—flexibility and rigidity.

The Glass Transition Temperature

An important parameter of most thermoplastic polymers is the **glass transition temperature (T_g)**. Above this temperature, the polymer is rubbery and tough;

We can make use of the T_g concept in everyday life. For example, we can remove chewing gum from clothing by applying a piece of ice to lower the temperature of the gum below the T_g of the polyvinyl acetate resin that makes up the bulk of the gum. The cold, brittle resin then crumbles readily and can be removed.

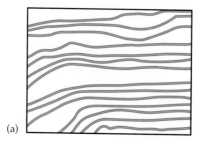
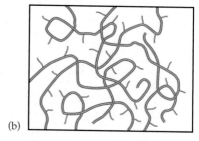

(a)   (b)

▶ **Figure 10.6** Organization of polymer molecules. (a) Crystalline arrangement. (b) Amorphous arrangement.

below it, the polymer is like glass: hard, stiff, and brittle. Each polymer has a characteristic T_g. We want automobile tires to be tough and elastic, and so we use materials with low T_g values. On the other hand, we want plastic substitutes for glass to be glassy; thus, they have T_g values well above room temperature.

Fiber-Forming Properties

Not all synthetic polymers can be converted to useful fibers, but those that can often have properties superior to those of natural fibers. This is why the majority of the fibers and fabrics used today in the United States are synthetic.

Silk fabrics are beautiful and have a luxurious "feel," but nylon fabrics are also beautiful and feel much like silk. Moreover, nylon fabrics wear longer, are easier to care for, and are less expensive than silk.

Polyesters such as PET can substitute for either cotton or silk, but they outperform the natural fibers in many ways. Polyesters are not subject to mildew as cotton is, and many polyester fabrics do not need ironing. Fabric made by blending polyester fibers with 35–50% cotton combines the comfort of cotton with the no-iron easy care of polyester fabrics.

Acrylic fibers, made from polyacrylonitrile, can be spun into yarns that look like wool. Acrylic sweaters have the beauty and warmth of wool, but they do not shrink in hot water, are not attacked by moths, and do not cause the allergic skin reactions caused by wool in some people.

▲ Formation of fibers by extrusion through a spinneret. A melted polymer is forced through the tiny holes to make fibers that solidify as they cool.

10.9 DISPOSAL OF PLASTICS

An advantage of plastics is that they are durable and resistant to many things in the environment. Perhaps some of them are too good in this respect; they last almost forever. Once they are dumped, they do not go away. You see them littering our parks, our sidewalks, and our highways; and if you should go out to the middle of the ocean, you would see them there, too. Small fish have been found dead with their digestive tracts clogged by bits of plastic foam ingested with their food.

Landfills

Plastics make up about 11% by mass of solid waste in the United States, but by volume they make more than 20%. This has created a problem because more than half of all solid waste goes into landfills, and it is increasingly difficult to find suitable landfill space.

Solid wastes in general are discussed in Chapter 11.

Incineration

Another way to dispose of discarded plastics is to burn them. Most plastics have a high fuel value. For example, a pound of polyethylene has about the same energy content as a pound of fuel oil. Some communities actually generate electricity with the heat from garbage incinerators. Some utility companies burn powdered coal mixed with a few percent of ground-up rubber tires. They not only obtain extra energy from the tires but also help solve the problem of tire disposal.

On the other hand, the burning of plastics and rubber can lead to some new problems. For example, PVC produces toxic hydrogen chloride gas when it burns, and burning automobile tires give off soot and a stinking smoke. Incinerators are corroded by acidic fumes and clogged by materials that are not readily burned.

Degradable Plastics

About half of our waste plastic is from packaging. One approach to the plastics disposal problem is to make plastic packages that are biodegradable or photodegradable (broken down in the presence of bacteria or light). Of course, it is important that the package remain intact and not start to decompose while it is still being used. For now, many people seem reluctant to pay extra for garbage bags that are designed to fall apart. However, recently there are more and more examples of

▲ Biodegradable plastic products such as those made from polylactic acid (PLA) are now widely available. Estimates see demand for degradable plastics experiencing double-digit growth annually through 2008 as prices and properties become more competitive with conventional polymers.

degradable, more environmentally friendly plastics for sale, including plastic tableware, cups, bottles, and delicatessen food containers.

Recycling

Recycling is perhaps the best way to handle waste plastics. The plastics must be collected, sorted, chopped, melted, and then remolded. Collection works well when there is strong community cooperation.

The separation step is simplified by code numbers (see the *Green Chemistry MediaLab* for this chapter) stamped on plastic containers. Once the plastics have been separated, they can be chopped into flakes, melted, and remolded or spun into fiber.

At present, the only plastic items being recycled on a large scale are those bearing Codes 1 (PET) and 2 (HDPE). For example, PET bottles are made into fiber, mainly for carpets, and HDPE containers are remolded into detergent bottles. Recycling is good news for green chemistry, because recycling keeps plastics out of landfills and prevents the use of petroleum-derived monomers or other hazardous chemicals.

▲ Recycled plastic can be used for many things. This dollhouse is built from "lumber" made from recycled milk jugs.

10.10 PLASTICS AND FIRE HAZARDS

The accidental ignition of fabrics, synthetic or otherwise, has caused untold human misery. The U.S. Department of Health and Human Services estimates that fires involving flammable fabrics kill several thousand people annually and injure as many as 150,000–200,000.

Research has led to a variety of flame-retardant fabrics. Many incorporate chlorine and bromine atoms within the polymeric fiber. Federal regulations require that children's sleepwear, in particular, be made of such flame-retardant materials.

Another synthetic fabric, meta-aramid, or Nomex (from DuPont), has such heat resistance that it is used for protective clothing for firefighters and race car drivers. The fibers don't ignite or melt when exposed to flames or high heat. Nomex is also used in electric insulation and for machine parts exposed to high heat.

Burning plastics often produce toxic gases. Hydrogen cyanide is formed in large quantities when polyacrylonitrile and other nitrogen-containing polymers burn. Lethal amounts of cyanide found in the bodies of plane crash victims have been traced to burned plastics. Firefighters often refuse to enter burning buildings without gas masks for fear of being overcome by fumes from burning plastics. Smoldering fires also produce lethal quantities of carbon monoxide. To make plastics safer when burned, green chemists are making new kinds of polymers that don't burn or that don't generate toxic chemicals when burned.

10.11 PLASTICIZERS AND POLLUTION

Chemicals used in plastics manufacture can also present problems. Plasticizers are an important example. Some plastics, particularly vinyl polymers, are hard and brittle and thus difficult to process. A **plasticizer** can make them more flexible and less brittle by lowering the glass transition temperature, T_g. Unplasticized PVC is rigid and is used for water pipes. Thin sheets of pure PVC crack and break easily, but plasticizers make them soft and pliable. Plastic raincoats, garden hoses, and seat covers for automobiles can be made from plasticized PVC. Plasticizers are liquids of low volatility and are generally lost by diffusion and evaporation as a plastic article ages. The plastic becomes brittle and then cracks and breaks.

Once used widely as plasticizers, but now banned, polychlorinated biphenyls (PCBs) are derived from a biphenyl ($C_{12}H_{10}$), a hydrocarbon that has two benzene rings joined at a corner. In PCBs, some of the hydrogen atoms of biphenyl are replaced with chlorine atoms (Figure 10.7). Note that PCBs are structurally similar to the insecticide DDT.

PCBs were also widely used as insulating materials in electric equipment (transformers, condensers, and other apparatus) because of their high electric re-

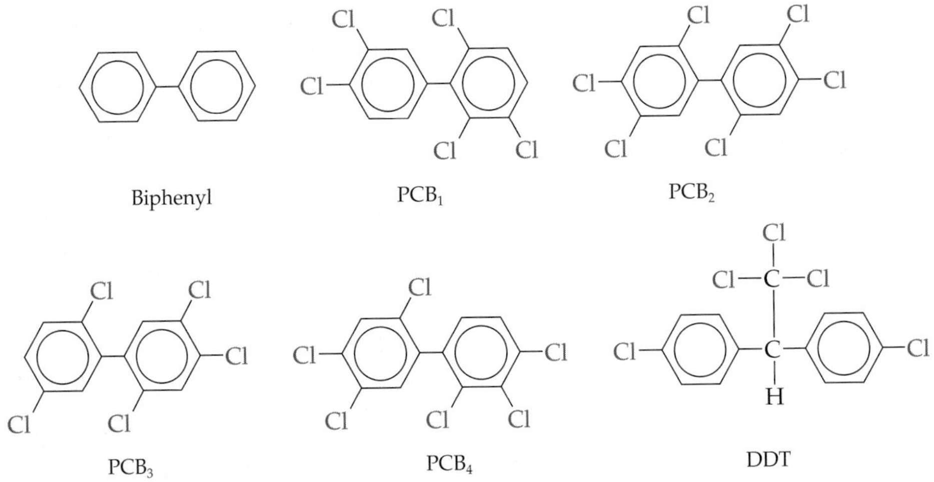

◀ Figure 10.7 Biphenyl and some of the PCBs derived from it. These are but a few of the hundreds of possible PCBs. DDT is shown for comparison.

sistance. The same properties that made PCBs so desirable as industrial chemicals cause them to be an environmental hazard. They degrade slowly in nature, and their solubility in nonpolar media—animal fat as well as vinyl plastics—leads to their concentration in the food chain. PCB residues have been found in fish, birds, water, and sediments. The physiologic effect of PCBs is similar to that of DDT. Monsanto Corporation, the only company in the United States that produced PCBs, discontinued production in 1977, but they still remain in the environment.

Today the most widely used plasticizers for vinyl plastics are phthalate esters, a group of diesters derived from phthalic acid (1,2-benzenedicarboxylic acid). Phthalate plasticizers (Figure 10.8) have low acute toxicity, and although some studies have suggested possible harm to young children, the FDA has recognized these plasticizers as being generally safe. They pose little threat to the environment because they degrade fairly rapidly. Another approach that green chemists adopt is to make plastics that have the right amount of flexibility and do not require any plasticizers to be used.

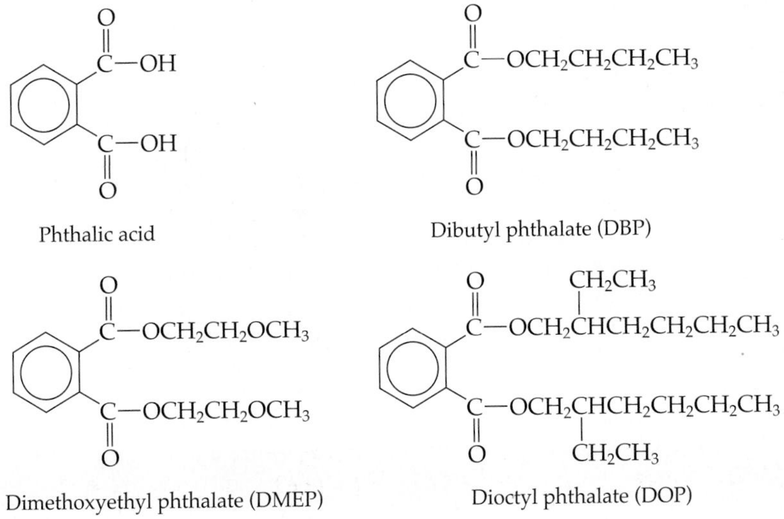

◀ Figure 10.8 Phthalic acid and some esters derived from it. Dioctyl phthalate is also called di-2-ethylhexyl phthalate.

10.12 PLASTICS AND THE FUTURE

Widely used today, synthetic polymers are the materials of the future. There will be new kinds of polymers and even wider use in the future. We already have polymers that conduct electricity, amazing new adhesives, and synthetic materials that are stronger than steel but much lighter in weight. Plastics present problems, but they have become such an important part of our daily lives that we would find it difficult to live without them.

In medicine, body replacement parts made from polymers have become common. Worldwide in 2003 there were about 800,000 hip replacements and 700,000 knee replacements, about half of each in the United States. In the future we will have artificial bones that can stimulate new bone growth so as to knit the new pieces more securely in place. We can also expect to have artificial lungs as well as artificial hearts.

Home construction today uses PVC water pipes, siding, and window frames, plastic foam insulation, and polymeric surface coatings. Tomorrow's homes may also contain lumber and wall panels of artificial wood, perhaps made from recycled plastics.

Synthetic polymers are used extensively in airplane interiors, and the bodies and wings of some planes are made of lightweight composite materials. Many automobiles now have bodies made from plastic composites. Electrically conducting polymers will aid in making lightweight batteries for electric automobiles. The electronics industry will use increasing amounts of electrically conducting thermoplastics in their miniaturized circuits.

But here is something to think about. Most synthetic polymers are made from petroleum or natural gas. Both these natural resources are nonrenewable, and our supplies are limited. We are likely to run out of petroleum during this century. You might suppose that we would be actively conserving this valuable resource, but unfortunately, this is not the case. We are taking petroleum out of the ground at a rapid rate, converting most of it to gasoline and other fuels, and then simply burning it. There are other sources of energy, but is there anything that can replace petroleum as the raw material for making plastics? Yes! Today, several new types of plastics, such as polylactic acid and polyhydroxybutyrates, are made from renewable resources such as corn, soybeans, and sugarcane. Polymers are an active area of green chemistry research.

Petroleum

That viscous, tarry liquid nature laid beneath the ground
A hundred million years before man came upon the scene
Can be transformed to marvelous new products we have found—
Like nylon, orlon, polyesters, polyethylene,
Synthetic rubber, plastics, films, adhesives, drugs, and dyes,
And other things, some that we now can only dream about.
Who knows what wondrous products man might some day synthesize
From oil! Except, alas, that our supplies are running out.
The time is near when Earth's prodigious flow of oil may stop
(We've taken so much from the ground, with no way to return it.)
Meanwhile, we strive to find and draw out every precious drop,
And then … incredibly … we take the bulk of it and burn it.

Critical Thinking Exercises

Apply knowledge that you have gained in this chapter and one or more of the FLaReS principles (Chapter 1) to evaluate the following statements or claims.

10.1 An environmental activist states that all synthetic plastics should be banned and replaced by natural materials.

10.2 A news report states that incinerators that burn plastics are a major source of chlorine-containing toxic compounds called dioxins (Chapter 16).

10.3 We know that vinyl chloride is a carcinogen. Parents fear that their children will get cancer from playing with toys made of PVC.

SUMMARY

Sections 10.1–10.2—A **polymer** is a giant molecule made from smaller molecules. The small "building block" molecules that make a polymer are called **monomers**. There are many natural polymers, including starch, cellulose, nucleic acids, and proteins.

Sections 10.3–10.4—**Celluloid** was the first semisynthetic polymer, made from natural cellulose modified with nitric acid. The simplest, least expensive, and highest-volume synthetic polymer is polyethylene, made from the monomer ethylene. High-density polyethylene molecules can pack closely together for a rigid, strong structure. Low-density polyethylene molecules have many side chains and the product is more flexible. Linear low-density polyethylene is a **copolymer**, formed from two (or more) different monomers.

Polyethylene is one of many **thermoplastic polymers** that can be softened and reshaped with heat and pressure. **Thermosetting resins** are polymers that decompose rather than softening when heated, and cannot be reshaped.

Section 10.5—In **addition polymerization** the monomer molecules add to one another; the polymer contains all atoms of the monomer(s). Addition polymers, such as polyethylene, polypropylene, polystyrene, and PVC, are made from monomers containing double bonds. Polypropylene is tough and resists moisture, oils, and solvents. Polystyrene is used primarily to make Styrofoam. PVC or polyvinyl chloride is used for plumbing, artificial leather, and flexible tubing. PTFE or Teflon has great nonstick properties and is chemically inert. Thermoplastic polymers can be molded in many different ways.

Section 10.6—Natural rubber is an **elastomer**, a polymer that can be stretched yet returns to its shape. The monomer of natural rubber, isoprene, can be made artificially and polymerized. Polyisoprene is soft and tacky when hot. In a process called **vulcanization**, polyisoprene is reacted with sulfur. The sulfur cross-links the hydrocarbon chains together, making the rubber harder, stronger, and more elastic.

Other elastomers are similar in structure to polyisoprene, including polybutadiene, polychloroprene, and styrene-butadiene rubber (SBR). All of these have molecules that can coil and uncoil.

Section 10.7—In **condensation polymerization**, parts of the monomer molecules are not incorporated in the product. Small molecules such as water are split out as by-products. Nylon is a condensation polymer that is a **polyamide**, with amide [—(CO)NH—] linkages joining the monomers. A **polyester** is a condensation polymer made from monomers with alcohol and carboxylic acid functional groups. Polyethylene terephthalate is the most common polyester. Bakelite, the first fully synthetic polymer, condenses from phenol and formaldehyde. Other similar resins can be prepared using urea or melamine instead of phenol. Other types of condensation polymers include polycarbonates (bullet-proof windows), polyurethanes (tough rubber), epoxies (glues), and **silicones**, which have silicon and oxygen atoms instead of carbon atoms. Combining a polymer with high-strength fibers gives a composite material that exploits the best properties of both materials.

Section 10.8—Properties of polymers are very different from those of their monomers. Polymer strength arises partly because the molecules entangle one another and partly because the large molecules have relatively strong dispersion forces. Polymers may be crystalline or amorphous. Above the **glass transition temperature** (T_g) a polymer is rubbery and tough; below it, the polymer is brittle. Strong fibers are made from crystalline polymers that have molecules neatly aligned with one another. Polymer fibers often have advantages over their natural counterparts.

Section 10.9—The longevity of plastics can be a problem. They make up a large fraction of the content of landfills. Proper incineration is one way of disposing of discarded plastics. Some polymers are engineered to be degradable. Recycling of plastics requires that they be collected and sorted according to the code numbers. Recycled plastics are seeing increasing use.

Sections 10.10–10.12—Polymers used for clothing often are made of flame-retardant materials that incorporate chlorine or bromine atoms. The chemicals used in manufacture can present problems. A **plasticizer** makes a polymer more flexible. Polychlorinated biphenyls (PCBs) were once used as plasticizers but are now banned. Phthalates are the most common plasticizers today.

Synthetic polymers are the materials of the future. They will be used for body replacement parts, construction, automobiles, batteries, and electronics. However, most synthetic polymers come from petroleum, which is a limited and nonrenewable natural resource. This is another important reason for recycling plastics.

REVIEW QUESTIONS

1. Define the following terms.
   a. macromolecule   b. monomer
   c. polymer   d. elastomer
   e. copolymer   f. plasticizer

2. What is celluloid? How is it made? Why is it no longer used to make movie film?

3. How does the structure of PVC differ from that of polyethylene? List several uses of PVC.

4. What is addition polymerization? What structural feature usually characterizes molecules used as monomers in addition polymerization?

5. What is Teflon? What unique property does it have? What are some of its uses?

6. What polymer is used to make clear, brittle, disposable drinking cups? From what monomer are disposable foamed plastic coffee cups made?

7. What plastic is used to make
   a. gallon milk jugs?
   b. 2-L soda pop bottles?

8. What type of polymers is used in paints? What is the function of the polymers?

9. What is condensation polymerization? Give some examples.

10. What is Bakelite? From what monomers is it made?

11. What is a thermosetting polymer? Give an example.

12. What type of fibers, natural or synthetic, is used most in the United States? Why?

13. What are silicones? Give an example.

14. What are PCBs? Why are they no longer used as plasticizers?

15. What problems arise when plastics are
    a. discarded into the environment?
    b. disposed of in landfills?
    c. disposed of by incineration?

16. What steps must be taken in order to recycle plastics?

17. Discuss plastics as fire hazards.

18. What elements are incorporated into polymers to make flame-retardant fabrics?

PROBLEMS

Polyethylene

19. Describe the structure of high-density polyethylene (HDPE). How does this structure explain the properties of this polymer?

20. Describe the structure of low-density polyethylene (LDPE). How does this structure explain the properties of this polymer?

21. Describe the structure of linear low-density polyethylene (LLDPE). How does its structure differ from that of HDPE? from LDPE?

22. What is a thermoplastic polymer? Give an example.

Addition Polymerization

23. Give the structure of the monomer from which each of the following polymers is made.
    a. polyvinyl chloride
    b. polystyrene

24. Give the structure of the monomer from which each of the following polymers is made.
    a. PTFE (Teflon)
    b. polyacrylonitrile

25. Give the structure of a segment of polymer chain that is at least eight carbon atoms long for each of the following.
    a. polyethylene
    b. polypropylene

26. Give the structure of a segment of polymer chain that is at least four monomer units long for polymers formed from the following monomers.
    a. vinylidene fluoride ($CH_2{=}CF_2$)
    b. methyl methacrylate

$$CH_2{=}\overset{\overset{\textstyle CH_3}{\textstyle |}}{C}COOCH_3$$

27. Give the structure of the polymer made from each of the following. Show at least four repeat units.
    a. acrylonitrile ($CH_2{=}CH{-}C{\equiv}N$)
    b. vinyl acetate ($CH_2{=}CH{-}O{-}CO{-}CH_3$)
    c. methyl acrylate ($CH_2{=}CHCOOCH_3$)

28. Give the structure of the polymer made from each of the following. Show at least four repeat units.
    a. trifluoroethylene ($CF_2{=}CHF$)
    b. 1-pentene ($CH_2{=}CHCH_2CH_2CH_3$)
    c. methyl cyanoacrylate

$$CH_2{=}\overset{\overset{\textstyle C{\equiv}N}{\textstyle |}}{C}COOCH_3$$

Rubber and Other Elastomers

29. Give the structure of isoprene, the monomer of natural rubber.

30. Give the structure of the monomer from which polybutadiene is made.

31. Describe the process of vulcanization. How does vulcanization change the properties of rubber?

32. Explain the elasticity of rubber.

33. How does polychloroprene (Neoprene) differ from natural rubber in properties? In structure?

34. How does polybutadiene differ from natural rubber in properties? In structure?

35. Name three synthetic elastomers and the monomer(s) from which they are made.

36. How is SBR made? From what monomer(s) is it made?

Condensation Polymerization

37. Nylon 46 is made from the monomers $H_2NCH_2CH_2CH_2CH_2NH_2$ and $HOOCCH_2CH_2CH_2CH_2COOH$.

    Give the structure of nylon 46 showing at least two units from each monomer.

38. Give the structure of a polymer made from glycolic acid (hydroxyacetic acid, $HOCH_2COOH$). Show at least four repeating units. (*Hint:* Compare with nylon 6 in Section 10.7.)

39. Kodel is a polyester fiber. The monomers are terephthalic acid (page 284) and 1,4-cyclohexanedimethanol (in the following figure). Write a condensed (bracketed) repeating structure of the Kodel polyester molecule.

    $HOCH_2$—⟨ ⟩—CH_2OH

    1,4-Cyclohexanedimethanol

40. Kevlar, a polyamide used to make bullet-proof vests, is made from terephthalic acid (page 284) and *para*-phenylenediamine (1,4-benzenediamine). Write a condensed (bracketed) repeating structure of the Kevlar molecule.

    H_2N—⟨○⟩—NH_2

    para-Phenylenediamine

Properties of Polymers

41. What three factors give polymers properties different from materials made up of small molecules?

42. What does the word plastic mean
    a. in everyday life and
    b. in chemistry?

43. What is the glass transition temperature (T_g) of a polymer? For what uses do we want polymers with a low T_g? With a high T_g?

44. How do plasticizers make polymers less brittle?

ADDITIONAL PROBLEMS

45. In the following equation, identify the parts labeled a, b, and c as monomer, polymer, and repeat unit. What type of polymerization (addition or condensation) is represented?

    $$\overbrace{n\ CH_2{=}CHF}^{a} \longrightarrow$$

    $$\underbrace{{\sim}CH_2{-}CHF{-}\overbrace{CH_2{-}CHF}^{b}{-}CH_2{-}CHF{-}CH_2{-}CHF{\sim}}_{c}$$

46. Is nylon 46 (Problem 37) an addition polymer or a condensation polymer? Explain.

47. From what monomers could the following copolymer, called poly(styrene-co-acrylonitrile) (SAN), be made?

    $${\sim}CH_2CH{-}CH_2CH{-}CH_2CH{\sim}$$
    $$\quad\ \ C{\equiv}N \qquad ○ \qquad C{\equiv}N$$

48. One type of Saran has the structure shown. Give the structures of the two monomers from which it is made.

    $${\sim}CH_2CCl_2{-}CH_2CHCl{-}CH_2CCl_2{-}CH_2CHCl{\sim}$$

49. Cyanoacrylates, such as those made from methyl cyanoacrylate (Problem 28c) are used in instant-setting "super" glues. However, poly(methyl cyanoacrylate) can irritate tissues. Cyanoacrylates with longer alkyl ester groups are less harsh and can be used in surgery in place of sutures. A good example is poly(octyl cyanoacrylate). Write a polymerization reaction for the formation of poly(octyl cyanoacrylate) from octyl cyanoacrylate, using a condensed (bracketed) repeat structure for the polymer.

50. Polyethylene terephthalate has a high T_g. How can the T_g be lowered so that a manufacturer can permanently crease a pair of polyethylene terephthalate slacks?

51. Isobutylene [$CH_2{=}C(CH_3)_2$] polymerizes to form polyisobutylene, a sticky polymer used as an adhesive. Give the structure of polyisobutylene. Show at least four repeat units.

52. Copolymerized with isoprene, isobutylene (Problem 51) forms butyl rubber. Give the structure of butyl rubber. Show at least three isobutylene repeat units and one isoprene repeat unit.

53. The bacteria *Alcalgenes eutrophus* produce a polymer called polyhydroxybutyrate with the structure shown. Give the structure of the hydroxy acid from which this polymer could be made.

    $$\begin{array}{cc} CH_3 & O \\ | & \parallel \\ {+}O{-}CH{-}CH_2{-}C{+}_n \end{array}$$

54. The highly heat-resistant polyamide Nomex is a polymer of *meta*-phenylenediamine (1,3-benzenediamine) and *meta*-phthalic acid (1,3-benzenedicarboxylic acid). Write a condensed (bracketed) repeat structure of the Nomex polyester molecule.

    ⟨○⟩–NH_2 ⟨○⟩–NH_2
    NH_2 NH_2

    meta-phenylenediamine   *meta*-phenylenediamine

55. Based on the condensed repeating structural formula of poly(ethylene naphthalate) (PEN) given,
    a. give the structures of the monomer(s) and
    b. state whether the polymer is a polyester or a polyamide.

    $$\left[O{-}\overset{O}{\overset{\parallel}{C}}{-}⟨○○⟩{-}\overset{O}{\overset{\parallel}{C}}{-}O{-}CH_2{-}CH_2 \right]_n$$

56. A student writes the formula for polypropylene as shown. Describe the error(s) in the formula.

$$\left[CH_2 - CH - CH_3 \right]_n$$

57. What kind of polymer is involved in the following phenomena? A rubber ball dropped from 100 cm will bounce back up to about 60 cm. Balls made from polybutadiene, called Superballs, will bounce back up to about 85 cm. Golf balls, made from cross-linked polybutadiene, bounce back up to 89 cm. Which kind of ball exhibits the phenomenon to the greatest degree?

58. Draw a structure of the likely product(s) if ethanol is substituted for ethylene glycol in the reaction with terephthalic acid (page 284).

59. Shell Chemical developed Carilon® polymers, which are highly versatile and have been used to make machine parts, fibers, laboratory equipment, specialty medical supplies, electrical connectors, hoses, and bearings. What functional group is present in the polymer chain? Carilon polymers are copolymers of carbon monoxide and an alkene. What alkene serves as a monomer in the following molecule?

$$\sim CH_2 - CH_2 - \overset{\displaystyle O}{\overset{\|}{C}} - CH_2 - CH_2 - \overset{\displaystyle O}{\overset{\|}{C}} - CH_2 - CH_2 - \overset{\displaystyle O}{\overset{\|}{C}} \sim$$

60. On page 283 we showed nylon 6 being formed from 6-aminohexanoic acid by condensation polymerization. In practice, however, nylon 6 is usually made from a cyclic compound called ε-caprolactam (see equation). Is this an addition polymerization or a condensation polymerization? Can you see why some chemists prefer the labels *chain-reaction polymerization* for those processes involving monomers with double bonds and *step-reaction polymerization* for those that do not?

$$\underset{\text{ε-caprolactam}}{\text{structure}} \longrightarrow \left[CH_2 - CH_2 - CH_2 - CH_2 - CH_2 - \overset{\displaystyle O}{\overset{\|}{C}} - NH \right]_n$$

nylon 6

COLLABORATIVE GROUP PROJECTS

Prepare a PowerPoint, poster, or other presentation (as directed by your instructor) for presentation to the class.

61. Prepare a brief report on possible sources of plastics when Earth's supplies of coal and petroleum become too expensive.

62. Prepare a brief biographical report on one of the following and share it with your group.
    a. Charles Goodyear
    b. Wallace Carothers
    c. Stephanie Kwolek
    d. Roy Plunkett
    e. Leo Baekeland

63. Make a list of plastic objects (or parts of objects) that you encounter in your daily life. Try to identify a few of the kinds of polymers used in making these items. Compare your list with those of some of your classmates.

64. Do a risk–benefit analysis (Section 1.6) for the use of synthetic polymers as one or more of the following.
    a. grocery bags
    b. building materials
    c. clothing
    d. carpets
    e. food packaging
    f. picnic coolers
    g. automobile tires
    h. artificial hip sockets

65. To what extent are plastics a litter problem in your community? Survey one city block (or other area as directed by your instructor) and inventory the litter found. (You might as well pick it up while you are at it.) What proportion of the trash is plastics? What proportion of it is fast-food containers? To what extent are plastics recycled in your community? What factors limit further recycling?

66. Prepare a brief report on one of the following polymers and share it with your group. List sources and commercial uses.
    a. polybutadiene
    b. ethylene–propylene rubber
    c. polyurethanes
    d. epoxy resins

67. Do some research on Kevlar. Where is it used in addition to bullet-resistant vests? What makes it so strong? What is the Kevlar Survivors' Club, jointly sponsored by the International Chiefs of Police and DuPont?

68. Compare the treatment of environmental questions about plastics and other polymers at a website such as that of the American Plastics Council, an industry group, and at an environmental site such as Greenpeace.

REFERENCES AND READINGS

1. Ashley, Steven. "Artificial Muscles." *Scientific American*, October 2003, pp. 52–59.

2. Bren, Linda. "Joint Replacement: An Inside Look." *FDA Consumer*, March–April 2004, pp. 12–19.

3. Campbell, Ian M. *Introduction to Synthetic Polymers*, 2nd edition. New York: Oxford University Press, 2000.

4. Cook, Perry A., Sue Hall, and Jill Donahue. "Pondering Packing Peanut Polymers." *Journal of Chemical Education*, November 2003, pp. 1288A–1288B.

5. Deanin, Rudolph D. "The Chemistry of Plastics." *Journal of Chemical Education*, January 1987, pp. 45–47.

6. "Facts and Figures for the Chemical Industry." *Chemical and Engineering News*, July 5, 2004, pp. 23–63. Plastics are reported on p. 56. These figures are updated annually.

7. Friedel, Robert. "The Accidental Inventor." *Discover*, October 1996, pp. 58–69. Roy Plunkett and the invention of Teflon.

8. Jacoby, Mitch. "Composite Materials." *Chemical and Engineering News*, August 30, 2004, pp. 34–39.

9. LaJeunesse, Sara. "Plastic Bags." *Chemical and Engineering News*, September 20, 2004, p. 54.

10. Marvel, C. S. "The Development of Polymer Chemistry in America—The Early Days." *Journal of Chemical Education*, July 1981, pp. 535–539.

11. McCurdy, Patrick P. "Better Things for Better Living—How? Through Chemistry!" *Today's Chemist at Work*, July–August 1996, p. 84. Stephanie Kwolek and the Kevlar story.

12. Murphy, Marina. "Middle Earth's Monster Makers." *Chemistry and Industry*, December 15, 2003, pp. 10–13.

13. *Polymers and People.* Philadelphia: Beckman Center for the History of Chemistry, 1990.

14. Seymour, Raymond B., and George B. Kauffman. "Elastomers II: Synthetic Rubbers." *Journal of Chemical Education*, March 1991, pp. 217–220.

GREEN CHEMISTRY

Polymers: Giants Among Molecules

Jennifer L. Young, ACS Green Chemistry Institute

In this chapter, you have learned about many polymers that you already use in your daily life. You have seen polypropylene bags, polystyrene cups, polyethylene terephthalate soda bottles, nylon clothing and carpeting, styrene-butadiene rubber tires, and many more examples. Today, almost all polymers are made from petroleum and other fossil fuels. Many kinds of polymers you use everyday are recyclable. However, after the polymer products have been used, they often end up discarded into landfills. But times are changing, and polymers are becoming more environmentally friendly or "greener." Today, many new polymers are coming to market. These greener polymers are made from renewable resources (such as corn), contain less toxic components, and are designed to biodegrade. You will learn more about these greener polymer topics in the following Web Investigations.

WEB INVESTIGATIONS

Investigation 1
Recycling
By recycling, you reduce the amount of plastic that ends up in landfills and reduce the quantity of new plastic that is made from fossil fuels. It is an easy way for people to feel that they are actively helping to protect the environment. Did you ever notice that many plastic objects that you use on a regular basis are labeled with the type of plastic (a number symbol)? Some of these plastics, particularly polyethylene terephthalate (PET or PETE) and high-density polyethylene (HDPE), are commonly recycled through recycling programs. Use a search engine to conduct a Web search to learn about resin identification codes and the common plastic materials made from each coded plastic type. Read more about a recy-

cling topic of your choice from the websites you visit. Now that you know which plastic codes are recyclable, which of the plastics you used today were recyclable? Did you recycle them?

Investigation 2
Biodegradable Plastics
What does it mean for a plastic to be biodegradable? Visit an online dictionary or encyclopedia for the definition. You will see that the definition isn't straightforward. Biodegradability is related to how much of a polymer degrades into carbon dioxide and water, over what period of time, and under what conditions (temperature, moisture, sunlight). It also depends on which official body makes the definition: for instance, the European Standard EN13432, the Japanese Greenpla Standard, and the USA ASTM D6400-99 Standards all are different in scope and in methods of enforcement. According to these three standards, how should plastic be disposed of in order to biodegrade? Should it be thrown in the trash/landfill, composted, or something else? What are some classes or types of biodegradable plastics? The Japanese Biodegradable Plastics Society's homepage is an excellent resource to frame further research.

Investigation 3
Plastic from Corn
Some polymers are made from renewable resources, such as corn, soybeans, grass, and sugar cane. One example is polylactic acid made from corn. Another example is polyhydroxyalkanoates (PHAs), made from plant sugars and oils. Use these two polymer names as keywords to find more Web-based information on polylactic acid and PHAs.

▲ For every ten pounds of material in landfills, over a pound is plastic that can be recycled or reused.

Making plastics from corn and other crops raises several ethical questions. Will there be enough land to support the needs for both food and plastics, particularly in developing countries? Is it safe to use genetically modified organisms (GMOs) to make plastics? Choose one of these ethical topics and use a search engine to find more information.

Investigation 4
Nontoxic Plastic
Recent studies have linked detrimental health effects to several components in polymers. Included are phthalates —plasticizers used in polyvinyl chloride (PVC)—and bisphenol A—a monomer commonly used in polycarbonate plastics. What is the latest news related to either of these polymer components? Use a search engine or other online news sites to learn more. What common materials (such as bottles, plastic wraps, children's toys, and so on) contain phthalates or bisphenol A? Can you find information about some safer replacement materials (safer plasticizers, replacements for polyvinyl chloride or for bisphenol A polycarbonate)?

COMMUNICATE YOUR RESULTS

Exercise 1
Recycling
Design a flier to post around school or around town to encourage recycling. Use what you have learned already or find more information on the Internet to design a convincing flier.

Exercise 2
Biodegradable Plastics
A large market for biodegradable plastics is for serviceware: disposable plates, cups, and eating utensils. Places where biodegradable plastics could be used at your school are in the cafeteria, at sporting events, or at social events in sororities and fraternities. Does your school use any biodegradable plastics? What other opportunities can you think of for using biodegradable plastics at your school?

Exercise 3
Plastic from Corn
Based on what you have read in Web Investigation 3, what is your recommendation on the issue of farmland or GMO use by the plastics industry? Are you in favor of a plastics industry based on corn and other renewable resources? Are you in favor of a plastics industry based on genetically modified organisms? Choose a side of one of these issues and write a one-page paper defending your position. Defend your position in a class debate.

Exercise 4
Nontoxic Plastic
Write an article for a school publication, such as a school newspaper or newsletter, to make other students aware that components of some plastics may be toxic (such as some water bottles made from polycarbonate) and what nontoxic alternatives are available.

Cooperative Exercise 1
If your school doesn't have a recycling program, find out why not. With some or all of the students from class, develop a plan for starting a recycling program at your school. Look for useful websites that can help you start a recycling program. Get other students and teachers/administrators involved to start a recycling program.

Cooperative Exercise 2
Can you convince your school to use (or expand the use of) biodegradable plastics in any of the locations just mentioned? Can you convince your school to collect and compost biodegradable plastics rather than throwing them into the trash? Establish an argument in favor of using biodegradable plastics and work as a class to implement the use of biodegradable plastics in your school. There are some disadvantages to biodegradable plastics, and you may want to address and counter those points in your argument.

CHAPTER 11

Chemistry of Earth

Earth is a storehouse of chemicals—minerals, metals, and much more. Materials we extract from Earth, such as copper from this open-pit copper mine in Arizona, are the basis of much of our modern civilization. People use chemistry to modify many of these materials to make them more useful.

Metals and Minerals

This wondrous world of ours is a fertile sphere blanketed in air, with about three-fourths of its surface covered by water. Although it is but a tiny, blue-green jewel in the vastness of space, our Spaceship Earth is about 40,000 km in circumference, with a surface area of 500 million km^2.

Human astronauts have walked on the barren surface of Earth's airless moon. Our space probes have explored the desolation of Mars, the crushing pressure and hellish heat of Venus with its hurricane clouds of sulfuric acid, and Saturn's moon Titan with its liquid hydrocarbon seas and water-ice rocks. Probes have also given us close-up portraits of dry, pockmarked Mercury, and of Jupiter and Saturn with their horrendous lightning storms and their turbulent atmospheres of hydrogen and helium. Earth is but a small island in the inhospitable immensity of space, a tiny oasis uniquely suited to the life that inhabits it.

From earliest times, people have used Earth's resources. Primitive people used ordinary stones for hunting and as tools; modern people excavate ores and coal and drill for oil and gas. Spaceship Earth carries 6.5 billion passengers, and their numbers are increasing rapidly. What kinds of materials do we have aboard this spaceship? Are they sufficient for this enormous load of passengers? Let us begin by looking at the composition of Earth.

11.1 SPACESHIP EARTH: THE MATERIALS MANIFEST

Earth is divided into three main regions: the core, the mantle, and the crust (Figure 11.1). The *core* is thought to consist largely of iron with some nickel. Because the core is not accessible and does not seem likely to become so, we won't consider it as a source of materials. Nor shall we consider the Moon, Mars, or asteroids as a source; access to any such resources is decades, if not centuries, in the future.

The *mantle* is mostly silicates—compounds of silicon and oxygen with a variety of metals. Although the mantle may be reached eventually, it probably has few useful materials that are not available in the relatively very thin but much more accessible crust.

The *crust* is the outer solid shell of Earth, often called the *lithosphere*. The watery part, made up of the oceans, seas, lakes, rivers, and so on, is the *hydrosphere*. The *atmosphere* is the air that surrounds the planet. We discuss the atmosphere in Chapter 12 and the hydrosphere in Chapter 13. In this chapter we focus mainly on the lithosphere.

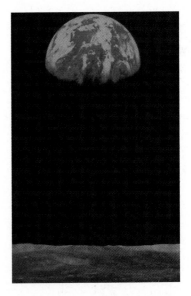

▲ Spaceship Earth is a blue jewel in the blackness of space. The desolate moon is in the foreground.

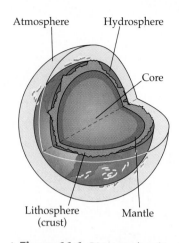

Atmosphere   Hydrosphere

Core

Lithosphere   Mantle
(crust)

▲ **Figure 11.1** Diagram showing the structural regions of Earth. The drawing is not to scale.

TABLE 11.1 Elemental Composition of the Earth's Surface			
Element	Number of Atoms in a Sample of 10,000 Atoms	Atom Percent	Percent by Mass
Oxygen	5,330	53.3	49.5
Silicon	1,590	15.9	25.7
Hydrogen	1,510	15.1	0.9
Aluminum	480	4.8	7.5
Sodium	180	1.8	2.6
Iron	150	1.5	4.7
Calcium	150	1.5	3.4
Magnesium	140	1.4	1.9
Potassium	100	1.0	2.4
All others	370	3.7	1.4
Total	10,000	100.0	100.0

The lithosphere is about 35 km thick under the continents and about 10 km thick under the oceans. Through extensive sampling of the atmosphere, hydrosphere, and lithosphere, scientists have been able to estimate the elemental composition of the outer portion of Earth. Let's consider a random sample of 10,000 atoms. The composition, featuring the nine most abundant elements, is summarized in Table 11.1.

Note that more than half the atoms in our 10,000-atom sample are oxygen. This element occurs in the atmosphere as molecular oxygen (O_2), in the hydrosphere in combination with hydrogen as water, and in the lithosphere in combination with silicon (pure sand is largely SiO_2) and various other elements. Silicon is the second most abundant element in the sample with 1590. Hydrogen is third with 1510 atoms, most of them in combination with oxygen in water. Hydrogen is such a light element—the lightest of all—that it makes up only 0.9% of Earth's crust by mass. Just three elements—oxygen, silicon, and hydrogen—make up 8430 of the atoms in the outer portion of Earth. The top nine elements account for 9630 of the atoms, leaving only 370 atoms of all the other elements. However, these minor constituents include some elements very important to life, such as carbon, nitrogen, and phosphorus.

11.2 THE LITHOSPHERE: ORGANIC AND INORGANIC

Carbonate Ion

The lithosphere is mainly rocks and minerals. Prominent among these are *silicate minerals* (compounds of metals with silicon and oxygen), *carbonate minerals* (metals combined with carbon and oxygen), *oxide minerals* (metals combined with oxygen only), and *sulfide minerals* (metals combined with sulfur only). Thousands of these mineral compounds make up the *inorganic* portion of the solid crust. We discuss some silicate minerals in Section 11.4. Some typical nonsilicate minerals are listed in Table 11.2.

TABLE 11.2 Some Nonsilicate Minerals of Economic Importance			
Mineral Type	Name	Chemical Formula	Use
Oxide	Hematite	Fe_2O_3	Ore of iron; pigment
	Magnetite	Fe_3O_4	Ore of iron
	Corundum	Al_2O_3	Gemstone; abrasive
Sulfide	Galena	PbS	Ore of lead
	Chalcopyrite	$CuFeS_2$	Ore of copper
	Cinnabar	HgS	Ore of mercury
	Sphalerite	ZnS	Ore of zinc
Carbonate	Calcite	$CaCO_3$	Cement; lime

EXAMPLE 11.1   **Writing Equations**

You are told by a geoscientist that hematite (Table 11.2) can be dissolved in hydrochloric acid to form $FeCl_3$ and water. Write a balanced equation for the process.

Solution

From Table 11.2, we see that hematite is Fe_2O_3. From Chapter 7, we know that hydrochloric acid is HCl. We are told in the example that the products are $FeCl_3$ and H_2O so we now have enough information to write an unbalanced equation.

$$Fe_2O_3 + HCl \longrightarrow FeCl_3 + H_2O \quad \text{(not balanced)}$$

To balance the equation, we see that there are two Fe atoms on the left and only one on the right, so we add a coefficient of 2 to $FeCl_3$.

$$Fe_2O_3 + HCl \longrightarrow 2\,FeCl_3 + H_2O \quad \text{(not balanced)}$$

Now there are six Cl atoms on the right, so a coefficient of 6 is needed for HCl.

$$Fe_2O_3 + 6\,HCl \longrightarrow 2\,FeCl_3 + H_2O \quad \text{(not balanced)}$$

Finally, there are three O atoms and six H atoms on the left, so a 3 in front of the H_2O will complete the task.

$$Fe_2O_3 + 6\,HCl \longrightarrow 2\,FeCl_3 + 3\,H_2O \quad \text{(balanced)}$$

▶ **Exercise 11.1A**
Quartz dissolves in hydrofluoric acid to give silicon tetrafluoride and water. Write and balance an equation for this process.

▶ **Exercise 11.1B**
What mass in grams of HF is needed to dissolve 12 g of quartz?

Although much, much smaller in quantity, the *organic* portion of Earth's outer layers includes all living creatures, their waste and decomposition products, and fossilized materials (such as coal, natural gas, petroleum, and oil shale) that once were living organisms. This organic material always contains the element carbon, nearly always has combined hydrogen, and often contains oxygen, nitrogen, and other elements.

11.3 MEETING OUR NEEDS: FROM STICKS TO BRICKS

People long have been able to modify nature's materials to help satisfy their needs and wants. For centuries, technological developments were based largely on trial and error. By developing an understanding of the structures of materials, modern science has greatly increased the human ability to modify natural materials. Thus, technological advances have been greatly accelerated by scientific knowledge.

The needs of early people were few. They obtained food by hunting and gathering. The skins of animals provided clothing when it was needed, and shelter was found in a convenient cave or constructed from available sticks and stones and mud.

After the agricultural revolution (about 10,000 years ago), people no longer were forced to search for food. Domesticated animals and plants supplied their needs for food and clothing. A reasonably assured food supply enabled them to live in villages and created a demand for more sophisticated building materials. People learned to convert natural materials into other products with superior properties. Adobe bricks, which serve well in arid areas, could be made simply by drying a mixture of clay and straw in the sun.

Calcium Carbonate: Limestone

One of the more common rocks on Earth is limestone, which is calcium carbonate ($CaCO_3$). In nature, calcium carbonate is found in many different physical forms; marble, seashells, eggshells, calcite, travertine, aragonite, coral, pearls, and chalk all are calcium carbonate. Striking examples of calcium carbonate also are found in the stalagmite and stalactite formations in caves, such as those in the Temple of the Sun, Carlsbad Cavern, New Mexico (Figure 11.2).

◀ **Figure 11.2** Limestone formations in the Carlsbad Cavern. The stalactites hanging from the ceiling, and the stalagmites growing from the floor, are the result of centuries of buildup of calcium carbonate.

The cave formations are produced over centuries. Carbon dioxide in the air dissolves in rainwater to form a weakly acidic solution. As this acidic water seeps through limestone formations, insoluble calcium carbonate is converted to soluble calcium hydrogen carbonate.

$$CaCO_3(s) + H_2O(l) + CO_2(g) \rightleftharpoons Ca(HCO_3)_2(aq)$$

Over time this action produces large caves. The reaction is reversible, however. As dripping water—saturated with calcium hydrogen carbonate—evaporates, stalactites and stalagmites of insoluble calcium carbonate are formed.

Limestone is used as a building stone. It is also used to make lime (Chapter 7) and in the manufacture of cement (page 308) and steel (page 309). Its role in the alleviation of air pollution from coal-burning power plants is discussed in Chapter 12.

Along the way people discovered fire, and they learned quite early that cooking improved the flavor and digestibility of meat and grains. Fire was one of the earliest agents of chemical change. With it people learned to heat their adobe bricks to make much stronger ceramic bricks. A similar process of firing clay at high temperatures produced ceramic pots, which made cooking and storage of food much easier.

With access to high temperatures, people later learned to make glass and to extract metals from ores. They were able to make tools from bronze, and later from iron.

11.4 SILICATES AND THE SHAPES OF THINGS

There are thousands of different minerals in the lithosphere. We will consider only a few representative ones here. First, let's look at some silicates. The basic unit of silicate structure is an SiO_4 tetrahedron (Figure 11.3). As shown in Table 11.3, silicate tetrahedra can exist singly as silicate anions or can be joined in a variety of ways.

Quartz is pure silicon dioxide (SiO_2). The ratio of silicon to oxygen atoms is 1:2. However, because each silicon atom is surrounded by *four* oxygen atoms, the basic unit of quartz is the SiO_4 tetrahedron. These tetrahedra are arranged in a complex three-dimensional structure. Crystals of pure quartz (rock crystal) are colorless, but various impurities produce a variety of quartz crystals sometimes used as gems (Figure 11.4).

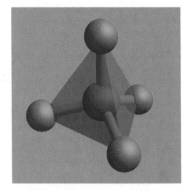

▲ **Figure 11.3** The silicate tetrahedron has an Si atom at the center and an O atom at each of the four corners. The shaded area (green) shows how the SiO_4 tetrahedron is typically represented in a mineral structure.

TABLE 11.3 SiO₄ Tetrahedra in Some Silicate Minerals			
Mineral(s)	**SiO₄ Arrangement**	**Formula**	**Uses**
Zircon	Simple anion (SiO_4^{4-})	$ZrSiO_4$	Ceramics; gemstones
Spudomene	Long chains of tetrahedra	$LiAl(SiO_3)_2$	Source of lithium and its compounds
Chrysotile asbestos	Double chains of tetrahedra	$Mg_3(Si_2O_5)(OH)_4$	Fireproofing (now banned)
Muscovite mica	Sheets of tetrahedra	$KAl_2(AlSi_3O_{10})(OH)_4$	Insulation; fancy paints; packing (vermiculite)
Quartz	Three-dimensional array of tetrahedra	SiO_2	Making glass (sand); gemstones (amethyst, agate, citrine)

Micas are composed of SiO₄ tetrahedra arranged in two-dimensional sheetlike arrays. Micas are easily cleaved into thin, transparent sheets (Figure 11.5). Pieces of mica once were used as panels for lanterns and as windows in the doors of stoves. Today, micas are used as insulators for electronics and in high-temperature furnaces. *Vermiculite* is a form of mica that has been "puffed" by heat, like popcorn. It is used as insulation, as a soil additive to hold moisture, and as lightweight, absorbent packing material.

Asbestos is a generic term for a variety of fibrous silicates. Perhaps the best known of these is *chrysotile*, a magnesium silicate (Figure 11.6). Note that chrysotile is a double chain of SiO₄ tetrahedra. The oxygen atoms that have only one covalent bond also bear a negative charge. Magnesium ions (Mg^{2+}) are associated with these negative charges.

It might appear that the four oxygen atoms surrounding each silicon atom contradict the SiO_2 formula for silica, but remember that the four O atoms are also shared by neighboring Si atoms.

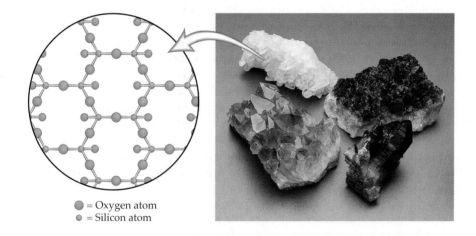

● = Oxygen atom
● = Silicon atom

◀ **Figure 11.4** A variety of quartz crystals. Counterclockwise from upper right: citrine (yellowish quartz), colorless quartz, amethyst (purple quartz), and smoky quartz. The chemical structure of quartz shows that each silicon atom is bonded to four oxygen atoms, and each oxygen atom is bonded to two silicon atoms.

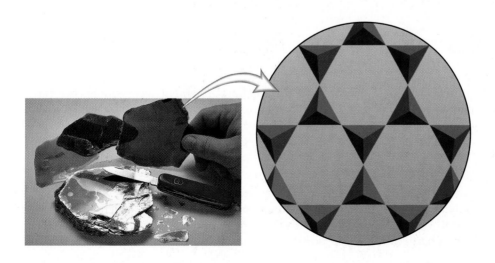

◀ **Figure 11.5** A sample of mica, showing cleavage into thin, transparent sheets. The chemical structure of mica shows sheets of SiO₄ tetrahedra. The sheets are bound together by cations, principally Al^{3+} (not shown). Mica is used as transparent "window" material in industrial furnaces.

Silicon Carbide

▲ **Figure 11.6** A sample of chrysotile asbestos. The chemical structure of chrysotile shows double chains of SiO_4 tetrahedra. The two chains are joined to each other through oxygen atoms. The double chains in turn are bound to each other by cations, principally Mg^{2+} (not shown).

▲ Vermiculite is a lightweight material made from mica. It is a useful soil additive, holding moisture much better than soil alone.

Asbestos: Risks and Benefits

Asbestos is an excellent thermal insulator. It has been used widely to insulate furnaces, heating ducts, and steam pipes. Protective clothing for firefighters and others who are exposed to flames and high temperatures once was made of asbestos. Asbestos was also used in brake linings for automobiles.

The health hazards to those who work with asbestos are well known. Inhalation of fibers 5–50 μm long over a period of 10–20 years causes asbestosis, and after 30–45 years, some asbestos workers contract lung cancer. Others get mesothelioma, a rare and incurable cancer of the linings of body cavities.

Long-term occupational exposure to asbestos increases the risk of lung cancer by a factor of two. Cigarette smoking causes a tenfold increase in the risk of lung cancer. We would expect asbestos workers who smoke to have a risk 20 times that of nonsmokers who do not work with asbestos, but instead their risk is increased by a factor of 90. Cigarette smoke and asbestos fibers act in such a way that each enhances the action of the other. Such a joint action is called a **synergistic effect**.

The harmful effects of asbestos are due mainly to a relatively rare form called crocidolite. Chrysotile, which makes up 95% of the asbestos used in the United States, appears not to be nearly as dangerous. Government regulations do not distinguish between the two types.

Public fears of developing cancer from asbestos have led to regulations that require its removal from schools and other public buildings. Costs of removal will total billions of dollars, and benefits are uncertain.

▲ Clothing made of asbestos fibers was once widely used to protect workers from high temperatures and hot materials, shown in this old photo of a firefighter on the USS Ranger. Asbestos for this purpose has been largely replaced by synthetic polymers such as Nomex (Chapter 10).

11.5 MODIFIED SILICATES: CERAMICS, GLASS, AND CEMENT

Pottery work, glassmaking, and formulation of cement were among the earliest technologies developed. People learned to modify natural materials such as sand, clay, and limestone, mainly by mixing and heating, to make much more useful products.

Ceramics

Clays are exceedingly complex and their compositions vary widely, but they are basically aluminum silicates. Early potters used natural clays, which they hardened by heat. When clay is mixed with water, it can be molded into any shape. Firing leaves a hard, durable (but porous) product. Bricks and tile are made in this manner. When porosity is not desirable—as in a cooking pot or a water jug—the pottery can be glazed by adding various salts to the surface. Heat then converts the entire surface to a glasslike matrix. Bricks and pottery are examples of **ceramics**, inorganic materials made by heating clay or other mineral matter to a high temperature at which the particles partially melt and fuse together.

Ceramic research has led to the development of some amazing new ceramic materials. Some have such high heat resistance that they can withstand the extreme temperatures of rocket reentry into the atmosphere. They are used to make rocket nose cones, surface tiles, and exhaust nozzles. Other ceramics have magnetic properties and can serve as memory elements in computers. Still others have such exceptional electric conductivity at the temperature of liquid nitrogen that they are known as "superconductors." Such superconducting materials are needed to build the powerful electromagnets used in particle accelerators and magnetically levitated ("maglev") trains.

$YBa_2Cu_3O_7$

▲ Ceramics are durable. This pottery jar from the Naqada period of Egypt is over five thousand years old.

Glass

Glass, a noncrystalline solid, is another technological development of ancient times. The first glass probably was made in ancient Egypt about 5000 years ago by heating a mixture of sand, sodium carbonate (Na_2CO_3), and limestone (calcium carbonate, $CaCO_3$). As the mixture melts, it becomes a homogeneous liquid. When the liquid cools, it becomes hard and transparent.

Crystalline materials have a regular, repeated arrangement of atoms, ions, or molecules. The forces among these unit particles are the same throughout. When heated, crystals melt over a narrow temperature range. For example, sodium hydroxide, an ionic crystal, melts sharply at 318 °C, and crystalline aspirin, a molecular substance, melts at 135 °C. Glass is different; when heated, it gradually softens. While soft, it can be blown, rolled, pressed, or molded into almost any shape. The properties of glass result from an irregular arrangement, in three dimensions, of SiO_4 tetrahedra. The chemical bonds in this arrangement are not all equivalent. Thus, when glass is heated, the weaker bonds break first and the glass softens gradually.

The basic ingredients in glass can be used in different proportions. Oxides of various metals can be substituted in whole or in part for the lime, sodium carbonate ("soda"), or sand. Thus, many special types of glass can be made. Some examples are given in Table 11.4.

No vital raw materials are involved in the manufacture of ordinary glass, but the furnaces used to melt and shape glass require energy. Disposal is a potential

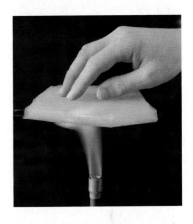

▲ Aerogels are specially prepared ceramic materials that are over 95% air. They are the best solid insulators in the world.

TABLE 11.4	**Compositions and Properties of Various Glasses**	
Type	**Composition**	**Special Properties and Uses**
Soda–lime glass	Sodium and calcium silicates	Ordinary glass (for windows, bottles, and so on) (sand plus soda plus lime)
Borosilicate glass	Boron oxide (instead of lime)	Heat-resistant (for laboratory ware and ovenware) (Pyrex, Kimax)
Aluminosilicate glass	Aluminum oxide (instead of soda)	More highly heat-resistant (for top-of-stove cookware and fiberglass)
Lead glass	Lead oxide (instead of lime)	Highly refractive (for optical glass, art glass, table crystal)
Alabaster glass	Sodium chloride (salt) added	White, opaque ("milk glass")
Colored glass	Selenium compounds added	Red color (ruby glass)
	Cobalt compounds added	Blue color (cobalt glass)
	Chromium compounds added	Green color
	Manganese compounds added	Violet color
	Cadmium sulfide added	Yellow color
	Carbon and iron oxide added	Brown color (amber glass)
Photochromic glass	Silver chloride or bromide added	Light sensitive; darkens when exposed to light (for sunglasses, hospital windows)
Laser glass	Contains neodymium	Powerful lasers
Frosted glass	Etched with hydrofluoric acid (HF)	Satiny frosted surface

problem because glass is one of the most permanent materials known. It doesn't rot when discarded. Glass is easily recycled, however, as long as different kinds are separated. It can be melted and formed into new objects at considerable energy savings compared with the manufacture of new glass. In fact, it is standard practice to add *cullet* (broken glass) to each batch of sand, soda, and limestone to be melted. The cullet is usually scrap glass from previous melts. This practice not only gets rid of waste glass but also actually improves the melting process.

▲ Heat-softened glass can be shaped by blowing or molding.

▲ These optical fibers have become the preferred transmission medium for telecommunications. The glass fibers are generally protected with plastic coatings to preserve strength and to make them easier to handle.

▲ Limestone is mixed with clay and sand in a long rotary kiln to produce a common kind of cement known as portland cement.

Optical Fibers

Modern glass research has resulted in many wonderful new products, but one of the most remarkable developments is a product that is practically invisible. Pure threads of glass, no thicker than a human hair, have replaced many old telephone lines. A bundle of these optical fibers can carry several hundred times as many messages as a copper cable of the same size.

Optical fibers carry messages as intermittent bursts of light. When you speak into a telephone, the sound is translated into an electric signal, which then pulses a laser to give off light messages that are carried by the glass fibers. At the end of the line, the light pulses are converted back to sound waves.

Optical fibers are also used in computer interfacing. They are employed by medical personnel to view inside blood vessels and the digestive system without surgery, and to guide the use of instruments during laparoscopic surgery. Some chemical instruments use optical fibers to measure absorbance of light, which can then be used to determine the composition of an unknown mixture. Future applications of optical fibers can only be imagined.

Cement and Concrete

Cement, a complex mixture of calcium and aluminum silicates, is another ancient technological development. The Romans used a type of cement to construct roads, aqueducts, and the famous Roman baths. The raw materials for the production of cement are limestone (calcium carbonate, $CaCO_3$) and clay (aluminum silicates). The materials are finely ground, mixed, and roasted at about 1500 °C in a rotary kiln heated by burning natural gas or powdered coal. The finished product is mixed with sand, gravel, and water to form *concrete*.

Our understanding of the complex chemistry of cement is still imperfect. Nevertheless, extensive research has made various special cements available, including a fast-setting cement with high early strength, a white cement, a waterproof cement, and a cement that sets at high temperatures.

Concrete is used widely in the construction of buildings and roads because it is inexpensive, strong, chemically inert (and thus nonpolluting in itself), durable, and tolerant of a wide range of temperatures. However, its production involves extensive mining, with entire mountains being torn down for limestone rocks. The rotary kiln process consumes fossil fuels. Particulate matter from the crushing operations and smoke and sulfur dioxide from the burning of fossil fuels make air pollution from cement plants especially serious. Large expanses of concrete now cover what once were green acres. In cities, areas paved with asphalt and concrete are sufficiently extensive to change the climate (temperature and rainfall). Runoff of rainfall from paved areas is especially rapid and contributes to flash flooding. However, concrete can be broken up and used as rock fill; it is also sometimes used to encase hazardous waste for disposal.

11.6 METALS AND ORES

Mineral Sources of Metal Ores

Human progress through the ages is often described in terms of the materials used for making tools. We speak of the Stone Age, the Bronze Age, and the Iron Age. Ancient peoples knew about gold and silver because they are often found in the free state in nature, but these metals are too soft to use as tools. Metal tools were not possible until artisans learned to extract certain metals from ores by smelting.

Copper and Bronze

Copper is sometimes found in the *native* or uncombined state, and it was probably the first metal to be freed from its ore by early smelting techniques. Ancient Egyptian records show that copper was known 5500 years ago. The Egyptians probably isolated copper by heating copper carbonate ($CuCO_3$) or copper sulfide (Cu_2S) ore.

$$Cu_2S(s) + O_2(g) \longrightarrow 2\,Cu(s) + SO_2(g)$$

Many copper ores are blue to green in color and easy to identify in the ground. The ancient Egyptians produced thousands of tons of copper metal. Today copper is important mainly because of its excellent electric conductivity. It is used primarily for electric wiring.

Even more important than copper was **bronze**, a copper alloy containing about 10% tin. Bronze is harder than copper, and it could therefore be made into many useful tools. In some societies, bronze remained the most widely used metal for about 2000 years.

Iron and Steel

The technology for iron and steel greatly lagged behind that for copper and bronze for two reasons. Iron reacts so readily with oxygen and sulfur that it is not found uncombined in nature, and it melts at a much higher temperature than copper or bronze. To produce iron from ore, carbon (coke) is employed as the reducing agent. First, the carbon is converted to carbon monoxide.

$$2\,C(s) + O_2(g) \longrightarrow 2\,CO(g)$$

Then the carbon monoxide reduces the iron oxide to iron metal.

$$Fe_2O_3(s) + 3\,CO(g) \longrightarrow 2\,Fe(l) + 3\,CO_2(g)$$

Iron can be made from ore in a huge chimneylike vessel called a *blast furnace* (Figure 11.7). The raw materials fed into the furnace are iron ore, *coke*, and limestone. The coke (which is mainly carbon) is made by heating coal in the absence of air to drive off coal oil and tar. The limestone is added to combine with silicate impurities to form a molten **slag** that floats on top of the iron and is drawn off. Molten iron is drawn off at the bottom of the furnace. The product of the blast furnace, called *pig iron*, has a fairly low melting point, and so it is easily cast into molds. Hence, it is also known as cast iron.

Because cast iron is brittle and has many impurities, most iron from blast furnaces is *alloyed*. An **alloy** is a mixture of two or more elements, at least one of which is a metal. Alloys have metallic properties. The principal alloy of iron is **steel**, an alloy with carbon. Many varieties of steel also contain other elements. In a steel furnace, pressurized oxygen reacts with impurities, such as phosphorus, silicon, and excess carbon. The properties of steel can be varied over a wide

▲ This double-headed bronze axe from about 1200 B.C.E. was recovered from a shipwreck excavation off the coast of Turkey. The Bronze Age occurred at different times in different cultures. Some cultures seem to have skipped it altogether, going directly from the Stone Age to the Iron Age.

Ore, limestone, coke

Waste gases

400 °C

1000 °C

1500 °C

Air

Pig iron   Slag   Slag out

Molten iron   Molten iron

▲ **Figure 11.7** A modern blast furnace. Iron ore, coke, and limestone are added at the top, and hot air is injected at the bottom.

Question: What is the role of each of the three substances added at the top of the furnace?

About half of all steel made today in the United States comes from "mini-mills," which make steel from scrap iron instead of ore. Almost all automobiles today are recycled.

range by adjusting the amount of carbon in it. High-carbon steel is hard and strong, whereas low-carbon steel is ductile and malleable.

Steel is commonly alloyed with other metals to give it special properties. For example, manganese imparts hardness, tungsten gives high-temperature strength, and additions of chromium and nickel produce stainless steel. Because it is so abundant and can be made into so many different alloys, iron is the most useful of the metals.

Modern Steelmaking

Most steel producers now use the basic oxygen process (Figure 11.8). The furnace is first charged with pig iron. Powdered limestone is added just above the molten iron, and oxygen gas at about 10 atm pressure is injected there also. Metallic impurities in the pig iron are converted to oxides, which react with SiO_2 to form a slag that floats above the liquid iron and is poured off. Then any desired alloying elements are added, often by adding scrap steel of a particular composition.

Chemistry is employed extensively during the process. As the melt progresses, samples are drawn from the furnace and rushed to an on-site laboratory for analysis of the alloying metals and nonmetals. If, for example, the steel contains too much carbon, more oxygen is used to "burn off" the excess, or low-carbon scrap may be added. In this way the impurities are removed and the desired composition of the steel is achieved at the same time.

Some companies now make iron directly from iron ore in a single-step continuous process. They avoid the great energy requirements of the blast furnace by using temperatures below the melting points of any of the raw materials. The ore is reduced by H_2 and CO, which in turn are made from natural gas. This *direct reduction* method is used mainly in developing countries that do not already have large iron and steel industries in place.

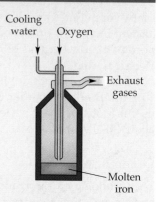

▲ **Figure 11.8** A basic oxygen furnace. The steel vessel is lined with dolomite, a mixed calcium and magnesium carbonate ($CaCO_3 \cdot MgCO_3$).

The method we still use to produce aluminum was discovered more than 100 years ago by a college student named Charles Martin Hall. Interestingly, the same process was discovered almost simultaneously in France by another college student named Paul Héroult. Both men were born in 1863, and they developed the process for manufacturing aluminum in 1886. The coincidence doesn't end there; both died in 1914 at the age of 51.

Aluminum: Abundant and Light

Aluminum, the most abundant metal in Earth's crust, has replaced iron for many purposes. However, the aluminum in nature is tightly bound in compounds, and considerable energy is required to extract the metal from its ores. The principal ore of aluminum is *bauxite,* impure aluminum oxide. The Al_2O_3 is extracted from the impurities with a strong base, and then the melted oxide is reduced by passing electricity through it.

$$2\ Al_2O_3 \xrightarrow{\text{electricity}} 4\ Al + 3\ O_2$$

It takes 2 metric tons (t) of aluminum oxide and 17,000 kilowatt hours (kWh) of electricity to produce 1 t of aluminum.

Aluminum is light and strong. An aluminum object weighs only one-third as much as a steel article the same size. Although it is considerably more active than iron, aluminum corrodes much more slowly. Freshly prepared aluminum metal reacts with oxygen to form a hard, transparent film of aluminum oxide (Al_2O_3) over its surface. The film then protects the metal from further oxidation. Iron, on the other hand, forms an oxide coating that is porous and flaky. Instead of protecting the metal, this coating flakes off, allowing further oxidation.

The Environmental Costs of Iron and Aluminum

Steel and aluminum play vital roles in the modern industrial world, and it would be hard to overemphasize their economic value. But shouldn't we include the environmental cost of producing, using, and discarding these materials?

Both steel mills and aluminum plants produce considerable waste. In days gone by, steel mills discharged lime, acids, grease, oil, and iron salts into waterways. They released carbon monoxide, nitrogen oxides, particulate matter, and

Metal	Symbol	Important Ore	Selected Properties	Typical Uses
Chromium	Cr	$FeCr_2O_4$	Shiny, resists corrosion	Chrome plating, stainless steel
Gold	Au	Au	Yellow metal, soft, dense	Coinage, jewelry, dentistry, electrical contacts
Lead	Pb	PbS	Low melting, dense, soft	Plumbing, batteries
Magnesium	Mg	$MgCl_2$	Light, strong	Auto wheels, luggage
Mercury	Hg	HgS	Dense liquid	Thermometers, barometers
Nickel	Ni	NiS	Resists corrosion	Coinage, alloy for stainless steel
Platinum	Pt	Pt	Inert, high melting	Catalyst, instruments
Silver	Ag	Ag_2S	Excellent electric conductor	Electric contacts, mirrors, jewelry, coins
Sodium	Na	NaCl	Reactive, soft	Heat transfer medium, reducing agent
Tin	Sn	SnO_2	Resists corrosion	Coating for steel cans
Tungsten	W	$CaWO_4$	Very high melting	Light-bulb filaments
Uranium	U	U_3O_8	Fissionable	Energy source
Zinc	Zn	ZnS	Forms protective coating	Galvanizing coating

TABLE 11.5 Some Other Important Metals

other pollutants into the air. Aluminum plants discharged iron, aluminum, and other metal oxides into waterways, and particulate matter, fluorides, and other pollutants into the air. Modern treatment methods have reduced the quantity of pollutants released.

Scientists estimate that it takes 15 times as much fuel energy to produce aluminum as it does to produce a comparable weight of steel. Aluminum is less dense than steel, but making an aluminum can requires 6.3 times as much energy as making a steel can. On the other hand, when aluminum is recycled, the energy needed is only a fraction of that required to make the metal from ore. Further, the low density of aluminum means energy savings down the road; an airplane made largely from aluminum is so much lighter than a similar craft made entirely from steel that it takes a good deal less energy to operate it.

Other Important Metals

There are many other important metals. Some of them, with typical ores, properties, and uses, are listed in Table 11.5.

Metals and minerals are vital to a vigorous economy, but no industrial nation is 100% self-sufficient in all the vital metals and minerals. The United States has depleted much of its high-grade ores and is now more than 90% dependent on imports for chromium, molybdenum, and tungsten—elements vital for steelmaking and other industries—and for platinum (Pt) and palladium (Pd)—elements used in making catalysts, including those needed to reduce pollution from automobile exhaust emissions. Metals and minerals are essential components of the modern global economy.

How can we ever run short of a metal? Aren't atoms conserved? Yes, there is as much iron on Earth as there was 100 years ago, but by using it we scatter the metal throughout the environment. Gathering it back to a factory to make new objects requires energy, just as obtaining metals from low-grade ores requires more energy than extracting them from high-grade ores.

11.7 RUNNING OUT OF EVERYTHING: EARTH'S DWINDLING RESOURCES

Dramatic fuel shortages during the 1970s brought important changes in government policies and in industrial and individual practices. But now many of the conservation efforts begun at that time have largely been forgotten. We remain vulnerable to future fuel shortages.

We also face potential shortages of metals and minerals. The United States has exhausted its high-grade ores of metals such as copper and iron. When the Europeans first came to North America, the native Americans were mining nearly pure

▲ These crystals of native copper are from Houghton, Michigan.

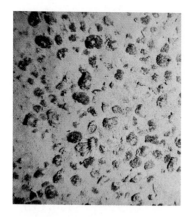

▲ Manganese-rich nodules on the ocean floor. They also contain iron and traces of cobalt, copper, and nickel.

In 1974 the U.S. Central Intelligence Agency (CIA) used a putative search for manganese nodules as a cover for an attempt to raise the sunken Soviet submarine *K129* from the depths of the Pacific Ocean. The recovery ship *Glomar Explorer* hoisted aboard a 100-ft section of the *K129* containing two nuclear-tipped torpedoes and the bodies of six Russian sailors. Just what intelligence information the CIA obtained is still secret. The sailors were buried at sea in a solemn ceremony filmed by the CIA.

▲ **Figure 11.9** Waste cyanide solution from the leaching of gold from ore is stored in a large containment pond. This facility is part of a gold mining operation in the Mojave Desert of California.

copper in Michigan. Now we mine ore with less than 0.5% copper. For much of the nineteenth century, we mined high-grade iron ore from the Mesabi range in Minnesota. Now we mine taconite, a hard rock that contains only small, dispersed amounts of iron oxide. Australia has vast deposits of high-grade iron ore, but importing this ore is costly. Similarly, miners once found nuggets of pure gold and silver, but the "glory holes" of western North America are long gone. Now we mine ores with a fraction of an ounce of gold in a ton of gold ore (see Problem 62). Indeed, there are few high-grade deposits of any metal ores left that are readily accessible to the industrialized nations.

What's wrong with low-grade ores? Because more material must be mined for the same final amount of metal, there is more environmental disruption (Figure 11.9), and more energy is required to concentrate the ores. Both environmental cleanup and energy cost money. Consider an analogy. A bag of popcorn is useful. You can pop it and eat it. The same popcorn, scattered all over your room, would be less useful. Of course, you could gather it all up and then pop it, but that would take a lot of energy—perhaps more energy than you would care to expend and maybe more energy than you would get back when you popped the corn and ate it.

We also must import quite a few metals because their ores are not found in the United States. For example, we get tin from Bolivia, chromium from Rhodesia, and platinum from South Africa and Russia. But the metal reserves in other countries are being depleted, too.

Where will we get metals in the future? The sea is one possible source. Nodules rich in manganese cover vast areas of the ocean floor. These nodules also contain copper, nickel, and cobalt. Questions of who owns them and how to mine them without major environmental disruption remain to be resolved.

11.8 LAND POLLUTION: SOLID WASTES

Productivity and creativity have enabled people in industrialized nations to have a seemingly endless variety of consumer goods in enormous quantities. These goods often are packaged in paper or plastic. Their wrappers litter the environment and fill our disposal sites. As the goods break, wear out, or merely become obsolete, they too add to our disposal problems. The approximate composition of municipal solid wastes (MSWs)—commonly called trash or garbage—in the United States is given in Figure 11.10.

The U.S. Environmental Protection Agency (EPA) ranks strategies for MSW as follows: (1) Source reduction (including reuse) is best, (2) recycling and composting is next best, and (3) disposal in combustion facilities and landfills is a last resort.

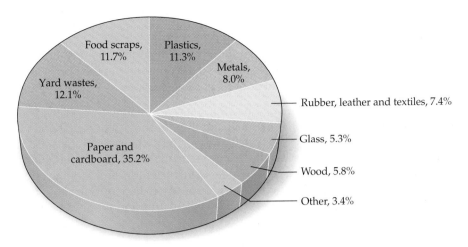

Food scraps, 11.7%
Plastics, 11.3%
Metals, 8.0%
Yard wastes, 12.1%
Rubber, leather and textiles, 7.4%
Paper and cardboard, 35.2%
Glass, 5.3%
Wood, 5.8%
Other, 3.4%

◀ **Figure 11.10** Municipal wastes in the United States are largely paper, cardboard, and yard wastes. Paper and cardboard can be recycled. Yard wastes and food wastes can be composted.

Currently in the United States, 30% of MSW is recovered and recycled or composted, 14% is burned at combustion facilities, and 56% is disposed in *sanitary landfills*.

In the past, most solid wastes were simply discarded in open dumps. This led to infestation by rats, flies, and other pests that often spread to nearby areas. Open burning led to offensive and unhealthful air pollution. These open dumps have now been largely phased out.

When disposed in a sanitary landfill, garbage and trash are piled into a trench, compacted, and covered over. This eliminates the problems of rats, flies, and odors. However, some landfills leak and contribute to groundwater contamination. Furthermore, materials in landfills decompose slowly. Newspapers are still readable after being entombed for 20 years. Without water and oxygen, microorganisms are unable to carry out the normal decay processes.

Disposal of solid wastes by *incineration* accounts for about 14% of today's MSW. When carried out in properly designed incinerators, air pollution is minimal. However, many incinerators are old, or poorly designed, and they generate considerable smoke and odor. Building new incinerators tends to be unpopular with local citizens.

Research in solid waste disposal has led to new processes. The U.S. Bureau of Mines has developed a method for converting garbage to oil with an overall yield of 25%. Ground-up rubber tires can be mixed with powdered coal and used as fuel. Waste glass and rubber tires have both been added to road-paving mixtures for highway construction.

A tin can—really steel with a thin coating of tin—rusts when you throw it away, eventually disintegrating. Scientists at Pennsylvania State University estimate that it would take 500 years for an aluminum can to degrade completely.

Some cities not only burn their combustible solid wastes but also use the heat from the incinerators to warm buildings, to generate power, or both. A pound of polyethylene produces about 20,000 British thermal units (Btu) of energy, about the same as a pound of no. 2 fuel oil.

11.9 THE THREE R's OF GARBAGE: REDUCE, REUSE, RECYCLE

There are several ways to deal with our garbage problem. Not all are equally desirable. We should choose wisely among them in order to best save energy and protect the environment.

The best way to deal with our solid waste problem is to *reduce* the amount of throwaway materials produced. Decreasing the volume of materials produced and sold saves resources and energy and minimizes the disposal problem.

Next best is the *reuse* of materials. When possible, items should be made durable enough to withstand repeated use, rather than designed for a single use and disposal. Consider a glass beverage bottle. It takes about 6300 kJ (1500 kcal) of energy to make a nonreturnable half-liter bottle. It takes a little more energy, 8300 kJ (2000 kcal), to make a more durable, returnable bottle of the same capacity. However, the returnable bottle is used an average of 12.5 times before it is broken, which means it consumes only about 10% of the energy used to make the single-use nonreturnable bottle. Of course, there are issues with reuse that complicate the situation. For example, a returnable bottle must be cleaned thoroughly before refilling, which increases water usage. All considerations must be examined when determining the best overall solution to a reuse problem.

EXAMPLE 11.2   **Energy Calculations**

Based on the data in the previous paragraph, calculate the quantity of energy needed to produce 10,000 returnable bottles, and compare it to the energy needed to produce the equivalent number of disposable bottles.

Solution

We haven't yet discussed energy in detail, but there is enough information in the paragraph to permit us to do the calculation using dimensional analysis. The paragraph notes that 6300 kJ of energy is needed to make a single disposable bottle, that 8300 kJ is needed to make a single returnable bottle, and that the returnable is used an average of 12.5 times. That means that one returnable is the equivalent of 12.5 disposables. We begin by calculating the number of disposables equivalent to 10,000 returnables.

$$10{,}000 \text{ returnables} \times \frac{12.5 \text{ disposables}}{1 \text{ returnable}} = 125{,}000 \text{ disposables}$$

Now we can calculate the energy cost of the disposables, using the factor 6300 kJ = 1 disposable.

$$125{,}000 \text{ disposables} \times \frac{6300 \text{ kJ}}{1 \text{ disposable}} = 7.88 \times 10^8 \text{ kJ}$$

Then we can calculate the energy cost of the returnables in a similar way.

$$10{,}000 \text{ returnables} \times \frac{8300 \text{ kJ}}{1 \text{ returnable}} = 8.3 \times 10^7 \text{ kJ}$$

Notice that we can do this calculation without knowing just what a joule is!

▶ **Exercise 11.2**

When one mole (16.0 g) of methane is burned in air, 890 kJ of energy is produced. What mass of methane is equivalent to the energy difference between the returnables and the disposables in Example 11.2?

About 12 million automobiles are junked each year in the United States, and 300 million tires are thrown away. Nearly all the automobiles are recycled for their steel and other components. About 80% of the tires are recycled or reused in some way.

Every day in the United States, each person discards an average of about 4.5 lb of trash, up from 2.7 lb per person per day in 1970.

Recycling a 1-m stack of newspapers saves the equivalent of a 10-m pine tree. Making paper from scrap instead of virgin wood pulp yields a 50% savings of water and energy.

According to the Steel Recycling Institute, about 5740 kJ of energy is conserved for each pound of steel recycled. That is enough energy to light a 60-watt (W) bulb for more than 26 hr.

The third way to reduce the volume of wastes is to *recycle* them. Recycling requires energy, and some material is unavoidably lost, but recycling saves materials. Metals such as iron and steel, aluminum, copper, and lead are largely recycled. More than 57% of steel cans and about 55% of beer and soft drink aluminum cans are recycled. About 35% of plastic soft drink bottles and 26% of glass containers are recycled. About 45% of paper and cardboard are recycled. Lawn trimmings and food wastes can be composted, converting them to soil. About 57% of yard trimmings are composted. In the future, as raw materials become scarcer, these proportions are bound to increase.

Recycling also saves energy. It takes only 5% as much energy to make new aluminum cans from old ones as it does to make them from aluminum ore. Using a ton of scrap to make new iron and steel saves 1.5 tons of iron ore and 0.3 ton of coal. It also results in a 74% savings in energy, an 86% reduction in air pollution, and a 76% reduction in water pollution. However, recycling is "energy intensive" in another way: Many items to be recycled must be sorted by hand or carefully segregated. When a batch of polyethylene plastic is contaminated with paper, wood, or metal chips, the batch can no longer be recycled easily. Used motor oil contaminated with paint requires expensive processing to make it useful again. Some cities with curbside recycling have separate bins so that the individual can sort items quickly and easily. Others collect mixed recyclables that are separated in processing plants.

By following the three R's—reduce, reuse, and recycle—we can protect the environment at the same time that we save materials, energy, and money.

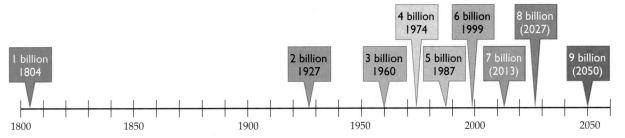

▲ **Figure 11.11** Two thousand years ago, Earth's population was about 300 million. It changed little over the next thousand years and is estimated to have been only 310 million at the end of the first millennium. It reached 1 billion about 1804. This graph shows the intervals in which an additional billion people were added to Earth's population, with projections to 2050. (*Source:* United Nations, *World Population Prospects: The 2004 Revision.* New York: 2005.)

11.10 HOW CROWDED IS OUR SPACESHIP?

Each day, people die and other people are born, but the number born is about 200,000 more than the number that die. Every five days our world population goes up by a million people, and it reached 6.5 billion in 2006.

Figure 11.11 is a graphic picture of population growth. For tens of thousands of years, human population was never more than a few hundred million. It did not reach the 1 billion mark until about 1800. Population growth peaked in the 1960s at about 2.2% a year. The present rate of growth is about 1.3% per year. Overall, the world's population quadrupled during the twentieth century. The Population Division of the United Nations Department of Economic and Social Affairs estimates a world population of 9.0 billion by the middle of the twenty-first century.

How many people can Spaceship Earth accommodate? Many believe that we have already gone beyond the optimum population for this planet. They predict increasing conflicts over resources and increasing pollution of the environment.

Why did Earth's population start increasing so rapidly after centuries of little change? Population growth depends on both the birth rate and the death rate. If they are equal, population growth is zero. After being almost equal for thousands of years, scientific progress in the 1800s led to a decline in the death rate in developed countries. Yet the birth rate changed little and the population grew rapidly. These changes reached the less developed countries in the 1950s, greatly lowering the death rate. However, the birth rate remained high and the population exploded. Population growth has slowed considerably in recent years, especially in developed countries. Scientific progress continues to cause a decline in death rates, but birth rates have also declined, especially in Europe. The decrease in birth rates in the less developed countries has come more slowly. With more people comes more waste to be processed. We must continue to develop technologies and methods to deal with an ever-increasing waste problem.

Population and the food supply are discussed in Chapter 16.

Critical Thinking Exercises

Apply knowledge that you have gained in this chapter and one or more of the FLaReS principles (Chapter 1) to evaluate the following statements or claims.

11.1 An economist has said that we need not worry about running out of copper because it can be made from other metals.

11.2 A citizen testifies against establishing a landfill near his home, claiming that the landfill will leak substances into the groundwater and contaminate his water well.

11.3 A citizen lobbies against establishing an incinerator near her home, claiming that plastics burned in the incinerator will release hydrogen chloride into the air.

11.4 An environmental activist claims that we could recycle all goods, leaving no need for the use of raw materials to make new ones.

11.5 A concerned parent demands that all the insulation on the steam pipes in the schools in her district be removed and replaced with new insulation. Her contention is that the pipe insulation may contain asbestos and therefore may cause cancer in the students.

SUMMARY

Sections 11.1–11.3—Earth is divided into the core, the mantle, and the crust. The core is thought to be mainly iron; the mantle is mostly silicate minerals. The crust consists of the solid lithosphere, the liquid (water) hydrosphere, and the gaseous atmosphere. Nine elements make up 96% of the atoms in the crust. The lithosphere is made up largely of minerals such as silicates, carbonates, oxides, and sulfides. The organic part of the lithosphere consists of all things once and currently living. People have manipulated the lithosphere to meet their wants and needs for about 10,000 years.

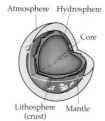

Section 11.4—Silicates contain SiO_4 tetrahedra in various arrangements. **Quartz** is a crystalline mineral that consists of a three-dimensional network of these tetrahedra. Sand is mainly quartz. The fibrous mineral **asbestos** is made of chains of SiO_4 tetrahedra. Asbestos is implicated in lung cancer in asbestos workers. The incidence of lung cancer among asbestos workers who smoke is much greater than the simple combination of asbestos exposure and smoking might indicate; this is called a **synergistic effect**. The sheet-like mineral **mica** consists of sheets of SiO_4 tetrahedra.

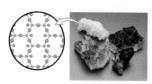

Section 11.5—Ceramics, glass, and cement are modified silicates. **Ceramics** are made by heating clay and other inorganic minerals to fuse them partially. New kinds of ceramics have been developed that have high heat resistance, low electrical conductivity, and increased durability. **Glass** is a noncrystalline solid that does not have a fixed melting point. Most glass is made from sand, with various additives to impart desired properties such as heat resistance, color, and light sensitivity. Glass does not degrade in the environment but it is easy to recycle; recycled glass aids the melting process for newly manufactured glass. **Cement** is a complex mixture of calcium and aluminum silicates made from limestone and clay. Cement with gravel, sand, water, and additives is used to make concrete.

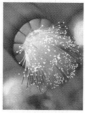

Section 11.6—Copper was one of the first metals to be obtained from its ore but is too soft to be used for tools. **Bronze** is a mixture of copper and tin that is much harder than pure copper. Bronze implements were used for thousands of years. Iron technology came about much later. Iron is prepared in a blast furnace, from iron ore, limestone, and coke. The limestone combines with impurities to produce **slag** that floats on top of the iron and is removed. Cast iron is brittle, so it is usually further processed with oxygen to burn off additional impurities. This changes the iron to **steel**. Steel is an **alloy**, which is a mixture of a metal and at least one other element (carbon, in the case of steel). Various elements can be added to steel to give it different properties. Aluminum is less dense than steel and does not rust the way steel does. Aluminum was first isolated from its ore, bauxite, about 100 years ago, by an electrical process. There are many other useful metals and alloys.

Section 11.7—We face potential shortages of many metals and minerals because most of the high-grade ores have been used up. The United States now imports some metals and ores that it used to mine because of these shortages. It takes much less energy to recycle most metals than it does to produce them from ores. However, there are other possible sources of some metals, notably manganese nodules at the bottom of some oceans.

Sections 11.8–11.10—The United States produces large amounts of waste, most of which is paper, cardboard, and yard wastes. Municipal solid waste (MSW) used to be discarded in open dumps, but now it is buried in a sanitary landfill, burned, or otherwise processed to reduce the amount of MSW. Several new uses for MSW have been developed. The three R's of garbage are as follows: Reduce the amount of garbage, reuse items such as reusable bottles when possible, recycle when possible. Steel, aluminum, plastic, and paper are easily recycled, and yard waste can be composted. Although birth rates have declined somewhat, Spaceship Earth is crowded and is going to become more so in years to come. We must come to terms with the amount and type of waste we produce.

REVIEW QUESTIONS

1. What are the three main regions of Earth? From which region do we get most of the materials we use?

2. What are silicates? List three types.

3. What is a synergistic effect? Give an example.

4. What environmental problems are associated with the manufacture of cement?

5. What is concrete?

6. Why was copper one of the first metals used?

7. What is bronze? How is it superior to copper?

8. Why did the Bronze Age come about before the Iron Age?

9. What environmental problems are associated with steel mills?

10. What environmental problems are associated with aluminum production?

11. If matter is conserved, how can we ever run out of a metal?

12. List three methods of solid waste disposal. Give the advantages and disadvantages of each.

13. What are the environmental three R's of the waste we produce?

14. Explain why metals can be recycled fairly easily. What factors limit the recycling of metals?

PROBLEMS

Composition of Earth

15. Define *lithosphere*, *hydrosphere*, and *atmosphere*.

16. What element makes up most of the core of Earth?

17. What is the most abundant element in the outer portion of Earth?

18. Name four kinds of minerals found in Earth's crust and give an example of each.

19. What materials make up the organic portion of the lithosphere?

20. How did the discovery of fire change the types of materials available to humans?

Silicate Minerals

21. What is the basic structural unit of silicate minerals? Draw it.

22. What is the chemical composition of quartz? How are the basic structural units of quartz arranged?

23. Why does mica occur in sheets? How are the basic structural units of mica arranged?

24. Why does asbestos occur as fibers? How are the basic structural units of chrysotile asbestos arranged?

Modified Silicates

25. How does the structure of glass differ from that of crystalline silicates?

26. How do the properties of glass differ from those of crystalline substances? Why?

27. What are the three principal raw materials used for making glass?

28. How is the basic recipe for glass modified to make glasses with special properties?

29. What is cement?

30. What are the two basic raw materials required for making cement?

Metals and Ores

31. What are the principal raw materials used in modern iron production? What is the purpose of each?

32. What is coke? What is slag?

33. By what kind of chemical process is a metal obtained from its ore? Give an example.

34. By what chemical process does a metal corrode? Give an example.

35. Both aluminum and iron combine (slowly) with oxygen at room temperature. Why is that a problem for iron but not for aluminum?

36. What is an alloy? Give an example.

37. What functions are served by a blast furnace in the metallurgy of iron?

38. What is the purpose of limestone in iron production?

39. What is pig iron? What are its principal impurities?

40. What is steel? How is it made from cast iron?

41. What is the most abundant metal in Earth's crust? Why is it not the most widely used?

42. Which metal is the most widely used? Why?

Oxidation and Reduction

For Problems 43–48, answer the following questions. (These problems require some knowledge of material in Chapter 8.) **(a)** What substance is reduced? **(b)** What is the reducing agent? **(c)** What substance is oxidized? **(d)** What is the oxidizing agent?

43. Manganese metal may be prepared by reacting manganese(IV) oxide with silicon.
$$Si + MnO_2 \longrightarrow SiO_2 + Mn$$

44. The Hunter method for the production of titanium uses the reaction
$$TiCl_4 + 4\,Na \longrightarrow Ti + 4\,NaCl$$

45. Thorium metal is prepared by reacting thorium(IV) oxide with calcium.
$$ThO_2 + 2\,Ca \longrightarrow Th + 2\,CaO$$

46. Potassium metal is prepared by reacting molten potassium chloride with sodium metal.
$$KCl + Na \longrightarrow K + NaCl$$

47. The Toth Aluminum Company produces aluminum from kaolin. One step converts aluminum oxide to aluminum chloride by the following reaction.
$$Al_2O_3 + 3\,C + 3\,Cl_2 \longrightarrow 2\,AlCl_3 + 3\,CO$$

48. Copper(I) oxide, heated in a stream of hydrogen gas, is converted to copper metal.
$$Cu_2O + H_2 \longrightarrow 2\,Cu + H_2O$$

Mass Relationships

Problems 49–52 require some knowledge of material in Chapter 6.

49. Copper metal is prepared by blowing air through molten copper(I) sulfide.
$$Cu_2S + O_2 \longrightarrow 2\,Cu + SO_2$$
How much copper is obtained from 143 g of Cu_2S?

50. What mass in grams of potassium is produced from 1.73 g of KCl? Use the equation in Problem 46.

51. What mass in grams of calcium chloride will be produced from 0.529 g of cesium chloride? The reaction is
$$2\,CsCl + Ca \longrightarrow 2\,Cs + CaCl_2$$

52. What mass in grams of aluminum is required to reduce 33.2 g of calcium oxide? The reaction is
$$3\,CaO + 2\,Al \longrightarrow 3\,Ca + Al_2O_3$$

ADDITIONAL PROBLEMS

53. An aluminum plant produces 65 million kg of aluminum per year. How much aluminum oxide is required? How much bauxite is required? (It takes 2.1 kg of crude bauxite to produce 1.0 kg of aluminum oxide.)

54. It takes 17 kWh of electricity to produce 1.0 kg of aluminum. How much electricity does the plant in Problem 53 use for aluminum production in 1 year?

55. An ore deposit at Crandon, Wisconsin, is estimated at 75 million tons and assays at 5.0% zinc, 0.4% lead, and 1.1% copper. What mass of each metal is contained in the deposit?

56. The Parc mine at Llanrwst, North Wales, United Kingdom, was closed in 1954, leaving behind a mound of about 250,000 tons of tailings. The tailings assayed at 3.22% Zn, 0.82% Pb, and 0.023% Cd. Erosion has washed away 13,000 tons of the tailings. How much of each of the three elements has been washed away?

57. Approximately 83,000 troy ounces of gold are mined worldwide each year. What size cube of gold would this make if it were all in one piece? The density of gold is 19.3 g/cm^3 and 1 troy oz = 31.10 g.

58. The mass of gem-quality rough diamonds mined each year is about 10 million carats. What is the mass of these diamonds in kilograms? (1 carat = 200 mg.)

59. A blast furnace produces 1.0×10^7 kg of pig iron per day. Assume the pig iron is 95% Fe by mass. How many kilograms of iron ore are consumed in this furnace per day? Assume that the ore is 82% by mass hematite (Fe_2O_3).

60. The blast furnace described in Problem 59 produces 0.50 kg of slag per kilogram of pig iron. What is the minimum daily requirement of limestone in this furnace? Assume that the limestone is 91% $CaCO_3$, that the slag is exclusively $CaSiO_3$, and that the limestone is the only source of calcium in the slag.

61. The note on page 314 gives the amount of trash produced per person per day in the United States. If the trash has an average density of 85 lb per cubic foot, how many cubic feet of trash are produced each year by a small town of 14,000 people? If a sanitary landfill is constructed that is 100 feet wide and 100 feet long, how deep must it be to contain this trash?

62. The equation for the reaction by which a cyanide solution dissolves gold from its ore is

$$4\,Au + 8\,NaCN + 2\,H_2O + O_2 \longrightarrow$$
$$4\,NaAu(CN)_2 + 4\,NaOH$$

What is the minimum mass of sodium cyanide required to dissolve the 10 g of Au in a ton of gold ore?

63. Refer back to Figure 11.1 and to the data on the lithosphere in Section 11.1. The Earth is 12756 km in diameter. Measure the diameter of the Earth in the figure, and calculate approximately how thick (in mm) the lithosphere should be on that scale. The atmosphere is about 100 km thick; can the atmosphere be shown to scale?

COLLABORATIVE GROUP PROJECTS

Prepare a PowerPoint, poster, or other presentation (as directed by your instructor) for presentation to the class.

64. Prepare a brief report on one of the metals listed in Table 11.5 and share it with your group. List principal ores and describe how the metal is obtained from one of its ores. Describe some of the uses of the metal and explain how its properties make it suitable for those uses.

65. Prepare a brief report on recycling one of the following types of municipal solid wastes. List advantages and problems involved in the process.
    a. glass
    b. plastics
    c. yard wastes
    d. paper
    e. lead
    f. aluminum

66. Compare disposal in landfills, incineration, and recycling as methods of dealing with each of the types of municipal solid wastes listed in Problem 64. List advantages and disadvantages of each method for each type of waste.

67. Search the Internet for recent mine waste disasters or long-term mine waste problems. Prepare a brief report on one of them.

68. Use the Internet to find the current population of the United States and of the world. What is the current rate of growth, in percent per year in **(a)** Germany, **(b)** Kenya, **(c)** India, **(d)** United States, and **(e)** the world?

REFERENCES AND READINGS

1. Bjornerud, Marcia. *Reading the Rocks: The Autobiography of the Earth.* Cambridge, MA: Westview, 2005.

2. Hill, John W., Ralph H. Petrucci, Terry W. McCreary, and Scott S. Perry. *General Chemistry*, 4th edition. Upper Saddle River, NJ: Prentice Hall, 2005. Chapter 24, "Chemistry of Materials," deals with some of the topics discussed in this chapter in more detail.

3. Kolb, Kenneth E., and Doris K. Kolb "Glass—Sand + Imagination." *Journal of Chemical Education,* July 2000, 812–816.

4. Lesney, Mark. "Eyeing the Glass Past," *Today's Chemist at Work*, February 2004, pp. 55–56.

5. McCoy, Michael. "Electronic Chemicals." *Chemical and Engineering News,* June 28, 2004, pp. 18–24.

6. McGuire, Nancy K. "Ceramics: Beyond the Coffee Mug," *Today's Chemist at Work*, May 2002, pp. 32–36.

7. Ohashi, Nobuo. "Modern Steelmaking." *American Scientist*, November–December 1992, pp. 540–555.

8. Tarbuck, Edward J., Frederick K. Lutgens, and Dennis Tasa. *Earth Science*, 11th edition. Upper Saddle River, NJ: Prentice Hall, 2006.

GREEN CHEMISTRY

Catalysts

Denyce K. Wicht, Suffolk University

In this chapter you learned about the metals and minerals that are extracted from Earth's crust and further refined for material use. The unique properties of these materials, such as corrosion resistance, electrical conductivity, toughness, chemical inertness—just to name a few—make them valuable to humans for a variety of purposes. For example, the ordinary glass used for windows is comprised primarily of sand, or silicon dioxide. The sand is processed with sodium carbonate and calcium carbonate (limestone), to produce a material primarily composed of silicon and oxygen, with smaller amounts of sodium and calcium. The properties that make this material useful for windows include transparency, toughness, and absorbance of ultraviolet light. Certainly everyone appreciates being able to see out from within a building, protected from the weather and potentially damaging sunlight. Rooms would be less desirable to occupy if nontransparent material were used for windows. Glass manufacturing continues on a relatively large scale, producing massive quantities of this material. The high demand for glass is due to the physical properties of this combination of elements in bulk quantities.

Metals and minerals are also useful for the properties they exhibit on the atomic and molecular level. In Chapter 8 you learned about *catalysts*, substances that increase the rate of a chemical reaction without themselves being consumed or destroyed. Catalysts are so useful because they aid in the cleavage and formation of chemical bonds at the molecular level. You may think of them as "assistants" to the movement of electrons from one atom or molecule to another during a chemical reaction. Using catalysts to drive a chemical reaction is desirable because catalysts lower the activation energy of the reaction, making it easier to start the reaction and keep it going. As you learned in the Green Chemistry activity for Chapter 1, the ninth principle of green chemistry promotes the use of catalysts, as they are typically used in small amounts. Each catalyst molecule can react with other molecules in the reaction many times. Catalyst use is preferable to simply "adding more reactant" because the reactants typically are used in excess and only react one time with other molecules during the reaction. Many metals and minerals are regarded as valuable materials due to the catalytic properties they exhibit. In these series of Green Chemistry activities you will learn about some useful chemical reactions that use metals or minerals as catalysts.

WEB INVESTIGATIONS

Investigation 1
Zeolite Catalysts

Zeolites are crystalline solids comprised of silicon, aluminum, and oxygen and are of interest from a catalytic perspective because of the cavities that are formed within the three-dimensional arrangement of atoms and bonds in space. A wide variety of different sizes and shapes are possible for the cavities within different zeolites. Within these cavities, molecules can be held or "sequestered" while electrons, atoms, or other molecules are shuttled to and from the reactant molecule as necessary. Do a keyword search on "zeolites" to learn about the many uses of this interesting technology. Pay particular attention to the industry that relies heavily on the organic reactions catalyzed by zeolites.

▲ This model of a zeolite shows the many sizes and shapes of cavities within the mineral. These cavities can act as sites for catalytic action.

Investigation 2
Catalytic Converters

The burning of fossil fuels for energy is a very effective way to run automotive engines, allowing the operation of cars that transport people from one place to another. Consider for a moment how reliant our society is on automobiles for the purposes of work, school, and travel. Burning fossil fuels in a typical combustion engine, however, can produce undesirable emission of by-products that must be converted to innocuous species before they are emitted from the tailpipe of a car. This is where the catalytic converter comes into play; it is placed between the engine and the muffler in cars to convert harmful emissions to more benign species like nitrogen, oxygen, carbon dioxide, and water. While catalytic converters prevent the release of harmful emissions to the atmosphere, this pollution prevention treatment is considered "end-of-pipe"; it is treating pollution after it has already been formed. As you work through this Web exercise, consider what new green transportation technologies might prevent the production of pollution in the first place. Use the keywords "catalytic converter chemistry" in a Web search to learn more about the important chemistry involved in this process and the metals and minerals that act as catalysts in these systems.

COMMUNICATE YOUR RESULTS

Exercise 1
Zeolites

What major industry relies on the use of zeolites as catalysts for organic reactions? Can you draw a connection between this industry and how it relates to the need for catalytic converters? Metal catalysts are also useful for reduction chemistry; you learned about this general type of reaction in Chapter 8 of this textbook. How does reduction chemistry fit? Based on the information you have learned in this Web exercise, write a two- to three-page analysis of any societal problems you foresee, based on the assumption that the rate of reduction reactions is slow relative to the rate of petroleum refining and combustion reactions.

Exercise 2
Catalytic Converters

What are the common emissions produced by running a car engine? What are the common harmful emissions or by-products produced by a car engine? What are the specific metals and/or minerals commonly used in automotive catalytic converters? Write the balanced chemical transformations with which the catalysts used in catalytic converters assist in the removal of undesirable by-products. What environmental problems would become exacerbated if the natural sources of these metals become scarce? What technological advances can you think of that may make the need for catalytic converters obsolete? Capture your responses to these questions in a two- to three-page report; use a factual, journalistic style.

CHAPTER 12

Air

Earth's atmosphere is a thin, transparent skin of molecules that protects us from harmful ultraviolet rays, provides vital oxygen to living organisms, and warms the planet so that life can flourish. The human impacts on the atmosphere can be damaging, as shown in this 2005 satellite of southern Europe and northern Italy. On March 17, the haze of pollution that often builds up against the mountains of northern Italy extended east over the Adriatic Sea, Slovenia, and Croatia. The smog is so thick that in many places the surface features are barely visible.

The Breath of Life

We could live about a month without food. We could even live for several days without water. But without air we cannot live more than a few minutes. Every breath we take is automatic; we seldom think about it. Under normal conditions, we cannot hold our breath for very long, even if we want to.

We may run short of food, and we may run short of fresh water, but we are not likely to run out of air. On the other hand, we might foul the air so badly in some places that it could become unfit to breathe. In some areas the air is already so bad that people become sick from breathing it, and some even die because of it. The World Health Organization estimates that a billion people live in places where the air is substandard (mostly in urban areas) and that air pollution kills 8000 people a day worldwide. About 90% of the deaths occur in developing countries. We begin with a discussion of the natural chemistry of the atmosphere and then present some of the changes that human intervention is having on that natural balance and some possible solutions.

12.1 EARTH'S ATMOSPHERE: DIVISIONS AND COMPOSITION

We passengers on Spaceship Earth live under a thin blanket of air called the **atmosphere**. It is difficult to measure exactly how deep the atmosphere is. It does not end abruptly but gradually gets thinner as it fades into space. But we know that 99% of the atmosphere lies within 30 km of Earth's surface. This is a thin air layer indeed (a bit like the skin on a large apple, but even thinner, relatively speaking) when compared to the diameter of the Earth.

The atmosphere is divided into layers (Figure 12.1) depending on the temperature variation with altitude within each layer. The layer nearest Earth, the *troposphere*, harbors nearly all living things and nearly all human activity. Although other planets in our solar system have atmospheres, Earth's atmosphere is unique in its ability to support higher forms of life. We normally think that the temperature continually decreases as the altitude increases. For the troposphere this is true, but the opposite trend occurs in the *stratosphere*, when the temperature actually increases as you go higher. This chapter focuses on the troposphere, but we also examine some threats to the ozone layer in the stratosphere. We will not be concerned with higher regions of the atmosphere.

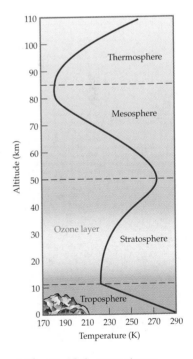

▲ **Figure 12.1** Approximate altitudes and temperature variations of the several layers of the atmosphere. The height of the troposphere varies from about 8 km at the poles to 16 km at the equator.

TABLE 12.1	Composition of Clean Dry Air Near Sea Level
Component	**Number of Molecules**[*]
Nitrogen (N_2)	78,083
Oxygen (O_2)	20,945
Argon (Ar)	934
Carbon dioxide (CO_2)	38
Trace substances[†]	<1

[*]Values are number of molecules per 100,000 molecules of air.

[†]Total less than 1 molecule in 100,000: neon (Ne), helium (He), methane (CH_4), krypton (Kr), hydrogen (H_2), dinitrogen monoxide (N_2O), xenon (Xe), ozone (O_3), sulfur dioxide (SO_2), nitrogen dioxide (NO_2), ammonia (NH_3), carbon monoxide (CO), and iodine (I_2).

Oxygen

Carbon Dioxide

Air is a mixture of gases. Dry air is (by volume) about 78% nitrogen (N_2), 21% oxygen (O_2), and 1% argon (Ar). Damp air can contain up to about 4% water vapor. There are a number of minor constituents, the most important of which is carbon dioxide (CO_2). The concentration of carbon dioxide in the atmosphere increased from about 280 parts per million (ppm) in the preindustrial world to its present value of 380 ppm. The composition of the atmosphere is summarized in Table 12.1.

12.2 CHEMISTRY OF THE ATMOSPHERE

Nitrogen

Nitrogen is a component of many organic chemicals (Chapter 9), and of proteins and nucleic acids vital for living organisms. Although nitrogen makes up 78% of the atmosphere, N_2 molecules can't be used directly by animals or by most plants. Recall that the Lewis structure of N_2 shows that the two nitrogen atoms are held together by a strong triple bond. Therefore, organisms first must *fix nitrogen*—that is, combine nitrogen with another element.

Several natural phenomena convert nitrogen gas to more usable forms. Lightning fixes nitrogen by causing it to combine with oxygen. The equations show that nitrogen monoxide (NO) and nitrogen dioxide (NO_2) are formed.

Nitrogen Dioxide

$$N_2 + O_2 + \text{energy (lightning)} \longrightarrow 2\,NO$$
$$2\,NO + O_2 \longrightarrow 2\,NO_2$$

Nitrogen dioxide reacts with water to form nitric acid (HNO_3).

$$3\,NO_2 + H_2O \longrightarrow 2\,HNO_3 + NO$$

Nitric Acid

The nitric acid falls in rainwater, adding to the supply of available nitrates in the oceans and the soil, but also acidifying streams and lakes (Chapter 13).

Nitrogen is fixed industrially by combining nitrogen with hydrogen to form ammonia, a procedure called the Haber–Bosch process (Chapter 16).

$$N_2 + 3\,H_2 \longrightarrow 2\,NH_3$$

This technology has greatly increased our food supply because the availability of fixed nitrogen is often the limiting factor in the production of food. Not all the consequences of this intervention have been favorable, however; excessive runoff of nitrogen fertilizer has led to serious water pollution problems in some areas (Chapter 13).

Most nitrogen fixation is performed by bacteria in the roots of legumes (peas, beans, clover, and the like). Certain types of bacteria oxidize N_2 to nitrates, while others undertake the reduction of N_2 to ammonia. Other plants are then able to take up the nitrates and ammonia from the soil, and then animal species can get their required nitrogen from the plants. The **nitrogen cycle** (Figure 12.2) is completed by the action of microbes, which can use nitrate as their oxygen source in the decomposition of organic matter and release N_2 gas back to the atmosphere.

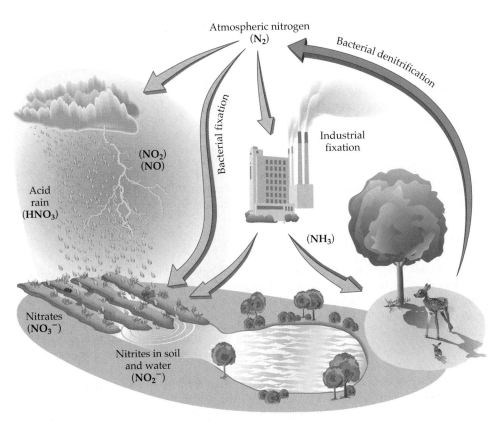

◀ **Figure 12.2** The nitrogen cycle.

The Oxygen Cycle

Oxygen makes up 21% of Earth's atmosphere. As with nitrogen, a balance is set up between consumption and production of oxygen in the troposphere. Both plants and animals use oxygen in the metabolism of foods. The decay and combustion of plant and animal materials consume oxygen and produce carbon dioxide, as does the burning of fossil fuels like coal, natural gas, and oil. The rusting of metals and the weathering of rocks also consume oxygen. A simplified **oxygen cycle**, illustrated in Figure 12.3, shows the influence of green plants, including one-celled organisms

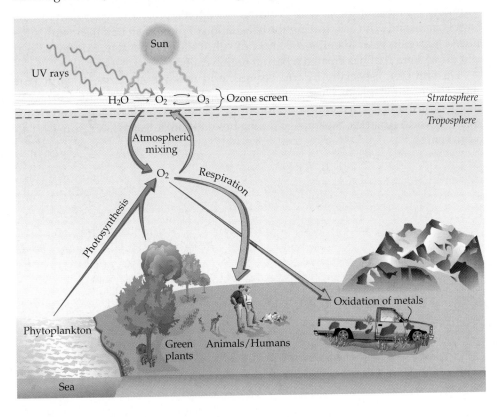

◀ **Figure 12.3** The oxygen cycle.

(phytoplankton) in the sea, which constantly consume carbon dioxide in photosynthesis and also replenish the oxygen supply.

$$6\,CO_2 + 6\,H_2O \longrightarrow C_6H_{12}O_6 + 6\,O_2$$

Another vital balancing act occurs in the stratosphere. There, oxygen is formed by the action of ultraviolet (UV) rays on water. Perhaps more important, some oxygen is converted to ozone (Section 12.12).

$$3\,O_2(g) + \text{energy (UV radiation)} \longrightarrow 2\,O_3(g)$$

This ozone absorbs high-energy UV radiation that might otherwise make higher forms of life on Earth impossible. We discuss the atmosphere in more detail in subsequent sections, paying particular attention to changes wrought by human activities.

Ozone

12.3 TEMPERATURE INVERSION

Normally, air is warmest near the ground and gets cooler with increased altitude. On clear nights, the ground cools rapidly, but the wind usually mixes the cooler air near the ground with the warmer air above it. Occasionally the air is still, and the lower layer of cold air becomes trapped by the layer of warm air above it. This condition is known as a **temperature inversion** (or a thermal inversion). Pollutants in the cooler air are trapped near the ground, and the air can become quite seriously polluted in a short time (Figure 12.4).

A temperature inversion can also occur when a warm front collides with a cold front. The less dense warm air mass slides over the cold air mass, producing a condition of atmospheric stability. Because there is no vertical air movement, the air stagnates and air pollutants accumulate.

12.4 NATURAL POLLUTION

Air pollution has always been with us. Wildfires, windblown dust, and volcanic eruptions added pollutants to the atmosphere long before humans existed. Volcanoes spew ash and poisonous gases into the atmosphere. In 2005, several volcanic explosions in Colima, Mexico, sent dust and ash to altitudes of 20,000 feet. The 1995 eruptions of the Soufriere Hills volcano on Montserrat covered much of the Caribbean island with ash, forcing the evacuation of more than half the country. Kilauea volcano in Hawaii emits 200–300 t of sulfur dioxide per day, leading to acid rain downwind that has created a barren region called the Kau Desert.

Dust storms, especially in arid regions, add massive amounts of particulate matter to the atmosphere. Dust from the Sahara Desert often reaches the Caribbean and South America. Dust and pollution from China blanket Japan and even reach Western North America. Swamps and marshes emit noxious gases. Nature isn't always benign.

▲ The 1991 eruption of Mount Pinatubo in the Philippines deposited ash up to 33 cm deep, and 10 cm of ash covered an area of 2000 km^2. From 200 to 800 people (accounts vary) died during the eruption, most as heavy wet ash collapsed the roofs of their houses. Mount Pinatubo also ejected between 15 and 30 million tons of sulfur dioxide gas, which reacted with water and oxygen in the atmosphere to form an aerosol cloud of sulfuric acid. This eruption was 10 times larger than that of Mount St. Helens in 1980.

▶ **Figure 12.4** Ordinarily, the air gets colder as the altitude increases (a). During an atmospheric (thermal) inversion (b), a cold layer near the surface lies beneath a warmer layer.

(a)

(b)

A Closed Ecosystem?

If you think for a moment about the oxygen and nitrogen cycles, you may realize that a *closed ecosystem* should be possible. In such a system, the plant and animal life would be balanced. The O_2 given off by the plants would sustain the animal life, and the CO_2 given off by the animals would be sufficient to keep the plants alive. Other animal wastes would supply the nitrogen needed by the plants, and the animals would partially consume the plants or one another for food. Trace elements would need to be present in sufficient quantity. Sunlight would provide the energy necessary to "run" the ecosystem; nothing else goes in and nothing comes out.

Earth is itself a closed ecosystem. The planet absorbs sunlight energy, but no significant amount of matter enters or leaves. But can such a system be constructed artificially? The answer is of great interest in a number of fields, including space exploration. A long space voyage would require that huge amounts of resources be carried along—at enormous expense. The required resources could be minimized by carrying along plants to provide oxygen and consume carbon dioxide and human wastes.

Tiny ecosystems are available commercially as desktop novelties. These systems have a limited lifetime of a few years. *Biosphere II* was a large-scale experiment involving humans and a semiclosed ecosystem.

▲ The Ecosphere® is a sealed, self-contained ecosystem in which plants and animals can live for a year or more, with sunlight as the only "input."

The experiment was cut short because of several problems, including rising levels of nitrogen oxides, water pollution, unforeseen low levels of carbon dioxide, and other difficulties that could only be examined in such an experiment. Although additional studies and experiments are slated, Earth remains the only long-term successful closed ecosystem.

12.5 THE AIR OUR ANCESTORS BREATHED

People have always altered their environment. The discovery of tools and the use of fire forever changed the balance between people and their environment. The problems of air pollution probably first occurred when fires were built in poorly ventilated caves or other dwellings. People cleared land, making larger dust storms possible. They built cities, and the soot from their hearths and the stench from their wastes filled the air. The Roman author Seneca wrote in C.E. 61 of the stink, soot, and "heavy air" of the imperial city. In 1257 the queen of England had to move away from the city of Nottingham because she could not endure the heavy smoke. The industrial revolution brought even worse air pollution. People burned coal to power factories and to heat homes. Soot, smoke, and sulfur dioxide filled the air. The good old days? Not in factory towns. But there were large rural areas that were relatively unaffected by air pollution.

Air pollution today is much more complex than that our ancestors worried about. When society changes its activities, it often changes the nature of its waste materials, including the ones it dumps into the atmosphere. Earth has gone through many changes. Change is not new, but today's rapid rate of change is unprecedented. The present pace of change is at a level that our environment may not be able to absorb.

12.6 POLLUTION GOES GLOBAL

Air pollution problems are no longer just local concerns. Problems are worldwide. Winds and precipitation do purify the air, but we are pouring more pollutants into the atmosphere than it can readily handle. Air pollution knows no

Air Pollution in China

As China strives to become an industrial power, its people are paying a heavy price in pollution. Coal burning supplies about three-fourths of China's commercial energy needs. China now accounts for 27% of the world's steel output, consuming huge amounts of coal and power. The coal has a high sulfur content, and emission controls are often inadequate. As a result, levels of sulfur dioxide and particulate matter (ash, soot, and dust) are among the highest in the world. More than half of 500 Chinese cities failed to meet national air quality standards in 2004. More than 80% of these cities had sulfur dioxide or nitrogen oxide levels above maximum guideline levels set by the World Health Organization (WHO). Many of the cities had sulfur dioxide emissions more than double the standard. Many also exceeded WHO guidelines for particulate matter. According to World Bank estimates, some 590,000 people a year will suffer premature deaths due to urban air pollution in China in the first two decades of the twenty-first century.

China's fleet of motor vehicles will grow from 5 million in 2005 to 140 million by 2020, according to the Chinese Ministry of Communications. Most vehicles are operated in large cities, and because few have effective emission controls, they contribute heavily to the smog in these cities.

China is beginning to attack the problem by closing heavily polluting factories in some larger metro-

▲ The polluted air of Beijing, China, shrouds the capital in a nasty haze on November 5, 2005. People were warned to stay indoors as the pollution index hit its highest level for the second day.

politan areas. The Chinese government has also invested in gas and in cleaner, more efficient briquettes as replacements for raw coal as a fuel for domestic cooking and heating.

political boundaries: Pollution originating in the United States contributes to acid rain in Canada and to strained relationships between the two countries. Contaminants from England and Germany pollute the snow in Norway. Satellites orbiting at 150 km and higher can trace plumes of pollution from China to North America. Within the United States, Los Angeles smog drifts to Colorado and beyond. Pollution from Midwestern power plants leads to acid rain in the Northeast.

As population increases, the world is becoming more urban. Especially in developing countries, industrialization takes precedence over the environment. Cities in China, Iran, Mexico, Indonesia, and many other countries have experienced frightening episodes of air pollution. Large metropolitan areas are most afflicted by air pollution, but rural areas are affected, too. Neighborhoods around smoky factories have seen evidence of increased rates of spontaneous abortion and poor wool quality in sheep, decreased egg production and high mortality in chickens, and increased feed and care requirements for cattle. Plants are stunted, deformed, and even killed.

What is a **pollutant**? It is too much of any substance in the wrong place or at the wrong time. A chemical may be a pollutant in one place and helpful in another. For example, ozone is a natural and important constituent of the stratosphere, where it shields Earth from life-destroying UV radiation. In the troposphere, however, ozone is a dangerous pollutant (Section 12.12).

12.7 COAL + FIRE → INDUSTRIAL SMOG

The word *smog* is a contraction of the words *smoke* and *fog*. There are two basic types of smog. Polluted air associated with industrial activities is often called **industrial smog**. (*Photochemical* smog is discussed in Section 12.9.) It is characterized by the

An Air Pollution Episode: London, England

So common a problem in London, industrial smog was once called *London smog*. A notorious episode of this type of smog began in London on Thursday, December 4, 1952. A large, cold air mass moved into the valley of the Thames River. A temperature inversion placed a blanket of warm air over the cold air. With nightfall, a dense fog and below-freezing temperatures caused the people of London to heap sulfur-rich coal into their stoves. These fires and coal-fired power plants poured sulfur dioxide and soot into the air. The next day (Friday), schools closed and transportation was disrupted. People continued to burn coal because the temperature remained below freezing.

Saturday was a day of darkness. For 20 mi around London, no light came through the smog. A performance of the opera *La Traviata* had to be abandoned because smog obscured the stage. The air remained cold and still, and the coal fires continued to burn throughout the weekend. On Monday, December 8, more than 100 people died of respiratory and related conditions. People tried to sleep sitting up in chairs in order to breathe a little easier. The city's hospitals overflowed with patients with pneumonia and bronchitis.

By the time a breeze cleared the air on Tuesday, December 9, more than 4000 deaths had been attributed to the smog. Many others became ill, some of whom died later. The total premature deaths before the death rate returned to normal the following summer are now estimated to be as many as 12,000 people. That is more people than were ever killed in any single tornado, mine disaster, shipwreck, or airplane crash. A second incident occurred in December of 1991, in which 160 people died. Air pollution episodes may not be as dramatic as other disasters, but they can be just as deadly.

▲ The great London smog episode of 1952.

presence of smoke, fog, sulfur dioxide, and particulate matter such as ash and soot. The burning of coal, especially high-sulfur coal such as that found in the eastern United States, China, and eastern Europe, causes most industrial smog.

Chemistry of Industrial Smog

The chemistry of industrial smog is fairly simple. High-grade coal is a complex combination of organic materials and inorganic materials. The organics are mainly carbon and burn when the coal is combusted. The inorganic materials—minerals—wind up as ash. Some coal can contain 10% or more sulfur. When burned, the carbon in coal is oxidized to carbon dioxide and heat is given off.

$$C(s) + O_2(g) \longrightarrow CO_2(g) + \text{heat}$$

Not all the carbon is completely oxidized, and some of it winds up as carbon monoxide.

$$2\,C(s) + O_2(g) \longrightarrow 2\,CO(g)$$

Still other carbon, essentially unburned, ends up as soot.

Sulfur Oxides

The sulfur in coal also burns, forming sulfur dioxide, a choking, acrid gas.

$$S(s) + O_2(g) \longrightarrow SO_2(g)$$

Sulfur dioxide is readily absorbed in the respiratory system. It is a powerful irritant and is known to aggravate the symptoms of people who suffer from asthma, bronchitis, emphysema, and other lung diseases.

And things get worse. Some of the sulfur dioxide reacts further with oxygen in the air to form sulfur trioxide.

$$2\,SO_2(g) + O_2(g) \longrightarrow 2\,SO_3(g)$$

▲ Burning of coal in electric power plants and other industrial operations is the main source of industrial smog characterized by high levels of sulfur dioxide, particulate matter, and sometimes carbon monoxide. Industrial smog can be alleviated by the use of electrostatic precipitators, scrubbers, and other devices.

Sulfur Trioxide

The two oxides of sulfur are often designated collectively as SO_x. Sulfur trioxide then reacts with water to form sulfuric acid.

$$SO_3(g) + H_2O(l) \longrightarrow H_2SO_4(l)$$

Fine droplets of this acid form an aerosol, a suspension of very small particles of a solid or droplets of a liquid dispersed in a gas. Sulfuric acid is a strong acid, and this aerosol is even more irritating to the respiratory tract than sulfur dioxide.

Particulate Matter

Usually, industrial smog also has high levels of **particulate matter (PM)**, solid and liquid particles of greater than molecular size. The largest particles are often visible in the air as dust and smoke. PM consists mainly of the mineral matter that occurs in coal and of *soot* (unburned carbon). Minerals do not burn, even in the roaring fire of a huge factory or power plant boiler. Some mineral matter is left behind as *bottom ash*, but much of it is carried aloft in the tremendous draft created by the fire. This *fly ash* settles over the surrounding area, covering everything with dust. It is also inhaled, contributing to respiratory problems in animals and humans. Small particles—those less than 10 μm in diameter (PM10 particulates)—are believed to be especially harmful. They contribute to respiratory and heart disease. The EPA also monitors fine particles, those less than 2.5 μm in diameter (PM2.5). EPA studies indicate that as many as 40,000 premature deaths a year are linked to PM and that the majority of them are due to PM2.5. PM concentrations have been lowered significantly since nationwide monitoring began in 1999. However, 62 million people lived in counties with particle pollution levels higher in 2003 than EPA's standards for PM2.5, PM10, or both.

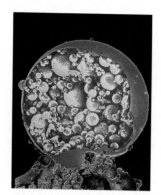

▲ False-color scanning electron micrograph of PM from a coal-burning power plant.

Health and Environmental Effects of Industrial Smog

Minute droplets of liquid sulfuric acid and smaller particulates (PM10 and smaller) are easily trapped in the lungs. Interaction may considerably magnify the harmful effects of pollutants. A certain level of sulfur dioxide, without the presence of PM, might be reasonably safe. A particular level of PM might be fairly harmless without sulfur dioxide around, but the effect of both together might well be deadly. *Synergistic effects* such as this are quite common whenever chemicals act together.

When the pollutants in industrial smog come into contact with the alveoli of the lungs, the cells are broken down. The alveoli lose their resilience, making it difficult for them to expel carbon dioxide. Such lung damage contributes to pulmonary emphysema, a condition characterized by increasing shortness of breath.

The oxides of sulfur and the aerosol mists of sulfuric acid also damage plants. Leaves become bleached and splotchy when exposed to SO_x. The yield and quality of farm crops can be severely affected. These compounds are also major ingredients in the production of acid rain.

What to Do About Industrial Smog

Much effort goes into the prevention and alleviation of industrial smog. PM can be removed from smokestack gases in several ways:

Fly ash particle

50,000 volts

Stack gas flow

▲ **Figure 12.5** A cross section of a cylindrical electrostatic precipitator for removing particulate matter from smokestack gases. Electrons from the negatively charged discharge electrode (in the center) become attached to the particles of fly ash, giving them a negative charge. The charged particles are then attracted to and deposited on the positively charged collector plate.

- An **electrostatic precipitator** (Figure 12.5) induces electric charges on the particles, which are then attracted to oppositely charged plates and deposited.
- *Bag filtration* works much like the bag in a vacuum cleaner. Particle-laden gases pass through filters in a bag house. These filters can be cleaned by shaking and by periodically blowing air through them in the opposite direction.
- A *cyclone separator* is arranged so that the stack gases spiral upward with a circular motion. The particles hit the outer walls, settle out, and are collected at the bottom.
- A **wet scrubber** removes PM by passing stack gases through water. The water is usually sprayed in as a fine mist. The wastewater has to be treated to remove the particulates, which adds to the cost of this method.

The choice of device depends on the type of coal being burned, the size of the power plant, and other factors. All require energy—electrostatic precipitators use 10% of the plant's output—and the collected ash has to be put somewhere. Some of the ash is used to make concrete, as a substitute for aggregate in road base, as a soil modifier, and for backfilling mines. The rest has to be stored; most of it goes into ponds, and the rest into landfills.

It is harder to remove sulfur dioxide than PM. Sulfur can be removed by processing coal before burning, but both the flotation method and gasification or liquefaction processes (Chapter 14) are expensive. Another way to get rid of sulfur is to scrub sulfur dioxide out of stack gases after the coal has been burned. The most common scrubber uses the limestone–dolomite process. Limestone ($CaCO_3$) and dolomite (a mixed calcium–magnesium carbonate) are pulverized and heated. Heat drives off carbon dioxide to form calcium oxide (lime), a basic oxide that reacts with sulfur dioxide to form solid calcium sulfite ($CaSO_3$).

$$CaCO_3(s) + heat \longrightarrow CaO(s) + CO_2(g)$$
$$CaO(s) + SO_2(g) \longrightarrow CaSO_3(s)$$

This by-product presents a sizable disposal problem. Removal of 1 t of sulfur dioxide produces almost 2 t of solids. Some modified scrubbers oxidize the calcium sulfite to calcium sulfate, $CaSO_4$.

$$2\,CaSO_3(s) + O_2(g) \longrightarrow 2\,CaSO_4(s)$$

Calcium sulfate is much more useful than calcium sulfite because the sulfate is used to make commercial products such as plasterboard. In nature, calcium sulfate occurs as gypsum ($CaSO_4 \cdot 2\,H_2O$), which has the same composition as hardened plaster.

12.8 AUTOMOBILE EMISSIONS

When a hydrocarbon burns in sufficient oxygen, the products are carbon dioxide and water. For example, consider the combustion of octane, one of the hundreds of hydrocarbons that make up the mixture we call gasoline:

$$2\,C_8H_{18}(l) + 25\,O_2(g) \longrightarrow 18\,H_2O(g) + 16\,CO_2(g)$$

Because both of the products are normal constituents of air, they are not generally considered pollutants. Unfortunately, internal combustion engines do not burn fuel with perfect efficiency, and there are compounds other than hydrocarbons in gasoline. The next sections describe several types of pollutants that result.

Carbon Monoxide: The Quiet Killer

When insufficient oxygen is present during combustion, carbon monoxide (CO) is formed. Millions of metric tons of this invisible but deadly gas are poured into the atmosphere each year. In the United States, carbon monoxide makes up more than 60% (by mass) of all air pollutants entering the atmosphere, with more than three-fourths of all CO emissions coming from transportation sources.

In urban areas, the motor vehicle contribution to carbon monoxide pollution can be 85 to 95%. Other sources of CO include industrial processes such as metals processing, residential wood burning, and natural sources such as forest fires. The U.S. EPA has set danger levels of 9 ppm carbon monoxide (average) over 8 h and 35 ppm (average) over 1 h. On streets, danger levels are exceeded much of the time. Even in off-street urban areas, levels often average 7–8 ppm. Such levels do not cause immediate death, but over a long period exposure can cause physical and mental impairment.

Synergistic interactions of asbestos and cigarette smoke were discussed in Chapter 11, and synergistic interactions of drugs are discussed in Chapter 19.

The incidence of asthma has increased dramatically in recent years. It is unlikely that air pollution is the direct cause of the increase. Rather, medical studies show that pollutants at moderate levels can exert a small but measurable effect on the lung function and level of symptoms of asthmatic individuals. It seems more likely that pollutants and allergens interact to enhance the severity of asthma attacks.

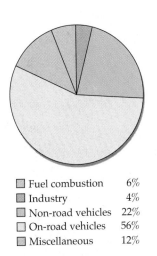

☐ Fuel combustion	6%
☐ Industry	4%
☐ Non-road vehicles	22%
☐ On-road vehicles	56%
☐ Miscellaneous	12%

▲ Sources of carbon monoxide in the United States.

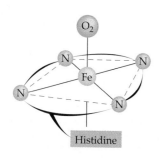

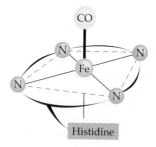

▲ Figure 12.6 Schematic representations of a portion of the hemoglobin molecule. Histidine is an amino acid. Carbon monoxide bonds much more tightly than oxygen, as indicated by the heavier bond line.

 Heme Group

Carbon monoxide is a nonirritating, invisible, odorless, tasteless gas. We can't tell that it is around except by using a CO detector or test reagents. Drowsiness is usually the only symptom, and drowsiness is not always unpleasant. How many auto accidents are caused by drowsiness or sleep induced by carbon monoxide? No one knows for sure.

Carbon monoxide exerts its insidious effect by tying up the hemoglobin in the blood. The normal function of hemoglobin is to transport oxygen (Figure 12.6). Carbon monoxide binds to hemoglobin so strongly that the hemoglobin is hindered in carrying oxygen. The symptoms of carbon monoxide poisoning are therefore those of oxygen deprivation. All except the most severe cases of acute carbon monoxide poisoning are reversible, but prolonged hospital stays with oxygen therapy are sometimes necessary.

In chronic carbon monoxide poisoning, the heart has to work harder to supply oxygen to the tissues, and this may lead to an increased chance of a heart attack. People already suffering from heart or lung disease may be severely affected. There are thousands of cases of carbon monoxide poisoning in the United States each year, and hundreds die. Nonlethal levels of carbon monoxide have been shown to impair time-interval discrimination. In severe cases, performance on psychomotor tests is impaired.

Nitrogen Oxides: Some Chemistry of Amber Air

Nitrogen oxides are formed when N_2 reacts with O_2 at high temperatures. Because air is 78% N_2 and 21% O_2, nitrogen oxides can be formed during the combustion of any fuel whether or not nitrogen is present in the fuel. Automobile exhaust and power plants that burn fossil fuels are major sources of nitrogen oxides. When nitrogen and oxygen combine, the main product is nitrogen monoxide (NO or nitric oxide).

$$N_2(g) + O_2(g) \longrightarrow 2\,NO(g)$$

The NO is then oxidized to nitrogen dioxide.

$$2\,NO(g) + O_2(g) \longrightarrow 2\,NO_2(g)$$

Nitrogen dioxide is an amber-colored gas that causes eye irritation and a brownish haze. The two nitrogen oxides are often designated collectively as NO_x; they play a vital (villain's) role in atmospheric chemistry.

At present atmospheric levels, nitrogen oxides don't seem particularly dangerous in themselves. Nitric oxide at high concentrations reacts with hemoglobin; as in carbon monoxide poisoning, this leads to oxygen deprivation. Such high levels seldom, if ever, result from ordinary air pollution, but they might be reached in areas close to industrial sources. Nitrogen dioxide is an irritant to the eyes and the respiratory system. Tests with laboratory animals indicate that chronic exposure in the range of 10–25 ppm might lead to emphysema and other degenerative diseases of the lungs.

The most serious environmental effect of nitrogen oxides is that they lead to smog formation. These gases also contribute to the fading and discoloration of fabrics, and, by forming nitric acid, contribute to acid rain (Section 12.10). They also contribute to crop damage, although their specific effects are difficult to separate from those of sulfur dioxide and other pollutants.

Volatile Organic Compounds

Substances called **volatile organic compounds (VOCs)** are major contributors to smog formation. There are many sources of VOCs, including gasoline vapors, combustion products of fuels, and consumer products such as paints and aerosol sprays. Many VOCs are hydrocarbons.

Hydrocarbons can act as pollutants even when they aren't burned. They are released from a variety of natural sources, such as decay in swamps. Only about 15%

of all hydrocarbons found in the atmosphere are put there by people. In most urban areas, however, the processing and use of gasoline are the major sources of hydrocarbons in the environment. Gasoline can evaporate anywhere along the line, contributing substantially to the total amount of hydrocarbons in urban air. The automobile's internal combustion engine also contributes by exhausting unburned and partially burned hydrocarbons.

Certain hydrocarbons, particularly alkenes, combine with oxygen atoms or ozone molecules to form aldehydes. As a class, aldehydes have foul, irritating odors. Another series of reactions involving hydrocarbons, oxygen, and nitrogen dioxide leads to the formation of peroxyacetyl nitrate (PAN).

$$\underset{\text{PAN}}{CH_3\overset{\overset{\displaystyle O}{\|}}{C}-O-ONO_2}$$

Ozone, aldehydes, and PAN are responsible for much of the destruction wrought by smog. They make breathing difficult and cause the eyes to smart and itch. People who already have respiratory ailments may be severely affected, and the very young and the very old are particularly vulnerable.

Hydrocarbons from automobiles come largely from those with no pollution control devices or those with devices that are not operating properly. Significant amounts enter the atmosphere from spillage during automobile fueling.

12.9 PHOTOCHEMICAL SMOG: MAKING HAZE WHILE THE SUN SHINES

Individually ozone, aldehydes, and PAN are worrisome and have dramatic environmental impacts, including acid rain. Collectively however, and with the inclusion of sunlight, these can form a complex series of reactions that produce **photochemical smog**, visible as an amber haze.

Unlike industrial smog, which accompanies cold, damp air, photochemical smog usually occurs during dry, sunny weather. The warm, sunny climate that has drawn so many people to the Los Angeles area is also the perfect setting for photochemical smog at any time of year. The principal culprits are unburned hydrocarbons (HC) and nitrogen oxides (NO_x) from automobiles.

The chemistry of photochemical smog is exceedingly complex (Figure 12.7). In the initial step, NO_2 absorbs a photon of sunlight and breaks apart to NO and O. The oxygen atoms are quite reactive; they combine with other components of automobile exhaust and the atmosphere to produce a variety of irritating and toxic chemicals, including ozone.

▲ Photochemical smog results from the action of sunlight on oxides of nitrogen emitted from automobiles and other high-temperature combustion sources. An amber haze like that shown here over Paris in June 2003, characterizes this type of smog.

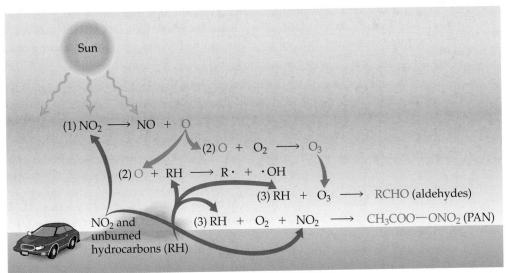

◀ **Figure 12.7** Some chemical processes in the formation of photochemical smog. Highly reactive oxygen atoms are formed in the initiating steps. Many other reactive intermediates have been omitted from this simplified scheme.

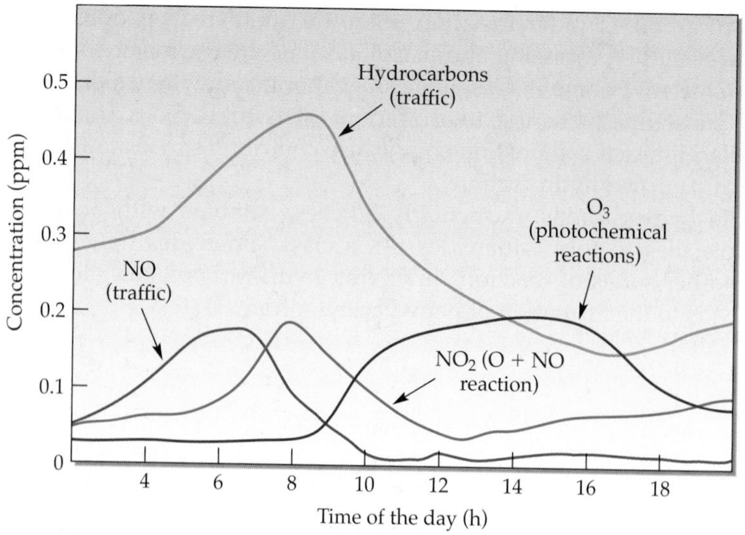

▶ **Figure 12.8** Concentrations of several urban air pollutants at different times during a typical sunny day. Hydrocarbons and NO are produced first. As they interact with sunlight, NO₂ and then ozone forms.

The development of air pollutants on a typical sunny summer day is shown in Figure 12.8.

Solutions to Photochemical Smog

Photochemical smog requires the action of sunlight on nitrogen oxides and hydrocarbons. Reducing the quantity of any of these would diminish the amount of smog. It isn't likely that we would want to reduce the amount of sunlight, so let's focus on the other two.

Hydrocarbons have many important uses as solvents. If we could reduce the quantity of hydrocarbons entering the atmosphere, the amounts of aldehydes and PAN formed would also be reduced. Improved design of storage and dispensing systems has decreased hydrocarbon emissions from gasoline stations. Similarly, modified gas tanks and crankcase ventilation systems have reduced evaporative emissions from automobiles. The main approach, however, has been through the use of **catalytic converters** to reduce hydrocarbon and carbon monoxide emissions in automotive exhausts. The oxidation catalyst in these converters is a precious metal such as platinum (Pt) and/or palladium (Pd). Hydrocarbons and carbon monoxide react rapidly with oxygen on the surface of the metal, forming water and carbon dioxide.

Reducing the quantity of nitrogen oxides is more difficult. Whereas carbon monoxide and hydrocarbons are removed by oxidation, NO must be reduced to N_2. Lowering the operating temperature of an engine helps, but the engine then becomes less efficient. Running an engine on a richer (more fuel, less air) mixture lowers NO_x emissions but tends to raise carbon monoxide and hydrocarbon emissions. With a separate reduction catalyst, such as rhodium (Rh) or palladium (Pd), some of the carbon monoxide can be used to reduce the quantity of nitric oxide.

$$2\,NO(g) + 2\,CO(g) \longrightarrow N_2(g) + 2\,CO_2(g)$$

The remaining carbon monoxide and the hydrocarbons are then oxidized over the platinum catalyst. Over the past 30 years, the use of catalytic converters has prevented more than 12 billion tons of harmful exhaust gases from entering Earth's atmosphere.

Several states have mandated increases in the fraction of *hybrid vehicles* sold. A hybrid vehicle uses both an electric motor and a gasoline engine. The small gasoline engine drives the vehicle most of the time and recharges the batteries that run the electric motor. The electric motor provides additional power when it is needed (for

Catalytic Converters and Nanotechnology

For maximum effectiveness, the catalysts inside a catalytic converter must have a large surface area because the reactions take place on the surface of the precious metal. To achieve this, the catalyst is thinly coated on a collection of small ceramic beads or honeycomb matrix. One liter of ceramic pellets can have up to 500,000 m^2 of support surface. A honeycomb mesh support can have a surface area of about 23,000 m^2. The support is coated with 1 or 2 g of precious metals such as palladium, rhodium, and platinum. (These metals really are precious; gold costs about $18 per gram, but rhodium costs about $100 per gram.)

Catalytic converters have used nanotechnology for several decades. A converter only uses about 4–5 g of precious metal to coat the honeycomb structure or pellets. This exposes the exhaust stream of the vehicle to a catalyst area equal to that of several football fields.

Researchers at the University of California, Davis, are studying iridium (Ir) catalysts just four atoms in size. Catalysts often are coated onto a support material (such as ceramic), but these researchers found that the catalytic nanoclusters are actually chemically bonded to the support. Previously the support material was thought to be inert; these workers found that modifying the support material under the nanoparticles actually changed the efficiency of the catalyst up to tenfold, possibly opening the way for a new class of nano-sized catalysts. This work could lead to the ability to design new catalysts for specific functions.

Experiments on cerium(IV) oxide nanoparticles carried out at Brookhaven National Laboratory may lead to catalytic converters that are better at cleaning up auto exhaust. Researchers there used a novel technique to synthesize the CeO_2 nanoparticles and then impregnated them with zirconium (Zr) in one case and with gold (Au) in another. The zirconium-doped nanoparticles store or release oxygen, depending on the engine conditions. This made the catalyst more efficient in converting CO and NO to CO_2 and N_2.

Other experiments have shown that gold-doped nanoparticles act as active catalysts in helping CO combine with O_2 to make CO_2. Gold is not catalytic in its bulk form, but when made into particles of less than 6 nm, it is a very effective catalyst.

Nanoscale catalysts open the way for many process innovations that will make various chemical processes more efficient and thus save resources.

example, when climbing a hill). Hybrid vehicles are significantly more efficient than conventional gasoline-engined vehicles, and several manufacturers now market hybrids.

12.10 ACID RAIN: AIR POLLUTION → WATER POLLUTION

We have seen how sulfur oxides are converted to sulfuric acid (Section 12.7) and nitrogen oxides to nitric acid (Section 12.2). These acids fall on Earth as acid rain or acid snow or are deposited from acid fog or adsorbed on PM. **Acid rain** is defined as rain having a pH less than 5.6. Rain with a pH as low as 2.1 and fog with a pH of 1.8 have been reported (Figure 12.9). These values are lower than the pH of vinegar or lemon juice.

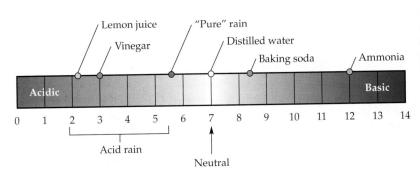

◄ **Figure 12.9** The pH of acid rain in relation to that of other familiar substances. Normal rainwater, if saturated with dissolved CO_2, has a pH of 5.6. During thunderstorms, the pH of rainwater can be much lower because of nitric acid formed by lightning. Recall (Chapter 7) that a solution with a pH value one unit lower than that of another is 10 times as acidic. A decrease in pH from 5.6 to 4.6, for example, means an increase in acidity by a factor of 10. Two pH units lower means 100 times as acidic; and three pH units lower means 1000 times as acidic.

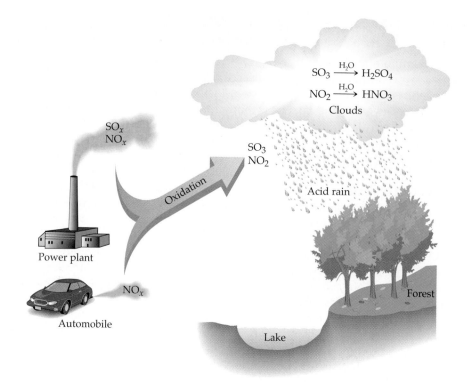

▶ **Figure 12.10** Acid rain. The two principal sources of acid rain are sulfur dioxide (SO_2) from power plants, and nitric oxide (NO) from power plants and automobiles. These oxides are converted to SO_3 and NO_2, which then react with water to form H_2SO_4 and HNO_3. The acids fall in rain, often hundreds of kilometers from their sources.

▲ Marble statues are slowly eroded by the action of acid rain on marble (calcium carbonate).

Acid rain comes mainly from sulfur oxides emitted from power plants and smelters and from nitrogen oxides discharged from power plants and automobiles. These acids are often carried far before falling as rain or snow (Figure 12.10).

$$Fe(s) + H_2SO_4(aq) \longrightarrow FeSO_4(aq) + H_2(g)$$

Iron (in steel) ⟶ A soluble salt

Acids corrode metals and can even erode stone buildings and statues. Sulfuric acid eats away metal to form a soluble salt and hydrogen gas.

$$CaCO_3(s) + H_2SO_4(aq) \longrightarrow CaSO_4(aq) + H_2O + CO_2(g)$$

Marble or limestone

The reaction shown here is oversimplified. For example, in the presence of water and oxygen (air), the iron is converted to rust (Fe_2O_3). Marble buildings and statues are disintegrated by sulfuric acid in a similar reaction that forms calcium sulfate, a slightly soluble, crumbly compound.

12.11 THE INSIDE STORY: INDOOR AIR POLLUTION

Indoor air pollution can pose a major health concern. It can be bad in the United States but is often much worse in developing countries where people cook and heat using wood, coal, or even dung as fuel.

EPA studies on human exposure to air pollutants indicate that indoor air levels of many pollutants may be 2–5 times higher, and on occasion more than 100 times higher, than outdoor levels. Near high-traffic areas, the air indoors has the same carbon monoxide level as the air outside. In some areas, office buildings, airport terminals, and apartments have indoor CO levels that exceed government safety stan-

Wood Smoke

Many people like the smell of smoke from burning wood or leaves. However, wood smoke is far from benign. Each kilogram of wood burned yields 80–370 g of carbon monoxide, 7–27 g of VOCs, 0.6–5.4 g of aldehydes, 1.8–2.4 g of acetic acid, and 7–30 g of particulates. The smoke also contains hydrocarbons such as methane (14–25 g/kg wood), benzene (0.6–4.0 g), naphthalene (0.24–1.6 g), and a variety of alkylbenzenes (1–6 g) and other familiar pollutants such as nitrogen oxides (0.2–0.9 g) and sulfur oxides (0.16–0.24 g).

Aldehydes irritate the eyes and nasal passages, accounting in part for the effect of smoke on those gathered around a campfire. But wood smoke is more than an irritant; it contains a wide variety of carcinogenic polycyclic aromatic hydrocarbons (PAHs), chlorinated dioxins, and related compounds. The odor of wood smoke may evoke memories of autumns past or singing around a campfire, but it can also lead to asthma attacks and possible long-term harm to the lungs.

dards. Woodstoves, gas stoves, cigarette smoke, and unvented gas and kerosene space heaters are also sources of CO indoors.

The Home: No Haven from Air Pollution

We try to save energy by installing better insulation and by sealing air leaks around windows and doors, but in doing so we often make indoor pollution worse by trapping pollutants inside. A kitchen with a gas range often has levels of nitrogen oxides above U.S. government standards. Freestanding kerosene stoves produce levels of NO_x up to 20 times greater than those permitted by federal regulations for outdoor air. (These regulations do not apply to indoor air.)

Cigarettes and Secondhand Smoke

Another indoor polluter is tobacco smoke. Only about 15% of cigarette smoke is inhaled by the smoker; the rest remains in the air for others to breathe. More than 40 carcinogens have been identified among the 4000 or so chemical compounds found in cigarette smoke. The EPA classifies secondhand smoke as a Class A carcinogen, a substance known to cause cancer in humans.

The carcinogens in tobacco smoke are similar to those in wood smoke. Tobacco smoke has PAHs, aromatic amines, nitrosamines, and radioactive polonium-210.

The health effects of smoking on the smoker, widely publicized in a series of reports released by the Surgeon General of the United States, are well known, including its role as the principal cause of emphysema (Chapter 18). The first report, published in 1964, led to a ban on television advertisements for cigarettes. Reports have also dealt with the effects of tobacco smoke as an air pollutant. Air quality in smoke-filled rooms is poor. Smoke from just one burning cigarette can raise the level of particulate matter above government standards. Even with good room ventilation, the level of carbon monoxide is often equal to, or greater than, the legal limits for ambient air.

The risk to nonsmokers from cigarette smoke is well established. Medical research has shown that nonsmokers who regularly breathe secondhand smoke suffer many of the same diseases as active smokers, including lung cancer and heart disease. Women who have never smoked but live with a smoker have a 91% greater risk of heart disease and twice the risk of dying from lung cancer. Women who are exposed to secondhand smoke during pregnancy have a higher rate of miscarriages and stillbirths. Their children have decreased lung function and a greater risk of sudden infant death syndrome (SIDS). Children exposed to secondhand smoke are more likely to experience middle ear and sinus infections. They also have an increased frequency of asthma, colds, bronchitis, pneumonia, and other lung diseases.

These and other problems have caused many states and municipalities to ban smoking in many public places.

Radon and Her Dirty Daughters

A most enigmatic indoor air pollutant is radon. A noble gas, radon is colorless, odorless, tasteless, and unreactive chemically. Radon is, however, radioactive. The decay products of radon, called **daughter isotopes**, are the problem. Radon-222 decays by alpha emission (Chapter 4) with a half-life of 3.8 days.

$$^{222}_{86}\text{Rn} \longrightarrow \text{}^{218}_{84}\text{Po} + \text{}^{4}_{2}\text{He}$$

When radon is inhaled, the polonium-218 and other daughters, including lead-214 and bismuth-214, are trapped in the lungs, where their decay damages the tissues.

Radon is released naturally from soils and rocks, particularly granite and shale, and from minerals such as phosphate ores and pitchblende. The ultimate source is uranium atoms found in these materials. Radon is only one of several radioactive materials formed during the multistep decay of uranium. However, radon is unique in that it is a gas and escapes. The others are solids and remain in the soil and rock.

Outdoors, radon dissipates and presents no problems. However, a house built on a solid concrete slab or on a basement can trap the gas inside. Radon levels build up, sometimes reaching several times the maximum safe level established by the EPA (0.15 disintegration per second per liter of air). At five times this level, the hazard is thought to equal that of smoking two packs of cigarettes a day. Scientists at the National Cancer Institute estimate that radon causes 15,000 lung cancer deaths a year; however, the exact threat isn't known. The "safe level" was set for people who work in uranium mines. These workers have high rates of lung cancer, but they also breathe radioactive dust and many smoke cigarettes. These risk factors may be synergistic and are difficult to separate.

Other Indoor Pollutants

Many homes use kerosene heaters or unvented natural gas heaters to help reduce heating costs in the winter. Heating only the occupied areas of a home seems a good idea. However, when such heaters are not properly adjusted, carbon monoxide can be generated. Carbon monoxide from gas stoves, gas furnaces, gasoline generators, even automobile exhaust from an attached garage can also contribute to the high levels seen in some houses. The EPA recommends a maximum level of CO of 9 ppm over an eight-hour period. Carbon monoxide levels from a poorly adjusted gas heater can exceed 30 ppm—continuously. Technology is now available to detect the problem. Commercial carbon monoxide detectors are reliable and easy to install. Proper ventilation is key to preventing high levels of carbon monoxide.

Mold is an often-ignored pollutant in many homes. Leaky pipes or cool areas where water vapor condenses can lead to moisture problems, and mold will grow where there is moisture. Mold spores are a form of PM that can worsen asthma, bronchitis, and other lung diseases. Generally, mold is controlled by controlling moisture.

Advertisements for electronic "air cleaners" are seen almost every day. It sounds like a great idea; just plug in the cleaner and let it remove all the bad pollutants. Unfortunately, some of these cleaners actually *produce* pollution. In particular, some cleaners generate *ozone*. As we shall see in the next section, ozone is itself a pollutant. Moreover, ozone is quite reactive. It can indeed reduce the levels of some pollutants by reacting with them. But in doing so, it can generate additional pollutants such as formaldehyde. The EPA's recommendation is, "The public is advised to use proven methods of controlling indoor air pollution. These methods include eliminating or controlling pollutant sources, increasing outdoor air ventilation, and using proven methods of air cleaning."

12.12 OZONE: THE DOUBLE-EDGED SWORD

The ordinary oxygen in the air we breathe is made up of O_2 molecules. Ozone is a form of oxygen that consists of O_3 molecules. Ozone (O_3) and oxygen (O_2) are **allotropes**: two different forms of the same element. Ozone is a familiar constituent of photochemical smog. Inhaled, it is a toxic, dangerous chemical. Ozone is also an im-

2005 Rank	Metropolitan Area	2004 Rank	2003 Rank
	TABLE 12.2 Metropolitan Areas with the Worst Ozone Air Pollution		
1	Los Angeles–Long Beach–Riverside, CA	1	1
2	Bakersfield, CA	3	3
3	Fresno, CA	2	2
4	Visalia–Porterville, CA	4	4
5	Merced, CA	6	7
6	Houston–Baytown–Huntsville, TX	5	5
7	Sacramento–Arden–Arcade–Truckee, CA, NV	7	6
8	Dallas–Fort Worth, TX	10	12
9	New York–Newark–Bridgeport, NY–NJ–CT–PA	13	14
10	Philadelphia–Camden–Vineland, PA–NJ–DE–MD	12	13

Source: Data from the American Lung Association www.lungusa.org

portant natural component of the stratosphere, where it shields Earth from life-destroying UV radiation. Ozone is a good example of just what a pollutant is: a chemical substance out of place in the environment. In the stratosphere, it helps make life possible. In the lower troposphere—the part we breathe—it makes life difficult.

Ozone as an Air Pollutant

Most U.S. urban areas have experienced ozone alerts. Table 12.2 shows the 2005 rankings of the 10 worst areas for ozone pollution. The American Lung Association estimates that over 152 million Americans live in counties that exceed recommended concentrations of ozone.

Ozone is a powerful oxidizing agent and is highly reactive. This makes it a severe irritant to the respiratory system. Episodes of high ozone levels correlate with increased hospital admissions and emergency room visits for respiratory problems. At low levels, it causes eye irritation. Repeated exposure can make people more susceptible to respiratory infection, cause lung inflammation, aggravate asthma, decrease lung function, and increase chest pain and coughing. Ozone is particularly damaging to children. The WHO estimates that 80% of all deaths attributed to air pollution are among children.

Ozone causes economic damage beyond its adverse effect on health. It is a strong oxidizing agent that causes rubber to harden and crack, shortening the life of automobile tires and other rubber items. Ozone also causes extensive damage to crops, especially tobacco and tomatoes.

▲ Ozone causes rubber to harden and crack. Tire manufacturers incorporate paraffin wax into the rubber to protect tires from ozone. The alkanes that make up the wax are among the few substances that are resistant to attack by ozone. As tires age, the paraffin wax migrates to the surface to replace that worn off.

The Stratospheric Ozone Shield

In the mesosphere (Figure 12.1), some ordinary oxygen molecules are split into oxygen atoms by short-wavelength, high-energy UV radiation.

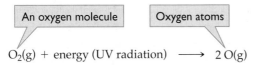

$$O_2(g) + energy\ (UV\ radiation) \longrightarrow 2\ O(g)$$

Some of these highly reactive atoms diffuse down to the stratosphere, where they react with O_2 molecules to form ozone.

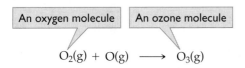

$$O_2(g) + O(g) \longrightarrow O_3(g)$$

The ozone in turn absorbs longer-wavelength, but still lethal, UV rays, thus shielding us from this harmful radiation. In absorbing the rays, ozone is converted back to oxygen molecules and oxygen atoms in a reversal of the previous reaction.

$$O_3(g) + energy\ (UV\ radiation) \longrightarrow O_2(g) + O(g)$$

Undisturbed, the concentration of ozone in the stratosphere is kept fairly constant; over 300 billion tons of ozone are destroyed and created every day by this cyclic process. However, because of the reduced pressure in the stratosphere, if the entire ozone layer were compressed to normal sea level pressures, it would be only 3 mm thick! This delicate barrier protects all living things from harmful UV radiation. In recent decades, human activity has threatened to upset the balance and deplete some of the protective qualities of the ozone layer.

Chlorofluorocarbons and the Ozone Hole

The less ozone there is in the stratosphere, the more harmful UV radiation reaches Earth's surface. For humans, increased exposure to UV radiation can lead to more cases of skin cancer, cataracts, and impaired immune systems. The U.S. National Research Council predicts a 2–5% increase in skin cancer for each 1% depletion of the ozone layer. Even a 1% increase in UV radiation can lead to a 5% increase in melanoma, a deadly form of skin cancer. Increased UV also damages many crops resulting in lower yields. A depletion of stratospheric ozone can also cause a decreased population of phytoplankton in the ocean, disrupting the web of life in the sea and disrupting the oxygen cycle (page 325).

The thickness of the ozone layer changes with latitude and also with season. The cycling of ozone concentration is most prevalent over Antarctica and is related to the fact that the chemical reactions responsible for ozone destruction occur much more rapidly on the surfaces of ice crystals. During the Antarctic winter (June–August) more clouds containing ice crystals are formed and the levels of ozone decrease. In the early 1970s, serious concerns were raised because the ozone layer was not recovering properly during the warm months. In 1974, Mario Molina and F. Sherwood Rowland proposed a mechanism for the enhanced ozone depletion that implicated the presence of chlorofluorocarbons (CFCs) in the stratosphere. Their work was awarded the Nobel Prize in chemistry in 1995.

CFCs were used as the dispersing gases in aerosol cans, as foaming agents for plastics, and as refrigerants. At room temperature, CFCs are either gases or liquids with low boiling points. They are essentially insoluble in water and inert toward most other substances. These properties make them ideal for many uses. However, once released into the atmosphere, this lack of reactivity allows them to persist and diffuse to the higher regions of the stratosphere. There, above the protective ozone layer, energetic UV radiation cleaves one of the C—Cl bonds and initiates a series of reactions.

▲ In 1986 Susan Solomon, a chemist at the Oceanic and Atmospheric Administration in Boulder, Colorado, proposed a mechanism for ozone depletion over the poles by CFCs. Later experiments confirmed the theory and led to an international ban on the use of CFCs. Solomon was awarded a 1999 National Medal of Science for her work.

$$CF_2Cl_2 + energy\ (UV\ light) \longrightarrow CF_2Cl \cdot + Cl \cdot$$
$$Cl \cdot + O_3 \longrightarrow ClO \cdot + O_2$$
$$\cdot ClO + O \longrightarrow Cl \cdot + O_2$$

Both products of the initial reaction are highly reactive free radicals (Chapter 5). The chlorine atoms then react with ozone molecules producing molecular oxygen. Note that the last step results in the formation of another chlorine atom that can break down another molecule of ozone. The second and third steps are repeated many times; the decomposition of one CFC molecule can result in the destruction of *thousands* of molecules of ozone.

International Cooperation

The United Nations has addressed the problem of ozone depletion through the 1987 Montreal Protocol, international agreements enforcing the reduction and eventual elimination of the production and use of ozone-depleting substances. As a result, CFCs have been banned in the United States and many other countries, and the search for effective substitutes has already been successful. CFCs and other chlorine- and bromine-containing compounds that might diffuse into the stratosphere have been replaced for the most part by more benign substances. Hydrofluorocarbons (HFCs) such as CH_2FCF_3 are one alternative. The C—H and C—F bonds are stronger than C—Cl bonds, and therefore the HFCs are not susceptible to photochemical breakdown in the stratosphere. Other substitutes are hydrochlorofluorocarbons (HCFCs) such as $CHCl_2CF_3$. Although these compounds contain chlorine atoms, most break down before reaching the stratosphere. For every solution, it seems there is a problem; CFCs, HFCs, and HCFCs are all greenhouse gases. (Section 12.13).

UV light sometimes is used to sterilize tools in the hairdresser's shop and safety glasses in laboratories to prevent the transfer of infectious microorganisms from one person to another. We do not want the entire planet rendered sterile, however, for we are part of the life that would be harmed.

EXAMPLE 12.1 Lewis Structures

Draw Lewis structures for (a) the hydrofluorocarbon CH_2FCF_3; (b) the hydrochlorofluorocarbon $CHCl_2CF_3$.

Solution

Although these molecules both look complicated, it isn't hard to draw skeletal structures and Lewis structures. Each molecule has two carbon atoms, and all the remaining atoms tend to have a single bond each. That means that the two carbon atoms are connected, and each carbon atom has three more atoms connected to it.

Now we count the valence electrons: H = 1, C = 4, Cl = 7, F = 7.   Subtract a pair of electrons for each bond in the skeletal structure to find the number of electrons remaining.

$$38 - 14 = 24 \qquad 44 - 14 = 30$$

In structure (a) the hydrogen atoms and the carbon atom have filled shells. The four fluorine atoms need six more electrons each, so we place 24 electrons on the fluorine atoms (six each). In structure (b) the three fluorine and two chlorine atoms each need six more electrons, so we place six electrons on each.

▶ Exercise 12.1A
Draw a Lewis structure for Freon-12, which has the formula CCl_2F_2.

▶ Exercise 12.1B
Freon-12 is unreactive in part because it is almost nonpolar. Explain the low polarity of the Freon-12 molecule. (*Hint*: What are the electronegativities of fluorine and chlorine?)

The Montreal Protocol seems to be having the desired effects. The concentration of CFCs in the stratosphere has peaked. Chlorine concentrations leveled off in 1998 and will likely continue to decline slowly for decades. There is some indication that the ozone hole over Antarctica is *getting smaller* (Figure 12.11), but its size varies with temperature and other meteorological factors. It may be several years before we can be sure that the problem is being solved.

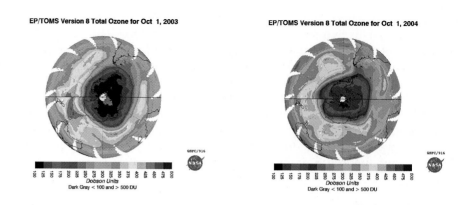

▶ **Figure 12.11** The ozone hole over Antarctica. Notice that the black area, which indicates a very low ozone concentration, is much smaller in 2004 than in 2003. Also, the overall area of reduced concentration appears to have decreased between 2003 and 2004. Global chlorine concentrations are also on the decline.

12.13 CARBON DIOXIDE AND THE GREENHOUSE EFFECT

No matter how clean an engine or a factory is, as long as it burns coal or petroleum products it produces carbon dioxide. The concentration of carbon dioxide in the atmosphere increased about 31% since 1750, which scientists attribute to global industrialization and the burning of fossil fuels. We generally don't even consider carbon dioxide a pollutant because it is a natural component of the environment and not toxic. Certainly its immediate effect on us is slight, but what about its long-term effects?

The sun radiates many different types of radiation of which visible, infrared, and ultraviolet are most prominent. About half of this energy is either reflected or absorbed by the atmosphere; the light that gets through (mostly visible) acts to heat the surface of Earth. Carbon dioxide and other gases produce a **greenhouse effect** (Figure 12.12). They let in the sun's visible light to warm the surface, but when the Earth radiates infrared energy back toward space, these greenhouse gases absorb and trap the energy. A similar thing happens in a car left in the sun with the windows closed. Visible light entering through the glass heats the interior, but the heat cannot efficiently escape back through the windows. The temperature inside the car

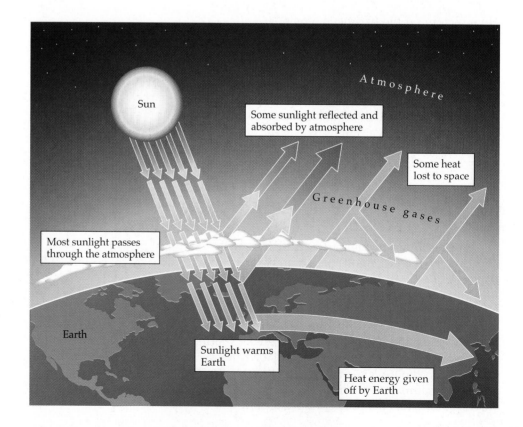

▶ **Figure 12.12** The greenhouse effect. Sunlight passing through the atmosphere is absorbed, warming Earth's surface. The warm surface emits infrared radiation. Some of this radiation is absorbed by CO_2, H_2O, CH_4, and other gases and retained in the atmosphere as heat energy.

can rise significantly above the outside temperature. Some degree of the greenhouse effect is necessary to life. Without an atmosphere, all of the radiated heat would be lost to outer space, and the Earth would be much colder than it is. In fact, just based on the distance from the sun, it is estimated that the Earth's average temperature would be a cold $-18\,°C$.

EXAMPLE 12.2 **Stoichiometry**

What mass of carbon dioxide (molar mass = 44.0 g/mol) is produced by the burning of 2670 g (1 gallon) of gasoline? Gasoline may be represented by the formula C_8H_{18} (molar mass = 114.0 g/mol).

$$2\,C_8H_{18}(l) + 25\,O_2(g) \longrightarrow 18\,H_2O(g) + 16\,CO_2(g)$$

Solution

The problem gives us a mass of a reactant, and we are to find the mass of a product, so this is a stoichiometry problem, as covered in Chapter 6 (see Example 6.13 for a similar example).

We begin by converting the mass of the given substance, C_8H_{18}, to moles.

$$2670\text{ g }C_8H_{18} \times \frac{1\text{ mol }C_8H_{18}}{114.0\text{ g }C_8H_{18}} = 23.4\text{ mol }C_8H_{18}$$

Now we use the relationship of 2 mol $C_8H_{18} \backsimeq 16$ mol CO_2 to find moles of CO_2.

$$23.4\text{ mol }C_8H_{18} \times \frac{16\text{ mol }CO_2}{2\text{ mol }C_8H_{18}} = 187\text{ mol }CO_2$$

The final step is to use the molar mass of CO_2 to determine the mass of CO_2.

$$187\text{ mol }CO_2 \times \frac{44.0\text{ g }CO_2}{1\text{ mol }CO_2} = 8240\text{ g }CO_2$$

The mass of CO_2 is actually greater than the mass of gasoline consumed, because the carbon of the gasoline is combined with oxygen from the air.

▶ **Exercise 12.2A**
Calculate the mass in grams of water vapor produced from burning 2670 g of gasoline.

▶ **Exercise 12.2B**
Calculate the mass in kilograms of carbon dioxide produced by a class of 60 students in one week, if each student uses 10 gallons of gasoline in a week.

Greenhouse Gases and Global Warming

What many fear is an *enhanced* greenhouse effect caused by increased concentrations of carbon dioxide and other greenhouse gases in the atmosphere. This enhancement could lead to a rise in Earth's average temperature, an effect called **global warming.** Indeed, periods of high CO_2 concentration in Earth's past have been correlated with increased global temperatures (Figure 12.13).

Human activities add 25 billion t of carbon dioxide to the atmosphere each year, of which 22 billion t comes from the burning of fossil fuels. About 15 billion t is removed by plants, the soil, and the oceans, leaving a net addition to the atmosphere of 10 billion t CO_2/year.

Methane, CFCs, and other trace gases also contribute to the greenhouse effect. The concentration of methane in the atmosphere has been increasing since 1977.

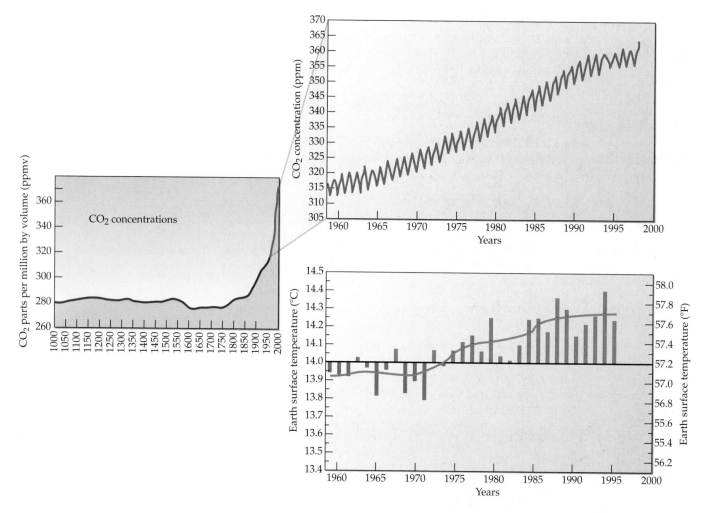

▲ **Figure 12.13** Carbon dioxide content of the atmosphere and surface temperature variations. The increase in carbon dioxide correlates with the beginning of the Industrial Revolution, around 1850.

Although present in much smaller amounts than carbon dioxide, these trace gases are much more efficient at trapping heat. Methane is 20–30 times, and CFCs 20,000 times, as effective as carbon dioxide at holding heat in Earth's atmosphere. Water is about 10 times less efficient at trapping the heat, though when present in significant amounts has a noticeable effect. For instance, cloudy winter nights tend to be warmer than clear winter nights due to the presence of significant amounts of water in the clouds, which acts to trap the heat close to the Earth's surface.

Predictions and Consequences

Many complex and interrelated natural variables (clouds, volcanoes, El Niño weather patterns, and such) as well as varying estimates of future greenhouse gas emissions make it difficult to predict the magnitude of future warming. Different scenarios (with different assumptions) have predicted a warming trend of 1.5–4.0 °C by the year 2100. Most predictions are made by computer models, but scientists still are not sure they have input all the needed data. Perhaps this temperature rise doesn't seem like much, but scientists are concerned (yet not completely certain) about the ramifications of such warming. Some fear that global warming will melt the polar ice caps and flood coastal cities. The ice caps may not need to melt for global warming to be problematic. When water warms, it expands. The oceans are now rising 3 mm/year. Again, this rise may seem trivial, but even slight rises in ocean level

The World Health Organization estimates that Earth's warming climate contributes to more than 150,000 deaths and 5 million illnesses each year. WHO predicts that this toll could double by 2030. According to data published in the journal *Nature,* a warmer climate is driving up rates of malaria, malnutrition, and diarrhea.

The Kyoto Conference

In 1997, leaders from over 100 nations gathered in Kyoto, Japan, for the largest environmental summit in history. At issue was the state of the atmosphere, specifically the concentration of greenhouse gases and the industrialization that most believe has contributed to its increase. The Kyoto Protocol aimed to reduce emissions of carbon dioxide, requiring that countries have a 2012 emission target 5.2% below their 1990 levels. The agreement would grant countries "pollution permits." If improvements were made, and fewer emissions produced, one country could sell or trade permits to another country, but the total amount of CO_2 emitted would be limited. The treaty took effect in 2005, but the United States has not signed the treaty, believing that it is not in its economic best interest.

will increase tides and result in higher, more damaging storm surges. Various models predict a rise of 0.09–0.88 m by 2100.

Are We Really Heating Up?

It seems clear that both the surface temperatures and atmospheric CO_2 concentration have risen since the middle of the twentieth century. The U.S. National Academy of Sciences and other leading world science organizations have expressed the consensus views that there is now strong evidence that significant global warming is occurring and that it is likely that most of the warming in recent decades can be attributed to human activities. However, some aspects of global warming remain controversial. The chemistry of the atmosphere, with the interconnected elemental cycles (oxygen, carbon, nitrogen) and the interactions of gas molecules with radiation, is an incredibly complex phenomenon. The immensity of the laboratory (the entire Earth) makes it difficult to do the science, and we lack the ability to perform controlled experiments. What can we do? It is unlikely that imposing drastic, economically disastrous limits on energy consumption will have the desired effects, yet the opposite extreme, denial of the global warming problem, seems equally unwise. Global climate change could have dramatic effects on human civilization.

12.14 THE ULTIMATE POLLUTANT: HEAT

Hybrid cars—vehicles with engines that run on a combination of gasoline and batteries—are a possible partial solution to air pollution. These cars could help to reduce air pollution in urban areas. But hybrid cars require electric power, and electric power requires power plants. Conventional power plants burn coal, gas, and oil. It might be argued that replacing cars that run on fossil fuel with ones that run partially on batteries would serve only to change the site of the pollution and perhaps spread it out a bit. Efficiency plays a role, however; if a hybrid vehicle is more efficient in its use of fuel, there can be a net benefit.

No matter what we do, however, there is one pollutant we cannot avoid: heat. According to the second law of thermodynamics, in any energy conversion some of the energy winds up as heat. All power plants dump residual heat into the environment as they produce electric energy, and this heat may be the ultimate pollutant. Even now human structures and activities often make urban areas "heat islands" that are 4–5 °C warmer than the surrounding rural areas.

12.15 WHO POLLUTES? HOW MUCH?

The U.S. EPA lists six "criteria" pollutants (Table 12.3), so called because they employ scientific criteria to measure the health effects. Since the first Earth Day in 1970, nationwide total emissions of the six criteria pollutants have declined 54%. These improvements in air quality have come while the U.S. population increased

TABLE 12.3	Ambient Air Quality Standards for Six Criteria Air Pollutants	
Pollutant	**Limit**	**% Change (1970–2003)**
Carbon monoxide		−53
8-h average	9 ppm	
1-h average	35 ppm	
Nitrogen dioxide		−24
Annual average	0.053 ppm	
Ozone[a]		−29[b]
8-h average	0.08 ppm	
1-h average	0.12 ppm	
Particulate matter		−91
<10 μm, 24-h average	150 μg/m^3	
<2.5 μm, 24-h average	65 μg/m^3	
Sulfur dioxide		−46
Annual average	0.03 ppm	
24-h average	0.14 ppm	
3-h average	0.50 ppm	
Lead		−99
3-month average	1.5 μg/m^3	
VOC[c]		−54

[a]Formed from precursor volatile organic compounds (VOC) and NO_x.

[b]1980–2003.

[c]Not a criteria pollutant, but monitored because it is an ozone precursor.

40%, the gross domestic product increased 187%, and vehicle miles traveled increased 171%.

Air pollution causes material damage by dirtying and destroying buildings, clothing, and other objects. It increases health hazards, especially for the very young, the old, and those already ill. It causes crop damage by stunting or killing green plants. Air pollution reduces visibility, thus increasing auto and air traffic accidents. With its ugly smoke plumes and unpleasant odors, it is even an aesthetic problem.

Who causes all this pollution? Table 12.4 summarizes the seven major air pollutants, their main sources, and their health and environmental effects. Note that motor vehicles are a prominent source. Indeed, they are the source of nearly one-half (by mass) of all air pollutants. Our transportation system accounts for about 80% of carbon monoxide emissions, 40% of hydrocarbon emissions, and 40% of nitrogen oxide emissions. Because most transportation in the United States is by private automobile, we can conclude that cars are a major source of pollution.

On the other hand, most PM comes from power plants (about 40%) and industrial processes (about 45%). Similarly, more than 80% of sulfur oxide emissions come from power plants, with an additional 15% coming from other industries. Power plants alone contribute about 55% of nitrogen oxide emissions. Who uses electricity from power plants? We all do. A 100-W bulb burning for 1 year uses the electricity generated by burning 275 kg of coal.

What is the worst pollutant? Carbon monoxide is produced in huge amounts and is quite toxic. Yet it is deadly only in concentrations approaching 4000 ppm. Its contribution to cardiovascular disease, by increasing stress on the heart, is difficult to measure. The WHO rates sulfur oxides as the worst pollutants. They are powerful irritants, and, according to WHO, people with respiratory illnesses are more likely to die from exposure to sulfur oxides than from exposure to any other kind of pollutant. Sulfur oxides and sulfuric acid formed from them have been linked to more than 50,000 deaths per year in the United States.

We all share the responsibility for pollution. In the words of Walt Kelly's comic-strip character Pogo, "We have met the enemy, and he is us." But we can all be part of the solution by conserving fuel and electricity. Many utility companies now offer suggestions for saving energy. Some even give free analyses of home energy use.

Pollutant	Formula or Symbol	Major Sources	Health Effects	Environmental Effects
Carbon monoxide	CO	Motor vehicles	Interferes with oxygen transport, causing dizziness, death; possibly contributes to heart disease	Slight
Hydrocarbons	C_nH_m	Motor vehicles, industry, solvents	Narcotic at high concentrations; some aromatics are carcinogens	Precursors of aldehydes, PAN
Sulfur oxides	SO_x	Power plants, smelters	Irritate respiratory system; aggravate lung and heart diseases	Reduce crop yields; precursors of acid rain, SO_4^{2-} particulates
Nitrogen oxides	NO_x	Power plants, motor vehicles	Irritate respiratory system	Reduce crop yields; precursors of ozone and acid rain; produce brown haze
Particulate matter	—	Industry, power plants, dust from farms and construction sites	Irritates respiratory system; synergistic with SO_2; contains adsorbed carcinogens, toxic metals	Impairs visibility
Ozone	O_3	Secondary pollutant from NO_2	Irritates respiratory system; aggravates lung and heart diseases	Reduces crop yields; kills trees (synergistic with SO_2); destroys rubber, paint
Lead	Pb	Motor vehicles, smelters	Toxic to nervous system and blood-forming system	Toxic to all living things

TABLE 12.4 The Seven Major Air Pollutants

12.16 PAYING THE PRICE

Air pollution costs us tens of billions of dollars each year. It wrecks our health by causing or aggravating bronchitis, asthma, emphysema, and lung cancer. It destroys crops and sickens and kills livestock. It corrodes machines and blights buildings.

Elimination of air pollution will be neither cheap nor easy. It is especially difficult to remove the last fractions of pollutants. Cost curves are exponential; costs soar toward infinity as pollutants approach zero. For example, if it costs $200 per car to reduce emissions by 50%, it usually costs about $400 to reduce them by 75%, $800 to reduce them by 87.5%, and so on. It would be extremely expensive to reduce emissions by 99%. We can have cleaner air, but how clean depends on how much we are willing to pay.

What would we gain by getting rid of air pollution? How much is it worth to see the clear blue sky or to see the stars at night? How much is it worth to breathe clean, fresh air?

Critical Thinking Exercises

Apply knowledge that you have gained in this chapter and one or more of the FLaReS principles (Chapter 1) to evaluate the following statements or claims.

12.1 A citizen claims that air pollution problems in the Los Angeles area could be solved by allowing only zero-emission vehicles to be sold and licensed in the area.

12.2 Ozone in aerosol cans was once used as a room deodorizer. An inventor proposes that such preparations be reintroduced because any ozone entering the environment would help to restore the protective ozone layer in the stratosphere.

12.3 Although Earth's surface has grown warmer during the last few decades, satellite data collected over an 18-year period show a cooling trend in the upper atmosphere. On the basis of these data, some people claim that global warming as a result of an enhanced greenhouse effect is a myth.

12.4 A notable scientist claims that we must continue to *increase* the amount of carbon dioxide in the atmosphere to prevent a coming ice age.

12.5 Some experts believe that much of global warming comes from methane produced by such natural sources as plant decay in swamps and cattle flatulence. They say that industrial activities that produce carbon dioxide raise the standard of living and should be allowed to continue without interference since they aren't the only source of global warming.

12.6 An entrepreneur promotes a new vehicle, which is an ordinary automobile adjusted to burn hydrogen. He claims that the only product of hydrogen combustion is water, and so the new vehicle is completely nonpolluting.

12.7 A scientist claims that the increase in carbon dioxide from 0.02 to 0.03% is not important in terms of global warming. His reasoning is that water vapor makes up about 2–3% of the atmosphere, and although water vapor is only about one-tenth as effective a greenhouse gas as CO_2, the 0.01% increase in CO_2 is equivalent to a trivial 0.1% increase in water vapor.

SUMMARY

Section 12.1—The thin blanket of air that is our **atmosphere** is made up of layers: the troposphere (nearest Earth), the stratosphere (which includes the ozone layer), the mesosphere, and the thermosphere. The atmosphere is a mixture of gases, about 78% nitrogen, 21% oxygen, and 1% argon by volume, plus trace gases and up to 4% water vapor.

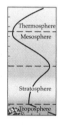

Section 12.2—Animals and most plants cannot use atmospheric nitrogen unless it has been fixed (combined with other elements). Nitrogen goes from the air into plants and animals and eventually back to the air via the **nitrogen cycle**. Oxygen from the air is involved in oxidation (of plant and animal materials) and is eventually returned to the air via the **oxygen cycle**. Oxygen is converted to ozone and then changed back to oxygen in the stratosphere. In this process, the ozone absorbs UV radiation that might otherwise make life impossible.

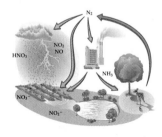

Sections 12.3–12.6—A **temperature inversion** occurs when a layer of cold, dirty air is trapped under a layer of warmer air. The cold air can become quite polluted. A **pollutant** is a chemical that is in the wrong place and causes problems there. Air pollution has existed for millions of years. Natural air pollution includes dust storms, noxious gases from swamps, and ash and sulfur dioxide from erupting volcanoes. Today air pollution is more complex than in the past. Air pollution is a global problem because pollutants from one geographic area often migrate to another. The primary air pollutants introduced by human activity are the

smoke and gases produced by the burning of fuels and the fumes and particulates emitted by factories.

Section 12.7—**Industrial smog** is the kind of smog produced in cold, damp air by excessive burning of fossil fuels. It consists of a combination of smoke and fog with sulfur dioxide, sulfuric acid, and **particulate matter (PM)**, which is solid or liquid particles of greater than molecular size. PM commonly takes the form of soot and fly ash from coal combustion. Carbon monoxide is a poisonous gas produced in most combustion that ties up hemoglobin in the blood so that it cannot transport oxygen to the cells. The pollutants that make up industrial smog can act synergistically to cause severe damage to living tissue. PM can be removed from smokestack gases using an **electrostatic precipitator**, which charges the particles, or with a **wet scrubber**, which passes the gases through water. Sulfur dioxide can be scrubbed using limestone, to form calcium sulfite, which can then be oxidized to useful calcium sulfate.

Section 12.8—Complete combustion of gasoline produces carbon dioxide and water, but combustion is always incomplete. Carbon monoxide is one product of incomplete combustion, and three-fourths of the carbon monoxide produced results from transportation. Carbon monoxide ties up hemoglobin so that it cannot carry oxygen to the body. Nitrogen oxides are present in automobile exhaust gases and in emissions from power plants that burn fossil fuels. The greatest problem with nitrogen oxides is their contribution to smog. The brown color of nitrogen dioxide causes the brownish haze often seen over certain large cities. **Volatile organic compounds (VOCs)** are major contributors to smog. VOCs come from gasoline vapor and other fuels, and from some consumer products. Hydrocarbons can be pollutants in themselves and can react to form other pollutants such as aldehydes and peroxyacetyl nitrate (PAN), both of which are very irritating to tissues.

Sections 12.9–12.10—**Photochemical smog** results when hydrocarbons, nitrogen oxides, and ozone are exposed to bright sunlight. A complex series of reactions produces a variety of pollutants. Catalytic converters in automobiles act to reduce photochemical smog by oxidizing unburned hydrocarbons. A separate catalyst reduces the quantity of nitrogen oxides. Hybrid vehicles are more efficient than conventional vehicles, with a reduction in emission of pollutants as well as photochemical smog. Nitrogen oxides (mainly from automobiles and power plants) and sulfur oxides (mainly from power plants) can dissolve in atmospheric moisture, forming acid rain. Acid rain often falls far from the original sources, and it can corrode metals and erode marble buildings and statues.

Sections 12.11–12.13—Indoor pollution can be as bad as, or worse than, pollution outdoors. Cigarette smoke can raise both PM and carbon monoxide above the allowed EPA levels. Because of the effects of secondhand smoke, most cities and states restrict tobacco smoking in public places. The radioactive gas radon can accumulate indoors in some cases, and inhaled radon produces daughter isotopes that also are radioactive and can accumulate in the lungs when breathed. The daughter isotopes can produce a significant risk of lung cancer.

Ozone, an allotrope of oxygen, is a harmful air pollutant in the troposphere but forms a protective layer in the stratosphere that shields Earth against UV radiation from the sun. Chlorofluorocarbons (CFCs) have been linked to the hole in the ozone layer and are now widely banned. It

$CH_3\overset{O}{\overset{\|}{C}}-O-ONO_2$
PAN

appears that the ozone hole may be shrinking because of this ban. Carbon dioxide and other gases contribute to the greenhouse effect, which occurs when the infrared radiation emitted from the Earth's surface cannot escape. The result appears to be global warming, an increase in the Earth's average temperature.

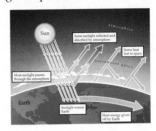

Sections 12.14–12.16—Although increased efficiency of automobiles helps to limit pollution, there is one pollutant that cannot be avoided: heat, which is always a by-product. The EPA has listed six "criteria" pollutants, of which all have been reduced in recent years. Motor vehicles are a prominent source of carbon monoxide, hydrocarbons, and nitrogen oxides, while most PM and sulfur oxides come from industrial processes. Carbon monoxide and sulfur oxides are the most problematic pollutants. We all share responsibility for pollution—and we share the responsibility for reducing it.

REVIEW QUESTIONS

1. How does the ozone layer protect the inhabitants of Earth's surface?

2. What is an atmospheric (temperature) inversion? How is it related to air pollution?

3. Is all air pollution the result of human activity? Explain.

4. List two (former) uses of CFCs.

5. List one potential replacement for CFCs. What problems do the replacements have?

6. Describe how (a) a bag filter and (b) a cyclone separator remove particulate matter from stack gases.

7. How does a wet scrubber remove particulate matter? What is a major disadvantage of wet scrubbers?

8. What is bottom ash? What is fly ash? Give two uses for fly ash.

9. What is smog?

10. What are the health effects of ozone in polluted air?

11. What is a greenhouse gas? Give two examples.

12. List two pollutants often found in air indoors.

13. Consider (a) the hole in the ozone layer and (b) the phenomenon of global warming. Are they the same thing? Are they related in any way?

14. What one pollutant will always be produced by human activity?

PROBLEMS

The Atmosphere: Composition and Cycles

15. Which layer of the atmosphere lies nearest Earth? Which contains the ozone layer?

16. List the three major components of dry air and give the approximate (nearest whole number) percentage by volume of each.

17. What are the two most important variable components of the atmosphere? Give the approximate concentration range of each.

18. What is meant by nitrogen fixation? Why is it important?

19. How has industrial fixation of nitrogen to make fertilizers affected the nitrogen cycle?

20. Give equations for the reactions by which lightning fixes nitrogen.

21. How is the oxygen supply of Earth's atmosphere replenished?

22. What is a pollutant? Which of the following are pollutants? Explain.
    a. NO_2 in the troposphere
    b. O_3 in the stratosphere
    c. N_2 in the stratosphere
    d. O_3 in the troposphere

Industrial Smog

23. What are the chemical components of industrial smog?

24. What weather conditions characterize industrial smog?

For Problems 25–28, give the equation for the indicated reaction.

25. Sulfur is oxidized to sulfur dioxide.

26. Sulfur dioxide is oxidized to sulfur trioxide.

27. Sulfur trioxide reacts with water to form sulfuric acid.

28. Sulfur dioxide reacts with hydrogen sulfide to form elemental sulfur and water.

29. Describe the synergistic action of sulfur dioxide and particulate matter.

30. What is an electrostatic precipitator? How does it work?

31. Describe how a limestone scrubber removes sulfur dioxide from stack gases. Give the chemical reaction(s) involved.

32. What is particulate matter? What is an aerosol?

Photochemical Smog

33. What are the chemical components of photochemical smog?

34. What weather conditions characterize photochemical smog?

35. Under what conditions do nitrogen and oxygen combine? Give the equation for the reaction.

36. Give the equation for the reaction of nitric oxide with oxygen to form nitrogen dioxide.

37. What happens to nitrogen dioxide in sunlight? Give the equation for the reaction.

38. What are VOCs?

39. What is the main source of hydrocarbons in polluted air?

40. What is PAN? From what is it formed? What are its health effects?

41. What are catalytic converters? How do they work?

42. Give two ways in which the level of nitrogen oxide emissions from an automobile can be reduced.

Carbon Monoxide

43. What is the main source of carbon monoxide in polluted air?

44. What are the health effects of carbon monoxide?

45. How might exposure to carbon monoxide contribute to heart disease?

46. Give the equation by which carbon monoxide reduces nitric oxide to nitrogen gas.

The Ozone Layer

For Problems 47–48, give the equation for the indicated reaction.

47. Oxygen is converted to ozone in the ozone layer.

48. CFCs destroy ozone.

49. What health effect might result from depletion of the ozone layer?

50. How are free radicals related to the ozone layer?

Acid Rain

51. What is acid rain? How is it formed?

52. How might acid rain be alleviated?

53. What is the effect of acid rain on iron? Give an equation.

54. What is the effect of acid rain on marble? Give the equation.

Indoor Air Pollution

55. List several indoor air pollutants. What is the source of each?

56. List some risks associated with secondhand cigarette smoke.

57. What are the health effects of radon gas? How can a radon problem in a building be alleviated?

58. How does better insulation of buildings make indoor air pollution worse?

59. When and how is carbon monoxide considered an indoor air pollutant?

60. What kind of particulate matter is found inside the home?

Global Warming

61. What is the greenhouse effect?

62. How can global warming be alleviated?

ADDITIONAL PROBLEMS

63. Why is zero pollution not possible?

64. Is the electric car a solution to air pollution on a nationwide basis? Within a given city?

65. The average person breathes about 20 m^3 of air a day. What weight of particulates would a person breathe in a day if the particulate level were 400 μg/m^3?

66. The atmosphere contains 5.2×10^{15} t of air. How much carbon dioxide (CO_2) is in the atmosphere if the concentration is 363 ppm?

67. The world's termite population is estimated to be 2.4×10^{17}. These termites produce an estimated 4.6×10^{16} g of carbon dioxide, 1.5×10^{14} g of methane, and 2×10^{14} g of hydrogen each year. How much of each gas is produced by each termite? What percentage increase in the total amount of carbon dioxide in the atmo-

sphere (see Problem 66) would the termites cause in one year if none of it were removed?

68. Since 1978 the giant ice cap that floats in the Arctic Ocean has lost half its average thickness and has decreased in size by an area larger than the state of Texas. The ice cap atop the Antarctic continent also appears to be melting at the edges, dropping huge icebergs into the ocean. Which occurrence, if either, has caused a rise in sea level? (*Hint:* Consider

▲ The number and size of the icebergs that break off the continental shelf in Antarctica seem to be increasing as a result of global warming.

whether the water level rises when an ice cube melts in a glass of water and when an ice cube is added to a glass of water.)

69. Water is a greenhouse gas that is present in significant concentrations in the Earth's atmosphere. Water is also a product of the combustion of fossil fuels. Why is there lit-tle concern over increases of water vapor and its contribution to global warming?

70. Review Example 12.2. If combustion of gasoline is 99% ef-ficient—that is, 99% of the gasoline burns to carbon dioxide—what mass of carbon *monoxide* is then produced when one gallon of gasoline is burned?

COLLABORATIVE GROUP PROJECTS

Prepare a PowerPoint, poster, or other presentation (as directed by your instructor) for presentation to the class.

71. What are the average and peak concentrations of each of the following in your community or in a nearby large city? How can you find out?
    a. carbon monoxide **b.** ozone **c.** nitrogen oxides
    d. sulfur dioxide **e.** particulate matter

72. Does your community have an air pollution problem? If so, describe it. How could it be solved?

73. Should students be allowed to smoke in school buildings? On school grounds? Defend your position.

74. Research the arguments put forth both for and against rat-ification of the Kyoto Agreement.

75. Using the Internet or books, compare the amount of sulfur dioxide from Mt. Kilauea in Hawaii with that from all U.S. industry. (Remember to use the same time span for both.)

76. There is still controversy about whether government in-tervention is warranted to totally ban the use of CFCs. Find pro and con arguments on the Internet. Discuss.

77. Substitutes for CFCs are more expensive than CFCs. What factors should you look at to do a cost–benefit analysis of this substitution? Should the government sub-sidize the switch?

78. Air quality standards are maximum allowable amounts of various pollutants. The federal government and some state governments issue these standards. Using the Inter-net, find the current standards issued by the EPA. Does your state issue such standards? (*Hint*: Try the EPA site and your own state's environmental protection site.)

79. Divide your class into several teams, each taking a dif-ferent view of the global warming problem. Possible po-sitions are (1) we must act now, (2) we need more study before we act, and (3) there is no global warming. Re-search your position and be prepared to state your case with data to support your arguments. Hold a debate with the other groups in your class, paying close atten-tion to the information presented by the other groups. Are their arguments persuasive? Are their sources reli-able? Are yours?

REFERENCES AND READINGS

1. Alley, Richard B. "Abrupt Climate Change." *Scientific American*, November 2004, pp. 62–69.

2. Appenzeller, Tim. "The Case of the Missing Carbon." *National Geographic*, February 2004, pp. 88–117.

3. Becker, Jasper. "China's Growing Pains." *National Geo-graphic*, March 2004, pp. 68–95.

4. Hanson, David. "Defusing the Global Warming Time Bomb." *Scientific American*, March 2004, pp. 68–77.

5. Herzog, Howard, Baldur Eliasson, and Olav Kaarstad. "Capturing Greenhouse Gases." *Scientific American*, Feb-ruary 2000, pp. 72–79.

6. Johnson, Jeff. "Putting a Lid on Carbon Dioxide: Carbon Sequestration, Clean-Coal Research Mark Government Response to Climate-Change Threat." *Chemical & Engi-neering News*, December 20, 2004, pp. 36–42.

7. Kiester, Edwin, Jr. "A Darkness in Donora." *Smithsonian*, November 1999, pp. 22–24.

8. Klemm, O. "Local and Regional Ozone: A Student Study Project," *Journal of Chemical Education*, vol. 78, 2001, pp. 1641–1646.

9. Kunzig, Robert. "Will the Methane Bubble Burst?" *Discover*, March 2004, pp. 34–41.

10. Michaels, Patrick J., and Robert C. Balling. *The Satanic Gases*. Cato Institute, Oxnard, CA, May 15, 2000.

11. Sarmiento, Jorge L., and N. Gruber. "Sinks for Anthro-pogenic Carbon." *Physics Today*, August 2002, pp. 30–36.

12. Sperling, Daniel. "The Case for Electric Vehicles." *Scientific American*, November 1996, pp. 54–59.

13. Speth, James Gustave. *Red Sky at Morning: America and the Crisis of the Global Environment*. New Haven: Yale Univer-sity Press, 2004.

14. Spiro, Thomas G., and William M. Stigliani. *Chemistry of the Environment*, 2nd ed. Upper Saddle River, NJ: Prentice Hall, 2003. Part II, The Atmosphere.

15. Stone, Richard. "Counting the Cost of London's Killer Smog." *Science*, December 13, 2002, pp. 2106–2107.

16. Sturm, Matthew, Donald K. Perovich, and Mark C. Ser-reze. "Meltdown in the North." *Scientific American*, Octo-ber 2003, pp. 60–67.

17. Tarbuck, Edward J., and Frederick K. Lutgens. *Earth Sci-ence*, 9th edition. Upper Saddle River, NJ: Prentice Hall, 2000. Part 3, The Atmosphere.

18. Wilson, Elizabeth. "Ozone Hole Recovery May Be De-layed." *Chemical & Engineering News*, December 12, 2005, p. 9.

19. Wuethric, Bernice. "When Permafrost Isn't." *Smithsonian*, February 2000, pp. 32–33.

GREEN CHEMISTRY

Air Toxics: Gone but Not Forgotten

Irvin Levy, Gordon College

If you have ever painted a room in your home, you probably recall the odor of the solvents in the paint that permeated the house. Perhaps you opened some windows to "air out" the room. After a little fresh air, the room begins to seem like normal again. From the point of view of the person in the house, the unpleasant odor has been eliminated—or has it? In fact, we have only relocated the vapors that annoyed us by mixing them with the huge container of gases that is our atmosphere. Consequently, airing out foul or hazardous vapors from a small space simply moves the problem materials to another space so vast that we can easily forget that we have even made the deposit of these gases into our air.

Many of the materials that we vent into the atmosphere are compounds that have not existed on our planet prior to the past century. Some of those materials may be fairly harmless to us when we are exposed to them for a short period of time, yet may still pose serious atmospheric threats because of the reactions that they can catalyze in our upper atmosphere, upsetting natural balances that have existed for millions of years in the history of Earth.

Other materials that we put into the environment are natural compounds that do not bother us when they are in the room with us. For example, the carbon dioxide that we exhale with every breath is essential to the normal biological cycles between plants and animals, but when we pro-

◀ Both termites and man produce prodigious quantities of carbon dioxide, a greenhouse gas. However, much of the CO_2 produced by human activities comes from fossil fuels, which have not been part of the biosphere for millions of years. The net result is an increase in CO_2 concentration in the atmosphere.

duce quantities far more rapidly than the biosphere can absorb, natural balances can be upset in alarming ways.

In this Web Investigation you will examine several specific examples of gases that pose a threat when released into the air. In each case, you will explore further to find ways that the green chemistry principles have been used to develop strategies to combat these threats, sometimes leading to solutions that have received recognition through the Presidental Green Chemistry Challenge (PGCC) awards. In Communicate Your Results, you will share information about the PGCC, find practical ways to support the reduction of air toxics in your own community, and become more aware of the complex policy decisions that are forged in national and international communities to protect our air.

WEB INVESTIGATIONS

Investigation 1
Air Toxics, Greenhouse Gases, and Ozone-Depleting Compounds

Using your favorite search engine, investigate these three subjects: "air toxics", "greenhouse gases", and "ozone CFC." The "air toxics" search should provide general background about toxic air pollutants. Greenhouse gases are substances that may lead to an overall warming of the climate on earth. As described in the chapter, ozone is a necessary component of our upper atmosphere, protecting us from much of the ultraviolet radiation from the sun. Chlorofluorocarbons (CFCs) were the first class of molecules suspected to be responsible for a global reduction in ozone.

Investigation 2
Ozone Depletion Potential

Although many molecules are believed to be responsible for the depletion of the ozone layer, some substances have the potential to deplete ozone even more than others when they are released into the atmosphere. A measure of this ozone depletion potential is referred to as the ODP for the substance. Search the EPA website for a glossary of terms related to ozone depletion. Also, find the EPA's list of substances with ODP of 0.2 or greater, and review that list. What is the designation for such substances? Since ozone depletion poses a serious risk, international agreements have been forged to regulate the use of these chemicals. Search the Web to examine the details of The Montreal Protocol on Substances that Deplete the Ozone Layer.

Investigation 3
Creative Solutions 1, Replacements for Air Toxics

Carbon dioxide is the best-known greenhouse gas. As described in this chapter, burning fossil fuels produces huge quantities of CO_2. Interestingly, creative uses for CO_2 have been discovered in recent years, allowing for the replacement of some commonly encountered air toxics, such as perchloroethylene (PERC). Search the Web using the keywords "perc alternative" to learn about creative solutions for the elimination of PERC.

Investigation 4
Creative Solutions 2, Replacements for Ozone-Depleting Substances

Chemicals are designed for the useful properties that they possess; however, other properties (sometimes undesirable) often accompany those desired. Green chemistry principles guide us toward the development of solutions with the least negative impact on human health and the environment. Conduct a Web search using the keywords "polystyrene foam CFC" to examine the technology used to produce this material without any CFCs.

On the list of ozone depleting substances (see Investigation 2), Halon has the largest ODP value, indicating that, on a mass basis, it is the most aggressive of these substances. Search for the product called "Pyrocool" to explore award-winning technology that targets the reduction of Halon.

COMMUNICATE YOUR RESULTS

Exercise 1
Presidential Green Chemistry Challenge

Consider the Presidential Green Chemistry Challenge program. What are the goals, focus areas, and award categories in the program? What are the criteria for winning the award? In 2005, several of the awards related to methods that reduce the emission of volatile organic compounds (VOCs) into the atmosphere. Discuss these methods as examples of award winners. Present your answer as a slide show for a brief in-class presentation.

Exercise 2
Cleaner Cleaners

Using the Web as a search tool, identify several dry-cleaning companies that have switched to a non-PERC method of dry cleaning. If one of these companies has a store in your community, visit the store to learn more about its methods. Write an informative essay comparing and contrasting these establishments to traditional dry-cleaning stores. Include a discussion of the challenges and benefits associated with converting from older technology to a new method.

Exercise 3
International Agreements for Improvement

In your Web investigation of the Montreal Protocol, you may have noticed that there are many tables embedded in the main document. These tables refer to the ozone-depleting molecules CFC-11, CFC-12, CFC-13, CFC-114, and CFC-115. Prepare an informative poster about the Montreal Protocol. Your poster should include the current uses and principal substitutes for each of the CFCs. Include the Lewis (electron-dot) structures (see Chapter 5) of these molecules. Use the Web as a search tool to locate the chemical structures.

Cooperative Exercise 1
First on the Agenda

Consider the hazards associated with air toxics, greenhouse gases, and ozone-depleting substances. In your opinion, which of these three categories demands the most immediate remedy? Adopt a position, organize a group of several students who share that position, and then write a persuasive argument. Find another group in the class that adopts an alternate position, and then have an informal debate about the topic.

CHAPTER 13

Water

Water is truly a unique substance. Liquid water is readily converted to steam by heating on a kitchen stove or in a geyser as seen here. Ice floats on liquid water because when water freezes it becomes less dense. Water is also known as the "universal solvent." Water's solvent power makes it incredibly useful; it dissolves many of the substances needed for proper function of plant and animal life. It also makes it easy to contaminate and difficult to purify.

Rivers of Life; Seas of Sorrows

Earth is a water world; most of its surface is covered with oceans and seas. People contain a lot of water, too: About two-thirds of our body weight is water. The water in our blood is quite similar to water in the ocean, containing a great variety of dissolved ions. You might even say that we are walking sacks of seawater.

The presence of large quantities of water makes our planet unique in the solar system, probably the only one capable of supporting higher forms of life. Water is the only substance on Earth that commonly exists in large amounts in all three physical states. Go outside on a snowy day and you are likely to "see" all three forms at once. Gaseous water vapor is invisible (and is all around us), but tiny droplets of liquid water form clouds. Solid water falls to the ground as snow, and melts into puddles of liquid water.

The Ionian philosopher Thales (ca. sixth century B.C.E.)—whom some consider the first "scientist"—held water in the highest regard, believing that it was the "primordial substance", that from which all other things are made. Little wonder from a man from an island nation. Water's unique nature makes it essential to life, and the nature of life as we know it makes it dependent on water. Our search for life on Mars is based to a large extent on evidence that the planet once had vast quantities of water and that relics of that life might still exist below the dry surface.

The properties that make water able to support life also make it easy to pollute and therefore potentially hazardous. Many substances are soluble in water and are therefore easily dispersed throughout the environment. Once they are there, it is not easy to remove dissolved substances from the water. Furthermore, many strains of bacteria and other microorganisms thrive in water but are harmful to human beings.

We start this chapter with a discussion of some of the unusual properties of water. Then we look at water pollution and water treatment. We should not underestimate the importance of clean drinking water nor take it for granted. We could live for several weeks without food, but without water we would last a few days at most.

13.1 WATER: SOME UNIQUE PROPERTIES

Although water is quite familiar, it is a most unusual compound. It is the only common liquid on the surface of our planet. For most substances, the solid form is more dense than the liquid. However, the solid form of water (ice) is less dense than the liquid. This means water expands when it freezes. The consequences of this peculiar characteristic are essential for life on Earth. Ice forms on the surfaces of lakes and insulates the lower layers of water. This enables fish and other aquatic organisms to survive winter in the temperate zones. If ice were denser than liquid water, it would sink to the bottom as it formed, and even the deeper lakes of the northern United States would freeze solid in winter.

The same property—ice being less dense than liquid water—has dangerous consequences for living cells. When living tissues freeze, ice crystals are formed, and the expansion ruptures and kills cells. The slower the cooling, the larger the crystals of ice and the more damage there is to the cell. Frozen-food manufacturers make "flash-frozen" products by freezing foods so rapidly that the ice crystals are kept very small and thus do minimal damage to the cellular structure of the food.

Water also is more dense than most other familiar liquids. As a consequence, liquids less dense than water and insoluble in water float on its surface. The massive oil spills that occur when a tanker ruptures or when an offshore well gets out of control are a fairly common problem. The oil, floating on the surface of the water, washes onto beaches, where it does considerable ecological and aesthetic damage. If water were less dense than oil, the problem would certainly be different, although not necessarily less serious.

Another unusual property of water is its high specific heat. Different substances have different capacities for storing energy absorbed as heat. The **heat capacity** of a substance is the quantity of heat required to raise the temperature of the substance by 1 °C. **Specific heat** is the heat capacity of a 1-g sample; that is, it is the quantity of heat required to change the temperature of 1 g of the substance by 1 °C. Table 13.1 gives the specific heats of several familiar substances in the old metric units—calories per gram per degree (cal/g °C)—and in the SI units—joules per gram per kelvin (J/g K). Note that it takes almost 10 times as much heat to raise the temperature of 1 g of water 1 °C as to raise the temperature of 1 g of iron by the same amount. Conversely, to cause even a small drop in water temperature, a relatively large quantity of heat must be removed. Imagine heating two iron kettles on the stove, one empty and one filled with water. It takes much longer (and much more energy) to heat the water. Because the water stores heat so well, it will stay hot a lot longer when the stove is turned off. The vast amounts of water on the surface of Earth thus act as a giant heat reservoir to moderate daily temperature variations. You need only consider the extreme temperature changes on the surface of the waterless moon, ranging from 100 °C at noon to −173 °C at night, to appreciate this important property of water.

▶ Oil coats the water of Prince William Sound, Alaska, where the *Exxon Valdez* ran aground on Bligh Reef in 1989, spilling 42 million L of oil. One liter of oil can create a slick 2.5 hectares (6.2 acres) in size.

TABLE 13.1 Specific Heats of Some Familiar Substances at 25 °C		
Substance	Specific Heat	
	(cal/g °C)	(J/g K)
Aluminum (Al)	0.216	0.902
Copper (Cu)	0.0920	0.385
Ethanol (CH_3CH_2OH)	0.588	2.46
Iron (Fe)	0.107	0.449
Lead (Pb)	0.0306	0.128
Silver (Ag)	0.0562	0.235
Sulfur (S)	0.169	0.706
Water	1.00*	4.180

*Note that in cal/g °C, the specific heat of water is 1.00. As the metric system was established, the properties of water were often taken as the standard.

Water also is unique in that it has a high **heat of vaporization**; that is, a large amount of heat is required to evaporate a small amount of water. This is of enormous importance to us because large amounts of body heat can be dissipated by the evaporation of small amounts of water (perspiration) from the skin. This effect also accounts in part for the climate-modifying property of lakes and oceans. A large portion of the heat that would otherwise heat up the land instead vaporizes water from the surface of lakes or seas. Thus, in summer it is cooler near a large body of water than in interior land areas.

All these fascinating properties of water depend on the unique structure of the highly polar water molecule (Chapter 5). In the liquid state, the molecules are tumbling over one another but always are associated with one another through strong hydrogen bonds. The tumbling becomes more and more violent as the temperature increases until the boiling point is reached. Under those conditions, a large fraction of the molecules have enough energy to break all of the hydrogen bonds with the other molecules surrounding them and they fly out of the liquid into the gas phase.

Conversely, when water freezes, its molecules take on a more ordered arrangement, forming four hydrogen bonds per molecule. The solid ice contains large hexagonal holes (Figure 13.1). This three-dimensional structure extends out for billions and billions of molecules. The holes collapse when the ice melts, accounting for the fact that ice is less dense than liquid water.

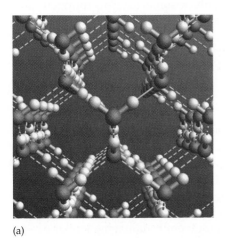

(a)

(b)

▲ **Figure 13.1** Hydrogen bonds in ice. (a) Oxygen atoms are arranged in layers of distorted hexagonal rings. Hydrogen atoms lie between pairs of oxygen atoms, closer to one (covalent bond) than to the other. The yellow dashed lines indicate the hydrogen bonds. The structure has large "holes." (b) At the macroscopic level, the hexagonal arrangement of water molecules in the crystal structure of ice is revealed in the hexagonal shapes of snowflakes.

▲ Three-fourths of Earth's surface is covered with water, making the planet appear mostly blue when viewed from space. The world appears to be mainly water, but in fact only about 1/4000 of Earth's mass is water. If Earth were the size of a basketball and you could hold it in your hands, you would notice that it was wet on the surface, but the sphere would seem quite solid otherwise.

▲ An adequate supply of water is essential to life. This photo shows a community water well in Malawi.

13.2 WATER, WATER, EVERYWHERE

Three-fourths of the surface of Earth is covered with water, but nearly 98% of it is salty seawater—unfit for drinking and not suitable for most industrial purposes. Water, due to its polar nature, readily dissolves most ionic substances. This solvent power of water accounts for the saltiness of the sea. Rainwater dissolves the ions of minerals, and these ions are carried by streams and rivers to the sea. There the heat of the sun evaporates part of the water, leaving the salts behind. Are the oceans growing saltier as the years go by? Apparently not. The rates of addition appear to be balanced by precipitation of minerals and absorption of ions onto clays.

Rain falls on Earth in enormous amounts, but most of it falls into the sea or on areas otherwise inaccessible. About 2% of Earth's water is frozen in the polar ice caps, leaving less than 1% available as fresh water, most of that underground. Lakes and streams account for only 0.01% of the fresh water on the planet.

Although potable (suitable for drinking) water is generally readily available in the United States, in many nations the situation is quite different. About 3.6 million people die each year because of contaminated water or lack of adequate sanitation. They die mainly of bacterial, viral, and parasitic infections. More than a billion people around the world have inadequate amounts of fresh water. According to the Pacific Institute, Oakland, California, that number is expected to double by 2015 and triple by 2050.

13.3 THE WATER CYCLE AND NATURAL CONTAMINANTS

Although the percentages of water apportioned to the oceans, ice caps, and freshwater rivers, lakes, and streams remain fairly constant, water is dynamically cycled among these various repositories (Figure 13.2). It constantly evaporates from both water and land surfaces, and water vapor condenses into clouds and returns to Earth as rain, sleet, and snow. This fresh water becomes part of the ice caps, runs off in streams and rivers, and fills lakes and underground pools of water in rocks and sand called *aquifers*.

Gases

Rainwater is not pure H_2O. It carries down dust particles, and it dissolves some oxygen, nitrogen, and carbon dioxide as it falls through the atmosphere. The carbon dioxide makes natural rainwater slightly acidic because it reacts with water to form carbonic acid (H_2CO_3):

$$CO_2(g) + H_2O(l) \longrightarrow H_2CO_3(aq)$$

Lightning causes nitrogen, oxygen, and water vapor to combine to form nitric acid, which dissolves in rainwater as well.

Groundwater also contains the naturally occurring gas radon, a product of the decay of radioactive uranium and thorium. Although radon emits damaging ionizing radiation and produces a variety of hazardous daughter nuclei, it is only slightly soluble in water. Thus, the water used for showering, washing dishes, and cooking generally contributes only a small proportion (about 1 to 2%) of the total radon in indoor air.

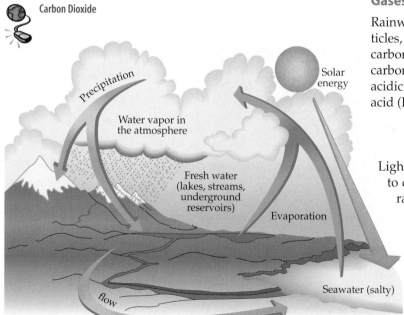

▲ **Figure 13.2** The hydrological (water) cycle.

TABLE 13.2	Some Substances Found in Natural Waters	
Substance	**Formula**	**Source**
Carbon dioxide	CO_2	Atmosphere
Dust	—	Atmosphere
Nitrogen	N_2	Atmosphere
Oxygen	O_2	Atmosphere
Nitric acid (thunderstorms)	HNO_3	Atmosphere
Sand and soil particles	—	Soil and rocks
Sodium ions	Na^+	Soil and rocks
Potassium ions	K^+	Soil and rocks
Calcium ions	Ca^{2+}	Limestone rocks
Magnesium ions	Mg^{2+}	Dolomite rocks
Iron(II) ions	Fe^{2+}	Soil and rocks
Chloride ions	Cl^-	Soil and rocks
Sulfate ions	SO_4^{2-}	Soil and rocks
Bicarbonate ions	HCO_3^-	Soil and rocks
Radon	Rn	Radioactive decay

Dissolved Minerals

As water moves along or beneath the surface of Earth, it dissolves minerals from rocks and soil. Recall that minerals (salts) are ionic and that ions are either positively charged (cations) or negatively charged (anions). The principal cations in natural water are ions of sodium (Na^+), potassium (K^+), calcium (Ca^{2+}), magnesium (Mg^{2+}), and sometimes iron (Fe^{2+} or Fe^{3+}). The anions are usually sulfate (SO_4^{2-}), bicarbonate (HCO_3^-), and chloride (Cl^-). Table 13.2 provides a summary of substances found in natural waters.

Water containing calcium, magnesium, or iron salts is called **hard water**. The positive ions react with the negative ions in soap to form a scum that clings to clothes and bathroom fixtures and leaves them dingy looking (Chapter 17). *Soft water* may contain ions, such as Na^+ or K^+, but these ions do not form insoluble scum with soap.

The water cycle replenishes our supply of fresh water. When water evaporates from the sea, salts are left behind. When water moves through the ground, impurities are trapped in the rock, gravel, sand, and clay. This capacity to purify is not infinite, however.

Some Biblical Chemistry

According to the Biblical story of Moses, getting a dependable supply of fresh water has long been a problem. When Moses led the Israelites out of Egypt into the wilderness, he encountered a desert area where potable water was scarce. Many people know the Biblical account of how Moses struck the rock to bring forth water. But an incident at Marah, where the Israelites couldn't drink the water because it was bitter, seems less well known. According to Exodus, God commanded Moses to throw a certain tree into the water to sweeten (neutralize) it.

Basic (alkaline) solutions are bitter. The pool at Marah was probably basic; such *alkaline* waters are common in desert areas. Attempts have been made to give a chemical explanation of Moses's purification of the brackish water. The tree was probably dead, its cellulose bleached by the desert sun, with the alcohol groups in cellulose oxidized to carboxylic acid groups.

These acidic groups neutralize the alkali in the water. (Tannins, found in plant material and composed of polyphenols and other acidic compounds, also may have contributed to the neutralization.) Moses didn't have to understand the science of water purification in order to apply the appropriate technology.

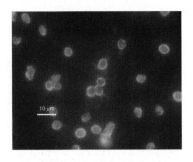

▲ *Cryptosporidium*, shown here in a fluorescent image, sickened 400,000 people and killed about 100 in the Milwaukee area in 1993. Since then, utilities have generally improved their treatment systems.

▲ Contaminated beaches are closed to swimming and other recreational activities.

▲ Acid water draining from an old mine.

Organic Matter

Rainwater dissolves matter from decaying plants and animals. In small quantities as part of the natural cycle in a forest, this enriches the soil, but in large quantities it contaminates soil and water. Bacteria, microorganisms, and animal wastes are all potential contaminants of natural waters.

13.4 CHEMICAL AND BIOLOGICAL CONTAMINATION

Early people did little to pollute the water and the air, if only because their numbers were so few. Most pollutants at that time were natural: hard-water ions, alkaline contaminants, and animal wastes. The coming of the agricultural revolution and the rise of cities brought together enough *Homo sapiens* to pollute the environment seriously. Even then, pollution was mostly local and largely biological. Human wastes were dumped on the ground or into the nearest stream. Disease organisms were transmitted through food, water, and direct contact.

Waterborne Disease

Contamination of water supplies by *pathogenic* (disease-causing) microorganisms from human wastes was a severe problem throughout the world until about 100 years ago. During the 1830s, severe epidemics of cholera swept the Western world. Typhoid fever and dysentery were common. In 1900, for example, there were more than 35,000 deaths from typhoid in the United States. Today, as a result of chemical treatment, municipal water supplies in developed nations are generally safe. Waterborne diseases are still quite common in much of Asia, Africa, and Latin America, where even today there are occasional epidemics of cholera and typhoid. Dysentery is rampant much of the time. An estimated 80% of all the world's sickness is caused by contaminated water.

The threat of biological contamination has not been totally eliminated from developed nations. The EPA estimates that 30 million people in the United States are at risk because of bacterial contamination of drinking water. One ongoing threat is *cryptosporidium*, a protozoa excreted in human and animal feces that resists standard chemical disinfection. Utilities are constantly improving their treatment systems to deal with this and other problems.

Biological contamination also lessens the recreational value of water, when swimming, fishing, and other recreational activities become hazardous.

Acid Rain

Acids formed from sulfur oxides (SO_x) and nitrogen oxides (NO_x) come down from the sky as acid rain, fog, and snow. These acids corrode metals, dissolve limestone and marble, and even ruin the finishes on our automobiles. Acids also flow into streams from abandoned mines.

Acid rain has been somewhat alleviated over the last few decades by the reduction of SO_x emissions (Chapter 12), but thousands of bodies of water in eastern North America are acidified, and thousands more have only a limited ability to neutralize the acids that enter them. The acids are presumed to originate mainly in the Ohio River Valley and Great Lakes regions. The SO_x and NO_x—mainly from coal-fired power plants, but also from other industries and automobiles—travel hundreds of kilometers downwind and fall as sulfuric and nitric acids.

Acid water is detrimental to life in lakes and streams. It has been linked to declining crop and forest yields. Acid waters probably affect living organisms in many ways. One way is that they cause the release of toxic ions from rocks and soil. For example, aluminum ions, which are tightly bound in clays and other minerals, are released by acid. Aluminum ions have low toxicity for humans but they can be deadly to young fish. Many dying lakes have only old fish because none of the young survive. Ironically, lakes destroyed by excess acidity are often quite beautiful. The water is clear and sparkling—quite a contrast to those in which fish are killed by oxygen depletion following algal blooms.

Acids are no threat to lakes and streams in areas where the rock is limestone (calcium carbonate), which can neutralize excess acid.

$$CaCO_3(s) + 2 H^+(aq) \longrightarrow Ca^{2+}(aq) + CO_2(g) + H_2O$$

Limestone — Acid

Where rock is principally granite, however, no such neutralization occurs.

Sewage and Dying Lakes

Pathogenic microorganisms are not the only problem caused by the dumping of human sewage into our waterways. The breakdown of organic matter by bacteria depletes **dissolved oxygen** in the water and enriches the water with plant nutrients. A stream can handle a small amount of waste without difficulty, but when massive amounts of raw sewage are dumped into a waterway, undesirable changes occur.

Most organic material can be degraded (broken down) by microorganisms. Biodegradation can be either *aerobic* or *anaerobic*. **Aerobic oxidation** occurs in the presence of dissolved oxygen. A measure of the amount of oxygen needed for this degradation is the **biochemical oxygen demand (BOD)**. The greater the quantity of degradable organic wastes, the higher the BOD. If the BOD is high enough, dissolved oxygen is depleted and no life (other than odor-producing anaerobic microorganisms) can survive in the lake or stream. Flowing streams can regenerate themselves; those with rapids soon come alive again as the swirling water dissolves oxygen. Lakes with little or no flow can remain dead for years.

With adequate dissolved oxygen, aerobic bacteria (those that require oxygen) oxidize the organic matter to carbon dioxide, water, and a variety of inorganic ions (Table 13.3). The water is relatively clean, but the ions, particularly the nitrates and phosphates, may serve as nutrients for the growth of algae, which also cause problems. When the algae die, they become organic waste and increase the BOD. This process is called **eutrophication**. The eutrophication of a lake is a natural process, but the action can be greatly accelerated by human wastes, phosphates from detergents, and the runoff of fertilizers from farms and lawns which stimulates algal bloom and die-off. Streams and lakes die because nature cannot purify them nearly as quickly as we can pollute them.

When too much organic matter depletes the dissolved oxygen in a body of water—whether from sewage, dying algae, or other sources—**anaerobic decay** processes take over. Instead of oxidizing the organic matter, anaerobic bacteria reduce it. Methane (CH_4) is formed. Sulfur is converted to hydrogen sulfide (H_2S) and foul-smelling organic compounds. Nitrogen is reduced to ammonia and odorous amines (Chapter 9). The foul odors are a good indication that the water is overloaded with organic wastes. Only anaerobic microorganisms can survive in such water.

▲ Common sewage from homes and businesses depletes the dissolved oxygen in water.

TABLE 13.3 Some Substances Formed in Water by the Breakdown of Organic Matter	
Substance	**Formula**
Aerobic conditions	
Carbon dioxide	CO_2
Nitrate ions	NO_3^-
Phosphate ions	PO_4^{3-}
Sulfate ions	SO_4^{2-}
Bicarbonate ions	HCO_3^-
Anaerobic conditions	
Methane	CH_4
Ammonia	NH_3
Amines	RNH_2*
Hydrogen sulfide	H_2S
Methanethiol	CH_3SH

*See Chapter 9.

▶ The eutrophication of a lake is a natural process, but the action can be greatly accelerated by human wastes and the runoff from farms, lawns, and golf courses.

Dumping our sewage into waterways isn't the only way we can foul up an ecological cycle. Fertilizer runoff from farm fields, golf courses, and lawns, and seepage from feedlots all add inorganic nutrients to the cycle, and an algal bloom can lead to oxygen depletion and death for the fish. Perhaps the most enigmatic influence of all comes from the introduction of new substances into the ecological water cycle: pesticides, radioisotopes, detergents, toxic metals, and industrial chemicals.

The Industrial Revolution added a new dimension to our water pollution problems. Factories were often built on the banks of streams, and wastes were dumped into the water to be carried away. The rise of modern agriculture has led to increased contamination as fertilizers and pesticides have found their way into the water system. Transportation of petroleum results in oil spills in oceans, estuaries, and rivers. Acids enter waterways from mines and factories and from acid precipitation. Household chemicals also contribute to water pollution when detergents, solvents, and other chemicals are dumped down drains.

By creating new materials, chemists have brought on a variety of new ecological problems. But chemists and chemistry are also necessary to any understanding of our pollution problems. Through the development of sophisticated analytical methods and instruments, we have become aware of some of our ecological problems. A stinking lake is obvious to everyone, but only by using advanced analytical techniques can scientists determine the level of dangerous, yet invisible, materials in the water we drink.

13.5 AN EXAMPLE OF INDUSTRIAL POLLUTION: BUILDING A CAR

It takes several hundred kilograms of steel to produce a typical automobile. To make a metric ton of steel requires about 100 t of water. About 4 t of water is lost through evaporation. The remainder is contaminated with acids, grease and oil, lime, and iron salts. This polluted water can be cleaned up, and most of it is recycled.

Chrome plating on bumpers, grills, and ornaments was once a serious source of pollution. Waste chromium, in the form of chromate ions (CrO_4^{2-}), and cyanide ions (CN^-) are products of this process. In the past, these toxic substances were dumped into waterways. Today, chemical treatment generally removes a large amount.

Cyanide is treated with chlorine and a base to form nitrogen gas, bicarbonate ions, and chloride ions.

$$10\,OH^- + 2\,CN^- + 5\,Cl_2 \longrightarrow N_2 + 2\,HCO_3^- + 10\,Cl^- + 4\,H_2O$$

The products are much less toxic than cyanide. The chromate is removed by reduction with sulfur dioxide to Cr^{3+} ion, and the sulfur dioxide is oxidized to sulfate.

$$2\,CrO_4^{2-} + 3\,SO_2 + 2\,H_2O \longrightarrow 2\,Cr^{3+} + 3\,SO_4^{2-} + 4\,OH^-$$

TABLE 13.4	Water Required to Produce Various Materials		
Industrial Products	**Water Required[a]**	**Consumer Products**	**Water Required[b]**
Steel	100	Laptop computer	10,600
Paper	20	1 kg flour	77
Copper	400	1 bowl rice	525
Rayon	800	1 L red wine	720
Aluminum	1280	1 cup coffee	140
Synthetic rubber	2400	1 XL cotton tee shirt	30,300

[a]In cubic meters per metric ton. A cubic meter of water weighs 1000 kg, or 1 t.
[b]In liters.

Sulfate is generally not a serious pollutant, and Cr^{3+} is relatively insoluble in alkaline solution but is soluble enough in acidic media to constitute a problem.

With the environmental and economic costs of elastomers for tires, fabrics for upholstery, glass for windows, electricity, and so on, it is easy to see that the private automobile is an ecological problem even before it hits the road. And once on the road, it is a major contributor to air pollution (Chapter 12).

Most other industries also contribute to water pollution. Table 13.4 lists the water required (per metric ton) for the production of a variety of materials. Much of this water is cleaned up and recycled, but the need for clean water is still enormous. Industries in the United States have substantially reduced their contribution to water pollution. Most are in compliance with the Water Pollution Control Act, which requires that they use the best practicable technology. We examine here only a few examples of industrial pollution and some ways to alleviate it.

Water Pollution in Russia

Until the Soviet Union collapsed in 1991, their leaders covered up the environmental problems of Eastern Europe. Let's consider the case of one of the most famous rivers in the world, the mighty Volga in Russia (Figure 13.3). So many dams have been built along the Volga that its flow has been slowed to a crawl. It used to take 50 days for water to travel the 2300 mi from the source to the mouth of the river, but now it takes 18 months. The sluggish flow plus the pollution from all the factories along the way have created an ecological catastrophe. Tons of industrial waste (cleaning fluids, fertilizers, pesticides, heavy metals, toxic chemicals, radioactive waste, and waste from pulp and paper mills) pour into the Caspian Sea, which appears to be dying. In some places you can see yellow, red, and black streams of water carrying sulfur, iron oxide, and various oils. The Volga once teemed with caviar-producing sturgeon, but pollution and poaching now threaten the sturgeon with extinction.

◄ **Figure 13.3** The Soviet Union built so many dams and factories along the Volga River that it is no longer the mighty, vigorous waterway it once was. Much of the damage was hidden from the rest of the world until the USSR collapsed in 1991.

13.6 GROUNDWATER CONTAMINATION → TAINTED TAP WATER

About half the people in the United States drink surface water (from streams and lakes). The other half get their drinking water from groundwater via wells or springs. In rural areas, 97% of the population drink groundwater. Because rocks have different porosity and permeability, water does not move around the same way in all rocks. An aquifer forms when water-bearing rocks readily transmit water to wells and springs. Wells are drilled into an aquifer, and water is pumped out. Precipitation recharges the aquifer. For many aquifers, the rate of pumping water is much greater than the rate of recharge, and the water table drops. The dropping water table in most parts of the country requires that wells be drilled deeper and deeper.

Well water can be contaminated. Toxic chemicals have been found in the groundwater in some areas. For example,

- Many wells throughout the United States are contaminated with the gasoline additive methyl *tert*-butyl ether (MTBE) (Chapter 14), making the water undrinkable due to its offensive taste and odor.
- Wells around the Rocky Mountain Arsenal, near Denver, are contaminated by wastes from the production of pesticides.
- Wells near Minneapolis, Minnesota, are contaminated with creosote, a chemical used as a wood preservative.
- Wells on Long Island and in parts of Wisconsin have been contaminated with aldicarb, a pesticide used on potato crops.
- Some community water supplies in New Jersey have been shut down because of contamination with industrial wastes.

Groundwater contamination is particularly alarming because, once contaminated, an underground aquifer may remain unusable for decades or longer. There is no easy way to remove the contaminants. Pumping out the water and purifying it could take years and cost billions of dollars. We will examine a few important sources of groundwater contamination in a bit more detail.

> Half of Chinese cities have significantly polluted groundwater and some cities are facing a water crisis. More than 400 Chinese cities are threatened with water shortage, with 136 of them experiencing severe shortages.

Nitrates

In many agricultural areas, well water is contaminated with nitrate ions (NO_3^-). Excessive nitrates are especially dangerous to infants. In an infant's digestive tract, nitrate ion is reduced to nitrite ion. The result is *methemoglobinemia*, or blue baby syndrome. The baby turns blue and can die after drinking the water if not treated. In some farming areas, such as California's Imperial Valley, some parents have to buy bottled water for their babies.

Nitrates in the groundwater (Figure 13.4) come from fertilizers used on farms and lawns, from the decomposition of organic wastes in sewage treatment, and from runoff from animal feed lots. They are highly soluble and thus difficult to remove from water, requiring expensive treatment.

Volatile Organic Chemicals

We mentioned volatile organic compounds (VOCs) as air pollutants (Chapter 12). VOCs are also water pollutants and add an undesirable odor to water. Many VOCs are suspected carcinogens. VOCs are used as solvents, cleaners, and fuels, and they are components of gasoline, spot removers, oil-based paints, thinners, and some drain cleaners. Common VOCs are hydrocarbon solvents, such as benzene and toluene, and chlorinated hydrocarbons, such as carbon tetrachloride (CCl_4), chloroform ($CHCl_3$), and methylene chloride (CH_2Cl_2). Especially common is trichloroethylene ($CCl_2{=}CHCl$), widely used as a dry-cleaning solvent and as a degreasing compound. When spilled or discarded, VOCs enter the soil and eventually get into the groundwater.

Probability of Nitrate Contamination of Shallow Ground Water

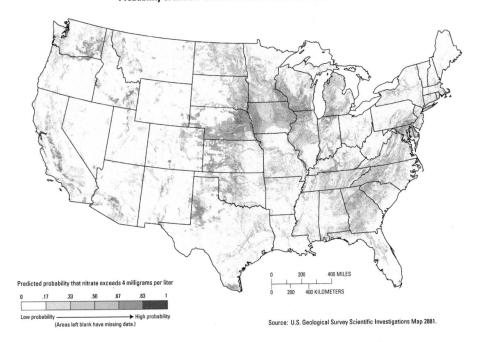

Predicted probability that nitrate exceeds 4 milligrams per liter

| 0 | .17 | .33 | .50 | .67 | .83 | 1 |

Low probability ⟶ High probability
(Areas left blank have missing data.)

0   200   400 MILES
0   200   400 KILOMETERS

Source: U.S. Geological Survey Scientific Investigations Map 2881.

◀ **Figure 13.4** Areas of the contiguous United States with groundwater likely to be contaminated by nitrates.

Chemicals buried in dumps—often years ago, before there was much awareness of environmental problems—have now infiltrated groundwater supplies. Often, as at the Love Canal site in Niagara Falls, New York, people built schools and houses on or near old dump sites. VOCs generally are only slightly soluble in water, with solubilities often in the parts-per-million or parts-per-billion range. These trace amounts are found in groundwater. Many VOCs are characterized by a lack of reactivity. They react so slowly that they are likely to be around for a long time.

Leaking Underground Storage Tanks

A major source of groundwater contamination is underground storage tanks (USTs) that contain petroleum or hazardous chemicals. These buried steel tanks last an average of about 15 years before they rust through and begin to leak. In the 1980s the U.S. EPA estimated there were about 2.5 million such tanks. More than 370,000 leaks from USTs have been detected, with petroleum products such as gasoline and certain gasoline additives (Chapter 14) the most prevalent contaminants found in nearby wells. Laws now require replacement of old gasoline tanks and proper cleanup of any contaminated ground. About 1.6 million of the tanks subject to regulation have been closed, and the frequency and severity of leaks from UST systems have been reduced greatly. However, there were still 7850 newly confirmed releases in fiscal year 2004.

13.7 MAKING WATER FIT TO DRINK

The per capita use of water in the United States is nearly 2 million L per year, and this number is increasing rapidly. This value includes that used for industrial, agricultural, and other purposes and is significantly larger than consumption in other nations. Although often taken for granted, significant energy and resources are spent every year to ensure the safety and quality of drinking water supplied by the over 170,000 public water systems in the United States.

Safe Drinking Water Act

The U.S. Safe Drinking Water Act was passed in 1974 and amended in 1986 and 1996. The Act gives the EPA power to set, monitor, and enforce national

Homes and a school were built on the site of an old chemical dump in the Love Canal section of Niagara Falls, New York. A contaminated area around the old dump was fenced off to keep people out. Over 250 different industrial contaminants were found on the site. Cleanup was completed in the 1990s and the evacuated neighborhood was repopulated.

TABLE 13.5	U.S. Environmental Protection Agency Drinking Water Standards for Selected Substances*	
Substance		**Maximum Contaminant Level (mg/L)**
Primary standards: inorganic compounds		
Arsenic		0.010
Barium		2
Copper		1.3
Cyanide		0.2
Fluoride		4.0
Lead		0.015
Nitrate		10[†]
Primary standards: organic compounds		
Atrazine		0.003
Benzene		0.005
p-Dichlorobenzene		0.075
Dichloromethane		0.005
Heptachlor		0.0004
Lindane		0.0002
Toluene		1
Trichloroethylene		0.005
Secondary standards (nonenforceable)		
Chloride		250
Iron		0.3
Manganese		0.05
Silver		0.10
Sulfate		250
Total dissolved solids		500
Zinc		5

*A more detailed list and a more detailed explanation of the rules can be found on the EPA website.
[†]Measured as N. Measured as nitrate ion, the level is 45 mg/L.

health-based standards for a variety of contaminants in municipal water supplies. As modern analytical techniques continue to improve, our ability to identify smaller and smaller concentrations of potentially harmful substances also advances. As a result, the number of regulated substances has also increased from 22 in 1976 to 90 in 2004 and includes metals, microorganisms, and other hazardous chemicals (Table 13.5). Many of these contaminants are present in minute amounts, and chemists use some special units to describe their concentrations quantitatively.

Calculations of Parts per Million and Parts per Billion

We discussed solution concentrations in Chapter 6. For solutions that are extremely dilute, we often express concentrations in ppm, ppb, or even ppt (parts per trillion). For example, in fluoridated drinking water, the fluoride ion concentration is maintained at about 1 ppm. A typical level of the contaminant chloroform ($CHCl_3$) in municipal drinking water taken from the lower Mississippi River is 8 ppb.

For aqueous solutions, ppm, ppb, and ppt are generally based on mass. Thus, 1 ppm of solute in a solution is the same as 1 g solute per 1×10^6 g (1 million grams) solution, and 1 ppb is 1 g solute per 10^9 g (1 billion grams) solution. Or

$$1 \text{ ppm} = \frac{1 \text{ g solute}}{10^6 \text{ g solution}} \qquad 1 \text{ ppb} = \frac{1 \text{ g solute}}{10^9 \text{ g solution}}$$

Some Comparisons that Put the Figures in Perspective
1 Part per Million (ppm) Is
    1 inch in 16 miles
    1 minute in 2 years
    1 cent in $10,000
1 Part per Billion (ppb) Is
    1 inch in 16,000 miles
    1 second in 32 years
    1 cent in $10 million
1 Part per Trillion (ppt) Is
    1 inch in 16 million miles
    1 second in 320 centuries
    1 cent in $10 billion

Calculations are much like those for percent by mass (Section 6.7). In Example 13.1, we introduce a useful relationship for aqueous solutions.

EXAMPLE 13.1 **Contaminant Concentrations**

The maximum allowable level of fluoride ion in drinking water set by the EPA is 4 mg F⁻ per liter. What is this level expressed in ppm?

Solution

The density of water, even if it contains traces of dissolved substances, is essentially 1.00 g/mL. One liter of water has a mass of 1000 g. So we can express the allowable fluoride level as the ratio

$$\frac{4 \text{ mg F}^-}{1000 \text{ g water}}$$

To convert this fluoride level to ppm, we need to have the numerator and denominator in the same units. By using milligrams, we make the denominator 1 million mg. The numerator then expresses the ppm of solute, that is, ppm F⁻.

$$\frac{4 \text{ mg F}^-}{1000 \text{ g water}} \times \frac{1 \text{ g water}}{1000 \text{ mg water}} = \frac{4 \text{ mg F}^-}{1,000,000 \text{ mg water}} = 4 \text{ ppm F}^-$$

▶ **Exercise 13.1A**

What is the concentration in **(a)** ppb and **(b)** ppt corresponding to a maximum allowable level in water of 0.1 μg/L of the gasoline additive MTBE (methyl *tert*-butyl ether)?

▶ **Exercise 13.1B**

What is the molarity of the solution in Exercise 13.1A?

But Not a Drop to Drink

Clean drinking water is scarce in Bangladesh, one of the poorest countries in the world. Ravaged frequently by floods, the surface waters that supplied drinking water for most of the country until about 35 years ago were often contaminated by disease-causing microorganisms. During the 1970s, UNICEF and the World Bank provided funds for the construction of tubewells to provide a consistent supply of clean drinking water. The project involved constructing hundreds of thousands of tubewells; a well within 100 m of every family. The project halved the infant mortality rate during the 1980s.

Much to everyone's surprise, the wells brought a new risk: chronic arsenic poisoning. Arsenic occurs naturally in the mineral formations of the region, and it now contaminates the water collecting in many of the wells. These tubewells are the main source of potable water in 59 out of Bangladesh's 64 districts, and arsenic poisoning threatens 75 million of Bangladesh's 120 million people, mostly in the north and west of the country. The initial stages of arsenic poisoning are characterized by skin pigmentation, warts, diarrhea, and ulcers. Arsenic also damages the lungs, kidneys, and other internal organs. In the most severe cases, arsenic poisoning causes liver and renal deficiencies or cancer that can lead to death. Remediation could take dozens of years and hundreds of millions of dollars.

▲ A village tubewell in Bangladesh.

13.8 WATER TREATMENT PLANTS

About half the population of the United States gets its drinking water from surface water (lakes, reservoirs) while the other half gets theirs from groundwater supplies (wells and aquifers). Contamination can threaten either of these sources. Most cities in developed nations treat their water supply before it flows into homes (Figure 13.5). The water to be purified is usually placed in a settling basin where it is treated with slaked lime and a flocculent such as "alum." These materials react to form a gelatinous mass of aluminum hydroxide.

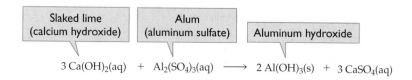

Slaked lime (calcium hydroxide)   Alum (aluminum sulfate)   Aluminum hydroxide

$$3\,Ca(OH)_2(aq) \;+\; Al_2(SO_4)_3(aq) \;\longrightarrow\; 2\,Al(OH)_3(s) \;+\; 3\,CaSO_4(aq)$$

The aluminum hydroxide carries down dirt particles and bacteria. The water is then filtered through sand and gravel.

Usually water is treated by *aeration*; it is sprayed into the air to remove odors and improve its taste (water without dissolved air tastes flat). Sometimes it is filtered through charcoal to remove colored and odorous compounds.

Chemical Disinfection

In the final step, chlorine is added to kill any remaining bacteria. In some communities that use river water, a lot of chlorine is needed to kill all the bacteria, and you can taste the chlorine in the water. Some people question the use of chlorine because it converts dissolved organic compounds into chlorinated hydrocarbons. Analyses of the drinking water of several cities that take their water from rivers have found chlorinated hydrocarbons, including such known carcinogens as chlor-

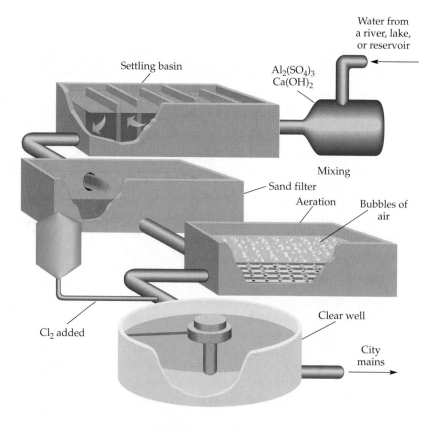

▶ **Figure 13.5** A diagram of a municipal water purification plant.

oform and carbon tetrachloride. The concentration is in the parts-per-billion range, probably posing only a small threat, but it is worrisome nonetheless. (It is not nearly so worrisome, however, as the waterborne diseases that prevail in much of the world where adequate water treatment is not available.)

Chlorination is not the only way to disinfect drinking water. Ozone (O_3) is used widely in Europe and increasingly in the United States. Ozone is more expensive than chlorine, but less of it is needed. An added advantage is that ozone kills viruses on which chlorine has little, if any, effect. For example, ozone is 100 times as effective as chlorine in killing polioviruses.

Ozone acts by transferring its "extra" oxygen atom to the contaminant. Oxidized contaminants are generally less toxic than chlorinated ones. In addition, ozone imparts no chemical taste to water. Unlike chlorine, however, ozone does not provide residual protection against microorganisms. Some systems therefore use a combination of disinfectants: ozone for initial treatment and subsequent chlorine addition to provide residual protection.

Other Technologies

Water contaminated with a variety of microorganisms can be purified by irradiation with ultraviolet (UV) light. UV works rapidly, and it can be cost effective in small-scale applications. No chemical generation, storage, or handling is required. UV is effective against *Cryptosporidium*, and there are no known by-products formed at levels that cause concern. The disadvantages include no residual protection for drinking water and no taste and odor control. UV has limited effectiveness in turbid water and generally costs more than chlorine treatment.

Fluorides

Dental caries (tooth decay) was once considered the leading chronic disease of childhood. That this is no longer true is attributed mainly to fluoride toothpastes (Chapter 17) and the addition of fluoride to municipal water supplies. The hardness of tooth enamel can be correlated with the amount of fluoride present. Tooth enamel is a complex calcium phosphate called hydroxyapatite. Fluoride ions replace some of the hydroxide ions, forming a harder mineral called fluorapatite.

Hydroxyapatite   Fluorapatite

$$Ca_5(PO_4)_3OH(s) + F^-(aq) \longrightarrow Ca_5(PO_4)_3F(s) + OH^-(aq)$$

The fluoride concentration of the drinking water of many communities has been adjusted to 0.7–1.0 ppm (by mass) by adding fluoride, usually as H_2SiF_6 or Na_2SiF_6. Early studies showed reduction in the incidence of dental caries by 50 to 70%. More recent studies show a much smaller effect, perhaps because of fluoride toothpastes and the presence of fluorides in food products prepared with fluoridated water.

Some people object to water fluoridation. Fluoride salts are acute poisons in moderate to high concentrations. Indeed, sodium fluoride (NaF) is used as a poison for roaches and rats. Small amounts of fluoride ion, however, seem to contribute to our well-being through strengthening bones and teeth. There is some concern about the cumulative effects of consuming fluorides in drinking water, in the diet, in toothpaste, and from other sources. Excessive fluoride consumption during early childhood can cause mottling of tooth enamel. The enamel becomes brittle in certain areas and gradually discolors. Fluorides in high doses also interfere with calcium metabolism, with kidney action, with thyroid function, and with the actions of other glands and organs. Although there is little or no evidence that optimal fluoridation causes problems such as these, the fluoridation of public water supplies will most likely remain a subject of controversy.

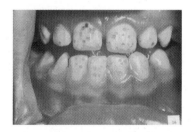

▲ Excessive fluoride consumption in early childhood can cause mottling of tooth enamel. The severe case shown here was caused by continuous use during childhood of a water supply that had an excessive natural concentration of fluoride.

13.9 FROM WASTEWATER TO DRINKING WATER

The water that half the U.S. population drinks comes from reservoirs, lakes, and rivers. Many cities depend on water that has been used by other municipalities upstream. Cities have to treat the used water before discharging it back into the environment. In this section we discuss the treatment and purification of wastewater that renders it suitable for returning it to the water cycle.

Wastewater Treatment Plants

For decades, most communities treated sewage simply by holding it in settling ponds for a while before discharging it into a stream, lake, or ocean, a process now called **primary sewage treatment** (Figure 13.6). Primary treatment removes some of the solids as *sludge*. The effluent still contains a lot of dissolved and suspended organic matter and has a huge BOD. Often all the dissolved oxygen in the pond is used up, and anaerobic decomposition—with its resulting odors—takes over.

A **secondary sewage treatment** plant passes effluent from the primary treatment facility through sand and gravel filters. There is some aeration in this step, and aerobic bacteria convert much of the organic matter to inorganic materials. In the **activated sludge method** (Figure 13.7), a combination of primary and secondary treatment methods, the sewage is placed in tanks and aerated with large blowers. This causes the formation of large, porous clumps called *flocs*, which filter and absorb contaminants. The aerobic bacteria further convert the organic material to sludge. Part of the sludge is recycled to keep the process going, but huge quantities must be removed for disposal. This sludge is stored on land (where it requires large areas), dumped at sea (where it pollutes the ocean), or burned in incinerators (where it requires energy—such as natural gas—and can contribute to air pollution). Some of the sludge is used as fertilizer.

Secondary treatment of wastewater often is inadequate. To an increasing extent, federal mandates require **advanced treatment** (sometimes called *tertiary treatment*). Several advanced processes are in use, and most are quite costly.

- In **charcoal filtration**, charcoal adsorbs organic molecules that are difficult to remove by any other method. The organic molecules are adsorbed on the surface of the charcoal and thus removed from the water. After a period of time, the charcoal becomes saturated and is no longer effective. It can be regenerated by heating to drive off the adsorbed substances.

- In **reverse osmosis**, pressure forces water through a semipermeable membrane, leaving impurities behind.

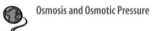 Osmosis and Osmotic Pressure

Finding the money to finance adequate sewage treatment will be a major political problem for years to come. Table 13.6 summarizes wastewater treatment methods.

The effluent from sewage plants is usually treated with chlorine to kill any remaining pathogenic microorganisms before it is returned to a waterway. Chlorination has been quite effective in preventing the spread of waterborne infectious

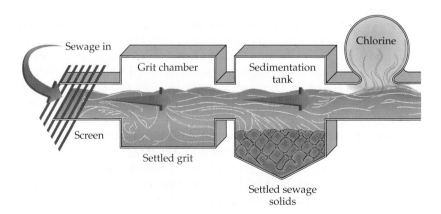

▶ **Figure 13.6** Diagram of a primary sewage treatment plant.

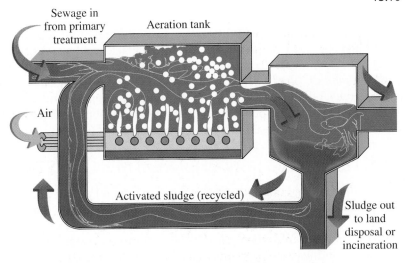

Sewage in from primary treatment

Aeration tank

Air

Activated sludge (recycled)

Sludge out to land disposal or incineration

◀ **Figure 13.7** Diagram of a secondary sewage treatment plant that uses the activated sludge method of treatment.

TABLE 13.6	Summary of Wastewater Treatment Methods*		
Method	**Cost**	**Material Removed**	**Amount Removed (%)**
Primary			
Sedimentation	Low	Dissolved organics	25–40
		Suspended solids	40–70
Secondary			
Trickling filters	Moderate	Dissolved organics	80–95
		Suspended solids	70–92
Activated sludge	Moderate	Dissolved organics	85–95
		Suspended solids	85–95
Advanced (tertiary)			
Carbon bed with regeneration	Moderate	Dissolved organics	90–98
Ion exchange	High	Nitrates and phosphates	80–92
Chemical precipitation	Moderate	Phosphates	88–95
Filtration	Low	Suspended solids	50–90
Reverse osmosis	Very high	Dissolved solids	65–95
Electrodialysis	Very high	Dissolved solids	10–40
Distillation	Extremely high	Dissolved solids	90–98

*See References and Readings for a discussion of methods not covered in the text.

diseases such as typhoid fever. Further, some chlorine remains in the water, providing residual protection against pathogenic bacteria. However, chlorination is not effective against viruses such as those that cause hepatitis.

13.10 THE NEWEST SOFT DRINK: BOTTLED WATER

Bottled water is the fastest growing and most profitable segment of the beverage industry. In 2004, worldwide sales of bottled water was 156 billion L at a cost that exceeded $50 billion, and the industry was growing at 10% a year. In the United States, bottled water is the second most popular drink, exceeded only by carbonated soft drinks. More than half of all Americans drink bottled water, even though it costs 240–10,000 times as much per liter as tap water. Per capita consumption is 90 L per year. The trend is based largely on the misconception that bottled water is safer or healthier than tap water.

    The federal government considers bottled water a food, and it is therefore regulated by the Food and Drug Administration (FDA) rather than the EPA. Although the FDA has adopted EPA standards for tap water as standards for bottled water, testing of bottled waters is typically less rigorous than for municipal water.

    Because municipal water supplies are usually chlorinated, tap water can have a chlorine taste. This might cause some to buy bottled water, but beware: About 25% of bottled water sold in the United States comes from municipal water supplies.

Some bottled water is disinfected with ozone, just as some municipal water supplies are, but just because it comes in a bottle with a glacier and a mountain spring on the label does not necessarily mean water is "more pure." Mineral water in fact is likely to have *more dissolved ions* (typically Ca^{2+} and Mg^{2+}) than normal tap water. Fortunately, in the United States in general, both bottled water and municipal water are quite safe. However, some people will continue to buy bottled water for its convenience, taste, and falsely presumed health benefits.

13.11 ALTERNATIVE SEWAGE TREATMENT SYSTEMS

For most of human history, our body wastes were an integral part of Earth's natural recycling system. The wastes provided food for microorganisms that degraded them, returning nutrients to soil and water. But with the growth of cities, human waste was disconnected from the cycle. We now dump our wastes in waterways, fouling the water and wasting nutrients that could fertilize the land.

Many communities dry and sterilize sludge and then transport it to farmlands for return of nutrients to the soil. Others pump wet, suspended sludge directly to the fields. The water in the mixture irrigates the crops and the sludge provides nutrients and humus. One concern is that sludge is often contaminated with toxic metals that could be taken up by plants and eventually end up in our food.

A few communities treat their sewage by first holding it in sedimentation tanks where the solids settle out as a sludge that is removed and processed for use as fertilizer. The effluent is then diverted into a marshy area where the marsh plants filter the water and use nutrients from the sewage as fertilizer. Some plants even remove toxic metals. Effluent from the marsh is often as clean as the water in municipal reservoirs.

Other solutions are possible. Toilets have been developed that compost wastes and use no energy or water. The mild heat of composting drives off water from the wastes. The system is ventilated to keep the process aerobic, and no odors enter the house from a properly installed system. Dried waste is removed about once a year. The initial cost is much higher than that of a flushed toilet, however. For each person in the United States, we flush about 35,000 L of drinking-quality water annually. Perhaps we should consider an alternative to flushing our wastes into the water we drink.

13.12 WE'RE THE SOLUTION TO WATER POLLUTION

We generally take our drinking water for granted. Perhaps we shouldn't. There are between 4000 and 40,000 cases of waterborne illnesses in the United States each year. Twenty million people have no running water at all, and many more obtain water from suspect sources: 10% of public water supplies do not meet one or more of the EPA standards. Another 30 million tap individual wells or springs that are often uncontrolled and of unknown quality. Much remains to be done before we can all be assured of safe drinking water.

How much water does one really need? Only about 1.5–2.0 L/day for drinking. In the United States each day, we use about 7 L per person for drinking and cooking, but we use directly a total of about 380 L. We use much more water *indirectly* in agriculture and industry (Table 13.7) to produce food and other materials; it takes 800 L of water to produce 1 kg of vegetables and 13,000 L of water to produce a

TABLE 13.7 Use of Water in the United States	
Use	**Percentage**
Coolant electric power plants	48
Irrigation	34
Public water supplies	11
Industrial	5
Miscellaneous*	2

*Includes mining, livestock, aquaculture, and other uses.

steak. We also use water for recreation (for example, swimming, boating, and fishing). For most of these purposes, we need water that is free of bacteria, viruses, and parasitic organisms.

The Clean Water Act has drastically reduced water pollution from industrial sources. However, we hope it is clear that complete elimination of pollution is not possible; to use water is to pollute it. All of us can do our share by conserving water and by minimizing our use of products that require vast amounts of water to make. As our population grows, it will cost a lot just to maintain the present water quality. To clean up our water, and then keep it clean, will cost even more. However, the cost of unclean water is even higher—discomfort, loss of recreation, illness, and even death.

Critical Thinking Exercises

Apply knowledge that you have gained in this chapter and one or more of the FLaReS principles (Chapter 1) to evaluate the following statements or claims.

13.1 An activist group claims that all chlorine compounds should be banned because they are toxic.

13.2 A mother discovers that tests have found 1.0 ppb of trichloroethylene in the well water the family uses for drinking, bathing, cooking, and so on. The family has used the well for two years. The mother decides that the family members should be tested for cancer.

13.3 A group of citizens want to ban fluoridation of the community water supply because fluorides are toxic.

13.4 A company claims that its "super-oxygenated water can boost athletic performance." (Note that under pressure, there might be at most 0.3 g O_2/L H_2O, the amount in a 1-L breath.)

13.5 A company claims that "oxygenated and structured water is a water that has smaller molecules of water so it can be penetrated quicker into the cells, and therefore hydrate your body faster and more efficiently."

SUMMARY

Section 13.1—Water has unusual properties: It is the only common liquid on Earth; its solid form is less dense than the liquid; and it is more dense than most other common liquids.   Water also has a high **heat capacity**, the amount of heat needed to raise its temperature by 1 °C. **Specific heat** is heat capacity per gram; for water it is 1 cal/(g °C). Water also has a high **heat of vaporization**, the amount of heat needed to vaporize a fixed amount of water. These properties arise because of strong hydrogen bonding between water molecules.

Sections 13.2–13.3—Water covers three-fourths of Earth's surface, but only about 1% of it is available as fresh water. Water dissolves many ionic substances, which is why the sea is  salty. Potable water is readily available in most parts of the United States but many other countries do not have sufficient fresh water. In the water cycle, water evaporates from oceans and lakes, condenses into clouds, and returns to Earth as rain or snow. During this cycle it can be contaminated by various chemicals and microorganisms, both natural and from human activities. **Hard water** is formed when minerals (salts of calcium, magnesium, and/or iron) dissolve in groundwater.

Section 13.4—Waterborne diseases such as cholera, typhoid fever, and dysentery were a severe problem until about 100 years ago and are still common in many parts of the world. Acid rain from sulfur oxides and nitrogen oxides, which in turn arise from burning coal and other fuels, has caused acidification of thousands of bodies of water in North America. In addition to its corrosive effects, acid rain is detrimental to both plant and animal life. When sewage is dumped into waterways, that sewage is biodegraded by microorganisms. **Aerobic oxidation** occurs in the presence of **dissolved oxygen**, and the **biochemical oxygen demand (BOD)** is a measure of the amount of oxygen needed. Sewage increases the BOD. If the BOD is high enough, only **anaerobic decay**, or decay in the absence of oxygen, can occur. Sewage, phosphates, and fertilizers can accelerate **eutrophication**, in which algae grow and die, thereby increasing the BOD of the water. Other new substances introduced into the water cycle may cause new problems.

Sections 13.5–13.6—Building a car takes a great deal of water, which is polluted in the process and must be cleaned up and recycled. Chromium and cyanide were once serious sources of pollution but various chemical treatments now remove them. About half the U.S. population drinks surface water and half drinks groundwater; the latter is especially common in rural areas. The aquifers of drilled wells are being consumed faster than they are being refilled from precipitation, and deeper wells are now required in many places. Well water may be contaminated with various chemicals including nitrates and VOCs from underground storage tanks.

Sections 13.7–13.8—The U.S. Safe Water Drinking Act sets and enforces standards for water quality. In many cases, contaminants must be expressed in tiny units such as parts per million (ppm) or parts per billion (ppb).

$$1 \text{ ppm} = \frac{1 \text{ g solute}}{10^6 \text{ g solution}} \qquad 1 \text{ ppb} = \frac{1 \text{ g solute}}{10^9 \text{ g solution}}$$

Prior to drinking, water treatment usually includes settling, filtration, aeration, and chemical disinfection with chlorine or ozone. Fluorides may also be added to prevent tooth decay.

Section 13.9—Treatment of wastewater (used water) begins with primary sewage treatment, allowing the wastes to settle and removing sludge before discharging into a stream, lake, or ocean. In secondary sewage treatment the effluent is filtered through sand and gravel filtration. Also, the activated sludge method may be used, in which the sewage is aerated and the sludge is removed. Advanced (tertiary) treatment may involve charcoal filtration to adsorb organic compounds or reverse osmosis, in

which water is forced through a semipermeable membrane, leaving contaminants behind. Increasing treatment is increasingly expensive.

Sections 13.10–13.12—Bottled water is a popular but relatively expensive beverage. It is widely but incorrectly perceived that bottled water is safer or healthier than tap water. Bottled water often comes from municipal water supplies and may have more dissolved ions than normal tap water.

Alternative methods for sewage treatment are being investigated. Much more water is used indirectly by consumers than directly. Maintaining water quality is expensive and will become more so, but not maintaining the quality of our water will cost much more in terms of our health and comfort.

REVIEW QUESTIONS

1. What proportion of Earth's water is fresh liquid water?
2. How is our supply of fresh water replenished by natural processes?
3. What is BOD? Why is a high BOD undesirable?
4. List some waterborne diseases. Why are these diseases no longer common in developed countries?
5. What is eutrophication?
6. What are pathogenic microorganisms?
7. What problems do leaking underground storage tanks cause?
8. List some ways in which groundwater is contaminated.
9. Why do chlorinated hydrocarbons remain in groundwater for such a long time?
10. List some common industrial contaminants of groundwater.

PROBLEMS

Properties of Water

11. Why is ice less dense than liquid water? What consequences does this property have for life in northern lakes?
12. Why should foods be flash-frozen rather than frozen slowly?
13. A barge filled with gasoline sinks and breaks open. Will the gasoline dissolve in the water? Will it float or sink?
14. An offshore oil well leaks from a broken pipe. Will the oil float or sink? Explain.
15. Define heat capacity. Why is the high heat capacity of water important to planet Earth?
16. Why is the high heat of vaporization of water important to our bodies?
17. Why is it cooler near a lake than inland during the summer?
18. Why are seas salty? Are they getting saltier?

Natural Waters

19. What impurities are present in rainwater?
20. What is hard water? Why is it sometimes undesirable?
21. List four cations and three anions present in groundwater.
22. What are the products of the breakdown of organic matter by aerobic bacteria? List some of the products of anaerobic decay.

Acidic Waters

23. List two ways in which lakes and streams have become acidic.
24. Why is acidic water especially harmful to fish?
25. What kind of rocks neutralize acidic waters?
26. List several ways in which the acidity of rain can be reduced.

27. How can we restore (at least temporarily) lakes that are too acidic?

28. List two toxic compounds found in wastes from the chrome plating process. How is each removed?

Wastewater Treatment

29. Describe a primary sewage treatment plant. What impurities does it remove?

30. Describe a secondary sewage treatment plant. What impurities does it remove?

31. What substances remain in wastewater after effective secondary treatment?

32. Describe the activated sludge method of sewage treatment.

33. Why is wastewater chlorinated before it is returned to a waterway?

34. What is meant by advanced treatment of wastewater? List two methods of advanced treatment.

35. What kinds of substances are removed from wastewater by charcoal filtration?

36. What are the advantages and disadvantages of spreading sewage sludge on farmlands?

Municipal Water Supplies

37. Why are municipal water supplies treated with aluminum sulfate and slaked lime?

38. List the advantages and disadvantages of chlorination of drinking water.

39. Why are municipal water supplies aerated?

40. List the advantages and disadvantages of ozone as a disinfectant of drinking water.

41. Describe a nonchemical disinfecting treatment for drinking water.

42. How does fluoride strengthen tooth enamel?

43. What is the optimal level of fluoride in drinking water?

44. What are some health effects of too much fluoride in the diet?

45. What is the source of nitrate ions in well water?

46. What is methemoglobinemia? How is it caused?

Chemical Equations

47. Write the balanced equation for the neutralization of acidic rain (assume HNO_3 is dissolved in water) by limestone (calcium carbonate).

48. Write the balanced equation for the reaction of slaked lime (calcium hydroxide) with alum (aluminum sulfate) to form aluminum hydroxide.

Parts per Million and Parts per Billion

49. Express the following aqueous concentrations in the unit indicated.
    a. 6 μg benzene per liter of water, as ppb benzene
    b. 0.0145% $CaCl_2$, by mass, as ppm $CaCl_2$

50. Express the following aqueous concentrations in the units indicated.
    a. 25 μg trichloroethylene in 11 L water, as ppb trichloroethylene
    b. 38 g Cl_2 in 1.00×10^4 L water, as ppm of Cl_2

ADDITIONAL PROBLEMS

51. What types of substances are most effectively removed by each of the following methods of water purification? Which of the list of substances—$CHCl_3$, NaCl, phosphate ion, sand—is effectively removed from water by each?
    a. activated carbon bed
    b. distillation
    c. filtration
    d. ion exchange
    e. precipitation with Al^{3+}

52. Describe an alternative to the flush toilet.

53. Which takes more energy per gram: melting ice (solid at 0 °C to liquid at 0 °C), or boiling water (liquid at 100 °C to gas at 100 °C). Explain briefly.

54. Express the following aqueous concentrations in the unit indicated.
    a. 2.4 ppm F^-, as molarity of fluoride ion, $[F^-]$
    b. 45 ppm NO_3^-, as molarity of nitrate ion, $[NO_3^-]$
    c. Why do some scientists prefer the ppm and ppb units over molarity with water contaminants?

55. Wastewater disinfected with chlorine must be dechlorinated before it is returned to sensitive bodies of water. The dechlorinating agent is often sulfur dioxide. The reaction is

$$Cl_2 + SO_2 + 2\,H_2O \longrightarrow 2\,Cl^- + SO_4^{2-} + 4\,H^+$$

Is the chlorine oxidized or reduced? Identify the oxidizing agent and reducing agent in the reaction.

56. Radioactive aluminum-26 is used to study the mobilization of aluminum by acidic waters. When it decays, the isotope is transmuted into magnesium-26. What kind of particle is emitted by aluminum-26?

57. The text indicates that radioactive radon gas is slightly soluble in water. What are the intermolecular forces responsible for the solubility?

58. Two iron teakettles are heated on a stove. One has 2 L of water; the other is empty. Which gets hotter? Explain.

COLLABORATIVE GROUP PROJECTS

Prepare a PowerPoint, poster, or other presentation (as directed by your instructor) for presentation to the class.

59. Consult the references or other sources and prepare a report on the desalination of seawater by one of the following methods.
    a. distillation
    b. freezing
    c. electrodialysis
    d. reverse osmosis
    e. ion exchange

60. What sort of wastewater treatment is used in your community? Is it adequate?

61. Where does your drinking water come from? What steps are used in purifying it?

62. Call your water utility office or consult its website and obtain a chemical analysis of your drinking water. What substances are monitored? Are any of these substances considered problems? (If you use water from a private well, has the water been analyzed? If so, what were the results?)

63. The website of the United States EPA's Office of Wastewater Management (OWM) has a menu of water topics. First, notice how many water-related topics there are. Then choose one and write a few paragraphs about that aspect of water management.

64. Take a poll of your class regarding their consumption of bottled water. Why do people drink it? Do they have any concerns regarding its purity?

65. Using the EPA's OWM site or your state's environmental website, see what you can learn about the water quality where you live.

66. Find a website related to water quality and write a few paragraphs about its sponsoring organization.

67. Using the website of the Natural Resources Defense Council (NRDC) or other environmental group, investigate and write about pollution from urban stormwater runoff, which some say rivals sewage plants and factories as a source of water contamination.

REFERENCES AND READINGS

1. Ball, Philip. *Life's Matrix: A Biography of Water*. New York: Farrar, Straus & Giroux, 2000.

2. Dalton, Louisa. "Two Sides to CO_2 Rise." *Chemical & Engineering News*, July 19, 2004, p. 6.

3. Ertle, James. "Using 'Bugs' to Treat Wastewater." *Environmental Technology*, January/February 2000, pp. 13–14.

4. Freemantle, Michael. "Chemistry for Water." *Chemical & Engineering News,* July 19, 2004, pp. 25–30.

5. Gleick, P. H. "Making Every Drop Count." *Scientific American*, February 2001, pp. 40–45.

6. Glennon, Robert. *Water Follies: Groundwater Pumping and the Fate of America's Fresh Waters*. Washington, DC: Island Press, 2002. The misuse of America's aquifers is creating a crisis for a vital natural resource.

7. Gorman, Jessica. "Tums of the Sea." *Science News*, Aug. 7, 2002, pp. 104–105. Dissolved CO_2 is neutralized by shells floating in the sea.

8. Luoma, Jon R. "Water for Profit: The Price of Water." *Mother Jones*, November–December 2002, pp. 34–37, 88.

9. Marzuola, Carol. "Ocean View." *Science News*, Dec. 7, 2002, pp. 362–364.

10. Postel, Sandra, and Amy Vickers. "Boosting Water Productivity." In *State of the World 2004*. New York: W. W. Norton, 2004, pp. 46–67.

11. Sampat, P. "The Polluting of the World's Major Fresh Water Sources." *World Watch*, January–February 2000, pp. 10–22.

12. Spiro, Thomas G., and William M. Stigliani. *Chemistry of the Environment*, 2d ed. Upper Saddle River, NJ: Prentice Hall, 2002. Part III, Hydrosphere/Lithosphere.

13. Tarbuck, Edward J., and Frederick K. Lutgens. *Earth Science*, 11th edition. Upper Saddle River, NJ: Prentice Hall, 2006. Chapter 5, "Running Water and Groundwater," and Chapter 14, "Ocean Water and Ocean Life."

14. Tullo, A. H. "Turning on the Tap." *Chemical & Engineering News*, November, 2002. p. 37.

15. Van der Ryn, Sim. *The Toilet Papers: Recycling Waste and Conserving Water*. White River Junction, VT: Chelsea Green Publishing Company, 2000.

16. *Water for People—Water for Life: The United Nations World Water Development Report*. Paris: UNESCO Publishing, 2003.

17. Wolverton B. C., and John D. Wolverton. *Growing Clean Water: Nature's Solution to Water Pollution*. Picayune, MS: Wolverton Environmental Services, 2001.

GREEN CHEMISTRY

Antifouling in Rivers of Life; Seas of Sorrow

Denyce K. Wicht, Suffolk University

In this chapter, you are learning about the unique properties of water and how important an adequate supply of potable water is to life on this planet. Chemical contamination of both seawater and freshwater supplies can disrupt the delicate balance that exists between dissolved salts and gases in water and the animal and plant life that thrives in these environments. Ensuring that our water supply remains pristine requires careful consideration and a thorough investigation into how the introduction of possibly harmful chemicals can be minimized, if not eliminated. It also requires a clear understanding of how chemicals introduced into a water system biodegrade over time, so as to produce minimal disruption to the marine ecosystem.

Often chemists and chemical engineers face a challenge in which a chemical or chemical process initially designed to solve a problem effectually creates a new problem due to some unforeseen deleterious property. As you learned in the Green Chemistry investigations in Chapter 1, one of the 12 principles of green chemistry is the design of safer chemicals, which states that chemicals should be designed to carry out their desired function while minimizing their toxicity. In the Web Investigations you will explore the concept of biofouling, an interesting and complex problem, and learn how one compound, known as Sea-Nine®, was developed as a safer alternative to a chemical that, at one time, was thought to be an extremely effective chemical solution to the biofouling problem.

Biofouling is defined as a gradual accumulation of water-borne microorganisms, plants, and animals on the surfaces of marine-based structures, such as the hulls of boats. Excessive amounts of this type of accumulation are undesirable, as the organisms add weight to the hull and cause greater hull friction, leading to increased drag and a subsequent increased fuel demand. Historically, com-

pounds containing tin were shown to be quite effective at preventing unwanted growth, and therefore tin-based antifouling agents were investigated, manufactured, and incorporated into the paint that was coated on the hulls of vessels. However, it was found that, over time, the tin-based compounds leached from the paint into the water and were being taken up by marine life. This bioaccumulation of the tin-based chemicals caused a myriad of problems for some members of the marine ecosystem. In response, the Organotin Antifouling Paint Control Act of 1988 prohibited the use of antifouling paints containing tin-based chemicals on vessels less than 25 meters in length. In addition, this legislation further mandated that the Environmental Protection Agency must certify that any antifouling paint that does contain tin-based chemicals does not release more than 4.0 micrograms per square centimeter per day into the water system.

▲ Barnacles are one form of biofouling. Barnacles attach themselves to the hull of a boat, increasing friction and increasing the fuel used by the boat. Treatment of the hull is necessary to avoid this kind of biofouling.

Investigation 1
1996 Presidential Green Chemistry Award Winner
Go to the EPA website and review the Presidential Green Chemistry Challenge. Select "Award Winners" and scroll down to find the document titled *Summary of 1996 Award Entries and Recipients* and read the content related to the Designing Safer Chemicals Award.

Investigation 2
The Sea-Nine® Website
Use the keywords "antifouling sea nine" in your favorite search engine to learn more about the Sea-Nine® application process and how it works to prevent fouling on a ship's hull.

COMMUNICATE YOUR RESULTS

Exercise 1
Opposing Opinions
Drawing on what you've just learned, write two op-ed articles for a well-recognized newspaper, one supporting the need for antifouling materials on the hulls of vessels and the other supporting the elimination of antifouling materials. Use concrete evidence from your reading to back up each argument. Are there any conflicting aspects of these two opinions? Explain.

Exercise 2
Life Cycles
Use pictures or a slide presentation to communicate the "life" cycle of a chemical antifouling agent. Begin with the preparation of the antifouling agent at a plant production site, assuming that the chemical is assembled there from the appropriate elemental material. Draw the process by which this agent could be incorporated into a large water supply as discussed in this Green Chemistry exercise. Be sure to close the loop of this life cycle and estimate a time line.

CHAPTER 14

Energy

Our society is driven by energy, produced mainly from the burning of fossil fuels such as oil, coal, and natural gas. Producing and using that energy forms pollutants that cause adverse health effects, economic damage, and global climate change. Renewable, cleaner alternative energy sources (such as solar and wind power) will become more and more important in the years ahead.

A Fuels Paradise

We were tempted to call this chapter Fire, making the titles of Chapters 11, 12, 13, and 14 Earth, Air, Water, and Fire—the four elements of the ancient world.

Actually, fire could be an appropriate title for this chapter. We obtain most of our energy by burning fuels, which certainly involves fire. However, many sources of energy do not involve burning of anything, and we will depend more heavily on these other sources as our reserves of fossil fuels are depleted.

Energy is not matter; **energy** is the ability to do work. Everything we consume or use—our food and clothes, our homes and their contents, our cars and roads—requires energy to produce, package, distribute, operate, and dispose of. The United States, with less than 5% of the world's population, uses one-fourth of all the energy currently generated on the planet. Abundant energy has enabled the United States to provide its people with tremendous benefits and conveniences, adding up to a high standard of living, although generally lower than that of Canada, Australia, Japan, and the Scandinavian countries, who achieve their higher ranking with lower per capita use of energy. This energy use also has enormous costs: to our health, to the environment, and to our national security. It affects our foreign debt and the stability of the Middle East. It pollutes the air we breathe and the water we drink.

Industry uses about 33% of all energy produced in the United States. This energy is used to convert raw materials to the many products our society seems to demand. Transportation uses about 27%, to power automobiles, trucks, trains, airplanes, and buses. Private homes and commercial spaces use about 40%. Utilities use about 35% of the nation's energy production, primarily to generate electricity.

Energy lights our homes, heats and cools our living spaces, and makes us the most mobile society in the history of the human race. It powers the factories that provide us with abundant material goods. Indeed, energy is the basis of modern civilization.

These figures add up to more than 100% because the electricity is used in industry and homes and commercial spaces and some of it is therefore counted twice.

14.1 HEAVENLY SUNLIGHT FLOODING EARTH WITH ENERGY

All living things on Earth—us included—depend on nuclear energy for survival. Most of the energy available to us on this planet comes from the giant nuclear reactor we call the sun. Although it is about 150 million km away, the sun has been supplying Earth with most of its energy for billions of years, and it will likely continue to do so for billions more (Table 14.1).

Energy and power are related in the same way that distance and speed are related. Speed is the *rate* at which distance is covered, and power is the *rate* at which energy is used.

TABLE 14.1 Earth's Energy Ledger (Rough Estimates)

Item	Energy (TW)	Approximate Percent
Energy in		
Solar radiation	173,000	99+
Internal heat	32	0.02
Tides	3	0.002
Energy out		
Direct reflection	52,000	30
Direct heating*	81,000	47
Water cycle*	40,000	23
Winds*	370	0.2
Photosynthesis*	40	0.02

Source: M. King Hubbert, "The Energy Resources of the Earth," *Scientific American*, September 1971.

*This energy is eventually returned to space by means of long-wave radiation (heat).

▲ The sun fuses more than 600 million t of hydrogen into helium each second. It has enough hydrogen to last another five billion years. The solar flare shown here is over twenty times larger than Earth.

The SI unit of energy is the joule (J); 1 J = 0.2388 cal. A watt (W), the SI unit of power, is 1 joule per second (J/s).

$$1 \text{ W} = 1 \frac{J}{s}$$

A unit of more convenient size is the kilowatt: 1 kW = 1000 W. To measure energy consumption, the watt is combined with a unit of time. A familiar example is the *kilowatt hour (kWh)*, the quantity of energy used by a 1-kW device in one hour; 1 kWh = 3600 kJ.

Recall (Chapter 4) that the sun is a nuclear fusion reactor that steadily converts hydrogen to helium. The sun has a power output of 4×10^{26} W. Earth receives about 1.73×10^{17} W [173,000 terawatts (TW); 1 TW = 10^{12} W] from the sun—an amount equivalent to the output of 115 million nuclear power plants. In three days, Earth receives energy from the sun equivalent to all our fossil fuel reserves. Yet this is only about 1 part in 50 billion of the sun's output!

EXAMPLE 14.1   Power and Energy Conversion

How much electrical energy, in joules, is consumed by a 75-W bulb burning for 1.0 hr?

Solution

From the definition of a watt, 1 W = 1 J/s, we know that 75 W = 75 J/s. The bulb burns for

$$1 \text{ h} \times \frac{60 \text{ min}}{1 \text{ h}} \times \frac{60 \text{ s}}{1 \text{ min}} = 3600 \text{ s}$$

$$3600 \text{ s} \times \frac{75 \text{ J}}{1 \text{ s}} = 270,000 \text{ J}$$

▶ **Exercise 14.1A**

How much electrical energy, in joules, does a 650-W microwave oven consume in heating a cup of coffee for 1.5 min?

▶ **Exercise 14.1B**

(a) How much electrical energy, in kilojoules, does a 1200-W space heater consume when it runs for 8.0 h? **(b)** How much energy is that in kilowatt hours?

TABLE 14.2 Some Examples of Potential and Kinetic Energy		
Potential Energy		
Energy stored by position		Water at the top of a waterfall
		A child at the top of a sliding board
		A skier poised at the top of a mountain slope
		A ball ready to roll down a hill
		A swimmer ready to dive
		A baseball player poised to swing his bat
		A golfer at the top of her backswing ready to tee off
Energy stored in chemical bonds		Fuel (coal, gasoline, natural gas)
		Food (carbohydrates, fats, proteins)
		Explosives (nitroglycerin)
Energy stored in bound nuclear particles		Nuclear energy (power plants, bombs)
Energy stored by compression		A compressed spring
		A squeezed rubber ball
Kinetic Energy		
Energy of any moving object		A rolling freight train
		A spinning waterwheel
		A rolling bowling ball
		A moving molecule
		A sailboat skimming across a lake
		An eagle soaring above a mountain
		An exploding firecracker
		A pair of figure skaters
		A baseball hurtling toward home plate
		A football passing through the goal posts
		Horses racing along a track
		The explosion of a volcano
		The explosion of a building being demolished
		A nuclear explosion

Kinetic and Potential Energy

Energy exists in two main forms. Energy due to position or arrangement is called *potential* energy. The water at the top of a dam has potential energy due to gravitational attraction. When the water is allowed to flow through a turbine to a lower level, the potential energy is converted to *kinetic* energy (the energy of motion). As the water falls, it moves faster. Its kinetic energy becomes greater as its potential energy decreases. The turbine can convert part of the kinetic energy of the water into electrical energy. The electricity thus produced can be carried by wires to homes and factories where it can be converted to light energy, to heat, or to mechanical energy. Table 14.2 gives several examples each of kinetic and potential energy.

Energy and the Life-Support System

The *biosphere* is the thin (about 15 km thick) film of air, water, and soil in which all life exists. Only a small fraction of the energy the biosphere receives is used to support life. About 30% of incident radiation is immediately reflected back into space as short-wave radiation (ultraviolet and visible light). Nearly half is converted to heat, making the third planet a warm and habitable place. About 23% of solar radiation powers the water cycle (Chapter 13), evaporating water from land and seas. The radiant energy of the sun is converted to the potential energy of water vapor, water droplets, and ice crystals in the atmosphere. This potential energy is converted to the kinetic energy of falling rain and snow and of flowing rivers.

A tiny but most important fraction—less than 0.02%—of solar energy is absorbed by green plants, which use it to power **photosynthesis**. In the presence of

Chlorophyll-a

green chlorophyll pigments, this energy converts carbon dioxide and water to glucose, a simple sugar rich in energy.

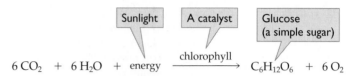

$$6\ CO_2\ +\ 6\ H_2O\ +\ \text{energy}\ \xrightarrow{\text{chlorophyll}}\ C_6H_{12}O_6\ +\ 6\ O_2$$

Photosynthesis also replenishes oxygen in the atmosphere. Glucose can be stored, or it can be converted to more complex foods and structural materials. All animals depend on the stored energy of green plants for survival.

Energy: Some Scientific Laws

A study of chemistry is incomplete without a discussion of energy. In calculations based on chemical equations, we can include quantities of energy. In addition, we find some scientific laws to be of considerable help in understanding chemical processes. We examine some of these ideas in the following sections.

Catalysis

Methane

14.2 ENERGY AND CHEMICAL REACTIONS

How fast—or how slowly—chemical reactions take place depends on several factors. One factor is *temperature*. Reactions generally take place faster at higher temperatures. For example, coal (carbon) reacts so slowly with oxygen (from the air) at room temperature that the change is imperceptible. However, if coal is heated to several hundred degrees, it reacts rapidly. The heat evolved in the reaction keeps the coal burning smoothly.

We use the kinetic–molecular theory (Chapter 6) to explain the effect of temperature on the rates of chemical reactions. At high temperatures, molecules move more rapidly. Thus, they collide more frequently, increasing the chance for reaction. The increase in temperature also supplies more energy for the breaking of chemical bonds—a condition necessary for most reactions. We make use of our knowledge of the effect of temperature on chemical reactions in our daily lives. For example, we freeze foods to retard chemical reactions that lead to spoilage. When we want to speed up the reactions involved in cooking food, we turn up the heat.

Another factor affecting the rate of a chemical reaction is the *concentration of reactants*: The more molecules there are in a given volume of space, the more likely they are to collide. With more collisions, there are more reactions. For example, if you light a wood splint and then blow out the flame, the splint continues to glow as the wood reacts slowly with oxygen in the air. When the glowing splint is placed in pure oxygen, the splint bursts into flame, indicating a much more rapid reaction. We use the concentration of oxygen to interpret the phenomenon: Air is about 21% oxygen, and so the concentration of O_2 molecules in 100% oxygen is almost five times as great as that in air.

Catalysts (Chapter 8) also affect the rates of chemical reactions. Catalysts are of great importance in the chemical industry and in biology. Appropriate catalysts increase the rate of reactions that otherwise would be so slow as to be impractical.

▲ When liquid oxygen—much more concentrated than oxygen gas—is poured on a lit cigarette, the cigarette sparks, bursts into flame, and quickly disintegrates.

Catalysts are even more important in living organisms. Biological catalysts, called *enzymes*, mediate nearly all the chemical reactions that take place in living systems (Chapter 15).

The two main units of energy are the joule (J) and the calorie (cal). To convert between them we can use the equality 1 cal = 4.184 J. In this chapter, we will use joules (or kilojoules) as the unit of energy. In Chapter 16, when the topic shifts to nutrition, we will use calories (and kilocalories).

Energy Changes and Chemical Reactions

The energy changes associated with chemical reactions are quantitatively related to the *amounts* of chemicals involved. For example, burning 1.00 mol (16.0 g) of methane to form carbon dioxide and water releases 803 kJ (192 kcal) of energy as heat. We can list the amount of heat evolved as a product of the reaction.

$$CH_4(g)\ +\ 2\ O_2(g)\ \longrightarrow\ CO_2(g)\ +\ 2\ H_2O(g)\ +\ 803\ kJ\ (192\ kcal)$$

Burning 2.00 mol (32.0 g) of methane produces twice as much heat, 1606 kJ.

(a)

(b)

◀ The action of a catalyst is illustrated. Hydrogen peroxide decomposes slowly to water and oxygen (a). When platinum metal is inserted into a solution of hydrogen peroxide (b), the reaction proceeds rapidly. The heat released during the process produces steam, and the solution froths as oxygen gas is evolved.

EXAMPLE 14.2 **Energies of Chemical Reactions**

Burning 1.00 mol of propane releases 2201 kJ of energy.

$$C_3H_8(g) + 5\,O_2(g) \longrightarrow 3\,CO_2(g) + 4\,H_2O(g) + 2201\ kJ$$

How much energy is released when 15.0 mol of propane is burned?

Solution
We start with 15.0 mol C_3H_8 and use the balanced equation to form a conversion factor, just as we did with chemical conversions in Chapter 6.

$$15.0\ \text{mol}\ C_3H_8 \times \frac{2201\ kJ}{1\ \text{mol}\ C_3H_8} = 33{,}000\ kJ$$

▶ **Exercise 14.2A**
The reaction of nitrogen and oxygen to form nitrogen monoxide (nitric oxide) requires an input of energy.

$$N_2(g) + O_2(g) + 18.07\ kJ \longrightarrow 2\,NO(g)$$

How much energy is absorbed when 5.05 mol of N_2 reacts with oxygen to form NO?

▶ **Exercise 14.2B**
Photosynthesis is an endothermic reaction in which carbon dioxide and water are converted to sugars in a reaction that requires 15,000 kJ of energy for each 1.00 kg of glucose produced. How much energy, in kilojoules, is required to make 1.00 mol of glucose?

Chemical reactions that result in the release of heat to the surroundings are **exothermic** reactions. The burning of methane, gasoline, and coal (Figure 14.1) are all exothermic reactions. In each case, chemical energy is converted to heat energy.

In other reactions, such as the decomposition of water, energy must be supplied to the reactants from the surroundings. We can write the amount of heat required for this process as a reactant in the chemical equation:

$$2\,H_2O(g) + 573\ kJ \longrightarrow 2\,H_2(g) + O_2(g)$$

Thermite Reaction

▶ **Figure 14.1** Coal burns in a highly exothermic reaction. The heat released can convert water to steam that can turn a turbine to generate electricity.

In an **endothermic** reaction, energy is supplied as heat. It takes 573 kJ of energy to decompose 36.0 g (2.00 mol) of water into hydrogen and oxygen. It should be noted that exactly the same amount of energy is released when enough hydrogen is burned to form 36.0 g of water.

$$2\,H_2(g) \,+\, O_2(g) \longrightarrow 2\,H_2O(g) \,+\, 573 \text{ kJ}$$

Physical processes can also be either exothermic or endothermic. Table 14.3 lists several examples of physical and chemical processes and classifies them as exothermic or endothermic.

▶ **Figure 14.2** A striking endothermic reaction occurs when stoichiometric amounts of barium hydroxide octahydrate react with ammonium thiocyanate to produce barium thiocyanate, ammonia gas, and water.

$$\text{Heat} \,+\, Ba(OH)_2 \cdot 8\,H_2O(s) \,+\, 2\,NH_4SCN(s) \longrightarrow Ba(SCN)_2(s) \,+\, 2\,NH_3(g) \,+\, 10\,H_2O(l)$$

When the reaction is carried out in a beaker placed on a wet board, the temperature drops well below the freezing point of water, thus freezing the beaker to the board.

TABLE 14.3   Some Exothermic and Endothermic Processes

Exothermic Processes	Endothermic Processes
Freezing of water	Melting of ice
Condensation of water vapor	Evaporation of water
Burning of wood	Cooking of food
Metabolism in animals	Photosynthesis in plants
Making chemical bonds	Breaking chemical bonds
Discharging of a battery	Charging a battery
Oxidation of Mg to MgO	Decomposition of HgO
Mixing CaO (or H_2SO_4) with water	Mixing NH_4Cl with water
Explosion of dynamite	Evaporation of a chlorofluorocarbon

EXAMPLE 14.3   **Energy Changes in Chemical Reactions**

How much energy is released when 225 g of propane (see Example 14.2) is burned?

Solution

The formula mass of propane (C_3H_8) is

$$(3 \times C) + (8 \times H) = (3 \times 12.0\ u) + (8 \times 1.0\ u) = 36.0 + 8.0\ u = 44.0\ u$$

The molar mass is therefore 44.0 g. Next, we use the molar mass to convert grams of propane to moles of propane.

$$225\ \text{g } C_3H_8 \times \frac{1\ \text{mol } C_3H_8}{44.0\ \text{g } C_3H_8} = 5.11\ \text{mol } C_3H_8$$

Now, proceeding as in Example 14.2,

$$5.11\ \text{mol } C_3H_8 \times \frac{2201\ \text{kJ}}{1\ \text{mol } C_3H_8} = 11{,}200\ \text{kJ}$$

▶ **Exercise 14.3A**

How much energy, in kilocalories, is released when 4.42 g of methane is burned?

$$CH_4 + 2\,O_2 \longrightarrow CO_2 + 2\,H_2O + 803\ \text{kJ}$$

▶ **Exercise 14.3B**

How much energy, in kilojoules, is absorbed when 0.528 g of N_2 is converted to NO (see Exercise 14.2A)?

14.3 ENERGY AND THE FIRST LAW: ENERGY IS CONSERVED

We will take a look at our present energy sources and our future energy prospects shortly. To do so scientifically, we first examine some natural laws. Recall that natural laws merely summarize the results of many experiments. We won't recount all those experiments here; we will merely state the laws and some of their consequences.

The **first law of thermodynamics** (*thermo* refers to heat; *dynamics* to motion) grew out of a variety of experiments conducted during the early 1800s. By 1840 it was clear that, although energy can be changed from one form to another, it is neither created nor destroyed. This law (also called the **law of conservation of energy**) has been restated in several ways, including, "You can't get something for nothing" and "There is no such thing as a free lunch." Energy can't be made from nothing. Neither does it just disappear, although it may go someplace else.

From the first law of thermodynamics we can conclude that we can't "win." We can't come out ahead by making a machine that produces more energy than it takes in. From the first law alone, however, we might conclude that we can't possibly run out of energy because energy is conserved. This is true enough, but it doesn't mean that we don't have problems. There is another long-armed law from which we cannot escape.

14.4 ENERGY AND THE SECOND LAW: THINGS ARE GOING TO GET WORSE

Despite innumerable attempts (and even several granted patents), no one has ever built a successful "perpetual-motion" machine. You can't make a machine that produces more energy than it consumes. Even if an engine isn't doing any work, it loses energy (as heat) because of the friction of its moving parts. In fact, in any real engine, not only is it impossible to get as much useful energy out as you put in ... you can't even break even.

▲ **Figure 14.3** Energy always flows spontaneously from a hot object to a cold one, never the reverse. It flows from a hot fire to cold hands. (A spontaneous event is one that occurs without outside influence.)

The energy to run an engine usually comes from the concentrated chemical energy inside molecules of oil or coal. In biochemistry, food molecules are the concentrated energy source. In either case, energy is spread out during the process. The spread-out energy in the product gases—CO_2 and H_2O in each case—is less useful. A car won't run on exhaust gases, nor can organisms obtain energy for life processes from respiratory products.

Estimation of Entropy Changes

Using Energy

If energy is neither created nor destroyed, why do we always need more? Won't the energy we have now last forever? The answer lies in the facts that

- energy can be changed from one form to another,
- not all forms are equally useful, and
- high-grade (more useful) forms of energy are constantly being degraded into low-grade (less useful) forms.

Energy flows downhill. Mechanical energy is eventually changed into heat energy. Hot objects cool off by transferring their heat to cooler objects. There is a tendency toward an even distribution of energy. Energy always flows from a hot object to a cooler one (Figure 14.3). The reverse does not occur spontaneously.

Observations of heat flow led to formulation of the **second law of thermodynamics**. In one form (of many), this law states that energy does not flow spontaneously from a cold object to a hot one. It is true that we can make energy flow from a cold region to a hot one—that's what refrigerators are all about—but we cannot do so without producing changes elsewhere. We can reverse a natural process only at a price. The price, in the case of refrigerators, is the consumption of electricity; that is, you must use energy to cause energy flow from a cold space to a warmer one.

When we change energy from one form to another, we can't concentrate all the energy in a particular source to do the job we want it to do. For example, we use energy in a fuel to push a piston in a car engine or use water rushing down from the top of a dam to run dynamos to generate electricity. In either case—indeed in all cases—some of the energy is wasted, mainly as unusable heat released to the environment. (*Wasted* isn't always the proper word. We use the energy from food for all our activities, but the waste heat helps us to maintain our body temperature at about 37 °C.)

Entropy

Another way to look at the second law is in terms of entropy. Scientists use the term **entropy** as a measure of the dispersal of energy. The more the energy of a system is spread out, the higher is its entropy. Natural processes tend toward greater entropy, or are exothermic, or both.

As we mentioned in Chapter 6, molecules move about constantly. Their motion in solids is limited to rapid but tiny back-and-forth movements much like vibra-

Entropy and Disorder

Entropy is best expressed quantitatively by way of mathematical relationships that are beyond the scope of this text. Strictly speaking, the concept of entropy applies only to the behavior of internally energetic, mobile molecules and atoms. The energy of these particles indeed tends to be as spread out as possible. Macro objects, such as books and stones, stay exactly where they are unless some energy flow from outside forces them to move. Macro objects have no inherent tendency toward disorder. Rather the energy flow that moves them causes the disorder. There are simply many, many more "disorderly" arrangements than "orderly" ones.

This tendency toward the spreading of energy helps to explain why it is relatively easy to pollute air or water and so difficult to clean it up once it is polluted. For example, it takes little energy to dump a ton of a chlorofluorocarbon (CFC) into the air. The rapidly moving CFC molecules would quickly disperse among the nitrogen, oxygen, and other molecules in the air. Once the CFC molecules were scattered, it would take a lot of energy to concentrate them from the air. It costs a lot to clean up polluted water, soil, or air. Prevention of pollution is by far the better alternative.

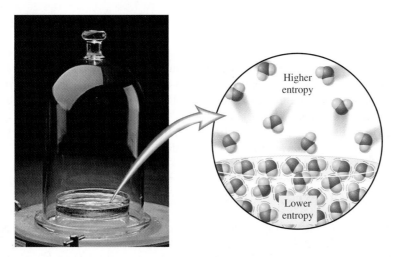

▲ **Figure 14.4** The photograph depicts the vaporization of water; a sample of liquid water [$H_2O(l)$] at room temperature spontaneously changes to $H_2O(g)$ through the process of evaporation. At the macroscopic level, nothing appears to be taking place. However, in the molecular view, we see that the molecules are in motion and are much more widely spaced in the vapor state than in the liquid state. Vaporization is spontaneous because vapor has greater *entropy* than the liquid; molecules in the vapor can be "arranged" in many more ways in their spread-out spacing than can the molecules of liquid. We can make liquid water from water vapor, but only by compressing the gas and lowering its temperature.

tions about a nearly fixed point. The molecules in liquids move more freely but still over only short distances. Those in gases move still more freely and over much greater distances.

In Chapter 3 we saw that the energy of light occurs in little packets called photons. Similarly, the energy of molecular motion is measured in tiny bundles that are called *microstates*. In comparing the states of matter, it is easy to see that a given sample of matter has the fewest energy microstates available when in the solid state because the molecules are limited mainly to vibrational motion. In the liquid state, the molecules can move more freely, and thus the sample has many more energy microstates available. When converted to a gas, the molecules can move still more freely and for much greater distances. This means that matter in the gaseous state has even more microstates available to its molecules. As any form of matter is heated, the energy in it is spread out among more accessible microstates. This concept is illustrated in Figure 14.4. Of course, we can often reverse the tendency toward greater entropy—but only through an input of energy.

14.5 PEOPLE POWER: EARLY USES OF ENERGY

Early people obtained their energy (food and fuel) by collecting wild plants and hunting wild animals. They expended this energy in hunting and gathering. Domestication of horses and oxen increased the availability of energy only slightly. The raw materials used by these work animals were natural, replaceable plant materials.

Plant materials were also the first fuels. These combustible materials kept early fires burning. As late as 1760, wood was almost the only fuel being used. Even today, wood and dried dung remain the primary fuels for about one-third of the world's people.

One of the first mechanical devices used to convert energy to useful work was the waterwheel. The Egyptians first used waterpower about 2000 years ago, primarily for grinding grain. Later on, waterpower was used for sawmills, textile mills, and other small factories. Windmills were introduced into western Europe during the Middle Ages, primarily for pumping water and grinding grain. More recently, wind power has been used to generate electricity.

▲ Waterwheels at Hama, Syria, are 2000 years old. Some are more than 20 m high. Several are still used to lift water into aqueducts. Waterwheels were rarely used in Europe, however, until the Middle Ages. The Romans used human slaves instead.

Windmills and waterwheels are fairly simple devices for converting the kinetic energy of blowing wind and flowing water to mechanical energy. They were sufficient to power the early part of the Industrial Revolution, but development of the steam engine freed factories from locations along waterways. Since 1850, turbines turned by water, steam, and gas, the internal-combustion engine, and a variety of other energy-conversion devices have boosted the energy available for use by an estimated factor of 10,000.

Let's turn our attention now to the fossils that fueled the Industrial Revolution and still serve as the basis for modern civilization.

Fossil Fuels

The combustion of long-buried fossils initiated and sustain our modern industrial civilization. More than 90% of the energy used to support our way of life comes from **fossil fuels**—coal, petroleum, and natural gas. In the following sections, we consider the origin and chemical nature of these fuels and how they are burned to release energy that plants of ages past captured from rays of sunlight.

A **fuel** is a substance that burns readily with the release of significant amounts of energy. Fuels are *reduced* forms of matter, and the burning process is oxidation (Chapter 8). If an atom already has its maximum number of bonds to oxygen (or to other electronegative atoms such as chlorine or bromine), the atom cannot serve as a fuel. Indeed, such substances can be used to put out fires. Figure 14.5 shows some representative fuels and nonfuels.

14.6 RESERVES AND CONSUMPTION RATES OF FOSSIL FUELS

Earth has only a limited supply of fossil fuels. Estimated U.S. and world reserves and annual U.S. and world consumption are given in Table 14.4. Estimates of reserves vary greatly, depending on the assumptions made. Even the most optimistic estimates, however, lead to the conclusion that nonrenewable energy resources are being depleted rapidly. Indeed, in just a century, we will have used up more than half the fossil fuels that were formed over the ages. In only a few hundred more years, we will have removed from Earth and burned virtually all the remaining recoverable fossil fuels. Of all that ever existed, about 90% will have been used in a period of 300 years. Within the lifetime of today's 18-year-old, natural gas and petroleum will likely become so scarce and so expensive that they won't be used much as fuels. At the current rate of production, U.S. reserves of petroleum and natural gas will be substantially depleted sometime during this century. Coal reserves should last perhaps 300 years. However, the rate of use of all fossil fuels is increasing, especially in developing nations.

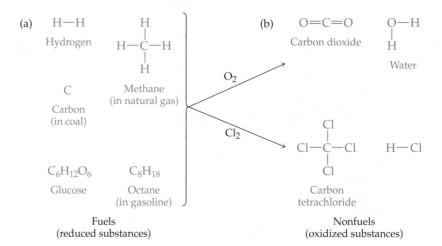

▶ **Figure 14.5** Some fuels (a). Fuels are reduced forms of matter that release relatively large quantities of heat when burned. Some nonfuels (b). These compounds are oxidized forms of matter.

TABLE 14.4 Estimated U.S. and World Reserves of Economically Recoverable Fuels and Annual Consumption of Fossil Fuels*

Fuel	Reserves		Consumption	
	United States	World	United States	World
Coal	121,962	501,171	545	2,342
Petroleum	4,184	156,700	904	3,507
Natural gas	4,711	158,198	517	2,122
Total	130,857	816,069	1,966	7,971

Source: World Resources Institute, Washington, DC. Website (http://www.wri.org/) includes assumptions made in making these estimates.
*Expressed in millions of metric tons of oil equivalents (Mtoe) for ready comparison of energy content.

Conversion factors:

1 metric ton oil equivalent (toe) = 7.8 barrels oil

1 toe = 1270 m^3 of natural gas

1 toe = 2.3 metric ton (t) of coal

Presumably, fossil fuels are still being formed in nature, a process that is perhaps most evident in peat bogs. The rate of formation is extremely slow, estimated at only one fifty-thousandth the rate we are using them.

14.7 COAL: THE CARBON ROCK OF AGES

People probably have used small amounts of coal since prehistoric times. After the steam engine came into widespread use (by about 1850), the industrial revolution was powered largely by coal. By 1900 about 95% of the world's energy production came from the burning of coal.

Coal is a complex combination of organic materials that burn and inorganic materials that produce ash. Its main element is carbon, but it also contains small percentages of other elements. The quality of coal as an energy source is based on its carbon content. Complete combustion of the carbon produces carbon dioxide.

> By energy *production*, we mean conversion of some form of energy into a more useful form. For example, production of petroleum means pumping, and transporting, and refining it. Energy *consumption* means using it in a way that changes it to a less useful form. In either case, we are neither making energy nor destroying it.

Carbon (from coal) Oxygen (from air)

$$C(s) + O_2(g) \longrightarrow CO_2(g)$$

In limited quantities of air, however, carbon monoxide and soot are formed.

$$2\,C(s) + O_2(g) \longrightarrow 2\,CO(g)$$

Soot is mostly unburned carbon.

Coal is ranked by carbon content, from low-grade peat and lignite to high-grade anthracite (Table 14.5). The energy obtained from coal is roughly proportional to its carbon content. Soft (bituminous) coal is much more plentiful than hard coal (anthracite). Lignite and peat have become increasingly important as the supplies of higher grades of coal have been depleted.

TABLE 14.5 Approximate Composition (Percent by Mass) and Energy Content of Typical Grades of Coal (Dry Basis)

Grade of Coal	Carbon	Hydrogen	Oxygen	Nitrogen	Energy Content (MJ/kg)
Wood (for comparison)	50	6	43	1	—
Peat	60	6	33	2	14.7
Lignite (brown coal)	71	5	25	1	23
Bituminous (soft) coal	86–91	5	5–15	1	36
Anthracite (hard coal)	95	2–3	2–3	Trace	35.2

Source: Diessel, C. F. K. *Coal-Bearing Depositional Systems.* New York: Springer-Verlag, 1992.

▲ Giant ferns, reeds, and grasses grew during the Pennsylvanian period 300 million years ago. These plants were buried and through the ages were converted to the coal we burn today.

Coal deposits that exist today are less than 600 million years old. For millions of years, Earth was much warmer than it is now, and plant life flourished. Most plants lived, died, and decayed—playing their normal role in the carbon cycle (Figure 14.6). But some plant material became buried under mud and water. There, in the absence of oxygen, it decayed only partially. The structural material of plants is largely cellulose—a compound of carbon, hydrogen, and oxygen. Under increasing pressure, as the material was buried more deeply, the cellulose molecules broke down. Small molecules rich in hydrogen and oxygen escaped, leaving behind a material increasingly rich in carbon. Thus, peat is a young coal, only partly converted, with plant stems and leaves clearly visible. Anthracite, on the other hand, has been almost completely carbonized.

Abundant But Inconvenient Fuel

Coal is by far the most plentiful fossil fuel. The United States has about a quarter of the world's proven reserves. Electric utilities in the United States burn nearly 900 million metric tons of coal each year, generating 2000 billion kWh of electricity (51% of the total). Unfortunately, this coal use is associated with serious environmental problems.

Coal, a solid, is an inconvenient fuel to use and very hazardous to obtain. Coal mining is one of the most dangerous occupations in the world, accounting for over 100,000 deaths in the United States in the twentieth century alone. Devastating strip mining is usually used to remove it. Most coal is hauled in trains, barges, and trucks. We continue to pay a variety of great costs to extract coal from the ground and transport it to the power plant or factory where it is to be used.

Source of Pollution

Unlike natural gas and liquid petroleum, solid coal contains minerals that are left as ash when the coal is burned. As we noted in Chapter 12, some minerals enter the air as particulate matter, constituting a major pollution problem.

The flotation method removes most inorganic sulfur compounds, such as pyrite, from coal. It does not remove sulfur atoms covalently bonded to the carbon atoms in the coal. This bound sulfur still contributes to pollution.

Perhaps even worse, much of our remaining coal is high in sulfur. When it burns, the stack gases must be scrubbed, or choking sulfur dioxide pours into the atmosphere. The SO_2 reacts with oxygen and moisture in the air to form sulfuric acid. This acid slowly damages steel and aluminum structures, marble buildings and statues, and human lungs.

Some coal is cleaned before it is burned. A flotation method makes use of the different densities of coal and its major impurities. Coal has a density of about

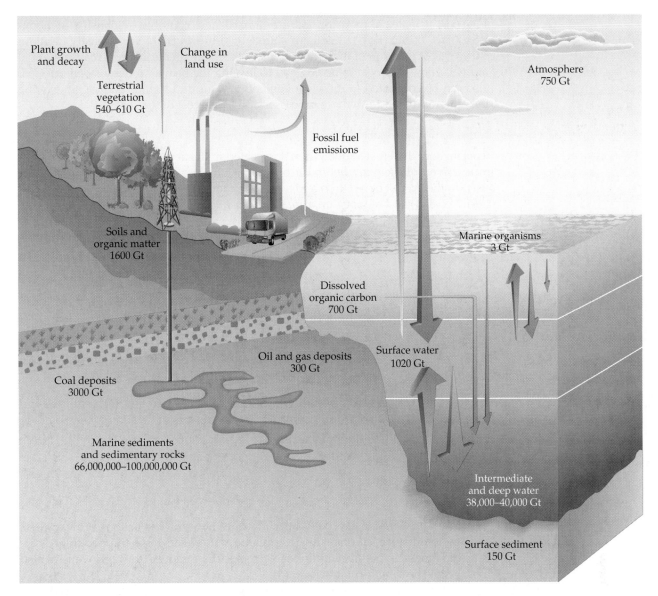

▲ **Figure 14.6** The carbon cycle. The numbers indicate the quantity of carbon, in gigatons (10^9 tons), in each location.

▲ Strip mining bares vast areas of vegetation. The exposed soil washes away, filling streams with mud and silt. U.S. laws now require the (expensive) restoration of most stripped areas. Unfortunately, much damage was done before these laws were passed, and many consider the present laws inadequate. These photographs show the site of a strip mine in Morgan County, Tennessee. (a) The abandoned mine in 1963, and (b) the area after it was reclaimed in a demonstration project in 1971.

1.3 g/cm³. Shale, a rock formed from hardened clay, has a density of about 2.5 g/cm³. Pyrite (FeS_2), the major source of sulfur in coal, has a density of about 5.0 g/cm³. A solution of the proper density, with detergents added to cause the coal to float, allows the coal to be floated off, leaving the heavier minerals behind. Coal also can be converted to more convenient gaseous and liquid fuels, again leaving the minerals behind. All these processes add to the cost of the coal.

Source of Chemicals

Coal is more than just a fuel. When it is heated in the absence of air, volatile material is driven off, leaving behind a product—mostly carbon—called *coke*, which is used in the production of iron and steel. The more volatile material is condensed to liquid coal oil and a gooey mixture called *coal tar*, both of which are sources of organic chemicals for medical and industrial use.

14.8 NATURAL GAS: MOSTLY METHANE

Natural gas, composed principally of methane, is the cleanest of the fossil fuels. Minor components of this gaseous fossil fuel vary greatly. In North America, a pipeline natural gas supply might contain about 82% methane, 6% ethane, 2% propane, and smaller amounts of butanes and pentanes. The gas burns with a relatively clean flame, and its products are mainly carbon dioxide and water.

$$CH_4(g) + 2 O_2(g) \longrightarrow CO_2(g) + 2 H_2O(g) + heat$$

The gas, as it comes from the ground, often contains nitrogen (N_2), sulfur compounds, and other substances as impurities. As in any combustion in air, some nitrogen oxides are formed. When natural gas is burned in limited air, carbon monoxide and elemental carbon (soot) can be major products.

$$2 CH_4(g) + 3 O_2(g) \longrightarrow 2 CO(g) + 4 H_2O(g)$$
$$CH_4(g) + O_2(g) \longrightarrow C(s) + 2 H_2O(g)$$

Natural gas was most likely formed—ages ago—by the actions of heat, pressure, and perhaps bacteria on buried organic matter. The gas is trapped in geological formations capped by impermeable rock. It is removed through wells drilled into the gas-bearing formations.

Most natural gas is used as fuel, but it is also an important raw material. In North America, some higher alkanes are separated out of natural gas. Ethane and propane are cracked—decomposed by heat in the absence of air—to form ethylene and propylene. These alkenes are intermediates in the synthesis of plastics (Chapter 10) and many other useful commodities. Natural gas is also the raw material from which methanol and many other organic compounds are made. Like other fossil fuels, its supply is not unlimited (Table 14.4).

14.9 PETROLEUM: LIQUID HYDROCARBONS

With the development of the internal combustion engine, petroleum became increasingly important and by 1950 had replaced coal as the principal fuel. **Petroleum** is a complex liquid mixture of organic compounds—as many as 17,000 were separated from a sample of Brazilian crude oil. Most are hydrocarbons: alkanes, cycloalkanes, and aromatic compounds. As with natural gas, complete combustion of these substances yields mainly carbon dioxide and water. A representative reaction is that of an octane.

$$2 C_8H_{18}(l) + 25 O_2(g) \longrightarrow 16 CO_2(g) + 18 H_2O(g)$$

Combustion in air also produces nitrogen oxides. Incomplete burning yields carbon monoxide and soot. Petroleum usually contains small amounts of sulfur compounds that produce sulfur dioxide when burned.

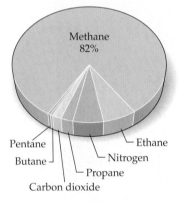

▲ Composition of typical natural gas. Natural gas is mostly methane, with small amounts of other hydrocarbons, nitrogen, and a few other gases.

▲ Natural gas burns with a relatively clean flame.

While coal is primarily of plant origin, petroleum is thought to be of animal origin. Most likely it is formed primarily from the fats (Chapter 15) of ocean-dwelling, microscopic animals because it is nearly always found in rocks of oceanic origin. Fats are made up mainly of compounds of carbon and hydrogen, with a little oxygen (for example, $C_{57}H_{110}O_6$). Removal of the oxygen and slight rearrangement of the carbon and hydrogen atoms in these fats form typical petroleum hydrocarbon molecules.

Efficiently burned, petroleum products are rather clean fuels. Fuel oil, used for heating homes and to produce electricity, can be burned efficiently and thus contributes only moderately to air pollution. Gasoline, however, the major fraction of petroleum, is used to power automobiles; because the internal combustion engines in most automobiles are rather inefficient, the combustion of gasoline contributes greatly to air pollution (Chapter 12). Petroleum-derived fuels are generally dirtier than natural gas because they contain more impurities.

Air pollution isn't the only problem: Burning up our petroleum reserves will leave us without a ready source of many familiar materials. Most industrial organic chemicals come from petroleum (others are derived from coal and natural gas); and plastics, synthetic fibers, solvents, and many other consumer products are made from these chemicals.

Worldwide, petroleum still appears to be quite abundant (Table 14.4). However, U.S. petroleum reserves have been declining since 1972. Further tapping of offshore deposits and deeper drilling would produce only a few years' supply, and would require a great deal more energy (and money) than drilling on land.

The developed countries of North America, western Europe, and Japan depend heavily on oil imports from developing nations, some of which are politically unstable. This dependence makes the industrial nations vulnerable economically and politically.

Our dependence on imported oil leads to economic and political problems. Political events in the oil-rich Middle East have caused huge oil price increases several times. Continued military involvement in Iraq, disastrous hurricanes in the Gulf of Mexico, and increased demand in developing countries, especially China, caused crude oil prices to hit record highs in 2005 and 2006.

Obtaining and Refining Petroleum

Crude oil is liquid, a convenient form for its transportation. Petroleum is pumped easily through pipelines or hauled across the oceans in giant tankers. Pumping petroleum from the ground requires little energy. As domestic supplies diminish, however, we will have to expend an increasing fraction of the energy content of petroleum to transport it to where it is needed. Also, as petroleum is pumped out of a well, it becomes increasingly difficult to extract the remainder. The oil that is left behind is more scattered, and more energy is required to collect it.

As it comes from the ground, crude oil is of limited use. To make it better suit our needs, we separate it into fractions by boiling it in a distillation column (Figure 14.7). Petroleum deposits nearly always have associated natural gas, part or all of which is sent through a separation process. The lighter hydrocarbon molecules come off at the top of the column, and the heavier ones at the bottom (Table 14.6).

Gasoline is usually the fraction of petroleum most in demand, and fractions that boil at higher temperatures are often converted to gasoline by heating in the

TABLE 14.6 Typical Petroleum Fractions			
Fraction	Typical Range of Hydrocarbons	Approximate Range of Boiling Points (°C)	Typical Uses
Gas	CH_4 to C_4H_{10}	Less than 40	Fuel, starting materials for plastics
Gasoline	C_5H_{12} to $C_{12}H_{26}$	40–200	Fuel, solvents
Kerosene	$C_{12}H_{26}$ to $C_{16}H_{34}$	175–275	Diesel fuel, jet fuel, home heating; cracking to gasoline
Heating oil	$C_{15}H_{32}$ to $C_{18}H_{38}$	250–400	Industrial heating, cracking to gasoline
Lubricating oil	$C_{17}H_{36}$ and up	Above 300	Lubricants
Residue	$C_{20}H_{42}$ and up	Above 350 (some decomposition)	Paraffin, asphalt

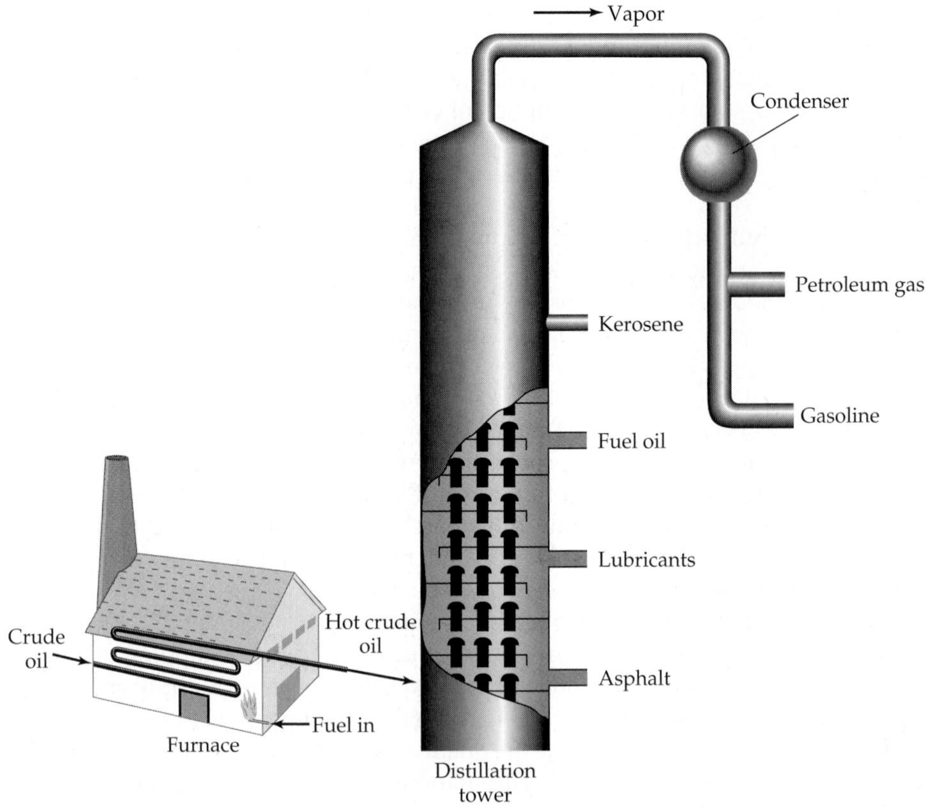

Vapor

Condenser

Petroleum gas

Kerosene

Gasoline

Fuel oil

Crude oil

Hot crude oil

Lubricants

Asphalt

Fuel in

Furnace

Distillation tower

▶ **Figure 14.7** The fractional distillation of petroleum. Crude oil is vaporized, and the column separates the components according to their boiling points. The lower-boiling constituents reach the top of the column, and the higher-boiling components come off lower in the column. A nonvolatile residue collects at the bottom.

absence of air. Called *cracking*, this process breaks down bigger molecules into smaller ones. The effect of this process, with $C_{14}H_{30}$ as an example, is illustrated in Figure 14.8. Cracking not only converts some molecules to those in the gasoline range (C_5H_{12} through $C_{12}H_{26}$), but it also produces a variety of useful by-products. The unsaturated hydrocarbons are starting materials for the manufacture of a host of petrochemicals.

The cracking process illustrates the way chemists modify nature's materials to meet human needs and desires. Starting with petroleum or coal tar, chemists can create a dazzling array of substances with a wide variety of properties. They can make plastics, pesticides, herbicides, perfumes, preservatives, painkillers, antibiotics, stimulants, depressants, dyes, and detergents.

The ability to modify hydrocarbon molecules enables the petroleum industry to shift production to whatever fraction is desired. The industry can, on demand, increase the proportion of gasoline in summer, or of fuel oil in winter, from a given supply of petroleum. It can even make gasoline from coal. It can't, however, increase the amount of fossil fuels aboard Spaceship Earth. Scientists are seeking new sources of energy that do not depend on petroleum. Perhaps we can soon stop this profligate waste of resources. Spaceship Earth has aboard it all the supplies it will ever have. We must use them wisely.

▶ **Figure 14.8** Formulas of a few of the possible products formed when $C_{14}H_{30}$, a typical molecule in kerosene, is cracked. In practice, a wide variety of hydrocarbons (most of which have fewer than 14 carbon atoms), hydrogen gas, and char (mostly elemental carbon) are formed. Cyclic and branched-chain hydrocarbons are also produced.

$$CH_3CH_2CH_2CH_2CH_2CH_2CH_2CH_2CH_2CH_2CH_2CH_2CH_2CH_3 \xrightarrow[\text{catalyst}]{\text{heat}}$$

$$CH_3CH_2CH_2CH_2CH_2CH_2CH_2CH_2CH_2CH_2CH_2CH_3 \ + \ CH_2{=}CH_2$$

and

$$CH_3CH_2CH_2CH_2CH_2CH_2CH_2CH_2CH_2CH_2CH_3 \ + \ CH_3CH{=}CH_2$$

and

$$CH_3CH_2CH_2CH_2CH_2CH_2CH_2CH_3 \ + \ CH_3CH_2CH_2CH_2CH{=}CH_2$$

and so on

Gasoline

Gasoline, like the petroleum from which it is derived, is mainly a mixture of hydrocarbons. A commercial gasoline typically contains more than 150 different compounds, but up to 1000 have been identified in some blends. Among the hydrocarbons, a typical gasoline sample, by volume, might have 4–8% straight-chain alkanes, 25–40% branched-chain alkanes, 2–5% alkenes, 3–7% cycloalkanes, 1–4% cycloalkenes, and 20–50% aromatic hydrocarbons (0.5–2.5% benzene). The gasoline also has a variety of additives, including antiknock agents, antioxidants, antirust agents, antiicing agents, upper-cylinder lubricants, detergents, and dyes. Typical alkanes in gasoline range from C_5H_{12} to $C_{12}H_{26}$. There are also small amounts of some sulfur- and nitrogen-containing compounds.

The gasoline fraction of petroleum as it comes from a distillation column is called *straight-run gasoline*. It doesn't burn very well in modern high-compression automobile engines, but chemists are able to modify it to make it burn more smoothly.

The Octane Ratings of Gasolines

In an internal combustion engine, the gasoline–air mixture sometimes ignites before the spark plug "fires." This is called *knocking* and can damage the engine. Early on, scientists learned that some types of hydrocarbons, especially those with branched structures, burned more evenly and were less likely to cause knocking than others. An arbitrary performance standard, called the **octane rating**, was established in 1927. Isooctane was assigned a value of 100 octane. An unbranched-chain compound, heptane, was given an octane rating of 0. A gasoline rated 90 octane was one that performed the same as a mixture that was 90% isooctane and 10% heptane.

$$CH_3-\underset{\underset{CH_3}{|}}{\overset{\overset{CH_3}{|}}{C}}-CH_2-\underset{}{\overset{\overset{CH_3}{|}}{CH}}-CH_3 \qquad CH_3CH_2CH_2CH_2CH_2CH_2CH_3$$

Isooctane   Heptane

During the 1930s, chemists discovered that the octane rating of gasoline could be improved by heating it in the presence of a catalyst such as sulfuric acid (H_2SO_4) or aluminum chloride ($AlCl_3$). This converted (*isomerized*) part of the unbranched structures to highly branched molecules. For example, heptane molecules can be isomerized to branched structures:

$$CH_3CH_2CH_2CH_2CH_2CH_2CH_3 \xrightarrow[\text{heat}]{H_2SO_4} CH_3-CH_2-\underset{\underset{CH_3}{|}}{CH}-\overset{\overset{CH_3}{|}}{CH}-CH_3$$

Chemists also can combine small hydrocarbon molecules (below the gasoline range) into larger ones more suitable for use as fuel. This process is called *alkylation*. In a typical alkylation reaction, shown below, isobutylene is reacted with propane.

$$CH_2=\underset{\underset{CH_3}{|}}{\overset{\overset{CH_3}{|}}{C}} \;+\; \underset{\underset{CH_3}{|}}{CH_2}-CH_3 \longrightarrow CH_3-\underset{\underset{CH_3}{|}}{\overset{\overset{CH_3}{|}}{C}}-\underset{\underset{CH_3}{|}}{CH}-CH_3$$

The product molecules are in the right size range for gasoline, and the highly branched molecules are high in octane number.

Certain additives substantially improve the antiknock quality of gasoline. Tetraethyllead, $Pb(CH_2CH_3)_4$, was found to be especially effective. As little as 1 mL tetraethyllead per liter of gasoline (one part per thousand) increases the octane rating by 10 or more.

▲ Fuel with the octane booster tetraethyllead, designated "Ethyl" gasoline, was available at many gas stations before leaded gasoline was phased out, beginning in the 1970s.

Lead fouls the catalytic converters used in modern automobiles. More important, lead is especially toxic to the brain. Even small amounts can lead to learning disabilities in children. In the United States unleaded gasoline became available in 1974, and all leaded gasoline now has been phased out.

Scientists have found other ways to get high octane ratings in unleaded fuels. For example, petroleum refineries use **catalytic reforming** to convert low-octane alkanes to high-octane aromatic compounds. Hexane (with an octane number of 25) is converted to benzene (octane number, 106).

$$CH_3CH_2CH_2CH_2CH_2CH_3 \xrightarrow[\text{heat}]{\text{catalyst}} \bigcirc + 4\,H_2$$

(C_6H_{14}) (C_6H_6)

Octane boosters to replace tetraethyllead include ethanol, methanol, *tert*-butyl alcohol, and methyl *tert*-butyl ether (MTBE) (Chapter 9). None of these is nearly as effective as tetraethyllead in boosting the octane rating. They must therefore be used in fairly large quantities. The amount that can be used in gasoline is limited by solubility problems. For example, ethanol in excess of 10% tends to separate from the gasoline, especially if moisture gets into the fuel. In the United States ethanol is usually made by the fermentation of corn.

Unlike gasoline, which is made up of hydrocarbons, the various alcohols and their derivatives all contain oxygen and therefore are sometimes called *oxygenates*. Not only do these additives improve the octane rating, but they also decrease the amount of carbon monoxide in auto exhaust gas. A disadvantage of the oxygenates is that their energy content is lower than pure hydrocarbon fuels, and so the distance one can travel per tankful is somewhat shorter.

MTBE, like ethers in general (Chapter 9), is rather unreactive chemically, but it is soluble in water to the extent of 4.8 g per 100 g of water. When gasoline spills or leaks from storage tanks, MTBE enters the groundwater, leading to widespread contamination. MTBE is listed as a hazardous substance under the Federal Superfund law and is considered a potential human carcinogen by the EPA. The exact threat to human health is far from clear, but some states have banned the use of MTBE in gasoline.

Alternative Fuels

An automobile engine can be made to run on nearly any liquid or gaseous fuel. There are cars on the road today powered by natural gas, by propane, by diesel fuel, by fuel cells, and even by used fast-food restaurant grease. There are electric cars that can be plugged into the power grid and hybrid cars that can switch from electricity to gasoline as needed.

Diesel fuel for automobiles overlaps the kerosene fraction of petroleum (Table 14.6); it consists mainly of C_9 to C_{20} hydrocarbons with a boiling range of about 250 °C to 350 °C. Diesel fuel has a greater proportion of straight-chain alkanes than gasoline does. The standard for performance, called the *cetane number*, is based on hexadecane ($C_{16}H_{34}$) once known as cetane. A renewable fuel called *biodiesel* can be used in unmodified diesel engines. Biodiesel is made by reacting ethanol with vegetable oils and animal fats. The triacylglycerol esters are converted to ethyl esters of the fatty acids in the oils and fats.

Brazil makes extensive use of ethanol, made by fermentation of sucrose from a major local crop, sugar cane. A fuel that is 85% ethanol and only 15% gasoline, called E85, is available in some parts of the United States. More than 4 million American cars and trucks have the ability to run on E85 right now, but most people who own them are not aware of that. Ethanol alone is not likely to be an answer to our energy problems; widespread use would require huge tracts of farmland to be converted to corn production.

▲ Only about 500 of the 180,000 gasoline stations in the United States provide E85 fuel. Most are in corn-growing states.

Energy Return on Energy Invested

When evaluating an energy source, we need to consider the *energy return on energy invested* (EROEI) of exploiting the source. EROEIs are difficult to evaluate but can be quite important because if the EROEI of a resource is 1 or less, it becomes an energy sink and is no longer a primary source of energy.

In the early days of the petroleum industry, the EROEI for Texas crude oil may have been as much as 100 to 1; investing the equivalent of 1 barrel of oil could produce 100 barrels. The EROEI for oil from the Middle East is still about 30 to 1. Oil from ever more remote places or ever deeper offshore wells will have ever lower EROEIs.

The EROEIs of alternative energy sources are often quite low. Canada has vast deposits of tar sands, but the EROEI is only about 1.5 to 1. Colorado has huge deposits of oil shale, but the EROEI is likely to be quite low.

The EROEI for ethanol is the subject of much debate, ranging from a low of 0.78 to 1 to the highest estimate for ethanol from corn of 1.67 to 1. Two reasons for the low EROEI for ethanol are as follows: Much energy is needed to distill or evaporate the ethanol after fermentation, and ethanol has a lower energy content than petroleum fuels.

Hydrogen has an EROEI of less than 1 to 1 because it cannot be extracted from the ground nor generated by a biosystem. Hydrogen can be generated by electricity or by heating steam to very high (2000 °C) temperatures. For now, those measures require more energy than the hydrogen produces when burned.

In addition to the EROEI, we need to consider the environmental consequences of exploiting an energy source. For example, getting oil from tar sands or oil shale is a messy process. Whatever else, our quest for plentiful energy will be costly and controversial.

14.10 CONVENIENT ENERGY: ELECTRICITY

The convenience of a fossil fuel depends on its physical state: gases are most convenient, liquids next, and solids least convenient. Perhaps the most convenient form of energy of all is electricity (Table 14.7). With electricity we can have light and hot water, and we can run motors of all sorts. We can use it to heat and cool our homes and workplaces. When looking at future energy sources, then, we look mainly at ways of generating electricity.

Any fuel can be burned to boil water, and the steam produced can turn a turbine to generate electricity. Figure 14.9 shows a coal-fired steam power plant. At present, about 51% of U.S. electric energy comes from coal-burning plants (Figure 14.10). Such facilities are at best only about 40% efficient; 60% of the energy of the fossil fuel is wasted as heat. Some power installations use the waste heat generated to warm buildings, a technique called *cogeneration*.

Coal Gasification and Liquefaction

Coal can be converted to gas or oil. When we run short of gas and petroleum, why not make them from coal? The technology has been around for years. Gasification and liquefaction do have some advantages: Gases and liquids are easy to transport,

TABLE 14.7 Convenience of Fuels in Various Physical States

Physical State	Extraction	Transportation to Cities	Distribution Within a City	In Use Convenience	In Use Cleanliness
Solids (coal, wood)	Shovels, borers, blasting	Trucks, trains, barges (slurry with water in pipe)	Trucks, buckets	Least	Dirtiest
Liquids (gasoline, fuel oil)	Pumps	Pipelines, tankers, barges, trucks	Trucks		
Gases (natural gas)	Pumps	Pipeline	Pipes		
Electricity* (electron flow)	—	Wires	Wires	Most	Cleanest

*Produced by burning any of the primary fuels, and included for comparison.

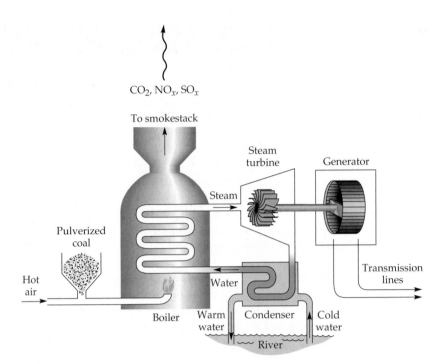

▶ **Figure 14.9** A diagram of a coal-burning power plant for generating electricity.

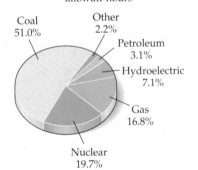

Industry total = 3800 billion kilowatt-hours

Coal 51.0%
Other 2.2%
Petroleum 3.1%
Hydroelectric 7.1%
Gas 16.8%
Nuclear 19.7%

▲ **Figure 14.10** Percentages of electric power generation in the United States in 2003 from various energy sources.

and the process of conversion leaves much of the sulfur and minerals behind, thus overcoming a serious disadvantage of coal as a fuel.

The basic process for converting coal to a synthetic gaseous fuel is a reduction of carbon by hydrogen. Passing steam over hot charcoal produces *synthesis gas*, a mixture of hydrogen and carbon monoxide.

$$C(s) + H_2O(g) \longrightarrow CO(g) + H_2(g)$$

The hydrogen can be used to reduce the carbon in coal or other substance to form methane.

$$C(s) + 2\,H_2(g) \longrightarrow CH_4(g)$$

Coal can also be converted to methanol by way of synthesis gas. Part of the CO in synthesis gas is reacted with water to form more H_2.

$$CO(g) + H_2O(g) \longrightarrow CO_2(g) + H_2(g)$$

The remaining CO is combined with H_2 to form methanol.

$$CO(g) + 2\,H_2(g) \longrightarrow CH_3OH(l)$$

The methanol can be used directly as fuel or converted, in turn, to gasoline-like hydrocarbons that we represent as C_nH_m.

$$n\,CH_3OH(l) \longrightarrow C_nH_m(l) + x\,H_2O(l)$$

Each step requires an appropriate catalyst.

Both gasification and liquefaction of coal require a lot of energy: Up to one-third of the energy content of the coal is lost in the conversion. Liquid fuels from coal are high in unsaturated hydrocarbons and in sulfur, nitrogen, and arsenic compounds, and their combustion products are high in particulate matter. Coal conversions also require large amounts of water, yet the large coal deposits on which they would be based are in arid regions. Further, the conversions are messy. Without stringent safeguards, plants would seriously pollute both air and water.

Other processes that have been used to make liquid fuels from coal are the Bergius and Fischer–Tropsch methods. In the Bergius process, coal is reacted with hydrogen. In the Fischer–Tropsch method, coal is reacted with steam to make a mixture of carbon monoxide and hydrogen, which is then reacted over an iron catalyst to produce a mixture of hydrocarbons. Both processes were used in Germany during World War II. Since 1955, the Sasol process, which uses the Fischer–Tropsch method, has provided a significant portion of liquid fuels for South Africa.

Nuclear Energy

In Chapter 4 we explored some nuclear reactions and noted that both nuclear fission and nuclear fusion can be used in bombs. Nuclear fission can also be controlled to generate power. So far nuclear fusion reactions cannot be controlled to provide energy; they can be employed only in bombs.

14.11 NUCLEAR FISSION

Nuclear fission reactions can be controlled in a **nuclear reactor**. The energy released during fission can be used to generate steam, which can turn a turbine to generate electricity (Figure 14.11).

At the dawn of the nuclear age, some people envisioned nuclear power to be destined to fulfill the biblical prophecy of a fiery end to our world. Others saw nuclear power as a source of unlimited energy. During the late 1940s, some claimed that electricity from nuclear plants would become so cheap that it eventually would not have to be metered. Nuclear power has not yet brought us either paradise or perdition, but it has become a most controversial issue. Our great demand for energy indicates that the controversy will continue for years to come.

At present, nearly 20% of U.S. electricity comes from nuclear power plants. The eastern seaboard and upper midwestern states, many of which have minimal fossil fuel reserves, are heavily dependent on nuclear power for electricity.

We could rely more on nuclear power as other nations do (Table 14.8), but the public is quite fearful of nuclear power plants. This apprehension was exacerbated by the accident at Chernobyl, Ukraine. The United States has 104 operating nuclear reactors, but no new plants have been ordered since 1978. It takes about 10 years to build a nuclear power plant. There will have to be a dramatic change in public attitude if nuclear power is to play a big role in our future.

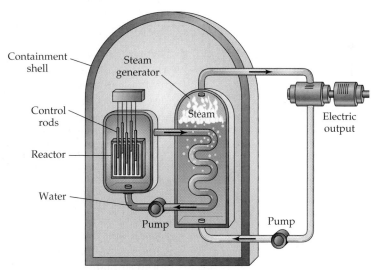

▲ **Figure 14.11** A diagram of a nuclear power plant for generating electricity. The nuclear reactors are housed in containment buildings made of steel and reinforced concrete. The structures are designed to withstand nuclear accidents without releasing radioactive substances into the environment.

TABLE 14.8	Electricity Generation Using Nuclear Power Plants (Selected Countries)		
Country	Total Electricity from Nuclear Power (%)	Number of Operating Nuclear Power Plants	Number of Nuclear Power Plants Under Construction
Belgium	55	7	0
China	2	9	2
France	78	59	0
Germany	32	17	0
India	3	15	8
Japan	29	55	2
Korean Republic	38	20	0
Lithuania	72	1	0
Russian Federation	16	31	4
Slovak Republic	55	6	0
Sweden	52	10	0
Switzerland	40	5	0
Ukraine	51	15	2
United States	20	104	0
World total	11	442	24

Source: International Atomic Energy Agency, *PRIS Database.* Vienna: 2005.

Types of Nuclear Power Plants

There are several types of nuclear power plants, but we won't attempt to discuss all the possible types here. The one illustrated in Figure 14.11 is a pressurized water reactor. Earlier models were mainly boiling water reactors in which the steam from the reactor was used to power the turbine directly.

Nuclear power plants use the same fission reactions employed in nuclear bombs, but nuclear power plants cannot blow up like bombs. The uranium used in power plants is enriched to only 3–4% uranium-235. A bomb requires about 90% uranium-235.

In a nuclear power plant, a *moderator* is used to slow down the fission neutrons so that they can be absorbed by U-235 atoms. The reaction is controlled by the insertion of boron steel or cadmium *control rods.* Boron and cadmium absorb neutrons readily, preventing them from participating in the chain reaction. These rods are installed when the reactor is built. Removing them part way starts the chain reaction; pushing them in all the way stops the reaction.

The Nuclear Advantage: Minimal Air Pollution

The main advantage of nuclear power plants over those that burn fossil fuels is in what they do not do. Unlike fossil-fuel-burning plants, nuclear power plants produce no carbon dioxide to add to the greenhouse effect, and they add no sulfur oxides, nitrogen oxides, soot, or fly ash to the atmosphere. They contribute almost nothing to global warming, air pollution, or acid rain. They reduce our dependence on foreign oil and lower our trade deficit. If the 104 nuclear plants in the United States were replaced by coal-burning plants, airborne pollutants would increase by 18,000 tons/day!

Problems with Nuclear Power

Nuclear power plants have some disadvantages. Elaborate and expensive safety precautions must be taken to protect plant workers and the inhabitants of surrounding areas from radiation. The reactor must be heavily shielded and housed inside a containment building of metal and reinforced concrete. Because loss of coolant water can result in a meltdown of the reactor core, backup emergency cooling systems are required. Despite what proponents call utmost precautions, some opponents of nuclear power still fear a runaway nuclear reaction in which the containment building is breached and massive amounts of radioactivity escape into the

◀ Mildly radioactive tailings are a by-product of the processing of uranium ores.

environment. The chance of such an accident is probably exceedingly small, but if one did occur, thousands of people could be killed and large areas rendered uninhabitable for centuries. The benefit of nuclear power—abundant electric energy—is clear, but the small probability of an accident causes scientists and others to endlessly debate its desirability.

Another problem is that the fission products are highly radioactive and must be isolated from the environment for centuries. Again, scientists disagree about the feasibility of nuclear waste disposal. Proponents of nuclear power say that such wastes can be safely stored in old salt mines or other geologic formations. Opponents fear that the wastes may arise from their "graves" and eventually contaminate the groundwater. It is impossible to do a million-year experiment in a few years to determine who is right.

Mining and processing uranium ore produces wastes called tailings. Over 200 million tons of tailings now plague ten western states. These tailings are mildly radioactive, giving off radon gas and gamma radiation. Dust from these tailings carries problems to surrounding areas.

One problem, thermal pollution (Chapter 13), is unavoidable. As the energy from any material is converted to heat to generate electricity, some of the energy is released into the environment as waste heat. Nuclear power plants generate more thermal pollution than plants that burn fossil fuels, but the difference is perhaps not as important as the other problems we have mentioned.

Most of the nuclear waste currently awaiting disposal is from nuclear weapons production. High-level waste from nuclear power plant operation in the United States totals about 3000 metric tons per year.

Yucca Mountain

For many, the solution to the problem of ever-accumulating radioactive spent nuclear fuel and high-level radioactive waste is situated 300 m underground about 160 km northwest of Las Vegas, Nevada. Yucca Mountain is slated to become the United States' first high-level radioactive waste depository. Currently, spent fuel rods and other radioactive materials are stored at 126 sites throughout the country.

Since 1978 scientists and engineers have been studying the suitability of Yucca Mountain for the

▶ Yucca Mountain, site of the U.S. proposed nuclear waste depository.

depository site. They need not only to understand the natural characteristics of the mountain that could affect a potential repository's safety, but also to engineer repository systems that further isolate the waste. By 2002 costs had reached $7 billion. In 2002 the U.S. Department of Energy (DOE) was authorized to prepare an application to obtain the Nuclear Regulatory Commission license to proceed with construction of the repository.

Few argue against the need for a long-term solution to a growing problem, but there are concerns, especially over the transportation of radioactive material to Nevada. Estimates are that 19,200 rail cars and 93,000 trucks would be required to relocate the radioactive material—mostly in the form of heavy solid waste pellets—moving it through 43 states. Once on site, the pellets would be placed in casks and maneuvered by rail into tunnels deep in Yucca Mountain. Despite the protests from the citizens of Nevada, the DOE plans to begin entombing 70,000 metric tons of spent nuclear fuel in 2010. Most of the high-level wastes will decay away in a few decades, but the site would remain radioactive for 10,000 years or more.

A 20-year study of 70,000 nuclear shipyard workers found that they had 24% *lower* mortality than non-nuclear workers in the general population. Those with the highest chronic radiation exposure had the *lowest* mortality from all causes—including cancer. Could it be that small doses of radiation are beneficial?

Nuclear Accidents: Real and Imagined Risks

In 1979 a loss-of-coolant accident at the Three Mile Island nuclear power plant near Harrisburg, Pennsylvania, released a tiny amount of radioactivity into the environment. Although no one was killed or seriously injured, the accident whetted public fear of nuclear power. A 1986 accident at Chernobyl, Ukraine, was much more frightening. There a reactor core meltdown killed several people outright. Others died from radiation sickness in the following weeks and months, and 135,000 people were evacuated. The Ukraine Radiological Institute estimates the accident caused more than 2500 deaths overall. A large area will remain contaminated for decades. Radioactive fallout spread across much of Europe. The Ukraine thyroid cancer rate increased tenfold. Thousands of others have an increased risk of cancer because of exposure to radiation. At Three Mile Island, a containment building kept most of the radioactive material inside. The Chernobyl plant had no such protective structure.

There is considerable controversy over most aspects of nuclear power. Although scientists may be able to agree on the results of laboratory experiments, they don't always agree on what is best for society.

Breeder Reactors: Making More Fuel Than They Burn

The supply of fissionable uranium-235 isotope is limited, making up less than 1% of naturally occurring uranium. Separation of uranium-235 leaves behind large quantities of uranium-238, which is not fissionable. However, uranium-238 can be converted to fissile plutonium-239 by bombardment with neutrons. Unstable uranium-239 is formed initially, but it rapidly decays to neptunium-239.

$$^{238}_{92}U + ^{1}_{0}n \longrightarrow ^{239}_{92}U \longrightarrow ^{239}_{93}Np + ^{0}_{-1}e$$

Neptunium-239 then decays to plutonium:

$$^{239}_{93}Np \longrightarrow ^{239}_{94}Pu + ^{0}_{-1}e$$

If a reactor is built with a core of fissionable plutonium surrounded by uranium-238, neutrons from the fission of plutonium convert the uranium-238 shield to more plutonium. In this way, the reactor, called a **breeder reactor**, produces more fuel than it consumes. There is enough uranium-238 to last several centuries, and so one of the disadvantages of nuclear plants could be overcome by the use of breeder reactors.

Breeder reactors have some problems of their own, however. Plutonium is fairly low melting (640 °C), and a plant is therefore limited to fairly cool, inefficient operation. Water is not adequate as a coolant, and therefore molten sodium metal is used in the primary loop. These reactors are often called *liquid-metal fast breeder reactors*. If an accident occurred in such a breeder, the sodium could react violently with both the water and the air.

▲ An accident at this nuclear power plant at Chernobyl, Ukraine, increased public fears of nuclear power. It released considerable radioactivity because it had no reinforced containment building. The plant used graphite as a moderator (U.S. plants use water), and the burning carbon hindered efforts to tame the runaway reaction.

Because plutonium is low melting, a failure of the cooling system could cause the reactor's core to melt. All reactors are required to have an emergency backup core-cooling system. Whether these systems work or not is a principal area of controversy.

Plutonium is highly toxic and has a half-life of about 24,000 years. It emits alpha particles, making it especially dangerous if ingested. An estimated 1 μg in the lungs of a human is enough to induce lung cancer. Yet another problem is that plutonium is chemically different from uranium-238. That means it is much easier to separate it from U-238, for illicit nuclear weapons use, than is uranium-235.

An alternate breeder reaction converts thorium-232 into fissile uranium-233.

$$^{232}_{90}\text{Th} + ^{1}_{0}\text{n} \longrightarrow ^{233}_{90}\text{Th} \longrightarrow ^{233}_{91}\text{Pa} + ^{0}_{-1}\text{e}$$

$$^{233}_{91}\text{Pa} \longrightarrow ^{233}_{90}\text{U} + ^{0}_{-1}\text{e}$$

Thorium-232 is especially attractive for a breeder reactor because it is quite abundant. There is about as much thorium in the earth's crust as there is lead.

Uranium-233 is like plutonium in that it emits damaging alpha particles, but it is thought to be rather difficult to make bombs from reactor-grade uranium-233.

Only 11 breeder reactors of more than 100 MW have ever been built, none in the United States. Only four were still operating in 2004, and two of those are scheduled to be shut down.

14.12 NUCLEAR FUSION: THE SUN IN A MAGNETIC BOTTLE

In Chapter 4, we discussed the thermonuclear reactions that power the sun and that occur in the explosion of hydrogen bombs. Control of these fusion reactions to produce electricity would give us nearly unlimited power. To date, fusion reactions have been useful only for making bombs, but research on the control of nuclear fusion is progressing (Figure 14.12).

Controlled fusion would have several advantages over nuclear fission reactors. The principal fuel, deuterium ($^{2}_{1}\text{H}$), is plentiful and is obtained by the fractional electrolysis (splitting apart by means of electricity) of water. (Only 1 hydrogen atom in 6500 is a deuterium atom, but we have oceans of water to work with.) The problem of radioactive wastes would be minimized. The end product, helium, is stable and biologically inert. Escape of tritium ($^{3}_{1}\text{H}$), which undergoes beta decay with a half-life of 12.3 years, might be a problem because this hydrogen isotope would be readily incorporated into organisms. Also, neutrons are emitted in most fusion reactions, and neutrons can convert stable isotopes into radioactive ones. Finally, we would still be concerned with thermal pollution: the unavoidable loss of part of the energy as heat.

Great technical difficulties have to be overcome before a controlled fusion reaction can be used to produce energy. A sustainable fusion reaction would require

- attainment of a critical ignition temperature of 100 million to 200 million °C,
- a confinement time of 1–2 s, and
- a sufficient ion density of about 2–3 × 10^{20} ions per cubic meter.

In practice, the required conditions are measured by a product of these three values.

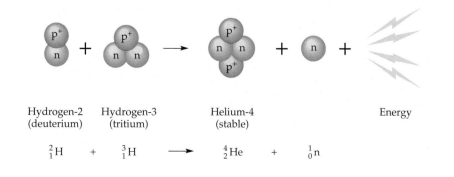

Hydrogen-2 (deuterium)   Hydrogen-3 (tritium)   Helium-4 (stable)   Energy

$$^{2}_{1}\text{H} + ^{3}_{1}\text{H} \longrightarrow ^{4}_{2}\text{He} + ^{1}_{0}\text{n}$$

◀ **Figure 14.12** The most promising fusion reaction is the deuterium–tritium reaction. A hydrogen-2 (deuterium) nucleus fuses with a hydrogen-3 (tritium) nucleus to form a helium-4 nucleus. A neutron is released along with a considerable quantity of energy.

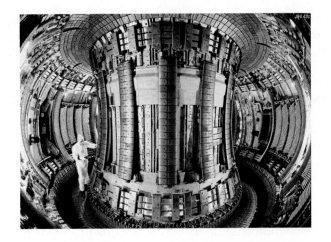

► **Figure 14.13** A giant, donut-shaped electromagnet called a *tokamak* is designed to confine plasma at the extremely high temperatures and pressures required for nuclear fusion.

No molecule could hold together at the fusion temperature; no material on Earth can withstand more than a few thousand degrees. Even atoms are unstable under these conditions; atoms are stripped of their electrons, and the nuclei and free electrons form a mixture called a *plasma*. The plasma, made of charged particles (nuclei and electrons), can be contained by a strong magnetic field (Figure 14.13).

Nuclear fusion may well be our best hope for relatively clean, abundant energy in the future, but much work remains to be done. Even when controlled fusion is achieved in the laboratory, it will still be decades before it becomes a practical source of energy.

Renewable Energy Sources

Burning fossil fuels leads to air pollution and to the depletion of vital resources. Using nuclear power also presents problems, nuclear fuel is not unlimited, and the waste problems remain unsolved. Are there no renewable energy resources? There are indeed, and we shall devote the remainder of this chapter to them.

14.13 HARNESSING THE SUN: SOLAR ENERGY

At the beginning of this chapter, we saw that nearly all the energy on Earth comes from the sun. With all that energy from our celestial power plant, why do we need fossil fuels or nuclear power? The answer lies in the fact that this solar energy is thinly spread out and difficult to concentrate.

Solar Heating

Diffuse energy is not very useful. As it arrives on the surface of Earth, 30% of solar energy is simply reflected back into space. About half is converted to heat. We can increase the efficiency of this conversion rather easily. A black surface absorbs radiation better than a light-colored one. To make a simple solar collector, we need only cover a black surface with a glass plate. The glass is transparent to the incoming solar radiation, but it partially prevents the heat from escaping back into space. The hot surface is used to heat water or other liquids, and the hot liquids are usually stored in an insulated reservoir.

Water heated this way can be used directly for bathing, dish washing, and laundry, or it can be used to heat a building. Air is passed around the warm reservoir, and the warmed air is then circulated through the building (Figure 14.14). Even in cold northern climates, solar collectors could meet about 50% of home heating requirements. These installations are expensive but could pay for themselves, through fuel savings, in a few years.

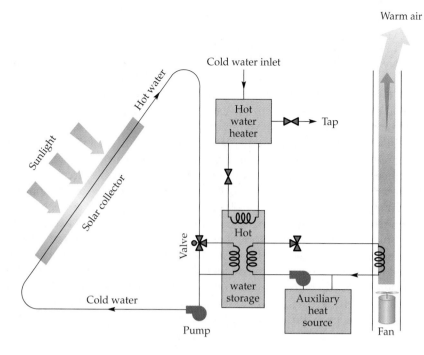

◀ **Figure 14.14** Energy in sunlight is absorbed by solar collectors and used to heat water. The hot water can be used directly, or it can be circulated to partially heat the building. This diagram shows how a solar collector furnishes hot water and warm air for heating the building.

Solar Cells: Electricity from Sunlight

Sunlight also can be converted directly to electricity by devices called **photovoltaic cells** or *solar cells*. These devices can be made from a variety of substances, but most are made from elemental silicon. In a crystal of pure silicon, each silicon atom has four valence electrons and is covalently bonded to four other silicon atoms (Figure 14.15). To make a solar cell, extremely pure silicon is doped with small amounts of specific impurities and formed into single crystals.

One type of crystal has about 1 ppm of arsenic added. Arsenic atoms have five valence electrons, four of which are used to form bonds to silicon atoms. The fifth electron is relatively free to move around. Because this material has extra electrons, and electrons are negatively charged, it is called an *n-type semiconductor* (*n* = negative). Adding about 1 ppm of boron to silicon forms a different type of material. Boron has three valence electrons, producing a shortage of one electron and leaving a *positive hole* in the crystal. This boron-doped silicon is called a *p-type semiconductor*.

Joining the two types of crystals (*n*-type and *p*-type) forms a photovoltaic cell (Figure 14.16); electrons flow from the *n*-type region, which has a high concentration, to the *p*-type region. However, the holes near the junction are quickly filled by nearby mobile electrons, and the flow ceases.

When sunlight hits the photovoltaic cell, an electric current is generated. The energetic photons knock electrons out of the Si—Si bonds, creating more mobile electrons and more positive holes. Because of the barrier at the junction between the

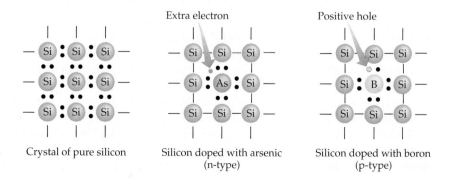

◀ **Figure 14.15** Models of silicon crystals. Crystals doped with impurities such as arsenic and boron are more conductive than pure silicon, and are used in solar cells.

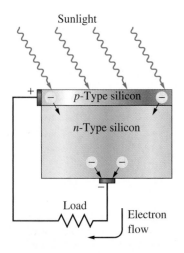

Sunlight

+

− p-Type silicon −

n-Type silicon

− −

−

Load

Electron flow

▲ **Figure 14.16** Schematic diagram of the operation of a solar cell. Electrons flow from the n-type (upper layer) to the p-type (lower layer) through the external circuit.

Hemp, the fiber crop from which rope is made, is a good energy source. Its woody stalks are 77% cellulose, and hemp can produce 10 t of biomass per acre in just 4 months. The oil from hemp seeds can be used as diesel fuel. The problem is that it is illegal to raise hemp in the United States because the leaves and flowers constitute the drug marijuana.

▲ These willow trees are grown specifically as fuel. The biomass is harvested by cutting the trees to near ground level. This stimulates regrowth. Willow is especially suitable for biomass because of its rapid growth.

two semiconductors, electrons cannot move through the interface. When an external circuit connects the two crystals, electrons flow from the n-type region around the circuit to the p-type region and that current can be used directly.

An array of solar cells, combined to form a solar battery, can produce about 100 W/m^2 of surface; it takes 1 m^2 of cells to power one 100-W light bulb. Solar batteries have been used for years to power spaceships. They are now widely used to power small devices such as electronic calculators. They are also used to provide electricity for weather instruments in remote areas.

Solar cells are not very efficient (usually about 10%). Much of the sunlight striking them is reflected back into space or converted to heat. The generation of enough energy to meet a significant portion of our demands would require covering vast areas of desert land with solar cells. It would require 2000 hectares (about 5000 acres) of cloud-free desert land to produce as much energy as one nuclear power plant. Research has led to more efficient solar cells, and their cost is decreasing.

Using solar energy requires storage of energy for use at night and on cloudy days. The technology is now widely available for the use of solar energy for space heating and for providing hot water. Several companies now sell home-sized solar electric systems that can be connected to the normal house supply, and can reduce electric bills to nearly zero. However, they are quite expensive, and widespread use of solar electricity is probably still some years away.

14.14 BIOMASS: PHOTOSYNTHESIS FOR FUEL

Why bother with solar collectors and photovoltaic cells to capture energy from the sun when green plants do it every day? Indeed, dry plant material, called **biomass** when used as a fuel, burns quite well. It could be used to fuel a power plant for the generation of electricity. Biomass burns cleanly; the emissions are almost entirely water vapor and the carbon dioxide the plant took from the air in the first place. "Energy plantations" could grow plants for use as fuel. Biomass is a renewable resource whose production is powered by the sun.

Unfortunately, there are several disadvantages to this scheme, too. Most available land is needed for the production of food. Even where productive land is available, plants have to be planted, harvested, and transported to the power plant. Often, the land is far from where the energy is needed. However, some hybrid species of trees grow quickly and densely, reducing the needed growing area. The overall efficiency of biomass use is even less than that of solar cells—only about 3% at best. Nevertheless, there is considerable research activity in the area of biomass; nature "constructs" plants much more easily and more quickly than we can construct industrial plants.

We don't have to burn plant material directly.

- Starches and sugars from plants can be fermented to form ethanol. Wood can be distilled in the absence of air to produce methanol. Both alcohols are liquids and convenient to transport and are excellent fuels that burn quite cleanly.
- Bacterial breakdown of plant material produces methane. Under proper conditions, this process can be controlled to produce a clean-burning fuel similar to natural gas.

Each of these conversions, however, results in the loss of a portion of the useful energy. The laws of thermodynamics tell us that we would get the most energy by burning the biomass directly rather than converting it to a more convenient liquid or gaseous fuel.

The shortage of land probably means that we will never obtain a major portion of our energy from biomass. We could, however, supplement our other sources by burning agricultural wastes, fermenting some to ethanol, producing methanol from wood where wood is plentiful, and fermenting human and animal wastes to produce methane. The technology for all these processes is readily available. Each has been used in the past, and all are now being used on a limited scale.

14.15 HYDROGEN: LIGHT AND POWERFUL

Just as natural gas is sent through pipes to wherever it is needed, so could other fuel gases be sent through these same pipes. One such gas is hydrogen.

When hydrogen burns, it produces water and gives off energy.

$$2 H_2(g) + O_2(g) \longrightarrow 2 H_2O(l) + 572 \text{ kJ}$$

On a mass basis, hydrogen has more energy than any other chemical fuel. It is also a clean fuel, yielding only water as a chemical product. Although hydrogen is the most abundant element in the universe, as elemental hydrogen (H_2) it is almost nonexistent on Earth. There is a lot of hydrogen on Earth, but it is tied up in chemical compounds, mainly water, and releasing it requires more energy than the hydrogen produces when it is burned. Because hydrogen can be made from seawater, the supply is almost inexhaustible, but finding a way to make H_2 from water economically is a complicated matter.

Fuel Cells

A **fuel cell** is a device in which fuel is oxidized in an electrochemical cell (Chapter 8) so as to produce electricity directly. Fuel cells differ from the usual electrochemical cell in two ways:

- The fuel and oxygen are fed continuously. As long as fuel is supplied, current is generated.
- The electrodes are made of an inert material such as platinum that does not react during the process.

Most of today's fuel cells use hydrogen and oxygen (Figure 14.17). At the platinum anode, H_2 is oxidized forming hydrogen ions and electrons:

$$2 H_2(g) \longrightarrow 4 H^+ + 4 e^-$$

The electrons produced at the anode travel through the external circuit and arrive at the cathode, where they combine with the hydrogen ions and oxygen molecules to form water.

$$4 e^- + O_2(g) + 4 H^+ \longrightarrow 2 H_2O(g)$$

The overall reaction is identical to combustion, but not all the chemical energy is converted to heat. About 40–55% of the chemical energy is converted directly to electricity, making fuel cells much more efficient than internal combustion engines.

$$2 H_2(g) + O_2(g) \longrightarrow 2 H_2O(l)$$

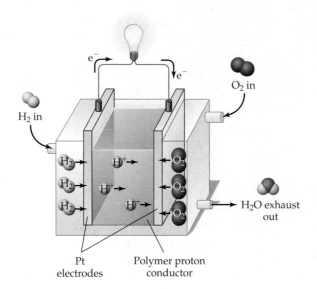

Pt electrodes   Polymer proton conductor

◀ **Figure 14.17** A hydrogen–oxygen fuel cell. The fuel cell continues to provide electricity as long as reactants are supplied.

Hydrogen in Your Future (Car)

Hydrogen can be used as a fuel for cars, either directly or in fuel cells. In his 2003 State of the Union address, President George W. Bush announced a $1.2 billion federal initiative to support hydrogen fuel cell research with the hope that it will lead to the widespread use of hydrogen-powered cars. Among the advantages of hydrogen as a fuel are that

- water vapor is the only exhaust, which cuts down on urban air pollution and the production of greenhouse gases,
- hydrogen can be produced from renewable sources, and
- this could reduce our reliance on foreign oil, and its associated economic and political costs.

There are however, significant obstacles to this transition to hydrogen; the roads will not be filled with hydrogen cars any time soon. Among the problems are the following:

- Hydrogen is not a *source* of energy; there are no hydrogen "mines" or "reservoirs."
- The hydrogen-powered cars available now are expensive novelties, costing $100,000 or more each.
- There is no distribution system of hydrogen fueling stations.
- Hydrogen is a low-density gas that liquefies at −253 °C (20 K). Use of the gas requires large,

heavy tanks. Use as a liquid requires that it be maintained at a temperature only a few degrees above absolute zero. Therefore, storage, transportation and dispensing of the fuel is problematic.

Progress has been made in several areas. California has a dozen or so pilot stations run by utilities and carmakers. Governor Arnold Schwarzenegger's "hydrogen blueprint" calls for up to 2000 hydrogen vehicles and 100 refueling stations by 2010 at an estimated cost of $54 million.

Fuel cell technology is becoming more efficient and costs are coming down, but there is still the problem of producing hydrogen gas economically. Hydrogen has to be made using another energy source: fossil fuels, nuclear power, or—preferably—some renewal source such as solar energy or wind power. Some experimental cells make hydrogen on-board from methanol or natural gas by a high-temperature process called "reforming."

$$CH_4(g) + 2 H_2O(g) \longrightarrow CO_2(g) + 4 H_2(g)$$

The drawback is that reforming emits carbon dioxide, negating one of the reasons for using fuel cells in the first place.

Even liquid H_2, with a density of 0.07 g/cm^3, requires a large fuel tank, and the tank must be well insulated. It may be possible to store hydrogen in other ways: Scientists have found that various metals can absorb up to a thousand times their own volume of hydrogen gas. Special forms of carbon, such as ultra-thin carbon nanotubes, may also hold large quantities.

The cost of establishing pipelines and filling stations to handle hydrogen will be immense. It takes many years for revolutionary technologies to come to the market place. The increased emphasis on research into hydrogen fuel cells appears to be a step in the right direction.

▲ Hydrogen-powered Shelby Cobras start at $149,000. As modified by the Hydrogen Car Company, the hydrogen tank in the Shelby Cobra takes up most of the trunk space.

Fuel cells are used on spacecraft to produce electricity. Fuel cells overall have a weight advantage over storage batteries—an important consideration when launching a spaceship—and the water produced can be used for drinking. Because they have no moving parts, fuel cells are tough and reliable.

On Earth, research is underway to reduce cost and design long-lasting cells. Perhaps someday fuel cells will provide electricity to meet peak needs in large power plants. Unlike huge boilers and nuclear reactors, they can be started and stopped simply by turning the fuel on or off.

Some people are afraid of using hydrogen as fuel for home heating because of the possibility of explosion, but natural gas leaks can also lead to serious explosions. Hydrogen escapes more readily than natural gas or gasoline because of its smaller molecular size, but it also dissipates more readily, rather than forming a flammable low-lying vapor as gasoline sometimes does.

14.16 OTHER RENEWABLE ENERGY SOURCES

Modern civilization requires tremendous amounts of energy and will require more and more in the coming years. We are depleting our reserves of fossil fuels, and many see significant disadvantages to nuclear power. Using the sun's energy seems to be a good idea, but it would be difficult to meet all our needs with solar energy. In this section we look at other kinds of energy sources, some of which are already in use.

Wind and Water

The sun, by heating Earth, causes winds to blow and water to evaporate and rise into the air, later to fall as rain. The kinetic energy of blowing wind and flowing water can be used as sources of energy—as they have been for centuries.

Why not use wind power and waterpower to help solve the energy crisis? We do. Waterpower provides more than 7% of our current electricity production, most of it in the mountainous western United States. In a modern hydroelectric plant, water is held behind huge dams. Some of it is released through penstocks against the blades of water turbines. The potential energy of the stored water is converted to the kinetic energy of flowing water. The moving water imparts mechanical energy to the turbine. The turbine drives a generator that converts mechanical energy to electrical energy.

Hydroelectric plants are relatively clean, but most of the good dam sites in the United States have already been used. To obtain more hydroelectric energy, we would have to dam up scenic rivers and flood valuable cropland and recreational areas. Reservoirs silt up over the years, and sometimes dams break, causing catastrophic floods. Even renewable hydroelectric power has its problems.

We could use more wind power. The kinetic energy of moving air is readily converted to mechanical energy to pump water, grind grain, or turn turbines and generate electricity. Wind power amounts to only about 0.5% of U.S. energy production but is now the fastest growing energy source, increasing at an annual rate of 20 to 30%.

Wind is clean, free, and abundant. However, the wind does not always blow, and some means of energy storage or an alternative source of energy is needed. Land use might become a problem if wind power were used widely, but land under windmills can be used for farming or grazing.

Geothermal Energy

The interior of Earth is heated by immense gravitational forces and by natural radioactivity. This heat comes to the surface in some areas through geysers and volcanoes. **Geothermal energy** has long been used in Iceland, New Zealand, Japan, and Italy. It has some potential in the United States—it is being used now in California (Figure 14.18)—but in the near future this potential could be realized only in areas

What is the connection between waterpower and wind power and chemistry? Chemists produce new materials such as metal alloys, reinforced plastics, and lubricants vital to the construction and operation of dams, windmills, and turbines. Chemists also monitor water quality above and below dams, and they participate in other activities vital to protection of the environment around generating facilities.

▲ Hoover Dam, on the Colorado River, was completed in 1936. It generates 4 billion kWh a year—enough to serve 1.3 million people.

▶ Wind turbines, such as these in an offshore wind farm, supply about 20% of Denmark's electricity. By the end of 2003, over 39,000 MW of generating capacity were operating worldwide, producing some 90 billion kWh each year.

where steam or hot water is at or near the surface. One drawback of geothermal energy is that the wastewater is quite salty; its disposal could be a problem.

Oceans of Energy

The oceans that cover three-fourths of Earth's surface are an enormous reservoir of potential energy. The use of *ocean thermal energy* was first proposed in 1881 and was shown to be workable in the 1930s. The difference in temperature between the surface and the depths is 20 °C or more, enough to evaporate a liquid and use the vapor to drive a turbine. The liquid is condensed by the cold from the ocean depths and the cycle is repeated.

(a)

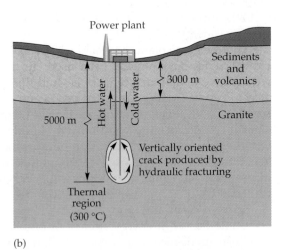

(b)

▲ **Figure 14.18** Geothermal energy contributes to energy supplies in some areas. (a) A geothermal field at Wairakei, New Zealand. (b) Energy is extracted from dry, hot rock 5 km below the surface by pumping water into the rock to be heated.

Other ways of using ocean energy have been tested. In some areas of the world, the daily rise and fall of the tide can be harnessed, in much the same way as a hydroelectric plant. At high tide, water fills a reservoir or bay. At low tide, the water escapes through a turbine to generate electricity. Even the energy of the waves crashing to shore can be used with the appropriate technology.

14.17 ENERGY: HOW MUCH IS TOO MUCH?

Our energy future is quite uncertain. World petroleum production will peak soon—some claim it has peaked already. Developing nations will increasingly compete for fuels on the world market. Science and technology will be asked to provide an ever-growing supply of energy. The Energy Information Administration of the U.S. Department of Energy projects that world energy consumption will increase by 57% from 2002 to 2025, most of it in developing countries.

Each of the energy sources discussed in this chapter has advantages and disadvantages. How do we choose the best way to deal with energy problems? Wise choices require informed citizens who examine the process from beginning to end. We must know what is involved in the construction of power plants, the production of fuels, and the ultimate use of energy in our homes and factories. We must know that energy is wasted (as heat) at every step in the process.

Will our profligate consumption of energy affect Earth's climate? Our activities have already modified the climate in and around metropolitan areas. The worldwide effects of our expanding energy consumption are harder to estimate.

What can we do as individuals? We can conserve. We can walk more and use cars less. We can reduce our wasteful use of electricity. We can buy more efficient appliances, and we can avoid purchasing energy-intensive products.

Over the past three decades we have made significant progress in energy conservation. Appliances are more energy efficient. New furnaces are 90–95% efficient, as compared with the 60% efficiency of 20-year-old furnaces. New refrigerators and air conditioners use about 35% less energy than earlier models. More people use fluorescent lamps, which are at least four times as energy efficient as incandescent lamps. Overall, American industry has reduced its energy consumption per product by almost 30%.

We could significantly reduce energy consumption in the United States by greater use of public transportation. In Europe, where there is much wider use of public transit, per capita energy use is considerably less. So far, few people used to the convenience of personal cars seem eager to start using buses and trains. In Europe the average price for gasoline is usually more than twice that in the United States. As the price of gasoline in the United States continues to rise, perhaps public transportation will seem more attractive.

The simple facts are that our population on Spaceship Earth is going up and our fuel resources are going down. People in developing countries want to raise their standard of living, and that will require more energy. We need to conserve energy and to look for new energy sources.

Critical Thinking Exercises

Apply knowledge that you have gained in this chapter and one or more of the FLaReS principles (Chapter 1) to evaluate the following statements or claims.

14.1 In 1989 chemists Stanley Pons and Martin Fleischman announced that they had achieved nuclear fusion in an ordinary laboratory at ordinary temperatures, a process known as "cold fusion." They had electrolyzed heavy water using palladium electrodes and claimed that much more energy was produced at the cathode than was used for the electrolysis. Many scientists have tried to repeat this work. Some seem to have had some success, but most have not. The system is unpredictable, and excess heat is not given off every time. In addition, Pons and Fleischman claim to have observed neutrons being emitted during the reaction. A few scientists have verified one or more of their claims, but most have failed in attempts to reproduce their experiments.

14.2 An inventor claims to have discovered a chemical that converts water to gasoline.

14.3 An inventor claims to have discovered a catalyst that speeds the conversion of linear alkanes, such as hexane, to branched-chain isomers.

14.4 While watching TV during the summer, you hear someone proclaim that a way to cool your kitchen is to keep the refrigerator door open.

14.5 A man claims that a battery pack of car batteries will supply your house with electricity because he has a battery charger that is 200% efficient. An inverter unit hooked to the batteries makes more power output than line power from the house power.

14.6 A website claims that magnets attached to the fuel line increases gas mileage for automobiles up to 300%.

14.7 A company claims that it can use a metal such as magnesium as a fuel for automobiles. A coil of the metal would be fed into a chamber where it would react with superheated steam to produce $H_2(g)$; the H_2 would then power a fuel cell that would run the car.

SUMMARY

Section 14.1—**Energy** is the ability to do work. The United States has about 5% of the world's population but uses about one-fourth of all energy generated. Most of the energy used on Earth originated from the sun. The SI unit of energy is the joule (J), and the unit of power is the watt (W), 1 joule per second. Energy can be classified as **potential energy** (energy of position) or **kinetic energy** (energy of motion). In the process called **photosynthesis**, solar energy is absorbed by plants and stored as glucose, and oxygen is generated.

Section 14.2—Rates of reactions are affected by temperature, by concentrations of the reactants, and by the presence of catalysts. A chemical reaction or a physical process may be either **exothermic** (giving off energy) or **endothermic** (requiring energy). When a given process is exothermic, the reverse process is endothermic to exactly the same extent. Reactions involving burning of fuels are all exothermic.

Sections 14.3–14.4—The **first law of thermodynamics (law of conservation of energy)** says that energy can be neither created nor destroyed. The **second law of thermodynamics** states that in a spontaneous process, energy is degraded from more useful forms to less useful forms. **Entropy** is the dispersal of energy among the possible states of a system. There is a tendency for a system to go toward greater entropy.

Section 14.5—Ancient societies used slave labor and animal power to do work. The primary fuel was wood. The first mechanical devices for doing work were the waterwheel and the windmill. Energy from water, steam, gas, the internal combustion engine, and other sources fueled the industrial revolution.

Sections 14.6–14.7—A **fuel** is that which will burn readily with release of energy. **Fossil fuels** are natural fuels derived from once-living plants and animals. They include coal, petroleum, and natural gas. Coal is a rock that is a fuel, and is also a source of chemicals. The main element in coal is carbon. Coal rank is determined by its carbon and energy content, with anthracite being highest in rank and peat lowest. Coal is the most plentiful fossil fuel but is inconvenient to use and produces more pollution than other fuels.

Sections 14.8–14.9—Natural gas is mainly methane and burns relatively cleanly. It is both an important fuel and a raw material for synthesis of plastics and other prod-ucts. Petroleum or crude oil is a thick liquid mixture of organic compounds, mainly hydrocarbons. Petroleum products can be burned fairly cleanly but inefficient combustion produces pollution. Petroleum is refined by separating it, by distillation, into different fractions. Gasoline is the fraction consisting of hydrocarbons from about C_5 to C_{12}. Branched-structure hydrocarbons such as isooctane burn more smoothly than straight-chain hydrocarbons. The octane rating of gasoline compares its anti-knock performance to pure isooctane. To raise the octane rating, oil refiners use iso-merization (to increase branching), catalytic reforming (to convert straight-chain hydrocarbons to aromatic rings), and alkylation (to convert gases to highly branched gasoline molecules). Octane boosters such as MTBE, methanol, ethanol, and *t*-butyl alcohol are added as well. Tetraethyllead was used to improve octane rating for more than 50 years but has been discontinued.

Distillation tower

Sections 14.10–14.12—Electricity is the most convenient form of energy. About half of the electricity in the United States is generated from coal-fired steam power plants. Coal can be converted to more-convenient gaseous or liquid fuels, but at an energy cost. Nuclear fission reactions can be controlled in a nuclear reactor, which uses a low concentration (3–5%) of uranium-235. The reaction is controlled with control rods of cadmium or boron steel. The energy released generates steam which turns a turbine to generate electricity. A nuclear power plant cannot explode like a nuclear bomb, it produces minimal air pollution, and its wastes are

collected rather than dispersed. But elaborate safety precautions are needed to prevent escape of radioactive material, the needed fuel is limited in availability, and ultimate disposal of radioactive waste is an ongoing problem. A breeder reactor converts other isotopes to useful nuclear fuel and can make more fuel than it uses. Nuclear fusion cannot yet be controlled, as there are great technical difficulties. Temperatures in the millions of degrees are needed to generate a plasma containing free electrons and nuclei, for controlled fusion. But fusion has very important potential advantages over most other energy sources.

Sections 14.13–14.15—Solar energy is rather diffuse and hard to use. However, simple solar collectors can produce hot water for heating. Solar cells or photovoltaic cells produce electricity directly from sunlight, using *n*-type and *p*-type semiconductor material. The cells are expensive, not very efficient, and require a storage system for night use. Plant biomass such as wood can be burned directly or can be converted to liquid or gaseous fuels, but biomass requires much land area and has a very low overall efficiency. Hydrogen is an energy-storage and transport method rather than an energy source. It burns cleanly but requires more energy to make than it produces. One advantage of hydrogen is that it can be used in fuel cells, which consume fuel and oxygen continuously to generate electricity efficiently.

Sections 14.16–14.17—Other renewable energy sources include wind, which can be used to power windmills, and hydroelectric power. Windmills are expensive and not terribly efficient, and hydroelectric power can be generated only in certain areas. Heat energy from the interior of Earth is a form of geothermal energy, which has been used in several countries. It can only be used in areas where steam or hot water is near the surface. We can also make use of the temperature difference between surface and deep waters, and tidal energy, and the energy of the waves. All energy sources have advantages, limitations, and consequences. We can improve the energy situation locally by using more efficient appliances and simply by using less energy when practical.

REVIEW QUESTIONS

1. What was the principal fuel used in the United States before 1800?

2. What is a fuel? List the three principal fossil fuels.

3. What kind of reaction powers the sun?

4. What was the principal fuel used in the United States from about 1850 to 1950? What has been our principal fuel since 1950?

5. What is the ultimate source of nearly all the energy on Earth?

6. List some advantages and disadvantages of hydrogen as a vehicle fuel.

PROBLEMS

Fuels and Combustion

7. Which of the following are fuels?
   a. sucrose ($C_{12}H_{22}O_{11}$)
   b. H_2
   c. acetylene (C_2H_2)
   d. H_2O_2

8. Which of the following are fuels?
   a. ethylene (C_2H_4)
   b. argon (Ar)
   c. C_4H_{10}
   d. iodine (I_2)

9. For each substance in Problem 7 that is a fuel, write a balanced equation for its complete combustion in oxygen gas (O_2).

10. For each substance in Problem 8 that is a fuel, write a balanced equation for its complete combustion in oxygen gas (O_2).

11. Give the equation for the complete combustion of coal, assuming that the coal is carbon.

12. Give the equation for the incomplete combustion of coal to form carbon monoxide, assuming that the coal is carbon.

13. Give the equation for the complete combustion of methane.

14. Give the equation for the incomplete combustion of methane to form carbon monoxide and water vapor.

15. *Water gas* is a mixture of H_2 and CO. Can water gas serve as a fuel? Explain.

16. Water gas can be made by reacting red hot charcoal with steam. Write the equation for the reaction.

Energy and Chemical Reactions

17. How does temperature affect the rate of a chemical reaction?

18. Why does wood burn more rapidly in pure oxygen than in air?

19. What is an exothermic reaction? Give an example.

20. What is an endothermic reaction? Give an example.

21. Burning 1.00 mol of methane releases 803 kJ of energy. How much energy is released by burning 5.25 mol of methane?

$$CH_4(g) + 2\,O_2(g) \longrightarrow CO_2(g) + 2\,H_2O(g) + 803\text{ kJ}$$

22. It takes 572 kJ of energy to decompose 2.00 mol of liquid water. How much energy does it take to decompose 55.5 mol of water?

$$2\,H_2O(l) + 572\text{ kJ} \longrightarrow 2\,H_2(g) + O_2(g)$$

23. When burned, 1.00 g of gasoline gives off 1030 cal. What is this quantity in joules?

24. How much heat, in kilojoules, is given off when 4.301 grams of $H_2(g)$ reacts with $O_2(g)$ to form steam [$H_2O(g)$] according to the following equation?

$$2\,H_2(g) + O_2(g) \longrightarrow 2\,H_2O(g) + 483.6\text{ kJ}$$

25. Review Problem 21, and determine how much energy in kJ must be supplied to convert one mole of gaseous CO_2 and two moles of water vapor into methane gas and oxygen gas.

26. Review Problem 24, and determine how much energy in kJ must be supplied to convert one mole of water vapor into hydrogen gas and oxygen gas.

Energy and the Laws of Thermodynamics

27. State the first law of thermodynamics.

28. State the second law of thermodynamics in terms of energy flow and in terms of degradation of energy.

29. What is entropy? Is entropy increased or decreased when a fossil fuel is burned?

30. Energy is conserved. How can we ever run out of energy?

Fossil Fuels

31. Which of the fossil fuels is most plentiful in the United States?

32. What are the advantages and disadvantages of coal as a fuel?

33. What is the physical state of each of the three fossil fuels? List some advantages of gaseous and liquid fuels over solid fuels.

34. Why did the United States shift from coal to petroleum and natural gas when we have much larger reserves of coal than of the two hydrocarbon fuels?

35. The estimated proven world oil reserves in 2005 were 1189 billion barrels. The world rate of annual use was 30.2 billion barrels. How long will these reserves last if this rate of use continues? (The actual rate of use is increasing more than 2% a year.)

36. The estimated proven U.S. oil reserves in 2005 were 21.9 billion barrels. **(a)** How long will these reserves last if there are no imports or exports and if the U.S. annual rate of use of 7.4 billion barrels continues? **(b)** Taking the most optimistic projection for oil reserves in the Arctic National Wildlife Refuge—10.4 billion barrels—how long would exploiting that resource extend our reserves at the current U.S. annual rate of use?

37. The estimated world natural gas reserves in 2005 were 6340 trillion ft^3. The world rate of annual use was 98.5 trillion ft^3. How long will these reserves last if this rate of use continues?

38. The U.S. proven natural gas reserves in 2005 were 189 trillion ft^3. How long will these reserves last if there are no imports or exports and if the U.S. annual rate of use of 23 trillion ft^3 continues?

Natural Gas

39. What is the main component of natural gas?

40. What are the advantages and disadvantages of natural gas as a fuel?

Petroleum

41. What is thought to be the origin of petroleum?

42. What are the advantages and disadvantages of petroleum as a source of fuels?

43. How is crude petroleum modified to better meet our needs and wants? What are the advantages and disadvantages of tetraethyllead as an octane booster?

44. What is meant by the octane rating of a gasoline?

Nuclear Power

45. What proportion of U.S. electricity is generated by nuclear power plants?

46. What proportion of electricity in France is generated by nuclear power plants?

47. Can a nuclear power plant explode like a nuclear bomb? Explain your answer.

48. Can nuclear bombs be made from reactor-grade uranium? Explain your answer.

49. How does a breeder reactor produce more fuel than it consumes?

50. List some advantages of nuclear power plants over coal-fired plants.

51. Can a nuclear bomb be made from reactor-grade plutonium?

52. What are some of the disadvantages of breeder reactors?

53. List some possible advantages of a nuclear fusion reactor over a fission reactor. What is plasma?

54. List some possible problems with nuclear fusion reactors.

55. Give nuclear equations showing how thorium-232 is converted to fissile uranium-233.

56. Give nuclear equations showing how uranium-238 is converted to fissile plutonium-239.

57. Give the nuclear equation that shows how deuterium and tritium fuse to form helium and a neutron.

58. Give the nuclear equation that shows how four protons fuse to form helium and two positrons.

Renewable Energy Sources

59. What is a photovoltaic cell?

60. What are some problems associated with the use of solar energy?

61. Review the material on solar cells in Section 14.13, and consider a house that requires about 4 kW of power for normal operation. What area in square meters and in square feet would a solar battery occupy, to provide this power?

62. Review the material on solar cells in Section 14.13, and consider a small factory that requires about 550 kW of power for normal operation. Will an area of solar cells that is 50 m (164 ft) square be sufficient to provide this power?

63. What is plant biomass?

64. List some advantages and disadvantages of the use of biomass as a source of energy.

65. List two ways that fuel cells differ from electrochemical cells.

66. List the advantages, disadvantages, and limitations of each of the following as an energy source.
    a. wind power
    b. geothermal power
    c. power from tides
    d. hydroelectric power

Synthetic and Converted Fuels

67. Give the chemical equation for the basic process by which coal is converted to methane.

68. Give the chemical equation for the conversion of coal (carbon) to carbon monoxide and hydrogen.

69. Give the chemical equation for the conversion of carbon monoxide and hydrogen to methanol.

70. Give the chemical equation for the reaction that occurs in a hydrogen–oxygen fuel cell.

ADDITIONAL PROBLEMS

71. A cold pack (photo) works by the dissolving of ammonium nitrate in water. Cold packs are carried by athletic trainers when transporting ice is not possible. Is this process endothermic or exothermic? Explain.

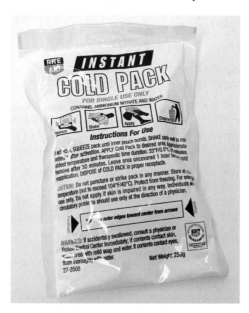

72. A hot pack hand warmer (photo) contains iron powder, water, salt, activated carbon, and vermiculite. It is activated by exposing the contents to air. Hot packs are used by hunters and other outdoor sportsmen and workers where access to heaters or fires is not possible. Is this process endothermic or exothermic? Explain.

73. A woman uses 1125 kcal in running 10.0 km in 39 min 18 s. Calculate her average power output in watts. (1.000 kcal = 4184 J.)

74. The energy requirements of the human brain are about 20% of the total body metabolism. Calculate the power output (in watts) of your brain if you use 2175 kcal of energy per day.

75. What mass of carbon dioxide is formed by the combustion of 1250 kg of coal that is 51.2% carbon?

76. Heats of combustion of several gaseous fuels, in kilojoules per mole, are given below. Which yields the most energy **(a)** per kilogram and **(b)** per liter when the volume is measured under the same conditions of temperature and pressure?

    Hydrogen (H_2), 286.6 kJ

    Isobutane [$CH_3CH(CH_3)_2$], 2868 kJ

    Neopentane [$CH_3C(CH_3)_3$], 3515 kJ

77. Indicate a natural process or processes by which carbon atoms are **(a)** removed from the atmosphere; **(b)** returned to the atmosphere; **(c)** effectively withdrawn from the carbon cycle.

78. The United States leads the developed world in per capita emissions of $CO_2(g)$ with 19.8 t per person per year (1 t = 1000 kg). What mass, in metric tons, of each of the following fuels would yield this quantity of CO_2?
    a. CH_4
    b. C_8H_{18}
    c. coal that is 94.1% C by mass

79. A large coal-fired electric plant burns 2500 tons of coal per day. The coal contains 0.65% S by mass. Assume that all the sulfur is converted to SO_2. What mass of SO_2 is formed? If a thermal inversion traps all this SO_2 in a parcel of air that is 45 km $\times$ 60 km $\times$ 0.40 km, will the level of SO_2 in the air exceed the primary national air quality standard of 365 μg SO_2/m^3 air?

80. The contribution of the combustion of various fuels to the buildup of CO_2 in the atmosphere can be assessed in different ways. One way relates the mass of CO_2 formed to the mass of fuel burned; another relates the mass of CO_2 to the quantity of heat evolved in the combustion. Which of the three fuels C(graphite), $CH_4(g)$, or $C_4H_{10}(g)$ produces the smallest mass of CO_2 **(a)** per gram of fuel and **(b)** per kilojoule of heat evolved? The heat released *per mole* of the three substances is C(graphite), 393.5 kJ; $CH_4(g)$, 803 kJ; and $C_4H_{10}(g)$, 2877 kJ.

COLLABORATIVE GROUP PROJECTS

Prepare a PowerPoint, poster, or other presentation (as directed by your instructor) for presentation to the class.

81. Which would you rather have in your neighborhood, a nuclear power plant or a coal-burning plant? Why? Explain your choice fully.

82. What characteristics should an ideal energy source have? Which of these characteristics do the different forms of energy have?

83. Social cost pricing takes into account all the costs of a product or activity that are external to market costs. For example, one of the factors in the cost of gasoline-based transportation is air pollution. Select three forms of energy and list the possible social costs that should be included.

84. Compare the quantity of electricity used for two appliances that are alike except for different efficiencies. Extend this comparison to CO_2 emissions from the appliances, assuming the electricity is generated by burning coal.

85. Compare qualitatively the efficiency of heating with natural gas versus electricity produced from burning natural gas at a power plant.

86. Compare qualitatively the efficiency of using solar panels on the roof versus photovoltaic cells in a "solar farm."

87. Which of the following is the best fuel for heating your home? What problems with supply, use, and waste products are involved in each case?
  a. natural gas   **b.** electricity
  c. coal   **d.** fuel oil

88. There was an accident at a uranium processing plant in Tokai, Japan, on September 30, 1999. Using your favorite search engine, find out what happened and compare this accident with the ones at Chernobyl and Three Mile Island. (*Hint:* Search on the name, followed by "nuclear accident.") How could these accidents have been avoided? Do these incidents prove that nuclear power plants should be phased out? Why or why not?

89. The use of appliances and heating equipment that employ alternative forms of energy (such as solar power) can help in individual energy conservation. Using internet sources, find such equipment. Do companies selling these products appear to be doing so for mostly idealistic—or mainly commercial—reasons? (A good place to begin is SolarAccess.com, which features solar industry directories.)

REFERENCES AND READINGS

1. Asimov, Isaac. "In Dancing Flames a Greek Saw the Basis of the Universe." *Smithsonian*, November 1971, pp. 52–57. Discusses heat, from the Greek "element" fire to the first law of thermodynamics. Vintage Asimov.

2. Bent, Henry A. "Haste Makes Waste: Pollution and Entropy." *Chemistry*, October 1971, pp. 6–15. A classic.

3. Dunn, Seth. *Micropower: The Next Electrical Era*. Washington, DC: Worldwatch Institute, July 2000. Worldwatch Paper 151.

4. Henry, Celia M. "A Fine Look at Crude Oil." *Chemical & Engineering News*, March 31, 2003, p. 39.

5. Hill, John W., Ralph H. Petrucci, Terry W. McCreary, and Scott S. Perry. *General Chemistry*, 4th edition. Upper Saddle River, NJ: Prentice Hall, 2005. Chapter 6, "Thermochemistry," and Chapter 17, "Thermodynamics."

6. Hoffert, M. I., et al. "Advanced Technology Paths to Global Climate Stability: Energy for a Greenhouse Planet." *Science*, November 1, 2002, pp. 981–987.

7. Jensen, William B. "Entropy and Constraint of Motion." *Journal of Chemical Education*, May 2004, pp. 639–640.

8. Johnson, Jeff. "Power from Moving Water." *Chemical & Engineering News*, October 4, 2004, pp. 23–30.

9. Kolb, Doris, and Kenneth E. Kolb. "Chemical Principles Revisited: Petroleum Chemistry." *Journal of Chemical Education*, July 1979, pp. 465–469.

10. Lambert, F. L. "Entropy Is Simple, Qualitatively." *Journal of Chemical Education*, October 2002, vol. 79, pp. 1241–1246.

11. Moreira, Naila. "Growing Expectations: New Technology Could Turn Fuel Into a Bumper Crop." *Science News*, October 1, 2005, pp. 218–220.

12. Ritter, Steve. "What's That Stuff: Gasoline." *Chemical & Engineering News*, February 21, 2005, p. 37.

13. Schor, Juliet B., and B. Taylor. *Sustainable Planet: Solutions for the 21st Century*. Boston: Beacon Press, 2002.

14. Scigliano, Eric. "Wave Energy." *Discover*, December 2005, pp. 42–45.

15. Walker, J. Samuel. *Three Mile Island: A Nuclear Crisis in Historical Perspective*. Berkeley: University of California Press, 2004.

GREEN CHEMISTRY

Energy: A Fuels Paradise; Alternative Energy

Scott Reed, Portland State University

Currently, oil, coal, and natural gas provide the world with most of its energy needs. These sources of fuel are not renewable on the timescale of a human life span, and there are negative environmental consequences associated with their combustion. Chemistry plays a role in the design of new energy sources and technologies. For example, chemists have improved the efficiency of fuel cells and batteries. Chemistry also helps create new ways of harnessing renewable sources such as sunlight. In the following investigation you will learn more about the energy sources currently available and the exploration of new greener energy options.

WEB INVESTIGATIONS

Investigation 1
Renewable Resources
In the United States, less than 10% of energy used comes from renewable sources. Renewable energy, unlike fossil fuel, is derived from sources that are not depleted over time. Using your favorite search engine, discover what renewable energies are currently used and to learn about new ones that scientists are developing. Consider which of these are suitable for heating homes or powering cars.

Investigation 2
Fuel Cells, Biomass, and Solar Energy
The development of alternate sources of energy has been a hot topic of political debate and is garnering more public interest as hybrid, electric, and biodiesel-fueled cars are becoming more and more common. Chemists are currently working on fuel cells that burn hydrogen gas (H_2) instead of oil. When H_2 is burned (a reaction with O_2), water is the only by-product. Plants, as well as plant products like soybean oil and canola oil, are other sources of renewable energy. Plants utilize sunlight through photosynthesis and can be harvested to provide energy-rich biomass. Solar energy is an abundant, clean, and renewable energy source. Chemistry has played a role in the design of solar cells that collect sunlight and convert it to electricity. Visit the United States Department of Energy website to find out more about the most recent discoveries for the use of each of these renewable resources by searching the site for "fuel cell," "biomass" and "solar."

▲ Fuel cells are highly efficient, and emit much less pollution than do ordinary internal-combustion engines.

420

Exercise 1
Alternative Energy Sources

In the preceding investigations, you learned about many alternative energy sources. Pick one to three alternative energy sources and design a pamphlet that describes the pros and cons of each.

Exercise 2
You're the Boss

Imagine you are the owner of a small taxicab company with a fleet of ten cars. Gasoline prices have risen to an all-time high and you are concerned about the consequences of consuming fossil fuels. It is time to invest in a new technology, but how do you decide which technology to go with? Analyze the options available to you: what would be the initial costs; what would be the long-term impact on the environment; what would be the impact on your bottom line?

Exercise 3
Energy Map

What areas of the United States have alternative energy potential and what resources are currently being used? Pick a region of the United States and write a report summarizing the energy use and the renewable resources that are available.

Exercise 4
Think Globally, Act Locally

Taking into account the resources available in your region, select an energy source that you think is a good choice for reducing use of fossil fuels. Describe a specific plan for reducing dependence on nonrenewable energy sources. Write your argument in the form of a letter to a local power company. Weigh both the environmental and economic impacts of your plan.

CHAPTER 15

Biochemistry

Earth has millions of forms of life. The "blueprint" for each is carried in a distinctive DNA molecule. Zebrafish (*Danio rerio*), common both in labs
and as pets, are normally black and gray striped. Scientists at the National University of Singapore genetically engineered the fish,
inserting a gene for a red fluorescent protein from sea anemones and coral. The original intent was to detect water pollution.
Fluorescent fish, called GloFish®, are now available as pets.

A Molecular View of Life

The human body is an incredible chemical factory, far more complex than any industrial plant. To stay in good condition and perform its varied tasks, it needs many specific chemical compounds, and it manufactures most of them in an exquisitely organized network of chemical production lines.

Every minute of every day, thousands of chemical reactions take place within each of the 100 trillion tiny cells in your body. The study of these reactions and the chemicals they produce is called *biochemistry*.

15.1 THE CELL

Biochemistry is the chemistry of living things and life processes. The structural unit of all living things is the *cell*. Every cell is enclosed in a *cell membrane* through which it gains nutrients and gets rid of wastes. Plant cells (Figure 15.1) also have

Membrane Structure and Function

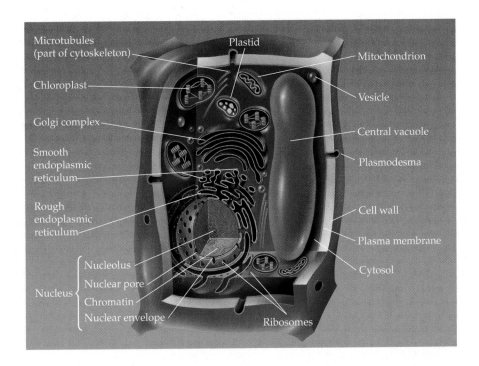

◄ **Figure 15.1** An idealized plant cell. Not all the structures shown here occur in every type of plant cell.

423

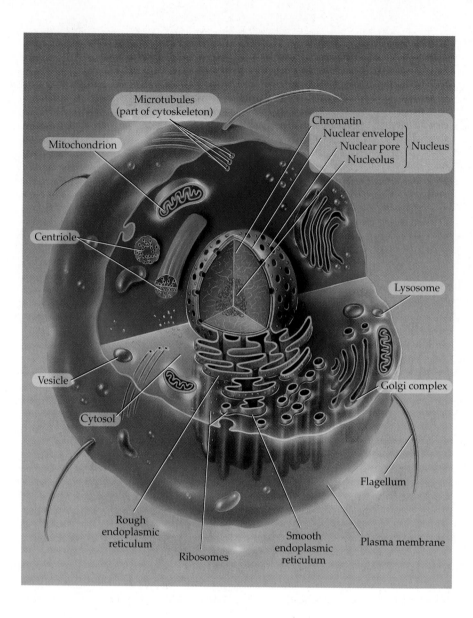

▶ **Figure 15.2** An animal cell. The entire range of structures shown here seldom occurs in a single cell, and we discuss only a few of them in this text. Each kind of plant or animal tissue has cells specific to the function of that tissue. Muscle cells differ from nerve cells, nerve cells differ from red blood cells, and so on.

walls made of cellulose. Animal cells (Figure 15.2) do not have cell walls. Cells have a variety of interior structures that serve a multiplicity of functions. We can consider only a few of them here.

The largest interior structure is usually the *cell nucleus*, which contains the material that controls heredity. Protein synthesis takes place in the *ribosomes*. The *mitochondria* are the cell "batteries" where energy is produced. Plant cells (but not animal cells) also contain *chloroplasts* in which energy from the sun is converted to chemical energy, which is stored in the plant in the form of carbohydrates.

15.2 ENERGY IN BIOLOGICAL SYSTEMS

Life requires energy. Living cells are inherently unstable and only a continued input of energy keeps them from falling apart. Living organisms are restricted to using certain forms of energy. Supplying a plant with heat energy by holding it in a flame will do little to prolong its life. On the other hand, a green plant is uniquely able to tap sunlight, the richest source of energy on Earth. Chloroplasts in green plant cells capture the radiant energy of the sun and convert it to chemical energy, which is then stored in carbohydrate molecules. The photosynthesis of glucose is represented by the equation

$$6\,CO_2 + 6\,H_2O \longrightarrow C_6H_{12}O_6 + 6\,O_2$$

Plant cells can also convert these carbohydrate molecules to fat molecules and, with the proper inorganic nutrients, to protein molecules.

Animals cannot directly use the energy of sunlight. They must get their energy by eating plants or by eating other animals that eat plants. Animals obtain energy from three major types of foods: carbohydrates, fats, and proteins.

Once digested and transported to a cell, a food molecule can be used as a building block to make new cell parts or to repair old ones, or it can be "burned" for energy. The entire series of coordinated chemical reactions that keep cells alive is called **metabolism**. In general, metabolic reactions are divided into two classes. The degrading of molecules to provide energy is called **catabolism**, and the process of building up or synthesizing the molecules of living systems is termed **anabolism**.

Carbohydrates, fats, and proteins as foods are discussed in Chapter 16. In this chapter, we deal mainly with the synthesis and structure of these vital materials.

15.3 CARBOHYDRATES: A STOREHOUSE OF ENERGY

It is difficult to give a simple formal definition of **carbohydrates**. Chemically, they are polyhydroxy aldehydes or ketones or compounds that can be hydrolyzed (split by water) to form such aldehydes and ketones. Composed of the elements carbon, hydrogen, and oxygen, carbohydrates include sugars, starches, and cellulose. Usually, the atoms of these elements are present in a ratio expressed by the formula $C_x(H_2O)_y$. Glucose, a simple sugar, has the formula $C_6H_{12}O_6$, which could be written as $C_6(H_2O)_6$. The term *carbohydrate* is derived from formulas such as these. We use the name even though carbohydrates are not actually hydrates of carbon.

Glucose

Some Simple Sugars

Sugars are sweet-tasting carbohydrates. The simplest of these are **monosaccharides**, carbohydrates that cannot be further hydrolyzed. Three familiar monosaccharides are shown in Figure 15.3. They are glucose (also called dextrose), fructose (fruit sugar), and galactose (a component of lactose, the sugar in milk). Glucose and galactose are **aldoses**, monosaccharides with an aldehyde functional group. Fructose is a **ketose**, a monosaccharide with a ketone functional group.

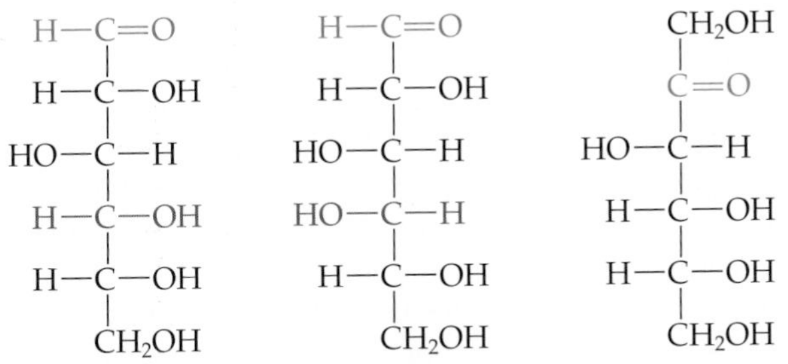

◄ Figure 15.3 Three common monosaccharides. All have hydroxyl groups. Glucose and galactose have aldehyde functions (red), and fructose has a ketone group (blue). Glucose and galactose differ only in the arrangement of the H and OH on the fourth carbon (green) from the top. Glucose is also called dextrose, fructose is fruit sugar, and galactose is a component of milk sugar.

QUESTION: What is the molecular formula of each of the three monosaccharides? How are the three compounds related?

CONCEPTUAL EXAMPLE 15.1   **Classification of Monosaccharides**

Shown below are structures of (left to right) erythrulose, gulose, mannose, and ribose.

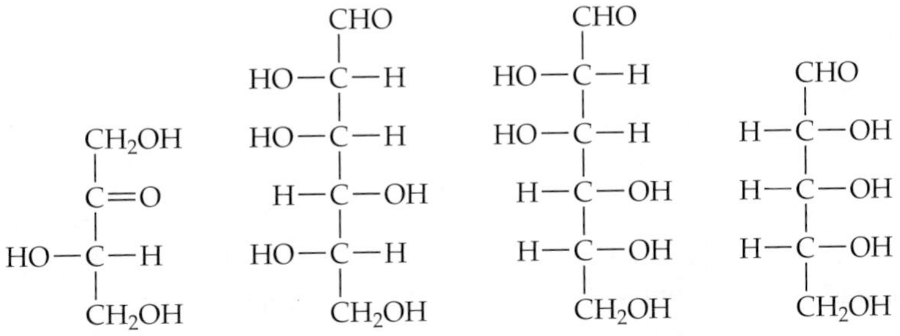

Classify erythrulose and gulose as an aldose or a ketose.

Solution

Erythrulose is a ketose. The carbonyl (C=O) group is on the second carbon atom from the top; it is between two other carbon atoms, a situation that defines a ketone. Gulose is an aldose. The carbonyl group is on an "end" carbon atom, as it is in all aldehydes.

▶ **Exercise 15.1A**

Classify mannose and ribose as an aldose or a ketose.

▶ **Exercise 15.1B**

How does the structure of gulose differ from that of glucose? How does the structure of mannose differ from that of glucose?

▶ **Figure 15.4** Cyclic structures for glucose, galactose, and fructose. A corner with no letter represents a carbon atom. Glucose and galactose are represented as six-membered rings. They differ only in the arrangement of the H and OH (green) on the fourth carbon. Fructose is shown as a five-membered ring. Some sugars exist in more than one cyclic form. For simplicity, we show only one form of each here.

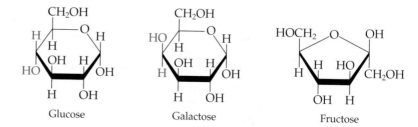

Glucose   Galactose   Fructose

These monosaccharides are represented in Figure 15.3 as open-chain compounds to show the aldehyde or ketone functional groups. However, these sugars exist mainly as cyclic molecules (Figure 15.4).

Sucrose and lactose are examples of **disaccharides**, carbohydrates consisting of molecules that can be hydrolyzed to two monosaccharide units (Figure 15.5). Sucrose is split into glucose and fructose. Hydrolysis of lactose gives glucose and galactose.

Although we have represented these cyclic monosaccharides as flat hexagons, they are actually three-dimensional. Most assume a conformation (shape) called a *chair conformation* because it outlines a structure that somewhat resembles a reclining chair.

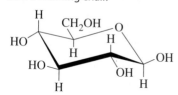

$$\text{Sucrose} + \text{H}_2\text{O} \longrightarrow \text{Glucose} + \text{Fructose}$$
$$\text{Lactose} + \text{H}_2\text{O} \longrightarrow \text{Glucose} + \text{Galactose}$$

Polysaccharides: Starch and Cellulose

Polysaccharides are composed of large molecules that yield many monosaccharide units on hydrolysis. Polysaccharides include starches, which comprise the main energy storage system of many plants, and cellulose, which is the structural material of plants. Figure 15.6 shows short segments of starch and cellulose molecules. Notice that both are polymers of glucose. Starch molecules generally have from 100 to about 6000 glucose units. Cellulose molecules are composed of 1800–3000 or more glucose units.

A crucial structural difference between starch and cellulose is in the way the glucose units are hooked together. Consider starch first. With the CH$_2$OH at the top as a reference, the oxygen atom joining the glucose units is pointed *down*. This arrangement is called an *alpha linkage*. In cellulose, again with the CH$_2$OH as a point of reference, the oxygen atom connecting the glucose segments is pointed *up*, an arrangement called a *beta linkage*. This subtle but important difference in link-

Sucrose

▶ **Figure 15.5** Sucrose and lactose are disaccharides. On hydrolysis, sucrose produces glucose and fructose, whereas lactose yields glucose and galactose. Sucrose is cane or beet sugar and lactose is milk sugar.

Question: What is the molecular formula of each of these disaccharides? How are the two compounds related? Label the two monosaccharide units in each as fructose, galactose, or glucose.

Sucrose   Lactose

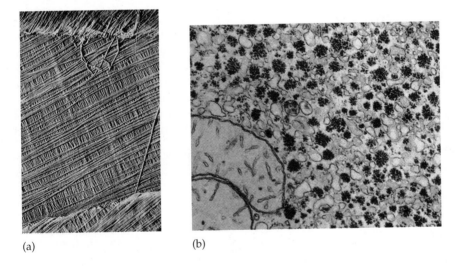

(a)

α-linkage

α-linkage

(b)

β-linkage

β-linkage

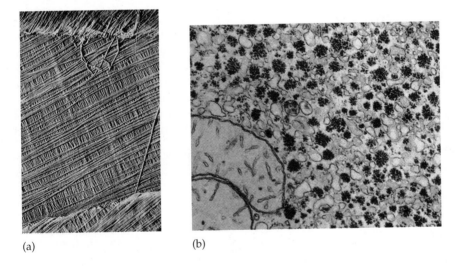

(a)

(b)

α-linkage

β-linkage

◄ **Figure 15.6** Both starch and cellulose are polymers of glucose. They differ in that the glucose units are joined by alpha linkages (blue) in starch and by beta linkages (red) in cellulose.

Question: The formulas for most polymers can be written in condensed form (page 274). Write a condensed formula for starch. How would the condensed formula for cellulose differ from that of starch?

ages determines whether the materials can be digested (Chapter 16). The different linkages also result in different three-dimensional forms for cellulose and starch. For example, cellulose in the cell walls of plants is arranged in *fibrils*, bundles of parallel chains. As shown in Figure 15.7(a), fibrils in turn lie parallel to each other in each layer of the cell wall. In alternate layers, the fibrils are perpendicular, an arrangement that imparts great strength to the wall.

There are two kinds of plant starch. One, called *amylose*, has glucose units joined in a continuous chain like beads on a string. The other kind, *amylopectin*, has branched chains of glucose units. These starches are perhaps best represented schematically, as in Figure 15.8, where each glucose unit is represented by the abbreviation Glc.

(a)

(b)

◄ **Figure 15.7** Cellulose molecules form fibers whereas starch (glycogen) forms granules. Electron micrographs of (a) the cell wall of an alga, made up of successive layers of cellulose fibers in parallel arrangement, and (b) glycogen granules in a liver cell of a rat.

. . . Glc-Glc-Glc-Glc-Glc-Glc-Glc-Glc-Glc-Glc-Glc-Glc-Glc-Glc . . .

Amylose

. . . Glc-Glc

. . . . Glc-Glc-Glc-Glc

. . . Glc-Glc-Glc-Glc-Glc-Glc-Glc-Glc-Glc

. . . Glc-Glc-Glc-Glc-Glc-Glc-Glc-Glc-Glc-Glc-Glc-Glc

. . . . Glc-Glc-Glc-Glc-Glc-Glc-Glc-Glc

Amylopectin

◄ **Figure 15.8** Schematic representations of amylose and amylopectin. Glc stands for a glucose unit. The structure of glycogen is similar to that of amylopectin.

When cornstarch is heated at about 400 °F for an hour, the starch polymers are broken down into shorter chains, but not all the way to glucose monomers. The product, called *dextrin*, has properties intermediate between sugars and starches. Mixed with water, dextrin makes a sticky and slightly thick mixture. Millions of pounds of dextrin are used each year as the base for adhesives including envelope "gum" and "lick and stick" labels.

Animal starch is called *glycogen*. Like amylopectin, it is composed of branched chains of glucose units. In contrast to cellulose, we see in Figure 15.7(b) that glycogen in muscle and liver tissue is arranged in granules, clusters of small particles. Plant starch, on the other hand, forms large granules. Granules of plant starch rupture in boiling water to form a paste, and on cooling, the paste gels. Potatoes and cereal grains form this type of starchy broth. All forms of starch are hydrolyzed to glucose during digestion.

15.4 FATS AND OTHER LIPIDS

Fats are the predominant forms of a class of compounds called *lipids*. These substances are not defined by functional groups, as are most other families of organic compounds. Rather, lipids have common solubility properties. A **lipid** is a cellular constituent that is soluble in organic solvents of low polarity such as hexane, diethyl ether, or carbon tetrachloride. Lipids are insoluble in water. In addition to fats, the lipid family includes **fatty acids** (long-chain carboxylic acids), steroids such as cholesterol and sex hormones (Chapter 18), fat-soluble vitamins (Chapter 16), and other substances. Figure 15.9 shows three representations of palmitic acid, a typical fatty acid.

$CH_3CH_2CH_2CH_2CH_2CH_2CH_2CH_2CH_2CH_2CH_2CH_2CH_2CH_2CH_2COOH$
(a)

(b)

(c)

▶ **Figure 15.9** Three representations of palmitic acid: (a) Condensed structural formula. (b) Line-angle formula, in which the lines denote bonds and each intersection and end of a line represents a carbon atom. (c) Wireframe model, in which the wires represent bonds, and the atoms are symbolized as points where the bonds meet.

A **fat** is an ester of fatty acids and the trihydroxy alcohol glycerol (Figure 15.10). A fat has three fatty acid chains joined to glycerol through ester linkages. Fats are often called *triglycerides* or triacylglycerols. Related compounds are also classified according to the number of fatty acid chains they contain: A *monoglyceride* has one fatty acid chain joined to glycerol, and a *diglyceride* has two.

Naturally occurring fatty acids nearly always have an even number of carbon atoms. Representative ones are listed in Table 15.1. Animal fats are generally rich in

Number of Carbon Atoms	Common Condensed Structure	Name	Source
4	$CH_3CH_2CH_2COOH$	Butyric acid	Butter
6	$CH_3(CH_2)_4COOH$	Caproic acid	Butter
8	$CH_3(CH_2)_6COOH$	Caprylic acid	Coconut oil
10	$CH_3(CH_2)_8COOH$	Capric acid	Coconut oil
12	$CH_3(CH_2)_{10}COOH$	Lauric acid	Palm kernel oil
14	$CH_3(CH_2)_{12}COOH$	Myristic acid	Oil of nutmeg
16	$CH_3(CH_2)_{14}COOH$	Palmitic acid	Palm oil
18	$CH_3(CH_2)_{16}COOH$	Stearic acid	Beef tallow
18	$CH_3(CH_2)_7CH=CH(CH_2)_7COOH$	Oleic acid	Olive oil
18	$CH_3(CH_2)_4CH=CHCH_2CH=CH(CH_2)_7COOH$	Linoleic acid	Soybean oil
18	$CH_3CH_2(CH=CHCH_2)_3(CH_2)_6COOH$	Linolenic acid	Fish oils
20	$CH_3(CH_2)_4(CH=CHCH_2)_4CH_2CH_2COOH$	Arachidonic acid	Liver

TABLE 15.1 Some Fatty Acids in Natural Fats

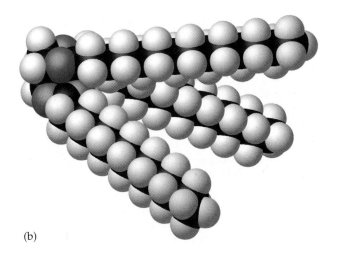

(a)

Glycerol   Fatty acids   A triglyceride

(b)

◀ **Figure 15.10** Triglycerides (triacylglycerols) are esters in which the trihydroxy (three OH groups) alcohol glycerol is esterified with three fatty acid groups. (a) The equation for the formation of a triglyceride. (b) Space-filling model of a triglyceride.

saturated fatty acids and have a smaller proportion of unsaturated fatty acids. At room temperature, most animal fats are solids. Liquid fats, called *oils*, are obtained principally from vegetable sources. Oils typically have a higher proportion of unsaturated fatty acid units than do fats.

Fats are often classified according to the degree of unsaturation of the fatty acids they incorporate. A *saturated fatty acid* contains no carbon-to-carbon double bonds, a *monounsaturated fatty acid* has one carbon-to-carbon double bond per molecule, and a *polyunsaturated fatty acid* molecule has two or more carbon-to-carbon double bonds. A *saturated fat* contains a high proportion of saturated fatty acids; these fat molecules have relatively few carbon-to-carbon double bonds. A *polyunsaturated fat* (oil) incorporates mainly unsaturated fatty acids; these fat molecules have many double bonds.

The iodine number usually measures the degree of unsaturation of a fat or oil. The **iodine number** is the number of grams of iodine that are consumed by 100 g of fat or oil. Iodine, like other halogens, undergoes an addition reaction with carbon-to-carbon double bonds.

$$\begin{array}{c} \diagdown \\ C \end{array}{=}\begin{array}{c} \diagup \\ C \\ \diagdown \end{array} + \ I_2 \longrightarrow \begin{array}{c} | \quad | \\ -C-C- \\ | \quad | \\ I \quad I \end{array}$$

The more double bonds a fat contains, the more iodine is required for the addition reaction; thus, a high iodine number means a high degree of unsaturation. Representative iodine numbers are listed in Table 15.2. Note the generally lower values for animal fats (butter, tallow, and lard) compared with those for vegetable oils. Coconut oil, which is highly saturated, and fish oils, which are relatively unsaturated, are notable exceptions to the general rule.

Fats and oils feel greasy. They are less dense than water and float on it.

▲ Many salad dressings are made of an oil and vinegar. The one shown here has olive oil floating on top of balsamic vinegar, an aqueous solution of acetic acid.

TABLE 15.2 Typical Iodine Numbers for Some Fats and Oils*

Fat or Oil	Iodine Number	Fat or Oil	Iodine Number
Coconut oil	8–10	Cottonseed oil	100–117
Butter	25–40	Corn oil	115–130
Beef tallow	30–45	Fish oils	120–180
Palm oil	37–54	Canola oil	125–135
Lard	45–70	Soybean oil	125–140
Olive oil	75–95	Safflower oil	130–140
Peanut oil	85–100	Sunflower oil	130–145

* Oils shown in blue are from plant sources. Three fats and one oil come from animals.

Proteins: The Stuff of Life

Proteins are vital components of all life. No living part of the human body—or of any other organism, for that matter—is completely without protein. There is protein in blood, muscles, brain, and even tooth enamel. The smallest cellular organisms—bacteria—contain protein. Viruses, so small that they make bacteria look like giants, are little more than proteins and nucleic acids. This combination of nucleic acids and proteins is found in all cells and is the stuff of life itself.

Each type of cell makes its own kinds of proteins. Proteins serve as the structural material of animals, much as cellulose does for plants. Muscle tissue is largely protein; so are skin and hair. Silk, wool, nails, claws, feathers, horns, and hooves are proteins. All proteins contain the elements carbon, hydrogen, oxygen, and nitrogen, and most also contain sulfur. The structure of a short segment of a typical protein molecule is shown in Figure 15.11.

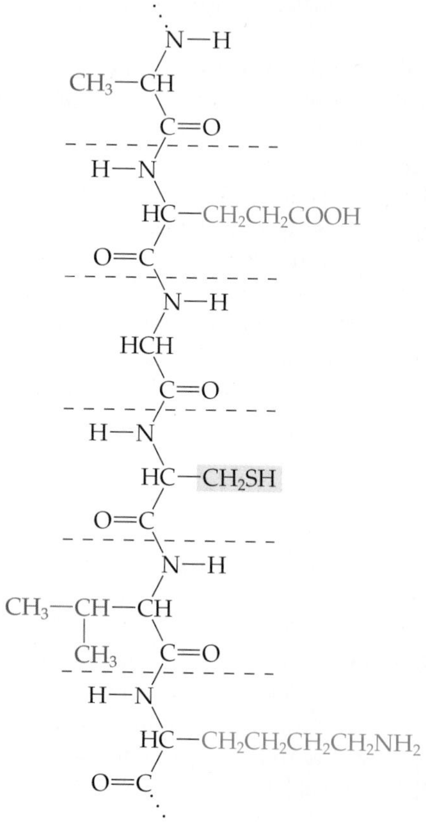

▶ **Figure 15.11** Structural formula (a) and spacefilling model (b) of a short segment of a protein molecule. In the structural formula, hydrocarbon side chains (green), an acidic side chain (red), a basic side chain (blue), and a sulfur-containing side chain (amber) are highlighted. The broken lines in (a) indicate where two amino acid units join (see Section 15.6).

Question: The formulas for most polymers can be written in condensed form (page 274). Those for natural proteins cannot. Explain.

(a)   (b)

15.5 PROTEINS: POLYMERS OF AMINO ACIDS

Like starch and cellulose, *proteins* are polymers. They differ from other polymers that we have studied in that the monomer units are about 20 different amino acids (Table 15.3). The amino acids differ in their side chains (blue). An **amino acid** has

TABLE 15.3   The 20 Amino Acids Specified by the Genetic Code

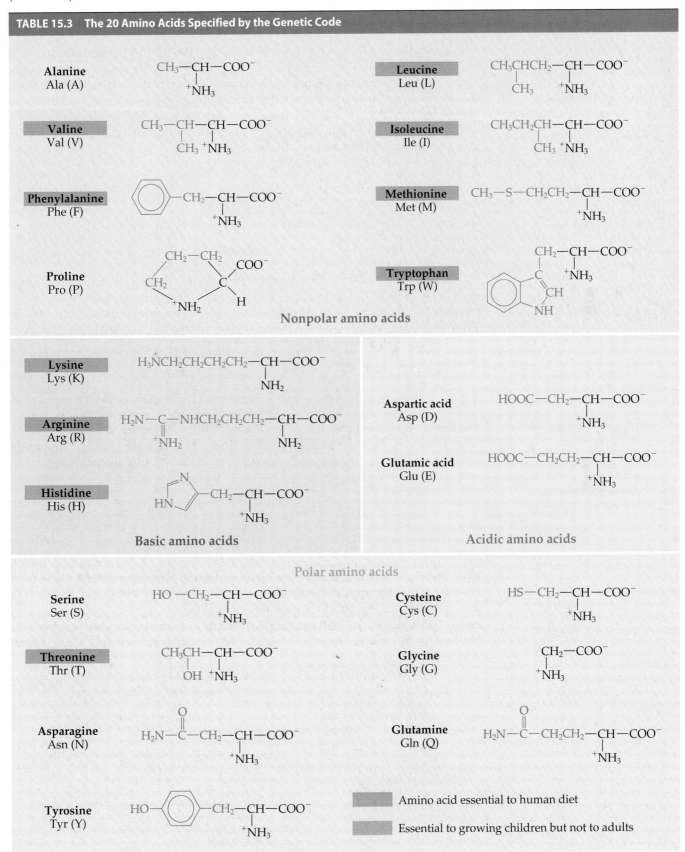

two functional groups: an amino group (—NH$_2$) and a carboxyl group (—COOH) attached to the same carbon atom (called the *alpha carbon*).

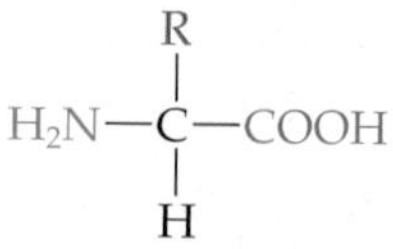

$$H_2N-\underset{\underset{H}{|}}{\overset{\overset{R}{|}}{C}}-COOH$$

In this formula, the R can be H (as it is in glycine, the simplest amino acid), or any of the groups shown in color in Table 15.3. The structure shows the proper placement of these groups, but it is not really correct. Acids react with bases to form salts; the carboxyl group is acidic, and the amino group is basic. The two functional groups interact, the acid transferring a proton to the base. The resulting product is an inner salt, or **zwitterion**, a compound in which the negative charge and the positive charge are on different parts of the same molecule.

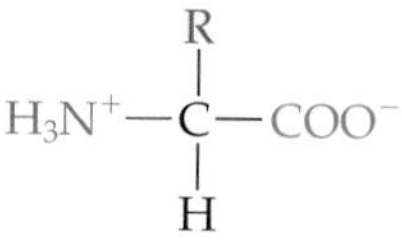

$$H_3N^+-\underset{\underset{H}{|}}{\overset{\overset{R}{|}}{C}}-COO^-$$

A zwitterion

Plants are able to synthesize proteins from carbon dioxide, water, and minerals such as nitrates (NO$_3^-$) and sulfates (SO$_4^{2-}$). Animals require proteins as one of the three major classes of foods. Humans can synthesize some of the amino acids in Table 15.3; others must be part of the proteins consumed in a normal diet. The latter are called *essential amino acids*.

15.6 THE PEPTIDE BOND: PEPTIDES AND PROTEINS

The human body contains tens of thousands of different proteins. Each of us has a tailor-made set. Proteins are polyamides. The amide linkage is called a **peptide bond** (shaded part of the structure below) when it joins two amino acid units.

Formation of a Peptide Bond

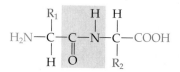

Note that there is still a reactive amino group on the left and a carboxyl group on the right. These groups can react further to join more amino acid units. This process can continue until thousands of units have joined to form a giant molecule—a polymer called a *protein*. We will examine the structure of proteins in the next section, but first let us look at some molecules called *peptides* that have only a few amino acid units.

When only two amino acids are joined, the product is a *dipeptide*.

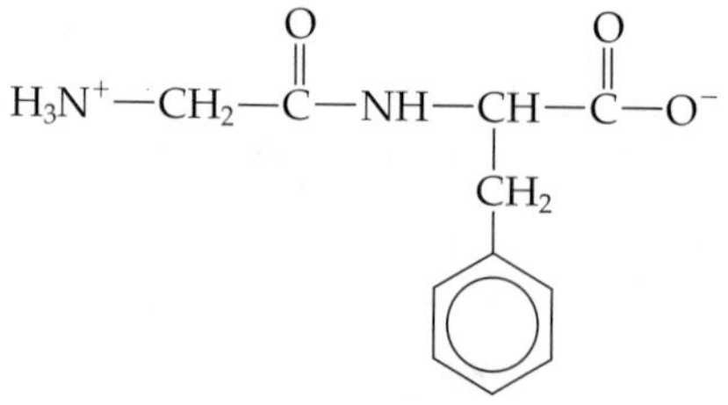

$$H_3N^+-CH_2-\overset{\overset{O}{\|}}{C}-NH-\underset{\underset{CH_2}{|}}{CH}-\overset{\overset{O}{\|}}{C}-O^-$$

Glycylphenylalanine
(a dipeptide)

Three amino acids combine to form a *tripeptide*.

$$H_3N^+-CH-\overset{\overset{\textstyle O}{\|}}{C}-NH-CH-\overset{\overset{\textstyle O}{\|}}{C}-NH-CH-\overset{\overset{\textstyle O}{\|}}{C}-O^-$$

$$\underset{CH_2OH}{|} \qquad \underset{CH_3}{|} \qquad \underset{CH_2SH}{|}$$

Serylalanylcysteine
(a tripeptide)

In describing peptides and proteins, scientists find it simpler to indicate the amino acids in a chain by using the abbreviations given in Table 15.3. The three-letter abbreviations are more common, but the alternate single-letter set is also used. Thus, glycylphenylalanine is written either Gly-Phe or G-F, and serylalanylcysteine is written as Ser-Ala-Cys or S-A-C. Example 15.2 further illustrates this naming.

EXAMPLE 15.2 **Names of Peptides**

Give the **(a)** alternate designation using one-letter abbreviations and the **(b)** full name for the pentapeptide Met-Gly-Phe-Ala-Cys. You may use Table 15.3.

Solution

a. M-G-F-A-C
b. The endings of the names for all except the last amino acid are changed from *ine* to *yl*. The name is therefore methionylglycylphenylalanylalanylcysteine.

▶ **Exercise 15.2A**
Give the **(a)** alternate designation using one-letter abbreviations and the **(b)** full name for the tetrapeptide His-Pro-Val-Ala.

▶ **Exercise 15.2B**
Use **(a)** three-letter abbreviations and **(b)** one-letter abbreviations to indicate the amino acids in the peptide threonylglycylalanylalanylleucine.

A molecule with more than 10 amino acid units is often simply called a **polypeptide**. When the molecular weight of a polypeptide exceeds about 10,000, it is called a **protein**. The distinctions are arbitrary and are not always precisely applied.

The Sequence of Amino Acids

For peptides and proteins to function properly, it is not enough that they incorporate certain *amounts* of specific amino acids. The order or *sequence* in which the amino acids are connected is also of critical importance. The sequence is written starting at the end with a free amino group (—NH$_2$), called the *N-terminal*, and continuing to the end with a free carboxyl group (—COOH), called the *C-terminal*.

Some protein molecules are enormous, with molecular weights in the tens of thousands. The molecular formula for hemoglobin, the oxygen-carrying protein in red blood cells, is $C_{3032}H_{4816}O_{780}N_{780}S_8Fe_4$, corresponding to a molar mass of 64,450 g/mol. Although they are huge compared with ordinary molecules, a billion average-sized protein molecules could still fit on the head of a pin.

$$H_3N^+CH\overset{\overset{\textstyle O}{\|}}{C}-NHCH\overset{\overset{\textstyle O}{\|}}{C}\left(-NHCH\overset{\overset{\textstyle O}{\|}}{C}\right)_n-NHCH\overset{\overset{\textstyle O}{\|}}{C}-NHCH\overset{\overset{\textstyle O}{\|}}{C}-O^-$$

$$\underset{R}{\uparrow} \qquad \underset{R}{|} \qquad \underset{R}{|} \qquad \underset{R}{|} \qquad \underset{R}{|}\uparrow$$

N-terminal   C-terminal

Glycylalanine is therefore different from alanylglycine. The sequence for glycylalanine is written Gly-Ala, and that for alanylglycine is Ala-Gly. Although the difference seems minor, the structures are different and the two substances behave differently in the body.

$$H_3N^+CH_2CO-NHCHCOO^- \qquad\qquad H_3N^+CHCO-NHCH_2COO^-$$

$$\qquad\qquad\underset{CH_3}{|} \qquad\qquad\qquad\qquad \underset{CH_3}{|}$$

Glycylalanine   Alanylglycine

As the length of a peptide chain increases, the number of possible sequential variations becomes enormous. Just as we can make millions of different words with our 26-letter English alphabet, we can make millions of different proteins with 20 amino acids. And just as we can write gibberish with the English alphabet, we can make nonfunctioning proteins by putting together the wrong sequence of amino acids.

EXAMPLE 15.3 **Numbers of Peptides**

How many different tripeptides can be made from the three amino acids methionine, valine, and phenylalanine? Write them, using the three-letter designations in Table 15.3.

Solution

Write out the various possibilities:

Met-Val-Phe	Val-Met-Phe	Phe-Val-Met
Met-Phe-Val	Val-Phe-Met	Phe-Met-Val

There are six possible tripeptides.

▶ **Exercise 15.3A**

How many different tripeptides can be made from two methionine units and one valine unit?

▶ **Exercise 15.3B**

How many different tetrapeptides can be made from the four amino acids methionine, valine, tyrosine, and phenylalanine?

Although the correct sequence is ordinarily of utmost importance, it is not always absolutely required. Just as you can sometimes make sense of incorrectly spelled English words, a protein with a small percentage of "incorrect" amino acids may continue to function. It may not function as well, however. And sometimes a seemingly minor change can have a disastrous effect. Some people have hemoglobin with one incorrect amino acid unit out of about 300. That "minor" error is responsible for sickle cell anemia, an inherited condition that ordinarily proves fatal.

15.7 STRUCTURE OF PROTEINS

The structure of proteins has four organizational levels:

- **Primary structure**: Amino acids are linked by peptide bonds to form polypeptide chains. The primary structure of a protein molecule is simply the order of its amino acids. By convention, this order is written from the amino end (N-terminal) to the carboxyl end (C-terminal).
- **Secondary structure**: Polypeptide chains can fold into regular structures such as the alpha helix and the beta sheet (described in this section).
- **Tertiary structure**: Protein folds yield spatial relationships of amino acid units that are relatively far apart in the protein chain.
- **Quaternary structure**: With more than one polypeptide chain, the chains can assemble into multiunit structures.

Primary Structure

To specify the primary structure, we write out the sequence of amino acids. For even a small protein molecule, this sequence can be quite long. For example, it takes about half a page, using one-letter abbreviations, to give the sequence of the 451 amino acid units in a protein called hexokinase, which aids in glucose metabolism. The primary structure of angiotensin II, a peptide that causes powerful constriction of blood vessels and is produced in the kidneys, is

Asp-Arg-Val-Tyr-Ile-His-Pro-Phe

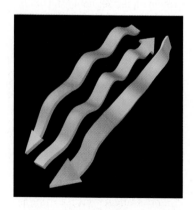

▲ The protein strands in a pleated sheet are often represented as ribbons.

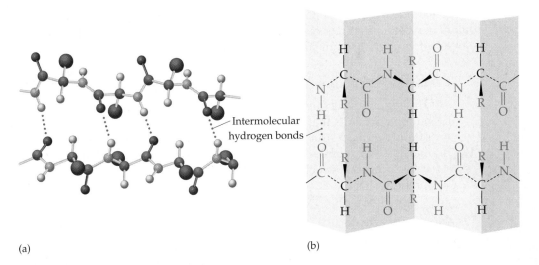

(a)

(b)

Intermolecular hydrogen bonds

◀ **Figure 15.12** Beta pleated sheet conformation of protein chains. (a) Ball-and-stick model. (b) Model emphasizing the pleats. The side chains extend above or below the sheet and alternate along the chain. The protein chains are held together by interchain hydrogen bonds.

This sequence specifies an octapeptide with aspartic acid at the N-terminal, joined to arginine, valine, tyrosine, isoleucine, histidine, and proline and ending with phenylalanine at the C-terminal.

Secondary Structure

Protein chains are held together in unique configurations. The secondary structure of a protein refers to the arrangement of chains about an axis. Common arrangements are a *pleated sheet*, as in silk, and a *helix*, as in wool.

In the pleated sheet conformation (Figure 15.12), protein chains exist in an extended zigzag arrangement. The molecules are stacked in extended arrays, with hydrogen bonds holding adjacent chains together. The appearance gives this type of secondary structure its name, the **beta pleated sheet** arrangement. This structure, with its multitude of hydrogen bonds, makes silk strong and flexible.

The protein molecules in wool, hair, and muscle contain large segments arranged in the form of a right-handed helix, or **alpha helix** (Figure 15.13). Each turn of the helix requires 3.6 amino acid units. The N—H groups in one turn form

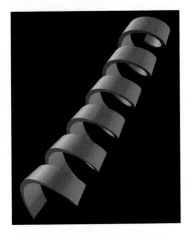

▲ The protein strands in an alpha helix are often represented as coiled ribbons.

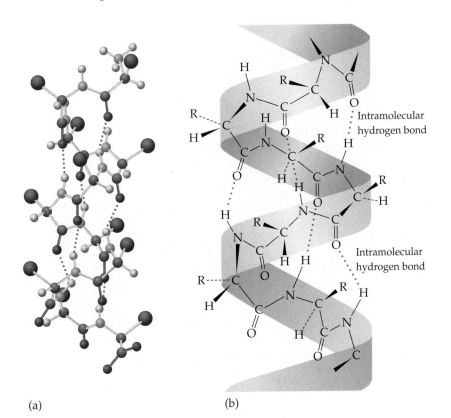

(a)

(b)

Intramolecular hydrogen bond

Intramolecular hydrogen bond

◀ **Figure 15.13** Two representations of the α-helical conformation of a protein chain. (a) Intrachain hydrogen bonding between turns of the helix is shown in the ball-and-stick model. (b) The skeletal representation best shows the helix.

hydrogen bonds to carbonyl (C=O) groups in the next turn. These helices wrap around one another in threes or sevens like strands of a rope. Unlike silk, wool can be stretched, much as we can stretch a spring by pulling the coils apart.

Tertiary Structure

The tertiary structure of a protein refers to the spatial relationships of amino acid units that are relatively far apart in the protein chain. In describing tertiary structure, we frequently talk about how the molecule is folded. An example is the protein chain in **globular proteins.** Figure 15.14 shows the structure of myoglobin, which is folded into a compact, spherical shape.

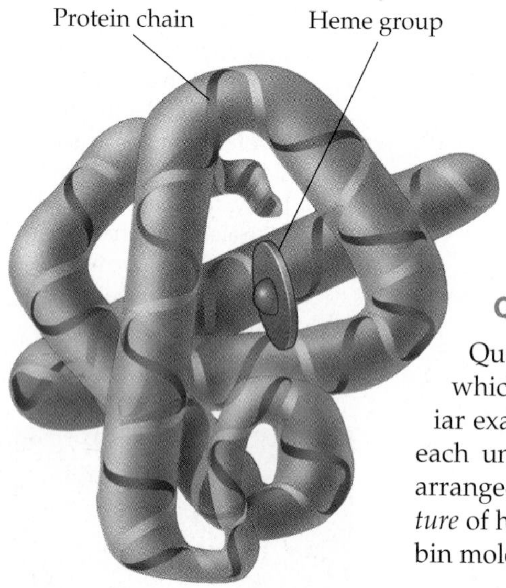

Quaternary Structure

Quaternary structures exist only if there is more than one polypeptide chain, in which case there can be an aggregate of subunits. Hemoglobin is the most familiar example. A single hemoglobin molecule contains four polypeptide units, and each unit is roughly comparable to a myoglobin molecule. The four units are arranged in a specific pattern (Figure 15.15). When we describe the *quaternary structure* of hemoglobin, we describe the way the four units are stacked in the hemoglobin molecule.

Four Ways to Link Protein Chains

As we noted earlier, the primary structure of a protein is the order of the amino acids, which are held together by peptide bonds. What forces determine the secondary, tertiary, and quaternary structures? There are four kinds of forces that operate between protein chains—hydrogen bonds, ionic bonds, disulfide linkages, and dispersion forces. These forces are illustrated in Figure 15.16.

The most important hydrogen bonding in a protein involves an interaction between the atoms of one peptide bond and those of another. Thus, the amide hydro-

Protein chain   Heme group

▲ **Figure 15.14** The tertiary structure of myoglobin. The protein chain is folded to form a globular structure, much as a string can be folded into a ball. The disk shape represents the heme group, which binds the oxygen carried by myoglobin (or hemoglobin).

Ionic and covalent bonds, hydrogen bonds, and dispersion forces, introduced in Chapter 5 as intermolecular forces, can be intramolecular as well, as indicated here in proteins.

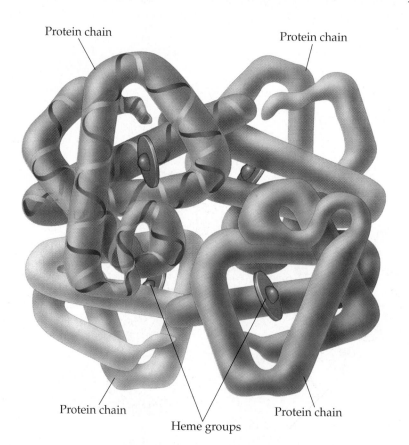

Protein chain   Protein chain

Protein chain   Protein chain

Heme groups

▶ **Figure 15.15** The quaternary structure of hemoglobin has four coiled chains, each analogous to myoglobin (Figure 15.14), stacked in a nearly tetrahedral arrangement.

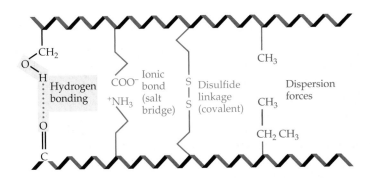

gen (N—H) of one peptide link can form a hydrogen bond to a carbonyl (C=O) oxygen located (a) some distance away on the same chain, an arrangement that occurs in the secondary structure of wool (alpha helix), or (b) on an entirely different chain, as in the secondary structure of silk (beta pleated sheet). In either case, there is usually a pattern of such interactions. Because peptide links are regularly spaced along the chain, the repetition of hydrogen bonds is also regular, both intramolecularly (in wool) and intermolecularly (in silk). Other types of hydrogen bonding can occur (such as with side chains of amino acids) but are less important.

Ionic bonds, sometimes called *salt bridges*, occur when an amino acid with a basic side chain appears opposite one with an acidic side chain. Proton transfer results in opposite charges, which then attract one another. These interactions can occur between relatively distant groups that happen to come in contact because of folding or coiling of a single chain. They also occur between chains.

A **disulfide linkage** is formed when two cysteine units (whether on the same chain or on two different chains) are oxidized. A disulfide linkage is a covalent bond, and thus much stronger than a hydrogen bond. Although far less numerous than hydrogen bonds, disulfide linkages are critically important in determining the shape of some proteins (for example, many of the proteins that act as enzymes) and the strength of others (such as the fibrous proteins in connective tissue and hair).

Dispersion forces are the only kind of forces that exist between nonpolar side chains. Recall that dispersion forces are relatively weak. These interactions can be important, however, when other types of interactions are missing or are minimized. They are made stronger by the cohesiveness of the water molecules surrounding the protein. Nonpolar side chains minimize their exposure to water by clustering together on the inside folds of the protein in close contact with one another (Figure 15.17). Dispersion forces become fairly significant in structures such as that of silk, in which a high proportion of amino acids in the protein have nonpolar side chains.

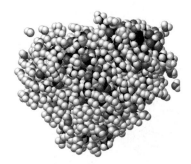

▲ **Figure 15.17** A protein chain often folds with nonpolar groups on the inside, where they are held together by dispersion forces. The outside has polar groups, visible in this space-filling model of a myoglobin molecule as oxygen atoms (red) and nitrogen atoms (blue). The carbon atoms (gray) are mainly on the inside.

Infectious Prions: Deadly Protein

In December 2003 and June 2005 the phrase *mad cow disease* cast a shadow over the beef industry, causing uncertainty and consumer concern over beef safety. On those dates two cows in the United States were found to have the disease, properly called bovine spongiform encephalopathy (BSE).

Just what is BSE? Although the agent responsible has yet to be fully characterized and understood, it is clear that BSE is not a bacterial disease like anthrax or cholera. Nor is it a viral disease like the common cold and influenza. Instead, BSE arises from an infectious *prion*, an abnormal form of a normal protein. Because they are simple proteins rather than living organisms, prions can remain unchanged under conditions of con-

siderable heat, disinfectants, and antibiotics used to destroy bacteria and viruses.

Much of the fear over BSE is unfounded. BSE is not contagious like bacterial and viral disease. It is generally spread by consumption of contaminated feed. U.S. Department of Agriculture restrictions on material that may be used in cattle feed have been quite effective. The incidence of prion-caused diseases is extremely low among people; the best-known example is *kuru* or laughing sickness, which seems to arise from cannibalistic practices in certain societies. Nonetheless, the U.S. government continues to be vigilant in avoiding any potential spread of BSE because of the near impossibility of treating the disease once contracted.

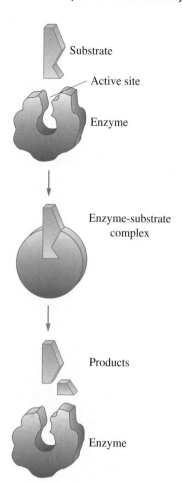

▲ **Figure 15.18** The induced-fit model of enzyme action. In this case, a single reactant molecule is broken into two product molecules.

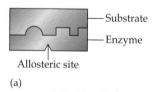

Allosteric site

(a)

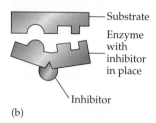

(b)

▲ **Figure 15.19** A model for the inhibition of an enzyme. (a) The enzyme and its substrate fit like a lock and key. (The allosteric site is where the inhibitor binds.) (b) With the inhibitor bound to the enzyme, the active site of the enzyme is distorted, and the enzyme cannot bind to its substrate.

15.8 ENZYMES: EXQUISITE PRECISION MACHINES

A highly specialized class of proteins, **enzymes,** are biological catalysts produced by cells. They have enormous catalytic power and are ordinarily highly specific, catalyzing only one reaction or a closely related group of reactions. Nearly all known enzymes are proteins.

Enzymes enable reactions to occur at much more rapid rates and lower temperatures than they otherwise would. They do this by changing the reaction path. Biochemists often use the model pictured in Figure 15.18, called the *induced-fit* model, to explain enzyme action. In this model, the shapes of the substrate and the active site are not perfectly complementary, but the active site adapts to fit the substrate, much like a glove molds to fit the hand that is inserted into it. The reacting substance, called the **substrate,** attaches to an area on the enzyme called the **active site,** to form an enzyme–substrate complex. The enzyme–substrate complex decomposes to form products, and the enzyme is regenerated. We can represent the process as follows:

Enzyme + Substrate ⟶ Enzyme–substrate complex ⇌ Enzyme + Products

Not only must the substrate shape fit the enzyme, but the complex is also usually held together by electrical attraction. This requires that certain charged groups on the enzyme complement certain charged or partially charged groups on the substrate. The formation of new bonds to the enzyme by the substrate weakens bonds within the substrate, and these weakened bonds can then be more easily broken to form products.

Portions of the enzyme molecule other than the active site may be involved in the catalytic process. Some interaction at a position remote from the active site can change the shape of the enzyme and thus change its effectiveness as a catalyst. In this way it is possible to slow or stop the catalytic action. Figure 15.19 offers a model for the inhibition of enzyme catalysis. An inhibitor molecule attaches to the enzyme at a position remote from the active site where the substrate is bound. The enzyme changes shape as it accommodates the inhibitor, and the substrate is no longer able to bind to the enzyme. This is one of the mechanisms by which cells "turn off" enzymes when their work is done.

Some enzymes consist entirely of protein chains. In others, another chemical component, called a **cofactor,** is necessary for proper function of the enzyme. This cofactor may be a metal ion such as zinc (Zn^{2+}), manganese (Mn^{2+}), magnesium (Mg^{2+}), iron(II) (Fe^{2+}), or copper(II) (Cu^{2+}). An organic cofactor is called a **coenzyme.** By definition, coenzymes are nonprotein. The pure protein part of an enzyme is called the **apoenzyme.** Both the coenzyme and the apoenzyme must be present for enzymatic activity to take place.

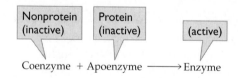

Coenzyme + Apoenzyme ⟶ Enzyme

Many coenzymes are vitamins or are derived from vitamin molecules. Enzymes are essential to the function of every living cell.

Enzymes in Medicine

Enzymes find many uses in medicine, industry, and everyday life. Diabetics use test strips coated with two enzymes and a dye to monitor their blood sugar. When a drop of blood is added to the strip, one enzyme catalyzes the oxidation of glucose, producing hydrogen peroxide as a by-product. The second enzyme catalyzes the breakdown of the hydrogen peroxide and in turn oxidizes a dye to produce a color change. The greater the amount of glucose, the greater the concentration of hydrogen peroxide produced and the more intense the color formed.

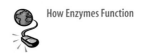

How Enzymes Function

Clinical analysis for enzymes in body fluids or tissues is a common diagnostic technique in medicine. For example, a diseased or damaged liver can leak enzymes normally found only in the liver into the bloodstream. Appearance of these enzymes in blood confirms liver damage. Analysis for specific enzymes in blood is a valuable diagnostic tool.

Enzymes are used to break up blood clots after a heart attack. This has increased survival rates for heart attack patients when these medicines are given as soon as possible after an attack. Other enzymes are used to do just the opposite: blood clotting factors are used to treat hemophilia, a disease in which the blood fails to clot normally.

Enzymes in Industry

Enzymes are widely used in a variety of industries, from baby foods to beer. Enzymes in intact microorganisms have long been used to make bread, beer, wine, yogurt, and cheese. Now the use of enzymes has been broadened to include such things as the following:

- Enzymes that act on proteins (called proteases) are used in the manufacture of baby foods to predigest complex proteins and make them easier for a baby to digest.
- Enzymes that act on starches (carbohydrases) are used to convert corn starch to corn syrup (mainly glucose) for use as a sweetener.
- An enzyme called isomerase is used to convert the glucose in corn syrup into fructose, which is sweeter than glucose and can be used in smaller amounts to give the same sweetness in diet foods. (Glucose and fructose are isomers; both have the molecular formula $C_6H_{12}O_6$.)
- Protease enzymes are used to make beer and fruit juices clear by breaking down the proteins that cause cloudiness.
- Enzymes that degrade cellulose (cellulases) are used in stonewashing blue jeans. These enzymes break down the cellulose polymers of cotton fibers. By carefully controlling their activity, manufacturers can get the desired effect without destroying the cotton material. Varying the cellulase enzymes creates different effects to make true "designer jeans."
- Protease and carbohydrases are added to animal feed to make nutrients more easily absorbed and improve the animal's digestion.

Gelatin dessert packages usually caution about adding fresh pineapple, kiwi, or papaya to the gelatin. Each of these fruits contains a protease enzyme that breaks down proteins in gelatin. The gelatin will not gel! When the fruits are cooked, the enzyme is inactivated, so canned pineapple is perfectly fine in gelatin.

Enzymes and Green Chemistry

Scientists are hard at work trying to find enzymes to carry out all kinds of chemical conversions. Enzymes are often more efficient than the usual chemical catalysts, and they are more specific in that they produce more of the desired product and fewer by-products that have to be disposed of. Further, processes involving enzymes do not require toxic solvents, thus producing less waste.

Companies spend hundreds of millions of dollars studying the special characteristics of the enzymes inside *extremophiles*, microorganisms that live in harsh and strange environments that would kill other creatures. Scientists hope to replace ordinary enzymes with enzymes from extremophiles. They are already finding uses in making specialty chemicals and new drugs. Many industrial processes can only be carried out at high temperatures and pressures. Enzymes make it possible to carry out some of these processes at fairly low temperatures and atmospheric pressure, thus reducing the energy and the need for expensive equipment.

Another area of enzyme use is for cleaning up pollution. For example, polychlorinated biphenyls (PCBs), notoriously persistent pollutants, are usually destroyed by incineration at 1200 °C. Microorganisms have been isolated that can degrade PCBs and thus have the potential to destroy these toxic pollutants. However, much research is still needed to make them cost-effective and able to compete economically with their chemical processes.

Enzymes in Everyday Life

Enzymes are important components of some detergents. Enzymes that act on fats (called lipases) and proteins (called proteases) break down stains by attacking the protein-type stains, such as blood, meat juice, dairy products, and the fats and grease that make up many stains, breaking them down into smaller, water-soluble substances. A protease enzyme is the active ingredient in meat tenderizers, making a tough piece of meat easier to eat. All told, the worldwide production of enzymes is worth more than $1 billion per year.

Nucleic Acids: The Chemistry of Heredity

Life on Earth has a fantastic range of forms, but all life arises from the same molecular ingredients: five nucleotides that serve as the building blocks for DNA and RNA (Section 15.9), and 20 amino acids (Section 15.5) that are the building blocks for proteins. These components limit the chemical reactions that can occur in cells and thus restrict what life is like.

15.9 NUCLEIC ACIDS: PARTS AND STRUCTURE

Nucleic acids serve as the information and control centers of the cell. There are two kinds of nucleic acids: *deoxyribonucleic acid (DNA)*, which serves as the blueprint for all the proteins of an organism, and *ribonucleic acid (RNA)*, which carries out protein assembly. DNA is a coiled threadlike molecule found primarily in the cell nucleus. RNA is found in all parts of the cell, where different forms do different jobs. Both DNA and RNA are long chains of repeating units called **nucleotides**. Each nucleotide in turn consists of three parts (Figure 15.20): a pentose (five-carbon) sugar, a phosphate unit, and a heterocyclic amine (Chapter 9) base. The sugar is either ribose (in RNA) or deoxyribose (in DNA). Looking at the sugar in the nucleotide, note that the hydroxyl group on the first carbon atom is replaced by one of five

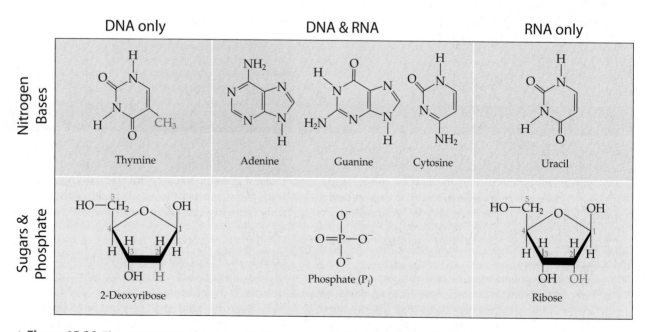

▲ **Figure 15.20** The components of nucleic acids. The sugars are 2-deoxyribose (found in DNA) and ribose (found in RNA). Note that deoxyribose differs from ribose in that it lacks an oxygen atom on the second carbon atom. *Deoxy* indicates that an oxygen atom is "missing." Phosphate units, often abbreviated as P$_i$, a compact designation of "inorganic phosphate." Heterocyclic bases found in nucleic acids. Adenine, guanine, and cytosine are found in both DNA and RNA. Thymine occurs only in DNA, and uracil only in RNA. Note that thymine has a methyl group (red) that is lacking in uracil.

bases. The bases with two fused rings, adenine and guanine, are classified as **purines**. The **pyrimidines** cytosine, thymine, and uracil have only one ring.

The hydroxyl group on the fifth carbon of the sugar unit is converted to a phosphate ester group. Adenosine monophosphate (AMP) is a representative nucleotide. In AMP, the base (blue) is adenine and the sugar (black) is ribose.

Adenosine monophosphate

Nucleotides can be represented schematically as in the following, where P_i is the biochemist's designation for "inorganic phosphate." A general representation is shown on the left and a specific schematic for AMP on the right.

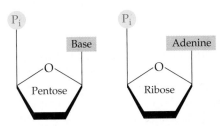

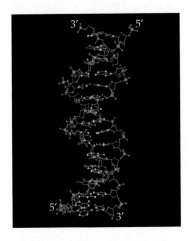

▲ **Figure 15.21** A DNA molecule. Each repeating unit is composed of a sugar, phosphate unit, and base. The sugar and phosphate units make a backbone on the outside. The bases are attached to the sugar units. Pairs of bases make "steps" on the inside of the spiral staircase.

Nucleotides are joined to one another through the phosphate group to form nucleic acid chains. The phosphate unit on one nucleotide forms an ester linkage to the hydroxyl group on the third carbon atom of the sugar unit in a second nucleotide. This unit is in turn joined to another nucleotide, and the process is repeated to build up a long nucleic acid chain (Figure 15.21). The backbone of the chain consists of alternating phosphate and sugar units, and the heterocyclic bases branch off this backbone.

~sugar-P_i-sugar-P_i-sugar-P_i-sugar-P_i-sugar-P_i-sugar-P_i~
    | | | | | |
  base   base   base   base   base   base

If this diagram represents DNA, the sugar is deoxyribose and the bases are adenine, guanine, cytosine, and thymine. In RNA the sugar is ribose and the bases are adenine, guanine, cytosine, and uracil.

EXAMPLE 15.4 **Nucleotides**

Consider the following nucleotide. Identify the sugar and the base. State whether it appears in DNA or in RNA.

▲ Francis H. C. Crick (1916–2004) (seated) and James D. Watson (1928–) used data obtained by British chemist Rosalind Franklin (1920–1958) to propose a double helix model of DNA in 1953. Franklin used a technique called X-ray diffraction, which showed DNA's helical structure. Without her permission, Franklin's colleague Maurice Wilkins shared her work with Watson and Crick. These data helped Watson and Crick decipher DNA's structure. Wilkins, Watson, and Crick were awarded the Nobel Prize for this discovery in 1962, a prize Franklin might have shared had she lived.

Solution

The sugar has two H atoms on the second C atom and therefore is deoxyribose. The base is a pyrimidine (one ring). The amino (NH_2) group helps identify it as cytosine.

▶ **Exercise 15.4A**

Consider the following nucleotide. Identify the sugar and the base. State whether it appears in DNA or in RNA.

▶ **Exercise 15.4B**

A nucleotide is composed of a phosphate unit, ribose, and thymine. Does it appear in DNA, RNA, or neither? Explain.

Base Sequence in Nucleic Acids

All the vast genetic information needed to build living organisms is stored in the sequence of the four bases along the nucleic acid strand. Not surprisingly, these molecules are huge, with molecular masses ranging into the billions for mammalian DNA. Along these chains, the four bases can be arranged in almost infinite variations. We will examine this aspect of nucleic acid chemistry shortly, but first we consider another important feature of nucleic acid structure.

The Double Helix

By 1950 experiments designed to probe the structure of DNA showed that the molar amount of adenine (A) in DNA corresponds to the molar amount of thymine (T). Similarly, the molar amount of guanine (G) is essentially the same as that of cytosine (C). To maintain this balance, the bases in DNA must be paired, A to T and G to C. But how? At the midpoint of the twentieth century, it was clear that whoever answered this question would win a Nobel Prize. Many illustrious scientists worked on the problem, but two who were relatively unknown announced in 1953 that they had worked out the structure of DNA. Using data that involved quite sophisticated chemistry, physics, and mathematics, and working with models not unlike a child's construction set, James D. Watson and Francis H. C. Crick determined that DNA must be composed of two helices wound about one another. The two strands are antiparallel; they run in opposite directions. The phosphate and sugar backbone of the polymer chains form the outside of the structure, which is rather like a spiral staircase. The heterocyclic amines are paired on the inside—with guanine always opposite cytosine and adenine always opposite thymine. In our spiral staircase analogy, these base pairs are the steps (Figure 15.22).

Why do the bases pair in this precise pattern, always A to T and T to A, and always G to C and C to G? The answer is hydrogen bonding and a truly elegant molecular arrangement. Figure 15.23 shows the two sets of base pairs. You should notice two things. First, a pyrimidine is paired with a purine in each case, and the long dimensions of both pairs are identical (1.085 nm).

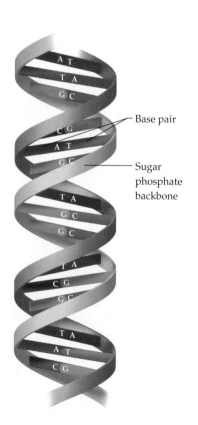

Base pair

Sugar phosphate backbone

◀ **Figure 15.22** A model of a portion of a DNA double helix is given in Figure 15.21. Here we show a schematic representation of the double helix with the sugar–phosphate backbones shown as ribbons and complementary base pairs shown as steps on the spiral staircase.

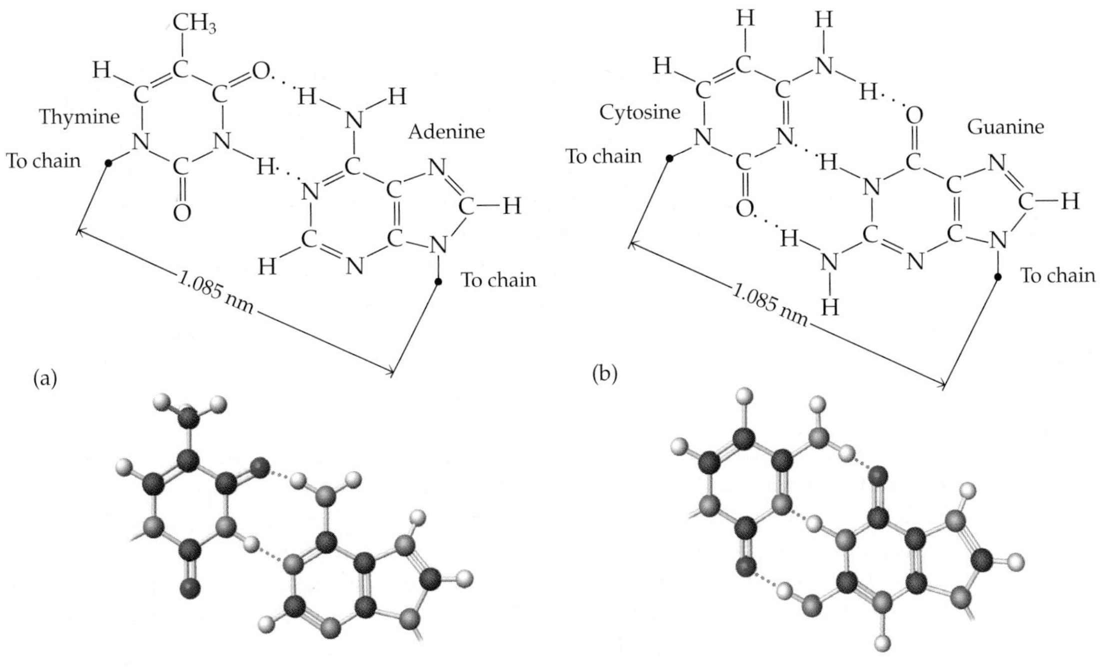

▲ Figure 15.23 Pairing of the complementary bases thymine and adenine (a) and cytosine and guanine (b). The pairing involves hydrogen bonding, as in DNA.

The second thing you should notice in Figure 15.23 is the hydrogen bonding between the bases in each pair. When guanine is paired with cytosine, three hydrogen bonds can be formed between the bases. No other pyrimidine–purine pairing permits such extensive interaction. Indeed, in the combination shown in the figure, each pair of bases fits like a lock and key.

Other scientists around the world quickly accepted the Watson–Crick structure because it answers so many crucial questions. It explains how cells are able to divide and go on functioning, how genetic data are passed on to new generations, and even how proteins are built to required specifications. It all depends on the base pairing.

Structure of RNA

RNA molecules consist of single strands of nucleic acid. Some internal (intramolecular) base pairing can occur in sections where the molecule folds back on itself. Portions of some RNA molecules exist in double-helical form (Figure 15.24).

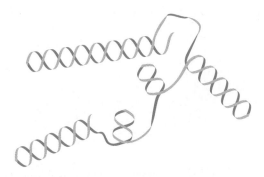

▲ Figure 15.24 RNA occurs as single strands that can form double-helical portions by internal pairing of bases.

DNA Replication Movie

15.10 DNA: SELF-REPLICATION

Cats have kittens that grow up to be cats. Birds lay eggs that hatch and grow up to be birds. How is it that each species reproduces its own kind? How does a fertilized egg "know" that it should develop into a kangaroo and not a koala?

The physical basis of heredity has been known for a long time. Most higher organisms reproduce sexually. A sperm cell from the male unites with an egg cell from the female. The fertilized egg so formed must carry all the information needed to make the various cells, tissues, and organs necessary for the functioning of a new individual. In addition, if the species is to survive, information must be passed along in germ cells—both sperms and eggs—for the production of new individuals.

Chromosomes and Genes

The hereditary material is found in the nuclei of all cells, concentrated in elongated, threadlike bodies called *chromosomes*. Chromosomes form compressed X-shaped structures when strands of DNA coil up tightly just before a cell divides. The number of chromosomes varies with the species. In sexual reproduction, chromosomes come in pairs, with one member of the pair from each parent. Each human inherits 23 pairs; thus our body cells have 46 chromosomes. Thus, the entire complement of chromosomes is achieved only when the egg's 23 chromosomes combine with a like number from the sperm.

Chromosomes are made of DNA and proteins. Arranged along the chromosomes are the basic units of heredity, the genes. Structurally, a **gene** is a section of the DNA molecule, although some viral genes contain only RNA. Genes control the synthesis of proteins, and in this way tell cells, organs, and organisms how to function in their surroundings. The environment helps determine which genes become active at a particular time. The complete set of genes of an organism is called its *genome*. When cell division occurs, each chromosome produces an exact duplicate of itself. Transmission of genetic information therefore requires the **replication** (copying or duplication) of DNA molecules.

The Watson–Crick double helix provides a ready model for the process of replication. If the two chains of the double helix are pulled apart, each chain can direct the synthesis of a new DNA chain using the nucleotides available in the cellular

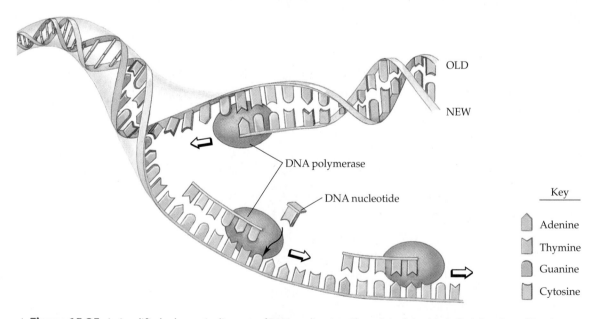

▲ **Figure 15.25** A simplified schematic diagram of DNA replication. The original double helix is "unzipped" and new nucleotides (as triphosphate derivatives) are brought into position by the enzyme DNA polymerase. Phosphate bridges are formed, restoring the original double helix configuration. Each newly formed double helix consists of one old strand and one new strand.

fluid surrounding the DNA. Each of the separating chains serves as a template, or pattern, for the formation of a new complementary chain. Synthesis begins with a base on a nucleotide pairing with its complementary base on the DNA strand—adenine with thymine and guanine with cytosine (Figure 15.25). Each base unit in the separated strand can pick up only a unit identical to the one with which it was paired before.

As the nucleotides align, enzymes connect them to form the sugar–phosphate backbone of the new chain. In this way, each strand of the original DNA molecule forms a duplicate of its former partner. Whatever information was encoded in the original DNA double helix is now contained in each of the replicates. When the cell divides, each daughter cell gets one of the DNA molecules and all the information that was available to the parent cell.

How is information stored in DNA? The code for the directions for building all the proteins that comprise an organism and enable it to function resides in the *sequence* of bases along the DNA chain. Just as *cat* means one thing in English and *act* means another, the sequence of bases CGT means one amino acid, and GCT means another. Although there are only four "letters"—the four bases—in the genetic codes of DNA, their sequence along the long strands can vary so widely that essentially unlimited information storage is available. Each cell carries in its DNA all the information it needs to determine all its hereditary characteristics.

Humans were long thought to have 80,000 or more active genes. Scientists completed a massive project, called the Human Genome Project, in 2000 and now think there are only about 20,000 to 25,000 genes in the human genome.

Genes have functional regions called *exons* interspersed with inactive portions called *introns*. During protein synthesis, introns are snipped out and not translated.

15.11 RNA: PROTEIN SYNTHESIS AND THE GENETIC CODE

Protein Synthesis

DNA carries a message that must somehow be relayed and acted on in a cell. Because DNA does not leave the cell nucleus, its information, or "blueprint," must be transported by something else, and it is. In the first step, called **transcription**, DNA transfers its information to a special RNA molecule called *messenger RNA (mRNA)*. The base sequence of DNA specifies the base sequence of mRNA. Thymine in DNA calls for adenine in mRNA, cytosine specifies guanine, guanine calls for cytosine, and adenine requires uracil (Table 15.4). Remember that in RNA molecules, uracil is used in place of DNA's thymine. Notice the similarity in the structure of these two bases (Figure 15.20).

The next step in creating a protein involves deciphering the code copied by mRNA and **translation** of that code into a specific protein structure. The decoding occurs when the mRNA travels from the nucleus and attaches itself to a ribosome in the cytoplasm of the cell. Ribosomes are constructed of RNA and proteins.

Another type of RNA molecule, called *transfer RNA (tRNA)*, carries amino acids from the cell fluid to the ribosomes. A tRNA molecule has the looped structure shown in Figure 15.26. At the head of the molecule is a set of three base units, a *base triplet* called the *anticodon*, that pairs with a set of three complementary bases on mRNA, called the codon. This triplet determines which amino acid is carried at the tail of the tRNA. To illustrate, the base triplet GUA on a segment of mRNA pairs with the base triplet CAU on a tRNA molecule. All tRNA molecules with the base triplet CAU always carry the amino acid valine. Once the tRNA has paired with the base triplet of mRNA, it releases its amino acid and returns to the cell fluid to pick up another amino acid molecule.

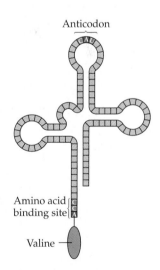

▲ **Figure 15.26** A given transfer RNA (tRNA) doubles back on itself, forming three loops with intermolecular hydrogen bonding between complementary bases. The anticodon triplet at the head of the molecule joins with a complementary codon triplet on mRNA. Here the base triplet CAU in the anticodon specifies the amino acid valine.

TABLE 15.4 DNA Bases and Their Complementary RNA Bases	
DNA Base	**Complementary RNA Base**
Adenine (A)	Uracil (U)
Thymine (T)	Adenine (A)
Cytosine (C)	Guanine (G)
Guanine (G)	Cytosine (C)

Why is a genetic code that must specify 20 different amino acids based on a triplet of bases? Four "letters" can be arranged in $4 \times 4 \times 4 = 64$ possible three-letter "words." A "doublet" code of four letters would have only $4 \times 4 = 16$ different words, and a "quadruplet" code of four letters can be arranged in $4 \times 4 \times 4 \times 4 = 256$ different ways. A triplet code can call for 20 different amino acids with some redundancy, a doublet code is inadequate, and a quadruplet code would provide too much redundancy.

TABLE 15.5 The Genetic Code

		SECOND BASE					
		U	**C**	**A**	**G**		THIRD BASE
FIRST BASE	**U**	UUU=Phe UUC=Phe UUA=Leu UUG=Leu	UCU=Ser UCC=Ser UCA=Ser UCG=Ser	UAU=Tyr UAC=Tyr UAA=Termination UAG=Termination	UGU=Cys UGC=Cys UGA=Termination UGG=Trp	U C A G	
	C	CUU=Leu CUC=Leu CUA=Leu CUG=Leu	CCU=Pro CCC=Pro CCA=Pro CCG=Pro	CAU=His CAC=His CAA=Gln CAG=Gln	CGU=Arg CGC=Arg CGA=Arg CGG=Arg	U C A G	
	A	AUU=Ile AUC=Ile AUA=Ile AUG=Met	ACU=Thr ACC=Thr ACA=Thr ACG=Thr	AAU=Asn AAC=Asn AAA=Lys AAG=Lys	AGU=Ser AGC=Ser AGA=Arg AGG=Arg	U C A G	
	G	GUU=Val GUC=Val GUA=Val GUG=Val	GCU=Ala GCC=Ala GCA=Ala GCG=Ala	GAU=Asp GAC=Asp GAA=Glu GAG=Glu	GGU=Gly GGC=Gly GGA=Gly GGG=Gly	U C A G	

Errors can occur at each step in the replication–transcription–translation process. In replication alone, each time a human cell divides, 4 billion bases are copied to make a new strand of DNA, and there may be up to 2000 errors. Most such errors are corrected, and others are unimportant, but some have terrible consequences: genetic disease or even death may result.

Each of the 61 different tRNA molecules carries a specific amino acid into place on a growing peptide chain. The protein chain gradually built up in this way is released from the tRNA as it is formed. A complete dictionary of the genetic code has now been compiled (Table 15.5). It shows which amino acids are specified by all the possible mRNA base triplets. There are 64 possible triplets and only 20 amino acids, and so there is some redundancy in the code. Three amino acids (serine, arginine, and leucine) are each specified by six different codons. Two others (tryptophan and methionine) have only one codon each. Three base triplets on mRNA are *stop* signals that call for termination of the protein chain. The codon AUG signals "start" as well as specifying methionine in the chain. Figure 15.27 provides an overall summary of protein synthesis.

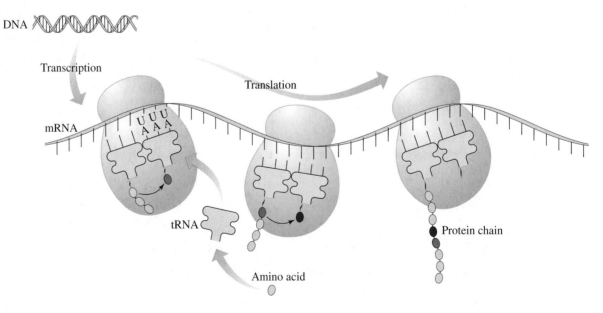

▲ **Figure 15.27** Protein synthesis visualized from DNA through mRNA and tRNA to a protein.

15.12 THE HUMAN GENOME

Many human diseases have clear genetic components. Some are directly caused by one defective gene, others have several genes involved. Some of these genes have been located on the human genome map, and the function of some of them has been identified. Once all the genes are identified, the ability to use this information to diagnose and cure genetic diseases will revolutionize medicine.

Genetic Testing

DNA is sequenced by using enzymes to cleave it into segments of a few to several hundred nucleotides each. A technique called the *polymerase chain reaction (PCR)* is used to duplicate and amplify each DNA sequence. PCR employs bacterial enzymes, called *DNA polymerases*, to multiply the small amount of DNA extracted from a sample into the large quantities needed for DNA sequencing. The fragments are then separated by length from longest to shortest, and a "print" is obtained. The print can be in the form of peaks on a chart or as a series of horizontal bars resembling the bar codes imprinted on packaged goods sold in supermarkets.

Patterns of DNA fragments are characteristic of certain families and can be used to look for inherited diseases. If the DNA pattern of a relative resembles that of a person with a genetic disease closely enough, that relative will probably develop the disease. It is thus possible to identify and predict the occurrence of a genetic disease.

DNA Fingerprinting

In 1985 the British scientist Alec Jeffreys invented a technique for which he coined the term *DNA fingerprinting*. Like fingerprints, each person's DNA is unique. Any cells—skin, blood, semen, saliva, and so on—can supply the necessary DNA sample.

DNA samples from evidence found at the crime scene are amplified by PCR and compared with the DNA obtained from a suspect. The DNA fingerprints are then compared to those of any suspects and of people known to have been at the scene.

DNA fingerprinting is a major advance in criminal investigation, and thousands of criminal cases have been solved with this technology. Also, because children inherit half their DNA from each parent, DNA fingerprinting has been used to establish the parentage of a child of contested origin. The odds in favor of being right in such cases are excellent—at least 100,000:1.

DNA fingerprinting has led to many criminal convictions, but perhaps more important, it can readily prove someone innocent. If the DNA does not match, the suspect could not have left the biological sample. Hundreds of people have been shown to be innocent and freed, some after spending years in prison. The technique is a major advance in the search for justice.

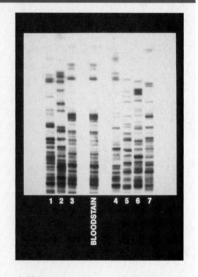

▲ This set compares a DNA segment from a bloodstain at the scene of a crime to the blood of seven suspects. Who did it?

Recombinant DNA: Using Organisms as Chemical Factories

All living organisms (except some viruses) have DNA as their hereditary material. *Recombinant DNA* is DNA that has been created artificially. DNA from two different sources is incorporated into a single recombinant DNA molecule. Scientists treat DNA from both sources with the same restriction endonuclease. This enzyme cuts the two DNA molecules at the same site. There are three different methods by which recombinant DNA is made. We describe one method here.

After scientists determine the base sequence of the gene that codes for a particular protein, they can isolate it and amplify it by PCR. The gene can then be spliced into a special kind of bacterial DNA called a *plasmid*. The recombined plasmid is then inserted into the host organism (Figure 15.28). Once inside the host, the plasmid replicates, making multiple exact copies of itself. Producing many identical copies of the same recombinant molecule is called *cloning*. As the engineered bacteria multiply, they become effective factories for producing the desired protein.

Insulin, a protein coded by DNA, is required for the proper use of glucose by cells. People with diabetes, an insulin-deficiency disease, formerly had to use insulin from pigs or cattle. Now human insulin is made using recombinant DNA technology. Scientists take the human gene for insulin production and paste it into the DNA of *Escherichia coli*, a bacterium commonly found in the human digestive tract. The bacterial cells multiply rapidly, making billions of copies of themselves,

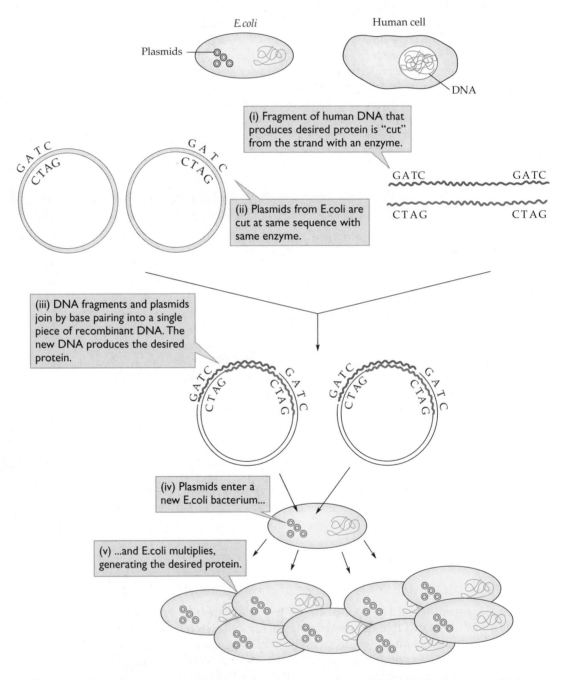

▲ **Figure 15.28** Recombinant DNA: cloning a human gene into a bacterial plasmid. The bacteria multiply and, in doing so, make multiple copies of the human gene. The gene can be one that codes for insulin, human growth hormone, or another valuable protein. Potentially, a gene from any organism—microbe, plant, or animal—can be incorporated into any other organism.

and each new *E. coli* cell carries in its DNA a replica of the gene for human insulin. Diabetics are no longer dependent on insulin from cows or pigs—by-products of the meat packing industry. The hope for the future is that a functioning gene for insulin can be incorporated directly into the cells of insulin-dependent diabetics.

In addition to insulin, many other valuable materials used in human therapy and difficult to obtain in any other way are now made using recombinant DNA technology. Some examples are

- human growth hormone (HGH), used to treat children who fail to grow properly
- erythropoietin (EPO) for treating anemia
- factor VIII for treating males with hemophilia A and factor IX for treating hemophilia B
- tissue plasminogen activator (TPA) for dissolving blood clots
- granulocyte-macrophage colony-stimulating factor (GM-CSF) for stimulating bone marrow after a bone marrow transplant
- parathyroid hormone
- angiostatin and endostatin, being tried as anticancer drugs
- interferons, promising anticancer agents
- leptin, a possible treatment for obesity
- epidermal growth factor, which stimulates the growth of skin cells and is used to speed the healing of burns and other skin wounds
- adenosine deaminase (ADA) for treating some forms of severe combined immunodeficiency (SCID), often called "bubble boy disease"

▲ Under a powerful microscope, a tiny pipet is used to introduce foreign genetic material into animal cells.

Gene Therapy

Gene therapy involves introducing a functioning gene into a person's cells to correct the action of a defective gene. Viruses are commonly used to carry DNA into cells. Current gene therapy is experimental, and human gene transplants so far have had only modest success:

- In 1999, ten children with SCID were injected with working copies of a gene that helps the immune system develop. Almost all the children got better and most are healthy today, but in 2002, two of the children developed leukemia.
- Five women had genetically modified cells injected into arthritic knuckles. The new cells blocked a chemical that causes joint irritation and swelling. The knuckles were in such bad shape that the women were scheduled to have joint replacement surgery. Because the joints were replaced, there is no way to know if the transplanted genes would have halted the disease.
- Genetically altered cells injected into the brain of patients with Alzheimer's disease appear to nourish ailing neurons and may slow the cognitive decline in these people.

Many problems remain, including getting the gene in the right place; and targeting it so it only inserts itself into the cells where it is needed. The stakes in gene therapy are huge. We all carry some defective genes. About one in ten people either has or will develop an inherited genetic disorder. Gene therapy could become an efficient way of curing disease.

Controversy and Promise in Genetic Engineering

Genes can now be transferred from nearly any organism to any other. Transgenic cows can produce proteins in their milk for the treatment of human ailments. New animals can be cloned from cells in adult animals. Genes for herbicide resistance are now incorporated into soybeans so that herbicide application will kill the weeds without killing the soybean plants. People worry that these genes will escape into wild plants (weeds), making them too resistant to herbicides. Genes for bacterial toxins are now incorporated into corn plants to kill caterpillars that would otherwise feed on the plants. People worry that these toxins will hasten the time when insects will become resistant to the bacterial toxin and butterflies and other desirable insects will also be killed. They also worry that genetically modified food plants will have proteins that will cause allergic reactions in some people. To protect against such developments, strict guidelines for recombinant DNA research have been instituted. Some people think these are not strict enough.

Molecular genetics has already resulted in some impressive achievements. Its possibilities are mind-boggling—elimination of genetic defects, a cure for cancer, and who knows what else? Knowledge gives power, but it does not necessarily give wisdom. Who will decide what sort of creature the human species should be? The greatest problem we are likely to face in our use of bioengineering is choosing who is to play God with the new "secret of life."

Critical Thinking Exercises

Apply knowledge that you have gained in this chapter and one or more of the FLaReS principles (Chapter 1) to evaluate the following statements or claims.

15.1 In a paternity case, a woman states that a certain man is the father of her child and that DNA fingerprinting will prove her case.

15.2 DNA is an essential component of every living cell. An advertisement claims that DNA is a useful diet supplement.

15.3 A restaurant claims that "Other restaurants use saturated beef fat for frying. We use only pure vegetable oil to fry our fish and french fries, for healthier eating."

15.4 The wrapper of an energy bar claims, in large print: "21,000 mg of amino acids in each bar!"

15.5 A dietician claims that a recommendation of 60 grams of protein per day for an adult is somewhat misleading. She claims that the right kinds of protein, with the proper amino acids in them, must be consumed, otherwise even a much larger amount of protein per day may be insufficient to maintain health.

15.6 A man convicted of rape claims he is innocent because the DNA fingerprinting of a semen sample taken from the rape victim shows that it contains DNA from another man.

SUMMARY

Sections 15.1–15.2—**Biochemistry** is the chemistry of life processes. Life requires chemical energy. Cell metabolism is the set of chemical reactions that keep an organism alive. Metabolism includes many processes of both **anabolism** (building up of molecules) and **catabolism** (breaking down of molecules). Food molecules include carbohydrates, fats, and proteins.

Section 15.3—A **carbohydrate** is a compound whose formula can be written as a hydrate of carbon. Sugars, starches, and cellulose are carbohydrates. The simplest sugars are **monosaccharides**, molecules that cannot be further hydrolyzed. A monosaccharide is either an **aldose** (aldehyde functional group) or a **ketose** (ketone functional group). Glucose and fructose are monosaccharides. A **disaccharide** can be hydrolyzed into two monosaccharide units. Sucrose and lactose are disaccharides. **Polysaccharides** contain many saccharide units linked together. Starches and cellulose are polysaccharides. Starches are polymers of glucose held together by alpha linkages; cellulose is a glucose polymer held together by beta linkages. There are two kinds of plant starches, amylose and amylopectin. Animal starch is called glycogen.

Glucose   Galactose   Fructose

Section 15.4—A **lipid** is a cellular component that is insoluble in water and soluble in nonpolar solvents. Lipids include solid fats and liquid oils; both are triglycerides (esters of fatty acids with glycerol). Lipids also include steroids, hormones, and **fatty acids**, long-chain carboxylic acids. A **fat** is an ester of fatty acids and glycerol. Fats may be saturated (all carbon-carbon bonds in the fatty acids are single bonds), monounsaturated (one C=C), or polyunsaturated (more than one C=C). The **iodine number** of a fat is the number of grams of iodine that reacts with 100 grams of fat. Oils tend to have higher iodine numbers than fats because there are more double bonds in oils.

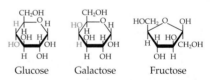

Sections 15.5–15.6—An **amino acid** contains an amino (—NH$_2$) group and a carboxyl (—COOH) group. An amino acid exists as a **zwitterion** in which a proton from the carboxyl group is transferred to the amino group. Proteins are polymers of amino acids, with 20 amino acids forming thousands of different proteins. Plants can synthesize proteins from CO$_2$, water, and other substances, while animals require proteins in their diets. The bond that joins two amino acid units in a protein is called a **peptide bond**. Two amino acids joined form a dipeptide, three form a tripeptide, and more than about 10 amino acids joined form a **polypeptide**. When the molecular weight of a polypeptide exceeds about 10,000, it is called a **protein**. The sequence of the amino acids in a protein is of great importance in its function. Three-letter or single-letter abbreviations are used to designate amino acids.

Section 15.7—Protein structure has four levels. The **primary structure** of a protein is simply the sequence of its amino acids. The **secondary structure** is the arrangement of protein chains about an axis. Two such arrangements are the **beta pleated sheet**, in which the arrays of chains form zigzag sheets, and the **alpha helix**, which is a coil of proteins. The **tertiary structure** of a protein is its folding pattern; **globular proteins** such as myoglobin are folded into compact shapes. Some proteins have a **quaternary structure** in which the protein contains subunits in a specific pattern.

There are four types of forces that generate secondary, tertiary, and quaternary structure in proteins. They are (a) hydrogen bonding; (b) ionic bonds formed where acidic and basic side chains meet; (c) **disulfide linkages**, which are S—S bonds formed between cysteine groups; and (d) dispersion forces between the nonpolar regions of the molecule.

Section 15.8—**Enzymes** are biological catalysts. In operation, the reacting substance or **substrate** attaches to the **active site** of the enzyme to form a complex, which then decomposes to the products. An enzyme is made up of an **apoenzyme** (protein) and a **cofactor** that is necessary for proper function. An organic cofactor (such as a vitamin) is called a **coenzyme**. Sometimes a cofactor is inorganic (such as a metal ion).

Section 15.9—**Nucleic acids** are the information and control centers of a cell. DNA and RNA are polymers of **nucleotides**. Each nucleotide contains a pentose sugar unit, a phosphate unit, and one of four amine bases. Bases with one ring are called **pyrimidines**, and those with two fused rings are **purines**. In DNA (deoxyribonucleic acid), the sugar is deoxyribose and the bases are adenine, thymine, guanine, and cytosine. In RNA (ribonucleic acid) the sugar is ribose and the nitrogen bases are the same as those in DNA, except that uracil replaces thymine. Nucleotides are joined through their phosphate groups to form nucleic acid chains. In DNA the bases thymine and adenine are always paired, as are the bases cytosine and guanine. Pairing occurs via hydrogen bonding between the bases, and gives rise to the double-helix structure of DNA. RNA consists of single strands of nucleic acids, some parts of which may hydrogen bond to form double helixes.

Sections 15.10–15.11—Chromosomes are made of DNA and proteins. A **gene** is a section of DNA. Genetic information is stored in the base sequence; a three-base sequence means a particular amino acid. Transmission of genetic information requires **replication** or copying of the DNA molecules. When the chains of DNA are pulled apart, each half can replicate from nucleotides in the cellular fluid. Each base unit picks up a unit identical to that with which it was paired before. Enzymes connect the nucleotides to form the new chain. When a cell divides, one DNA molecule goes to each daughter cell, with all the accompanying genetic information.

In **transcription** the DNA transfers its information to messenger RNA (mRNA). The next step is the **translation** of the code into a specific protein structure. Transfer RNA (tRNA) delivers amino acids to the growing protein chain. Each **base triplet** on mRNA codes for a specific amino acid.

Section 15.12—Many diseases have genetic components. In genetic testing, DNA is cleaved and duplicated. The patterns of DNA fragments are examined and compared to those of individuals with genetic diseases, to identify and predict the occurrence of such diseases. A similar technique is used to compare DNA from different sources in DNA fingerprinting. In recombinant DNA technology, the base sequence for the gene that codes for a desired protein is determined, and that new gene is spliced into a plasmid. The plasmid is inserted into a host organism which generates the protein as it reproduces. Gene therapy—replacement of defective genes—could be used to cure diseases. Genetic engineering holds great promise and carries with it great responsibility.

REVIEW QUESTIONS

1. What is biochemistry?

2. Briefly identify and state a function of each of the following parts of a cell.
   a. cell membrane   **b.** cell nucleus
   c. chloroplasts   **d.** mitochondria
   e. ribosomes

3. How do green plants obtain food? How do animals obtain food?

4. What are the three major types of foods from which animals obtain energy?

5. Define each of the following.
   a. anabolism   **b.** catabolism   **c.** metabolism

6. What are carbohydrates? What are the three principal elements in carbohydrate molecules?

7. What is a lipid? Name at least three kinds of lipids.

8. List some of the properties of fats and oils.

9. In what parts of the body are proteins found? What tissues are largely proteins?

10. What is the chemical nature of proteins?

11. How many different amino acids are incorporated in proteins by the genetic code?

12. How does the elemental composition of proteins differ from those of carbohydrates and fats?

13. What is a peptide bond?

14. Of what importance is the sequence of amino acids in a protein molecule?

15. What is the difference between a polypeptide and a protein?

16. Define and give an example of each of the following in relation to proteins.
    a. primary structure
    b. secondary structure
    c. tertiary structure
    d. quaternary structure

17. What is a globular protein?

18. Name the two kinds of nucleic acids. Which is found primarily in the nucleus of the cell?

19. What kind of intermolecular force is involved in base pairing?

20. Describe the process of replication.

21. How do DNA and RNA differ in structure?

22. What is the relationship among the cell parts called chromosomes, the units of heredity called genes, and the nucleic acid DNA?

23. What is the polymerase chain reaction (PCR)?

24. How do DNA fragment patterns indicate whether a person will probably develop a genetic disease?

25. List the steps in recombinant DNA technology.

26. How can a virus be used to replace a defective gene in a human?

PROBLEMS

Carbohydrates

27. What are monosaccharides? Name three common ones.

28. What are disaccharides? Name three common ones.

29. What is glycogen? How does it differ from amylose? From amylopectin?

30. In what way are amylose and cellulose similar? What is the main structural difference between starch and cellulose?

31. Which of the following are monosaccharides?
    a. amylose
    b. cellulose
    c. fructose
    d. galactose

32. Which of the following are monosaccharides?
    a. glucose
    b. glycogen
    c. lactose
    d. sucrose

33. Which of the carbohydrates in Problems 31 and 32 are disaccharides?

34. Which of the carbohydrates in Problems 31 and 32 are polysaccharides?

35. What are the hydrolysis products of each of the following?
    a. amylose   b. lactose   c. glycogen

36. What are the hydrolysis products of each of the following?
    a. amylopectin   b. sucrose   c. cellulose

37. What functional groups are present in the formula for the open-chain form of glucose?

38. Give the formula for the open-chain form of fructose. What functional groups are present?

39. What functional groups are present in the formula for the open-chain form of galactose? How does the structure of galactose differ from that of glucose?

40. Mannose differs from glucose only in that the H and OH on the second carbon atom are reversed. Give the formula for the open-chain form of mannose. What functional groups are present?

Lipids: Fats and Oils

41. To what class of organic compounds do fats belong?

42. Define monoglyceride, diglyceride, and triglyceride.

43. How do fats and oils differ in structure? In properties?

44. What does the iodine number of a fat mean? In the determination of the iodine number of a fat, what part of the fat molecule reacts with the reagent?

45. What is a saturated fat?

46. What is a polyunsaturated fat (oil)?

47. Which of the following fatty acids are saturated? Which are unsaturated?
    a. linolenic acid
    b. linoleic acid
    c. oleic acid
    d. palmitic acid
    e. stearic acid

48. Which of the fatty acids in Problem 47 are monounsaturated? Which are polyunsaturated?

49. How many carbon atoms are there in a molecule of each of the following fatty acids?
    a. linolenic acid   b. palmitic acid   c. stearic acid

50. How many carbon atoms are there in a molecule of each of the following fatty acids?
    a. linoleic acid   b. oleic acid   c. butyric acid

51. Which would you expect to have a higher iodine number—corn oil or beef tallow? Explain your reasoning.

52. Which would you expect to have a higher iodine number—lard or liquid margarine? Explain your reasoning.

Proteins

53. What functional groups are found on amino acid molecules? What is a zwitterion?

54. What is a dipeptide? A tripeptide? A polypeptide?

55. Is the dipeptide represented by Ser-Ala the same as the one represented by Ala-Ser? Explain.

56. A chemist is asked to make Phe-Ile-Leu. He makes Leu-Ile-Phe. Evaluate the chemist for possible job advancement.

57. Give structural formulas for the following amino acids.
    a. alanine   **b.** serine

58. Give structural formulas for the following amino acids.
    a. glycine   **b.** phenylalanine

59. Give structural formulas for the following dipeptides.
    a. glycylalanine   **b.** alanylserine

60. Give structural formulas for the following dipeptides.
    a. alanylglycine   **b.** phenylalanylserine

61. List the four different ways protein chains are bonded to one another.

62. Describe **(a)** the induced fit model of enzyme action and **(b)** how an inhibitor deactivates an enzyme.

Nucleic Acids

63. What is the sugar unit in RNA? What sugar is found in DNA?

64. List the heterocyclic bases found in DNA. List those found in RNA.

65. Which of the following nucleotides would occur in DNA, which in RNA, and which in neither?
    a. thymine
       |
       deoxyribose—P_i

    b. adenine
       |
       ribose—P_i

    c. cytosine
       |
       ribose—P_i

66. Which of the following nucleotides would occur in DNA, which in RNA, and which in neither?
    a. uracil
       |
       ribose—P_i

    b. adenine
       |
       deoxyribose—P_i

    c. cytosine
       |
       deoxyribose—P_i

67. Identify the sugar and the base in the following nucleotide.

68. Identify the sugar and the base in the following nucleotide.

69. In DNA, which base would be paired with the base listed?
    a. cytosine
    b. adenine
    c. guanine
    d. thymine

70. In an RNA molecule, which base would pair with the base listed?
    a. adenine
    b. guanine
    c. uracil
    d. cytosine

71. In replication, a parent DNA molecule produces two daughter molecules. What is the fate of each strand of the parent DNA double helix?

72. We say DNA controls protein synthesis, yet most DNA resides within the cell nucleus and protein synthesis occurs outside the nucleus. How does DNA exercise its control?

73. Explain the role of **(a)** mRNA, and **(b)** tRNA in protein synthesis.

74. Which nucleic acid or acids are involved in **(a)** the process referred to as transcription, and **(b)** the process referred to as translation?

75. The base sequence along one strand of DNA is AATTCG. What would be the sequence of the complementary strand of DNA?

76. What sequence of bases would appear in the mRNA molecule copied from the original DNA strand shown in Problem 75?

77. If the sequence of bases along a mRNA strand is UCC-GAU, what was the sequence along the DNA template?

78. What are the complementary triplets on tRNA for the following triplets on mRNA?
    a. UUU
    b. CAU
    c. AGC
    d. CCG

79. What are the complementary base triplets on mRNA for the following triplets on tRNA?
    a. UUG
    b. GAA

80. What are the complementary codons on mRNA for the following anticodons on tRNA?
    a. UCC
    b. CAC

ADDITIONAL PROBLEMS

81. In the schematic below, each circle represents a glucose unit. What substance is indicated?

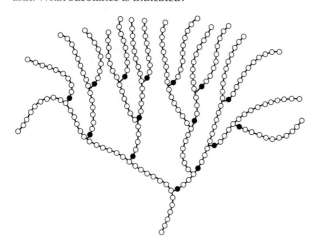

82. Which of the following is a carbohydrate? Which is a lipid? Which is a peptide?

(a)

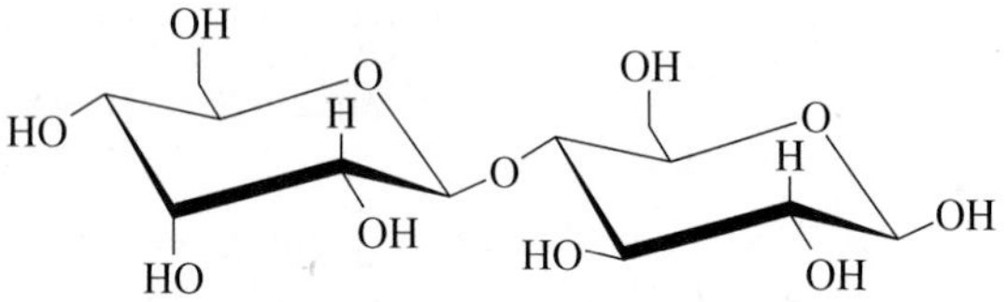

(b) $CH_3(CH_2)_7CH=CH(CH_2)_7C(O)O-CH_2$
$CH_3(CH_2)_7CH=CH(CH_2)_7C(O)O-CH_2$
$CH_3(CH_2)_{14}C(O)O-CH_2$

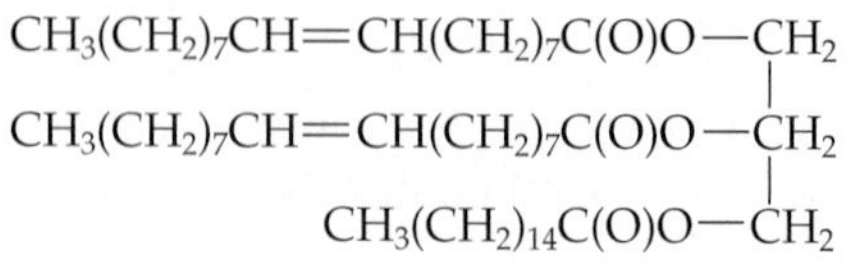

(c)

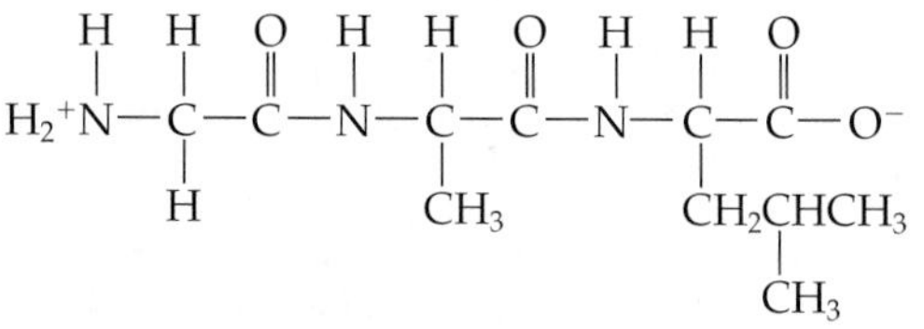

83. Human insulin molecules are made up of two chains with the following structure:

Chain A: G-I-V-E-Q-C-C-T-S-I-C-S-L-Y-Q-L-E-N-Y-C-N

Chain B: F-V-N-Q-H-L-C-G-D-H-L-V-E-A-L-Y-L-V-C-G-E-R-G-F-F-Y-T-P-K-T

The A chain has a loop formed by a disulfide linkage between the sixth and eleventh amino acids (designated A-6 to A-11) and is joined to the B chain by two disulfide linkages (A-7 to B-7 and (A-20 to B-19). What level of protein structure is described by (a) the one-letter abbreviations describing each chain and (b) by the description of the loop and chain linkages?

84. Synthetic polymers can be formed from single amino acids. Write the formula for a segment of (a) polyglycine and (b) polyserine. Show at least four repeating units of each.

85. Identify the bases in the structures in (a) Problem 67 and (b) Problem 68 as purines or pyrimidines.

86. Answer the following questions for the molecule shown.

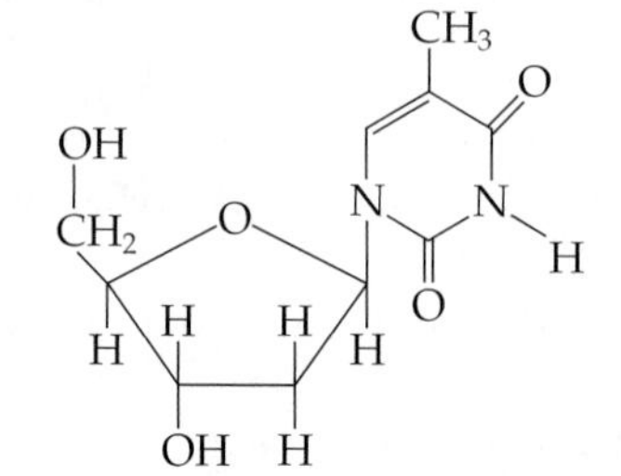

a. Is the base a purine or a pyrimidine?
b. Would the compound be incorporated in DNA or in RNA?

87. Answer the questions posed in Problem 86 for the following compound.

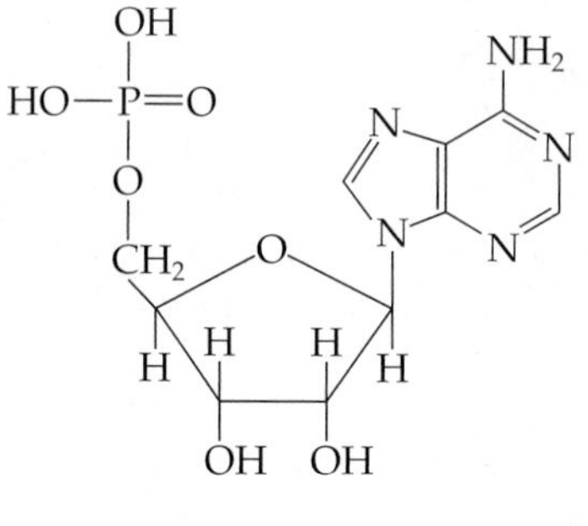

88. What amino acid is specified by the DNA triplet CTC? What amino acid is specified if a mutation changes the DNA triplet to CTT? To CTA? to CAC?

89. What happens if a mutation changes the DNA triplet CTC to ATC?

90. Write the (a) three-letter and (b) one-letter abbreviated versions of the following structural formula.

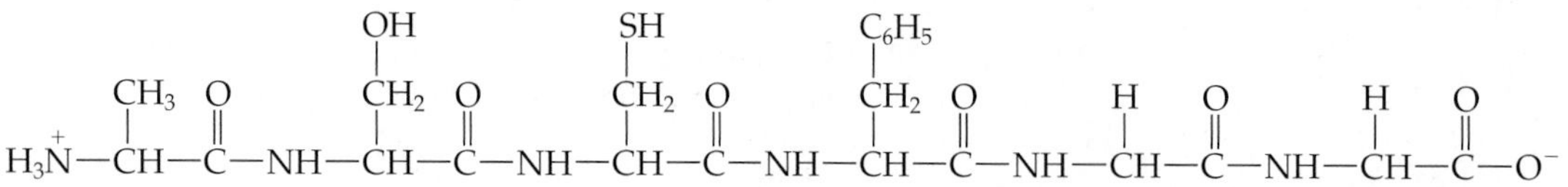

COLLABORATIVE GROUP PROJECTS

Prepare a PowerPoint, poster, or other presentation (as directed by your instructor) for presentation to the class.

91. Prepare a brief report on one of the following carbohydrates. List principal sources and uses of the carbohydrate.
   a. glycogen   **b.** maltose
   c. fructose   **d.** sucrose
   e. lactose   **f.** starch

92. Prepare a brief report on one of the following lipids. List principal sources and uses of the lipid.
   a. palmitic acid   **b.** oleic acid
   c. beef tallow   **d.** palm oil
   e. corn oil   **f.** lard

93. Prepare a brief report on one of the following applications of genetic engineering. List some advantages and disadvantages of the application.
   a. golden rice
   b. herbicide-resistant transgenic corn
   c. transgenic cotton that produces Bt toxin

   d. transgenic tobacco that produces human serum albumin
   e. transgenic tomatoes that have improved virus resistance
   f. transgenic rabbits that produce human interleukin-2, a protein that stimulates the production of T-lymphocytes that play a role in fighting selected cancers.

94. Genetic testing carries both promise and peril and is an important area of bioethics today. After some online research, write a brief essay on one of the following—or on a similar topic assigned by your instructor.
   a. In what cases might a person not want to be tested for a disease his or her parent had?
   b. When is genetic testing most valuable? What privacy issues need to be addressed in this area?
   c. Does testing negative for the breast cancer genes mean that a woman doesn't need mammograms?

REFERENCES AND READINGS

1. Dagani, Ron. "Gates Foundation Awards 43 Grants For Global Health Research." *Chemical & Engineering News*, July 4, 2005, p. 9. Has a link to The Grand Challenges in Global Health initiative.

2. Gibbs, W. Wayt. "The Unseen Genome: Gems Among the Junk." *Scientific American*, November 2003, pp. 46–53.

3. Hileman, Bette. "NRC Report Explores Potential Problems Posed By Biotech Animals." *Chemical & Engineering News*, August 26, 2002, p. 11.

4. Hopkins, Tom. "Chicken Genome 60% Shared with Humans." *Chemistry & Industry*, vol. 20 December 2004, p. 5.

5. Kaiser, Jocelyn. "Retroviral Vectors: A Double-Edged Sword." *Science*, June 17, 2005, pp. 1735–1736. One of several reports on gene therapy in this issue.

6. Lesney, Mark S. "Biocatalysts on Board." *Today's Chemist at Work*, December 2003, pp. 20–23.

7. Murphy, Marina. "DNA, Race and Disease Linked." *Chemistry & Industry*, February 21, 2005, p. 6.

8. Pöpping, Bert. "Are You Ready for [a] Roundup?—What Chemistry Has to Do with Genetic Modifications." *Journal of Chemical Education*, June 2001, pp. 752–756.

9. Roberts, Leslie. "The Gene Hunters." *U.S. News & World Report*, January 3–10, 2000, pp. 34–44. Includes attractive graphics in a centerfold primer.

10. Thayer, Ann. "Genomics Moves On." *Chemical & Engineering News*, October 14, 2002, pp. 25–36.

11. Watson, James D. *The Double Helix*. New York: New American Library, 1968. Watson's personal account of the discovery of the structure of DNA.

12. Welsh, Paul. "Two Peas in a Pod? A Case of Questionable Twins." *Journal of College Science Teaching*, February 2004, pp. 23–27.

13. Williams, R. J. P., and J. J. R. Fraústo da Silva. "The Trinity of Life: The Genome, the Proteome, and the Mineral Chemical Elements."*Journal of Chemical Education*, May 2004, pp. 738–741.

14. Yang, Jay, and Christopher L. Wu. "Gene Therapy for Pain." *American Scientist*, March–April 2001, pp. 126–135.

GREEN CHEMISTRY

Biocatalysts: Nature's Tools for Chemical Synthesis

Edward J. Brush, Bridgewater State College

Catalysts are amazing chemical materials that are used extensively in the chemical industry for the production of a wide range of consumer products. Green chemistry processes that avoid the use of toxic or hazardous reagents, have high atom efficiency, reduce energy consumption, and minimize or eliminate the production of hazardous waste all rely on catalysts. As consumer demand for new products continues to grow along with increasingly stringent environmental regulations, the chemical industry is becoming more reliant on green chemistry technologies, which in turn has led to an increased interest in the use of biocatalysts.

People have used natural catalysts for thousands of years. Wine-making is one of the oldest examples of industrial biotechnology; it involves the fermentation of sugar to alcohol by yeast. We now know that enzymes are responsible for catalyzing a wide variety of biological processes.

Thousands of enzymes have been identified from diverse biological sources, providing biocatalysts for a large number of industrial applications. The market demand for industrial enzymes is huge, reaching nearly $2 billion in 2005. Known enzymes catalyze a variety of chemical transformations on their usual substrates as well as on other compounds, providing consumers with a wide range of enzyme formulations in commercial products, production of fine chemicals, bioremediation of waste, and genetic engineering. Furthermore, enzyme catalysts can be recovered and reused, which adds to their economic and green chemistry viability.

▲ In wine-making and beer-making, the enzymes in yeast are biological catalysts that help to convert sugars to alcohol.

In this Web Investigation, you will examine the importance of enzymes as catalysts in the chemical industry for the production of a variety of consumer products used in your everyday life. You will build a body of knowledge that will allow you to effectively communicate your knowledge of biocatalysts as efficient "green catalysts" to representatives of the chemical industry and to general consumers. Finally, you will critically evaluate a variety of new biocatalytic processes and recommend one for the coveted Biocatalysis for Changing Times award based on how they follow green chemistry principles.

WEB INVESTIGATIONS

Investigation 1
Understanding Catalysts and Biocatalysts
In this activity you may have seen a number of terms that you may have heard before but are not completely familiar with. At the Chemical Industry Education Center website, click on the "Contents" tab, and read about a brief history of biocatalysis under the "Applied Catalysts" link. At the same website also visit "Principles of Catalysis," where you will read about how catalysts work and view simple animations on enzyme catalysis. Follow the links in the "Examples" box and read more about industrial biocatalysis and some specific applica-

tions. Additional information on catalysts and enzymes can be found at the Brown University educational website and Procter & Gamble website, respectively.

Investigation 2
Evaluating the Advantages and Disadvantages of Biocatalysts
Biocatalysts are currently available for any number of industrial applications. However, any new technology must be thoroughly evaluated and its advantages and limitations compared to the traditional processes. Read the article on biocatalysis from the May 21, 2001 issue of

Chemical & Engineering News. Identify advantages and disadvantages of the industrial use of biocatalysts, and list them side-by-side in a table. Using a critical eye, write two to three sentences evaluating this information, and make a clear statement as to the practicality of using enzymes for industrial applications.

Investigation 3
Better Living Through Biocatalysis
The market demand for biocatalysts is huge, thanks in part to nature's rich diversity of enzymatic reactions. Enzymes are involved in a multitude of industrial processes and impact many areas of our everyday lives. But what exactly are all these enzymes doing?

Read the article "Bio-Catalysis on Board," from the December 2003 issue of *Today's Chemist at Work*. Prepare a table with columns for "Type of Enzyme," "Type of Reaction," and "Industrial Application." In a fourth column, rate each industrial application on a scale of your (or your instructor's) choosing, in terms of its impact on your everyday life. A more extensive list of industrial enzymes and applications can be found at the Biotechnology Industry Organization's website.

COMMUNICATE YOUR RESULTS

Exercise 1
The Green Chemistry Appeal of Biocatalysts
You are employed at a major chemical company that produces consumer products but does not employ green chemistry principles or use biocatalysts. The Board of Directors has hired a new, progressive CEO, who has promoted you to Director of Industrial Research. You have been asked to prepare a one- to two-page brief that proposes switching to biocatalysts for all of your company's industrial processes and explains how these processes are consistent with the principles of green chemistry. Refer to the information you gathered in the previous investigations and the 12 principles of green chemistry to guide you in preparing this brief.

Exercise 2
Communicating Biocatalysis to the Consumer
You are a green chemist working for the Public Affairs Office of a major industrial supplier of the food additive aspartame. You have been asked to write a newspaper article explaining why aspartame is used in foods, how it is produced by the chemical industry, and whether the production technology follows green chemistry principles. It is critical to your company that the story is understood by nonscientific readers. Information on aspartame can be found at the Chemical Industry Education Center website. Click on the "Contents" tab, and go to "Principles of Catalysis: catalysis by enzymes."

Cooperative Exercise
The Biocatalysis for Changing Times Award
You are a member of an award committee composed of four to eight other students, charged with the task of identifying the recipient of the coveted Biocatalysis for Changing Times award. Your committee has been asked to review all EPA Presidential Green Chemistry awards and identify the awards that deal with biocatalysis. Your committee then breaks up into smaller groups, assigning one or two of the selected biocatalysis Green Chemistry Challenge awards to each group for assessment. Using the 12 principles of green chemistry as a guide, prepare a written evaluation as to how closely your assigned awards relate to the 12 principles.

All groups then meet (perhaps with the entire class) to debate and defend the technologies you evaluated as recipients of the EPA award. By class vote, decide which technology should receive the Biocatalysis for Changing Times award.

CHAPTER 16

Food

The food we eat supplies all the molecular building blocks from which our bodies are made and all the energy for our life activities. That energy comes ultimately from the sun through the remarkable process of photosynthesis.

Molecular Gastronomy

Chemistry has contributed much to the quality, variety, and abundance of our food supply.

FOOD! From holiday feasts to late-night snacks, many of life's joys involve food. Yet the prime purpose of food is not to give pleasure; it is to sustain life. Food supplies all the molecular building blocks from which our bodies are made and all the energy for our life activities. That energy comes ultimately from the sun through the remarkable process of photosynthesis.

In many places on Earth people never have enough food. More than 800 million people are always hungry. According to The Hunger Project, begun in 1977, an estimated 15,000,000 people died of hunger that year. Ten years later, that number had declined to 13,000,000. Now that estimate is 9,000,000 each year. Three-fourths are children under the age of five. Famine and wars make the news, but they cause just 10% of hunger deaths. Most result from chronic malnutrition. Because of extreme poverty, families simply cannot get enough to eat.

While millions starve, other locales have a surplus. Most of us in the Western world have an overabundance of food and often eat too much. Yet even in the United States, the U.S. Department of Agriculture estimates that 31 million people live in households that experience hunger or the risk of hunger. Nearly 9 million people, including more than 3.2 million children, actually experience hunger, frequently skipping meals or eating too little. These people often have lower quality diets, or they resort to seeking food from emergency sources.

Many of us also frequently eat the wrong kinds of food. Our diet, often too rich in saturated fats, sugar, and alcohol, has been linked in part to at least 5 of the 10 leading causes of death in the United States: heart disease, cancer, stroke, diabetes, and kidney disease. The 1999–2002 National Health and Nutrition Examination Survey estimated that more than 65% of the adults in the United States are overweight. This excess weight contributes to poor health, increased risk of diabetes, heart attacks, strokes, and some forms of cancer. It also leads to increased susceptibility to other diseases.

The human body is a conglomeration of chemicals. A chemical analysis of the human body would show that it is about two-thirds water. The body contains several minerals and thousands of other chemicals. Our foods are also made up of chemicals. They enable children to grow, they provide people with energy, and they supply the chemicals needed for the repair and replacement of body tissues.

In this chapter we discuss food—what it's made of and where it comes from.

Worldwide, six children under five years old die every minute from eating food or drinking water contaminated with microorganisms. One of every eight people on Earth suffers from malnutrition severe enough to stunt physical and mental growth and to shorten life.

The Food We Eat

The three main classes of foods are carbohydrates, fats, and proteins. These, too, are chemicals. For proper nutrition, our diet should include balanced proportions of these three foodstuffs, plus water, vitamins, minerals, and fiber.

16.1 CARBOHYDRATES IN THE DIET

Dietary carbohydrates include sugars and starches. Sugars are mainly monosaccharides and disaccharides; starches are polysaccharides.

Sweet Chemicals: Sugars

Sugars have been used for ages to make food sweeter. Two monosaccharides, *glucose* (or *dextrose*) and *fructose* (fruit sugar), and the disaccharide *sucrose* are the most common dietary sugars (Figure 16.1). Common table sugar is sucrose, usually obtained from sugar cane or sugar beets. Glucose is the sugar used by the cells of our bodies for energy. Because it is the sugar that circulates in the bloodstream, it is often called **blood sugar**. Fructose is found in honey and in some fruits, but much of it is made from glucose. Corn syrup, made from starch, is mainly glucose. *High-fructose corn syrup* is made by treating corn syrup with enzymes to convert much of the glucose to fructose. Fructose is sweeter than sucrose or glucose. Foods sweetened to the same degree with fructose have somewhat fewer calories than those sweetened with sucrose.

Since 1968, per-capita consumption of sugars in the United States has skyrocketed from 11 kg/year to 72 kg/year. Most are consumed in soft drinks, presweetened cereals, candy, and other highly processed foods with little or no other nutritive value. The sugars in sweetened foods provide empty calories and contribute to tooth decay and obesity.

Digestion and Metabolism of Carbohydrates

Glucose and fructose are absorbed directly into the bloodstream from the digestive tract. Sucrose is hydrolyzed ("split by water") during digestion to glucose and fructose.

$$\text{Sucrose} + H_2O \longrightarrow \text{Glucose} + \text{Fructose}$$

The disaccharide lactose occurs in milk. During digestion, it is hydrolyzed to two simpler sugars, glucose and galactose.

$$\text{Lactose} + H_2O \longrightarrow \text{Glucose} + \text{Galactose}$$

 Glucose

 Fructose

Some of the chemistry of carbohydrates is discussed in Section 15.3. The role of foods in health and fitness is discussed in Chapter 18.

▶ **Figure 16.1** The principal sugars in our diet are sucrose (cane or beet sugar), glucose (corn syrup), and fructose (fruit sugar, often in the form of high-fructose corn syrup).

Question: What is the molecular formula for glucose? For fructose? How are the two compounds related? How is sucrose related to glucose and fructose? *Hint:* You may refer to Chapter 15, if necessary.

Nearly all human babies have the enzyme necessary to accomplish this breakdown, but many adults do not. People who lack the enzyme get digestive upsets from drinking milk, a condition called *lactose intolerance*. When milk is cooked or fermented, the lactose is at least partially hydrolyzed. People with lactose intolerance may still be able to enjoy cheese, yogurt, or cooked foods containing milk with little or no discomfort. Lactose-free milk, made by treating milk with an enzyme that hydrolyzes lactose, is available in most grocery stores.

All monosaccharides are converted to glucose during metabolism. Some babies are born with *galactosemia*, a deficiency of the enzyme that catalyzes the conversion of galactose to glucose. For proper nutrition, they require a synthetic formula (Figure 16.2) in place of milk.

Complex Carbohydrates: Starch and Cellulose

Starch and cellulose are both polymers of glucose, but the connecting links between the glucose units are different (Figure 15.6). Humans can digest starch, but not cellulose. **Starch**, a polymer of alpha-glucose, is an important part of any balanced diet (Figure 16.3). **Cellulose**, a beta-glucose polymer, is an important component of dietary fiber.

When digested, starch is hydrolyzed to glucose, as represented by the following equation:

$$(C_6H_{10}O_5)_n + n\,H_2O \xrightarrow{\text{Carbohydrases}} n\,C_6H_{12}O_6$$

$$\text{Starch} \qquad\qquad\qquad\qquad\qquad \text{Glucose}$$

The body then metabolizes the glucose, using it as a source of energy. Glucose is broken down through a complex set of more than 50 chemical reactions to produce carbon dioxide and water, with the release of energy.

$$C_6H_{12}O_6 + 6\,O_2 \longrightarrow 6\,CO_2 + 6\,H_2O + \text{Energy}$$

The net reaction is essentially the reverse of photosynthesis. In this way, animal organisms are able to make use of the energy from the sun that was captured by plants in the process of photosynthesis.

Carbohydrates, which supply 4 kcal of energy per gram, are our bodies' preferred fuels. When we eat more than we can use, small amounts of carbohydrates can be stored in the liver and in muscle tissue as **glycogen** (animal starch), a highly branched polymer of alpha-glucose. Large excesses, however, are converted to fat for storage. Most health authorities recommend obtaining carbohydrates from a diet rich in whole grains, fruits, vegetables, and legumes (beans). We should minimize our intake of the simple sugars and refined starches found in many prepared foods.

Cellulose is the most abundant carbohydrate. It is present in all plants, forming their cell walls and other structural features. Wood is about 50% cellulose, and cotton is almost pure cellulose. Unlike starch, cellulose cannot be digested by humans or other meat-eating animals. We get no caloric value from dietary cellulose because its glucose units are joined by beta linkages, and most animals lack the enzymes needed to break this beta glycoside bonding. Certain bacteria that live in the gut of termites and in the digestive tract of grazing animals such as cows do have such enzymes, however, so that these animals can convert cellulose to glucose. Cellulose does serve as *dietary fiber* in humans (page 471).

16.2 FATS AND CHOLESTEROL

Fats—esters of fatty acids and glycerol—are high-energy foods, yielding about 9 kcal of energy per gram. Some fats are "burned" as fuel for our activities. Others are used to build and maintain important constituents of our cells, such as cell

▲ **Figure 16.2** Infants born with galactosemia can thrive on a milk-free substitute formula.

▲ **Figure 16.3** Bread, flour, cereals, and pasta are rich in starches.

► **Figure 16.4** Cream, butter, margarine, cooking oils, and foods fried in fat are rich in fats. The average American diet contains too much fat; about 34% of our calories come from fat.

membranes. The fat in our diet comes from many sources, some of which are shown in Figure 16.4.

Digestion and Metabolism of Fats

Dietary fats are mainly *triacylglycerols*, commonly called **triglycerides**. Fats are digested by enzymes called *lipases*, ultimately forming fatty acids and glycerol (Figure 16.5). Some fat molecules are hydrolyzed only to a monoglyceride (*monoacylglycerol*; glycerol combined with only one fatty acid) or a diglyceride (*diacylglycerol*; ester of two fatty acids and glycerol). Once absorbed, these products of fat digestion are reassembled into triglycerides, which are attached to proteins for transportation through the bloodstream.

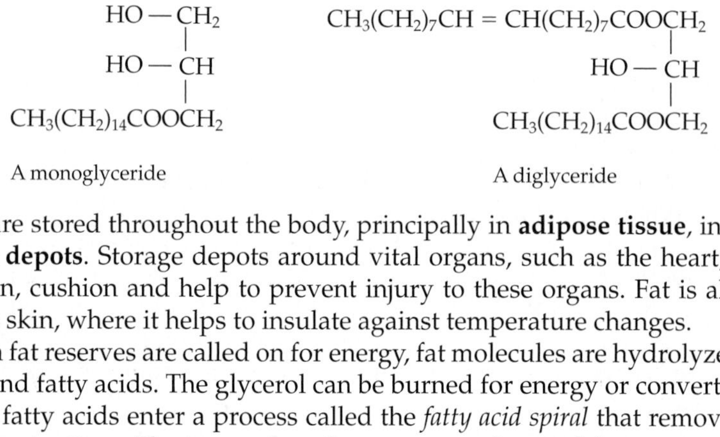

$$HO - CH_2$$
$$HO - CH$$
$$CH_3(CH_2)_{14}COOCH_2$$

A monoglyceride

$$CH_3(CH_2)_7CH = CH(CH_2)_7COOCH_2$$
$$HO - CH$$
$$CH_3(CH_2)_{14}COOCH_2$$

A diglyceride

Fats are stored throughout the body, principally in **adipose tissue**, in locations called **fat depots**. Storage depots around vital organs, such as the heart, kidneys, and spleen, cushion and help to prevent injury to these organs. Fat is also stored under the skin, where it helps to insulate against temperature changes.

When fat reserves are called on for energy, fat molecules are hydrolyzed back to glycerol and fatty acids. The glycerol can be burned for energy or converted to glucose. The fatty acids enter a process called the *fatty acid spiral* that removes carbon atoms two at a time. The two-carbon fragments can be used for energy or for the synthesis of new fatty acids.

Fats, Cholesterol, and Human Health

Dietary *saturated fats* and cholesterol have been implicated in *arteriosclerosis* ("hardening of the arteries"). (Recall from Section 15.5 that saturated fats are those made of a

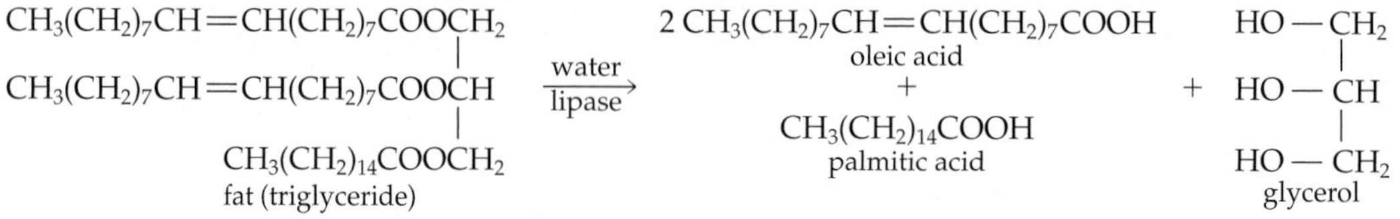

$$CH_3(CH_2)_7CH = CH(CH_2)_7COOCH_2$$
$$CH_3(CH_2)_7CH = CH(CH_2)_7COOCH \xrightarrow[\text{lipase}]{\text{water}}$$
$$CH_3(CH_2)_{14}COOCH_2$$
fat (triglyceride)

$$2\ CH_3(CH_2)_7CH = CH(CH_2)_7COOH$$
oleic acid
$$+$$
$$CH_3(CH_2)_{14}COOH$$
palmitic acid

$$HO - CH_2$$
$$+ \quad HO - CH$$
$$HO - CH_2$$
glycerol

▲ **Figure 16.5** In the digestion of fats, triglyceride is hydrolyzed to fatty acids and glycerol in a reaction catalyzed by the enzyme lipase.

Question: What diglyceride is formed by removal of the palmitic acid part of the triglyceride? What monoglyceride is formed by removal of both oleic acid parts? *Hint:* You may refer to Chapter 15, if necessary.

Emulsions

When oil and water are vigorously shaken together, the oil is broken up into tiny, microscopic droplets and dispersed throughout the water. Such a mixture is called an *emulsion*. Unless a third substance has been added, the emulsion usually breaks down rapidly, the oil droplets recombining and floating to the surface of the water.

Emulsions can be stabilized by adding certain types of gum, a soap, or a protein that can form a protective coating around the oil droplets and prevent them from coming together. Compounds called *bile salts* keep tiny fat droplets suspended in aqueous media during human digestion.

Many foods are emulsions. Milk is an emulsion of butterfat in water. The stabilizing agent is a protein

called *casein*. Mayonnaise is an emulsion of salad oil in water, stabilized by egg yolk.

large proportion of saturated fatty acids, such as palmitic and stearic.) Incidence of cardiovascular disease is strongly correlated with diets rich in saturated fats. As the disease develops, deposits form on the inner walls of arteries. Eventually these deposits harden, and the vessels lose their elasticity (Figure 16.6). Blood clots tend to lodge in the narrowed arteries, leading to a heart attack (if the blocked artery is in heart muscle) or a stroke (if the blockage occurs in an artery that supplies the brain).

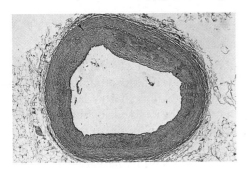
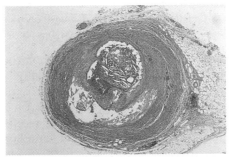

◀ **Figure 16.6**
Photomicrographs of cross sections of a normal artery and a hardened artery, showing deposits of plaque that contain cholesterol.

The plaque in clogged arteries is rich in cholesterol, a fatlike steroid alcohol found in animal tissues and various foods. Cholesterol is normally synthesized by the liver and is important as a constituent of cell membranes and a precursor to steroid hormones. High blood levels of cholesterol, like those of triglycerides, correlate closely with the risk of cardiovascular disease. Like fats, cholesterol is insoluble in water, as we can determine from its molecular formula ($C_{27}H_{45}OH$) and structure (margin). Cholesterol is transported in blood by water-soluble proteins. The cholesterol–protein combination is called a lipoprotein. A **lipoprotein** is any of a group of proteins combined with a lipid, such as cholesterol or a triglyceride.

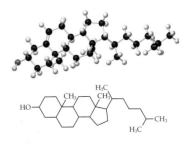

Lipoproteins are usually classified according to their density (Table 16.1). Very-low-density lipoproteins (VLDLs) serve mainly to transport triglycerides, whereas low-density lipoproteins (LDLs) are the main carriers of cholesterol. LDLs carry cholesterol to the cells for use, and these lipoproteins are the ones that deposit cholesterol in arteries, leading to cardiovascular disease. High-density lipoproteins (HDLs) also carry cholesterol, but they carry it to the liver for processing

TABLE 16.1	Lipoproteins in the Blood			
Class	**Abbreviation**	**Protein (%)**	**Density (g/mL)**	**Main Function**
Very low density	VLDL	5	1.006–1.019	Transport triglycerides
Low density	LDL	25	1.019–1.063	Transport cholesterol to the cells for use
High density	HDL	50	1.063–1.210	Transport cholesterol to the liver for processing for excretion

and excretion. Exercise is thought to increase the levels of HDL, the lipoprotein sometimes called "good" cholesterol. High levels of LDL, called "bad" cholesterol, increase the risk of heart attack and stroke. The American Heart Association recommends a maximum of 300 mg/day of cholesterol for the general population and 200 mg/day for people with heart disease or at risk for it.

Fats differ in their effect on blood cholesterol levels. Many nutritionists advise us to use olive oil and canola oil as our major sources of dietary lipids. These oils contain a high percentage of monounsaturated fatty acids, which have been shown to lower LDL cholesterol. There is also statistical evidence that fish oils can prevent heart disease. For example, Greenlanders who eat a lot of fish have a low risk of heart disease despite a diet that is high in total fat and cholesterol. The probable effective agents are polyunsaturated fatty acids such as eicosapentaenoic acid (EPA) and docosahexaenoic acid (DHA):

$$CH_3(CH_2CH{=}CH)_5(CH_2)_3COOH \qquad CH_3(CH_2CH{=}CH)_6(CH_2)_2COOH$$
$$\text{EPA} \qquad\qquad\qquad\qquad \text{DHA}$$

These fatty acids are known as *omega-3 fatty acids* because they have a carbon–carbon double bond that begins on the *third* carbon from the end opposite the COOH group—the *omega* end. Studies have also shown that diets with added omega-3 fatty acids lead to lower cholesterol and triglyceride levels in the blood.

Fats and oils containing double bonds can undergo hydrogenation. Hydrogenation of vegetable oils to produce semisolid fats is an important process in the food industry. The chemistry of this conversion process is essentially identical to the hydrogenation reaction described for alkenes in Chapter 9.

$$CH_3(CH_2)_7CH{=}CH(CH_2)_7COOH \xrightarrow[\text{Ni}]{H_2} CH_3(CH_2)_7CH_2CH_2(CH_2)_7COOH$$
$$\text{Oleic Acid (monounsaturated)} \qquad\qquad \text{Stearic acid (saturated)}$$

By properly controlling the reaction conditions, inexpensive vegetable oils (cottonseed, corn, soybean) can be partially hydrogenated, forming soft and pliable fats suitable for use in oleomargarine, or fully hydrogenated into harder fats like shortening. The consumer would get much greater unsaturation by using the oils directly, but most people would rather spread margarine than pour oil on their toast.

Concern about the role of saturated fats in raising blood cholesterol and clogging arteries caused many consumers to switch from butter to margarine. However, recently it was found that some of the unsaturated fats that remain after partial hydrogenation of vegetable oils have structures similar to saturated fats. These unsaturated fats raise cholesterol levels and increase the risk of coronary heart disease.

Figure 16.7 provides molecular models of three types of fatty acids. Most naturally occurring unsaturated fatty acids have a *cis* arrangement about the double bond. For example, oleic acid has the structure given by Figure 16.7(a). Recall from Chapter 9 that a *cis* arrangement about double-bonded carbon atoms has both hydrogen atoms on the same side of the double bond. During hydrogenation, in some of the molecules the arrangement is changed so that the hydrogen atoms of the double-bonded carbon atoms are on opposite sides of the double bond, a *trans* arrangement Figure 16.7(b). (Draw a straight line passing through the double-bonded carbon atoms. If the hydrogen atoms on those carbon atoms are on the same side of the line, the molecule represents a *cis* fatty acid. If they fall on opposite sides, it is a *trans* fatty acid.)

Note in Figures 16.7(a) and (b) that both saturated fatty acids (such as stearic acid) and *trans* fatty acids are more or less straight and can stack neatly together like logs. This maximizes the intermolecular attractive forces, making saturated and *trans* fatty acids both more likely to be solids than the *cis* fatty acids. On the other hand, *cis* fatty acids (Figure 16.7c) have a bend in their structure, fixed in position by the double bond. They have weaker intermolecular forces and are more likely to be liquids. Because *trans* fatty acids also resemble saturated fatty acids in their tendency to raise blood levels of LDL cholesterol, the FDA required manufacturers to label foods with the *trans* fat content, effective January 1, 2006.

Half of the total fat, three-fourths of the saturated fat, and all the cholesterol in a typical human diet come from animal products such as meat, milk, cheese, and eggs. Advertising that a vegetable oil (for example) contains no cholesterol is silly. No vegetable product contains cholesterol.

▲ Food labels now must include the trans fat content.

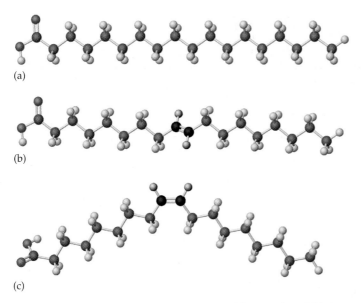

(a)

(b)

(c)

▲ **Figure 16.7** The structures of (a) stearic acid (a saturated fatty acid) and (b) the *trans* isomer of oleic acid (a *trans* unsaturated fatty acid) are very similar in shape and behave similarly in the body, while (c) oleic acid (a *cis* unsaturated fatty acid) has a very different shape.

A lot of Americans eat too much fat, especially saturated and *trans* fats. Health experts generally advise us to limit our intake of fats in general, and saturated and *trans* fats in particular. Many health professionals recommend that fat should not exceed 30% of total calories and no more than one-third of the fat should be saturated fat.

EXAMPLE 16.1   **Nutrient Calculations**

A single-serving pepperoni pan pizza has 38 g of fat and 780 total calories. (Recall that a food calorie is a kilocalorie.) Estimate the percentage of the total calories from fat.

Solution
First, calculate the calories from fat. Fat furnishes about 9 kcal/g, so 38 g of fat furnishes

$$38 \text{ g fat} \times \frac{9 \text{ kcal}}{1 \text{ g fat}} = 340 \text{ kcal}$$

Now divide the calories from fat by the total calories. Then multiply by 100% to get the percentage (parts per 100).

$$\% \text{ calories from fat} = \frac{340 \text{ kcal}}{780 \text{ kcal}} \times 100\% = 44\%$$

(This answer is only an estimate for two reasons: The value of 9 kcal/g is only approximate and the "38 g of fat" is known only to the nearest gram. A proper answer is "about 44%.")

▶ **Exercise 16.1A**
A serving of two fried chicken thighs has 51 g of fat and furnishes 720 kcal. What percentage of the total calories is from fat?

▶ **Exercise 16.1B**
A lunch consists of a regular hamburger with 12 g of fat and 275 kcal, a serving of french fried potatoes with 12 g of fat and 240 kcal, and a chocolate milk shake with 9 g of fat and 360 kcal. What percentage of the total calories is from fat?

EXAMPLE 16.2   **Fatty Acids**

Following are line-angle structures of three fatty acids. **(a)** Which is a saturated fatty acid? **(b)** Which is a monounsaturated fatty acid? **(c)** Which is an omega-3 fatty acid?

I.

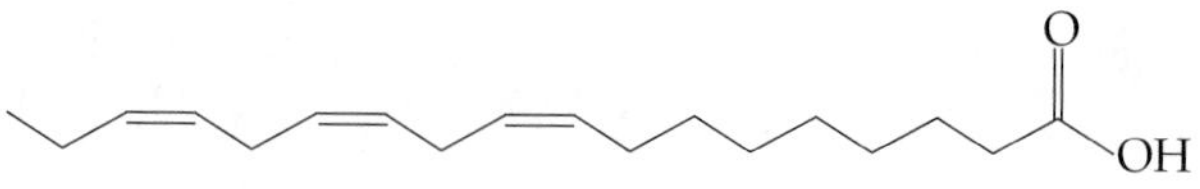

II.

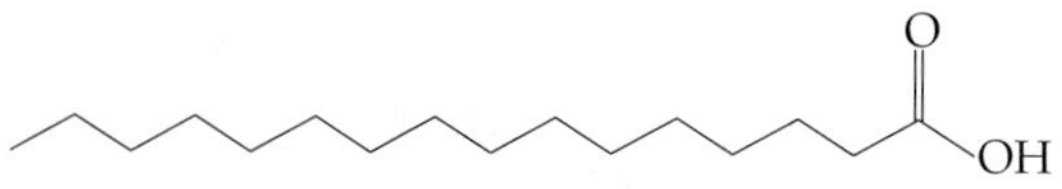

III.

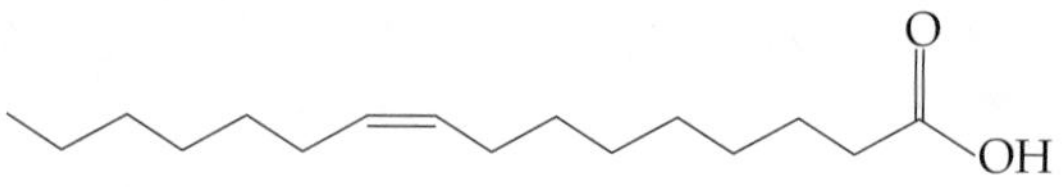

Solution

a. Fatty acid II has no carbon-to-carbon double bonds; it is a saturated fatty acid.
b. Fatty acid III has one carbon-to-carbon double bond; it is a monounsaturated fatty acid.
c. Fatty acid I has one of its carbon-to-carbon double bonds three carbons removed from the end farthest from the carboxyl group (the omega end); it is an omega-3 fatty acid.

▶ **Exercise 16.2A**

Which of the fatty acids in Example 16.2 is polyunsaturated?

▶ **Exercise 16.2B**

Is fatty acid III likely to be more similar physiologically to fatty acid II or to fatty acid I? Explain.

16.3 PROTEINS: MUSCLE AND MUCH MORE

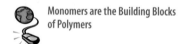 Monomers are the Building Blocks of Polymers

As we saw in Chapter 15, proteins are polymers of amino acids. A gene carries the blueprint for a specific protein, and each protein serves a particular purpose. We require protein in our diet in order to provide the amino acids needed to make muscles, hair, enzymes, and many other cellular components vital to life.

Protein Metabolism: Essential Amino Acids

Proteins are broken down in the digestive tract into their component amino acids.

$$\text{Proteins} + n\,\text{H}_2\text{O} \xrightarrow{\text{Proteases}} \text{Amino acids}$$

From these amino acids, our bodies synthesize proteins for growth and repair of tissues. When a diet contains more protein than is needed for the body's growth and repair, the excess protein is used as a source of energy, providing about 4 kcal per gram.

The adult human body can synthesize all but nine of the amino acids needed for making proteins. These nine are called **essential amino acids**—isoleucine, lysine, phenylalanine, tryptophan, leucine, methionine, threonine, and valine—(see Table 15.3 for structures) and must be included in our diet. Each of the essential amino acids is a **limiting reactant** in protein synthesis. When the body is deficient in one of them, it can't make proper proteins.

An *adequate* (or *complete*) *protein* supplies all the essential amino acids in the quantities needed for the growth and repair of body tissues. Most proteins from animal sources contain all the essential amino acids in adequate amounts. Lean meat,

The distinction between essential and nonessential amino acids is somewhat ambiguous, because some amino acids can be produced from others. For example, both methionine and cysteine contain sulfur, and although methionine cannot be synthesized from other precursors, cysteine can partially meet the need for methionine. Similarly, tyrosine can partially substitute for phenylalanine.

milk, fish, eggs, and cheese supply adequate protein. Gelatin is one of the few inadequate animal proteins. It contains almost no tryptophan and has only small amounts of threonine, methionine, and isoleucine.

In contrast, most plant proteins are deficient in one or more amino acids. Corn protein has insufficient lysine and tryptophan, and people who subsist chiefly on corn may suffer from malnutrition even though they get adequate calories. Protein from rice is short of lysine and threonine. Wheat protein lacks enough lysine. Even soy protein, one of the best nonanimal proteins, is deficient in the essential amino acid methionine.

Protein Deficiency in Young and Old

Our requirement for protein is about 0.8 g per kilogram of body weight. Diets with inadequate protein are common in some parts of the world. A protein-deficiency disease called *kwashiorkor* (Figure 16.8) is rare in the developed countries but is common during times of famine in parts of Africa where corn is the major food. In the United States, protein deficiency occurs mainly among the elderly persons in nursing homes.

Nutrition is especially important in a child's early years. This is readily apparent from the fact that the human brain reaches nearly full size by the age of two years. Early protein deficiency leads to both physical and mental retardation.

▲ **Figure 16.8** An extreme lack of proteins and vitamins causes a deficiency disease called kwashiorkor. The symptoms include retarded growth, discoloration of skin and hair, bloating, a swollen belly, and mental apathy.

Vegetarian Diets

Green plants trap a small fraction of the energy that reaches them from the sun. They use some of this energy to convert carbon dioxide, water, and mineral nutrients (including nitrates, phosphates, and sulfates) to proteins. Cattle eat plant protein, digest it, and convert a small portion of it to animal protein. It takes 100 g of protein feed to produce 4.7 g of edible beef or veal protein, an efficiency of only 4.7%. People eat this animal protein, digest it, and reassemble some of the amino acids into human protein. Some of the energy originally transformed by green plants is lost as heat at every step. If people ate the plant protein directly, one highly inefficient step would be skipped. A vegetarian diet conserves energy. Pork production, at a protein conversion efficiency of 12.1%, and chicken or turkey production (at 18.2%) are more efficient. Milk production (22.7%) and egg production (23.3%) are still more efficient but do not compare well with eating the protein directly.

Vegetarians generally are less likely than meat eaters to have high blood pressure. Vegetarian diets that are low in saturated fat can help us avoid or even reverse coronary artery disease. These diets also offer protection from some other diseases. However, although complete proteins can be obtained by eating a carefully selected mixture of vegetable foods, total (*vegan*) vegetarianism can be dangerous, especially for young children. Even when the diet includes a wide variety of plant materials, an all-vegetable diet is short in several nutrients, including vitamin B_{12} (a nutrient not found in plants), calcium, iron, riboflavin, and vitamin D (required by children not exposed to sunlight). A modified vegetarian (*ovolacto*) diet that includes eggs, milk, and milk products can provide excellent nutrition, with red meat totally excluded.

A variety of ethnic dishes supply relatively good protein by combining vegetable foods, usually a cereal grain with a legume (beans, peas, peanuts, and so on). The grain is deficient in tryptophan and lysine, but it has sufficient methionine. The legume is deficient in methionine, but it has enough tryptophan and lysine. A few such combinations are listed in Table 16.2. Peanut butter sandwiches are a popular American example of a legume–cereal grain combination.

TABLE 16.2	Ethnic Foods that Combine a Cereal Grain with a Legume
Group	**Food***
Mexicans	Corn tortillas and refried beans
Japanese	Rice and soybean curds (tofu)
English working classes	Baked beans on toasted bread
American Indians	Corn and beans (succotash)
Western Africans	Rice and peanuts (ground nuts)
Cajuns (Louisiana)	Red beans and rice

*Cereal grains are in red, and legumes are in blue.

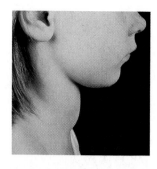

▲ **Figure 16.9** A person with goiter. The swollen thyroid gland in the neck results from a dietary deficiency of the trace element iodine.

Iron provides an example of a substance that can be both essential and toxic. The National Institutes of Health recommends daily intake of 8 mg for adult males, 18 mg for adult females. The Institute also points out that a 200 mg dose of iron in children can cause death.

16.4 MINERALS: INORGANIC CHEMICALS AND LIFE

Several inorganic substances, called **dietary minerals,** are vital to life. Minerals represent about 4% of the weight of the human body. Some of these minerals, such as chlorides (Cl^-), phosphates (PO_4^{3-}), bicarbonates (HCO_3^-), and sulfates (SO_4^{2-}), occur in the blood and other body fluids. Others, such as iron (as Fe^{2+}) in hemoglobin and phosphorus in nucleic acids (DNA and RNA) are constituents of complex organic compounds.

Table 16.3 lists the 30 elements known to be essential to one or more living organisms. The 11 that comprise the structural elements and the macrominerals make up more than 99% of all the atoms in the human body. The 19 trace elements include iron, zinc, copper, and 16 others called *ultratrace elements.*

Minerals serve a variety of functions; that of iodine is quite dramatic. A small amount of iodine is necessary for proper thyroid function. An iodine deficiency has dire effects, of which goiter is perhaps the best known (Figure 16.9). Iodine is available naturally in seafood, but, to guard against iodine deficiency, a small amount of potassium iodide (KI) is often added to table salt. The use of iodized salt has greatly reduced the incidence of goiter.

Iron(II) ions (Fe^{2+}) are necessary for proper function of the oxygen-transporting compound hemoglobin. Without sufficient iron, oxygen supply to body tissues is reduced, and anemia, a general weakening of the body, results. Foods especially rich in iron compounds include red meat and liver. It appears that most adult males need very little dietary iron because iron seems to be retained by the body and lost mainly through bleeding.

Calcium and phosphorus are necessary for the proper development of bones and teeth. Growing children need about 1.5 g of each mineral per day. These elements are available from milk. The calcium and phosphorus needs of adults are less widely known but are very real just the same. For example, calcium ions are necessary for the coagulation of blood (to stop bleeding) and for maintenance of the rhythm of the heartbeat. Phosphorus occurs as phosphate units in adenosine triphosphate (ATP), the "energy currency" of the body. Phosphorus is necessary for

TABLE 16.3 Elements Essential to Life

Element	Symbol	Form Used	Element	Symbol	Form Used
Bulk Structural Elements			**Ultratrace Elements**		
Hydrogen	H	Covalent	Manganese	Mn	Mn^{2+}
Carbon	C	Covalent	Molybdenum	Mo	Mo^{2+}
Oxygen	O	Covalent	Chromium	Cr	?
Nitrogen	N	Covalent	Cobalt	Co	Co^{2+}
Phosphorus*	P	Covalent	Vanadium	V	?
Sulfur*	S	Covalent	Nickel	Ni	Ni^{2+}
			Cadmium	Cd	Cd^{2+}
Macrominerals			Tin	Sn	Sn^{2+}
Sodium	Na	Na^+	Lead	Pb	Pb^{2+}
Potassium	K	K^+	Lithium	Li	Li^+
Calcium	Ca	Ca^{2+}	Fluorine	F	F^-
Magnesium	Mg	Mg^{2+}	Iodine	I	I^-
Chlorine	Cl	Cl^-	Selenium	Se	SeO_4^{2-}?
Phosphorus*	P	$H_2PO_4^-$	Silicon	Si	?
Sulfur*	S	SO_4^{2-}	Arsenic	As	?
			Boron	B	H_3BO_3
Trace Elements					
Iron	Fe	Fe^{2+}			
Copper	Cu	Cu^{2+}			
Zinc	Zn	Zn^{2+}			

*Note that phosphorus and sulfur each appear twice; they are structural elements and are also components of the macrominerals phosphate and sulfate.

the body to obtain, store, and use energy from foods. As phosphate units, phosphorus is an important part of the nucleic acid backbone (Section 15.9).

Sodium ions and chloride ions make up sodium chloride (salt). In moderate amounts, salt is essential to life. It is important in the exchange of fluids between cells and plasma, for example. The presence of salt increases water retention, however, and a high volume of retained fluids can cause swelling and high blood pressure (hypertension). Most physicians agree that our diets generally contain too much salt. About 65 million people in the United States suffer from hypertension, the leading risk factor for stroke, heart attack, kidney failure, and heart failure. Antihypertensives are among the most widely prescribed drugs in the United States.

Iron, copper, zinc, cobalt, manganese, molybdenum, calcium, and magnesium are essential to the proper functioning of metalloenzymes, which are life sustaining. A great deal remains to be learned about the role of inorganic chemicals in our bodies. Bioinorganic chemistry is a flourishing area of research.

16.5 THE VITAMINS: VITAL, BUT NOT ALL ARE AMINES

Why are British sailors called "limeys"? And what does this term have to do with food? Sailors on long voyages have been plagued since early times by *scurvy*. In 1747 Scottish navy surgeon James Lind showed that this disease could be prevented by including fresh fruit and vegetables in the diet. Fresh fruits conveniently carried on long voyages (there was no refrigeration) were limes, lemons, and oranges. British ships put to sea with barrels of limes aboard, and sailors ate a lime or two every day. That is how they came to be known as "lime eaters," or simply limeys.

In 1897 the Dutch scientist Christiaan Eijkman showed that polished rice lacked something found in the hull of whole-grain rice. Lack of that substance caused the disease *beriberi*, which was a serious problem in the Dutch East Indies at that time. A British scientist, F. G. Hopkins, found that rats fed a synthetic diet of carbohydrates, fats, proteins, and minerals were unable to sustain healthy growth. Again, something was missing.

In 1912 Casimir Funk, a Polish biochemist, coined the word *vitamine* (from the Latin word *vita*, meaning "life") for these missing factors. Funk thought all these factors contained an amine group. In the United States, the final *e* was dropped after it was found that not all the factors were amines. The generic term became *vitamin*. Eijkman and Hopkins shared the 1929 Nobel Prize in physiology and medicine for their discoveries relating to vitamins.

Vitamins are specific organic compounds that are required in the diet (in addition to the usual proteins, fats, carbohydrates, and minerals) to prevent specific diseases. Some vitamins, along with their sources and deficiency symptoms, are shown in Table 16.4. The role of vitamins in the prevention of deficiency diseases, such as those shown in Figure 16.10, has been well established.

Vitamins do not share a common chemical structure. They can, however, be divided into two broad categories: *fat-soluble vitamins* (A, D, E, and K) and *water-soluble vitamins* (B vitamins and vitamin C). The fat-soluble vitamins incorporate a high proportion of hydrocarbon elements. They contain one or two oxygen atoms but are only slightly polar as a whole.

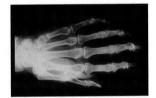
(a)

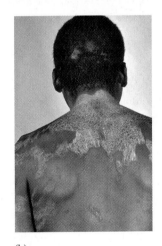
(b)

▲ **Figure 16.10** (a) This X-ray shows softened bones caused by a deficiency of vitamin D. (b) Inflammation and abnormal pigmentation characterize pellagra, caused by a deficiency of niacin.

Retinol

Vitamin D₂ (calciferol)

Fat-soluble vitamins

Vitamin*	Name	Sources	Deficiency Symptoms
TABLE 16.4 Some of the Vitamins			
Fat-Soluble Vitamins			
A	Retinol	Fish, liver, eggs, butter, cheese; also a vitamin precursor in carrots and other vegetables	Night blindness
D_2	Calciferol	Cod liver oil, irradiated ergosterol (milk supplement)	Rickets
E	α-Tocopherol	Wheat germ oil, green vegetables, egg yolks, meat	Sterility, muscular dystrophy
K_1	Phylloquinone	Spinach, other green leafy vegetables	Hemorrhage
Water-Soluble Vitamins			
B_1	Thiamine	Germ of cereal grains, legumes, nuts, milk, and brewer's yeast	Beriberi—polyneuritis resulting in muscle paralysis, enlargement of heart, and ultimately heart failure
B_2	Riboflavin	Milk, red meat, liver, egg white, green vegetables, whole wheat flour (or fortified white flour), and fish	Dermatitis, glossitis (tongue inflammation)
B_3	Niacin	Red meat, liver, collards, turnip greens, yeast, and tomato juice	Pellagra—skin lesions, swollen and discolored tongue, loss of appetite, diarrhea, various mental disorders (Figure 16.10)
B_6	Pyridoxine	Eggs, liver, yeast, peas, beans, and milk	Dermatitis, apathy, irritability, and increased susceptibility to infections; convulsions in infants
	Folic acid	Liver, kidney, mushrooms, yeast, and green leafy vegetables	Anemias (folic acid is used to treat megaloblastic anemia, a condition characterized by giant red blood cells); neural tube defects in fetuses of deficient mothers
B_{12}	Cyanocobalamin	Liver, meat, eggs and fish (not found in plants)	Pernicious anemia
C	Ascorbic acid	Citrus fruits, tomatoes, green peppers	Scurvy

*Some vitamins exist in more than one chemical form. We name a common form of each here.

In contrast, a water-soluble vitamin contains a high proportion of the electronegative atoms oxygen and nitrogen, which can form hydrogen bonds to water; therefore, the molecule as a whole is soluble in water.

Vitamin C (ascorbic acid)

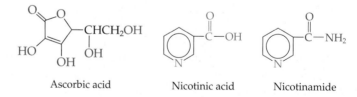

Ascorbic acid   Nicotinic acid   Nicotinamide

Water-soluble vitamins (sites for hydrogen bonding in color)

Fat-soluble vitamins dissolve in the fatty tissue of the body, where reserves can be stored for future use. An adult can store several years' supply of vitamin A. If the diet becomes deficient in vitamin A, these reserves are mobilized, and the adult remains free of the deficiency disease for quite a while. However, a small child who has not built up a store of the vitamin soon exhibits deficiency symptoms. Children in developing countries often are permanently blinded as a result of vitamin A deficiency.

Because they are efficiently stored in the body, overdoses of fat-soluble vitamins can have adverse effects. Large excesses of vitamin A cause irritability, dry skin, and a feeling of pressure inside the head. Too much vitamin D can cause pain in the bones, hard deposits in the joints, nausea, diarrhea, and weight loss. Vitamins E and K are also fat soluble, but they are metabolized and excreted. They are not stored to the extent that vitamins A and D are, and excesses seldom cause problems.

The body has a limited capacity to store water-soluble vitamins. It excretes anything over the amount that can be used immediately. Water-soluble vitamins are needed frequently, every day or so. Some foods lose their vitamin content when they are cooked in water and then drained. The water-soluble vitamins go down the drain with the water.

16.6 OTHER ESSENTIALS: FIBER AND WATER

We need carbohydrates, proteins, fats, minerals, and vitamins, but some other items in our diets are also important. One is *fiber;* another is *water.*

Dietary Fiber

Dietary fiber may be soluble or insoluble. The insoluble fiber is usually cellulose, while the soluble fibers are generally sticky materials called gums and pectins. High-fiber diets prevent constipation and are an aid to dieters. Fiber has no calories because the body doesn't absorb it. Therefore, high-fiber foods such as fruits and vegetables are low in fat and often low in calories. Fiber takes up space in the stomach, making us feel full, and therefore eat less food. Soluble fiber lowers cholesterol levels, perhaps by removing bile acids that digest fat. It may also help control blood sugar by delaying stomach emptying, thus slowing sugar absorption after a meal. This may reduce the amount of insulin needed by a diabetic. These properties make dietary fiber beneficial to people with high blood pressure, diabetes, heart disease, and diverticulitis.

It's a Drug! No, It's a Food! No, It's . . . a Dietary Supplement!

Almost every day we see advertisements for products that are claimed to promote rapid weight loss, improve stamina, aid memory, or provide other seemingly miraculous benefits. Although these products may appear to act as drugs, many are not classified as such. Legally, they are *dietary supplements.* The Dietary Supplement Health and Education Act, enacted by the U.S. Congress in 1994, changed the law so that dietary supplements are now regulated as foods and not as drugs. Manufacturers are permitted to describe some specific benefits that may be attributed to use of the supplement. However, they must also include a disclaimer, such as: "This statement has not been evaluated by the Food and Drug Administration. This product is not intended to diagnose, treat, cure, or prevent any disease."

Some supplements are simply combinations of various vitamins. Others contain minerals, amino acids, herbs or other plant materials, other nutrients, or ingredients and extracts of animal and plant origin.

Should you take a particular supplement? Maybe; maybe not. A supplement containing omega-3 oils might well aid in reducing LDL cholesterol. But a tablespoon a day of a special preparation of vitamins and minerals is not likely to reverse the effects of aging, no matter how brightly colored the bottle, no matter how

Supplement Facts

Serving size 1 Tablet

Amount Per Serving	%DV
Vitamin B-6 (as pyridoxine HCl)......5 mg	250%
Folic Acid....................................400 mcg	250%
Vitamin B-12............................1000 mcg	16666%

DIRECTIONS: As a dietary supplement, take 1 or 2 tablets daily. Allow tablet to dissolve under tongue. Conforms to USP <2091> for weight.

This statement has not been evaluated by the Food and Drug Administration. This product is not intended to diagnose, treat, cure, or prevent any disease.

▲ Most dietary supplements do not go through the rigorous testing that is required by law for all prescription drugs and over-the-counter drugs.

beautiful the person in the advertisement. Perhaps a judicious application of the FLaReS principles (see this chapter's Critical Thinking Exercises) to the manufacturer's claims will provide insight.

Water

Much of the food we eat is mainly water. It should come as no surprise that tomatoes are 90% water or that melons, oranges, and grapes are largely water. But water is one of the main ingredients in practically all foods, from roast beef and seafood to potatoes and onions.

In addition to the water we get in our food, we need to drink about 1.0–1.5 L of water each day. We could satisfy this need by drinking plain water, but often we choose other beverages—milk, coffee, tea, and soft drinks. Many people also drink beverages that contain ethyl alcohol. These are made by fermenting grains or fruit juices. Beer is usually made from malted barley, and wine from grape juice. Even these alcoholic beverages are mainly water.

16.7 STARVATION AND FASTING

When the human body is totally deprived of food, whether voluntarily or involuntarily, the condition is known as **starvation.** Involuntary starvation is a serious problem in much of the world. Although starvation is seldom the sole cause of death, those weakened by malnutrition succumb readily to disease. Even a seemingly minor disease, such as measles, can become life threatening.

Metabolic changes similar to those occurring in starvation take place during fasting. During total fasting, the body's glycogen stores are depleted in less than a day and the body calls on its fat reserves. Fat is first taken from around the kidneys and the heart. Then it is removed from other parts of the body, eventually even from the bone marrow.

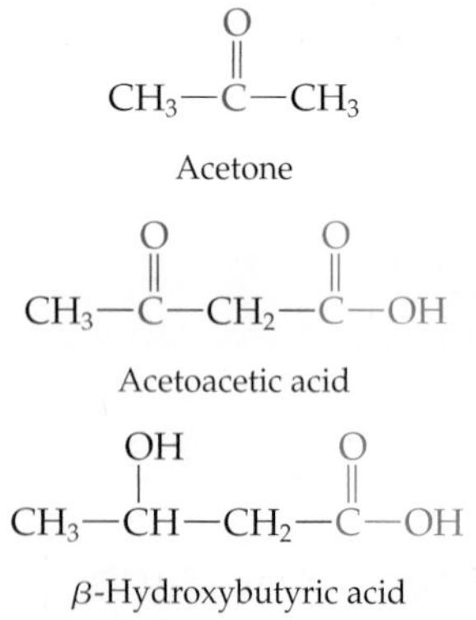

Increased dependence on stored fats as an energy source leads to *ketosis*, a condition characterized by the appearance of compounds called *ketone bodies* in the blood and urine (Figure 16.11). Ketosis rapidly develops into *acidosis*; the blood pH drops, and oxygen transport is hindered. Oxygen deprivation leads to depression and lethargy.

In the early stages of a *total* fast, body protein is metabolized at a relatively rapid rate. After several weeks, the rate of protein breakdown slows considerably as the brain adjusts to using the breakdown products of fatty acid metabolism for its energy source. When fat reserves are substantially depleted, the body must again draw heavily on its structural proteins for its energy requirements. The emaciated appearance of a starving individual is due to the depletion of muscle proteins.

Processed Food: Less Nutrition

Malnutrition need not be due to starvation or dieting. It can be the result of eating too much highly processed food.

Whole wheat is an excellent source of vitamin B_1 and other vitamins. To make white flour, the wheat germ and bran are removed from the grain. This greatly increases the storage life of the flour, but the remaining material has few minerals or vitamins and little fiber. We eat the starch and use much of the germ and bran for animal food. Our cattle and hogs often get better nutrition than we do. Similarly, polished rice has had most of its protein and minerals removed, and it has almost no vitamins. The disease beriberi became prevalent when polished rice was introduced into Southeast Asia.

When many fruits and vegetables are peeled, they lose most of their vitamins, minerals, and fiber. The peels are often dumped (directly or through a garbage disposal) into a sewage system, where they contribute to water pollution (Chapter 13). The heat used to cook food also destroys some vitamins. If water is used in cooking, some of the water-soluble vitamins (C and B complex) and some of the minerals are often drained off and discarded with the water.

It is estimated that 90% of the food budget of an average family in the United States goes to buy processed foods. A diet of hamburgers, potato chips, and colas is lacking in many essential nutrients. Highly processed convenience foods threaten

In 2002, worldwide consumption of beverages included an estimated 195 billion L of milk, 139 billion L of coffee, 185 billion L of soda pop, 143 billion L of beer, and 131 billion L of bottled water.

Weight loss through diet and exercise is discussed in Chapter 18.

$CH_3-\overset{\overset{O}{\|}}{C}-CH_3$

Acetone

$CH_3-\overset{\overset{O}{\|}}{C}-CH_2-\overset{\overset{O}{\|}}{C}-OH$

Acetoacetic acid

$CH_3-\overset{\overset{OH}{|}}{CH}-CH_2-\overset{\overset{O}{\|}}{C}-OH$

β-Hydroxybutyric acid

▲ **Figure 16.11** The three ketone bodies.

Question: Identify the functional groups in each of the compounds. Which ketone body is not a ketone?

to leave the people of developed nations obese but poorly nourished despite their abundance of food.

Food Additives

The label reads "egg whites, vegetable oils, nonfat dry milk, lecithin, mono- and diglycerides, propylene glycol monostearate, xanthan gums, sodium citrate, aluminum sulfate, artificial flavor, iron phosphate, niacin, riboflavin, and irradiated ergosterol." Ingredients on a food label are listed in decreasing order by weight, but you recognize only a few of the ingredients. Just what is in the food we eat? There are more than 3000 substances that together comprise a list, which can be found on the FDA's website, called *Everything Added to Food in the United States* (*EAFUS*). We examine some of those in the following sections.

16.8 ADDITIVES TO ENHANCE OUR FOOD

On the label of many processed foods, most of the substances listed on the label are **food additives**, substances other than basic foodstuffs that are present in food as a result of some aspect of production, processing, packaging, or storage. Because food processing removes certain essential food substances, some additives are included in prepared food to increase its nutritional value. Other substances are added to enhance color and flavor, to retard spoilage, to provide texture, to sanitize, to bleach, to ripen (or prevent ripening), to control moisture levels, or to control foaming. There are several thousand different additives. Sugar, salt, and corn syrup are used in the greatest amounts. These three, plus citric acid, baking soda, vegetable colors, mustard, and pepper, make up more than 98% (by weight) of all additives.

Some food additives, such as salt to preserve meat and fish and spices to flavor and preserve foods, have been used since ancient times. Throughout the centuries, other additives were found to be useful. Movement of the population from farms to cities has increased the necessity for using preservatives. More widespread consumption of convenience foods has also led to greater use of additives.

In the United States, food additives are regulated by the Food and Drug Administration (FDA). The original Food, Drug, and Cosmetic Act was passed by Congress in 1938. Under this act, the FDA had to prove that an additive was unsafe before its use could be prevented. The Food Additives Amendment of 1958 shifted the burden of proof to the food industry. A company that wishes to use a food additive must first furnish proof to the FDA that the additive is safe for the intended use. The FDA can also regulate the quantity of additives that can be used.

Some people express concern about the "chemicals in our food," but food itself consists of chemicals. Table 16.5 shows the chemical composition of a typical breakfast. Many of the chemicals in this breakfast might be harmful in large amounts but are harmless in the trace amounts that occur naturally in foods. Indeed, some make important contributions to delightful flavors and aromas.

Our bodies are also collections of chemicals. If broken down into its elements (Table 16.6), your body would be worth only a few dollars. It is the unique combination and arrangement of the elements in every human body that make you different from everyone else and make each individual's worth beyond measure. Since food is chemical and we are chemical, we shouldn't have to worry about chemicals in our food—except perhaps for some specific ones.

Additives That Improve Nutrition

The first nutrient supplement approved by the Bureau of Chemistry of the U.S. Department of Agriculture, potassium iodide (KI), was added to table salt in 1924 to reduce the incidence of goiter (Figure 16.9). (The Bureau of Chemistry later became the FDA.)

TABLE 16.5 Your Breakfast—As Seen by a Chemist[*]			

Chilled Melon

Starches	Anisyl propionate
Sugars	Amyl acetate
Cellulose	Ascorbic acid
Pectin	Vitamin A
Malic acid	Riboflavin
Citric acid	Thiamine
Succinic acid	

Scrambled Eggs

Ovalbumin	Lecithin
Conalbumin	Lipids (fats)
Ovomucoid	Fatty acids
Mucin	Butyric acid
Globulins	Acetic acid
Amino acids	Sodium chloride
Lipovitellin	Lutein
Livetin	Zeazanthine
Cholesterol	Vitamin A

Sugar-Cured Ham

Actomyosin	Adenosine triphosphate (ATP)
Myogen	Glucose
Nucleoproteins	Collagen
Peptides	Elastin
Amino acids	Creatine
Myoglobin	Pyroligneous acid
Lipids (fats)	Sodium chloride
Linoleic acid	Sodium nitrate
Oleic acid	Sodium nitrite
Lecithin	Sodium phosphate
Cholesterol	Sucrose

Coffee

Caffeine	Acetone
Essential oils	Methyl acetate
Methanol	Furan
Acetaldehyde	Diacetyl
Methyl formate	Butanol
Ethanol	Methylfuran
Dimethyl sulfide	Isoprene
Propionaldehyde	Methylbutanol

Cinnamon Apple Chips

Pectin	Propanol
Cellulose	Butanol
Starches	Pentanol
Sucrose	Hexanol
Glucose	Acetaldehyde
Fructose	Propionaldehyde
Malic acid	Acetone
Lactic acid	Methyl formate
Citric acid	Ethyl formate
Succinic acid	Ethyl acetate
Ascorbic acid	Butyl acetate
Cinnamyl alcohol	Butyl propionate
Cinnamic aldehyde	Amyl acetate
Ethanol	

Toast and Coffee Cake

Gluten	Mono- and diglycerides
Amylose	Methyl ethyl ketone
Amino acids	Niacin
Starches	Pantothenic acid
Dextrins	Vitamin D
Sucrose	Acetic acid
Pentosans	Propionic acid
Hexosans	Butyric acid
Triglycerides	Valeric acid
Sodium chloride	Caproic acid
Phosphates	Acetone
Calcium	Diacetyl
Iron	Maltol
Thiamine	Ethyl acetate
Riboflavin	Ethyl lactate

Tea

Caffeine	Phenyl ethyl alcohol
Tannin	Benzyl alcohol
Essential oils	Geraniol
Butyl alcohol	Hexyl alcohol
Isoamyl alcohol	

[*]The chemicals listed are those found normally in the foods. The chemical listings are not necessarily complete.

Source: Manufacturing Chemists Association, Washington, DC.

TABLE 16.6 Approximate Elemental Analysis of the Human Body	
Element	**Percent by Weight in Human Body**
Oxygen	65
Carbon	18
Hydrogen	10
Nitrogen	3
Calcium	1.5
Phosphorus	1
Potassium	0.35
Sulfur	0.25
Chlorine	0.15
Sodium	0.15
Magnesium	0.05
Iron	0.004
Trace elements to make 100%	

Several other chemicals are added to foods specifically to prevent deficiency diseases. Addition of vitamin B_1 (thiamine) to polished rice is essential in the Far East, where beriberi still is a problem. The replacement of the B vitamins thiamine, riboflavin, folic acid, and niacin (which are removed in processing) and the addition of iron, usually ferrous carbonate ($FeCO_3$), to flour is called **enrichment**. Enriched bread or pasta made from this flour still isn't as nutritious as bread made from whole wheat. It lacks vitamin B_6, pantothenic acid, zinc, magnesium, and fiber, nutrients usually provided by whole-grain flour. Despite these shortcomings, the enrichment of bread, corn meal, and cereals has essentially eliminated pellagra, a disease that once plagued the southern United States.

Vitamin C (ascorbic acid) is frequently added to fruit juices, flavored drinks, and beverages. Although our diets generally contain enough ascorbic acid to prevent scurvy, some scientists recommend a much larger intake than minimum daily requirements. Vitamin D is added to milk in developed countries, and this use of fortified milk has led to the almost total elimination of rickets. Similarly, vitamin A, which occurs naturally in butter, is added to margarine so that the substitute more nearly matches butter in nutritional quality.

If we ate a balanced diet of fresh foods, we probably wouldn't need nutritional supplements. For some people in cities, however, such a diet might be impossible, and people everywhere find that convenience foods are indeed convenient. With our usual diets rich in highly processed foods, we often need the nutrients provided by vitamin and mineral food additives.

Additives That Taste Good

Spice cake, soda pop, gingerbread, sausage, and many other foods depend on spices and other additives for most of their flavor. Cloves, ginger, cinnamon, and nutmeg are examples of natural spices, and basil, marjoram, thyme, and rosemary are widely used herbs. Natural flavors are also extracted from fruits and other plant materials.

Some esters that serve as flavors are listed in Table 9.7. Other flavor compounds are shown in Figure 16.12. Chemists can analyze natural flavors and then synthesize the components to make mixtures that resemble the natural products. Major components of natural and artificial flavors are often identical. For example, both vanilla extract and imitation vanilla owe their flavor mainly to vanillin. The natural flavor is often more complex because it contains a wider variety of chemicals than

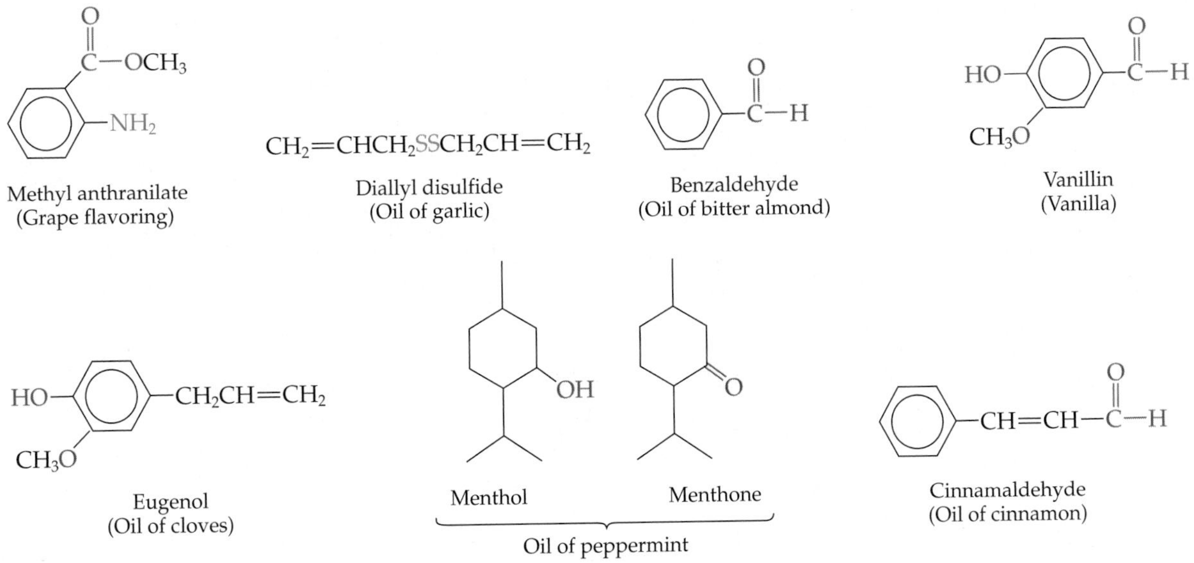

▲ **Figure 16.12** Some molecular flavorings. See Table 9.7 for others.

Question: Which of those shown here have aldehyde functional groups? Which are phenols? Ethers? Alkenes? Which is an ester? A ketone? An alcohol?

the imitation. Flavor additives, whether natural or synthetic, probably present little hazard when used in moderation, and they contribute considerably to our enjoyment of food.

Artificial Sweeteners

Obesity is a major problem in most developed countries. Presumably, we could reduce the intake of calories by replacing sugars with noncaloric sweeteners, but there is little evidence that artificial sweeteners are of value in controlling obesity. People eat more and more sugar, now averaging more than 150 lb per capita annually in the United States.

For many years, the major artificial sweeteners were saccharin and cyclamates. Cyclamates were banned in the United States in 1970 after studies showed they caused cancer in laboratory animals. (Subsequent studies have failed to confirm these findings, but the FDA has not lifted the ban.) In 1977 saccharin was shown to cause bladder cancer in laboratory animals. However, the move by the FDA to ban it was blocked by Congress because at that time saccharin was the only approved artificial sweetener. Its ban would have meant the end of diet soft drinks and low-calorie products.

There are now six FDA-approved sweeteners (Figure 16.13). Aspartame (the methyl ester of the dipeptide aspartylphenylalanine) was approved in 1981. A closely related compound, called neotame, has a 3,3-dimethylbutyl group $[CH_3C(CH_3)_2CH_2CH_2—]$ on the nitrogen atom of the aspartyl unit of the dipeptide. There are anecdotal reports of problems with aspartame, but repeated studies have shown it to be generally safe, except for people with phenylketonuria, an inherited condition in which phenylalanine cannot be metabolized properly. Other artificial sweeteners approved for use in the United States include acesulfame K (Sunette) and sucralose (Splenda), sweeteners that can survive the high temperatures of cooking processes, whereas aspartame is broken down by heat.

Table 16.7 compares the sweetness of a variety of substances. What makes a compound sweet? There is little structural similarity among the compounds. Most bear little resemblance to sugars.

Recall from Chapter 15 that sugars are polyhydroxy compounds. Like sugar, many compounds with hydroxyl groups on adjacent carbon atoms are sweet. Ethylene glycol ($HOCH_2CH_2OH$) is sweet, although it is quite toxic. Glycerol, obtained from the hydrolysis of fats (Section 16.2), is also sweet. It is used as a food additive,

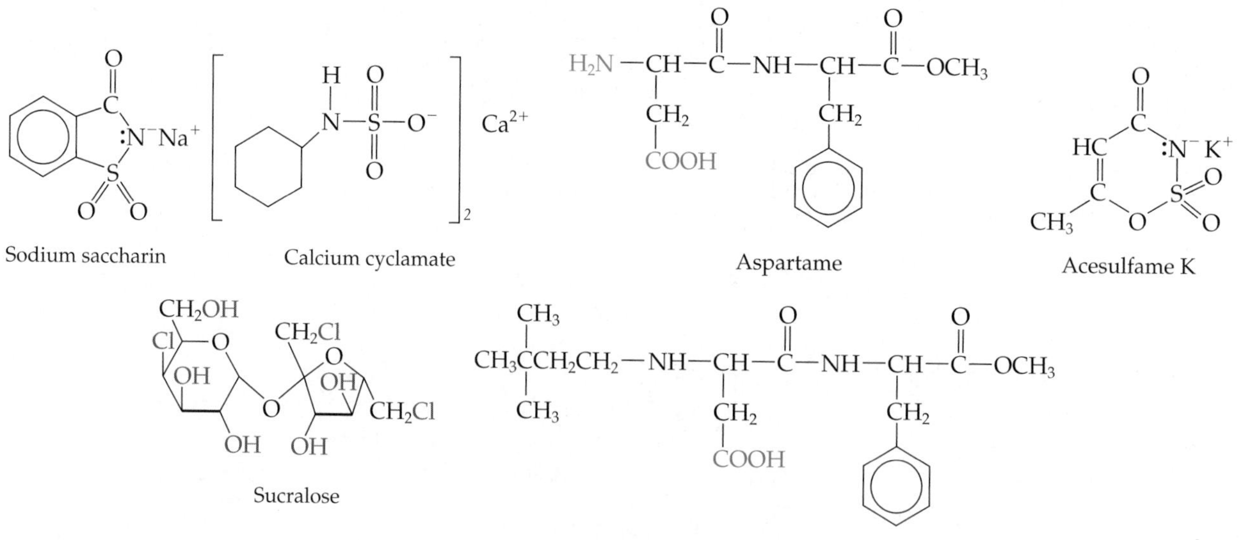

▲ **Figure 16.13** Six artificial sweeteners. Note that sucralose (Splenda) is a chlorinated derivative of sucrose. It is the only artificial sweetener actually made from sucrose.

TABLE 16.7 Approximate Sweetness of Some Compounds	
Compound	Relative Sweetness*
Lactose	0.16
Maltose	0.33
Glucose	0.74
Sucrose	1.00
Fructose	1.73
Cyclamate	45
Aspartame	180
Acesulfame K	200
Saccharin	300
Sucralose	600
Neotame	13,000

*Sweetness is relative to sucrose at a value of 1.

principally because of its properties as a **humectant** (moistening agent), and only incidentally as a sweetener.

Other polyhydroxy alcohols used as sweeteners are *sorbitol*, made by the reduction of glucose, and *xylitol*, which has five carbon atoms with a hydroxyl group on each. These compounds have an advantage over sugars in that they are not broken down in the mouth and thus do not contribute to tooth decay. This makes them useful in sugar-free chewing gums.

Several thousand sweet-tasting compounds have been discovered. They belong to more than 150 chemical classes. We now know that all the sweet substances act on a single taste receptor. (In contrast, we have more than 30 receptors for bitter substances.) Unlike any other known receptor, the sweet receptor has more than one area that can be activated by various molecules. The different areas have different affinities for the various molecules. For example, sucralose fits the receptor more tightly than sucrose does, partly because its chlorine atoms carry a stronger charge than the oxygen atoms in the OH groups they replaced. Neotame, approved by the FDA in 2002, fits the receptor so tightly it keeps the receptor firing repeatedly.

$$
\begin{array}{cc}
CH_2OH & \\
CHOH & CH_2OH \\
CHOH & CHOH \\
CHOH & CHOH \\
CHOH & CHOH \\
CH_2OH & CH_2OH \\
Sorbitol & Xylitol
\end{array}
$$

Flavor Enhancers

Some chemical substances, although not particularly flavorful themselves, are used to enhance other flavors. Common table salt (sodium chloride) is a familiar example. In addition to being a necessary nutrient, salt seems to increase sweetness and helps to mask bitterness and sourness.

Another popular flavor enhancer is monosodium glutamate (MSG). MSG is the sodium salt of glutamic acid, one of the 20 amino acids that occur naturally in proteins. It is used in many convenience foods.

$$
\begin{array}{cc}
& O \\
& \| \\
HOOC-CH_2CH_2CH-C-O^- & \\
| & \\
^+NH_3 & \\
\text{Glutamic acid} &
\end{array}
\qquad
\begin{array}{c}
O \\
\| \\
HOOC-CH_2CH_2CH-C-O^-\ Na^+ \\
| \\
NH_2 \\
\text{MSG}
\end{array}
$$

Although glutamates are found naturally in proteins, there is evidence that huge excesses can be harmful. MSG can numb portions of the brains of laboratory animals. It may also be teratogenic, causing birth defects when eaten in large amounts by women who are pregnant.

Spoilage Inhibitors

Food spoilage can result from the growth of molds, yeasts, or bacteria. Substances that prevent this are often called *antimicrobials*. Certain carboxylic acids and their salts are commonly used. Propionic acid and its sodium and calcium salts are used

CH₃CH₂COOH

Propionic acid

CH₃CH₂COO⁻ Na⁺

Sodium propionate

CH₃CH=CHCH=CHCOOH

Sorbic acid

CH₃CH=CHCH=CHCOO⁻ K⁺

Potassium sorbate

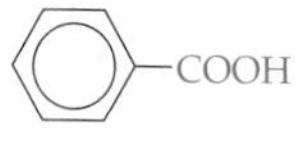

$\langle\bigcirc\rangle$—COOH

Benzoic acid

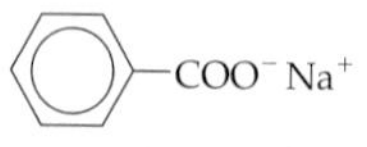

$\langle\bigcirc\rangle$—COO⁻ Na⁺

Sodium benzoate

▲ **Figure 16.14** Most spoilage inhibitors are carboxylic acids or salts of carboxylic acids.

In addition to sulfur dioxide, common sulfating agents include sodium sulfite (Na_2SO_3), potassium sulfite (K_2SO_3), sodium bisulfite ($NaHSO_3$), potassium bisulfite ($KHSO_3$), sodium metabisulfite ($Na_2S_2O_5$), and potassium metabisulfite ($K_2S_2O_5$).

to inhibit molding in bread and cheese. Sorbic acid, benzoic acid, and their salts are also used (Figure 16.14).

Some inorganic compounds are also added as spoilage inhibitors. Sodium nitrite ($NaNO_2$) is used in meat curing and to maintain the pink color of smoked hams, frankfurters, and bologna. It also contributes to the tangy flavor of processed meat products. Nitrites are particularly effective as inhibitors of *Clostridium botulinum*, the bacterium that produces botulism poisoning. However, only about 10% of the amount used to keep meat pink is needed to prevent botulism. Nitrites have been investigated as possible causes of cancer of the stomach. In the presence of the hydrochloric acid (HCl) in the stomach, nitrites are converted to nitrous acid,

$$NaNO_2 + HCl \longrightarrow HNO_2 + NaCl$$

which may then react with secondary amines (amines with two alkyl groups on nitrogen) to form nitroso compounds.

$$H-O-N=O + R-\underset{\underset{H}{|}}{N}-H \longrightarrow R-\underset{\underset{N=O}{|}}{N}-N=O + H_2O$$

Nitrous acid   A secondary amine   A nitroso compound

The R groups can be alkyl groups such as methyl (CH₃—) or ethyl (CH₃CH₂—), or they can be more complex. In any case, these nitroso compounds are among the most potent carcinogens known. The rate of stomach cancer is higher in countries that use prepared meats than in developing nations where people eat little or no cured meat. However, the incidence of stomach cancer is *decreasing* in the United States, perhaps because ascorbic acid (vitamin C) inhibits the reaction between nitrous acid and amines to form nitrosamines, and we often have orange juice with our breakfast bacon or sausage. Nevertheless, possible problems with nitrites have led the FDA to approve sodium hypophosphite (NaH_2PO_2) as a meat preservative.

Other inorganic food additives include sulfur dioxide and sulfite salts. A gas at room temperature, sulfur dioxide (SO_2) serves as a disinfectant and preservative, particularly for dried fruits such as peaches, apricots, and raisins. It is also used as a bleach to prevent the browning of wines, corn syrup, jelly, dehydrated potatoes, and other foods. Sulfur dioxide seems safe for most people when ingested with food, but it is a powerful irritant when inhaled and is a damaging ingredient of polluted air in some areas (Chapter 12). Sulfur dioxide and sulfite salts cause severe allergic reactions in some people. The FDA requires their use in foods to be indicated on the label.

Antioxidants: BHA and BHT

Antioxidants are preservatives that inhibit the chemical spoilage of food that occurs in the presence of oxygen. These substances are added to foods (or their packaging) to prevent fats and oils from forming rancid products that make the food unpalatable. Antioxidants also minimize the destruction of some essential amino acids and vitamins. Packaged foods that contain fats or oils (bread, potato chips, sausage, breakfast cereal) often have antioxidants added. Compounds commonly used include butylated hydroxytoluene (BHT), butylated hydroxyanisole (BHA), *tert*-butylhydroquinone, and propyl gallate (Figure 16.15).

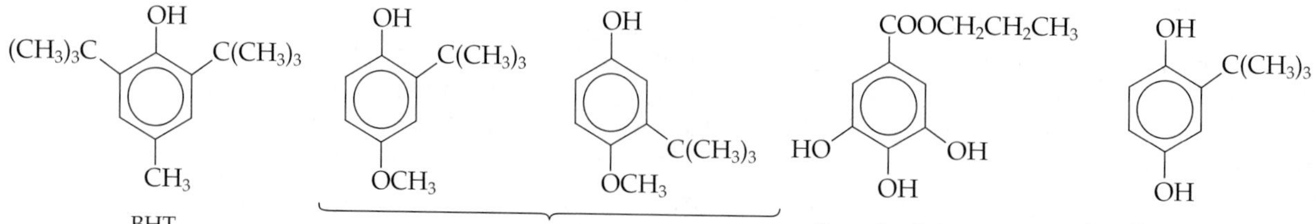

▲ **Figure 16.15** Four common antioxidants. BHA is a mixture of two isomers.

Question: What functional group is common to all four compounds? What other functional group is present in BHA? In propyl gallate?

Fats turn rancid, in part as a result of oxidation, a process that occurs through the formation of molecular fragments called **free radicals**, which have an unpaired electron as a distinguishing feature. (Recall from Chapter 5 that covalent bonds are shared *pairs* of electrons.) We need not concern ourselves with the details of the structures of radicals, but we can summarize the process. First, a fat molecule reacts with oxygen to form a free radical.

$$\text{Fat} + O_2 \longrightarrow \text{Free radical}$$

Then the radical reacts with another fat molecule to form a new free radical that can repeat the process. A reaction such as this, in which intermediates are formed that keep the reaction going, is called a **chain reaction**. One molecule of oxygen can lead to the decomposition of many fat molecules.

To preserve foods containing fats, processors package the products to exclude air, but it cannot be excluded completely, and so chemical antioxidants are used to stop the chain reaction by reacting with the free radicals.

BHT

The new radical formed from BHT is rather stable. The unpaired electron doesn't have to stay on the oxygen atom but can move around in the electron cloud of the benzene ring. The BHT radical doesn't react with fat molecules, and the chain is broken.

Why are the butyl groups important? Without them, the phenols would simply couple when exposed to an oxidizing agent.

With the bulky butyl groups aboard, the rings can't get close enough together for coupling. They are free, then, to trap free radicals formed by the oxidation of fats.

Many food additives have been criticized as being harmful, and BHA and BHT are no exceptions. They have been reported to cause allergic reactions in some people. In one study, pregnant mice fed diets containing 0.5% BHA or BHT gave birth to offspring with brain abnormalities. On the other hand, when relatively large amounts of BHT were fed to rats daily, their life spans were increased by a human equivalent of 20 years. One theory about aging is that it is caused in part by the formation of free radicals. BHT retards this chemical breakdown in cells in the same way that it retards spoilage in foods. BHA and BHT are synthetic chemicals, but there are also natural antioxidants, such as vitamin E. Vitamin E (like BHT) is a phenol, with several substituents on the benzene ring. Presumably, its action as an antioxidant is quite similar to that of BHT. However, one recent study suggests that large doses of vitamin E may also be somewhat harmful.

Food Colors

Some foods are naturally colored. For example, the yellow compound β-carotene (read "beta-carotene") occurs in carrots. β-Carotene (shown in the following line-angle formula) is used as a color additive in other foods such as butter and margarine. (Our bodies convert β-carotene to vitamin A by cutting the molecule in half at the center double bond; thus it is a vitamin additive as well as a color additive.)

Although too much vitamin A (from liver or supplements) can be toxic, if too much β-carotene is consumed, the body merely stores it in fat, eventually using or excreting it. Very large amounts of β-carotene have been known to cause the skin to turn orange! (The effect is temporary and harmless.)

Other natural food colors include beet juice, grape-hull extract, and saffron (from autumn-flowering crocus flowers).

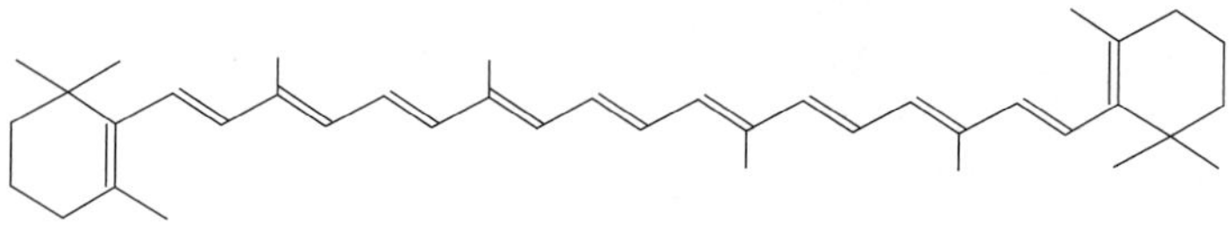

We expect many foods to have characteristic colors. To increase the attractiveness and acceptability of its products, the food industry has used synthetic food colors for decades. Since the Food and Drug Act of 1906, the FDA has regulated the use of these coloring chemicals and set limits on their concentrations. But the FDA is not infallible. Some colors once on the approved list were later shown to be harmful and were removed from the list. In 1950 a candy company used large amounts of Food, Drug, and Cosmetic (FD & C) Orange No. 1 in an attempt to give Halloween candy a bright orange color like that of pumpkins. Although safe in the smaller amounts previously used, the dye caused gastrointestinal upsets in several children and was then banned by the FDA.

In following years, other dyes were banned. FD & C Yellow Nos. 3 and 4 were found to contain small amounts of β-naphthylamine, a carcinogen that causes bladder cancer in laboratory animals. Furthermore, they reacted with stomach acids to produce more β-naphthylamine. FD & C Red No. 2 was also shown to be a weak carcinogen in laboratory animals. The structures of these dyes are given in Figure 16.16.

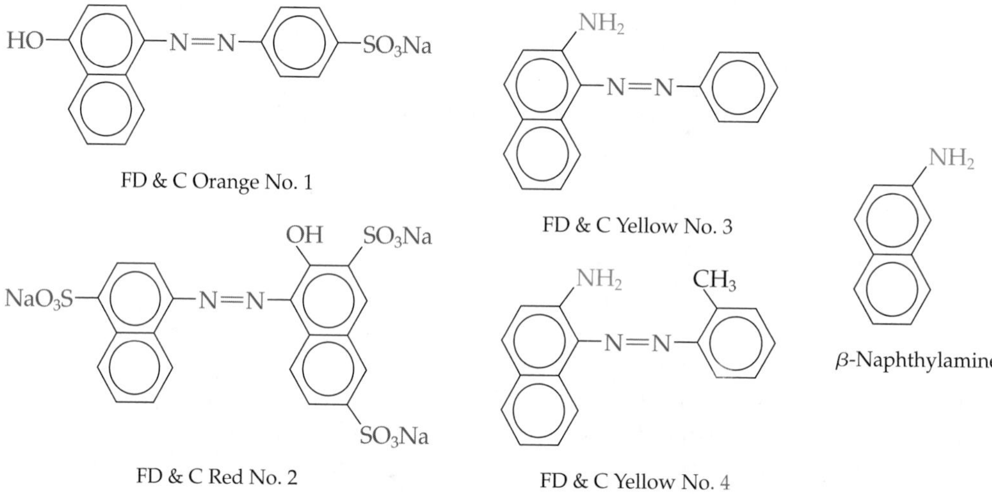

▶ **Figure 16.16** Four synthetic food colors that have been banned by the U.S. FDA. Note that the two yellow dyes are related to β-naphthylamine, a carcinogen.

Food colors, even those that have been banned, present little risk. They have been used for years with apparent safety and are normally used in tiny amounts. Although the risk is low, however, the benefit is largely aesthetic. Any foods that contain artificial colors must say so on the label, and so you can avoid them if you want to.

The GRAS List

Some food additives have been used for many years without apparent harmful effects. In 1958 the U.S. Congress established a list of additives *generally recognized as safe* (*GRAS*). Many of the **GRAS** substances are familiar spices, flavors, and nutrients. Some of the substances placed on the 1958 list have since been removed, including cyclamate sweeteners and some food colors. There were some deficiencies in the original testing procedures, and the FDA has reevaluated several of them based on new research findings—and greater consumer awareness. Improved instruments and better experimental designs have revealed possible harm, however slight, where none was thought to exist formerly. Most of the newer experiments involve feeding massive doses of additives to small laboratory animals, and these studies have been criticized for that reason. Some other substances have been added to the original list; about 7000 substances are now given GRAS status.

16.9 POISONS IN OUR FOOD

People have been trying to deal with poisons in their food for millennia. Early foragers learned—by the painful process of trial and error—that some plants and animals were poisonous. Rhubarb leaves contain toxic oxalic acid. Injured celery produces psoralens. These compounds are powerful mutagens and carcinogens. The Japanese relish a variety of puffer fish that contains deadly poison in its ovaries and liver. More than 100 Japanese die each year from improperly prepared puffers.

One of the most toxic substances known is the toxin botulin produced by the bacterium *Clostridium botulinum*. This organism grows in improperly canned food by a perfectly natural process. If the food isn't properly sterilized before it is sealed in jars or cans, the microorganism flourishes under anaerobic (without air) conditions. Botulin is so toxic that 1 g of it could kill more than a million people. The point is that a food is not inherently good simply because it is natural. Neither is it necessarily bad because a synthetic chemical substance has been added to it.

$$HO-\overset{\overset{\displaystyle O}{\|}}{C}-\overset{\overset{\displaystyle O}{\|}}{C}-OH$$

Oxalic acid

Botulin toxin (Botox®) is used in minute quantities for producing long-term (months) paralysis of muscles. Originally intended for the relief of unmanageable muscle spasms, Botox® is increasingly used for cosmetic purposes, to paralyze facial muscles to conceal wrinkles.

Carcinogens

There is little chance that we will suffer acute poisoning from approved food additives. But what about cancer? Could all these chemicals in our food increase our risk of cancer? The possibility exists, even though the risk is low.

Carcinogens occur naturally in food. A charcoal-broiled steak contains 3,4-benzpyrene, a carcinogen also found in cigarette smoke and automobile exhaust fumes. Cinnamon and nutmeg contain safrole, a carcinogen that has been banned as a flavoring in root beer.

3,4-Benzpyrene   Safrole

Among the most potent carcinogens are **aflatoxins**, compounds produced by molds growing on stored peanuts and grains (Figure 16.17). Aflatoxin B_1 is estimated to be 10 million times as potent a carcinogen as saccharin, and there is no way to completely keep it out of our food. The FDA sets a tolerance of 20 ppb for aflatoxins.

Aflatoxin B_1

◀ **Figure 16.17** Aflatoxins, toxic and carcinogenic compounds, are produced by molds that grow on peanuts and stored grains.

Scientists estimate that our consumption of natural carcinogens is 10,000 times that of synthetic carcinogens. Should we ban steaks, spices, peanuts, and grains because they contain naturally occurring carcinogens? Probably not. The risk is slight, and life is filled with more serious risks. Should we ban additives that have been shown to be carcinogenic? Certainly we should carefully weigh the risks against any benefits the additives might provide.

Incidental Additives

There are two major categories of food additives: *Intentional additives* are put in a product on purpose to perform a specific function; *incidental additives* get in accidentally during production, processing, packaging, or storage. Pesticide residues, insect parts, and antibiotics added to animal feeds are examples of incidental additives. There are about 3000 intentional additives, and perhaps 10,000 incidental additives. The incidental additives often receive wide publicity.

In 1989 the discovery of residues of daminozide (Alar) on apples caused great public concern. Daminozide is a plant growth regulator that causes apples to ripen at the same time and have a better appearance than untreated apples.

$$\underset{\text{Daminozide}}{\overset{\displaystyle O \qquad\qquad O}{\overset{\displaystyle \| \qquad\qquad \|}{HOCCH_2CH_2CNHN(CH_3)_2}}}$$

The compound itself appears to be harmless, but it breaks down to yield dimethylhydrazine $[(CH_3)_2NNH_2]$, a suspected carcinogen. To get a dose equivalent to that fed to laboratory animals, a person would have to eat 13,000 kg of apples a day for 70 years. To most scientists, the risk seemed vanishingly small, but many consumers were unwilling to assume the risk, however small. Falling sales of apples led to a withdrawal of the chemical from use by apple growers and removal of the compound from the market by manufacturers.

Alar was not the first incidental additive to cause concern. In 1959 the sale of cranberries was forbidden after some shipments were found to be contaminated by the herbicide aminotriazole, a compound shown to be carcinogenic in tests on laboratory animals. In 1969 coho salmon taken from Lake Michigan were shown to contain DDT above the tolerance level, and the sale of these fish was also banned.

Aminotriazole

Polychlorinated biphenyls (PCBs; Chapter 10) have been found in poultry and eggs. These products, too, have been seized and destroyed. Related compounds, polybrominated biphenyls (PBBs), meant to be used as fire retardants, were accidentally mixed with animal feed in western Michigan. Many farm animals were destroyed, and yet PBBs still got into the food supply.

Typical PBB compounds

Substances in animal feed often show up in the meat we eat. Antibiotics are added in low-level dosages to animal feed to promote weight gain. In fact, half or more of the 23 million kg of antibiotics produced annually in the United States goes into animal feeds. Residues of these antibiotics, sometimes found in meat, may result in the sensitization of individuals who eat the meat, thus hastening the development of allergies. Scientists fear the use of antibiotics in animal feeds will hasten the process by which bacteria become drug resistant (Chapter 19).

Diethylstilbestrol (DES), a synthetic female hormone, was once added to animal feeds to promote weight gain. It was banned after evidence showed that DES caused vaginal cancer in the daughters and testicular cancer in the sons of women who had taken it during pregnancy.

Note that the effect of DES did not show up for 15 years and that even then it appeared in the *offspring* of women who took the drug. This points out some of the problems involved in evaluating a chemical for its possible harmful effects.

Diethylstilbestrol

16.10 A WORLD WITHOUT FOOD ADDITIVES

Could we get along without food additives? Some of us could. But food spoilage might drastically reduce the food supply in some parts of an already hungry world, and diseases due to vitamin and mineral deficiencies might increase. Foods might cost more and be less nutritious. Food additives seem to be a necessary part of modern society. There are potential hazards associated with the use of some food additives, but the major problem with our food supply is still contamination by rodents, insects, and harmful microorganisms. Indeed, there are about 76 million illnesses, 325,000 hospitalizations, and 5000 deaths annually in the United States resulting from food poisoning caused by bacteria, viruses, and parasites. Few, if any, deaths associated with the use of intentional food additives have been documented.

What should we do about food additives? We should be sure that the FDA is staffed with qualified personnel to ensure the adequate testing of proposed food additives. People trained in chemistry are necessary for the control and monitoring of food additives and the detection of contaminants. Research on the analytical techniques necessary for the detection of trace quantities is vital to adequate consumer protection. We should demand laws adequate to prevent the unnecessary and excessive use of pesticides and other agricultural chemicals that might contaminate our food. Above all, we should be alert and informed about these problems that are so vital to our health and well-being.

Growing Our Food

The earliest humans obtained their food by hunting and gathering. The coming of the agricultural revolution about 10,000 years ago led to a rapid increase in population, which in turn led to a need for more productive ways to obtain food. We examine some of those ways in the following sections.

16.11 GREEN PLANTS: SUN-POWERED FOOD-MAKING MACHINES

Our food comes ultimately from green plants. All organisms can transform one type of food into another, but only green plants can use sunlight to convert carbon dioxide and water to the sugars that directly or indirectly fuel all living things.

$$6\,CO_2 + 6\,H_2O \xrightarrow{\text{sunlight}} C_6H_{12}O_6 + 6\,O_2$$

This reaction also replenishes the oxygen in the atmosphere and removes carbon dioxide (Chapter 12). With other nutrients, particularly compounds of nitrogen and phosphorus, plants can convert the sugars from photosynthesis to proteins, fats, and other chemicals that we use as food.

The structural elements of plants—carbon, hydrogen, and oxygen—are derived from air and water. Other plant nutrients are taken from the soil, and energy is supplied by the sun. In early agricultural societies, people grew plants for food and obtained energy from the food. Much of this energy was reinvested in the production

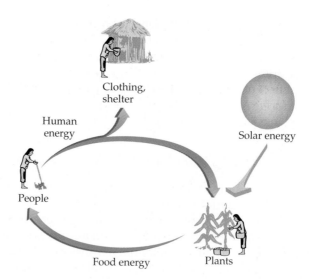

▶ **Figure 16.18** Energy flow in an early agricultural society.

Soil is rock broken down by weathering, plus various types of decaying plant and animal matter. To build even a few millimeters of soil from rocks takes many years. The United States loses 5 billion t of soil each year to erosion. Developing countries often suffer far more devastating losses as more land is cleared for farming in an attempt to feed an increasing population.

of food, although a portion went into making clothing and building shelter. Figure 16.18 shows a simplified diagram of the energy flow in such a society. A real society would be much more complicated than indicated here. Some plants, for example, might be fed to animals, and human energy would in turn be obtained from animal flesh or animal products (such as milk and eggs).

In early societies, nearly all the energy came from renewable resources. One unit of human work energy, supplemented liberally by energy from the sun, might produce 10 units of food energy. The surplus energy could be used to make clothing or to provide shelter. It also might be used in games or cultural activities.

The flow of nutrients is also rather simple. Unused portions of plants and human and animal wastes are returned to the soil. These are broken down by microorganisms to provide nutrients for the growth of new plants (Figure 16.19) and maintain the humus content of the soil. Properly practiced, this kind of agriculture could be continued for centuries without seriously depleting the soil. The only problem is that farming at this level supports relatively few people.

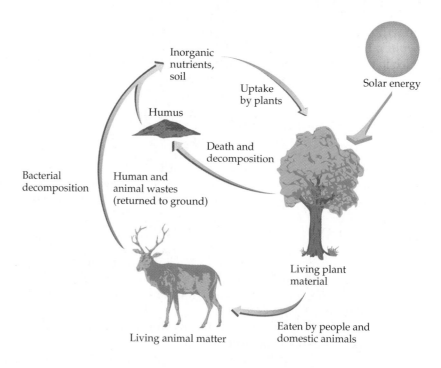

▶ **Figure 16.19** Flow of nutrients in a simple system.

16.12 FARMING WITH CHEMICALS: FERTILIZERS

To replace nutrients lost from the soil and to increase crop production, modern farmers use a variety of chemical fertilizers. The three **primary plant nutrients** are nitrogen, phosphorus, and potassium. Let's consider nitrogen first.

Nitrogen Fertilizers

Nitrogen, although present in air in the elemental form (N_2), is not generally available to plants. Some bacteria are able to *fix* nitrogen—that is, to convert it to a combined, soluble form. Colonies of bacteria that can perform this vital function grow in nodules on the roots of legumes (plants such as clovers and peas) (Figure 16.20). Thus, farmers are able to restore fertility to the soil by crop rotation. A nitrogen-fixing crop (such as clover) is alternated with a nitrogen-consuming crop (such as corn).

Legumes are still used to some extent to supply nitrogen to the soil. But the modern methods of high-yield production demand chemical fertilizers. Why get a corn crop every other year when you can have one every year?

Plants usually take up nitrogen in the form of nitrate ions (NO_3^-) or ammonium ions (NH_4^+). These ions are combined with carbon compounds from photosynthesis to form amino acids that join to form proteins, compounds essential to all life processes. For years, farmers were dependent on manure as a source of nitrates. The discovery of deposits of sodium nitrate (called Chile saltpeter) in the deserts of northern Chile led to use of this substance as a source of nitrogen.

A rapid rise in population growth during the late nineteenth and early twentieth centuries led to increasing pressure on the available food supply, and this pressure led to an increasing demand for nitrogen fertilizers. The atmosphere offered a seemingly inexhaustible supply of nitrogen—if only it could be converted to a form useful to humans. Every flash of lightning forms some nitric acid in the air (Figure 16.21). Lightning contributes about a billion tons of fixed nitrogen each year—more than is provided by nitrogen-fixing bacteria.

The first real breakthrough in nitrogen fixation came in Germany on the eve of World War I. A process developed by Fritz Haber and transformed into a large-scale method using a catalyst and high pressure by Carl Bosch made possible the combination of nitrogen and hydrogen to make ammonia.

$$3 H_2 + N_2 \longrightarrow 2 NH_3$$

By 1913 one nitrogen-fixation plant was in production and several more were under construction. The Germans were able to make ammonium nitrate (NH_4NO_3), an explosive, by oxidizing part of the ammonia to nitric acid.

$$NH_3 + 2 O_2 \longrightarrow HNO_3 + H_2O$$

The nitric acid was then reacted with ammonia to produce ammonium nitrate.

$$HNO_3 + NH_3 \longrightarrow NH_4NO_3$$

The Germans were interested in ammonium nitrate mainly as an explosive, but it turned out to be a valuable nitrogen fertilizer as well.

A gas at room temperature, ammonia is easily compressed into a liquid that can be stored and transported in tanks. In this form, called "anhydrous ammonia" (*anhydrous* means "without water"), it is applied directly to the soil as fertilizer (Figure 16.22). Some ammonia is reacted with carbon dioxide to form urea,

▲ **Figure 16.20** Bacteria in nodules on the roots of legumes fix atmospheric nitrogen by converting it to soluble compounds that plants can use as nutrients.

▲ **Figure 16.21** The intense heat in a lightning flash fixes nitrogen that then falls to Earth as nitric acid, enriching the soil.

Theft of anhydrous ammonia from farm storage tanks has become a major problem. The thieves use the potentially deadly chemical to convert the antihistamine pseudoephedrine in ordinary cold tablets into the illegal drug methamphetamine.

▲ **Figure 16.22** A farmer applies anhydrous ammonia, a source of the plant nutrient nitrogen, to his fields.

an organic compound that releases nitrogen into the soil slowly (rather than all at once as inorganic forms do).

$$2\,NH_3\ +\ CO_2\ \longrightarrow\ NH_2\!-\!\overset{\overset{\displaystyle O}{\|}}{C}\!-\!NH_2\ +\ H_2O$$
Urea

As we mentioned, some ammonia is converted to ammonium nitrate. This and some other conversions of ammonia to crystalline solids for application as fertilizers are listed in Table 16.8. These products can be applied separately or combined with other plant nutrients to make a more complete fertilizer.

The atmosphere has vast amounts of nitrogen, but current industrial methods of fixing it require a lot of energy and use nonrenewable resources. (The hydrogen needed for the synthesis of ammonia is made from natural gas.) Genes for nitrogen fixation have been transferred from bacteria into higher (nonlegume) plants, and perhaps someday soon corn and cotton will be able to produce their own nitrogen fertilizer the way clovers and peas do.

Phosphate Ion

Phosphorus Fertilizers

Availability of phosphorus is often the limiting factor in plant growth. In plants, phosphates are incorporated into DNA and RNA (Chapter 15) and other compounds essential to plant growth. Phosphates have probably been used as fertiliz-

Fritz Haber

Fritz Haber was awarded the Nobel Prize in chemistry in 1918. There are a number of ironies in this. Alfred Nobel, the Swedish inventor and chemist who died in 1896, endowed the Nobel Prize (including the peace prize) with a fortune derived from his own work with explosives. During his lifetime, he was bitterly disappointed that the explosives he had developed for excavation and mining were put to such destructive uses in war. And Haber, a man who helped his country during World War I to the extent of going to the front to supervise the release of chlorine in the first poison-gas attack, was exiled from his native land in 1933. He was Jewish, and Nazi racial laws forced him out of his position as director of the Kaiser Wilhelm Institute of

◄ Fritz Haber (1868–1934), a German chemist, invented a process for manufacturing ammonia from N_2 and H_2.

Physical Chemistry. He accepted a post at Cambridge University in England but died of a stroke less than a year later.

TABLE 16.8	Various Forms of Nitrogen Fertilizers Made from Ammonia	
Reagent	**Product Fertilizer**	**Formula**
None	Anhydrous ammonia	NH_3
Carbon dioxide	Urea	NH_2CONH_2
Sulfuric acid	Ammonium sulfate	$(NH_4)_2SO_4$
Nitric acid	Ammonium nitrate	NH_4NO_3
Phosphoric acid	Ammonium monohydrogen phosphate	$(NH_4)_2HPO_4$

ers since ancient times—in the form of bone, guano (bird droppings), or fish meal. However, phosphates as such were not recognized as plant nutrients until 1800. Following this discovery, the great battlefields of Europe were dug up and bones were shipped to chemical plants for processing into fertilizer.

Animal bones are rich in phosphorus, but the phosphorus is tightly bound and not readily available to plants. In 1831 the Austrian chemist Heinrich Wilhelm Kohler treated animal bones with sulfuric acid to convert them to a more soluble form called superphosphate. In 1843 John Lawes applied the same treatment to phosphate rock. The essential reaction for forming superphosphate is

$$Ca_3(PO_4)_2 \ + \ 2\,H_2SO_4 \ \longrightarrow \ \underbrace{Ca(H_2PO_4)_2 \ + \ 2\,CaSO_4}$$

Phosphate rock
or bone (insoluble)
　　　　　　　　　Superphosphate
　　　　　　　　　(more soluble)

Modern phosphate fertilizers are often produced by treating phosphate rock with phosphoric acid to make water-soluble calcium dihydrogen phosphate.

$$Ca_3(PO_4)_2 + 4\,H_3PO_4 \longrightarrow 3\,Ca(H_2PO_4)_2$$

More common today is the use of ammonium monohydrogen phosphate [$(NH_4)_2HPO_4$], which supplies both nitrogen and phosphorus.

Phosphates are common in the soil, but available forms are often at concentrations too low for adequate support of plant growth. Fortunately, there are more concentrated deposits. The presence of bones and teeth from early fish and other animals in these ores indicates that the deposits are largely the skeletal remains of sea creatures of ages past (Figure 16.23). Unfortunately, fluorides are often associated with phosphate ores, and they can cause a serious pollution problem during the production of phosphate fertilizers.

About 90% of all phosphates produced are used in agriculture. The United States is the leading producer and user, but the rich phosphate deposits in the United States may soon be depleted. Large offshore deposits have been discovered in the Atlantic Ocean off North Carolina. Worldwide, Morocco has about half of the reserves. All the high-grade phosphate reserves may be exhausted in 30 or 40 years. Our use of phosphates scatters them irretrievably throughout the environment.

Potassium Fertilizers

The third major element necessary for plant growth is potassium. Plants use it in the form of the simple ion K^+. Generally, potassium is abundant, and there are no problems with solubility. Potassium ions, along with Na^+ ions, are essential to the fluid

▲ **Figure 16.23** Phosphate fertilizers are obtained from an ore called rock phosphate, which is largely the skeletal remains of ancient sea creatures. This phosphate ore sample, from Morocco, contains a tooth from a shark of the Paleozoic era, between 200 and 500 million years ago.

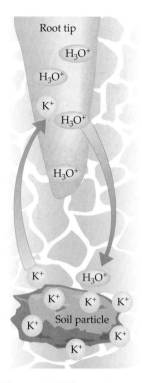

▲ **Figure 16.24** When a root tip takes up K⁺ from the soil, H_3O^+ ions are transferred to the soil. The uptake of potassium ions therefore tends to make the soil acidic.

balance of cells. They also seem to be involved in the formation and transport of carbohydrates. Also, K^+ may be necessary for the assembly of proteins from amino acids. Uptake of potassium ions from the soil leaves the soil acidic; each time one potassium ion enters the root tip, a hydronium ion must leave in order for the plant to maintain electrical neutrality (Figure 16.24).

The usual chemical form of potassium in commercial fertilizers is potassium chloride (KCl). Vast deposits of this salt occur in Stassfurt, Germany, and for years this source supplied nearly all the world's potassium fertilizer. With the coming of World War I, the United States sought supplies within its own borders. Deposits at Searles Lake, California, and Carlsbad, New Mexico, now supply much of the needs of the United States. Canada has vast deposits in Saskatchewan and Alberta. Beds of potassium chloride up to 200 m thick lie about 1.5 km below the Canadian prairies (Figure 16.25). Although reserves are large, potassium salts are a nonrenewable resource and we should use them wisely.

Other Essential Elements

In addition to the three major nutrients (nitrogen, phosphorus, and potassium), a variety of other elements are necessary for proper plant growth. Three **secondary plant nutrients**—magnesium, calcium, and sulfur—are needed in moderate amounts. Calcium, in the form of lime (calcium oxide), is used to neutralize acidic soils.

$$CaO(s) + 2 H^+(aq) \longrightarrow Ca^{2+}(aq) + H_2O(l)$$

Calcium ions are also necessary plant nutrients. Magnesium ions (Mg^{2+}) are incorporated into chlorophyll molecules and therefore are necessary for photosynthesis. Sulfur is a constituent of several amino acids, and it is necessary for protein synthesis.

Eight other elements, called **micronutrients**, are needed in small amounts. These elements are summarized in Table 16.9. Many soils contain these trace elements in sufficient quantities, but some are deficient in one or more, and their productivity can be markedly increased by adding small amounts of the needed elements.

Plants may also require other elements, including sodium, silicon, vanadium, chromium, selenium, cobalt, fluorine, and arsenic. Most of these elements are present in soil, but it is not known whether they are necessary for plant growth.

Fertilizers: A Mixed Bag

Farmers and gardeners often buy *complete fertilizers*, which, despite the name, usually contain only the three main nutrients. The three numbers—for example,

▶ **Figure 16.25** Mining potassium chloride about 1.5 km below the prairies of Saskatchewan, Canada.

TABLE 16.9   Eight Micronutrients Necessary for Proper Plant Growth

Element	Form Used by Plants	Function	Deficiency Symptoms
Boron	H_3BO_3	Required for protein synthesis; essential for reproduction and for carbohydrate metabolism	Death of growing points of stems, poor growth of roots, poor flower and seed production
Copper	Cu^{2+}	Constituent of enzymes; essential for reproduction and for chlorophyll production	Twig dieback, yellowing of newer leaves
Iron	Fe^{2+}	Constituent of enzymes; essential for chlorophyll production	Yellowing of leaves, particularly between veins
Manganese	Mn^{2+}	Essential for redox reactions and for the transformation of carbohydrates	Yellowing of leaves, brown streaks of dead tissue
Molybdenum	MoO_4^{2-}	Essential in nitrogen fixation by legumes and reduction of nitrates for protein synthesis	Stunting, pale-green or yellow leaves
Nickel	Ni^{2+}	Required for iron absorption; constituent of enzymes	Failure of seeds to germinate
Zinc	Zn^{2+}	Essential for early plant growth and maturing	Stunting, reduced seed and grain yields
Chlorine	Cl^-	Increases water content of plant tissue; involved in carbohydrate metabolism	Shriveling

5-10-5—on fertilizer bags (Figure 16.26) indicate the proportions of nitrogen, phosphorus, and potassium (NPK). The first number represents the percent of nitrogen (N); the second, the percent of phosphorus (calculated as P_2O_5); and the third, the percent of potassium (calculated as K_2O). So 5-10-5 means that a fertilizer contains 5% N, 10% P_2O_5, and 5% K_2O; the rest is inert material.

Fertilizers must be water soluble to be used by plants. When it rains, nutrients from fertilizers can be washed into streams and lakes, where they can stimulate blooms of algae. These chemicals, particularly nitrates, also enter the groundwater.

▲ **Figure 16.26** The numbers on this bag of fertilizer indicate that the fertilizer is 5% nitrogen (N), 10% P_2O_5, and 5% K_2O. There really is no K_2O or P_2O_5 in fertilizer; these formulas are used merely as a basis for calculation. The actual form of potassium is nearly always KCl, although any potassium salt would furnish the needed K^+ ion. Phosphorus is supplied as one of several salts.

EXAMPLE 16.3   **Fertilizer Production Equations**

Write an equation to show the formation of ammonium sulfate (Table 16.8) from ammonia.

Solution

We have seen that ammonium nitrate is made by reacting ammonia with nitric acid. Reacting ammonia with sulfuric acid should therefore give ammonium sulfate. In the form of an equation, we have

$$NH_3 + H_2SO_4 \longrightarrow (NH_4)_2SO_4 \quad \text{(not balanced)}$$

Balancing the equation requires the coefficient 2 for NH_3:

$$2\,NH_3 + H_2SO_4 \longrightarrow (NH_4)_2SO_4$$

▶ **Exercise 16.3A**

Write an equation to show the formation of ammonium monohydrogen phosphate (Table 16.10) from ammonia.

▶ **Exercise 16.3B**

The micronutrient zinc is often applied in the form of zinc sulfate. Write the equation for the formation of zinc sulfate from zinc oxide.

In 1912 American farmers produced an average of 26 bushels of corn per acre. In 2003 the yield per acre was 142 bushels. The fivefold increase is mainly due to the increased use of fertilizers.

16.13 THE WAR AGAINST PESTS

People have always been plagued by insect pests. Three of the ten plagues of Egypt (described in the Book of Exodus) were insect plagues—lice, flies, and locusts. The decline of Roman civilization has been attributed in part to malaria, a disease carried by mosquitoes, which destroys vigor and vitality when it does not kill. Bubonic plague carried by rats (and by fleas from rats to humans) swept through the Western world repeatedly during the Middle Ages. One such plague during the 1660s is estimated to have killed 25 million people, 25% of the population of Europe at that time. The first attempt to dig a canal across Panama (made by the French during the 1880s) was defeated by outbreaks of yellow fever and malaria.

The use of modern chemical **pesticides** (substances that kill organisms that we consider pests) may be the only thing that stands between us and some of these insect-borne plagues. Pesticides also prevent the consumption of a major portion of our food supply by insects and other pests.

In earlier days, people tried to control insect pests by draining swamps, pouring oil on ponds (to kill mosquito larvae), and using various chemicals. Most of these chemicals were compounds of arsenic. Lead arsenate $[Pb_3(AsO_4)_2]$ is a particularly effective poison because both the lead and the arsenic in it are toxic. A few pesticides, such as pyrethrum (used in mosquito control) and nicotine sulfate (Black Leaf 40), are obtained from plant matter.

Only a few insect species are harmful. Many are directly beneficial, and others play important roles in ecological systems and are indirectly beneficial. Most poisons are indiscriminate. They kill all insects, not just those we consider pests. Many are also toxic to humans and other animals. Some say we should call such poisons *biocides* (because they kill living things) rather than **insecticides** (substances that kill insects). Table 16.10 lists the toxicities of some insecticides.

TABLE 16.10 Toxicity of Insecticidal Preparations Administered Orally to Rats		
Pesticide	**LD_{50}***	
Pyrethrins[1]	1200	Least toxic
Malathion	1000 (1375)	
Lead arsenate	825	
Diazinon	285 (250)	
Carbaryl	250	
Nicotine[2]	230	
DDT[3]	118 (113)	
Lindane	91 (88)	
Methyl parathion	14 (24)	
Parathion	3.6 (13)	
Carbofuran[4]	2	
Aldicarb	1	Most toxic

*Dose in milligrams per kilogram of body weight that will kill 50% of test population (see Section 20.13). Values in parentheses are for male rats.

Notes:

(1) Active ingredients of pyrethrum.

(2) In mice. Nicotine is much more toxic by injection.

(3) Estimated LD_{50} for humans is 500 mg/kg.

(4) In mice.

Source: Susan Budavari (ed.), *The Merck Index*, 12th ed. Rahway, NJ: Merck and Co., 1996.

EXAMPLE 16.4   **Pesticide Toxicity**

What quantity of the pesticide lindane could lead to the death of a 15-kg (33-lb) child if the lethal dose is the same as it is for male rats (Table 16.10)?

Solution

The lethal dose is given as 88 mg of lindane per kilogram of body weight and the child's body weight is 15 kg.

$$\text{Lethal dose} = 15 \text{ kg} \times \frac{88 \text{ mg}}{1 \text{ kg}} = 1300 \text{ mg}$$

The lethal dose is 1300 mg or 1.3 g.

▶ **Exercise 16.4**

How much parathion would it take to kill a 75-kg (165-lb) farm worker if the lethal dose were the same as it is for rats (Table 16.10)?

DDT: The Dream Insecticide

Shortly before World War II, Swiss scientist Paul Müller (1899–1965) found that DDT (dichlorodiphenyltrichloroethane), a chlorinated hydrocarbon, is a potent insecticide. DDT was soon used effectively against grapevine pests and against a particularly severe potato beetle infestation.

When the war came, supplies of pyrethrum, a major insecticide of the time, were cut off by the Japanese occupation of Southeast Asia and the Dutch East Indies (now Indonesia). Lead, arsenic, and copper that went into insecticides were diverted for armaments and other military purposes.

The Allies, desperately needing an insecticide to protect soldiers from disease-bearing lice, ticks, and mosquitoes, obtained a small quantity of DDT and quickly tested it. Combined with talcum, DDT was an effective delousing powder. Clothing was impregnated with DDT, and it seemed to have no harmful effects even on those exposed to large doses. Allied soldiers were nearly free of lice, but German troops were heavily infested and many were sick with typhus. In wars before World War II, more soldiers probably died from typhus than from bullets.

DDT is easily synthesized from cheap, readily available chemicals. Chlorobenzene (C_6H_5Cl) and chloral hydrate [$Cl_3CCH(OH)_2$] are warmed in the presence of sulfuric acid. When the reaction is complete, the mixture is poured into water, and the DDT separates out because, like other chlorinated hydrocarbons, DDT is essentially insoluble in water.

DDT
(Dichlorodiphenyltrichloroethane)

Chlorobenzene   Chloral hydrate

DDT

A cheap insecticide effective against a variety of insect pests, DDT came into widespread use after the war (Figure 16.27). Other chlorinated hydrocarbons were synthesized, tested, and used in the war against insects. Although invaluable to farmers in the production of food and fiber, chlorinated hydrocarbons won their most dramatic victories in the field of public health. According to the World Health Organization, approximately 25 million lives have been saved and hundreds of millions of illnesses prevented by the use of DDT and related pesticides.

DDT seemed to be a dream come true. It would at last free the world from insect-borne diseases. It would protect crops from the ravages of insects and

▲ **Figure 16.27** Insecticides are often applied by aerial spraying. Winds can spread the pesticide over a wide area.

thus increase food production. In recognition of his discovery, Müller was awarded the Nobel Prize in physiology and medicine in 1948.

The Decline and Fall of DDT

Even before Müller received his prize, however, there were warnings that all was not well. Houseflies resistant to DDT were reported as early as 1946 and DDT's toxicity to fish by 1947. Such early warnings were largely ignored, and it was also assumed that the toxicity would disappear soon after the chemical was discharged into the environment. DDT was used extensively to protect crops and control mosquitoes and to try to prevent Dutch elm disease. By 1962, the year Rachel Carson's book *Silent Spring* appeared, U.S. production of DDT had reached 76 million kg/year.

Today, in developed countries, DDT is known mostly for its harmful environmental effects. Birds were threatened because DDT causes their eggs to have thin shells that are poorly formed and easily broken. Even a few parts per billion of DDT interferes with the growth of plankton and the reproduction of crustaceans such as shrimp. As bad as DDT sounds, however, it has probably saved the lives of more people than any other chemical substance.

Chlorinated hydrocarbons generally are unreactive. This lack of reactivity was a major advantage of DDT. Sprayed on a crop, DDT stayed there and killed insects for weeks. This *pesticide persistence* was also a major disadvantage: The substance did not break down readily in the environment. Although not very toxic to humans and other warm-blooded creatures, DDT is much more toxic to cold-blooded organisms, which include insects, of course, and fish. DDT's lack of reactivity toward oxygen, water, and components of the soil led to its build-up in the environment, where it threatened fish, birds, and other wildlife.

Biological Magnification: Concentration in Fatty Tissues

Chlorinated hydrocarbons are good solvents for fats, and fats are good solvents for chlorinated hydrocarbons such as DDT. When these compounds are ingested as contaminants in food or water, they are concentrated in fatty tissues. Their fat-soluble nature causes chlorinated hydrocarbons to be concentrated up the food chain. This *biological magnification* was graphically demonstrated in California in 1957. Clear Lake, about 100 miles north of San Francisco, was sprayed with DDT in an effort to control gnats. After spraying, the water contained only 0.02 ppm of DDT, but the microscopic plant and animal life contained 5 ppm—250 times as much. Fish feeding on these microorganisms contained up to 2000 ppm. Grebes, the diving birds that ate the fish, died by the hundreds (Figure 16.28).

DDT and other chlorinated hydrocarbons such as PCBs (Chapter 10) are nerve poisons. They are concentrated in the fatlike compounds that make up nerve sheaths. DDT also interferes with calcium metabolism essential to the formation of healthy bones and teeth. Although harm to humans has never been conclusively demonstrated, the disruption of calcium metabolism in birds was disastrous for some species. The shells of their eggs, composed mainly of calcium compounds,

Insect Resistance

In the six decades or so that pesticides have been widely used, hundreds of insect species have become resistant to one or more of the commonly used pesticides. This has resulted in increases in crop losses and in insect-transmitted diseases. Some strains of insect species such as the cotton bollworm, the Colorado potato beetle, malaria-transmitting *Anopheles* mosquitoes, and the German cockroach exhibit resistance to all commercially available pest control agents.

Insects develop resistance to a pesticide by natural selection. A genetic mutation for resistance can spread quickly: For example, if a farmer uses a new pesticide on a field with 10.0 billion pest insects, perhaps 9.9 billion are killed. Resistant insects survive and pass the gene for resistance to the next generation. The resistant insect group becomes a larger percentage of the whole. By repeating this process through several generations, the resistant insects become the predominant form and the pesticide is no longer effective.

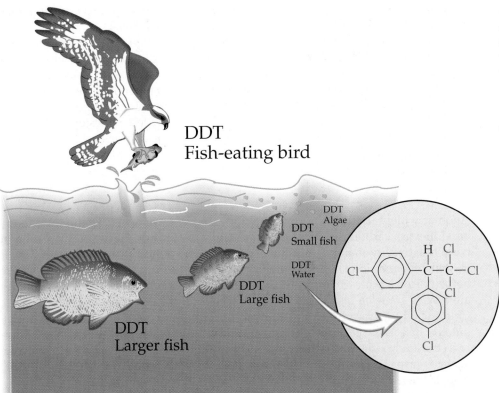

DDT
Fish-eating bird

◄ **Figure 16.28** Concentration of DDT up the food chain. Animals at the top of the food chain have the highest concentrations of the pesticide. DDT dissolves in the fatty membranes of cells. The wedge-shaped DDT molecules open channels in the membranes, causing them to leak.

were thin and poorly formed. The bald eagle, peregrine falcon, brown pelican, and other birds became endangered species but have made dramatic recoveries since DDT was banned in the United States and other industrialized countries in the 1970s.

DDT is still used in some countries where mosquito-borne malaria and typhus are great health problems. It is applied primarily inside buildings and by selective spraying. The environmental impact is much less than the widespread use of DDT in agriculture in earlier years. Malaria infects between 300 million and 500 million people and kills 1 million every year. About 90% of the deaths occur in Africa and most of them in children under 5. In 2000, more than 120 countries signed a treaty to phase out *persistent organic pollutants* (POPs), a group of chemicals, many of them chlorinated hydrocarbons, including DDT. This treaty has the goal of reducing and ultimately eliminating the use of DDT, but individual countries may continue to use it for controlling malaria.

Organic Phosphorus Compounds

Bans and restrictions on chlorinated hydrocarbon insecticides have led to increased use of organic phosphorus compounds such as malathion, diazinon, and parathion (Figure 16.29). More than two dozen of these insecticides are available commercially.

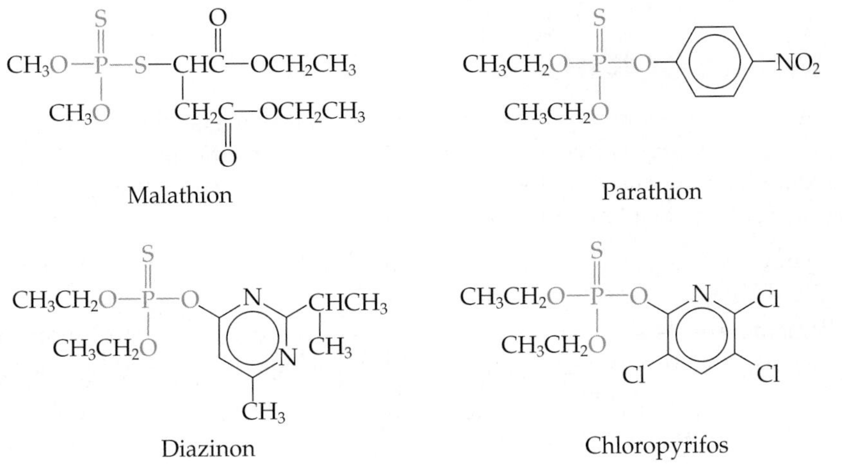

Malathion

Parathion

Diazinon

Chloropyrifos

◄ **Figure 16.29** Four organic phosphorus compounds used as insecticides.

They have been extensively studied and evaluated for effectiveness against insects and for toxicity to people, laboratory animals, and farm animals.

Organic phosphorus compounds are nerve poisons (Section 20.11); they interfere with the conduction of nerve signals of insects. In cases of high levels of exposure, they do the same in humans. Most are more toxic to mammals than chlorinated hydrocarbons (Table 16.10), but malathion is a notable exception; it is less toxic than DDT. Like chlorinated hydrocarbons, organic phosphorus pesticides concentrate in fatty tissues. Phosphorus compounds are less persistent in the environment; they break down in days or weeks, whereas chlorinated hydrocarbons often persist for years. Residues of organic phosphorus compounds are seldom found in food.

Carbamates

Another widely used family of insecticides are the carbamates. Examples are carbaryl (Sevin), carbofuran (Furadan), and aldicarb (Temik) (Figure 16.30). Carbaryl has a low toxicity to mammals, but carbofuran and aldicarb have toxicities similar to that of parathion. Like the organic phosphorus compounds, carbamates are nerve poisons, but carbamates have a shorter duration of action. Most carbamates are *narrow-spectrum insecticides* directed specifically at one, or a few, insect pests. Chlorinated hydrocarbons and organic phosphorus compounds that kill many kinds of insects are called *broad-spectrum insecticides*. Carbamates break down rapidly in the environment. Carbaryl, a widely used carbamate, is particularly effective against wasps but unfortunately is also quite toxic to honeybees. Millions of these valuable insects have been wiped out by spraying crops with carbaryl.

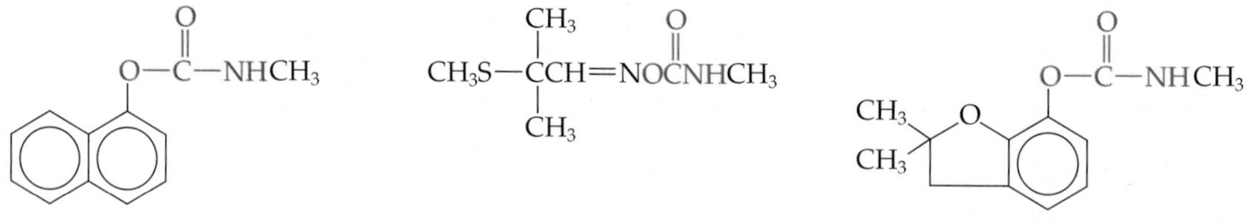

▶ **Figure 16.30** Three carbamate insecticides.

Carbaryl (Sevin)   Aldicarb (Temik)   Carbofuran (Furadan)

Carbamates are characterized by the presence of the carbamate group:

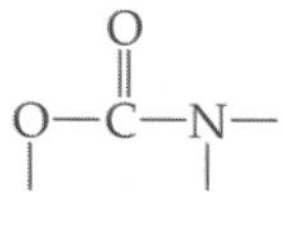

Many variations can be obtained by changing the groups attached to the O and the N. Generally, carbamates break down fairly rapidly in the environment and do not accumulate in fatty tissue, which is an advantage of these insecticides over chlorinated hydrocarbons and organic phosphorus compounds.

A carbamate insecticide was being made at a chemical plant in Bhopal, India, in December 1984 when an explosion released large amounts of the toxic gas methyl isocyanate.

16.14 BIOLOGICAL INSECT CONTROLS

The use of natural enemies is another method to control pests. These biological controls include predatory insects, mites, and mollusks; parasitic insects; and microbial controls, which are insect pathogens such as bacteria, viruses, fungi, and nematodes. Following are a few examples:

- Predatory insects such as praying mantises and ladybugs are sold commercially and are used to destroy garden pests.
- Parasitic insects such as *Trichogramma* wasps are used to control moth larvae such as tomato hornworm, corn earworm, cabbage looper, codling moth, cutworm, and armyworm.

- *Bacillus thuringiensis (Bt)*, sold under trade names such as Dipel, is widely used by home gardeners against cabbage loopers, hornworms, and other moth caterpillars. Bt is also used against Colorado potato beetles, mosquito larvae, black flies, European corn borers, grape leaf rollers, and gypsy moths.
- Naturally occurring nuclear polyhedrosis virus (NPV) is present at low levels in many insect populations. These viruses can be grown in culture and applied to crops. NPV is highly selective, affecting only *Heliothis* caterpillars. Natural viral pesticides appear to be harmless to humans, wildlife, and beneficial insects. They are completely biodegradable, but their production is generally quite expensive.

Microbial controls such as Bt are host specific: They don't harm organisms other than the targeted ones. The gene for the toxin produced by Bt has been inserted into cotton, corn, potato, and other plants. Such genetically modified (GM) cotton plants are protected against the cotton bollworm. These GM plants have had some success, but insects are already developing resistance to the toxin. Some people fear the introduced genes will spread to wild relatives of cultivated species, leading to the development of "super weeds." People also fear that proteins coded for in GM food plants will cause allergic reactions in susceptible individuals. For example, introduction of a gene for a peanut protein into another food plant could cause a dangerous reaction in people allergic to peanuts.

A highly successful biological approach is the breeding of insect- and fungus-resistant plants. Yields of corn (maize), wheat, rice, and other grains have been increased substantially in this manner. Not all such research is successful, however. Plant breeders have produced a potato that is insect resistant, but it had to be taken off the market because it also is toxic to people. Public fears of food from genetically modified plants have slowed developments, particularly in Europe.

> To protect themselves from being eaten, some plants produce their own pesticides, often making up 5–10% of the dry weight of the plants. Nicotine protects tobacco plants. Pyrethrins are also plant-produced pesticides. Physostigmine is a naturally occurring carbamate. Our consumption of natural pesticides is probably 10,000 times as great as that of synthetic ones.

Sterile Insect Technique

A sterile insect technique (SIT) method of insect control involves rearing large numbers of males; sterilizing them with radiation, chemicals, or cross-breeding; and then releasing them in areas of infestation. These sterile males, which far outnumber the local fertile males, mate with wild females. If a female mates with a sterile male, no offspring are produced, thus reducing their reproductive potential. This leads to control of the insect pest population. Successful SIT programs have been conducted against screwworm, a serious pest that affects cattle, in the southern United States, Mexico and Central America; and also in Libya. SIT programs have also been employed against the Mediterranean fruit fly (medfly) in Latin America and against the codling moth in Canada. The great expense and limited applicability of this process probably mean that it will not become a major method of insect control.

Pheromones: The Sex Trap

Substances called *pheromones* are increasingly important in insect control. A **pheromone** is a chemical that is secreted externally by an insect to mark a trail, send an alarm, or attract a mate. Insect **sex attractants** are usually secreted by females to attract males (Figure 16.31). Chemical research has identified pheromones of hundreds of insect species. Most are blends of two or more chemicals that must be present in exactly the right proportions to be biologically active. Chemists can then synthesize these compounds, and workers can use them to lure male insects into traps to determine what pests are present and the level of infestation. Then they can undertake measures, including the use of conventional pesticides, to minimize damage to the crop. If the attractant is exceptionally powerful and the insect population level is very low, a pheromone trap technique called "attract and kill" might achieve sufficient control. Alternatively, the attractant can be used in quantities sufficient to confuse and disorient males, who detect a female in

▲ **Figure 16.31** A male gypsy moth uses its large antennae to detect pheromones from a female.

every direction but can't find one to mate with. Mating disruption has been used successfully in controlling some insect pests. Many grape growers in Germany and Switzerland use this technique, allowing them to produce wine without using conventional insecticides.

Some sex attractants have relatively simple structures. The sex attractant for the codling moth, which infests apples, is an alcohol with a straight chain of 12 carbon atoms and two double bonds.

However, most attractants have complicated structures, and all are secreted in extremely tiny amounts. This can make research on sex attractants difficult and tedious. For example, a team of U.S. Department of Agriculture researchers had to use the tips of 87,000 female gypsy moths to isolate a minute amount of a powerful sex attractant.

Pheromones are effective at extremely low concentrations. A male silkworm moth can detect as few as 40 molecules per second. If a female releases as little as 0.01 mg, she can attract every male within 1 km. Gypsy moth larvae have defoliated great forests, mainly in the northeastern United States, and have now spread over much of the country. The gypsy moth pheromone has been used mainly in traps to monitor insect populations.

Pheromones are usually too expensive to play a huge role in insect control, though recombinant DNA methods (Section 15.12) hold promise for future work. For now the method is costly, and research is painstaking and time-consuming. Workers must be careful not to get the attractants on their clothes. Who wants to be attacked on a warm summer night by a million sex-crazed gypsy moths?

Juvenile Hormones

Some insects can be controlled by the use of *juvenile hormones*. Hormones are the chemical messengers that control many life functions in plants and animals, and minute quantities produce profound physiological changes. In the insect world, a **juvenile hormone** controls the rate of development of the young. Normally, production of the hormone is shut off at the appropriate time to allow proper maturation to the adult stage.

Chemists have been able to isolate insect juvenile hormones and determine their structures. With knowledge of the structure, they can synthesize the hormone or an analog. The application of juvenile hormones to ponds where mosquitoes breed keeps mosquitoes in the harmless preadult stage. Because only adult insects can reproduce, juvenile hormones appear to be a nearly perfect method of mosquito control.

A natural juvenile hormone

Methoprene, a juvenile hormone analog, is approved by the EPA for use against mosquitoes and fleas.

Methoprene

The synthesis of juvenile hormones is difficult and expensive, and they are only for use against insects that are pests at the adult stage. Little would be gained by keeping a moth or a butterfly in the caterpillar stage for a longer period of time; caterpillars have voracious appetites and do a lot of damage to crops.

16.15 HERBICIDES AND DEFOLIANTS

The United States produces about 900 million kg of pesticides annually, about half of which is **herbicides**. Herbicides, four of which are shown in Figure 16.32, are chemicals used to kill weeds. Closely related are **defoliants**, which cause leaves to fall off plants.

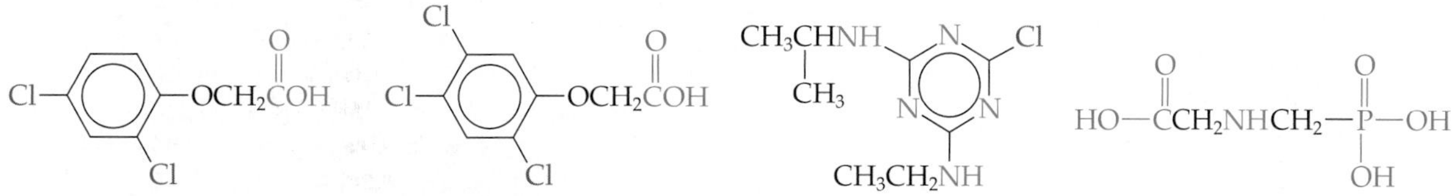

2,4-Dichlorophenoxyacetic acid (2,4-D)

2,4,5-Trichlorophenoxyacetic acid (2,4,5-T)

Atrazine

Glyphosate

2,4-D and 2,4,5-T

Crops that have no competition from weeds produce more abundant harvests. Weeds cause estimated crop losses of $6 billion a year in the United States. Removing weeds by hand and hoe is tedious, backbreaking work, and so chemical herbicides have been employed for a number of years to kill unwanted plants. Early herbicides included solutions of copper salts, sulfuric acid, and sodium chlorate ($NaClO_3$), but it wasn't until the introduction of 2,4-D (2,4-dichlorophenoxyacetic acid or one of its derivatives) in 1945 that the use of herbicides became common. These chemicals are growth regulator herbicides and are especially effective against newly emerged, rapidly growing broad-leaved plants. A relative of 2,4-D, called 2,4,5-T (2,4,5-trichlorophenoxyacetic acid), is especially effective against woody plants; it works by causing leaves to fall off plants (defoliation). Combined in a formulation called **Agent Orange**, 2,4-D and 2,4,5-T were used extensively in Vietnam to remove enemy cover and to destroy crops that maintained enemy armies. Agent Orange caused vast ecological damage, and is suspected of causing birth defects in children born to both American soldiers and Vietnamese exposed to the herbicides. Laboratory studies show that 2,4-D and 2,4,5-T, when pure, do not cause abnormalities in fetuses of laboratory animals. Extensive birth defects are caused, however, by contaminants called **dioxins**, once frequently found in the herbicides. The most toxic dioxin is 2,3,7,8-tetrachlorodibenzo-*para*-dioxin, abbreviated as 2,3,7,8-TCDD, or just TCDD. Continuing concern about dioxin contamination led the EPA to ban 2,4,5-T in 1985.

Atrazine and Glyphosate

The most widely used herbicides in the United States are atrazine and glyphosate. Atrazine binds to a protein in chloroplasts in plant cells, shutting off the electron transfer reactions of photosynthesis. Atrazine is often used on corn crops. Corn plants deactivate atrazine by removing the chlorine atom. Weeds cannot deactivate the compound and are killed. Some studies indicate that atrazine interferes with the hormone system, disrupting estrogen function. Atrazine has been linked to sexual abnormalities in frogs.

Glyphosate, a derivative of the amino acid glycine, is used to control perennial grasses. It is not selective, however, and kills all vegetation by inhibiting the function of a certain plant enzyme. Glyphosate is metabolized by bacteria in the soil, so that other plants can be sown or transplanted into treated areas shortly after spraying. It is sold under several trade names, including Round-Up®.

Paraquat: A Preemergent Herbicide

A **preemergent herbicide** is one used to kill weed plants before crop seedlings emerge. Paraquat, an ionic compound toxic to most plants but that is rapidly broken down in the soil, is a notable example. Paraquat inhibits photosynthesis by accepting the electrons that otherwise would reduce carbon dioxide.

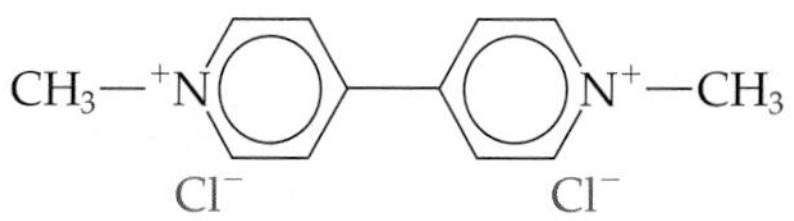

▲ **Figure 16.32** Four common herbicides. (a) 2,4-dichlorophenoxyacetic acid (2,4-D), (b) 2,4,5-trichlorophenoxyacetic acid (2,4,5-T), (c) atrazine, and (d) glyphosate.
Question: What alkyl groups are on the atrazine molecule?

One of the earliest widely used defoliants was calcium cyanamide (CaNCN), which causes cotton plants to lose their leaves when the cotton bolls become mature. This defoliant makes it possible to use mechanical cotton pickers to harvest cotton. Leaves left on the plants would be crushed by the machinery and would stain the cotton.

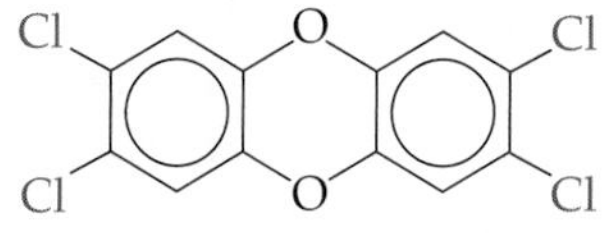

2,3,7,8-Tetrachlorodibenzo-para-dioxin (a "dioxin")

Paraquat

Pesticides—Risks and Benefits

The U.S. Environmental Protection Agency estimates that 10,000–20,000 physician-diagnosed pesticide poisonings occur each year among agricultural workers. Many agricultural poisoning cases are never reported, and many other incidents occur around the home, lawn, and garden. The World Health Organization estimates that pesticide poisonings kill some 40,000 agricultural workers yearly worldwide and adversely affect the health of 2 million to 5 million more.

Pesticides are usually poisonous because their purpose is to kill—harmful insects, perhaps, or troublesome weeds. There are about 50,000 different pesticide products on the market, and they contain about 1400 active ingredients, most of them quite toxic. About 67 million pounds of pesticides are used each year on lawns in the United States, and lawns and gardens usually receive heavier pesticide applications per acre than agricultural land.

Do we really need to use pesticides? Insect pests cause billions of dollars of damage each year, and weeds can greatly reduce the yields of agricultural products, but can we really justify adding so much toxic material to our environment in order to get rid of them? Consider the case of DDT, which was banned because it had done so much environmental damage. It was used to kill mosquitoes, but it also killed birds. It is true that DDT killed mosquitoes that spread malaria, but is malaria really a health problem today? If you consider the whole world, it is. Worldwide at least 100 million people have the disease, and each

▲ Many pesticides are used in homes against insect infestations and on gardens and lawns to prevent plant damage and to kill weeds.

year it causes more than a million deaths, mostly among young children.

What about herbicides? Except for problems with dioxin contamination in 2,4,5-T, herbicides seem to have a better safety record than insecticides. Roadsides, vacant lots, and industrial areas have been kept free of weeds, and the value of agricultural crops has been increased by billions of dollars. People who suffer from allergies to ragweed, poison ivy, or other noxious weeds have been helped greatly by the use of herbicides. But we still don't know the ultimate effect on the environment of their long-term use.

16.16 SUSTAINABLE AGRICULTURE

Problems with pesticides have led to calls for alternatives to conventional farming, which uses pesticides and fertilizers from chemical plants. The Board of Agriculture of the National Research Council has urged farmers to consider other ways to deal with pests and provide plant nutrients. Among the Board's suggestions were crop diversification, integrated pest management (using a mixture of biological controls and synthetic chemicals), disease prevention by careful crop management, and genetic improvement of crops. Much of this advice is similar to that given to organic farmers and gardeners for generations.

Modern agriculture is also energy intensive. Nonrenewable petroleum energy is required for the production of fertilizers, pesticides, and farm machinery. Energy is also required to run the machinery needed to till, harvest, dry, and transport the crops and to process and package the food (Figure 16.33).

A 21-year Swiss study compared conventional plots that used mineral fertilizers and synthetic pesticides to organic plots that were fertilized with manure and treated only occasionally with a copper fungicide. Identical crops of potatoes, winter wheat, grass clover, barley, and beets were grown under otherwise identical conditions. The organic systems were claimed to be more efficient, with yields only 20% less, although the nutrient input was reduced by 50%. The organic method was said to use 20 to 56% less energy when energy required for production of fertilizers and pesticides was taken into account.

One-third of the cost of food in the United States goes to pay for transportation. We spend $6 million and use 3.6 million L of fuel each year just to transport broccoli from California to New York.

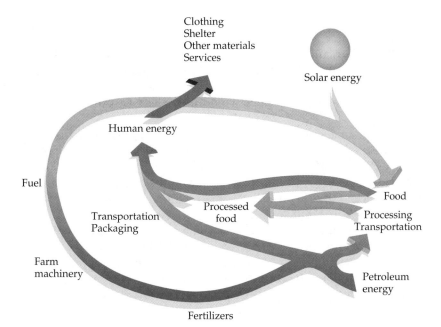

Clothing
Shelter
Other materials
Services

Solar energy

Human energy

Fuel

Transportation
Packaging

Processed
food

Food

Processing
Transportation

Farm
machinery

Petroleum
energy

Fertilizers

Figure 16.33 Energy flow in modern agriculture. Petroleum energy has largely supplanted human energy.

Organic farming is carried out without the use of synthetic fertilizers or pesticides. Organic farmers (and gardeners) use manure from farm animals for fertilizer, and they rotate other crops with legumes to restore nitrogen to the soil. They control insects by planting a variety of crops, alternating the use of fields. (A corn pest has a hard time surviving during the year that its home field is planted in soybeans.) Organic farming also is less energy intensive. According to a study by the Center for the Biology of Natural Systems at Washington University, comparable conventional farms used 2.3 times as much energy as organic farms. Production on organic farms was 10% lower, but costs were lower by a comparable percent.

Nanoscience and the Food Industry

Nanoscience and nanotechnology involve the production and use of materials at a scale of billionths of a meter—nanometers—and exploitation of the fact that materials have different properties at this scale from those at a larger scale. This has the potential to revolutionize many aspects of food quality, production, and packaging in the future. Developments in nanoscience and nanotechnology may well produce food that can adjust its color, flavor, or nutrient content to accommodate each customer's taste or health condition. It may also produce packaging that can sense when food inside is spoiling and alert the user. Some other possible future developments include the following:

- **Nano foods:** Several companies are using nanotechnology to change the structure of food, creating drinks containing nanocapsules that can change color and flavor, and making ice cream with nanoparticle emulsions to improve texture.

- **Nano packaging:** Some companies are developing new nanomaterials that extend food shelf life and change color to signal when the food spoils. Nanoparticles in packaging film could improve the transport of some gases through the plastic

film, perhaps removing carbon dioxide or blocking out gases such as oxygen and ethylene that shorten the shelf life of food.

- **Nanoparticle pesticides:** Industries are developing pesticides made up of nanoparticles. These pesticides are more readily taken up by plants. They can also be designed in "time-release" forms.

- **Nano animal feed:** Researchers are feeding bioactive polystyrene nanoparticles that bind with bacteria to chickens as a substitute for antibiotics in poultry production. Others are adding nanoparticle vaccines to trout ponds to be taken up by fish.

- **Nano security devices:** Nanosensors are being developed for pathogen and contaminant detection, and nanodevices are being investigated that would allow the tracking of individual shipments.

Nanotechnology centers around the world are developing an impressive body of research. Scientists will build many new possibilities for food production on this research. Economists estimate that the global market for nanotechnology could be $1 trillion within a decade. Nanotechnology offers enormous potential benefit to the economy and to society.

Organic farms require 12% more labor than conventional ones, but human labor is a renewable resource, whereas petroleum energy is not.

Conventional agriculture can result in severe soil erosion and is the source of considerable water pollution. No doubt we should practice organic farming to the limit of our ability to do so. But we should not delude ourselves. Abrupt banning of synthetic fertilizers and pesticides would likely lead to a drastic drop in food production.

As far as human energy is concerned, U.S. agriculture is enormously efficient. Each farm worker produces enough food for about 80 people. But this productivity is based on fossil fuels; about 10 units of petroleum energy are required to produce 1 unit of food energy. If we consider production per hectare, modern farming is marvelously efficient. If we consider the energy used in relation to the energy produced, it is remarkably inefficient. It should be noted, however, that in an energy-efficient early agricultural society, nearly all human energy went into food production. In modern societies, only about 10% of human energy is devoted to producing food. The other 90% is used to provide the materials and services that are so much a part of our civilization. We should try to make our food production more energy efficient, but it is unlikely that we will want to return to an outmoded way of life.

16.17 SOME MALTHUSIAN MATHEMATICS

In 1830 Thomas Robert Malthus, an English clergyman and political economist, made the statement that population increases faster than the food supply. Unless the birth rate was controlled, he said, poverty and war would have to serve as restrictions on the increase.

Malthus's predictions were based on simple mathematics: Population grows geometrically, while the food supply increases arithmetically. In **arithmetic growth**, a constant amount is added during each growth period. As an example, consider a cookie jar savings account. The first week, a child puts in the 25¢ she received for her birthday. Each week thereafter, she adds 25¢. The growth of the savings is arithmetic; it increases by a constant amount (25¢) each week. At the end of the first week, she will have 25¢; at the end of the second, 50¢; the third, 75¢; the fourth, $1.00; the fifth, $1.25; the sixth, $1.50; and so on.

In **geometric growth**, the increment increases in size for each growth period. Again, let's use a child's bank as an example. The first week she puts in 25¢. The second week she puts in another 25¢ to double the amount (to 50¢). The third week she puts in 50¢ to double this amount again. Each week, she puts in an amount equal to what is already there and doubles the amount in the bank. At the end of the first week, she has 25¢; the second, 50¢; the third, $1.00; the fourth, $2.00; the fifth, $4.00; the sixth, $8.00; and so on. Before long, she will have to start robbing banks to keep up her geometrically growing deposits.

Table 16.11 compares arithmetic and geometric growth through ten growth periods. These data are shown graphically in Figure 16.34. Note that arithmetic growth is slow and steady; geometric growth starts slowly and then shoots up like a rocket.

For a population growing geometrically, we can calculate the **doubling time** from the **rule of 72**. Simply divide 72 by the percent of annual growth. For example, Earth's population was 6.46 billion in 2005 and growing 1.14% per year. If it continues to grow at this rate, it will double to nearly 13 billion in 63 years.

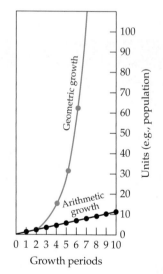

▲ **Figure 16.34** Arithmetic growth and geometric growth through 10 periods, each starting with one unit.

TABLE 16.11	Arithmetic and Geometric Growth Through Ten Periods[*]										
	Growth Period										
	0	1	2	3	4	5	6	7	8	9	10
Arithmetic growth	1	2	3	4	5	6	7	8	9	10	11
Geometric growth	1	2	4	8	16	32	64	128	256	512	1024

[*]Starting with one unit.

EXAMPLE 16.5 **Population Growth**

The population of the United States was 300 million in 2005 and was growing at a rate of 0.92% per year. If growth continues at the same rate, when will the population have doubled to 580 million?

Solution
Divide the annual growth rate into 72.

$$\text{Doubling time} = \frac{72}{0.92/\text{y}} = 78 \text{ y}$$

If growth continues at 0.92% per year, U.S. population will double to 600 million people in 78 years; that is, by 2083. Many factors will affect the growth rate. It will change as the birth rate changes and as immigration laws change. An epidemic or other catastrophe might also change the death rate.

▶ **Exercise 16.5A**
The population of Albania was estimated to be 3.56 million in 2005 and was growing at a rate of 0.52% per year. If growth continues at the same rate, when will the population have doubled to 7.12 million?

▶ **Exercise 16.5B**
The population of Bulgaria was estimated to be 7.45 million in 2005 and was declining at a rate of 0.89% per year. If the decline continues at this rate, when will the population have been halved to 3.73 million?

Earth's population has grown enormously since the time of Malthus, reaching almost 6.5 billion in 2005. Famine has brought death to millions in war-ravaged areas around the world (Figure 16.35), but modern farming has piled up surpluses in developed countries. Even developing nations have made great progress in food production, many of them becoming self-sufficient. Despite surpluses in some parts of the world, an estimated 800 million people are seriously malnourished. Half are children under five years old, who will carry the physical and mental scars of this deprivation for the rest of their lives.

Scientific developments, such as those outlined in this chapter, have brought abundance to many, but millions still go hungry.

◀ **Figure 16.35** In developed countries, science has thwarted the Malthusian prediction of hunger. But starvation is still a fact of life for many people in developing countries. In 2005, up to 150,000 people faced starvation in Niger because the international community was too slow to respond to the country's food crisis.

16.18 CAN WE FEED A HUNGRY WORLD?

If Earth's population continues at its present growth rate, it is expected to reach 9 billion by midcentury. Can we feed all those people?

An alliance between science and agriculture has brought an increase in food supplies beyond the imagination of people of a few generations ago. Through the use of irrigation, synthetic fertilizers, pesticides, and improved genetic varieties of plants and animals, most of the people of the world have abundant food.

We may be able to increase food production even more through genetic engineering (Chapter 15). Scientists can design plants that produce more food, that are resistant to disease and insect pests, and that grow well in hostile environments. They can design animals that grow larger and produce more meat and milk. They may someday provide us with food in unimaginable abundance. But the hungry are with us now, and 75 million more people come to dinner each year.

Even by quadrupling present food production, we could meet our needs for only a few decades. Virtually all the world's available arable land is now under cultivation, and we lose farm land every day to housing, roads, erosion, encroaching deserts, and the increasing salt content of irrigated soils. It will be difficult to keep food production ahead of the rate of population growth, particularly in developing countries. Fortunately, the rate of growth has slowed in many of these nations.

We seem to be hooked on a high-energy form of agriculture that uses synthetic fertilizers, pesticides, and herbicides and depends on the power of machinery that burns fossil fuels. Ultimately, the only solution to the problem is population stabilization. Obviously, that will come someday. The only questions are when and how. Population control could come through decreasing the birthrate. This is happening in almost all the developed world and in many developing nations. A few nations still have populations growing at explosive rates that far outstrip their food supplies. Birth control isn't the only way to limit populations. Another possibility is an increase in the death rate—through catastrophic war, famine, pestilence, or the poisoning of the environment by the wastes of an ever-expanding population. The ghost of Thomas Malthus haunts us yet.

Critical Thinking Exercises

Apply knowledge that you have gained in this chapter and one or more of the FLaReS principles (Chapter 1) to evaluate the following statements or claims.

16.1 The author of a new diet book claims that he and many other people have been able to lose a substantial amount of weight by following a diet very low in carbohydrates (such as bread, cereals, pasta) but high in proteins (such as beef and lamb).

16.2 An advertisement claims that a certain organically grown cereal provides a "life energy" that does not exist in ordinary processed cereals.

16.3 A physician noted that the blood cholesterol of one of his patients had gone down dramatically. On the advice of a friend, the patient had been taking daily supplements of niacin, one of the B vitamins. Although he was skeptical, the doctor found six other patients who were willing to try the niacin treatment without any further medication. A year later he was surprised to find that five of these patients had lowered their cholesterol levels significantly. The doctor says he believes that niacin can be used to lower blood cholesterol.

16.4 Information on a website states that a chemical from marijuana, delta 1-tetrahydrocannabinol (Delta 1-THC), is a fat-soluble vitamin they call "vitamin M." They say they can make the claim "because many of the properties of delta 1-THC are similar to those of the fat-soluble vitamins."

16.5 Information on a website states that natural vitamins may be worth the extra cost because vitamins in nature occur within a family of "things like trace elements, enzymes, co-factors and other unknown factors that help them absorb into the human body and function to their full potential" whereas "synthetic vitamins are usually isolated into that one pure chemical, thereby missing out on some of the additional benefits that nature intended."

16.6 A friend claims that it is fine to eat all the ground beef she likes, as long as it is labeled "85% lean." She shows you the label, which says that a 100-gram portion of cooked meat contains 15 grams of fat and has 210 calories.

SUMMARY

Section 16.1—Foods are chemicals too. The three main classes of foods are carbohydrates, fats, and proteins. We also need vitamins, minerals, fiber, and water in our diets. Carbohydrates include sugars, starches, and cellulose.  Glucose, also called blood sugar, is the sugar used directly by cells for energy. Other sugars include fructose, sucrose (table sugar) and lactose. Sucrose and lactose are broken down during digestion. People with *lactose intolerance* do not have the enzyme to accomplish this. Starch is a polymer of alpha-glucose, and cellulose is a beta-glucose polymer. The body hydrolyzes starch to glucose, but cellulose is not hydrolyzed, instead serving as dietary fiber. We store small amounts of carbohydrates as glycogen or animal starch.

Section 16.2—Dietary fats are esters called triglycerides, digested by enzymes to form fatty acids and 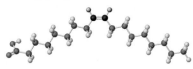 glycerol. Fats are stored in adipose tissue located in fat depots and are hydrolyzed back to glycerol and fatty acids when needed. A lipoprotein is a cholesterol– or triglyceride–protein combination transported in the blood. Low-density lipoproteins (LDLs, "bad cholesterol") carry cholesterol to the cells for use and increase the risk of heart disease, while high-density lipoproteins (HDLs, "good cholesterol") carry it to the liver for excretion. Polyunsaturated fats (mostly from plants), *cis*-fatty acids, and omega-3 fatty acids tend to lower LDL levels. Saturated fats (mostly from animals) and *trans*-fatty acids from hydrogenation of polyunsaturated fats have been implicated in cardiovascular disease.

Section 16.3—Proteins are polymers of amino acids, and are broken down to those amino acids during digestion. The essential amino acids must be included in the diet. Each essential amino acid is a limiting reactant in protein synthesis; deficiencies cause an inability to make the proper proteins. Animal proteins usually have these essential amino acids; plant proteins usually are deficient and must be combined with other plant proteins for a complete diet. Protein deficiency is unusual in the United States but is common in other parts of the world.

Sections 16.4–16.7—Dietary minerals are inorganic substances vital to life. Besides the bulk structural elements such as carbon, hydrogen, and oxygen, there are macrominerals such as sodium, potassium, and calcium.  There are also trace elements needed, such as iron, and ultratrace elements including manganese, chromium, and vanadium. Minerals serve a variety of functions: Iodine is needed by the thyroid; iron is part of hemoglobin; and calcium and phosphorus are important in the bones and teeth. Vitamins are organic compounds that our bodies need but cannot make. Most are needed as coenzymes. The B vitamins and vitamin C are water soluble and are needed frequently, whereas A, D, E, and K are all fat soluble and can be stored to some extent. Dietary fiber may be soluble (gums and pectins) or insoluble (cellulose) and prevents digestion problems such as constipation. Drinking enough water is important, but it need not be plain water. Most drinks are mainly water. Starvation is the deprivation of food and causes metabolic changes. The body first depletes its glycogen, then metabolizes fat and muscle tissue. Malnutrition can result from starvation, dieting, or even the lack of a proper diet.

Section 16.8—Food additives are substances other than basic foodstuffs that are added to food to: aid nutrition; inhibit spoilage; add color, flavor, or sweetness; bleach; or provide texture. There are thousands of additives. The FDA regulates which additives, and how much, can be used. In enrichment, chemicals are added to foods to prevent deficiency diseases: B vitamins in grains, vitamin C in drinks, vitamin D in milk. Natural herbs and spices, or their synthetic components, add flavor. Artificial sweeteners reduce sugar intake but do not appear to aid in controlling obesity. Glycerol is a sweetener that is used primarily as a humectant or moisturizing agent. Salt and MSG are flavor enhancers. Propionates, sorbates, benzoates, and nitrites inhibit spoilage and bacterial growth. Fats may turn rancid from formation of free radicals with unpaired electrons. One free radical can cause a chain reaction that leads to decomposition of many fat molecules. BHT and BHA are antioxidants that react with free radicals and prevent rancidity. Natural food colorings such as β-carotene, and human-made ones, improve the appearance of food. Some additives have been found to be harmful; additives are usually banned if they are shown to cause cancer in laboratory animals. The list of additives generally recognized as safe is the GRAS list, which is constantly reevaluated.

Sections 16.9–16.10—Toxic substances found in food are sometimes natural ingredients, but sometimes they are added inadvertently to our food in the form of pesticide residues or animal feed additives. Charcoal-broiled steak, cinnamon, and nutmeg contain carcinogens. Aflatoxins are produced by molds on peanuts and grain and are potent carcinogens. We may consume about 10,000 times as much natural carcinogens as synthetic carcinogens. Some additives are intentional, while others are incidental. Incidental additives include Alar, PCBs, PBBs, and DES. Although we may be able to get along without food additives, spoilage and disease would be likely to cause serious problems worldwide. Additives continue to be carefully scrutinized and regulated.

Sections 16.11–16.12—All food originates in plants. Photosynthesis in green plants stores energy from the sun, providing fuel for all living things. The three **primary plant nutrients** are nitrogen, phosphorus, and potassium. Atmospheric nitrogen is fixed naturally by soil bacteria and by lightning. Artificial fixation of nitrogen produces ammonia and nitrates which can be added to the soil. Phosphorus can be made available from phosphate rock. Potassium, as potassium chloride, is usually mined. The **secondary plant nutrients** are calcium, magnesium, and sulfur. Eight other **micronutrients** are needed in small amounts. The three numbers on commercial fertilizers indicate the % N, % P_2O_5, and % K_2O in the fertilizer.

Section 16.13—Pests can cause plagues and ruin the food supply. **Pesticides** are used to kill organisms that we consider pests, including insects, rodents, diseases, and weeds. Most **insecticides** kill all insects, though only a few insects are harmful. The insecticide DDT saved millions of lives during and after World War II. However, its persistence in the environment caused problems. The fat-soluble pesticide was concentrated up the food chain and interfered with calcium metabolism in birds. DDT was eventually banned, though it continues to be used in developing countries for malaria control. Organic phosphorus compounds are nerve poisons. Although they accumulate in fatty tissues, they are less persistent in the environment than chlorinated compounds like DDT. Carbamates are narrow-spectrum insecticides, harmful to just a few insect pests. They neither accumulate in fatty tissue nor persist in the environment.

Sections 16.14–16.15—Some pests can be controlled naturally, using praying mantises, ladybugs, and pest-specific bacteria and viruses. Some insects can be controlled by sterilization and release of males. A **pheromone** is a chemical secreted by an insect to mark a trail, send an alarm, or, in the case of a **sex attractant**, attract a mate. **Juvenile hormones** can be used to keep adult pests in an immature stage, in which they cannot reproduce. Pheromones and juvenile hormones have proved useful to some extent in monitoring and control of insect pests but are difficult and expensive to use.

Herbicides kill plant pests, and **defoliants** cause leaves to fall off plants. The herbicide 2,4-D and the defoliant 2,4,5-T were combined in **Agent Orange**, used to remove ground cover in Vietnam. Highly toxic **dioxins** were contaminants in this mixture and caused birth defects, resulting in a ban on 2,4,5-T. Atrazine and glyphosate are the most widely used herbicides in the United States. Paraquat is a **preemergent herbicide** that kills weed plants before crop seedlings emerge.

2,3,7,8-Tetrachlorodibenzo-para-dioxin (a "dioxin")

Sections 16.16–16.18—**Organic farming** avoids the use of synthetic fertilizers and pesticides. It is less energy intensive and less productive than conventional farming but costs less and uses renewable resources. According to Thomas Malthus, food production sees a steady increase called **arithmetic growth**. Because population increase is a **geometric growth**, with the rate of increase itself increasing, food production cannot keep up with population. The **doubling time** can be predicted from the **rule of 72**; doubling time equals 72 divided by the percent annual growth. Additional scientific and technical advances may help increase food production or control population in years to come, but Malthusian predictions still hang over us.

REVIEW QUESTIONS

1. List the three major types of food.

2. What is the role of carbohydrates in the diet?

3. What functions do fats serve?

4. What is the role of proteins in the diet?

5. In general, what problems are associated with a strict vegetarian diet?

6. A diet high in meat products makes less efficient use of the energy originally captured by plants through photosynthesis than a vegetarian diet does. Explain why.

7. Vitamins and minerals are discussed in this chapter. Which are organic and which are inorganic?

8. Is an excess of a water-soluble vitamin or an excess of a fat-soluble vitamin more likely to be dangerous? Why?

9. What is starvation?

10. In fasting, which stores are depleted first, fats or glycogen?

11. What is a food additive?

12. What are the two major categories of food additives? Give an example of each.

13. What is enriched bread? Is it equal in nutritional value to bread made from whole grain?

14. What is MSG?

15. What is botulism? What are aflatoxins?

16. What vitamin serves as a fat-soluble antioxidant?

17. Name three artificial sweeteners. Which are approved for current use in the United States?

18. What is the chemical nature of aspartame?

19. What is the GRAS list?

20. List some incidental additives that have been found in foods.

21. What is the usual fate of a food additive shown to cause cancer in laboratory animals?

22. What United States government agency regulates the use of food additives? What must be done before a food company can use a new food additive?

23. List five functions of food additives.

24. Where does most of the matter of a growing plant come from?

25. What is a preemergent herbicide?

26. List the four main features of alternative agriculture.

27. Compare organic farming with conventional farming. Consider energy requirements, labor, profitability, and crop yields.

28. What did Thomas Malthus predict in his famous 1830 statement? Why has Malthus's prediction not come true?

PROBLEMS

Carbohydrates

29. What are the chemical names of the following sugars?
 a. blood sugar   **b.** table sugar   **c.** fruit sugar

30. What is the principal sugar in corn syrup? How is high-fructose corn syrup made?

31. What is the dietary difference between starch and cellulose?

32. What sugar is formed when starch is digested? What type of chemical reaction is involved?

33. What are the ultimate products formed when cells metabolize glucose?

34. How much energy is supplied by 1 g of carbohydrates?

Fats and Cholesterol

35. What products are formed when a fat is digested? What type of chemical reaction is involved?

36. Where is fat stored in the body?

37. How do animal fats differ from vegetable oils?

38. What is a trans fat? How are trans fats made?

39. How much energy is supplied by 1 g of fat?

40. What is the maximum percentage of calories in our diet that should come from fats? From saturated fats?

Problems 41–44 refer to the following line-angle structures.

I.

II.

III.

41. Which is a saturated fatty acid?

42. Which is a monounsaturated fatty acid?

43. Which is an omega-3 fatty acid?

44. Which is linoleic acid?

Proteins

45. What products are formed when a protein is digested? What type of chemical reaction is involved?

46. What are essential amino acids? What is an adequate protein? List some foods that contain adequate proteins.

47. Which essential amino acids are likely to be lacking in corn? In beans?

48. What is a limiting reactant?

49. What is kwashiorkor? What age group in developed countries suffer a similar condition?

50. A new type of bread is made by adding pea flour to wheat flour. Will the bread provide adequate protein? Why or why not?

Minerals

51. Indicate a biological function for each of the following dietary minerals.
 a. iodine
 b. iron
 c. calcium
 d. phosphorus

52. Which of the following minerals would you expect to find in relatively large amounts in the human body? Explain your reasoning.
 a. Ca   **b.** Cl
 c. Co   **d.** Mo
 e. Na   **f.** P
 g. Zn

Vitamins

53. Classify the following vitamins as water soluble or fat soluble.

Pantothenic acid

Phylloquinone

Biotin

54. Coenzyme Q_{10} (CoQ_{10}) is claimed by some to have a vitaminlike activity: It prevents LDL oxidation and may act as an antihypertensive, perhaps reducing the risk of cardiovascular disease. CoQ_{10} belongs to a family of substances called ubiquinones and is synthesized in the body. Is CoQ_{10} (below) **(a)** a vitamin? **(b)** Fat soluble or water soluble?

55. Which of the following are B vitamins?
    a. folic acid   **b.** riboflavin
    c. β-hydroxybutyric acid **d.** thiamine
    e. niacin

56. Match the compound with its designation as a vitamin.

Compound	Designation
Ascorbic acid	Vitamin A
Calciferol	Vitamin B_{12}
Cyanocobalamin	Vitamin C
Retinol	Vitamin D
Tocopherol	Vitamin E

57. Identify the vitamin associated with each of the following deficiency diseases.
    a. scurvy
    b. rickets
    c. night blindness

58. In each case, identify the deficiency disease associated with a diet lacking in the indicated vitamin.
    a. vitamin B_1 (thiamine)
    b. niacin
    c. vitamin B_{12} (cyanocobalamin)

59. Identify each of the following vitamins as water soluble or fat soluble.
    a. vitamin A
    b. vitamin B_6
    c. vitamin B_{12}
    d. vitamin C
    e. vitamin K

60. Identify each of the following vitamins as water soluble or fat soluble.
    a. calciferol
    b. niacin
    c. riboflavin
    d. tocopherol

Food Additives

61. What is the function of each of the following food additives?
    a. potassium iodide
    b. vanillin
    c. MSG
    d. sodium nitrite

62. What is the function of each of the following food additives?
    a. $FeCO_3$
    b. SO_2
    c. potassium sorbate

63. What is the purpose of each of the following food additives?
    a. BHA
    b. FD & C Yellow No. 5
    c. saccharin

64. What is the purpose of each of the following food additives?
    a. aspartame
    b. vitamin D
    c. sodium hypophosphite

65. Which of the following is true about a dietary supplement labeled "natural"? It is
    a. mild acting
    b. without risk of side effects
    c. always safe to use with other medications
    d. none of the above

66. Propyl gallate is made by esterification of gallic acid, obtained by the hydrolysis of tannins from Tara pods. Is propyl gallate a natural antioxidant?

67. What are antioxidants? Name a natural antioxidant and two synthetic antioxidants.

68. What is a chain reaction? What is the action of an antioxidant on a chain reaction?

Fertilizers

69. List the three structural elements of a plant.

70. In what form is nitrogen used by plants? What is the role of nitrogen in plant nutrition?

71. How is ammonia made? What raw materials are required for ammonia synthesis?

72. List several ways that nitrogen can be fixed.

73. What is urea? From what components is synthetic urea made?

74. What is anhydrous ammonia? How is it used?

75. What is the source of phosphate fertilizers? What is the role of phosphorus in plant nutrition?

76. Why does soil become acidic when potassium ions are absorbed by plants?

Pesticides

77. List the advantages and disadvantages of DDT as an insecticide.

78. Why is DDT especially harmful to birds?

79. Describe how chlorinated hydrocarbons become concentrated in a food chain.

80. Define and give an example of **(a)** a narrow-spectrum insecticide and **(b)** a broad-spectrum insecticide?

81. What are pheromones? How are they used in insect control?

82. What are juvenile hormones? How are they used against insect pests?

83. Describe the technique of sterilization as a method for controlling insects. Why is it not more widely used?

84. Define and give an example of a **(a)** a herbicide and **(b)** a defoliant?

Calorie and Nutrient Calculations

85. The label on a can of soup indicates that each half-cup portion supplies 60 kcal and has 3 g of protein, 8 g of carbohydrate, and 1 g of fat. Calculate the percent of calories each from carbohydrate, fat, and protein.

86. The label on a can of milk substitute indicates that each half-cup serving supplies 150 kcal and has 8 g of protein, 12 g of carbohydrate, and 8 g of fat. Calculate the percent of calories each from carbohydrate, fat, and protein.

87. A cup of lowfat cottage cheese provides 31 g protein, 4 g fat, and 8 g of carbohydrate. About how many kilocalories does the cottage cheese provide? What percentage of the calories are from fat?

88. A 6-oz portion of baked orange roughy fish provides 148 kcal, 32 g protein, and 2 g fat. How much carbohydrate does the fish provide?

89. A male teenager needs about 3000 kcal per day. If the guideline of 30% or less calories from fat is to be followed, how much fat, in grams, is permitted?

90. An adult female goes on a diet that provides 1200 kcal per day with no more than 25% from fat and no more than 30% of that fat being saturated fat. How much saturated fat, in grams, is permitted in this diet?

Toxicity

91. How much DDT would it take to kill a person weighing 60 kg if the lethal dose is 0.50 g per kilogram of body weight?

92. How much parathion would it take to kill the person in Problem 91 if the lethal dose is 5 mg per kilogram of body weight?

Growth

93. What is arithmetic growth? Give an example.

94. What is geometric growth? Give an example.

95. You start a stamp collection, planning to buy two stamps each week. How many weeks will it take to acquire 100 stamps? Does the collection grow arithmetically or geometrically?

96. You are raising rabbits. Starting with two, you have four at the end of two months, eight at the end of four months, and so on; the rabbit population doubles every two months. How many rabbits will you have at the end of 24 months? Is the rabbit population growing arithmetically or geometrically?

97. The population of India was estimated to be 1.1 billion in 2005. It is growing at 1.4% a year. At this rate, how many years of growth will it take for the population to double? What factors are likely to change this rate of growth?

98. The population of China, now 1.3 billion, is growing at an annual rate of 0.58%. At this rate, how many years of growth will it take for the population to double to 2.6 billion? What factors are likely to change this rate of growth?

ADDITIONAL PROBLEMS

99. Give the chemical equation for the photosynthesis reaction.

100. Give the chemical equation for the synthesis of ammonia.

101. The compound ammonium monohydrogen phosphate is frequently used as a fertilizer. **(a)** What two plant nutrients does it supply? **(b)** Write the equation for the formation of the compound from ammonia.

102. The micronutrient copper is often applied in the form of copper sulfate. Write the equation for the formation of copper sulfate from copper oxide.

103. Omega-6 fatty acids are unsaturated fatty acids in which the double bond closest to the methyl (omega) end of the molecule occurs at the sixth carbon from that end. Which of the following are omega-6 fatty acids? Refer to Table 15.1 for structures not given here.
    a. palmitoleic acid, $CH_3(CH_2)_5CH=CH(CH_2)_7CO_2H$
    b. oleic acid
    c. linoleic acid
    d. linolenic acid
    e. arachidonic acid

104. Following are line-angle formulas for two fats. **(a)** Which is a saturated fat and which is a polyunsaturated fat? **(b)** Identify the fatty acid units in each.

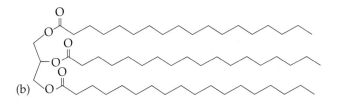

105. Referring to Section 16.14, identify the functional group(s) in the **(a)** sex attractant for the codling moth, **(b)** juvenile hormone molecule, and **(c)** methoprene molecule.

106. The sex attractant of the female tiger moth (*Homomelia imaculata*) is a 2-methyl branched-chain alkane with a molecular formula of $C_{18}H_{38}$; the straight-chain isomer is not attractive to males. Recalling the structural difference between butane and isobutane (2-methylpropane) in Chapter 9, draw the line-angle formula of this pheromone.

COLLABORATIVE GROUP PROJECTS

Prepare a PowerPoint, poster, or other presentation (as directed by your instructor) for presentation to the class.

107. Several fat substitutes have been developed. Prepare a brief report on one of the following fake fats. List principal advantages and disadvantages of each.
    a. Simplesse, made by NutraSweet
    b. Olestra, made by Procter & Gamble
    c. Leanesse (Oatrim), a ConAgra product

108. Prepare a brief report on one of the minerals listed in Table 16.3. List principal dietary sources and uses of the mineral.

109. Prepare a brief report on one of the vitamins listed in Table 16.4. List principal dietary sources and uses of the vitamin.

110. Prepare a brief report on one of the insecticides listed in Table 16.10. List principal advantages and disadvantages of each.

111. Prepare a brief report on one of the following alternative methods of insect control. List principal advantages and disadvantages of each.
    a. Bt
    b. a viral agent
    c. sterilization
    d. sex attractants
    e. a juvenile hormone

112. Examine the label on a sample of each of the following.
    a. a can of soft drink
    b. a can of beer
    c. a dried soup mix
    d. a can of soup
    e. a can of fruit drink
    f. a cake mix

    Make a list of the food additives in each. Try to determine the function of each additive. Do all the products offer this information?

113. Examine the label on a package of yard or garden pesticide. List its trade name and active ingredients.

114. Examine the label on a bag of fertilizer. What is its composition?

115. Prepare a brief report on one of the following scientists and share it with your group.
    a. Fritz Haber
    b. F. G. Hopkins
    c. Justus von Liebig
    d. Ellen Swallow Richards

116. Look for several websites with different opinions about the use of fertilizers and pesticides. Take one side of an issue involving the use of these substances and defend it.

117. Use data from Problems 93 and 94 to predict when the population of India will be greater than that of China.

REFERENCES AND READINGS

1. Brown, Lester R., Gary Gardner, and Brian Halweil. *Beyond Malthus: Nineteen Dimensions of the Population Challenge.* New York: W. W. Norton, 1999.

2. Charles, Daniel. *Lords of the Harvest: Biotech, Big Money and the Future of Food.* Cambridge, MA: Perseus, 2000.

3. Cohen, Joel E. "Ten Myths of Population." *Discover*, April 1996, pp. 42–47.

4. Dalton, Louisa. "What's That Stuff: Food Preservatives." *Chemical & Engineering News*, November 11, 2002, p. 40.

5. D'Mello, J. P. F. (ed.). *Food Safety: Contaminants and Toxins.* Cambridge, MA: CABI Publishing, 2003.

6. McGee, Harold. *On Food and Cooking: The Science and Lore of the Kitchen.* New York: Scribner, 2004. (Complete revision and updating of the 1984 edition.)

7. McWilliams, Margaret. *Food Around the World: A Cultural Perspective.* Upper Saddle River, NJ: Prentice Hall, 2003.

8. O'Driscoll, Cath. "Good Chemists Make Great Chefs." *Chemistry and Industry*, May 3, 2004, pp. 12–13.

9. Selim, Jocelyn. "The Chemistry of . . . Artificial Sweeteners." *Discover*, August 2005, pp. 18–19.

10. Vroom, Jay J. "Another View Of Pesticide Use." *Chemical & Engineering News*, November 4, 2002, p. 3.

11. Wolke, Robert L. *What Einstein Told His Cook: Kitchen Science Explained.* New York: W. W. Norton, 2002.

12. Zurer, Pamela S. "Organic Farming." *Chemical & Engineering News*, June 3, 2002, p. 8. Compared with conventional methods, organic farming uses less energy and leads to healthier soils.

13. Zurer, Pamela S. "Ridding the World of Unwanted Chemicals." *Chemical & Engineering News*, September 2, 2002, pp. 34–35.

GREEN CHEMISTRY

Renewable Feedstocks and Food Production

Scott Reed, Portland State University

The tools of synthetic chemistry have long been applied to understanding and altering food. From a chemical perspective, food can be thought of as a raw material from which we can produce new chemicals and materials. The use of corn to produce ethanol is one example. Another example is the use of naturally occurring polymers such as cellulose, which can be used in materials science applications after chemical modification.

The world's growing population requires more and more food to be produced from limited resources. Fertilizers and pesticides were designed to reduce disease and increase food production; however, their use may lead to detrimental side effects. These investigations will have you look at food production and the role of green chemistry in addressing the challenge of producing healthy food with minimal environmental impact.

WEB INVESTIGATIONS

Investigation 1
Trans Fatty Acids
Recent research has shown that trans fatty acids may increase the risk of heart disease. As a result, the FDA now requires food to be labeled with the trans fat content. A recent green chemistry award winner used the enzyme lipase to prevent the formation of trans fatty acids in vegetable oils. Visit the EPA's Green Chemistry Award site to read more about this work.

Investigation 2
Greener Pesticides
Early pesticides were successful at controlling pests, but this came with unintended environmental consequences. These pesticides were often chlorinated hydrocarbons that persist in the environment. Learning from this,

chemists have worked toward replacements that reduce unwanted side effects. Go to the National Integrated Pest Management Network and search their site for information about a green chemistry process that eliminates many of the side effects of previous pesticides.

Investigation 3
Biomass
Agricultural products can be used in the production of energy and other useful materials. The benefits of this renewable resource must be balanced against the costs of producing, transporting, and transforming food into a usable form. Search the site for the North American Commission for Environmental Cooperation, using the keyword "biomass." Select and read one of the articles on the uses of biomass.

COMMUNICATE YOUR RESULTS

Exercise 1
Good Fats, Bad Fats
Partially hydrogenated vegetable oils can be produced by catalytic hydrogenation or by the enzymatic process you learned about in Investigation 1. Compare the structures of fats produced by these different processes. Prepare a report that discusses the known health effects of saturated, unsaturated, and trans fats. The American Heart Association's site will have a wealth of information to get you started.

▲ Solid shortening tends to be higher in saturated fats and trans-fatty acids than is liquid cooking oil.

Exercise 2
Old and New Pesticides
Write a report that describes the differences between the pesticides you learned about in Investigation 2 and older pesticides, such as DDT. Relate the chemical structures of the pesticides to their side effects. Start by describing the properties you would want in an ideal pesticide.

Exercise 3
Regional Biomass
The ideal source of biomass varies from region to region. Using what you learned in Investigation 3, write a report that describes the biomass feedstocks available in your region. Describe the energy input that is needed to transform these feedstocks into usable energy.

CHAPTER 17

Household Chemicals

The household chemicals sold in greatest volume are cleaning products—soaps, detergents, and various special-purpose and multipurpose cleaning mixtures.

Helps and Hazards

If someone offered you a job in a place where poisonous chemicals were used every day, where toxic vapors and harmful dusts and molds were common, where corrosive acids and alkalis were often used, and where highly flammable liquids and vapors posed a fire hazard, would you take the job? You probably have, many times. This is a description of a typical American home.

People tend to be careful in a chemistry laboratory because they know the place can be dangerous. They often are much less cautious in their homes, even though some of the chemicals on the shelves can be just as toxic and hazardous as those in a laboratory—and many are exactly the same substances. There are perhaps half a million chemical products available for use in the American home. They include waxes, wax removers, paints, paint removers, bleaches, insecticides, rodenticides, spot removers, solvents, disinfectants, detergents, toothpaste, shampoo, perfumes, lotions, shaving cream, deodorants, hair spray, and many others (Figure 17.1). Some are quite harmless, but other products contain corrosive or toxic chemicals or present fire hazards. Some cause environmental problems when they are used or discarded.

Extensive use of chemicals in the home has led to many accidents. Studies show that chemicals are often used without regard to the directions or precautions given on their labels. Frequently the labels aren't read at all, and *misuse* of household chemicals can sometimes end in tragedy.

We discussed the chemical composition of food and the chemicals that are used in producing it in Chapter 16. Some of the agricultural chemicals are also used around the home, especially in the yard and garden. We discussed polymers present in our clothing and home furnishings in Chapter 10 and fuels used in our furnaces and automobiles in Chapter 14. We discussed the chemistry of our own bodies in Chapter 15. But there are many other chemicals around the house—cleaning products, personal care supplies, and an assortment of other things. In this chapter we look at some household chemicals. Let's begin with cleaning agents, which make up the largest volume of household chemicals.

▲ **Figure 17.1** A modern home is stocked with a variety of chemical products.

A typical U.S. neighborhood supermarket offers about 5000 different consumer products. An online superstore may offer a million or more.

17.1 A HISTORY OF CLEANING

In developing societies, even today in some places, clothes are cleaned by beating them with rocks in the nearest stream. Sometimes plants, such as the soapworts of Europe or the soapberries of tropical America, are used as cleansing

513

Personal Cleanliness

The history of cleanliness of body and clothing is rather spotty. The Romans, with their great public baths, probably did not use any sort of soap. They covered their bodies with oil, worked up a sweat in a steam bath, and then had the oil wiped off by a slave. A dip in a pool of fresh water completed the "cleansing." And the slaves? They probably didn't bathe at all.

During the Middle Ages body cleanliness was prized in some cultures—if not always attained. Twelfth-century Paris, with a population of about 100,000, had many public bathhouses. The Renaissance, from the fourteenth to the seventeenth centuries, was noted for a revival of learning and art. However, it was not noted for cleanliness. Queen Elizabeth I of England (1558–1603) bathed once a month, a habit that caused many to think her overly fastidious. A common approach to unpleasant body odor back then was the liberal use of perfume. Today? Perhaps we have gone too far in the other direction. With soaps, detergents, body wash, shampoo, conditioners, antitangle sprays, deodorants, antiperspirants, aftershave, colognes, and perfumes,

▲ Queen Isabella of Castille (1474–1504), who supported the 1492 voyage of Columbus to the New World, is reported to have bathed only twice in her life.

it's quite possible that we add more after a shower than we remove in the shower.

agents. The leaves of soapworts and soapberries contain *saponins*, chemical compounds that produce a soapy lather. These saponins may have been the first detergents.

Ashes of plants contain potassium carbonate (K_2CO_3) and sodium carbonate (Na_2CO_3). The carbonate ion present in both these compounds reacts with water to form an alkaline solution that has detergent properties. These alkaline plant ashes were used as cleansing agents by the Babylonians at least 4000 years ago. Europeans used plant ashes to wash their clothes as recently as 100 years ago. Sodium carbonate is still sold today as washing soda.

Although soap has been known for hundreds of years, it was first used mainly as a medicine. The discovery of disease-causing microorganisms, and subsequent public health practices, increased interest in cleanliness during the eighteenth century. By the middle of the nineteenth century, soap was in common use.

17.2 FAT + LYE → SOAP

The first written record of soap is found in the writings of Pliny the Elder, the Roman who described the Phoenicians' synthesis of soap using goat tallow and ashes. By the second century C.E., sodium carbonate (produced by the evaporation of alkaline water) was heated with lime (from limestone or seashells) to produce sodium hydroxide (lye).

$$Na_2CO_3 + Ca(OH)_2 \longrightarrow 2\,NaOH + CaCO_3$$

The sodium hydroxide was heated with animal fats or vegetable oils to produce soap (Figure 17.2). Note that a **soap** is a salt of a long-chain carboxylic acid (Chapter 9). American pioneers made soap in much the same manner. Lye was added to animal

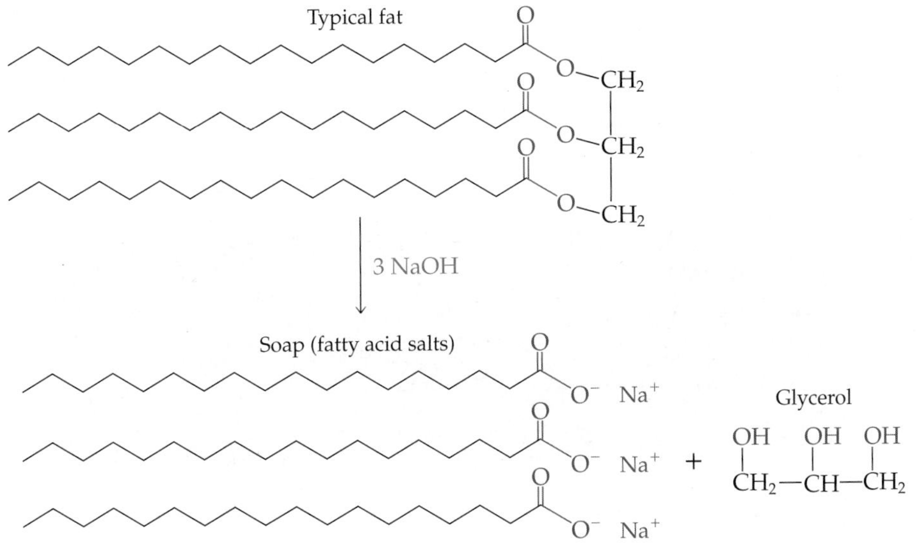

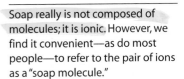
Triglyceride

Typical fat

3 NaOH

Soap (fatty acid salts)

O⁻ Na⁺

O⁻ Na⁺ +

O⁻ Na⁺

Glycerol

OH   OH   OH
| | |
CH_2—CH—CH_2

▲ **Figure 17.2** Soap can be made by reacting animal fat or vegetable oil with sodium hydroxide. Vegetable oils, with unsaturated carbon chains, generally produce softer soaps. Coconut oils, with shorter carbon chains, yield soaps that are more soluble in water.

Question: Write the structure of triolein, the oil in which all three fatty acid residues are from oleic acid.

fat in a huge iron kettle, and the mixture was cooked over a wood fire for several hours. The soap rose to the surface and, on cooling, solidified. The glycerol remained as a liquid on the bottom of the pot. Both the glycerol and the soap often contained unreacted alkali, which eroded the skin. Grandma's lye soap is not just a myth.

In modern commercial soapmaking, fats and oils are often hydrolyzed with superheated steam. The fatty acids are then neutralized to make soap. Toilet soaps usually contain additives such as dyes, perfumes, creams, and oils. Scouring soaps contain abrasives such as silica and pumice. Some soaps have air blown in before they solidify to lower their density so that they float. Some bath bars contain synthetic detergents. Their action is similar to that of soap, and consumers seldom make a distinction between soap and synthetic detergents.

Potassium soaps are softer than sodium soaps, and they produce a finer lather. They are used alone, or in combination with sodium soaps, in liquid soaps and shaving creams. However, most liquids for hand cleaning contain synthetic detergents. Soaps are also made by reacting fatty acids with triethanolamine. These substances are used in shampoos and other cosmetics.

How Soap Works

Dirt and grime usually adhere to skin, clothing, and other surfaces because they are combined with greases and oils—body oils, cooking fats, lubricating greases, and other similar substances—that act a little like sticky glues. Because oils are not miscible with water, washing with water alone does little good.

A soap molecule has a dual nature. One end is ionic and therefore polar and **hydrophilic** (water attracting). The rest of the molecule is hydrocarbon-like and therefore nonpolar and **hydrophobic** (water repelling). Figure 17.3 shows a typical soap. The hydrophilic "head" dissolves in water, while the hydrophobic "tail" dissolves in nonpolar substances, such as oils.

Soap really is not composed of molecules; it is ionic. However, we find it convenient—as do most people—to refer to the pair of ions as a "soap molecule."

Sodium Stearate

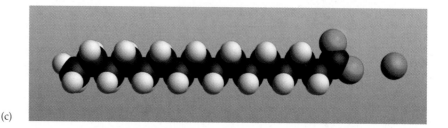

(a)

$$CH_3CH_2CH_2CH_2CH_2CH_2CH_2CH_2CH_2CH_2CH_2CH_2CH_2CH_2CH_2COO^- \ Na^+$$

Hydrocarbon tail   Ionic head

(b)

(c)

(d)

▶ **Figure 17.3** Representations of sodium palmitate, a soap. (a) Structural formula. (b) Line-angle formula. (c) Space-filling model. (d) Schematic representation of the palmitate anion.

The cleansing action of soap is illustrated in Figure 17.4. A spherical collection of molecules like this is called a **micelle**. The hydrocarbon tails stick in the oil, with the ionic heads remaining in the aqueous phase. In this manner, the oil is broken into tiny droplets and dispersed throughout the solution. The droplets don't coalesce because of the repulsion of the charged groups (the carboxyl anions) on their surfaces. The oil and water form an *emulsion*, with soap acting as the *emulsifying agent*. With the oil no longer "gluing" it to the surface, the dirt can be removed easily. Any agent, including soap, that stabilizes the suspension of non-polar substances—such as oil and grease—in water is called a **surface-active agent** (or **surfactant**). If not mixed into the water, soap tends to act at the liquid surface, the ionic heads interacting with the water, while the hydrocarbon tails stick up out of the water.

When we mix oil and vinegar to make mayonnaise, we need an emulsifying agent. Because soaps are not noted for their flavor, we use instead the proteins found in an egg yolk.

A Disadvantage of Soap

A major disadvantage of soap is that it doesn't work well in hard water. Hard water is water that contains certain metal ions, particularly calcium, magnesium,

Surfactant Molecules

▶ **Figure 17.4** The cleaning action of soap is visualized in this diagram of a soap micelle. A tiny oil droplet is suspended in water by having the hydrophobic hydrocarbon tails of soap molecules immersed in the oil, while their hydrophilic ionic heads extend into the water. Attraction between the water and the ionic ends of the soap molecules carries the oil droplet into the water.

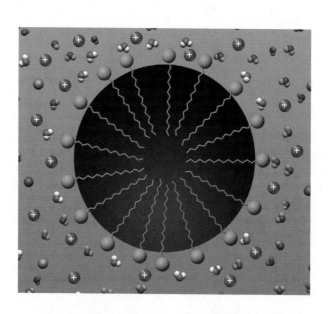

Soap curd in hard water

and iron ions. Soap anions react with these metal ions to form greasy, insoluble curds (Figure 17.5). With calcium ions, the reaction is

$$2 \, CH_3CH_2CH_2CH_2CH_2CH_2CH_2CH_2CH_2CH_2CH_2CH_2CH_2CH_2CH_2COO^- \, Na^+(aq) + Ca^{2+}(aq) \longrightarrow$$
A sodium soap (soluble)

$$(CH_3CH_2CH_2CH_2CH_2CH_2CH_2CH_2CH_2CH_2CH_2CH_2CH_2CH_2CH_2COO^-)_2Ca^{2+}(s) + 2 \, Na^+(aq)$$
A calcium soap (insoluble)

These deposits make up the familiar ring around the bathtub. They leave freshly washed hair sticky and are responsible for "tattletale gray" in the family laundry. For cleaning clothes and for many other purposes, soap has been largely replaced by synthetic detergents.

Water Softeners

To aid the action of soaps, various water-softening agents and devices are used. Water softeners are used to remove ions of calcium, magnesium, and iron, the ions that make water "hard." An effective water softener is washing soda, sodium carbonate ($Na_2CO_3 \cdot 10 \, H_2O$). The carbonate ion makes the water basic (preventing the precipitation of fatty acids) by reacting with water to raise the pH.

$$CO_3^{2-}(aq) + H_2O(l) \longrightarrow HCO_3^-(aq) + OH^-(aq)$$

The carbonate ion also reacts with the ions that cause hard water and removes them as insoluble salts.

$$Mg^{2+}(aq) + CO_3^{2-}(aq) \longrightarrow MgCO_3(s)$$
$$Ca^{2+}(aq) + CO_3^{2-}(aq) \longrightarrow CaCO_3(s)$$

Trisodium phosphate (Na_3PO_4) is another water-softening agent. Like washing soda, it makes water basic and precipitates calcium and magnesium ions.

$$PO_4^{3-}(aq) + H_2O(l) \longrightarrow HPO_4^{2-}(aq) + OH^-(aq)$$
$$2 \, PO_4^{3-}(aq) + 3 \, Mg^{2+}(aq) \longrightarrow Mg_3(PO_4)_2(s)$$

Phosphates also seem to aid in the cleaning process in some other way that is not yet well understood.

Water-softening tanks are also available for use in homes and businesses. These tanks contain an insoluble polymeric material that attracts and holds calcium, magnesium, and iron ions to its surface, replacing them with sodium ions and thus

A substance that produces hydroxide ions in water is a base (Chapter 7). Carbonate and phosphate ions are bases because they react with water to form hydroxide ions. Bases are also defined as proton acceptors, and carbonate and phosphate ions accept protons from water molecules.

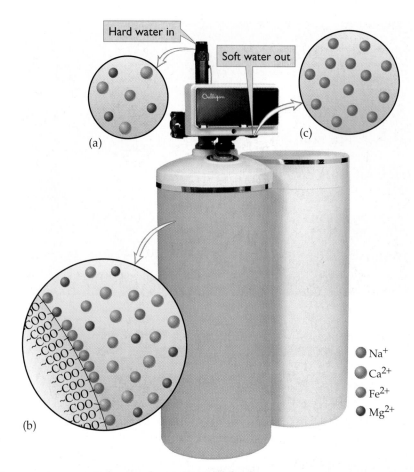

▶ **Figure 17.6** Water softeners work by *ion exchange*. Hard water going into the softener contains doubly charged cations— Ca^{2+}, Mg^{2+}, and Fe^{2+}. Those ions attach to the ion-exchange granules, and Na^+ (violet) is released. The water leaving the softener contains mostly Na^+ ions.

Question: If one mole each of Ca^{2+}, Mg^{2+}, and Fe^{2+} is present in the hard water, how many moles of Na^+ will be released?

softening the water (Figure 17.6). After a period of use, the polymer becomes saturated and must be regenerated by adding salt (NaCl) to replace the resin's sodium ions.

Before leaving the subject of soap, let's mention that soap has some advantages. It is an excellent cleanser in soft water, it is relatively nontoxic, it is derived from renewable resources (animal fats and vegetable oils), and it is biodegradable.

17.3 SYNTHETIC DETERGENTS

A second technological approach to the problems associated with soap was to develop a new synthetic detergent. Molecules of synthetic detergents were enough like those of soap to have the same cleaning action but different enough to resist the effects of acids and hard water. The raw materials for soap manufacture were scarce and expensive during and immediately after World War II. The synthetic detergent industry developed rapidly in the postwar period.

ABS Detergents: Nonbiodegradable

Within a few years, cheap synthetic detergents, made from petroleum products, were widely available. Alkylbenzenesulfonate (ABS) detergents were made from propylene (CH_2=$CHCH_3$), benzene (C_6H_6), and sulfuric acid (H_2SO_4). The resulting sulfonic acid (RSO_3H) was neutralized with a base, usually sodium carbonate, to yield the final product.

$$\text{---SO}_3^- \text{ Na}^+$$

Sales of ABS detergents soared. For a decade or more, nearly everyone was happy, but suds began to accumulate in sewage treatment plants. Foam piled high in rivers, and in some areas a head of foam even appeared on drinking water (Figure 17.7). The branched-chain structure of ABS molecules was not readily broken down by the microorganisms in sewage treatment plants, and the

(a)   (b)

◀ **Figure 17.7** ABS detergents do not degrade in nature. They led to foam on rivers such as that on the Bogota River in Colombia (a). They also caused contaminated water from a groundwater source to foam as it came from the tap (b).

groundwater supply was threatened. Public outcries caused laws to be passed and industries to change their processes. Biodegradable detergents were quickly put on the market, and nonbiodegradable detergents were banned.

LAS Detergents: Biodegradable

Biodegradable detergents called linear alkylsulfonates (LAS) have linear chains of carbon atoms.

$$\text{---SO}_3^- \text{ Na}^+$$

Microorganisms can break down LAS molecules by producing enzymes that degrade the molecule two (and only two) carbon atoms at a time (Figure 17.8). The branched chain of ABS molecules blocks this enzyme action, preventing their degradation. Thus, technology has solved the problem of foaming rivers.

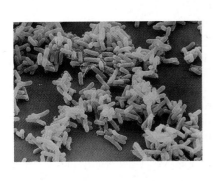

$$\text{CH}_3\text{CH}_2\text{CH}_2\text{CH}_2\text{CH}_2\text{CH}_2\text{CH}_2\text{CH}_2\text{CH}_2\text{CH}_2\text{CH}\text{---SO}_3^-\text{Na}^+$$
$$\underset{\text{CH}_3}{|}$$

$$\text{CH}_3\text{CHCH}_2\text{CHCH}_2\text{CHCH}_2\text{CH}\text{---SO}_3^-\text{Na}^+$$
$$\underset{\text{CH}_3}{|}\quad\underset{\text{CH}_3}{|}\quad\underset{\text{CH}_3}{|}\quad\underset{\text{CH}_3}{|}$$

(a)   (b)

▲ **Figure 17.8** Microorganisms such as *Escherichia coli*, shown in (a) magnified 42,500 times by scanning electron microscopy, are able to degrade LAS detergents. They can readily metabolize LAS detergents (b), removing two carbon atoms at a time from the molecule. It takes them much longer to break down the branched chains of the ABS molecules.

The cleansing action of synthetic detergents is quite similar to that of soaps. However, detergents work better than soap in acidic solution and in hard water. Their calcium and magnesium salts, unlike those of soap, are soluble and do not separate out, even in extremely hard water. As Figure 17.5 shows, the cleansing action of a synthetic detergent is little affected by hard water.

17.4 LAUNDRY DETERGENT FORMULATIONS

Detergent products used in homes and commercial laundries usually contain a variety of ingredients. The main component is a surfactant, but other substances are added to provide a variety of functions, such as increasing cleaning performance for specific soils or surfaces. Some additives ensure product stability. Others are intended to supply a unique identity to a particular product. Common additives include builders, brighteners, fabric softeners, and other substances to lessen the redeposition of dirt or simply to reduce cost.

Surfactants

Surface active agents (surfactants) enable the cleaning solution to wet the surface to be cleaned. They loosen and remove soil, usually with the aid of mechanical action. Surfactants also emulsify oily soils and keep them dispersed and suspended so they do not settle back on the surface (Figure 17.4). Many cleaning products include two or more surfactants.

Surfactants are generally classified by the ionic charge, if any, borne by the working part of the material. An ionic detergent is always associated with a small ion of opposite charge, called a *counter ion*. Soap, ABS, and LAS are all **anionic surfactants**; they have a negative charge on the active part. Other common anionic surfactant categories include the alkyl sulfates, such as sodium dodecyl sulfate, and the alcohol ethoxysulfates (AES). The counter ion for most anionic detergents is often Na^+.

$$CH_3(CH_2)_nOSO_3^- \ Na^+$$
An alkyl sulfate
$n = 11$ to 13

$$CH_3(CH_2)_mO(CH_2CH_2O)_nSO_3^- \ Na^+$$
An alcohol ethoxysulfate (AES)
$m = 6$ to 13 $n = 7$ to 13

Anionic surfactants are used in laundry and hand dishwashing detergents, household cleaners, and personal cleansing products. They have excellent cleaning properties and generally make a lot of suds.

As the name indicates, **nonionic surfactants** have no electrical charge. They are low sudsing and resistant to water hardness. They work well on most soils and are typically used in laundry and automatic dishwasher formulations. The most widely used nonionic surfactants are the alcohol ethoxylates.

$$CH_3(CH_2)_mO(CH_2CH_2O)_nH$$
An alcohol ethoxylate
$m = 6$ to 13 $n = 7$ to 13

Cationic surfactants have a positive charge on the active part. These cationics are not particularly good detergents, but they have a germicidal action and are therefore used as cleansers and disinfectants in the food and dairy industries and as the sanitizing ingredient in some household cleaners. The most common cationic surfactants are *quaternary ammonium salts*, so called because they have four hydrocarbon groups attached to a nitrogen atom that bears a positive charge.

$$CH_3(CH_2)_nCH_2N^+(CH_3)_3 \ Cl^-$$
A quaternary ammonium salt
$n = 10$ to 16

Sometimes cationic surfactants are used along with nonionic surfactants. Cationic surfactants are seldom used with anionic ones because the ions of opposite charge tend to clump together and precipitate from solution, destroying the detergent action of both.

Amphoteric surfactants carry both a positive and a negative charge. They react with both acids and bases and are noted for their mildness, sudsing, and stability. They are used in personal cleansing products such as shampoos for babies and in household cleaning products such as liquid handwashing soaps. Typical amphoterics include compounds called betaines.

$$CH_3(CH_2)_nCH_2NH_2^+ \ CH_2COO^-$$
A betaine
$n = 10$ to 16

Surfactants are used in thousands of products, and not just in cleaning. A small amount of lecithin, an edible surfactant, is often added to chocolate in the manufacturing process. The lecithin helps the chocolate disperse on the tongue more easily. The chocolate tastes "more chocolatey."

Classification of Surfactants

Classify each of the following surfactants as anionic, cationic, nonionic, or amphoteric.

a. $CH_3(CH_2)_{14}CH_2N^+(CH_3)_3\ Cl^-$

b. $CH_3(CH_2)_{13}O(CH_2CH_2O)_7SO_3^-\ Na^+$

c. $CH_3(CH_2)_{12}CH_2NH_2^+CH_2COO^-$

d. $CH_3(CH_2)_{12}CON(CH_2CH_2OH)_2$

Solution

a. The N atom of the large, active part bears a positive charge; this is a cationic surfactant. (The counter ion is Cl^-.)

b. There is a negative charge on the SO_3 group of the active part, making this an anionic surfactant. (The counter ion is Na^+.)

c. This substance has both a positive charge (on the N atom) and a negative charge (on the COO^- group), making this an amphoteric detergent.

d. This substance has neither a positive charge nor a negative charge; it is a nonionic surfactant.

▶ **Exercise 17.1**

Classify each of the following surfactants as anionic, cationic, nonionic, or amphoteric.

a. $CH_3(CH_2)_8-C_6H_4-O(CH_2CH_2O)_9H$

b. $CH_3(CH_2)_{14}COO^-K^+$

c. $CH_3(CH_2)_9-C_6H_4-SO_3^-\ Na^+$

d. $CH_3(CH_2)_{12}CH_2N^+(CH_3)_3\ Cl^-$

Builders

Any substance added to a surfactant to increase its detergency is called a **builder**. The main function of builders is to reduce water hardness. Sodium citrate ($Na_3C_6H_5O_7$) and complex phosphates such as sodium tripolyphosphate ($Na_5P_3O_{10}$) and sodium hexametaphosphate [$(NaPO_3)_6$] tie up Ca^{2+} and Mg^{2+} in soluble complexes, a process called *sequestration*. The complex phosphates also enhance detergency by a synergistic process. For example, if a wash solution has 1.00 g/L of an alkyl sulfate detergent, it will have a certain cleaning action. Similarly, a wash solution with 1.00 g/L of sodium tripolyphosphate will have a certain detergency. But if a solution has 0.50 g/L of alkyl sulfate detergent and 0.50 g/L of sodium tripolyphosphate, clothes washed in it will be cleaner and brighter than if washed in either of the first two.

Phosphates speed the eutrophication of lakes (Chapter 13). In some areas, phosphates from detergents contribute somewhat to this process. Several state and local governments have banned the sale of detergents containing phosphates.

Sodium carbonate acts by reacting with the hard water ions to form an insoluble substance (Section 17.2). Automatic washing machines can be harmed by the $CaCO_3$ and $MgCO_3$ precipitates. These builders also produce a mild alkalinity, which assists cleaning, especially of acidic soils.

Complex aluminosilicates called *zeolites* are often used as builders. Zeolite anions act by *ion exchange* (Section 17.2), trapping calcium ions by exchanging them for their own sodium ions.

$$Ca^{2+}(aq) + Na_2Al_2Si_2O_8(s) \longrightarrow 2\ Na^+(aq) + CaAl_2Si_2O_8(s)$$

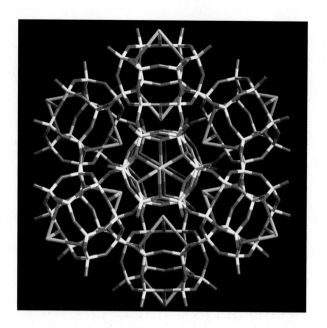

▶ Zeolites used in detergent formulations trap positive ions such as Ca^{2+} and Mg^{2+} in the large, cagelike structures and hold them in suspension in the wash water.

Calcium and magnesium ions are held in suspension by the zeolites (rather than being precipitated).

Brighteners

Almost all detergent formulations include fluorescent white dyes or **optical brighteners**. After white clothing is worn a few times, it often looks yellowish and drab. Optical brighteners absorb ultraviolet rays in sunlight and release them as blue rays. These blue rays then camouflage the yellowish color and give the garment the appearance of being "whiter than white" (Figure 17.9). Clothes treated with an optical brightener on the surface may be dirty underneath, but they look "whiter and brighter than new." Brighteners are also used in cosmetics, paper, soap, plastics, and other products.

Optical brighteners appear to have low toxicity to humans, although they cause skin rashes in some people. They may cause developmental and reproductive effects, but additional testing is needed to confirm or refute this. Their effect on lakes and streams is largely unknown, despite the large amounts entering our waterways. The only benefit of these compounds is cosmetic.

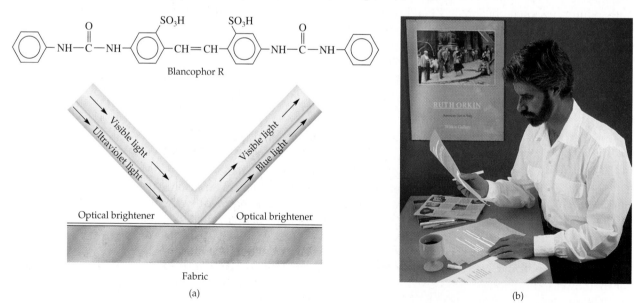

▲ **Figure 17.9** (a) An optical brightener, called a *blancophor* (or colorless dye), converts invisible ultraviolet light to visible blue light, making the fabric look brighter and masking any yellowish color. (b) Clothes washed in a detergent formulation that has an optical brightener glow in the light from an ultraviolet lamp (black light).

Liquid Laundry Detergents

Early laundry detergent preparations were powders, but liquid laundry detergent formulations now command much of the market. In general, powders are more concentrated and clean better, but many consumers find liquids to be more convenient. Some liquid laundry detergents are "built" with sodium citrate, sodium carbonate, or zeolites. Unbuilt formulations are high in surfactants but contain no builders.

LAS surfactants are the cheapest surfactants in liquid laundry detergents. In unbuilt formulations, they are usually used as the sodium salt (page 519) or triethanolamine [$N(CH_2CH_2OH)_3$] salt.

$$CH_3(CH_2)_9CH(CH_3)-C_6H_4-SO_3^-\ HN^+(CH_2CH_2OH)_3$$
LAS, triethanolamine salt

In built varieties, LAS are often present as the potassium salt.

$$CH_3(CH_2)_9CH(CH_3)-C_6H_4-SO_3^-\ K^+$$
LAS, potassium salt

Other popular surfactants for liquid formulations are AES. AES have a hydrocarbon portion derived from an alcohol (or alkylphenol), a polar portion derived from ethylene oxide (margin), and a sulfate salt portion. AES surfactants are efficient but more expensive than LAS.

Both AES and LAS are anionic surfactants. Some liquid detergents also contain nonionic surfactants, such as alcohol ethoxylates and alkylphenol ethoxylates. The oxygen atoms of the nonionic surfactants hydrogen bond with water molecules, making the ends of the molecules water soluble, just like the ionic end of an anionic surfactant. Nonionic surfactants effectively remove oily soil from fabrics. They are not as good as anionic surfactants at keeping dirt particles in suspension. Alcohol ethoxylates have the unusual property of being more soluble in cold water than in hot, which makes them particularly suitable for cold-water laundering.

Like powdered formulations, liquid laundry detergents can have a variety of other ingredients: bleaches, fragrances, fabric softeners, fluorescent whiteners, foam stabilizers, and more. Liquid products may also have a solvent such as ethanol, isopropyl alcohol, or propylene glycol to help prevent separation of the other components and to help dissolve greasy soils.

Some detergent formulations include enzymes such as lipases (enzymes that work on lipids) to help remove fats and oils and proteases (enzymes that work on proteins) to aid in the removal of protein stains such as blood.

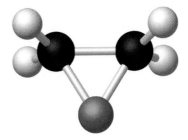
▲ Ethylene oxide.

17.5 DISHWASHING DETERGENTS

Liquid detergents for washing dishes by hand generally contain one or more surfactants as the main active ingredients. As with liquid laundry detergents, the surfactants in these formulations include LAS as the sodium and/or the triethanolamine salt. Some use nonionic surfactants such as cocamido DEA, an amide made from a fatty acid and diethanolamine [$HN(CH_2CH_2OH)_2$].

$$CH_3(CH_2)_nCON(CH_2CH_2OH)_2$$
Cocamido DEA
$n = 8$ to 12

Few dishwashing liquids contain builders; those that do usually have only small amounts. They may also contain enzymes to help remove greases and protein stains, fragrances, preservatives, solvents, bleaches, and other ingredients. Most liquid dishwashing detergents differ significantly only in the concentration or effectiveness of the surfactant.

Detergents for automatic dishwashers are quite another matter. They often are strongly alkaline and should never be used for hand dishwashing. They contain sodium tripolyphosphate ($Na_5P_3O_{10}$), sodium carbonate, sodium metasilicate (Na_2SiO_3), sodium sulfate, a chlorine or oxygen bleach, and only a small amount of surfactant, usually a nonionic type. Some contain sodium hydroxide. They depend mainly on their strong alkalis and the vigorous agitation of the machine for cleaning. Dishes that are washed repeatedly in an automatic dishwasher may be permanently etched by the strong alkali of the detergent, which is why delicate crystal stemware is often washed by hand.

17.6 FABRIC SOFTENERS: QUATERNARY AMMONIUM SALTS

In Section 17.4 we mentioned cationic surfactants, in which the working part is a positive ion, and many of which are quaternary ammonium salts. The cations have four hydrocarbon groups attached to a nitrogen atom that bears a positive charge. One of the hydrocarbon groups is a long chain with 12 to 18 carbon atoms. The other three are small, usually methyl groups. An example of such a cationic surfactant is octadecyltrimethylammonium chloride.

$$CH_3CH_2CH_2CH_2CH_2CH_2CH_2CH_2CH_2CH_2CH_2CH_2CH_2CH_2CH_2CH_2CH_2CH_2\overset{\overset{\displaystyle CH_3}{|}}{\underset{\underset{\displaystyle CH_3}{|}}{N^+}}-CH_3$$

Cationics make up only a small portion of the surfactant market, but another kind of quaternary salt with *two* long carbon chains and two smaller groups on nitrogen is used as a fabric softener. An example is dioctadecyldimethylammonium chloride. These compounds are strongly adsorbed by the fabric, forming a film one molecule thick on the surface. The long hydrocarbon chains lubricate the fibers, imparting increased flexibility and softness to the fabric.

$$CH_3CH_2CH_2CH_2CH_2CH_2CH_2CH_2CH_2CH_2CH_2CH_2CH_2CH_2CH_2CH_2CH_2CH_2-\overset{\overset{\displaystyle CH_3}{|}}{\underset{\underset{\displaystyle CH_3CH_2CH_2CH_2CH_2CH_2CH_2CH_2CH_2CH_2CH_2CH_2CH_2CH_2CH_2CH_2CH_2CH_2}{|}}{N^+}}-CH_3$$

17.7 LAUNDRY BLEACHES: WHITER WHITES

Bleaches are oxidizing agents (Chapter 8) that remove colored stains from fabrics. Most familiar liquid "chlorine" laundry bleaches (such as Clorox and Purex) are 5% sodium hypochlorite (NaOCl) solutions that differ primarily in price. The newer "ultra" chlorine bleaches have a higher concentration of NaOCl. Hypochlorite bleaches release chlorine rapidly, and high concentrations of chlorine can be quite damaging to fabrics. These bleaches do not work well on polyester fabrics, often causing yellowing rather than the desired whitening.

Other bleaches are available in solid forms that release chlorine slowly in water so as to minimize damage to fabrics. Symclosene is an example of a cyanurate-type bleach.

Oxygen-releasing bleaches usually contain sodium percarbonate ($2\,Na_2CO_3 \cdot 3\,H_2O_2$) or sodium perborate ($NaBO_2 \cdot H_2O_2$). As indicated by the formulas, these compounds are complexes of $NaBO_2$ or Na_2CO_3 and hydrogen peroxide (H_2O_2). In hot water, the hydrogen peroxide is liberated and acts as a bleach.

Borates are somewhat toxic. Perborate bleaches are less active than chlorine bleaches and require higher temperatures, higher alkalinity, and higher concentrations to do an equivalent job. They are used mainly for bleaching white, resin-treated polyester–cotton fabrics. These fabrics last much longer with oxygen bleaching than with chlorine bleaching. Percarbonate bleaches act like a combination of hydrogen peroxide and sodium carbonate. Less toxic than the perborate bleaches, percarbonate bleaches are rapidly becoming the predominant forms of oxygen bleaches. Properly used, oxygen bleaches make fabrics whiter than do chlorine bleaches.

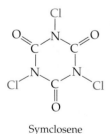

Symclosene

Bleaches act on certain light-absorbing chemical groups, called *chromophores*, that cause a substance to be colored. If we represent a chromophore as $-CH=CH-C(=O)-CH_2$, we can show its oxidation as

$$-CH=CH-\overset{\overset{O}{\|}}{C}-CH_2- \xrightarrow{[O]} -\overset{\overset{O}{\|}}{C}-OH + HO-\overset{\overset{O}{\|}}{C}-\overset{\overset{O}{\|}}{C}-CH_2-$$

Chromophore (colored)   Fragments (colorless)

Thus the bleach attacks the color-producing groups, not the entire colored soil.

> The physical and chemical processes that occur in cleaning a surface take some time. An often-overlooked technique in cleaning is: allow the surface to soak. Plain water or dilute soap solution is an amazingly effective, cheap, and nonpolluting cleaner if it is simply sprayed on a surface and allowed to stand for a few minutes.

17.8 ALL-PURPOSE CLEANING PRODUCTS

Various all-purpose cleaning products are available for use on walls, floors, countertops, appliances, and other tough, durable surfaces. Those for use in water solution may contain surfactants, sodium carbonate, ammonia, solvent-type grease cutters, disinfectants, bleaches, deodorants, and other ingredients. Some are great for certain jobs but not as good for others. They damage some surfaces but work especially well on others. Most important, they may be harmful when used improperly. Reading the labels is important.

Household ammonia solutions, straight from the bottle, are good for loosening baked-on grease or burned-on food. Diluted with water, they clean mirrors, windows, and other glass surfaces. Mixed with detergent, ammonia rapidly removes wax from vinyl floor coverings. Ammonia vapors are highly irritating, and this cleanser should not be used in a closed room. Ammonia should not be used on asphalt tile, wood surfaces, or aluminum because it may stain, pit, or erode these materials.

Baking soda (sodium bicarbonate, $NaHCO_3$) straight from the box, is a mild abrasive cleanser. It absorbs food odors readily, making it good for cleaning the inside of a refrigerator. Vinegar (acetic acid) cuts grease film. It should not be used on marble, because it reacts with the marble, pitting the surface.

$$CaCO_3(s) + 2\ CH_3COOH(aq) \longrightarrow Ca^{2+}(aq) + 2\ CH_3COO^-(aq) + CO_2(g) + H_2O(l)$$

Marble   Vinegar

▲ Vinegar reacts with marble, pitting the surface. Carbon dioxide gas is given off, forming bubbles.

Hazards of Mixing Cleaners

Mixing bleach with other household chemicals can be quite dangerous. For example, mixing a hypochlorite bleach with hydrochloric acid produces poisonous chlorine gas.

$$2\ HCl(aq) + ClO^-(aq) \rightarrow Cl^-(aq) + H_2O(l) + Cl_2(g)$$

Mixing bleach with toilet bowl cleaners that contain HCl is especially dangerous. Most bathrooms are small and poorly ventilated, and generating chlorine in such a limited space is hazardous. Chlorine can do enormous damage to the throat and the entire respiratory tract. If the concentration of chlorine is high enough, it can kill.

Mixing bleach with ammonia is also extremely hazardous. Two of the gases produced are chloramine (NH_2Cl) and hydrazine (NH_2NH_2) both of which are quite toxic. *Never* mix bleach with other chemicals without specific directions to do so. In fact, it is a good rule not to mix any chemicals unless you know exactly what you are doing.

17.9 SPECIAL-PURPOSE CLEANERS

There are many highly specialized cleaning products on the market. For metals there are various kinds of chrome cleaners, brass cleaners, copper cleaners, and silver cleaners. There are lime removers, rust removers, and grease removers. There

H₂C—COOH

are cleaners specifically for wood surfaces, for vinyl floor coverings, and for ceramic tile. Let us look at just a few special purpose cleaners that you would find in almost any home.

Toilet Bowl Cleaners

$$H_2C-COOH$$
$$HO-C-COOH$$
$$H_2C-COOH$$

Citric acid

The "lime" build-up that forms in toilet bowls is mainly calcium carbonate ($CaCO_3$), deposited from hard water, and often discolored by such things as iron compounds and fungal growth. Because calcium carbonate is readily dissolved by acid, toilet bowl cleaners tend to be strongly acidic. The solid crystalline products usually contain sodium bisulfate ($NaHSO_4$), and the liquid cleaners are hydrochloric acid (HCl), citric acid, or some other acidic material.

Scouring Powder

Most powdered cleansers contain an abrasive such as silica (SiO_2) that "scrapes" stains from hard surfaces. They also usually contain a surfactant to dissolve grease, and some feature a bleach. These cleansers are mainly intended for removing stains from porcelain tubs and sinks. Such abrasive cleansers may scratch the finish on appliances, countertops, and metal utensils. They may even scratch the surfaces of sinks, toilet bowls, and bathtubs. Dirt gets into the scratches and makes cleaning even more difficult.

Glass Cleaners

Cleaners for windowpanes and mirrors are volatile liquids that evaporate without leaving a residue. A common glass cleaner is simply isopropyl alcohol (rubbing alcohol) diluted with water. Sometimes ammonia or vinegar is added for greater cleaning power.

Drain Cleaners

When kitchen drains become clogged, it is usually because the pipes have become clogged with grease. Drain cleaners often contain sodium hydroxide, either in the solid form or as a concentrated liquid. The sodium hydroxide reacts with the water in the pipe to generate heat, which melts much of the grease. The sodium hydroxide then reacts with some of the fat, converting it to soap, which helps clean out more of the grease in the pipe.

In some products there are also bits of aluminum metal that react with the sodium hydroxide solution to form hydrogen gas, which bubbles out of the clogged area of the drain, creating a stirring action.

Many liquid drain cleaners contain bleach (NaOCl) as well as concentrated sodium hydroxide. Shower drains are often clogged with hair; the bleach degrades the hair, helping to unplug the drain. The best drain cleaner is prevention: Don't pour grease down the kitchen sink and try to keep hair out of the drain. If a drain does get plugged, a mechanical device such as a plumber's snake is often a better choice than a chemical drain cleaner.

Oven Cleaners

Most oven cleaners also contain sodium hydroxide. Several popular products dispense it as an aerosol foam. The greasy deposits on oven walls are converted to soaps when they react with the sodium hydroxide. The resulting mixture can then be washed off with a wet sponge. (Wear rubber gloves; sodium hydroxide is extremely caustic and hard on the skin.)

17.10 ORGANIC SOLVENTS IN THE HOME

Solvents are used in the home to remove paint, varnish, adhesives, waxes, and other materials. Petroleum solvents are also added to some all-purpose cleansers as grease cutters (Figure 17.10). They dissolve grease readily but, like gasoline, are

▲ **Figure 17.10** Cleansers that contain petroleum distillates warn of combustibility and of the hazard of swallowing them.

highly flammable and deadly when swallowed. The lungs become saturated with hydrocarbon vapors, fill with fluid, and fail to function.

Most organic solvents used around the home are volatile and flammable. Many have toxic fumes, and nearly all are narcotic at high concentrations. Such solvents should be used only with adequate ventilation and never used around a flame. Be sure to read—and heed—all precautions before you use any solvent. Gasoline should not be used for cleaning; it is too hazardous in too many ways.

A troubling problem connected with household solvents is the practice of inhaling fumes to get "high." The popularity of solvent sniffing seems to be related mainly to peer pressure, especially among young teenagers. Long-term sniffing can cause permanent damage to vital organs, especially the lungs. There can be irreversible brain damage, and some have died of heart failure while sniffing solvents.

17.11 PAINTS

Paint is a broad term that covers a wide variety of products—lacquers, enamels, varnishes, oil-base coatings, and a number of different water-base finishes. Any or all of these materials can be found in any home. A paint contains three basic ingredients: a pigment, a binder, and a solvent. The universal pigment today is titanium dioxide (TiO_2), which has taken the place of "white lead," the poisonous white pigment that was finally banned in 1977. White lead is basic lead carbonate, $2\ PbCO_3 \cdot Pb(OH)_2$.

Titanium dioxide is a brilliant white pigment with great stability and excellent hiding power. All ordinary paints are pigmented with titanium dioxide. For colored paints, small amounts of colored pigments or dyes are added to the white base mixture. The binder, or film former, is the substance that binds the pigment particles together and holds them on the painted surface. In oil paints the binder is usually tung oil or linseed oil. In water-base paints it is a polymer of some kind. Most interior paints have polyvinyl acetate as the binder. Exterior water-base paints use acrylic resins as binders. Acrylic latex paints are much more resistant to rain and sunlight. The solvent is added in order to keep the paint fluid until it is applied to a surface. The solvent might be an alcohol, a hydrocarbon, an ester (or some mixture thereof), or water.

The paint may also contain additives: a drier (or activator) to make the paint dry faster, a fungicide to act as a preservative, a thickener to increase the paint's viscosity, an antiskinning agent to keep the paint from forming a skin inside the can, and perhaps a surfactant to stabilize the mixture.

▲ Addition of colored pigments or dyes to a white base mixture can produce a great variety of colored paints.

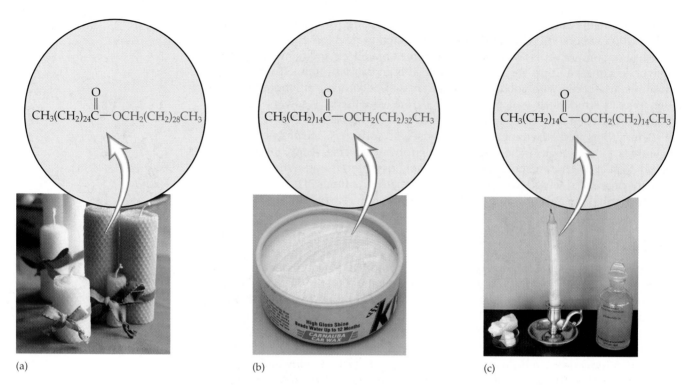

$$CH_3(CH_2)_{24}\overset{O}{\underset{}{C}}-OCH_2(CH_2)_{28}CH_3$$

$$CH_3(CH_2)_{14}\overset{O}{\underset{}{C}}-OCH_2(CH_2)_{32}CH_3$$

$$CH_3(CH_2)_{14}\overset{O}{\underset{}{C}}-OCH_2(CH_2)_{14}CH_3$$

(a) (b) (c)

▲ **Figure 17.11** The molecular views show the esters that are typical components of (a) beeswax, (b) carnauba wax, and (c) spermaceti.

17.12 WAXES

Chemically, a **wax** is an ester of a long-chain organic acid (fatty acid) with a long-chain alcohol. Waxes are produced by plants and animals mainly as protective coatings. (These compounds are not to be confused with paraffin wax, which is made up of hydrocarbons.)

Three typical waxes are shown in Figure 17.11. Beeswax is the material from which bees build honeycombs and is used in such household products as candles and shoe polish. Carnauba wax is a coating that forms on the leaves of certain palm trees in Brazil. It is a mixture of esters similar to those in beeswax and is used in making automobile wax, floor wax, and furniture polish. Spermaceti wax, which is extracted from the head of a sperm whale, is largely cetyl palmitate. Once widely used in making cosmetics and other products, spermaceti wax is now in short supply because the sperm whale has been hunted almost to extinction.

Lanolin, the grease in sheep's wool, is also a wax. Because it forms stable emulsions with water, it is useful in making various skin creams and lotions.

Many natural waxes have been replaced by synthetic polymers such as silicones (Chapter 10), which are often cheaper and more effective.

17.13 COSMETICS: PERSONAL CARE CHEMICALS

Ages ago, people used materials from nature for cleansing, beautifying, and otherwise altering their appearance. Evidence indicates that 7000 years ago, Egyptians used powdered antimony and the green copper ore malachite as eye shadow. Egyptian pharaohs used perfumed hair oils as far back as 3500 B.C.E. Claudius Galen, a Greek physician of the second century C.E. is said to have invented cold cream. Dandy gentlemen of seventeenth-century Europe used cosmetics lavishly, often to cover the fact that they seldom bathed. Ladies of eighteenth-century Europe whitened their faces with lead carbonate ($PbCO_3$) and many died from lead poisoning.

The use of cosmetics has a long and interesting history, but nothing in the past comes close to the amounts and varieties of cosmetics used by people in the modern industrial world. Each year we spend billions of dollars on everything from hair sprays to toenail polishes, from mouthwashes to foot powders. Combined sales of the world's 100 largest cosmetics companies were more than $89 billion in 2001.

What is a cosmetic? The United States Food, Drug, and Cosmetic Act of 1938 defined **cosmetics** as "articles intended to be rubbed, poured, sprinkled or sprayed on, introduced into, or otherwise applied to the human body or any part thereof, for cleansing, beautifying, promoting attractiveness or altering the appearance...." Soap, although obviously used for cleansing, is specifically excluded from coverage by the law. Also excluded are substances that affect the body's structure or functions. Antiperspirants, products that reduce perspiration, are legally classified as drugs. So are antidandruff shampoos. The main difference between drugs and cosmetics is that drugs must be proven "safe and effective" before they are marketed; cosmetics generally do not have to be tested before they're marketed. Most brands of a given type of cosmetic contain the same (or quite similar) active ingredients. Thus, advertising is usually geared toward selling a name, a container, or a fragrance rather than the actual product itself.

The Food, Drug, and Cosmetic Act states that if a product has drug properties, it must be approved as a drug. Nevertheless, the cosmetics industry uses the term *cosmeceutical* to refer to cosmetic products that allegedly have medicinal or druglike benefits. We will discuss several such products in the following sections.

Skin Creams and Lotions

Skin is the body's largest organ with an area of about 18 ft². It encloses the body, forming a barrier to keep harmful substances out and moisture and nutrients in (Figure 17.12). The outer layer of skin is called the *epidermis*. The epidermis in turn is divided into two parts: dead cells on the outside (the corneal layer), and living cells on the inside, continually replacing corneal cells, which are then sloughed off.

The corneal layer is composed mainly of a tough, fibrous protein called **keratin**. Keratin has a moisture content of about 10%. Below 10% moisture, the human skin is dry and flaky. Above 10%, conditions are ideal for the growth of harmful microorganisms. Skin is protected from loss of moisture by **sebum**, an oily secretion of the sebaceous glands. Exposure to sun and wind can leave the skin dry and scaly, and too-frequent washing also removes natural skin oils.

Cosmetics are applied to the dead cells of the corneal layer. Most of the preparations applied to the skin consist mainly of lotions and creams. A **lotion** is an emulsion of tiny oil droplets dispersed in water. A **cream** is the opposite; tiny water

About 8000 different chemicals are used in cosmetics. They are sold in tens of thousands of different combinations.

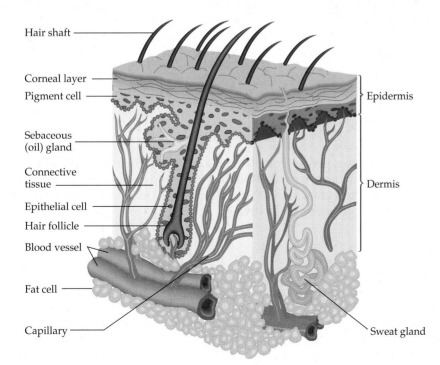

Hair shaft
Corneal layer
Pigment cell
Epidermis
Sebaceous (oil) gland
Connective tissue
Dermis
Epithelial cell
Hair follicle
Blood vessel
Fat cell
Capillary
Sweat gland

◄ **Figure 17.12** Cross section of an area of skin. Skin has two main layers: Each square inch of the dermis contains about 1 m of blood vessels, 4 m of nerves, 100 sweat glands, several oil glands, more than 3 million cells, and the active part of hair follicles. The epidermis, just over the dermis, has a thin layer of cells that divide continuously. These cells move upward to the corneal layer, dying as they do so. Cosmetics generally affect only the outer corneal layer of dead cells.

(a)

(b)

▲ **Figure 17.13** A lotion is an emulsion of an oil in water (a). A cream is an emulsion of water in an oil (b). A lotion feels cool because evaporating water removes heat from the skin. A cream feels greasy.

droplets are dispersed in oil (Figure 17.13). The essential ingredient of each is a fatty or oily substance that forms a protective film over the skin and helps hold moisture in. Typical substances used are mineral oil and/or petroleum jelly, both mixtures of alkanes obtained from petroleum. Others include natural fats and oils, propylene glycol, glycerol, perfumes, waxes, water, and emulsifiers (compounds that keep the oily portions from separating from the water). Lanolin, a fat obtained from sheep's wool, aloe vera gel, olive oil, and palm oil are also used, and beeswax is often added to harden the product.

Some creams have been formulated with hormones, vitamins, and other strange ingredients. Few, if any, have been found to confer any particular benefit. Creams and lotions protect the skin by coating and softening it. Such skin softeners are called **emollients**. Petroleum jelly (such as Vaseline®) or a good grade of white mineral oil (baby oil) works as well as fancy creams.

It may seem strange that gasoline, a mixture of alkanes, dries out the skin while the higher alkanes in mineral oil and petroleum jelly soften it. Keep in mind, however, that gasoline is a thin, free-flowing liquid. It dissolves natural skin oils and carries them away. Higher alkanes are viscous, staying right on the skin and serving as emollients.

Moisturizers hold moisture to the skin. They work best when applied while the skin is still wet from a bath or shower. Moisturizers don't actually add moisture to skin. Rather, they form a physical barrier that hinders evaporation of water from the skin. Moisturizers are often substances such as lanolin. Collagen (the protein in connective tissue) is an effective moisturizer in skin lotions.

Some cosmetics contain *humectants* such as glycerol, lactic acid, or urea. Glycerol, with its three hydroxyl groups, holds water by hydrogen bonding. Urea actually binds water below the surface of the epidermis, right down to the skin-building cells. Urea and lactic acid also cause the old dead cells to fall off more rapidly, leaving new, fresher-looking skin exposed.

Sunscreen Lotions

Ultraviolet rays in sunlight turn light skin darker by triggering production of the pigment **melanin**. The dark melanin then protects the deeper layers of the skin from damage. Excessive exposure to ultraviolet radiation causes premature aging of the skin and leads to skin cancer. Our quest for tanned skin has resulted in an epidemic of skin cancer. Shorter-wavelength (UV-B) rays are more energetic and are especially harmful. Most **sunscreen lotions** block UV-B radiation while letting through the less energetic long-wave UV-A rays that promote tanning. For many years the active ingredient in many of these preparations has been *para*-aminobenzoic acid or one of its esters.

$$H_2N-\underset{}{\bigcirc}-\overset{\overset{O}{\|}}{C}-OH$$

para-Aminobenzoic acid

PABA derivatives generally are water soluble and thus wash off when swimming. They provide only partial UV-B and no UV-A protection and are therefore now used in very few sunscreens. Octyl methoxycinnamate (OMC) is now used as a UV-B filter in most sunscreen lotions.

$$CH_3O-\underset{}{\bigcirc}-CH=CHCOOCH_2CH(CH_2CH_3)CH_2CH_2CH_2CH_3$$

Octyl methoxycinnamate (OMC)
(2-ethylhexyl 4'-methoxycinnamate)

The eight-carbon alkyl group helps make this compound insoluble in water and less likely to wash off while swimming.

Antiaging Creams and Lotions

Some cosmetics are claimed to reduce wrinkles, spots, and other signs of aging skin. Prominent among them are products containing alpha hydroxy acids (AHAs), derived from fruit and milk sugars. An alpha hydroxy acid is a carboxylic acid that has a hydroxyl group on the carbon atom next to the carboxyl carbon. Lactic acid is a typical example. Consumers spend billions of dollars annually for cosmetics containing AHAs. There is some scientific evidence that they may work.

$$CH_3CH \overset{\displaystyle OH}{\underset{\displaystyle |}{}} - \overset{\displaystyle O}{\underset{\displaystyle \|}{C}} - OH$$

Lactic acid

AHAs and related substances act as exfoliating agents; they cause the old dead skin cells to be sloughed off, to be replaced by newer, fresher looking skin. They seem to be relatively safe. They do cause redness, swelling, burning, blistering, bleeding, rash, itching, and skin discoloration in some people. People who use AHA-containing products have greater sensitivity to sun, perhaps putting them at greater risk of skin cancer.

The best treatment for wrinkling of the skin is prevention, accomplished largely by avoiding excessive exposure to the sun and not smoking cigarettes.

Various concentrations of ultraviolet-absorbing substances in a lotion provide the **skin protection factor (SPF)** ratings. SPF values vary from 2 to 35 or more. An SPF of 15, for example, is supposed to mean that you can stay in the sun without burning 15 times as long as you could with unprotected skin.

Physical sunscreens, or sunblocks, block all UV radiation. They contain zinc oxide or titanium dioxide powders, both of which absorb UV strongly.

Cigarette smoking also leads to premature aging of the skin. Nicotine causes constriction of the tiny blood vessels that feed the skin. Repeated constrictions over the years cause the skin to lose its elasticity and become wrinkled.

Lipstick

Lipsticks and lip balms are quite similar to skin creams in composition and function. They are made of an oil and a wax, with a higher proportion of wax than in creams. Because it has little in the way of protective oils, the skin of the lips easily dries out, leading to chapped lips. Lipsticks and lip balms prevent moisture loss and soften and brighten the lips. The oil is frequently castor oil, sesame oil, or mineral oils. Waxes often employed are beeswax, carnauba, and candelilla. Dyes and pigments provide color to lipsticks. Perfumes are added to cover up the unpleasant fatty odor of the oil, and antioxidants are used to retard rancidity. Bromo acid dyes such as tetrabromofluorescein, a bluish-red compound, are responsible for the color of most modern lipsticks.

Tetrabromofluorescein

Eye Makeup

Various chemicals are used to decorate the eyes. Mascara, to darken eyelashes, has a base of soap, oils, fats, and waxes. Mascara is colored brown by iron oxide pigments, black by carbon (lampblack), green by chromium(III) oxide (Cr_2O_3), or blue by ultramarine (a silicate that contains some sulfide ions). A typical composition is 40% wax, 50% soap, 5% lanolin, and 5% coloring matter. Eyebrow pencils have about the same ingredients.

Eye shadow has a base of petroleum jelly with the usual fats, oils, and waxes. It is colored by dyes or made white by zinc oxide (ZnO) or titanium dioxide (TiO_2) pigments. A typical composition is 60% petroleum jelly, 10% fats and waxes, 6% lanolin, and the remainder dyes, pigments, or both.

Some people have allergic reactions to ingredients in eye makeup. A more serious problem is eye infection caused by bacterial contamination. It is recommended that eye makeup be discarded after three months. Mascara is of special concern because of the wand applicator. A slip of the hand and it can scratch the cornea.

Triclosan

Methyl *para*-hydroxybenzoate
(Methylparaben)

Deodorants and Antiperspirants

Deodorants are products that have perfume to mask body odor and a germicide to kill odor-causing bacteria. Bacteria act on perspiration residue and sebum (natural body oil) to produce malodorous compounds such as short-chain fatty acids and amines. The germicide is usually a long-chain quaternary ammonium salt (Section 17.7) or a phenol such as triclosan (left).

Antiperspirants are usually deodorants as well, but they also retard perspiration. The active antiperspirant ingredient is a variable complex of chlorides and hydroxides of aluminum and zirconium. Aluminum chlorohydrate $[Al_2(OH)_5Cl \cdot 2\ H_2O]$ is a typical compound found in many antiperspirants. Zirconium and aluminum chlorides and hydroxides function as **astringents**, which constrict the openings of the sweat glands, thus restricting the amount of perspiration that can escape. Antiperspirants can be formulated into creams or lotions, or they can be dissolved in alcohol and applied as sprays. Sweating is a natural, healthy body process, and to stop it on a routine basis is probably not wise. Regular bathing and changing of clothes make antiperspirants unnecessary for most people. But if you still feel the need for underarm protection, a simpler deodorant product should suffice.

17.14 TOOTHPASTE: SOAP WITH GRIT AND FLAVOR

After soap (which doesn't count, because the law says it isn't a cosmetic), toothpaste is the most important cosmetic product. Its only essential components are a detergent and an abrasive. Soap and sodium bicarbonate could do the job quite well but would be rather unpalatable. The ideal abrasive should be hard enough to clean teeth but not hard enough to damage tooth enamel. Abrasives frequently used in toothpaste are listed in Table 17.1. Some have been criticized as being too harsh.

Typical detergents are alkyl sulfates, such as sodium dodecyl sulfate (sodium lauryl sulfate).

$$CH_3CH_2CH_2CH_2CH_2CH_2CH_2CH_2CH_2CH_2CH_2CH_2OSO_3^-\ Na^+$$

Any pharmaceutical grade of soap or detergent probably would work satisfactorily. Most toothpastes today are full of minty flavors, colors, aromas, and sweet tastes. Ingredients include sweeteners such as sorbitol, glycerol (glycerin), and saccharin (Chapter 16); flavors such as wintergreen and peppermint; thickeners such as cellulose gum and polyethylene glycols; and preservatives such as methyl *para*-hydroxybenzoate. Table 17.2 gives a typical recipe for toothpaste.

Tooth decay is caused primarily by bacteria that convert sugars to sticky dextrans or plaque, and to acids such as lactic acid ($CH_3CHOHCOOH$). Acids dissolve tooth enamel. Brushing and flossing remove plaque and thus prevent decay. Decay can be minimized by eating sugars only at meals rather than in snacks and by brushing immediately after eating. Acids such as lactic acid in beer and phosphoric acid (H_3PO_4) in soda pop may also erode the tooth enamel of people who consume these beverages in large amounts.

TABLE 17.1 Abrasives Commonly Used in Toothpaste	
Name	**Chemical Formula**
Precipitated calcium carbonate	$CaCO_3$
Insoluble sodium metaphosphate	$(NaPO_3)_n$
Calcium hydrogen phosphate	$CaHPO_4$
Titanium dioxide	TiO_2
Tricalcium phosphate	$Ca_3(PO_4)_2$
Calcium pyrophosphate	$Ca_2P_2O_7$
Hydrated alumina	$Al_2O_3 \cdot nH_2O$
Hydrated silica	$SiO_2 \cdot nH_2O$

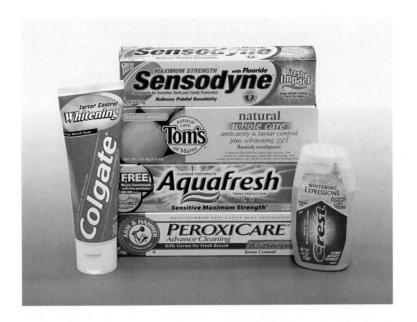

TABLE 17.2 A Typical Recipe for Toothpaste		
Ingredient	**Function**	**Amount**
Precipitated calcium carbonate	Abrasive	46 g
Castile soap or sodium dodecyl sulfate	Detergent	4 g
Glycerol (glycerin)	Sweetener	20 g
Gum tragacanth or gum cellulose	Thickener	1 g
Oil of peppermint (or peppermint extract)	Flavoring	1 mL
Water		28 mL

Many modern toothpastes contain fluorine compounds, such as stannous fluoride (SnF_2), shown to be effective in reducing the incidence of tooth decay. The enamel of teeth is composed mainly of hydroxyapatite [$Ca_5(PO_4)_3OH$]. Fluoride from toothpaste (or drinking water) converts part of the enamel to fluorapatite [$Ca_5(PO_4)_3F$]. Fluorapatite is stronger and more resistant to decay than hydroxyapatite.

The main cause of tooth loss in adults is gum disease. Toothpaste formulations with ingredients such as baking soda ($NaHCO_3$) and hydrogen peroxide (H_2O_2) are purported to prevent gum disease, but there seems little evidence that they really do.

Popular teeth whiteners often include hydrogen peroxide as a bleaching agent. The peroxide works on colored compounds in teeth in much the same way as bleaches work on stained clothing. However, the action of the H_2O_2 is slow. In toothpaste, the peroxide is not in contact with the teeth long enough to have a significant effect. Most of the successful teeth whiteners require application for 30 minutes or more. Dentist's preparations often use higher concentrations of peroxide along with ultraviolet light, which speeds the reaction.

17.15 PERFUMES, COLOGNES, AND AFTERSHAVES

Perfumes are among the most ancient and widely used cosmetics. Originally, all perfumes were extracted from natural sources such as fragrant plants. Their chemistry is exceedingly complex, but chemists have identified many of the components of perfumes and synthesized them in the laboratory, and so there are many synthetic perfumes. The best ones are probably still made from natural materials because chemists have so far been unable to identify all the many important, but minor, ingredients.

A good perfume may have a hundred or more constituents. Often the components are divided into three categories, called *notes*, based on differences in volatility.

TABLE 17.3	Compounds with Flowery and Fruity Odors Used in Perfumes	
Name	**Structure**[*]	**Odor**
Citral		Lemon
Irone		Violet
Jasmone		Jasmine
Phenylacetaldehyde		Lilac, hyacinth
2-Phenylethanol		Rose

[*]Note that each compound has only 1 oxygen atom for 10 or more carbon atoms.

The most volatile fraction (that which vaporizes most readily) is called the **top note**. This fraction, made up of relatively small molecules, is responsible for the odor when a perfume is first applied. The **middle note** is intermediate in volatility. It is responsible for the lingering aroma after most of the top-note compounds have vaporized. Top-note and end-note components generally have floral odors (Table 17.3). Some of these are synthesized in large quantities for use in perfumes. A typical top note ingredient is phenylacetaldehyde, which has a lilac odor. 2-Phenylethanol, which has the aroma of roses, is a typical middle note component.

The **end-note** fraction has low volatility and is made up of compounds with large molecules, often with musky odors. Odors vary with dilution. A concentrated solution may be unpleasant, yet a dilute solution of the same compound may have a pleasant aroma. Musks (Table 17.4) are often added to moderate the odor of the flowery or fruity top and middle note components.

Musks and similar compounds have extremely disagreeable odors when concentrated but are often pleasant in extreme dilution. The Ethiopian civet cat, from

Note that fragrance molecules (Table 17.3) are fat soluble. They dissolve in skin oils and can be retained for hours.

TABLE 17.4	Unpleasant Compounds Used to Fix Delicate Odors in Perfumes	
Compound	**Structure**	**Natural Source**
Civetone		Civet cat
Muscone		Musk deer
Indole		Feces

which the compound civetone is obtained, is a skunklike animal. Its secretion, like that of the skunk, is a defensive weapon. The secretion from musk deer is probably a pheromone (Chapter 16) that serves as a sex attractant for the deer.

Are there sex attractants for humans? The evidence is scanty, but some perfume makers add α-androstenol to their products. α-Androstenol is a steroid that occurs naturally in human hair and urine. Some studies hint that it may act as a sex attractant for human females, but this seems more likely to be just an advertising ploy. Even if there are human sex attractants, it is doubtful that they have much influence on human behavior because we probably have overriding cultural constraints. Anyway, we seem to prefer the sex attractant of the musk deer.

A perfume usually consists of 10–25% fragrant compounds and fixatives dissolved in ethyl alcohol. **Colognes** are perfumes diluted with ethyl alcohol or an alcohol–water mixture. They are only about 10% as strong as perfumes, usually containing only about 1 or 2% of perfume essence.

Aftershave lotions are similar to colognes. Most are about 50–70% ethanol, the remainder being water, perfume, and food coloring. Some have menthol added for a cooling effect on the skin; others have an added emollient of some sort to soothe chapped skin.

Perfumes are the source of many allergic reactions associated with cosmetics and other consumer products. **Hypoallergenic cosmetics** are those that purport to cause fewer allergic reactions than regular products. The term has no legal meaning, but most "hypoallergenic" cosmetics do not contain perfume.

α-Androstenol

Menthol

Aromatherapy

Aromatherapy is the practice of using volatile plant oils—*essential oils*—as a purported way to promote health and well-being. The aromas from these oils are supposed to have an effect on the nervous system. Aromatherapy includes the use of soap, bath additives, shower gels, and cosmetic products containing essential oils. It also includes the use of accessories such as car scenters, candles, room fragrances, oil burners, and sprays for bed linens.

Commonly used essential oils are those from marigold, lemon, orange, grapefruit, geranium, lavender, jasmine, and bergamot. The essential oils are diluted by carrier oils, which are vegetable oils such as sweet almond oil, apricot kernel oil, and grapeseed oil, before being applied to the skin.

Some aromas seem to have a positive psychological effect on the body. It may be as simple as a pleasant feeling caused when the scent reminds one of a pleasant experience. The United States aromatherapy market is valued at $800 million a year and accounts for most of worldwide sales. Home fragrance products account for nearly half the aromatherapy sales in the United States. It is difficult to evaluate the claims of the proponents by scientific experiments. Most of the reported effects are highly subjective.

17.16 SOME HAIRY CHEMISTRY

Like skin, hair is composed primarily of the fibrous protein keratin. Recall (Chapter 15) that protein molecules are made up of amino acid chains held together by four types of forces: hydrogen bonds, salt bridges, disulfide linkages, and dispersion forces. Of these, hydrogen bonds and salt bridges are important to our understanding of the actions of shampoos and conditioners. Hydrogen bonds between protein chains are disrupted by water (Figure 17.14), salt bridges are destroyed by changes in pH (Figure 17.15), and disulfide linkages are broken and restored when hair is permanently waved or straightened.

Shampoo

The keratin of hair has five or six times as many disulfide linkages as the keratin of skin. When hair is washed, the keratin absorbs water and is softened and made more stretchable. The water disrupts hydrogen bonds and some of the salt bridges. Acids and bases are particularly disruptive to salt bridges, making control of pH important in hair care. The number of salt bridges is maximized at a pH of 4.1.

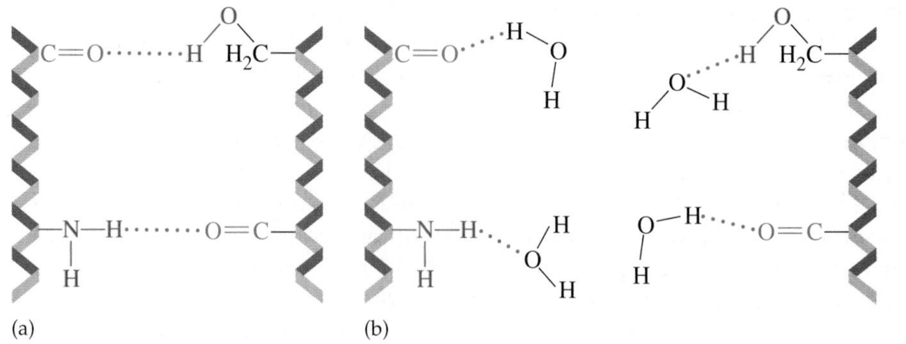

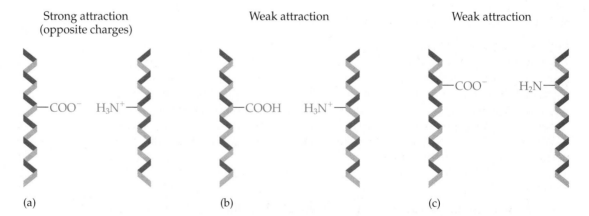

▶ **Figure 17.14** In hair, adjacent protein chains are held together by hydrogen bonds that link the carbonyl group of one chain to the amide group of another (a). When hair is wet, these groups can hydrogen bond with water rather than with each other (b), disrupting the hydrogen bonds between chains.

(a)   (b)

The visible portion of hair is dead; only the root is alive. The hair shaft is lubricated by sebum. Washing the hair removes this oil and any dirt adhering to it.

Before World War II, the cleansing agent in shampoos was soap. Soap-based shampoos worked well in soft water but left a dulling film on hair in hard water. People often removed the film by using a rinse containing vinegar or lemon juice; however, such rinses are usually not needed with today's products.

Modern shampoos use a synthetic detergent as a cleansing agent. In shampoos for adults, the detergent is often an anionic type, such as sodium dodecyl sulfate, the same detergent used in many toothpastes. In shampoos used for babies and children, the detergent is often an amphoteric surfactant (Section 17.4) that is less irritating to the eyes.

Amphoteric detergents react with both acids and bases. In an acidic solution, it can accept a proton on the negatively charged oxygen. In a basic solution, it can give up one of the protons from the nitrogen.

The only essential ingredient in shampoo is a detergent of some sort. What, then, is all the advertising about? You can buy shampoos that are fruit- or herb-scented, protein enriched, pH balanced, or made especially for oily or dry hair. Shampoos for oily, normal, and dry hair seem to differ primarily in the concentration of the detergent. Shampoo for oily hair is more concentrated; dry hair shampoo is more dilute.

Because hair is protein, a protein-enriched shampoo does give it more body. The protein (usually keratin or collagen) coats the hair and literally glues split ends together. Protein is often added to conditioners, too. Conditioners are mainly long-chain alcohols or long-chain quaternary ammonium salts. "Quats" are similar to the compounds used in fabric softeners (Section 17.7), and they work in much the same way to coat hair fibers.

There are acidic and basic groups on the protein chains in hair. It therefore stands to reason that the acidity or basicity of a shampoo affects hair. Hair and skin

Strong attraction (opposite charges)   Weak attraction   Weak attraction

$-COO^-$ H_3N^+- $-COOH$ H_3N^+- $-COO^-$ H_2N-

(a)   (b)   (c)

▲ **Figure 17.15** Strong electrostatic forces hold a protein chain with an ionized carboxyl group to another chain with an ionized amino function (a). A change in pH can disrupt this salt bridge. If the pH drops, the carboxyl group takes on a proton and loses its charge (b). If the pH rises, a proton is removed from the amino group and it is left uncharged (c). In either case, the forces between the protein chains are weakened considerably.

are both slightly acidic. Highly basic (high-pH) or strongly acidic (low-pH) shampoos would damage hair, and such products would also irritate the skin and eyes. Most shampoos, however, have pH values between 5 and 8, close to neutral or very slightly acidic.

How about all those flavors and fragrances? Ample evidence indicates that such "natural" ingredients as milk, honey, strawberries, herbs, cucumbers, and lemons add nothing to the usefulness of shampoos or other cosmetics. Why are they there? Smells sell, and there is an appeal to those interested in the back-to-nature movement. There is one hazard in the use of such fragrances: bees, mosquitoes, and other insects like certain fruit and flower odors, too. Using such products before going on a picnic or a hike could lead to a bee in your bonnet.

Hair Coloring

The color of hair and skin is determined by the relative amounts of two pigments: *melanin*, a brownish-black pigment, and *phaeomelanin*, a red-brown pigment that colors the hair and skin of redheads. Brunettes have lots of melanin in the hair, but blondes have little of either pigment. Brunettes who would like to become blondes can do so by oxidizing the colored pigments in their hair to colorless products. Hydrogen peroxide is the oxidizing agent usually used to bleach hair.

Hair dyeing is more complicated than bleaching. The color may be temporary, from water-soluble dyes that can be washed out, or more permanent, from dyes that penetrate the hair and remain there. These dyes often are used in the form of a water-soluble, often colorless, precursor that soaks into the hair and then is later oxidized by hydrogen peroxide to a colored compound.

▲ Shampoos are available in many forms and colors and under many brand names. The only essential ingredient in any shampoo is a detergent.

▲ Brown and black hair color is determined by the brownish-black pigment melanin; red hair by the red-brown pigment phaeomelanin; and blonde hair color by scarcity of both pigments.

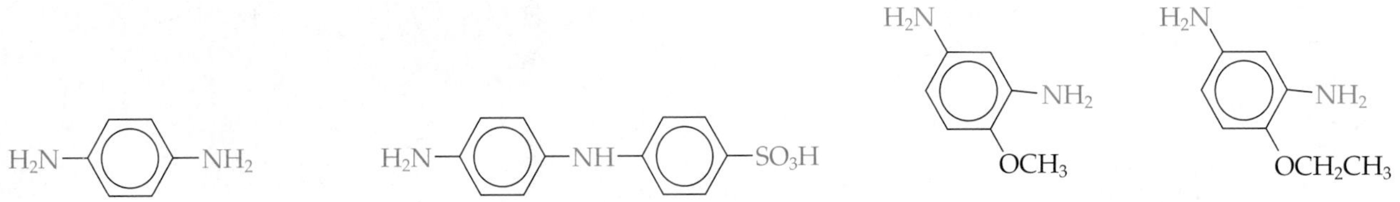

para-Phenylenediamine (a) | *para*-Aminodiphenylaminesulfonic acid (b) | MMPD (c) | EMPD (d)

▲ **Figure 17.16** Most hair dyes are derivatives of *para*-phenylenediamine (a). These include *para*-aminodiphenylaminesulfonic acid (b), *para*-methoxy-*meta*-phenylenediamine (MMPD) (c), and *para*-ethoxy-*meta*-phenylenediamine (EMPD) (d).

Permanent dyes are often derivatives of an aromatic amine called *para*-phenylenediamine (Figure 17.16). Variations in color can be obtained by placing substituents on this molecule. The *para*-phenylenediamine itself produces a black color, and the derivative *para*-aminodiphenylaminesulfonic acid is used in blonde formulations. Intermediate colors can be obtained by the use of other derivatives. One derivative, *para*-methoxy-*meta*-phenylenediamine (MMPD), has been shown to be carcinogenic when fed to rats and mice, but the hazard to those who use it as hair dye is not yet known. It is interesting to note that one substitute for MMPD was its homolog, *para*-ethoxy-*meta*-phenylenediamine (EMPD). Screening revealed that EMPD causes mutations in bacteria. Such mutagens often are also carcinogens.

The FDA cannot ban a cosmetic without proof of harm, but testing for carcinogenicity (Chapter 20) would cost at least $500,000 and take two years.

The chemistry of colored oxidation products is quite complex. These products probably include quinones and nitro compounds, among others, and are well-known products of the oxidation of aromatic amines.

Permanent dyes affect only the dead outer portion of the hair shaft. New hair, as it grows from the scalp, has its natural color.

Hair treatments, such as Grecian Formula®, that develop color gradually use rather simple chemistry. A solution containing colorless lead acetate

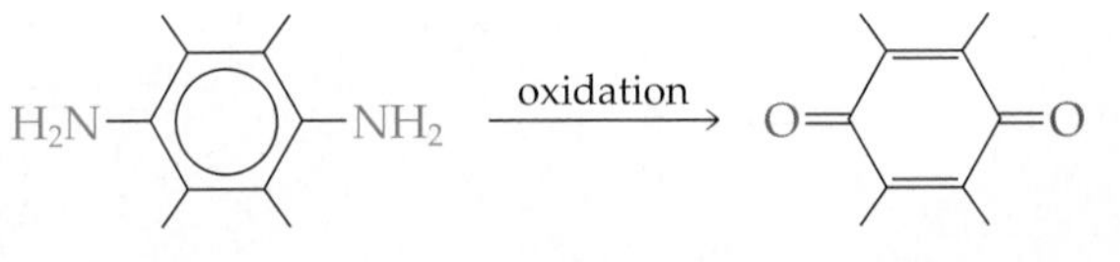

A diamine | A quinone

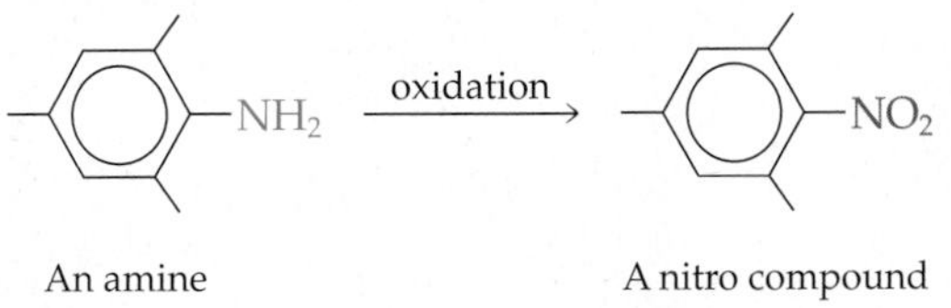

An amine | A nitro compound

[$Pb(CH_3COO)_2$] is rubbed on the hair. As it penetrates the hair shaft, Pb^{2+} ions react with sulfur atoms in the hair to form black lead sulfide (PbS). Repeated applications produce darker colors as more lead sulfide is formed.

Permanent Waving: Chemistry to Curl Your Hair

The chemistry of curly hair is interesting. Hair is protein, and adjacent protein chains are held together by disulfide linkages. Permanent wave lotion contains a reducing agent such as thioglycolic acid ($HSCH_2COOH$). This wave lotion ruptures the disulfide linkages (Figure 17.17), allowing the protein chains to be pulled apart as the hair is held in a curled position on rollers. The hair is then treated with a mild oxidizing agent such as hydrogen peroxide. Disulfide linkages are formed in new positions to give shape to the hair.

The same chemical process can be used to straighten naturally curly hair. The change in curliness depends only on how you arrange the hair after the disulfide bonds have been reduced and before the linkages have been restored. As in the case of permanent dyes, permanent curls grow out as new hair is formed.

Hair Sprays

Hair can be held in place by using **resins**, solid or semisolid organic materials that form a sticky film on the hair. Common resins used on hair are polyvinylpyrrolidone (PVP) and its copolymers.

The resin is dissolved in a solvent and sprayed on the hair where the solvent evaporates. The propellants usually used are volatile hydrocarbons, which are also flammable.

Holding resins are also available as mousses. A **mousse** is simply a foam or froth. The active ingredients, like those of hair sprays, are resins such as PVP. Coloring agents and conditioners are also available as mousses. Silicone polymers are often used to impart a sheen to the hair.

Hair Removers

Chemicals that remove unwanted hair are called **depilatories**. Most of them contain a soluble sulfur compound, such as sodium sulfide or calcium thioglycolate $[Ca(OOCCH_2SH)_2]$, formulated into a cream or lotion. These strongly basic mixtures destroy some of the peptide bonds in the hair so that it can be washed off. Remember that skin is made of protein, too, and so any chemical that attacks the hair can also damage the skin.

Hair Restorers

For years, men have been searching for a way to make hair grow on bald spots. Women suffer from baldness, too, but seem to consider wigs more acceptable than men do.

Minoxidil was first introduced as a drug for treating high blood pressure. It acts by dilating the blood vessels. When people who were taking the drug started growing hair on various parts of their bodies, it was applied to the scalps of people who were becoming bald. Minoxidil can produce a growth of fine hair anyplace on the skin where there are hair follicles. Minoxidil is sold under the trade name Rogaine. To be effective it must be used on a continuous basis, and the cost can be $1000 per year or more.

17.17 THE WELL-INFORMED CONSUMER

We cannot even attempt to tell you everything about the many chemical products you have in your home. Books have been written about some of them, and there are many volumes in the library that can help you if you want to know more. There is also much information available on the World Wide Web, but much of it is little more than advertising hype; use it with caution. We hope that the knowledge you have gained here and the information you read on product labels will make you a better-informed consumer.

When you look over the many different brands of a product at the supermarket, remember that the most expensive one is not necessarily better than the others. Consider the soap and detergent industry, which does a lot of advertising. Each laundry powder and shampoo claims to be superior to all its competitors, but most brands are rather similar. Advertising is expensive, and so the brands that are most widely advertised probably cost the most, too. Of course, it sometimes happens that one brand really is somewhat better than another, but the better product might be the less expensive one.

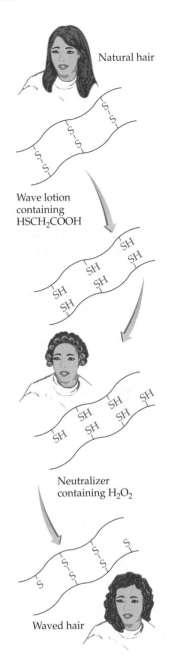

Natural hair

Wave lotion containing $HSCH_2COOH$

Neutralizer containing H_2O_2

Waved hair

▲ **Figure 17.17** Permanent waving of hair is accomplished by breaking disulfide bonds between protein chains and then reforming them in new positions.

Question: Write a balanced equation for the reaction of a thiol, RSH, with hydrogen peroxide to form a disulfide.

If you are unhappy with a product you have bought, tell the company that made it. The cosmetic industry gets approximately 5000 complaints from customers every year. In addition, you can complain to the FDA. There are many reports each year to the FDA regarding adverse effects of cosmetic products: makeup products, skin care products, fragrances, and various kinds of hair care products.

Most cosmetics are made from inexpensive ingredients, and yet many highly advertised cosmetics have extremely high price tags. Are they worth it? This is a judgment that lies beyond the realm of chemistry. The product inside that fancy little $100 bottle with the famous label may have cost the manufacturer less than $5 to produce. On the other hand, if the product makes you look better or feel better, perhaps the price is not important. Only you can decide.

Critical Thinking Exercises

Apply knowledge that you have gained in this chapter and one or more of the FLaReS principles (Chapter 1) to evaluate the following advertising claims.

17.1 A laundry detergent "leaves clothes 50% brighter 50% whiter [and is] 100% new & improved."

17.2 A laundry detergent contains "allergen fighters."

17.3 A shampoo "make[s] hair 10 times stronger."

17.4 A shampoo promises "to revitalize, rejuvenate or repair hair."

17.5 A new skin cream "reduces wrinkles and prevents or reverses damage caused by aging and sun exposure."

17.6 A vitamin in a skin cream "nourishes the skin."

17.7 "[P]roducts that contain artificial ingredients do not provide true aromatherapy benefits."

17.8 "Laundry balls contain no chemicals [and yet are] reusable indefinitely [in the washing machine to] clean, deodorize, sterilize, bleach and soften clothes."

SUMMARY

Sections 17.1–17.2—Some plants contain saponins, which produce lather. Plant ashes contain K_2CO_3 and Na_2CO_3, which are alkaline and have detergent properties. A **soap** is a salt of a long-chain carboxylic acid. Traditionally, soap was made from fat by a reaction with NaOH. Soap has been known for centuries, but has only come into common use since the mid-1800s. A soap molecule has a **hydrophobic** hydrocarbon tail (soluble in oil or grease) and a **hydrophilic** polar head (soluble in water). Soap molecules disperse oil or grease in water by forming **micelles**, tiny oil droplets surrounded by soap molecules. Agents that stabilize suspensions in this way are called **surface-active agents (surfactants)**. Soaps do not work well in hard water (water containing Ca^{2+}, Fe^{2+}, or Mg^{2+} ions) because they form insoluble salts. Carbonates and phosphates act as water softeners, removing as insoluble salts the ions that otherwise cause water to be hard. Home water softeners remove the unwanted ions by replacing them with sodium or potassium ions (Na^+, K^+).

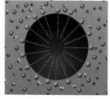

Sections 17.3–17.5—A synthetic detergent usually has a large hydrocarbon tail and a polar or ionic head. ABS detergents have branched-chain structures and are not easily biodegraded. LAS detergents have linear chains that can be broken down by microorganisms. Soap, ABS detergents, and

LAS detergents are **anionic surfactants** that have negative charges on the active part. A **nonionic surfactant** has no charge. A **cationic surfactant** such as a quaternary ammonium salt has a positive charge. An **amphoteric surfactant** carries both positive and negative charges. Detergent formulations often include **builders** such as complex phosphates to increase detergency. **Optical brighteners**, which are fluorescent whitening agents, are often added to make whites "whiter." Other components such as bleaches, fragrances, and fabric softeners may be included. Hand-dishwashing detergents contain mainly surfactants as the active ingredients. Automatic dishwasher detergents are often strongly alkaline and are not to be used for hand dishwashing.

Sections 17.6–17.7—Cationic surfactants (alkylammonium salts) are useful as disinfectants. Those with two long hydrocarbon chains are used as fabric softeners. **Bleaches** are oxidizing agents. Hypochlorite bleaches release chlorine. Oxygen-releasing bleaches usually contain perborates or percarbonates. Bleaches act by oxidizing chromophores (colored compounds) to colorless fragments.

Sections 17.8–17.9—All-purpose cleaners may contain surfactants, ammonia, disinfectants, and other ingredients. Reading the label is imperative, so that the cleaner may be used properly and safely. Household ammonia, sodium bicarbonate (baking soda), and vinegar are good cleaners for many purposes. Special-purpose cleaners each have a specific purpose. Toilet bowl cleaners usually contain an acid. Scouring powders contain abrasives. Glass cleaners contain ammonia. Drain cleaners contain NaOH or bleach. Oven cleaners contain NaOH.

Sections 17.10–17.12—Solvents are used to remove paint and other materials. Many organic solvents are volatile and flammable. **Paint** contains a pigment, a binder, and a solvent. Titanium dioxide is the most common pigment. Water-based paints often use a polymer binder; oil-based paints use tung oil or linseed oil. A **wax** is an ester of a fatty acid with a long-chain alcohol. Beeswax and carnauba wax are commonly used waxes. **Lanolin**, from sheep's wool, is also a wax, useful for skin creams and lotions.

Section 17.13—A **cosmetic** is something applied to the body for cleansing, promoting attractiveness or for similar purposes, but cosmetics do not have to be proven safe and effective like drugs. Toothpaste and soaps legally are not cosmetics. Skin is made of a tough, fibrous protein called **keratin**. **Sebum**, an oily secretion, protects skin from loss of moisture. A **lotion** is an emulsion of oil droplets in water; a **cream** is an emulsion of water droplets in oil. Each is a **moisturizer** that forms a protective physical barrier over the skin to hold in moisture. Creams and lotions also soften skin; such skin softeners are called **emollients**. Humectants such as glycerol hydrogen bond to water and hold it to the skin. Ultraviolet rays trigger the production of the pigment **melanin**, which darkens the skin. Most **sunscreen lotions** block shorter wavelength UV-B while letting through the longer

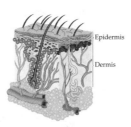

Epidermis

Dermis

wavelength, less-energetic UV-A rays that promote tanning. PABA and OMC are the most common ingredients in sunscreens. The **skin protection factor (SPF)** is the factor by which the UV light is reduced. Eye makeup and lipstick are mainly blends of oil, waxes, and pigments. **Deodorants** mask body odor and kill odor-causing bacteria. **Antiperspirants** also retard perspiration. The most common antiperspirants act as **astringents**, constricting the openings of the sweat glands.

Sections 17.14–17.15—Toothpaste contains a detergent, an abrasive, a thickener, flavoring, and often an added fluoride compound. The fragrance of a **perfume** can usually be split into three basic fractions: the **top note** (which is the most volatile), the **middle note** (of intermediate volatility), and the **end note** (of lowest volatility). **Colognes** are diluted perfumes. **Hypoallergenic cosmetics** purport to cause fewer allergic reactions than regular products, but the term has no legal meaning.

Section 17.16–17.17—Shampoos are synthetic detergents blended with perfume ingredients and sometimes mixed with protein or other conditioners to give the hair more body. Dark hair contains the pigment melanin, red hair contains phaeomelanin, and blonde hair contains very little of either pigment. Many hair dyes are based on *para*-phenylenediamine. Permanent waving of the hair uses a reducing agent to break the disulfide linkages and then an oxidizing neutralizer to reform the disulfide bonds in different places. Hair sprays contain **resins**, organic materials that leave a sticky film on the hair. Such resins are also available in foam form as **mousses**. Chemicals that remove hair are called **depilatories**, and most such chemicals can also damage the skin. The most expensive brand of a given product is not necessarily the best.

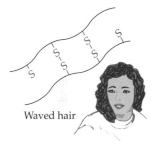

Waved hair

REVIEW QUESTIONS

1. What are saponins?

2. How do carbonate ions make water basic?

3. What are some advantages of soap?

4. What are some advantages of synthetic detergents?

5. What is an optical brightener? How do these compounds work?

6. What ingredients are used in liquid dishwashing detergents (for hand dishwashing)? How do detergents for automatic dishwashers differ from those used for hand dishwashing?

7. List some uses of diluted household ammonia. What safety precautions should be followed when using ammonia?

8. What sorts of surfaces should not be cleaned with ammonia? Why?

9. List two properties of baking soda that make it useful as a cleaning agent.

10. Mention a use of vinegar in cleaning. Why should marble surfaces not be cleaned with vinegar?

11. What is the legal definition of a cosmetic? Is soap a cosmetic?

12. What is the principal material of the corneal layer of the skin? What is sebum?

13. What is a deodorant?

14. What is an antiperspirant?

PROBLEMS

Soaps

15. What is a soap? Give an example.

16. List some ingredients found in toilet soaps.

17. How does a potassium soap differ from a sodium soap?

18. What kind of ingredients are used in scouring soaps?

19. What is hard water?

20. How does soap (or detergent) clean a dirty surface?

21. Why does soap fail to work in acidic water?

22. How does hard water affect the action of soap?

23. What products are formed when tripalmitin reacts with lye (sodium hydroxide)?

24. Give the structural formula for each of the following.
    a. potassium stearate   b. sodium palmitate

25. Give the structures of the products formed in the following reaction.

$$CH_2O-\overset{\overset{\displaystyle O}{\|}}{C}(CH_2)_{10}CH_3$$
$$|$$
$$CH-O-\overset{\overset{\displaystyle O}{\|}}{C}(CH_2)_{10}CH_3 \quad + \quad 3\,NaOH \quad \longrightarrow$$
$$|$$
$$CH_2O-\overset{\overset{\displaystyle O}{\|}}{C}(CH_2)_{10}CH_3$$

26. Give the structures of the products formed in the following reaction.

$$CH_2O-\overset{\overset{\displaystyle O}{\|}}{C}(CH_2)_{6}CH_3$$
$$|$$
$$CH-O-\overset{\overset{\displaystyle O}{\|}}{C}(CH_2)_{4}CH_3 \quad + \quad 3\,NaOH \quad \longrightarrow$$
$$|$$
$$CH_2O-\overset{\overset{\displaystyle O}{\|}}{C}(CH_2)_{8}CH_3$$

Water Softeners

27. List two ways by which trisodium phosphate softens water.

28. How does a water-softening tank work?

29. Write the equation that shows how phosphate ion reacts with water to form a basic solution.

30. Write the equation that shows how carbonate ion reacts with water to form a basic solution.

31. Write the equation that shows how carbonate ions precipitate hard-water ions. Use M^{2+} to represent the hard-water ion.

32. Write the equation that shows how phosphate ions precipitate hard-water ions. Use M^{2+} to represent the hard-water ion.

Builders

33. What is a (detergent) builder?

34. How does sodium tripolyphosphate aid the cleaning action of a soap or detergent?

35. How do zeolites aid the cleaning action of a soap or detergent?

36. What advantages do zeolites have over carbonates as builders?

Detergents: Structures and Properties

37. What is an LAS detergent?

38. What is a surface-active agent (surfactant)?

39. What is an anionic surfactant?

40. What is a cationic surfactant?

41. What is a nonionic surfactant?

42. What is an amphoteric surfactant?

43. Cationic surfactants are often used along with nonionic surfactants but are seldom used with anionic surfactants. Why not?

44. Cationic surfactants are not particularly good detergents, yet they are widely used, especially in the food industry. Why?

Problems 45 through 48 refer to compounds represented by the following structures.

45. Which are biodegradable?

46. Which is an LAS detergent?

47. Which is an alkyl sulfate?

48. Which are anionic detergents? Which is a cationic detergent?

Fabric Softeners

49. Give an example of a compound that acts as a fabric softener.

50. How do fabric softeners work?

Bleaches

51. How do cyanurate bleaches (such as Symclosene) work?

52. How do sodium percarbonate and sodium perborate work as a bleaches?

Organic Solvents

53. List some uses of organic solvents in the home.

54. List two hazards of cleansers that contain petroleum distillates.

55. What are the main hazards of using solvents in the home?

56. Should gasoline be used as a cleaning solvent? Why or why not?

Creams and Lotions

57. What is an emollient?

58. What is the ideal moisture content of skin?

59. What are the principal components of creams and lotions?

60. What is a skin moisturizer?

Lipsticks

61. How does lipstick differ from skin cream?

62. What materials are used to color lipsticks?

Toothpastes

63. What are the essential ingredients in a toothpaste?

64. What is the principal cause of tooth decay?

65. How do fluorides strengthen tooth enamel?

66. What are the best ways to prevent tooth decay?

Perfumes and Colognes

67. What is a perfume?

68. What are the three fractions of a perfume? How do these fractions differ at the molecular level?

69. What is musk? Why are musks added to perfumes?

70. What are colognes?

71. What are the principal ingredients of aftershave lotion?

72. What is the function of menthol in some aftershave lotions?

Hair and Hair Care

73. What is the only essential ingredient of shampoos?

74. What type of detergent is used in baby shampoos? Why?

75. What chemical compound is used most often to bleach hair?

76. What chemical substances determine the color of hair and skin?

77. What is the difference between a temporary and a permanent hair dye? Why is a permanent dye not really permanent?

78. What are resins? How are resins used in hair care?

79. What kind of chemical reagent is used to break disulfide linkages in hair? What kind of chemical reaction is involved?

80. What kind of chemical reagent is used to restore disulfide linkages in hair? What kind of chemical reaction is involved?

81. What kind of chemical reaction is involved when a colorless hair dye compound is converted to a colored dye?

82. What kind of chemical reaction is involved in the bleaching of hair?

ADDITIONAL PROBLEMS

Problems 83 through 86 refer to structures I through IV below.

83. Which are synthetic detergents?

84. Which are soaps?

85. Which is an amphoteric detergent?

86. Which is a principal ingredient in baby shampoo?

87. Write the structure for isopropyl palmitate, an ingredient in skin creams.

88. The hardness of water is reported as milligrams of calcium carbonate per liter of water. Calculate the quantity of sodium carbonate needed to soften 10.0 L of water with a hardness of 295 mg/L. (*Hint:* How much Ca^{2+} is present in the water?)

I. $CH_3CH_2CH_2CH_2CH_2CH_2CH_2CH_2CH_2CH_2CH_2CH_2CH_2CH_2CH_2CH_2\overset{\overset{\displaystyle H}{|}}{\underset{\underset{\displaystyle H}{|}}{N^+}}-CH_2\overset{\overset{\displaystyle O}{\|}}{C}-O^-$

II. $CH_3CH_2CH_2CH_2CH_2CH_2CH_2CH_2CH_2CH_2CH_2COO^-\ Na^+$

III. $CH_3CH_2CH_2CH_2CH_2CH_2CH_2CH_2CH_2CH_2\underset{\underset{\displaystyle CH_3}{|}}{CH}-\langle\bigcirc\rangle-SO_3^-\ Na^+$

IV. $CH_3CH_2CH_2CH_2CH_2CH_2CH_2CH_2CH_2CH_2CH_2CH_2-\langle\bigcirc\rangle-\overset{\overset{\displaystyle CH_3}{|}}{\underset{\underset{\displaystyle CH_3}{|}}{N^+}}-CH_3\quad Cl^-$

COLLABORATIVE GROUP PROJECTS

Prepare a PowerPoint, poster, or other presentation (as directed by your instructor) for presentation to the class.

89. Examine the label on a sample of each of the following household products. Insofar as you can, list the essential ingredients in each product. Use a reference such as *The Merck Index* (Reference 1) to determine the formula and function of each ingredient.
    a. scouring cleaner
    b. window cleaner
    c. oven cleaner
    d. toilet bowl cleaner
    e. liquid drain cleaner
    f. solid drain cleaner
    g. paint remover

90. Read the labels on several brands of toothpaste. Insofar as you can, list the essential ingredients in each product. Use a reference such as *The Merck Index* (Reference 1) to determine the formula and function of each ingredient.

91. Read the labels on five brands of lotions or creams (designed for use on the skin—face, hands, or body). Insofar as you can, list the essential ingredients in each product. Use a reference such as *The Merck Index* (Reference 1) to determine the formula and function of each ingredient.

92. Read the labels on five brands of shampoos. Insofar as you can, list the essential ingredients in each product. Use a reference such as *The Merck Index* (Reference 1) to determine the formula and function of each ingredient.

93. The Clearwater Chemical Company has announced the development of a biodegradable detergent. Discuss with your group what tests, if any, should be made before the detergent is marketed. Should Clearwater be allowed to market the detergent until it is proven harmful? Can a product ever be proven safe?

94. Maxisuds, Inc., has announced a new replacement for phosphate builders in detergents. Discuss with your group what tests, if any, should be made before the builder is used in detergent formulations. Should Maxisuds be allowed to market the builder until it is proven harmful? Can a product ever be proven safe?

95. Consult an industry-sponsored website about soaps and detergents. Then find a site sponsored by an environmental organization that talks about these materials in the environment. Compare the two viewpoints.

96. A product called Febreze claims to be an odor-removing spray. Search the Web to see what you can find out about the composition and mode of action of this product.

REFERENCES AND READINGS

1. Budavari, S., M. J. O'Neill, and A. Smith (eds.). *The Merck Index*, 13th ed. Rahway, NJ: Merck & Co., 2001.

2. "Chemistry of Colors and Curls." *Science News*, August 25, 2001, pp. 124–126. A chemist looks at the coloring, perming, and styling of hair.

3. "Coming Clean on Bath Soap." *Consumer Reports*, October 2001, pp. 32–33.

4. Delaney, Niamh. "Fluoride: The Magic Ingredient." *Chemistry and Industry*, September 16, 2002, pp. 8–9.

5. "The Dirt on Detergents." *Consumer Reports*, February 2000, pp. 35–37.

6. "Dishwashing Liquids." *Consumer Reports*, August 2002, pp. 32–33.

7. Emsley, John. *Vanity, Vitality, and Virility: The Science Behind the Products You Love To Buy*. Oxford, UK: Oxford University Press, 2004.

8. Fortineau, Anne-Dominique. "Chemistry Perfumes Your Daily Life." *Journal of Chemical Education*, January 2004, pp. 45–50.

9. Giroux, Robin. "What's That Stuff? Shampoo." *Chemical & Engineering News*, April 15, 2002, p. 42.

10. Holton, Sarah. "Natural Skin Science." *Chemistry & Industry*, April 19, 2004, pp. 16–17. Discusses cosmeceuticals—cosmetics with medical claims.

11. Johnson, Rita. "What's That Stuff? Lipstick." *Chemical & Engineering News*, July 12, 1999, p. 31.

12. McCoy, Michael. "Soaps and Detergents." *Chemical & Engineering News*, January 26, 2004, pp. 23–28. A survey of the industry. One of a series of such articles published each January.

13. McCoy, Mike. "What's That Stuff? Fluoride." *Chemical & Engineering News*, April 16, 2001, p. 42.

14. Raber, Linda. "What's That Stuff? Hair Coloring." *Chemical & Engineering News*, March 13, 2000, p. 52.

15. Rakita, Philip E. "Dentifrice Fluoride." *Journal of Chemical Education*, May 2004, pp. 677–680.

16. Reisch, Marc. "What's That Stuff? Sunscreens." *Chemical & Engineering News*, June 24, 2002, p. 38.

17. Rouhi, Maureen. "What's That Stuff? Shower Cleaners." *Chemical & Engineering News*, December 3, 2001, p. 39.

18. "Sunscreens." *Consumer Reports*, June 2001, pp. 27–29.

19. "What's That Stuff? Self-Tanners." *Chemical & Engineering News*, June 12, 2000, p. 46.

20. "Which Dishwashing Liquids Are Clear Winners?" *Consumer Reports*, August 2002, pp. 32–33.

21. Yarnell, Amanda. "What's That Stuff? Teeth Whiteners." *Chemical & Engineering News*, February 10, 2003, p. 29.

GREEN CHEMISTRY

Greening Dry Cleaning

By Liz Gron, Hendrix College

Dry cleaning is a common practice to protect delicate fabrics; however, many people do not understand this process or its impact on our environment.

The term *dry cleaning* encompasses a wide variety of nonaqueous cleaning methods. Since the 1960s, the most common process has used perchloroethylene (PERC).[1] As you can see from the structure below, PERC is completely nonpolar, making it an excellent solvent for hydrophobic grease. Unfortunately, PERC has significant long-term health consequences. Exposure through inhalation and skin contact can lead to neurological problems.[1] Due to the increase in certain cancers among workers, the National Institute for Occupational Safety and Health (NIOSH) has designated PERC as a "potential occupational carcinogen." Evaporation of PERC into the atmosphere also has environmental impacts. The chlorocarbon structure leads to ozone depletion. At present, 87% of modern dry cleaners use PERC. Although significant improvements have been made by reducing the amount of solvent used and capturing fugitive emissions, the underlying risk of the solvent remains.

Perchloroethylene (PERC)

Chemists and engineers are working to reduce the risks associated with dry cleaning by creating greener processes. In general terms, we can consider chemical risk as a function of hazard and exposure[2]:

$$\text{Risk} = \text{Hazard} \times \text{Exposure}$$

Common sense and principles of green chemistry[3] lead us to understand that reducing exposure to toxic substances decreases risk, but eliminating the toxic chemicals entirely would be superior.

A surprising alternative to the PERC process is dry cleaning with supercritical carbon dioxide ($SCCO_2$). We are all familiar with CO_2 as a component of our own breath, a natural result of cellular respiration. Although CO_2 is a gas at room temperature (25 °C), at higher temperatures and pressures ($T > 31$ °C, $P > 72.8$ atm), CO_2 can be compressed into an unusual phase that has properties of both a gas and of a liquid simultaneously, without actually being either.[4] This is known as the *supercritical* phase. $SCCO_2$ is nonpolar, as you can see below, and is able to dissolve small nonpolar molecules, but it is not able to dissolve larger molecules that make up oil and grease. This problem can be solved by the use of cleverly designed detergents.

$$\ddot{O}=C=\ddot{O}$$

Carbon dioxide

As you have learned in this chapter, detergents bridge solubility differences between a solvent and the solute, $SCCO_2$ and oily grease in this case. Detergents for this system contain a "CO_2-philic" area (usually containing fluorine atoms) that prefers to associate with CO_2 and a "CO_2-phobic" area that prefers to associate with hydrophobic oils. Analogous to surfactants in water, these detergents form micelles with an oil or grease droplet at the center, surrounded by the nonpolar, "CO_2-phobic" end of the detergent with the other "CO_2-philic" end sticking out into the $SCCO_2$ solvent. This work earned Professor Joseph DeSimone of University of North Carolina-Chapel Hill a Presidential Green Chemistry Award in 1997. The green dry-cleaning process is presently in commercial use under the Hangers® trademark.

REFERENCES

1. USEPA 1998 Cleaner Technologies Substitutes Assessment: Professional Fabricare Processes, EPA 774-B-98-001.

2. Poliakoff, M., J. M. Fitzpatrick, T. R. Farren, and P. T. Anastas. "Green Chemistry: Science and Politics of Change." *Science*, vol. 297, 2002, 297, pp. 807–810.

3. Anastas, P. T., and J. K. Warner. *Green Chemistry: Theory and Practice*. New York: Oxford University Press, 1998, p. 30.

4. DeSimone, J. M. "Practical Approaches to Green Solvents." *Science*, vol. 297, 2002, pp. 799–802.

5. McClain, J. B., D. E. Betts, D. A. Canelas, E. T. Samulski, J. M. DeSimone, J. D. Londono, H. D. Cochran, G. D. Wignall, D. Chillura-Martino, and R. Triolo. "Design of Nonionic Surfactants for Supercritical Carbon Dioxide." *Science*, vol. 274, pp. 2049–2052.

6. Presidential Green Chemistry Award 1997 abstract on EPA website. http://www.epa.gov/greenchemistry/docs/award_entries_and_recipients1997.pdf

WEB INVESTIGATIONS

Investigation 1
Identifying Health Hazards of Chemicals
Exposure limits for chemicals used in manufacturing can be found on the materials safety data sheet (MSDS) associated with each chemical. For dry cleaning, inhalation is the most likely route of exposure for the dry-cleaning chemicals. The units for inhalation limits are usually given as TLV or PEL using a STEL or TWA. Use a Web browser to find definitions of these abbreviations. Be sure to type "define" as part of your search. What do these abbreviations stand for, and what do the terms mean?

Investigation 2
Relative Risks of PERC Versus CO_2
Using your browser, search for an MSDS from any manufacturer of these chemicals, PERC and CO_2. In the exposure limits section, copy the TLV in ppm, listed as TWA for PERC versus CO_2. Notice the differences in these limits.

Investigation 3
Supercritical CO_2—What Is It and What Can It Do?
Use your favorite search engine to find a phase diagram of carbon dioxide, showing temperature and pressure, that includes the supercritical region. An image search may be useful. Using the phase diagram for temperature and pressure, what is the phase of CO_2 at normal room temperature (25 °C + 273 = 298 K) and pressure (about 1 atm or 1 bar)? What pressure is necessary to keep CO_2 as a liquid at room temperature? Can the pressure be high enough to maintain CO_2 as a solid at room temperature? Sketch the phase diagram and label the phases as gas, liquid, and supercritical CO_2.

Investigation 4
Soaps and Cleaning
Search the Web using the keyword "micelle" to see in more detail how a micelle is formed and works to dissolve unlike materials.

COMMUNICATE YOUR RESULTS

Exercise 1
Chemical Safety Information—Skepticism and Insight
Visit the DHMO Web page to find the MSDS of hazards of dihydrogen monoxide.

- Discuss what this website says about the toxicity of this material with your partner or group. Together, write a paragraph about your conclusions.

- Before you proceed further, write the formula for dihydrogen monoxide, have a drink of water, and reevaluate the "report" you read.

- Together, write a two-page response to this report. Were the statements made about dihydrogen monoxide untrue or merely misleading? What can you do to assure yourself of the validity of information on the Web?

Exercise 2
Chemical Safety Information—PERC
While we are fairly well versed in the dangers of CO_2, more information on PERC might be in order. Another method for evaluating chemical safety is to look up the rules supplied by OSHA, the Occupational Safety and Health Administration. OSHA regulates chemical exposure for workers. Search the OSHA website and answer the following questions in a one- to two-page paper.

- What are the known health hazards to humans from PERC?

- What is the primary health concern about long-term exposure?

- What is perchloroethylene used for besides dry cleaning?

Exercise 3
Chemical Safety Information—From Where?
Investigate the health hazards of PERC from several sites. Three contrasting sources are listed on the Companion website. You should notice that while the underlying science is the same, the level of concern changes from site to site. Write a two-page paper about the contrasting claims. Read the text carefully. Compare and contrast the reported health hazards. How do the recommendations differ? Which recommendations do you think are reasonable, and why?

CHAPTER 18

Fitness and Health

Exercise is essential to good health. Through chemistry, we gain a better understanding of the effects of exercise on our bodies and minds.

Some Chemical Connections

For much of human history, the main concern of most people was obtaining enough food to stay alive. How the world has changed! Especially in developed countries, people now have scores of labor-saving devices and food is abundant. Most people earn their daily bread with little physical exertion. So much food is available, and so little physical effort required to obtain it, that many people eat more than they should and get much less exercise than they need. In 2005 the National Center for Health Statistics estimated that 64% of U.S. adults, 15% of adolescents age 12–19 years, and 15% of children age 6–11 years were overweight. The rate of obesity is rising rapidly. In 1991 the percentage of Americans who were obese was 12%. By 2005 that figure had increased to 31%. Obesity has become a major epidemic.

How do we stay healthy and physically fit? By some special diet or kind of exercise? In this chapter we examine nutrition, electrolyte balance, muscle action, drugs, and other matters of concern to those interested in their own well-being and physical fitness.

Nutrition in Health and Fitness

A nutritious diet is essential to good health. Good nutrition includes sufficient carbohydrates, fats, and proteins; the essential minerals and vitamins in proper proportions; and adequate dietary fiber and water (Chapter 16).

18.1 CALORIES: QUANTITY AND QUALITY

Even today for many people around the world, it is a daily struggle to obtain enough food to survive. At the same time, in developed countries, too much food that is too rich in fats and calories makes obesity a major health problem. Research shows that undereating is healthier than overeating. Large studies with mice have shown that when given 40% less food than a control group chooses to eat, the mice live longer and have fewer tumors. Humans also seem to stay healthier when they eat a bit too little than when they eat too much.

▶ **Figure 18.1** A new daily food guide pyramid. One diet does not fit all: The actual recommendations are based on an individual's age, sex, activity level, and general health condition. These colors represent the food groups.

- Orange for grains
- Green for vegetables
- Red for fruits
- Yellow for oils
- Blue for milk
- Purple for meat and beans

Total calories consumed is important, but the distribution of calories is even more important. The U.S. Department of Agriculture (USDA) *Dietary Guidelines for Americans 2005* include the key recommendations that

- we consume "a variety of nutrient-dense foods and beverages within and among the basic food groups while choosing foods that limit the intake of saturated and trans fats, cholesterol, added sugars, salt, and alcohol;" and
- we "meet recommended intakes within energy needs by adopting a balanced eating pattern, such as the USDA Food Guide or the DASH Dietary Approaches to Stop Hypertension, Eating Plan" (Reference 19).

A generalized food pyramid (Figure 18.1) can be helpful, with details of a food plan then tailored to the individual. For a 2000-calorie diet, the guidelines suggest

- Four daily servings (2 cups total) from the *fruit group*.
- Five daily servings $\left(2\frac{1}{2}\text{ cups total}\right)$ from the *vegetable group*.
- Six servings (6 oz total), half of which should be whole grains, from the *grain group*.
- Five and a half daily servings (5.5 oz total) from the *meat and beans group*.
- Three servings (3 cups total) from the *milk group*.
- Six teaspoons or 24 g from the *oils group*.
- After other requirements have been met with nutrient-dense foods, up to 267 calories in *discretionary calories* can be consumed in solid fats or added sugars.

Less Total Fats and Less Saturated Fats and Trans Fats

A key recommendation is to keep fat intake between 20 and 35% of calories from fat. Most fats should be made up of polyunsaturated and monounsaturated fatty acids. Less than 10% of calories should come from saturated fatty acids, and trans fatty acids should be kept to a minimum. In addition, a person should take in less than 300 mg/day of cholesterol.

The "good" fats come from sources such as fish, nuts, and vegetable oils. Olive oil and canola oil are rich in monounsaturated fatty acids, and fish oils contain beneficial omega-3 fatty acids. A study of people living on the Greek island of Crete showed that in spite of their diet, which averaged about 40% fat—much of it from olive oil—the people had amazingly little cardiovascular disease.

More than half the fats in the American diet are animal fats, and 70% of saturated fat comes from animal products. Saturated fats and cholesterol are implicated in artery clogging. They are found in animal products such as butter, lard, red meats, and fast-food hamburgers. We should choose meat, poultry, milk, and milk products that are lean, low fat, or fat free. Trans fats, produced when vegetable oils are hydrogenated (for example, to produce margarine), are high in trans fatty acids (Chapter 16). Trans fatty acids contain double bonds, just as the fatty acids in vegetable oils do, but these double bonds have a trans configuration, so that they behave more like saturated fatty acids (Figure 16.7). Unless margarine has been specially treated to remove the trans fat, it has about the same health effects as butter.

The *percentage* of fat in the American diet has dropped from 40% in 1990 to about 34% today, but the absolute quantity of fat actually increased because Americans are eating more total calories. The 34% is an average figure; for some people the figure runs as high as 45%.

A healthy diet can help us avoid disease. Harvard Medical School studies indicate that correct dietary choices (as outlined previously and spelled out in more detail in the *Dietary Guidelines for Americans 2005*), regular exercise, and avoiding smoking would prevent about 82% of heart attacks, about 70% of strokes, over 90% of type 2 diabetes, and over 70% of colon cancer. In contrast, the most effective drugs, called statins, can reduce heart attacks by about 20 or 30%.

Nutrition and the Athlete

The recommended ranges of energy nutrients for most people are 20–35% of calories from fat, 45–65% of calories from carbohydrates, and 10–35% of calories from protein. Athletes generally need more calories because they expend more than the average sedentary individual. Those extra calories should come mainly from carbohydrates. Foods rich in carbohydrates, especially starches, are the preferred source of energy for the healthy body.

Fatty and protein-rich foods also supply calories, but protein metabolism produces more toxic wastes that tax the liver and kidneys. The pregame steak dinner consumed by some athletes in the past was based on a myth that protein builds muscle. It doesn't. Although athletes do need the **Dietary Reference Intake (DRI)** quantity of protein (0.8 grams per kilogram of body weight), with few exceptions, they do not need an excess. DRIs are nutrient-based reference values established by the Food and Nutrition Board of the U.S. National Academy of Sciences for use in planning and assessing diets; they replace the Recommended Dietary Allowances (RDAs) that have been published since 1941 by the National Academy of Sciences. (An RDA is an intake that meets the nutrient need of almost all [97 to 98%] of the healthy individuals in a specific age and gender group.) Protein consumed in amounts greater than needed for synthesis and repair of tissue only makes the athlete fatter (as a result of excessive calorie intake)—not more muscular.

Muscles are built through exercise, not through eating excess protein. When a muscle contracts against a resistance, an amino acid called *creatine* is released. Creatine stimulates production of the protein myosin, thus building more muscle tissue. If the exercise stops, the muscle begins to shrink after about two days. After about two months without exercise, muscle built through an exercise program is almost completely gone. (The muscle does *not* turn to fat, as some athletes believe. Former athletes often do get fat, however, because they continue to take in the same number of calories while expending fewer.)

18.2 VITAMINS AND MINERALS

Vitamins are organic substances that the human body needs but cannot manufacture. Unlike hormones and enzymes, which the body can synthesize, vitamins must be included in the diet. Minerals, inorganic elements that the body needs, must also be present in the diet. Table 18.1 lists DRIs of vitamins and minerals for young adult women and men. A more complete list for various ages and with a number of qualifications is available from the Food and Nutrition Board.

Many medical professionals say that vitamin and mineral supplements are unnecessary when the diet is well balanced and includes the DRI for each nutrient. However, for those who choose to take such supplements, reasonable doses of vitamin and mineral tablets and capsules are not likely to cause harm. In some cases they can be beneficial. There are times when the body has an increased demand for vitamins and minerals, as during periods of rapid growth, during pregnancy and lactation, and during periods of trauma and recovery from disease. A study of retired people aged 65 years and older showed that a daily dose of 18 vitamins and minerals seems to strengthen the immune system.

Extreme-endurance athletes, such as triathletes and ultramarathoners, may need a bit more than the DRI for protein. Because nearly all Americans eat 50% more protein than they need, even these athletes seldom need protein supplements. Weightlifters and bodybuilders need no extra protein.

$$NH_2-\underset{\underset{NH}{\parallel}}{C}-\underset{\underset{CH_3}{\mid}}{N}-CH_2COOH$$

Creatine

Vitamins and minerals are discussed in Chapter 16. See Table 16.3 for a list of essential minerals and trace minerals. Table 16.4 lists most vitamins along with food sources and deficiency symptoms.

TABLE 18.1	Dietary Reference Intakes of Vitamins and Minerals for Young Adults	
Nutrient	**Females**	**Males**
Vitamin A	700 μg[‡]	900 μg[‡]
Vitamin C	75 mg[*]	90 mg[*]
Vitamin D	200 IU[†]	200 IU[†]
Vitamin E	30 IU[†]	30 IU[†]
Thiamine, B_1	1.1 mg	1.2 mg
Riboflavin, B_2	1.1 mg	1.3 mg
Niacin	14 mg	16 mg
Pyridoxine, B_6	1.5 mg	1.7 mg
Cyanocobalamin, B_{12}	3 μg	3 μg
Folacin	400 μg	400 μg
Pantothenic acid	5 mg	5 mg
Biotin	100–200 μg	100–200 μg
Calcium	1200 mg	1200 mg
Phosphorus	700 mg	700 mg
Magnesium	320 mg	420 mg
Iron	15 mg	10 mg
Zinc	12 mg	15 mg
Iodine	150 μg	150 μg
Fluoride	3 mg	4 mg
Selenium	55 μg	70 μg
Potassium	2 g	2 g

[*]Smokers add 35 mg.

† IU = International unit.

‡ To the extent that the vitamin A requirement is met by ingested β-carotene, multiply these by 6.

The number of infections in the study group was 50% less than that in a control group, which received no supplements.

The DRI values for vitamins and minerals set by the Food and Nutrition Board are quite modest, and they are readily supplied by any well-balanced diet. The DRI values for vitamins are the amounts that will prevent deficiency diseases such as beriberi, pellagra, and scurvy. Some scientists believe that *optimum* intakes of some vitamins are somewhat higher than their DRI values. For example, they think that DRIs for vitamins C, D, and E are too low. If you eat a well-balanced diet and you are in good health, you probably do not need vitamin supplements. Small doses of vitamins are not toxic; however, some of the fat-soluble vitamins are toxic in extremely large doses, so caution is warranted.

Vitamin A

Vitamin A is essential for good vision, bone development, and skin maintenance. There is some evidence that it may confer resistance to certain kinds of cancer. This may help to explain the anticancer activity that has been noted for cruciferous vegetables such as broccoli, cauliflower, Brussels sprouts, and cabbage. Most cruciferous vegetables are rich in β-carotene, which is converted to vitamin A in the body.

Vitamin A is a fat-soluble vitamin stored in the fatty tissues of the body and especially in the liver. Large doses can be toxic. The recommended upper limit for vitamin A is 3000 μg/day. Studies have shown that in affluent countries ingesting vitamin A from supplements and in fortified foods at levels even slightly above the DRI leads to an increased risk of bone fractures later in life.

Larger quantities of β-carotene can be taken safely because excess β-carotene is not converted to vitamin A. It may have important functions other than as a precursor of vitamin A. Some nutritionists argue that β-carotene is needed in quantities greater than those needed to meet the vitamin A requirement. Ingesting β-carotene is probably the safest way to obtain the proper quantity of vitamin A.

B Vitamins

There are eight members of the vitamin B family. Because they are all water soluble, excess intake is excreted in the urine. There appears to be no toxicity connected with B vitamins, with the possible exception of vitamin B_6, which apparently can cause neurological damage in some people if taken in extremely large daily doses. As a group, B vitamins are important in maintaining the skin and the nervous system.

In addition to preventing the skin lesions of pellagra, niacin (vitamin B_3) offers some relief from arthritis. It also helps in lowering the blood cholesterol level. Very large daily doses (5–30 g) have been taken for years by schizophrenic patients without apparent toxic effects.

Vitamin B_6, pyridoxine, has been found to help people with arthritis by shrinking the synovial membranes that line the joints. Vitamin B_6 is a coenzyme for more than 100 different enzymes.

Vitamin B_{12} is not found in plants, and vegetarians are apt to be deficient in this vitamin. That deficiency can lead to pernicious anemia. The structure of vitamin B_{12}, also called cyanocobalamin, is quite complicated. Its molecular formula is $C_{63}H_{88}CoN_{14}O_{14}P$ and its molecular mass is 1355.38 u. Its structure was determined by Dorothy Hodgkin of Oxford University in 1956 using X-ray crystallography. No one had ever attempted to establish the structure of a molecule of this size and complexity before. Hodgkin received the 1964 Nobel Prize in chemistry for this work.

Another B vitamin is folic acid, which is critical in the development of the nervous system of a fetus. Its presence in a pregnant woman's diet prevents spina bifida in her baby. Folic acid also helps prevent cardiovascular disease. An upper limit of 1000 μg has been set; more than that can cause nerve damage.

Vitamin C

Vitamin C is ascorbic acid, the component in citrus fruits that acts as an antiscurvy factor. About 60–80 mg daily will prevent scurvy, but vitamin C does more than that. It promotes the healing of wounds, burns, and lesions such as gastric ulcers. It also seems to play an important role in maintaining the body's collagen supply. Like vitamin E, it is an antioxidant, and these two vitamins, along with β-carotene, are included in many antioxidant formulations. Antioxidants may act as anticarcinogens. About 200 mg of vitamin C per day is probably optimal, and an upper limit of 2000 mg/day has been set for vitamin C.

Vitamin C seems to be essential for efficient functioning of the immune system. *Interferons*, large molecules formed by the action of viruses on their host cells, are agents in the immune system. By producing an interferon, a virus can interfere with the growth of another virus. An increased level of vitamin C has been shown to increase the body's production of interferons.

Vitamin D

Vitamin D is a steroid-type vitamin that protects children against rickets. It promotes the absorption of calcium and phosphorus from foods to produce and maintain healthy bones. Too much vitamin D can lead to excessive calcium and phosphorus absorption, with subsequent formation of calcium deposits in various soft body tissues, including those of the heart. The DRI for vitamin D increases with age, from 200 international units (IU)[1] for adults from age 20 to 50, and reaching 600 IU for those over 70. The recommended upper limit is 2000 IU.

[1]An international unit (IU) measures the activity of many vitamins and drugs. For each substance to which the IU applies, an international agreement specifies the biological effect expected with a dose of 1 IU. Other quantities of the substance are then expressed as multiples of this standard. For example, 1 IU represents 25 nanograms for vitamin D. For many substances, there is no definite conversion between IUs and mass units because preparations of those substances vary in activity.

▲ Dorothy Crowfoot Hodgkin (1910–1994) used X-rays to determine the structures of several important organic compounds, including vitamin B_{12}, penicillin, and cholesterol.

  Vitamin C

People with low levels of vitamin C are more likely to develop cataracts and glaucoma, as well as gingivitis and periodontal disease. Some suggest that very large doses of vitamin C may cure or prevent colds, but there is little evidence for this.

There is good evidence that eating lots of fruits and vegetables protects against cancer, but there is little evidence that pills containing antioxidants do so.

Vitamin E

Vitamin E is a mixture of tocopherols (Table 16.4). Its antioxidant activity may have value in maintaining the cardiovascular system, and it has been used to treat coronary heart disease, angina, rheumatic heart disease, high blood pressure, arteriosclerosis, varicose veins, and a number of other cardiovascular problems. As an antioxidant, vitamin E can inactivate free radicals. It is generally believed that much of the physiological damage from aging is a result of the production of free radicals. Indeed, vitamin E has been called the *antiaging vitamin.* Studies of these effects are often contradictory and far from conclusive. Vitamin E is also an anticoagulant that has been useful in preventing blood clots after surgery.

Rats deprived of vitamin E become sterile. For this reason, vitamin E is sometimes called the antisterility vitamin. Vitamin E deficiency can also lead to muscular dystrophy, a disease of the skeletal muscles. A lack of vitamin E can lead to a deficiency in vitamin A. (Vitamin A can be oxidized to an inactive form when vitamin E is no longer present to act as an antioxidant.) Vitamin E also collaborates with vitamin C in protecting blood vessels and other tissues against oxidation. Vitamin E is the fat-soluble and vitamin C the water-soluble antioxidant vitamin. Oxidation of unsaturated fatty acids in the cell membrane can be prevented or reversed by vitamin E, which is itself oxidized in the process. Vitamin C can then restore vitamin E to its unoxidized form. The upper limit for vitamin E is 800 IU/day or 400 IU/day depending on side effects. Higher levels bring the risk of excessive bleeding.

LDL cholesterol is oxidized before it is deposited in arteries, and vitamin E prevents the oxidation of cholesterol. Some think this is the mechanism by which it helps to protect against cardiovascular disease.

Electrolytes and Nonelectrolytes

18.3 BODY FLUIDS AND ELECTROLYTES

Another aspect of the relationship of chemistry and nutrition is the balance between fluid intake and electrolyte intake. An **electrolyte** is a substance that conducts electricity when dissolved in water. In the body, electrolytes are ions required by cells to control the electric charge and thus the flow of water molecules across the cell membrane. The main electrolytes are sodium ions (Na^+), potassium ions (K^+), and chloride ions (Cl^-). Others include calcium (Ca^{2+}), magnesium (Mg^{2+}), sulfate (SO_4^{2-}), hydrogen phosphate (HPO_4^{2-}), and bicarbonate ions (HCO_3^-).

Water is an essential nutrient, a fact obvious to anyone who has been deprived of it. Many people seem to prefer to meet the body's need for water by consuming soda pop, coffee, beer, fruit juice, and other beverages. Some of these products actually impair the body's use of the water in these beverages. Alcoholic drinks promote water loss by blocking action of the *antidiuretic hormone (ADH).* Caffeine has a diuretic effect on the kidneys, promoting urine formation and consequent water loss.

The best way to replace water lost through sweat, tears, respiration, and urination is to drink water. Unfortunately, thirst is often a *delayed* response to water loss, and it may be masked by such symptoms of dehydration as exhaustion, confusion, headache, and nausea. Sweat is about 99% water, but a liter of sweat typically contains 1.15 g Na^+ and 1.48 g Cl^- (the ions of table salt), 0.02 g Ca^{2+}, 0.23 g K^+, 0.05 g Mg^{2+}, plus minute amounts of urea, lactic acid, and body oils. (This composition varies from person to person.) The typical American diet probably contains too much sodium chloride (NaCl) and only traces of the other electrolytes are lost. Thus, it makes the most sense to replace the water component of lost sweat with pure water itself. Commercial "sports drinks" usually contain more sugars than electrolytes. Nonetheless, they are quite popular with both serious and weekend athletes, but because they are so concentrated (a hazard that can lead to diarrhea), they are of marginal value except for endurance athletes (Figure 18.2).

How do you know if you are drinking enough water and are in a proper state of hydration? The urine is a good indicator of hydration. When you are dehydrated, your urine gets cloudy or dark because your kidneys are trying to conserve water so as to keep the blood volume from shrinking and to prevent shock.

▲ **Figure 18.2** The best replacement for fluids lost during exercise generally is plain water. Sports drinks help little, except when engaging in endurance activities lasting for 2 hr or more. At that point, carbohydrates in the drinks delay the onset of exhaustion by a few minutes.

Chemistry and Athletic Performance

Chemistry has transformed athletic performance in recent decades. New materials have provided protective gear such as plastic helmets and foam padding for football, ice hockey, skateboarding, and other sports.

New equipment has changed the nature of most sports events. Baseball, football, and other sports are quite different when played on artificial turf or under a dome than when played outdoors on grass. Traction changes, the ball bounces differently, and so on. Skateboards of the 1950s had metal rollerskate wheels on a flat wooden board. Today's skateboards have wide urethane wheels, specially shaped composite-construction boards, and shock-absorbing wheel

mounts that make incredible stunts almost commonplace. A few decades ago, running shoes had thin leather soles and little or no foot support. Compare this to the thick, shock-absorbing soles of modern shoes. One of the most startling changes came in pole vaulting, a sport that was revolutionized by the replacement of wooden poles with fiberglass-reinforced plastic ones in the 1960s.

Biochemistry has brought a better understanding of the human body and the way it works. Chemistry has helped to provide better nutrition and drugs that keep athletes healthier. Today's athletes are bigger, faster, and stronger than ever before.

◄ Modern plastic helmets and foam padding protect skateboarders and other athletes from injury.

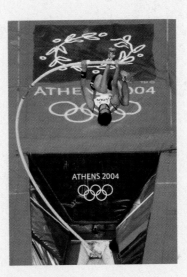

◄ Pole vaulting was transformed by the replacement of wooden poles with fiberglass-reinforced plastic ones in the 1960s. Plastic foam padding protects the landing vaulter from injury.

TABLE 18.2 Effects of Varying Degrees of Dehydration	
Percent Body Weight Lost as Sweat	**Physiological Effect**
2%	Performance impaired
4%	Muscular work capacity declines
5%	Heat exhaustion
7%	Hallucinations
10%	Circulatory collapse and heat stroke

Dehydration can be quite serious, even deadly. We become thirsty when total body fluid volume decreases by 0.5–1.0%. Beyond that, things get worse (Table 18.2). Muscles tire and cramp. Dizziness and fainting may follow, and brain cells shrink, resulting in mental confusion. Finally, the heat regulatory system fails, causing *heat stroke*, which can be fatal without prompt medical attention.

18.4 WEIGHT-LOSS DIETS

With obesity an epidemic and so many people overweight, dieting is a major U.S. industry. It is possible to lose weight through dieting. One pound of adipose (fatty) tissue stores about 3500 kcal of energy. If you reduce your intake by 100 kcal/day and keep your activity constant, you will burn off a pound of fat in 35 days. Unfortunately, people are seldom so patient and they resort to more stringent diets. To achieve their goals more rapidly, they exclude certain foods and reduce the amounts of others. Such diets can be harmful. Diets with fewer than 1200 kcal/day are likely to be deficient in necessary nutrients, particularly in B vitamins and iron. Further, dieting slows down metabolism. Weight lost through dieting is quickly regained when the dieter resumes old eating habits.

EXAMPLE 18.1 **Weight Loss Through Diet**

If you ordinarily expend 2000 kcal/day and you go on a diet of 1500 kcal/day, about how long will it take to diet off 1.0 lb of fat?

Solution

You will use 2000 − 1500 = 500 kcal/day more than you consume. There are about 3500 kcal in 1.0 lb of fat, and so it will take

$$3500 \text{ kcal} \times \frac{1 \text{ day}}{500 \text{ kcal}} = 7 \text{ days}$$

(Keep in mind, however, that your weight loss will not be all fat. You will probably lose more than 1 lb, but it will be mostly water with some protein and a little glycogen.)

▶ **Exercise 18.1A**

A person who expends 1800 kcal/day goes on a diet of 1200 kcal/day without a change in activities. Estimate how much fat she will lose if she stays on this diet for three weeks.

▶ **Exercise 18.1B**

A person who expends 2200 kcal/day goes on a diet of 1800 kcal/day and adds exercise activities that use 180 kcal/day. Estimate how much fat he will lose if he stays on this program for three weeks.

Biochemistry of Hunger

Scientists are just beginning to understand the complex biochemistry of hunger mechanisms. Several molecular signals have been identified that regulate body weight. Some determine whether we want to eat now or stop eating. Other molecules monitor long-term fat balance. Much of what we now know is disappointing for people who want to lose weight: The mechanisms protect against weight loss and favor weight gain.

Two peptide hormones, known as ghrelin and peptide YY (PYY), are produced by the digestive tract and are linked to short-term eating behaviors. Ghrelin, a modified peptide, is an appetite stimulant produced by the stomach. PYY acts as an appetite suppressant. Studies at Imperial College, London, found that people ate about 30% less after they were given a dose of PYY. The research also found that obese people had lower natural levels of PYY, which may explain why the obese are hungrier and overeat.

The hormone insulin and a substance called leptin, which is produced by fat cells, determine longer term weight balance. When insulin levels go up, glucose levels go down and we experience hunger. Conversely, when glucose levels are high, the activity of brain cells sensitive to glucose is lessened, and we feel satiated.

Leptin, a protein with 146 amino acid units that is produced by fat cells, produces weight loss in mice by decreasing their appetite and increasing their metabolic rates. It signals the hypothalamus gland about how much fat is in the body. Humans also produce leptin, and scientists hoped it would be the route to a cure for our ever-growing obesity problem. Those hopes are largely unrealized. Only a few cases of severe human obesity, caused by defects in leptin production, have been helped by leptin treatment. As it turns out, most obese humans have *higher* than normal blood levels of leptin and are resistant to its actions. Leptin's main role seems to be to protect against weight loss in times of scarcity rather than against weight gain in times of plentiful food.

Other substances involved in weight control include cholecystokinin, a peptide formed in the intestine that signals that we have had enough to eat, and a class of compounds called melanocortins, which act on the brain to regulate food intake.

It remains to be seen whether or not this new knowledge about the body's weight-control systems will pay off in better antiobesity treatments, in part because many social factors are also involved in eating behavior. Some people are motivated by environmental cues to eat. A family gathering, the sight or smell of food, and stressful situations can all trigger the hunger mechanism.

Crash Diets: Quick = Quack

Any weight-loss program that promises a loss of more than a pound or two a week is likely to be dangerous quackery. Most quick-weight-loss diets depend on factors other than fat metabolism to hook a prospective customer. The diets often include a **diuretic**, such as caffeine, to increase the output of urine. Weight loss is water loss, and weight is regained when the body is rehydrated.

Other such diets depend on depleting the body's stores of glycogen. On a low-carbohydrate diet the body draws on its glycogen reserves, depleting them in about 24 hr. Recall (Chapter 15) that glycogen is a polymer of glucose. Glycogen molecules have lots of hydroxyl (OH) groups that can form hydrogen bonds to water molecules. Each pound of glycogen carries about 3 lb of water held to it by these hydrogen bonds. Depleting the pound or so of glycogen results in a weight loss of about 4 lb (1 lb glycogen + 3 lb water). No fat is lost, and the weight is quickly regained when the dieter resumes eating carbohydrates.

One pound of fatty tissue stores about 3500 kcal of energy. If your normal energy expenditure is 2400 kcal/day, the most fat you can lose by *total fasting* for a day is 0.69 lb (2400 kcal/day divided by 3500 kcal/lb adipose tissue). This assumes that your body burns nothing but fat. It doesn't. The brain runs on glucose; if that glucose isn't supplied in the diet, it is obtained from protein. Any diet that restricts carbohydrate intake results in a loss of muscle mass as well as fat. When you gain the weight back (as 90% of all dieters do), you gain mostly fat. People who diet without exercising and then gain back the lost weight replace metabolically active tissue (muscle) with inactive fat. Weight loss becomes harder with each subsequent attempted diet.

Many crash diets are also deficient in minerals such as iron, calcium, and potassium. A deficiency of these minerals can interrupt the smooth function of nerve impulse transmission to muscle, which impairs athletic performance. Impulse transmission to vital organs may also be impaired in cases of severe restriction, and death can result from cardiac arrest.

There is little evidence that commercial weight-loss programs are effective in helping people drop excess pounds. Almost no rigorous studies of the programs have been carried out, and U.S. Federal Trade Commission officials say that companies are unwilling to conduct such studies.

Exercise for Health and Fitness

Studies consistently show that people who exercise regularly live longer. They are sick less often and have fewer signs of depression. They can move faster, and they have stronger bones and muscles. Although there are dozens of good reasons for regular exercise, many people begin exercise programs for one simple reason: They want to lose weight.

18.5 EXERCISE FOR WEIGHT LOSS

People who do not increase their food intake when they begin an exercise program lose weight. Contrary to a common myth, exercise (up to 1 hr/day) does not cause an increase in appetite. Most of the weight loss from exercise results from an increase in metabolic rate during the activity, but the increased metabolic rate continues for several hours after completion of the exercise. Exercise helps us maintain both fitness and proper body weight.

EXAMPLE 18.2 | **Weight Loss Through Exercise**

A 135-lb person doing high-impact aerobics burns off about 6.9 kcal/min. How long would the person have to exercise to burn off 1.0 lb of adipose (fat) tissue?

Solution

One pound of adipose tissue stores 3500 kcal of energy. To burn it off at 6.9 kcal/min requires

$$3500 \text{ kcal} \times \frac{1 \text{ min}}{6.9 \text{ kcal}} = 500 \text{ min}$$

It takes about 500 min (more than 8 hr) to burn off 1 lb of fat, even doing high-impact aerobics. (However, you don't have to do it all in one day.)

▶ **Exercise 18.2A**

Walking a mile burns off about 100 kcal. About how far do you have to walk to burn off 1.0 lb of fat?

▶ **Exercise 18.2B**

A moderately active person can calculate the calories needed each day to maintain proper weight by multiplying the desired weight (in pounds) by 15 kcal/lb. How many calories per day do you need to maintain a weight of 150 lb?

On a weight-loss diet (without exercise), about 65% of the weight lost is fat and about 11% is protein (muscle tissue). The rest is water and a little glycogen.

The most sensible approach to weight loss is to adhere to a balanced low-calorie diet that meets the DRI for essential nutrients and to engage in a reasonable, consistent, individualized exercise program. In this way the principles of weight loss are applied by decreasing intake and increasing output.

Weight loss or gain is based on the law of conservation of energy (Chapter 14). When we take in more calories than we use up, the excess calories are stored as fat. When we take in fewer calories than we need for our activities, our bodies burn some of the stored fat to make up for the deficit. One pound of adipose tissue requires 200 mi of blood capillaries to serve its cells. Excess fat therefore puts extra strain on the heart.

Fad Diets

Weight loss diets are often lacking in balanced nutrition and can be harmful to one's health. In the more extreme low-carbohydrate diets, ketosis (Chapter 16) is deliberately induced, and possible side effects include depression and lethargy. In the early stages of a diet deficient in carbohydrates, the body converts amino acids to glucose. The brain must have glucose; it can't obtain sufficient energy from fats. If there are enough adequate proteins in the diet, tissue proteins are spared.

Even low-carbohydrate diets high in adequate proteins are hard on the body, which must rid itself of the nitrogen compounds—ammonia and urea—formed by the breakdown of proteins. This puts an increased stress on the liver, where the waste products are formed, and the kidneys, where they are excreted.

Contrary to a popular notion, fasting does not cleanse the body. Indeed, quite the reverse occurs. A shift to fat metabolism produces ketone bodies, and protein breakdown produces ammonia, urea, and other wastes. You can lose weight by fasting, but the process should be carefully monitored by a physician.

18.6 MEASURING FITNESS

To measure fitness is usually to measure fatness. How much fat is enough? The male body requires about 3% essential body fat, and the average female body needs 10–12%. It is not easy to measure percent body fat accurately. Weight alone does not indicate degree of fitness. A tall 180-lb man may be much more fit than a 150-lb man with a smaller frame. Skinfold calipers are quite inaccurate and measure water retention as well as fat.

One way to estimate body fat is by measuring a person's density. Mass is determined by weighing, and volume is calculated with the help of a dunk tank like the one shown in Figure 18.3. The difference between a person's weight in air and when submerged in water corresponds to the *mass* of the displaced water. When this mass is divided by the density of water and then corrected for air in the lungs, it gives a reasonable value for the person's volume. Mass divided by volume yields the person's density, although results can vary with the amount of air in the lungs. Fat is less dense (0.903 g/mL) than the water (1.000 g/mL) that makes up most of the mass of our bodies. The higher the proportion of body fat a person has, the lower the density and the more buoyant that person is in water.

A simpler way to estimate degree of fatness is by making measurements of the waist and hips. The waist should be measured at the narrowest point and the hips should be measured where the circumference is the largest. The waist measurement should then be divided by the hip measurement. For men the ratio should be less than 1, and for women it should be 0.8 or less. (Because the units cancel out, measurements can be made in inches or centimeters, as long as the same units are used for both.)

Some modern scales provide a measure of body fat content, using *bioelectric impedance analysis*. The person stands on the scale in bare feet, and a small electric current is sent through the body. Fat has a greater impedance (resistance to varying current) than does muscle. By measuring the impedance of the body, the percentage of body fat can be calculated based on height and weight. However, there are many other variables, such as bone density, water content, and location of fat. Such scales are not very accurate, and are useful mostly for monitoring *changes* in body fat content.

Body Mass Index

Body mass index (BMI) is a commonly used measure of fatness. It is defined as weight (in kilograms) divided by the square of the height (in meters). For a person who is 1.8 m tall and weighs 82 kg, the body mass index is $82/(1.8)^2 = 25$. Average

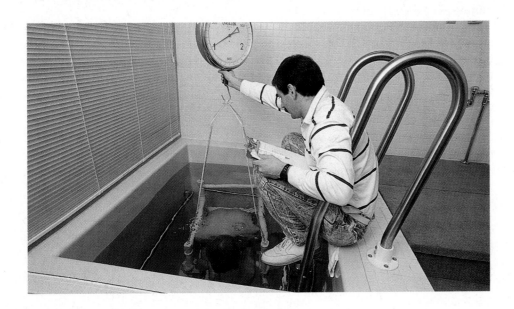

◀ **Figure 18.3** When submerged in water, a body displaces its own volume of water. The difference between a person's weight in air and when submerged in water corresponds to the *mass* of the displaced water. Because the density of water is 1.00 g/mL, the mass of water displaced (in grams) equals the person's volume (in milliliters).

Question: What is the volume, in liters, of a person who displaces 48.5 kg of water?

In Europe during the Middle Ages, people accused of crimes were often tried by some kind of ordeal. In trial by water, the innocent sank and the guilty floated. Suspected witches were tied hand to foot and thrown into the water. The guilty floated, were fished out, dried off, and burned at the stake. In those days, body density was *really* important!

BMI values for adults are between 18.5 and 24.9. A BMI of 25–29.9 indicates that a person is overweight, and a BMI of 30 or greater indicates obesity. When measurements are in pounds and inches, the equation is

$$BMI = \frac{705 \times \text{body weight (lb)}}{[\text{height (in.)}]^2}$$

In other words, a person who is 5 ft 10 in. tall (70 in.) and weighs 180 lb has a BMI of

$$\frac{705 \times 180}{70 \times 70} = 26$$

which is a little on the high side.

EXAMPLE 18.3 **Body Mass Index**

What is the BMI for a person who is 6 ft 1 in. tall and weighs 205 lb?

Solution
The person's height is $(6 \times 12) + 1 = 72 + 1 = 73$ in.

$$BMI = \frac{705 \times 205}{73 \times 73} = 27$$

The body mass index is 27.

▶ **Exercise 18.3A**
What is the BMI for a person who is 5 ft tall and weighs 120 lb?

▶ **Exercise 18.3B**
What is the maximum weight, in pounds, that a person 5 feet 10 inches tall can maintain and have a BMI that does not exceed 25.0?

V_{O_2} Max: A Measure of Fitness

As we increase exercise intensity, our uptake of oxygen must also increase. For example, the faster we run, the more oxygen we need to sustain the pace. However, we reach a point at which our body simply cannot increase the amount of oxygen it consumes even if we increase the intensity of exercise. That point, called the V_{O_2} max, is the maximum amount of oxygen, in milliliters, a person can use in one minute per kilogram of body weight.

V_{O_2} max is therefore a measure of fitness. The higher the V_{O_2} max, the more fit an athlete is. A high V_{O_2} max means the person can exercise more intensely than those who are not as fit. The V_{O_2} max can be increased by working out at an intensity that raises the heart rate to 65 to 85% of its maximum for at least 20 minutes three to five times a week. Limitations on V_{O_2} max include the ability of muscle cells to use oxygen in metabolizing fuels and the ability of the cardiovascular system and lungs to transport oxygen to the muscles.

Direct testing of V_{O_2} max requires expensive equipment, including a gas analyzer to measure O_2 taken in and exhaled. Percent V_{O_2} max can be estimated indirectly from percent maximum heart rate (%MHR).

$$\%MHR = (0.64 \times \%V_{O_2} \text{ max}) + 37$$

The relationship holds quite well for both males and females of all ages and activity levels. For example, 80% MHR corresponds to a $\%V_{O_2}$ max of

$$80 = (0.64 \times \%V_{O_2} \text{ max}) + 37$$
$$\%V_{O_2} \text{ max} = 67$$

Equations for determining MHR and tables for evaluating fitness levels can be found in exercise physiology text books and on the Web (see Collaborative Group Project 80).

18.7 SOME CHEMISTRY OF MUSCLES

The human body has about 600 muscles. Exercise makes these muscles larger, more flexible, and more efficient in their use of oxygen. Exercise helps to protect against heart disease because the heart is an organ comprised mainly of muscle. A strong heart is a healthy heart. With regular exercise, resting pulse and blood pressure usually decline. After several months of an effective exercise program, pulse and blood pressure remain lower even *during* exercise. The net result, called the **training effect**, is that a person who exercises regularly is able to do more physical work with less strain.

Exercise is an art, but it is also increasingly a science—a science in which chemistry plays a vital role.

Energy for Muscle Contraction: ATP

When cells metabolize glucose or fatty acids, only part of the chemical energy in these substances is converted to heat. Some of the energy is stored in the high-energy phosphate bonds of adenosine triphosphate (ATP) molecules.

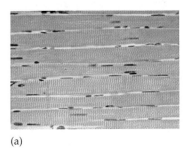

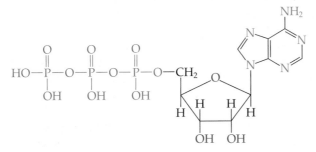

Adenosine triphosphate

The stimulation of muscles causes them to contract. This contraction is *work*, and it requires energy that the muscles get from the molecules of ATP. The energy stored in ATP powers the physical movement of muscle tissue. Two proteins, actin and myosin, play important roles in this process. Together they form a loose complex called *actomyosin*, the contractile protein of which muscles are made (Figure 18.4). When ATP is added to actomyosin in the laboratory, the protein fibers contract. It seems likely that the same process occurs in the muscles of living animals. Not only does myosin serve as part of the structural complex in muscles, it also acts as an enzyme for the removal of a phosphate group from ATP. Thus, it is directly involved in liberating the energy required for the contraction.

In a resting person, muscle activity (including that of the heart muscle) accounts for only about 15–30% of the energy requirements of the body. Other activities, such as cell repair, transmission of nerve impulses, and maintenance of body temperature, account for the remaining energy needs. During intense physical activity, the energy requirements of muscle may be more than 200 times the resting level.

Aerobic Exercise: Plenty of Oxygen

The ATP in muscle tissue is sufficient for activities lasting for at most a few seconds. Fortunately, muscles have a more extensive energy supply in the form of glycogen, a starchlike stored form of dietary carbohydrate that has been ingested, digested to glucose, and absorbed.

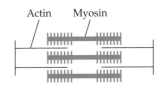

Actin   Myosin

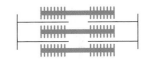

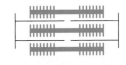

(b)

▲ **Figure 18.4** Skeletal muscle tissue has a banded or *striated* appearance, shown here in a micrograph at 180X magnification (a). The diagram of the actomyosin complex in muscle (b) shows extended muscle fibers (top), resting fibers (middle), and partially contracted fibers (bottom).

 Oxygen

▲ **Figure 18.5** Aerobic exercise is performed at a pace that allows us to get enough oxygen to our muscle cells to oxidize pyruvic acid to carbon dioxide and water. Aerobic dance is a popular form of aerobic exercise.

When muscle contraction begins, glycogen is converted to pyruvic acid by muscle cells in a series of steps.

$$(C_6H_{10}O_5)_n \longrightarrow 2n \ CH_3C{-}COH$$

Glycogen   Pyruvic acid

Then, if sufficient oxygen and other factors are readily available, the pyruvic acid is oxidized to carbon dioxide and water in another series of steps.

$$2 \ CH_3C{-}COH \ + \ 5 \ O_2 \longrightarrow 6 \ CO_2 \ + \ 4 \ H_2O$$

Muscle contractions that occur under these circumstances—that is, in the presence of oxygen—constitute **aerobic exercise** (Figure 18.5).

Anaerobic Exercise and Oxygen Debt

When sufficient oxygen is not available, pyruvic acid is reduced to lactic acid.

$$CH_3COCOOH + [2 \ H] \longrightarrow CH_3CHOHCOOH$$

(The [2 H] represent hydrogen atoms from one of several reducing agents.) If this **anaerobic exercise** persists, an excess of lactic acid builds up in the muscle cells. This lactic acid ionizes, forming lactate ions and hydronium ions.

$$CH_3CHOHCOOH + H_2O \longrightarrow CH_3CHOHCOO^- + H_3O^+$$

Muscle fatigue correlates well with lactate levels, but fatigue has many sources (Reference 1) and may well be mainly attributable to changes in V_{O_2} max. Several other products of muscle metabolism, including phosphocreatine (the phosphorylated form of creatine, page 551); ATP; and ions such as Ca^{2+}, Na^+, K^+, and $H_2PO_4^-$, change during fatigue. Each of these may affect proteins that regulate muscle contraction. Even though other substances are no doubt involved, exercise scientists find it useful to assess an athlete's **lactate threshold,** the point at which lactic acid concentration builds up rapidly in the blood and muscles, and the athlete suffers severe muscle fatigue. Lactate threshold assessment is used for determining endurance training adaptations.

$$H_2O_3PNH{-}C{-}N{-}CH_2COOH$$
$$NH \quad CH_3$$

Phosphocreatine

Well-trained athletes can continue for a while after reaching their lactate threshold, but most people quit. At this point the **oxygen debt** is repaid (Figure 18.6). After exercise ends, the cells' demand for oxygen decreases, making more oxygen available to oxidize the lactic acid resulting from anaerobic metabolism back to pyruvic acid. This acid is then converted to carbon dioxide, water, and energy.

Most athletes emphasize one type of training (anaerobic or aerobic) over the other. For example, an athlete training for a 60-m dash does mainly anaerobic work, but one planning to run a 10-km race does mainly aerobic training. Sprinting and weightlifting are largely anaerobic activities; a marathon run is largely aerobic. During a marathon, athletes must set a pace to run for more than 2 hr. Their muscle cells depend on slow, steady aerobic conversion of carbohydrates to energy. During anaerobic activities, however, muscle cells use almost no oxygen. Rather, they need the quick energy provided by anaerobic metabolism.

▲ **Figure 18.6** After a race, the metabolism that fueled the effort continues. The athlete gulps air to repay her oxygen debt.

After glycogen stores are depleted, muscle cells can switch over to fat metabolism. Fats are the main source of energy for sustained activity of low or moderate intensity, such as the last part of a marathon run (42 km).

Muscle Fibers: Twitch Kind Do You Have?

Muscles are tools with which we do work. The quality and type of muscle fibers that an athlete has affect his or her athletic performance, just as the quality of machinery influences how work is done. For example, we can remove snow from a driveway in more than one way. We can eliminate it in 10 min with a snowblower, or we can spend 50 min shoveling. Muscles are classified according to the speed and effort required to accomplish this work. The two classes of muscle fibers are **fast-twitch fibers** (those stronger and larger and most suited for anaerobic activity) and **slow-twitch fibers** (those best for aerobic work). Table 18.3 lists some characteristics of these two types of muscle fibers.

Type I (slow-twitch) fibers are called upon during activity of light or moderate intensity. The respiratory capacity of these fibers is high, which means that they can provide a large amount of energy via aerobic pathways. The myoglobin level is also high. Aerobic oxidation requires oxygen, and muscle tissue rich in slow-twitch fibers is geared to supplying high levels of oxygen.

Type I muscle fibers have a low capacity to use glycogen and thus are not geared to anaerobic generation of energy. Their action does not require the hydrolysis of glycogen. The catalytic activity of the actomyosin complex is low. Remember that actomyosin not only is the structural unit in muscle that actually undergoes contraction but also is responsible for catalyzing the hydrolysis of ATP to provide energy for the contraction. Low catalytic activity means that the energy is parceled out more slowly. This is not good if you want to lift 200 kg, but it is great for a 15-km run.

Type IIB (fast-twitch) fibers have characteristics opposite those of type I fibers. Low respiratory capacity and low myoglobin levels do not bode well for aerobic oxidation. A high capacity for glycogen use and a high catalytic activity of actomyosin allow tissue rich in fast-twitch fibers to generate ATP rapidly and also to hydrolyze this ATP rapidly during intense muscle activity. Thus, this type of muscle tissue provides the capacity to do short bursts of vigorous work. We say "bursts" because this type of muscle fatigues rather quickly. A period of recovery, during which lactic acid is cleared from the muscle, is required between brief periods of activity.

Building Muscles

Endurance exercise increases myoglobin levels in skeletal muscles. This provides for faster oxygen transport and increased respiratory capacity (Figure 18.7), and these changes usually are apparent within two weeks. Endurance training does not necessarily increase the size of muscles.

Myoglobin is the heme-containing protein in muscle that stores oxygen (similar to the way hemoglobin stores and transports oxygen in the blood).

▲ **Figure 18.7** The New York City Marathon attracts more than 30,000 runners each year.

Question: What is the predominant muscle type in the elite runners in this race? Do their muscles have high or low aerobic capacity?

TABLE 18.3 A Comparison of Two Types of Muscle Fibers		
Characteristic	**Type I**	**Type IIB**[*]
Category	Slow twitch	Fast twitch
Color	Red	White
Respiratory capacity	High	Low
Myoglobin level	High	Low
Catalytic activity of actomyosin	Low	High
Capacity for glycogen use	Low	High

[*]A type IIA fiber exists that resembles type I in some respects and type IIB in others. We will discuss only the two types described in this table.

▲ **Figure 18.8** Weightlifting is a popular activity in many gyms and fitness centers.

Question: What is the predominant muscle type in elite weightlifters? Do their muscles have high or low aerobic capacity?

If you want larger muscles, try weightlifting. Weight training (Figure 18.8) develops fast-twitch muscle fibers. These fibers increase in size and strength with repeated anaerobic exercise. Weight training does *not* increase respiratory capacity, however.

Muscle-fiber type seems to be inherited. Research shows that world-class marathon runners may possess up to 80–90% slow-twitch fibers, compared with championship sprinters who may have up to 70% fast-twitch muscle fibers. Some exceptions have been noted, however, and factors such as training and body composition still are important in athletics. So are other factors, including nutrition, fluid and electrolyte balance, and drug use or misuse.

Drugs, Athletic Performance, and the Brain

Muscle chemistry and metabolism, aerobic and anaerobic exercise, and nutrition and fluid balance all relate to normal internal processes. It would seem that hard work and proper nutrition are the best answers to improved performance. However, many athletes turn instead to drugs, such as the stimulants amphetamine and cocaine, and to anabolic steroids to build muscles.

18.8 DRUGS AND THE ATHLETE

Like most other people, athletes use drugs to alleviate pain or soreness that results from overuse and to treat injuries. These **restorative drugs** include analgesics (painkillers), such as aspirin and acetaminophen (Chapter 19), and anti-inflammatory drugs, such as aspirin, ibuprofen, ketoprofen, and cortisone. Cortisone derivatives are often injected to reduce swelling in damaged joints and tissues. Relief is often transitory, however, and side effects can be severe. Prolonged use of these drugs can cause fluid retention, hypertension, ulcers, disturbance of the sex hormone balance, and other problems. Another substance, one that can be bought over the counter and applied externally, is methyl salicylate, or oil of wintergreen. This substance causes a mild burning sensation when applied to the skin, thus serving as a counterirritant for sore muscles. An aspirin derivative, it also acts as an analgesic.

Stimulant Drugs

Some athletes use *stimulant drugs* (Chapter 19) to try to improve performance. The caffeine in a strong cup of coffee may be of some benefit. Caffeine triggers the release of fatty acids. Metabolism of these compounds may conserve glycogen, but any such effect is small. Caffeine also increases the heart rate, speeds metabolism, and increases urine output. The latter could lead to dehydration.

Other athletes resort to stronger stimulants, including amphetamines and cocaine. A multipurpose drug, cocaine was first used in the late 1800s as a local anesthetic to deaden pain during dental work or minor surgery. It is also a powerful stimulant, which is the reason for its current popularity. Cocaine stimulates the central nervous system, increasing alertness, respiration, blood pressure, muscle tension, and heart rate. Both stimulants mask symptoms of fatigue and give an athlete a sense of increased stamina. The cocaine intoxication deaths of several thousand people, including prominent professional athletes, brought this problem to public attention. Other athletes have been arrested, suspended, and banned for cocaine use.

Anabolic Steroids

Many strenuous athletic performances depend on well-developed muscles. Muscle mass depends on the level of the male hormone *testosterone* (Chapter 19). When boys reach puberty, testosterone levels rise, and the boys become more muscular if they exercise. Men generally have larger muscles than women because they have more testosterone.

Testosterone and some of its semisynthetic derivatives are taken by athletes in an attempt to build muscle mass more rapidly. Steroid hormones used to increase muscle mass are called **anabolic steroids.** These chemicals aid in the building (anabolism) of body proteins and thus of muscle tissue.

There are no reputable controlled studies demonstrating the effectiveness of anabolic steroids. They do seem to work—at least for some people—but the side effects are many. In males, side effects include testicular atrophy and loss of function, impotence, acne, liver damage that may lead to cancer, edema (swelling), elevated cholesterol levels, and growth of breasts.

Anabolic steroids act as male hormones (androgens), making women more masculine. They help women build larger muscles, but they also result in balding, extra body hair, a deep voice, and menstrual irregularities. What price will an athlete pay for improved performance?

Drugs, Athletic Performance, and Drug Screening

The use of drugs to increase athletic performance is not only illegal and physically dangerous but can also be psychologically damaging. Stimulant drugs give the user a false sense of confidence by affecting the person's ego, giving the delusion of invincibility. Such "greatness" often ends as the competition begins, however, because the stimulant effect is short lived (it lasts for only half an hour to an hour for cocaine). Exhaustion and extreme depression usually follow. More stimulants are needed to combat these "down" feelings, and the user then runs the risk of addiction or overdose.

Because drug use for the enhancement or improvement of performance is illegal, chemists have another role to play in sports: screening blood and urine samples for illegal drugs. Using sophisticated instruments, chemists can detect minute amounts of illegal drugs. Drug testing has been used at the Olympic Games since 1968, and it is rapidly becoming standard practice for athletes in college and professional sports.

Drugs are used at great risk of health (and legal!) problems. The use of illegal drugs violates the spirit of fair athletic competition. Let us work toward a world in which events and medals are won by drug-free athletes who are highly motivated, well-nourished, and well-trained.

18.9 EXERCISE AND THE BRAIN

Hard, consistent training improves athletic performance. Few drugs help, but our own bodies manufacture a variety of substances that aid performance. Examples are the pain-relieving **endorphins** (Section 19.17).

After vigorous activity, athletes have an increased level of endorphins in their blood. Stimulated by exercise, deep sensory nerves release endorphins to block the pain message. Exercise extended over a long period of time, such as distance running, sometimes causes in an athlete many of the same symptoms experienced by opiate users. Athletes get a euphoric high during or after a hard run. Unfortunately, both natural endorphins and their analogs, such as morphine, seem to be addictive. Athletes, especially runners, tend to suffer from withdrawal: They feel bad when they don't get the vigorous exercise of a long, hard run. This can be positive because exercise is important to the maintenance of good health. It can be negative if the person exercises when injured or to the exclusion of family or work obligations. Athletes also seem to develop a tolerance to their own endorphins: They have to run farther and farther to get that euphoric feeling.

Exercise also directly affects the brain by increasing its blood supply and by producing **neurotrophins**, substances that enhance the growth of brain cells. Exercises that involve complex motions, such as dance movements, lead to an increase in the connections between brain cells. Both our muscles and our brains work better when we exercise regularly.

18.10 NO SMOKING

Tobacco use and smoking are dangerous addictions that cause a wide variety of diseases, including cancer, heart disease, and emphysema, and often result in death. Most tobacco users are hooked when they are children. If you want to stay healthy and fit, one thing you certainly should *not* do is use tobacco. According to the U.S. Centers for Disease Control and Prevention, cigarette smoking is associated with death from 24 different diseases, including heart disease, stroke, and lung diseases such as cancer, stroke, emphysema, and pneumonia. Smoking causes diseases of the gums and mouth, chronic hoarseness, vocal cord polyps, and premature aging and wrinkling of the skin. Lung cancer is not the only malignancy caused by cigarette smoking. Cancers of the pancreas, bladder, breast, kidney, mouth and throat, stomach, larynx, and cervix are also associated with smoking.

We fear cancer, but heart attack and stroke are the major causes of disability and death of smokers. Cigarette smokers have a 70% higher risk of heart attack than nonsmokers. When they are also overweight, with high blood pressure and high cholesterol, their risk is 200% higher. At every age the death rate is higher among smokers than among nonsmokers.

Women who smoke during pregnancy have more stillborn babies, more premature babies, and more babies who die before they are a month old. Women who take birth control pills and also smoke have an abnormally high risk of stroke. Women who smoke are advised to use some other method of birth control.

One of the gases in cigarette smoke is carbon monoxide, which ties up about 8% of the hemoglobin in the blood. Smokers therefore breathe less efficiently. They work harder to breathe, but less oxygen reaches their cells. It is no wonder that fatigue is a common problem for smokers. Some chemicals that have been identified in cigarette smoke are listed in Table 18.4.

Two million people die each year in the United States, about 1 million of them prematurely. Tobacco use contributes to about half of these early deaths. More than 5 million children alive today will die prematurely because they decide to smoke cigarettes.

More than 4000 chemical compounds have been identified in cigarette smoke, including many that are quite poisonous and more than 63 that are known or suspected carcinogens.

TABLE 18.4 A Few of the Many Chemicals that Have Been Identified in Cigarette Smoke*

Acetaldehyde	Chromium	Isoprene	**Nitrosamines**
Acetic acid	**Chrysene**	Lead	**Nitrosonomicotine**
Acetone	Copper	Limonine	Nomicotine
Acetylene	**Crotonaldehyde**	Linoleic acid	Palmitic acid
Acrolein	Cyclotenes	Linolenic acid	Phenanthrenes
Acrylonitrile	DDT	Mercury	Phenol
Aluminum	**Dibenz(a,h)acridine**	Methane	Picolines
Aminobiphenyl	**Dibenz(a,h)anthracene**	Methanol	**Polonium-210**
Ammonia	**Dibenz(a,j)acridine**	Methyl formate	Propionic acid
Anabasine	**Dibenzo(a,l)pyrene**	Methylamine	Pyrenes
Anatabine	**Dibenzo(c,g)carbazole**	NNK	Pyrrolidine
Aniline	Dieldrin	Methylpyrrolidine	**Quinoline**
Anthracenes	**Dimethylhydrazine**	**N-Nitrosoanabasine**	Quinones
Argon	Ethanol	**N-Nitrosodiethanolamine**	Scopoletin
Arsenic	**Ethylcarbamate**	**N-Nitrosodiethylamine**	Sitosterol
Benz(a)anthracene	Fluoranthenes	**N-Nitrosodimethylamine**	Skatole
Benzene	Fluorenes	N-Nitrosoethylmethylamine	Solanesol
Benzo(a)pyrene	**Formaldehyde**	**N-Nitrosomorpholine**	Stearic acid
Benzo(b)fluoranthene	Formic Acid	**N-Nitrosopyrrolidine**	Stigmasterol
Benzo(j)fluoranthene	Furan	Naphthalene	Styrene
Butadiene	Glycerol	**Naphthylamine**	Titanium
Butane	Hexamine	Neophytadienes	Toluene
Cadmium	**Hydrazine**	**Nickel**	**Toluidine**
Campesterol	Hydrogen cyanide	Nicotine	**Urethane**
Carbon monoxide	Hydrogen sulfide	Nitric oxide	**Vinyl chloride**
Carbon disulfide	**Indeno(1,2,3-c,d)pyrene**	Nitrobenzene	Vinylpyridine
Catechol	Indole	**Nitropropane**	

* Those proven or suspected of causing cancer are in boldface type. Those proven or suspected of causing birth defects are in green type.

In the linings of the lungs there are cilia (tiny, hairlike processes) that help to sweep out particulate matter breathed into the lungs. But in smokers the tar from cigarette smoke forms a sticky coating that keeps the cilia from moving freely. The reduced action of the cilia may explain the smoker's cough. Pathologists claim that during an autopsy it is usually easy to tell if the body was that of a smoker. Lungs are normally pink; a smoker's lungs are black.

Unfortunately, cigarette smoke also harms nonsmokers. Exposure to second-hand smoke causes an estimated 3000 premature deaths each year in the United States and 300,000 children to suffer from lower respiratory tract infections. Children of smokers miss twice as much school time because of respiratory infections as do children of nonsmokers.

Efforts to lessen indoor pollution caused by smoking have led many restaurants and government buildings to be declared nonsmoking areas, and further bans are in the works. Smoking has never been a smart or a healthy thing to do, but now it isn't even fashionable.

A study of military veterans, conducted at Harvard University, concluded that smokers are 4.3 times as likely as nonsmokers to develop Alzheimer's disease.

Chemistry of Sports Materials

The athlete's world is shaped by chemicals. The ability to perform reflects biochemistry within the athlete. Chemicals taken by athletes may hinder or enhance their performance. Yet another chemical dimension to the sports world involves the chemicals that provide athletic clothing and equipment (Table 18.5). Athletic clothing depends on chemicals for its shape and protective qualities.

Swimsuits, ski pants, and elastic supports stretch because of synthetic fibers. Undergarments made of polypropylene or polyester fibers wick perspiration away from the skin and keep the athlete dry even during vigorous exercise. Joggers run in rain and winds in suits made of materials such as Gore-Tex that keep them from getting cold and wet. Gore-Tex is a thin, membranous material made by stretching fibers of polytetrafluoroethylene (Teflon). The material has billions of tiny holes that are too small to pass drops of water but readily pass water vapor. Raindrops falling on the outside are held together by surface tension and run off rather than penetrating the fabric. Water vapor from sweating skin, however, can pass from the warm area inside the suit where vapor concentration is high to the cooler area outside the suit where the vapor

TABLE 18.5 A Few of the Many Sports and Recreational Items Made from Petroleum			
Wet suits	Motorcycle helmets	Darts	Ice buckets
Parachutes	Dune-buggy bodies	Tents	Fishing nets
Card tables	Checkers	Stadium cushions	Hiking boots
Golf-cart bodies	Chess boards	Fingerpaints	Frisbees
Warmup suits	Shorts	Foul-weather gear	Fishing boots
Ping-Pong paddles	Tennis shoes	Foot pads	Diving masks
Rafts	Paddles	Visors	Guitar picks
Uniforms	Decoys	Swimming-pool liners	Beach balls
Phonographs	Volley balls	Vinyl tops for cars	Sunglasses
Racks	Sleeping bags	Ice chests	Dog leashes
Track shoes	Sports-car bodies	Life jackets	Dice
Dominoes	Tennis balls	Audio tapes	Pole-vaulting poles
Windbreakers	Reclining chairs	Model planes	Aquariums
Sails	Insect repellents	In-line skates	Golf club impact surfaces
Football helmets	Canoes and kayaks	Artificial turf	Basketballs
Swimsuits	Bicycle tires, seats, and handles		

concentration is lower. The runner stays relatively dry (Figure 18.9).

Joggers are not the only athletes pampered by equipment made of materials developed by the chemical industry. Football, hockey, and baseball players are protected by polycarbonate helmets and protective pads of synthetic foamed rubbers. Brightly dyed nylon uniforms with synthetic colors add to the glamour of amateur and professional teams. Many sports events are played on artificial turf, a carpet of nylon or polypropylene with an underpad of synthetic foamed polymers. Stadiums are protected from the weather by Teflon covers reinforced with glass fibers and held aloft by air pressure (Figure 18.10).

The dramatic effect of new materials is perhaps best illustrated in pole vaulting. Between 1940 and 1960 the world-record height for this event increased only 23.5 cm. After the development of plastic poles reinforced with glass fibers, however, the record rapidly rose another 100 cm. Other sports have been affected similarly, if less dramatically. From the soles of sports shoes and the wax used on cross-country skis to titanium golf clubs and tennis sweaters that stretch and yet retain their shape, the sports world is thoroughly immersed in chemicals.

(a)

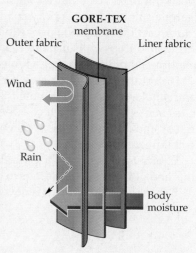
(b)

▲ **Figure 18.9** Imaginative use of fibers and other materials makes it possible to exercise in relative comfort even in inclement weather (a). Gore-Tex fabrics (b) repel wind and raindrops but allow body moisture to escape.

◀ **Figure 18.10** "Baseball played on an artificial surface inside a domed stadium is quite different from baseball played outside on grass."

Critical Thinking Exercises

Apply knowledge that you have gained in this chapter and one or more of the FLaReS principles (Chapter 1) to evaluate the following statements or claims.

18.1 An advertisement in an American sports magazine claims that an amino acid supplement helps build muscle mass.

18.2 An athlete claims that a certain brand of shoe helps him jump higher than any other kind of shoe.

18.3 An athlete claims that a piece of tape with a certain shape applied across the bridge of the nose helps her breathe better and improves her athletic performance.

18.4 The various ways by which performance can be improved are called *ergogenic aids*. These include mechanical, physiological, psychological, pharmacological, and nutritional aids. Unfortunately, many of the claims for these aids have not been substantiated by rigorous scientific trials. Evaluate the following claims:

a. A creatine supplement "makes muscles... much less prone to fatigue and the capacity to undertake strenuous exercise is increased."

b. Sodium bicarbonate reduces the acidity of the blood and "may be able to draw more of the acid produced within the muscle cells out into the blood and thus reduce the level of acidity within the muscle cells themselves. This could delay the onset of fatigue."

   c. Two factors that limit prolonged exercise are depletion of the body's carbohydrate stores and dehydration. Consuming sports drinks containing carbohydrates and electrolytes before, during, and after exercise will help prevent blood glucose levels falling too low, help maintain the body's glycogen stores, and replace electrolytes lost through perspiration.

18.5 An advertisement for a weight-loss pill claims that you can "Eat all you want and still lose weight (the pill does all the work)."

18.6 An advertisement for a weight-loss pill claims that the substance in the pill "melts fat from your body."

SUMMARY

Section 18.1—Good health and fitness require a nutritious diet and regular exercise. Today's dietary guidelines emphasize nutrient-dense foods and a balanced eating program. A balanced diet consists largely of fruits, vegetables, whole grains, and dairy products in the proper proportions. Total fats should not exceed 35% of calories; trans and saturated fats (mostly of animal origin) less than 10% of calories. Athletes generally need more calories, which should come mostly from carbohydrates—especially starches. Exercise, not extra protein, builds muscle tissue. The Dietary Reference Intake (DRI) for protein and other food items is established by the U.S. Academy of Sciences for planning and assessing diets.

Sections 18.2–18.3—The body needs organic substances (vitamins) and inorganic substances (minerals) that it cannot produce in the diet. Vitamin and mineral supplements may not be necessary in a well-balanced diet, but they do no harm and may be beneficial. Vitamin A is essential for vision and skin maintenance. There are several B vitamins; B_3 and B_6 may help arthritis patients. B_{12} is not found in plants, and a lack can cause anemia. Vitamin C prevents scurvy, promotes healing, is needed for the immune system, and is an antioxidant. Vitamin D promotes absorption of calcium and phosphorus. Vitamin E is an antioxidant and an anticoagulant, and deficiency can lead to muscular dystrophy.

   An electrolyte is a substance that conducts electricity in water solution. The main electrolytes in the body are ions of sodium, potassium, chlorine, calcium, and magnesium, as well as sulfate, hydrogen phosphate, and bicarbonate ions. Water is an essential nutrient. Most athletes probably should rely on water rather than sports drinks for hydration.

Sections 18.4–18.5—Drastic diets are unlikely to be permanently successful, and the weight lost is usually regained. One pound of adipose tissue stores about 3500 kcal of energy. A variety of chemical substances that are produced in the body are appetite regulators. Quick-loss diets often include a diuretic to increase urine output, which results in temporary weight loss. Low-carbohydrate diets cause quick weight loss by depleting glycogen and water. Exercise aids in weight loss, because the increased metabolic rate during exercise continues after exercise ends. The most sensible approach to weight loss involves a balanced low-calorie diet and reasonable exercise.

Sections 18.6–18.7—The male body requires about 3% body fat; the female body, 10–12%. Body fat can be estimated by several methods. Density measurement is probably the most accurate. Body mass index also expresses fatness, and V_{O_2} max is a measure of fitness.

   Exercise makes muscles larger, more flexible, and more efficient. A person who exercises regularly is able to do more work with less strain; this is called the training effect. Energy for muscular work comes partly from ATP. A protein complex called *actomyosin* contracts when ATP is added to it. ATP provides energy for just a few seconds, after which glycogen (animal starch) is metabolized. If metabolism occurs with plenty of oxygen—during aerobic exercise—CO_2 and H_2O are formed. With insufficient oxygen—during anaerobic exercise—lactic acid is formed and muscle fatigue occurs. An athlete's lactate threshold is the point at which lactic acid concentration builds up rapidly, resulting in muscle fatigue. When exercise ends, the oxygen debt is repaid; oxygen demand of cells decreases and the lactic acid can be oxidized to CO_2 and H_2O. There are two classes of muscle fibers. Fast-twitch fibers are larger and stronger and good for rapid bursts of power, while slow-twitch fibers are geared for steady exercise of long endurance. Weight training develops fast-twitch muscles but does not increase respiratory capacity. Long-distance running develops slow-twitch muscles and increases respiratory capacity.

Sections 18.8–18.10—Athletes sometimes use restorative drugs, such as aspirin, ibuprofen, or cortisone, to alleviate pain and treat injuries. Occasionally they use stimulant drugs such as caffeine and cocaine. Anabolic steroids appear to increase muscle mass, but they have many bad side effects, including impotence and liver cancer. Our bodies can manufacture pain-relieving endorphins that may help performance, but it appears that some athletes can become addicted to them. Exercise also directly affects the brain by producing neurotrophins that enhance the growth of brain cells. Smoking reduces athletic performance and contributes to disease and death, and athletes should avoid it. From modern sports clothing and equipment to artificial turf, chemistry has contributed greatly to the world of sport.

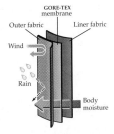

REVIEW QUESTIONS

1. What are the two major requirements for good health and fitness?

2. Why is the 2005 food guide pyramid (Figure 18.2) not the same for everyone?

3. The recommended diet contains what percent of its calories as fat? As saturated fat?

4. What kind of products are the main dietary source of saturated fats?

5. How might you lose 5 lb in just 1 week? Would it be a useful way to lose weight and keep it off?

6. Do you think it is likely that obesity is inherited?

7. List three ways that chemistry has had an impact on sports.

8. Describe the training effect.

9. How do the nutritional needs of an athlete differ from those of a sedentary individual? How is this extra need best met?

10. List two ways to determine percent body fat. Describe a limitation of each method.

11. How many calories of energy are stored in 1 lb of adipose tissue?

12. List some problems that result from low-calorie diets.

13. How much glycogen can the average human body store?

14. Why does excess body fat put a strain on the heart?

15. What are restorative drugs?

16. What are the effects of anabolic steroids on females?

PROBLEMS

Nutrition

17. Mention two good reasons for not going on a high-protein diet.

18. Do athletes need more protein than other people? Explain.

Vitamins and Minerals

19. What are Dietary Reference Intake (DRI) values?

20. Do we ever need more of a given vitamin than its DRI?

21. What are some of the benefits of vitamin A?

22. In addition to preventing pellagra, what does vitamin B_3 (niacin) do?

23. What kinds of problems are helped by vitamin B_6 (pyridoxine)?

24. What benefit is derived from vitamin B_{12}?

25. What are the benefits of vitamin D? Is taking a daily megadose of vitamin D a good idea?

26. Vitamins C and E are antioxidants. What does this mean?

Electrolytes

27. List the three major electrolytes essential for proper cellular function.

28. What is a diuretic? Describe the function of antidiuretic hormone (ADH).

29. What fluid is best for replacing water lost during exercise? Why are beer and cola drinks not recommended for fluid replacement?

30. Is thirst a good indicator of dehydration? Are you always thirsty when dehydrated?

31. How is the appearance of urine related to dehydration?

32. What are some of the effects of dehydration? What is heat stroke?

Diet and Exercise

33. Why does a diet that restricts carbohydrate intake lead to loss of muscle mass as well as fat?

34. List two ways in which fad diets lead to "quick weight loss." Why is this weight rapidly regained?

35. A giant double-decker hamburger provides 600 kcal of energy. How long would a 130-lb person have to walk to burn off this energy if 1 hr of walking uses about 210 kcal?

36. One kilogram of fat tissue stores about 7700 kcal of energy. An average person burns about 40 kcal in walking 1 km. If such a person walks 5 km/day, how much fat will be burned in one year?

37. How long would a 155-lb person have to run to burn off the 110 kcal in one glass of beer if 1 hr of running burns off 565 kcal?

38. How far would you have to run to burn off 5 kg of fat if running burns off 100 kcal/km? (See Problem 36.)

39. The DRI for protein is about 0.8 g/kg body weight. How much protein is required each day by a 125-kg football player?

40. How much protein is required each day by a 50-kg gymnast? (See Problem 39.)

41. A 70-kg man can store about 2000 kcal as glycogen. How far can such a man run on this stored starch if he expends 100 kcal/km while running and the glycogen is his only source of energy?

42. A 70-kg man can store about 100,000 kcal of energy as fat. Estimate how far such a man could run on this stored fat if he expends 100 kcal/km while running and the fat was his only source of energy.

43. Describe the role of each of the following substances involved in the control of body weight.
    a. insulin   **b.** ghrelin

44. Describe the role of each of the following substances involved in the control of body weight.
    a. leptin   **b.** PYY   **c.** cholecystokinin

Body Mass Index

45. What is the body mass index for a 6-ft baseball player who weighs 185 lb?

46. What is the body mass index for a 6-ft 6-in. basketball player who weighs 250 lb?

Muscles

47. Muscles are made of protein. Does eating more protein result in larger muscles?

48. How many muscles do humans have?

49. What is the immediate source of energy for muscle contraction?

50. What is aerobic exercise? What is anaerobic exercise?

51. What two proteins make up the actomyosin protein complex?

52. What are the two functions of the actomyosin protein complex?

53. Which type of metabolism (aerobic or anaerobic) is primarily responsible for providing energy for intense bursts of vigorous activity?

54. Which type of metabolism (aerobic or anaerobic) is primarily responsible for providing energy for prolonged low levels of activity?

55. What is meant by oxygen debt?

56. What is the lactate threshold?

57. Identify type I and type IIB muscle fibers as
    a. fast twitch or slow twitch.
    b. suited to aerobic oxidation or to anaerobic use of glycogen.

58. Explain why high levels of myoglobin are appropriate for muscle tissue geared to aerobic oxidation.

59. Why does the high catalytic activity of actomyosin in type IIB fibers suggest that these are the muscle fibers engaged in brief, intense physical activity?

60. Why can the muscle tissue that uses anaerobic glycogen metabolism for its primary source of energy be called on only for brief periods of intense activity?

61. Which type of muscle fiber is most affected by endurance training exercises? What changes occur in the muscle tissue?

62. Birds use large, well-developed breast muscles for flying. Pheasants can fly 80 km/hr, but only for short distances. Great blue herons can fly only about 35 km/hr but can cruise great distances. What kind of fibers do each have in their breast muscles?

63. Muscle is protein. Does an athlete need extra protein (above the DRI) to build muscles? What is the only way to build muscles?

64. Describe the biochemical process by which muscles are built.

Smoking and Health

65. What are some of the health problems related to smoking?

66. Why is it important for pregnant women not to smoke?

Drugs

67. How are cortisone derivatives used in sports medicine? What are some of the side effects of its use?

68. What are anabolic steroids? List some side effects of the use of anabolic steroids in males.

69. Does cocaine enhance athletic performance? Explain fully.

70. What happens when the effect of cocaine wears off?

71. Give a biochemical explanation of the "runner's high."

72. What is the role of chemists in the control of drug use by athletes?

73. List the evidence indicating that long-distance running is addictive.

74. Give a biochemical explanation of addiction to long-distance running.

ADDITIONAL PROBLEMS

75. If you are moderately active and want to maintain a weight of 150 lb, about how many calories do you need each day?

76. An athlete can run a 400-m race in 45 s. Her maximum oxygen intake is 4 L/min, yet working muscles at their maximum exertion requires about 0.2 L of oxygen gas per minute for each kilogram of body weight. If an athlete weighs 50 kg, what oxygen debt will she incur?

77. Fat tissue has a density of about 0.9 g/mL, and lean tissue a density of about 1.1 g/mL. Calculate the density of a person with a body volume of 80 L who weighs 85 kg. Is the person fat or lean?

78. Following is a table of data used to determine an individual's percent body fat.

Body Density (g/cm^3)	% Body Fat
1.010	38.3
1.030	29.5
1.050	21.0
1.070	12.9
1.090	5.07

Make a graph of the data, and then estimate the percent body fat of the following.
a. Person A, who has a body density of 1.037 g/cm^3.
b. Person B, who weighs 165 lb in air and 14 lb when submerged in water.
c. Which person is more likely to be a well-trained athlete?
d. A 20-year-old male weighs 89.34 kg on land and 5.30 kg underwater. The water has a density of 0.9951 kg/L. The male's residual lung volume is 1.57 L, and his maximum lung volume is 6.51 L. Calculate his body density. Why does a person float in spite of the fact that his or her body density is greater than that of water?

COLLABORATIVE GROUP PROJECTS

Prepare a PowerPoint, poster, or other presentation (as directed by your instructor) for presentation to the class.

79. Several popular diets are based on books, including
   a. *Dr. Atkins' New Diet Revolution*, by Robert C. Atkins,
   b. *The Zone*, by Barry Sears,
   c. *The McDougall Program for Maximum Weight Loss*, by John McDougall,
   d. *Eat More, Weigh Less*, by Dean Ornish, and
   e. *The Carbohydrate Addict's Lifespan Program*, by Richard Heller and Rachael Heller.

   For each of these, search the Web or print literature for pro and con views of each. Prepare a brief report, pro or con, on one of the diets.

80. Equations for determining maximum heart rate (MHR) and tables for evaluating fitness levels can be found in exercise physiology text books and on the Web. Use these resources to calculate your V_{O_2} max. How does your fitness level compare with that of your classmates? With that of the following athletes? (a) endurance runners and bicyclists: >75 mL/kg/min; (b) volleyball (female): 50 mL/kg/min; (c) football (male): 60–65 mL/kg/min; and (d) baseball (male): 50 mL/kg/min.

81. Choose one or more of the chemicals in Table 18.4. Use a reference such as *The Merck Index* (Reference 2) to try to determine the formula and properties of the substance.

82. Choose one or more of the sports or recreational items in Table 18.5 and try to determine what synthetic materials are used in it.

83. Only three Nobel Prizes in chemistry have been awarded to women: Marie Curie (1911), her daughter Iréne Joliot-Curie (1935), and Dorothy Crowfoot Hodgkin (1964). Write a brief essay about Hodgkin based on information sources from the library or the Internet.

84. Search the Internet or sources from the library for information on the various drugs that athletes use to enhance their performance. Use a reference such as *The Merck Index* (Reference 2) to determine the formula and function of one or more such drugs.

REFERENCES AND READINGS

1. Allen, David, and Håkan Westerblad. "Lactic Acid—The Latest Performance-Enhancing Drug." *Science*, August 20, 2004, pp. 1112–1113.

2. Budavari, S., M. J. O'Neill, and A. Smith (eds.). *The Merck Index*, 13th ed. Rahway, NJ: Merck & Co., 2001.

3. Butler, Richard. "Pushing Materials." *Chemistry & Industry*, October 6, 2003, pp. 19–20. How the search for high-performance sports materials drives innovation.

4. Gately, Iain. *Tobacco, A Cultural History of How an Exotic Plant Seduced Civilization*. New York: Grove, 2001. A fascinating look at 18,000 years of the various uses of tobacco.

5. Halford, Bethany. "Chemistry in Concert." *Chemical & Engineering News*, November 22, 2004, pp. 57–63. Polymeric materials used in musical instruments play a major role in making beautiful music.

6. Marx, Jean. "Cellular Warriors at the Battle of the Bulge." *Science*, February 7, 2003, pp. 846–849. Provides new information about molecular signals that regulate body weight.

7. Martindale, Diane. "Muscle Twitch Switch." *Scientific American*, December 2004, pp. 22–24.

8. Micheli, Lyle J. *Sports Medicine Bible for Young Athletes*. Naperville, IL: Sourcebooks, 2001.

9. Nash, J. Madeline, "Cracking the Fat Riddle." *Time*, September 2, 2002, pp. 46–55. Should you count calories or carbohydrates? Is dietary fat your biggest enemy?

10. Olshansky, S. Jay, and Bruce A. Carnes. *The Quest for Immortality: Science at the Frontiers of Aging*. New York: W. W. Norton. 2002.

11. Plowman, Sharon A., and Denise Smith. *Exercise Physiology for Health, Fitness and Performance*, 2d ed. San Francisco: Benjamin Cummings, 2003.

12. Reader's Digest Editors. *Medical Breakthroughs 2003*. Pleasantville, NY: Reader's Digest, 2003.

13. "Saving Our Children from Tobacco." *FDA Consumer*, October 1996, pp. 7–8. Each day 3000 young people in the United States become regular smokers, and 1000 of them will die prematurely from smoking-related diseases.

14. "Secondhand Smoke: Is It a Hazard?" *Consumer Reports*, January 1995, pp. 27–33.

15. Shell, Ellen Ruppel. *The Hungry Gene: The Science of Fat and the Future of Thin*. New York: Atlantic Monthly Press, 2002.

16. Smith, Wilder D. "Dietary Fats: Good or Evil?" *Today's Chemist*, November 2002, pp. 35–36.

17. "Smoking Leaves Fingerprints on DNA." *Science News*, November 2, 1996, p. 284. The definitive link between cigarette smoking and lung cancer is described.

18. United States Department of Health and Human Services, National Institutes of Health, National Heart, Lung, and Blood Institute. NIH Publication No. 03-4082, Facts about the DASH Eating Plan, Karanja N. M. et al. *Journal of the American Dietetic Association (JADA)* 8:S19–27, 1999. Revised May 2003. NIH Publication No. 03-4082.

19. U.S. Department of Health and Human Services and U.S. Department of Agriculture. *Dietary Guidelines for Americans 2005*, 6th ed. Washington DC: U.S. Government Printing Office, 2005.

20. Willett, Walter C., and Meir J. Stampfer. "Rebuilding the Food Pyramid." *Scientific American*, January 2003, pp. 64–71.

21. Willett, Walter C., P. J. Skerrett, and Edward L. Giovannucci. *Eat, Drink, and Be Healthy: The Harvard Medical School Guide to Healthy Eating*. New York: Simon & Shuster, 2001.

GREEN CHEMISTRY

Choosing a Healthy Lifestyle and Environment

Julie Manley, ACS Green Chemistry Institute

Exercise is a key component of a healthy lifestyle. It is your choice if you will exercise, how much you will exercise, what you eat, what you drink, and what you wear. We all know someone who runs a few miles and then returns to eat chocolate cake and ice cream for lunch. Did the exercise improve the person's health? The same philosophy could be applied to our impact on the environment. Hypothetically, people could wear running shorts and shoes manufactured by a process with minimal consideration for the environment. Then they could run a few miles along a busy roadway while being exposed to ground-level ozone. Upon returning from exercising, they could eat fruit containing chemical pesticides. Are we cognizant of the impact we have on the environment and, conversely, the impact the environment has on us? Are these environmental exposures acceptable to us or do we want to reduce them?

This Green Chemistry activity is intended to illustrate how your choices impact your healthy lifestyle and the environment. In Communicate Your Results, you will consider how green chemistry principles affect air quality, water sustainability, food contaminants, and clothing. The Green Chemistry culminates in an exercise to evaluate how your choices affect the environment.

WEB INVESTIGATIONS

Investigation 1
The Air We Breathe
If you live in a city and go for a run, do you choose to run down the sidewalk? Or do you find a park or school jogging track? Do you drive a fuel-efficient car or a gas guzzler? Motor vehicle exhaust and gasoline vapors contribute to the formation of ground-level ozone, which can be harmful to your health and the environment. Search using the keywords "ground level ozone" to learn more about ground-level ozone and its formation. Then search under the keywords "hybrid automobile" to learn more about the design of hybrid electric vehicles.

Investigation 2
The Water We Use
Water is used for so many different reasons: to drink, to cook, to clean, to manufacture products, and so on. Have you ever considered the impact of our water use on the environment? It is estimated that 40% of the world will live in water-scarce regions by 2025.[1] Perform a search on "water sustainability" and look for examples of businesses that have implemented such a program.

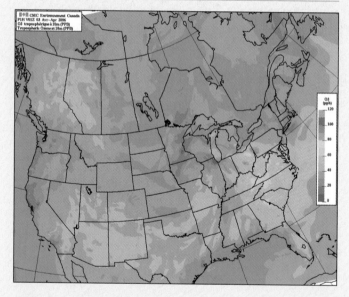

▲ The concentration of ozone at ground level varies tremendously across the United States.

Investigation 3
The Food We Eat
Pesticides are designed and utilized to kill pests while protecting the growth of a crop. However, there is a risk associated with human exposure to chemical pesticides. Perform a search on "Agraquest" to review the 2003

---

[1]World Resources Institute, United Nations Environment Programme, and the World Business Council for Sustainable Development, 2002. "Tomorrow's Markets, Global Trends, and Their Implications for Business." Paris, France.

Presidential Green Chemistry Challenge Award winner that developed a green alternative to chemical pesticides.

Investigation 4
The Clothes We Wear

To continue our discussion of pesticides, did you know that growing and harvesting cotton accounts for 25% of all the pesticides used in the United States?[2] Search using the keywords "organic cotton" to understand the opportunities and challenges associated with a market for certified organically grown and sustainable cotton. Narrow the search by including the keyword "Nike" to learn more about one company's strategy for environmental sustainability and associated new product line.

[2]Sustainable Cotton Project, http://www.sustainablecotton.org/html/who_we_are.html

COMMUNICATE YOUR RESULTS

Exercise 1
The Air We Breathe

From your research in Web Investigation 1 and information you have learned from previous chapters, determine how your actions currently contribute to the generation of ground level ozone. Additionally, consider ways you could reduce your contribution to and exposure to ground-level ozone. Summarize your conclusions in the form of a letter to the editor of a local newspaper to share your insight with the general public.

Exercise 2
The Water We Use

From your review of the business case studies on water sustainability and the information you learned in previous chapters (especially Chapter 1), choose two cases that incorporate green chemistry principles. Write a one-page assessment identifying the green chemistry principle used in each case and explaining how the principle improves water sustainability.

Exercise 3
The Food We Eat

From your evaluation of pesticide impact on human health and an example of the bio-based alternatives, decide if you accept the risk posed by exposure to pesticides as regulated by the EPA. Write a one-page summary of your decision and justify your conclusion with data from your investigation.

Exercise 4
Your Healthy Choices

After considering what you have learned from the Web Investigations and Communications in this chapter, determine if any of the information has changed your perspective on a healthy lifestyle. Will any of your decisions about water usage, food choices, or clothing manufacturers change as a result of this knowledge? How do your choices affect the environment? Summarize your conclusions in a paper, not to exceed two pages in length.

CHAPTER 19

Drugs

In the 1960s, the U.S. National Cancer Institute (NCI) started a program of screening extracts taken from a variety of natural sources for biological activity. A compound, designated Dolastatin-10, isolated from the sea hare *Dolabella auricularia*, is in clinical trials for treatment of various types of cancer. Dolastatin-10 has the molecular formula $C_{42}H_{68}N_6O_6S$ and molecular mass 785.

Chemical Cures, Comforts, and Cautions

The word *drug* originally meant a dried plant (or plant part) used as a medicine, either directly or after extracting active ingredients as a tea. Today the word has a negative connotation for many people because it brings up the idea of illegal narcotics, back-street dope dealers, all kinds of substance abuse, and jail sentences. On the other hand, many people literally owe their lives to drugs. Drugs can kill bacteria, lower blood pressure, prevent seizures, relieve allergies, and do any number of remarkable things.

A **drug** is a chemical substance that affects the functioning of living things and is used to relieve pain, to treat an illness, or to improve one's state of health or well-being. Many drugs are still obtained from plants and other natural sources, including the following obtained from plants.

- Quinine, the antimalarial drug, from the bark of *Cinchona* species;
- Morphine, the analgesic, from the opium poppy;
- Digoxin, for heart disorders, from *Digitalis purpurea*;
- Vinblastine and vincristine, anticancer agents isolated from the Madagascar periwinkle, *Catharanthus roseus*;
- Reserpine, the antihypertensive agent, from *Rauwolfia serpentina*, traditionally used for snakebites and other ailments;
- The anticancer drug paclitaxel (Taxol) initially isolated from the bark of the Pacific yew tree *Taxus brevifolia*;
- Ephedrine, an antiasthma agent, from *Ephredra sinica*; and
- Tubocurarine, the muscle relaxant, from *Chondrodendron* and *Curarea species*, used by Amazon natives in the arrow poison curare.

Many other drugs are obtained from microorganisms. Some important ones are

- Antibacterial agents (penicillins) and cholesterol-lowering agents (statins) from *Penicillium* species;
- Immunosuppressants, such as cyclosporins, from *Streptomyces* species;
- A potential new antidiabetic agent from a *Pseudomassaria* fungal species found in the African rainforest.

▲ Opium poppy.

▲ Madagascar periwinkle.

▲ Pacific yew tree *Taxus brevifolia*.

577

▲ Caribbean gorgonian coral *Pseudopterogorgia elisabethae.*

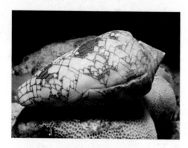

▲ Cone snail *Conus textile.*

▲ Some cults in ancient Greece worshiped opium. This Greek coin from that period shows an opium poppy capsule at the nape of the warrior's neck.

Drug industry sales worldwide in 2004 were $550 billion, with $250 billion in North America. Americans had 3.4 billion prescriptions filled in 2004, an average of 12 per person. They spent about $221 billion for prescription drugs.

Another vast resource for new drugs is marine organisms. Some drugs from the sea are

- The anticancer drug dolastatin-10 from a sea hare (page 576);
- Analgesic and anti-inflammatory drugs called pseudopterosins, from the Caribbean gorgonian *Pseudopterogorgia elisabethae*;
- An anti-inflammatory agent, manoalide, from the sponge *Luffarriella variabilis*; and
- Ziconotide, a drug for treatment of severe pain, derived from cone snail venom.

Many other drugs are *semisynthetic*, made by a chemical modification of a substance from a natural source. Examples include

- Salicylic acid (from willow bark) is converted to acetylsalicylic acid (aspirin).
- Morphine (from opium poppies) is converted to heroin.
- Lysergic acid (from ergot fungus) is converted to LSD.

Today there are many completely synthetic drugs on the market. These substances are found in nearly all categories, from antibacterial drugs to sleeping pills to narcotics such as methadone and fentanyl.

Drugs are used to treat, diagnose, and prevent diseases. People use drugs for many reasons—to relieve pain, to fight infections, to stay awake, to get to sleep, to calm anxiety, to prevent conception, and to treat various conditions from arthritis to heart disease to pneumonia to cancer. There are also some drugs that are used for the wrong reasons, and they can sometimes hurt much more than they help.

The use of drugs by humans dates to prehistoric times and extends to all cultures. Most societies used alcohol, and the use of marijuana goes back to at least 3000 B.C.E. The narcotic effect of the opium poppy was known to the Greeks in the third century B.C.E, and the Indians of the Andes Mountains have long chewed leaves of the coca plant because of the stimulating effect of the cocaine.

19.1 SCIENTIFIC DRUG DESIGN

In 1904 Paul Ehrlich (1854–1915), a German chemist, found that certain dyes used to stain bacteria to make them more visible under a microscope could also be used to kill the bacteria. He used dyes against the organism that causes African sleeping sickness. He also prepared an arsenic compound that proved effective against the organism that causes syphilis. Ehrlich realized that certain chemicals were more toxic to disease organisms than to human cells and could therefore be used to control or cure infectious diseases. He coined the term **chemotherapy** (from "chemical therapy"). Ehrlich was awarded the Nobel Prize in physiology or medicine in 1908.

The use of drugs is not new, but never before has there been such a vast array of them. There are dozens of pharmaceutical companies producing thousands of drugs. Vast research laboratories develop new drugs each year. Many drugs are sold under generic names and several different brand names, resulting in thousands of prescription medicines. A few hundred medicines are available over the counter (OTC), but there are many brand names and various combinations. Some of these drugs are vital to the lives and good health of the people who take them. Other drugs are mainly recreational and can be quite destructive. Many of these are addictive, and most are also illegal. In this chapter we look at some of the many drugs that are available today.

19.2 PAIN RELIEVERS: NONSTEROIDAL ANTI-INFLAMMATORY DRUGS (NSAIDS)

Acetylsalicylic acid, commonly called aspirin in the United States, is the most widely used drug in the world with 50 billion tablets consumed annually. Its history goes back at least 2400 years to the time of Hippocrates in ancient Greece, who

knew that sick people could ease their pain and lower their fever by chewing willow leaves. Even earlier, the Old Testament book of Leviticus (Chapter 23, Verse 40)
tells the Israelites to " take you on the first day the boughs of goodly trees, … and
willows of the brook." It was not until 1835 that chemists isolated the active ingredient, salicylic acid (Chapter 9), an effective **analgesic** (pain reliever), **antipyretic**
(fever reducer), and **anti-inflammatory** drug, but it also caused considerable stomach distress and bleeding. Aspirin was introduced in 1893 by the Baeyer company
in Germany. Acetylsalicylic acid had the desirable medicinal properties of salicylic
acid, but it was gentler to the stomach.

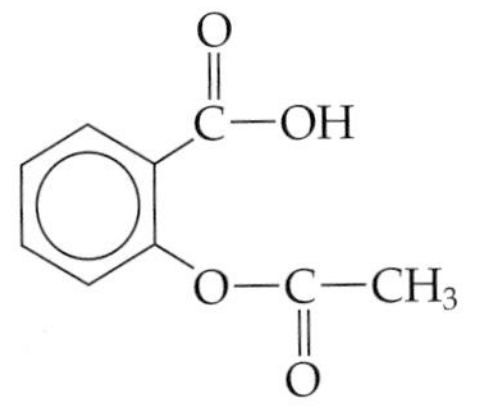

Aspirin tablets are sold under a variety of trade names and store brands. A typical aspirin tablet contains 325 mg of acetylsalicylic acid held together with an inert
binder (usually starch). Tablets with 81 mg aspirin are used by some people at risk for
stroke or heart attack to prevent the formation of blood clots and thus reduce the risk
of stroke and protect against heart attacks. "Extra strength" formulations usually
contain 500 mg. "Buffered" aspirin contains added ingredients to prevent the stomach irritation that sometimes occurs when aspirin is taken on an empty stomach.

Aspirin is one of several substances called **nonsteroidal anti-inflammatory
drugs (NSAIDs)**, a designation that distinguishes these drugs from the more potent
steroidal anti-inflammatory drugs such as cortisone and prednisone (Section 19.7).
Other NSAIDs include ibuprofen, ketoprofen, and naproxen (Figure 19.1). Acetaminophen, like aspirin, relieves minor aches and reduces fever but is not anti-
inflammatory and is therefore not an NSAID.

NSAIDs act to relieve pain and reduce inflammation by inhibiting the production of *prostaglandins*, compounds involved in sending pain messages to the brain
(Section 19.7). They don't cure whatever is causing the pain; they merely "kill the
messenger." Inflammation is caused by an overproduction of prostaglandin derivatives, and inhibition of their synthesis reduces the inflammatory process.

Regrettably, for those who take NSAIDs on a regular basis, prostaglandins also
have an effect on the blood platelets, the kidneys, and the stomach lining. Suppression of these functions can lead to the side effects most often associated with taking
NSAIDs, excessive bleeding and stomach pains. About 1.5% of patients with
rheumatoid arthritis taking NSAIDs for one year had a serious stomach complication. This may seem to be a small proportion of patients, but it is a large number of
people because of the vast quantities of NSAIDs used. This **anticoagulant** action
(inhibition of the clotting of blood) may be harmful when NSAIDs are used for
treatment of arthritis, a disease characterized by the inflammation of joints and connective tissues. It can also be helpful for prevention of blood clots in people at risk
for heart disease and stroke.

Aspirin for Heart Attack and Stroke Prevention

Because NSAIDs act as anticoagulants, they should not be used by people facing
surgery, childbirth, or some other hazard involving the possible loss of blood
(NSAIDs should be withheld for at least a week before the event). On the other

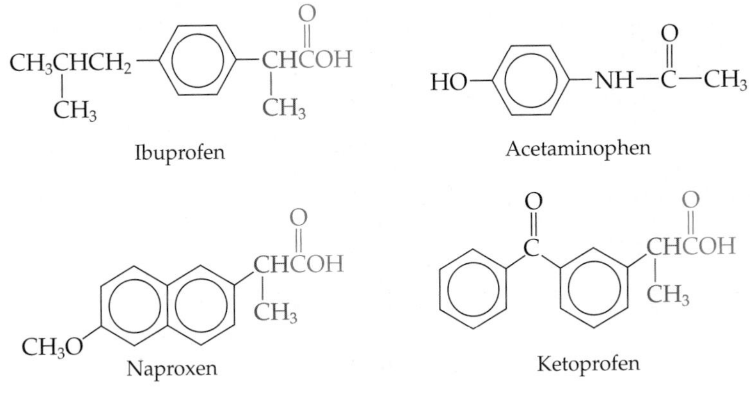

▲ Acetylsalicylic acid (aspirin).

Aspirin

Drugs such as NSAIDs are often
available in several brand names,
generic forms, and in combinations. Ibuprofen brands include
Motrin, Advil, and Nuprin; naproxen
brands are Aleve, Anaprox,
Naprelan, and Naprosyn; and ketoprofen brands are Orudis KT, Oru-
vail, and Actron. Acetaminophen is
sold under 50 brand names, including the best known, Tylenol.

Ibuprofen

Acetaminophen

Naproxen

Ketoprofen

◀ **Figure 19.1** Three NSAIDs and
acetaminophen. Ibuprofen, naprox-
en, and ketoprofen are NSAIDs; all
three are derivatives of propionic
acid (CH_3CH_2COOH). Acetaminophen (Tylenol®) is not an
NSAID.

hand, small daily doses seem to lower the risk of heart attack and stroke, presumably by the same anticoagulant action that causes bleeding in the stomach. Adult low-strength (usually 81 mg) tablets are now routinely used by people at risk for heart attack and stroke.

Reye's Syndrome

The use of aspirin to treat fevers in children suffering from the flu or chicken-pox is associated with *Reye's syndrome*. This syndrome is characterized by vomiting, lethargy, confusion, and irritability. Fatty degeneration of the liver and other organs can lead to death unless treatment is begun promptly. Just how aspirin is involved in the onset of Reye's syndrome is not known, but the correlation is quite strong. Aspirin products carry a warning indicating that they should not be used to treat children with fevers.

NSAIDs and Fever Reduction

Fevers are induced by substances called *pyrogens*. These compounds are produced by and released from leukocytes and other circulating cells. Pyrogens usually use prostaglandins as secondary mediators. Fevers therefore can be reduced by aspirin and other prostaglandin inhibitors.

Pyrogens that do not work through prostaglandins are not affected by aspirin.

Our bodies try to fight off infections by elevating the temperature. Mild fevers in adults (those below 39 °C or 102 °F) are usually best left untreated. High fevers, however, can cause brain damage and require immediate treatment.

How NSAIDs Work

Prostaglandins are produced in the body from a cell membrane component called arachidonic acid (Section 19.7). Scientists have identified several types of enzymes, called cyclooxygenases (COX), that catalyze the conversion. Two of the forms are called COX-1, found in stomach and kidney tissues where NSAID side effects occur, and COX-2, found in tissues where inflammation occurs. The older NSAIDs, such as aspirin, ibuprofen, ketoprofen, and naproxen, inhibit both COX-1 and COX-2, providing relief from pain and inflammation but also producing undesirable side effects. Newer drugs preferentially inhibit COX-2. Three such drugs, celecoxib (Celebrex), rofecoxib (Vioxx), and valdecoxib (Bextra) are anti-inflammatory agents claimed to have fewer NSAID-type side effects than the older drugs (Figure 19.2). However, Vioxx and Bextra were removed from the market because of a higher risk of stroke and heart attacks. Celebrex remains on the market but carries a new warning label, called a "Black Box" warning because it is the strongest warning that a medicine can get on its label. The U.S. Food and Drug Administration (FDA) now requires labels on all prescription NSAIDs to include a boxed warning highlighting the potential for increased risk of heart problems, as well as information regarding allergic reactions and internal bleeding.

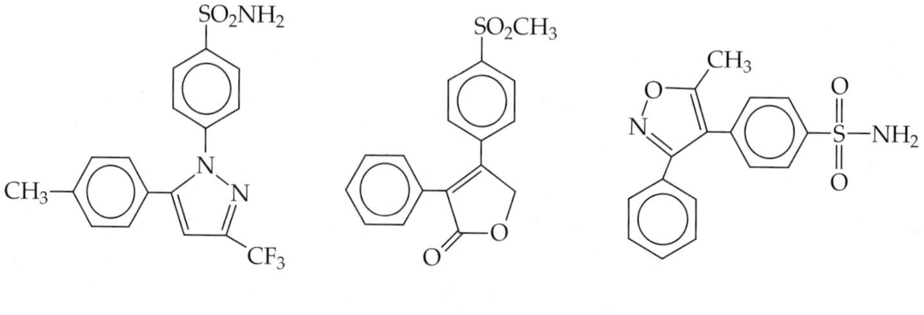

▶ **Figure 19.2** Three COX-2 inhibitors.

Celecoxib (Celebrex)   Rofecoxib (Vioxx)   Valdecoxib (Bextra)

Acetaminophen

Acetaminophen gives relief of pain and reduction of fever comparable to that of aspirin. However, unlike the NSAIDS, acetaminophen is not effective against inflammation, and it is not an anticoagulant. People allergic to aspirin or susceptible to bleeding can safely take acetaminophen, but it is of limited use to people with arthritis. Acetaminophen is often used to relieve the pain that follows minor surgery. Regular acetaminophen tablets are 325 mg, and extra-strength forms are 500 mg. Overuse of acetaminophen, especially when combined with alcohol, has been linked to liver and kidney damage. As little as 4 g (eight extra-strength acetaminophen tablets) in 24 hours, coupled with recent alcohol use, can cause severe liver toxicity.

Acetaminophen acts on a third variant of the cyclooxygenase enzyme, COX-3. Inhibition of COX-3 may represent a mechanism by which acetaminophen decreases pain and fever. However, COX-3 appears to have no role in inflammation.

Combination Pain Relievers

Many analgesic products are some combination of one or more NSAIDs, acetaminophen, caffeine, antihistamines (Section 19.3), or other drugs. Acetaminophen, for example, is found in more than 70 combination products. These combinations are available under various brand names and as store brands. Familiar brands include Excedrin, which can can be purchased in five different combinations, and Anacin, which is available in four different formulations.

Many other combination pain relievers are available. Most claim some advantage over regular aspirin. However, such advantages are marginal at best. For occasional use by most people, plain aspirin may be the cheapest, safest, and most effective product.

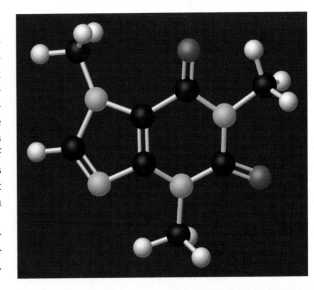

▲ Caffeine, $C_8H_{10}N_4O_2$, is a stimulant (Section 19.17) and is added to NSAIDs in some combination pain reliever formulations.

 Caffeine

"Children's" NSAIDs are typically one-quarter of the normal adult dose per tablet, with fewer tablets per bottle than their adult counterparts. This reduces significantly the possibility of an accidental overdose.

19.3 CHEMISTRY, ALLERGIES, AND THE COMMON COLD

From giant warehouse-type stores to small convenience stores, we find shelves stocked with a variety of cold and allergy medicines. They contain compounds that act as antihistamines, cough suppressants, expectorants, bronchodilators, or nasal decongestants. None of the products cure the common cold. Colds are caused by as many as 200 related viruses, and no antibiotic or other drug provides an effective cure. Most cold remedies treat just the symptoms. An FDA advisory panel reviews these substances periodically for safety and effectiveness.

Antihistamines

Many cold medicines contain an **antihistamine**, such as chlorpheniramine, diphenhydramine, and triprolidine (Figure 19.3). Antihistamines relieve the symptoms of allergies: sneezing, itchy eyes, and runny nose. They don't cure colds, but they can temporarily relieve some of the symptoms, especially those caused by allergies.

When an **allergen** (a substance that triggers an allergic reaction) binds to the surfaces of certain cells, it triggers the release of *histamine*, which causes the redness, swelling, and itching associated with allergies. Antihistamines inhibit the release of histamine, but many over-the-counter ones also enter the brain and act on the cells controlling sleep, thus making patients drowsy; loratadine (Claritin®) is an exception among the over-the-counter drugs. Prescription drugs, such as fexofenadine (Allegra®) and desloratadine (Clarinex®), inhibit the release of histamine but do not enter the brain and so do not cause drowsiness. Cetirizine (Zyrtec®), also a prescription drug, is somewhat more sedating than fexofenadine and desloratadine.

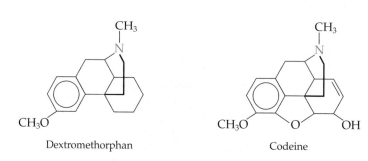

(a) Histamine

(b) Chlorpheniramine

(c) Diphenhydramine

(d) Triprolidine

▶ **Figure 19.3** Histamine (a) and three antihistamines: (b) chlorpheniramine (Chlor-Trimeton®), (c) diphenhydramine (Benadryl®), and (d) triprolidine (Actifed®). Diphenhydramine is also used as a cough suppressant.

Question: To what family of organic compounds do these three substances belong? How does brompheniramine differ from chlorpheniramine?

A 2004 study at Pennsylvania State University questioned the effectiveness of the OTC antitussives for children. The study found that diphenhydramine and dextromethorphan were no better than placebo (Section 19.20) for cough suppression in children aged 2 to 18.

Cough Suppressants

The two *antitussives* (cough suppressants) available OTC are diphenhydramine and dextromethorphan. Codeine and hydrocodone (Table 19.5) are effective antitussives but are available only by prescription; both are narcotics. The question arises, though: Should a cough really be suppressed? Ordinarily, a cough is functional; the respiratory tract uses the cough mechanism to rid itself of congestion. When a cough is dry or interferes with needed rest, however, temporary cough suppression may be advisable.

Dextromethorphan

Codeine

Expectorants and Decongestants

An *expectorant* is a substance that helps bring up mucus from the bronchial passages. The only expectorant rated safe and effective is glyceryl guaiacolate (guaifenesin), and it is only marginally effective.

Nasal *decongestants* seem to be safe and effective for occasional use, although repeated use leads to a rebound effect in which the nasal passages swell and make congestion seem worse. Examples of long-acting nasal decongestants are oxymetazoline and xylometazoline. These are effective in a few minutes and last for 6 to 12 hours. Short-acting nasal decongestants include naphazoline and phenylephrine.

Guaifenesin

Oxymetazoline

Xylometazoline

Naphazoline

Phenylephrine

How should you treat a common cold? First, you should drink plenty of liquids and get lots of rest. When you are in good physical condition and have a strong immune system, colds seem to strike less often. Frequent washing of the hands, especially when colds are prevalent, can dramatically reduce transmission of the virus. Many people claim to get cold relief by using some favorite over-the-counter drug. And, of course, there are those who still swear by chicken soup!

19.4 ANTIBACTERIAL DRUGS

A century ago, infectious diseases were the principal cause of death in the United States (Figure 19.4). At the time of the American Civil War, thousands of wounded soldiers required amputations of wounded arms or legs, and nearly 80% of the patients died after coming to a hospital, mostly from infections. Today infectious disease categories have moved toward the bottom of the top 10 list of leading causes of death. (These top 10 causes account for well over 90% of all causes of death in the United States.) Today many diseases have been brought under control by the use of *antibacterial drugs*.

Sulfa Drugs

The first antibacterial drugs were *sulfa drugs*, the prototype of which was discovered in 1935 by the German chemist Gerhard Domagk (1895–1964). Sulfa drugs were used extensively during World War II to prevent wound infections. Many soldiers lived who would have died in earlier wars.

Sulfanilamide, the simplest sulfa drug, was one of the first to have its action understood at the molecular level. Its effectiveness is based on a case of mistaken identity. Bacteria need *para*-aminobenzoic acid (PABA) to make folic acid, which is essential for the formation of certain compounds the bacteria require for proper growth. But bacterial enzymes can't tell the difference between sulfanilamide and PABA because the substances are so similar in structure.

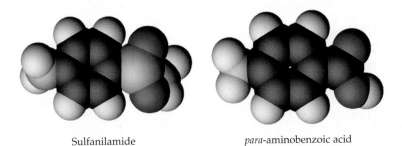

Sulfanilamide   *para*-aminobenzoic acid

When sulfanilamide is applied to an infection in large amounts, bacteria incorporate it into pseudo–folic acid molecules that cannot act normally; hence, the bacteria cease to grow. Of the thousands of sulfanilamide analogs that have been developed and tested, only a few are used today. The structures of two common ones are given in the margin. Some sulfa drugs tend to cause kidney damage or other problems.

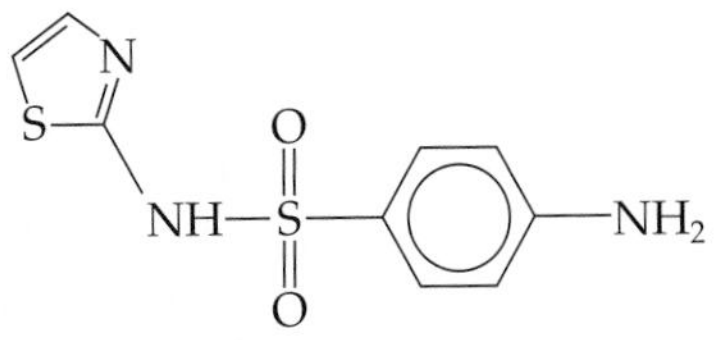

Sulfathiazole

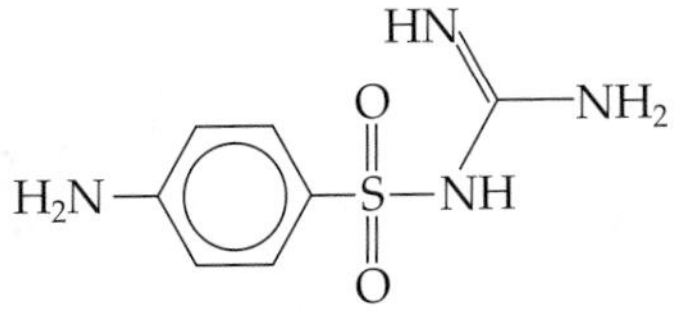

Sulfaguanidine

Penicillins

The next important discovery was that of penicillin, an antibiotic. An **antibiotic** is a soluble substance (derived from molds or bacteria) that inhibits the growth of other microorganisms. Penicillin was first discovered in 1928 but was not tried on humans until 1941. Alexander Fleming (1881–1955), a Scottish microbiologist then working at the University of London, first observed the antibacterial action of a

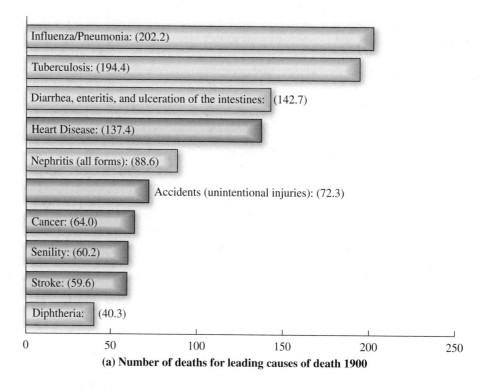

(a) Number of deaths for leading causes of death 1900

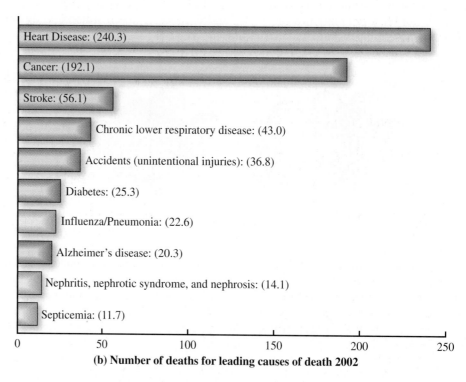

(b) Number of deaths for leading causes of death 2002

▲ **Figure 19.4** (a) In 1900 five of the ten leading causes of death—and three-fifths of all deaths—in the United States were infectious diseases. (b) The leading causes of death in 2002 were so-called lifestyle diseases: heart disease, stroke, and diabetes are related in part to our diets; cancer and lung diseases to cigarette smoking; and many accidents to excessive consumption of alcohol.

mold, *Penicillium notatum*. Fleming was studying an infectious bacterium, *Staphylococcus aureus*, and one of his cultures became contaminated with a blue mold. Contaminated cultures generally are useless. Most investigators probably would have destroyed the culture and started over, but Fleming noted that bacterial colonies had been destroyed in the vicinity of the mold.

Fleming was able to make crude extracts of the active substance. This material, later called *penicillin*, was further purified and improved by Howard Florey (1898–1968), an Australian, and Ernst Boris Chain (1906–1979), a refugee from Nazi Germany, both working at Oxford University. Fleming, Florey, and Chain shared the 1945 Nobel Prize in physiology or medicine for their work on penicillin.

Penicillin is not a single substance but a group of related compounds (Figure 19.5). By designing molecules with different structures, chemists can change the properties of the drugs. The penicillins now produced vary in effectiveness. Some can be taken orally; others must be injected. Bacteria resistant to one penicillin may be killed by another. Amoxicillin has a broad spectrum of activity against many types of microorganisms. It is among the top 10 most prescribed drugs with more than 40 million prescriptions in 2004. Nine million prescriptions were written for penicillin V.

Penicillin works by inhibiting enzymes that the bacteria use to make their cell walls. The bacterial cell walls are made up of *mucoproteins*, polymers in which amino sugars are combined with protein molecules. Penicillin prevents cross-linking between these large molecules. This leaves holes in their cell walls, and the bacteria swell and rupture. Cells of higher animals have only external membranes; they do not have mucoprotein walls and are therefore not affected by penicillin. Thus, penicillin can destroy bacteria without harming human cells. However, many people can't take penicillin because they are allergic to it.

In their early days, antibiotics were known as miracle drugs. The number of deaths from blood poisoning (septicemia), pneumonia, and other infectious diseases was reduced substantially by the use of antibiotics. Prior to 1941 a person with a major bacterial infection almost always died. Today, such deaths are rare except for those ill with other conditions such as AIDS. Six decades ago, pneumonia was a dreaded killer of people of all ages. Today, it kills mainly the elderly and those with AIDS. Antibiotics have indeed worked miracles in our time, but even miracle drugs are not without problems. It wasn't long after these drugs were first used that disease organisms began to develop strains resistant to them. One strain can spread antibiotic resistance to another by sharing genes, and the emergence of more and more resistant strains of bacteria is a serious threat to world health. For example, tuberculosis was once almost completely wiped out in developed countries. Now drug-resistant tuberculosis is now rampant in Russia and among AIDS patients everywhere.

Another example involves erythromycin, an antibiotic obtained from *Streptomyces erythreus*. As long as it had only limited use, erythromycin could handle all strains of staphylococci. After it was put into extensive use, resistant strains

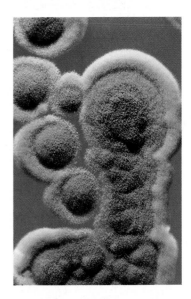

▲ Penicillin molds. These symmetrical colonies of mold are *Penicillium chrysogenum*, a mutant form of which now produces nearly all commercial penicillin.

Penicillin G

(a) General formula

(b) Penicillin G

(c) Amoxicillin

▲ **Figure 19.5** Penicillins. A general formula where R is a variable side chain (a). Penicillin G (b) is generally considered the most effective penicillin but can only be given by injection. Amoxicillin (c) can be administered orally.

began to appear. Staph infections are now a serious problem in hospitals. People who go to a hospital to be cured sometimes develop serious bacterial infections instead. Some even die of staph infections picked up at hospitals. A Northwestern Memorial Hospital (Chicago) study estimates staph infections cause 12,000 deaths a year and cost U.S. hospitals $9.5 billion.

Cephalosporins

In a race to stay ahead of resistant strains of bacteria, scientists continue to seek new antibiotics. Penicillins have been partially displaced by related compounds called *cephalosporins* such as cephalexin (Keflex). Unfortunately, some strains of bacteria are resistant to cephalosporins.

Cephalexin

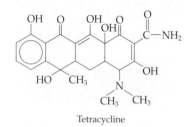

Tetracycline

Aureomycin
(chlortetracycline)

Terramycin
(oxytetracycline)

▲ **Figure 19.6** Three tetracycline antibiotics.

Tetracyclines

Another important development in the field of antibiotics was the discovery of a group of compounds called *tetracyclines*. The first of these four-ring compounds, aureomycin, was isolated in 1948 by Benjamin Duggor (1872–1956) from *Streptomyces aureofaciens*. Scientists at Pfizer Laboratories isolated terramycin from *Streptomyces rimosus* in 1950, and both drugs were later found to be derivatives of tetracycline, a compound now obtained from *Streptomyces viridifaciens*. All three compounds (Figure 19.6) are **broad-spectrum antibiotics**, so called because they are effective against a wide variety of bacteria.

Tetracyclines bind to bacterial ribosomes, inhibiting bacterial protein synthesis and blocking growth of the bacteria. Tetracyclines do not bind to mammalian ribosomes and thus do not affect protein synthesis in host cells. Several disease-causing organisms have developed strains resistant to tetracyclines.

When given to young children, tetracyclines can cause the discoloration of permanent teeth, even though the teeth may not appear until several years later. Women are usually told to avoid tetracyclines during pregnancy for the same reason. This probably results from the interaction of tetracyclines with calcium during the period of tooth development. Calcium ions in milk and other foods combine with hydroxyl groups on tetracycline molecules.

Fluoroquinolones

Fluoroquinolone antibiotics were first introduced in 1986. They act against a broad spectrum of bacteria and seemingly have few side effects. A major use is against bacteria with penicillin resistance. Fluoroquinolones act by inhibiting bacterial DNA replication through interference with the action of an enzyme called DNA gyrase. Humans do not have this enzyme so fluoroquinolones do not harm human cells. Because their mechanism of action is different from that of other antibiotics, it was hoped that resistance to these drugs would be slow to emerge. However, resistance to fluoroquinolones is already being seen. Ciprofloxacin (Cipro®) is the leading fluoroquinolone, with more than 9 million prescriptions written in 2004 in the United States.

Ciprofloxacin (Cipro®)

19.5 VIRUSES AND ANTIVIRAL DRUGS

For most of us, antibiotics have taken the terror out of bacterial infections such as pneumonia and diphtheria. We worry about resistant strains of bacteria, but generally these problems are not insurmountable. However, viral diseases cannot be

Structure–Function Relationships

Fluoroquinolones are of particular interest to chemists and pharmacologists because many of their structural features are correlated with their activity, making possible a rational, logical approach to synthesis of new drugs in this class.

The basic skeletal structure for fluoroquinolones is

The fluorine atom at position 6 is essential for broad antimicrobial activity when taken orally. Effectiveness depends on a carboxyl group at position 3 and a carbonyl oxygen at position 4. These are responsible for binding to the bacterial DNA complex. A nitrogen-containing ring structure at position 7 or a methoxy group at position 8 broadens the range of bacteria affected. If position 7 has a bulky substituent, the drug has fewer side effects on the nervous system.

cured by antibiotics, and viral infections from colds and influenza to herpes and **acquired immune deficiency syndrome (AIDS)** still plague us. Some viral infections—such as poliomyelitis, mumps, measles, and smallpox—can be prevented by vaccination. Influenza vaccines are quite effective against common recurrent strains, but there are many different strains of flu viruses, and new ones appear periodically.

DNA Viruses and RNA Viruses

Viruses are composed of nucleic acids and proteins (Figure 19.7). They have an external coat with a repetitive pattern of protein molecules. Some coats also include a lipid membrane, and others have sugar–protein combinations called glycoproteins. The genetic material of a virus is either DNA or RNA. A *DNA virus* enters a host cell where the DNA is replicated, and it directs the host cell to produce viral proteins. The viral proteins and viral DNA assemble into new viruses that are released by the host cell. These new viruses can then invade other cells and continue the process.

Most *RNA viruses* use their nucleic acids in much the same way. The virus penetrates a host cell, where the RNA strands are replicated and induce the synthesis of viral proteins. The new RNA strands and viral proteins are then assembled into new viruses. Some RNA viruses, called **retroviruses**, synthesize DNA in the host cell. This process is the opposite of the transcription of a DNA code into RNA (Chapter 15) that normally occurs in cells. The synthesis of DNA from an RNA template is catalyzed by an enzyme called *reverse transcriptase*. The human immunodeficiency virus (HIV) that causes AIDS is a retrovirus. HIV invades and eventually destroys *T cells*, white blood cells that normally help protect the body from infections. With the T cells destroyed, the AIDS victim often succumbs to pneumonia or some other infection.

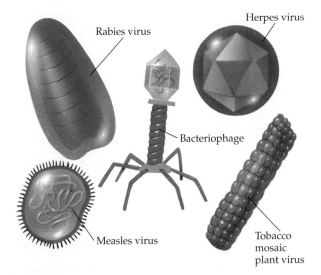

▲ **Figure 19.7** Viruses come in a variety of shapes, which are determined by their protein coats.

Worldwide, as many as 50 million people are living with AIDS. More than 20 million are already dead from AIDS, and 3 million more die each year. Some 5 million people become infected with HIV each year. In the United States, about 900,000 people have AIDS, and an estimated 1–2 million persons have HIV infection that has not yet developed into AIDS. Another 40,000 new HIV infections occur each year.

Antiviral Drugs

Scientists have developed a variety of drugs that are effective against some viruses. None provide cures. Structures of some common antiviral agents are given in Figure 19.8. Amantadine (Symmetrel®) and is used to prevent or treat certain kinds of influenza (flu). It is given either alone or in combination with flu shots. Acyclovir (Zovirax®) is used to treat chickenpox, shingles, cold sores, and the symptoms of genital herpes.

Antiretroviral drugs prevent the reproduction of retroviruses such as HIV and are used against AIDS. Three important kinds are

- *Nucleoside analogs,* or nucleoside reverse transcriptase inhibitors (NRTIs), such as didanosine (ddI, Videx®), lamivudine (3TC, Epivir®), stavudine (d4T, Zerit®), zalcitabine (ddC, Hivid®), and zidovudine (AZT, Retrovir®). Substitution of an analog for a nucleoside in the viral DNA cripples it, slowing down its replication.

- *Nonnucleoside reverse transcriptase inhibitors (NNRTIs)* stop the reverse transcriptase from working properly to make more virus. NNRTIs include delavirdine (Rescriptor®), loviride, and nevirapine (Viramune®).

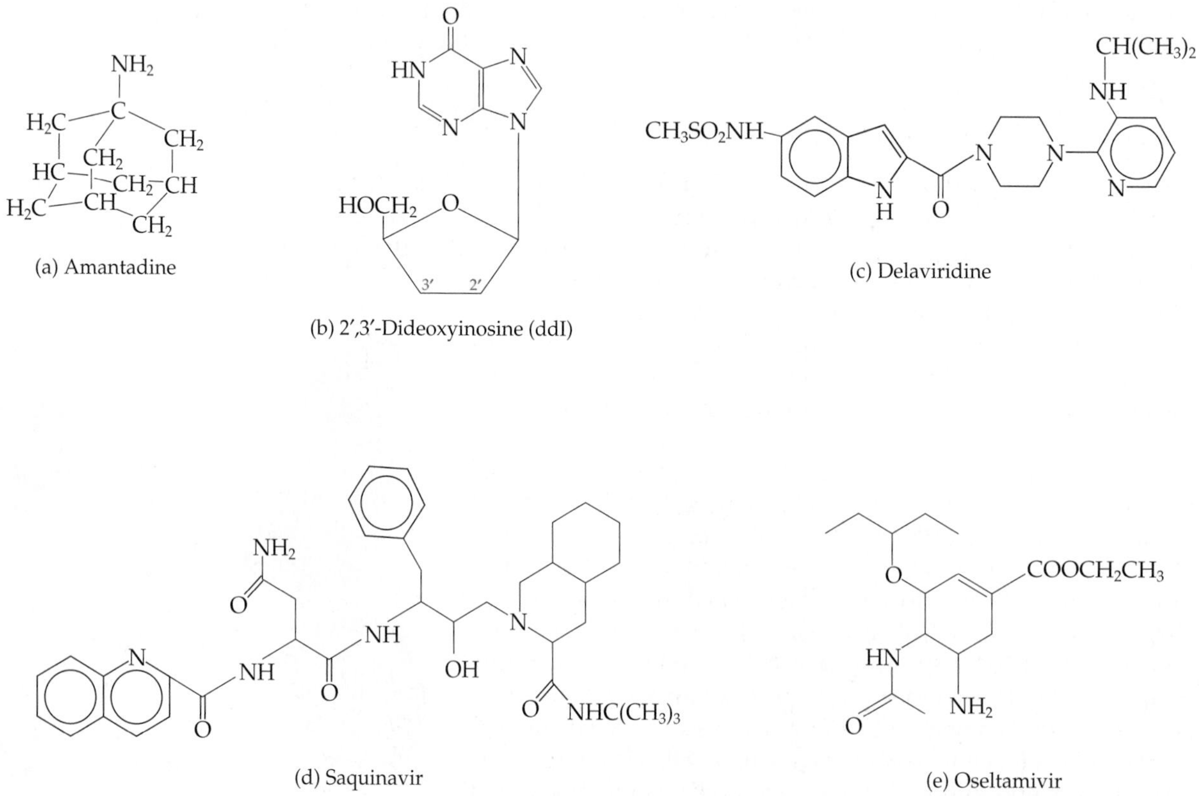

(a) Amantadine

(b) 2′,3′-Dideoxyinosine (ddI)

(c) Delaviridine

(d) Saquinavir

(e) Oseltamivir

▲ **Figure 19.8** Some antiviral drugs. Amantadine (a) is completely synthetic. Didanosine (ddI) (b) is a nucleoside analog or nucleoside reverse transcriptase inhibitors (NRTI). Delavirdine (c) is a nonnucleoside reverse transcriptase inhibitor (NNRTIs). Saquinavir (d) is a protease inhibitor. Oseltamivir (e) (Tamiflu) can prevent influenza if it is given in early stages.

- *Protease inhibitors* block the enzyme protease so that new copies of the virus can't infect new cells. Protease inhibitors include ritonavir (Norvir®), indinavir (Crixivan®), nelfinavir (Viracept®), and saquinavir (Invirase®).

Combinations of AIDS drugs seem to be more effective than any one of them alone. For example, Trizivir® contains abacavir, lamivudine (3TC), and zidovudine (azidothymidine). Unfortunately, these drugs are too toxic to use in quantities sufficient to stop virus replication completely.

Basic Research and Drug Development

Scientists had to understand the normal biochemistry of cells before they could develop drugs to treat diseases caused by viruses. Gertrude Elion and George Hitchings of Burroughs Wellcome Research Laboratories in North Carolina and James Black of Kings College in London did much of the basic biochemical research that led to the development of antiviral drugs and many anticancer drugs (Section 19.8). They determined the shapes of cell membrane receptors, and they learned how normal cells work. They and other scientists were then able to design drugs to block receptors in infected cells. Elion, Hitchings, and Black shared the 1988 Nobel Prize in physiology or medicine for this work.

Drug design in its early days was often a hit-or-miss procedure. Today scientists use powerful computers to design molecules to fit receptors, and drug design is becoming a more precise science.

Combinatorial Chemistry

For a century now chemists have synthesized one substance at a time and then tested it for biological activity. Then they made a new compound by varying the structure a bit, hoping to obtain a substance with more desirable properties and less undesirable attributes. Now many chemists use *combinatorial chemistry*, a technique in which a set of starting chemicals are reacted in all possible combinations. For example, a set of 10 compounds reacted with 10 different reagents gives 100 possible products. The products are then tested for biological activity in much the same manner. The process is highly automated.

These robotic techniques yield enormous numbers of different compounds whose properties can vary widely. Most hold little promise and are discarded or filed away. Promising substances are tested further. The very few with the most potential eventually start the long, expensive route to human testing and possible FDA approval.

19.6 CHEMICALS AGAINST CANCER

Despite decades of "war" on cancer, it remains a dread disease. Each year in the United States, more than 1.3 million new cases are diagnosed and cancer kills more than half a million people. Scientists have made progress, but much remains to be done. A major problem is that the drugs that kill cancer cells damage normal cells as well, and a primary aim of cancer research is to find a way to kill cancerous tissue without killing too many normal cells. Treatment with drugs, radiation, and surgery yields a high rate of cure for some kinds of cancer. For example, thyroid cancer has an overall five-year survival rate of 96%. For other types, such as lung cancer, the rate of cure is still quite low: The overall five-year survival rate is only 15%, mainly because the lung cancer is often advanced at the time of diagnosis. The five-year survival rate for all cancers is 63%. Dozens of anticancer drugs are used widely, and this number will no doubt increase as our understanding of basic cell chemistry increases.

Cancer chemotherapy affects any body cells that undergo rapid replacement, not just cancer cells. Included are the cells that line the digestive tract and those that produce hair. Side effects of chemotherapy therefore include nausea and loss of hair.

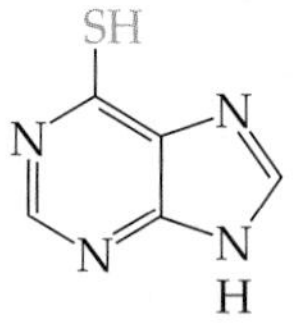

6-Mercaptopurine

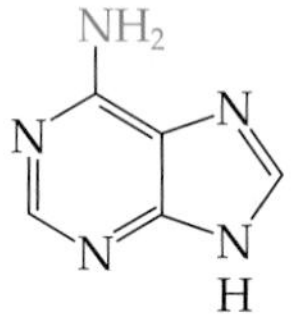

Adenine

Antimetabolites: Inhibition of Nucleic Acid Synthesis

An **antimetabolite** is a compound that closely resembles a substance essential to normal body metabolism and therefore interferes with physiological reactions involving it. Rapidly dividing cells, characteristic of cancer, require an abundance of DNA. Cancer antimetabolites block DNA synthesis and therefore block the increase in the number of cancer cells. Because cancer cells are undergoing rapid growth and cell division, they are generally affected to a greater extent than normal cells.

Gertrude Elion and George Hitchings patented 6-mercaptopurine (6-MP) in 1954. Prior to 6-MP, half of all children with acute leukemia died within a few months. Combined with other medications, 6-MP was able to cure approximately 80% of child leukemia patients. This compound can substitute for adenine in a phosphate–sugar–base unit called a *nucleotide* (Chapter 15), and the pseudonucleotide then inhibits the synthesis of nucleotides incorporating adenine and guanine. This slows DNA synthesis and cell division, thus inhibiting the multiplication of cancer cells.

Two other prominent antimetabolites are 5-fluorouracil and its deoxyribose derivative, 5-fluorodeoxyuridine. In the body, both these compounds are incorporated into a nucleotide, and the fluorine-containing nucleotide inhibits the formation of thymine-containing nucleotides required for DNA synthesis. Thus, both compounds slow the division of cancer cells. They have been employed against a variety of cancers, especially those of the breast and the digestive tract.

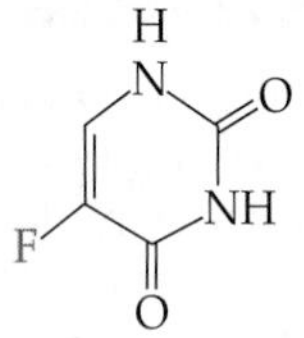

5-Fluorouracil 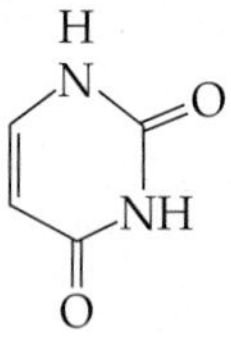   Uracil 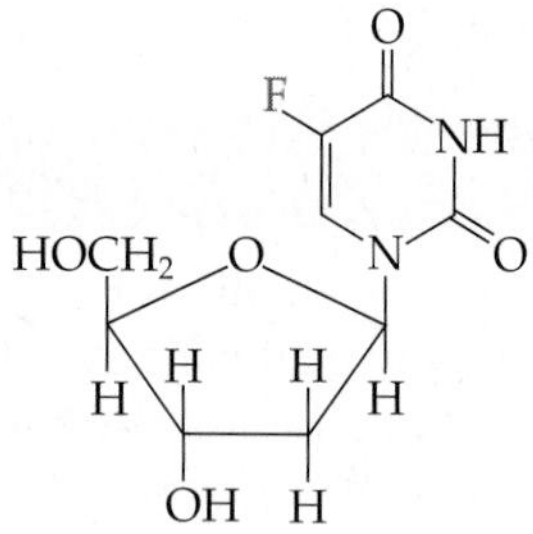   5-Fluorodeoxyuridine

The antimetabolite methotrexate acts in a somewhat different manner. Note the similarity between its structure and that of folic acid. Like the pseudofolic acid formed from sulfanilamide, methotrexate competes successfully with folic acid for an enzyme but cannot perform the growth-enhancing function of folic acid. Again, cell division is slowed and cancer growth is retarded. Methotrexate is used frequently against leukemia.

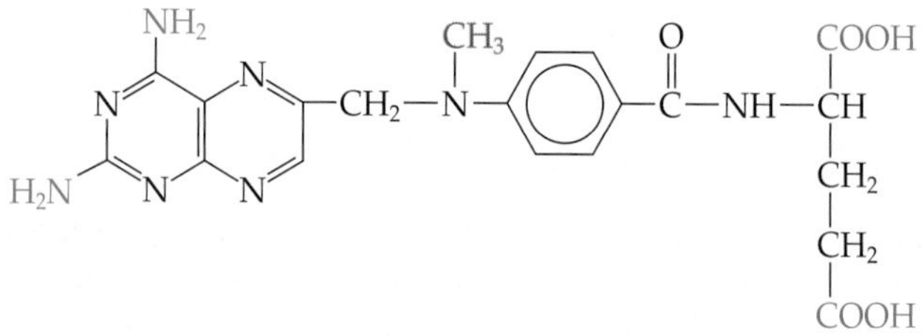

Methotrexate

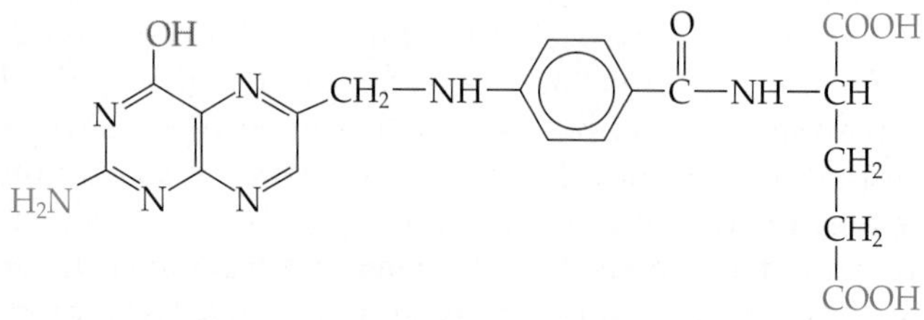

Folic acid

Cisplatin: The Platinum Standard of Cancer Treatment

The platinum-containing compound cisplatin, $PtCl_2(NH_3)_2$, is a prominent anticancer drug. Cisplatin was first synthesized in 1844. Its biological effect was discovered by accident in 1965 by Barnett Rosenberg at Michigan State University. Rosenberg was using inert platinum electrodes in an experiment designed to measure the effect of electrical currents on cell division. He found that *E. coli* were growing in length but failing to divide, reaching 300 times their normal length. After much further study, Rosenberg and his team found that the electric current had caused a chemical reaction between the platinum in the electrodes and nutrients in the solution containing the bacteria. A new compound, now called cisplatin, was produced. Because the compound inhibited cell division, Rosenberg reasoned that it might be effective as an anticancer drug. It is. Cisplatin binds to DNA and blocks its replication. It is widely used to treat cancers such as those of the ovaries, testes, uterus, head, neck, breast, lung, and advanced bladder cancers.

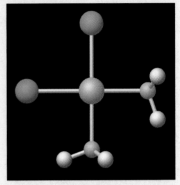

Cisplatin

Alkylating Agents: Turning War Gases on Cancer

Alkylating agents are highly reactive compounds that can transfer alkyl groups to compounds of biological importance. These foreign alkyl groups then block the usual action of the biological molecules. Some alkylating agents are used against cancer. Typical among these are nitrogen mustards, compounds that arose out of chemical warfare research.

The original "mustard gas" was a sulfur-containing blister agent used during World War I. Contact with either the liquid or the vapor causes blisters that are painful and slow to heal. It is easily detected, however, by its garlic or horseradish odor. Mustard gas is denoted by the military symbol H.

$$Cl-CH_2CH_2-S-CH_2CH_2-Cl$$

Mustard gas
H

Nitrogen mustards (symbol HN) were developed about 1935. Though not quite as effective overall as mustard gas, nitrogen mustards produce greater eye damage and don't have an obvious odor. Structurally, nitrogen mustards are chlorinated amines.

$$CH_3CH_2-N{\overset{\displaystyle CH_2CH_2-Cl}{\underset{\displaystyle CH_2CH_2-Cl}{}}} \qquad CH_3-N{\overset{\displaystyle CH_2CH_2-Cl}{\underset{\displaystyle CH_2CH_2-Cl}{}}} \qquad Cl-CH_2CH_2-N{\overset{\displaystyle CH_2CH_2-Cl}{\underset{\displaystyle CH_2CH_2-Cl}{}}}$$

HN_1 HN_2 HN_3

In cancer treatment, nitrogen mustards, which are bifunctional alkylating agents, act by cross-linking two DNA strands. This prevents or hinders replication and thus impedes growth of a cancer. Knowledge gained through science is neither good nor evil. The same knowledge—in this case, the ability to make nitrogen mustards—can be used either for our benefit or for our destruction.

The nitrogen mustard of choice for cancer therapy nowadays is often a compound called cyclophosphamide (Cytoxan). It is used to treat Hodgkin's disease, lymphomas, leukemias, and other cancers.

Alkylating agents can cause cancer as well as cure it. For example, the nitrogen mustard HN_2 causes lung, mammary, and liver tumors when injected into mice; yet it can be used with some success in the management of certain human tumors. There is still a lot of mystery—and seeming contradiction—regarding the causes and cures of cancer.

Cyclophosphamide

Chemical Warfare

Chemical agents were used extensively during World War I. More than 30 such substances were employed, killing 91,000 and wounding 1.2 million (many of them for life). Fritz Haber supervised the release of chlorine gas in the first attack by the Germans. Adolf Hitler was among those wounded when the British retaliated with phosgene a few days later. Haber considered gas warfare to be "a higher form of killing." Fortunately, his views have not become widely accepted.

The use of chemical warfare agents was largely avoided during World War II. However, they have been used in smaller wars, such as by Iraq against Iran in the 1980s, and by the Iraqi government against Kurdish rebels. They remain today as one of the *weapons of mass destruction (WMDs)* that can kill huge numbers of people and might be used by a rogue nation or a terrorist group. (The other WMDs are nuclear devices and biological agents such as anthrax and botulism.)

Miscellaneous Anticancer Agents

There are many other anticancer agents. Alkaloids from vinca plants have been shown to be effective against leukemia and Hodgkin's disease. Paclitaxel (Taxol), obtained from the Pacific yew tree, is effective against cancers of the breast, ovary, and cervix. Several antibiotics have been found to kill cancer cells as well as bacteria. Actinomycin, obtained from the molds *Streptomyces antibioticus* and *S. parvus*, is used against Hodgkin's disease and other types of cancer. It is quite effective but extremely toxic. Actinomycin acts by binding to the double helix of DNA, thus blocking the replication of RNA on the DNA template. Protein synthesis is inhibited.

Chemotherapy is only part of the treatment for cancer. Surgical removal of tumors and radiation treatment remain major weapons in the war on cancer. It is unlikely that a single agent will be found to cure all cancers. Perhaps a greater hope lies in the prevention of cancer. Deaths from smoking-related cancers still account for 80% of all cancer deaths. Much active research is underway on the mechanisms of carcinogenesis. The more we learn about what causes cancer, the better equipped we will be in trying to prevent it.

19.7 HORMONES: THE REGULATORS

Before we can discuss the next group of drugs, we need to take a brief look at the human endocrine system and some of the chemical compounds, called hormones, this system manufactures. A **hormone** is a chemical messenger produced in the endocrine glands (Figure 19.9). Those released in one part of the body send signals for profound physiological changes in other parts of the body. By causing reactions to speed up or slow down, hormones control growth, metabolism, reproduction, and many other functions of body and mind. Some of the important human hormones and their physiological effects are listed in Table 19.1 (p. 594). Still other hormones are formed in tissues of the heart, liver, kidney, gut, and placenta.

Prostaglandins: Hormone Mediators

A **prostaglandin** is a hormonelike lipid compound derived from a fatty acid. Each prostaglandin molecule has 20 carbon atoms, including a five-carbon ring. Prostaglandins act much like hormones in that they act on target cells. However, they differ from hormones in that they (1) act near the site where they are produced, (2) can have different effects in different tissues, and (3) are rapidly metabolized. Prostaglandins are synthesized in the body from arachidonic acid (Figure 19.10). There are six primary prostaglandins, and these potent biological chemicals are widely distributed throughout the body. Extremely small doses can elicit marked changes. Many others have been identified, and hundreds of synthetic analogs have been produced.

Major Endocrine Glands

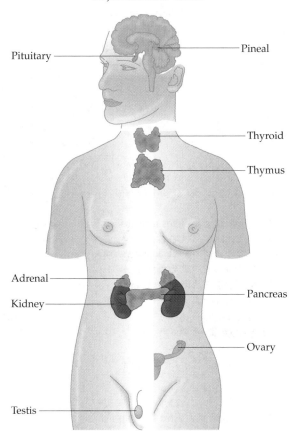

◀ **Figure 19.9** The approximate locations of the eight major endocrine glands in the human body.

Prostaglandins act as mediators of hormone action. They regulate such things as smooth muscle activity, blood flow, and secretion of various substances. This range of physiological activity has led to the synthesis of hundreds of prostaglandins and their analogs. Prostaglandin E_2 (PGE$_2$), also known as dinoprostone (Cervidil®), is used to induce labor. PGE$_1$ was used to treat erectile dysfunction, but it had to be injected into the penis prior to intended sexual intercourse. It has largely been replaced for this purpose by Viagra, an oral medication. Other prostaglandins can be used clinically to lower or increase blood pressure, to inhibit stomach secretions, to relieve nasal congestion, to provide relief from asthma, to treat glaucoma, to treat pulmonary hypertension, and to prevent formation of the blood clots associated with heart attacks and strokes.

PGF$_{2\alpha}$ is used in cattle breeding. A prize cow is treated with a hormone and with PGF$_{2\alpha}$ to induce the release of many ova. The ova are then fertilized with sperm from a champion bull, and the developing embryos are implanted in less valuable cows. This procedure enables a farmer to get several calves a year from one outstanding cow.

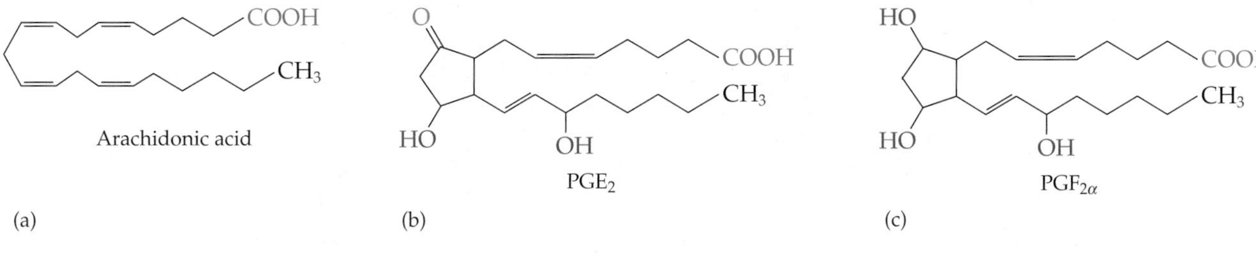

(a) Arachidonic acid

(b) PGE$_2$

(c) PGF$_{2\alpha}$

▲ **Figure 19.10** Prostaglandins are derived from arachidonic acid, an unsaturated carboxylic acid with 20 carbon atoms (a). Two representative prostaglandins are shown here: (b) prostaglandin E_2 and (c) prostaglandin $F_{2\alpha}$.

TABLE 19.1 Some Human Hormones and Their Physiological Effect

Name	Chemical Nature	Function(s)
Pituitary Hormones		
Vasopressin	Nonapeptide	Stimulates contractions of smooth muscle; regulates water uptake by the kidneys
Oxytocin	Nonapeptide	Stimulates contraction of the smooth muscle of the uterus; stimulates secretion of milk
Growth hormone (GH) (or somatotropin)	Protein	Controls general body growth; controls bone growth
Thyroid-stimulating hormone (TSH)	Protein	Stimulates growth of the thyroid gland and production of thyroxin
Adrenocorticotrophic hormone (ACTH)	Protein	Stimulates growth of the adrenal cortex and production of cortical hormones
Follicle-stimulating hormone (FSH)	Protein	Stimulates growth of follicles in ovaries of females and of sperm cells in testes of males
Luteinizing hormone (LH)	Protein	Controls production and release of estrogens and progesterone from ovaries and testosterone from testes
Prolactin	Protein	Maintains the production of estrogens and progesterone; stimulates the formation of milk
Hypothalamic Hormones		
Corticotropin-releasing factor (CRF)	Protein	Acts on corticotrope to release ACTH and β-endorphin (lipotropin)
Gonadotropin-releasing factor (GnRF)	Decapeptide	Acts on gonadotrope to release LH and FSH
Somatostatin release inhibiting factor (SIF)	Polypeptide	Inhibits GH and TSH secretion
Thyrotropin-releasing factor (TRF)	Tripeptide	Stimulates TSH and prolactin secretion
Thyroid Hormone		
Thyroxine	Amino acid derivative	Increases rate of cellular metabolism
Parathyroid Hormone		
Parathyroid	Protein	Controls metabolism of phosphorus and calcium
Pancreatic Hormones		
Insulin	Protein	Increases cell usage of glucose; increases glycogen storage
Glucagon	Protein	Stimulates conversion of liver glycogen to glucose
Somatostatin	Polypeptide	Inhibits insulin and glucagon secretion from pancreas and release of numerous gut peptides
Adrenal Cortical Hormones		
Cortisol	Steroid	Stimulates conversion of proteins to carbohydrates
Aldosterone	Steroid	Regulates salt metabolism; stimulates kidneys to retain Na^+ and excrete K^+
Adrenal Medullary Hormones		
Epinephrine (adrenaline)	Amino acid derivative	Stimulates a variety of mechanisms to prepare the body for emergency action including the conversion of glycogen to glucose
Norepinephrine (noradrenaline)	Amino acid derivative	Stimulates sympathetic nervous system; constricts blood vessels; stimulates other glands
Gonadal Hormones		
Estradiol	Steroid	Stimulates female sex characteristics; regulates changes during menstrual cycle
Progesterone	Steroid	Regulates menstrual cycle; maintains pregnancy
Testosterone	Steroid	Stimulates and maintains male sex characteristics
Pineal Hormone		
Melatonin	Amine derivative	Regulation of circadian rhythms

Steroids

A **steroid** is a compound with the skeletal four-ring structure shown in the margin. Steroids occur widely in living organisms, and not all are hormones (Figure 19.11). Cholesterol is a common component of all animal tissues. About 10% of the brain is cholesterol, which makes up a vital part of the membranes of nerve cells. Cholesterol is a major component of certain types of gallstones and is also found in deposits in hardened arteries. Cortisol is an adrenal hormone (Table 19.1). Prednisone is a synthetic steroid used to reduce inflammation in arthritis sufferers and to treat injuries.

Many drugs, both natural and synthetic, are based on steroids. Some of these are the anabolic steroids (Section 18.8) taken illegally by many athletes to improve their performance. Some are anti-inflammatory drugs used to treat such conditions as arthritis, bronchial asthma, dermatitis, and eye infections. And some are sex hormones used in formulating birth control pills (Section 19.8).

Serendipity often plays a role in the discovery of new drugs. Percy Julian, a grandson of slaves, was a chemist at the Glidden Paint Company doing research on soybeans when his work led to the development of new steroid-based drugs. It often happens that new drugs are discovered by chemists who are not really looking for them.

Sex Hormones

Sex hormones are steroids (Figure 19.12). Note that male sex hormones differ only slightly in structure from female sex hormones. In fact, the female hormone progesterone can be converted to the male hormone testosterone by a simple biochemical reaction. The physiological actions of these structurally similar compounds, however, are markedly different.

An **androgen** is any natural or synthetic compound that stimulates or controls the development and maintenance of masculine characteristics. These male sex hormones are secreted by the testes. In males, the pituitary hormones FSH and LH (Table 19.1) are required for the continued production of sperm and of androgens.

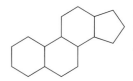

▲ Steroid skeletal structure.

▲ Percy Lavon Julian (1899–1975) was involved in the synthesis of a variety of steroids, including physostigmine, a drug used to treat glaucoma. Untreated, glaucoma causes blindness. Julian is shown here on a 1992 U.S. postage stamp.

(a)   Cholesterol

(b)   Cortisol

(c)   Prednisone

(d)   Methandrostenolone

◀ **Figure 19.11** Line-angle formulas of four steroids. (a). Cholesterol is an essential component of all animal cells. Cortisol (b) is a hormone secreted by the adrenal glands in response to physical or psychological stress. Prednisone (c) is a synthetic anti-inflammatory substance. Methandrostenolone (d), also called Dianabol®, is an anabolic steroid (Chapter 18). Note that all have the same basic four-ring structure.

Testosterone

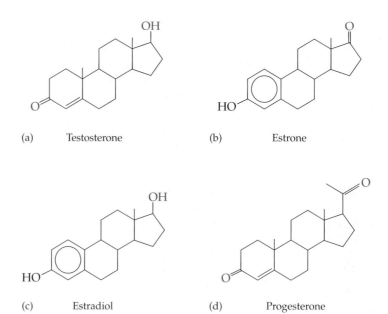

(a)   Testosterone   (b)   Estrone

(c)   Estradiol   (d)   Progesterone

▶ **Figure 19.12** Line-angle formulas of the principal sex hormones. Testosterone (a) is the main male sex hormone (androgen). Estrone (b) and estradiol (c) are female sex hormones (estrogens). Progesterone (d), also produced by females, is essential to the maintenance of pregnancy.

These hormones are responsible for development of the sex organs and for secondary sexual characteristics, such as voice and hair distribution. The most important androgen is testosterone.

An **estrogen** is a compound that controls female sexual functions, such as the menstrual cycle, the development of breasts, and other secondary sexual characteristics. The two important estrogens are estradiol and estrone. These female sex hormones are produced mainly in the ovaries. Another female sex hormone is progesterone, which prepares the uterus for pregnancy and prevents the further release of eggs from the ovaries during pregnancy. A related type of compound, called a **progestin**, is any steroid hormone that has the effect of progesterone. In females, the pituitary hormones FSH and LH (Table 19.1) are required for the production of ova and of estrogens and progesterone.

Sex hormones—both natural and synthetic—are sometimes used therapeutically. For example, a woman who has passed menopause may be given hormones to compensate for those no longer being produced by her ovaries. Hormone replacement therapy provide both risks and benefits, and its use is quite controversial (see Collaborative Group Project 90).

19.8 CHEMISTRY AND SOCIAL REVOLUTION: THE PILL

When administered by injection, progesterone serves as an effective birth control drug; it fools the body into acting as if it were already pregnant. The structure of progesterone was determined in 1934 by Adolf Butenandt (1903–1995), who received the Nobel Prize in chemistry in 1939. Other chemists began to try to design a contraceptive that would be effective when taken orally.

In 1938 Hans Inhoffen (1906–1992) synthesized the first oral contraceptive, ethisterone, which has an ethynyl group ($-C\equiv CH$). (The ethynyl group is derived from ethyne, $HC\equiv CH$, also called acetylene.) However, ethisterone had to be taken in large doses to be effective and was not widely used.

In 1951, Carl Djerassi synthesized 19-norprogesterone, which is simply progesterone with one of its methyl groups missing. It was four to eight times as effective as progesterone as a birth control agent. However, like progesterone itself, 19-norprogesterone had to be given by injection, an undesirable property. Djerassi then combined the two ideas: removal of a methyl group to make the drug more effective, and addition of the ethynyl group to allow oral administration. Djerassi synthesized norethindrone (Norlutin®), patented in 1956, which proved effective when taken in small doses (Figure 19.13).

▲ Carl Djerassi (1923–), professor of chemistry at Stanford University and president of Zoecon Corporation, Palo Alto, California, synthesized Norlutin, a progestin, in 1951.

(a) Ethisterone

(b) Norethindrone (Norlutin)

(c) Norethynodrel

(d) Mestranol

◀ **Figure 19.13** Line-angle formulas of some synthetic steroids that act as sex hormones: ethisterone (a), norethindrone (b), norethynodrel (c), and mestranol (d). In 1960, G. D. Searle's Enovid became the first birth control pill approved by the FDA in the United States. An Enovid pill contained 9.85 mg of norethynodrel and 150 mg of mestranol. Djerassi's norethindrone and Colton's norethynodrel differ only in the position of a double bond.

The progestins—norethynodrel, norethindrone, and related compounds—mimic the action of progesterone. Mestranol is a synthetic estrogen added to regulate the menstrual cycle. Progestin acts by establishing a state of false pregnancy. A woman does not ovulate when she is pregnant (or in the state of false pregnancy established by the progestin), and because she does not ovulate, she cannot conceive.

Emergency Contraceptives

Several products, called emergency contraceptives, are available for use to prevent pregnancy after unprotected intercourse. There are two kinds of emergency contraception pills (ECPs). Some ECPs are combination pills with both estrogen and progestin—synthetic analogs of the natural substances. Others contain progestin only. These are given in high doses of the same medications used in birth control pills. This disrupts hormone patterns needed for pregnancy, stopping development of the uterine lining and inhibiting ovulation or fertilization. This makes pregnancy less likely.

One brand of ECP, called Previn®, contains both an estrogen, ethinyl estradiol, and levonorgestrel, a progestin. Another ECP, called Plan B®, contains only levonorgestrel. ECPs are taken as two doses, 12 hours apart. They are 75–89% effective in preventing pregnancy after unprotected sexual intercourse, and they work best if taken as soon as possible after intercourse. Only one woman out of 100 will become pregnant after taking progestin-only ECPs.

An intrauterine device (IUD), the Copper T 380A IUD (ParaGard®), is also used for emergency contraception. Emergency IUD insertion reduces the risk of pregnancy by 99.9% if inserted within three to five days of unprotected intercourse. Copper(II) ions in the IUD interfere with sperm transport and fertilization. The copper IUD probably causes an inflammation that makes the endometrium unsuitable for implantation. Only one out of 1000 women will become pregnant after emergency IUD insertion.

▲ Working at about the same time as Djerassi, Frank Colton (1923–) synthesized norethynodrel, another progestin, for which G. D. Searle was awarded patents in 1954 and 1955. Norethynodrel, a progestin, was used in Enovid, the first birth control pill approved by the FDA (1960).

▲ The Copper T 380A IUD (ParaGard®).

(a) Levonorgestrel

(b) Ethinyl Estradiol

Chemical Treasure in the Rainforest

In 1940 progesterone was one of the most expensive compounds in the world. Obtained with great difficulty from animals, it sold for $200 per gram—equivalent to about $1600 in today's dollars. American chemist Russell Marker (1902–1995) wondered if progesterone could be made from a plant steroid. He took diosgenin, which occurs in tiny amounts in lily roots, and converted it to progesterone.

He then looked for a more abundant source of this plant steroid, examining more than 400 plants from all over the world. He finally identified a Mexican wild yam, *Dioscorea villosa*, that contained a significant quantity of diosgenin in its roots. He left his faculty post at Pennsylvania State University to look for more of these vines in the tropical forests of Mexico. Eventually he found a profuse growth of them and within a few months synthesized almost a kilogram of progesterone. Marker had found a commercial source not only for progesterone but for all the other steroid hormones as well. (Wild yam extracts are sold with claims that they may relieve symptoms associated with menopause, but the yam as such does not contain progesterone nor anything else that would act like progesterone.)

Tropical rainforests are warm, moist, and fertile areas where there are more different kinds of plants and animals than any place else on Earth. These forests are living chemical factories that contain valuable and irreplaceable natural resources. Unfortunately, the world's rainforests are rapidly disappearing.

▲ The Mexican wild yam is a source of diosgenin, from which progesterone is made.

ECPs are not the same as the so-called medical abortion pills such as mifepristone (also called RU-486). Mifepristone works after a woman becomes pregnant and the fertilized egg has attached to the wall of the uterus. It causes the uterus to expel the egg, ending the pregnancy. ECPs prevent pregnancy after sexual intercourse, while mifepristone ends an unwanted pregnancy at an early stage.

In Chapter 16, we noted the use of diethylstilbestrol (DES) as a growth promoter in cattle. Its use for that purpose was banned in 1979. It has been used like mifepristone as a missed-period pill. DES is not a steroid, but its molecular shape is similar to that of estradiol (Figure 19.13). By its action, DES is classified as a synthetic estrogen.

DES

Mifepristone

Mifepristone inhibits the action of progesterone. Because progesterone is essential for maintaining pregnancy, if its action is blocked, pregnancy cannot be established or maintained. RU-486 has replaced a substantial number of surgical abortions. A woman who wishes to abort a pregnancy takes three mifepristone tablets, followed in a few days by an injection of a prostaglandin (Section 19.9). The lining of the uterine wall and the implanted fertilized egg are sloughed off, and the pregnancy is terminated.

People who believe that human life begins when the ovum is fertilized by the sperm equate the use of mifepristone with abortion and oppose the use of the drug. Others consider the drug just another method of birth control and see its use as being safer than a surgical abortion.

Risks of Taking Birth Control Pills

Oral contraceptives have been used by millions of women in the United States since 1960. They are safe in most cases, but some women experience hypertension, acne, or abnormal bleeding. These pills increase the risk of blood clotting in some women, but so does pregnancy. Blood clots can clog arteries and cause death by stroke or heart attack. The death rate associated with birth control pills is about 3 in 100,000, only one-tenth of that associated with childbirth. For smokers, however, the risks are much higher. For women over 40 who smoke 15 cigarettes per day, the risk of death from stroke or heart attack is 1 in 5000. The FDA advises all women who smoke, especially those over 40, to use some other method of contraception.

The Minipill

Because most of the side effects of oral contraceptives are associated with the estrogen component, the amount of estrogen in these pills has been greatly reduced over the years. Today's pills contain only a fraction of a milligram of estrogen. "Minipills" are now available that contain only small amounts of progestin and no estrogen at all. Minipills are not quite as effective as the combination pills, but they have fewer side effects.

A Pill for Males?

Why should females have to bear all the responsibility for contraception? Why not a pill for males? Many men are willing to share the risks and the responsibility of contraception, but condoms and vasectomies are the only effective forms of contraception available. There are biological reasons for females to bear the burden. Women are the ones who get pregnant when contraception fails, and in females contraception has to interfere with only one monthly event: ovulation. Males produce sperm continuously. Daily injection of testosterone can reduce or shut down sperm production, but the side effects of acne and weight gain make this approach undesirable for most males.

In males, the pituitary hormones FSH and LH (Table 19.1) are required for the continued production of sperm and of the male hormone testosterone. A new method tested in mice combines two hormone analogs, cyproterone, a progestin, and testerone undecanoate, an androgen. The cyproterone inhibits the pituitary gland's release of FSH and LH. Low levels of FSH result in reduced sperm production, down to zero in some test subjects. But low levels of LH result in reduced production of testosterone, which reduces the male libido. The testerone undecanoate is added to replace the testosterone. Although it has not yet been tested in humans, in mice the drug combination seems to be an ideal male birth control pill. It is effective orally and acts quickly, and its effects are readily reversible.

It may be years before a safe, effective contraceptive for men is available.

19.9 DRUGS FOR THE HEART

The heart is a muscle that beats virtually every second of every day for as long as we live. Diseases of the cardiovascular system are responsible for one-third of the deaths in the United States and more than one-fifth of deaths worldwide. Drugs that treat these diseases are a most important set of life-saving compounds. These drugs prolong life and improve its quality.

Major diseases of the heart and blood vessels include ischemic ("lacking oxygen") coronary artery disease, heart arrhythmias (abnormal heartbeat), hypertension (high blood pressure), and congestive heart failure. Atherosclerosis (fatty deposit build-up in the lining of the arteries) is the primary cause of coronary artery

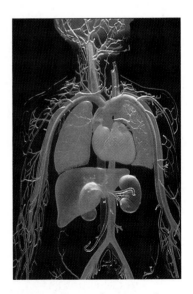

▲ The cardiovascular system. Over the course of 24 hr, the heart pumps more than 8000 L of blood through nearly 100,000 km of blood vessels.

disease, which in turn causes myocardial infarction (heart attack). Thus, most drug treatments for the heart aim to increase its supply of blood (and oxygen), to normalize its rhythm, to lower blood pressure, or to prevent accumulation of lipid plaque deposits in blood vessels.

Lowering Blood Pressure

Hypertension, or high blood pressure, is defined as pressure above 140/90. Normal blood pressure is defined as less than 120/80, and the range 120–139/80–89 is called prehypertension. Hypertension is the most common cardiovascular disease. Because it normally does not produce symptoms, many people have hypertension without realizing it. Four major drug categories for lowering blood pressure are

- *diuretics*, which cause the kidneys to excrete more water and thus lower the blood volume;
- *beta blockers*, which slow the heart rate and reduce the force of the heartbeat,
- *calcium channel blockers*, which are powerful vasodilators, inducing muscles around the blood vessels to relax; and
- *angiotensin-converting enzyme (ACE) inhibitors*, which inhibit the action of an enzyme that causes blood vessels to contract.

Studies show that diuretics, the oldest, simplest, and cheapest of the blood-pressure-lowering drugs, may also be the most effective drugs for treating hypertension in most people. However, both a diuretic and one of the other medications often are prescribed. Two common beta blockers are propranolol and metaprolol (Section 19.12).

Normalizing Heart Rhythm

An *arrhythmia* is an abnormal heartbeat. Some arrhythmias exhibit no symptoms and are discovered only during a physical examination, but others can be life threatening. The electrical properties of nerves and muscles arise from the flow of ions across cell membranes, and drugs for arrhythmia alter this flow. There are several types of such drugs, with different mechanisms of action. These drugs tend to have a narrow margin between the therapeutic dose and a harmful amount. A change in the heart's rhythm can have serious consequences. The too-rapid heartbeat, known as *fibrillation*, is so common that defibrillator devices are now available on many airplanes and in various other public places.

Treating Coronary Artery Disease

A common symptom of coronary artery disease is chest pain called *angina pectoris*. This pain is caused by the heart getting less oxygen than it needs, usually due to partial blockage of the coronary arteries by lipid-containing plaque (arteriosclerosis). When the blockage is complete, a heart attack occurs, and some of the heart muscle dies. Medical treatment for coronary artery disease usually involves dilation (widening) of the blood vessels to the heart to increase the blood flow and slowing of the heart rate to decrease its work load and its demand for oxygen. Some of the drugs used are the same as those used to lower high blood pressure (beta blockers, for example). Several organic nitro compounds, especially amyl nitrite ($CH_3CH_2CH_2CH_2CH_2ONO$) and nitroglycerin have a long history in the treatment of angina pectoris. These compounds act by releasing nitric oxide (NO) which relaxes the constricted vessels that are reducing the supply of blood and oxygen to the heart.

Although many new drugs are now being used to treat heart failure, one that has been used for centuries still plays an important role. Digitalis, from the foxglove plant, was used by the ancient Egyptians and Romans and is still used to treat patients with heart failure. Digitalis is a mixture of glycosides that yield carbohydrates and steroids on hydrolysis. Hydrolysis of digitoxin (one of the digitalis glycosides) yields the steroid digitoxigenin.

$$H_2C-O-NO_2$$
$$HC-O-NO_2$$
$$H_2C-O-NO_2$$

Nitroglycerin

NO—A Messenger Molecule

Less than two decades ago, the simple molecule NO, notorious as an air pollutant (Chapter 12), was found to act as a *messenger molecule* that carries signals between cells in the body. All previously known messenger molecules were complex substances such as norepinephrine and serotonin (Section 19.12) that act by fitting specific receptors in cell membranes. Nitrogen monoxide, commonly called nitric oxide, is essential to maintaining blood pressure and establishing long-term memory. It also aids in the immune response to foreign invaders in the body and mediates the relaxation phase of intestinal contractions in the digestion of food.

NO is formed in cells from arginine, a nitrogen-rich amino acid (Chapter 15), in a reaction catalyzed by an enzyme. NO kills invading microorganisms, probably by deactivating iron-containing enzymes in much the same way that carbon monoxide destroys the oxygen-carrying capacity of hemoglobin (Chapter 12).

Louis Ignarro, Robert F. Furchgott, and Ferid Murad, who discovered the physiological role of NO, were awarded a 1998 Nobel Prize. This award has a fortuitous link back to Alfred Nobel, whose invention of dynamite provided the financial basis of the Nobel Prizes. Nitroglycerin, the explosive ingredient of dynamite, relieves the chest pain of heart disease. In his later years, Nobel refused to take nitroglycerin for his own heart disease because it causes headaches, and he did not think it would relieve his chest pain. Ferid Murad showed that nitroglycerin acts by releasing NO.

NO also dilates the blood vessels that allow blood flow into the penis to cause an erection. Research on this role of NO led to the development of the anti-impotence drug Viagra. One of the physiological effects of Viagra is the production of small quantities of NO in the bloodstream. Related research has led to drugs for treating shock and a drug for treating high blood pressure in newborn babies.

19.10 DRUGS AND THE MIND

A **psychotropic drug** is one that affects the human mind. Probably the first drugs used by primitive peoples were mind-altering substances. Alcohol, marijuana, opium, cocaine, peyote, and other plant materials have been known for thousands of years. Generally, only one or two of these were used in any one society. However, today thousands of drugs that affect the mind are readily available. Some still come from plants, but most are synthetic.

There is no clear distinction between drugs that affect the mind and those that affect the body. Most drugs probably affect our minds as well as our bodies. It can still be useful, however, to distinguish between drugs that act primarily on the body and drugs that affect primarily the mind.

Psychotropic drugs that affect the mind are divided into three classes.

- A **stimulant drug**, such as cocaine and amphetamines, increases alertness, speeds mental processes, and generally elevates the mood.
- A **depressant drug**, such as alcohol, most anesthetics, opiates, barbiturates, and minor tranquilizers, reduces the level of consciousness and the intensity of reactions to environmental stimuli. In general, depressants dull emotional responses.
- A **hallucinogenic drug**, such as lysergic acid diethylamide (LSD) and marijuana, alters qualitatively the way we perceive things.

We will examine some of the major drugs in each category, but first we need to look at some chemistry of the nervous system.

19.11 SOME CHEMISTRY OF THE NERVOUS SYSTEM

The nervous system is made up of about 12 billion **neurons** (nerve cells) with 10^{13} connections between them. The brain operates with a power output of about 25 W and has the capacity to handle about 10 trillion bits of information. Nerve cells vary a great deal in shape and size. One type is shown in Figure 19.14. The essential parts of each cell are the cell body, the axon, and the dendrites. We discuss here only the nerves that make up the involuntary (autonomic) nervous system. These nerves carry messages between organs and glands that act involuntarily (such as the heart, the digestive organs, and the lungs) and the brain and spinal column.

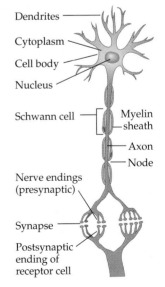

▲ **Figure 19.14** Diagram of a human nerve cell.

TABLE 19.2	Selected Neurotransmitters*	
Transmitter Molecule	**Derived From**	**Site of Synthesis**
Dopamine	Tyrosine	Central nervous system (CNS)
Norpinephrine	Tyrosine	CNS
Epinephrine	Tyrosine	Adrenal medulla, some CNS cells
Serotonin [5-Hydroxytryptamine (5-HT)]	Tryptophan	CNS, gut
Acetylcholine	Choline	CNS, parasympathetic nerves
GABA	Glutamate	CNS
Glutamate		CNS
Aspartate		CNS
Glycine		Spinal cord
Histamine	Histidine	Hypothalamus
Adenosine	ATP	CNS, peripheral nerves
ATP		Sympathetic, sensory and enteric nerves
Nitric oxide, NO	Arginine	CNS, gastrointestinal tract

*Not all of these are discussed in this text.

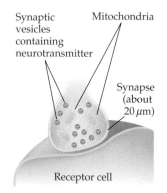

Synaptic vesicles containing neurotransmitter

Mitochondria

Synapse (about 20 μm)

Receptor cell

▲ **Figure 19.15** Diagram of a synapse. When an electrical signal reaches the presynaptic nerve ending, neurotransmitter molecules are released from the vesicles. They migrate across the synapse to the receptor cell, where they fit specific receptor sites.

Axons on a nerve cell may be up to 60 cm long, but there is no continuous pathway from an organ to the central nervous system. Messages must be transmitted across tiny, fluid-filled gaps, or **synapses** (Figure 19.15). When an electrical signal from a nerve cell in the brain reaches the end of an axon of a presynaptic nerve cell, chemicals called **neurotransmitters** are released from tiny vesicles into the synapse. Receptors on the postsynaptic cell bind the neurotransmitter, causing changes in the receiving cell. After a brief interval, the neurotransmitter is released from the receptor and carried back to the presynaptic cell by substances called transporters.

There are many neurotransmitters (Table 19.2), and each has a specific function. Messages are carried to other nerve cells, to muscles, and to the endocrine glands (such as the adrenal glands). Each neurotransmitter fits one or more receptor sites on the receptor cell. Many drugs (and some poisons) act by mimicking the action of the neurotransmitter. Others act by blocking the receptor and preventing the neurotransmitter from acting on it. Several neurotransmitters are amino acids. Others are amines, as are some of the drugs that affect the chemistry of our brains. Still others are peptides.

19.12 BRAIN AMINES: DEPRESSION AND MANIA

We all have our ups and downs in life. These moods probably result from multiple causes, but it is likely that a variety of chemical compounds formed in the brain are involved. Before we consider these ups and downs, however, let's take a look at epinephrine, an amine formed in the adrenal glands and some central nervous system cells (Figure 19.16).

Often called adrenaline, epinephrine is secreted by the adrenal glands when a person is under stress or is frightened. A tiny amount of epinephrine causes a great increase in blood pressure, and its flow prepares the body for fight or flight. Because culturally imposed inhibitions prevent fighting or fleeing in most modern situations, the adrenaline-induced supercharge is not used. This sort of frustration has been implicated in some forms of mental illness.

Biochemical Theories of Mental Illness

Biochemical theories of mental illness involve brain amines, one of which is norepinephrine (NE), a relative of epinephrine. NE is a neurotransmitter formed in the brain (Figure 19.16). When produced in excess, NE causes a person to be elated—perhaps even hyperactive. In large excess, NE induces a manic state. A deficiency of NE, on the other hand, can cause depression.

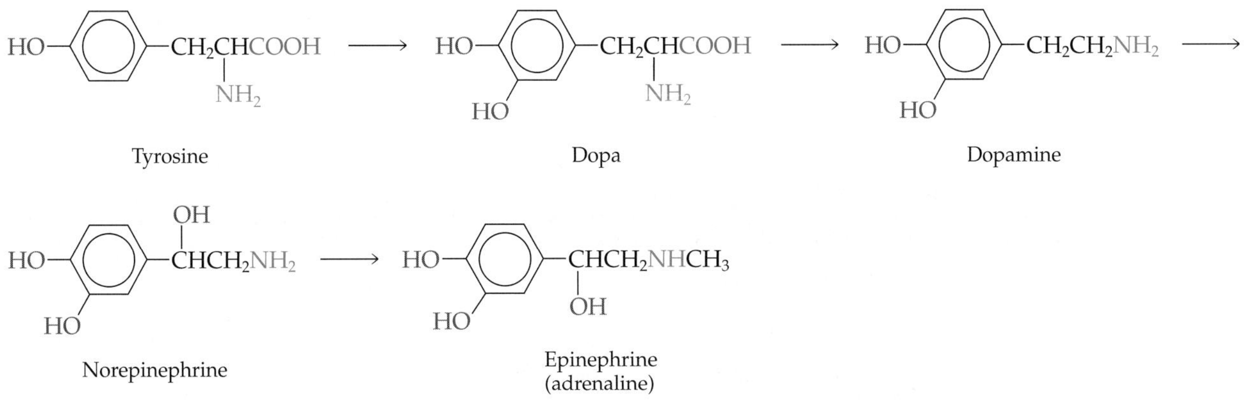

HO—⬡—CH₂CHCOOH ⟶ HO—⬡—CH₂CHCOOH ⟶ HO—⬡—CH₂CH₂NH₂ ⟶

Tyrosine   Dopa   Dopamine

HO—⬡—CHCH₂NH₂ ⟶ HO—⬡—CHCH₂NHCH₃

Norepinephrine   Epinephrine (adrenaline)

▲ **Figure 19.16** The biosynthesis of epinephrine and norepinephrine from tyrosine. L-Dopa, the lefthanded form of the compound (see "Right- and Left-handed Molecules" on page 604), is used to treat Parkinson's disease, which is characterized by rigidity and stiffness of muscles. Parkinson's results from inadequate dopamine production, but dopamine cannot be used directly to treat the disease because it is not absorbed into the brain. Dopamine has been used to treat hypertension. It also is a neurotransmitter; schizophrenia has been attributed to an overabundance of dopamine in nerve cells.

Another brain amine is the neurotransmitter *serotonin*. Serotonin is involved in sleep, sensory perception, pleasure sensations, and regulation of body temperature. A metabolite of serotonin, 5-hydroxyindoleacetic acid (5-HIAA), is found in unusually *low* levels in the spinal fluid of violent suicide victims. This indicates that abnormal serotonin metabolism plays a role in depression. Recent research suggests that a reduced flow of serotonin through the synapses in the frontal lobe of the brain causes depression.

Levels of 5-HIAA are also low in murderers and other violent offenders. However, they are higher than normal in people with obsessive–compulsive disorders, sociopaths, and people with guilt complexes.

Brain Amine Agonists and Antagonists in Medicine

Cells have several different variations of the receptor that are activated by NE and related compounds. NE **agonists** (drugs that enhance or mimic its action) are stimulants. NE **antagonists** (drugs that block the action of NE) slow down various processes. Drugs called beta blockers (Figure 19.17) reduce the stimulant action of epinephrine and NE on various kinds of cells. Propranolol (Inderal) is used to treat cardiac arrhythmias, angina, and hypertension by slightly lessening the force of the heartbeat. Unfortunately, it also causes lethargy and depression. Metoprolol (Lopressor) acts selectively on the cells of the heart. It can be used by hypertensive patients who have asthma because it does not act on receptors in the bronchi.

Serotonin agonists are used to treat depression, anxiety, and obsessive–compulsive disorder. Serotonin antagonists are used to treat migraine headaches and to relieve the nausea caused by cancer chemotherapy.

Brain Amines and Diet: You Feel What You Eat

Richard Wurtman of the Massachusetts Institute of Technology has demonstrated a relationship between diet and serotonin levels in the brain. As shown in Figure 19.19, serotonin is produced in the body from the amino acid tryptophan. Wurtman found that diets high in carbohydrates lead to high levels of serotonin. High levels of protein lower the serotonin concentration.

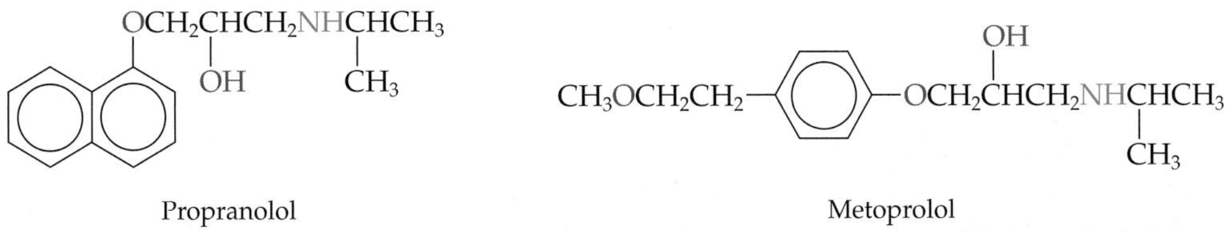

Propranolol   Metoprolol

▲ **Figure 19.17** Propranolol and metoprolol are beta blockers, widely used to treat high blood pressure.

Norepinephrine is synthesized in the body from the amino acid tyrosine. As noted in Figure 19.16, the synthesis is complex and proceeds through several intermediates. Because tyrosine is also a component of our diets, it may well be that our mental state depends to a fair degree on our diet.

Right- and Left-handed Molecules

Like people, molecules can be either right- or lefthanded. Using models, we can show two structures for the amino acid alanine [$CH_3CH(NH_2)COOH$] (Figure 19.18). These structures are **stereoisomers**, isomers having the same structural formula but differing in the arrangement of atoms or groups of atoms in three-dimensional space. Models of the two stereoisomers of alanine are as alike—and as different—as are a pair of gloves. If you place one of the models in front of a mirror, the image in the mirror will be identical to the second stereoisomer. Molecules that are nonsuperimposable (nonidentical) mirror images of each other are stereoisomers of a specific type called **enantiomers** (Greek *enantios*, "opposite").

Enantiomers have a carbon atom to which four different groups are attached. For example, the central carbon of alanine has an H, a CH_3, a COOH, and an NH_2 joined to it. A carbon atom that has four different groups attached is a **chiral carbon**. If a molecule contains one or more chiral carbons, it is likely to exist as two or more stereoisomers. In contrast, propane ($CH_3CH_2CH_3$) does not contain a chiral carbon and thus does not exist as a pair of stereoisomers.

Stereoisomerism is quite common in organic chemicals of biological importance. The simple sugars (Chapter 15) are all righthanded; 19 of the 20 amino acids that make up proteins are lefthanded. The other, glycine (H_2NCH_2COOH), has no handedness.

A righthand glove doesn't fit on a left hand, or vice versa. Similarly, enantiomers fit enzymes differently; thus they have different effects. As an example, consider Lipitor®, a statin (lipid-lowering) drug widely used to lower blood cholesterol. Initial research in 1989 was promising but not outstanding. Later it was found that only the lefthanded isomer is active and that it is much more effective when separated from its righthanded form. The development of drugs that contain only one enantiomer is becoming increasingly important in the pharmaceutical industry.

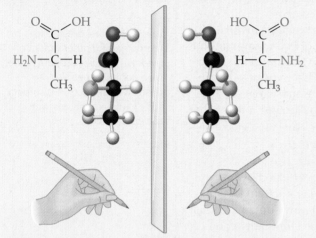

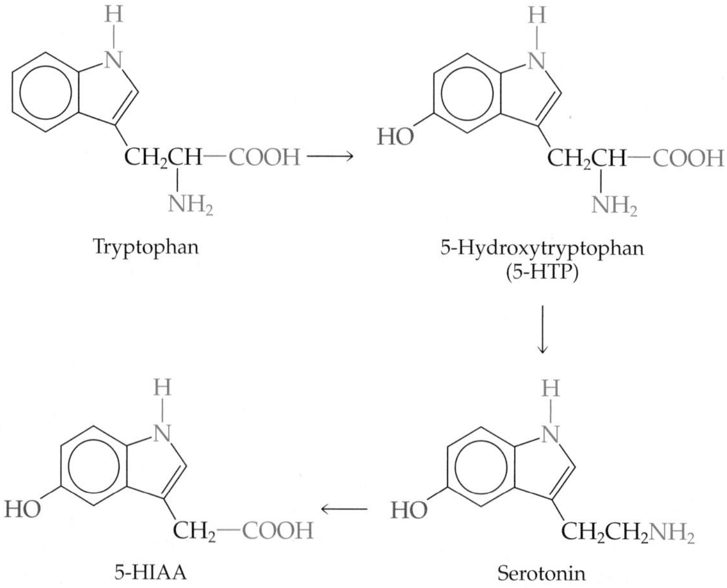

▲ **Figure 19.18** Ball-and-stick models and structural formulas of D- and L-alanine. The molecules are *enantiomers*, mirror images of one another. Like a right and left hand, enantiomers cannot be superimposed on one another.

▶ **Figure 19.19** Serotonin is produced in the brain from the amino acid tryptophan. The synthesis involves several steps; some are omitted here for simplicity. 5-HIAA is a metabolite of serotonin.

Tryptophan

5-Hydroxytryptophan (5-HTP)

Serotonin

5-HIAA

Nearly 1 out of every 10 people in the United States suffers from mental illness. Over half the patients in hospitals are there because of mental problems. When the biochemistry of the brain is more fully understood, mental illness may be cured (or at least alleviated) by the administration of drugs—or by adjusting the diet.

Some Chemistry of Love, Trust, and Sexual Fidelity

The notion that such emotions as love and trust might be chemical in origin is unsettling. However, emotions that trigger romantic relationships seem to be governed in part by a substance called β-phenylethylamine (PEA). PEA functions as a neurotransmitter in the human brain. It appears to create excited, alert feelings and aggressive moods. Increased levels of PEA produce a feeling much like the feeling of "being in love." The chemical structure of PEA resembles that of norepinephrine.

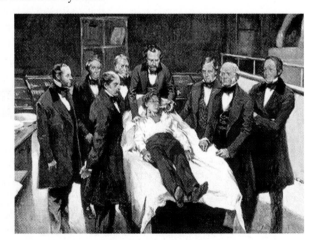

$CH_2CH_2NH_2$

β-Phenylethylamine

CH_2-COOH

Phenylacetic acid

How much PEA does it take to get back that old feeling? Levels of PEA in the brain can be estimated by measuring levels of its metabolite, phenylacetic acid, in the urine. Low levels of urinary phenylacetic acid correlate with depression. There are no good food sources of PEA. Chocolate contains small quantities of PEA, but most of the PEA in chocolate is metabolized before it reaches the brain. PEA releases dopamine, producing an antidepressant effect. Some research has shown that oral doses of PEA relieve depression in some people.

Oxytocin, sometimes called the "cuddling chemical," increases the bond between lovers. It also induces contractions during childbirth and lactation afterward (Table 19.1). Swiss scientists have shown that oxytocin increases trust in people. People playing a trust game were more trusting after taking one sniff of oxytocin.

If oxytocin equals love, vasopressin may keep couples together. Few mammals are monogamous, mating and bonding with one partner for life. The prairie vole is one such. Montane voles, a closely related species, are polygamous. Scientists have found that within 24 hours after mating, the male prairie vole is hooked for life. Postcoital production of vasopressin in the male is responsible for this monogamous behavior. Male montane voles injected with vasopressin did not react significantly to the female vole. Montane voles have a different vasopressin receptor pattern than the prairie vole. When given a compound to suppress the effect of vasopressin, the prairie voles lose their devotion to each other. Perhaps the solution to our high divorce rate is a few sniffs of vasopressin.

19.13 ANESTHETICS

An **anesthetic** is a substance that causes lack of feeling or awareness. A **general anesthetic** acts on the brain to produce unconsciousness and a general insensitivity to pain. A local anesthetic causes loss of feeling in a part of the body. Diethyl ether ($CH_3CH_2OCH_2CH_3$), the first general anesthetic, was introduced into surgical practice in 1846 by a Boston dentist, William Morton. Inhalation of ether vapor produces unconsciousness by depressing the activity of the central nervous system. Ether is relatively safe, since there is a fairly wide gap between the dose that produces an effective level of anesthesia and the lethal dose. Its disadvantages are high flammability and an undesirable side effect, nausea.

Nitrous oxide, or laughing gas (N_2O), was tried by Morton without success before he tried ether. Nitrous oxide was discovered by Joseph Priestley in 1772, and its narcotic effect was soon noted. Mixed with oxygen, nitrous oxide is used in modern anesthesia. It is quick acting but not very potent. Concentrations of 50% or greater must be used for it to be effective. When nitrous oxide is mixed with ordinary air instead of oxygen, not enough oxygen gets into the patient's blood, and permanent brain damage can result.

Chloroform ($CHCl_3$) was introduced as a general anesthetic in 1847. It quickly became popular after Queen Victoria gave birth to her eighth child while anesthetized by chloroform in 1853, and it was used widely for years. It is nonflammable and

▲ This painting shows an operation in Boston in 1846 during which ether was used as an anesthetic.

TABLE 19.3	Inhaled Anesthetic Agents			
Generic or Chemical Name	Formula	Commercial Name	Year Introduced	Currently in Use?
Diethyl ether	$CH_3CH_2OCH_2CH_3$	Ether	1842	No
Nitrous oxide	N_2O	Nitrous oxide	1844	Yes
Chloroform	$CHCl_3$	Chloroform	1847	No
Halothane	$CHBrClCF_3$	Fluothane®	1956	Yes
Enflurane	CHF_2OCF_2CHClF	Ethrane®	1974	Yes
Isoflurane	$CHF_2OCHClCF_3$	Forane®	1980	Yes
Desflurane	$CHF_2OCHFCF_3$	Suprane®	1992	Yes
Sevoflurane	$CHF_2OCH(CF_3)_2$	Ultane®	1995	Yes

produces effective anesthesia, but it has serious drawbacks. For one, it has a narrow safety margin; the effective dose is close to the lethal dose. It also causes liver damage, is a suspected carcinogen, and must be protected from oxygen during storage to prevent the formation of deadly phosgene gas.

Modern Anesthesia

Modern inhalant anesthetics include fluorine-containing compounds (Table 19.3). These compounds are nonflammable and relatively safe for patients. Their safety for operating room personnel, however, has been questioned. For example, female operating room workers suffer a higher rate of miscarriages than women in the general population.

Modern surgical practice usually makes use of a variety of drugs. Generally, a patient is given

- a tranquilizer such as a benzodiazepine (Valium or Versed; Section 19.16) to decrease anxiety,
- an intravenous anesthetic such as thiopental (Section 19.14) to produce unconsciousness quickly,
- a narcotic pain killer (such as fentanyl) to block pain,
- an inhalant anesthetic (Table 19.3) to provide insensitivity to pain and keep the patient unconscious, often combined with oxygen and nitrous oxide to support life, and
- a relaxant such as succinylcholine chloride to relax the muscles make it easier to insert the breathing tube.

General anesthetics reduce nerve transmission at the synapses by affecting the response of receptors to neurotransmitters. The anesthetics bind only weakly to their sites of action, and it is difficult to determine their exact mode of action. They also seem to work in part by dissolving in the fatlike membranes of nerve cells, changing the membrane permeability to sodium ions and thus depressing the conductivity of the neurons.

Solvent Sniffing: Self-Administered Anesthesia

Nearly all gaseous and volatile liquid organic compounds exhibit anesthetic action. This action of organic solvents leads to their abuse. The sniffing of glue solvents, gasoline, aerosol propellants, and other inhalants is perhaps the deadliest form of drug abuse. The dose required for intoxication often is not far from that which will stop the heart. And it is difficult to measure the dose inhaled from the plastic or paper bag normally used in glue sniffing. Also, as with nitrous oxide, sublethal doses can cause permanent brain damage by cutting down the oxygen supply to the brain.

TABLE 19.4 Common Substances Used as Local Anesthetics	
Substance	**Duration of Action**
Procaine (Novocain)	Short
Lidocaine (Xylocaine)	Medium (30–60 min)
Mepivacaine (Carbocaine, Polocaine)	Fast (6–10 min)
Bupivacaine (Marcaine, Sensorcaine)	Moderate (8–12 min)
Prilocaine	Medium (30–90 min)
Chloroprocaine (Nesacaine)	Short (15–30 min)
Cocaine	Medium
Etidocaine (Duranest)	Long (120–180 min)

Local Anesthetics

Local anesthetics are used to render a part of the body insensitive to pain. They block nerve conduction by reducing the permeability of the nerve cell membrane to sodium ions. The patient remains conscious during dental work or minor surgery.

The first local anesthetic to be used successfully was cocaine, a drug first isolated in 1860 from the leaves of the coca plant (Figure 19.20). Its structure was determined by Richard Willstätter in 1898. Scientists have made many attempts to develop synthetic compounds with similar properties. Cocaine is a powerful stimulant. Its abuse is discussed in Section 19.17.

Certain esters of *para*-aminobenzoic acid (PABA) act as local anesthetics (Figure 19.21). The ethyl and butyl esters are used to relieve the pain of burns and open wounds. These esters are applied as ointments, usually in the form of picrate salts.

More powerful in their anesthetic effects are a series of derivatives with a second nitrogen atom in the alkyl group of the ester (Table 19.4). Perhaps the best known of these is procaine (Novocaine), first synthesized in 1905 by Alfred Einhorn (1865–1917), who had worked with Willstätter on the structure of cocaine. Procaine can be injected as a local anesthetic or injected into the spinal column to deaden the entire lower portion of the body. A spinal block prevents messages of pain from the lower parts of the body from reaching the brain.

The local anesthetic of choice nowadays is often lidocaine or bupivacaine. These compounds are not derivatives of PABA, but they share some structural features with compounds that are. They are highly effective but have fairly low toxicity compared to cocaine and procaine (see Collaborative Group Project 93).

▲ **Figure 19.20** Coca leaves contain the alkaloid cocaine, a local anesthetic and stimulant.

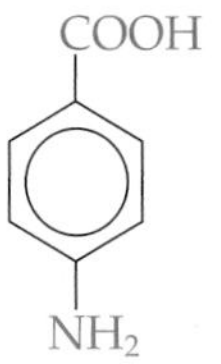

para-Aminobenzoic acid

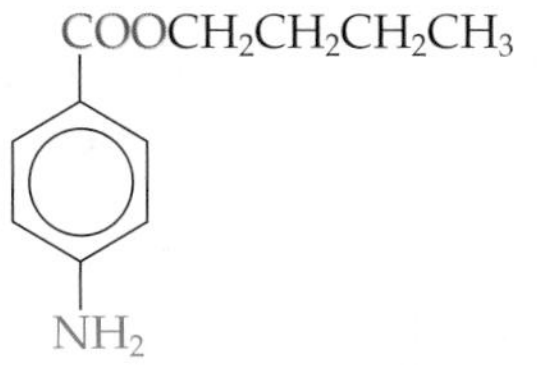

Butyl *para*-aminobenzoate
(Butesin)

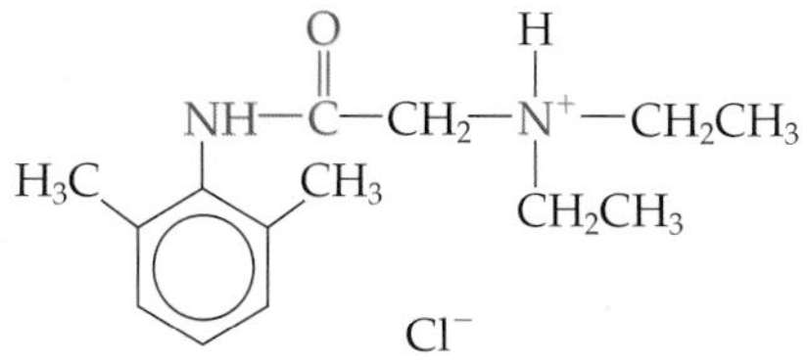

Lidocaine (Xylocaine)

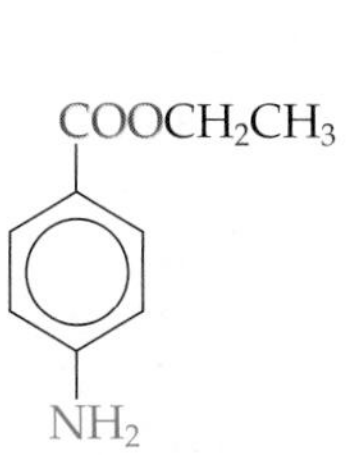

Ethyl *para*-aminobenzoate
(Benzocaine)

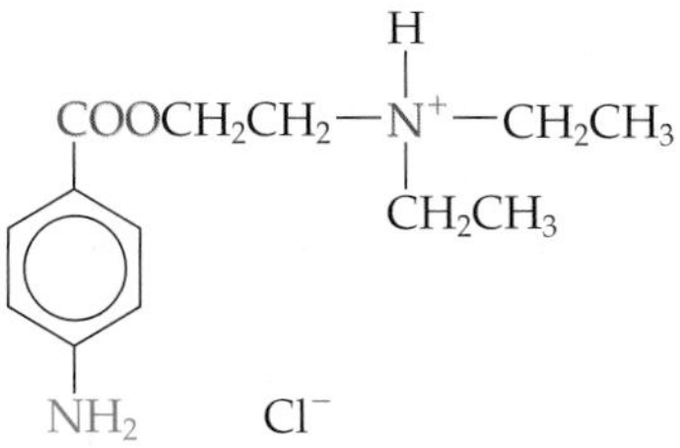

Procaine
(Novocaine)

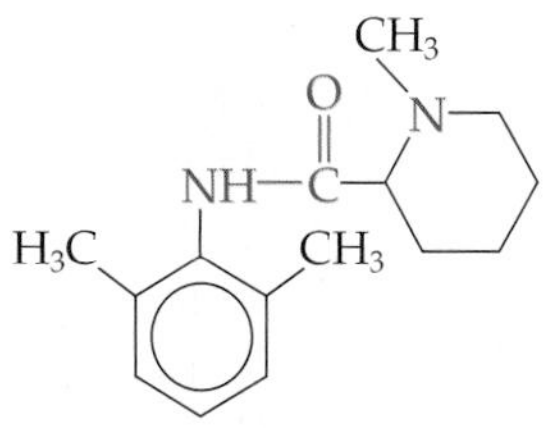

Mepivacaine

▲ **Figure 19.21** Some local anesthetics. They are often used in the form of a hydrochloride or picrate salt, which is more soluble in water than the free base.

Question: Which are derived from *para*-aminobenzoic acid? Which have amide functional groups? Which are esters?

Dissociative Anesthetics: Ketamine and PCP

Ketamine, an intravenous anesthetic, is called a *dissociative anesthetic* because it dissociates a person's perception from his or her sensations. The person appears awake but is unaware of his or her surroundings. Ketamine induces hallucinations similar to those reported by people who have had near-death experiences. They seem to remember observing their rescuers from a vantage point above the scene, or moving through a dark tunnel toward a bright light. Little is known about the action of ketamine at the molecular level. If it acts by binding to receptors in the body, we can assume that our bodies also produce chemicals that fit these receptors. These compounds may be synthesized or released only in extreme circumstances—such as in near-death experiences.

Ketamine is used widely in veterinary medicine. Because it suppresses breathing much less than most other available anesthetics, ketamine is used in human medicine as an anesthetic for victims with unknown medical history and for children and persons of poor health.

Phencyclidine, commonly called PCP, is closely related to ketamine. PCP was formerly used as an animal tranquilizer and for a brief time as a general anesthetic for humans. It is now fairly common on the illegal drug scene.

PCP, which is soluble in fat and has no appreciable water solubility, is stored in fatty tissue and released when the fat is metabolized, accounting for the "flashbacks" commonly experienced by users. Many users experience bad "trips" with PCP, and about 1 in 1000 develops a severe form of schizophrenia. High doses cause illusions and hallucinations and can also cause seizures, coma, and death. However, death of PCP users more often results from an accident or suicide during a trip.

Ketamine

Phencyclidine
(PCP)

19.14 DEPRESSANT DRUGS

In this section we explore several familiar depressant drugs. We discussed ethyl alcohol in some detail in Chapter 9, but we start with it again here because it is by far the most used and abused depressant drug in the world.

Alcohol

In nature, sugars in fruit are often fermented into alcohol through contact with airborne yeasts, but alcohol use would not have begun in earnest until people began to farm. The first brew was probably date palm wine, originating in Mesopotamia. As early as 3700 B.C.E., the Egyptians fermented fruit to make wine and the Babylonians made beer from barley. People have been fermenting fruits and grains ever since, but relatively pure alcohol was not available until the invention of distillation during the Middle Ages.

People often think they are stimulated ("get high") when they drink alcohol, but ethanol is actually a depressant. It slows down both physical and mental activity. Nearly two-thirds of adults in the United States drink alcohol, and nearly one-third have five or more drinks on at least one day each year. There are at least 10 million alcoholics, and alcoholism is the third leading health problem, right after cancer and heart disease. Alcohol is a factor in about 20,000 deaths each year, not counting accidents and homicides. For 15- to 24-year-olds, the three leading causes of death are automobile accidents, homicides, and suicides, and alcohol is a leading factor in all three.

There is a positive side to drinking alcohol. Longevity studies indicate that those who use alcohol only moderately (no more than a drink or two a day) live longer than nondrinkers. This is probably a result of the relaxing effect of the alcohol. Heavier drinking, however, can cause many health problems and shortens the life span.

Some Biochemistry of Ethanol

Alcoholic beverages are generally high in calories. Pure ethanol furnishes about 7 kcal/g. Ethanol is metabolized by oxidation to acetaldehyde at a rate of about 1 oz

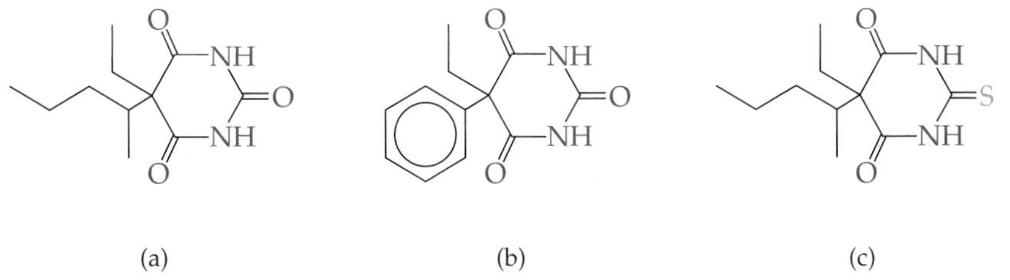

◀ **Figure 19.22** Line-angle formulas of some barbiturate drugs. (a) Pentobarbital (Nembutal), (b) phenobarbital (Luminal), and (c) thiopental (Pentothal). These drugs, derived from barbituric acid, are often used in the form of their sodium salts; for example, thiopental is used as sodium pentothal.

(28 mL)/hr, and a build-up of ethanol concentration in the blood caused by drinking more than this amount results in intoxication.

Oxidation of ethanol uses up a substance called NAD^+, the oxidized form of nicotinamide adenine dinucleotide, represented in the reduced form as NADH, where the H is a hydrogen atom.

$$CH_3CH_2OH + NAD^+ \longrightarrow CH_3CHO + NADH + H^+$$

Ethanol   Acetaldehyde

Because NAD^+ is usually employed in the oxidation of fats, when one drinks too much alcohol, the fat is deposited—mainly in the abdomen—rather than metabolized. This results in the familiar "beer gut" of many heavy drinkers.

Just how ethanol intoxicates is still somewhat of a mystery. Researchers have found that ethanol disrupts receptors for two neurotransmitters (Table 19.2): *gamma*-aminobutyric acid (GABA), which depresses nerve cell activity, and glutamate, which excites certain nerve cells. Ethanol thus disrupts the brain's control over muscle activity, causing the drunk to stagger and fall.

There are several theories about the cause of alcoholism. Studies on twins raised separately show that there is undoubtedly a genetic component, but we still have much to learn about this ancient drug.

Barbiturates

A family of related depressants, the barbiturates, display a wide range of properties. They can be employed to produce mild sedation, deep sleep, or even death. Several thousand barbiturates have been synthesized over the years, but only a few have found widespread use in medicine (Figure 19.22).

- Pentobarbital is employed as a short-acting hypnotic drug. Before the discovery of modern tranquilizers, it was used widely to calm anxiety.
- Phenobarbital is a long-acting drug. It is employed widely as an anticonvulsant for epileptics and brain-damaged people.
- Thiopental, which differs from pentobarbital only in that an oxygen atom on the ring has been replaced by a sulfur atom, is used widely as an anesthetic.

Barbiturates were once used in small doses as sedatives, the dosage generally being a few milligrams. In larger dosages (about 100 mg), barbiturates induce sleep. They were once the sleeping pills of choice but now have been replaced by drugs such as a benzodiazepine (Section 19.16). Barbiturates are quite toxic; the lethal dose is about 1500 mg (1.5 g).

Barbituric acid, from which the barbiturates are derived, was first synthesized in 1864 by Adolph von Baeyer. He made it from urea, which occurs in urine, and malonic acid, which is made from apples.

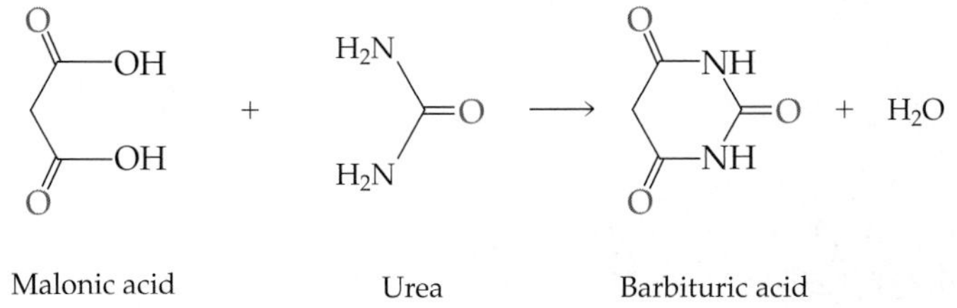

Malonic acid   Urea   Barbituric acid

The term *barbiturates*, according to Willstätter, came about because at the time of the discovery von Baeyer was infatuated with a woman named Barbara. The word comes from *Barbara* and *urea*.

Synergism: Barbiturates and Alcohol

Barbiturates are especially dangerous when ingested along with ethyl alcohol. This combination produces an effect much more drastic than the sum of the effects of two depressants. The effects of barbiturates are enhanced by a factor of up to 200 when they are taken with alcoholic beverages. This effect, in which two chemicals bring about an effect greater than that of each chemical individually, is called a **synergistic effect**. The presence of one chemical enhances the effects of the second. Synergistic effects can be deadly, and they are not limited to alcohol–barbiturate combinations. Two drugs should never be taken at the same time without competent medical supervision.

Barbiturates are strongly addictive. Habitual use leads to the development of a tolerance to the drugs, and ever-larger doses are required to produce the same degree of intoxication. The side effects of barbiturates are similar to those of alcohol: hangovers, drowsiness, dizziness, and headaches. Withdrawal symptoms are often severe, accompanied by convulsions and delirium, and can cause death.

Barbiturates are cyclic amides. The mechanism of their action is not known. The most likely site of action is on a GABA receptor.

19.15 NARCOTICS

A **narcotic** is a drug that produces narcosis (stupor or general anesthesia) and analgesia (relief of pain). Many drugs produce these effects, but in the United States only those that are also *addictive* are legally classified as narcotics. Their use is regulated by federal law.

Opium and Morphine

Opium is the dried, resinous juice of the unripe seeds of the oriental poppy (*Papaver somniferum*) (Figure 19.23). It is a complex mixture of about 20 nitrogen-containing organic bases (alkaloids), sugars, resins, waxes, and water. The principal alkaloid, morphine, makes up about 10% of the weight of raw opium. Opium was used in many patent medicines during the nineteenth century.

Morphine was first isolated in 1805 by Friedrich Sertürner, a German pharmacist. With the invention of the hypodermic syringe in the 1850s, a new method of administration became available. Injection of morphine directly into the bloodstream was more effective for the relief of pain, but this method also seriously escalated the problem of addiction.

Morphine was used widely during the American Civil War (1861–1865) for relief of pain from battle wounds. One side effect of morphine use is constipation.

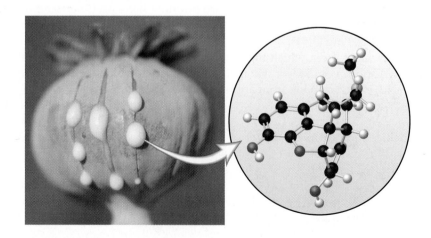

▶ **Figure 19.23** Opium poppy seed pod. The slits on the seed pod exude the resinous juice, which when dried is called opium. About 10% of the weight of raw opium is morphine.

Noting this, soldiers came to use morphine as a treatment for that other common malady of men on the battlefront—dysentery. More than 100,000 soldiers became addicted to morphine while serving in the war. The affliction was so common among veterans that it came to be known as "soldier's disease."

Morphine and other narcotics were placed under control of the federal government by the Harrison Act of 1914. Morphine is still used by prescription for the relief of severe pain. It also induces lethargy, drowsiness, confusion, euphoria, chronic constipation, and depression of the respiratory system. Morphine is addictive, and this becomes a problem when it is administered in amounts greater than the prescribed dose or for a period longer than the prescribed time.

▲ Trade card advertising Mrs. Winslow's Soothing Syrup, an opium-containing patent medicine.

Codeine and Heroin

Changes in the morphine molecule produce altered physiological properties. Replacement of the phenolic OH group by a methoxy (OCH_3) group produces codeine (see Additional Problem 76). Codeine is present in opium to an extent of about 0.7 to 2.5%, but it is usually synthesized by methylating the more abundant morphine molecules. Codeine is similar to morphine in its action, but it is less potent, has less tendency to induce sleep, and is less addictive. For relief of moderate pain, codeine is often combined with acetaminophen (for example, in Tylenol 3®) or with aspirin (in Empirin®). Codeine for cough suppression is used in liquid preparations that are regulated less stringently than stronger narcotics.

Conversion of both the OH groups of the morphine molecule to acetate ester groups produces heroin (see Additional Problem 77). This semisynthetic morphine derivative was first prepared by chemists at the Baeyer Company of Germany in 1874. It received little attention until 1890, when it was proposed as an antidote for morphine addiction. Shortly thereafter, Baeyer widely advertised heroin as a sedative for coughs, often in the same ads describing aspirin. However, it soon was found that heroin induced addiction more quickly than morphine and that heroin addiction was harder to cure.

▲ A Civil War–era morphine kit with an early hypodermic syringe.

The physiological action of heroin is similar to that of morphine. Heroin is less polar than morphine; it enters the fatty tissues of the brain more rapidly and seems to produce a stronger feeling of euphoria than morphine. Heroin is not legal in the United States, even by prescription. It has, however, been advocated for use in the relief of pain in terminal cancer patients and has been so used in Britain.

Addiction probably has three components: emotional dependence, physical dependence, and tolerance. Psychological dependence is evident in the uncontrollable desire for the drug. Physical dependence is shown by acute withdrawal symptoms such as convulsions. Tolerance for the drug is evidenced by the increasing dosages required to produce in the addict the same degree of narcosis and analgesia.

Deaths from heroin are usually attributed to overdoses, but the situation is not always clear. The problem often seems to be a matter of quality control. Street drugs vary considerably in potency, and a particular dose can be much higher in heroin than expected.

Morphine

Synthetic Narcotics

Much research has gone into developing a drug that would be as effective as morphine for the relief of pain but would not be addictive. Several synthetic narcotics are available. Some common ones are listed in Table 19.5. Structures of some of these and other narcotics are given in Figure 19.24.

The synthetic narcotic methadone is widely used to treat heroin addiction. Like heroin, methadone is highly addictive. However, when taken orally, it does not induce the sleepy stupor characteristic of heroin intoxication. Unlike a heroin addict, a person on methadone maintenance is usually able to hold a productive job. In

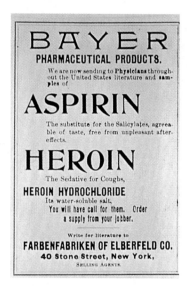

Heroin was regarded as a safe medicine in 1900; it was widely used as a cough suppressant. It was also thought to be a nonaddictive substitute for morphine.

TABLE 19.5 Narcotic Pain Medications

Substance	Duration of Action (h)	Approximate Dosage Relative to Morphine at 30 mg[a]
Codeine	4–6	200
Meperidine (Demerol)	2–4	300
Hydrocodone	4–8	20
Fentanyl	1–2[b]	(b)
Oxycodone (Oxycontin)	4–6[d]	20
Hydromorphone (Dilaudid)	4–5	7.5
Methadone	4–6[c]	(c)

[a]Oral doses for equivalent pain relief.
[b]Used in transdermal patches for chronic pain.
[c]Used for treatment of heroin addiction; not usually given for pain relief.
[d]Often dispensed in time-release form that allows 12 h between doses.

Meperidine (Demerol) (a)

Methadone (b)

Pentazocine (Talwin) (c)

Naloxone (d)

▶ **Figure 19.24** Line-angle formulas of some synthetic narcotics: (a) meperidine, (b) methadone, (c) pentazocine (Talwin®), and (d) naloxone.

some places methadone is available free in clinics. If an addict who has been taking methadone reverts to heroin, the methadone in his or her system effectively blocks the euphoric rush normally given by heroin and so reduces the addict's temptation to use heroin.

When injected into the body, methadone produces an effect similar to that of heroin, and methadone has been diverted for illegal use in this manner. An addict on methadone is still an addict. All the problems of tolerance—and cross-tolerance with heroin and morphine—still exist.

Morphine Agonists and Antagonists

Chemists have synthesized thousands of morphine analogs, but relatively few have shown significant analgesic activity, and most are addictive. Morphine acts by binding to receptors in the brain. A molecule that has morphinelike action is called a morphine **agonist**. A morphine **antagonist** inhibits the action of morphine by blocking the receptors. Some molecules have both agonist and antagonist effects. An example is pentazocine, Figure 19.24(c), which is less addictive than morphine and is effective for the relief of pain.

Pure antagonists such as naloxone, Figure 19.24(d), are of value in treating opiate addicts. An addict who has overdosed can be brought back from death's door by an injection of naloxone.

Natural Opiates

Morphine acts by binding to specific receptor sites in the brain. Why should the human brain have receptors for a plant-derived drug such as morphine? Scientists concluded that the body must produce its own substances that fit these receptors. Actually, it produces several morphinelike substances, called **endorphins** (endogenous morphines). Each is a short peptide chain composed of amino acid units. Those with five amino acid units are called *enkephalins*. There are two enkephalins, which differ only in the amino acid at the end of the chain. *Leu*-enkephalin has the sequence Tyr-Gly-Gly-Phe-Leu, and *Met*-enkephalin is Tyr-Gly-Gly-Phe-Met. Other endorphins have chains of up to 30 or more amino acids.

Some enkephalins have been synthesized and shown to be potent pain relievers. Their use in medicine is quite limited, however, because after being injected they are rapidly broken down by the enzymes that hydrolyze proteins. Researchers have sought to make analogs more resistant to hydrolysis that can be employed as morphine substitutes for the relief of pain. Unfortunately, both natural enkephalins and their analogs, such as morphine, seem to be addictive.

It appears that endorphins are released as a response to pain deep in the body. Some evidence indicates that acupuncture anesthetizes somewhat by stimulating the release of brain "opiates." The long needles stimulate deep sensory nerves that cause the release of peptides that then block the pain signals.

Endorphin release has also been used to explain other phenomena once thought to be largely psychological. A soldier, wounded in battle, may feel no pain until the skirmish is over. His body has secreted its own painkiller.

19.16 ANTIANXIETY AGENTS

The hectic pace of life in the modern world causes some people to seek rest and relaxation in chemicals. Ethyl alcohol is undoubtedly the most widely used tranquilizer. The drink before dinner—to "unwind" from the tensions of the day—is a part of the way of life for many people. Others, however, seek relief in other chemical forms.

One class of widely used antianxiety drugs are the benzodiazepines, compounds that feature seven-member heterocyclic rings (Figure 19.25). More than 50 benzodiazepines have been marketed in more than 100 different preparations. Three common ones are diazepam, a classic antianxiety agent; clonazepam, an anticonvulsant as well as antianxiety drug; and temazepam, used to treat insomnia. Antianxiety agents make people feel better simply by making them feel dull and insensitive, but they do not solve any of the underlying problems that cause anxiety. They are thought to act on a GABA receptor, dampening neuron activity in higher function parts of the brain.

Antianxiety agents are sometimes called *minor tranquilizers*.

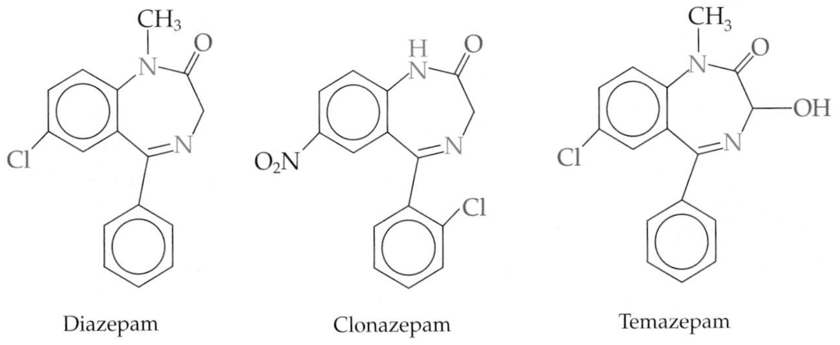

Diazepam · Clonazepam · Temazepam

◀ **Figure 19.25** Three benzodiazepines: (a) diazepam (Valium®), (b) clonazepam (Klonopin®), and (c) temazepam (Restoril®).

What price tranquility? After 20 years of use, benzodiazepines were found to be addictive. People trying to go off these drugs after prolonged use experience painful withdrawal.

Antipsychotic Agents

For centuries, the people of India used the snakeroot plant, *Rauwolfia serpentina*, to treat a variety of ailments including fever, snakebite, and other poisonings, and—most importantly—to treat maniacal forms of mental illness. Western scientists became interested in the plant near the middle of the twentieth century—after dismissing such remedies as quackery for generations.

In 1952 rauwolfia was introduced into American medical practice as an antihypertensive agent by Robert Wilkins of Massachusetts General Hospital, the same year that Swiss chemist Emil Schlittler isolated an active alkaloid from rauwolfia and named it reserpine. Reserpine was found not only to reduce blood pressure but also to bring about sedation. The latter finding attracted the interest of psychiatrists, who found it so effective that by 1953 it had replaced electroshock therapy for 90% of psychotic patients.

Antipsychotic agents are often referred to as *major tranquilizers*.

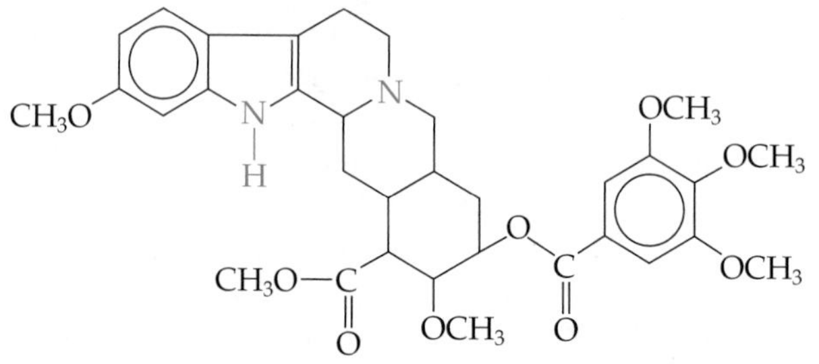

Reserpine

Phenothiazines

Also in 1952 chlorpromazine (Thorazine®) was administered as a tranquilizer to psychotic patients in the United States. The drug had been tested in France as an antihistamine, and medical workers there had noted that it calmed mentally ill patients being treated for allergies. Chlorpromazine was found to be helpful in controlling the symptoms of schizophrenia and truly revolutionized mental illness therapy.

Chlorpromazine is one of several related compounds called phenothiazines. Several of these compounds are used in medicine. Promazine (chlorpromazine without the chlorine atom) is also a tranquilizer, but it is much less potent than chlorpromazine.

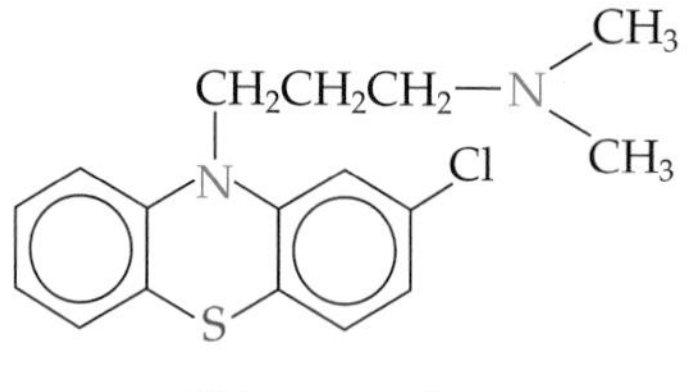

Chlorpromazine (Thorazine)

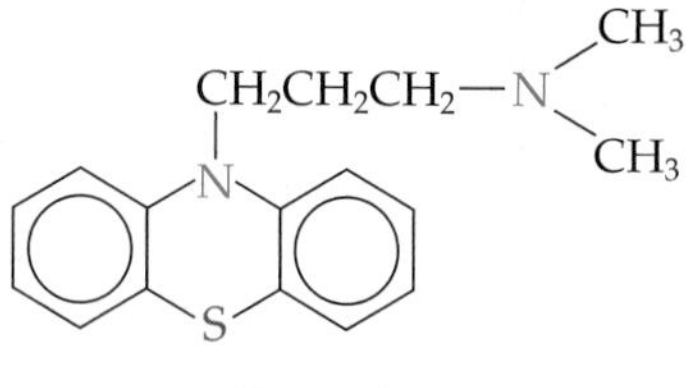

Promazine

Phenothiazines act in part as dopamine antagonists; they block postsynaptic receptors for dopamine, a neurotransmitter important in the control of detailed motion (such as grasping small objects), in memory and emotions, and in exciting the cells of the brain. Some researchers think schizophrenic patients produce too much dopamine, whereas others think that they have too many dopamine receptors. In either case, blocking the action of dopamine relieves the symptoms of schizophrenia.

Newer so-called atypical antipsychotics are now available. They are as effective as the older drugs for reducing the positive psychotic symptoms such as hallucinations and delusions, but may be better than the older medications at relieving the negative symptoms, such as apathy, withdrawal, and problems with thinking. Atypical antipsychotics include aripiprazole (Abilify®), risperidone (Risperdal®), clozapine (Clozaril®), and olanzapine (Zyprexa®). These drugs are thought to act on a type of serotonin receptor that loosens the binding of dopamine receptors, allowing more dopamine to reach the neurons.

Antipsychotic drugs have served to reduce greatly the number of patients confined to mental hospitals. They do not cure schizophrenia, but control its symptoms to the extent that 95% of all schizophrenics no longer need hospitalization. Patients who go off their medication relapse. Many homeless people are mentally ill, and there is no good way to get them to take medication without incarceration.

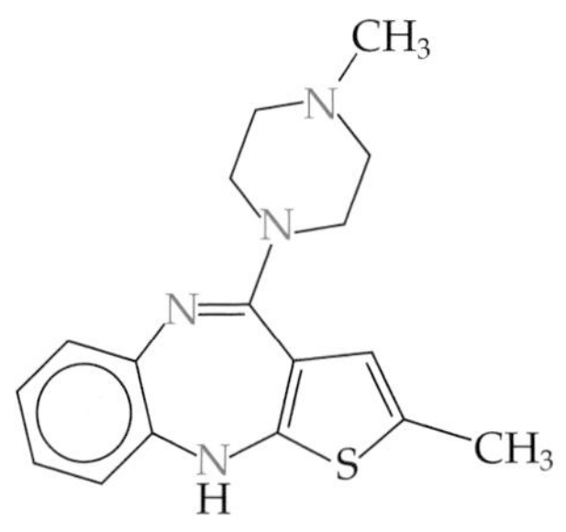

Olanzapine (Zyprexa®)

Antidepressant Drugs

Slight changes in the structure of a molecule can result in profound changes in properties. Promazine is not very potent, but replacing the sulfur atom of promazine with a CH_2CH_2 group produces imipramine (Tofranil®), a fairly potent tricyclic (three-ring) antidepressant. Another tricyclic antidepressant drug is amitriptylene (Elavil®), in which the ring nitrogen atom also is replaced by a carbon atom. Tricyclic antidepressants have been available since the 1950s, but they have serious side effects and have been largely replaced by newer drugs.

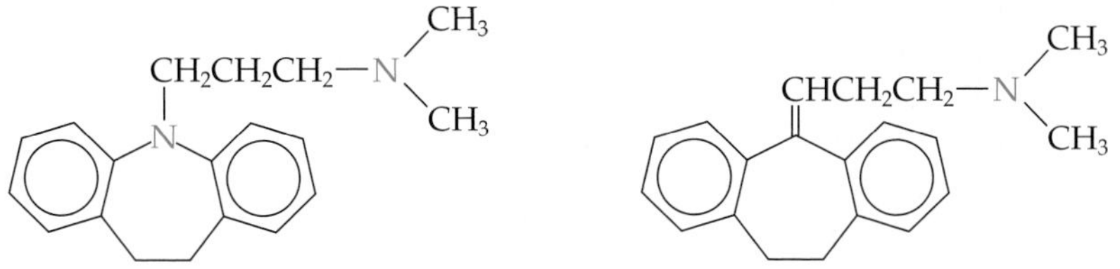

Imipramine
(Tofranil)

Amitriptylene
(Elavil)

Common antidepressants today include drugs called *selective serotonin reuptake inhibitors* (SSRIs) Three major ones are shown in Figure 19.26. Doctors prescribe these drugs to help people cope with gambling problems, obesity, fear of public speaking, and premenstrual syndrome (PMS). The drugs work by enhancing the effect of serotonin, blocking its reabsorption ("reuptake") by nerve cells. They seem to be safer than tricyclic antidepressants and more easily tolerated.

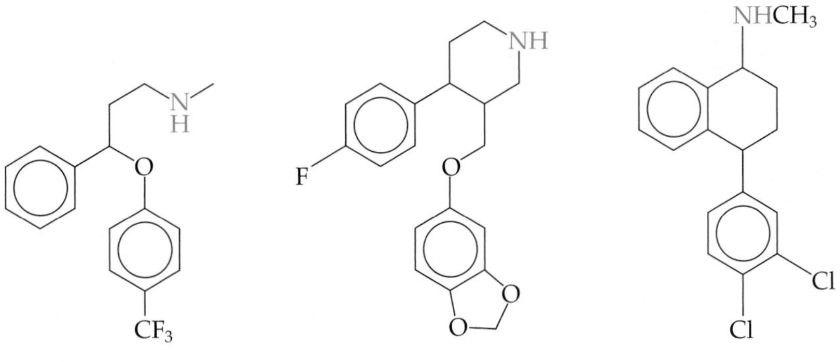

Fluoxetine

Paroxetine

Sertraline

▲ **Figure 19.26** Three selective serotonin reuptake inhibitors (SSRIs): Fluoxetine (Prozac®), paroxetine (Paxil®), and sertraline (Zoloft®).

19.17 STIMULANT DRUGS

Among the more widely known stimulant drugs are a variety of synthetic amines related to β-phenylethylamine (Figure 19.27). These drugs, called amphetamines, are similar in structure to epinephrine and norepinephrine (Section 19.12). They probably act as stimulants by mimicking natural brain amines such as norepinephrine.

Amphetamines

Amphetamine and methamphetamine are inexpensive and have been widely abused. **Amphetamine** was once—but is no longer—used as a diet drug, and it has also been employed in treating mild depression and narcolepsy, a rare form of sleeping sickness. Amphetamine induces excitability, restlessness, tremors, insomnia, dilated pupils, increased pulse rate and blood pressure, hallucinations, and psychoses.

Methamphetamine has a more pronounced psychological effect than amphetamine. Illegal "meth labs" have been found in all parts of the country. Methamphetamine is made readily from an antihistamine and ammonia. Several states have instituted restrictions in the availability of the ingredients, hoping to combat methamphetamine manufacture. Like other amine drugs, amphetamines are usually available as hydrochloride salts. A free-base form of methamphetamine (like crack cocaine) is used for smoking. It is called "ice" because of its crystalline appearance.

Phenylpropanolamine, like amphetamine, was widely used as an over-the-counter appetite suppressant. Like its relatives, this compound is a stimulant. Studies show that it is at best marginally effective as a diet aid, and it poses a threat to people with hypertension.

One controversial use of amphetamines is their employment in the treatment of attention deficit disorder (ADD) in children. The drug of choice is often methylphenidate (Ritalin®). Although it is a stimulant, the drug seems to calm children who otherwise can't sit still. This use has been criticized as "leading to drug abuse" and as "solving the teacher's problem, not the kid's."

Like many other drugs, amphetamine exists as mirror image isomers (enantiomers). Benzedrine® is the trade name for a mixture of the two isomers in equal amounts. The dextro (righthanded) isomer is a stronger stimulant than the levo (lefthanded) isomer. Dexedrine® is the trade name for dextroamphetamine, a drug made of the pure dextro isomer.

Dexedrine is two to four times as active as Benzedrine. Many current drugs have dextro and levo isomers, and most of them are sold as mixtures of the isomers. The development of drugs that contain only one enantiomer is becoming increasingly important in the pharmaceutical industry. Worldwide sales of single-enantiomer drugs is almost $200 billion annually.

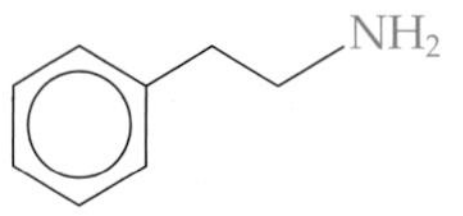

(a) 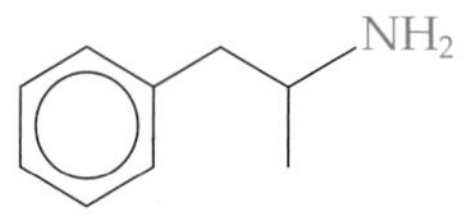   (b) 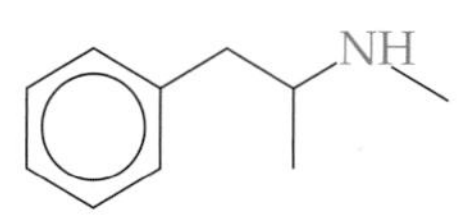   (c)

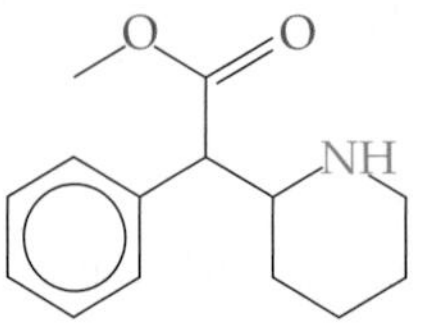

(d) 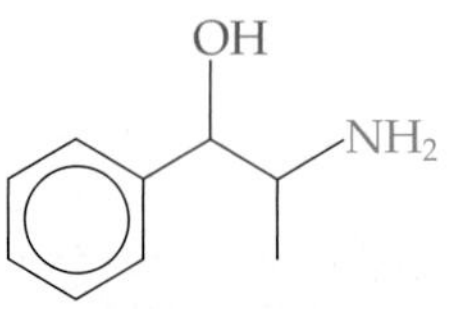   (d)

▶ **Figure 19.27** β-Phenylethylamine (a) and related compounds. (b) Amphetamine, (c) methamphetamine, (d) methylphenidate, and (e) phenylpropanolamine.

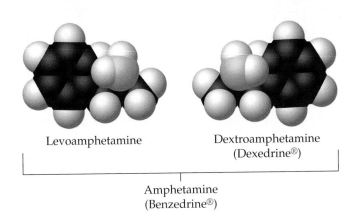

Levoamphetamine   Dextroamphetamine
(Dexedrine®)

Amphetamine
(Benzedrine®)

◀ Benzedrine is a mixture of Dexedrine and its enantiomer, levoamphetamine. As is often the case, the two isomers differ greatly in their biological activity.

Cocaine

Cocaine, first used as a local anesthetic (Section 19.14), also serves as a powerful stimulant. The drug is obtained from the leaves of a shrub that grows almost exclusively on the eastern slopes of the Andes Mountains. Many of the Indians living in and around the area of cultivation chew coca leaves—mixed with lime and ashes—because of their stimulant effect. Cocaine used to arrive in the United States as the salt cocaine hydrochloride, but now much of it comes in the form of broken lumps of the free base, a form called *crack cocaine.*

Cocaine hydrochloride is readily absorbed through the watery mucous membrane of the nose, and this is the form used by those who "snort" cocaine. Those who smoke cocaine use the free base (crack), which readily vaporizes at the temperature of a burning cigarette. When smoked, cocaine reaches the brain in 15 s. It acts by preventing the neurotransmitter dopamine from being reabsorbed after it is released by nerve cells. High levels of dopamine are therefore available to stimulate the pleasure centers of the brain. After the binge, dopamine is depleted in less than an hour, leaving the user in a pleasureless state and (often) craving more cocaine.

The use of cocaine increases stamina and reduces fatigue, but the effect is short lived. Stimulation is followed by depression. Once quite expensive and limited to use mainly by the wealthy, cocaine is now available in cheap, potent forms. Hundreds, including several well-known athletes, have died from cocaine overdose.

Cocaine

TABLE 19.6	Caffeine in Soft Drinks
Brand	**Caffeine***
Barq's Root Beer	22
Diet Coke	45
Mello Yello	51
Diet Mr. Pibb	40
Vanilla Coke	34
Dr Pepper	41
Sun Drop	63
Sunkist Orange	41
Mountain Dew	55
Pepsi-Cola	37
Royal Crown Cola	43

Source: American Beverage Association.

* Milligrams per 12-oz serving.

Caffeine: Coffee, Tea, or Cola

Coffee, tea, and cola soft drinks naturally contain the mild stimulant caffeine. Caffeine is also added to many other soft drinks (Table 19.6). An effective dose of caffeine is about 200 mg. This is about the amount in one cup of strong coffee or three to five cups of tea. Caffeine is also available in tablet form as a stay-awake or keep-alert type of drug. Two well known brands are No-Doz and Vivarin.

Is caffeine addictive? The "morning grouch" syndrome suggests that it is mildly so. There is also evidence that large doses of caffeine may be involved in chromosome damage. To be safe, people in their childbearing years should avoid large quantities of caffeine. Overall, the hazards of caffeine ingestion seem to be slight.

Caffeine

Nicotine: Going Up in Smoke

Another common stimulant is nicotine. This drug is taken by smoking or chewing tobacco. Nicotine is highly toxic to animals and has been used in agriculture as a contact insecticide. It is especially deadly when injected; the lethal dose for a human is estimated to be about 50 mg. Nicotine seems to have a rather transient effect as a stimulant. This initial response is followed by depression, but smokers generally keep a near-constant level of nicotine in their bloodstream by indulging frequently.

Is nicotine addictive? Casual observation of a person trying to quit smoking seems to indicate that it is. Consider the 1972 memorandum from a Philip Morris scientist who noted that "no one has ever become a cigarette smoker by smoking cigarettes without nicotine." He suggested that the company "think of the cigarette as a dispenser for a dose unit of nicotine."

Nicotine

Clonidine, a drug used to treat high blood pressure, reduces nicotine withdrawal symptoms.

19.18 PSYCHEDELIC DRUGS

Psychedelic drugs are consciousness-altering substances that induce changes in sensory perception. Sometimes called "mindbenders" because they qualitatively change the way we perceive things, common psychedelic drugs include LSD, psilocybin, mescaline, and dimethyltryptamine (DMT). Marijuana is sometimes considered a mild psychedelic.

Lysergic Acid Diethylamide (LSD)

Lysergic acid diethylamide (LSD), a semisynthetic drug, is a powerful psychedelic. Its physiological properties were discovered quite accidentally by Swiss chemist Albert Hofmann in 1943. Hofmann unintentionally ingested some LSD. He later took 250 mg, which he considered a small dose, to verify that LSD had caused the symptoms he had experienced. Hofmann had a rough time for the next few hours, experiencing such symptoms as visual disturbances and schizophrenic behavior.

LSD can create a feeling of lack of self-control and sometimes of extreme terror. The exact mechanism by which LSD exerts its psychic effects is still unknown, but it is thought to act on serotonin receptor sites in the central nervous system.

Lysergic acid is obtained from ergot, a fungus that grows on rye. This carboxylic acid is converted to its diethylamide derivative.

Several medical drugs are obtained from the ergot fungus. Because ergotamine shrinks blood vessels in the brain, it is used to treat migraine headaches. Ergonovine induces uterine contractions and can reduce bleeding after childbirth. Like LSD, both these compounds are lysergic acid amides.

Lysergic acid diethylamide
(LSD)

The potency of LSD is indicated by the small amount required for a person to experience its fantastic effects. The usual dose is probably about 10–100 μg. No wonder Hofmann had a bad time with 250 mg. To give you an idea of how small 10 μg is, let's compare that amount of LSD with the amount of aspirin in one tablet—one aspirin tablet contains 325,000 μg of aspirin.

Psilocybin and Mescaline

A variety of plants produce psychedelic compounds:

- Psilocybin, a psychedelic alkaloid of the tryptamine family, is found in several species of mushrooms, including *Psilocybe cubensis*. Its effects are similar to those of LSD but are of shorter duration.
- Mescaline (3,4,5-trimethoxyphenylethylamine) is a psychedelic drug related to phenylethylamine (compare to Figure 19.27). It is found in the peyote cactus, *Lophophora williamsii*, and other species. It is also made synthetically. Peyote is used in the religious ceremonies of some Indian tribes. The effects last for up to 12 hours.

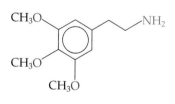

Mescaline

Marijuana

Books have been written about marijuana, yet all we know for certain about this drug would fill only a few pages. *Cannabis sativa* has long been useful. The stems yield tough fibers for making ropes. *Cannabis* has been used as a drug in tribal religious rituals and also has a long history as a medicine, particularly in India. In the United States, marijuana is second only to alcohol in popularity as an intoxicant.

The term **marijuana** refers to a preparation made by gathering the leaves, flowers, seeds, and small stems of the plant (Figure 19.28), which are generally dried

and smoked. They contain various chemical substances, many of them still uniden-
tified. The principal active ingredient, however, is tetrahydrocannabinol (THC). Although there are several active cannabinoids in marijuana, only one is shown here.

$$CH_3$$

Tetrahydrocannabinol
(THC)

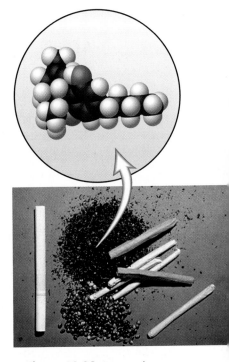

▲ **Figure 19.28** Prepared
marijuana.

Marijuana plants vary considerably in THC potency, depending on their genetic variety. Wild plants native to the United States have a low THC content, usually
about 0.1%, but some marijuana sold in North America now has a THC content approaching 6%.

The effects of marijuana are difficult to measure, partly because of the variable
amount of THC in different samples. A variety with standard potency is now
grown and used in controlled clinical studies. With this product, some of the effects
of marijuana can be measured in reproducible experiments. Smoking *Cannabis* increases the pulse rate, distorts the sense of time, and impairs some complex motor
functions. Other possible effects include a euphoric floating sensation, a feeling of
anxiety, a heightened enjoyment of food, and a false impression of brilliance. Although studies have shown no mind-expanding effects, users sometimes experience hallucinations.

The long-term effects of marijuana use are more difficult to evaluate. There is
some evidence that long-term marijuana use causes brain damage. People who use
marijuana heavily are often lazy, passive, and mentally sluggish, but it is difficult to
prove that marijuana is the cause. Even if it is, the damage is less extensive than that
caused by heavy use of alcohol. Some people claim that excessive use of marijuana
leads to psychoses. It has long been known to induce short-term psychotic episodes
in those already predisposed and in others who take excessive amounts. Long-term
psychoses, however, occur at about the same rate among regular marijuana users as
among members of the general population.

Some studies have shown that marijuana has a feminizing effect on male users,
causing loss of libido and a growth of breasts. Some evidence suggests that it stimulates the body to convert testoterone to estrogen, the female hormone. There are
about 9000 breast reduction surgeries on males each year, but just how many—if
any—result from marijuana use is uncertain. Experiments with rats show that
crude marijuana extract and condensed marijuana smoke compete with estradiol
for binding to the estrogen receptors. Pure THC, however, did not interact with the
estrogen receptor, nor did 10 of its metabolites. Of several other cannabinoids tested, only cannabidiol showed any estrogen receptor binding.

Scientists have learned how THC acts in the brain to produce its various effects.
When a person smokes marijuana, THC rapidly passes from the lungs into the
bloodstream and thus to the brain, where it connects to specific sites called cannabi-
noid receptors on nerve cells. Some regions of the brain have many such receptors;
other areas have few or none. The parts of the brain that influence pleasure, memory, thought, concentration, judgment, sensory and time perception, and regulate
movement are rich in receptors.

As THC enters the brain, it activates the brain's reward system in the same way
that food and drink do. Like nearly all drugs of abuse, THC causes a euphoric feeling by stimulating the release of dopamine.

The location of receptors in movement control centers in the brain explains the
loss of coordination seen in those intoxicated by the drug. The presence of receptors in memory and cognition areas of the brain explains why marijuana users do
poorly on tests. That few receptors are present in the brainstem where breathing

and heartbeat are controlled is probably why it is hard to get a lethal dose of pure marijuana.

If THC binds to receptors in the brain, the brain must produce a THC-like substance. This substance is thought to be anandamide derived from arachidonic acid, which is the precursor of prostaglandins.

Anandamide

Marijuana has some legitimate medical uses. It reduces eye pressure in people who have glaucoma. If not treated, this increasing pressure eventually causes blindness. Marijuana also relieves the nausea that afflicts cancer patients undergoing radiation treatment and chemotherapy.

19.19 DRUG PROBLEMS

The existence of illegal drugs causes enormous problems. Their marketing is covert, with sellers failing to report their income or pay taxes on it. This makes the illicit drug business enormously lucrative. Annual revenues for drug dealers amount to at least $400 billion, about half of which comes from the United States. Worldwide, people spend more money for illegal drugs than for food. So much unlawful wealth makes illegal drugs a dangerous business in which crimes are frequent, murders are common, and corrupt politicians abound. **Drug abuse** (using drugs for their intoxicating effects) is a serious problem, not only for the abusers but for society in general.

The illegal drug user is the biggest loser. Street drugs are expensive, and most are addictive. Some addicted users have to steal in order to pay for drugs and eventually end up in jail; one study found that 16% of convicted jail inmates admitted that they committed their offense to get money for drugs. In the workplace 50–80% of all accidents and personal injuries are drug related. Drug users are absent from work over twice as often as nonusers, and they are five times as likely to file claims for compensation. They are a drain on their employers' income, and they often have trouble keeping a job.

A recurring problem is that illegal drugs are not always what they are supposed to be. Buyers simply have to trust the information sellers give them about the identity and quality of the products they buy. In fact, crime labs have generally found nearly two-thirds of all drugs (other than marijuana) brought in for analysis to be something other than what the dealers said they were.

Of course, there can be difficulties even with legal drugs that have been approved and tested. Problems range from faulty prescriptions written by physicians, to pharmacy errors in filling prescriptions, to patient mistakes in taking medications. Many drugs have undesirable side effects that are worse than the condition being treated. There are drugs (both prescription and over the counter) that have such similar names that they are easily confused. **Drug misuse** (for example, using penicillin, which has no effect on viruses, to treat a viral infection) is all too common. Such overuse of penicillin has led to strains of bacteria that are now immune to penicillin.

Another problem is that the cost of many drugs makes them unaffordable for the people who need them. Pharmaceutical research and development is expensive, and companies must often charge high prices for the products they develop, at least for a while. After the patent rights expire, other companies can make "generic" versions of the same products at lower prices because they have incurred no research costs. Generic drugs are usually chemically identical to the original products.

Chemistry has provided many drugs of enormous benefit to society, but sometimes these drugs can create problems.

19.20 THE PLACEBO EFFECT

An interesting phenomenon associated with the testing of drugs is the **placebo effect**. A placebo is an inactive substance given in the form of medication to a patient.

A common way to evaluate a new drug is to administer it to one group of patients and to give placebos to a similar "control" group. In the "double-blind" study of the sort most accepted by the medical community, neither the patients nor the doctors know who is receiving the real drug and who is receiving the placebo.

Sometimes people who think they are receiving a certain drug expect positive results, and may actually experience such results, even though they have not been given the actual drug. In one study of a new tranquilizer, 40% of the patients who had been given placebos reported that they felt much better and rated the drug as highly effective. The psychological effect of a placebo can be very powerful.

19.21 NEW USES FOR OLD DRUGS

The average cost of developing a new drug is now estimated at more than $800 million. One of the most expensive and time-consuming steps in the process is scientific testing, especially human testing. If an old drug happens to be found successful in treating an ailment different from the one for which it is commonly used, the pharmaceutical company, and ultimately the consumer, saves a lot of money. The drug has already been tested for safety and has been approved for use in humans. Following are a few examples of some old drugs for which new uses have been found.

- The drug topirimate (Topamax®), used by epilepsy patients to prevent seizures, is also useful in reducing migraine headaches.
- The rheumatoid arthritis drug infliximab (Remicade®) seems also to be useful against psoriasis, greatly reducing its skin lesions. (The cause of psoriasis is not known, but it is accompanied by high levels of inflammatory molecules, as is also the case with arthritis.) This drug also shows promise as a treatment for Crohn's disease, which involves bowel inflammation. In some cases the drug has caused Crohn's disease to go into remission.
- Buproprion hydrochloride (Wellbutrin®, Zyban®), an antidepressant used to help people stop smoking, also seems to be an effective painkiller.
- Erythropoietin, long used to treat anemia, is also used to boost endurance in athletes. It may also protect against nerve damage in victims of stroke and spinal cord injuries and against glaucoma.
- Sildenafil is a drug originally developed by scientists to treat heart disease. It is now well known as the anti-impotence drug Viagra®.
- The Botox® injections that are given to reduce facial wrinkling also seem to reduce the incidence of migraine headaches. Botox is a dilute solution of botulin (Chapter 20), which is the deadly poison that causes botulism.
- Finally there is aspirin, the most widely used of all drugs. It relieves headaches and other pains, reduces fevers, and thins the blood helping to prevent strokes and heart attacks. One of the cheapest and safest of drugs, it may also reduce the risk of some cancers in men (but not in women). Studies show that it may also delay the onset of Alzheimer's disease.

Critical Thinking Exercises

Apply knowledge that you have gained in this chapter and one or more of the FLaReS principles (Chapter 1) to evaluate the following statements or claims.

19.1 An actor in a television advertisement states, "For aches and pain I take Tylenol, but for headaches I take Excedrin. Excedrin works better for headaches."

19.2 St. John's wort, an herb derived from a plant, *Hypericum perforatum*, is recommended for treatment of depression.

19.3 A nurse claims to be able to cure disease by "therapeutic touch," a technique that she says allows her to sense "human energy fields" by simply moving her hands above the patient's body.

19.4 A television advertisement shows an actor taking a medicine that inhibits the production of stomach acid, then eating a large meal of rich, spicy food. Does the ad prove the claim of heartburn prevention?

19.5 A television advertisement shows an actor taking an antihistamine and then joyously walking though a grassy, flower-filled meadow without experiencing the usual allergy symptoms. What, if anything, does the ad prove about alleviation of allergies?

19.6 A Web article recommends *S*-adenosylmethionine (SAMe) for treatment of depression, arthritis, and liver disease.

SUMMARY

Sections 19.1–19.2—A **drug** is a substance that relieves pain, treats illness, or improves one's health. Many drugs are obtained from natural sources. Paul Ehrlich founded **chemotherapy**, based on the fact that some chemicals were more toxic to disease organisms than to human cells. Aspirin is a **nonsteroidal anti-inflammatory drug (NSAID)** to distinguish it from more powerful steroids. It is an **analgesic** (pain reliever), **antipyretic** (fever reducer), and **anti-inflammatory** (reduces inflammation). NSAIDs inhibit the production of prostaglandins that send pain messages to the brain. Ibuprofen, ketoprofen, and naproxen are also NSAIDs. Acetaminophen is an analgesic and antipyretic but does not reduce inflammation. Many analgesic products are combinations of NSAIDs and other drugs.

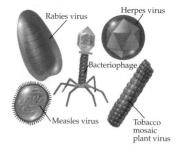

Acetylsalicylic acid (aspirin)

Sections 19.3–19.4—Cold symptoms are often treated with an **antihistamine** to relieve symptoms caused by an **allergen**. The allergen triggers the release of histamine, causing sneezing, itchy eyes, and runny nose. Colds are also treated with a cough suppressant, a nasal decongestant, and/or an expectorant to loosen mucus. None of these treatments cure a cold; they simply treat the symptoms.

Sulfanilamide

Sulfa drugs were the first antibacterial drugs; they mimic compounds essential for bacterial growth. An **antibiotic** is a soluble substance derived from mold or bacteria that inhibits microorganism growth. Penicillins prevent bacteria from forming their mucoprotein cell walls but do not harm animal cells which do not have mucoprotein walls. More modern antibiotics include cephalosporins, tetracyclines, and fluoroquinolines. These are **broad-spectrum antibiotics** that are effective against a wide variety of bacteria. Unfortunately, antibiotic-resistant bacteria often develop after an antibiotic has been used for a while.

Sections 19.5–19.6—Viral diseases cannot be cured with antibiotics. Some, such as measles, mumps, and smallpox, are usually prevented by vaccination, but vaccines are not available for all viral diseases. Herpes and **acquired immune deficiency syndrome (AIDS)** are viral infections that have no cure. Viruses are composed of nucleic acids—either DNA or RNA—and proteins. A **retrovirus** such as the AIDS virus is an RNA virus that synthesizes DNA in the host cell. Antiviral agents are used to treat some viral diseases. Drug design was often hit-or-miss in its early days, but now drugs can be designed to fit receptors. Most anticancer drugs are **antimetabolites**, which inhibit DNA synthesis, slowing the growth of cancer cells. Alkylating agents transfer alkyl groups to biomolecules, blocking their usual action. Nitrogen mustards, which arose from chemical warfare research, are alkylating agents.

Rabies virus
Herpes virus
Bacteriophage
Measles virus
Tobacco mosaic plant virus

Sections 19.7–19.8—A **hormone** is a chemical messenger, sending signals for physiological changes in other parts of the body. A **prostaglandin** is a hormonelike lipid that acts as a mediator of hormone action. There are six primary prostaglandins. Some hormones and prostaglandins are **steroids**, having a characteristic four-ring structure. Cholesterol and cortisol are steroids, as are sex hormones. An **androgen** stimulates or controls masculine charac-

Major Endocrine Glands

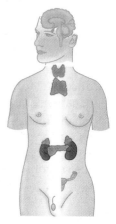

teristics, and an estrogen controls female sexual functions. Estradiol, estrone, and progesterone are estrogens. A progestin is a steroid hormone that has the effect of progesterone. Birth control pills are mixtures of an estrogen and a progestin, which act by creating a state of pseudopregnancy, preventing ovulation and conception. Birth control pills are safer than childbirth for nonsmokers but much less safe for smokers.

Section 19.9—Diseases of the circulatory system include coronary artery disease, arrhythmia, hypertension, and congestive heart failure. Drugs for hypertension include diuretics, beta blockers, calcium channel blockers, and ACE inhibitors. Drugs for arrhythmia alter the flow of ions across membranes. Coronary artery disease treatment is often the same as for hypertension; however, organic nitro compounds are also used.

Section 19.10—Psychotropic drugs are drugs that affect the mind. They can be divided into three classes: stimulant drugs such as cocaine and amphetamines; depressant drugs such as alcohol, anesthetics, opiates, and tranquilizers; and hallucinogenic drugs such as LSD.

Sections 19.11–19.12—The neurons (nerve cells) of the nervous system have tiny gaps called synapses between them. Neurotransmitters are chemicals released into the synapses during nerve transmissions which cause changes in the receiving cell. Mental illness is thought to involve brain amines such as norepinephrine (NE) and serotonin. NE agonists enhance or mimic the action of NE. NE antagonists block its action and slow down various processes. Agonist and antagonist molecules must have the proper shape. Many compounds exist as stereoisomers, with the same formula but different arrangements of atoms. Stereoisomers that are not superimposable on their mirror images are called enantiomers and usually have one or more chiral carbon atoms (have four different groups attached). Some brain amines appear to be affected by diet.

Synapse

Sections 19.13–19.14—An anesthetic is a depressant that causes lack of feeling or awareness. A general anesthetic produces unconsciousness and general insensitivity to pain. A local anesthetic causes loss of feeling in a part of the body. Older general anesthetics include diethyl ether, chloroform, and nitrous oxide. Modern inhalant anesthetics include fluorine-containing compounds. A variety of other drugs are used for anesthesia in surgical practice. Local anesthetics include procaine and lidocaine, which share structural features of *para*-aminobenzoic acid. A dissociative anesthetic dissociates perception from sensation. Ketamine and PCP are dissociative anesthetics, but PCP is abused, causing illusions and hallucinations.

Alcohol is the most common depressant used worldwide. It appears to disrupt two neurotransmitters that control muscle activity. Barbiturates such as pentobarbital and thiopental are used as sedatives and anesthetics. Alcohol and barbiturates together produce a synergistic effect, with the action of one enhancing the other; the result can be deadly.

Sections 19.15–19.16—A narcotic relieves pain and produces stupor or anesthesia but is addictive. Opium from the oriental poppy contains morphine, a natural narcotic. Codeine is made from morphine; it is less potent and less addictive. Codeine is often combined with other analgesics for pain relief, and it suppresses coughs. Heroin is a derivative of morphine that is much more potent and highly addictive. Several synthetic narcotics have been prepared. A morphine agonist has morphinelike action; a morphine antagonist blocks morphine receptors. The body produces morphine agonists called endorphins, in response to extreme pain.

Tranquilizers are depressants that some people use to relax. Some are addictive. Phenothiazines are one type of tranquilizer that act as dopamine antagonists and relieve symptoms of schizophrenia. Antidepressants are sometimes prescribed which enhance the effect of serotonin.

Sections 19.17–19.18—Amphetamines are stimulants, similar in structure to epinephrine. Amphetamine exists as enantiomers, one of which is much more potent than the other. Methamphetamine is even more potent than amphetamine and is easy to make. Ritalin is an amphetamine used to treat attention deficit disorder (ADD). Cocaine is obtained from the coca plant. It is a local anesthetic as well as a powerful stimulant and is very addictive. Caffeine in coffee, tea, and soft drinks and nicotine in tobacco are common legal stimulants. Psychedelic drugs induce changes in sensory perception. The best-known psychedelic drugs are LSD and marijuana. Psilocybin from certain mushrooms, and mescaline from peyote cactus, are natural psychedelics.

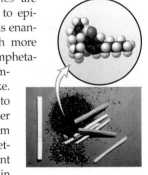

Sections 19.19–19.21—Drug abuse—using drugs for their intoxicating effects—is a serious problem for abusers and society. Drug misuse, involving legal drugs used incorrectly or by mistake, is also a problem. Drugs must be tested properly to avoid the placebo effect, in which an inactive substance, given as medication, produces results in a patient for psychological reasons. The testing procedure for drugs is expensive and time-consuming. Sometimes existing drugs are found to have new uses, which ultimately benefits both the pharmaceutical companies and the consumer.

REVIEW QUESTIONS

1. What is an antibiotic?
2. What is a hormone?
3. What is cortisol? Prednisone?
4. How do birth control pills work?
5. What are prostaglandins? How do prostaglandins differ from hormones?
6. Describe two medical uses of barbiturates and the relative dosage level of each.
7. What are the hazards of long-term barbiturate use?

8. Describe a synergistic reaction involving drugs.
9. Name two dissociative anesthetics.
10. Describe modern anesthesiology.
11. What is a narcotic?
12. What is opium?
13. How may our mental state be related to our diet?
14. Define or identify each of the following.
    a. neuron   b. synapse
    c. neurotransmitter

15. Are tranquilizers a cure for schizophrenia?
16. What are psychotropic drugs?
17. List the three classes of drugs that affect the mind.
18. What are the effects of a hallucinogenic drug?
19. What is a placebo?
20. What is the difference between drug misuse and drug abuse? Give an example of each.

PROBLEMS

NSAIDs and Related Drugs

21. What is an analgesic? How does an analgesic work?
22. What is an antipyretic? How does an antipyretic work?
23. What is the chemical name for aspirin?
24. How do NSAIDs suppress inflammation?
25. In what ways do aspirin and acetaminophen act similarly?
26. Why is acetaminophen chosen over aspirin for the relief of pain associated with surgical procedures?
27. Why is aspirin chosen over acetaminophen for the treatment of arthritis?
28. What is a COX-2 inhibitor?
29. What two functional groups are present in an aspirin (acetylsalicylic acid) molecule?
30. What two functional groups are present in an acetaminophen molecule?

Antibacterial Drugs

31. What are penicillins?
32. What are sulfa drugs? How do they work?
33. Looking at the molecular models of sulfanilamide and para-aminobenzoic acid on page 583, write structural formulas for the two compounds.
34. How can a drug such as penicillin kill bacterial cells without killing human cells?
35. Tetracyclines are broad-spectrum antibiotics. What does this mean?
36. What are fluoroquinolones?

Antiviral Drugs

37. Do antiviral drugs cure viral diseases? What is the best way to deal with viral diseases?
38. What are the three classes of antiviral drugs? How does each work?

Anticancer Drugs

39. List two major classes of anticancer drugs.
40. Do anticancer drugs sometimes cause cancer? Explain.

Hormones and Birth Control Drugs

41. What is an estrogen?

42. What is an androgen?
43. How does mifepristone (RU-486) work?
44. What are emergency contraceptives? How do they work?

Drugs and the Mind

45. What are the effects of a depressant drug?
46. What are the effects of a stimulant drug?
47. What is a psychotropic drug?
48. What is a hallucinogenic drug?

Anesthetics

49. What is a general anesthetic?
50. What is a local anesthetic?
51. Which of these anesthetics is dangerous because of flammability?
    a. diethyl ether
    b. halothane
    c. chloroform
52. For each of the following anesthetics, describe a disadvantage associated with its use. Do not include flammability.
    a. nitrous oxide
    b. halothane
    c. diethyl ether
53. From what carboxylic acid are many local anesthetics derived?
54. What structural feature characterizes the more powerful local anesthetics?

Narcotics

55. How does codeine differ from morphine in its physiological effects? In what ways are the two drugs similar?
56. How does heroin differ from morphine in its physiological effects? In what ways are the two drugs similar?
57. How does methadone maintenance work?
58. How are endorphins related to (a) the anesthetic effect of acupuncture and (b) the absence of pain in a wounded soldier?
59. What is an agonist?
60. What is an antagonist?

ADDITIONAL PROBLEMS

61. Which of the following compounds are classified as steroids?
- **a.** tristearin
- **b.** cholesterol
- **c.** testosterone
- **d.** prostaglandins

62. What structural feature makes a birth control steroid effective orally?

63. Review the "Structure–Function Relationships" essay (page 587), and then draw the structure for a fluoroquinoline drug, which should be similar to Ciprofloxacin in action but does not have the large nitrogen-containing ring at position 7.

64. A patient was treated with the antibiotic ampicillin for a urinary tract infection. The daily dose was 40mg/kg divided into three doses per day. **(a)** What was the daily dose, in grams, for the 140-lb man? **(b)** The ampicillin capsules each contain 500 mg of active ingredient. How many capsules did the patient have to take every eight hours?

65. What structural feature is shared by all steroids?

66. What structural feature is shared by all tetracyclines?

67. Diphenhydramine (Benadryl®) has the chemical formula $C_{17}H_{21}NO$. Is it water soluble or fat soluble? Explain your answer.

68. Ephedrine hydrochloride has the chemical formula $C_{10}H_{16}NO^+Cl^-$. Is the compound water soluble or fat soluble? Explain your answer.

69. What is the basic structure common to all barbiturate molecules? How is this structure modified to change the properties of individual barbiturate drugs?

70. The dissociative anesthetics ketamine and PCP are classified as derivatives of cyclohexylamine. Give the structural formula for cyclohexylamine.

Problems 71–74 use the concept of an LD_{50} value. The LD_{50} of a substance is the dosage that would be lethal to 50% of a population of test animals.

71. When administered orally to rats, ketoprofen and naproxen have LD_{50} values of 101 mg/kg and 534 mg/kg, respectively. Which is more toxic? Explain.

72. When administered intravenously to rats, procaine and cocaine have LD_{50} values of 50 mg/kg and 17.5 mg/kg, respectively. Which is more toxic? Explain.

73. When administered orally to rats, aspirin has an LD_{50} value of 1500 mg/kg of body weight. If the toxicity is the same for humans, how many 325 mg tablets would kill the typical 10.0-kg child?

74. When administered intravenously to rats, the LD_{50} of procaine is 50 mg/kg body weight, and that of nicotine, 1.0 mg/kg body weight. Which drug is more toxic? What is the approximate lethal dose of each for a 50-kg human?

75. If the minimum lethal dose (MLD) of amphetamine is 5 mg/kg, what is the MLD for a 70-kg person? Can toxicity studies on animals always be extrapolated to humans?

76. As noted in Section 19.15, replacement of the phenolic OH group by a methoxy (OCH_3) group produces codeine. Draw the structure of the codeine molecule.

77. As noted in Section 19.15, conversion of both the OH groups of the morphine molecule to acetate esters produces heroin. Draw the structure of the heroin molecule.

78. *gamma*-Hydroxybutyric acid (GHB) is a drug used illicitly by athletes and bodybuilders, who believe that it causes release of a growth hormone that contributes to anabolism. Draw the structure of the GHB molecule.

COLLABORATIVE GROUP PROJECTS

Prepare a PowerPoint, poster, or other presentation (as directed by your instructor) for presentation to the class.

79. Prepare a brief report on one of the following combination products. Make a list of the ingredients in each and give the chemical structure, medical use, toxicity, and (if possible) side effects of each.
- **a.** Advil® Cold & Sinus
- **b.** Aleve® Cold & Sinus
- **c.** Dimetapp® Cold & Flu
- **d.** Theraflu®
- **e.** Excedrin PM®
- **f.** Alka-Seltzer® Plus
- **g.** DayQuil®
- **h.** NyQuil®
- **i.** Excedrin® Extra Strength

For items 80–86, follow these general directions: Give the chemical structure, medical use, toxicity, and (if possible) side effects of each.

80. Anticancer drugs:
- **a.** bleomycin
- **b.** fludarabine
- **c.** hydroxyurea
- **d.** levamisole
- **e.** procarbazine
- **f.** tamoxifen
- **g.** vinblastine

81. Antihistamines:
- **a.** azatadine
- **b.** fexofenadine
- **c.** cyproheptadine
- **d.** cetirizine
- **e.** dexchlorpheniramine
- **f.** loratadine
- **g.** promethazine
- **h.** tripelennamine

82. Antibiotics:
- **a.** vancomycin
- **b.** gramicidin
- **c.** erythromycin
- **d.** ciprofloxacin
- **e.** trimethoprim
- **f.** cefoxitin
- **g.** cefoperazone
- **h.** azithromycin

83. Antiviral drugs:
- **a.** delavirdine
- **b.** famciclovir
- **c.** ganciclovir
- **d.** indinavir
- **e.** ribavirin
- **f.** rimantadine
- **g.** ritonavir
- **h.** valacyclovir

84. Opioid narcotics:
- **a.** propoxyphene
- **b.** oxycodone
- **c.** butorphanol
- **d.** oxymorphone
- **e.** hydromorphone
- **f.** MPTP
- **g.** fentanyl
- **h.** levorphanol

85. Club drugs (drugs used by young adults at all-night dance parties called "raves").
    a. MDMA
    b. *gamma*-hydroxybutyrate
    c. rohypnol
    d. ketamine
    e. methamphetamine
    f. LSD

86. Antidepressants:
    a. sertraline
    b. paroxetine
    c. citalopram
    d. desipramine
    e. isocarboxazid
    f. phenelzine
    g. fluvoxamine
    h. nortriptyline

87. Examine the labels of three over-the-counter sleeping pills (such as Nytol, Sominex, and Sleep-eze). Make a list of the ingredients in each. Look up the properties (medical uses, dosages, side effects, toxicities) of each in a reference work such as *The Merck Index*.

88. Do a cost analysis of five brands of plain aspirin. Calculate the cost per gram of each. Compare the cost per gram of an extra-strength aspirin formulation with that of plain aspirin.

89. Discuss some of the problems involved in proving a drug safe or proving it harmful. Which is easier? Why?

90. Search the Internet for information on the use of hormone replacement therapy (HRT) in postmenopausal women. List the risks and benefits of HRT.

91. Do an Internet search to learn more about drug trials so that you can answer the following questions.
    a. Why might a person who is gravely ill not want to participate in a placebo-controlled drug study?
    b. What are the advantages and disadvantages of participating in such a drug study?
    c. Why is it important that studies use placebos as well as the test drug?
    d. Sometimes the control group receives the standard treatment for the disease instead of a placebo. Why does this happen?

92. Search the Web for information on any new drugs for treating one of the following diseases.
    a. pneumonia
    b. arthritis
    c. cancer
    d. diabetes

93. Use the *The Merck Index* or similar reference to look up the toxicities of cocaine, procaine, lidocaine, mepivicaine, and bupivacaine. Is it always accurate to compare toxicities in different animals and extrapolate animal toxicities to humans? Does the method of administration have an effect on the observed toxicity?

94. Prepare a brief report on one of the following neurotransmitters. Describe its function and site of action.
    a. GABA
    b. glutamate
    c. aspartate
    d. glycine
    e. histamine
    f. adenosine

95. Prepare a brief report on thalidomide; its structure, original intended use, problems that arose, and proposed new uses. Identify the chiral carbon(s) on thalidomide that were the source of the problems.

96. Some states are attempting to reduce the incidence of illegal methamphetamine "labs" by restricting the sale or availability of some of the starting materials. Prepare a report that lists those states that have such laws, what items are restricted and how they are restricted, and the apparent effect of these restrictions.

REFERENCES AND READINGS

1. Budavari, S., M. J. O'Neill, and A. Smith. *The Merck Index: An Encyclopedia of Chemicals, Drugs, and Biologicals,* 13th ed., Rahway, NJ: Merck & Co., 2001.

2. Christensen, Damaris, "Old Drug, New Uses?" *Science News,* November 9, 2002, pp. 296–298.

3. Connor, J. T. H. "The Victorian Revolution in Surgery." *Science,* April 2, 2004, pp. 54–55.

4. Djerassi, Carl. *This Man's Pill: Reflections on the 50th Birthday of the Pill.* New York: Oxford University Press, 2001.

5. Ezzell, Carol. "Why? The Neuroscience of Suicide." *Scientific American,* February 2003, pp. 44–51.

6. Henry, Celia. "COX-3 Found: New Variant of Cyclooxygenase May Explain Acetaminophen Mechanism." *Chemical & Engineering News,* September 23, 2002, p. 16.

7. Hogg R. C., and D. Bertrand. "What Genes Tell Us About Nicotine Addiction." *Science,* November 5, 2004, pp. 983–985.

8. Landau, Ralph. *Pharmaceutical Innovation.* Philadelphia: Chemical Heritage Foundation, 1999.

9. Lipman, Marvin M., and the editors of *Consumer Reports. The Best of Health.* Yonkers, NY: Consumers Union, 2004. Answers to questions that people might want to ask their doctor.

10. Liska, Ken. *Drugs and the Human Body: With Implications for Society,* 7th ed. Upper Saddle River, NJ: Prentice Hall, 2004.

11. Marzuola, C. "Male Pill on the Horizon." *Science News,* September 14, 2002, pp. 373–374.

12. *The Merck Manual of Medical Information—Second Home Edition.* Rahway, NJ: Merck & Co., 2005.

13. Musto, D. F. "Opium, Cocaine, and Marijuana in American History." *Scientific American,* July 1991, pp. 40–47.

14. Nestler, Eric J., and Robert C. Malenka. "The Addicted Brain." *Scientific American,* March 2004, pp. 78–85.

15. Perkins, Bill. "How Does Anesthesia Work?" *Scientific American,* May 2005, p. 102.

16. *Physicians' Desk Reference,* 59th ed. Montvale, NJ: Medical Economics Company, 2005.

17. Shermer, Michael. "What's the Harm? Alternative Medicine Is Not Everything to Gain and Nothing to Lose." *Scientific American,* December 2003, p. 50.

18. Shnayerson, Michael, and Mark J. Plotkin. "Supergerm Warfare." *Smithsonian,* October 2002, pp. 114–126.

19. Stachulski, Andrew V., and Martin S. Lennard. "Drug Metabolism: The Body's Defense Against Chemical Attack." *Journal of Chemical Education,* March 2000, pp. 349–353.

GREEN CHEMISTRY

Can Pharmaceuticals Be Green?

Julie Manley, ACS Green Chemistry Institute

The objective of the pharmaceutical industry is to invent medicines that allow patients to live longer, healthier, and more productive lives. This objective has a strong focus on environmental and social sustainability. How does the pharmaceutical industry address the environmental and social impact of their business? Can drugs be developed and manufactured to minimize the impact on the environment? Research has shown that small quantities of pharmaceuticals, cosmetics, and other household compounds are being deposited into our soil and water. How can the industry develop pharmaceuticals to prevent the deposition and accumulation of their products in the environment?

In the Web Investigations you will examine how the pharmaceutical industry has incorporated green chemistry into their business. In Communicate Your Results you will be placed into a variety of roles: a corporate officer of a pharmaceutical corporation, a journalist, and an educated consumer. By putting yourself in these positions, you will learn and communicate unique perspectives on the application of green chemistry in the pharmaceutical industry.

▲ Sophisticated analytical methods such as gas chromatography (shown here) are used to determine the concentrations of pharmaceuticals in soil, water, foodstuffs. Most modern industries must prepare environmental impact statements that estimate the impact of their manufacturing operations on the environment.

WEB INVESTIGATIONS

Investigation 1
Award-Winning Green Chemistry

Many pharmaceutical case studies on green chemistry have been recognized with Presidential Green Chemistry Challenge Awards administered by the U.S. Environmental Protection Agency. Use your favorite search engine to review the winners of the Alternative Synthetic Pathways Award for the years 1997 through the present to discover the contribution the pharmaceutical industry has made to green chemistry and to the Presidential Green Chemistry Award Program.

Investigation 2
Communicating Success

Stakeholder engagement and communication are important for publicly traded organizations. What programs are considered important enough to communicate to the public? How do companies communicate their green chemistry programs and success? Is green chemistry mentioned on their websites? Review the websites of at least four of the following companies: Pfizer, Merck, Eli Lilly, Sanofiaventis, Abbott, Glaxo Smith-Kline, and Schering-Plough. In the "search" function in each website, enter the words "green chemistry" and review the contents of the search results to learn how companies communicate green chemistry to their public.

Investigation 3
Pharmaceuticals in the Environment: Perception or Reality?

In recent years there has been a growing concern regarding studies that have identified various concentrations of pharmaceuticals in the environment (PIE). How do the pharmaceuticals get into the water? Are they harmful to the public and/or aquatic animals at the levels detected? Use your favorite search engine and appropriate keywords to investigate the PIE debate.

COMMUNICATE YOUR RESULTS

Exercise 1
Award-Winning Green Chemistry

From your investigation of the 2005 Presidential Green Chemistry Awards, write a one-page summary describing the participation and impact the pharmaceutical industry has made on the Awards program and the environment. Be sure to include some numerical metrics to describe the impact.

Exercise 2
Company Green Chemistry Policy

From your evaluation of company websites and their communication of green chemistry, determine how you would communicate the importance of green chemistry in a pharmaceutical company. As a corporate officer of a major pharmaceutical company, write a one-page corporate policy on green chemistry's role in your company.

Exercise 3
Design for Degradation

From your research on the topic of PIE in Web Investigation 3 and the information you learned in previous chapters, write an article for a newspaper to describe how green chemistry principles could be used to address the PIE debate.

Exercise 4
Are You an Educated Consumer?

As an educated consumer, how would a product's green profile or the company's dedication to green chemistry affect your decision to choose one prescription drug over another competitor? Would you ask your doctor for a comparable drug from another company? Use all of the Web Investigations for this chapter as well as any other information you have learned from previous chapters. Summarize your decision in a one-page essay.

Poisons

The fly agaric mushroom (*Amanita muscaria*) contains the poison L-muscarine, the poison that was the focus of Dorothy L. Sayers' mystery, *The Documents in the Case*. Poisoning from wild mushrooms is quite common, usually occurring when a person mistakes a poisonous variety for an edible species. The LD$_{50}$ (Section 20.7) for muscarine is 0.23 mg/kg of body weight when administered intravenously in mice.

Chemical Toxicology

What is a poison? Perhaps a better question is: How much is a poison? A substance may be harmless in one amount but injurious—or even deadly—in another. It was probably Paracelsus who first suggested around C.E. 1500 that "the dose makes the poison." Even something as common as table salt or sugar can be poisonous if eaten in abnormally large amounts. A poison is a substance that causes injury or illness or death of a living organism, but we usually think of a substance as poisonous if it can kill or injure at a low dosage. The toxic dose need not be the same for everyone. An amount of salt that is safe for an adult could be deadly for a tiny infant. An amount of sugar that is all right for a healthy teenager might be quite harmful for her diabetic father.

Of course, some compounds are much more toxic than others. It would take a massive dose of salt to kill a healthy adult, whereas a few micrograms of botulin could be fatal. How the substance is administered also matters. Nicotine given intravenously is more than 50 times as toxic as when taken orally. Pure fresh water that is delightful to drink can be deadly if inhaled in large quantity.

In this chapter we take a look at poisons. **Toxicology** is the study of the effects of poisons, their detection and identification, and their antidotes. In our discussion here, we cover only a few of the many toxic substances that are known, giving priority to those that are more likely to be encountered in everyday life.

Recall that we noted in Chapter 16 that vitamin A is essential to health but that it can build to toxic levels in the body's fatty tissues.

20.1 NATURAL POISONS

When Socrates was accused of corrupting the youth of Athens in 399 B.C.E. and was given the choice of exile or death, he chose death by drinking a cup of hemlock. Poisons from plant and animal sources were well known in the ancient world. In fact, poisoning was an official means of execution in many ancient societies; hemlock was the official Athenian state poison. Snake and insect venoms as well as plant alkaloids were widely used. Today many aboriginal tribes still use various natural poisons for hunting and warfare. Curare, used by certain South American tribes to poison their arrows, is a notorious example.

Within the past 150 years we have learned much more about these natural poisons. The hemlock taken by Socrates was probably prepared from the unripe fruit of *Conium maclatum* (poison hemlock), dried and brewed into a tea (Figure 20.1). Hemlock contains coniine, which causes, nausea, weakness, paralysis, and—as in the case of Socrates—death.

631

▶ **Figure 20.1** Poison hemlock (*Conium maculatum*), the plant from which the hemlock taken by Socrates was probably prepared from the unripe but full-grown fruit of this plant. The principal alkaloid in poison hemlock is coniine.

Alkaloids are heterocyclic amines that occur naturally in plants. In addition to coniine and the toxins in curare, alkaloids include caffeine, nicotine, morphine, and cocaine (Section 19.16). One of the most poisonous alkaloids is strychnine, which occurs in the seeds of a tree grown in India. A dose of only 30 mg can be fatal.

In addition to poisons produced by plants, there are many toxins produced by animals such as snakes. Many insects and lots of microorganisms also produce poisons. For example, each year some 20,000 Americans get *E. coli* infections, and approximately 1% of them die. Perhaps the most poisonous bacterial toxin of all is botulin, the nerve poison that causes botulism (Section 20.6).

Renaissance Poisoners

Despite a lack of historical evidence, Lucrezia Borgia has become legendary in the annals of poisoning. The illegitimate daughter of Pope Alexander VI, this Italian Renaissance woman is considered one of the most evil women to have ever lived. She has been portrayed in that role in many works of art, novels, and films. She is accused of poisoning husbands and lovers with a mixture of arsenic, phosphorus, and copper, prepared in the abdomen of a decaying hog. Lucrezia is also thought to have worn a hollow ring that carried poison used in victims' drinks. Her reputation for evil may be undeserved, based in part on propaganda from Borgia's enemies and on the very real crimes of her brother Cesare, who is known to have killed several foes.

Other Renaissance Italians were also efficient poisoners. According to records of contracts still preserved, fifteenth-century Venice had a "Council of Ten" that poisoned people for a fee. And there were schools in Rome and Venice that taught how to murder with poison.

Buffalo Bill of America's Wild West named his gun "Lucretia Borgia" because it was deadly.

▲ Lucrezia Borgia (1480–1519).

Poisons in the Garden and on the Farm

Because many natural poisons are alkaloids that occur in plants, it is not surprising that poisons are found in gardens and on farms or ranches. In addition to the toxic pesticides that might be found on a shelf (Chapter 16), some of the plants themselves are toxic. For example, those beautiful iris, azaleas, and hydrangeas are all poisonous. So are holly berries, wisteria seeds, and the leaves and berries of privet hedges. Indoor plants can also be poisonous. Philodendron, one of the most popular of all, is quite toxic.

Farmers and ranchers have to deal with plants that poison their animals. Particularly notorious is the locoweed of western parts of the United States and Canada. Ranchers use the name *locoweed* for several different plants. In the 1950s and 1960s, the annual losses of beef cattle from poisonous plants were estimated to be over $17 million. Cattle graze on these plants in early spring before nutritious grasses emerge or in times of drought when grasses are scarce.

◀ White locoweed (*Oxytropis sericea*).

Perhaps typical is the white locoweed (*Oxytropis sericea*), a plant that causes loss of coordination and nervousness under stress in cattle, horses, and sheep. It can cause congestive heart failure, especially at higher altitudes. The active poison is *swainsonine*, an alkaloid. *Locoism* is a chronic disease that can cause pregnant animals to abort or give birth to young with congenital deformities, and it can cause death.

20.2 CORROSIVE POISONS: A CLOSER LOOK

In Chapter 7 we noted the corrosive effects of strong acids and strong bases on human tissue. Substances such as these indiscriminately destroy living cells. Corrosive chemicals in lesser concentrations exert more subtle effects.

Strong Acids and Bases

Both acids and bases catalyze the hydrolysis of amides, including proteins (polyamides).

$$\text{Intact protein molecule} + H_2O \xrightarrow{H^+ \text{ or } OH^-} \text{Fragments}$$

Shape is vital to the function of a protein, and the fragments formed by hydrolysis are not able to carry out the functions of the original protein. For example, if the protein is an enzyme, it is deactivated by hydrolysis. In cases of severe exposure, fragmentation continues until the tissue is completely destroyed.

Acids in the lungs are particularly destructive. In Chapter 12, we saw how sulfuric acid is formed when sulfur-containing coal is burned. Acids are also formed when certain plastics and other wastes are burned. These acid pollutants cause the breakdown of lung tissue.

Oxidizing Agents

Other air pollutants also damage living cells. Ozone, peroxyacetyl nitrate (PAN), and the other oxidizing components of photochemical smog probably do their main damage through the deactivation of enzymes. The active sites of enzymes often incorporate the sulfur-containing amino acids cysteine and methionine. Cysteine is readily oxidized by ozone to cysteic acid.

$$HS-CH_2-\underset{\underset{NH_2}{|}}{\overset{\overset{H}{|}}{C}}-COOH + O_3 \longrightarrow HO-\underset{\underset{O}{\|}}{\overset{\overset{O}{\|}}{S}}-CH_2-\underset{\underset{NH_2}{|}}{\overset{\overset{H}{|}}{C}}-COOH$$

Cysteine   Cysteic acid

Methionine is oxidized to methionine sulfoxide.

$$CH_3-S-CH_2CH_2-\overset{\overset{\displaystyle H}{|}}{\underset{\underset{\displaystyle NH_2}{|}}{C}}-COOH \;+\; O_3 \;\longrightarrow\; CH_3-\overset{\overset{\displaystyle O}{\|}}{S}-CH_2CH_2-\overset{\overset{\displaystyle H}{|}}{\underset{\underset{\displaystyle NH_2}{|}}{C}}-COOH$$

Methionine   Methionine sulfoxide

Tryptophan also reacts with ozone. Tryptophan, which does not contain sulfur, undergoes a ring-opening oxidation at the double bond.

Tryptophan   Oxidation product

No doubt oxidizing agents can break bonds in many other chemical substances in a cell. Such powerful agents as ozone are more likely to make an indiscriminate attack than to react in a highly specific way.

20.3 POISONS AFFECTING OXYGEN TRANSPORT AND OXIDATIVE PROCESSES

Certain chemical substances prevent cellular oxidation of metabolites by blocking the transport of oxygen in the bloodstream or by interfering with oxidative processes in the cells. These chemicals act on the iron atoms in complex protein molecules.

Blood Agents

Carbon Monoxide

Probably the best known of these metabolic poisons is carbon monoxide. Recall (Chapter 12) that CO blocks the transport of oxygen by binding tightly to the iron atom in hemoglobin.

Nitrate ions, found in dangerous amounts in the groundwater in some agricultural areas (Chapter 13), also diminish the ability of hemoglobin to carry oxygen. Microorganisms in the digestive tract reduce nitrates to nitrites. We can write the process as a reduction half-reaction (Chapter 8).

$$2\,H^+(aq) \;+\; \underset{\text{Nitrate ion}}{NO_3^-(aq)} \;+\; 2e^- \;\longrightarrow\; \underset{\text{Nitrite ion}}{NO_2^-(aq)} \;+\; H_2O$$

The nitrite ions oxidize the Fe^{2+} in hemoglobin to Fe^{3+}, forming *methemoglobin*, which is incapable of carrying oxygen. The resulting oxygen deficiency disease is called *methemoglobinemia*, known in infants as "blue-baby syndrome."

Cyanides: Agents of Death

Cyanides, compounds that contain a $C\equiv N$ group, are among the most notorious poisons in both fact and fiction. They include salts such as sodium cyanide (NaCN) and the covalent compound hydrogen cyanide ($H-C\equiv N$). Cyanides act quickly and are powerful. It takes only about 50 mg of HCN or 200–300 mg of a cyanide salt to kill. Cyanide exposure occurs frequently in patients with smoke inhalation from house fires. Experts use HCN gas to exterminate insects and rodents in ships, warehouses, and railway cars, and on citrus and other fruit trees. Sodium cyanide (NaCN) is employed to extract gold and silver from ores and electroplating baths.

Hemoglobin is bright red and is responsible for the red color of blood. Methemoglobin is brown. During cooking, red meat turns brown because hemoglobin is oxidized to methemoglobin. Dried bloodstains turn brown for the same reason.

Deaths caused by house fires are often actually caused by toxic gases. Smoldering fires produce carbon monoxide, but hydrogen cyanide (formed by burning plastics and fabrics, Chapter 10) is also important. Some fire department rescue crews carry spring-loaded hypodermic syringes filled with sodium thiosulfate solution for emergency treatment of victims of smoke inhalation.

Hydrogen cyanide is generated from the sodium salt by treatment with an acid.

$$NaCN(s) + H_2SO_4(aq) \longrightarrow HCN(g) + NaHSO_4(aq)$$

Cyanide acts by blocking the oxidation of glucose inside the cell; it forms a stable complex with iron(III) ions in oxidative enzymes called *cytochrome oxidases*. These enzymes normally act by providing electrons for the reduction of oxygen in the cell. Cyanide blocks this action and brings an abrupt end to cellular respiration, causing death in minutes. Cyanides are sometimes used to commit suicide, particularly by health care and laboratory workers.

Any antidote for cyanide poisoning must be administered quickly. Providing 100% oxygen to support respiration can sometimes help. Sodium nitrite is often given intravenously to oxidize iron atoms in the enzymes back to the active Fe^{3+} form. Sodium thiosulfate ("hypo" used in developing photographic film) is then used if time permits. The thiosulfate ion transfers a sulfur atom to the cyanide ion, converting it to the relatively innocuous thiocyanate ion.

Cyanide ion		Thiosulfate ion			Thiocyanate ion		Sulfite ion
$CN^-(aq)$	+	$S_2O_3{}^{2-}(aq)$	$\longrightarrow$		$SCN^-(aq)$	+	$SO_3{}^{2-}(aq)$

Unfortunately, few victims of cyanide poisoning survive long enough to be treated.

20.4 MAKE YOUR OWN POISON: FLUOROACETIC ACID

The body generally acts to detoxify poisons (Section 20.8), but it can also convert an essentially harmless chemical to a deadly poison. Fluoroacetic acid (FCH_2COOH) is one such compound. Our cells use acetic acid to produce citric acid, which is then broken down in a series of steps, most of which release energy. When fluoroacetic acid is ingested, it is incorporated into fluorocitric acid. The latter effectively blocks the citric acid cycle by tying up the enzyme that acts on citric acid. Thus, the energy-producing mechanism of the cell is shut off and death comes quickly.

Sodium fluoroacetate (FCH_2COONa), often known as *Compound 1080*, is used to poison rodents and predatory animals. It is not selective, making it dangerous to humans, pets, and other animals.

Fluoroacetic acid occurs in nature in an extremely poisonous South African plant called *gifblaar*. Natives have used this plant to poison the tips of their arrows.

20.5 HEAVY METAL POISONS

Metals with densities at least five times that of water are called *heavy metals*. Many heavy metals are toxic, but lead, mercury, and cadmium are especially notable. Of course, metals need not be dense to be toxic. One especially toxic metal is beryllium, a very light metal. All beryllium compounds are poisonous and can cause a painful and usually fatal disease known as *berylliosis*.

People have long used a variety of metals in industry and in agriculture and around the home. Most metals and their compounds show some toxicity when ingested in large amounts. Even essential mineral nutrients can be toxic when taken in excessive amounts. In many cases, too little of a metal ion (a deficiency) can be as dangerous as too much (toxicity). This effect is illustrated in Figure 20.2.

For an example involving humans, the average adult requires 10–18 mg of iron every day. When less is taken in, the person suffers from anemia. Yet an overdose can cause vomiting, diarrhea, shock, coma, and even death. As few as 10–15 tablets containing 325 mg each of iron (as $FeSO_4$) have been fatal to children.

In the World War II Nazi death camps, both carbon monoxide and hydrogen cyanide were used to murder 6 million Jews, gypsies, Poles, and others. Treblinka, Belzec, and Sobibor used CO from internal combustion engines in closed chambers. Auschwitz/Birkenau, Stutthof, and other camps used HCN in the form of *Zyklon-B* (HCN absorbed in a carrier, typically wood pulp or diatomaceous earth).

 Acetic Acid Citric Acid

$$\begin{array}{c} CH_2-COOH \\ | \\ HO-C-COOH \\ | \\ CH_2-COOH \end{array}$$

Citric acid

$$\begin{array}{c} F-CH-COOH \\ | \\ HO-C-COOH \\ | \\ CH_2-COOH \end{array}$$

Fluorocitric acid

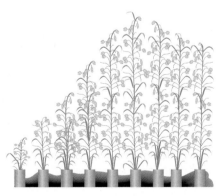

▲ **Figure 20.2** The effect of copper ions on the height of oat seedlings. From left to right, the concentrations of Cu^{2+} are 0, 3, 6, 10, 20, 100, 500, 2000, and 3000 $\mu g/L$. The plants on the left show varying degrees of deficiency; those on the right show copper ion toxicity. The optimum level of Cu^{2+} for oat seedlings is therefore about 100 $\mu g/L$.

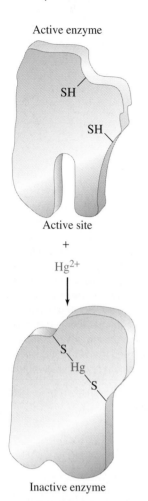

Active enzyme

Active site

+

Hg^{2+}

Inactive enzyme

▲ **Figure 20.3** Mercury poisoning. Enzymes catalyze reactions by binding a reactant molecule at the active site (Chapter 15). Mercury ions react with sulfhydryl groups to change the shape of the enzyme and destroy the active site.

We don't know exactly how iron poisoning works. Other heavy metals exert their action primarily by inactivating enzymes. In simple chemical reactions, heavy metal ions react with hydrogen sulfide to form insoluble sulfides.

$$Pb^{2+}(aq) + H_2S(g) \longrightarrow PbS(s) + 2\,H^+(aq)$$
$$Hg^{2+}(aq) + H_2S(g) \longrightarrow HgS(g) + 2\,H^+(aq)$$

Most enzymes have amino acids with sulfhydryl (—SH) groups. Heavy metal ions tie up these groups much as they react with H_2S, rendering the enzymes inactive (Figure 20.3).

Quicksilver → Slow Death

Mercury (Hg) is a most unusual metal. It is the only common metal that is a liquid at room temperature. People have long been fascinated by this bright, silvery, dense liquid, once known as *quicksilver*. Children sometimes play with the mercury from a broken thermometer.

Mercury metal has many uses. Dentists use it to make amalgams for filling teeth. It is used as an electrical contact in switches. Farmers use seeds treated with compounds of mercury.

Mercury presents a hazard to those who work with it because its vapor is toxic. An open container of mercury or a spill on the floor can release enough mercury vapor into the air to exceed the established maximum safe level by a factor of 200. Because mercury is a cumulative poison (it takes the body about 70 days to rid itself of *half* of a given dose), chronic poisoning is a threat to those continually exposed.

Fortunately, there are antidotes for mercury poisoning. British scientists, searching for an antidote for the arsenic-containing war gas lewisite, came up with a compound effective for heavy metal poisoning as well. The compound, a derivative of glycerol, is called British antilewisite (BAL). It acts by *chelating* (from the Greek *chela* meaning "claw") Hg^{2+} ions, surrounding the ions so that they are tied up and cannot attack vital enzymes.

$$CH_2-CH-CH_2$$
$$OH \quad SH \quad SH$$

BAL

Mercury atom chelated by two BAL molecules

The bad news is that the effects of mercury poisoning may not show up for several weeks. By the time the symptoms—loss of equilibrium, sight, feeling, and hearing—are recognizable, extensive damage has already been done to the brain and the nervous system. Such damage is largely irreversible. The BAL antidote is effective only when a person knows that he or she has been poisoned and seeks treatment right away.

Metallic mercury does not seem to be very toxic when ingested (swallowed). Most of it passes through the system unchanged. Indeed, there are numerous reports of mercury being given orally in the eighteenth and nineteenth centuries as a remedy for obstruction of the bowels. Doses varied from a few ounces to a pound or more.

Mercury vapor is quite hazardous when inhaled, particularly when exposure takes place over a long period of time. Such chronic exposure usually occurs with regular occupational use of the metal, such as in mining or extraction. By some as yet unknown mechanism, the body converts the inhaled mercury, to Hg^{2+} ions. All compounds of mercury, except those that are essentially insoluble in water, are poisonous no matter how they are administered.

Lead in the Environment

Lead is widespread in the environment, reflecting the many uses we have for this soft, dense, corrosion-resistant metal and its compounds. Lead (as Pb^{2+}) is present in many foods, generally in concentrations of less than 0.3 ppm. Lead (again as Pb^{2+}) also gets into our drinking water (up to 0.1 ppm) from lead-sealed pipes. Lead compounds were once widely used in house paints, and tetraethyllead was used in most gasolines. Exposure to lead has decreased dramatically since these two uses were banned. In 2004, 2.2% of children ages one to five had blood lead levels above $10\ \mu g/dL$, the amount set by the Centers for Disease Control as an unacceptable health risk. This was down from 4.4% in the early 1990s.

Lead and its compounds are quite toxic. Metallic lead is generally converted to Pb^{2+} in the body. Lead can damage the brain, liver, and kidneys. Extreme cases can be fatal.

Lead poisoning is especially harmful to children. Some children develop a craving that causes them to eat unusual things, and children with this syndrome (called *pica*) eat chips of peeling lead-based paints. These children probably also pick up lead compounds from the streets, where they were deposited by automobile exhausts. They also ingest lead from food and other sources. Large amounts of Pb^{2+} in a child's blood can cause mental retardation, behavior problems, anemia, hearing loss, developmental delays, and other physical and mental problems.

Adults can excrete about 2 mg of lead per day. Most people take in less than that from air, food, and water, and thus generally do not accumulate toxic levels. If intake exceeds excretion, however, lead builds up in the body and chronic irreversible lead poisoning results.

Lead poisoning is usually treated with a combination of BAL and another chelating agent called *ethylenediaminetetraacetic acid* (*EDTA*).

$$\text{HOOC}-CH_2 \diagdown \qquad\qquad CH_2-\text{COOH}$$
$$\text{N}-CH_2CH_2-\text{N}$$
$$\text{HOOC}-CH_2 \diagup \qquad\qquad CH_2-\text{COOH}$$

EDTA

The calcium salt of EDTA is administered intravenously. In the body, calcium ions are displaced by lead ions, which the chelate binds more tightly.

$$\text{CaEDTA}^{2-} + Pb^{2+} \longrightarrow \text{PbEDTA}^{2-} + Ca^{2+}$$

The lead–EDTA complex is then excreted.

As in mercury poisoning, the neurological damage done by lead compounds is essentially irreversible. Treatment must be begun early to be effective.

Cadmium: The "Ouch-Ouch" Disease

Cadmium is used in alloys, in the electronics industry, in nickel–cadmium rechargeable batteries, and in many other applications. Like mercury and lead, cadmium (as Cd^{2+} ions) is quite toxic. Cadmium poisoning leads to loss of calcium ions (Ca^{2+}) from the bones, leaving them brittle and easily broken. It also causes severe abdominal pain, vomiting, diarrhea, and a choking sensation.

The most notable cases of cadmium poisoning occurred along the upper Zintsu River in Japan, where cadmium ions entered the water in milling wastes from a mine. Downstream, farm families used the water to irrigate their rice fields and for drinking, cooking, and other household purposes. Soon these farm folk began to suffer from a strange, painful malady that became known as *itai-itai*, the "ouch-ouch" disease. Over 200 people died, and thousands were disabled.

Following the ban on leaded paints and leaded gasoline, the number of cases of children in the United States with low-level lead poisoning dropped from 14.9 million in the 1970s to about 900,000 today. Low-income children are most at risk of lead poisoning due to lead-based paint in older housing.

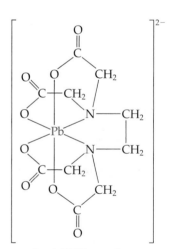

Lead–EDTA complex

Ethylenediaminetetraacetate ion-Co[EDTA]⁻

Arsenic Poisoning

Arsenic is not a metal, but it has some metallic properties. In commercial poisons, arsenic is usually found as arsenate (AsO_4^{3-}) or arsenite (AsO_3^{3-}) ions. Like heavy metal ions, these ions render enzymes inactive by tying up sulfhydryl groups.

Organic compounds containing arsenic are well known. One such compound, *arsphenamine*, was the first antibacterial agent and was once used widely in the treatment of syphilis.

Another arsenic compound, *lewisite*, first synthesized by (and named for) W. Lee Lewis, was developed as a blister agent for use in chemical warfare. The United States started large-scale production of lewisite in 1918, but fortunately World War I came to an end before this gas could be employed.

20.6 MORE CHEMISTRY OF THE NERVOUS SYSTEM

Some of the most toxic substances known act on the nervous system. Signals are shuttled across synapses between cells by *neurotransmitters* (Chapter 19). Neurotoxins can disrupt the action of a neurotransmitter in several ways: by interfering with its synthesis or transport, by occupying the transmitter's receptor site, or by blocking degradation of the transmitter.

One such chemical messenger is *acetylcholine* (*ACh*). It activates the postsynaptic cell by fitting a specific receptor and thus changing the permeability of the cell membrane to certain ions. Once ACh has carried the impulse across the synapse, it is rapidly hydrolyzed to acetic acid and relatively inactive choline in a reaction catalyzed by an enzyme, acetylcholinesterase.

People with bipolar (manic-depressive) disorder are overly sensitive to acetylcholine; they seem to have too many receptors, which may account for their wild swings in mood.

People with Alzheimer's disease are deficient in the enzyme acetylase. They produce too little acetylcholine for proper brain function.

The receptor cell releases the hydrolysis products and is then ready to receive further impulses. Other enzymes, such as acetylase, convert the acetic acid and choline back to acetylcholine, completing the cycle (Figure 20.4).

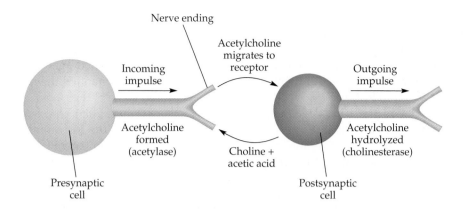

◀ **Figure 20.4** The acetylcholine cycle.

Nerve Poisons and the Acetylcholine Cycle

Various substances disrupt the acetylcholine cycle at three different points, as illustrated by the following examples.

- Botulin, the deadly toxin produced by *Clostridium botulinum* (an anaerobic bacterium) found in improperly processed canned food, is a powerful ACh antagonist; it blocks the synthesis of ACh. With no messenger formed, no messages are carried. Paralysis sets in and death occurs, usually from respiratory failure.

- Curare, atropine, and some local anesthetics act by blocking receptor sites. In this case, the message is sent but not received. In the case of local anesthetics, this can be good for pain relief in a limited area, but these drugs, too, can be lethal in sufficient quantity.

- Anticholinesterase poisons act by inhibiting the enzyme cholinesterase.

Atropine

Organophosphorus Compounds as Insecticides and Weapons of War

Organic phosphorus insecticides (Chapter 16) are well-known nerve poisons. The phosphorus–oxygen linkage is thought to bond tightly to acetylcholinesterase, blocking the breakdown of acetylcholine (Figure 20.5). Acetylcholine therefore builds up, causing receptor nerves to fire repeatedly. This over stimulates the muscles, glands, and organs. The heart beats wildly and irregularly, and the victim goes into convulsions and dies quickly.

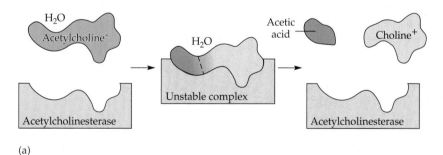

(a)

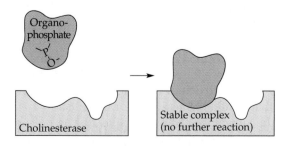

(b)

◀ **Figure 20.5** (a) Acetylcholinesterase catalyzes the hydrolysis of acetylcholine to acetic acid and choline. (b) An organophosphate ties up acetylcholinesterase, preventing it from breaking down acetylcholine.

Question: What functional group splits in the hydrolysis of acetylcholine?

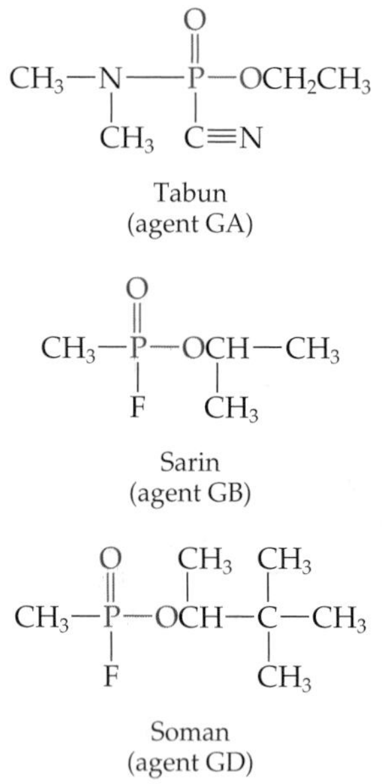

Tabun
(agent GA)

Sarin
(agent GB)

Soman
(agent GD)

▲ **Figure 20.6** Three organophosphorus compounds used as nerve poisons in chemical warfare. Compare these structures with those of the insecticides malathion and parathion (Chapter 16).

While doing research on organophosphorus compounds for use as insecticides during World War II, German scientists discovered some extremely toxic compounds with frightening potential for use in warfare. The Russians captured a German plant that manufactured a compound called *tabun* (designated agent GA by the U.S. Army), a compound with a fruity odor. The Soviets dismantled the factory and moved it to Russia; however, the U.S. Army captured some of the stock.

The United States has developed other nerve poisons (Figure 20.6). One is called *sarin* (agent GB) and is four times as toxic as tabun. It also has the "advantage" of being odorless. Another organophosphorus nerve poison is *soman* (agent GD). It is moderately persistent, whereas tabun and sarin are generally nonpersistent.

Note the variation in structure from one compound to the next. The approach in developing chemical warfare agents is much the same as the approach in developing drugs—find one that works and then synthesize and test structural variations.

These nerve poisons are among the most toxic synthetic chemicals known (although still not nearly as toxic as the natural toxin botulin). They are inhaled or absorbed through the skin, causing a complete loss of muscular coordination and subsequent death by cessation of breathing. The usual antidote is atropine injection and artificial respiration. Without an antidote, death may occur in 2–10 min.

The phosphorus-based insecticides malathion and parathion (Chapter 16) are similar to the warfare nerve agents but far less toxic. Nonetheless, they and other insecticides like them should be used with great caution.

Nerve poisons have helped us gain an understanding of the chemistry of the nervous system. This knowledge enables scientists to design antidotes for nerve poisons, to better understand diseases of the nervous system, and to develop new drugs against pain and diseases such as Alzheimer's disease and Parkinson's disease.

Chemistry and Counterterrorism

When people think of terrorism, they often think of the negative role of chemistry; they associate chemistry with chemical weapons and warfare. However, there are many applications of chemistry in counterterrorism. Chemistry plays an important role in countering the proliferation of chemical weapons and in detecting and defending against them in case of an attack. Knowledge of chemistry is essential in the evaluation of a threat. Chemistry can also help develop and use nonlethal devices. Counterterrorism efforts are interdisciplinary, involving toxicology, molecular biology, veterinary science, pathology, clinical medicine, and other fields of expertise that require knowledge of chemistry.

20.7 THE LETHAL DOSE

What do we mean by "most toxic"? Some substances are much more poisonous than others. In order to quantify toxicity, scientists use the term **LD$_{50}$** (lethal dose for 50%) to indicate a dosage that kills 50% of a population of test animals. This statistic is used because animals vary in strength and in their susceptibility to toxins. The dose that kills 50% is the average lethal dose.

Usually LD$_{50}$ values are given in terms of mass of poison per unit body weight of the test animal (for example, milligrams poison/kilograms animal). It is assumed that the material is given orally unless specified otherwise. (Values for intravenous administration can differ greatly from those for oral ingestion.) Although LD$_{50}$ values are useful in comparing the relative toxicities of various substances, they are not necessarily the same values that would be measured for humans.

We don't usually think of the substances listed in Table 20.1 as poisons, yet there is a dose of each one that can kill half of a given population of test animals.

TABLE 20.1 LD$_{50}$ Values for Some Common Substances

Substance	Test Animal	Oral LD$_{50}$ (g/kg)
Ethyl alcohol	Rat	10.3
Vitamin B$_1$	Mouse	8.2
Sodium chloride	Rat	3.75
Aspirin	Mouse	1.5
Acetaminophen	Mouse	0.34
Nicotine	Mouse	0.23
Caffeine	Mouse	0.13

TABLE 20.2 Estimated LD$_{50}$ Values for Some More Lethal Poisons

Substance	Oral LD$_{50}$ (mg/kg)
Sodium cyanide (NaCN)	15
Arsenic trioxide (As$_2$O$_3$)	15
Aflatoxin B	10
Rotenone	3
Strychnine	0.5
Muscarine	0.2
Tetanus toxin	0.000005
Botulin toxin	0.0000002

The larger the LD$_{50}$ value, the less toxic the substance. Nicotine and caffeine are the most toxic substances listed in Table 20.1 because they have the smallest LD$_{50}$ values.

The substances in Table 20.2 are highly lethal poisons. Notice that the doses are given in milligrams per kilogram of body weight, whereas the values in Table 20.1 are in grams per kilogram. The most poisonous substance of all is botulin toxin.

The Botox Enigma

Botulin is the most toxic substance known, with a median lethal dose of only about 0.2 ng/kg body weight. Yet botulin, in a commercial formulation called Botox®, is used in medicine to treat intractable muscle spasms and is widely used in cosmetic surgery to remove wrinkles. When a muscle contracts over a long period of time without relaxing, it can cause severe pain. Tiny amounts of Botox can be injected into the muscle, where Botox binds to receptors on nerve endings. This stops the release of acetylcholine, and the signal for the muscle to contract is blocked. The muscle fibers that were pulling too hard are paralyzed, providing relief from the spasm.

Botox is also used to treat afflictions such as uncontrollable blinking, crossed eyes, and Parkinson's disease. In the case of crossed eyes, for example, the optic muscles cause the eyeballs to be drawn to focus inward. Treated with botulin, the muscles are hindered from doing so, and repeated treatments correct the problem.

Botox curbs migraine headache pain in some people, probably by blocking nerves that carry pain messages to the brain and by relaxing muscles, making them less sensitive to pain. It is also used to treat excessive sweating, most likely by blocking the release of acetylcholine, the chemical that stimulates the sweat glands. Botox injections are used to treat some types of bladder problems, such as those in which the patients were suffering from involuntary contractions of the bladder muscle. These contractions can cause incontinence or make it impossible to completely empty the bladder. Botox blocks the contractions.

In cosmetic surgery, small, carefully delivered amounts of Botox inactivate small facial muscles, markedly reducing "surprise wrinkles" in the forehead, frown lines between the eyebrows, and "crow's feet" wrinkles at the corners of the eyes. Botox treatments wear off in a few months.

20.8 THE LIVER AS A DETOX FACILITY

A traditional antidote for methanol and ethylene glycol poisoning is ethanol, administered intravenously. This "loads up" the liver enzymes with ethanol, blocking oxidation of methanol until the compound can be excreted. A safer antidote, the drug fomepizole, inhibits the enzyme that catalyzes the oxidation of alcohols.

The human body can handle moderate amounts of some poisons. The liver is able to detoxify some compounds by oxidation, reduction, or coupling with amino acids or other normal body chemicals.

Perhaps the most common route is oxidation. Ethanol (Chapter 9) is detoxified by oxidation to acetaldehyde, which in turn is oxidized to acetic acid and then to carbon dioxide and water.

$$CH_3CH_2OH \longrightarrow CH_3CHO \longrightarrow CH_3COOH \longrightarrow CO_2 + H_2O$$

Highly toxic nicotine from tobacco is detoxified by oxidation to cotinine.

Nicotine   Cotinine

Cotinine is less toxic than nicotine, and the added oxygen atom makes it more water soluble, and thus more readily excreted in the urine, than nicotine.

The liver is equipped with a system of enzymes called *P-450* that oxidize fat-soluble substances (which are otherwise likely to be retained in the body) into water-soluble ones that are readily excreted. P-450 can also conjugate compounds with amino acids. For example, toluene is essentially insoluble in water. P-450 enzymes oxidize toluene to more soluble benzoic acid. The latter is then coupled with the amino acid glycine to form hippuric acid, which is even more soluble and is readily excreted.

Toluene   Benzoic acid   Hippuric acid

Note that liver enzymes simply oxidize, reduce, or conjugate. The end product is not always less toxic. For example, as we noted in Chapter 9, methanol is oxidized to more toxic formaldehyde, which then reacts with proteins in the cells to cause blindness, convulsions, respiratory failure, and death.

The same enzymes that oxidize alcohols deactivate the male hormone testosterone. Build-up of these enzymes in a chronic alcoholic leads to a more rapid destruction of testosterone. Thus, we have the mechanism for alcoholic impotence, one of the well-known characteristics of the disease.

Benzene, because of its general inertness in the body, is not acted upon until it reaches the liver. There it is slowly oxidized to an epoxide.

The epoxide is a highly reactive molecule that can attack certain key proteins. The damage done by this epoxide sometimes results in leukemia.

Carbon tetrachloride (CCl_4), is also quite inert in the body. But when it reaches the liver, it is converted to the reactive trichloromethyl free radical ($Cl_3C \cdot$), which in turn attacks the unsaturated fatty acids in the body. This action can trigger cancer.

Getting Rid of "Toxins"

Two thousand years ago Greek physicians used the theory of the four *humors*—blood, phlegm, black bile, and yellow bile—to diagnose illnesses. They held that people in good health had a balance of these fluids and that too much or too little of one or more of them caused disease. A person who is positive and happy is still said to be "in a good humor." For a patient short of a humor, the treatment included bed rest, a change in diet, or medicines. For those with too much of a humor, the practice included sweating, purging, and bloodletting. In some circles today, people believe that sweating gets rid of "toxins" and that laxatives somehow purify the body. Erroneous beliefs often persist for centuries.

▲ In the Middle Ages, bloodletting was often the treatment for inflammation from infection.

20.9 CHEMICAL CARCINOGENS: SLOW POISONS

A **carcinogen** is something that causes the growth of tumors. A tumor is an abnormal growth of new tissue and can be either benign or malignant. *Benign tumors* are characterized by slow growth; they often regress spontaneously, and they do not invade neighboring tissues. *Malignant tumors*, often called cancers, can grow slowly or rapidly and can invade and destroy neighboring tissues. Actually, cancer is not a single disease. Rather, it is a catch-all term for about 200 different afflictions, many of them not even closely related to each other. We worry about carcinogens because half of the men and one-third of the women in the United States will develop cancer at some point in life.

What Causes Cancer?

Most people seem to believe that synthetic chemicals are a major cause of cancer, but the facts indicate otherwise. Like overall deaths (Chapter 19), most cancers are caused by lifestyle factors. Nearly two-thirds of all cancer deaths in the United States are linked to tobacco, diet, or lack of exercise with resulting obesity (Figure 20.7).

The U.S. Occupational Safety and Health Administration (OSHA) has a National Toxicology Program (NTP) that compiles ongoing lists of carcinogens in two categories. The 2004 list has 58 entries on the list of "Known Human Carcinogens." For these, the NTP thinks there is sufficient evidence of carcinogenicity from studies in humans to establish a causal relationship between exposure to the agent, substance, or mixture, and human cancer. Another 188 are "Reasonably Anticipated to be Human Carcinogens." These suspected carcinogens often are structurally related to substances that are known to be carcinogens. Or there is limited evidence of carcinogenicity from studies in humans or in experimental animals.

Some carcinogens, such as sunlight, radon, and safrole in sassafras, occur naturally. Some scientists estimate that 99.99% of all carcinogens that we ingest are natural ones. Plants produce compounds to protect themselves from fungi, insects, and higher animals, including humans. Some of these compounds are carcinogens

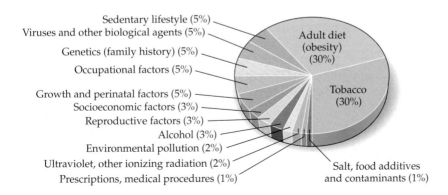

Sedentary lifestyle (5%)
Viruses and other biological agents (5%)
Genetics (family history) (5%)
Occupational factors (5%)
Growth and perinatal factors (5%)
Socioeconomic factors (3%)
Reproductive factors (3%)
Alcohol (3%)
Environmental pollution (2%)
Ultraviolet, other ionizing radiation (2%)
Prescriptions, medical procedures (1%)
Adult diet (obesity) (30%)
Tobacco (30%)
Salt, food additives and contaminants (1%)

◀ **Figure 20.7** Ranking cancer risks. About 30% of all cancer deaths are attributed to tobacco and another 30% or so to a diet high in fat and calories and low in fruits and vegetables. Environmental pollution and food additives rank high in public perception of causes, but are only relatively minor contributors to cancer deaths. (Based on data from Harvard University School of Public Health.)

Chapter 20 Poisons

found in mushrooms, basil, celery, figs, mustard, pepper, fennel, parsnips, and citrus oils—almost every place that a curious chemist looks. Carcinogens are also produced during cooking and as products of normal metabolism. Because carcinogens are so widespread, we must have some way of protecting ourselves from them.

How Cancers Develop

How do chemicals and physical factors cause cancer? Their mechanisms of action are probably quite varied. Some carcinogens chemically modify DNA, thus scrambling the code for replication and for the synthesis of proteins. For example, aflatoxin B, produced by molds on foods, is known to bind to guanine residues in DNA. Just how this initiates cancer, however, has not yet been determined.

Genetics plays a role in the development of many forms of cancer. Certain genes, called *oncogenes*, seem to trigger or sustain the processes that convert normal cells to cancerous ones. There are about 100 known oncogenes. They arise from ordinary genes that regulate cell growth and cell division. Chemical carcinogens, radiation, or perhaps some viruses can activate oncogenes. It seems that more than one oncogene must be turned on, perhaps at different stages of the process, before a cancer develops. We also have *suppressor genes* that ordinarily prevent the development of cancers. These genes must be inactivated before a cancer develops. Suppressor gene inactivation can occur through mutation, alteration, or loss. In all, several mutations may be required in a cell before it turns cancerous.

Chemical Carcinogens

A variety of widely different chemical compounds are carcinogenic. We will concentrate only on a few major classes here.

Polycyclic aromatic hydrocarbons, of which 3,4-benzpyrene (Chapter 12) is perhaps the best known, are among the more notorious carcinogens. These hydrocarbons are formed during the incomplete burning of nearly any organic material. They have been found in charcoal-grilled meats, cigarette smoke, automobile exhausts, coffee, burnt sugar, and many other materials. Not all polycyclic aromatic hydrocarbons are carcinogenic. There are strong correlations between carcinogenicity and certain molecular sizes and shapes.

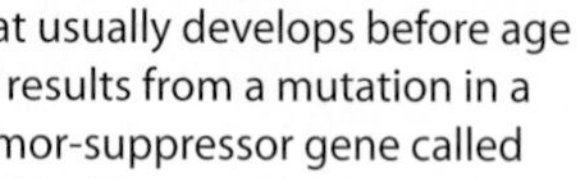

Aromatic amines make up another important class of carcinogens. Two prominent ones are β-naphthylamine (Chapter 16) and benzidine. These compounds, once widely used in the dye industry, were responsible for a high incidence of bladder cancer among workers whose jobs brought them into prolonged contact with these substances.

Not all carcinogens are aromatic. Two prominent aliphatic (nonaromatic) ones are dimethylnitrosamine (Chapter 16) and vinyl chloride (Chapter 10). Others include three- and four-membered heterocyclic rings containing nitrogen or oxygen (Figure 20.8). The epoxides and derivatives of ethyleneimine are examples. Others are cyclic esters called lactones.

This list is far from all inclusive. Its purpose is to give you an idea of some of the kinds of compounds that have tumor-inducing properties.

Anticarcinogens

If the food we eat has many natural carcinogens, why don't we all get cancer? Probably because other substances in our food act as **anticarcinogens**. Antioxidant vitamins are believed to protect against some forms of cancer, and the food additive butylated hydroxytoluene (BHT) may give protection against stomach cancer. Certain vitamins also have been shown to have anticarcinogenic effects.

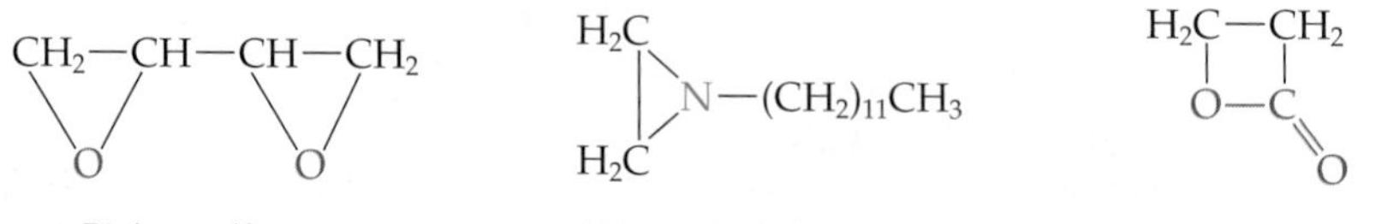

A hereditary form of breast cancer that usually develops before age 50 results from a mutation in a tumor-suppressor gene called *BRCA1*. Women who have a mutant form of *BRCA1* have an 82% risk of developing breast cancer, compared with most women in the United States, who have about a 10% risk.

Benzidine

Vinyl Chloride

▶ **Figure 20.8** Three small-ring heterocyclic carcinogens.

Question: To what family of compounds does each compound belong?

Bis(epoxy)butane

N-Laurylethyleneimine

β-Propiolactone

How Cigarette Smoking Causes Cancer

The association between cigarette smoking and cancer has been known for decades, but the precise mechanism was not determined until 1996. The research scientists who found the link focused on a tumor-suppressor gene called *P53*, a gene that is mutated in about 60% of all lung cancers, and on a metabolite of benzpyrene, a carcinogen found in tobacco smoke.

In the body, benzpyrene is oxidized to an active carcinogen (an epoxide) that binds to specific nucleotides in the gene—sites called "hot spots"—where mutations frequently occur. It seems quite likely that the benzpyrene metabolite causes many of these mutations.

Antioxidant vitamins include vitamin C, vitamin E, and β-carotene, a precursor of vitamin A. A diet rich in cruciferous vegetables (cabbage, broccoli, Brussels sprouts, kale, and cauliflower) has been shown to reduce the incidence of cancer both in animals and in human population groups. The evidence for the anticarcinogenicity of the vitamins taken as supplements is less conclusive; and in the case of β-carotene, even contradictory.

There are probably many other anticarcinogens in our food that have not yet been identified.

20.10 THREE WAYS TO TEST FOR CARCINOGENS

How do we know that a chemical causes cancer? Obviously, we can't experiment on humans to see what happens. That leaves us with no way to prove absolutely that a chemical does or does not cause cancer in humans. Three ways to gain evidence against a compound are: bacterial screening for mutagenesis, animal tests, and epidemiological studies.

The Ames Test: Bacterial Screening

The quickest and cheapest way to find out whether or not a substance may be carcinogenic is to use a screening test such as that developed by Bruce N. Ames of the University of California at Berkeley. The **Ames test** is a simple laboratory procedure that can be carried out in a petri dish (Figure 20.9). It assumes that most carcinogens are also mutagens, altering genes in some way. (This usually seems to be the case. About 90% of the chemicals that appear on a list of either mutagens or carcinogens are found on the other list as well.)

The Ames test uses a special strain of *Salmonella* bacteria that have been modified so that they require histidine as an essential amino acid. The bacteria are placed in an agar medium containing all nutrients except histidine. Incubating the mixture in the presence of a mutagenic chemical causes the bacteria to mutate so that they no longer require histidine and can grow like normal bacteria. Growth of bacterial colonies in the petri dish means that the chemical added was a mutagen, and probably also a carcinogen.

Animal Testing

Chemicals suspected of being carcinogens can be tested on animals. Tests involving low dosages and millions of rats would cost too much, and so tests are usually done by using large doses and 30 or so rats. An equal number of rats serve as controls, which are exposed to the same diet and environment but without the suspected carcinogen. A higher incidence of cancer in the experimental animals than in the controls indicates that the compound is carcinogenic.

Animal tests are not conclusive. Humans are not usually exposed to comparable doses; there may be a threshold below which a compound is not carcinogenic.

Mutant Induced Revertants

Spontaneous Revertants

▲ **Figure 20.9** The Ames test. The top petri dish, to which a mutagenic chemical has been added, shows the growth of several colonies of *Salmonella* bacteria, indicating that mutations have occurred. The bottom dish is a control that shows only a few spontaneous mutations.

Further, human metabolism is somewhat different from that of the test animals. A carcinogen might be active in rats but not in humans (or vice versa). There is only a 70% correlation between the carcinogenesis of a chemical in rats and that in mice. The correlation between carcinogenesis in either rodent and that in humans is probably less. Animal studies are expensive; in cancer studies, animal tests of a single substance may take four to eight years and cost $400,000 or more. Further, animal testing is controversial; supporters and opponents argue over both the ethics and the value of the tests.

Epidemiological Studies

The best evidence that a substance causes cancer in humans comes from *epidemiological* studies. A population that has a higher than normal rate of a particular kind of cancer is studied for common factors in their background. Studies of this sort showed that cigarette smoking causes lung cancer, that vinyl chloride causes a rare form of liver cancer, and that asbestos causes cancer of the lining of the pleural cavity (the body cavity containing the lungs). These studies sometimes require sophisticated mathematical analyses, and there is always the chance that some other (unknown) factor is involved in the carcinogenesis.

More recently, epidemiological studies were carried out to determine whether or not electromagnetic fields, such as those surrounding high-voltage power lines, certain electric appliances, and cellular phones could cause cancer. The results so far indicate that if there is such an effect, it is exceedingly small.

20.11 BIRTH DEFECTS: TERATOGENS

A **teratogen** is a substance that causes birth defects. Perhaps the most notable teratogen is the tranquilizer *thalidomide*. Fifty years ago thalidomide was considered so safe, based on laboratory studies, that it was often prescribed for pregnant women. In Germany it was available without a prescription. It took several years for the human population to provide evidence that laboratory animals had not provided. The drug had a disastrous effect on developing human embryos. About 12,000 women who had taken the drug during the first 12 weeks of pregnancy had babies who suffered from *phocomelia*, a condition characterized by shortened or absent arms and legs and other physical defects. The drug was used widely in Germany and Great Britain, and these two countries bore the brunt of the tragedy. The United States escaped relatively unscathed because Frances O. Kelsey of the FDA believed there was evidence to doubt the drug's safety and therefore did not approve it for use in the United States. Kelsey received the Distinguished Federal Civilian Service Award from President John F. Kennedy in 1962.

Other chemicals that act as teratogens include isotretinoin (Accutane), a prescription medication approved for use in treating severe acne. When taken by women during the first trimester of pregnancy, it can cause multiple major malformations. Educational materials provided to physicians and to patients have prevented all but a few tragedies associated with isotretinoin.

Thalidomide

▲ Above: Structure of thalidomide. Below: Frances O. Kelsey of the FDA receives a Distinguished Federal Civilian Service Award from President John Kennedy for her work in refusing to approve thalidomide.

Thalidomide is back in use for treating certain conditions, with strict restrictions in place to prevent its use by pregnant women. It has a unique anti-inflammatory action against a debilitating skin condition that often occurs with leprosy. Thalidomide is also being studied for use in the treatment of *wasting*, the severe weight-loss condition that often accompanies AIDS. Scientists are also studying thalidomide derivatives, hoping to enhance its beneficial effects while reducing its harmful properties.

Isotretinoin
(13-*cis*-retinoic acid)

By far the most hazardous teratogen, in terms of the number of babies born with birth defects, is ethyl alcohol, which causes fetal alcohol syndrome (Chapter 9).

What Kills You? What Makes You Sick?

There are many media reports of "toxic chemicals." You might think that chemicals are a leading cause of death and injury. It is true that some chemicals are quite toxic. Misused, they can make you sick or even kill you. However, when you look once more at Figure 19.4, it is difficult to associate any of the leading causes of death directly with chemicals. In developed countries, most of the causes of premature death are related to lifestyle. Nearly half are due to cardiovascular diseases, and another 20% or so are due to cancer. In these countries, it is "sociologicals," not "chemicals," that kill us.

Accidents and suicides often involve automobile crashes and guns. More than 20,000 people are murdered each year in the United States, most with guns, knives, or clubs of some sort. The physical force of a moving projectile, not its chemical nature, does the damage. These "physicals" kill us.

Worldwide, the main causes of premature death are infectious and parasitic diseases, "biologicals" that account for nearly one-third of the total. Every day in some of the world's poorest countries,

- 8000 people die from AIDS;
- 3900 people die from lack of access to clean water and sanitation; and

- almost 30,000 children die from illnesses that could easily be prevented, such as diarrheal dehydration, acute respiratory infections, measles, and malaria.

Even in the United States, food poisoning, caused by *Escherichia coli*, *Cyclospora*, *Listeria*, *Salmonella*, and other microorganisms sicken thousands each year. Chicken, meat, and eggs are sometimes contaminated with *Salmonella*. *E. coli* is a common inhabitant of cow intestines; people who eat beef that is carelessly slaughtered or eat foods exposed to water contaminated with cow manure are often infected.

In all these cases, of course, chemistry is involved at some level. We can identify the toxins produced by bacteria and work out the molecular mechanisms by which viruses invade and destroy cells. Nicotine is the chemical—a natural one, but a chemical no less—that makes tobacco addictive. Perfectly natural ethanol is the most dangerous chemical in alcoholic beverages.

Chemicals can be dangerous, but with care we can use them to our advantage. As far as being threats to our lives and health, though, chemicals are far down the list. We are even more likely to die of "geologicals"—earthquakes, floods, storms, and extremely hot or cold weather.

20.12 HAZARDOUS WASTES

Toxic substances, carcinogens, teratogens—the public has become increasingly concerned in recent years about issues involving hazardous wastes in the environment. Problems created by chemical dumps have made household words out of Love Canal in New York and "Valley of the Drums" in Kentucky. Although often overblown in the news media, serious problems do exist. Hazardous wastes can cause fires or explosions. They can pollute the air, and they can contaminate our food and water. Occasionally they poison by direct contact. As long as we want the products that our industries produce, however, we will have to deal with the problems of hazardous wastes (Table 20.3).

The first step in dealing with any problem is to understand what the problem is. A **hazardous waste** is one that can cause or contribute to death or illness or that

TABLE 20.3	Industrial Products and Hazardous Waste By-Products
Product	**Associated Waste**
Plastics	Organic chlorine compounds
Pesticides	Organic chlorine compounds, organophosphate compounds
Medicines	Organic solvents and residues, heavy metals (for example, mercury and zinc)
Paints	Heavy metals (for example, lead), pigments, solvents, organic residues
Oil, gasoline	Oil, phenols and other organic compounds, heavy metals, ammonium salts, acids, strong bases
Metals	Heavy metals, fluorides, cyanides, acid and alkaline cleaners, solvents, pigments, abrasives, plating salts, oils, phenols
Leather	Heavy metals, organic solvents
Textiles	Heavy metals, dyes, organic chlorine compounds, solvents

threatens human health or the environment when improperly managed. For convenience, hazardous wastes are divided into four types: reactive, flammable, toxic, and corrosive.

A **reactive waste** tends to react spontaneously or to react vigorously with air or water. They can generate toxic gases, such as hydrogen cyanide (HCN) or hydrogen sulfide (H_2S), or explode when exposed to shock or heat. Explosives such as trinitrotoluene (TNT) and nitroglycerin obviously are reactive wastes. Examples of household wastes that can explode include improperly handled propane tanks and aerosol cans.

Another example of a reactive waste is sodium metal. Wastes containing sodium caused explosions at Malkins Bank in Great Britain. Sodium reacted with water to form hydrogen gas.

$$2\,Na(s) + 2\,H_2O(l) \longrightarrow 2\,NaOH(aq) + H_2(g)$$

The hydrogen then exploded when ignited in air.

$$2\,H_2(g) + O_2(g) \longrightarrow 2\,H_2O(g)$$

Reactive wastes usually can be deactivated before disposal. Propane tanks and aerosol cans can be thoroughly emptied and properly discarded. Sodium can be treated with isopropyl alcohol, with which it reacts slowly, rather than being dumped without treatment.

A **flammable waste** is one that burns readily on ignition, presenting a fire hazard. An example is hexane, a hydrocarbon solvent. In one case, hexane (presumably dumped accidentally by Ralston-Purina) was ignited in the sewers of Louisville, Kentucky.

$$2\,C_6H_{14}(l) + 19\,O_2(g) \longrightarrow 12\,CO_2(g) + 14\,H_2O(g)$$

Explosions blew up several blocks of streets.

Examples of household flammable wastes include gasoline, lighter fluid, many solvents, and nail polish.

A **toxic waste** contains or releases toxic substances in quantities sufficient to pose a hazard to human health or to the environment. Most of the toxic substances discussed in this chapter would qualify as toxic wastes if they were improperly dumped in the environment. Some toxic wastes can be incinerated safely. For example, PCBs (Chapter 10) are burned at high temperatures to form carbon dioxide, water, and hydrogen chloride. If the hydrogen chloride is removed by scrubbing or is safely diluted and dispersed, this is a satisfactory way to dispose of PCBs. Some toxic wastes cannot be incinerated, however, and must be contained and monitored for years. Examples of household toxic wastes include paint, pesticides, motor oil, medicines, and cleansers.

A **corrosive waste** is one that requires special containers because it corrodes conventional container materials. Acids cannot be stored in steel drums because they react with and dissolve the iron.

$$Fe(s) + 2\,H^+(aq) \longrightarrow Fe^{2+}(aq) + H_2(aq)$$

Acid wastes can be neutralized (Chapter 7) before disposal, and lime (CaO) serves as cheap base for neutralization.

$$2\,H^+(aq) + CaO(s) \longrightarrow Ca^{2+}(aq) + H_2O(l)$$

Examples of household corrosive wastes include battery acid, drain cleaners, and oven cleaners.

The best way to handle hazardous wastes is not to produce them in the first place. This is a major focus of green chemistry. Many industries have modified manufacturing processes to minimize their wastes, and some wastes can be reprocessed to recover energy or materials. Hydrocarbon solvents such as hexane can be purified and reused or burned as fuels. Often the waste from one industry can

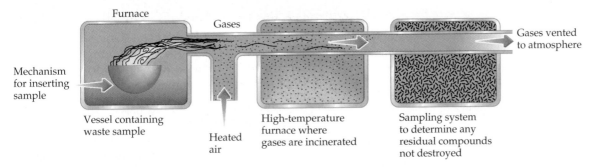

▲ **Figure 20.10** A schematic diagram of an incinerator for hazardous wastes.
Question: PCBs are burned in air at high temperatures to carbon dioxide, water, and hydrogen chloride. Write the equation for the incineration of the PCB component $C_6H_2Cl_3$—$C_6H_3Cl_2$.

become raw material for another industry. For example, waste nitric acid from the metals industry can be converted to fertilizer.

If a hazardous waste cannot be used or incinerated or treated to render it less hazardous, it must be stored in a secure landfill. Unfortunately, landfills often leak, contaminating the groundwater. We clean up one toxic waste dump and move the materials to another, playing a rather macabre shell game. The best technology at present for treating organic wastes, including chlorinated compounds, is incineration (Figure 20.10). At 1260 °C, 99.9999+% destruction is achieved. Biodegradation is another method that shows great promise. Scientists have identified microorganisms that degrade hydrocarbons such as those in gasoline. Other bacteria, when provided with proper nutrients, can degrade chlorinated hydrocarbons. Certain bacteria in the soil can even break down TNT and nitroglycerin. Through genetic engineering, scientists are developing new strains of bacteria that can decompose a great variety of wastes.

Often the trouble with incineration is finding a place to build the incinerator. No one wants an incinerator nearby.

20.13 WHAT PRICE POISONS?

We use so many poisons in and around our homes and workplaces that accidents are bound to happen. Poison control centers have been established in cities to help physicians deal with emergency poisonings. Are our insecticides, drugs, cleansers, and other chemicals worth the price we pay in terms of accidental poisonings? That is for you to decide. Generally, it is the misuse of these chemicals that leads to tragedy.

Perhaps it is easy to be negative about chemists and chemistry when you think of such horrors as nerve gases, carcinogens, and teratogens. But keep in mind that many toxic chemicals are of enormous benefit to us and that they can be used safely despite their hazardous nature. The plastics industry was able to control vinyl chloride emissions once the hazard was known. We are still able to have valuable vinyl plastics even though the vinyl chloride from which they are made causes cancer.

Increasingly, we have to decide whether the benefits we gain from hazardous substances are worth the risks we assume by using them. Many issues involving toxic chemicals are emotional; most of the decisions regarding them are political; but possible solutions to such problems lie mainly in the field of chemistry.

We hope that the chemistry you have learned here will help you make wise decisions. Most of all, we hope that you will continue to learn more about chemistry throughout the rest of your life, because chemistry affects nearly everything you do. We wish you success and happiness, and may the joy of learning go with you always.

Critical Thinking Exercises

Apply knowledge that you have gained in this chapter and one or more of the FLaReS principles (Chapter 1) to evaluate the following statements or claims.

20.1 An advertisement implies that rotenone is safer than competitive pesticides because it is obtained from natural sources.

20.2 A book claims that all chlorine compounds are toxic and should be banned.

20.3 A natural foods enthusiast contends that no food that has a carcinogen in it should be allowed on the market.

20.4 The advertisement claims "Organic Natural Chemical Free Mattresses." The company claims to "have to offer: No toxins, formaldehyde, VOCs, pesticides, synthetic or chemical foams."

20.5 The newspaper article suggests, "Try out tobacco 'sun tea' as a nontoxic pesticide." In the article is the claim, "This special tea can help you eliminate most garden pests without any unwanted toxicity."

SUMMARY

Sections 20.1–20.2—Toxicology is the branch of pharmacology that deals with poisons. "The dose makes the poison" means that anything can be a poison if the dose is large enough. Many natural poisons exist, including curare, strychnine, and botulin. Many plants, including hemlock, holly berries, iris, azalea, and hydrangea, are toxic. Strong acids and strong bases are toxic because they are corrosive. They cause hydrolysis of proteins. Strong oxidizing agents such as ozone and PAN damage largely by deactivation of enzymes.

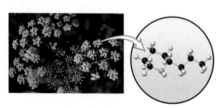

Sections 20.3–20.4—Carbon monoxide and nitrite ions are toxic because they interfere with the transport of oxygen by the blood. Cyanide is a poison because it shuts down cell respiration. Several substances, including fluoroacetic acid, are not highly toxic of themselves, but in the body they form substances that are toxic.

$$CH_2-COOH$$
$$HO-C-COOH$$
$$CH_2-COOH$$
Citric acid

$$F-CH-COOH$$
$$HO-C-COOH$$
$$CH_2-COOH$$
Fluorocitric acid

Sections 20.5–20.6—Heavy metal poisons such as lead and mercury inactivate enzymes by tying up their SH groups. Cadmium leads to loss of calcium ions from bones. Such poisons often are treated with chelating agents that tie up the metal ions so that they can be excreted. Nerve poisons, such as organophosphates, interfere with the acetylcholine cycle. Nerve poisons are among the most toxic synthetic chemicals known.

Active enzyme

Active site
+
Hg2+

Inactive enzyme

Sections 20.7–20.8—The term LD_{50} indicates a dosage that kills 50% of a population of test animals. Many substances not ordinarily thought of as poisons do have an LD_{50}. Botulism toxin has the lowest LD_{50} and is the most toxic substance known. The liver can detoxify substances by oxidizing or reducing them, or by coupling them with other body chemicals. The liver's P-450 enzyme system carries out many such detoxifying reactions. Sometimes this converts a substance less toxic, such as benzene or CCl_4, into one that is more toxic.

Sections 20.9–20.10—A carcinogen causes the growth of tumors. A benign tumor grows slowly and does not invade neighboring tissues. A malignant tumor or cancer can invade and destroy neighboring tissues. Some carcinogens are natural, but most cancer deaths are linked to tobacco, diet, and obesity. Oncogenes appear to trigger processes that convert normal cells to cancer, and suppressor genes ordinarily prevent the development of cancers. Some substances in our food, including antioxidant vitamins, act as anticarcinogens. The Ames test screens substances for carcinogenicity by examining bacterial growth. Animal testing is often necessary but is not conclusive. Epidemiological studies appear to be most useful.

Adult diet (obesity) (30%)

Tobacco (30%)

Sections 20.11–20.13—A teratogen causes birth defects. Hazardous wastes can be classified as reactive waste that tends to react spontaneously or vigorously; flammable waste that burns readily on ignition; toxic waste that contains or releases toxic substances; and corrosive waste that destroys tissue. The best way to handle hazardous wastes is not to produce them in the first place. Wastes may be stored in a landfill or incinerated, and biological degradation shows promise. Chemistry creates—and solves—many problems involving wastes.

Furnace
Gases

Heated air

REVIEW QUESTIONS

1. What is toxicology? What is a poison?

2. Is sodium chloride (table salt) poisonous? Explain your answer fully.

3. How can a substance be more harmful to one person than to another? Give an example.

4. How does the toxicity of a substance depends on the route of administration?

5. How does fluoroacetic acid exert its toxic effect?

6. List some sources of mercury poisoning and of lead poisoning.

7. What is the single leading cause of cancer?

8. Name several natural carcinogens.

9. What is a tumor? How are benign and malignant tumors different?

10. How does cigarette smoking cause cancer?

PROBLEMS

Corrosive Poisons

11. Which of the following are corrosive poisons?
    a. hexane
    b. hydrochloric acid
    c. nitroglycerin
    d. potassium hydroxide

12. How do dilute solutions of acids and bases damage living cells?

13. What kind of reaction is involved when acids break down protein molecules? What is the role of the acid in the reaction?

14. How does ozone damage living cells?

Poisons Affecting Oxygen Transport and Oxidative Processes

15. What is a blood agent? List two such poisons.

16. What is methemoglobin? How is it formed?

17. How does sodium thiosulfate act as an antidote for cyanide poisoning?

18. How do cyanides exert their toxic effect?

Heavy Metal Poisons

19. Iron (as Fe^{2+}) is a necessary nutrient. What are the effects of too little Fe^{2+}? Of too much?

20. Which of the following are heavy metals?
    a. beryllium
    b. mercury
    c. potassium
    d. thallium

21. How does mercury (as Hg^{2+}) exert its toxic effect?

22. How does BAL act as an antidote for mercury poisoning?

23. How does EDTA act as an antidote for lead poisoning?

24. How does cadmium (as Cd^{2+}) exert its toxic effect? What is *itai-itai* disease?

Chemistry of the Nervous System

25. How do each of the following affect the acetylcholine cycle?
    a. botulin   b. curare
    c. atropine

26. What is acetylcholine? Describe its action.

27. List three nerve poisons developed for use in warfare.

28. What are anticholinesterase poisons? How do they act on the acetylcholine cycle?

29. How does atropine act as an antidote for poisoning by organophosphorus compounds?

30. Describe some uses of botulin in medicine.

Detoxification

31. How does the liver detoxify ethanol?

32. What is the P-450 system? What is its function?

33. List two ways that the conversion of nicotine to cotinine in the liver lessens the risk of nicotine poisoning.

34. List two steps in the detoxification of ingested toluene. What is the effect of these steps?

35. Does the P-450 system always detoxify foreign substances? Explain.

36. How does ethanol work as an antidote for methanol poisoning?

Lethal Dose

37. The oral (in rats) LD_{50} for methyl isocyanate, the substance that caused the Bhopal tragedy, is 140 mg/kg body weight. Estimate the lethal dose for a 125-lb human.

38. The LD_{50} for ketoprofen (Orudis KT®), orally in rats, is 101 mg/kg body weight. If this value could be extrapolated to humans, what would be the lethal dose for a 23-kg child?

Chemical Carcinogens

39. What are oncogenes? How are they involved in the development of cancer?

40. What are suppressor genes? How are they involved in the development of cancer?

41. List some conditions under which polycyclic hydrocarbons are formed.

42. Name (a) two aromatic amines and (b) two aliphatic compounds that are carcinogens.

43. What is an epidemiological study? Can such a study prove absolutely that a compound causes cancer?

44. List some of the limitations involved in testing compounds for carcinogenicity by using laboratory animals.

Mutagens and Teratogens

45. What is a mutagen? A teratogen?

46. Describe the Ames test for mutagenicity. What are its limitations as a screening test for carcinogens?

Hazardous Wastes

47. Define and give an example of a flammable waste.

48. Which of the following are reactive wastes?
    a. CaS, which reacts with water to form toxic H_2S
    b. Ca metal, which reacts with water to form flammable H_2 gas
    c. nitroglycerin, an explosive
    d. hydrochloric acid, which dissolves steel drums

49. Define and give some examples of a hazardous waste.

50. What is the best method for the disposal of toxic organic wastes? Justify fully the method you select.

ADDITIONAL PROBLEMS

51. Botulin toxin has an estimated LD_{50} of 200 pg/kg body weight and an estimated molar mass of 150,000 g/mol. About how many molecules comprise a lethal dose for a 55-kg human?

52. The median lethal dose of botulin is about 2 ng/kg body weight. How much botulin would it take to kill (a) a 60-kg person and (b) all 6.3 billion people on Earth, assuming an average body weight of 60 kg?

53. Absinthe, a wormwood flavored liqueur that was highly popular in France about 1880–1914, contains thujone, a neurotoxin, LD_{50} (orally in mice) 87.5 mg/kg. Absinthe was the subject of famous artworks by Degas, Van Gogh, and others. The European Union (EU) limits thujone levels in commercially distilled absinthe to 10 mg/L or less. Absinthe is not sold in the United States. (a) Determine the molecular formula of thujone from the structure below. (b) Estimate the lethal dose for a 55-kg person. (c) What is the maximum mass of thujone that a 0.50-L bottle of the liqueur can contain in an EU country?

Thujone

54. Phosphorus comes in two forms or *allotropes*. Red phosphorus is used in manufacture of matches and is only mildly toxic. White phosphorus, a waxy yellow solid, is highly toxic, with a typical orally ingested lethal dose of just 70 mg for a 55-kg person. (a) What is the LD_{50} (in milligrams/kilogram) for white phosphorus? (b) Where does white phosphorus fall in Table 20.2?

55. A British study more than 50 years ago involved 1357 people with lung cancer. Of these 1350 were smokers and 7 were nonsmokers. A control group without cancer, matched in age and other characteristics, consisted of 1296 smokers and 61 nonsmokers. (a) Calculate the percentage of cancer cases who smoked and the percentage of controls who smoked. What can you infer from these proportions? (b) Calculate the odds of smoking for cancer cases and the odds of smoking for controls. (c) Calculate the odds ratio. Interpret these data.

COLLABORATIVE GROUP PROJECTS

Prepare a PowerPoint, poster, or other presentation (as directed by your instructor) for presentation to the class.

56. Prepare a brief report on one of the following pesticides. Include chemical formula, uses, toxicity, and possible carcinogenicity.
- **a.** alachlor
- **b.** fenoxycarb
- **c.** propoxur
- **d.** propachlor
- **e.** tribufos
- **f.** trichlorfon

57. Propose a disposal method for each of the following wastes. Be as specific as possible and justify your choice.
- **a.** hydrochloric acid contaminated with iron salts
- **b.** picric acid (an explosive)
- **c.** soybean oil contaminated with PCBs

58. Choose one of the following topics and be prepared to support either side (for or against) of the issue.
- **a.** Carcinogens that occur naturally in foods should be subjected to the same tests used to evaluate synthetic pesticides.
- **b.** Substances should be tested for toxicity or carcinogenicity on laboratory animals.
- **c.** Nerve gases are less humane than bullets in warfare.

59. Search the Web for information on brownfields. What are they? What is being done about them? Are there any near you?

60. Most tooth fillings consist of amalgams containing mercury, and the safety of these amalgams is a recurring controversy. Using your favorite search methods, find some websites dealing with this issue and write a brief analysis of the opposing points of view.

61. Search the Web for information on the effects of mercury in the environment and do a risk–benefit analysis of mercury use.

REFERENCES AND READINGS

1. Bower, B. "Excess Lead Linked to Boys' Delinquency." *Science News*, February 10, 1996, p. 86.

2. Burkholder, JoAnn M. "The Lurking Perils of Pfisteria." *Scientific American,* August 1999, pp. 42–47. Almost every year there are massive fish kills due to some highly poisonous toxins produced by microorganisms such as pfisteria. They look like algae, but they are not really plants.

3. Casarett, Louis J., John Doull, and Curtis D. Klaasen (eds.). *Casarett & Doull's Toxicology: The Basic Science of Poisons,* 6th ed. New York: McGraw-Hill, 2001.

4. Emsley, John. *The Elements of Murder: A History of Poisons.* Oxford, UK: Oxford University Press, 2005.

5. Gately, Iain, *Tobacco: A Cultural History of How an Exotic Plant Seduced Civilization.* New York: Grove, 2001.

6. Lewis, Ricki. "The Return of Thalidomide: Once-Feared Drug Finds New Applications." *The Scientist,* January 22, 2001, p. 1.

7. Lykknes, Annette, and Lise Kvittingen. "Arsenic: Not So Evil After All?" *Journal of Chemical Education,* May 5, 2003, pp. 497–500.

8. Marshall, Eliot. "Solving Louisville's Friday-the-Thirteenth Explosion." *Science,* March 27, 1981, p. 1405.

9. McGinn, Anne Platt. "Phasing Out Persistent Organic Pollutants," in Lester R. Brown, Christopher Flavin, and Hilary French (eds.), *State of the World 2000.* New York: W. W. Norton, 2000.

10. Macinnis, Peter. *Poisons: From Hemlock to Botox to the Killer Bean of Calabar.* New York: Arcade Publishing, 2005.

11. Parshall, George W. "Chemical Warfare Agents: How Do We Rid Ourselves of Them?" *The Chemist,* May–June 1995, pp. 5–7. We need a safe way to destroy weapons no longer needed.

12. Pizzi, Richard A. "Pointing to Poisons." *Today's Chemist at Work,* September 2004, pp. 43–45.

13. Stokstad, Erik. "A Snapshot of the U.S. Chemical Burden." *Science,* June 25, 2004, p. 1893; and "Pollution Gets Personal." *Science,* June 25, 2004, pp. 1892–1894.

14. Trewavas, Anthony. "Are We Oversensitive to Risks from Toxins?" *Chemistry and Industry,* December 20, 2004, pp. 14–15.

GREEN CHEMISTRY

Preservation of Lumber with Less Toxic Materials

Michael C. Cann, University of Scranton

The United States lumber industry produces about 35 billion board feet (1 ft × 1 ft × 1 in.) of softwood lumber each year, using over a billion trees. Approximately 20% of this lumber is used for exterior purposes. If left untreated, exterior lumber rots in about three to five years and must be replaced. In this Web Investigation you will learn about different ways that have been used to treat lumber and how green chemistry has been used to create less toxic preservatives and applied to the protection of wood.

WEB INVESTIGATIONS

Investigation 1
Creosote
A number of preservatives have been used to prolong the life of exterior lumber, including creosote (a phenol-containing mixture from coal tar), which is routinely used on telephone poles. Search the Web using the keyword "creosote" to learn more about where this mixture of chemicals comes from, how it is used, the effects on the environment, and health issues related to exposure to creosote. Use a search engine to find more information about creosote applications and use.

Investigation 2
CCA Pressure-Treated Wood
Wood that is used to make exterior structures such as decks is generally constructed with pressure treated wood (PTW). The process for producing PTW involves placing the wood in a large metal cylinder, using a vacuum to dehydrate the wood cells, and then treating the wood under pressure with a solution of preservative, allowing the preservatives to infiltrate the cells of the wood. PTW lasts from 10 to 20 times longer and estimates indicate that this saves approximately 6.5 billion cubic board feet of wood per year. This translates into about 435,000 new homes or 226 million trees. Preserving wood thus makes both environmental and economic sense.

Traditionally the preservative has been a mixture of copper, chromium, and arsenic salts known as chromated copper arsenate or CCA. However, health concerns have developed related to the chromium, and especially the ar-

senic, in CCA. Both chromium and arsenic are poisons and arsenic is a carcinogen. In a 6-foot-long, 2 × 6-inch piece of PTW there is about 13 grams of arsenic, enough to kill 100 adults. Although the chromium and arsenic are chemically locked into the wood, some studies suggest that small quantities of these chemicals can be transferred to the surroundings by leaching and to humans by direct contact with the skin. Search using the keywords "CCA and exposure" to find out about concerns regarding contact with the toxins in PTW.

▲ Wood that has been pressure-treated with chromated copper arsenate (CCA) is very resistant to insect damage and rot. Recently there have been some concerns over CCA and its ability to leach into the environment.

Investigation 3
Nontoxic Wood Preservatives

As of January 1, 2003, consumer wood products in the United States can no longer be preserved with CCA. Chemists at Chemical Specialties Inc. (CSI) developed new preservatives that are not toxic to replace CCA in pressure-treated wood. These preservatives are quaternary ammonium compounds (ACQ) and are similar to some of the chemicals that are used in disinfectants and cleaners. CSI's development of ACQ won a Presidential Green Chemistry Challenge Award in 2002. Go to CSI's website to learn more about this preservative. New technologies often bring new challenges. What problems have arisen with these new wood preservatives and how are they being addressed?

COMMUNICATE YOUR RESULTS

Exercise 1
Creosote

What other applications, past and present, has creosote been used for in addition to wood preservation? List at least 10 uses and applications. Specify when use stopped, if it is no longer used.

Exercise 2
Exposure to Toxins

List five ways in which humans may be exposed to arsenic in PTW that has been treated with CCA. Would you choose to use PTW with CCA in your home? Why or why not?

Exercise 3
Nontoxic Wood Preservatives

Based on what you've learned in the previous Web Investigations, prepare a two-page report comparing and contrasting the use of ACQ and CCA for preserving wood. Be sure to include advantages and disadvantages for each in terms of human health, environmental effects, and economic costs. Based on the information you have gathered for your report, which preservative would you use? Why?

Exercise 4
Next-Generation Wood Preservatives

Throughout this course you've seen many examples of green chemistry and learned the power of chemistry to design safer products. Based on what you've learned, consider the following discussion questions:

- Does using ACQ eliminate all risks to human health and the environment?
- What would the ideal wood preservative be like? List 5 to 10 characteristics.
- How might green chemistry be used to develop an even better wood preservative?

Using the discussion questions as a guide, write a letter to a company involved in the PTW industry. You should explain your vision and provide a convincing argument for the company to invest in research and development to achieve the ideal wood preservative.

Exercise 5
Wood Preservatives in Your Community

Investigate the public spaces in your college and surrounding community. Are there decks, fences, play structures, or other things made from PTW? Find out what kind of PTW was used. Write a one-page letter to your college or local newspaper summarizing your findings and making recommendations based on what you've learned in this Web Investigation. If replacement is recommended, provide an estimate of the cost.

Appendix A

Review of Measurement and Mathematics

Accurate measurements are essential to science. Measurements can be made in a variety of units, but most scientists use the International System of Measurement (SI). The data they record often has to be converted from one kind of unit to another and otherwise manipulated mathematically. In this appendix, we extend the discussion of metric measurement that we began in Chapter 1 and review some of the mathematics that you may find useful in this course.

A.1 THE INTERNATIONAL SYSTEM OF MEASUREMENT

As noted in Chapter 1, the standard unit of length in the International System of Measurement is the *meter*. This distance was once defined as 0.0000001 of Earth's quadrant—that is, of the distance from the North Pole to the equator measured along a meridian. The quadrant proved difficult to measure accurately, and today the meter is defined precisely as the distance light travels in a vacuum during 1/299,792,458 of a second.

The primary unit of mass is the *kilogram* (1 kg = 1000 g). It is based on a standard platinum–iridium cylinder kept at the International Bureau of Weights and Measures. The *gram* is a more convenient unit for many chemical operations.

The derived SI unit of volume is the *cubic meter*. The units more frequently employed in chemistry, however, are the *liter* (1 L = 0.001 m^3) and the *milliliter* (1 mL = 0.001 L). Other SI units of length, mass, and volume are derived from these basic units. Table A.1 lists additional metric units of length, mass, and volume and illustrates the use of prefixes.

TABLE A.1   Some Metric Units of Length, Mass, and Volume

Length
1 kilometer (km) = 1000 meters (m)
1 meter (m) = 100 centimeters (cm)
1 centimeter (cm) = 10 millimeters (mm)
1 millimeter (mm) = 1000 micrometers (μm)

Mass
1 kilogram (kg) = 1000 grams (g)
1 gram (g) = 1000 milligrams (mg)
1 milligram (mg) = 1000 micrograms (μg)

Volume
1 liter (L) = 1000 milliliters (mL)
1 milliliter (mL) = 1000 microliters (μL)
1 milliliter (mL) = 1 cubic centimeter (cm^3)

A.2 EXPONENTIAL (SCIENTIFIC) NOTATION

Scientists often use numbers that are so large or so small that they boggle the mind. For example, light travels at about 300,000,000 m/s. There are 602,200,000,000,000,000,000,000 carbon atoms in 12.01 g of carbon. On the small side, the diameter of an atom is about 0.0000000001 m, and the diameter of an atomic nucleus is about 0.000000000000001 m. We find it difficult to keep track of the zeros in such quantities. Scientists find it convenient to express such numbers in exponential notation.

A number is in *exponential notation*—often called *scientific notation*—when it is written as the product of a coefficient (usually with a value between 1 and 10) and a power of 10. Two examples are

$$4.18 \times 10^3 \text{ and } 6.57 \times 10^{-4}$$

Expressing numbers in exponential form generally serves two purposes.

1. We can write very large or very small numbers in a minimum of printed space and with a reduced chance of typographical error.
2. We can convey explicit information about the precision of measurements: The number of significant figures (Section A.4) in a measured quantity is stated unambiguously.

In the expression 10^n, n is the exponent of 10, and the number 10 is said to be raised to the nth power. If n is a *positive quantity*, 10^n has a value *greater than 1*. If n is a *negative quantity*, 10^n has a value *less than 1*. We are particularly interested in cases where n is an integer. For example,

Positive Powers of 10	*Negative Powers of 10*
$10^0 = 1$	$10^0 = 1$
$10^1 = 10$	$10^{-1} = 1/10 = 0.1$
$10^2 = 10 \times 10 = 100$	$10^{-2} = 1/(10 \times 10) = 0.01$
$10^3 = 10 \times 10 \times 10 = 1000$	$10^{-3} = 1/(10 \times 10 \times 10) = 0.001$
and so on	and so on

The power of 10 determines the number of zeros that follow the digit 1.

The power of 10 determines the number of places to the right of the decimal point where the digit 1 appears.

We express 612,000 in exponential form as

$$612,000 = 6.12 \times 100,000 = 6.12 \times 10^5$$

We express 0.000505 in exponential form as

$$0.000505 = 5.05 \times 0.0001 = 5.05 \times 10^{-4}$$

We can use a more direct approach to converting numbers to the exponential form.

- Count the number of places a decimal point must be moved to produce a coefficient having a value between 1 and 10.
- The number of places counted then becomes the power of 10.
- The power of 10 is *positive* if the decimal point is moved to the *left*.

$$6\,1\,2\,0\,0\,0 = 6.12 \times 10^5$$

Move the decimal point (here understood) *five* places to the *left*.

The exponent is (positive) 5.

- The power of 10 is *negative* if the decimal point is moved to the *right*.

$$0.000505 = 5.05 \times 10^{-4}$$

| Move the decimal point *four* places to the *right*. | The exponent is **−4**. |

To convert a number from exponential form to the conventional form, move the decimal point in the opposite direction.

$$3.75 \times 10^6 = 3.750000$$

| The exponent is **6**. | Move the decimal point *six* places to the *right*. |

$$7.91 \times 10^{-7} = 0.0000007.91$$

| The exponent is **−7**. | Move the decimal point *seven* places to the *left*. |

It is easy to handle exponential numbers on most calculators. A typical procedure is to enter the number, followed by the key [EXP]. The keystrokes required for the number 2.85×10^7 are [2] [.] [8] [5] [EXP] [7], and the result displayed is 2.85^{07} (the 10 is understood).

For the number 1.67×10^{-5}, the keystrokes are [1] [.] [6] [7] [EXP] [5] [±], and the result displayed is 1.67^{-05}. Many calculators can be set to convert all numbers and calculated results to the exponential form, regardless of the form in which the numbers are entered. Generally, a calculator can also be set to display a fixed number of digits in the results.

The keystrokes on your calculator—and the appearance of the display—may be different from those shown here. Check the instructions in the manual that came with the calculator.

Addition and Subtraction

To add or subtract numbers by hand in exponential notation, it is necessary to express each quantity as the *same power of 10*. In calculations, this treats the power of 10 in the same way as a unit—it is simply "carried along." In the following, each quantity is expressed with the power 10^{-3}.

$$(3.22 \times 10^{-3}) + (7.3 \times 10^{-4}) - (4.8 \times 10^{-4}) = (3.22 \times 10^{-3})$$
$$+ (0.73 \times 10^{-3}) - (0.48 \times 10^{-3})$$
$$= (3.22 + 0.73 - 0.48) \times 10^{-3}$$
$$= 3.47 \times 10^{-3}$$

If a scientific or graphing calculator is used, it is not necessary to change each exponential quantity to the same power of 10.

Multiplication and Division

To multiply numbers expressed in exponential form, *multiply* all coefficients to obtain the coefficient of the result and *add* all exponents to obtain the power of 10 in the result.

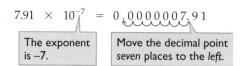

$$0.0803 \times 0.0077 \times 455 = (8.03 \times 10^{-2}) \times (7.7 \times 10^{-3}) \times (4.55 \times 10^2)$$

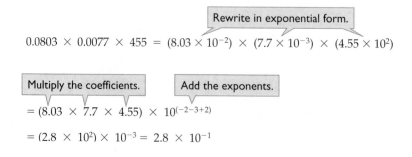

$$= (8.03 \times 7.7 \times 4.55) \times 10^{(-2-3+2)}$$
$$= (2.8 \times 10^2) \times 10^{-3} = 2.8 \times 10^{-1}$$

Generally, most calculators perform these operations automatically, and no intermediate results need be recorded.

To divide two numbers in exponential form, *divide* the coefficients to obtain the coefficient of the result and *subtract* the exponent in the denominator from the exponent in the numerator to obtain the power of 10. In the example below, multiplication and division are combined. First, the rule for multiplication is applied to the numerator and to the denominator, and then the rule for division is used.

> Rewrite in exponential form.

$$\frac{0.015 \times 0.0088 \times 822}{0.092 \times 0.48} = \frac{(1.5 \times 10^{-2})(8.8 \times 10^{-3})(8.22 \times 10^{2})}{(9.2 \times 10^{-2})(4.8 \times 10^{-1})}$$

> Apply the rule for multiplication to the numerator and denominator.

> Apply the rule for division.

$$= \frac{1.1 \times 10^{-1}}{4.4 \times 10^{-2}} = 0.25 \times 10^{-1-(-2)} = 0.25 \times 10^{1}$$

$$= 2.5 \times 10^{-1} \times 10^{1} = 2.5 \times 10^{0} = 2.5$$

Raising a Number to a Power and Extracting the Root of an Exponential Number

To raise an exponential number to a given power, raise the coefficient to that power and multiply the exponent by that power. For example, we can cube a number (that is, raise it to the *third* power) in the following manner.

> Rewrite in exponential form.
> Cube the coefficient.
> Multiply the exponent by 3.

$$(0.0066)^{3} = (6.6 \times 10^{-3})^{3} = (6.6)^{3} \times 10^{3 \times (-3)}$$

$$= (2.9 \times 10^{2}) \times 10^{-9} = 2.9 \times 10^{-7}$$

To extract the root of an exponential number, we raise the number to a fractional power—one-half power for a square root, one-third power for a cube root, and so on. Most calculators have keys designed for extracting square roots and cube roots. Thus, to extract the square root of 1.57×10^{5}, enter the number 1.57×10^{5} into a calculator, and use the $[\sqrt{\ }]$ key.

$$\sqrt{1.57 \times 10^{-5}} = 3.96 \times 10^{-3}$$

Some calculators allow you to extract roots by keying the root in as a fractional exponent. For example, we can take a cube root as follows.

$$(2.75 \times 10^{-9})^{1/3} = 1.40 \times 10^{-3}$$

PRACTICE PROBLEMS

1. Express each of the following numbers in exponential form.
   a. 0.000017   b. 19,000,000
   c. 0.0034   d. 96,500

2. Express each of the following numbers in exponential form.
   a. 4,500,000   b. 0.000108
   c. 0.0341   d. 406,000

3. Carry out the following operations. Express the answers in exponential form.
   a. $(4.5 \times 10^{13})(1.9 \times 10^{-5})$ b. $\sqrt{1.7 \times 10^{-5}}$
   c. $\dfrac{4.3 \times 10^{-7}}{7.6 \times 10^{22}}$ d. $\dfrac{9.3 \times 10^{9}}{3.7 \times 10^{27}}$

4. Carry out the following operations. Express the answers in exponential form.
   a. $(6.2 \times 10^{-5})(4.1 \times 10^{-12})$ b. $(2.1 \times 10^{-6})^2$
   c. $\dfrac{2.1 \times 10^{5}}{9.8 \times 10^{-7}}$ d. $\dfrac{4.6 \times 10^{-12}}{2.1 \times 10^{3}}$

A.3 UNIT CONVERSIONS

We live in a world in which almost all other countries use metric measurements. Suppose that an American applies for a job in another country and the job application asks for her height in centimeters. She knows her height in customary units is 66.0 in. Her actual height is the same, whether we express it in inches, feet, centimeters, or millimeters. Thus, when we measure something in one unit and then convert it to another one, we must not change the measured quantity in any fundamental way. We can use equivalencies such as those in Table A.2 to derive conversion factors.

TABLE A.2   Some Conversions Between Common and Metric Units
Length
1 mile (mi) = 1.61 kilometers (km)
1 yard (yd) = 0.914 meter (m)
1 inch (in.) = 2.54 centimeters (cm)
Mass
1 pound (lb) = 454 grams (g)
1 ounce (oz) = 28.4 grams (g)
1 pound (lb) = 0.454 kilogram (kg)
Volume
1 U.S. quart (qt) = 0.946 liter (L)
1 U.S. pint (pt) = 0.473 liter (L)
1 fluid ounce (fl oz) = 29.6 milliliters (mL)
1 gallon (gal) = 3.78 liters (L)

In mathematics, multiplying a quantity by 1 does not change its value. We can therefore use a factor equivalent to 1 to convert between inches and centimeters. We find our factor in the definition of the inch in Table A.2.

$$1 \text{ in.} = 2.54 \text{ cm}$$

Notice that if we divide both sides of this equation by 1 in., we obtain a ratio of quantities that is equal to 1.

$$1 = \frac{1 \text{ in.}}{1 \text{ in.}} = \frac{2.54 \text{ cm}}{1 \text{ in.}}$$

If we divide both sides of the equation by 2.54 cm, we also obtain a ratio of quantities that is equal to 1.

$$\frac{1 \text{ in.}}{2.54 \text{ cm}} = \frac{2.54 \text{ cm}}{2.54 \text{ cm}} = 1$$

The two ratios, one shown in red and the other in blue, are conversion factors. A *conversion factor* is a ratio of terms, equivalent to the number 1, used to change the unit in which a quantity is expressed. We call this process the *unit conversion method* of problem solving. Since we set up problems by examining the *dimensions* associated with the given quantity, those associated with the desired result, and those needed in the conversion factors, the process is sometimes called *dimensional analysis*.

What happens when we multiply a known quantity by a conversion factor? The original unit cancels out and is replaced by the desired unit. Thus, our general approach to using conversion factors is

Desired quantity and unit = given quantity and unit × conversion factors

Now let's return to the question about the woman's height, a measured quantity of 66.0 in. To get an answer in centimeters, we must use the appropriate conversion factor (in blue). Note that the desired unit (cm) is in the numerator and the unit to be replaced (in.) is in the denominator. Thus we can cancel the unit in. so that only the unit cm remains.

$$66.0 \text{ in.} \times \frac{2.54 \text{ cm}}{1 \text{ in.}} = 168 \text{ cm (rounded off from 167.64)}$$

You can see why the other conversion factor (in red) won't work. It gives a nonsensical unit.

$$66.0 \text{ in.} \times \frac{1 \text{ in.}}{2.54 \text{ cm}} = \frac{26.0 \text{ in.}^2}{\text{cm}}$$

Following are some examples and exercises to give you some practice in this method of problem solving.

Example A.1

a. Convert 0.742 kg to grams. **b.** Convert 0.615 lb to ounces.
c. Convert 135 lb to kilograms.

Solution
a. This is a conversion from one metric unit to another. We simply use a knowledge of prefixes to convert from meters to millimeters.

We start here. This converts kg to g. Our answer: the number and the unit.

$$0.742 \text{ kg} \times \frac{1000 \text{ g}}{1 \text{ kg}} = 742 \text{ g}$$

b. This is a conversion from one common unit to another. Here we use the fact that 1 lb = 16 oz to convert from pounds to ounces. Then we proceed as in part (a), arranging the conversion factor to cancel the unit lb.

$$0.165 \text{ lb} \times \frac{16 \text{ oz}}{1 \text{ lb}} = 9.48 \text{ oz}$$

c. This is a conversion from a common unit to a metric unit. We need data from Table A.2 to convert from pounds to kilograms. Then we proceed as in part (b), arranging the conversion factor to cancel the unit lb.

$$135 \text{ lb} \times \frac{0.454 \text{ kg}}{1 \text{ lb}} = 61.2 \text{ kg}$$

(We give our answers with the proper number of significant figures. If you do not understand significant figures and wish to do so, they are discussed in Section A.4. In this text we simply use three significant figures for most calculations.)

▶ **Exercise A.1**
a. Convert 16.3 mg to grams. **b.** Convert 24.5 oz to pounds.
c. Convert 12.5 fl oz to milliliters.

Quite often, to get the desired unit, we must use more than one conversion factor. We can do so by arranging all the necessary conversion factors in a single setup that yields the final answer in the desired unit.

Example A.2

What is the length, in millimeters, of a 3.25-ft piece of tubing?

Solution

No relationship between feet and millimeters is given in Table A.2, and so we need more than one conversion factor. We can think of the problem as a series of three conversions.

1. Use the fact that 1 ft = 12 in. to convert from feet to inches.
2. Use data from Table A.2 to convert from inches to centimeters.
3. Use a knowledge of prefixes to convert from centimeters to millimeters.

We could solve this problem in three distinct steps by making one conversion in each step, but it is just as easy to combine three conversion factors into a single setup. Then we proceed as indicated below.

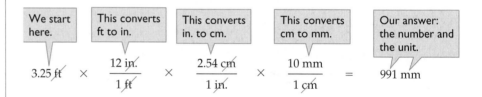

▶ **Exercise A.2**
Carry out the following conversions.
a. 90.3 mm to meters
b. 729.9 ft to kilometers
c. 1.17 gal to fluid ounces

Sometimes we need to convert two or more units in the measured quantity. We can do this by arranging all the necessary conversion factors in a single setup so that the starting units cancel and the conversions yield the final answer in the desired units.

Example A.3

A saline solution has 1.00 lb of salt in 1.00 gal of solution. Calculate the quantities in grams per liter of solution.

Solution

First let's identify the measured quantities. They can be expressed in the form of a ratio of mass of salt in pounds to a volume in gallons.

$$\frac{1.00 \text{ lb (salt)}}{1.00 \text{ gal (solution)}}$$

This ratio must be converted to one expressed in grams per liter. We must convert from pounds to grams in the numerator and from gallons to liters in the denominator. The following set of equivalent values from Table A.2 can be used to formulate conversion factors.

$$1 \text{ lb} = 453.6 \text{ g} \qquad 1 \text{ gal} = 3.75 \text{ L}$$

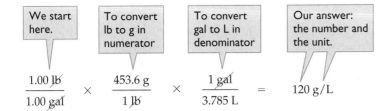

$$\frac{1.00 \text{ lb}}{1.00 \text{ gal}} \times \frac{453.6 \text{ g}}{1 \text{ lb}} \times \frac{1 \text{ gal}}{3.785 \text{ L}} = 120 \text{ g/L}$$

Note that we could also do the conversions in the numerator and denominator separately and then divide the numerator by the denominator.

Numerator: $1.00 \text{ lb} \times \dfrac{453.6 \text{ g}}{1 \text{ lb}} = 454 \text{ g}$

Denominator: $1.00 \text{ gal} \times \dfrac{3.785 \text{ L}}{1 \text{ gal}} = 3.785 \text{ L}$

Division: $\dfrac{454 \text{ g}}{3.785 \text{ L}} = 120 \text{ g/L}$

Both methods give the same answer to three significant figures.

▶ **Exercise A.3**
Carry out the following conversions.
a. 88.0 km/h to meters per second
b. 1.22 ft/s to kilometers per hour
c. 4.07 g/L to ounces per quart

A.4 PRECISION, ACCURACY, AND SIGNIFICANT FIGURES

Counting can give exact numbers. For example, we can count exactly 24 students in a room. Measurements, on the other hand, are subject to error. One source of error is in the measuring instruments themselves. For example, an incorrectly calibrated thermometer may consistently yield a result that is 0.2 °C too low. Other errors may result from the experimenter's lack of skill or care in using measuring instruments.

Precision and Accuracy

Suppose you were one of five students asked to measure a person's height using a meter stick marked off in millimeters. The five measurements are recorded in Table A.3. The *precision* of a set of measurements refers to how closely individual measurements agree with one another. We say the precision is good if each of the measurements is close to the average, and poor if there is a wide deviation from the average value.

How would you describe the precision of the data in Table A.3? Examine the individual data, note the average value, and determine how much the individual data differ from the average. Because the maximum deviation from the average value is 0.003 m, the precision is quite good.

The *accuracy* of a set of measurements refers to the closeness of the average of the set to the "correct" or most probable value. Measurements of high precision are more likely to be accurate than are those of poor precision, but even highly precise measurements are sometimes inaccurate. For example, what if the meter sticks used to obtain the data in Table A.3 were actually 1005 mm long but still had 1000-mm markings? The accuracy of the measurements would be rather poor, even though the precision would remain good.

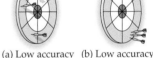

(a) Low accuracy   (b) Low accuracy
    Low precision   High precision

(c) High accuracy (d) High accuracy
    Low precision   High precision

▲ Comparing precision and accuracy: a dartboard analogy. (a) The darts are both scattered (low precision) and off-center (low accuracy). (b) The darts are in a tight cluster (high precision) but still off-center (low accuracy). (c) The darts are somewhat scattered (low precision) but evenly distributed about the center (high accuracy). (d) The darts are in a tight cluster (high precision) and well centered (high accuracy).

TABLE A.3 Five Measurements of a Person's Height	
Student	**Height, m**
1	1.827
2	1.824
3	1.826
4	1.828
5	1.829
Average	1.827

Sampling Errors

No matter how accurate an analysis, it will not mean much unless it is performed on valid, representative samples. Consider determining the level of glucose in the blood of a patient. High glucose levels are associated with diabetes, a debilitating disease. Results vary, depending on several factors such as the time of day and what and when the person last ate. Glucose levels are much higher soon after a meal high in sugars. They also depend on other factors, such as stress. Medical doctors usually take blood for analysis after a night of fasting. These fasting glucose levels tend to be more reliable, but physicians often repeat the analysis when high or otherwise suspicious results are obtained. Therefore, there is no one true value for the level of glucose in a person's blood. Repeated samplings provide an average level. Similar results from several measurements give much more confidence in the findings. For example, a diagnosis of diabetes is usually based on two consecutive fasting blood glucose levels above 120 mg/dL.

Significant Figures

Look again at Table A.3. Notice that the five measurements of height agree in the first three digits (1.82); they differ only in the fourth digit. We say that the fourth digit is uncertain. All digits known with certainty, plus the first uncertain one, are called *significant figures*. The precision of a measurement is reflected in the number of significant figures—the more significant figures, the more precise the measurement. The measurements in Table A.3 have four significant figures. In other words, we are quite sure that the person's height is between 1.82 m and 1.83 m. Our best estimate of the average value, including the uncertain digit, is 1.827 m.

The number 1.827 has four digits; we say it has four significant figures. In any properly reported measurement, all nonzero digits are significant. Zeros, however, may or may not be significant because they can be used in two ways: as a part of the measured value or to position a decimal point.

- Zeros between two other significant digits are significant. *Examples*: 4807 (four significant figures); 70.004 (five).
- We use a lone zero preceding a decimal point for aesthetic purposes; it is not significant. *Example*: 0.352 (three significant figures).
- Zeros that precede the first nonzero digit are also not significant. *Examples*: 0.000819 (three significant figures); 0.03307 (four).
- Zeros at the end of a number are significant if they are to the right of the decimal point. *Examples*: 0.2000 (four significant figures); 0.050120 (five).

We can summarize these four situations with a general rule: When we read a number from left to right, all the digits starting with the first nonzero digit are significant. Numbers without a decimal point that end in zeros are a special case, however.

- Zeros at the end of a number may or may not be significant if the number is written without a decimal point. *Example*: 700. We do not know whether the number 700 was measured to the nearest unit, ten, or hundred. To avoid this confusion, we can use exponential notation (Section A.2). In exponential notation, 700 is recorded as 7×10^2 or 7.0×10^2 or 7.00×10^2 to indicate one, two, or three significant figures, respectively. The only significant digits are those in the coefficient, not in the power of 10.

We use significant figures only with measurements—quantities subject to error. The concept does not apply to a quantity that is

1. inherently an integer, such as 3 sides to a triangle or 12 items in a dozen,
2. inherently a fraction, such as the radius of a circle equals one-half the diameter,

3. obtained by an accurate count, such as 18 students in a class,
4. a defined quantity, such as 1 km = 1000 m.

In these contexts, the numbers 3, 12, $\frac{1}{2}$, 18, and 1000 can have as many significant figures as we want. More properly, we say that each is an exact value.

Example A.4

In everyday life, common units are often used with metric units in parentheses (or vice versa), but significant figures are not always considered. Which of the following have followed proper significant figure usage?

a. A popular science magazine reports that *"T. rex ...* gained as much as 4.6 pounds (2 kilograms) daily."
b. A urinal is rated at "1.0 gpf (gallons per flush) or 3.8 lpf (liters per flush)."

Solution

a. The quantity "4.6 pounds" has two significant figures but "2 kilograms" has only one; significant figure rules were not followed.
b. Each quantity has two significant figures; significant figure rules were followed.

Significant Figures in Calculations: Multiplication and Division

If we measure a sheet of notepaper and find it to be 14.5 cm wide and 21.7 cm long, we can find the area of the paper by multiplying the two quantities. A calculator gives the answer as 314.65. Can we conclude that the area is 314.65 cm^2? That is, can we know the area to the nearest hundredth of a square centimeter when we know the width and length only to the nearest tenth of a centimeter? It just doesn't seem reasonable—and it isn't. A calculated quantity can be no more precise than the data used in the calculation, and the reported result should reflect this fact.

A strict application of this principle involves a fairly complicated statistical analysis that we will not attempt here, but we can do a fairly good job by using a practical rule involving significant figures:

> In multiplication and division, the reported result should have no more significant figures than the factor with the fewest significant figures.

In other words, a calculation is only as precise as the least precise measurement that enters into the calculation.

To obtain a numerical answer with the proper number of significant figures often requires that we round off numbers. In rounding, we drop all digits that are not significant and, if necessary, adjust the last reported digit. We use the following rules in rounding.

- If the leftmost digit to be dropped is *4 or less*, drop it and all following digits. *Example*: If we need four significant figures, 69.744 rounds to 69.74, and to 69.7 if we need three significant figures.
- If the leftmost digit to be dropped is *5 or greater*, increase the final retained digit by 1. *Example*: 538.76 rounds to 538.8 if we need four significant figures. Similarly, 74.397 rounds to 74.40 if we need four significant figures, and to 74.4 if we need three.

Example A.5

What is the area, in square centimeters, of a rectangular gauze bandage that is 2.54 cm wide and 12.42 cm long? Use the correct number of significant figures in your answer.

Solution

The area of a rectangle is the product of its length and width. In the result, we can show only as many significant figures as there are in the least precisely stated dimension, the width, which has three significant figures.

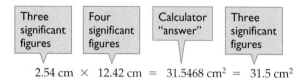

| Three significant figures | Four significant figures | Calculator "answer" | Three significant figures |

$$2.54 \text{ cm} \times 12.42 \text{ cm} = 31.5468 \text{ cm}^2 = 31.5 \text{ cm}^2$$

We use the rules for rounding off numbers as the basis for dropping the digits 468.

▶ **Exercise A.5**

Calculate the volume, in cubic meters, of a rectangular block of foamed plastic that is 1.827 m long, 1.04 m wide, and 0.064 m thick. Use the correct number of significant figures.

Example A.6

For a laboratory experiment, a teacher wants to divide all of a 226.8-g sample of glucose equally among the 18 members of her class. How many grams of glucose should each student receive?

Solution

The number 18 is a counted number, that is, an exact number that is not subject to significant figure rules. The answer should carry four significant figures, the same as in 226.8 g.

$$\frac{226.8 \text{ g}}{18} = 12.60 \text{ g}$$

In this calculation a calculator displays the result 12.6. We add the digit 0 to emphasize that the result is precise to four significant figures.

▶ **Exercise A.6**

A dozen eggs has a mass of 681 g. What is the average mass of one of the eggs, expressed with the appropriate number of significant figures?

Significant Figures in Calculations: Addition and Subtraction

In addition or subtraction, we are concerned not with the number of significant figures but with the number of digits to the right of the decimal point. When we add or subtract quantities with varying numbers of digits to the right of the decimal point, we need to note the one with the fewest such digits. The result should contain the same number of digits to the right of its decimal point. For example, if you are adding several masses and one of them is measured only to the nearest gram, the total mass cannot be stated to the nearest milligram no matter how precise the other measurements are.

We apply this idea in Example A.7. Note that in a calculation involving several steps, we need round off only the final result.

Example A.7

Perform the following calculation and round off the answer to the correct number of significant figures.

$$2.146 \text{ g} + 72.1 \text{ g} - 9.1434 \text{ g}$$

Solution

In this calculation, we add two numbers and subtract a third from the sum of the first two.

2.146 g ◁ three decimal places

+ 72.1 g ◁ one decimal place

74.246 g

− 9.1434 g ◁ four decimal places

65.1026 g = 65.1 g ◁ one decimal place

Note that we do not round off the intermediate result (74.246). When using a calculator, we generally don't need to write down an intermediate result.

▶ **Exercise A.7**

Perform the indicated operations and give answers with the proper number of significant figures. Note that in addition and subtraction all terms must be expressed in the same unit.

a. 48.2 m + 3.82 m + 48.4394 m
b. 148 g + 2.39 g + 0.0124 g
c. 15.436 L + 5.3 L − 6.24 L − 8.177 L
d. (51.5 m + 2.67 m) × (33.42 m − 0.124 m)

e. $\dfrac{125.1 \text{ g} - 1.22 \text{ g}}{52.5 \text{ mL} + 0.63 \text{ mL}}$

f. $\dfrac{0.307 \text{ g} - 14.2 \text{ mg} - 3.52 \text{ mg}}{(1.22 \text{ cm} - 0.28 \text{ mm}) \times 0.752 \text{ cm} \times 0.51 \text{ cm}}$

A.5 CALCULATIONS INVOLVING TEMPERATURE AND HEAT

We sometimes need to make conversions from one temperature scale to another or from one energy unit to another.

Temperature

On the Fahrenheit scale, the freezing point of water is 32 °F and the boiling point is 212 °F, whereas on the Celsius scale the freezing point of water is 0 °C and the boiling point is 100 °C. A 100° temperature interval (100 − 0) on the Celsius scale therefore equals a 180° interval (212 − 32) on the Fahrenheit scale. From these facts we can derive two equations that relate temperatures on the two scales. One of these requires multiplying the degrees of Celsius temperature by the factor 1.8 (that is, 180/100) to obtain the degrees of Fahrenheit temperature, followed by adding 32 to account for the fact that 0 °C = 32 °F.

$$°F = (1.8 \times °C) + 32$$

In the other equation, we subtract 32 from the Fahrenheit temperature to get the number of degrees Fahrenheit above the freezing point of water. Then this quantity is divided by 1.8.

$$°C = \dfrac{°F - 32}{1.8}$$

The SI unit of temperature is the kelvin (Chapters 1 and 6). The kelvin (K) is the same size as a Celsius degree, but the two scales have different starting points. The freezing point of water on the Kelvin scale is 273 K. To convert from °C to K, we simply add 273. And to convert from K to °C, we subtract 273.

$$K = °C + 273 \quad \text{and} \quad °C = K - 273$$

Example A.8 illustrates a practical situation where conversion between Celsius and Fahrenheit temperatures is necessary.

Example A.8

At home you keep your room thermostat set at 68 °F. When traveling, you have a room with a thermostat that uses the Celsius scale. What Celsius temperature will give you the same temperature as at home?

Solution

$$°C = \frac{°F - 32}{1.8} = \frac{68 - 32}{1.8} = 20 \ °C$$

▶ **Exercise A.8**
a. Convert 85.0 °C to degrees Fahrenheit.
b. Convert −12.2 °C to degrees Fahrenheit.
c. Convert 355 °F to degrees Celsius.
d. Convert −20.8 °F to degrees Celsius.

Heat

Scientists usually measure quantities of heat energy in *joules*. The joule is the SI unit of heat. A joule (J) is the work done by a force of 1 newton[1] acting over a distance of 1 meter. We also use the more familiar calorie.

$$1 \text{ cal} = 4.184 \text{ J}$$
$$1 \text{ kcal} = 1000 \text{ cal} = 4184 \text{ J}$$

A calorie (cal) is the amount of heat required to raise the temperature of 1 g of water 1 °C. (This quantity varies slightly with temperature; a calorie is defined more precisely as the amount of heat required to raise the temperature of 1 g of water from 14.5° to 15.5 °C.)

Some substances gain or lose heat more readily than others (Table A.4). The **specific heat** of a substance is the amount of heat required to raise the temperature of 1 g of the substance 1 °C. The definition of a calorie indicates that the specific heat of water is 1.00 cal/(g · °C). In SI, the specific heat of water is 4.184 J/(g · °C). Note that metals all have much lower values of specific heat than does water. This means that a sample of water must absorb much more heat to raise its temperature than a metal sample of similar mass. The metal sample may become red hot after absorbing a quantity of heat that makes the water sample only lukewarm.

We can use the following equation, in which ΔT is the change in temperature (in either °C or kelvins), to calculate the quantity of heat absorbed or released by a system.

$$\text{Heat absorbed or released} = \text{mass} \times \text{specific heat} \times \Delta T$$

[1] A *newton* (N) is the basic SI unit of force. A newton is the force required to give a 1-kg mass an acceleration of 1 m/s². That is, 1 N = 1 kg m/s². Therefore 1 J = 1 N m = 1 kg m²/s².

TABLE A.4   Specific Heats of Selected Substances

Substance	Specific Heat	
	cal/g °C	J/g °C
Aluminum (Al)	0.216	0.902
Copper (Cu)	0.0921	0.385
Ethyl alcohol (CH_3CH_2OH)	0.588	2.46
Iron (Fe)	0.106	0.443
Ethylene glycol ($HOCH_2CH_2OH$)	0.561	2.35
Magnesium (Mg)	0.245	1.025
Mercury (Hg)	0.0332	0.139
Sulfur (S)	0.169	0.706
Water (H_2O)	1.000	4.182

Example A.9

How much heat, in calories, kilocalories, and kilojoules, does it take to raise the temperature of 225 g of water from 25.0 °C to 100.0 °C?

Solution

Let's list the quantities we need for the calculations.

$$\text{Mass of water} = 225 \text{ g}$$
$$\text{Specific heat of water} = 1.00 \text{ cal}/(g \, °C)$$
$$\text{Temperature change} = (100.0 - 25.0) \, °C = 75.0 \, °C$$

Then we use the equation

$$\text{Heat absorbed} = \text{mass} \times \text{specific heat} \times \Delta T$$
$$= 225 \text{ g} \times 1.00 \text{ cal}/(g \, °C) \times 75.0 \, °C$$
$$= 16,900 \text{ cal}$$

We can then convert the unit cal to the units kcal and kJ.

$$16,900 \text{ cal} \times \frac{1 \text{ kcal}}{1000 \text{ cal}} = 16.9 \text{ kcal}$$

$$16.9 \text{ kcal} \times \frac{4.184 \text{ kJ}}{1 \text{ kcal}} = 70.7 \text{ kJ}$$

▶ **Exercise A.9**

How much heat, in calories, kilocalories, and kilojoules, is released by 975 g of water as it cools from 100.0 °C to 18.0 °C?

TABLE A.5   Some Conversion Units for Energy

1 calorie (cal) = 4.184 joules (J)
1 British thermal unit (Btu) = 1055 joules (J) = 252 calories (cal)
1 food Calorie = 1 kilocalorie (kcal) = 1000 calories (cal) = 4184 joules (J)

PRACTICE PROBLEMS

You may need data from Table A.4 and/or Table A.5 for some of these problems.

1. Perform the following conversions.
   a. 25 °C to kelvins
   b. 273 K to degrees Celsius
   c. 301 K to degrees Celsius
   d. 473 °C to kelvins

2. How many calories are there in 0.82 kcal?

3. How many kilocalories are there in 65,500 cal?

4. For each of the following, indicate which is the larger unit.
   a. °C or °F
   b. cal or Cal

5. Order the following temperatures from coldest to hottest: 0 K, 0 °C, and 0 °F.

6. Perform the following conversions.
   a. 37.0 °C to degrees Fahrenheit
   b. 5.5 °F to degrees Celsius
   c. 273 °C to degrees Fahrenheit
   d. 98.2 °F to degrees Celsius
   e. 2175 °C to degrees Fahrenheit
   f. 25.0 °F to degrees Celsius

7. Perform the following conversions.
   a. 0.741 kcal to joules
   b. 8.63 kJ to calories
   c. 1.36 kcal to kilojoules
   d. 345 cal to joules
   e. 873 kJ to kilocalories

8. How many calories are required to raise the temperature of 50.0 g of water from 20.0 °C to 50.0 °C?

9. How many kilojoules are required to raise the temperature of 131 g of iron from 15.0 °C to 95.0 °C?

Appendix B

Glossary

acid A substance that, when added to water, produces an excess of hydronium ion; a proton donor.

acid–base indicator A substance that is one color in acid and another color in base.

acidic anhydride A substance, such as a nonmetal oxide, that forms an acid on addition of water.

acid rain Rain having a pH less than 5.6.

acquired immune deficiency syndrome (AIDS) A disease caused by a retrovirus (HIV) that weakens the immune system.

activated sludge method A combination of primary and secondary sewage treatment methods in which some sludge is recycled.

activation energy The minimum quantity of energy that must be available before a chemical reaction can take place.

active site The spot on a molecule (usually on an enzyme or catalyst) at which reaction occurs.

addition polymerization A polymerization reaction in which all the atoms of the monomer molecules are included in the polymer.

addition reaction A reaction in which the single product contains all the atoms of two reactant molecules.

adipose tissue Connective tissue where fat is stored.

advanced sewage treatment Sewage treatment designed to remove phosphates, nitrates, and other soluble impurities.

aerobic oxidation An oxidation process occurring in the presence of oxygen.

aerobic exercise Exercise in which muscle contractions occur in the presence of oxygen.

aerosol Particles of 1 μm diameter, or less, dispersed in air.

aflatoxins Toxins produced by molds growing on stored peanuts and grains.

Agent Orange A combination of 2,4-D and 2,4,5-T used extensively in Vietnam to remove enemy cover and destroy crops that maintained enemy armies.

agonist A molecule that fits and activates a specific receptor.

AIDS See **acquired immune deficiency syndrome**.

alchemy A mystical blend of chemistry, magic, and religion that flourished in Europe during the Middle Ages (C.E. 500 to 1500).

alcohol (ROH) A compound composed of an alkyl group and a hydroxyl group.

aldehyde (RCHO) An organic molecule with a carbonyl group that has at least one hydrogen atom attached to the carbonyl carbon.

aldose A monosaccharide with an aldehyde functional group.

alkali metal A metal in group 1A of the periodic table.

alkaline earth metal An element in group 2A of the periodic table.

alkaloid A physiologically active nitrogen-containing organic compound obtained from plants.

alkalosis A physiological condition in which the pH of the blood rises to a life-threatening level.

alkane A hydrocarbon with only single bonds; a saturated hydrocarbon.

alkene A hydrocarbon containing one or more double bonds.

alkyl group (—R) The group of atoms that results when a hydrogen atom is removed from an alkane.

alkyne A hydrocarbon containing one or more triple bonds.

allergen A substance that triggers an allergic reaction.

allotropes Different forms of the same element in the same physical state.

alloy A mixture of two or more elements, at least one of which is a metal; an alloy has metallic properties.

alpha helix A secondary structure of a protein molecule in which the molecule has a spiral arrangement.

alpha (α) particle A cluster of two protons and two neutrons; a helium nucleus.

Ames test A laboratory test for mutagens, which are usually also carcinogens.

amide An organic compound having the functional group—CON—in which the carbon is double bonded to the oxygen atom and single-bonded to the nitrogen atom.

amine A compound that contains the elements carbon, hydrogen, and nitrogen; derived from ammonia by replacing one, two, or three of the hydrogen atoms with alkyl group(s).

amino acid An organic compound that contains both an amino group and a carboxylic acid group; amino acids combine to produce proteins.

amino group (—NH₂) A substituent group comprised of a nitrogen atom bonded to two hydrogen atoms.

amphetamines stimulant drugs related to β-phenylethylamine.

amphoteric surfactant A surfactant with a hydrocarbon tail and a water-soluble head that bears both a negative charge and a positive charge.

anabolic steroid A drug that aids in the building (anabolism) of body proteins and thus of muscle tissue.

anabolism The buildup of body tissues from simpler molecules.

anaerobic decay Decomposition in the absence of oxygen.

anaerobic exercise Exercise involving muscle contractions without sufficient amounts of oxygen.

analgesic A pain reliever.

androgen A male sex hormone.

anesthetic A substance that causes loss of feeling and blocks pain.

anion A negatively charged ion.

anionic surfactant A surfactant with a hydrocarbon tail and a water-soluble head that bears a negative charge.

anode The electrode at which oxidation occurs.

antagonist A drug that blocks the action of an agonist by blocking the receptors.

antibiotic A soluble substance, produced by a mold or bacterium, that inhibits growth of other microorganisms.

anticarcinogen A substance that inhibits the formation of cancer.

anticholinergic A drug that acts on nerves using acetylcholine as a neurotransmitter.

anticoagulant A substance that inhibits the clotting of blood.

anticodon The sequence of three adjacent nucleotides in a tRNA molecule that is complementary to a codon on mRNA.

antihistamine A substance that relieves the symptoms of allergies: sneezing, itchy eyes, and runny nose.

anti-inflammatory A substance that inhibits inflammation.

antimetabolite A compound that inhibits the synthesis of nucleic acids.

antioxidant A chemical that is so easily oxidized that it protects other substances from oxidation; a reducing agent.

antiperspirant A formulation that retards perspiration by constricting the openings of sweat glands.

antipyretic A fever-reducing substance.

apoenzyme The pure protein part of an enzyme.

applied research An investigation aimed at the solution of a particular problem.

aqueous solution A solution in which the solvent is water.

arithmetic growth A process in which a constant amount is added during each growth period.

aromatic compound A compound that has special bonding and properties like those of benzene.

asbestos A group of related fibrous silicates.

astringent A substance that constricts the openings of the sweat glands, thus reducing the amount of perspiration that escapes.

atmosphere The gaseous mass surrounding Earth.

atmosphere (atm) A unit of pressure equal to 760 mmHg.

atmospheric inversion A warm layer of air above a cool, stagnant lower layer.

atom The smallest characteristic particle of an element.

atomic mass unit (u) The unit of relative atomic weights, $\frac{1}{12}$ the mass of a carbon-12 atom.

atomic number (Z) The number of protons in the nucleus of an atom of an element.

atomic theory A model that explains the law of multiple proportions and the law of constant composition by stating that all elements are composed of atoms.

Avogadro's hypothesis Equal volumes of gases, regardless of their compositions, contain equal numbers of molecules under the same conditions of temperature and pressure.

Avogadro's number The number of atoms (6.022×10^{23}) in exactly 12 g of pure carbon-12.

background radiation Ever-present radiation from cosmic rays and from natural radioactive isotopes in air, water, soil, and rocks.

base A substance that, when added to water, produces an excess of hydroxide ions; a proton acceptor.

base triplet The sequence of three bases on a tRNA molecule that determine which amino acid it can carry.

basic anhydride A substance, such as a metal oxide, that forms a base on addition of water.

basic research The search for knowledge for its own sake.

battery A series of electrochemical cells.

beta (β) particle An electron emitted by a radioactive atom.

binary compound A compound consisting of two elements.

binding energy Energy derived from the conversion of mass to energy when neutrons and protons are combined to form nuclei.

biochemical oxygen demand (BOD) The quantity of oxygen required by microorganisms to remove organic matter from water.

biochemistry A study of the chemistry of living systems.

biomass The total mass of plants and animals; in energy studies, usually means plant material used as fuel.

bitumen A hydrocarbon mixture obtained from tar sands by heating.

bleach A substance used to remove unwanted color from fabrics, hair, or other materials.

blood sugar Glucose, a simple sugar, circulated in the bloodstream.

boiling point The temperature at which a substance can change state from a liquid to a gas throughout the bulk of the liquid.

bond See **chemical bond**.

bonding pair (BP) A pair of electrons that comprises a chemical bond.

Boyle's law For a given mass of gas at constant temperature, the volume varies inversely with the pressure.

breeder reactor A nuclear reactor that converts nonfissile isotopes to fissile isotopes.

broad-spectrum antibiotic An antibiotic that is effective against a wide variety of microorganisms.

bronze An alloy of copper and tin.

buffer A mixture that reacts with either acid or base to keep the pH of a solution essentially constant.

builder (in detergent formulations) Any substance added to a surfactant to increase its detergency.

calorie (cal) The quantity of heat required to raise the temperature of 1 g of water by 1 °C.

carbohydrate A compound consisting of carbon, hydrogen, and oxygen; a starch or sugar.

carbon-14 dating A technique for determining the age of artifacts based on the half-life of carbon-14.

carbonyl group (C=O) A carbon atom double bonded to an oxygen atom.

carboxyl group (COOH) A carbon atom double bonded to one oxygen atom and singly bonded to a second oxygen atom which in turn is bonded to a hydrogen atom; the functional group of carboxylic acids.

carboxylic acid (RCOOH) An organic compound that contains the COOH functional group.

carcinogen A substance or physical entity that produces tumors.

catabolism The metabolic process in which complex compounds are broken down into simpler substances.

catalyst A substance that increases the rate of a chemical reaction without itself being used up.

catalytic converter A device containing catalysts for oxidizing carbon monoxide and hydrocarbons to carbon dioxide and reduction of nitrogen oxides to nitrogen gas.

catalytic reforming A process that converts low-octane alkanes to high-octane aromatic compounds.

cathode The electrode at which reduction occurs.

cathode ray A stream of high-speed electrons emitted from a cathode in an evacuated tube.

cation A positively charged ion.

cationic surfactant A surfactant with a hydrocarbon tail and a water-soluble head that bears a positive charge.

celluloid Cellulose nitrate, a synthetic material derived from natural cellulose by reaction with nitric acid.

cellulose A polymer comprised of glucose units joined through beta linkages.

Celsius scale A temperature scale on which water freezes at 0° and boils at 100°.

cement (Portland cement) A mixture made from limestone, clay, and sand; when mixed with water it hardens like stone.

ceramic A hard, solid product made from clay or similar materials.

chain reaction A self-sustaining change in which one or more products of one event cause one or more new events.

charcoal filtration Filtration of water through charcoal to adsorb organic compounds.

Charles's law For a given mass of gas at constant pressure, the volume varies directly with the absolute temperature.

chemical bond The force of attraction that holds atoms or ions together in compounds.

chemical change A change in chemical composition.

chemical equation A before-and-after description in which chemical formulas and coefficients represent a chemical reaction.

chemical property A characteristic of a substance that describes the way in which the substance reacts with another substance to change its composition.

chemical symbol An abbreviation, consisting of one or two letters, that stands for an element.

chemistry The study of matter and the changes it undergoes.

chemotherapy The use of chemicals to control or cure diseases.

chiral carbon A carbon atom that has four different groups attached.

coal A fossilized black rock rich in carbon.

codon A sequence of three adjacent nucleotides in mRNA that specifies one amino acid.

coenzyme An organic molecule (often a vitamin) that combines with an apoenzyme to make a complete, functioning enzyme.

cofactor An inorganic component that combines with an apoenzyme to make a complete, functioning enzyme.

cologne A diluted perfume.

compound A pure substance made up of two or more elements combined in fixed proportions.

concentrated solution A solution that has a relatively large amount of solute per unit volume of solution.

condensation The reverse of vaporization; a change from the gaseous state to the liquid state.

condensation polymerization A reaction in which not all the atoms in the starting monomers are incorporated in the polymer because water (or other small) molecules are split out as the polymer is formed.

condensed structural formula An organic chemical formula that shows the atoms of hydrogen right next to the carbon atoms to which they are attached.

copolymer A polymer formed by the combination of two or more different monomer units.

corrosive waste A waste that requires a special container because it corrodes conventional container materials.

cosmetics Substances defined in the 1938 U.S. Food, Drug, and Cosmetic Act as "articles intended to be rubbed, poured, sprinkled or sprayed on, introduced into, or otherwise applied to the human body or any part thereof, for cleaning, beautifying, promoting attractiveness or altering the appearance."

cosmic rays Extremely high-energy rays from outer space.

covalent bond A bond formed by a shared pair of electrons.

cream An emulsion of tiny water droplets in oil.

critical mass The mass of an isotope above which a self-sustaining chain reaction can occur.

crystal A solid with plane surfaces at definite angles.

cyclic hydrocarbon A ring-containing hydrocarbon.

daughter isotope An isotope formed by the radioactive decay of another isotope.

defoliant A substance that causes premature dropping of leaves by plants.

Delaney Amendment A 1958 amendment to the U.S. Food, Drug, and Cosmetic Act that automatically bans from food any chemical shown to induce cancer in laboratory animals.

density The quantity of mass per unit volume.

deodorant A product that uses perfume to mask body odor; some claim to prevent body odor by killing odor-causing bacteria.

deoxyribonucleic acid (DNA) The type of nucleic acid found primarily in the nuclei of cells.

depilatory A hair remover.

depressant drug A drug that slows both physical and mental activity.

deuterium An isotope of hydrogen with a proton and a neutron in the nucleus (mass of 2 u).

dextro isomer A "right-handed" isomer.

dietary mineral A mineral required in the diet for proper health and well-being.

Dietary Reference Intake (DRI) A set of reference values for food and other nutrient intake: Estimated Average Requirements (EAR), Recommended Dietary Allowances (RDA), Adequate Intakes (AI), and Tolerable Upper Intake Levels (UL).

dilute solution A solution that has a relatively small amount of solute per unit volume of solution.

dioxins Highly toxic chlorinated cyclic compounds produced by burning wastes containing chlorinated compounds; once found as contaminants in herbicides.

dipole A molecule that has a positive end and a negative end.

dipole forces The attractive forces that exist among polar covalent molecules.

disaccharide A sugar that on hydrolysis yields two monosaccharide molecules per molecule of disaccharide.

dispersion forces The momentary, usually weak, attractive forces between molecules resulting from synchronized electron motion.

dissociative anesthetic A substance that causes gross personality disorders, including hallucinations similar to those in near-death experiences.

dissolved oxygen Oxygen dissolved in water; a measure of that water's ability to support fish and other aquatic life.

disulfide linkage A covalent linkage of two amino acid units through two sulfur atoms.

diuretic A substance that increases the output of urine.

double bond Two shared pairs of electrons.

doubling time The time it takes a population to double in size.

drug A chemical substance that affects the functioning of living things; used to relieve pain, to treat an illness, or to improve one's state of health or well-being.

drug abuse The use of a drug for its intoxicating effect.

drug misuse The use of a drug in a manner other than its intended use.

elastomer A synthetic polymer with rubberlike properties.

electrochemical cell A device that produces electricity by means of a chemical reaction.

electrode A device, such as a carbon rod or metal strip inserted into an electrochemical cell, at which oxidation or reduction occurs.

electrolysis The process of causing a chemical reaction by means of electricity.

electrolyte A compound that, in water solution, conducts an electric current.

electron The subatomic particle that bears a unit of negative charge.

electron capture (EC) A type of radioactive decay in which a nucleus absorbs an electron from the first or second electronic shell.

electron configuration The arrangement of an atom's electrons in space.

electron-dot (Lewis) structure The structural formula of a molecule in which valence electrons of all the atoms are indicated by dots.

electron-dot (Lewis) symbol The symbol of an element surrounded by dots representing the atom's outermost electrons.

electronegativity The attraction of an atom in a compound for a pair of shared electrons in a chemical bond.

electrostatic precipitator A device that removes particulate matter from smokestack gases by forming an electric charge on the particles, which are then removed by attraction to a surface of opposite charge.

element A fundamental substance in which all atoms have the same number of protons.

emollient An oil or grease used as a skin softener.

emulsion A suspension of submicroscopic particles of fat or oil in water.

enantiomers Nonsuperimposable (nonidentical) mirror image isomers.

end note The portion of perfume that has low volatility; composed of large molecules.

endorphins Naturally occurring peptides that bond to the same receptor sites as opiate drugs.

endothermic A process that absorbs energy from its surroundings.

energy The capacity for doing work.

energy levels (shells) The specific, quantized values of energy that an electron can have in an atom.

enrichment (food) Replacement of nutrients lost from a food during processing.

enrichment (isotope) The process by which the proportion of one isotope of an element is increased relative to those of the others.

entropy A measure of the distribution of elemental species among energy states.

enzyme A biological catalyst.

essential amino acid An amino acid not produced in the body that must be included in the human diet.

ester (RCOOR′) A compound derived from a carboxylic acid and an alcohol; the —OH of the acid is replaced by an —OR group.

estrogen A female sex hormone.

ether (ROR′) A molecule with two hydrocarbon groups attached to the same oxygen atom.

eutrophication The excessive growth of plants in a body of water that causes some of the plants to die because of a lack of light; the water becomes choked with vegetation, depleted of oxygen, and useless as a fish habitat or for recreation.

excited state A state in which an atom is supplied energy and an electron is moved from a lower to a higher energy level.

exothermic A process that releases heat to the surroundings.

fast-twitch fibers The stronger, larger kind of muscle fibers that are suited for anaerobic work.

fat A compound formed by the reaction of glycerol with three fatty acid units; a triglyceride or triacylglycerol.

fat depots Storage places for fats in the body.

fatty acid A carboxylic acid that contains 4 to 20 or more carbon atoms in a chain.

first law of thermodynamics Energy is neither created nor destroyed.

flammable waste A waste that burns readily on ignition, presenting a fire hazard.

FLaReS An acronym for the rules used to test a claim: falsifiability, logic, replicability, and sufficiency.

food additive Any substance other than basic foodstuffs that is present in food as a result of some aspect of production, processing, packaging, or storage.

formula A representation of a chemical substance in which the component chemical elements are represented by their symbols.

formula mass The sum of the atomic masses of a chemical compound as indicated by the formula, expressed in atomic mass units (u).

fossil fuels Coal, petroleum, and natural gas.

free radical A reactive neutral chemical species that contains an unpaired electron.

freezing The reverse of melting; a change from the liquid to the solid state.

fuel A substance that burns readily with the release of significant amounts of energy.

fuel cell A device that produces electricity directly from fuels and oxygen.

functional group The atom or group of atoms that confers characteristic properties on an organic molecule.

fundamental particle An electron, proton, or neutron.

gamma (γ) rays Rays similar to X-rays that are emitted from radioactive substances; have higher energy and are more penetrating than X-rays.

gas The state of matter in which the substance maintains neither shape nor volume.

gasoline The fraction of petroleum containing C_5 to C_{12} hydrocarbons, mainly alkanes, used as automotive fuel.

gene The segment of a nucleic acid molecule that contains the information necessary to produce a protein; the smallest unit of hereditary information.

general anesthetic A depressant that acts on the brain to produce unconsciousness as well as insensitivity to pain.

geometric growth A doubling in number for each growth period.

geothermal energy Energy derived from the heat of Earth's interior.

glass A noncrystalline material obtained by melting sand with soda, lime, and various other metal oxides.

glass transition temperature (T_g) The temperature above which a polymer is rubbery and tough, and below which the polymer is brittle.

global warming An increase in the Earth's average temperature.

globular protein A protein whose molecules fold into roughly spherical or ovoid shapes that can be dispersed in water.

glycogen A polymer of glucose with alpha linkages and branched chains; a storage form of starch in animals.

GRAS list A list, established by the U.S. Congress in 1958, of food additives generally recognized as safe.

greenhouse effect The retention of the sun's heat energy by Earth as a result of excess carbon dioxide or other substances in the atmosphere; causes an increase in Earth's average atmospheric and surface temperatures.

ground state The state of an atom in which all electrons are in the lowest possible energy levels.

group A vertical column in the periodic table; a family of elements.

half-life The length of time required for one-half of the radioactive nuclei in a sample to decay.

hallucinogenic drug A drug that produces visions and sensations that are not part of reality.

halogen An element in group 7A of the periodic table.

hard water Water containing ions of calcium, magnesium, and/or iron.

hazardous waste A waste that, when improperly managed, can cause or contribute to death or illness or threaten human health or the environment.

heat Energy transfer that occurs as a result of a temperature difference

heat capacity The quantity of heat needed to change the temperature of an object or substance by 1 °C.

heat of vaporization (of a substance) The amount of heat involved in the evaporation or condensation of 1 g of the substance.

heat stroke A failure of the body's heat regulatory system; unless the victim is treated promptly, the rapid rise in body temperature will cause brain damage or death.

herbicide A material used to kill plants.

heterocyclic compound A cyclic compound in which one or more atoms in the ring is not carbon.

homogeneous The same throughout; property of a sample that has the same composition in all parts.

homologous series A series of compounds in which adjacent members of the series differ by a fixed unit of structure.

hormone A chemical messenger secreted into the blood by an endocrine gland.

humectant A moistening agent.

hydrocarbon An organic compound that contains only carbon and hydrogen.

hydrogen bomb A bomb based on the nuclear fusion of isotopes of hydrogen.

hydrogen bond A type of intermolecular force in which a hydrogen atom covalently bonded in one molecule is simultaneously attracted to a nonmetal atom in a neighboring molecule; both the atom to which the hydrogen atom is bonded and the one to which it is attracted must be small atoms of high electronegativity, usually N, O, or F.

hydrolysis The reaction of a substance with water; literally, a splitting by water.

hydrophilic Attracted to polar solvents such as water.

hydrophobic Not attracted to water; associated with other nonpolar entities.

hypoallergenic cosmetics Cosmetics claimed to cause fewer allergic reactions than regular products.

hypothesis A reasoned guess that can be tested by experiment.

ideal gas law The volume of a gas is proportional to the amount of gas and its Kelvin temperature and inversely proportional to its pressure.

induced radioactivity Radioactivity caused by bombarding a stable isotope with elemental particles, forming a radioactive isotope.

industrial smog Polluted air associated with industrial activities, usually characterized by sulfur oxides and particulate matter.

inorganic chemistry The study of the compounds of all elements other than carbon.

insecticide A substance that kills insects.

iodine number The number of grams of iodine consumed by 100 g of fat or oil; an indication of the degree of unsaturation.

ion A charged atom or group of atoms.

ionic bond The chemical bond that results when electrons are transferred from a metal to a nonmetal; the electrostatic attraction between ions of opposite charge.

ionizing radiation Radiation that produces ions as it passes through matter.

isoelectronic Has the same number of electrons.

isomers Compounds that have the same molecular formula but different structural formulas and properties.

isotopes Atoms that have the same number of protons but different numbers of neutrons.

joule (J) The SI unit of energy (1 J = 0.239 cal).

juvenile hormone A hormone that controls the rate of development of the young; used to prevent insects from maturing.

kelvin (K) The SI unit of temperature. Zero on the Kelvin scale is absolute zero.

keratin The tough, fibrous protein that comprises most of the outermost layer of the epidermis.

kerogen The complex material found in oil shale; has an approximate composition of $(C_6H_8O)_n$, where n is a large number.

ketone (RCOR′) An organic compound with a carbonyl group between two carbon atoms.

ketose A monosaccharide with a ketone functional group.

kilocalorie A unit of energy equal to 1000 cal; one food calorie.

kilogram (kg) The SI unit of mass, a quantity equal to about 2.2 lb.

kinetic energy The energy of motion.

kinetic–molecular theory A model that uses the motion of molecules to explain the behavior of the three states of matter.

lactate threshold The upper limit of lactic acid concentration in muscle tissue above which the muscle is too fatigued to contract.

lanolin A natural wax obtained from sheep's wool.

law of combining volumes The volumes of gaseous reactants and products are in a small whole-number ratio when all measurements are made at the same temperature and pressure.

law of conservation of energy The quantity of energy within the universe is constant; energy cannot be created or destroyed, only transformed.

law of conservation of mass Matter is neither created nor destroyed during a chemical change.

law of definite proportions A compound always contains elements in certain definite proportions, never in any other combination; also called the law of constant composition.

law of multiple proportions Elements may combine in more than one proportion to form more than one compound—for example, CO and CO_2.

LD$_{50}$ The dosage that would be lethal to 50% of a population of test animals.

levo isomer A "lefthanded" isomer.

Lewis (electron-dot) structure The structural formula of a molecule in which valence electrons of all the atoms are indicated by dots.

Lewis (electron-dot) symbol The symbol of an element surrounded by dots representing the atom's outermost electrons.

limiting reactant The reactant that is used up first in a reaction, after which the reaction ceases no matter how much remains of the other reactants.

line spectrum The pattern of colored lines emitted by an element.

lipid A substance from animal or plant cells that is soluble in nonpolar solvents and insoluble in water.

lipoprotein A protein combined with a lipid, such as a triglyceride or cholesterol.

liquid The state of matter in which the substance assumes the shape of its container, flows readily, and maintains a fairly constant volume.

liter (L) A unit of volume equal to a cubic decimeter.

local anesthetic A substance that renders part of the body insensitive to pain while leaving the patient conscious.

lone pair (LP) A pair of electrons in the valence shell of an atom not involved in a bond; also called a nonbonding pair (NBP).

lotion An emulsion of submicroscopic fat or oil droplets dispersed in water.

main group element An element in the A groups of the periodic table (customary U.S. arrangement) and in groups 1, 2, and 13 to 18 in the periodic table recommended by IUPAC.

marijuana A preparation made from the leaves, flowers, seeds, and small stems of the *Cannabis* plant.

mass A measure of the quantity of matter.

mass–energy equation Einstein's equation $E = mc^2$, in which E is energy, m is mass, and c is the speed of light.

mass number (nucleon number) The sum of the numbers of protons and of neutrons in the nucleus of an atom.

matter The stuff of which all materials are made; anything that has mass and occupies space.

melanin A brownish-black pigment that determines the color of the skin and hair.

melting point The temperature at which a substance changes from the solid to the liquid state.

messenger RNA (mRNA) The type of RNA that contains the codons for a protein; travels from the nucleus of the cell to a ribosome.

metabolism The sum of all the chemical reactions by which the protoplasm of an organism grows, is maintained, obtains energy, and is degraded.

metalloid An element with properties intermediate between those of metals and those of nonmetals.

metals The group of elements to the left of the heavy, stepped, diagonal line in the periodic table.

meter (m) The SI unit of length, slightly longer than a yard.

mica A mineral composed of SiO_4 tetrahedra arranged in a two-dimensional, sheetlike array.

micelle A spherical cluster of surfactant molecules arranged so that their hydrophilic ends all lie along the outer surface.

micronutrient A substance needed by the body in only tiny amounts.

middle note The portion of perfume intermediate in volatility; responsible for the lingering aroma after most top-note compounds have vaporized.

minerals (dietary) The inorganic substances required in the diet for good health.

mixture Matter with a variable composition.

moisturizer (skin) A substance that adds moisture to the skin or acts to retain moisture in the skin.

molarity (M) The concentration of a solution in moles of solute per liter of solution.

molar mass The formula mass of a substance expressed in grams.

molar volume The volume occupied by 1 mol of a substance (usually a gas) under specified conditions.

mole (mol) The amount of a chemical substance that contains 6.02×10^{23} formula units of the substance.

molecular mass The mass of a molecule of a substance; the sum of the atomic masses as indicated by the molecular formula, expressed in atomic mass units (u).

molecule Two or more atoms joined together by covalent bonds; the smallest fundamental unit of a molecular substance.

monomer A substance of relatively low molecular mass. Monomer molecules combine to make a polymer.

monosaccharide A carbohydrate that cannot be hydrolyzed into simpler sugars.

mousse A foam or froth; a hair care product composed of resins used to hold hair in place.

mutagen Any entity that causes changes in genes without destroying the genetic material.

narcotic A depressant, analgesic drug that induces narcosis (sleep).

natural gas A mixture of gases, mainly methane, found in many underground deposits.

natural philosophy Philosophical speculation about nature.

neuron A nerve cell.

neurotransmitter A chemical that carries an impulse across a synapse from one nerve cell to the next.

neurotrophin A substance produced during exercise that promotes the growth of brain cells.

neutralization The reaction of an acid and a base to produce a salt and water.

neutron A fundamental particle with a mass of approximately 1 u and no electric charge.

nitrogen cycle The various processes by which nitrogen is cycled among the atmosphere, soil, water, and living organisms.

noble gases Generally unreactive gases that appear in the far right column (group 8A) of the periodic table.

nonbonding pair (NBP) A pair of electrons in the valence shell of an atom not involved in a bond; also called a lone pair (LP).

nonionic surfactant A surfactant with a hydrocarbon tail and a polar head whose oxygen atoms attract water molecules and make the head water soluble; bears no ionic charge.

nonmetals The group of elements to the right of the heavy, stepped, diagonal line in the periodic table.

nonpolar covalent bond A covalent bond in which there is an equal sharing of electrons.

nonsteroidal anti-inflammatory drug (NSAID) A milder anti-inflammatory drug as distinguished from the more potent steroidal anti-inflammatory drugs such as cortisone and prednisone.

nuclear fission The splitting of an atomic nucleus into two large fragments.

nuclear fusion The combination of two small atomic nuclei to produce one larger nucleus.

nuclear reactor A power plant that produces energy by nuclear fission.

nucleic acid A nucleotide polymer, DNA or RNA.

nucleon A proton or neutron in an atomic nucleus.

nucleon number (A) The total number of protons and neutrons in an atom; the mass number.

nucleotide A combination of a heterocyclic amine, a pentose sugar, and phosphoric acid; the monomer unit of nucleic acids.

nucleus Concentrated, positively charged matter at the center of an atom; composed of protons and neutrons.

octane rating The antiknock quality of a gasoline as compared with mixtures of isooctane (with a rating of 100) and heptane (with a rating of 0).

octet rule Atoms seek an arrangement that will surround them with eight electrons in the outermost energy level.

oil (food) A substance formed from glycerol and fatty acids, which is liquid at room temperature.

oil shale Fossil rock containing kerogen from which oil can be obtained at high cost by distillation.

optical brightener A compound that absorbs the invisible ultraviolet component of sunlight and reemits it as visible light at the blue end of the spectrum.

orbital A region of space in an atom occupied by one or two electrons.

organic chemistry The study of the compounds of carbon.

organic farming Farming without synthetic fertilizers or pesticides.

oxidation An increase in oxidation number; combination of an element or compound with oxygen; loss of hydrogen; loss of electrons.

oxidizing agent A substance that causes oxidation and is itself reduced.

oxygen cycle The various processes by which oxygen is cycled among the atmosphere, soil, water, and living organisms.

oxygen debt The demand for oxygen in muscle cells during anaerobic exercise.

ozone layer The layer of the stratosphere that contains ozone and shields living creatures on Earth from deadly ultraviolet radiation from the sun.

paint A surface coating that contains a pigment, a binder, and a solvent.

particulate matter (PM) A pollutant composed of solid and liquid particles of greater than molecular size.

peptide bond The amide linkage that bonds amino acids in chains of peptides, polypeptides, and proteins.

percent by mass The concentration of a solution, expressed as (mass of solute ÷ mass of solution) × 100%.

percent by volume The concentration of a solution, expressed as (volume of solute ÷ volume of solution) × 100%.

perfume A fragrant mixture of plant extracts and other chemicals dissolved in alcohol.

period A horizontal row of the periodic table.

periodic table A systematic arrangement of the elements in columns and rows; elements in a given column have similar properties.

pesticide A substance that kills some kind of pest (weeds, insects, rodents, and so on).

petroleum A dark, oily mixture of (mostly) hydrocarbons occurring in various deposits around the world.

pH The negative logarithm of hydronium ion concentration.

pharmacology The study of the response of living organisms to drugs.

phenol A compound with an OH group attached to a benzene ring.

pheromone A natural chemical secreted by an organism to mark a trail, send out an alarm, or attract a mate.

photochemical smog Smog created by the action of sunlight on unburned hydrocarbons and nitrogen oxides, mainly from automobiles.

photon A unit particle of energy.

photosynthesis The chemical process used by green plants to convert solar energy into chemical energy by reducing carbon dioxide.

photovoltaic cell A solar cell; a cell that converts sunlight directly to electric energy.

physical change A change in physical state or form.

physical property A quality of a substance that can be demonstrated without changing the composition of the substance.

placebo A substance that appears to be a real drug but has no active ingredients.

plasma A state of matter similar to a gas but composed of isolated electrons and nuclei rather than discrete whole atoms or molecules.

plasticizer A chemical substance added to some plastics, such as vinyl, to make them more flexible and easier to work with.

pleated sheet A secondary protein structure characterized by antiparallel molecules with a zigzag structure.

polar covalent bond A covalent bond in which more than half of the bond's negative charge is concentrated around one of the two atoms.

polar molecule A molecule that has a dipole moment.

pollutant A chemical that causes undesirable effects by being in the wrong place and/or in the wrong concentration.

polyamide A polymer that has structural units joined by amide linkages.

polyatomic ion An ion consisting of two or more atoms bonded together.

polyester A polymer made from a dicarboxylic acid and a dialcohol.

polymer A molecule with a large molecular mass; a chain formed of repeating smaller units.

polymerase chain reaction (PCR) A process that reproduces many copies of a DNA fragment.

polypeptide A polymer of amino acids, usually of lower molecular mass than a protein.

polysaccharide A carbohydrate, such as starch or cellulose, one molecule of which yields many molecules of monosaccharide(s) on hydrolysis.

polyunsaturated fat A fat containing fatty acid units with two or more carbon–carbon double bonds.

positron (β^+) A positively charged particle with the mass of an electron.

potential energy Energy by virtue of position or composition.

preemergent herbicide A herbicide that is rapidly broken down in the soil and can therefore be used to kill weed plants before crop seedlings emerge.

primary plant nutrients Nitrogen, phosphorus, and potassium.

primary sewage treatment Treatment of sewage in a plant with a holding pond intended to remove some of the sewage solids as sludge by settling.

primary structure The amino acid sequence in a protein or of nucleotides in a nucleic acid.

product A substance produced by a chemical reaction; product formulas follow the arrow in a chemical equation.

progestin A compound that mimics the action of progesterone.

prostaglandin A hormone-like compound derived from arachidonic acid that is involved in increased blood pressure, the contractions of smooth muscle, and other physiological processes.

protein An amino acid polymer.

proton The unit of positive charge in the nucleus of an atom; the hydrogen nucleus in acid–base chemistry.

psychotropic drugs Drugs that affect the mind.

pure substance Also known as **substance**; a sample of matter that always has the same composition, no matter how it is made or found.

purine A base with two fused rings, found in nucleic acids.

pyrimidine A base with one ring, found in nucleic acids.

quantum An energy packet of specific size; one photon of energy.

quartz A compound composed of SiO_4 tetrahedra arranged in a three-dimensional array.

quaternary structure An arrangement of protein subunits in a particular pattern.

radioactive decay The disintegration of an unstable atomic nucleus with spontaneous emission of radiation.

radioactivity The spontaneous emission of alpha, beta, or gamma rays by disintegration of the nuclei of atoms.

radioisotopes Radioactive isotopes.

reactant A starting material or original substance in a chemical change; reactant formulas precede the arrow in a chemical equation.

reactive wastes Wastes that tend to react spontaneously or to react vigorously with air or water.

Recommended Daily Allowance (RDA) The recommended level of a nutrient necessary for a balanced diet.

reducing agent A substance that causes reduction and is itself oxidized.

reduction A decrease in oxidation number; a gain of electrons; a loss of oxygen; a gain of hydrogen.

replication Copying or duplication; the process by which DNA reproduces itself.

resin A polymeric material, usually a sticky solid or semisolid organic material.

restorative drug A drug used to relieve the pain and reduce the inflammation resulting from overuse of muscles.

retrovirus An RNA virus that synthesizes DNA in the host cells.

reverse osmosis A method of pressure filtration through a semipermeable membrane; water flows from an area of high salt concentration to an area of low salt concentration.

ribonucleic acid (RNA) The form of nucleic acid found mainly in the cytoplasm, but also present in all other parts of the cell.

risk–benefit analysis An approach that estimates a desirability quotient by dividing the benefits by the risks.

rule of 72 A mathematical formula that gives the doubling time for a population growing geometrically; 72 divided by the annual rate equals the doubling time.

salt An ionic compound produced by the reaction of an acid with a base.

saponins Natural chemical compounds that produce a soapy lather.

saturated fat A fat composed of a large proportion of saturated fatty acids esterified with glycerol.

saturated hydrocarbon An alkane; a compound of carbon and hydrogen with only single bonds.

science A branch of knowledge based on the laws of nature.

scientific law A summary of experimental data; often expressed in the form of a mathematical equation.

scientific model A representation that serves to explain a scientific phenomenon.

sebum An oily secretion of the body that protects the skin from moisture loss.

second law of thermodynamics The entropy of the universe increases in any spontaneous process.

secondary plant nutrients Magnesium, calcium, and sulfur.

secondary sewage treatment Passing effluent from a primary treatment plant through gravel and sand filters to aerate the water and remove suspended solids.

secondary structure The arrangement of polypeptide chains in a protein—for example, helix or pleated sheet.

set-point theory An explanation of the difficulty of losing weight by dieting that holds that each person has an individual level of circulating fatty acids below which he or she is constantly hungry.

sex attractant A substance or mixture of substances released by an organism to attract members of the opposite sex of the same species for mating.

shell An electron energy level; the specific, quantized energy levels that an electron can have in an atom.

significant figures Those measured digits that are known with certainty plus one uncertain digit.

silicone A polymer with a base chain of alternating silicon and oxygen atoms.

single bond A pair of electrons shared between two atoms.

SI units (International System of Units) A measuring system used by scientists worldwide; it is based on seven base quantities and their multiples and submultiples.

skin protection factor (SPF) The rating of a sunscreen's ability to limit the penetration of ultraviolet radiation.

slag A relatively low-melting product of the reaction of limestone with silicate impurities in iron ore.

slow-twitch fibers Muscle fibers suited for aerobic work.

smog The combination of smoke and fog; polluted air.

soap A mixture of salts (usually sodium salts) of long-chain carboxylic acids.

solar cell A device used for converting sunlight to electricity; a photo-electric cell.

solid A state of matter in which the substance maintains its shape and volume.

solute The substance that is dissolved in another substance (solvent) to form a solution; usually present in a smaller amount than the solvent.

solution A homogeneous mixture of two or more substances.

solvent The substance that dissolves another substance (solute) to form a solution; usually present in a larger amount than the solute.

specific heat (of a substance) The amount of heat required to raise the temperature of 1 g of the substance by 1 °C.

standard temperature and pressure (STP) Conditions of 0 °C and 1 atm pressure.

starch A polymer of glucose units joined through alpha linkages; a complex carbohydrate.

starvation The withholding of nutrition from the body, whether voluntary or involuntary.

steel An alloy of iron containing small amounts of carbon and usually containing other metals such as manganese, nickel, and chromium.

stereoisomers Isomers having the same structural formula but differing in the arrangement of atoms or groups of atoms in three-dimensional space.

steroid A molecule that has a four-ring skeletal structure, with one cyclopentane and three cyclohexane fused rings.

stimulant drug A drug that increases alertness, speeds up mental processes, and generally elevates the mood.

stoichiometric factor A factor that relates the amounts of two substances through their coefficients in a chemical equation.

stoichiometry Quantity relationships between reactants and products in a chemical reaction.

strong acid An acid that ionizes completely in water; a potent proton donor.

strong base A base that dissociates completely in water; a potent proton acceptor.

structural formula A chemical formula that shows how the atoms of a molecule are arranged, to which other atom(s) they are bonded, and the kinds of bonds.

sublevel See **subshell**.

sublimation Conversion of a solid directly to the gaseous state without going through the liquid state.

subshell A subdivision of electron energy levels in an atom; also called a sublevel.

substance See **pure substance**.

substrate The substance that bonds to the active site of an enzyme; the substance acted upon.

surface-active agent (surfactant) Any agent that stabilizes the suspension in water of a nonpolar substance such as oil.

synapse A tiny gap between nerve fibers.

synergistic effect An effect greater than the sum of the expected effects.

tar sands Sands that contain bitumen, a thick hydrocarbon material.

technology The sum total of processes by which humans modify the materials of nature to better satisfy their needs and wants.

temperature A measure of heat intensity, or how energetic the particles of a sample are.

temperature inversion A warm layer of air above a cool, stagnant lower layer.

teratogen A substance that causes birth defects when introduced into the body of a pregnant female.

tertiary structure The folds, bends, and twists in protein or nucleic acid structure.

tetracyclines Antibacterial drugs with four fused rings.

theory A detailed explanation of the behavior of matter based on experiments; may be revised if new data warrant.

thermonuclear reactions Nuclear fusion reactions that require extremely high temperatures and pressures.

thermoplastic polymer A kind of polymer that can be heated and re-shaped.

thermosetting polymer A kind of polymer that cannot be softened and remolded.

top note The portion of perfume that vaporizes most quickly; composed of relatively small molecules; responsible for odor when perfume is first applied.

toxicology The division of pharmacology that deals with the effects of poisons on the body, their identification and detection, and remedies for them.

toxic waste A waste that contains or releases poisonous substances in amounts large enough to threaten human health or the environment.

tracers Radioactive isotopes used to trace movement or locate the sites of radioactivity in physical, chemical, and biological systems.

training effect The net effect, acquired through repeated exercise, of being able to do more physical work with less strain.

transcription The process by which DNA directs the synthesis of an mRNA molecule during protein synthesis.

transfer RNA (tRNA) A small molecule that contains anticodon nucleotides; the RNA molecule that bonds to an amino acid.

transition elements Metallic elements situated in the center portion of the periodic table, in the B groups (customary United States arrangement), and in groups 3 to 12 in the periodic table recommended by IUPAC.

translation The process by which the information contained in the codon of an mRNA molecule is converted to a protein structure.

transmutation The changing of one element into another.

triglyceride An ester of glycerol with three fatty acid units; also called a triacylglycerol.

triple bond The sharing of three pairs of electrons between two atoms.

tritium A radioactive isotope of hydrogen with two neutrons and one proton in the nucleus (hydrogen-3).

unsaturated hydrocarbon An alkene or alkyne; a hydrocarbon containing one or more double or triple bonds.

valence electrons Electrons in the outermost shell of an atom.

valence shell electron pair repulsion theory (VSEPR) A theory of chemical bonding useful in determining the shapes of molecules; it states that valence shell electron pairs locate themselves as far apart as possible.

vaporization The process by which a substance changes from the liquid to the gaseous (vapor) state.

variable A factor that changes during an experiment.

vitamin An organic compound that the body cannot produce in the quantity required for good health.

volatile organic compounds (VOCs) Compounds that cause pollution because they vaporize readily.

VSEPR theory See **valence shell electron pair repulsion theory**.

vulcanization The process of making naturally soft rubber harder by reaction with sulfur.

wax An ester of a long-chain fatty acid with a long-chain alcohol.

weak acid An acid that ionizes only slightly in water; a poor proton donor.

weak base A base that ionizes only slightly in water; a poor proton acceptor.

weight A measure of the force of attraction of Earth for an object.

wet scrubber A pollution control device that uses water or solutions to remove pollutants from smokestack gases.

X-rays Radiation similar to visible light but of much higher energy and much more penetrating.

zwitterion A molecule that contains both a positive charge and a negative charge; a dipolar ion.

Appendix C

Brief Answers to Selected Problems

Answers are provided for all in-chapter exercises. Brief answers are given for selected review questions; more complete answers can be obtained by reviewing the text. Answers are provided for all the odd-numbered problems in the matched sets, and for selected additional problems.

NOTE: For numerical problems, your answer may vary slightly from ours because of rounding and the use of significant figures (Appendix A).

Chapter 1

1.1   **A. a.** DQ would probably be small.
     b. DQ would probably be large.
     B. a. DQ would be uncertain.
     b. DQ would probably be large.
     C. a. DQ would probably be low.
     b. DQ would probably be large for children worldwide.

1.2   **A. a.** 1.00 kg   **b.** 179 lb
     B. a. 52.5 kg   **b.** 496 lb

1.3   **A.** physical: a, c; chemical: b
     B. physical: b, d; chemical: a, c

1.4   **A.** elements: He, No, Os; compounds: CuO, NO, KI
     B. 8

1.5   **a.** 7.24 kg   **b.** 4.29 μm   **c.** 7.91 ms
     d. 2.29 cg   **e.** 7.90 Mm

1.6   **A. a.** 7.45×10^{-9} m   **b.** 5.25×10^{-3} s
     c. 1.415×10^3 m   **d.** 2.06×10^{-3} m
     B. a. 2.84×10^{-7} m   **b.** 1.19×10^{-1} s
     c. 7.54×10^5 m   **d.** 6.19×10^3 m

1.7   **A.** 500 cm^2 **B.** 1600 cm^3

1.8   **a.** 7.5×10^3 mm   **b.** 2.056 L
     c. 2.06×10^6 mm   **d.** 7.38 mm

1.9   **A.** ice   **B.** Padouk floats, ebony sinks.

1.10   **A.** 1.11 g/mL   **B.** 11.2 g/cm^3

1.11   **A.** 253 g   **B.** 819 g

1.12   **A.** 486 cm^3 **B.** 341 mL

1.13   **A.** 373 K   **B.** 195 K

1.14   **A.** 1798 kcal   **B.** 8.6 kcal

1. Chemistry is the study of matter and the changes it undergoes. A chemical is a type of matter; a substance.

3. Science is testable, reproducible, explanatory, predictive, and tentative. Testability best distinguishes science.

5. Alchemy is a mystical, experimental chemistry that flourished during the Middle Ages.

7. Bacon's dream was that science would solve all the world's problems, increasing happiness and prosperity.

9. Technological progress has enabled food production to keep pace with population growth in most of the world.

11. A scientific law is a summary of observations and data about a given subject.

13. These problems usually involve too many variables to be treated by scientific methods.

15. Risk–benefit analysis compares benefits of an action to risks of that action.

17. A desirability quotient (DQ) is benefits divided by risks. A large DQ means that risks are minimal compared to benefits.

19. Physical properties are those that can be observed without reference to any other substance; chemical properties describe how one substance reacts with other substances.

21. Gases expand to fill the volume and take the shape of their container. Liquids take the shape, but do not fill the volume of their container. Solids have definite shape and volume.

27. Energy is the ability to do work.

29. Penicillin has saved thousands of lives, causing harm to a very few. The DQ for penicillin is large.

31. Food producers benefit through increased profits. The consumer assumes the greatest risk.

33. The DQ is uncertain; it depends on dose, potential for accidents, and many other factors.

35. matter: a, c, d

37. 70 kg

39. yes

41. 250 mL

43. 1.46×10^9 km^3

45. physical change: a; chemical change: b, c, d

47. chemical property: b; physical property: a, c, d

49. substance: a, b; mixture: c, d, e, f

51. homogeneous: a, d; heterogeneous: b, c

53. a substance; the law of definite proportions

55. elements: a, b, d; compound: c

57. **a.** C   **b.** Cl   **c.** Fe

59. **a.** hydrogen   **b.** oxygen   **c.** sodium

61. f

63. **a.** 8.01 μg   **b.** 7.9 mL   **c.** 1.05 km

65. **a.** 0.0374 L   **b.** 1.55×10^5 m   **c.** 198 mg
     d. 1.19×10^4 cm^2 **e.** 0.078 ms

67. **a.** cm   **b.** kg   **c.** dL

69. 1.00 mL; 15.3 mL

71. **a.** 1.17 g/mL   **b.** 1.26 g/mL

73. **a.** 120 g   **b.** 490 g

75. **a.** 53.1 cm^3 **b.** 18.7 mL

77. 310 K

79. 38.5 kcal

81. The brass weight

83. cabbage (1.65 kg) < potatoes (~2.3 kg) < sugar (2.5 kg)

85. 5.23 g/cm^3

87. 597 g

89. 6.73×10^{-6} cm

91. 0.908 metric tons

93. 4.03 km

Chapter 2

2.1   **A.** 1.19 g nitrogen   **B.** 63.0 g nitrogen

2.2   3 atoms hydrogen/1 atom arsenic

   3. discrete: c, d, e; continuous: a, b, f

   5. When a chemical change occurs, matter is neither created nor destroyed. The sum of the masses of the Fe and the S will equal the mass of the FeS formed.

   7. A compound contains elements in certain definite proportions and in no other combinations. All samples of ZnS will have 2.04 parts Zn to 1 part S by mass.

11. law of definite proportions

13. law of multiple proportions

15. 11.00 g; law of definite proportions

17. That water is not an element; it is a compound of hydrogen and oxygen.

19. No. The helium atoms have escaped into the air through the pores in the balloon.

21. No. They have become a part of the gas carbon dioxide.

23. Yes. Total mass before and after the reaction is the same, 1.2000 g.

25. (b)

27. 70 g hydrogen

29. 260 g carbon

31. (a)

33. **a.** Yes; Dalton assumed that atoms of different elements had different masses.

   b. No; Dalton assumed that atoms of one element differ (in mass) from atoms of any other element.

35. Contradict. Dalton regarded atoms as indivisible.

37. (d)

39. No; the ratio of carbon to hydrogen is different for the different samples.

41. (a)

43. Most of the wood is converted into gases that escape into the atmosphere.

45. He didn't know to take into account exchanges of matter between the tree and the atmosphere.

47. conservation of mass

49. 108.0 g of mercury oxide

Chapter 3

3.1   **A.** 78 neutrons   **B.** 40; potassium-40

3.2   **A.** 58 neutrons   **B.** $^{240}_{92}$U

3.3   **A.** $^{90}_{37}$X and $^{88}_{37}$X; $^{88}_{38}$X and $^{93}_{38}$X

   B. three

3.4   **A.** 32 electrons   **B.** 3

3.5   **a.** (Be) 2 2

   b. (Mg) 2 8 2; both have the same number of outer electrons.

3.6   **A. a.** $1s^2 2s^2 2p^5$

   b. $1s^2 2s^2 2p^6 3s^2 3p^5$ (Both have the valence configuration $ns^2 np^5$.)

   B. a. $1s^2 2s^2 2p^6 3s^2 3p^6 3d^2 4s^2$ **b.** $1s^2 2s^2 2p^6 3s^2 3p^6 3d^{10} 4s^2 4p^1$

3.7   **A. a.** (Rb) $5s^1$ **b.** (Se) $4s^2 4p^4$ **c.** (Ge) $4s^2 4p^2$

   B. a. (Ga) $4s^2 4p^1$ **b.** (In) $5s^2 5p^1$

   3. Radioactivity is spontaneous radiation from an atomic nucleus. Dalton's atomic theory stated that atoms were indestructible.

   5. Isotopes are atoms of the same element with different masses (different numbers of neutrons).

   7. A and B, no; A and C, no; A and D, yes; B and C, yes

   9. The tiny core of an atom with all the positive charge and most of the mass; protons and neutrons

11. Proton: mass ~1 u, charge +1; neutron: mass ~1 u, charge 0; electron: very little mass, charge −1

13. Electrons

15. Cf; californium; 251 u

17. Number of protons + neutrons

19. Protium: $^1_1$H; deuterium: $^2_1$H; tritium: $^3_1$H

21. Electrons

23. Emitted

25. **a.** 3   **b.** 4

27. **a.** 2   **b.** 11   **c.** 17   **d.** 8   **e.** 12   **f.** 16

29. **a.** $^{12}_5$B, boron-12   **b.** $^{125}_{53}$I; iodine-125

31. **a.** $^{69}_{31}$Ga   **b.** $^{60}_{27}$Co

33. **a.** 30 p, 32 n   **b.** 94 p, 147 n

35. (b)

37. 32

39. **a.** (He) 2   **b.** (Na) 2 8 1   **c.** (Cl) 2 8 7

   d. (O) 2 6   **e.** (Mg) 2 8 2   **f.** (S) 2 8 6

41. 2 electrons; spherical; 1 orbital

43. $1s^2 2s^2 2p^5$

45. **a.** 4   **b.** 7   **c.** 13

47. **a.** The outer two electrons should be in the 2p subshell, which fills before the 3s.

   b. The outer electron should be in the 2p subshell, which can hold up to 6 electrons and fills before the 3s.

   c. There is no d subshell in the second main energy level.

49. Sulfur has four 3p electrons; chlorine has five 3p electrons.

51. Argon

53. Group 7A (halogens)

55. Metal

57. F and Cl each have the valence configuration $ns^2 np^5$, but Cl has one more electron shell than F.

59. Metals are shiny, malleable, ductile, and conduct electricity. They form positive ions.

61. Metals: b, d; nonmetals: a, c

63. **a.** Cl, period 3   **b.** Os, period 6

   c. H, period 1   **d.** Li, period 2

65. Lack of chemical reactivity

67. (a) only

69. (a) and (b)

71. 2

73. **a.** No. An element cannot have fractional number of protons.
   b. Yes. It is possible to synthesize larger atoms.

75. **a.** 2   **b.** 3

77. They thought the nucleus had 19 protons, 10 of which were neutralized by 10 electrons.

79. **a.** An s subshell   **b.** A p subshell

81. **a.** False. Neutrons have mass, but are neutral.

   b. True. All atoms of an element have the same number of protons, but may differ in number of neutrons.

   c. False. They have the same nucleon numbers, but different elements have different atomic numbers.

83. From the line spectra, Ba (5540 A) and Na (5890 and 5900 A). The sodium in salt is part of our normal diet, so barium is more likely to be the culprit.

Chapter 4

4.1   **a.** $^{226}_{88}$Ra $\longrightarrow$ $^4_2$He + $^{222}_{86}$Rn

   b. $^{24}_{11}$Na $\longrightarrow$ $^0_{-1}$e + $^{24}_{12}$Mg

   c. $^{188}_{79}$Au $\longrightarrow$ $^0_{+1}$e + $^{188}_{78}$Pt

   d. $^{37}_{18}$Ar + $^0_{-1}$e $\longrightarrow$ $^{37}_{17}$Cl

4.2   A neutron ($_{0}^{1}$n)

4.3   **A.** 0.03825 mg   **B.** 6.25%

4.4   **A.** 0.500 mg   **B.** 420,000 y

4.5   **A.** 22,920 y   **B.** 142.5 y

4.6   **A.** A neutron ($_{0}^{1}$n)   **B.** Thorium-232 ($_{90}^{232}$Th)

4.7   Silicon-30 ($_{14}^{30}$Si)

4.8   Higher; the gas can be absorbed by inhalation, and decay in the body.

5. $_{1}^{1}$H, $_{1}^{2}$H, $_{1}^{3}$H

7. $_{35}^{83}$Br

9.   **a.** $_{31}^{69}$Ga   **b.** $_{42}^{98}$Mo   **c.** $_{42}^{99}$Mo   **d.** $_{43}^{98}$Tc

11.   (b) and (c)

13.   157

15.   Lithium-7

17.   **a.** $_{2}^{4}$He   **b.** $_{-1}^{0}$e   **c.** $_{0}^{1}$n   **d.** $_{+1}^{0}$e

19.   **a.** Atomic number decreases by 2; nucleon number decreases by 4.

   b. Neither the atomic number nor the nucleon number changes.

   c. Atomic number decreases by 1; nucleon number is unchanged.

21.   Iodine-131

23.   It produces no alpha or beta particles, and it has a short half-life.

25.   Gamma rays

27.   Move away from the source; use shielding.

29.   X-rays

31.   by mass

33.   **a.** $_{47}^{108}$Ag $\rightarrow$ $_{48}^{108}$Cd + $_{-1}^{0}$e
   b. $_{83}^{210}$Bi $\rightarrow$ $_{2}^{4}$He + $_{81}^{206}$Tl
   c. $_{29}^{64}$Cu $\rightarrow$ $_{28}^{64}$Ni + $_{+1}^{0}$e

35.   **a.** $_{2}^{4}$He   **b.** $_{-1}^{0}$e   **c.** $_{1}^{1}$H

37. $_{42}^{99}$Mo $\rightarrow$ $_{43}^{99m}$Tc + $_{-1}^{0}$e

39. $_{12}^{24}$Mg + $_{0}^{1}$n $\rightarrow$ $_{1}^{1}$H + $_{11}^{24}$Na; sodium-24

41. $_{85}^{215}$At $\rightarrow$ $_{2}^{4}$He + $_{83}^{211}$Bi; astatine-215

43. $_{48}^{110}$Cd

45.   **a.** 1500   **b.** 750

47.   6.25 mg

49.   **a.** 3 μCi   **b.** 0.19 μCi

51.   186 s

53.   5730 y

55.   5730 y; 11,460 y

57.   **a.** $_{51}^{121}$Sb + $_{2}^{4}$He $\rightarrow$ $_{53}^{124}$I + $_{0}^{1}$n   **b.** $_{53}^{124}$I $\rightarrow$ $_{52}^{124}$Te + $_{+1}^{0}$e

59.   1

61. $_{88}^{223}$Ra $\rightarrow$ $_{86}^{219}$Rn + $_{2}^{4}$He
   $_{88}^{223}$Ra $\rightarrow$ $_{82}^{209}$Pb + $_{6}^{14}$C

63.   853 cm³; 131 cm³; smaller than a baseball (~200 cm³)

65.   **a.** $_{5}^{10}$B + $_{0}^{1}$n $\rightarrow$ $_{1}^{1}$H + $_{4}^{10}$Be
   b. $_{51}^{121}$Sb + $_{1}^{1}$H $\rightarrow$ $_{0}^{1}$n + $_{52}^{121}$Te
   c. $_{27}^{59}$Co + $_{0}^{1}$n $\rightarrow$ $_{25}^{56}$Mn + $_{2}^{4}$He

67.   3 neutrons ($_{0}^{1}$n). The next set of fissions produce 9 neutrons, then 27, then 81, etc.

69.   No, the bottle of wine is somewhere between 3 and 4 half-lives (37 and 49 years) old!

71.   10.8 u

Chapter 5

5.1   a. $:\overset{..}{\underset{..}{Ar}}:$ b. $\cdot Ca \cdot$ c. $:\overset{..}{\underset{..}{F}}\cdot$ d. $\cdot\overset{..}{N}\cdot$ e. $K\cdot$ f. $:\overset{..}{\underset{.}{S}}\cdot$

5.2   **A.** $Li\cdot + :\overset{..}{\underset{..}{F}}\cdot \longrightarrow Li^{+} + :\overset{..}{\underset{..}{F}}:^{-}$

   B. $Rb\cdot + :\overset{..}{\underset{..}{I}}\cdot \longrightarrow Rb^{+} + :\overset{..}{\underset{..}{I}}:^{-}$

5.3 $2\cdot Al\cdot + 3:\overset{..}{\underset{.}{O}}\cdot \longrightarrow 2\,Al^{3+} + 3:\overset{..}{\underset{..}{O}}:^{2-}$

5.4   **A.** **a.** CaF_2 **b.** Li_2O
   B. $FeCl_2$ and $FeCl_3$

5.5   **a.** K_2O **b.** Ca_3N_2 **c.** CaS

5.6   **a.** calcium fluoride   **b.** copper(II) bromide

5.7   **A.** **a.** bromine trifluoride   **b.** bromine pentafluoride
   B. **a.** dinitrogen monoxide   **b.** dinitrogen pentoxide

5.8   **A.** **a.** PCl_3 **b.** Cl_2O_7
   B. **a.** NI_3 **b.** S_2Cl_2

5.9   a. $:\overset{..}{\underset{..}{Br}}\cdot + \cdot\overset{..}{\underset{..}{Br}}: \longrightarrow :\overset{..}{\underset{..}{Br}}:\overset{..}{\underset{..}{Br}}:$

   b. $H\cdot + \cdot\overset{..}{\underset{..}{Br}}: \longrightarrow H:\overset{..}{\underset{..}{Br}}:$

   c. $:\overset{..}{\underset{..}{I}}\cdot + \cdot\overset{..}{\underset{..}{Cl}}: \longrightarrow :\overset{..}{\underset{..}{I}}:\overset{..}{\underset{..}{Cl}}:$

5.10   **A.** **a.** polar covalent   **b.** ionic   **c.** nonpolar covalent
   B. **a.** polar covalent   **b.** polar covalent
   c. nonpolar covalent

5.11   **A.** **a.** $Ca(C_2H_3O_2)_2$ or $Ca(CH_3CO_2)_2$
   b. NH_4NO_3 **c.** $KMnO_4$
   B. **a.** 2   **b.** 4

5.12   **A.** **a.** calcium carbonate   **b.** magnesium phosphate
   B. **a.** potassium chromate   **b.** ammonium dichromate

5.13   **A.** **a.** $:\overset{..}{\underset{..}{F}} - \overset{..}{\underset{..}{O}} - \overset{..}{\underset{..}{F}}:$ **b.** $H - \overset{H}{\underset{H}{\overset{|}{\underset{|}{C}}}} - \overset{..}{\underset{..}{Cl}}:$

   B. **a.** $\left[:\overset{..}{\underset{..}{N}}=N=\overset{..}{\underset{..}{N}}:\right]^{-}$ **b.** $:\overset{..}{\underset{..}{O}}::N:\overset{..}{\underset{..}{O}}:\overset{..}{\underset{..}{F}}:$

5.14   **A.** **a.** pyramidal   **b.** triangular
   B. **a.** triangular   **b.** bent (at both O)

1.   The noble gases

3.   Sodium metal is quite reactive; sodium ions are quite unreactive.

7.   **a.** group 2A   **b.** group 5A   **c.** group 7A

9.   O, N, and C

11.   Similar: particles close together; different: liquid particles move randomly, solid particles virtually fixed.

17.   a. $:\overset{..}{\underset{.}{F}}\cdot$ b. $\cdot Ca\cdot$ c. $:\overset{.}{\underset{.}{N}}\cdot$ d. $\cdot\overset{.}{\underset{.}{C}}\cdot$

19.   a. $:\overset{..}{\underset{..}{I}}:^{-}$ b. Sr^{2+} c. $:\overset{..}{\underset{..}{N}}:^{3-}$

21.   a. $e^{-} + :\overset{..}{\underset{.}{Cl}}\cdot \longrightarrow :\overset{..}{\underset{..}{Cl}}:^{-}$ b. $\cdot Mg\cdot \longrightarrow Mg^{2+} + 2e^{-}$

23. $\cdot Mg\cdot + 2:\overset{..}{\underset{..}{Br}}\cdot \longrightarrow Mg^{2+} + 2\left[:\overset{..}{\underset{..}{Br}}:\right]^{-}$

25. $3 \cdot Ca \cdot + 2 \, \overset{\cdot}{\underset{\cdot}{:N \cdot}} \longrightarrow 3 \, Ca^{2+} + 2 \left[:\overset{\cdot\cdot}{\underset{\cdot\cdot}{N}}: \right]^{3-}$

27. **a.** $\cdot \overset{\cdot}{Al} \cdot$ and Al^{3+}

   b. $:\overset{\cdot\cdot}{Br} \cdot$ and $:\overset{\cdot\cdot}{\underset{\cdot\cdot}{Br}}:^{-}$

   c. $:\overset{\cdot\cdot}{\underset{\cdot\cdot}{O}}:^{2-}$ and $:\overset{\cdot\cdot}{\underset{\cdot\cdot}{Ne}}:$

29. **a.** Al: $1s^2 2s^2 2p^6 3s^2 3p^1$; Al^{3+}: $1s^2 2s^2 2p^6$
   b. Br: $1s^2 2s^2 2p^6 3s^2 3p^6 3d^{10} 4s^2 4p^5$;
      Br^-: $1s^2 2s^2 2p^6 3s^2 3p^6 3d^{10} 4s^2 4p^6$
   c. O^{2-}: $1s^2 2s^2 2p^6$; Ne: $1s^2 2s^2 2p^6$

31. **a.** $Mg^{2+} \; 2 \, :\overset{\cdot\cdot}{\underset{\cdot\cdot}{F}}:^{-}$ **b.** $Ca^{2+} \; 2 \, :\overset{\cdot\cdot}{\underset{\cdot\cdot}{Cl}}:^{-}$

   c. $2 \, Na^+ \; :\overset{\cdot\cdot}{\underset{\cdot\cdot}{O}}:^{2-}$ **d.** $2 \, K^+ \; :\overset{\cdot\cdot}{\underset{\cdot\cdot}{S}}:^{2-}$

33. **a.** $Mg^{2+} \; :\overset{\cdot\cdot}{\underset{\cdot\cdot}{O}}:^{2-}$ **b.** $Al^{3+} \; :\overset{\cdot\cdot}{\underset{\cdot\cdot}{N}}:^{3-}$ **c.** $2 \, Al^{3+} \; 3 \, :\overset{\cdot\cdot}{\underset{\cdot\cdot}{S}}:^{2-}$

35. **a.** potassium ion   **b.** calcium ion   **c.** zinc ion
   d. bromide ion   **e.** lithium ion   **f.** sulfide ion

37. **a.** iron(II) ion   **b.** copper(I) ion   **c.** iodide ion

39. **a.** Na^+ **b.** Al^{3+} **c.** O^{2-} **d.** Cu^{2+}

41. **a.** potassium iodide   **b.** calcium fluoride
   c. magnesium chloride   **d.** iron(II) chloride
   e. sodium sulfide   **f.** copper(II) oxide

43. **a.** CaS   **b.** Al_2O_3 **c.** FeS   **d.** CuCl

45. **a.** carbonate ion   **b.** hydrogen phosphate ion
   c. permanganate ion   **d.** hydroxide ion

47. **a.** NH_4^+ **b.** HSO_4^- **c.** CN^- **d.** NO_2^-

49. **a.** LiOH   **b.** Na_2CO_3 **c.** CaC_2O_4 **d.** $Zn(NO_2)_2$

51. **a.** $KMnO_4$ **b.** $AgHSO_4$ **c.** $Fe(OH)_3$ **d.** $CuSO_4$

53. **a.** potassium nitrite   **b.** lithium cyanide
   c. ammonium iodide   **d.** sodium nitrate
   e. potassium permanganate   **f.** calcium sulfate

55. $:\overset{\cdot\cdot}{\underset{\cdot\cdot}{I}} \cdot + \cdot \overset{\cdot\cdot}{\underset{\cdot\cdot}{I}}: \longrightarrow$
  (BP)
  $\left[:\overset{\cdot\cdot}{\underset{\cdot\cdot}{I}}:\overset{\cdot\cdot}{\underset{\cdot\cdot}{I}}: \right]$
  (LP)

57. $\cdot \overset{\cdot}{\underset{\cdot}{P}} \cdot + 3 \, H \cdot \longrightarrow H:\overset{\cdot\cdot}{\underset{H}{P}}:H$

59. $\cdot \overset{\cdot}{\underset{\cdot}{C}} \cdot + 4 \, :\overset{\cdot\cdot}{\underset{\cdot\cdot}{F}} \cdot \longrightarrow :\overset{\overset{\displaystyle :\overset{\cdot\cdot}{F}:}{}}{\underset{\underset{\displaystyle :\overset{\cdot\cdot}{F}:}{}}{F}}:\overset{\cdot\cdot}{\underset{\cdot\cdot}{C}}:\overset{\cdot\cdot}{\underset{\cdot\cdot}{F}}:$

61. **a.** N_2O_4 **b.** $BrCl_3$ **c.** NI_3 **d.** S_2F_2

63. **a.** carbon disulfide   **b.** dinitrogen tetrasulfide
   c. phosphorus pentafluoride   **d.** disulfur decafluoride

65. ClO_2

67. **a.** $:\overset{\overset{\displaystyle :\overset{\cdot\cdot}{F}:}{|}}{\underset{\underset{\displaystyle :\overset{\cdot\cdot}{F}:}{|}}{F}}-\overset{}{\underset{}{Si}}-\overset{\cdot\cdot}{\underset{\cdot\cdot}{F}}:$ **b.** $:\overset{\overset{\displaystyle H}{|}}{\underset{\underset{\displaystyle H}{|}}{N}}-\overset{\overset{\displaystyle H}{|}}{\underset{\underset{\displaystyle H}{|}}{N}}:$

   c. $H-\overset{\overset{\displaystyle H}{|}}{\underset{\underset{\displaystyle H}{|}}{C}}-\overset{\overset{\displaystyle H}{|}}{\underset{\underset{\displaystyle H}{|}}{N}}:$ **d.** $:\overset{\overset{\displaystyle H}{|}}{\underset{}{N}}-\overset{\cdot\cdot}{\underset{\cdot\cdot}{O}}:$

69. **a.** $:\overset{\cdot\cdot}{\underset{\cdot\cdot}{O}}=C-\overset{\cdot\cdot}{\underset{\cdot\cdot}{F}}:$ **b.** $:\overset{\cdot\cdot}{\underset{\cdot\cdot}{Cl}}-\overset{\overset{\displaystyle :\overset{\cdot\cdot}{Cl}:}{}}{\underset{\underset{\displaystyle :\overset{\cdot\cdot}{\underset{\cdot\cdot}{Cl}}:}{|}}{P}}-\overset{\cdot\cdot}{\underset{\cdot\cdot}{Cl}}:$

   c. $H-\overset{\cdot\cdot}{\underset{\cdot\cdot}{O}}-\overset{\overset{\displaystyle :\overset{\cdot\cdot}{F}:}{|}}{\underset{\underset{\displaystyle \overset{\cdot\cdot}{\underset{\cdot\cdot}{O}}-H}{|}}{P}}-\overset{\cdot\cdot}{\underset{\cdot\cdot}{O}}-H$ **d.** $H-C\equiv N:$

71. **a.** $\left[:\overset{\cdot\cdot}{\underset{\cdot\cdot}{Cl}}-\overset{\cdot\cdot}{\underset{\cdot\cdot}{O}}: \right]^{-}$ **b.** $\left[H-\overset{\cdot\cdot}{\underset{\cdot\cdot}{O}}-\overset{\overset{\displaystyle :\overset{\cdot\cdot}{O}:}{|}}{\underset{\underset{\displaystyle :\overset{\cdot\cdot}{\underset{\cdot\cdot}{O}}:}{|}}{P}}-\overset{\cdot\cdot}{\underset{\cdot\cdot}{O}}: \right]^{2-}$

   c. $\left[:\overset{\cdot\cdot}{\underset{\cdot\cdot}{O}}-\overset{\cdot\cdot}{\underset{\cdot\cdot}{Cl}}-\overset{\cdot\cdot}{\underset{\cdot\cdot}{O}}: \right]^{-}$ **d.** $\left[:\overset{\cdot\cdot}{\underset{\cdot\cdot}{O}}-\overset{\overset{}{}}{\underset{\underset{\displaystyle :\overset{\cdot\cdot}{\underset{\cdot\cdot}{O}}:}{|}}{Br}}-\overset{\cdot\cdot}{\underset{\cdot\cdot}{O}}: \right]^{-}$

73. **a.** polar   **b.** nonpolar   **c.** polar

75. **a.** $\overset{\longrightarrow}{H-O}$ **c.** $\overset{\longrightarrow}{B-F}$ **b.** is not polar

77. **a.** $\overset{\delta+ \quad \delta-}{Si-Cl}$ **c.** $\overset{\delta+ \quad \delta-}{O-F}$ **b.** is not polar

79. **a.** ionic   **b.** nonpolar covalent   **c.** ionic

81. **a.** nonpolar covalent   **b.** nonpolar covalent
   c. polar covalent

83. b, d

85. HCl

87. **a.** tetrahedral   **b.** bent   **c.** pyramidal

89. **a.** bent   **b.** tetrahedral   **c.** pyramidal

91. nonpolar; the two polar Be—F bonds cancel

93. **a.** nonpolar bonds; 109.5°; nonpolar molecule
   b. nonpolar bonds; 109.5°; nonpolar molecule
   c. nonpolar bonds; 109.5°; nonpolar molecule

95. **a.** polar bonds; 109.5°; polar molecule
   b. polar bonds; 109.5°; nonpolar molecule
   c. polar bonds; 109.5°; polar molecule

97. a, c

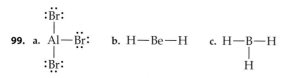

99. a.   **b.** H—Be—H   **c.** H—B—H
                                                                                  H

101. solvent: alcohol; solute: water

103. No. Benzene is nonpolar; it has no dipole to attract the Na^+ and Cl^- ions of NaCl.

105. Gasoline is the solvent; two-cycle motor oil is the solute.

107. Neon has an octet of valence electrons (a full outer shell).

109. AlP; Mg_3P_2

111. polar covalent

113. No; a potassium salt

115. a. Ba^{2+}: $1s^2 2s^2 2p^6 3s^2 3p^6 3d^{10} 4s^2 4p^6 4d^{10} 5s^2 5p^6$
    b. K^+: $1s^2 2s^2 2p^6 3s^2 3p^6$
    c. Se^{2-}: $1s^2 2s^2 2p^6 3s^2 3p^6 3d^{10} 4s^2 4p^6$
    d. I^-: $1s^2 2s^2 2p^6 3s^2 3p^6 3d^{10} 4s^2 4p^6 4d^{10} 5s^2 5p^6$
    e. N^{3-}: $1s^2 2s^2 2p^6$
    f. Te^{2-}: $1s^2 2s^2 2p^6 3s^2 3p^6 3d^{10} 4s^2 4p^6 4d^{10} 5s^2 5p^6$

117. The halogens have seven valence electrons. Forming single bonds by sharing a bond with one other atom gives them an octet of electrons.

Chapter 6

6.1 A. $3 H_2 + N_2 \longrightarrow 2 NH_3$
    B. $2 Fe_2O_3 + 3 C \longrightarrow 3 CO_2 + 4 Fe$

6.2 A. $2 C_4H_{10} + 13 O_2 \longrightarrow 8 CO_2 + 10 H_2O$
    B. It takes 13 molecules O_2 to burn 2 molecules C_4H_{10}; only 4 molecules O_2 to burn 2 molecules CH_4. 1 molecule C_4H_{10} gives 4 molecules CO_2; 1 molecule CH_4 gives 1 molecule CO_2.

6.3 A. $1.48 L CO_2$
    B. Propane; 8 L

6.4 A. a. 65.0 u   **b.** 98.0 u
    B. a. 147.0 u   **b.** 234.1 u

6.5 A. a. 2.04 g Si   **b.** 1000 g H_2O **c.** 17.0 g $Ca(H_2PO_4)_2$

6.6 a. 0.0664 mol Fe   **b.** 2.84 mol C_4H_{10}
    c. 14.8 mol $Mg(NO_3)_2$

6.7 A. 0.179 g/L
    B. 1.29 g/L; more than 7 times that of He

6.8 A. 51.5 g/mol   **B.** C_2H_6

6.9 *Molecular:* 2 molecules H_2S react with 3 molecules O_2 to form 2 molecules SO_2 and 2 molecules H_2O
    Molar: 2 mol H_2S react with 3 mol O_2 to form 2 mol SO_2 and 2 mol H_2O
    Mass: 68.2 g H_2S react with 96.0 g O_2 to form 128.1 g SO_2 and 36.0 g H_2O

6.10 a. 1.59 mol CO_2 **b.** 305 mol H_2O **c.** 0.606 mol CO_2

6.11 A. 0.763 g O_2
    B. a. 2130 g CO_2 **b.** 2350 g CO_2

6.12 A. 39.0 g NH_3
    B. a. 0.967 g O_2 **b** 8.02 g P_4O_{10}

6.13 400 mm Hg

6.14 A. 1800 mL   **B.** 503 mm Hg

6.15 A. a. 2.98 L   **b.** 234 L
    B. −167 °C

6.16 A. a. 0.0471 atm   **b.** 4.99 L
    B. 73.4 L

6.17 A. 0.00870 M   **B.** 0.968 M

6.18 a. 9.00 M   **b.** 1.26 M   **c.** 0.274 M

6.19 A. 673 g KOH   **B.** 5.61 g KOH

6.20 A. 29.7 mL   **B.** 6.30×10^{-4} g HNO_3

6.21 A. 9.28% ethanol   **B.** 34.8% toluene

6.22 A. Take 315 mL of isopropyl alcohol and add enough water to make 450 mL of solution.
    B. Take 195 mL of acetic acid and add enough water to make 2.00 L of solution.

6.23 A. 2.81% H_2O_2 **B.** 51.3% NaOH

6.24 A. Dissolve 5.62 g glucose in enough water to make 125 g solution.
    B. Dissolve 15.6 g NaCl in enough water to make 1750 g solution.

2. The atomic mass (O = 15.9994 u) represents the mass of one atom, which can occur in many compounds. The formula mass (31.9988 u) represents the mass of a molecule of the element as it occurs in nature—as O_2.

5. For a fixed amount of gas at a constant T, the volume varies inversely with pressure (PV = constant).

6. For a fixed amount of gas at a constant P, the volume is directly proportional to temperature; $V \propto T$.

9. Pressure times volume is equal to the product of the number of moles, the gas constant, and the temperature of the gas. $PV = nRT$.

13. a. 4   **b.** 3   **c.** 8

15. 2 Fe; 12 C; 6 H; 24 O

17. a. 4 molecules ammonia react with 3 molecules oxygen to form 2 molecules nitrogen and 6 molecules water.
    b. 4 mol NH_3 react with 3 mol O_2 to form 2 mol N_2 and 6 mol H_2O.
    c. 68.1 g NH_3 react with 96.0 g O_2 to form 56.0 g N_2 and 108.1 g H_2O.

19. (a) and (b) are balanced.

21. a. $4 Al + 3 O_2 \longrightarrow 2 Al_2O_3$
    b. $2 H_2 + V_2O_5 \longrightarrow V_2O_3 + 2 H_2O$
    c. $H_2O + Cl_2O_5 \longrightarrow 2 HClO_3$

23. a. $4 Fe + 3 O_2 \longrightarrow 2 Fe_2O_3$
    b. $CaCO_3 + 2 HCl \longrightarrow CaCl_2 + CO_2 + H_2O$
    c. $C_7H_{16} + 11 O_2 \longrightarrow 7 CO_2 + 8 H_2O$

25. a. 22.4 L   **b.** 22.4 L   **c.** 44.8 L

27.

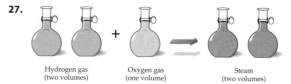

Hydrogen gas (two volumes)   Oxygen gas (one volume)   Steam (two volumes)

29. a. 17.2 L   **b.** 189 mL

31. 3.74 g/L

33. a. 47.5 g   **b.** 66.5 g

35. a. 6.02×10^{23} O_2 molecules   **b.** 1.20×10^{24} O atoms

37. (c)

39. a. 230.0 g   **b.** 142.0 g
    c. 74.1 g   **d.** 152.1 g

41. a. 1030 g $CaSO_4$ **b.** 3.17 g $CuCl_2$ **c.** 856 g $C_{12}H_{22}O_{11}$

43. a. 0.0195 mol Sb_2S_3 **b.** 0.133 mol MoO_3
    c. 3.56 mol $AlPO_4$ **d.** 0.0887 mol $Be(NO_3)_2$

45. a. 4.18 mol CO_2 **b.** 15.6 mol O_2

47. a. 2480 g NH_3 **b.** 193 g H_2

49. 266 g Mg_3P_2

51. **a.** 1450 mL **b.** 3470 mmHg

53. **a.** 9120 L **b.** 1140 min (19 h)

55. 117 mL

57. 433 K (160 °C)

59. **a.** decrease **b.** decrease **c.** increase

61. **a.** temperature decreases **b.** pressure decreases

63. **a.** 22.3 L **b.** 29.1 atm

65. 0.0781 mol Kr

67. **a.** 9.36 M HCl **b.** 0.277 M Li_2CO_3

69. **a.** 56.0 g NaOH **b.** 34.0 g $C_6H_{12}O_6$

71. **a.** 0.208 L **b.** 1.80 L

73. **a.** 4.83% by vol. **b.** 5.09% by vol

75. Add 277 g NaCl to 3098 g water.

77. Add 40.0 mL acetic acid to enough water to make 2.00 L of solution.

79. (b)

81. **a.** $Hg(NO_3)_2(s) \longrightarrow Hg(l) + 2\ NO_2(g) + O_2(g)$
b. $Na_2CO_3(aq) + 2\ HCl(aq) \longrightarrow H_2O(l) + CO_2(g) + 2\ NaCl(aq)$

83. **a.** yes **b.** no **c.** no

85. 16.6 g Fe_3O_4

87. (b), (c), and (d)

89. (c)

91. 3590 g nitric acid

93. **a.** 3.96% NaOH **b.** 7.31% ethanol

95. **a.** $NH_4NO_3 \longrightarrow N_2O + 2\ H_2O$

b. $:N\equiv N-\ddot{\underset{..}{O}}:$

c. 2.20 g N_2O

d. 2.24 L H_2O

97. 7.3 g XeF_6

99. 0.00825 L (8.25 mL)

101. **a.** 6.64×10^5 g Cu (664 kg Cu)
b. 2.35×10^5 L SO_2

103. 1.29 g/L; 0.804 g/L; Moist air is less dense than dry air.

Chapter 7

7.1 **A. a.** $HBr(aq) \longrightarrow H^+(aq) + Br^-(aq)$

b. $Ca(OH)_2(s) \xrightarrow{H_2O} Ca^{2+}(aq) + 2\ OH^-(aq)$

B. $CH_3COOH(aq) \longrightarrow H^+(aq) + CH_3COO^-(aq)$

7.2 **A.** $HBr(aq) + H_2O \longrightarrow H_3O^+(aq) + Br^-(aq)$
B. $HBr(aq) + CH_3OH \longrightarrow CH_3OH_2^+(aq) + Br^-(aq)$

7.3 **A.** H_2SeO_3 **B.** HNO_3

7.4 **A.** $Sr(OH)_2$ **B.** KOH

7.5 **A.** $LiOH(aq) + HC_2H_3O_2(aq) \longrightarrow LiC_2H_3O_2(aq) + H_2O$
B. $Ca(OH)_2(aq) + 2\ HCl(aq) \longrightarrow CaCl_2(aq) + 2\ H_2O$

7.6 **A.** pH = 11 **B.** pH = 3

7.7 **A.** 1.0×10^{-2} M (0.010 M) **B.** 1.0×10^{-3} M (0.0010 M)

7.8 **A.** (c)

1. **a.** Arrhenius: forms H^+ in water; Brønsted–Lowry: proton donor
b. Arrhenius: forms OH^- in water; Brønsted–Lowry: proton acceptor
c. Ionic compound from acid-base neutralization

2. No effect on litmus; no reaction with iron or zinc

4. A Brønsted–Lowry acid must have a hydrogen atom to donate a proton. A Brønsted–Lowry base must have a lone pair of electrons to accept a proton.

7. A strong acid ionizes to a much greater extent that does a weak acid.

8. In acid-base chemistry the proton is an H^+ ion. In nuclear chemistry the proton usually represents a subatomic particle within a nucleus.

11. $Mg(OH)_2$ is insoluble in water, forming few $OH^-(aq)$ ions.

13. A condition in which the blood is too alkaline.

14. H_2SO_4

16. All acids contain H (which they release in water as H^+). No, some bases (e.g., NH_3) and many neutral compounds (e.g., CH_4) contain H atoms.

17. Hydrogen ion (H^+) [or hydronium ion (H_3O^+)]

19. $HClO_4(aq) \longrightarrow H^+(aq) + ClO_4^-(aq)$

21. A proton donor; a reaction of the type
$HA(aq) + H_2O \longrightarrow H_3O^+(aq) + A^-(aq)$

23. **a.** base **b.** acid **c.** acid

25. $HCl(g) + H_2O \longrightarrow H_3O^+(aq) + Cl^-(aq)$; hydrochloric acid

27. $NH_3(aq) + H_2O \longrightarrow NH_4^+(aq) + OH^-(aq)$

29. **a.** HCl **b.** $Sr(OH)_2$
c. KOH **d.** H_3BO_3

31. **a.** nitric acid (an acid) **b.** cesium hydroxide (a base)
c. carbonic acid (an acid)

33. **a.** HNO_2 **b.** H_2SO_3

35. **a.** H_2SO_4 **b.** $Mg(OH)_2$

37. strong base

39. weak acid

41. **a.** weak acid **b.** strong base
c. salt **d.** strong acid

43. highest: a; lowest, b

45. **a.** $HI(aq) \longrightarrow H^+(aq) + I^-(aq)$

b. $LiOH(s) \xrightarrow{H_2O} Li^+(aq) + OH^-(aq)$

c. $HClO_2(aq) \longrightarrow H^+(aq) + ClO_2^-(aq)$

47. **a.** $HClO_2(aq) + H_2O \longrightarrow H_3O^+(aq) + ClO_2^-(aq)$
b. $HNO_2(aq) + H_2O \longrightarrow H_3O^+(aq) + NO_2^-(aq)$
c. $HCN(aq) + H_2O \longrightarrow H_3O^+(aq) + CN^-(aq)$

49. **a.** $KOH(aq) + HCl(aq) \longrightarrow KCl(aq) + H_2O$
b. $LiOH(aq) + HNO_3(aq) \longrightarrow LiNO_3(aq) + H_2O$

51. $H_3PO_4(aq) + 3\ NaOH(aq) \longrightarrow Na_3PO_4(aq) + 3\ H_2O$

53. **a.** acidic **b.** neutral **c.** acidic **d.** basic

55. pH = 5

57. 1.0×10^{-12} M H^+

59. (b)

61. $3\ HCl(aq) + Al(OH)_3(s) \longrightarrow AlCl_3(aq) + 3\ H_2O$
$2\ HCl(aq) + Mg(OH)_2(s) \longrightarrow MgCl_2(aq) + 2\ H_2O$

63. No. Covalent OH-containing compounds do not produce OH^- ions in water.

65. $SrCO_3(s) + 2\ HI(aq) \longrightarrow SrI_2(aq) + CO_2(g) + H_2O(l)$

67. $HPO_4^{2-}(aq) + H_2O \longrightarrow H_3O^+(aq) + PO_4^{3-}(aq)$
$HPO_4^{2-}(aq) + H_2O \longrightarrow H_2PO_4^-(aq) + OH^-(aq)$

69. **a.** pH = 11 **b.** pH = 12

71. 100 g stomach acid (500 mg HCl)

Chapter 8

8.1 oxidation: a, b, c, d

8.2 reduction: a; oxidation: b

8.3 reduction: a; oxidation: b, c, d

8.4 **a.** oxidizing agent: O_2; reducing agent: Se
b. oxidizing agent: CH_3CN; reducing agent: H_2
c. oxidizing agent: V_2O_5; reducing agent: H_2
d. oxidizing agent: Br_2; reducing agent: K

8.5   oxidation: $Al \longrightarrow Al^{3+} + 3\,e^-$; reduction:
$Br_2 + 2\,e^- \longrightarrow 2\,Br^-$

8.6   **A.** Half-reactions:
$Fe \longrightarrow Fe^{3+} + 3\,e^-$; $Mg^{2+} + 2\,e^- \longrightarrow Mg$
Overall: $2\,Fe + 3\,Mg^{2+} \longrightarrow 2\,Fe^{3+} + 3\,Mg$
B. Half-reactions:
$Pb \longrightarrow Pb^{2+} + 2\,e^-$; $Ag(NH_3)_2{}^+ + e^- \longrightarrow Ag + 2\,NH_3$
Overall: $Pb + 2\,Ag(NH_3)_2{}^+ \longrightarrow Pb^{2+} + 2\,Ag + 4\,NH_3$

8.7   **A.** $2\,Zn + O_2 \longrightarrow 2\,ZnO$
B. $Se + O_2 \longrightarrow SeO_2$

8.8   **A.** $2\,PbS + 3\,O_2 \longrightarrow 2\,PbO + 2\,SO_2$
B. $C_2H_5OH + 3\,O_2 \longrightarrow 2\,CO_2 + 3\,H_2O$

1. a. Oxidation: O atoms gained; reduction: O atoms lost
b. Oxidation: H atoms lost; reduction: H atoms gained
c. Oxidation: e^- lost; reduction: e^- gained

2. Oxidation: oxidation number increases; reduction: oxidation number decreases.

3. (a)

5. To allow ions to flow from one compartment to the other and thus keep the solutions electrically neutral

11. Iron is oxidized to iron(III) hydroxide; salt water acts as an electrolyte.

13. Silver is oxidized by hydrogen sulfide to (black) silver sulfide; use aluminum to reduce the silver sulfide to metallic silver.

19. a. oxidized (C_2H_4O gains O)
b. reduced (H_2O_2 loses O)
c. oxidized [$C_6H_4(OH)_2$ loses H]
d. reduced (Sn^{2+} gains e^-)
e. reduced (Cl_2 gains e^-)

21. a. $K \longrightarrow K^+ + e^-$ **b.** $Ca \longrightarrow Ca^{2+} + 2\,e^-$
c. $Zn \longrightarrow Zn^{2+} + 2\,e^-$

23. a. oxidizing agent: Fe_2O_3; reducing agent: C
b. oxidizing agent: O_2; reducing agent: P_4
c. oxidizing agent: H_2O; reducing agent: C
d. oxidizing agent: H_2SO_4; reducing agent: Zn

25. $Cr_2O_7{}^{2-}$ (goes from +6 to +3 state, forming Cr_2O_3)

27. a. oxidation: $Fe \longrightarrow Fe^{2+} + 2\,e^-$; reduction:
$2\,H^+ + 2\,e^- \longrightarrow H_2$
b. oxidation: $Al \longrightarrow Al^{3+} + 3\,e^-$; reduction:
$Cr^{2+} + 2\,e^- \longrightarrow Cr$

29. a. oxidation: $2\,I^- \longrightarrow I_2 + 2\,e^-$; reduction:
$Cl_2 + 2\,e^- \longrightarrow 2\,Cl^-$
overall: $2\,I^- + Cl_2 \longrightarrow I_2 + 2\,Cl^-$
b. oxidation: $SO_2 + 2\,H_2O \longrightarrow H_2SO_4 + 2\,H^+ + 2\,e^-$
reduction: $HNO_3 + H^+ + e^- \longrightarrow NO_2 + H_2O$
overall: $SO_2 + 2\,HNO_3 \longrightarrow 2\,NO_2 + H_2SO_4$

31. a. H_2CO is oxidized; H_2O_2 is the oxidizing agent
b. C_2H_6O is oxidized, $MnO_4{}^-$ is the oxidizing agent

33. Reduced; acetylene gains hydrogen.

35. MoO_3 is reduced; the reducing agent is H_2.

37. Zr was oxidized; water was the oxidizing agent

39. Nitrite ion is reduced; ascorbic acid is the reducing agent.

41. a. SO_2 **b.** H_2O
c. $CO_2 + H_2O$ **d.** $CO_2 + H_2O$

43. Indoxyl is oxidized; O_2 is the oxidizing agent.

45. To protect them from oxygen, preventing corrosion (oxidation)

47. $CuO + H_2 \longrightarrow Cu + H_2O$

49. 160 L O_2

51. Water is oxidized; CO_2 is the oxidizing agent. CO_2 is reduced; water is the reducing agent.

53. a. oxidized   **b.** neither   **c.** reduced

55. To act as oxidizing agents, metallic elements would have to form negative ions; they don't. Nonmetals can act as either oxidizing or reducing agents because they can take on positive or negative oxidation numbers.

57. $Cl_2 + H_2 \longrightarrow 2\,HCl$; Cl_2 is the oxidizing agent.

Chapter 9

9.1   **a.** molecular: C_6H_{14}; complete structural:

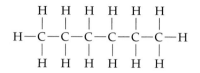

condensed structural: $CH_3CH_2CH_2CH_2CH_2CH_3$
b. molecular: C_8H_{18}; complete structural:

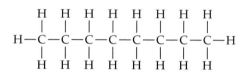

condensed structural: $CH_3CH_2CH_2CH_2CH_2CH_2CH_2CH_3$

9.2   **A.**

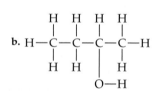

B. C_nH_{2n}

9.3   **a.** C_6H_{12} **b.** C_7H_{12}
9.4   **a.** $CHCl_3$ **b.** CCl_4

9.5   **a.** $CH_3CH_2CH_2CH_2OH$ **b.**

9.6   **a.** alcohol   **b.** ether   **c.** ether
d. phenol   **e.** ether
9.7   **A.** $CH_3OCH_2CH_2CH_3$ **B.** $CH_3CH_2OC(CH_3)_3$
9.8   **a.** ketone;   **b.** aldehyde   **c.** aldehyde
9.9   **A. a.** $CH_3CH_2CH_2COOH$ **b.** CH_3CHO
c. $CH_3CH_2COCH_2CH_3$
B. a. $CH_3CH_2CH_2CH_2CH_2COOH$
b. $CH_3CH_2COCH_2CH_2CH_2CH_3$
c. $CH_3CH_2CH_2CH_2CH_2CH_2CHO$
9.10   **a.** $CH_3CH_2CH_2CH_2NH_2$
b. $CH_3CH_2NHCH_2CH_3$
c. $CH_3NHCH_2CH_2CH_3$ **d.** $(CH_3)_2CHNHCH_3$
9.11   **a.** amine (NH)   **b.** amide (CONH)
c. amine (N)   **d.** both (NH and $CONH_2$)
9.12   heterocyclic compounds: (a) and (d)

1. The study of carbon compounds

2. Carbon atoms can bond strongly to each other; can bond strongly to other elements; can form chains, rings, and other kinds of structures

4. A series of compounds differing by a CH_2 unit; the straight-chain alkanes

5. Isomers have the same molecular formula but different structural formulas.

6. A hydrocarbon with the maximum number of hydrogen atoms, an alkane. A hydrocarbon with a double bond (alkene) or triple bond (alkyne).

7. A set of six delocalized electrons

9. 1–4 C, gases; 5–16 C, liquids; >18 C, solids

10. Alkanes are less dense than water; a layer of hexane over water.

11. **a.** ethanol **b.** 2-propanol (isopropyl alcohol)
c. methanol

13. Historical use: anesthetic; modern use: solvent

14. Antiseptics

15. An aldehyde has a hydrogen atom attached to the carbonyl carbon atom.

17. **a.** alcohol **b.** ketone **c.** ester
d. ether **e.** carboxylic acid **f.** aldehyde

19. **a.** organic **b.** organic
c. inorganic **d.** inorganic

21. **a.** 4 **b.** 8 **c.** 7 **d.** 5

23. **a.** propane **b.** acetylene (ethyne) **c.** ethylene (ethene)

25. **a.** molecular: C_5H_{12}; structural: $CH_3CH_2CH_2CH_2CH_3$
b. molecular: C_7H_{16}; structural:
$CH_3CH_2CH_2CH_2CH_2CH_2CH_3$

27. **a.** CH_3CH_2- **b.** $(CH_3)CH-$

29. **a.** methyl **b.** *sec*-butyl

31. **a.** ethanol **b.** butyl alcohol (1-butanol)

33. **a.** CH_3OH **b.** $CH_3CH_2CH_2OH$

35. —OH

37. **a.** $CH_3CH_2OCH_2CH_3$ **b.** $CH_3CH_2CH_2CH_2OCH_3$

39. **a.** CH_3CHO **b.** $HCHO$

41. **a.** propionaldehyde (propanal)
b. methyl propyl ketone (2-pentanone)

43. **a.** formic acid (methanoic acid)
b. propionic acid (propanoic acid)

45. **a.** $CH_3CH_2CH_2CH_2CH_2CH_2COOH$
b. $CH_3CH_2CH_2CH_2CH_2CH_2CH_2CH_2CH_2COOH$

47. **a.** CH_3COOH **b.** $CH_3CH_2CH_2CH_2COOH$

49. **a.** $CH_3COOCH_2CH_3$ **b.** $CH_3CH_2CH_2COOCH_3$

51. **a.** $CH_3CH_2NH_2$ **b.** CH_3NHCH_3

53. **a.** propylamine **b.** diethylamine

55. **a.** same **b.** same **c.** isomers

57. **a.** homologs **b.** none of these

59. **a.** unsaturated; alkene **b.** saturated; alkane

61.
$$\underset{\Vert}{-\overset{O}{C}}- \qquad \underset{\Vert}{-\overset{O}{C}}-OH$$

63. **a.** ester **b.** aldehyde **c.** amine
d. ether **e.** ketone **f.** carboxylic acid

65. **a.** not heterocyclic, cycloalkane **b.** heterocyclic, amine

67. C_9H_8O

69. **a.** $CH_3C{\equiv}CCH(CH_3)CH_2CH_2CH_2CH_3 + 2\,H_2 \xrightarrow{\text{Ni}}$
$\qquad\qquad CH_3CH_2CH_2CH(CH_3)CH_2CH_2CH_2CH_3$
b. $CH_2{=}C(CH_3)CH_2CH_2CH_3 + H_2 \xrightarrow{\text{Ni}}$
$\qquad\qquad (CH_3)_2CHCH_2CH_2CH_3$

71. **a.** $CH_3CH_2CH_2CH_2OH$ **b.** $CH_3CH_2CH(OH)CH_2CH_3$
c. $(CH_3)_2CHCH(OH)CH_3$

d. —OH

72. Phenols; methyl, isopropyl

73. Homology

74. Isomerism

75. $2\,CH_3OH + 3\,O_2 \longrightarrow 2\,CO_2 + 4\,H_2O$; 1070 g

Chapter 10

10.1 $CH_2{=}CH-C{\equiv}N$

10.2 $-[CH_2CH(OCH_3)CH_2CH(OCH_3)CH_2CH(OCH_3)$
$\qquad\qquad\qquad -CH_2CH(OCH_3)]-$

B. $-[CH_2CH(OCOCH_3)CH_2CH(OCOCH_3)$
$\qquad\qquad\qquad -CH_2CH(OCOCH_3)CH_2CH(OCOCH_3)]-$

10.3 **a.** $-[OCH_2CH_2COOCH_2CH_2COOCH_2$
$\qquad\qquad\qquad -CH_2COOCH_2CH_2CO]-$
b. $-[OCH_2CH_2(C{=}O)]_n$

2. A semisynthetic material derived from cellulose; made by treating cellulose with nitric acid; celluloid is very flammable.

3. PVC has a Cl atom on alternate C atoms.

4. Polymerization in which all the monomer atoms are incorporated in the polymer; a double bond.

5. Teflon is like PE but with all H atoms replaced by F atoms.

6. Polystyrene; styrene

7. **a.** HDPE **b.** PET

8. Acrylic polymers; acts as a binder and hardens to form a continuous surface coating.

12. Synthetic fibers; they are cheaper and have a wider range of properties.

18. Cl, Br

19. HDPE has linear molecules; a closely packed, fairly crystalline structure gives it greater rigidity and higher tensile strength than LDPE.

21. LLDPE is a copolymer of ethylene and a higher alkene; it has branched molecules similar to LDPE but branching is more controlled.

23. **a.** $CH_2{=}CHCl$ **b.** $CH_2{=}CH$—

25. **a.** $-[CH_2CH_2CH_2CH_2CH_2CH_2CH_2CH_2]-$
b. $-[CH_2CH(CH_3)CH_2CH(CH_3)CH_2CH(CH_3)$
$\qquad\qquad\qquad -CH_2CH(CH_3)]-$

27. **a.** $-[CH_2CH(C{\equiv}N)CH_2CH(C{\equiv}N)CH_2CH(C{\equiv}N)$
$\qquad\qquad\qquad -CH_2CH(C{\equiv}N)]-$
b. $-[CH_2CH(OCOCH_3)CH_2CH(OCOCH_3)$
$\qquad\qquad\qquad -CH_2CH(OCOCH_3)CH_2CH(OCOCH_3)]-$
c. $-[CH_2CH(COOCH_3)CH_2CH(COOCH_3)$
$\qquad\qquad\qquad -CH_2CH(COOCH_3)CH_2CH(COOCH_3)]-$

29. $CH_2{=}\underset{\underset{CH_3}{|}}{C}-CH{=}CH_2$

31. Hydrocarbon chains of natural rubber are cross-linked by sulfur, making the natural rubber harder; by crosslinking.

33. Neoprene is more resistant to oil and gasoline than other elastomers; a chlorine atom replaces the methyl group on isoprene.

35. Polybutadiene (made from butadiene), polychloroprene (from chloroprene), styrene-butadiene rubber (from styrene and butadiene).

37. $-$CO(CH$_2$)$_4$CONH(CH$_2$)$_4$NHCO(CH$_2$)$_4$CONH(CH$_2$)$_4$NH$-$

39. $-$[CH$_2$(C$_6$H$_{10}$)CH$_2$OOC(C$_6$H$_4$)$-$
$\quad$ $-$COOCH$_2$(C$_6$H$_{10}$)CH$_2$OOC(C$_6$H$_4$)COO$-$ (where C$_6$H$_{10}$ is a cyclohexane ring and C$_6$H$_4$ is a benzene ring)

41. The long chains can be entangled with one another; intermolecular forces are greatly multiplied in large molecules; large polymer molecules move more slowly than do small molecules.

43. The temperature at which the properties of the polymer change from hard, stiff, and brittle to rubbery and tough; rubbery materials such as automobile tires should have a low T_g; glass substitutes, a high T_g.

45. Monomer: a; repeating unit: b; polymer: c. Addition polymerization.

47. acrylonitrile (CH$_2$=CH$-$C≡N) and styrene (CH$_2$=CH$-$C$_6$H$_5$)

49.
$$\left[\begin{array}{c} C{\equiv}N \\ | \\ -CH_2-C- \\ | \\ O{=}COC_8H_{17} \end{array} \right]_n$$

51. ~CH$_2$C(CH$_3$)$_2$CH$_2$(CH$_3$)$_2$CH$_2$(CH$_3$)$_2$CH$_2$(CH$_3$)$_2$~

53. CH$_3$CH(OH)CH$_2$COOH

55. a.

HOOC—(naphthalene ring)—COOH   and   HOCH$_2$—CH$_2$OH

b. polyester

57. Elasticity; the golf ball

59. Carbonyl group; ethylene

Chapter 11

11.1   **A.** SiO$_2$ + 4 HF $\longrightarrow$ SiF$_4$ + 2 H$_2$O
$\quad$ **B.** 16 g HF

11.2 1.27×10^7 g CH$_4$

1. Crust, mantle, core; from the crust

2. Compounds of metals with silicon and oxygen; quartz, micas, asbestos

3. One substance enhances the effect of another; asbestos fibers and cigarette smoke.

4. Mining operations, particulate matter from crushing, high energy consumption

5. Cement mixed with sand and gravel

6. Found free in nature; easily shaped

7. A copper-tin alloy; harder than copper

8. Copper and tin are easier to obtain from ores than is iron.

11. The problem is keeping the metal in a usable form and not scattered in the environment.

13. Reduce; reuse; recycle

15. Lithosphere: solid portion of Earth; hydrosphere: watery portion; atmosphere: gaseous mass surrounding Earth

17. Oxygen

19. Living and once-living matter

21. The SiO$_4$ tetrahedron

23. The SiO$_4$ tetrahedra are arranged in flat two-dimensional arrays.

25. The SiO$_4$ tetrahedra in glass are arranged in irregular fashion rather than in an ordered array.

27. Sand (silica), sodium carbonate, and limestone (calcium carbonate)

29. A mixture of calcium and aluminum silicates

31. Iron ore (source of iron), limestone (removes impurities), and coke or coal (reducing agent).

33. Reduction; Fe$_2$O$_3$ + 3 CO $\rightarrow$ 2 Fe + 3 CO$_2$.

35. Aluminum forms an impervious coat of Al$_2$O$_3$; iron rust flakes off exposing fresh surface to corrosion.

37. The furnace heats the iron ore, melts impurities as slag, and reduces iron oxides to metallic iron.

39. Iron drawn off a blast furnace; phosphorus, silicon, and excess carbon.

41. Aluminum; it is difficult to obtain the metal from its ores.

43. a. MnO$_2$ **b.** Si   **c.** Si   **d.** MnO$_2$

45. a. ThO$_2$ **b.** Ca   **c.** Ca   **d.** ThO$_2$

47. a. Cl$_2$ **b.** C   **c.** C   **d.** Cl$_2$

49. 114 g Cu

51. 0.174 g CaCl$_2$

53. 120 million kg Al$_2$O$_3$; 250 million kg bauxite

55. 3.8 million t Zn; 0.3 million t Pb; 0.83 million t Cu

57. 51 cm

59. 1.7×10^7 kg ore per day

61. 282,000 cubic feet, 28 feet deep

63. In the figure, 33 mm in diameter, the lithosphere would be 0.09 mm thick, atmosphere would be about 0.26 mm thick; neither could be shown to scale properly.

Chapter 12

12.1   **A.**
$$\begin{array}{ccc} & \ddot{\text{:}}\overset{\cdot\cdot}{F}\text{:} & \\ & | & \\ \ddot{\text{:}}\overset{\cdot\cdot}{F} - & C - & \overset{\cdot\cdot}{C}l\ddot{\text{:}} \\ & | & \\ & \overset{\cdot\cdot}{\text{:}}\overset{}{C}l\text{:} & \end{array}$$

B. Both F and Cl are quite electronegative; the bond dipoles nearly cancel.

12.2   **A.** 3790 g H$_2$O   **B.** 4940 kg CO$_2$

1. By absorbing harmful UV radiation, thus shielding living creatures from it

2. A lower level of cold air trapped by layer of warm air above it; pollutants are trapped in cool air near the ground.

3. No; wildfires, dust storms, volcanoes, etc., also contribute.

4. Refrigerant; molding of plastic foams

5. HFCs or HCFCs; they contribute to the greenhouse effect.

9. A combination of smoke and fog

12. CO, NO$_2$, dust, molds

13. No, they are different phenomena.

15. Troposphere; stratosphere.

17. Water vapor: 4%; carbon dioxide: 0.4%

19. Causing water pollution problems by making more N compounds available to plants

21. By plants through photosynthesis

23. Particulate matter, sulfur oxides, carbon monoxide

25. S + O$_2$ $\longrightarrow$ SO$_2$

27. SO$_3$ + H$_2$O $\longrightarrow$ H$_2$SO$_4$

29. The two interact to give an effect greater than the sum of the individual effects on lung tissue and plants.

31. The lime reacts with and thus removes the SO$_2$;
$\quad$ CaO(s) + SO$_2$(g) $\longrightarrow$ CaSO$_3$(s).

33. Hydrocarbons, NO$_x$, O$_3$, aldehydes, PAN

35. At high temperatures; N$_2$ + O$_2$ $\longrightarrow$ 2 NO

37. It is broken down into nitric oxide (NO) and reactive oxygen atoms; NO$_2$ $\longrightarrow$ NO + O.

39. Automobiles

41. Devices containing catalysts for oxidation of CO and hydrocarbons to CO_2 and reduction of NO_x to N_2

43. Automobiles

45. By hindering O_2 transport and thus adding to the workload of the heart

47. $O_2 + O \longrightarrow O_3$

49. Increase in skin cancer

51. Rain with a pH below 5.6; SO_x and NO_x are oxidized and react with water to form acids.

53. Acids dissolve iron;
$Fe(s) + H_2SO_4(aq) \longrightarrow H_2(g) + FeSO_4(aq)$.

55. Carbon monoxide from poorly ventilated heaters; nitrogen oxides from gas ranges; radon from the ground; paticulates from cigarette smoke

57. Increased risk for lung cancer; with proper ventilation

59. When CO levels exceed a maximum of 9 ppm over an eight-hour period

61. The slow warming of Earth caused by gases absorbing infrared radiation

63. Some waste heat is formed in every process (second law of thermodynamics).

65. 8000 μg (or 8 mg)

67. 0.19 g CO_2, 6.3×10^{-4} g CH_4, 8.3×10^{-4} g H_2; 2.7%

69. Water vapor will form clouds and might even lead to cooling by reflecting sunlight back to space.

Chapter 13

13.1 **A. a.** 0.1 ppb   **b.** 100 ppt
     B. 1×10^{-9} M

1. Less than 1%

3. Biochemical oxygen demand; a high BOD can cause dissolved oxygen to be depleted.

4. Cholera, typhoid fever, and dysentery; modern water treatment facilities

5. The increased BOD that occurs when algae bloom and then die, becoming organic waste and leading to dead streams and lakes

6. Microorganisms that cause disease

7. They can leak contaminants such as gasoline and oils into underground water supplies.

9. Chlorinated hydrocarbons are unreactive and do not break down readily.

10. Dyes, acids, oils, metal salts, sulfur compounds

11. Water expands when it freezes because it forms a rigid structure with relatively large holes. The ice floats, protecting the deeper water from freezing.

13. The gasoline would not dissolve; it would float.

15. Heat capacity is the quantity of heat required to raise the temperature of an object 1 °C. The high specific heat capacity of the large amount of water moderates Earth's temperature fluctuations.

17. The high heat capacity of water allows the lake to absorb heat during the day, keeping the surrounding land cooler.

19. Dust, dissolved atmospheric gases such as carbon dioxide and oxygen, nitric acid from lightning storms; various pollutants.

21. Cations: Ca^{2+}, Mg^{2+}, Na^+, K^+; anions: SO_4^{2-}, HCO_3^-, Cl^-

23. Acid rain and mine runoff

25. Limestone (calcium carbonate)

27. Neutralize with lime or limestone.

29. Sewage is collected in a pond; it removes solids that are allowed to settle.

31. Organic molecules and inorganic ions such as nitrates and phosphates

33. To kill pathogenic microorganisms

35. Organic substances that are attracted to charcoal

37. The two react to produce a gelatinous aluminum hydroxide precipitate that drags down suspended solids and bacteria from the water.

39. To remove odors and improve the taste

41. Ultraviolet radiation kills microorganisms.

43. About 1 ppm

45. Fertilizer; drainage from feedlots

47. $2 HNO_3(aq) + CaCO_3(s) \longrightarrow$
$\qquad\qquad Ca^{2+}(aq) + 2 NO_3^-(aq) + CO_2(g) + H_2O(l)$

49. **a.** 6 ppb   **b.** 145 ppm

51. **a.** $CHCl_3$ **b.** NaCl, phosphate ion, sand
     c. sand   **d.** NaCl and phosphate ion   **e.** phosphate ion

53. Boiling water; molecules must be separated to relatively large distances.

55. Cl_2 is reduced; it is the oxidizing agent. SO_2 is the reducing agent.

57. Dispersion forces

Chapter 14

14.1 **A.** 58,500 J
     B. a. 34,600 kJ   **b.** 9.6 kWh

14.2 **A.** 91.2 kJ   **B.** 2700 kJ

14.3 **A.** 53.0 kcal   **B.** 81.5 cal

1. Wood

2. A material easily burned as a source of energy; coal, petroleum, natural gas

3. Nuclear fusion

4. Coal; petroleum

5. The sun

7. fuels: a, b, c

9. **a.** $2 C_2H_2(g) + 5 O_2(g) \longrightarrow 4 CO_2(g) + 2 H_2O(l)$
     b. $2 H_2(g) + O_2(g) \longrightarrow 2 H_2O(l)$
     c. $C_{12}H_{22}O_{11}(s) + 12 O_2(g) \longrightarrow 12 CO_2(g) + 11 H_2O(l)$

11. $C(s) + O_2(g) \longrightarrow CO_2(g)$

13. $CH_4 + 2 O_2 \longrightarrow CO_2(g) + 2 H_2O(l)$

15. Yes. Both H_2 and CO are readily oxidized (burn).

17. A reaction goes faster at a higher temperature.

19. A reaction that gives off energy; burning of methane.

21. 4220 kJ

23. 4310 J

25. 1606 kJ

27. Energy is conserved.

29. A measure of the degree of distribution of energy in a system; increased.

31. Coal

33. Coal, solid; petroleum, liquid; natural gas, gas

35. 39.4 years

37. 64.4 years

39. Methane

41. Ancient marine animals

43. By distillation, cracking, reforming; tetraethyllead is a highly effective octane booster but is quite toxic.

45. About 20%

47. No; the uranium is enriched to only about 3% uranium-235; to explode, it would have to be enriched to about 90%.

49. By converting nonfissile uranium-238 to fissile plutonium-239

51. Yes

53. Plentiful fuel, cleaner. Plasma is a mixture of atomic nuclei and electrons at temperatures such as that on the sun.

55. $^{232}_{90}\text{Th} + ^{1}_{0}\text{n} \longrightarrow ^{233}_{90}\text{Th}$

$^{233}_{90}\text{Th} \longrightarrow ^{0}_{-1}\text{e} + ^{233}_{91}\text{Pa}$

$^{233}_{91}\text{Pa} \longrightarrow ^{0}_{-1}\text{e} + ^{233}_{92}\text{U}$

57. $^{2}_{1}\text{H} + ^{3}_{1}\text{H} \longrightarrow ^{1}_{0}\text{n} + ^{4}_{2}\text{He}$

59. Cells that produce electricity directly from sunlight; solar cells.

61. 40 m^2, 430 ft^2

63. Plant material used as fuel.

65. Fuel is fed into a fuel cell continuously; the electrodes serve as a surface for the chemical reaction but are not used up.

67. $\text{C(s)} + 2\,\text{H}_2\text{(g)} \longrightarrow \text{CH}_4\text{(g)}$

69. $\text{CO(g)} + 2\,\text{H}_2\text{(g)} \longrightarrow \text{CH}_3\text{OH(l)}$

71. Endothermic; the reaction absorbs heat, cooling the treated area.

73. 2000 watts

75. $2.35 \times 10^6 \text{ g CO}_2$

77. **a.** Removal by photosynthesis
b. Animal respiration
c. Making plastics

79. 32 t SO_2; no, there would be 27 μg SO_2/m^3

Chapter 15

15.1 **A.** Both are aldoses.
B. Gulose differs from glucose in configuration about C-5 and C-2. Mannose differs from glucose in configuration about C-2 only.

15.2 **A. a.** H-P-V-A   **b.** histidylprolylvalylalanine
B. a. Thr-Gly-Ala-Ala-Leu   **b.** T-G-A-A-L

15.3 **A.** 3   **B.** 24

15.4 **A.** Sugar: ribose; base: uracil; RNA
B. Neither; thymine occurs only in DNA, ribose only in RNA.

1. The chemistry of living things and life processes

3. Plants make food by photosynthesis; animals eat plants or other animals.

4. Carbohydrates, fats, and proteins

6. Polyhydroxy aldehydes or ketones or compounds that can be hydrolyzed to form such compounds; starches and sugars; C, H, O

7. Tissue substance insoluble in water but soluble in nonpolar solvents: fats, fatty acids, steroids

9. In every cell; muscles, skin, hair, nails

10. They are polyamides.

11. 20

12. All have C, H, and O; proteins also have N and possibly S.

13. An amide bond that joins two amino acid units

15. If the molar mass of a polypeptide exceeds about 10,000 g, it is called a protein.

17. A protein with molecules that fold into a spheroid or ovoid shape

18. Deoxyribonucleic acid (DNA) and ribonucleic acid (RNA); DNA

19. Hydrogen bonds

21. DNA is a double helix; RNA is a single helix with some loops.

27. Carbohydrates that cannot be further hydrolyzed; glucose, fructose, and galactose

29. Glycogen: animal starch with highly branched molecules. Amylose: plant starch with continuous-chain molecules. Amylopectin: plant starch with branched molecules.

31. Monosaccharides: c, d

33. 31: none; 32: c, d

35. a. glucose   **b.** galactose and glucose   **c.** glucose

37. Aldehyde and alcohol (hydroxyl)

39. Aldehyde and alcohol (hydroxyl); in configuration about C-4.

41. Esters

43. At room temperature, fats are solids, oils are liquid. Structurally, oils have more C-to-C double bonds than fats have.

45. One that contains a high proportion of saturated (all C-to-C single bonds) fatty acid units.

47. saturated: d, e; unsaturated: a, b, c

49. a. 18   **b.** 16   **c.** 18

51. Corn oil; oils have more C-to-C double bonds than do fats.

53. An amino group and a carboxyl group; a zwitterion is a molecule that carries both a positive and a negative charge.

55. No; their N-terminal and C-terminal ends are opposite.

57. a. $\text{H}_3\text{N}^+\text{CH(CH}_3\text{)COO}^-$ **b.** $\text{H}_3\text{N}^+\text{CH(CH}_2\text{OH)COO}^-$

59. a. $\text{H}_3\text{N}^+\text{CH}_2\text{CO}{-}\text{NHCHCOO}^-$
$\qquad\qquad\qquad\qquad\quad |$
$\qquad\qquad\qquad\qquad\ \text{CH}_3$

b. $\text{H}_3\text{N}^+\text{CHCO}{-}\text{NHCHCOO}^-$
$\qquad\quad\ |\qquad\qquad\quad |$
$\qquad\quad \text{CH}_3\qquad\quad\ \text{CH}_2\text{OH}$

61. Hydrogen bonds, ionic bonds, disulfide linkages and dispersion forces

63. RNA: ribose; DNA: deoxyribose

65. DNA: a; RNA: b, c

67. ribose; uracil

69. a. guanine   **b.** thymine
c.   cytosine   **d.** adenine

71. One strand will go with one daughter cell nucleus, the other stand will go with the other daughter cell nucleus.

73. mRNA takes the information on DNA to a ribosome outside of the nucleus; tRNA is responsible for translating the specific base sequence of mRNA into a specific amino acid sequence in a protein.

75. TTAAGC

77. AGGCTA

79. a. AAC   **b.** CUU

81. Glycogen or amylopectin.

83. a. primary   **b.** secondary

85. a. pyrimidine   **b.** pyrimidine

87. a. purine   **b.** RNA

89. Instead of the amino acid glycine, a stop codon is specified.

Chapter 16

16.1 **A.** About 64%   **B.** About 34%

16.2 **A.** I
B. Fatty acid I, because fatty acid III is a cis fatty acid and not a trans fatty acid

16.3 **A.** $2\,\text{NH}_3 + \text{H}_3\text{PO}_4 \longrightarrow (\text{NH}_4)_2\text{HPO}_4$
B. $\text{ZnO} + \text{H}_2\text{SO}_4 \longrightarrow \text{ZnSO}_4 + \text{H}_2\text{O}$

16.4 270 mg

16.5 **A.** By 2143   **B.** By 2086

1. Carbohydrates, fats, proteins

3. An energy source, thermal insulation, and protects organs

7. Vitamins are organic; minerals are inorganic.

8. Fat soluble; an excess is stored and accumulated; excess water-soluble vitamins are excreted.

10. Glycogen

13. Bread with some vitamins and iron added; no

14. Monosodium glutamate, a flavor enhancer

16. Vitamin E

18. The methyl ester of the dipeptide aspartylphenylalanine
19. Additives generally recognized as safe
21. It is banned.
23. Add nutritional value, enhance flavor or color, retard spoilage, provide texture, sanitize (and others).
25. A herbicide used to kill weeds before crop seedlings emerge
27. Conventional farming uses more energy than organic farming. Production is 10% lower and 12% more labor is required on organic farms than conventional farms.
29. **a.** glucose   **b.** sucrose   **c.** fructose
31. Cellulose cannot be digested by humans; starch can.
33. CO_2 and H_2O
35. Fatty acids, glycerol, mono- and diglycerides
37. Most animal fats are solids; most vegetable oils are liquids; animal fats generally have fewer C-to-C double bonds than vegetable oils.
39. About 9 kcal
41. I
43. one
45. amino acids; hydrolysis
47. **a.** lysine   **b.** methionine
49. Protein deficiency; children
51. **a.** thyroid gland (regulation of metabolism)
    b. hemoglobin (oxygen transport)
    c. bones, teeth, blood clotting, heartbeat rhythm
    d. nucleic acids, ATP (obtaining, storing and using energy from foods), bones, teeth
53. Water soluble: pantothenic acid, biotin; fat soluble: phylloquinone
55. a, b, d, e
57. **a.** vitamin C   **b.** vitamin D   **c.** vitamin A
59. Water soluble: b, c, d; fat soluble: a, e
61. **a.** improves nutrition (thyroid function)
    b. adds flavor   **c.** flavor enhancer
    d. inhibit spoilage
63. **a.** antioxidant   **b.** color   **c.** artificial sweetener
65. (d)
67. Reducing agents, usually free-radical scavengers added to foods to prevent fats and oils from becoming rancid; vitamin E; BHT and BHA
69. C, H, O
71. The Haber process; hydrogen and nitrogen
73. An organic nitrogen fertilizer, H_2NCONH_2; ammonia and carbon dioxide
75. Phosphate rock is treated with phosphoric acid to make calcium dihydrogen phosphate; plays a role in the energy transfer processes of photosynthesis.
77. DDT is effective against many insects, but it remains toxic long after initial use.
79. They are fat soluble and become concentrated in fats moving up the food chain.
81. Chemicals secreted to mark a trail, send an alarm, or attract a mate; to attract male insects and determine when to use a pesticide.
83. Male insects are sterilized by radiation, chemicals, or cross breeding, then released for nonproductive breeding. Expensive and time consuming.
85. 53% from carbohydrates; 15% from fat; 20% from protein
87. 192 kcal, about 19% from fat
89. About 100 g fat
91. 30 g DDT

93. Addition of a constant amount each growth period; adding $10 to a savings account each week
95. 50 weeks; arithmetic
97. 51 years; changes in birth rate or death rate (war, famine, . . .)
99. $6\,CO_2 + 6\,H_2O \longrightarrow C_6H_{12}O_6 + 6\,O_2$
101. **a.** N, P   **b.** $2\,NH_3 + H_3PO_4 \longrightarrow (NH_4)_2HPO_4$
103. (a), (c), (e)
105. **a.** alkene, alcohol   **b.** ester, alkene, ether (epoxide)
     c. ether, alkene, ester

Chapter 17

17.1 **a.** nonionic   **b.** anionic
      c. anionic   **d.** cationic
1. Chemical compounds from plants that produce a soapy lather
3. An excellent cleanser in soft water, relatively nontoxic, derived from renewable sources, biodegradable
5. A compound that makes clothes appear brighter by absorbing ultraviolet light and emitting visible (blue) light
7. Loosens baked-on grease and burned-on food, excellent glass cleaner; vapors are irritating, toxic; don't mix with chlorine bleaches
9. Mildly abrasive and absorbs odors
10. Cuts grease film; vinegar (acetic acid) reacts with marble (calcium carbonate), pitting the surface
12. Keratin; an oily skin secretion
13. A product that uses a perfume to mask body odor; some have a substance that kills bacteria that cause odor
15. A salt of a long-chain carboxylic acid; $CH_3(CH_2)_{14}COO^-Na^+$
17. It is softer and produces a finer lather.
19. Water containing Ca^{2+}, Mg^{2+}, or Fe^{2+} ions
21. Acid neutralizes the ionic "head."
23. Sodium palmitate and glycerol
25. $3\,CH_3(CH_2)_{10}COO^-Na^+ \;+\;$
$$\begin{array}{c} CH_2{-}CH{-}CH_2 \\ \;|\qquad\;|\qquad\;\; | \\ OH\quad OH\quad OH \end{array}\; \text{(glycerol)}$$
27. Precipitates hard water ions; makes the water more alkaline
29. $PO_4^{3-} + H_2O \longrightarrow HPO_4^{2-} + OH^-$
31. $CO_3^{2-} + M^{2+} \longrightarrow MCO_3$
33. A substance added to a surfactant to increase its detergency
35. By binding Ca^{2+} and Mg^{2+} in soluble complexes
37. A linear alkylsulfonate; an anionic surfactant with a linear hydrocarbon "tail"
39. A surfactant molecule with a negative charge on its water-soluble head
41. A surfactant molecule with no charge on its water-soluble head
43. The positive end of the cationic surfactant would interact with the negative end of the anionic surfactant, destroying detergent action.
45. All three
47. II
49. Hexadecyltrimethylammonium chloride
51. They slowly release chlorine in water
53. To remove paint, varnish, adhesives, waxes, etc.
55. Flammability, toxic fumes
57. A skin softener
59. An oil and water
61. Lipstick is harder and contains more wax
63. Detergent and abrasive
65. Enamel is converted from hydroxyapatite to fluorapatite, a harder substance.

67. A complex mixture of compounds used as a fragrance
69. The sex attractant of musk deer or synthetic analog; to moderate the sweet, flowery odor
71. Alcohol, perfume, coloring
73. A detergent
75. Hydrogen peroxide
77. Temporary dyes are water soluble and can be washed out; permanent dyes last until the hair is cut off or falls out.
79. A reducing agent; redox reaction
81. Oxidation
83. I, III, IV
85. I
87. $CH_3(CH_2)_{14}COOCH(CH_3)_2$

Chapter 18

18.1   **A.** 3.6 lb   **B.** 3.5 lb
18.2   **A.** 35 mi   **B.** 2250 kcal
18.3   **A.** 23.5   **B.** 174 lb

1. Regular exercise and a healthy diet
3. At most 30%; at most 10%
5. Deplete glycogen stores; dehydrate yourself; no, it would be mostly water loss
7. Better understanding of muscle physiology, drugs, improved equipment
8. A person who exercises regularly is able to do more physical work with less strain
9. Athletes need more calories; by eating more carbohydrates
11. About 3500 kcal
12. Usually deficient in B vitamins, iron, and other nutrients; slows metabolism, making future dieting more difficult and weight gain easier.
13. About 1 lb
15. Drugs that alleviate pain or soreness that results from overtraining or to treat injuries
16. Masculinization (balding, extra body hair, deep voices) and menstrual irregularities
17. More toxic wastes that tax the liver and kidneys; such diets are also high in fat (artery clogging, cardiovascular disease).
19. Nutrient-based reference values established by the Food and Nutrition Board of the U.S. National Academy of Sciences for use in planning and assessing diets
21. Good vision, bone development, skin maintenance
23. Arthritis; B_6 is a cofactor for more than 100 enzymes
25. Promotes absorption of calcium and phosphorus; the upper limit is 2000 IU; more than that can be toxic.
27. Na^+, K^+, and Cl^-
29. Water; alcohol and caffeine are diuretics.
31. Cloudy, highly colored urine
33. Muscle protein is converted to glucose.
35. 2.9 h
37. 0.19 h (12 min)
39. 100 g
41. 20 km
43. **a.** When insulin levels go up, glucose levels go down and we experience hunger.
    b. Ghrelin is an appetite stimulant.
45. 25.2
47. No
49. ATP

51. Actin and myosin
53. Anaerobic
55. Lack of oxygen resulting from anaerobic exercise
57. **a.** Type I is slow twitch; Type IIB is fast twitch.
    b. Type I, aerobic; Type IIB, anaerobic.
59. ATP can be generated rapidly and hydrolyzed rapidly.
61. Type I; increase in myoglobin, faster oxygen transport, increased respiratory capacity
63. No, the only way to build muscles is by exercise.
65. Cancer, heart disease, stroke, emphysema, bronchitis, etc.
67. Anti-inflammatory; fluid retention, hypertension, ulcers, disturbance of sex hormone balance
69. Yes, briefly, by increasing alertness, respiration, blood pressure, muscle tension and heart rate; it masks fatigue and gives an athlete a sense of increased stamina.
71. The body releases endorphins that have much the same effect as some narcotics.
73. The "runner's high," feeling down when unable to run, having to run farther and farther to get the same high
75. 2250 kcal
77. 1.06 g/mL; lean

Chapter 19

1. A drug that kills or slows the growth of bacteria; originally limited to formulations derived from living organisms.
3. Cortisol: a hormone released in the body during stressed or agitated states; prednisone: a synthetic hormone similar to cortisone
5. Mediators of hormone action; prostaglandins act near where they are produced; hormones act throughout the body.
6. Mild sedation (small doses) and sleep (larger doses)
7. Addiction and tolerance
9. Ketamine, PCP
11. A drug that produces stupor and relief of pain
12. The dried, resinous juice of the unripe seeds of the opium poppy (*Papaver somniferum*).
13. Brain amines are made from dietary amino acids; for example, high-carbohydrate diets produce high serotonin levels in the brain.
15. No; they help control some symptoms.
17. Stimulant, depressant, hallucinogenic
19. An inactive substance given in place of a real drug; may have a psychological effect
20. Drug abuse: using drugs for their intoxicating effects (e.g., using cocaine to get high); drug misuse: using a drug improperly (e.g., using an antibiotic to treat a viral infection)
21. A pain reliever; it inhibits prostaglandin synthesis and thus blocks pain messages to the brain.
23. Acetylsalicylic acid
25. Both are analgesics and antipyretics.
27. Aspirin is anti-inflammatory; acetaminophen is not.
29. Carboxylic acid, ester
31. β-lactam antibiotics obtained from *Penicillium* molds.
33.
35. They are effective against a wide variety of bacteria.
37. No; vaccination
39. Antimetabolites and alkylating agents
41. A female sex hormone

43. It blocks pregnancy from being established by inhibiting the action of progesterone.

45. It reduces the level of consciousness and the intensity of reactions to environmental stimuli.

47. One that changes the way we perceive things

49. A drug that acts on the brain to produce unconsciousness and insensitivity to pain.

51. (a)

53. *para*-aminobenzoic acid

55. Codeine is a less potent analgesic and less addictive; both are analgesics and addictive.

57. It eases withdrawal symptoms without the stupor that accompanies heroin use.

59. An agonist mimics the action of a drug.

61. b, c

63. **a.** $C_6H_5—CH_2CH_2NH_2$
   b. $C_6H_5—CH_2CH(CH_3)NH_2$
   c. $C_6H_5—CH(CH_3)NHCH_3$
   d. $C_6H_5—CH(OH)CH(CH_3)NH_2$

65.

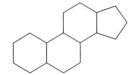

Steroid skeletal structure

67. Fat soluble; 17 C atoms with only 1 O atom and 1 N atom

69. Pyrimidine-like ring; by substituting various groups on the ring or replacing one of the carbonyl O atoms by an S atom

71. Ketoprofen, because it takes a smaller amount to kill the rat

73. 46 tablets

75. 350 mg; no

77.

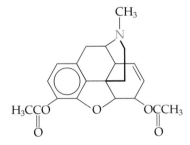

Chapter 20

1. Toxicology is a science that deals with the effects of poisons on the body. A poison is a substance that causes injury, illness, or death to an organism.

2. It can be, in extremely large amounts.

3. People have different metabolisms and different health conditions (e.g., sugar can be more harmful to a diabetic than to a nondiabetic).

5. It is converted to fluorocitric acid that inhibits the citric acid cycle by tying up the enzyme that acts on citric acid.

7. Cigarette smoking

9. An abnormal growth of new tissue; benign tumors grow slowly and do not invade neighboring tissue, while malignant tumors invade and destroy neighboring tissues.

11. b, d

13. Hydrolysis of amides; catalyst

15. A substance that blocks the transport of oxygen in the bloodstream; CO, nitrate ions

17. By converting toxic cyanide to harmless thiocyanate

19. Too little: anemia; too much: vomiting, shock, coma, and death

21. By tying up sulfhydryl groups, thus deactivating enzymes

23. By chelating Pb^{2+} ions, thus enhancing their excretion

25. **a.** blocks the synthesis of acetylcholine
   b. blocks acetylcholine receptor sites
   c. blocks acetylcholine receptor sites

27. Sarin, tabun, soman

29. It blocks acetylcholine receptor sites.

31. By oxidizing ethanol to acetaldehyde, then to acetic acid, and finally to carbon dioxide and water.

33. Cotinine is less toxic and is more water soluble (more readily excreted) than nicotine.

35. No. Some are oxidized to more toxic substances.

37. 7950 mg (7.95 g)

39. Genes that regulate cell growth; they sustain the abnormal growth characteristic of cancer

41. Incomplete burning of almost any organic material

43. A study of a human population to try to link human health effects (e.g., cancer) to a cause (e.g., exposure to a specific chemical)

45. A mutagen causes mutations; a teratogen causes birth defects.

47. A waste that burns readily on ignition, presenting a fire hazard; hexane, gasoline.

49. Any waste that can cause or contribute to death or illness or that threatens human health or the environment when improperly managed

51. 6.6×10^{10} molecules

53. **a.** $C_{10}H_{16}O$ **b.** 4800 mg   **c.** 5 mg

55. **a.** 99.48%; % smokers in control group: 95.50%
   b. 0.9948:1 for cancer cases; 0.9550:1 for controls
   c. 0.96:1

Appendix A

A.1 **a.** 0.0163 g   **b.** 1.53 lb   **c.** 370 mL

A.2 **a.** 0.0903 m   **b.** 0.2224 km   **c.** 150 fl oz

A.3 **a.** 24.4 m/s   **b.** 1.34 km/h   **c.** 0.136 oz/qt

A.4 **a.** No.   **b.** Yes

A.5 0.12 m^3

A.6 56.8 g

A.7 **a.** 100.5 m   **b.** 150 g   **c.** 6.3 L
   d. $1800 \text{ m}^2 (1.80 \times 10^3 \text{ m}^2)$ **e.** 2.33 g/mL
   f. 0.634 g/cm^3

A.8 **a.** 185 °F   **b.** 10 °F   **c.** 179 °C   **d.** −29 °C

A.9 80,000 cal; 80.0 kcal; 334 kJ

1. **a.** 298 K   **b.** 0 °C   **c.** 28 °C   **d.** 200 K

3. 65.5 kcal

5. (coldest) 0 K < 0 °F < 0 °C

7. **a.** 3100 J   **b.** 2060 cal   **c.** 5.69 kJ
   d. 1440 J   **e.** 209 kcal

9. 4.64 kJ

Photo Credits

Index

Some Metric Units of Length, Mass, and Volume

Length

1 kilometer (km)	= 1000 meters (m)
1 meter (m)	= 100 centimeters (cm)
1 centimeter (cm)	= 10 millimeters (mm)
1 millimeter (mm)	= 1000 micrometers (μm)

Mass

1 kilogram (kg)	= 1000 grams (g)
1 gram (g)	= 1000 milligrams (mg)
1 milligram (mg)	= 1000 micrograms (μg)

Volume

1 liter (L)	= 1000 milliliters (mL)
1 milliliter (mL)	= 1000 microliters (μL)
1 milliliter (mL)	= 1 cubic centimeter (cm^3)

Some Conversions Between Common and Metric Units

Length

1 mile (mi)	= 1.61 kilometers (km)
1 yard (yd)	= 0.914 meter (m)
1 inch (in.)	= 2.54 centimeters (cm)

Mass

1 pound (lb)	= 454 grams (g)
1 ounce (oz)	= 28.4 grams (g)
1 pound (lb)	= 0.454 kilograms (kg)

Volume

1 U.S. quart (qt)	= 0.946 liter (L)
1 U.S. pint (pt)	= 0.473 liter (L)
1 fluid ounce (fl oz)	= 29.6 milliliters (mL)
1 gallon (gal)	= 3.78 liters (L)

Some Conversion Units for Energy

1 calorie (cal) = 4.184 joules (J)

1 British thermal unit (Btu) = 1055 joules (J) = 252 calories (cal)

1 food Calorie = 1 kilocalorie (kcal) = 1000 calories (cal) = 4184 joules (J)

Some Common Polyatomic Ions

Charge	Name	Formula
1+	Ammonium ion	NH_4^+
	Hydronium ion	H_3O^+
1−	Hydrogen carbonate (bicarbonate) ion	HCO_3^-
	Hydrogen sulfate (bisulfate) ion	HSO_4^-
	Acetate ion	$CH_3CO_2^-$ (or $C_2H_3O_2^-$)
	Nitrite ion	NO_2^-
	Nitrate ion	NO_3^-
	Cyanide ion	CN^-
	Hydroxide ion	OH^-
	Dihydrogen phosphate ion	$H_2PO_4^-$
	Permanganate ion	MnO_4^-
2−	Carbonate ion	CO_3^{2-}
	Sulfate ion	SO_4^{2-}
	Chromate ion	CrO_4^{2-}
	Monohydrogen phosphate ion	HPO_4^{2-}
	Oxalate ion	$C_2O_4^{2-}$
	Dichromate ion	$Cr_2O_7^{2-}$
3−	Phosphate ion	PO_4^{3-}